AF572310

Ecology of the Southern California Bight

Ecology of the Southern California Bight

A Synthesis and Interpretation

Edited by
Murray D. Dailey
Donald J. Reish
Jack W. Anderson

UNIVERSITY OF CALIFORNIA PRESS
Berkeley / Los Angeles / London

University of California Press
Berkeley and Los Angeles, California

University of California Press
London, England

Library of Congress Cataloging-in-Publication Data

Ecology of the Southern California Bight: a synthesis and interpretation/edited by Murray D. Dailey, Donald J. Reish, Jack W. Anderson.
p. cm.
Includes bibliographical references and index.
ISBN 0–520–07578–1 (alk. paper)
1. Marine ecology—Southern California Bight (Calif. and Mexico) 2. Southern California Bight (Calif. and Mexico) I. Dailey, Murray D. II. Reish, Donald J. III. Anderson, Jack W.
QH95.45.S68E25 1993
574.5′2636′091641—dc20 93–34699
CIP

Printed in the United States of America

1 2 3 4 5 6 7 8 9

The paper used in this publication meets the minimum requirements of American National Standard for Information Sciences—Permanence of Paper for Printed Library Materials, ANSI Z39.48–1984

Reuben Lasker

This book is dedicated to Reuben Lasker, an eminent scientist who worked in the Southern California Bight. His investigations in the physiological ecology of marine organisms began as a Sverdrup Post-Doctoral Fellow at Scripps Institution of Oceanography (SIO) in 1956. He was educated at Stanford where he received his doctorate under the tutelage of Arthur C. Giese. At his untimely death in 1988, Reuben was leader of the Coastal Fisheries Resources Division of the Southwest Fisheries Science Center of the National Marine Fisheries Service on the campus of SIO, with a staff of about 50 scientists and technicians under his supervision. He was also involved in the preparation of this book as a member of the Quality Review Board. Reuben was active in graduate education and for 15 years held an appointment as Adjunct Professor of Marine Biology at SIO.

Reuben was an especially gregarious scientist, and his knowledge and warm support of marine science attracted a long list of collaborators during his career. One example of this is that in a bibliography prepared for the MMS project, 260 of the citations were produced by Reuben and 50 of his associates. He adopted as colleagues children of his fellow staff associates, local high school students, and National Academy of Science/National Research Council Senior Post-Doctoral Fellows. Reuben operated in all arenas of marine science: experimental laboratory work, satellite remote sensing, work at sea, and mathematical modeling.

Reuben Lasker received much recognition for his work: the Meritorious Service Award of the U.S. Department of the Interior, 1970; the Distinguished Service Award of the U.S. Department of Commerce, 1974; the Huntsman Medal for Excellence in Biological Oceanography from the Canadian Government

Bedford Institute of Oceanography; and the posthumous Outstanding Achievement Award of the American Institute of Fishery Research Biologists, 1988.

It is sufficient to say here that future studies of the ecology of the SCB, and similar marine habitats around the world, will be affected by this gregarious scientist, his scientific accomplishments, and his influence on his students, friends, and colleagues.

Paul E. Smith
NOAA/NMFS Southwest Fisheries
Science Center
La Jolla, California

Contents

Foreword

Today's student of marine ecology is faced with an almost overwhelming burden of information. The rapid expansion of knowledge in the marine sciences and the increasingly specialized nature of academic pursuits is a challenge to the scientist's ability to look beyond his or her own specialized field and not lose the ability to perceive and encompass the proverbial "forest through the trees." Understanding the relationships among the various biotic and abiotic elements in the sea has become increasingly important as escalating demands of recreation, fishing, minerals extraction, shipping, and waste discharge compete for limited ocean space. Resource agencies struggle with an imperfect view of ecological processes ("big picture") in an attempt to achieve some balance among potentially conflicting uses of the ocean and the seabed.

It is important that we, as marine scientists and managers, pause occasionally from our rush to acquire new, specialized information and, instead, make an effort to summarize, evaluate, and reinterpret information collected in the past. We must look for links in our knowledge of interacting physical, chemical, and biological components of environment, links that may have been previously overlooked.

In preparing this book, we collected a large base of published and unpublished information and assembled a multidisciplinary team to review and synthesize the information. The authors were encouraged to read and discuss each other's material, across disciplines, and by collective insight, extend each other's understanding of how the marine environment off southern California functions, both naturally and as influenced by human activities. A series of meetings and periods of writing and rewriting occurred over a period of two years, finally culminating with a book that we believe fulfills our plan for an integrated synthesis and interpretations of the ecology of the Southern California Bight.

We hope this book is useful to a wide audience, including the advanced undergraduate or graduate student, just discovering the fascinating world of marine life lying at our front door; the marine scientist, trying to differentiate between potential anthropogenic effects and an inherently variable ecosystem; and the resource manager, making decisions that may weigh economic considerations against environmental protection.

Gary Brewer
Fred Piltz
Environmental Studies Section
Pacific OCS Region
Minerals Management Service
U.S. Department of the Interior

Preface

The Southern California Bight (SCB) encompasses the body of water lying between Point Conception on the Santa Barbara County coast and a point just south of the United States–Mexico border. The mild Mediterranean climate, broad sandy beaches, and rocky shores provide an ideal environment for a water-oriented playground used by millions of people who inhabit the shores of the SCB. However, the recreational and commercial fishing elements in the SCB are often in conflict with other elements, including the onshore industrial complex, offshore oil production, operations of one of the world's busiest seaports (Los Angeles–Long Beach harbors), and its use as receiving waters for municipal wastes.

Prior to World War II, oceanographic research in the SCB was largely exploratory and descriptive in nature. It was not until the mid-1950s that integrated studies in oceanography in the SCB were undertaken. The first of these large-scale studies was funded by the California Water Pollution Control Board (now the California Water Quality Control Board) during the period 1956–1961. This study focused on those areas of the SCB that were polluted, were expected to become polluted, and were expected to remain unpolluted. Two later studies were funded by the Bureau of Land Management (now the Minerals Management Service, MMS): the Baseline Study (1975–1976) and the Benchmark Study (1976–1977). These two studies were primarily concerned with obtaining background data from areas where offshore drilling for oil may occur in the future as well as areas that serve as reference sites. The data generated from these studies constituted an important source of information for this book.

Synthesis of the accumulated oceanographic data of the SCB is a monumental task, and until publication of this compendium, only two other such works had been completed. In 1960, K. O. Emery wrote *The Sea off Southern California,* which integrated the physical, chemical, and geological knowledge of the SCB. The staff of the Southern California Coastal Water Research Project (SCCWRP) wrote the second synthesis, *Ecology of the Southern California Bight: Implication for Water Quality Management.* This synthesis focused on waste disposal, especially municipal wastes, in the SCB.

The impetus for the present volume grew out of public workshops sponsored by MMS in 1982. A report issued by MMS on the workshops stressed the need for a comprehensive review of the oceanography of the SCB and recommended that the study should emphasize the ecology of the area. *Ecology of the Southern California Bight: A Synthesis and Interpretation* is the result of that recommendation.

The primary goal of this work can best be summarized by a statement in the contract between MMS and the Ocean Studies Institute:

To produce a comprehensive hard-cover text that will make a significant contribution to the understanding of marine ecological processes within the SCB. By supporting the publication of a book written by a team of recognized experts in marine ecology and oceanography, MMS hopes to obtain a highly credible reference that will benefit all scientists as well as the environmental staff of MMS.

The contract specified that the text was to be "a synthesis and interpretation of existing data on the ecology of the SCB" and that "historical data and recent findings need to be thoroughly reviewed and then applied to new descriptions of how the marine ecosystem functions, both naturally and under the influence of human activities."

This book is written with both the advanced undergraduate and graduate student in mind, as well as the professional in need of a reference source.

The editors could not have completed this volume without the assistance and cooperation of many people. We appreciate the time and effort that each of the contributing authors has given to this project. We especially thank Dr. Donald W. Hood, author of the chapter "Ecosystem Interrelationships." He brought to the task not only his years of experience but also the expertise he acquired as co-editor of two similar volumes, *The Eastern Bering Sea Shelf: Oceanography and Resources,* published in 1981, and *The Gulf of Alaska,* published in 1987.

Many scientists served as reviewers, providing an invaluable service to the editors and authors throughout the writing of this book. A Quality Review Board was selected to advise and assist in producing the compendium. The board met with the authors at the outset to discuss the overall coverage of the book to be written. Then, after completion of the initial draft, the board members met again with the authors to offer suggestions for improvement. After reading the final draft, they met for the last time with the editors to give additional suggestions before the manuscript was submitted to MMS and subsequently to University of California Press. We, the editors, wish to thank these board members for their help and encouragement:

Dr. Donald Boesch, Director, Center for Environmental and Estuarine Studies, University of Maryland, Cambridge, Maryland
Dr. Alfred W. Ebeling, Biological Sciences, University of California, Santa Barbara
Dr. Donn Gorsline, John and Doris Zinsmeyer Chair in Marine Studies, University of Southern California, Los Angeles
Dr. Isaac Kaplan, Geology and Geochemistry, University of California, Los Angeles
Dr. Paul Smith, Fishery Biologist, National Oceanic and Atmospheric Administration, Southwest Fisheries Center, La Jolla, California
Dr. Reuben Lasker, Chief, Coastal Fisheries Resources Division, National Oceanic and Atmospheric Administration, Southwest Fisheries Center, La Jolla, California. (Dr. Lasker served as a member until his death in March 1988.)

Other scientists acted as peer reviewers for chapters in their areas of expertise. They include

Dr. Alice Alldredge, Department of Biological Sciences, University of California, Santa Barbara (zooplankton)
Dr. Farooq Azam, Marine Research Biology Department, University of California, San Diego (microbiology)
Dr. Kenneth T. Briggs, University of California, Davis (birds)
Dr. Roy Carpenter, School of Oceanography, University of Washington, Seattle (chemical oceanography)
Dr. Daniel Costa, Institute of Marine Sciences, University of California, Santa Cruz (mammals)
Dr. Gilbert F. Jones, Department of Biological Sciences, University of Southern California, Los Angeles (benthic invertebrates)
Dr. Ronald Kolpack, Marine Processes Research, West Covina, California (physical oceanography)
Dr. Mark Littler, Curator of Botany, National Museum of Natural History, Smith-

sonian Institution, Washington, D.C. (benthic macrophytes)
- Dr. Alan J. Mearns, Ecologist, National Oceanic and Atmospheric Administration, Seattle (human inputs)
- Dr. David Menzel, Director, Skidaway Institute of Oceanography, Savannah, Georgia (ecosystem interrelationships)
- Dr. Raymond Riznyk, Department of Natural Resources, Alaska Pacific University, Anchorage, Alaska (phytoplankton)
- Dr. Robert F. Rooney, Department of Economics, California State University, Long Beach (world shipping in human inputs)
- Dr. Maynard Silva, Assistant Executive Director, Guadalupe Community Redevelopment Agency, Guadalupe, California (governance)
- Dr. John Stephens, Department of Biology, Occidental College, Los Angeles (fish)
- Dr. Peter M. Williams, Research Chemist, University of California at San Diego and Scripps Institution of Oceanography (chemical oceanography)

We wish to thank Elizabeth Knoll, Sponsoring Editor at the University of California Press, and Kathy Walker, Copy Editor, for helping in the publication of this book. We are pleased to acknowledge the contributions of the staff of the Ocean Studies Institute, Linda Ennis and Dorothy ("Dee Dee") Rypka, who participated in all phases of this project. The art work was done by William Dunton, Carol Lyon, Ann Salness, and Marsha Schindler.

We also thank Dr. Gary Brewer, who represented MMS as project manager, for his help and encouragement, and Dr. Fred Piltz of MMS for his interest in this project since its inception.

Donald J. Reisch
Jack W. Anderson
Murray D. Dailey

Contributors

Editors

Jack W. Anderson
Columbia Aquatic Sciences
6060 Corte del Cedro
Carlsbad, California 92008

Murray D. Dailey
P.O. Box 1171
Gunnison, Colorado 81230

Donald J. Reish
Department of Biology
California State University
Long Beach, California 90840

Contributors

Larry Allen
Department of Biological Sciences
California State University
Northridge, California 91330

Patricia Herron Baird
Department of Biology
California State University
Long Beach, California 90840

Michael L. Bonnell
Institute of Marine Sciences
University of California
Santa Cruz, California 95060

Michael E. Brady
9426 Flower Street
Bellflower, California 90706

Richard N. Bray
Department of Biology
California State University
Long Beach, California 90840

Jeffrey Cross
Southern California Coastal Water Research Project
7171 Fenwick Lane
Westminster, California 92683

John K. Dawson
Institute for Marine and Coastal Studies
University of Southern California
Los Angeles, California 90089

John Dixon
Ecosystems Management
2270 Camino Vida Roble
Carlsbad, CA 92009

Robert P. Eganhouse
U.S. Geological Survey
National Center MS 431
Reston, Virginia 22092

David W. Fischer
Graduate Center for Public Policy and Administration
California State University
Long Beach, California 90840

Gill G. Geesey
Montana State University
Engineering Research Center
409 Cobleigh Hall
Bozeman, Montana 59717

Donn S. Gorsline
John and Doris Zinsmeyer Chair in Marine Sciences
University of Southern California
Los Angeles, CA 90089

Jack Hardy
Huxley College of Environmental Studies, ES-539
Western Washington University
Bellingham, Washington 98225

Barbara M. Hickey
School of Oceanography
University of Washington
Seattle, Washington 98195

Donald W. Hood
P.O. Box 57
Friday Harbor, Washington 98250

Steven N. Murray
Department of Biological Sciences
California State University
Fullerton, California 92634

Richard E. Pieper
Institute for Marine and Coastal Studies
University of Southern California
Los Angeles, California 90089

Elbert W. Segelhorst
2828 East 1st Street
Long Beach, California 90803

Steven Schroeter
Ecosystems Management
2270 Camino Vida Roble
Carlsbad, CA 92009

Robert B. Spies
Environmental Sciences Division
Lawrence Livermore National Laboratory
Livermore, California 94550

Bruce Thompson
Aquatic Habitat Institute
180 Richmond Field Station
1301 S. 46th Street
Richmond, California 94804

Indira Venkatesan
Institute of Geophysics
University of California
Los Angeles, California 90024

Chapter 1

The Southern California Bight: Background and Setting

Murray D. Dailey, Jack W. Anderson, Donald J. Reish, Donn S. Gorsline

Introduction

This work represents a major new effort by experts in many disciplines to contribute to a better understanding of the Southern California Bight (SCB) ecosystem. No other book has attempted to encompass the available literature on the SCB in all the areas of expertise represented in this volume. Toward this goal, each chapter in the work stands alone as a thorough and valuable update on pertinent findings in the field and as a major contribution toward our understanding of the SCB ecosystem.

This book endeavors to provide marine scientists of all disciplines a basic review of recent information gathered in their fields of study from the SCB. Furthermore, the authors of each chapter were challenged to furnish a combined summary and prospectus that serves two valuable purposes. First, it identifies study areas within the field that require further investigation, thus leading the next generation of researchers toward productive and essential study topics. Second, in the final chapter, it supplies the reader with the information needed to understand the synthesis of ecosystem interactions at work in the SCB. We also try to present the material so that those responsible for environmental management within the region will find the information to be a useful tool in future decision making concerning growth and development of the SCB.

The SCB is one of the most studied areas

of the United States. Although the name *Southern California Bight* is a regional name that has not been defined in geologic terms (*bight* is defined as a bend or curve in the coastline), it is used in this book to describe the southern California continental borderland. The SCB includes an area of about 78,000 km^2 of the California borderland. It encompasses that body of water stretching from Point Conception, north of the Santa Barbara Channel, to a point just south of the border between the United States and Mexico (fig. 1.1). The SCB measures about 1000 km in length and has a maximum width from shore to the base of the Patton Escarpment of about 300 km (fig. 1.2). The basin floor depths range from 600 m to over 3000 m. Basin sills deepen progressively south and west to the area of the Santo Thomas Fault, where they then shoal to the south (Doyle and Gorsline 1977).

Exploration and Early Human Inhabitants of the SCB

The body of water making up the SCB, the Pacific Ocean, was first seen by Balboa in 1513. He defined the shoreline for what was the most heavily populated part of the North American continent at that time: California. During that period, the early 1500s, an estimated 700,000 Indians were thought to be living within the boundaries of the present state. Many of the native southern Californians were organized as loose family units, not as tribes. Those inhabiting the Channel Islands and southern coastal areas lived in "wikiups," small grass-covered huts, which were grouped in settlements called "rancheria" by the Spanish. Their diet consisted primarily of seafood, such as fish, abalone, and sea mammals, supplemented with acorn meal and small terrestrial animals such as birds, reptiles, and insects (Narlon 1913).

The last known Indian inhabitant of the Channel Islands was Juana Maria, called the "Lone Woman" of San Nicolas. Reported to have jumped overboard while being evacuated from San Nicolas by priests in 1835, she was discovered by otter hunters nearly 20 years later (1853). She was taken to Santa Barbara where she died a few weeks after her rescue (Smull 1989).

Early Spanish Explorers

The first European explorers to set foot in what is now southern California were a small company of Spanish adventurers commanded by Juan Rodríguez Cabrillo. Cabrillo and his men landed at what is now San Diego Bay on September 28, 1542 (Hartman 1968). When Cabrillo first encountered the San Pedro Bay coastline a week later on October 8, 1542, it offered little or no positive features potential for future use. In the words of Richard Henry Dana, who visited the same area nearly 300 years later (in 1835),

> The land was of a clayey quality and as far as the eye could reach, entirely bare of trees, not even a house to be seen. What brought us to such a place one could not conceive. We lay exposed to every wind that could blow, except the northerly ones. We found at the low tide rocks and stones, covered with kelp and seaweed, lying bare for the distance of nearly an eighth of a mile. We all agreed that it was the worst place of all we had seen. (Queenan 1983).

How could Cabrillo or Dana have imagined that one day this desolate site would become one of the largest and most successful manmade harbors in the world?

It was not until May 1602 that Don Sebastián Vizcaíno set sail from Acapulco to search the coast of California for suitable harbors in support of Spanish vessels trading with the Philippines (the Manila Galleon trade). However, Vizcaíno's voyage was of more permanent value than Cabrillo's journey because he kept a careful and detailed record of the voyage. He touched at San Diego, Avalon, San Pedro, and Monterey. Don Vizcaíno named the Coronado Islands as well as Santa Catalina and San Clemente (Guinn 1902).

More than 160 years passed before the next recorded vessel arrived at San Diego Bay. In

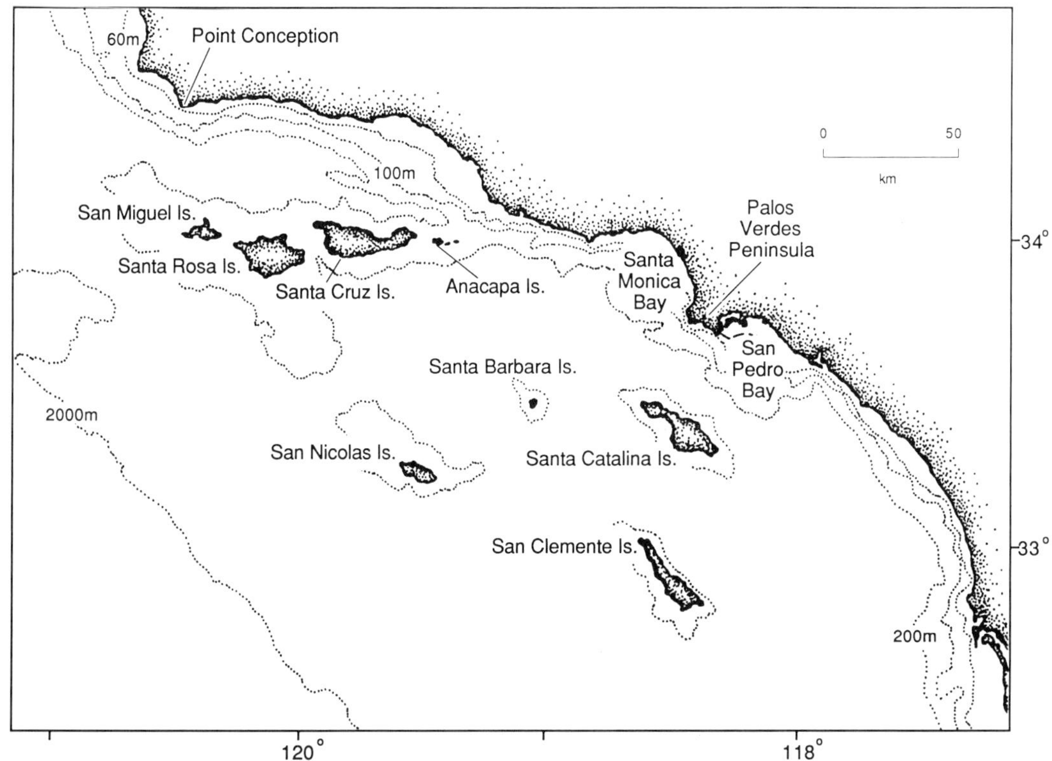

Figure 1.1. Index map of the Southern California Bight.

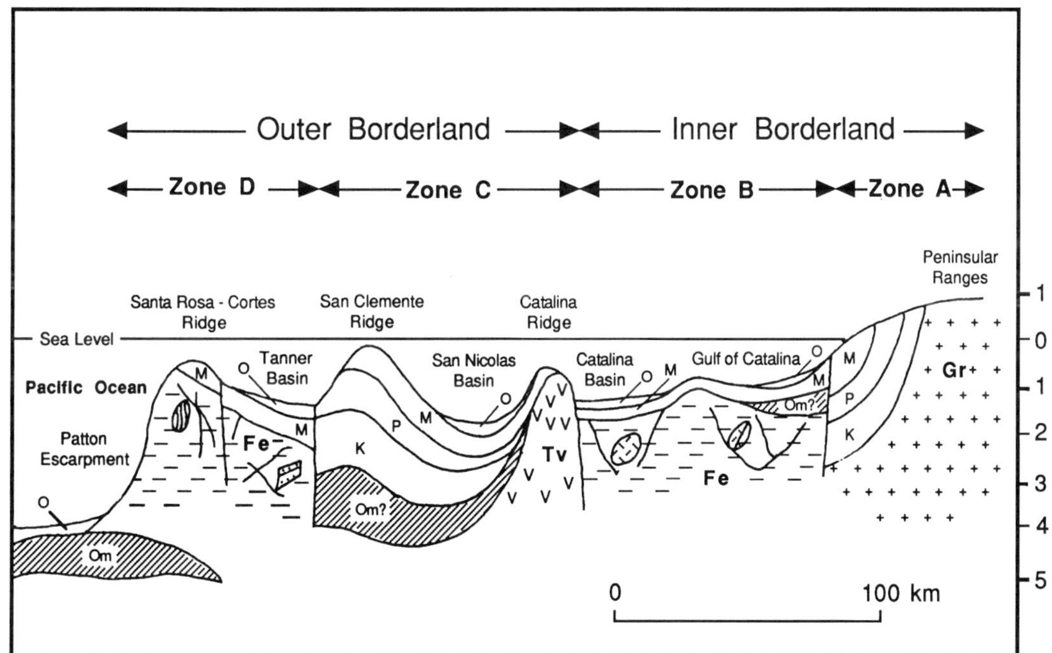

Figure 1.2. Geologic cross-section of the southern California borderland from the mainland shore to the Patton Escarpment. Numbers on right-hand side indicate 600-m intervals.

1769 the *San Antonio,* under the command of Juan Pérez, brought the first European settlers who came to make a permanent home in the SCB area. This was the first of a threefold "occupation" to be carried out by the Spanish government. This occupation incorporated a religious, military, and civil approach that was intended to stem the Russian occupation of Alta California, then encroaching from the north. The execution of this plan was entrusted to José de Galvez, the Royal Visitador of Mexico. The plan called for four military divisions, two to arrive by land and two by sea. Gaspar de Portolá was placed in command as military and civil governor, and Junípero Serra, as Father-President of the Franciscans, would oversee the religious mission (Lavender 1976). On Sunday, July 16, 1769, the first mission in California was founded by Father Serra in what is now known as Old Town in San Diego. Here Mission San Diego de Alcalá was formally dedicated.

During Serra's lifetime, there would be 9 missions established. By the end of the eighteenth century, there would be 18, and with the founding of San Francisco Solano in 1823, the total number reached 21. All 21 missions and their surrounding lands had been established in the southern California area under Spanish rule without expense to the royal treasury. The entire cost had been borne by private parties through what became known as the "Pious Fund." The Pious Fund consisted of money and property given by devout Catholics to the cause of proselytizing the California Indians. By 1768, the fund had reached over $1,273,000 and yielded an income of $50,000 per year to be invested almost entirely in land (Hartman 1968).

Spanish Colonization

At this time, the late 1700s, Spain was attempting to colonize the area bordering what is presently considered to be the SCB. Eleven families recruited from the Mexican provinces of Sonora and Sinaloa followed the trail northward for seven months to become the first settlers in El Pueblo de Nuestra Señora La Reina de Los Angeles de Porciúncula. The new town was founded on September 4, 1781, at a spot near the Indian village of Yang-na, later to become known as Pueblo de Los Angeles (Hartman 1968).

In 1784, the first of the vast California ranchos was established in the area around San Pedro Bay by Juan José Domínguez, a 65-year-old bachelor and veteran of the Portolá expedition. For his devoted service to Spain, Domínguez received a land grant of 74,000 acres, extending from what is now Redondo Beach south to include the entire Palos Verdes Peninsula and some distance eastward. Known as Rancho San Pedro, the original grant encompassed present-day Palos Verdes Estates, Rancho Palos Verdes, Rolling Hills Estates, San Pedro, Torrance, Gardena, Compton, Redondo Beach, Wilmington, Lomita, Harbor City, and Carson.

Other large land grants were doled out in the vicinity of the new pueblo of Los Angeles. Among the recipients was José María Verdugo, who received a large grassland area that covered the present-day city of Glendale and part of Burbank. Another huge grant, adjoining Rancho San Pedro and including present-day Long Beach and other nearby communities, went to Manuel Pérez Nieto. The founding fathers of Los Angeles were given title to their original holdings on the town's fifth anniversary in 1786. Along with the rancheros (those given land grants), these were the first private landholders in the pioneer province (Queenan 1983).

On May 13, 1846, the United States declared war on Mexico and moved American naval units into every port in California. Not long afterward, an uprising took place in Los Angeles when the pueblo inhabitants *pobladores* grew tired of petty tyrannies imposed by occupation leader Archibald Gillespie. The *pobladores* chased Gillespie and his soldiers out of Los Angeles and caused the resistance to spread throughout southern California. The resistance ended on December 6, 1846, near the Indian town of San Pasqual. By 1847 the *pobladores* had capitulated to the United States and the war in the West was over, although it would be another year before Mexico yielded and signed the peace treaty of Guadalupe Hidalgo (Hartman 1968).

Present Geologic, Climatic, and Oceanographic Setting

Geology

The Pacific margin along the western United States is a plate collision boundary that includes the typical narrow shelf (average width 25 km) fronting high-relief, coastal ranges. The shelf passes seaward to a steep slope and marginal trench. This morphology is characteristic of most of the Pacific rim except off southern California and northern Baja California, where continued large-scale overriding of the North American plate by the Pacific plate has produced movements along a major fault zone (San Andreas Fault System) (Teng 1985). The resulting Pacific margin is wide (up to 300 km) and is composed of a series of laterally shifted blocks that produce a roughly checkerboard pattern (Howell et al. 1980) (fig. 1.3).

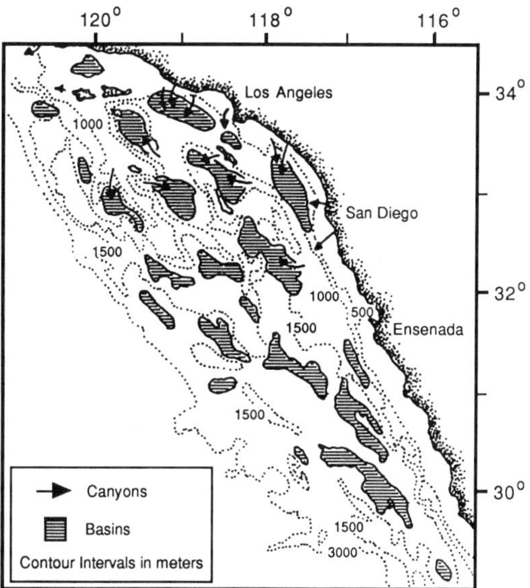

Figure 1.3. Submarine canyons and basins of the SCB and southward.

This checkerboard pattern is formed by basins that are arranged in rough rows trending northwest-southeast and converging to the south. Off California, the depressions are grouped as inner, central, and outer basins relative to the mainland. For the entire province, 23 depressions have been named (Emery 1960), 16 of which are located in the U.S. portion of the borderland. The actual continental slope at the seaward margin is the Patton Escarpment (Uchupi and Emery 1963).

Rivers along the Pacific coast of the United States typically drain tributary basins that are steep in gradient, are small in area, and produce a large amount of sand discharge (table 1.1). Southern California is noted for its mild temperatures, short wet winters, and long dry summers. There are only relatively small changes in these conditions between Santa Barbara to the north and San Diego to the south. Mean rainfall at Santa Barbara and Los Angeles is about 38.1 cm per year, while San Diego has a lower mean of 25.4. Mean annual air temperatures are also similar from north to south. They vary only from 17.8°C in Los Angeles to 15°C and 16.7°C in Santa Barbara and San Diego, respectively. The largest recorded river discharges occur at about 20- to 30-year intervals. The major southern California sediment discharge is delivered by the Santa Clara River in Ventura County, which has its sources in the San Gabriel Mountains at elevations of 2000 m (table 1.1).

This discharge is first sorted by wave action at the coast into coarser particles, usually sands and gravels, which move in traction or in short-term, near-bottom suspension. The coarse fraction travels along the shore within the beach and inshore zone, and offshore to the inner and central shelf at times of strong storm surging. Where submarine canyons cut into the nearshore, as at the ends of each coastal cell, they intercept much of this transport (Inman and Frautschy 1966). Silts and clays are transported as suspended load and

Table 1.1. *Discharges for Southern California and Baja California Streams*[a]

Drainage	Sediment Discharge ($\times 10^6$ t yr^{-1})	Suspension Discharge ($\times 10^6$ t yr^{-1})
Santa Ynez Mountains	0.70	0.52
Ventura River	0.93	0.58
Ventura area	0.02	0.01
Santa Clara River	3.72	2.44
Oxnard area	0.01	0.01
Calleguas Creek	0.26	0.16
Santa Monica Mountains	0.33	0.24
Los Angeles River	1.20	0.62
Long Beach area	0.01	0.01
San Gabriel River	1.20	0.62
Oceanside–San Diego area	3.80	?
Baja Coastal streams	1.00 ?	?
Total natural discharge	13.18	5–6 ?

[a] Average annual rates in metric tons per year (Brownlie and Taylor 1981; Schwalbach and Gorsline 1985). Estimates for southern streams are subject to error because of limited data on discharge. Natural discharge is based on estimates of controlled drainage of present streams and corrections for delivery from those areas in a natural, uncontrolled state (Brownlie and Taylor 1981).

follow the water circulation during their slow fall. The influx of fine sediment particulates is generally low except for times of winter runoff and the multi-year cycle of major flooding. Therefore, during much of the year and during dry years, the predominant suspension particulates are of biological origin (biogenic).

Biogenic components undergo extensive recycling before they reach the ocean bottom. Much of this material is probably aggregated in the form of pellets and aggregates of planktonic origin and may be degraded by bacterial action as they sink. Additional aggregation occurs from ingestion by benthic organisms and infaunal reworking. This component reaches the bottom principally by particle or aggregate settling. The process is continuous, but it occurs at varying rates related to the cycle of seasonal blooms. Biogenic particulates contribute about 20% of the total borderland sediment, which also includes carbonate, opaline silica, and other organic matter. During floods or wet seasons, the flux of terrigenous material dominates the deposited sediments.

We can simplify the description of the physical characteristics of the submarine canyons by considering them in relationship to the three environments that they cross: shelf, slope, and basin (trough floor). Emery (1960) states that there are 13 large named canyons and 19 smaller unnamed canyons in southern California. Of this total of 32 submarine canyons, 20 border the mainland, 10 border islands, and 2 are located off submarine banks. The canyons exert an influence on shelf water circulation because of the pumping of water by tide-driven flows up and down the canyon axis. This action draws some of the suspension load to the canyon circulation systems (Drake and Gorsline 1973; Shepard et al. 1979). Much also passes over the shelves in complex circulations and is ultimately concentrated in nepheloid plumes (Karl 1976). These are found in surface waters, in the water column, and as near-bottom turbid layers.

Sandy sediments initially deposited in nearshore canyon heads are progressively transferred downslope by mass movement processes and sediment gravity flows (Nardin et al. 1979). Fine sediments also initially accumulate in canyon walls and deeper canyon floors, where they are then incorporated and carried out of the canyons to submarine fans and basin floors (Shepard and Dill 1966). This process can be seen in the contours of the seafloor of the SCB (fig. 1.4).

The surface water circulation of the SCB tends to move fine suspended sediment into the Santa Barbara Basin from the California Current system to the west and through the Anacapa Passage from the southeast (Thornton 1981a, b). These conditions produce high rates of fine clay–silt sedimentation in Santa Barbara Basin. As clay content increases, organic carbon content increases. Oxygen demand for the decay of this material utilizes oxygen faster than the rate of recharge. This high-oxygen demand is superimposed on already low-oxygen content bottom waters that enter the basin over the western sill at about the depth of the core of the Pacific intermediate water (Reid 1965). This low-oxygen deep basin water can rapidly become dysaerobic (less than 0.3 ml l^{-1} dissolved oxygen), and where demand is increased, it can become anaerobic or anoxic (oxygen absent or below 0.1 ml l^{-1}). This factor is the major control on the basin floor benthic communities. In true anoxic environments, only anaerobic bacteria can flourish.

Since the central basin bottom waters of Santa Barbara Basin are presently anoxic, the seasonal sediment arrangements are preserved in the fine deposits. As shown by Emery and Hulsemann (1962) and Soutar and Crill (1977), these sediments preserve detailed annual records of biological and climatic events. These anoxic environments are widely viewed as the principal source conditions for petroleum generation (DeMaison and Moore 1980). Similar environments in the geologic past have produced the black shale facies found extensively in both deep ocean floor and continental stratigraphies.

In the recent past (about 15,000 years ago), other borderland basins have become anoxic. This has occurred during low sea level stands, which are typically periods of increased sedimentation rates and faster ocean circulation (and upwelling). These basins include Santa Monica and San Pedro basins, parts of Santa Catalina Basin, and possibly the San Diego Trough. In the past 200 years, the low-oxygen bottom waters of Santa Monica Basin have spread, possibly because of increasing anthropogenic influxes, and have formed a near-anoxic nonbioturbated bottom sediment layer. San Pedro Basin is presently dysaerobic, approaching anaerobic, and all other basins are in the low aerobic or dysaerobic state. These conditions permit bioturbation of bottom sediments to occur (Savrda et al. 1984).

Winds

A subtropical high-pressure system stationed offshore of the SCB produces a net weak southerly and onshore flow within the area (Dorman 1982). In general, the wind speeds can be classified as moderate-typical in the offshore region and are on the order of 10 km h^{-1}. The strength of the winds diminishes with proximity to the coast, averaging about half the speeds found offshore. Coastal wind speeds are approximately one-half those found off central and northern California (Hickey 1979). However, strong winds may occasionally accompany the passage of winter storm systems or, more rarely, the northward penetration of a tropical storm into the southern region of the SCB. Within 10–40 km of the coast, the diurnal land breeze becomes increasingly important, particularly during the summer, when a thermal low forms over the deserts to the east of the SCB.

On occasion, a high-pressure area can develop over the Great Basin area (the flat desert area to the east of the SCB), reversing the surface pressure gradient and generat-

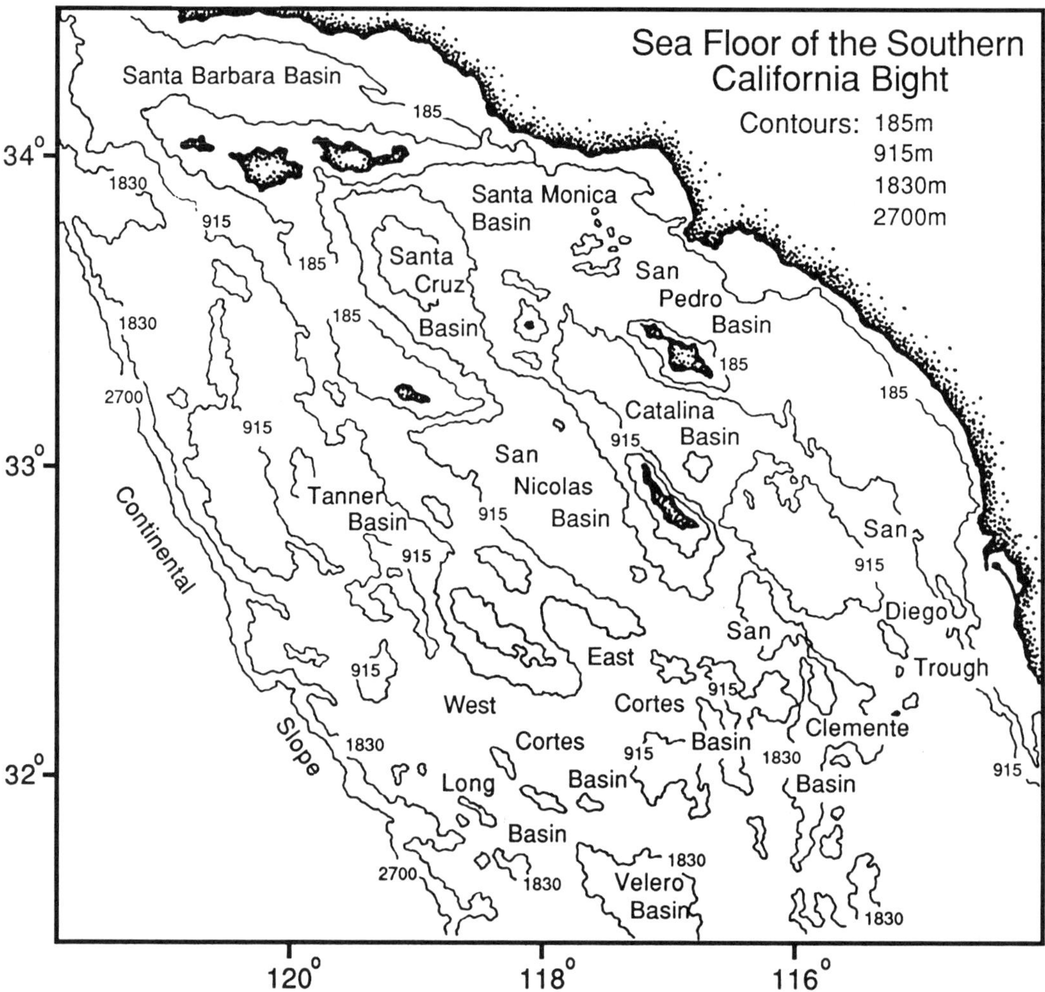

Figure 1.4. Seafloor of the SCB showing named submarine basins.

ing strong, dry, gusty offshore winds in the coastal area. These Santa Ana winds, as they are called, are most common in the late summer, but can occur during any time of the year.

The coastal mountain ranges and islands modify the strength and direction of the winds on a local and regional scale. San Miguel Island, for example, frequently blocks some of the wind flowing southeastward from Point Arguello, producing a zone of strong onshore flow along the north side of the northern Channel Islands. Strong Santa Ana winds can also produce strong flows below coastal canyons and valleys.

Currents and Water Properties

The primary surface current in this region is the California Current. This current flows southeastward off the central California coast with a maximum speed of about 10–15 cm s^{-1} (Pavlova 1966; Hickey 1979). Although the current is relatively slow moving, its broad width (approximately 300 km) results in a volume transport on the order of 1×10^7 m^3 s^{-1}. This rate, however, varies from year to year, with profound consequences for the properties of the water and the abundance and composition of the biota within the SCB.

Near Point Arguello and Point Concep-

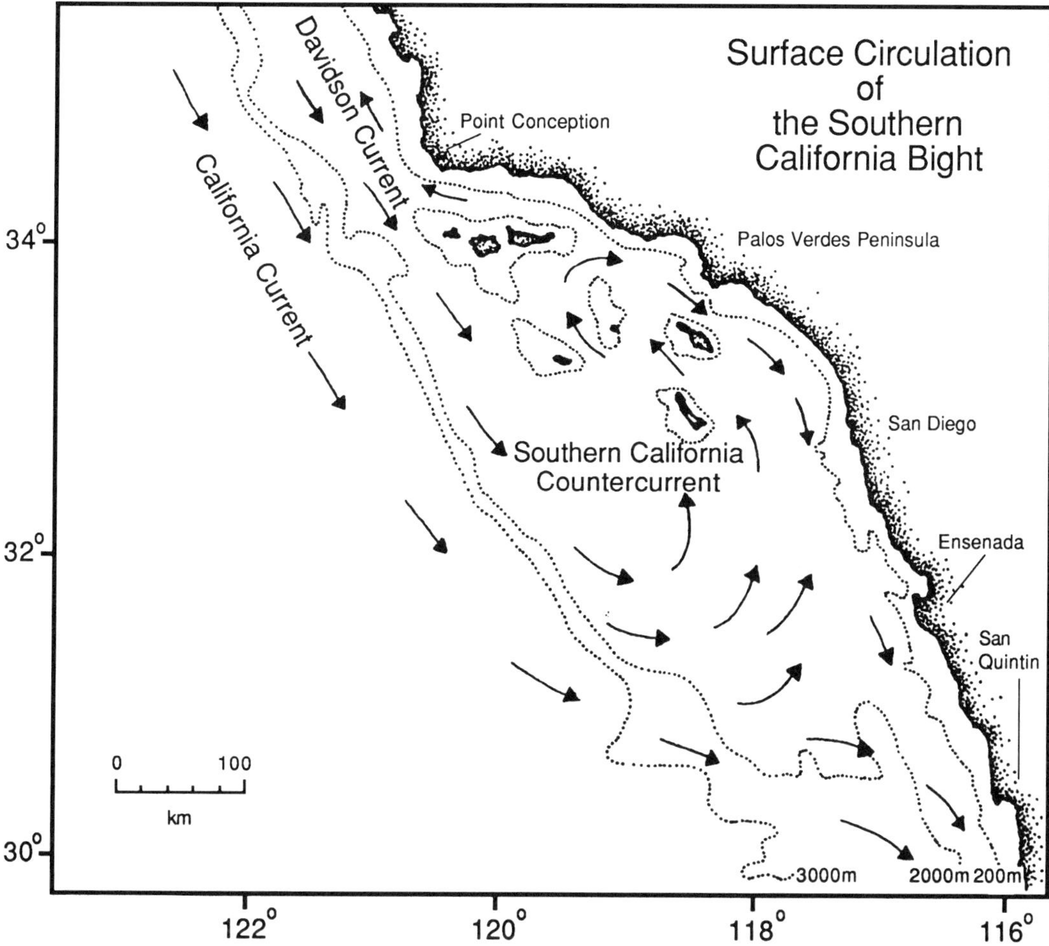

Figure 1.5. Surface circulation of the California Current and California Countercurrent in the SCB.

tion, the coast bends sharply to the east, forming the SCB. The outer edge of the continental borderland, however, continues southeastward before bending to the east near the border with Mexico. The inner edge of the California Current tends to flow along the edge of the borderland (except during spring), swinging inward toward the coast between San Diego, California, and Punta Colnett, Mexico, which is 210 km down the coast in Baja California (fig. 1.5).

In the central portion of the borderland, the average surface flow is northerly up the coast and is called the Southern California Countercurrent. This flow up the coast is substantially blocked by the northern Channel Islands, and the bulk of it is diverted to the west where it merges with the California Current (see fig. 1.5). This results in the formation of a counterclockwise-rotating gyre within the SCB (except during the spring). The remaining portion of the countercurrent flows into the Santa Barbara Channel. Large variations are observed in the strength of these currents, with time scales ranging from hours to months. Hickey (1992) has suggested that, on occasion, such as during extreme fluctuations, the entire SCB might be flushed within a few weeks.

The net flow beneath the California Current and the Southern California Countercurrent is poleward and is called the California

Undercurrent. Typical speeds are on the order of 10–20 cm s^{-1}, with maximum speeds occurring over slope areas (Hickey 1992). As for the surface currents, substantial variations in the undercurrent, frequently in excess of the net flow, occur over time scales of hours to weeks.

Flow over the shelves is generally weaker than over the basins and slopes. Also, equatorward mean flow is more common in the SCB than poleward flow, especially in the very near surface layers. Although shelf flows may be driven in part by the local wind field, they may also be related to larger scale basin flows.

The properties of the water in the SCB represent a mixture of subarctic water carried equatorward into the region by the California Current and equatorial waters carried poleward by the California Undercurrent. Differences between the two water masses have been used to estimate the relative percentages of each type of water within the SCB. For comparable densities, the equatorial waters have higher temperatures, higher salinities, higher phosphate concentrations, and lower dissolved oxygen. Between 200 m and 400 m deep, more than half of the water inshore of the Cortes Ridge is of southern origin (60–80%); offshore from the ridge, only about 20 to 30% of the water is of equatorial origin (Sverdrup et al. 1942).

The upper layer of the water column (40 m deep inshore and 100 m deep offshore) is relatively isosaline. Nearshore, salinities peak in July at about 33.6 ppt, decreasing to between 33.4 and 33.5 ppt in the late winter and early spring. Near the Cortes Ridge, salinities are 33.3 to 33.4 ppt. The upper portion of the water column is strongly thermally stratified from about May to October, reaching an average maximum surface temperature of approximately 19°C from July to September. During this period, the temperature changes by about 5°C over the upper 20 m of the water column. The minimum temperature (approximately 14.5°C) occurs in the late winter. Average concentrations of dissolved oxygen at the surface range from about 5.5 to 6.0 ml l^{-1} nearshore and slightly less at the outer edge of the SCB. Close to shore there are upward shifts in the isopleths of temperature, salinity, density, and dissolved oxygen for the period from April through July. These displacements occur to a depth of at least 400 m, average about 50–100 m, and are associated with wind-driven coastal upwelling. A similar displacement occurs offshore later in the year (June–August) and is associated with the large-scale California Current. Localized nearshore upwelling can also occur throughout the year in the vicinity of bathymetric features such as points of land.

El Niño

The phenomenon known as El Niño occurs yearly off the coast of Peru. The term *El Niño*, or "The Child," was apparently coined by the Peruvian fishermen because of its appearance around Christmas time. During the summer months, the Peruvian Current extends northward into the Northern Hemisphere and the Equatorial Countercurrent is displaced to the south. These waters are warmer and lower in salinity, and this current converges with the Peruvian Current. The combined currents then move southward, and the peak of this yearly event, the El Niño, is reached in February and March.

In some years, physical events occur that cause the Equatorial Countercurrent to extend farther south into the Southern Hemisphere. This change from the typical pattern can result in profound biological changes. Generally, the northwest winds of the South Pacific Ocean induce strong upwelling water movements along the coasts of Peru and Chile. However, in years of severe El Niño, physical changes occur that cause modifications in the environment. These include (1) a decrease in the strength of the upper westerly winds in the Pacific, (2) a difference in sea level between the two sides of the Pacific Ocean at the equator, and (3) differences in the transport of the Equatorial Countercurrent (Cushing and Dickson 1976).

These physical changes result in many alterations in the biological environment off the Peruvian coast. First, a decrease in phytoplankton productivity occurs, especially diatoms. Second, the anchovy *Engraulis ringens* moves into deeper, colder waters or migrates into Chilean waters. Third, a decrease in the recruitment of the anchovies results from either the failure to reach sexual maturity or the lack of sufficient food for the larvae. Fourth, whole populations of sea birds migrate to Chilean waters, especially the Peruvian cormorant (*Phalacrocorax bougainvillii*), the brown pelican (*Pelecanus occidentalis thagus*), and the Peruvian booby (*Sula nebouxi*). Finally, fish and other marine oganisms are killed as a result of the warmer waters (7°C higher) (Cushing and Dickson 1976; Schott 1913).

The decomposition of dead organisms leads to build-up of hydrogen sulfide, which can blacken the paint of boats. Peruvian fishermen call this discoloration of their boats "Aguaje" or "Callao Painter" (Sverdrup et al. 1942).

But how does the Peruvian El Niño affect the SCB? In some years, the south-moving California Current is weakened, resulting in an anomalous northern poleward movement of water. As a result of this change in current strength, some of the warmer Equatorial Countercurrent flows northward into the North Pacific Ocean. The marine waters off the SCB are elevated several degrees above normal. This condition in the Northern Hemisphere, also referred to as El Niño, does not occur yearly. It has been reported in the literature in 1911–1912, 1917, 1925, 1932, 1939–1942, 1953, 1957–1958, 1965, 1976–1977, and 1982–1983 (Vildosa 1974; Kerr 1988). Comparisons of the daily water temperatures recorded in 1983 to the 63-year mean indicated warmer waters than usual in the winter and fall months (Kerr 1988) (fig. 1.6). The 1982–1983 El Niño brought to the SCB not only increases in water temperature, but also severe storms along the coastline that winter, causing damage to marine structures and erosion to the beaches (Kerr 1988).

Quinn (1974) has developed a predictive technique for occurrences of El Niño in the Northern Hemisphere. He took into consideration the strength of the southeastern trade winds, basing his calculations on the differences in the sea level atmospheric pressure between Easter Island and Darwin, Australia. This difference results in an increased transport of warmer water to the east along the equator.

A contrasting phenomenon called La Niña ("The Girl") has been observed. La Niña is characterized by colder waters than normal for the SCB (Kerr 1988). Such events recur about every 4 years in the SCB, but the interval may be as great as 10 years.

The unique coastline of the SCB and the changing strengths of the California Current and the Equatorial Countercurrent result in recurring fluctuations in water temperatures in the area. Elevation of 7° to 10°C results in an alteration in the composition of the flora and fauna in the SCB. Species fluctuations are readily apparent in the plankton and shallow water biota, but it is not known if either the El Niño or La Niña affects the deeper water fauna. The measurable effects of El Niño on specific groups of plants and animals are addressed in later chapters.

Biological Setting

Its many habitats encourage a rich and varied marine life in the SCB. The mainland consists of a series of rocky shores, sandy beaches, and embayments of different types. Dredging and construction of harbors, marinas, jetties, and piers have increased the diversity of habitats. Eight major offshore islands, the Channel Islands, are distributed along the edge of the continental borderland of the SCB and provide additional habitats for marine organisms. They also serve as breeding grounds for marine birds and as

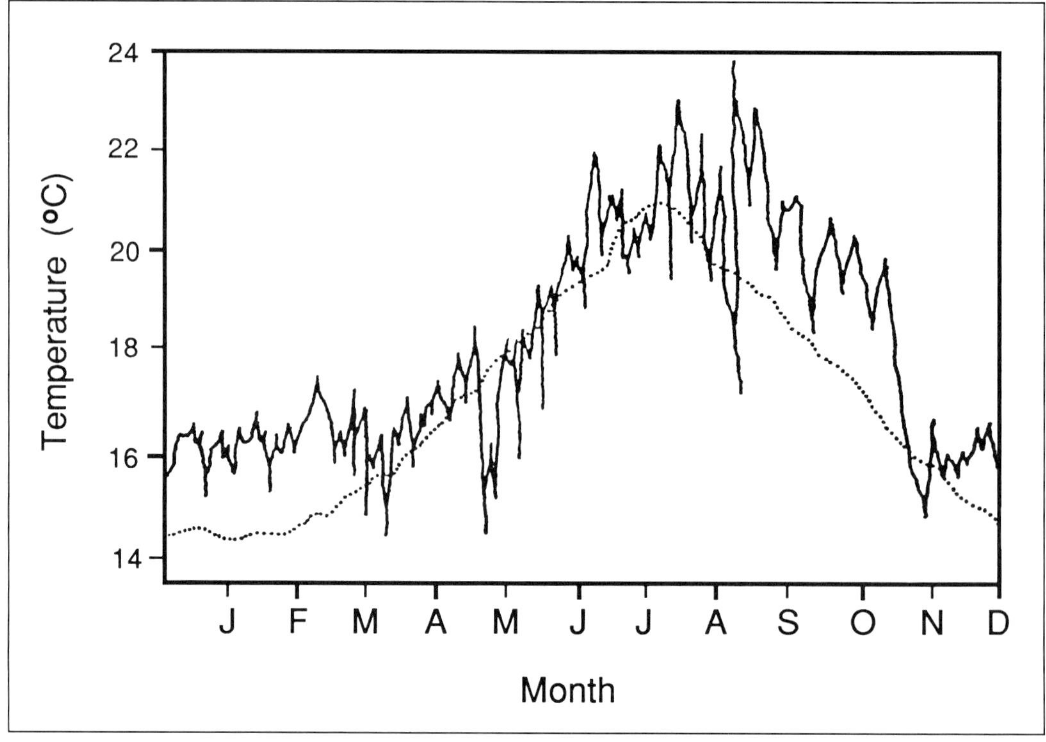

Figure 1.6. Annual water temperature profile for the SCB in 1983 (solid line). The dotted line shows the 63-year mean temperature profile for comparison.

protected shores for marine mammals. Since the Channel Islands are located some distance from the heavily populated mainland of the SCB, they represent the best examples of pristine marine environments in the southern California area. Distributed between the mainland and the Channel Islands (and beyond) are a series of submarine canyons, ridges, basins, and seamounts that provide unique habitats in the SCB.

As previously discussed, the SCB is subject to short-term and long-term temperature fluctuations, depending upon the strengths or weaknesses of the ocean current systems. The interplay of the physiography, current systems, and anthropogenic inputs also influences the richness of the marine life in the SCB.

Primary production in the SCB depends upon the source of the nutrients from storm runoff, aerial fallout, seasonal upwelling, and anthropogenic inputs coupled with long periods of sunshine. Emery (1960) calculated phytoplankton productivity to be about 500 g C dry wt m^{-2} yr^{-1} over the 78,000 km^2 area of the SCB. Productivity measurements over a 10-year period by Eppley and Holm-Hansen (1986) in the southern region of the SCB were 0.39 g C dry wt m^{-2} d^{-1} or 4.3×10^6 t of carbon per year for this 30,000 km^2 area.

Zooplankton displacement volumes were measured from 1949 to 1969 under the auspices of the California Cooperative Oceanic Fisheries Investigations (CalCOFI) project, which encompassed the SCB and adjacent areas to the north, south, and west of the SCB (Bernal 1979). The mean zooplankton volume in the SCB for this period was 243.7 ml/1000 m^3. This figure was about 60% of the volume measured for the area north of Point Conception, but it was greater than that for

Baja California and the open ocean. Monthly means ranged from a low of 101.2 ml m^{-3} in December to a high of 239.0 ml m^{-3} in February. The seasonal highs over the CalCOFI study area varied from year to year as to the month (spring, summer, or fall), but the lowest values were measured in the November–January period (see chap. 6).

Seventy percent of the known algal species from California occur in the SCB (see chap. 7). This high percentage is attributed to the extensive mainland shoreline of diverse habitats and the offshore Channel Islands. In the final analysis, this figure may be high, since many areas of the central California coast are difficult to reach and have not been studied extensively. Kelp beds form a unique shallow water community, which is not only important economically and recreationally but also provides a haven for a complex array of additional algal species, invertebrates, and fish. Productivity of individual algal species ranges from less than 0.1 to 11.2 mg C g^{-1} dry wt h^{-1} with the sheet species (such as *Ulva*) having the highest values and the crustose species (such as *Lithothamnium*) having the lowest values.

Over 5000 species of benthic marine invertebrates exist in the SCB. They inhabit all areas of the sea floor, from the high intertidal splash zone to the bottoms of the offshore basins (over 2500 m deep). The benthic fauna are so diverse because the region is an area of overlap between the northern Oregonian and southern Panamic biotic provinces and because of the wide range of habitats provided in the region (see chap. 8). Benthic invertebrates occur in rather discrete assemblages of species that also differ in diversity and biomass depending on water depth and substrate type. The ecology of shallow subtidal and intertidal assemblages has received considerable investigation, but there has been very little study of deep-water assemblages (see chap. 8).

The SCB supports a diverse and dynamic ichthyofauna. Of the 144 families and 554 species of California coastal marine fishes, 129 families and 481 species occur in the SCB. The SCB is the southern terminus of the ranges of many northern species and the northern terminus of the ranges of many southern species. Point Conception, a widely recognized faunal boundary, is more important as a barrier to southern species than to northern ones. Northern species cross Point Conception by moving into deeper water off southern California and by occupying upwelling areas on the southern side of headlands, especially off Baja California. Northward incursions of tropical fishes into the SCB during abnormally warm water years associated with El Niño demonstrate the dynamic nature of the southern California ichthyofauna. Less well known are the southerly incursions of northern fishes during cool years. Warm water and cool water events in the SCB affect fish recruitment and can alter the composition of fish assemblages for several years thereafter.

The complex bathymetry of the SCB offers a variety of habitats for fishes that live on (or are associated with) the bottom as well as fishes that live in the water column. The basins provide habitats for a significant number of mid-water and benthic deep sea fishes very near the coast. The nearshore waters contain a variety of habitats. Soft substrates, such as bays and estuaries, man-made harbors, exposed sandy beaches, shelves, and slopes, are abundant along the mainland and the offshore islands. Hard substrates, such as the rocky intertidal, shallow subtidal reefs, deep rock reefs, and kelp beds, are common along the mainland and abundant around the offshore islands (see chap. 9).

The SCB is the residence at least part of the year for over 195 species of coastal and offshore birds. Resident species constitute the greatest percentage of the bird population. The large population of birds in the SCB is attributed to the fact that this area is both the northern and southern limit of the breeding range for many species. Mainland wetlands and the Channel Islands are the principal breeding sites in the SCB. The conversion of two-thirds of the wetlands of California for other uses has destroyed valuable habitats, feeding areas, and breeding sites for many species of birds (see chap. 10). The establishment of marine sanctuaries in the SCB in recent years will preserve at least some of these wetlands for birds and other species (see chap. 13).

A total of 39 species of marine mammals have been sighted, some only rarely, in the SCB. Of these, 11 species are year-round residents, 22 are occasional or possible visitors, and 6 are migratory species (see chap. 11). While accurate numbers of marine mammals are difficult to obtain, as many as 300,000 individuals are believed to reside in or pass through the SCB within a year. The large population of mammals in the SCB is in part the result of a rich supply of food resources. Pinnipeds, of which many are year-round residents, are estimated to consume nearly 50% of the food eaten by all marine mammals. The migratory baleen whales account for nearly 20% of the food consumed by this group.

Human Impact and Marine Science

Population Growth

The growth of the SCB port cities has been very rapid. Los Angeles, a hamlet of 4385 people in 1860, possessed none of the natural advantages that enabled San Francisco to grow into one of the nation's major urban centers between 1847 and 1859. The southern city had no navigable rivers for opening easy transportation routes to inland mines or farming areas. Even San Diego's superlative harbor provided its citizens little, for the mountains and deserts behind the bay remained too undeveloped to support commerce.

However, after the completion of the Southern Pacific Railroad from the eastern United States in 1876, the Los Angeles and San Diego area populations grew rapidly. San Diego's population jumped from 5000 in 1884 to an estimated 32,000 in 3 years. Real estate promoters tried to hasten the growth of Los Angeles westward toward Santa Monica rather than southward to San Pedro, but eventually the victory went to San Pedro (fig. 1.7). In 1906, Los Angeles annexed a 20-mile strip of land that directly connected the city to the port. This allowed the municipal government, working with neighboring Long Beach, to turn their man-made harbor into one of the world's busiest ports. Within a few decades, the Los Angeles–Long Beach Harbor was able to surpass monetarily the San Francisco Harbor (Queenan 1983).

The growth of Santa Barbara, Ventura, Los Angeles, Orange, and San Diego counties ranked the area number two in growth in the nation during the 1950s and 1960s and number one during the 1970s and 1980s (U.S. Department of Commerce 1988). The combined population of the area increased from 6 million to 12 million during the 20-year period between 1960 and 1980. Between 1980 and 1985, the area grew by another 1.2 million people. This represents an increase of more than 100% during those 25 years. In the Los Angeles–Long Beach area, there are 1837 people per square mile of land (U.S. Department of Commerce 1988). According to a recent government report, it is projected that more than 300,000 new homes will be needed in Los Angeles County over the next 22 years to house the growing population (Southern California Association of Governments 1988). This prediction is based on a forecast that 5.5 million more people will live in southern California counties by the year 2010. This continued growth of an already impacted area will continue to exacerbate the already delicate balance found in the ecology of the SCB offshore and tideland waters.

Development of Marine Science

The history of oceanographic research in the SCB can be divided into three general peri-

Figure 1.7. Diagrammatic representation of early San Pedro Harbor. (Drawing courtesy of Los Angeles Maritime Museum, San Pedro.)

ods: the exploratory, descriptive, and experimental–analytical. These are not precise periods of time because exploratory oceanographic research is still being conducted today. Rather, it indicates the developmental process that has taken place in oceanography, which is characteristic of scientific research in general.

Early oceanographic explorations in the North Pacific Ocean have been summarized by Hood (1986), beginning with the voyage of the *St. Peter* in 1741. Captain George Vancouver is credited with the first known oceanographic work in 1793 during his first visit to the area. Between Vancouver's visit and California statehood in 1850, virtually nothing was added to the existing oceanographic information. These early expeditions were concerned with physical measurements, such as water temperature and salinity, which assisted scientists in the initial mapping of ocean currents. Biological investigations focused on vertebrate species that had potential for economic importance and could be easily exploited.

The descriptive phase of marine biology began elsewhere in the late eighteenth century, but did not begin in the SCB until in 1904 when subtidal collections were made from the U.S. Fish Commissioner's steamer, the *Albatross*. In the early part of the twentieth century, faculty and students from universities in the San Francisco Bay area began studying marine life in southern California. W. E. Ritter (who played a significant role in the formation of Scripps Institution of Oceanography) and others conducted investigations and instruction at several locales in southern California before establishing a permanent station in La Jolla. W. A. Setchell and N. L. Gardner published extensively on the algae of southern California and elsewhere.

D. S. Jordan and C. H. Gilbert contributed in a similar way to the knowledge of the fishes of the SCB. While much of the intertidal marine life of the SCB has been described, subtidal collections made under the auspices of the Bureau of Land Management (now the Minerals Management Service) and other public agencies concerned with the environment of the SCB have yielded hundreds of species of organisms that remain undescribed. The purely descriptive phase of marine biology in the SCB will continue well into the next century.

As the frontier of marine science advanced from the descriptive phase to a more experimental–analytical type of research, questions were directed to answer both academic and practical problems. The California Cooperative Oceanic Fisheries Investigations (CalCOFI) was established to investigate fishery productivity in the SCB and nearby areas. As part of its charge, CalCOFI has amassed one of the world's most extensive databases on the occurrence, distribution, and productivity of planktonic communities.

The rapid population growth of southern California has created many environmental problems. Since the ocean is the only accessible receptacle for municipal waste water, the sanitation districts in the SCB are required to monitor both their discharges and the receiving oceanic waters. The staffs of these agencies, with the assistance of personnel of the Southern California Coastal Water Research Project, have been collecting and analyzing vast amounts of chemical, physical, and biological data since the 1970s. Not only do these data assist public officials in evaluating environmental conditions, they also provide basic scientific information on the spatial and temporal variations of contaminants and organisms in the SCB.

Seashore-based laboratories make it possible to pursue studies that are otherwise difficult or impossible to conduct in a university setting, especially with living organisms. Scripps Institution of Oceanography was the first such station to be established in the SCB and is now recognized as one of the paramount oceanographic centers in the world. W. E. Ritter, a professor at the University of California, Berkeley, who had conducted summer marine biological programs in southern California (as noted earlier), became acquainted with E. W. and Ellen Scripps in 1903. With their financial backing, and that of others, the forerunner of the present Scripps was founded in La Jolla (Raitt and Moulton 1967). Later, two marine biological stations, neither in existence today, were administered by Pomona College in Laguna Beach and by the University of Southern California at Venice. The California Institute of Technology developed a small laboratory, initially under the direction of G. E. MacGinitie, near the mouth of Newport Bay.

More recently, the University of Southern California built an instructional and research facility at Santa Catalina Island and a research laboratory in Los Angeles Harbor. The California State University system and Occidental College currently operate marine research vessels for the collection of samples and basic instruction in the SCB. Many universities without shoreside facilities transport sea water to their campuses to provide limited opportunities to conduct marine research.

Literature Cited

Bernal, P. A., 1979. Large-scale biological events in the California Current. *Calif. Coop. Oceanic Fish. Invest. Rep.* 20:89–101.

Brownlie, W. R., and B. D. Taylor, 1981. Sediment management for southern California mountains, coastal plains and shoreline: Part C—Coastal sediment delivery by major rivers in southern California. *EQL Rep. 17C.* Prep. by California Institute of Technology, Pasadena, CA. 314pp.

Cushing, D. H., and E. R. Dickson, 1976. The biological response in the sea to climatic changes. In: F. S. Russell and M. Yonge, eds. *Advances in Marine Biology*. Academic Press, London. pp. 1–122.

DeMaison, G. J., and G. T. Moore, 1980. Anoxic

environments and oil source by genesis. *Am. Assoc. Pet. Geol. Bull.* 64:1179–1209.

Dorman, C. E., 1982. Winds between San Diego and San Clemente Island. *J. Geophys. Res.* 87(C12):9636–9646.

Doyle, L. J., and D. S. Gorsline, 1977. Marine geology of Baja California Borderland, Mexico. *Am. Assoc. Pet. Geol. Bull.* 61:903–917.

Drake, D. E., and D. S. Gorsline, 1973. Distribution and transport of suspended particulate matter in Hueneme, Redondo, Newport, and La Jolla submarine canyons, California. *Geol. Soc. Am. Bull.* 84:3949–3968.

Emery, K. O., 1960. *The Sea Off Southern California.* John Wiley & Sons, New York. 366pp.

Emery, K. O., and J. Hulsemann, 1962. The relationships of sediments, life, and water in a marine basin. *Deep-Sea Res.* 8:165–180.

Eppley, R. W., and O. Holm-Hansen, 1986. Primary production in the Southern California Bight. In: R. W. Eppley, ed. *Lecture Notes on Coastal and Estuarine Studies, Vol. 15. Plankton Dynamics of the Southern California Bight.* Springer–Verlag, Berlin. pp. 176–215.

Guinn, D. M., 1902. *Historical and Biographical Record of Southern California.* Chapman Publishing Co., Chicago. 295pp.

Hartman, D. N., 1968. *California and Man.* W. C. Brown Co., Dubuque, IA. 478pp.

Hickey, B. M., 1979. The California Current system—Hypotheses and facts. *Prog. Oceanogr.* 8(4):191–279.

Hickey, B. M., 1992. Circulation over the Santa Monica–San Pedro basin and shelf. *Prog. Oceanogr.* 30:37–115.

Hood, D. W., 1986. Physical setting and scientific history. In: D. W. Hood and S. T. Zimmerman, eds. *The Gulf of Alaska, Physical Environment and Biological Resources.* U.S. Department of Interior, Minerals Management Service, Washington, D.C. pp. 5–27.

Howell, D. G., J. K. Crouch, H. G. Greene, D. S. McCulloch, and I. G. Vedder, 1980. Basin development along the late Mesozoic and Cenozoic California margin: A plate tectonic margin of subduction, oblique subduction, and transform tectonics. *Int. Assoc. Sedimentol. Spec. Publ.* 4:43–62.

Inman, D. L., and J. D. Frautschy, 1966. Littoral processes and the development of shoreline. In: *Coastal Engineering Spec. Conf. Notes.* pp. 511–536.

Karl, H. A., 1976. Agents of sediment dispersal, San Pedro continental shelf, southern California. *EOS, Trans., Am. Geophys. Union.* 57(3): 150.

Kerr, R. A., 1988. La Niña's big chill replaces El Niño. *Science.* 241:1037–1038.

Lavender, D., 1976. *California, a Bicentennial History.* W. W. Norton and Co., Chicago. 243pp.

Nardin, T. R., F. D. Hein, D. S. Gorsline, and B. D. Edwards, 1979. A review of mass movement processes, sediment, and acoustic characteristics in slope and base-of-slope system versus canyon-fan-basin floor systems. In: L. Doyle and O. Pilkey, eds. *Geology of the Continental Slope. Soc. Econ. Paleontol. Mineral. Spec. Publ.* 27:61–73.

Narlon, H. K., 1913. *The Story of California from Earliest Days to Present.* A. C. McClurg and Co., Chicago. 390pp.

Pavlova, Y. V., 1966. Seasonal variations of the California Current (English translation). *Oceanology.* 6:806–814.

Queenan, C. F., 1983. *The Port of Los Angeles from Wilderness to World Port.* Los Angeles Harbor Department. 203pp.

Quinn, W. H., 1974. Outlook for El Niño-type conditions in 1975. *NORPAX Highlights.* 2(6): 2–3.

Raitt, H. J., and B. Moulton, 1967. *Scripps Institution of Oceanography.* Ward Ritchie Press, Los Angeles. 217pp.

Reid, J. L., Jr., 1965. *Intermediate Waters of the Pacific Ocean.* Johns Hopkins Oceanographic Studies, Baltimore. 85pp.

Savrda, C. E., D. J. Bottjer, and D. S. Gorsline, 1984. Development of a comprehensive oxygen-deficient marine biofacies model; evidence from Santa Monica, San Pedro, and Santa Barbara basins, California continental borderland. *Am. Assoc. Pet. Geol. Bull.* 68(9): 1179–1192.

Schott, G., 1913. Der Peru-strom und seine nordlichen Nachlichen Nachbargebiete in normaler und anormaler Ausbildung. *Ann. Hydrog. Mar. Meteorol.* 59:161–169, 200–213.

Schwalbach, J. R., and D. S. Gorsline, 1985. Holocene sediment budgets for the basins of the California continental borderland. *J. Sediment. Petrol.* 55:829–842.

Shepard, F. P., and R. F. Dill, 1966. *Submarine Canyons and Other Sea Valleys.* Rand-McNally and Co., Chicago. 381pp.

Shepard, F. P., N. F. Marshall, P. A. McLoughlin, and G. G. Sullivan, 1979. Currents in submarine canyons and other sea valleys. *Am. Assoc. Pet. Geol. Stud. Geol. No. 8.* 173pp.

Smull, J., 1989. *A Step into the Past: Island Dwellers of Southern California.* Museum of Anthropology, California State Univ., Fullerton, CA. 33pp.

Soutar, A., and P. A. Crill, 1977. Sedimentation and climatic patterns in the Santa Barbara Basin during the 19th and 20th centuries. *Geol. Soc. Am. Bull.* 88:1161–1172.

Southern California Association of Governments, 1988. Regional Housing Needs Assessment. South. Calif. Assoc. Gov., Los Angeles. 150pp.

Sverdrup, H. U., M. W. Johnson, and R. H. Fleming, 1942. *The Oceans, Their Physics, Chemistry, and General Biology.* Prentice-Hall, Englewood Cliffs, NJ. 1087pp.

Teng, L. S., 1985. Seismic stratigraphic study of the California continental borderland basins: Structure, stratigraphy, and sedimentation. Ph.D. Dissertation, Univ. of Southern California, Los Angeles. 197pp.

Thornton, S. E., 1981a. Suspended sediment transport in surface waters of the California Current off southern California; 1977–78 floods. *Geo-Mar. Lett.* 1(1):23–28.

Thornton, S. E., 1981b. Holocene stratigraphy and sedimentary process in Santa Barbara Basin: Influence of tectonics, ocean circulation, climate, and mass movement. Ph.D. Dissertation, Univ. of Southern California, Los Angeles. 351pp.

U.S. Department of Commerce, 1988. *Statistical Abstracts of the United States.* Bureau of Census. 943pp.

Uchupi, E., and K. O. Emery, 1963. The continental slope between San Francisco, California, and Cedros, Mexico. *Deep-Sea Res.* 10(4):397–447.

Vildosa, A. C., 1974. *Biological Aspects of the 1972–73 El Niño: IDOE Workshop on El Niño Phenomenon.* Guayaquil, Ecuador. 7pp.

Chapter 2

Physical Oceanography

Barbara M. Hickey

Introduction

The Southern California Bight (SCB) constitutes a unique physical environment within a major eastern boundary current system. A dramatic change in the angle of the coastline, coupled with the morphology of the southern California offshore coastal area (fig. 2.1), results in circulation patterns and forcing mechanisms that differ significantly from other locations on the west coast of the United States. In particular, because of the bend in the coastline, the coastal wind stress decreases by almost an order of magnitude between the central California coast and the SCB (Hickey 1979). Thus, we can expect the effects of local wind stress forcing, which predominate within approximately 40 km of the coast north of Point Conception, to be much reduced in the SCB.

The morphology of the SCB includes 12 major offshore basins (fig. 2.1). All of the basins are completely enclosed at some depth (for example, 740 m for the Santa Monica Basin) and semienclosed at shallower depths. Thus, the region includes time-variable circulations characteristic of enclosed basins as well as fluctuating flows over sills between the basins. Submarine canyons are also common in the area and may affect sediment movement as well as local water properties. The SCB includes both narrow (< 5 km) shelf regions as well as broader (approximately 20–40 km) shelves (Santa Monica and San Pedro). In contrast to coastal regions outside the SCB, the sections of shelf are virtually dis-

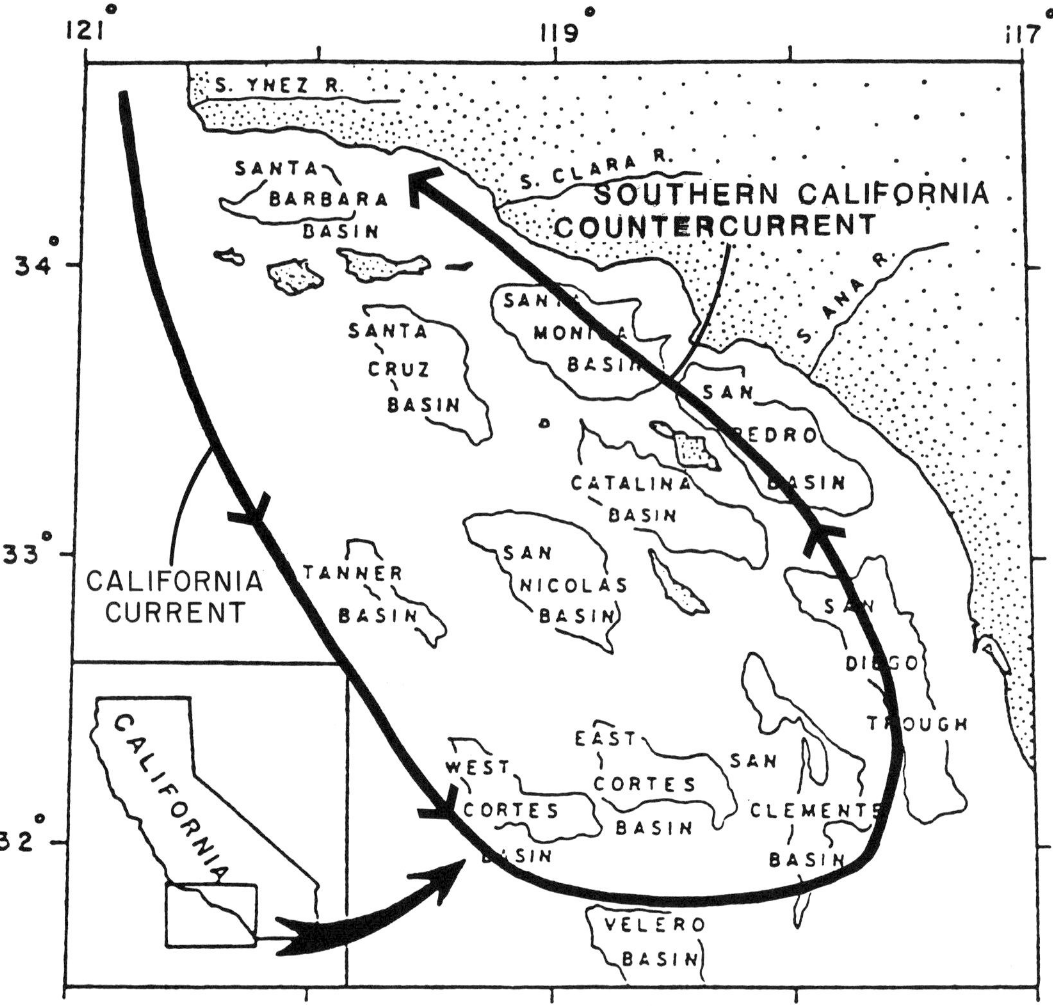

Figure 2.1. Schematic map of the coastal basins off southern California, showing the path of the large-scale California Current in the region.

connected by coastal promontories such as headlands and capes. Even the relatively long uninterrupted shelf from Newport Beach to San Diego is an order of magnitude shorter than the shelf north of Point Conception (100 km versus 1000 km).

The mean circulation in the SCB is dominated by the poleward-flowing Southern California Countercurrent, which may be thought of as a very large scale eddy in the California Current (Hickey 1979; Tsuchiya 1980) (fig. 2.1). The California Current is a broad, rather sluggish, equatorward flow that occurs off the west coast of the United States, with a seasonal mean speed maximum (about 10 cm s^{-1}) that occurs a few hundred kilometers offshore (Hickey 1979). The California Current has a seasonal speed maximum in late summer, as does the Southern California Countercurrent.

A poleward-flowing current, the California Undercurrent, occurs on the shoreward side and beneath the California Current all along the west coast. Flow in the undercurrent is generally concentrated over the continental slope. In the SCB, the undercurrent is concentrated over the nearshore continental slope (rather than the slope that occurs on the seaward side of the SCB) so that the California Current and Undercurrent are spatially separated. Also, in the SCB, where both the surface flow (the Southern California Countercurrent) and the subsurface flow (the California Undercurrent) are directed poleward, the presence of the undercurrent is demonstrated by a subsurface maximum in the current speed. The California Undercurrent, like the California Current, has a seasonal speed maximum in late summer.

Monthly mean winds blow equatorward off the southern California coast year round, with a seasonal maximum in spring (Hickey 1979). During much of the year (all but winter and early spring), winds in the nearshore half of the bight are almost an order of magnitude weaker than those offshore of the bight (Hickey 1992).

Emery (1960) was the first to address the physical oceanography of the SCB as a system. Emery's description was improved significantly by Jackson (1986), who included results from the California Cooperative Oceanic Fisheries Investigations (CalCOFI) program and from other then-available data. Several comprehensive data sets have recently become available. This chapter incorporates the results of these data sets to provide a further updated description of the currents and water properties of the region. The following section describes the principal data sets and their limitations. For simplicity, the discussion is organized according to the frequency band of the dominant fluctuations: seasonal, subtidal (approximately 2–30 days), high frequency (tidal and supratidal), and interannual bands. Discussion of basin-to-basin exchange and of the local wave climate follows the discussion of the several frequency bands.

Data Sources and Analyses

Several spatially and temporally comprehensive data sets are available for the SCB. These include CalCOFI hydrographic data and direct current measurements over the Santa Barbara Basin and shelf, the Santa Monica Basin and shelf, the San Pedro Basin, and the San Pedro and Newport Beach shelves, the shelf between Newport Beach and the entrance to San Diego Harbor, and the bight south of San Diego. Extensive conductivity and temperature versus depth (CTD) data are also available for the Santa Barbara Channel and the Santa Monica and San Pedro basins and shelves.

The CalCOFI data set consists of water property information available on a fixed station grid at roughly seasonal intervals since 1950 (Lynn and Simpson 1987) (fig. 2.2). Data coverage extends as far north as San Francisco and as far south as lower Baja California, although not on every cruise. These data provide important information on the large-scale current system and its seasonal and interannual variability. Data from over 16,000 stations are available for the period from 1950 to 1978. The sampling grid consists of parallel lines oriented roughly normal to the coast, spaced at approximately 74-km intervals along the coast. Stations are also separated along the lines by approximately 74 km, except near the coast, where the spacing is half or less. All stations are restricted to the upper water column (<500 m). Until 1964, data consisted of 18-bottle Nansen casts; subsequent data were obtained with a CTD. Seventeen standard stations are located within the SCB.

The Santa Barbara Channel data sets include both a 4-month pilot program and a year-long program (1984) of direct current measurements (SAIC data in fig. 2.3) as well as seasonal CTD measurements (Gunn et al. 1987). Seasonal CTD measurements were also made during 1969 (Kolpack 1971). The comprehensive 1984 measurement program was executed by Science Applications International Corporation (SAIC). An array of roughly 40 current meters included several moorings across both the east and west channel entrances as well as in all of the interisland passages. One mooring, the only mooring with surface current meters, was maintained in mid-basin. The CTD surveys

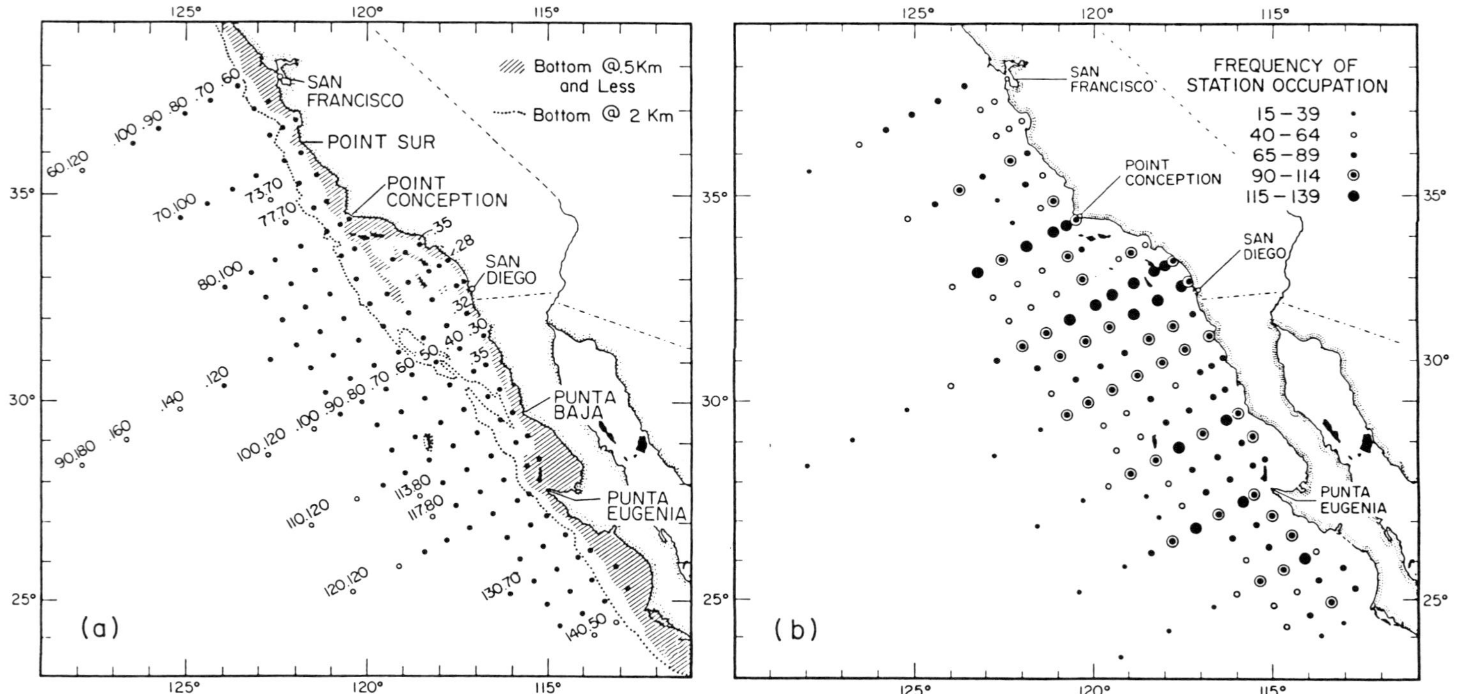

Figure 2.2. (a) CalCOFI station pattern. The 2000-m isobath is shown dotted. Bathymetry less than 500 m is hatched. (From Simpson et al. 1984.) (b) Record of station occupation. (From Simpson et al. 1984.)

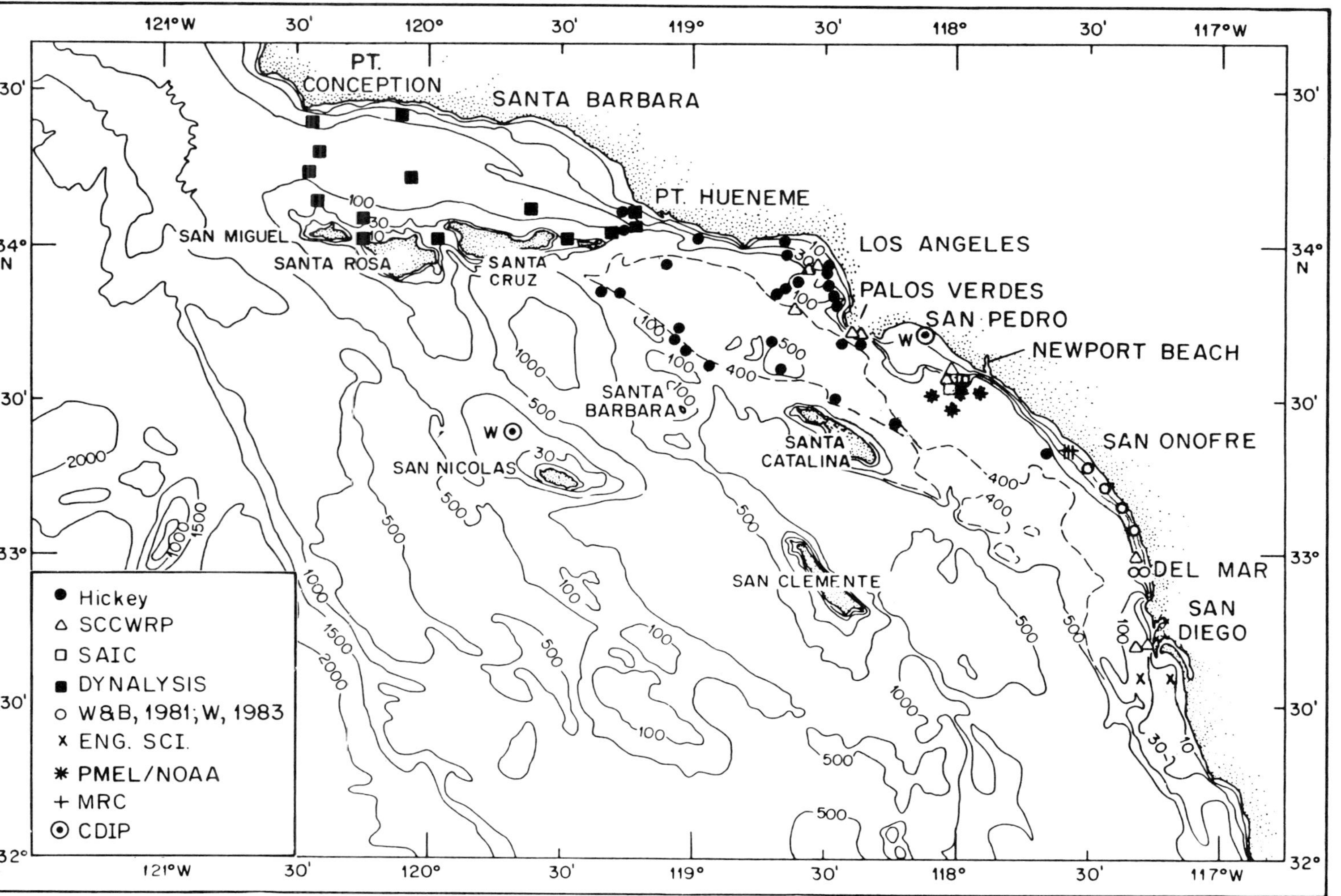

Figure 2.3. Locations of most comprehensive direct current records in the SCB. The location of offshore and nearshore wave monitoring sites is indicated by the symbol W. Depth is in fathoms.

consisted of quarterly surveys of approximately 50 stations along eight roughly north–south sections.

The comprehensive Santa Monica–San Pedro data set consisted of four separate experiments (Hickey 1991, 1992) (fig. 2.3). Each experiment included approximately 40 current meters distributed on 10–12 moorings over the basin and shelves in the area as well as intensive CTD surveys two or three times per year. Several transects were occupied on every CTD cruise. Additional lines were occupied along and across sills and in the vicinity of canyons, depending on the focus of the particular experiment. The first experiment (October 1985–February 1986), which was designed to establish cross-shore coherence scales and basic forcing mechanisms of the currents, consisted of a cross-shelf-slope-basin transect across the Santa Monica region. The second experiment (May–October 1986) was designed to establish slope-parallel and around-the-basin coherence scales and to determine basin circulation patterns below the depth of the deepest sill. The third experiment (April–October 1987) focused on exchange between Santa Monica and San Pedro basins and adjacent basins. The last experiment (February–October 1988) was designed to determine scales and patterns and forcing mechanisms of the shelf circulation in Santa Monica Bay as well as the effect of Redondo Canyon on the circulation. Only data from the first three experiments have been processed sufficiently to include in this synthesis.

The narrow southern California shelf also has been the subject of a focused set of three experiments (see fig. 2.3). The earliest experiments (Winant and Olson 1976) were performed on the Navy tower near San Diego during summer in a water depth of 18 m. This experiment was designed using closely spaced current meters to determine vertical scales of variability. The second set of experiments consisted of a cross-shelf array of 12 instruments distributed on moorings on the inner (15 m), middle (30 m), and outer (60 m) shelf at Del Mar, 30 km north of San Diego (Winant and Bratkovich 1981). This array was deployed for roughly 6-week periods in each of the four seasons to determine the cross-shelf structure of the shelf currents (see fig. 2.3). The third and last experiment was designed to determine longshelf coherence scales for current and temperature fluctuations during the summer season. Current meters were deployed from Del Mar to San Onofre along the 30-m isobath (two per mooring) at intervals as close as 2 km (Winant 1983).

The Southern California Coastal Water Research Project Authority (SCCWRP) has maintained moorings at selected sites near sewage outfalls along the southern California shelf since 1974 (Hendricks 1974, 1975, 1976, 1980, 1982, 1984). Most SCCWRP moorings were located on the outer shelf (in about 60 m of water) or on the upper slope (usually less than 200 m of water). Although as many as three shelf–slope sites have been occupied simultaneously for a month or more, the majority of SCCWRP records were obtained on individual mooring deployments. Measurements were also made in San Gabriel and Santa Monica submarine canyons. Only mooring data subsequent to 1979 are available in computer format and thus have been included quantitatively in this paper (fig. 2.3). The SCCWRP current meters, which are of a tiltmeter design specific to SCCWRP, have been intercompared successfully with an EG & G vector measuring current meter for nonwave-dominated regimes (Hendricks 1985).

Comprehensive current measurements have been made by SAIC off Newport Beach. The array varied from three to five moorings in bottom depths of 35–80 m. The array was maintained from June 1986 to July 1989; however, only the data from 1986 were available for this discussion. A comprehensive set of direct current measurements have been made by Engineering Science to study the small bight south of San Diego Harbor. A total of 15 tiltmeter-type current meters were deployed for a period of 13 months in water depths from 10 to 78 m (see fig. 2.3).

Other direct current measurements have been obtained in the region: in particular, off

Newport Beach and in the San Diego Trough by the Pacific Marine Environmental Laboratory of NOAA (G. Connor pers. comm.) and in mid Santa Monica Basin by the Navy (A. Bratkovich pers. comm.). However, these data are not yet available for public use. Extensive measurements were also made on the San Onofre shelf (water depths from 10 to 35 m) over a period of about 10 years (Erdman 1987) (see MRC data on fig. 2.3). Although the records are not continuous in time, they are unique in providing excellent coverage of shallow bottom depth and also near-surface regions of the water column. Vector-type current meters were used for all measurements. These data are now available in summary form (spatially averaged) and will be used in the discussion.

Direct measurements of current speed and direction are usually recorded at intervals of 20 minutes to 1 hour. The data are usually presented as north–south (v) and east–west (u) components of the velocity field. The data are generally filtered (smoothed) to hourly intervals and often are filtered or smoothed again to remove tidal and higher frequencies (a cut-off frequency of 40 hours is common, with a Lancotz-cosine filter). These data, which will appear to be much smoother than the original data, are useful for identifying current events that occur over periods of a few days or a season. Known as "subtidal data," these data are usually presented at 6-hour intervals as north–south or east–west velocity components or as a vector, as a function of time for several months. To identify possible seasonal variations, data are further averaged into monthly means for some presentations.

The now-standard technique of empirical orthogonal eigenfunction (EOF) analysis (Kundu and Allen 1976) is used in this chapter. This technique is simply a convenient way of describing the variability in data sets and is similar in many ways to the principal component analysis often used by biologists. Basically, correlations between stations are used in a least squares analysis to select the most efficient set of temporal patterns with which the data can be represented. Then both the temporal patterns and the maps of spatial amplitudes or weights are inspected to search for similarity to patterns expected for particular physical processes. The analysis will suggest several patterns (e.g., pattern a, b, c, . . .), each of which accounts for a certain amount of the variability in the observed data set. Thus, at any location, the total variability is made up of a specific amount of each pattern's amplitude (i.e., weights) (e.g., 0.1 × pattern a + 0.5 × pattern b + . . .).

The temporal pattern might be simply a seasonal variation. If one area has a strong seasonal variation, it might have a weight of 1.0, while another with a weak seasonal variation might have a weight of 0.1. A third might show a variation exactly opposite to the first: its weight would be –1.0. Thus, a map showing the weights over an area would show how the seasonal cycle varies from place to place. In this paper, the weights of the EOFs are denoted amplitudes, in keeping with common usage. A pattern need *not* be related to any physical process. However, in practice, at least the first pattern (or mode, as such patterns are called) which represents the most dominant variability can often be related to a physical process.

Currents at subtidal frequencies are generally oriented in the direction of local isobaths. In the SCB, the isobaths are extremely convoluted and are not usually oriented north–south. Flow directions are said to be poleward when they have an upcoast component and equatorward when they have a downcoast component. In keeping with common usage, both currents and wind stress direction are described by the direction *to* which they are directed. Wind is usually described by the direction *from* which it blows (e.g., southerly to designate wind *from* the south).

Seasonal Patterns

Currents

CalCOFI hydrographic data allow a comprehensive description of the large-scale eastern boundary currents that bathe the SCB. Syntheses and interpretation of the CalCOFI data by Hickey (1979), Tsuchiya (1980), Chelton (1980, 1984), and most recently, Lynn and Simpson (1987) provide descriptions of the seasonal and spatial structure of the California Current, the Southern California Countercurrent, and the California Undercurrent in this and other regions. However, the spatial resolution of the CalCOFI data set excludes regions shallower than 500 m, that is, much of the slope and all of the shelf, and provides no spatial details on scales of approximately 20–40 km. In the ensuing discussion, large-scale patterns are described using CalCOFI data; smaller scale details, in particular the shelf circulation, are described using available direct current measurements. Hydrographic data collected in the SCB between September 1974 and April 1977 with a much finer resolution than standard CalCOFI data furnish additional information on nearshore current structure (Tsuchiya 1980).

The California Current, which is fed by the West Wind Drift, is the eastern limb of the North Pacific gyre. The current flows equatorward along the west coast of the United States throughout the year, with maximum speeds approximately 300 km offshore (Hickey 1979; Lynn and Simpson 1987) (fig. 2.4). (Currents over the continental shelf are not generally considered to be part of the California Current and are discussed later.) The California Current is surface intensified, with maximum monthly mean speeds at the surface on the order of 10 cm s^{-1}, decreasing to about 2 cm s^{-1} at a depth of about 200 m (fig. 2.5). The California Current has a significant seasonal variation, with the seasonal maximum in summer (Hickey 1979; Lynn and Simpson 1987). The majority of the equatorward transport of the California Current occurs from 200 to 500 km offshore, seaward of the SCB. The total transport of the California Current along a section from the coast through the SCB to 1200 km offshore (fig. 2.5) varies from 5.8 Sv (Sverdrups) in January to 7.8 Sv in July (1 Sv = 10^6 m^3 s^{-1}).

The California Current turns shoreward near the southern U.S. border, and a branch of the current turns poleward into the SCB, where it is generally known as the Southern California Countercurrent. This countercurrent is strongest in summer when it is eddylike (that is, flow rejoins the California Current) and in winter when poleward flow through the Santa Barbara Channel can be continuous with the flow north of Point Conception (Hickey 1979). During spring, the countercurrent appears to be essentially absent; that is, flow enters the SCB, but turns equatorward rather than poleward (figs. 2.4 and 2.5).

A poleward-flowing undercurrent occurs beneath the shoreward side of the California Current in most seasons at all west coast locations (fig. 2.6). The undercurrent is distinguished from surface currents by its characteristic water properties, which are of southern or equatorial rather than subarctic origin (Tsuchiya 1980). The undercurrent at most locations is relatively narrow, having a high-speed core that is generally located over the continental slope (Hickey 1979; 1989a,b; 1992). The undercurrent at all west coast locations has a seasonal maximum in late summer (corresponding to that of the California Current) and a minimum in spring. The undercurrent has a second seasonal flow maximum in early winter at most locations, but often the winter maximum occurs at the sea surface and it is unclear whether the winter and summer poleward flows are dynamically the same (Hickey 1979; Chelton 1984; Lynn and Simpson 1987).

In the SCB, the continental slope is less well defined than in regions to the north and south because the area is populated with a number of islands and ridges. The large-scale CalCOFI data set, which has only four data points inside the SCB, suggests that the undercurrent occurs throughout the entire bight, with a width perhaps slightly broader than in regions outside the SCB (compare fig. 2 in Chelton 1984 with fig. 2.5 here). Direct current measurements over a much finer spa-

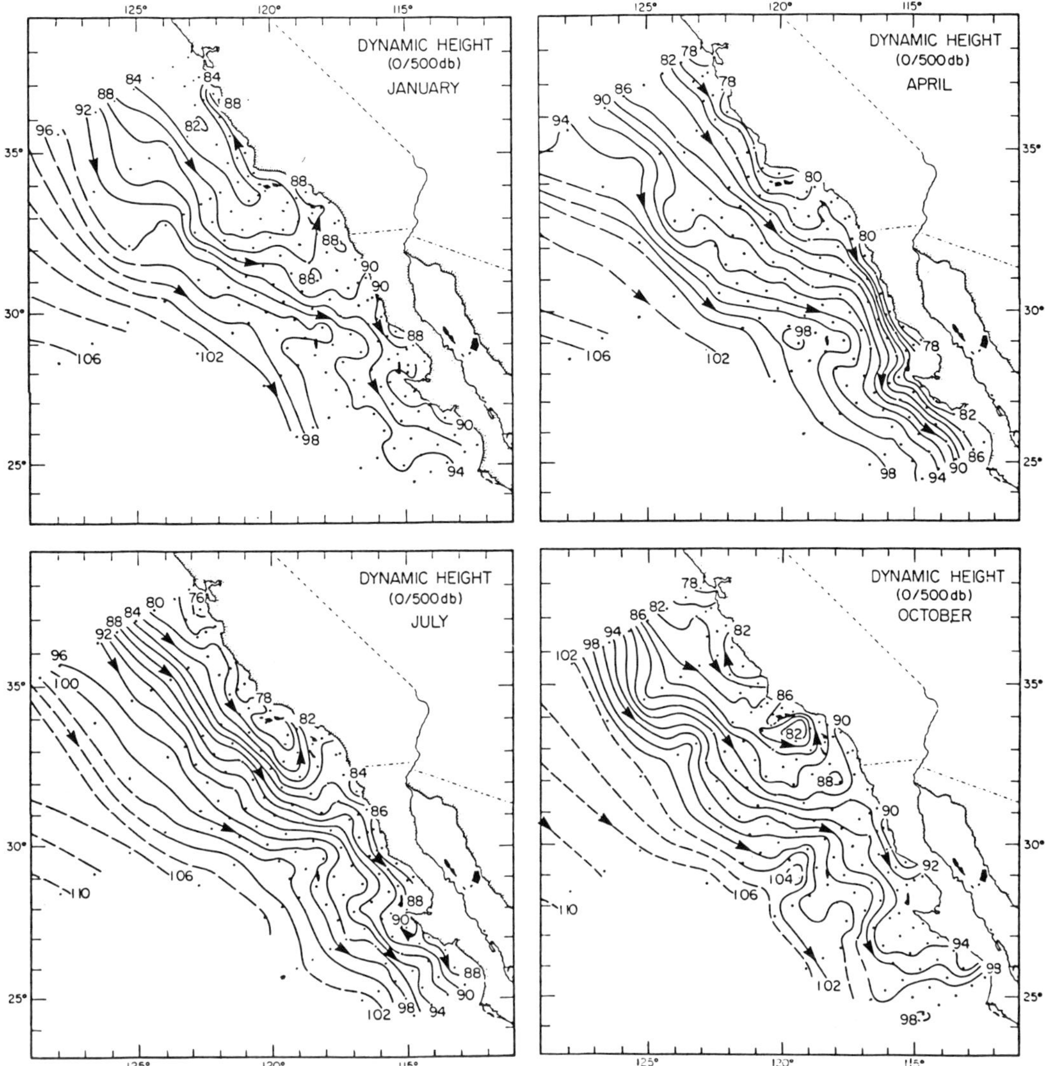

Figure 2.4. Mean dynamic height of the sea surface relative to 500 dbar (about 500 m) for four periods (of approximately 60 days) centered about January, April, July, and October. Contour interval is 2 dyn cm. (From Lynn and Simpson 1987.)

tial scale in the Santa Monica Basin suggest that the undercurrent is strongest over the nearshore continental slope (Hickey 1992). Of course, this Santa Monica undercurrent might represent only one of multiple cores. The exact spatial structure of the undercurrent within the SCB and its relationship to the upper water column California Countercurrent are the subjects of on-going research programs. If the poleward flow in the countercurrent and undercurrent are lumped together, we find that the SCB is flushed by a minimum transport of 0.8 Sv in April and by a maximum transport of 1.8 Sv in October.

A contoured section of geostrophic velocity across the SCB illustrates graphically that the islands in the SCB are not an impediment to the large-scale flow (fig. 2.5). This result reaffirms the large-scale nature of the flow field. The topography channels the

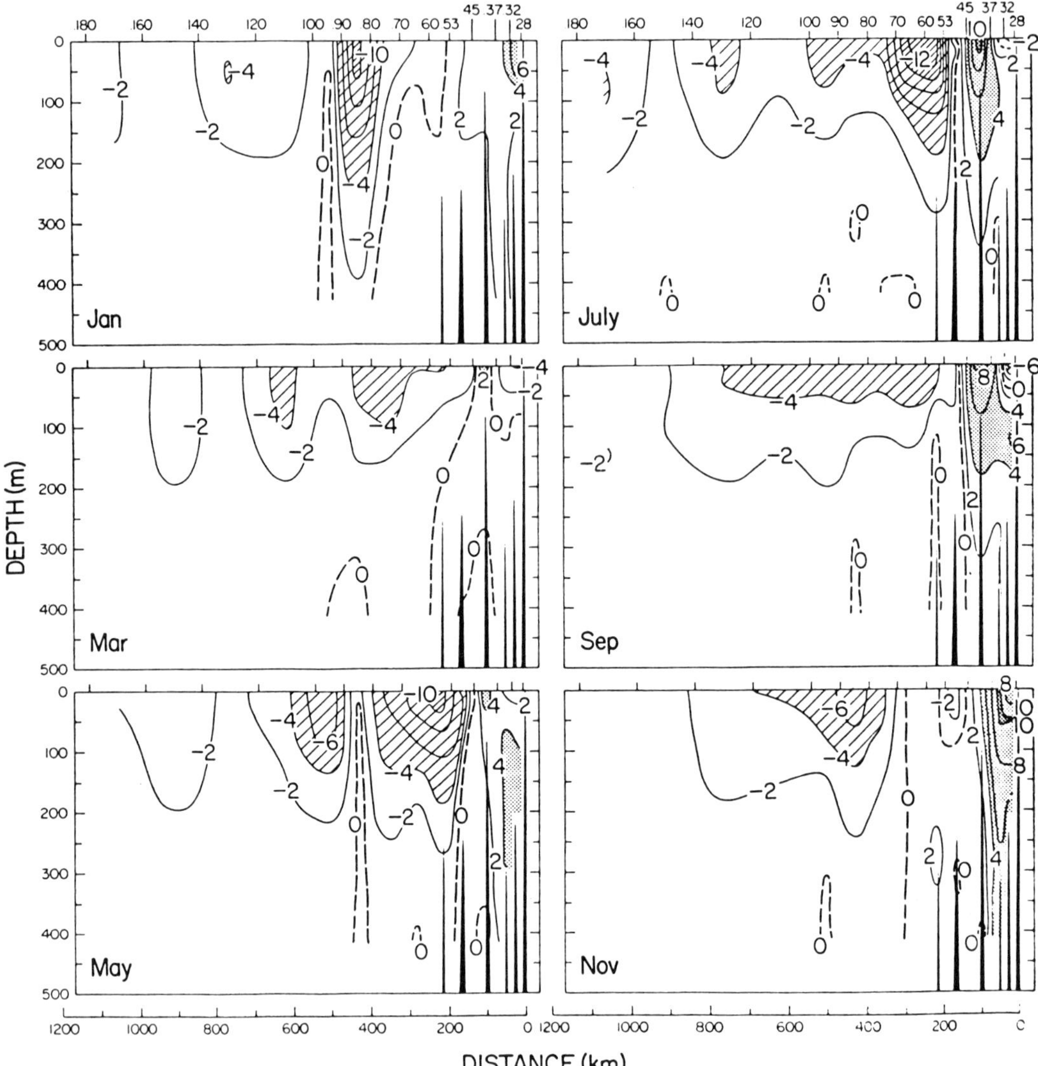

Figure 2.5. Geostrophic velocity (cm s^{-1}) relative to 500 dbar for odd months along CalCOFI line 90 (see fig. 2.2). Contour interval is 2 cm s^{-1}. Equatorward flow in excess of 4 cm s^{-1} is shaded. The Channel Islands and ridges are delineated by the black spikes emanating from the bottom of the graphs.

large-scale flow in two ways. First, the island chain on the southern side of the Santa Barbara Channel divides the Southern California Countercurrent into two branches, one that enters the Santa Barbara Channel and one that passes to the south of the island chain (Hickey 1992; Lynn and Simpson 1987). Second, the offshore banks of the borderland serve to divide the countercurrent from the California Current. Results from several small-scale hydrographic surveys in the Santa Monica–San Pedro–Santa Cruz basins demonstrate that the direction and strength of local currents are strongly affected by banks, ridges, and islands between and within these basins (Hickey 1992). Currents tend to follow ridges and go around islands. Narrow channels can also affect the local currents. For

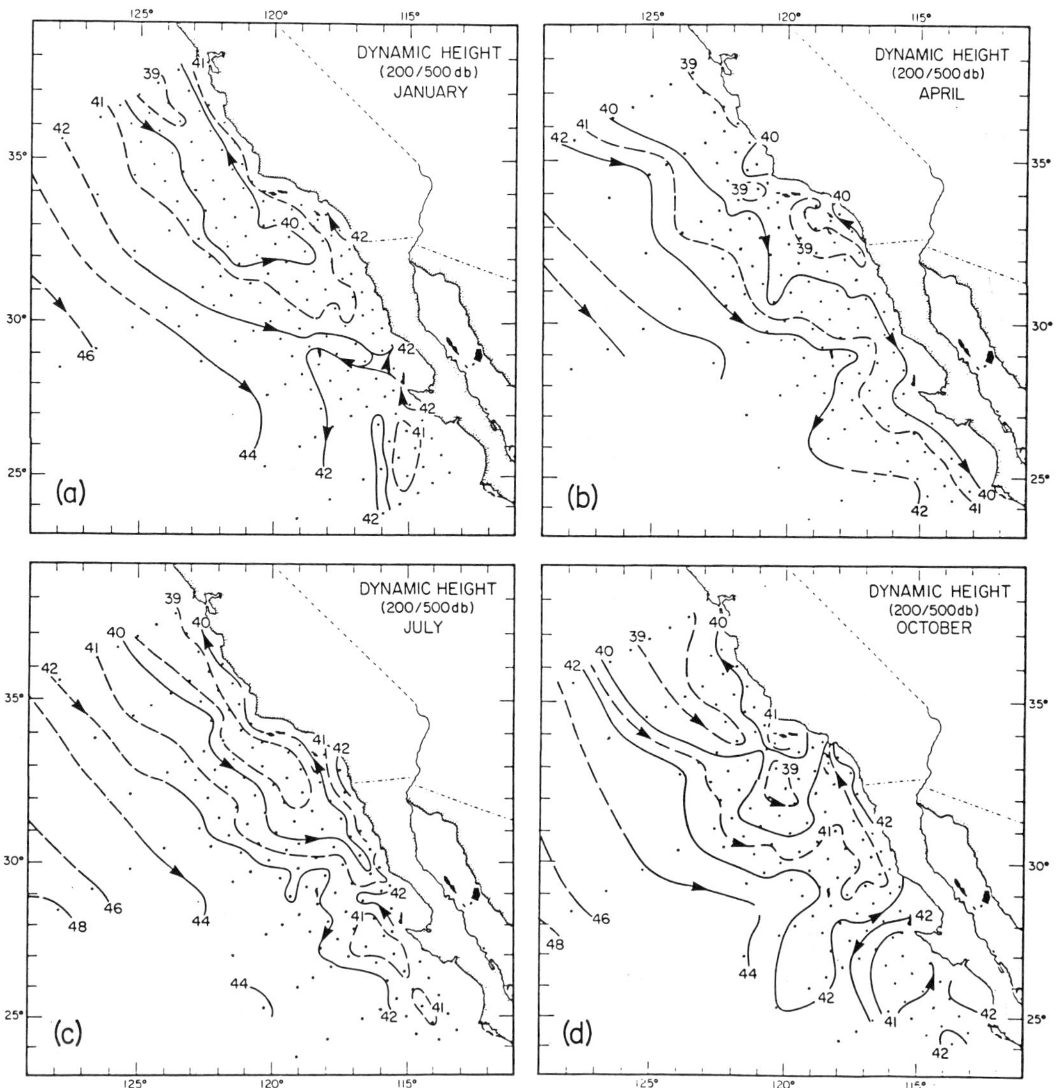

Figure 2.6. Mean dynamic height at 200 dbar (about 200 m) relative to 500 dbar (about 500 m) for four periods (of approximately 60 days) centered about January, April, July, and October. Contour interval is 1 dyn cm. (From Lynn and Simpson 1987.)

example, the channel between the mainland and Santa Catalina Island appears to significantly affect both the continuity and local direction and speed of poleward flow near the coast. At some depths, poleward flow is sometimes completely blocked; at other depths and times, it is swept from the east to the west side of the channel. Similar local perturbations to the flow field are expected in the vicinity of other ridges, islands, and channels in the SCB.

As already mentioned, direct current measurements in the SCB exhibit some differences from the CalCOFI results. In particular, direct measurements indicate that the subsurface maximum in poleward flow over the slope (the undercurrent) is both stronger and more continuous in time than indicated by the geostrophic data. For example, directly measured seasonal mean current speeds are on the order of 15–20 cm s^{-1} in comparison with 2–10 cm s^{-1} in the geostrophic data. The direct measurements in the Santa Monica region (Hickey 1992), as well as those in the Santa Barbara Channel region (Gunn et al. 1987), suggest that poleward flow persists beneath the surface layers, even during the spring, in contrast to the CalCOFI results. Also, direct current measurements in the Santa Monica Basin suggest that during at least one winter, equatorward flow occurs on the west side of the basin (Hickey 1989a) (fig. 2.7). This relatively small-scale feature is not resolved by the geostrophic large-scale data set.

Whereas the currents over the continental slope are predominantly poleward due to the presence of the large-scale countercurrent and undercurrent, currents over the mainland continental shelves in the SCB seaward of the very nearshore zone (bottom depth >10 m) appear to be predominantly equatorward, at least in the upper water column (0–15 m). As mentioned previously, the SCB shelf is not at all continuous; rather, several headlands indent the coastline in the northern SCB so that the shelves are almost disconnected from one another. On the wide Santa Monica shelf, measurements at depths of 5 and 10 m from the surface in 30 m of water at a location roughly one-third the distance from the southern end indicate that the seasonal mean flow is equatorward during fall and winter (Hickey 1992). On the narrow shelf north of San Diego, equatorward flow occurs in the upper layer (approximately 5–10 m) in 15-, 30-, and 60-m water depths in every season (Winant and Bratkovich 1981) (fig. 2.8). The strongest equatorward flow occurs in winter (60-m depth) or spring (15- and 30-m depth).

Equatorward flow has also been observed in several-year averages for winter and summer at all bottom depths (10–35 m) on the shelf off San Onofre (Erdman 1987). The strongest equatorward mean flows (up to 7 cm s^{-1}) occur in summer and in the upper layers (3–10 m) at this location. The magnitude of the equatorward flow and the thickness of the equatorward-flowing layer both increase in the offshore direction. In all cases, the magnitude of the mean equatorward flow on both narrow and wide shelves (approximately 5–10 cm s^{-1}) is much less than that of the poleward flow over the slope.

Poleward flow over the narrow shelf off Del Mar was observed in the lower half of the water column at the 30-m and 60-m sites, the signature of the poleward undercurrent discussed previously (Winant and Bratkovich 1981). Net poleward, subpycnocline flow has also been observed on the narrow shelf off Palos Verdes (Hendricks 1980, 1982) and on the somewhat wider shelf off San Diego (Hendricks 1976) and Newport Beach (Hendricks 1980, 1982).

Tsuchiya (1980) presents evidence from geostrophic (hydrographically derived) current data for the occurrence of equatorward flow over the upper slope from Newport Beach to just north of San Diego. Although his instantaneous data illustrate that equatorward flow can occur in all seasons, he makes the interesting suggestion that the strength of the equatorward flow is related to the strength of the poleward flow farther offshore, being strongest when the poleward flow is weakest. This conclusion would be consistent with the occurrence of the strongest equatorward flow nearshore during the spring, as suggested by the monthly averaged direct current measurements.

Although currents have been measured at many sites on the coastal side of the SCB, the measurements are not, for the most part, simultaneous. To attempt to develop more detailed seasonal flow patterns for the SCB, monthly mean data from all available data sets have been included in three seasonal maps, summer (August–September), winter (December–January), and spring (April–May) for 5–20 m, 30–50 m, and 80–120 m water column depths (fig. 2.9 a, b and c). Occasionally, data are included that exceed the depth range specified; the actual depths of

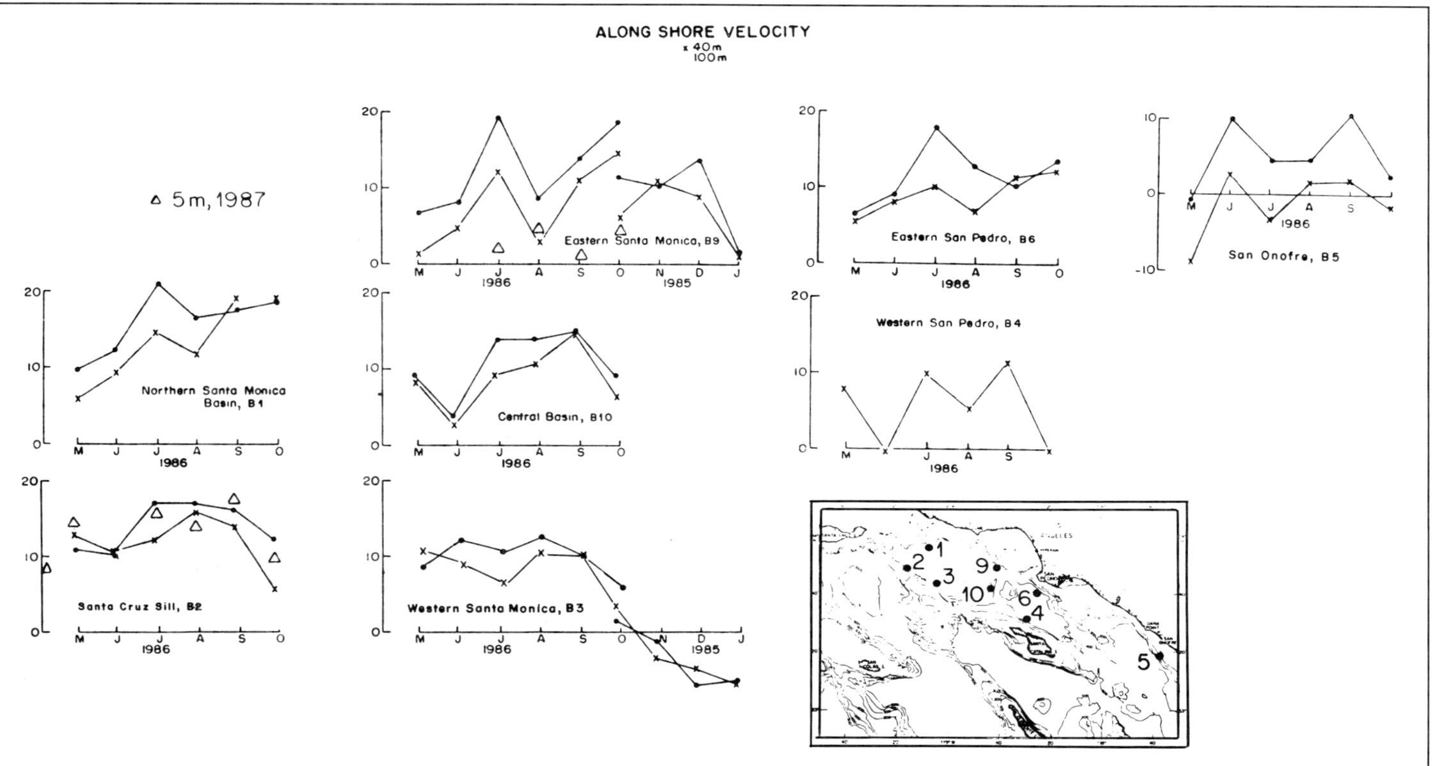

Figure 2.7. Monthly mean 40-m and 100-m current speeds (in cm s^{-1}) at selected locations in the SCB during 1985 and 1986. Because the velocity at these upper water column locations is strongly aligned either along the slope or across the sills, speeds very nearly represent the along-isobath or across-sill component of velocity. Positive speeds indicate an upcoast flow direction; negative speeds indicate a downcoast flow direction. Data from both 1985 and 1986 are available at some sites during October. To provide a qualitative estimate of an annual cycle, 1985 data have been added at the end of the 1986 data set.

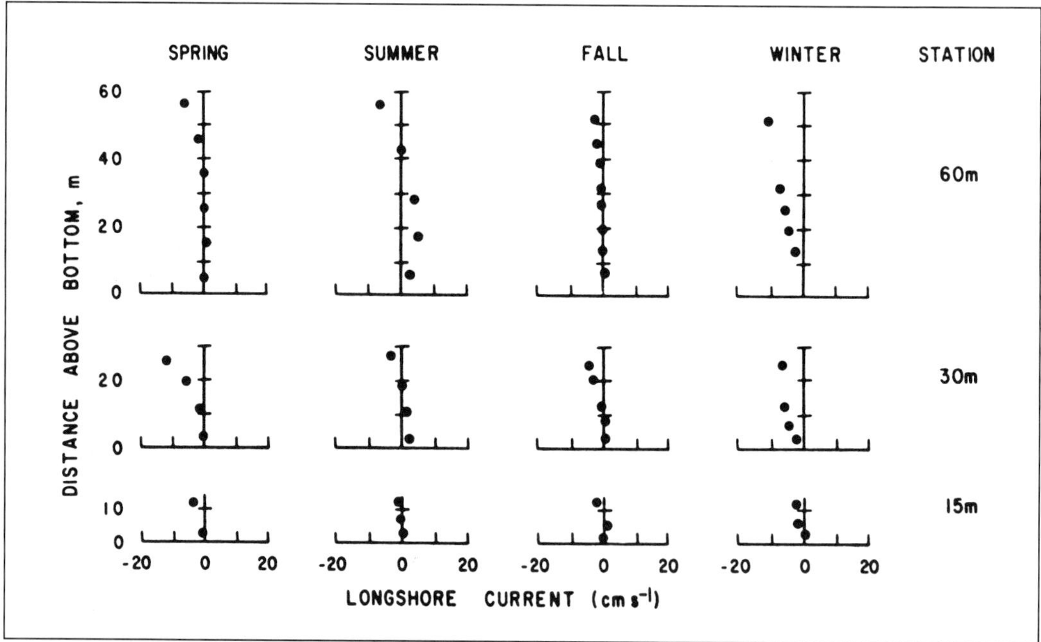

Figure 2.8. Distribution of mean longshore currents off Del Mar in bottom depths of 60, 30, and 15 m for the periods May 16–June 27, 1978 (spring), July 27–September 11, 1978 (summer), October 21–December 4, 1978 (fall), and December 21–March 26, 1979 (winter). Mooring locations are shown in figure 2.3.

such data are given in figure 2.9. In this depiction of seasonal current patterns, the effects of interannual variability (which can be significant) have been disregarded. However, note that most of the data from each source represents a single year (although, in each case, a different year). A second year of data for these sources is indicated with a dashed arrow in figure 2.9.

The seasonal maps illustrate clearly the extent to which the mean flow follows the direction of local, relatively shallow isobaths. For example, the flow on the San Pedro shelf is directed more westward (or eastward) than that off Palos Verdes or San Onofre. The majority of the flow vectors in the Santa Monica and Santa Barbara basins trend to the northwest. The flow vectors are directed more westward south of Santa Cruz Island as the flow is forced to bifurcate at the entrance to the Santa Barbara Channel.

The direction of flow at the eastern end of the Santa Barbara Channel appears to be extremely sensitive to the exact location, and possibly the angle, of the incident flow. If the flow is incident from the Santa Monica slope, the flow at the channel mouth might tend west-northwest; however, if the flow is incident from the southwest, the flow at the channel mouth would be expected to tend northeast. The most strongly northeastward flow at the mouth is observed in January when southeastward flow occurs on the seaward side of the Santa Monica Basin. (Remember, however, that data in the Santa Barbara Channel and Santa Monica Basin are from different years.) It is possible that flow across the Santa Cruz sill in winter is eastward rather than westward, as observed in the other seasons. No data are available with which to substantiate this hypothesis.

Data from the Santa Barbara Channel illustrate clearly net inflow through the island channels as well as the existence of an east-

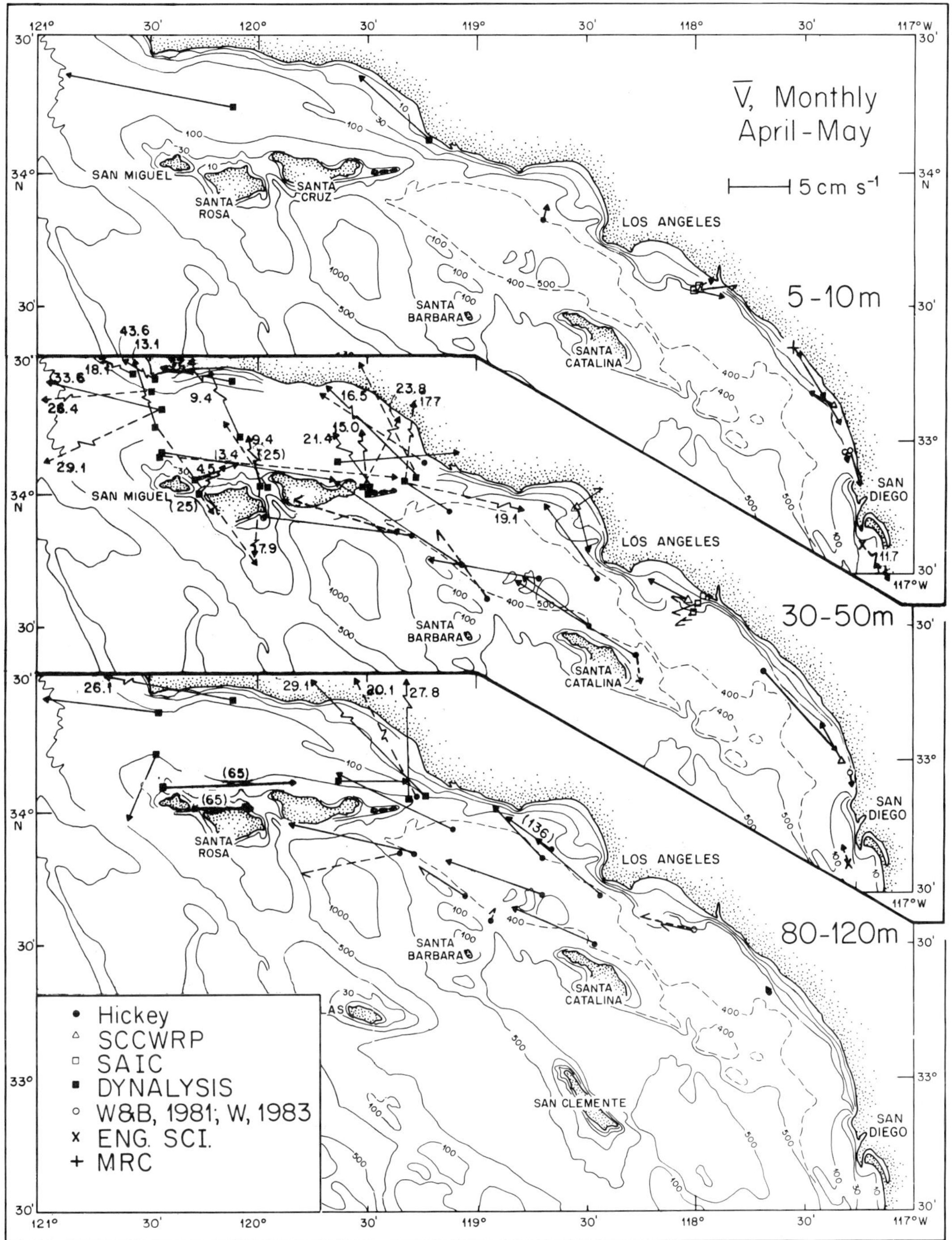

Figure 2.9. Monthly mean measured velocity vectors for available data in the SCB region for three depth intervals for (a) spring (April–May), (b) late summer (August–September), and (c) winter (December–January). The various data sets are described in the text. In general, the data sets are from different years and, thus, the figure ignores the effects of interannual variability. The vectors for the first month in each season are shown as one-sided arrowheads; the vectors for the second month are shown as two-sided arrowheads. A second year of data from the same source is shown as a dashed arrow. Numbers in parentheses indicate depths of data that were slightly outside the given range. Decimal numbers indicate magnitudes of vectors that exceeded map borders or that were shortened for clarity (shown by a zigzag in the vector).

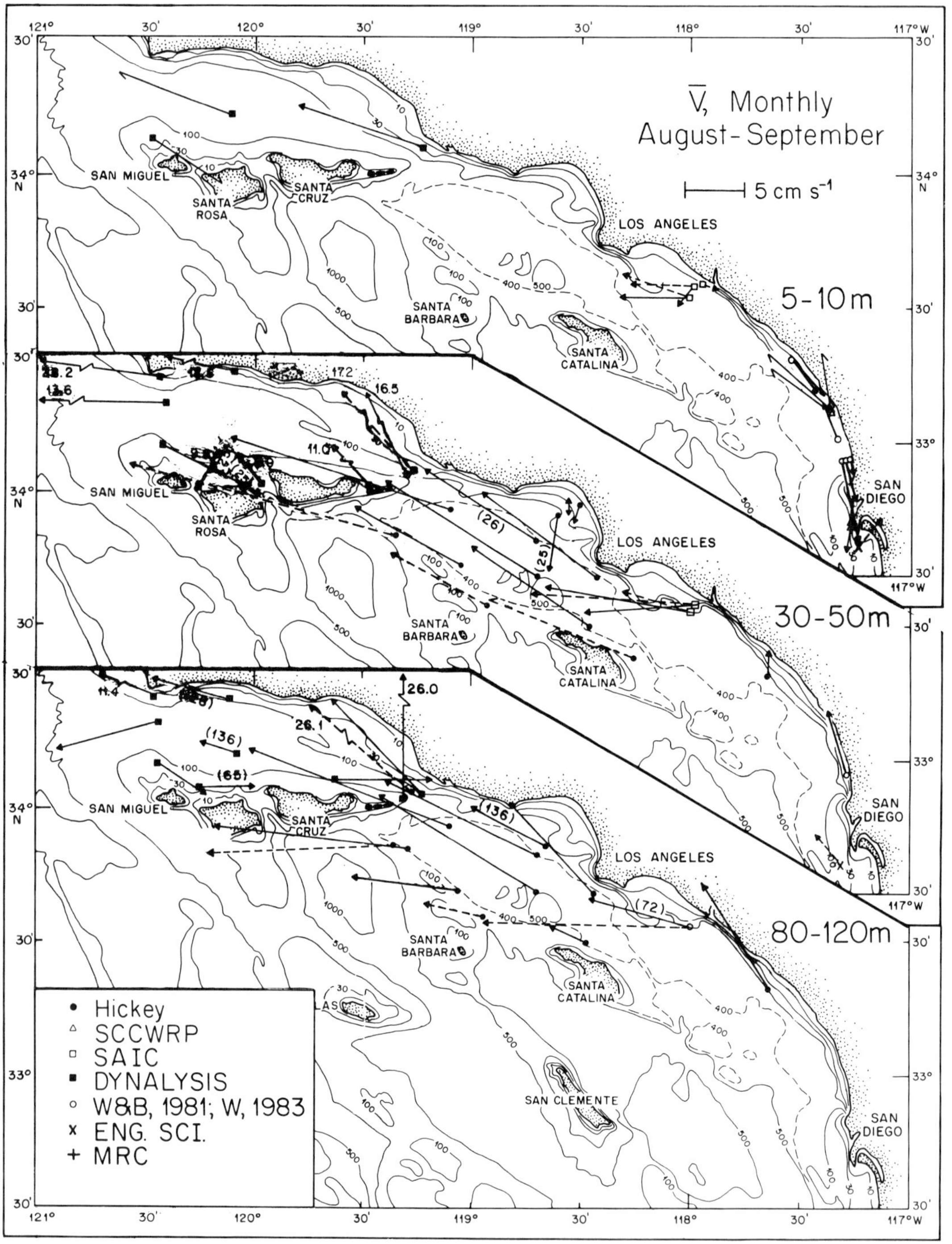

Figure 2.9.(b) (continued)

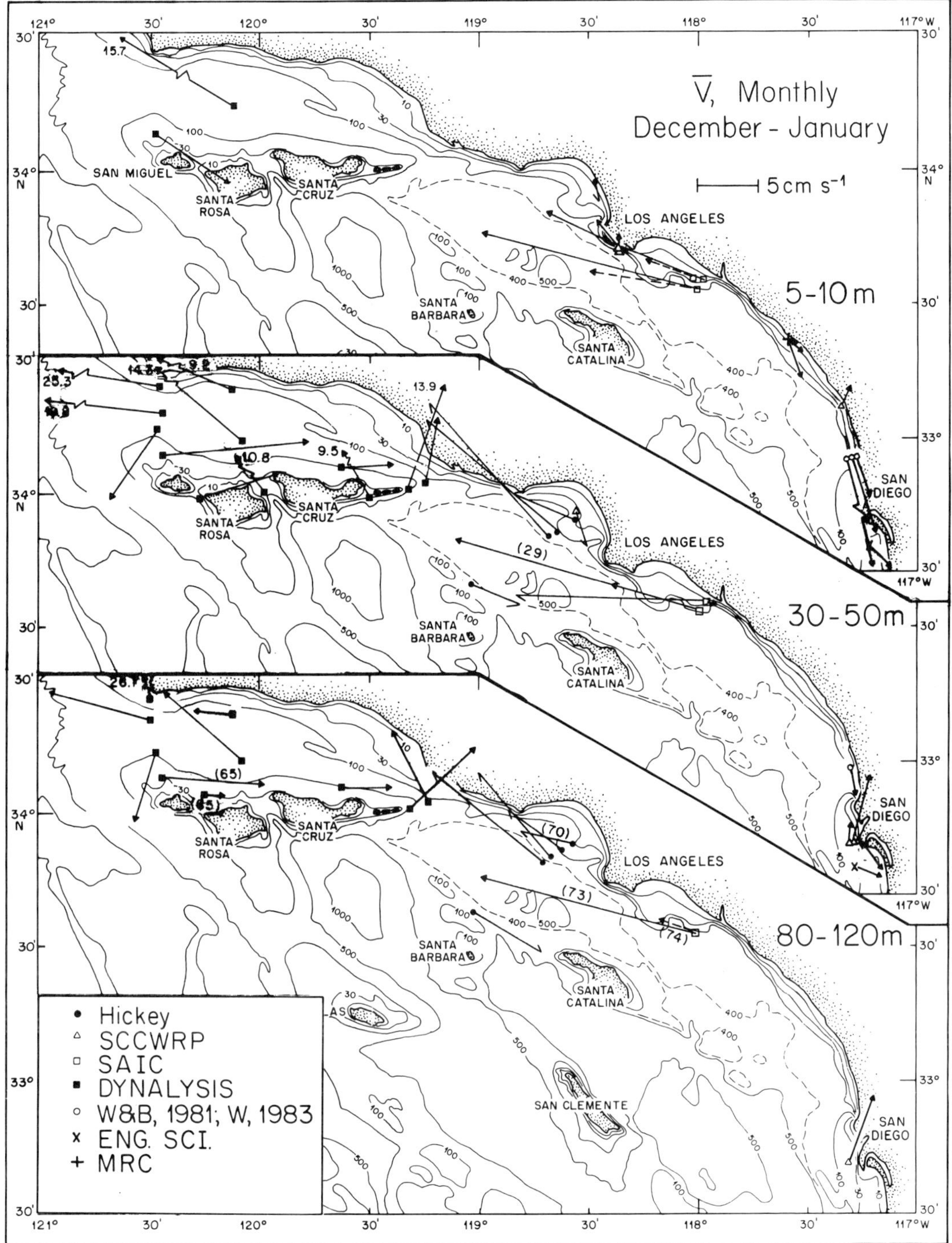

Figure 2.9.(c) (continued)

ward flow on the southern side of the basin. In the channel between Santa Rosa and Santa Cruz islands, measurements during the spring period for one year indicate outflow on the western side and inflow on the eastern side of the channel. The eastward flow along the southern side of the basin, which is strongest in spring, seems to be related to upwelling events off Point Conception (Brink and Muench 1986). One set of spring measurements suggests that outflow from the Santa Barbara Channel to the Santa Monica Basin can occur occasionally on the western side of the eastern channel entrance. Drifter studies as well as hydrographic surveys indicate that Santa Barbara Channel circulation consists of two counterclockwise gyres (Kolpack 1971; Brink and Muench 1986). The direction of the current meter data at mid-basin, which is more northward than the local isobath direction (fig. 2.9), is consistent with the existence of two (rather than one) gyres over the basin.

The composite maps provide additional information on flow direction over the shelf areas. Equatorward flow usually occurs in the top 5–10 m and is stronger at middle and outer shelf locations. Below the surface layers, the flow is likely to be poleward, especially during summer and fall. The direction of mean shelf currents may *appear* at first sight to be more variable than the direction of slope or basin currents. This is due in part to the smaller magnitude of the mean flow over the shelf, which may be more easily biased because of short-period current fluctuations. However, the majority of the apparent variability is most likely caused by the thinness of the layer of equatorward flow, which may be missed by a particular sampling scheme. Also, the location of the transition to the strong poleward flow that predominates over the slope can vary with time.

Water Properties

Lynn and Simpson (1987) recently updated the seasonal variation of the large-scale water properties of the California Current system, including the SCB. Earlier studies by Tibby (1941), Sverdrup and Fleming (1941), Wooster and Jones (1970), and others have also described the basic water mass characteristics. Water properties peculiar to the SCB have been described recently by Jackson (1986). The California Current system includes three distinctive water masses. The Pacific Subarctic water, which enters from the north, is characterized by relatively low temperature, low salinity, high dissolved oxygen, and high nutrients (Reid et al. 1958). North Pacific Central water, which enters from the west, is characterized by relatively warm temperature, high salinity, and low dissolved oxygen and nutrients (Reid et al. 1958). Pacific Equatorial water, which enters from the south, is characterized by relatively high temperature, high salinity, low dissolved oxygen, and high nutrients (Pickard 1964).

The equatorward advection of Pacific Subarctic water by the California Current is well illustrated in a map of salinity on the $\sigma_t = 25.0$ surface (Lynn et al. 1982; Lynn and Simpson 1987) (fig. 2.10). A tongue of low salinity occurs well offshore in the vicinity of the strongest equatorward flow. The tongue is bent shoreward and then poleward near the coast off San Diego as the California Current is drawn into the SCB, where it becomes the Southern California Countercurrent. North Pacific Central water is apparent as a high-salinity region offshore. This water does not generally enter the SCB directly, and it is consequently less readily identified in water properties in the SCB. Water in the SCB can be generally treated as a mixture of Subarctic and Equatorial water types (Tibby 1941).

The poleward advection of Equatorial-type water (alternately called "Southern" water) is illustrated by maps of salinity and oxygen on the $\sigma_t = 26.6$ surface (Lynn et al. 1982; Lynn and Simpson 1987) (fig. 2.11). This σ_t surface generally corresponds to a depth interval of 200–300 m, where strong poleward flow usually occurs. The jetlike nature of the poleward flow is confirmed by the narrow tongue of high-salinity, low-oxygen water adjacent to

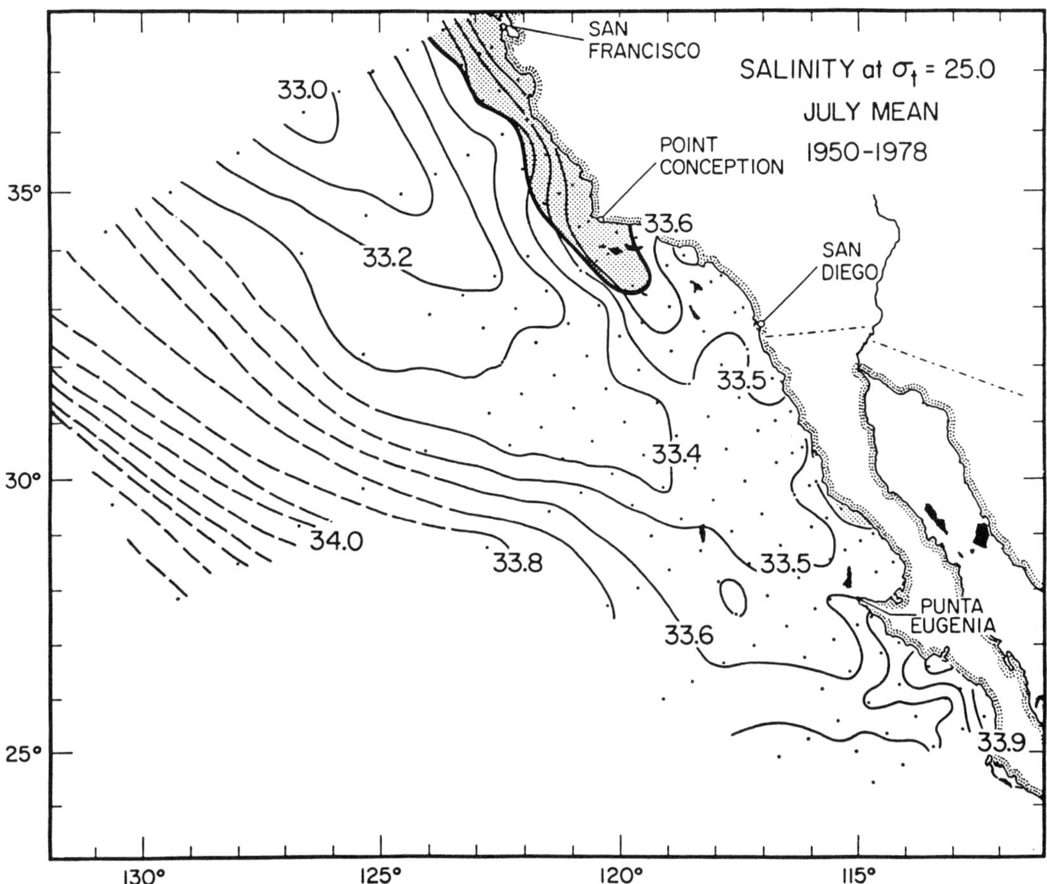

Figure 2.10. Salinity on the σ_t = 25.0 density surface during July, as derived from 1950–1978 averages of CalCOFI data. The intersection of this density surface with the sea surface is indicated by a bold line. Where density is less (shaded), surface salinity is used. Contour interval is 0.1 ppt. (From Lynn et al. 1982.)

the coast. As with the surface waters, spatial gradients in the water properties occur as the water is mixed laterally and vertically with surrounding waters.

Seasonal changes in the large-scale structure of the salinity, temperature, density, and oxygen fields in the SCB can be described from contoured sections of averages of data along CalCOFI line 90, which crosses the SCB (see fig. 2.2). Seasonal maps of all CalCOFI data averaged over the period 1950–1978 are presented in Lynn et al. (1982). Eber (1977) presents contoured depth–time charts of water properties for the period 1950–1966. The density field in all seasons except spring has the characteristic upward tilt toward shore and downward tilt inshore of the outer SCB below the upper 100 m that is indicative of the offshore equatorward flow and the nearshore poleward flow in the SCB (fig. 2.12a). The doming of the isopycnals is most pronounced during the summer when the currents are strongest. Isopycnals (as well as isotherms) move distances of 50–100 m in their seasonal march, with the maximum excursions occurring at depths of 50–300 m.

During spring in the upper ~200 m and summer in the upper ~100 m, the isopycnals adjacent to the coast tilt upward toward the coast. This upward tilt is indicative of the

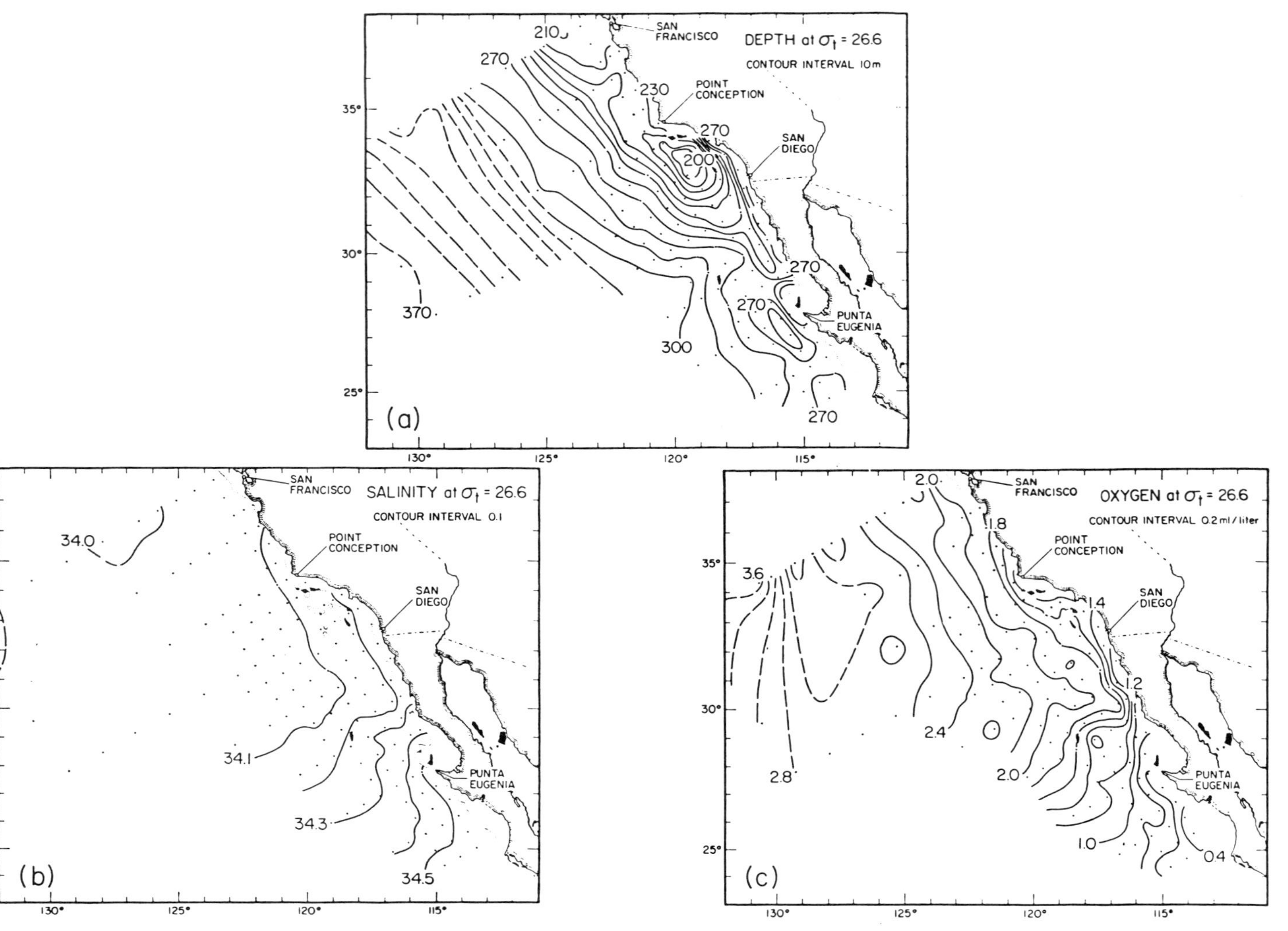

Figure 2.11. Salinity, oxygen, and depth on the σ_t = 26.6 density surface during July as derived from 1950–1978 averages of CalCOFI data. (From Lynn et al. 1982.)

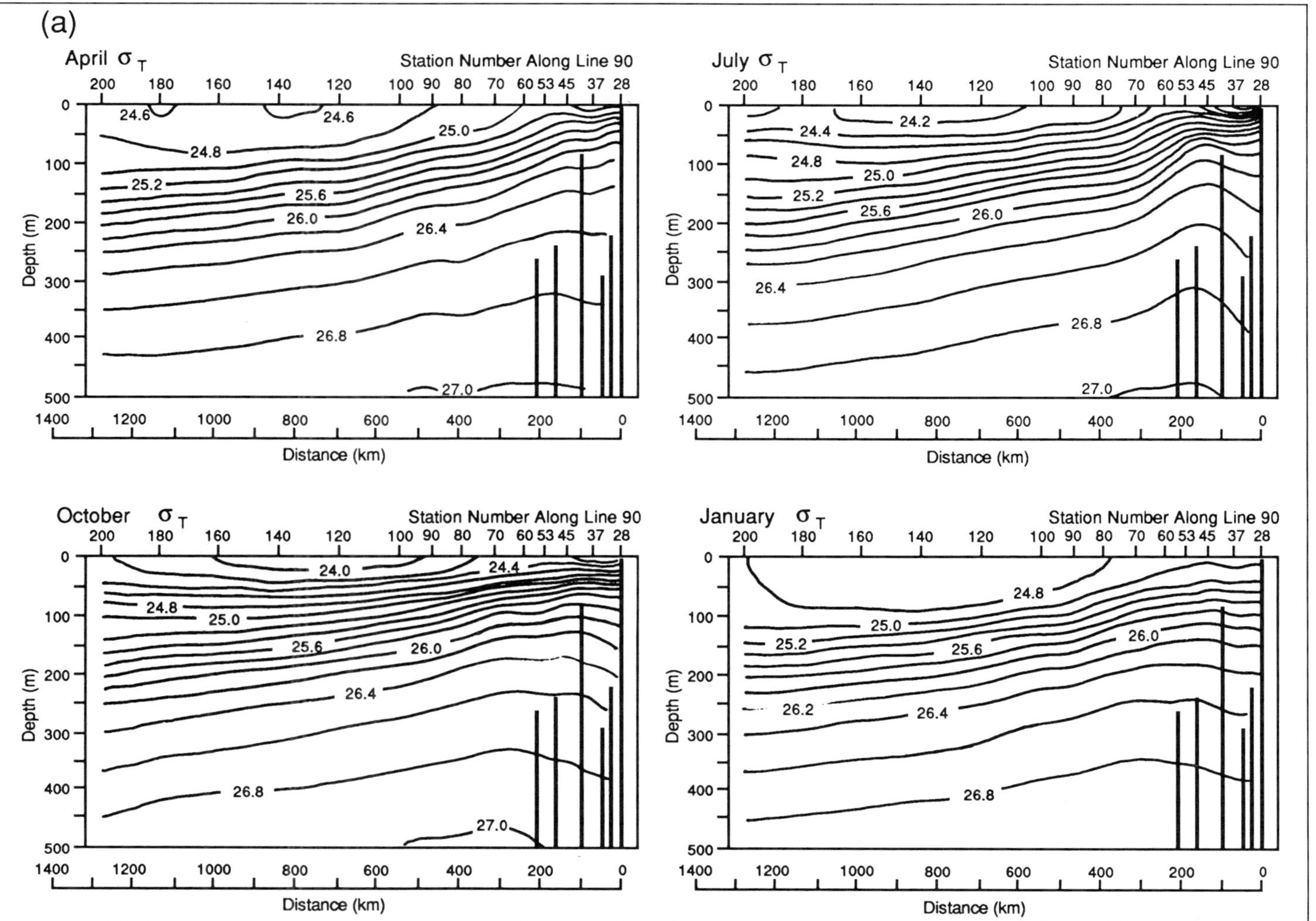

Figure 2.12. Contoured sections of seasonal averages of (a) density, (b) temperature, and (c) oxygen content along CalCOFI line 90 (see fig. 2.2). Data were averaged over about 60 days centered on the months given with each graph. The Channel Islands and ridges are delineated by the black spikes emanating from the bottom of the graphs. (From Lynn et al. 1982.)

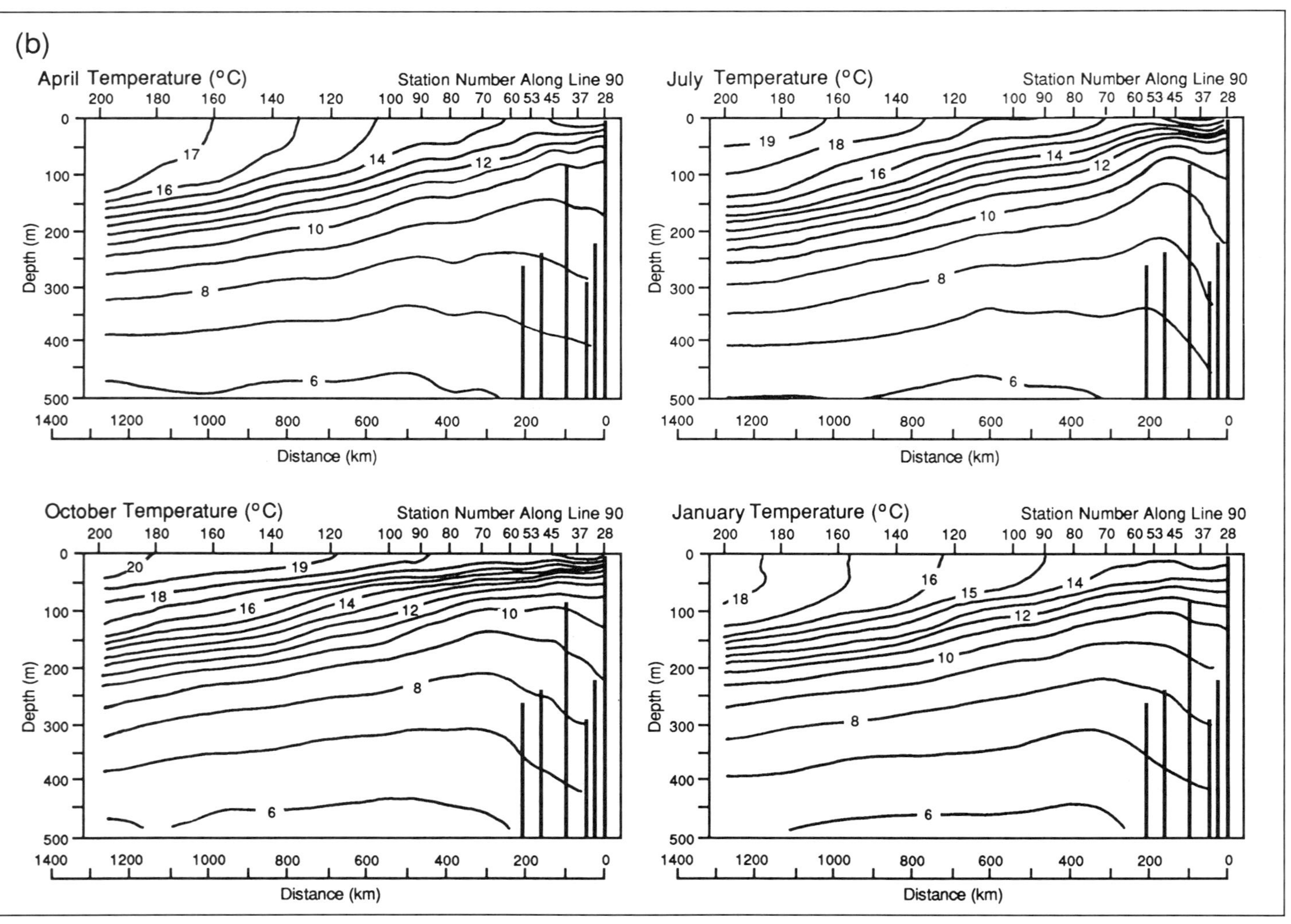

Figure 2.12. (continued)

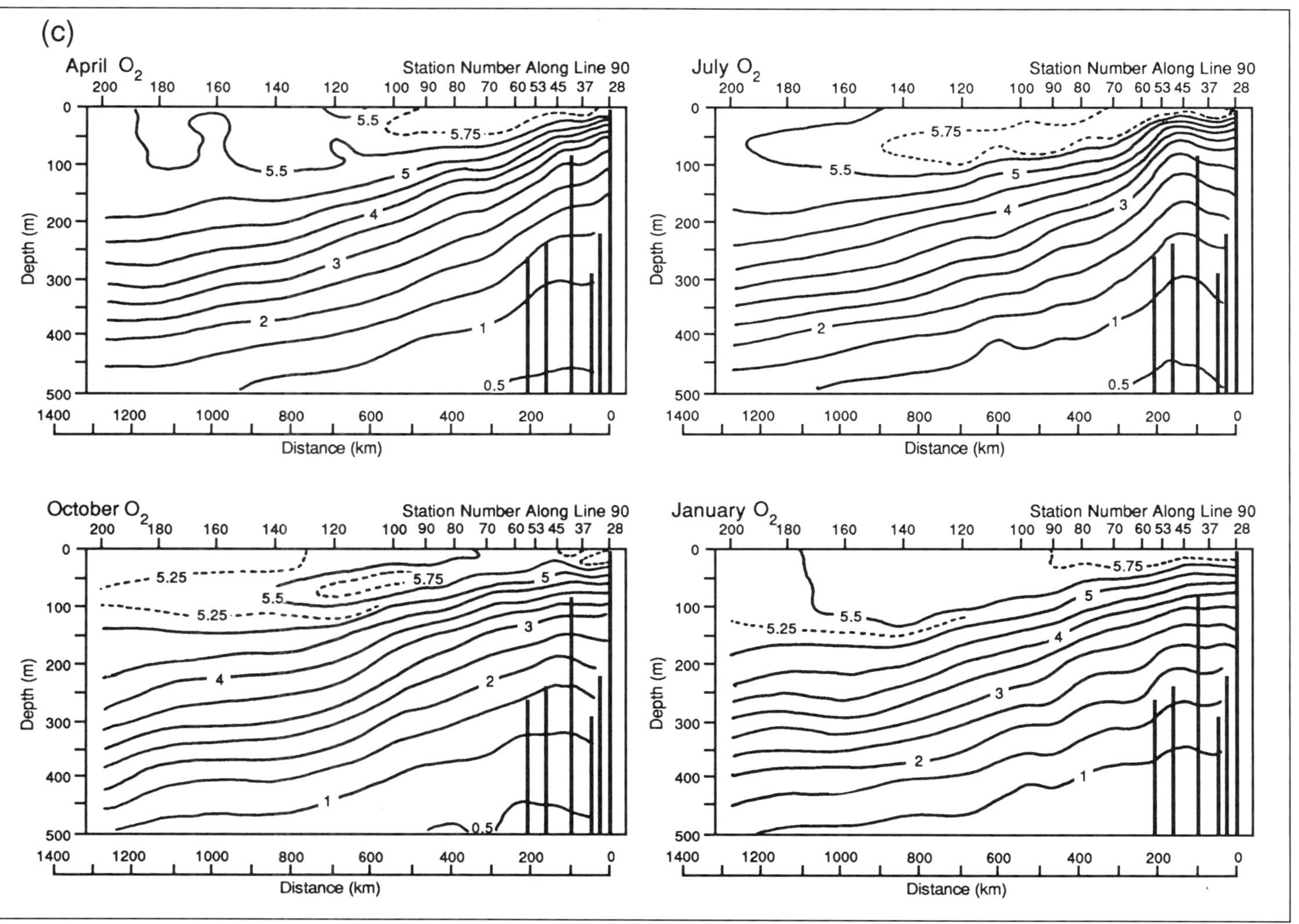

Figure 2.12. (continued)

equatorward flow and coastal upwelling usually observed near the coast. Because of the coastal upwelling, the maximum density (minimum temperature, maximum salinity) occurs a month or two earlier nearer the coast than farther offshore (Jackson 1986). The upward tilt of the isopycnals provides, via upwelling, a source of nutrients to the coastal waters. A strong pycnocline is apparent over the SCB during the summer. This pycnocline develops in response to seasonal warming (fig. 2.12b), and the strong stratification is maintained through early fall. Near-surface stratification is maintained in the SCB even during the winter season. Waters offshore of the SCB, which are subject to more intense wave action, do not develop as strong a seasonal thermocline or pycnocline.

The temperature field beneath the surface layer has roughly the same structure as the density field (compare figs. 2.12a and b). The thermocline tilts upward toward the coast and domes over or just offshore of the Channel Islands, particularly in summer and fall. Thus, water within the SCB at undercurrent depths (200–400 m) is typically at least 1°C warmer than that offshore at similar depths. Upwelling of cold, deep water is apparent next to the coast in spring and summer and is strongest during the spring. Warmest surface waters occur in early fall. The temperature in the surface waters increases in the offshore direction. In general, the surface waters inside the SCB are 2–4°C colder than those in the regions offshore of the SCB.

Oxygen isopleths, like those of temperature and density, tilt upward toward the coast (fig. 2.12c). Doming over the Channel Islands, although less dramatic than in the temperature field, is observed to be strongest in the summer. Upwelling of high-oxygen water is evident near the coast in the upper layers, particularly in the spring and summer. The high-oxygen water characteristic of the equatorward-flowing California Current is evident as a subsurface maximum in regions offshore of the SCB. The region of highest oxygen occurs closer to the coast in spring and winter.

Subtidal Fluctuations

Currents

The SCB is continuously bathed in current fluctuations of large amplitude (approximately 20–40 cm s^{-1}) with dominant long periods (20–30 days) (Hendricks 1976; Hickey 1992). This result may be surprising to investigators familiar with other west coast nearshore regions, such as the Coastal Ocean Dynamics Experiment (CODE) area off northern California (Winant et al. 1987) or the Oregon–Washington coast (Hickey 1989a). The circulation in these shelf regions is predominantly wind-driven and, consequently, has shorter time scales (~5–10 days). The relatively light winds in the SCB might lead one to expect the currents to be correspondingly small. However, large-amplitude fluctuations have been observed in the upper 200 m in and over the Santa Barbara (Brink and Muench 1986), Santa Monica, and San Pedro basins (Hickey 1992) and on the shoreward side of the San Diego Trough as far south as San Diego (Hendricks 1976). In the Santa Barbara Channel, the dominant fluctuations are coherent over the entire basin, including all the inter-island passages (fig. 2.13). The investigators concluded that these fluctuations were of a very large spatial scale. The fact that similar fluctuations also occur in the Santa Monica and San Pedro basins and at least on the shoreward edge of the San Diego Trough suggests that this is indeed the case.

The current fluctuations, like the mean flow, tend to follow the regional curvature of basin 200-m-depth isobaths, except in the very near surface and near-bottom frictional boundary layers (Hendricks 1976; Hickey 1992). Over the shoreward slope of the Santa Monica Basin, currents can attain speeds of more than 40 cm s^{-1} for one or more weeks (fig. 2.14). Such speeds would imply very short residence times in the SCB, if the currents were similar at all locations. As discussed later, however, the currents have considerable spatial variability, such as might be attributed to eddies or other physical pro-

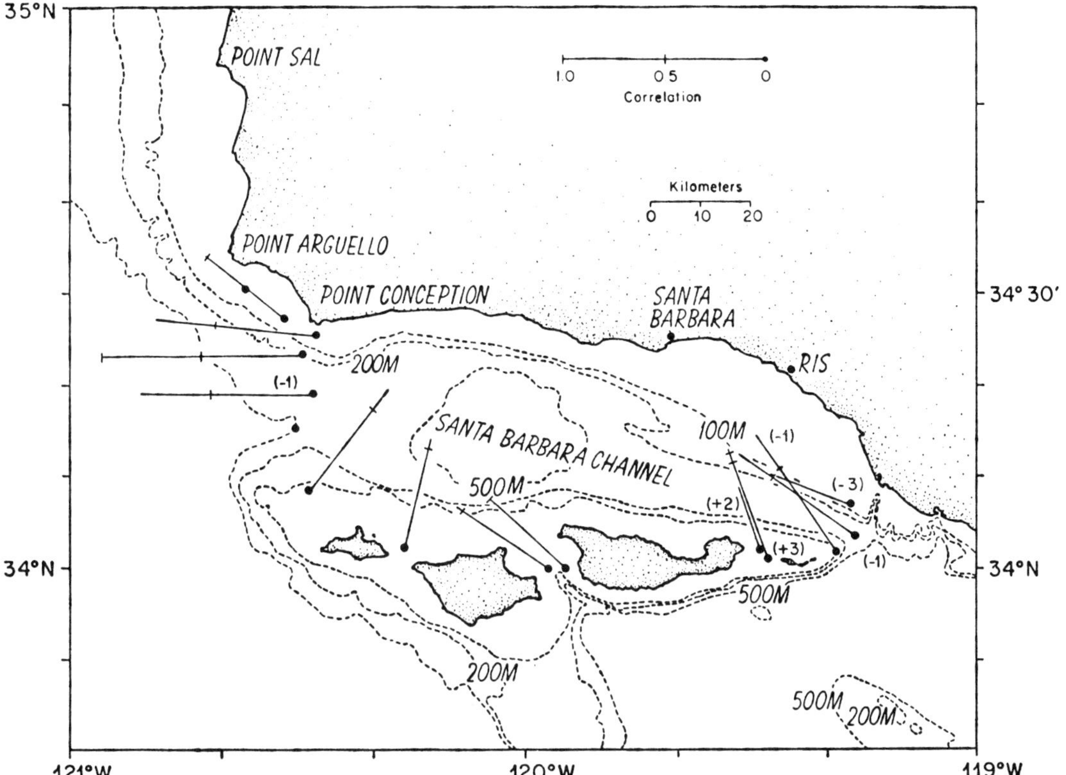

Figure 2.13. Correlations of first velocity EOF with velocity time series at a mid-channel location at the eastern entrance to the channel. The length of the arrow is proportional to correlation magnitude. Orientations are relative to having flow directly westward at a mid-channel location at the western entrance to the channel. Time lags (positive for the western entrance stations leading) are shown in parentheses for lag magnitudes greater than 0.75 day. (From Brink and Muench 1986.)

cesses, so that actual residence times could be much longer. This is particularly true in the near-surface layers where mean flows are much smaller and the eddy field is stronger (Hickey 1992). Surface turbidity data in Thornton (1981) clearly illustrate the mesoscale eddy structure in surface currents over these nearshore basins.

The longest period fluctuations have a subsurface maximum at about 100 m from the surface in the Santa Monica–San Pedro area (Hickey 1992), but may shoal toward the west in the Santa Barbara Channel (Blumberg et al. 1987). The amplitude of these fluctuations increases from a minimum in spring to a maximum in summer–fall and winter (Hickey 1992). Hickey (1992) uses the technique of EOF analysis to determine the spatial structure of the very long period (approximately 20–30 days) fluctuations over the Santa Monica and San Pedro basins. The long-period fluctuations shown in figure 2.14 are well described by spatial structures such as those shown in figure 2.15. The long-period fluctuations are stronger on the shoreward side of the Santa Monica and San Pedro basins during winter (fig. 2.14) and late summer to fall (fig. 2.15). This is not the case, however, during the early summer, when fluctuations are of relatively constant amplitude across the entire basin (fig. 2.15). A seasonal change in the alongshore scale of the fluctuations is also evident (fig. 2.16). During spring and early summer, it is difficult to see much similarity between adjacent pairs of records, whereas during late summer and fall, many spatially

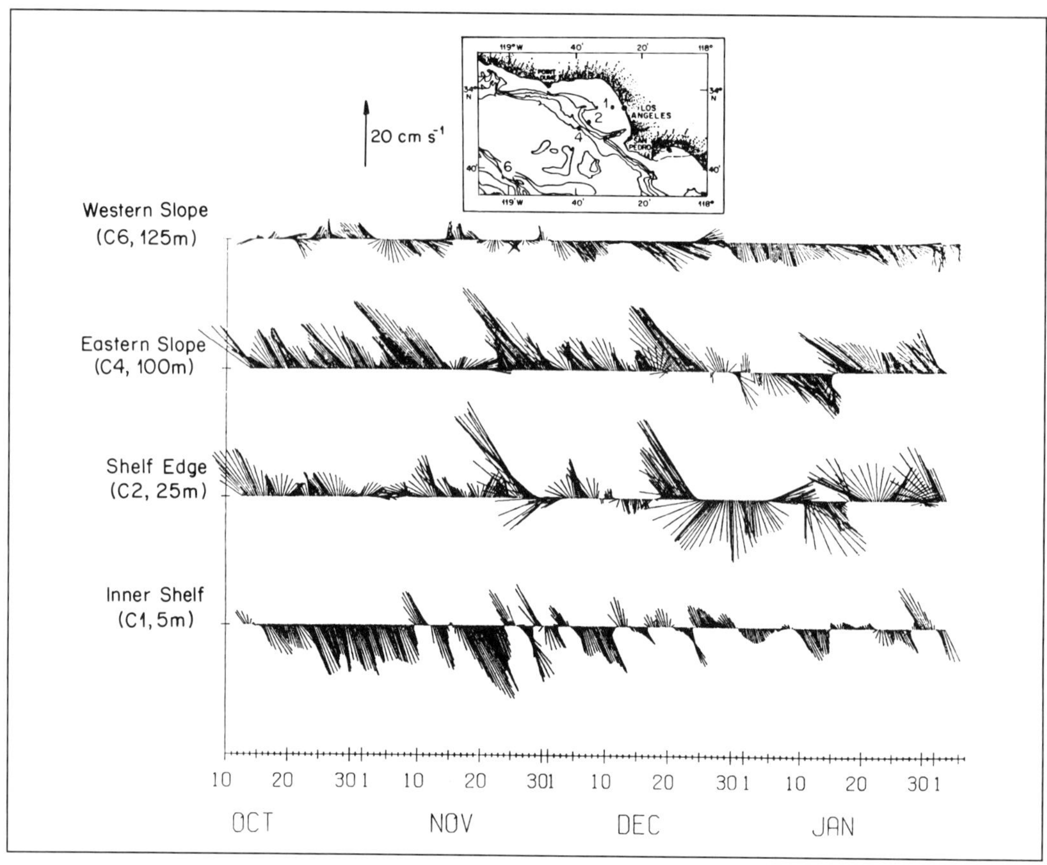

Figure 2.14. Six-hour velocity time series at sites on a section across the Santa Monica Basin and shelf from October 1985 to February 1986. (From Hickey 1992.)

coherent pulses are observed. Hickey (1992) shows that the pulses are also observed in coastal sea level records at locations 100 km and even 1000 km to the south.

Likely candidates to account for these basin-scale fluctuations include (1) freely propagating waves generated by remote wind forcing or some other mechanism and (2) eddy advection by the mean flow. The relatively weak wind stress makes local wind forcing seem highly unlikely, and Hickey (1992) and Brink and Muench (1986) find little evidence of strong correlation of basin currents with the local wind fields in the Santa Monica and Santa Barbara basins, respectively. Brink and Muench (1986) also demonstrate that remote wind forcing as far as 200 km upstream of the Santa Barbara Channel is not related to the observed variance. They find no support for poleward propagation of the signals from San Diego to the Santa Barbara Channel during the spring season investigated.

Halliwell and Allen (1984), however, demonstrate that sea level signals generated off Baja California during summer 1973 did propagate through the SCB in a manner consistent with coastal-trapped wave theory. Basically, this theory demonstrates that longshore wind stress off Baja California causes long (a few thousand kilometers) undulations in the sea surface with amplitudes of a few centimeters. These "waves," which hug the continental shelf and slope, travel poleward at speeds of a few hundred kilometers a day. As the waves pass by a given location, the currents are also affected. They are accelerated first poleward

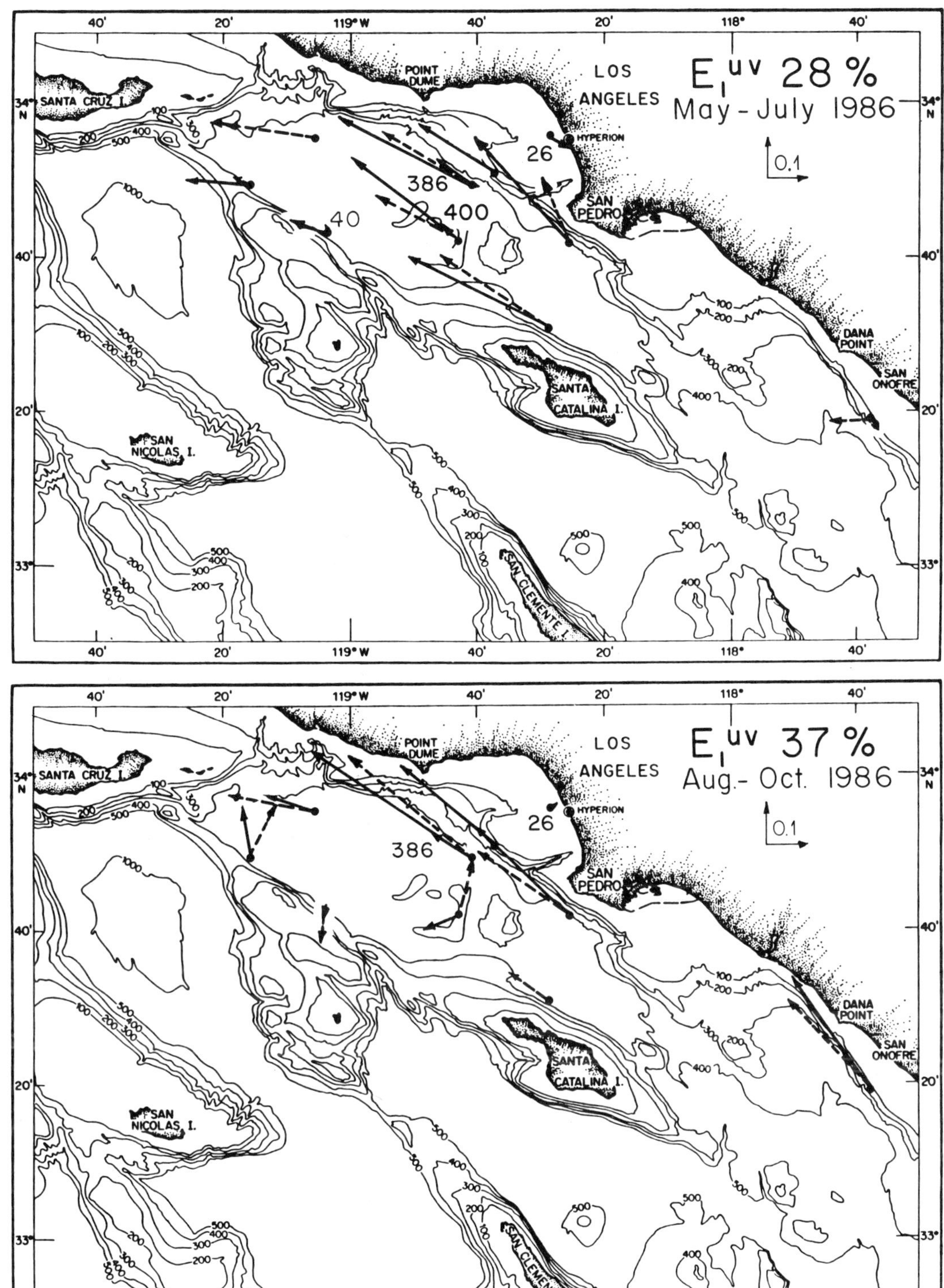

Figure 2.15. Amplitude of the first velocity EOF at upper water column locations in the Santa Monica and San Pedro basins and in the San Diego Trough during May–July 1986 (top) and August–October 1986 (bottom). Solid lines represent the 100-m depth, dashed lines the 40-m depth. Other depths are indicated by actual numbers. The percentage variance accounted for by the EOF is shown in the upper right corner. (From Hickey 1992.)

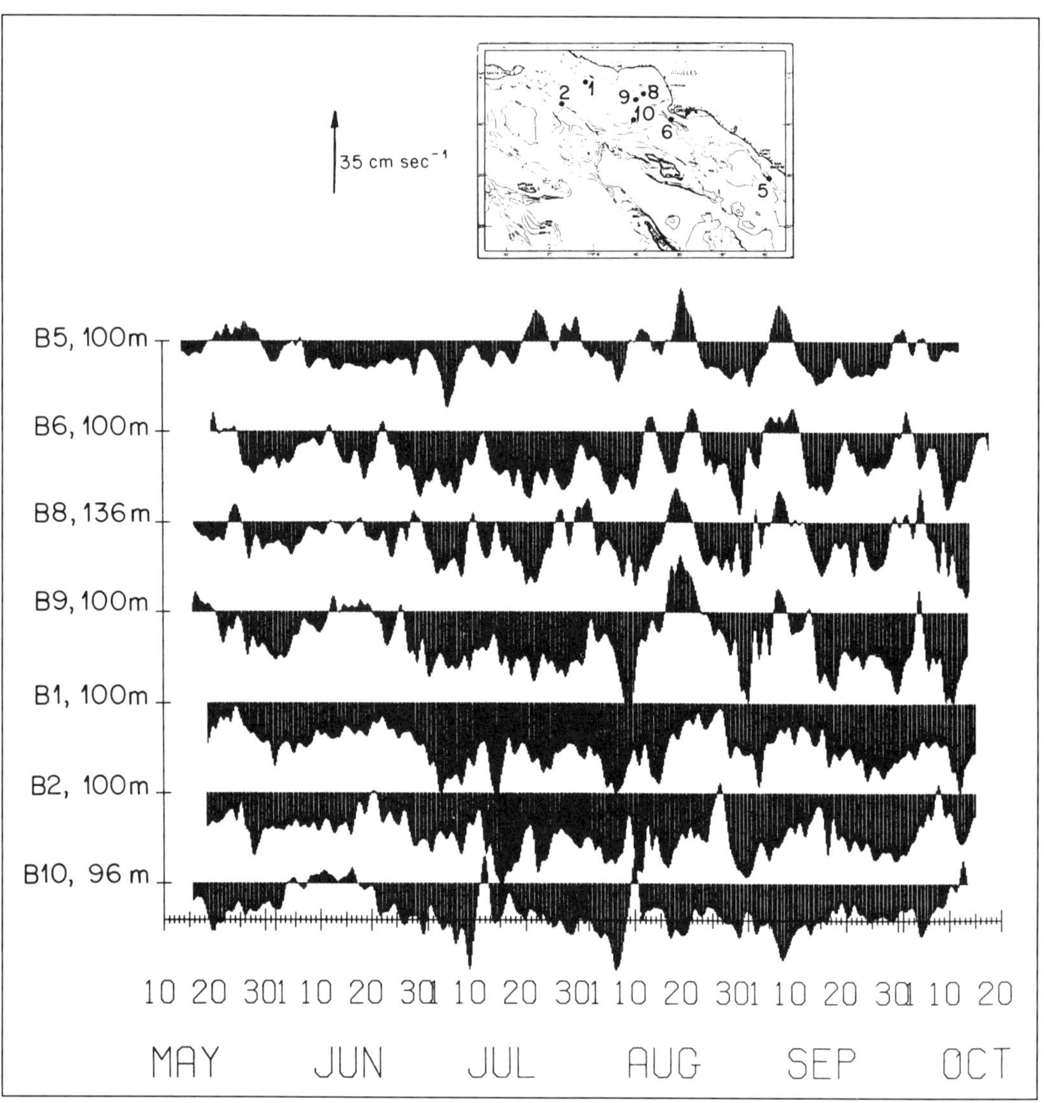

Figure 2.16. East-west component of velocity from May to October 1986. Onshore flow is positive. Station locations are shown on the inset map. (From Hickey 1992.)

and then equatorward (that is, one might see pulses of currents superimposed on any mean flow). The waves can be modified by local wind at a given location, the degree of change being dependent on the magnitude of the local longshore wind stress. Such waves are common over the Pacific northwest shelf (Hickey 1989a).

It appears likely that the late summer–fall and winter Santa Monica Basin fluctuations, which are more strongly trapped over the shoreward slope of the basin, may be the signature of freely propagating coastal-trapped waves. The time delay between the occurrence of signals at San Onofre and at San Pedro during winter 1985–1986 and also late summer–fall 1987 was consistent with poleward travel at roughly the speed of a first-mode shelf wave in this region (approximately 150–250 km d^{-1} at a period of 20 days)

(Hickey 1992). However, the speed decreased by an order of magnitude poleward of San Pedro. Hickey (1992) attributes the speed decrease to the changes in the shape of the coastal channel that occur poleward of the San Diego Trough. The long-period fluctuations over the Santa Monica and San Pedro basins during the spring do not demonstrate a significant relationship to fluctuations outside the basins. Hickey (1992) attributes these to more local effects such as would be caused by the passage of eddies being advected by the mean flow.

Fluctuations in currents over much of the relatively wide (approximately 20-km) semi-enclosed Santa Monica shelf are not directly related to the slope current fluctuations (fig. 2.14). Hickey (1992) demonstrates that although the outer shelf currents are directly related to the slope and basin currents, the inner shelf currents are driven in part by local wind stress, but primarily by the interaction of the slope currents with the semi-enclosed shelf: the cross-isobath pressure gradients associated with the slope currents drive flow in the opposite direction over the inner shelf.

On the relatively narrow (approximately 4-km) and open shelf from San Pedro to San Diego correlation with local wind stress is weak (Winant 1983; Winant and Bratkovich 1981) and a large percentage of the variance (as on the outer shelf in the wider shelf areas) is likely directly related to the large-scale basin and slope fluctuations. Indeed, dominant time scales in the Winant and Bratkovich data appear to be relatively long (approximately 20 days). The maximum energy occurs in fall and winter, and the energy in the fluctuations increases from the inner to the outer shelf in all seasons (Winant and Bratkovich 1981). However, the longshelf scales over which correlations are significant are relatively short (approximately 25 km) over the 50-day summer period of the observations (Winant 1983). It is likely that correlation scales for these long-period fluctuations were not well resolved by the relatively short data set.

Subtidal scale fluctuations have been observed at all depths throughout the water column in the Santa Monica and San Pedro basins and over their slopes (Hickey 1992). The amplitude of such fluctuations decreases dramatically below 200 m from the surface, to amplitudes of approximately 5 cm s^{-1} at depths of approximately 700 m from the surface. Small ($\leq$5 cm s^{-1}) fluctuations are observed even below the depth of the deepest basin sill (Hickey 1992). Similar fluctuations are expected to occur throughout the rest of the SCB. Much larger subtidal velocity fluctuations (up to 30 cm s^{-1}) with periods of 5–10 days have been observed in the bottom 100–200 m over the Santa Monica Basin sills and should be expected on the sills between other basins (Hickey 1993a).

Both near-surface and near-bottom flow can be substantially altered by the effects of friction. The retarding force of the bottom tends to both diminish the amplitude of the flow and change its direction in a layer whose thickness is on the order of 5–20 m over the shelf and slope. In general, the flow is rotated a few degrees to the left of its direction above the boundary layer. Thus, flow in the bottom boundary layer is directed offshore for poleward flow and onshore for equatorward flow. The cross-isobath flow can provide a mechanism for net offshore transport of sediment. In the near-surface layers, a local wind-driven component of flow occurs. This frictional current is superimposed on any large-scale currents and is particularly important over the shelf where the larger scale currents may be small. The thickness of the frictional surface layer varies from 5 to 40 m, often being delimited by the pycnocline. Over shallow shelf areas, surface and bottom frictional layers may overlap.

Submarine canyons occur frequently over the continental slopes in the SCB and may be important for the offshore dispersal of sediment to basin floors. For the most part, however, the canyons have a relatively small width (<5 km) and therefore can be expected to have a correspondingly minor effect on regional circulation. Indeed, recent current measurements over Redondo Canyon (just north of Palos Verdes) at a depth of 30 m suggest that the flow passes directly over the canyon with no apparent modification. SCCWRP measurements at a mooring within Santa Monica and San Gabriel canyons showed no significant correlation between canyon and shelf flows (Hendricks 1980, 1984). Below the lip of these canyons, subtidal fluctuations as well as net flows are weak, the currents being dominated by higher frequency fluctuations including diurnal and semidiurnal tides. Model studies of canyon circulation and the effect of canyons on regional circulation suggest that even very narrow canyons such as those in the SCB should have some effect on the overlying circulation, especially in the density fields (Klinck 1989). Thus, it is possible that (as has been demonstrated for wider canyons off Washington) a mean and fluctuating eddy (or both) is trapped within the canyon and that the density field is perturbed over the canyon walls (Hickey 1993b). Maximum velocities may also occur over the steep canyon walls where measurements are difficult to obtain. The flow in such submarine canyons and the effects of the canyons on the overlying circulation and on the density field are being examined as part of ongoing research programs.

Water Properties

Temperature–time series from a location just below the pycnocline over the Santa Monica slope illustrate that water properties undergo relatively large-amplitude fluctuations with roughly the same temporal scales as those of the velocity field (fig. 2.17). For example, water temperature at 40 m from the surface can change by as much as 2°C over a period of several days. Many of these changes are a result of vertical displacement of water masses that occur as the longshore current changes in speed. The density field adjusts to maintain geostrophic equilibrium. For this case, the changes in the density (temperature and salinity) field are coincident with those in the horizontal velocity field. Higher frequency (tidal and supratidal) fluctuations are also apparent in the record. These are due to vertical excursions related to the semidiurnal tide and to internal waves (see next section).

Satellite sea surface temperature (SST) data indicate that the SCB contains a rich mesoscale eddy field, at least in the very near-surface layers (e.g., Hickey 1992). The depth of such features appears to be at least the depth of the seasonal thermocline (Fiedler 1988; Hickey 1992). The scales of the many features appear to be relatively short (approximately 20–40 km) in comparison with the relatively large scales (approximately 200 km) associated with eddies in the offshore California Current (Simpson et al. 1984). This suggests that many of these features may be generated locally—either by horizontal or vertical current instabilities or by interaction of the mean flow with the highly irregular SCB bottom topography. The advection of such eddies by the mean flow will produce local changes in temperature and salinity as the eddy passes by a particular point. To the extent that the mesoscale eddy field is significant at a depth of approximately 40 m from the surface, some of the temperature fluctuations in the record shown in figure 2.17 could be attributed to horizontal eddy advection.

Patches of cold water that occur in response to local wind-driven upwelling are apparent in satellite SST maps. For example, such patches are frequently observed in Santa Monica Bay and over the San Pedro shelf and might be expected on the seaward side of islands in the SCB (Hickey 1992). Upwelling of cold water occurs during periods of equatorward winds when warmer surface waters are moved offshore and replaced by deeper water. Dramatic local upwelling events are only observed in winter and early spring, when nearshore winds within the bight are comparable in magnitude to those offshore (Hickey 1992). In summer and fall, winds are weak

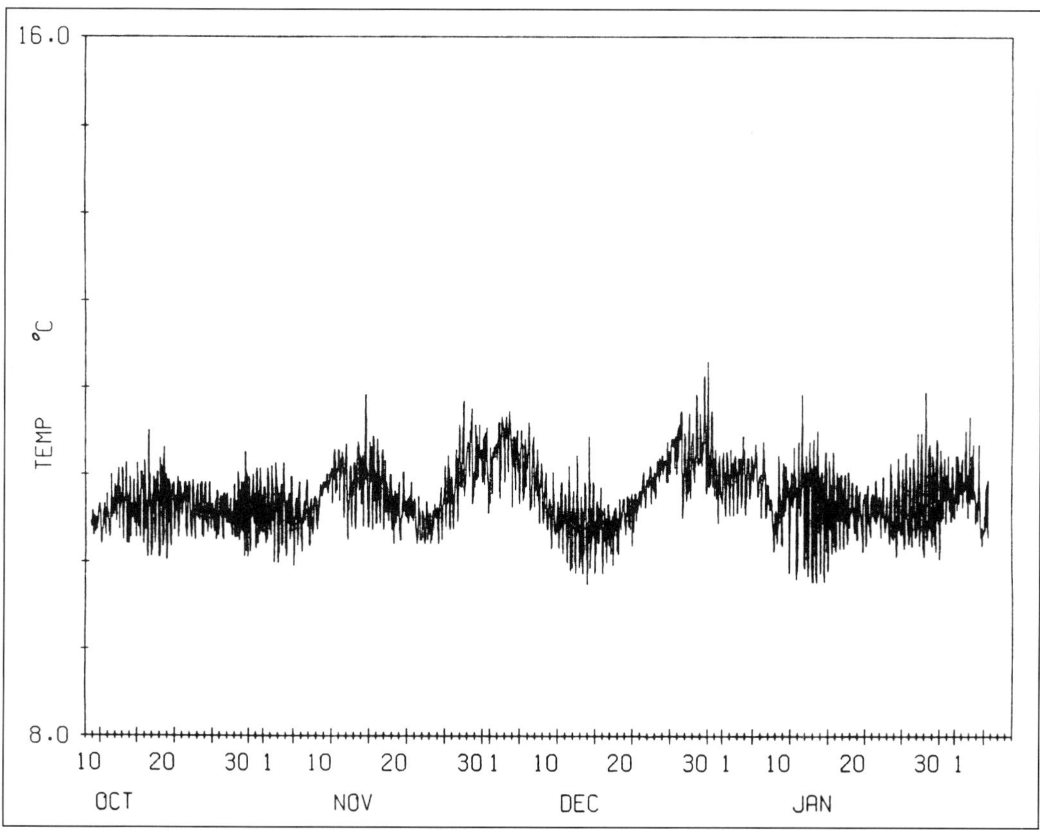

Figure 2.17. Time series of temperature at a location over the Santa Monica slope at a depth of approximately 100 m from the surface during winter 1985.

and local upwelling is rarely observed. Dorman and Palmer (1981) used sea surface temperature records in the SCB to demonstrate that significant upwelling events occur only a few times per year and last for 1 or 2 weeks in the region from Newport Beach to San Diego. Whether or not there are regions in which shelf waters move offshore in jets and squirts, such as is commonly observed off northern California (Kelly 1985), has not been addressed on a regional scale. However, a recent study utilizing direct current measurements, Lagrangian drifters, satellite sea surface temperature data, and hydrographic data demonstrate that strong offshelf flow, with resulting eddy-like features over the basin and its slope, occurs frequently over the southern half of the Santa Monica shelf (Hickey 1992). The reason for this offshelf flow is presently unknown and is the subject of ongoing studies in the area.

Upwelling and a major plume associated with it have been studied off Point Conception (Atkinson et al. 1986). During wind-driven upwelling events along the coast north of Point Conception, a southwest-directed plume of cold water emanates from the Point Conception area. The plume often turns eastward and enters the Santa Barbara Channel on its southern side, sometimes traveling all the way to the eastern entrance of the channel. The cold water plume is very evident in maps of satellite SST (Hickey 1992). It has a major effect on time-variable water properties in the Santa Barbara Channel, extending to depths of 50–70 m on the south side of the channel. The plume water appears to be mostly recirculated within the channel, that is, it does not generally affect water properties in nearshore basins farther south, such as Santa Monica. However, satellite data show that plume water often enters the SCB to the south of the Santa Barbara Channel, where it impacts water properties in the offshore basins (Atkinson et al. 1986). The contribution of the Point Conception upwelling plume to the SCB in terms of nutrient supply is likely to be an order of magnitude greater than that of local upwelling.

Tidal and Supratidal Fluctuations

Spatially comprehensive analyses of tidal and higher frequency current and water property fluctuations are available only for the southern California mainland shelf. The measurements were collected in three separate experiments by Winant (1983) and Winant and Bratkovich (1981). The first experiment provided detailed vertical resolution of currents at one site during the summer, the second described cross-shelf structure of current and temperature fluctuations and their seasonal variation, and the third focused on longshelf correlation of current fluctuations during the summer.

Results indicate that, at least for the shelf, the fluctuations can be divided roughly into a tidal band (periods of a few hours to one day) and a supratidal band (periods of a few hours to a few minutes) (Winant and Olson 1976; Winant and Bratkovich 1981). The tidal band includes not only tidally forced fluctuations, but also inertial and wind-forced fluctuations. The researchers made no attempt to separate these several forcing mechanisms. Most analyses were performed on the whole tidal band and are dominated by the semidiurnal and near-diurnal periods. At these latitudes the inertial period is approximately 22 hours, so that the diurnal tide and inertial peaks are indistinguishable in the standard spectral plots. At mid-shelf during the summer, the diurnal signal is much stronger near the surface than near the bottom, suggesting that inertial signals may be significant (Winant 1983).

A representative record from Winant and Olson (1976) illustrates that tidal band velocities over the shelf during the summer often exceed those of subtidal currents (fig. 2.18). In these records, which were obtained in a water depth of 17 m, semidiurnal cross-shelf tidal currents are on the order of 10 cm s^{-1} and are strongly baroclinic (depth and, presumably, density dependent). Maximum velocities occur not in the central water column but in the surface and bottom boundary layers. Vertical shear as large as 50 cm s^{-1} was commonly observed over the 12-m instrument span. Longshelf currents also exhibit semidiurnal periodicity. Longshelf velocities are on the order of 10 cm s^{-1}. In contrast to the cross-shelf currents, however, the longshelf currents are relatively barotropic (depth independent). The dominant semidiurnal longshelf fluctuations are not generally phase locked with the surface tide (Winant and Olson 1976; Winant and Bratkovich 1981), which is consistent with their baroclinic nature. This is evident in the cross-shelf currents shown in figure 2.18. The more barotropic longshelf currents are phase locked with the surface tide only on the shallow inner shelf (fig. 2.18).

EOF analyses at individual sites in 15-, 30-, and 60-m bottom depths (the inner, middle, and outer shelf, respectively) illustrate that these basic characteristics are similar across the shelf and in every season (fig. 2.19). However, details of the vertical structure of the fluctuations are a function of season and location, presumably due to stratification differences. For example, the zero crossing for the first mode for cross-shelf velocity occurs at 30–40 m near the shelf edge, at approximately 15 m over the mid-shelf, and at approximately 10 m over the inner shelf. Vertical shear in the cross-shelf currents is a maximum in summer and fall on the outer shelf, but in spring at mid-shelf. The amount of variance in cross-shelf currents in the tidal

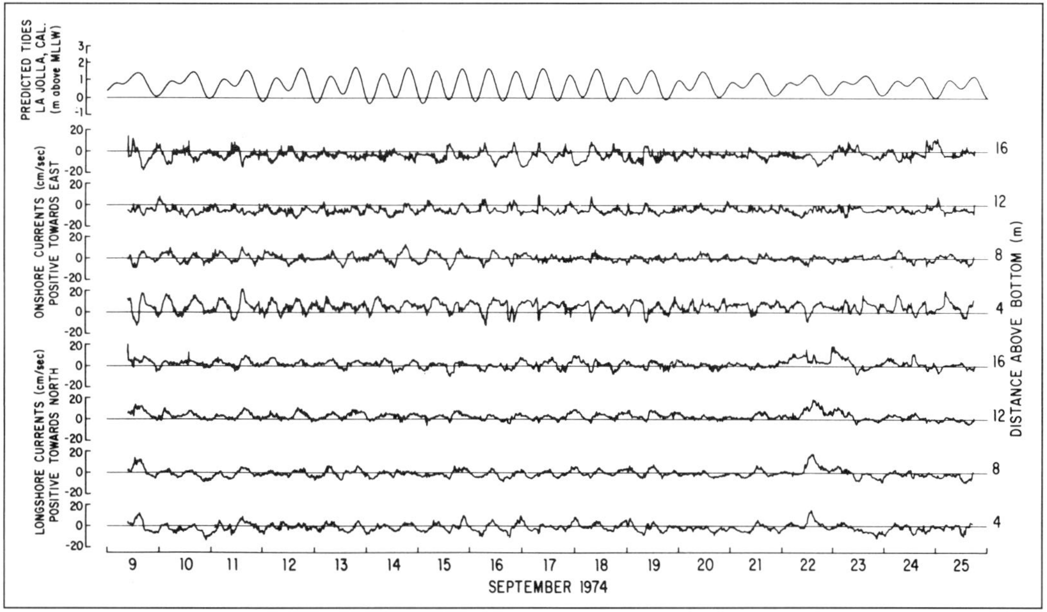

Figure 2.18. Current and tidal fluctuations, September 9–25, 1974. The bottom four traces represent the longshore currents, the next four traces represent the onshore currents, and the top trace is a prediction of tidal elevation for La Jolla, California. (From Winant and Olson 1976.)

band is also a strong function of season, decreasing from 24 $cm^2\ s^{-2}$ in spring to approximately 12 $cm^2 s^{-2}$ in summer and fall and to 7 $cm^2\ s^{-2}$ in winter. Variance in longshelf currents, in contrast, is relatively constant throughout the year (approximately 30 $cm^2\ s^{-2}$), consistent with the barotropic (not density related) nature of these fluctuations.

Cross-shelf tidal band currents decrease toward shore so that on the inner shelf, the energy of the cross-shelf currents is an order of magnitude less than that on the outer shelf (Winant and Bratkovich 1981). The relationship between the cross-shelf and longshelf velocity components also changes across the shelf. The two components are coherent, with an approximately 90° phase lag on the outer shelf, but are decoupled on the inner shelf. Cross-shelf currents and temperature are correlated at all shelf locations at tidal and higher frequencies, with a phase difference of approximately 90° near the bottom of the water column and approximately 270° near the surface. These phase differences suggest that the internal tide has the configuration of a standing wave in the cross-shelf plane (Winant and Bratkovich 1981).

Longshelf correlation scales for both cross-shelf and longshelf currents in the tidal band are, with one exception, very short ($\leq$5 km). Near the bottom at mid-shelf, some evidence suggests that cross-shelf tidal currents propagate poleward. The phase speed is on the order of 100 cm s^{-1}.

Supratidal current fluctuations are ubiquitous over the southern California shelf during periods of strong stratification (LaFond 1962; Cairns 1967). These internal waves progress toward the coast with phase speeds of 15–25 cm s^{-1} (Winant and Olson 1976; Winant and Bratkovich 1981). Such fluctuations are apparent in figure 2.17 as the jitter superimposed on the tidal currents. A blowup of part of the record in figure 2.18 illustrates the details of the internal wave currents and their associated temperature fluctuations (fig. 2.20). The current speeds are on the order of 3 cm s^{-1}, and as for the tidal band currents, the internal

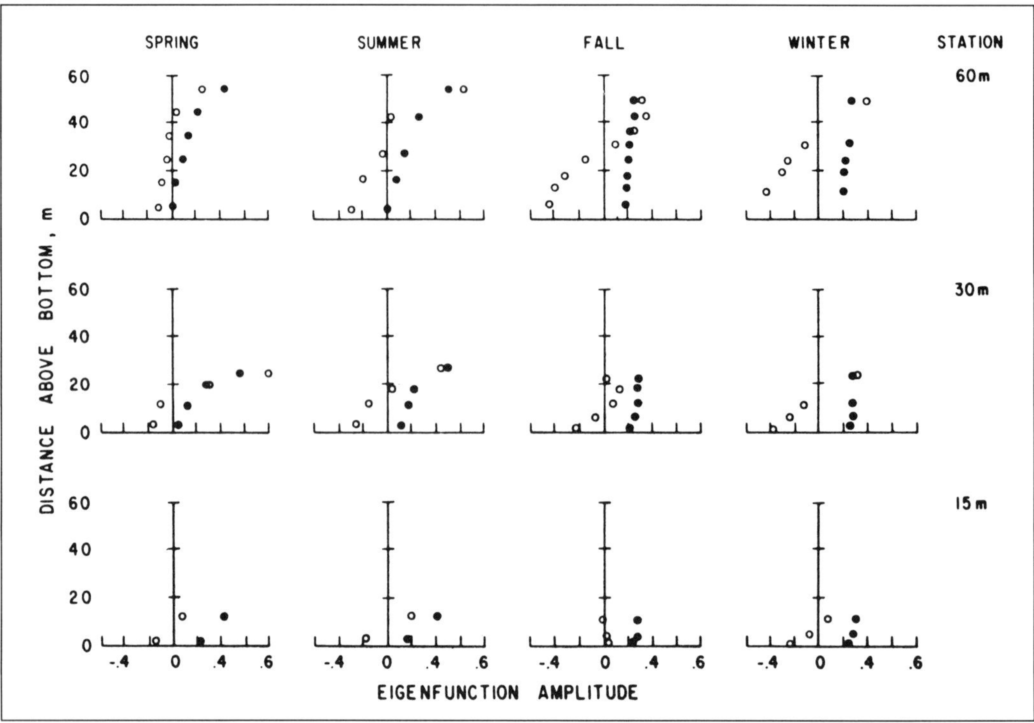

Figure 2.19. Seasonal distribution of the amplitude of the largest eigenvectors of the longshore (●) and cross-shelf (○) tidal band (periods between 4 and 36 hours) currents on a section across the shelf near Del Mar, California (see fig. 2.3 for locations). (From Winant and Bratkovich 1981.)

wave currents are larger in the upper and lower boundary layers than in the interior of the water column.

Winant and Olson (1976) illustrate that the observed velocity and temperature fluctuations are consistent with a three-layer model of the density field, with two well-mixed layers separated by a pycnocline in which density increases linearly with depth. The variance due to such high-frequency internal waves is a strong function of the density stratification and, hence, season. The variance associated with the high-frequency band is two to three times greater in spring and summer than in fall and winter at all locations on the shelf (Winant and Bratkovich 1981).

The broad details of these results for the tidal band should be applicable in all narrow shelf coastal areas in the SCB, for example, from Newport Beach to San Diego. Tidal currents around islands, in interisland passages, over sills, over the continental slope, and over wider shelves would be expected to have significantly different characteristics than those just described. In particular, for wider shelves, the internal tide might be expected to propagate toward shore, as observed off Oregon (Torgrimson and Hickey 1979), rather than forming a standing wave. The high-frequency internal wave results would be expected to have a broader applicability since they depend only on the density profile and not on the underlying bottom topography, except perhaps in very shallow areas. Thus, we would expect high-frequency internal waves to occur near the seasonal pycnocline over shelves, slopes, and basins in the SCB. The exact vertical structure and current amplitude would depend on the local density stratification.

Although detailed analyses of tidal currents have not been performed for wide shelf

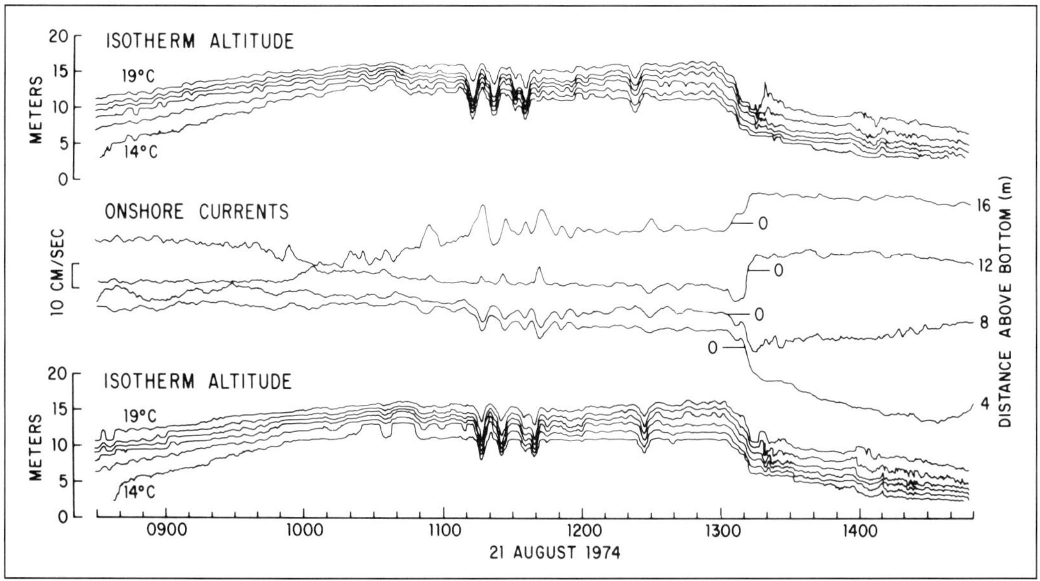

Figure 2.20. Temperature structure and onshore currents on August 21, 1974, offshore of San Diego. The top and bottom panels represent isotherm heights as a function of time as interpolated from measurements taken with vertical thermistor arrays located west (top) and east (bottom) of the current meter array. The thermistor arrays were 45 m apart and were located 130 m south of the current meter array. (From Winant and Olson 1976.)

or continental slope areas in the SCB, available hourly data can provide some insight into the spatial characteristics of the tidal band currents. In particular, data collected over the Santa Monica shelf and slope suggest that tidal currents on the inner shelf may be weaker than those on the narrower southern California shelf in the same water depth (approximately 5 cm s^{-1} amplitude) (fig. 2.21). Tidal band currents over the shelf edge and slope are larger than those over the shelf (approximately 15–20 cm s^{-1}) (fig. 2.21). They are also strongly baroclinic, decreasing by a factor of two by a depth of about 100 m and by a factor of four by approximately 300 m (fig. 2.21). Tidal band currents are also large in the upper 100 m over the open basin (not shown). However, they are consistently largest in the upper layers at locations near steep continental slopes. At depths well below the seasonal pycnocline, tidal band energy is much reduced (approximately 10 cm s^{-1}), except near regions of irregular topography. For example, tidal velocities exceeding 30 cm s^{-1} occur at depths of several hundred meters in the vicinity of the Santa Cruz sill (fig. 2.22). Analyses of tidal band currents for the Santa Monica–San Pedro region are part of an ongoing research program, from which results should be available in 1993.

Barotropic tidal currents within the Santa Barbara Channel basin and on the outer edge of the Santa Barbara shelf appear to be of the same order as those for the southern California shelf (approximately 10 cm s^{-1}) (Blumberg et al. 1987). Calculated ellipses for the semidiurnal lunar tide suggest that baroclinic tidal currents may be weaker in the upper 100 m of the water column than those on the southern California shelf. The semidiurnal tide is in phase throughout the Santa Barbara Channel, including the interisland passages. Tidal currents in the passages are approximately four times as strong as those over the basin, reaching speeds of 50 cm s^{-1} on strong ebbs (Blumberg et al. 1987).

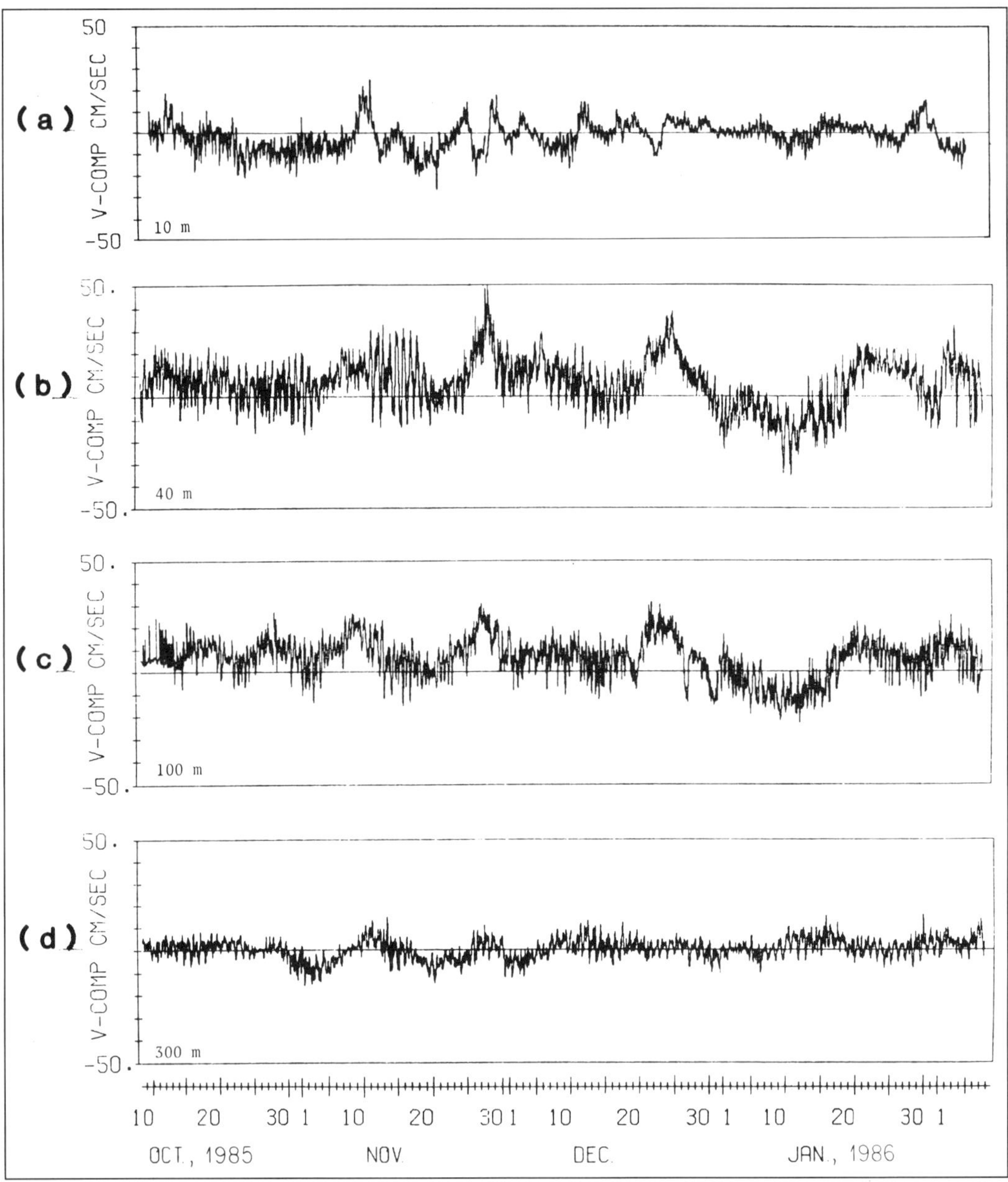

Figure 2.21. North–south velocity at selected depths over the Santa Monica inner shelf (a) and slope (b, c, and d).

As previously discussed, the water in the SCB is predominantly a mixture of two water masses: Subarctic, characterized by low temperature, low salinity, and high oxygen, and Equatorial, characterized by high temperature, high salinity, and low oxygen. Subarctic water is advected equatorward in the California Current and turns shoreward and poleward in the SCB. Equatorial water is advected from the south into the SCB by the California Undercurrent. For each water mass, the characteristic properties decrease in the

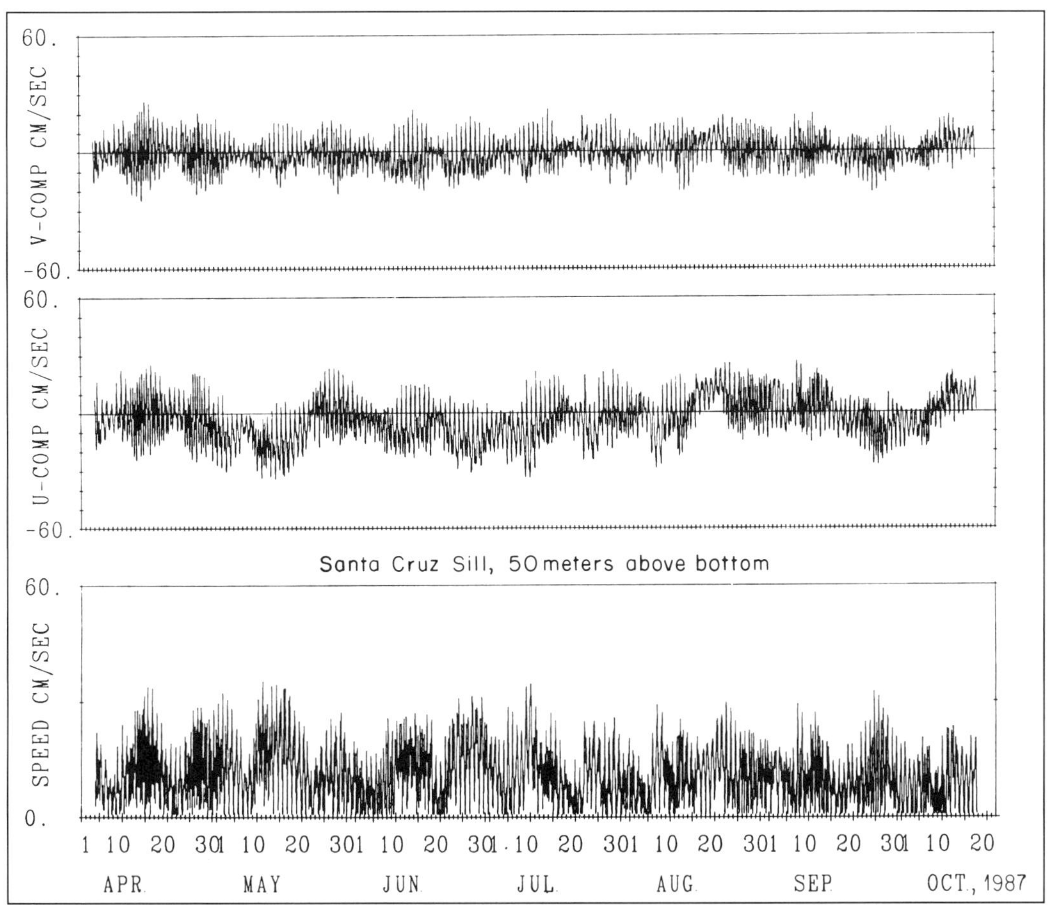

Figure 2.22. North–south (v) and east–west (u) components of velocity and speed at a location on the Santa Cruz sill, 50 m above the bottom and 623 m below the sea surface.

direction of the current pathway, indicating that mixing of the water mass has occurred (see figs. 2.10 and 2.11). Although the average seasonal patterns appear relatively smooth, the actual mixing is not smooth or continuous in either time or space. Rather, Gregg (1975) shows that the SCB is a region of spatially inhomogeneous and active turbulent mixing. Water profiles typically demonstrate 5–30 thick layers of the interleaving water masses. The intrusions are sometimes recognizable over distances of a few kilometers, but are sometimes not recognizable over distances as short as 2 km. The features disappear in a few hours, implying rapid mixing. Indeed, intense microstructure activity is observed at the vertical boundaries of the intrusions. Gregg (1975) attributes the mixing to shear instabilities and double diffusion phenomena.

The enclosed portions of Santa Monica and Santa Cruz basins have been the sites of dye experiments (SF_6) designed to estimate virtual eddy diffusivity both in the interior of the basins and near their boundaries. Away from the basin walls, diffusivity was about 0.25 $cm^2\ s^{-1}$ in Santa Monica Basin (Ledwell and Watson 1991). The spreading rate of the dye as well as changes in the heat content of the basin over several months demonstrated that diapycnal mixing in the basin is dominated by processes in the boundary regions (Led–

well and Hickey 1993). A reasonable overall diffusivity for the basin is about 1.1 $cm^2 s^{-1}$, more than four times greater than the interior value. Work in Santa Cruz Basin is still in progress (Ledwell and Bratkovich pers. comm.). The mixing rates determined from the dye experiment in Santa Monica Basin are somewhat higher than those found in the open ocean. Recently, Gregg and Kunze (1991) used shear and strain measurements to show that the elevated mixing rates are likely due to an elevated internal wave field within the basin.

Interannual Fluctuations

The California Current system, in which the SCB is embedded, undergoes significant year-to-year fluctuations (Chelton et al. 1982). Some of these fluctuations can be related to basin-wide processes such as occurrences of El Niño. In particular, the dominant EOF of the steric height of the sea surface relative to the 500-db pressure surface (a proxy, with certain assumptions, for actual sea surface height) for the southern California region for the period 1950–1980 indicates a weakening of the equatorward-flowing California Current during most El Niño periods and the reverse between El Niño events (fig. 2.23). The relationship between the California Current fluctuations as represented by the principal EOF and the El Niño events can be seen by comparing the northward transport associated with the EOF pattern with the eastern tropical Pacific sea surface temperature, which is often used as a proxy for the El Niño signal (fig. 2.24). This EOF accounts for roughly 30% of the interannual variability in steric height. The amplitude pattern of the EOF decreases offshore, indicating that the largest interannual effects occur nearshore. The interannual changes in currents are about 2–4 cm s^{-1} (Chelton et al. 1982).

Chelton et al. (1982) demonstrate that coastal sea level for the period 1950–1980 is highly correlated with fluctuations in the transport of the California Current (fig. 2.24). The authors use this relationship to extrapolate the time series for interannual fluctuations in the California Current backward in time to 1900 (fig. 2.25). The data suggest that significant interannual current fluctuations have occurred throughout this century, with a dominant period of 5–7 years. Related fluctuations also have been demonstrated for sea surface temperature and for zooplankton volume in the California Current (Chelton et al. 1982). During El Niño events, the water is warmer than between El Niño events and zooplankton biomass is reduced. Temperature fluctuations are typically on the order of 1°C. Interannual fluctuations in sea surface salinity also occur (Chelton et al. 1982). The amplitude of these fluctuations, however, is relatively small (approximately 0.1%), and the fluctuations do not demonstrate a fixed relationship to El Niño events. Interannual fluctuations in water properties throughout the water column are also expected. To our knowledge, however, these have not been examined.

The causes of the observed interannual variability in currents and water properties have not been unequivocally determined. Coastal sea level data suggest, however, that at least some of the variability may be due to the poleward propagation of coastal-trapped waves that originate in the tropics as a response to the El Niño phenomenon (Chelton and Davis 1982; Enfield and Allen 1983). The El Niño signal travels poleward at speeds of approximately 140 km d^{-1}. This conclusion has recently been validated with a two-layer model of the California Current system, which includes long waves of equatorial origin (that is, currents forced by the actual winds along the equator) as well as forcing by local winds along the coast (Pares-Sierra and O'Brien 1989). More than 75% of the interannual variability in sea level in the SCB was accounted for solely by the equatorial (remote) forcing.

Not only do the seasonal means of currents and water properties exhibit interannual variability but the statistics of the *fluctuations*

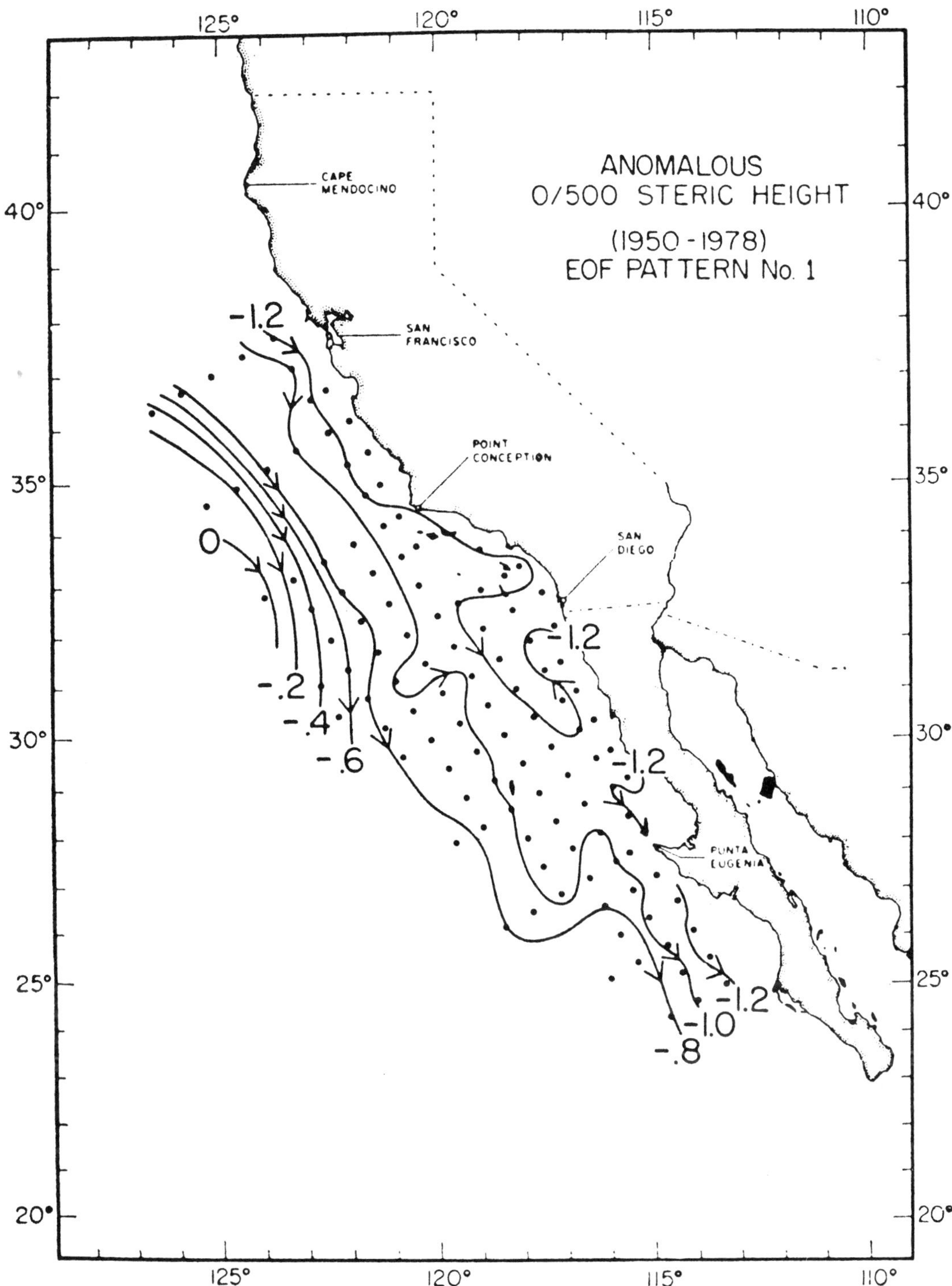

Figure 2.23. The principal spatial EOF of interannual 0/500-db steric height. The function values have been normalized to have a mean square value of 1. Arrows indicate direction of flow when the amplitude time series shown in figure 2.24a is negative. Positive values of the time series correspond to a reversal in the anomalous flow. (From Chelton et al. 1982.)

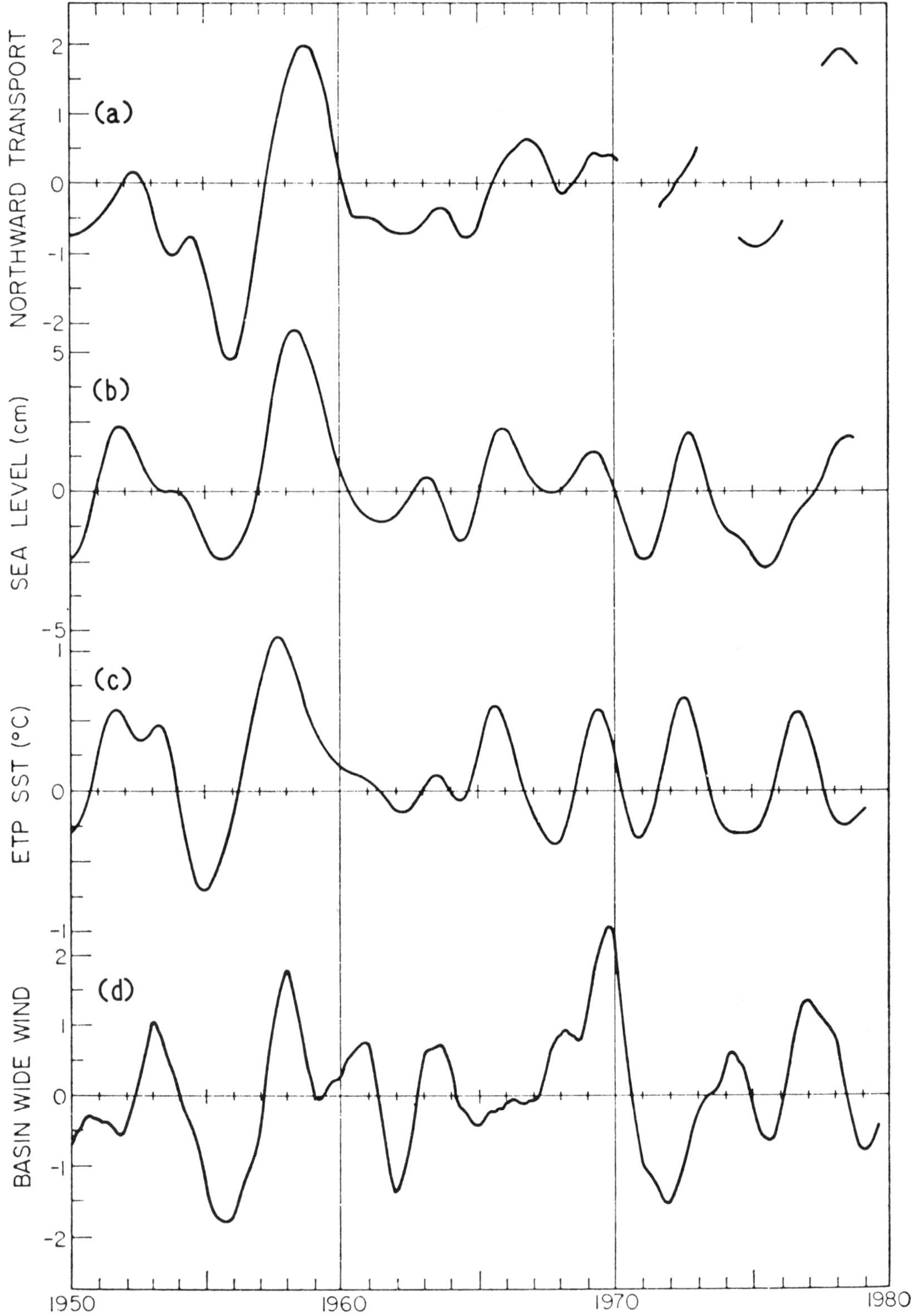

Figure 2.24. Low-frequency (double 13-month running mean) time series of (a) northward transport; (b) the average California sea level at San Francisco, Los Angeles, and San Diego; (c) eastern tropical Pacific sea surface temperature; and (d) the amplitude time series of the principal atmospheric pattern. (From Chelton et al. 1982.)

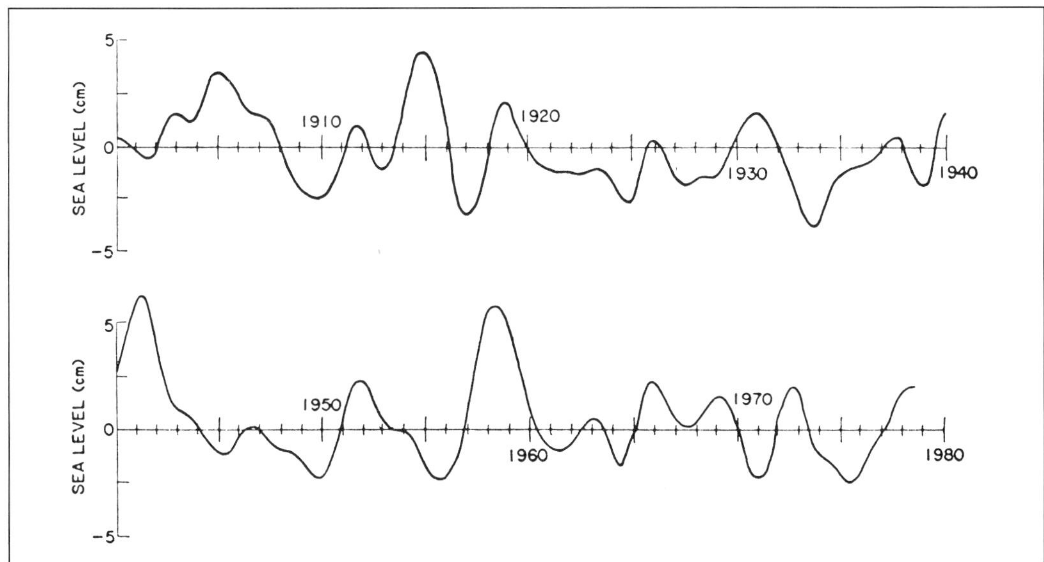

Figure 2.25. Low-frequency (double 13-month running mean) sea level from 1900 to 1979 averaged over San Francisco, Los Angeles, and San Diego. With some caution, this time series can be interpreted as an index of large-scale physical and biological variability over the past 80 years. Low sea level corresponds to above normal equatorward transport and zooplankton biomass and vice versa for high sea level. (From Chelton et al. 1982.)

can also vary considerably from year to year. It is well known that coastal weather patterns undergo significant interannual variability, some of which is related to the occurrence of El Niño. For example, winter storms (during El Niño) tend to be more vigorous and coastal winds are more poleward (Huyer and Smith 1984). These differences in the strength and direction of the local wind field can affect the direction and strength of currents, the depth of the mixed layer, and the location of the thermocline both over the basins and over the shelves for the several day scale. Moreover, year-to-year changes in the along coast structure of the wind field can affect the relative importance of local wind forcing and remote wind forcing at a particular site. For example, coastal-trapped waves generated off Baja California might make important contributions to the southern California current variance during one year but insignificant contributions the next year. Finally, the mesoscale eddy and meander jet field may be drastically altered by interannual changes in local wind and thermal forcing.

Basin Water Renewal

In each basin of the SCB the water is cut off from direct contact with other waters of the bight below the depth of the deepest sill of the basin; that is, horizontal advection of water is restricted by the basin walls. Renewal (or freshening, in the sense of an input of oxygen, which is depleted over time by organisms) occurs either by vertical mixing or by flow over the basin sills into the basins (Jackson 1986). Water flowing over the basin sills is generally denser than that at the same depth within the basin. Therefore, it sinks to some deeper depth (often the basin bottom) where it spreads laterally and presumably mixes vertically and laterally with the basin water. The "new" water flowing over the sill displaces "old" water from the basin.

Exact rates and mechanisms of renewal for the SCB are poorly understood. However, because the sills tend to shoal poleward, most of the basins are primarily renewed from the south (exceptions being the Santa Barbara Channel and Tanner Basin). This is illustrated by Emery (1960), who used water properties to trace the pathway of bottom water from basin to basin in the SCB (fig. 2.26). To our knowledge, this work has not generally been extended in the last 30 years, except in the Santa Monica and San Pedro basins. An array of current meters located around the perimeter of these basins demonstrates that the circulation below the deepest sill depth (>740 m) consists of a counterclockwise gyre (Hickey 1991). Net speeds are about 0.5 cm s^{-1}. In contrast with Emery (1960), Hickey's current data indicate net flow *from* Santa Monica Basin into San Pedro Basin on the westward side of the mid-basin knoll (fig. 2.26).

Analysis of current meter data as well as a 5-yr time series of hydrographic data in Santa Monica and San Pedro basins has provided the first insight into the spatial and temporal variability of both currents and water properties just above and well below the depth of the deepest sill of a coastal basin (Hickey 1992). Results indicate that neither the velocity field nor the temperature field is quiescent below the depth of the deepest sill on any time scale resolved by the data set (subtidal to interannual). The most energetic velocity fluctuations have some of the characteristics of topographic waves. At least below the depth of the deepest sill, the fluctuations propagate cyclonically around the basin perimeter and periods (10–20 days) are consistent with those expected for free topographic waves propagating around an elliptical basin. The amplitude of these fluctuations is ~4 cm s^{-1}. Relatively large-amplitude subtidal temperature fluctuations are also observed in the basins. The amplitudes of the fluctuations are ~0.1°C at 700 m and 0.05°C at 800 m. The characteristics of the temperature fluctuations suggest that much of their variance is related to the velocity fluctuations just described. The velocity fluctuations, which follow the topography *around* the lower basins, appear to be related to the regional circulation, which flows north–northwestward *over* the basin.

Seasonal variation of water properties occurs even below the depth of the deepest sill that connects to the regional flow (for example, 800 m in Santa Monica Basin versus a sill depth of 740 m). The amplitudes of the seasonal variation in temperature, salinity, and σ_t at mid-basin are roughly 0.1°C, 0.01‰, and 0.02 σ_t, respectively, 40 m above sill depth and roughly 0.04°C, 0.003‰, and 0.005 σ_t, respectively, 60 m below the sill. This seasonal variation is larger and extends deeper in the water column in the larger of the two basins considered. The seasonal variation is not uniform across the basins; the amplitude of the temperature signal can be a factor of two greater on the edges than at mid-basin.

Below the deepest sill in each basin, hydrographic properties are dominated by occasional intrusions, with renewal occurring at a frequency somewhat less than once per year. These results are consistent with estimates based on oxygen consumption rates in the sediment (approximately 500 days) (Jackson 1986). Two events were documented with Hickey's 5-yr data set, one in 1987 and one in 1988. The 1988 event was much stronger than the 1987 event and is comparable in magnitude to that reported by Berelson (1991) for 1984. In both the 1987 and 1988 renewal events, renewal is abrupt, occurring in a period on the order of 1 month or less and recovery to "normal" conditions is rapid, occurring in about 3 months. Both events, like those of Berelson, occur in spring. The renewal signal, as might be expected, is weaker and recovery may be faster farther downstream from the sill, that is, in Santa Monica Basin. An interesting water property study of San Pedro and San Nicolas basins for the period 1977–1988 demonstrated that renewal of some of the deep basins occurs simultaneously (Berelson 1991). This suggests that the renewal may be forced by changes in relatively large-scale processes such as the upper water column circulation over the basins. During a 7-month period in 1987, direct current and temperature measurements were made simultaneously on all of the Santa Monica and San Pedro basin sills and in the over-

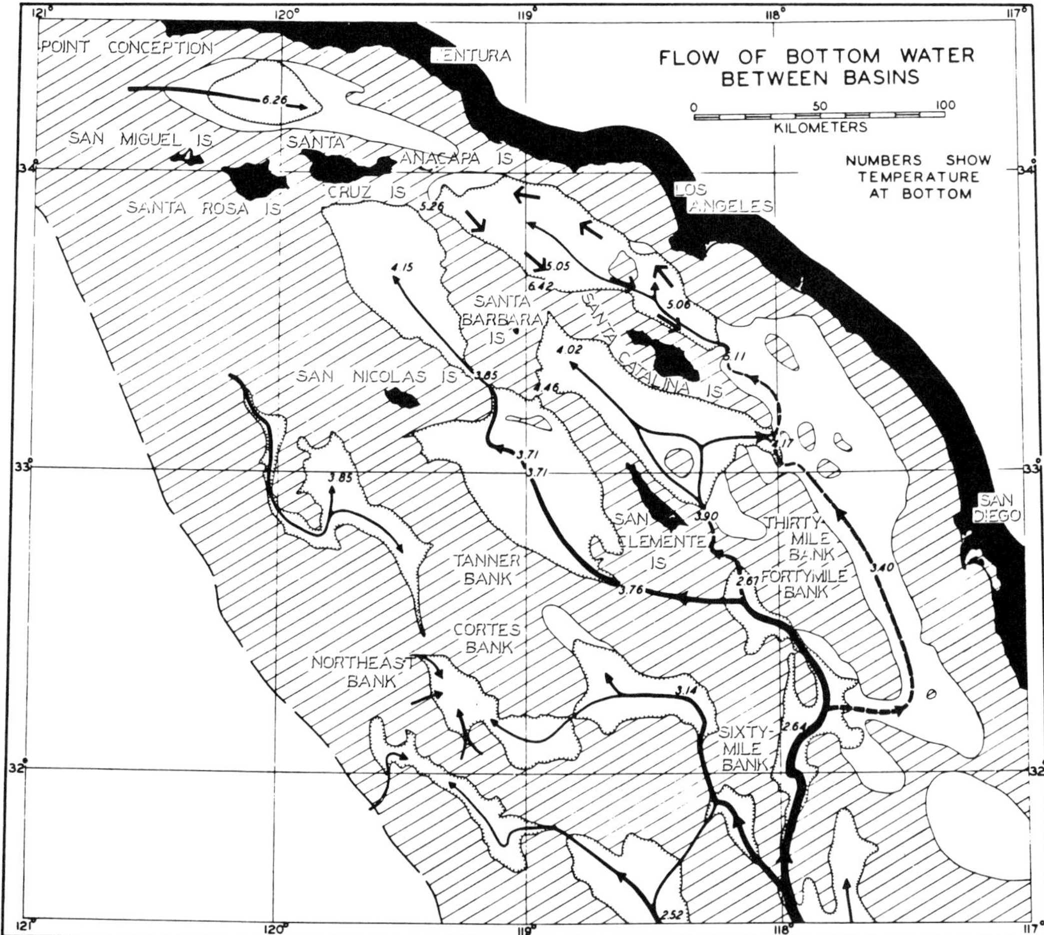

Figure 2.26. Paths followed by waters flowing from basin to basin in the southern Continental borderland as indicated by water properties. Width of lines is a rough indicator of transport volume. (From Emery 1960.) Note new short arrows in Santa Monica and San Pedro basins, derived from direct measurement of currents. Each short arrow represents data from one measurement site. (From Hickey 1991.)

lying water column. These data, as well as data from several hydrographic cruises, are presently being used to attempt to understand the mechanisms of basin water renewal in at least one deep coastal basin (Hickey 1993a).

Flushing rates of water in the Santa Barbara Channel were studied using water property analysis during 1969–1971 (Kolpack 1971; Sholkovitz and Gieskes 1971). For this relatively shallow (590 m) basin, rapid water renewal was observed as a result of an intense coastal upwelling event off Point Conception. No continuous current measurements were available to determine the exact timing and frequency of such events, nor has the time variable of circulation below sill depth been described. CalCOFI data indicate that the Santa Barbara Channel has overturned at least several times in the last 40 years, although not necessarily every year (Sholkovitz 1972). To our knowledge, water renewal rates and circulation below sill depth in basins other than the four just mentioned have not been studied.

Wave Climate

The best qualitative description of the seasonal wave climate in the SCB is given by Emery (1958). Data acquired during airplane flights were used to develop maps of wave and swell patterns for each season (fig. 2.27). Because the data were acquired using airplanes, there is an inherent "good weather" bias in the results, as we show later. The maps illustrate that the longer period swell is from the west–northwest in all seasons, while the intermediate period swell (near 3.5 seconds) is from the west or southwest. Shorter period (~2.0 seconds), southward traveling, locally generated wind waves are superimposed on the swell.

Since the mid-1970s, an array of wave measurement stations has been maintained along the California coast (Seymour and Sessions 1976). Data including significant wave height, wave period, and energy spectra are provided in monthly reports for each station. The data have been included in a database by the Minerals Management Service (Larson pers. comm.). Unfortunately, they have not, to our knowledge, been utilized to provide an updated qualitative summary of the wave climate in the SCB. Stations in the SCB are primarily located at the coast, where sheltering effects are significant. However, at least one station is maintained at a relatively offshore location in the SCB (Begg Rock, in fig. 2.3).

Data from monthly reports from wave measurement stations along the California coast dramatically illustrate the sheltering effects of the Channel Islands (Seymour and Sessions 1976). That is, the spectral amplitudes are an order of magnitude smaller at Sunset Beach than at Begg Rock (fig. 2.28a). The data also illustrate a seasonal change in the dominant swell period at Begg Rock from approximately 14–18 seconds in winter to 5–10 seconds in summer. The long-period winter swell is generated by North Pacific storms, whereas the shorter period waves are generated more locally. At Sunset Beach, the summer wave spectra are dominated by the very long period swell (16–18 seconds) coming from the Southern Hemisphere as described by Emery (1958). Begg Rock is sheltered from this long-period summer swell. Spectra during a major winter storm period are dramatically different from those during a typical winter period (fig. 2.28b). Severe waves such as these are usually generated from storms that develop between Hawaii and the Pacific coast. The wave amplitudes are an order of magnitude greater in all frequency bands. The dominant wave period is about 16 seconds. These waves are also felt at the coast, although their amplitude is significantly reduced.

The SCB is dotted with islands whose presence should alter the wave progression by reflection and refraction. In the Emery (1958) study, significant refraction was noted only near the mainland shores and near the islands. Cross-swell in the lee of islands was not detectable over large areas nor was reflection. However, shorter period swells and wind waves were influenced by island shadowing effects. Since Emery's (1958) work, research on the wave climate of the SCB has focused on the details of the effect of the islands on the wave climate. The offshore islands and ridges shelter the coast significantly from the effect of the deep ocean gravity waves. Much of the wave energy is dissipated in island surf zones or reflected back to the deep ocean. Behind each island there is a shadow zone, wherein energy of a particular wave might be expected to be zero. However, the wave energy can be spread into the lee of islands by wave refraction, scattering over shoals, diffraction, wave–current interactions, and nonlinear effects (Pawka et al. 1984). Wave refraction (Pawka et al. 1984) and nonlinear effects are thought to be the dominant processes (Vesecky et al. 1980).

The waves in the SCB are refracted to impinge on the coast more directly shoreward. The direction is not exactly shoreward, however, and the interaction of the waves with the topography in the SCB generates a net

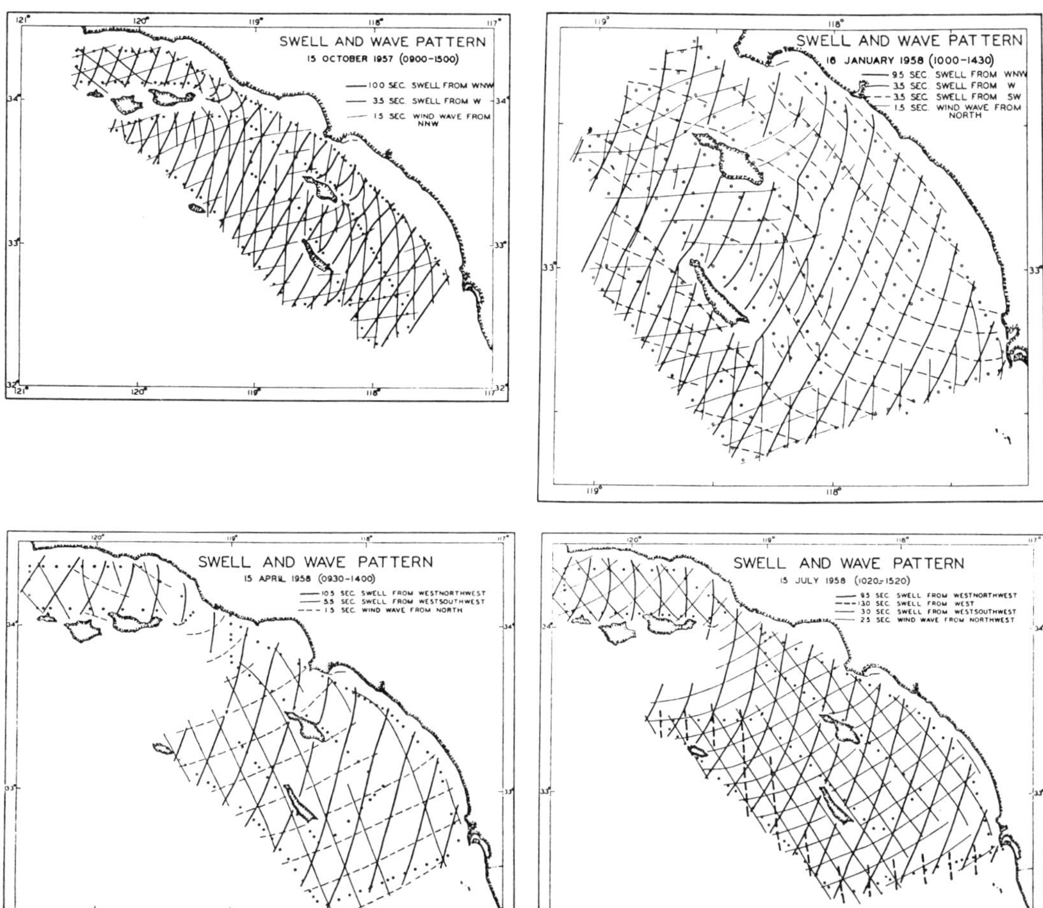

Figure 2.27. Seasonal swell and wave pattern for the SCB for several days in 1958. Dots show positions of observation points along airplane flight lines. The spacing of swell and wind wave crests is arbitrary and carries no implication regarding wave lengths. (From Emery 1958.)

southeastward longshore drift in the surf zone. This drift is responsible for much of the sediment movement along the coast (Emery 1960).

Summary and Prospectus for Future Research

The circulation over the SCB is much more complicated than that in other nearshore regions off the U.S. west coast. First, the wind over the basins is relatively weak in comparison to other west coast locations so that a large wind-driven signal is conspicuously lacking. The continental shelves are generally extremely narrow, but in some locations, shelves are both reasonably wide (~20 km) as well as very narrow (~3 km) within a few kilometers of each other. The topography is also very complicated over the basins. The channels over the basins become narrower as depth increases, so that the regional flow is blocked to a different degree at different depths and is completely blocked below the depth of the deepest sill. If coastal-trapped waves are incident on the SCB, they must squeeze through the passage between the mainland and Santa Catalina Island, then they must either break into two segments and/or shoal to enter the Santa Barbara Channel. The

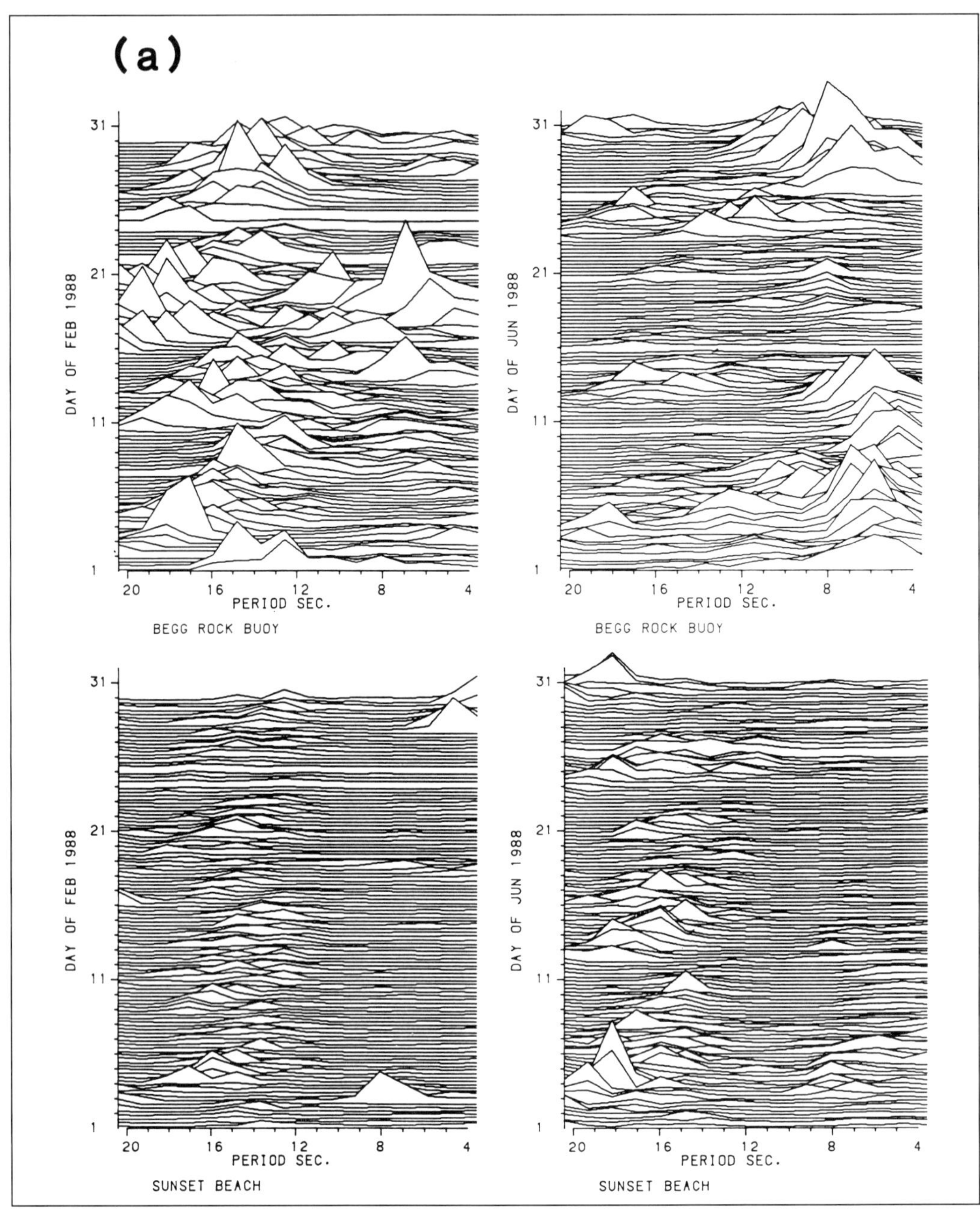

Figure 2.28. (a) Energy spectra of waves at Begg Rock and Sunset Beach during typical winter and summer months (see fig. 2.3 for locations). At 2041 on February 2 at Sunset Beach the total wave energy was 0.8×10^3 cm^2 and the significant wave height was 1.1 m. At the time of the largest peak at Begg Rock (0.8×10^3 on February 17), the total energy was 4.8×10^3 cm^2 and the significant wave height was 2.8 m. (b) Energy spectra of waves at Begg Rock and Sunset Beach during a major winter storm. At Sunset Beach at 0846 on January 17 the total wave energy was 2.5×10^3 cm^2 and the significant wave height was 2.0 m, whereas at Begg Rock at the peak of the severe storm at 1710 on the same day the total energy was 64.0×10^3 cm^2 and the significant wave height was 10.1 m. (Adapted from the Coastal Data Information Program 1988.)

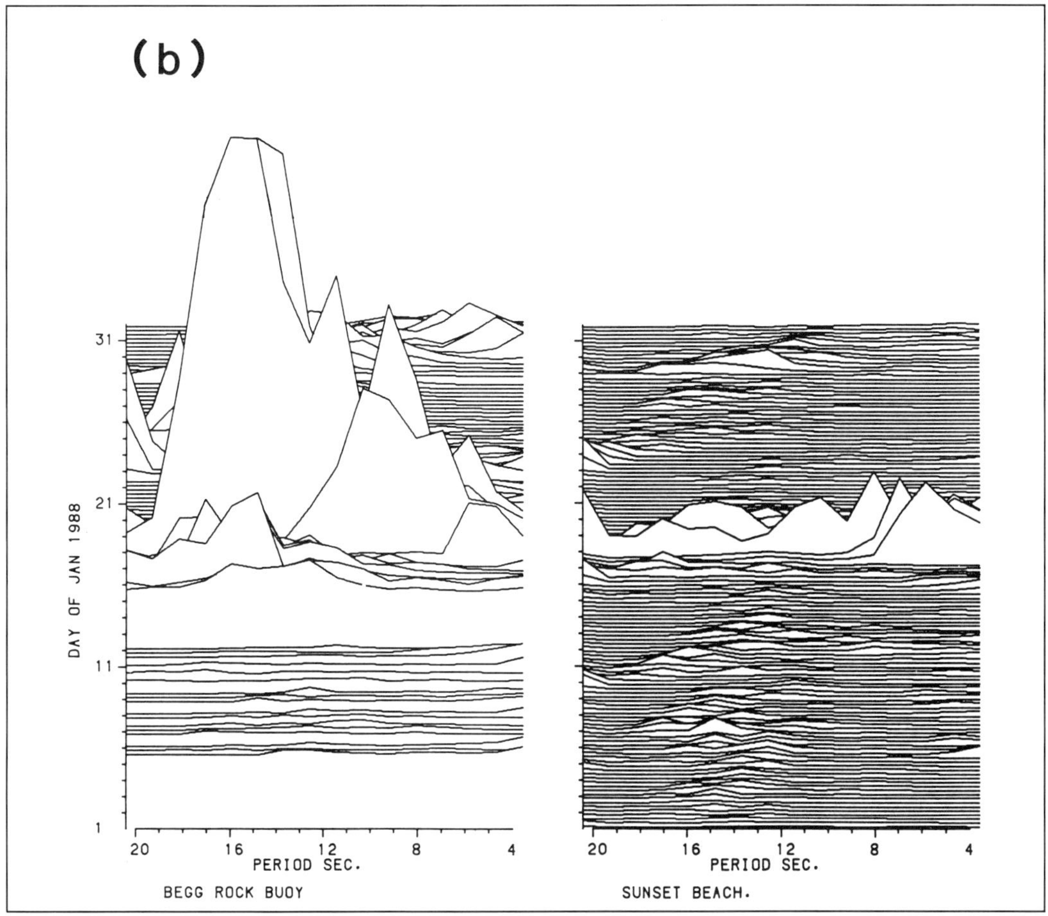

Figure 2.28. (continued)

coastal-trapped waves must also deal with the broadening of the shelf that occurs upstream of the San Pedro Basin. Additional complications are provided by the strong stratification that exists in the region much of the year. The stratification results in strong vertical shear in the currents. It is maintained not only because of the persistent heating that occurs but also because mixing induced by wind stress and/or surface waves are both inhibited in this comparatively protected area. The existing data base, while not adequate to describe the detailed circulation pattern in most areas, is sufficient to provide an overall characterization of the flow field.

Seasonal mean currents below the pycnocline over the slope are about 20–30 cm s^{-1} and are generally poleward inside the SCB, except perhaps in the immediate vicinity of local islands and banks. Above the pycnocline, it is likely that equatorward flow may occur, at least during spring, but this has not been substantiated with direct current measurements. Seasonal mean currents over the shelves are smaller than those over the slope and tend to be equatorward, particularly in the very near surface layers (<15 m) and over the middle and outer shelf. Poleward subpycnocline flow is usually observed during summer and fall on the shelves. Several day scale pulses of currents are typically 20–40 cm s^{-1} in the upper 200 m of the water column. Tidal and higher frequency currents can be as large as 10 cm s^{-1}. Interannual differences in the currents are in the range of a few centimeters per second. Thus, instantaneous currents below the pycnocline in the upper 200 m over the slope can occasionally attain speeds as large as 60 cm s^{-1} over the slope and the basins. Speeds over the shelves are substantially lower.

The circulation of the SCB is dominated by the effects of the large-scale Eastern Boundary Current system, the California Current, rather than by the effects of direct wind driving, as occurs in nearshore regions north of the SCB. The California Current flows equatorward offshore of the Channel Islands, roughly 200–500 km seaward of the coast. The current has a seasonal maximum in summer. It bends shoreward south of San Diego, then poleward, recycling water through the SCB in the large-scale Southern California Countercurrent. During spring, the countercurrent disappears, and flow at the surface tends to be equatorward throughout the SCB. A poleward undercurrent called the California Undercurrent occurs in the SCB within approximately 100 km of the coast. Although specific details of the seasonal mean flow differ throughout the SCB, the seasonal variation of the flow over the basins and their slopes (maxima in summer and winter), the existence of strong seasonal poleward means (except in surface layers during spring), and the general occurrence of subsurface maxima in the poleward flow (except in late winter and early spring) seem to be general and reproducible features of the flow field. Detailed hydrographic studies in Santa Monica–San Pedro Basin show that the flow is strongly affected by topographic features such as ridges, islands, and channels. The magnitude of the poleward transport through the SCB varies from 0.8 to 1.8 Sv with a maximum in the late summer–fall period.

The dominant water properties are also determined by the large-scale current system. Relatively low-temperature, low-salinity, high-oxygen, high-nutrient water enters the system from the north via the California Current. Relatively high-temperature, high-salinity, low-oxygen water enters the system from the south via the California Undercurrent. Isopleths of most water properties bow upward to the outer edge of the SCB and downward across the rest of the SCB to the coast. During spring, isopleths also tilt upward to the coast in response to wind-forced seasonal upwelling. Strong wind-driven local upwelling occurs in the SCB in winter and early spring. The upwelling occurs not only adjacent to the coast, but also near headlands and in the lee of islands. The strength and frequency of this upwelling is much less than that north of the SCB since the magnitude of the local wind is reduced in the bight by almost an order of magnitude in summer and fall, in comparison with regions offshore of and/or north of the bight. At least one location where offshelf jets of shelf water, such as are common north of the SCB, commonly occur has recently been documented on the southern Santa Monica shelf. Strong wind-driven upwelling at Point Conception itself causes dramatic modification of upper water column water properties in the SCB, particularly in the Santa Barbara Channel, but also over many of the offshore basins. The upwelled water is swept south of Point Conception and then onshore south of the Channel Islands.

Water properties vary seasonally at all locations as the strengths of the large-scale current systems vary (changing the slopes of isopleths) and the local upwelling varies. Subpycnocline isopleths rise and fall as much as 50–100 m annually, attaining their shallowest depth during summer just offshore of the SCB and during spring on the coastal side of the SCB. Water properties also fluctuate on subtidal and supratidal scales, primarily in response to vertical advection by the flow field.

Significant progress has been made over the past several years in understanding the magnitude and nature of subtidal (shorter than seasonal) current fluctuations in the SCB. In particular, results show that large-amplitude, long-period (20–30 day) pulses of currents are superimposed on the seasonal means. It is these pulses, rather than the seasonal means, that should be considered in estimating particle transit times for the SCB. Transit times could range from a week to a month below the near-surface layers (0–30 m) and perhaps several times that in the surface layers, where at least the mean flow is generally much smaller. Although such pulses have only been examined in detail in the Santa Barbara Channel and in the Santa Monica and San Pedro basins, it is likely that they occur throughout the SCB. The spatial scales of the pulses are shorter (≤20 km) in the spring than in the summer–fall, suggesting that many of the pulses may be the result of relatively small-scale eddies. Such eddies might be generated, for example, as the large-scale flow is forced past the various islands and banks in the SCB. It is clear that some of the current pulses, particularly during late summer, fall, and winter, are a result of poleward propagating, coastal-trapped waves. The origin of these waves may be located more than 1000 km south of the SCB. The speed of the waves decreases by an order of magnitude as they pass from San Pedro Basin into Santa Monica Basin, possibly a result of the dramatic topographic changes in the coastal channel that occur north of the San Diego trough. Current fluctuations on the shelves are relatively less energetic than those over the slopes and basins. The fluctuating shelf currents on the wider shelves (20–30 km) are at least partially due to local wind stress fluctuations, but, in the case of the semi-enclosed Santa Monica shelf, the fluctuations are primarily driven by (and opposite to) fluctuations in the slope currents.

The existence of a series of relatively deep basins beneath the upper water column circulation is one of the features that makes the SCB unique in the coastal margin of the U.S. west coast. Significant progress has been made in the last several years in understanding the variability in currents and water properties below the depth of the deepest sill in such basins. A recent study in Santa Monica and San Pedro basins illustrates that the circulation and water properties below the depth of the deepest sill in a coastal basin can be anything but quiescent. Long period (10–30 day) fluctuations occur in both the currents and water properties in these basins, and at least the dominant fluctuations are associated with waves that travel around the basin perimeter. The amplitude of associated velocity fluctuations is about 5 cm s^{-1}. The waves, which travel at speeds of about 20 km d^{-1}, appear to be forced by the upper water column currents as they pass over the deep basins. Water properties and currents in the deep basins can also have seasonal variations well below the depth of the deepest sill. At the deepest depths in the basins, fluctuations in water properties are predominantly interannual in nature, changes occurring only when new water upwells into the basin from an adjacent basin. Recent data from San Pedro, Santa Monica, and San Nicolas basins suggest that such renewal occurs on similar time scales in all three basins, roughly once every 2 or 3 years. Renewal is extremely rapid, occurring in a period somewhat less than 1 month.

The results summarized in this chapter demonstrate a clear difference between results obtained from large-scale hydrographic data and results obtained from direct current measurements. The direct measurements, in general, indicate much higher current speeds because of the increased sampling frequency and much richer spatial variability when sufficient stations are included. It is clear that if particle residence times are an issue, direct current measurements must be included in experimental programs.

Information on surface currents in the SCB is particularly limited as the result of the difficulty and expense of maintaining surface current measurement arrays. Any future studies should include direct measurements of the near-surface currents and their vertical structure—which implies that a combination of Eulerian (moorings) and Lagrangian (drifters) techniques should be employed.

Little is known about the effect of the irregular topography that occurs in the SCB on either the large-scale mean currents or on the current pulses, such as eddies or propagating coastal-trapped waves. However, recent studies in Santa Monica–San Pedro Basin suggest that effects of topography can be dramatic, such as significantly altering speeds of coastal-trapped waves and/or the local direction of large-scale currents. Models are now being developed that can resolve such effects as the scattering of shelf waves that occur near sharp bends in the topography or the generation of eddies by constrictions in the flow channel. The SCB might be an ideal candidate for testing models of this next generation. Such a modeling study should include a set of direct measurements for ground truth analysis.

Acknowledgments

This work was supported in part by the Minerals Management Service and in part by Department of Energy Grant #DE-FG05-85-ER60333.

Literature Cited

Atkinson, L. P., K. H. Brink, R. E. Davis, B. H. Jones, T. Paluszkiewicz, and D. W. Stuart, 1986. Mesoscale hydrographic variability in the vicinity of Points Conception and Arguello during April–May 1983: The OPUS 1983 experiment. *J. Geophys. Res.* 91(C11):12899–12918.

Berelson, W. M., 1991. The flushing of two deep sea basins, southern California borderland. *Limnol. Oceanogr.* 36(6):1150–1166.

Blumberg, A. F., D. E. Cover, J. T. Gunn, P. Hamilton, H. J. Herring, L. H. Kantha, G. L. Mellor, R. D. Muench, L. E. Piper, G. R. Stegen, and E. Waddell, 1987. *Santa Barbara Channel Circulation Model and Field Study.* Report 86. Dynalysis of Princeton, Princeton, NJ.

Brink, K. H., and R. D. Muench, 1986. Circulation in the Point Conception–Santa Barbara Channel region. *J. Geophys. Res.* 91(C1):877–895.

Cairns, S. L., 1967. Asymmetry of internal tidal waves in shallow coastal waters. *J. Geophys. Res.* 72:3563–3565.

Chelton, D. B., 1980. Low-frequency sea level variability along the west coast of North America. Ph.D. Dissertation, Scripps Institution of Oceanography, La Jolla, CA. 212pp.

Chelton, D. B., 1984. Seasonal variability of alongshore geostrophic velocity off central California. *J. Geophys. Res.* 89:3473–3486.

Chelton, D. B., and R. E. Davis, 1982. Monthly mean sea-level variability along the west coast of North America. *J. Phys. Oceanogr.* 12:757–784.

Chelton, D. B., P. A. Bernal, and J. A. McGowan, 1982. Large-scale interannual physical and biological interactions in the California current. *J. Mar. Res.* 40(4):1095–1125.

Coastal Data Information Program, 1988. Monthly summary reports. Institute of Marine Resources, Scripps Institution of Oceanography, La Jolla, CA.

Dorman, C. E., and D. P. Palmer, 1981. Southern California summer coastal upwelling. In: F. A. Richards, ed. *Coastal Upwelling.* American Geophysical Union, Washington, D.C. pp. 44–56.

Eber, L. E., 1977. Contoured depth–time charts (0 to 250 m, 1950–1966) of temperature, salinity, oxygen, and sigma-*t* at 23 CalCOFI stations in the California Current. *Calif. Coop. Oceanic Fish. Invest. Atlas No. 25.* 231 charts + 9pp.

Emery, K. O., 1958. Wave patterns off southern California. *J. Mar. Res.* 17:133–140.

Emery, K. O., 1960. *The Sea Off Southern California.* John Wiley & Sons, New York. 366pp.

Enfield, D. B., and J. S. Allen, 1983. The generation and propagation of sea level variability along the Pacific coast of Mexico. *J. Phys. Oceanogr.* 13:1012–1033.

Erdman, M. R., 1987. Inner-shelf circulation off San Onofre, CA, and the influence due to the operation of units 2 and 3 at the San Onofre nuclear generating system. Report to the Marine Review Committee, prepared by Eco-Systems Management Association.

Fiedler, P. C., 1988. Surface manifestations of subsurface thermal structure in the California Current. *J. Geophys. Res.* 93(C5):4975–4983.

Gregg, M. C., 1975. Microstructure and intrusions in the California Current. *J. Phys. Oceanogr.* 5:253–278.

Gregg, M. C., and E. Kunze, 1991. Shear and strain in Santa Monica Basin. *J. Geophys. Res.* 96(C9):16,709–16,719.

Gunn, J. T., P. Hamilton, H. S. Herring, L. H.

Kantha, G. S. Lagerloef, G. Mellor, R. D. Muench, and G. R. Stegen, 1987. *Santa Barbara Channel Circulation Model and Field Study*. Final Report. U.S. Department of Interior, Minerals Management Service, Washington, D.C. 400pp.

Halliwell, G. R., and S. S. Allen, 1984. Large-scale sea level response to atmospheric forcing along the west coast of North America, summer 1973. *J. Phys. Oceanogr.* 14:864–886.

Hendricks, T. J., 1974. Measurements of coastal currents. In: *1974 Annual Report*. South. Calif. Coastal Water Res. Proj., El Segundo, CA. pp.153–158.

Hendricks, T. J., 1975. Properties of nearshore currents: In: *1975 Annual Report*. South. Calif. Coastal Water Res. Proj., El Segundo, CA. pp. 167–172.

Hendricks, T. J., 1976. Measurements of subthermocline currents. In: *1976 Annual Report*. South. Calif. Coastal Water Res. Proj., Long Beach, CA. pp. 63–76.

Hendricks, T. J., 1980. Currents in the Los Angeles area. In: *Biennial Report 1979–1980*. South. Calif. Coastal Water Res. Proj., Long Beach, CA. pp. 243–256.

Hendricks, T. J., 1982. Shelf and slope currents off Newport Beach. In: *1982 Annual Report*. South. Calif. Coastal Water Res. Proj., Long Beach, CA. pp. 237–245.

Hendricks, T. J., 1984. Currents in San Gabriel Canyon. In: *Biennial Report 1983–1984*. South. Calif. Coastal Water Res. Proj., Long Beach, CA. pp. 143–153.

Hendricks, T. J., 1985. The use of inclinometer current meters in weak currents. In: *Oceans 1985 Conference Proceedings*. Marine Technology Society, Washington, D.C. pp. 742–748.

Hickey, B. M., 1979. The California Current system—hypotheses and facts. *Prog. Oceanogr.* 8(4):191–279.

Hickey, B. M., 1989a. Patterns and processes of circulation over the Washington continental shelf and slope. In: M. Landy and B. Hickey, eds. *Coastal Oceanography of Washington and Oregon*. Elsevier Science, Amsterdam. pp. 41–109.

Hickey, B. M., 1989b. Poleward flow near the northern and southern boundaries of the U.S. West Coast. In: S. J. Neshyba, Ch. N. K. Mooers, R. L. Smith, and R. T. Barber, eds. *Poleward Flows along Eastern Boundaries*. Coastal and Estuarine Studies Series No. 34. Springer-Verlag, New York. pp. 160–175.

Hickey, B. M., 1991. Variability in two deep coastal basins (Santa Monica and San Pedro) off southern California. *J. Geophys. Res.* 96(C9): 16,689–16,708.

Hickey, B. M., 1992. Circulation over the Santa Monica–San Pedro basin and shelf. *Prog. Oceanogr.* 30:37–115.

Hickey, B. M., 1993a. Flow over the sills of a deep basin off southern California (in prep.).

Hickey, B. M., 1993b. A topographically trapped, fluctuating vortex over Astoria submarine canyon (in prep.).

Huyer, A., and R. L. Smith, 1984. The signature of El Niño off Oregon, 1982–1983. *J. Geophys. Res.* 90(4):7133–7142.

Jackson, G. A., 1986. Physical oceanography of the Southern California Bight. In: R. W. Eppley, ed. *Lecture Notes on Coastal and Estuarine Studies, Vol. 15. Plankton Dynamics of the Southern California Bight*. Springer–Verlag, New York. pp. 13–52.

Kelly, K. A., 1985. The influence of winds and topography on the sea surface temperature patterns over the northern California slope. *J. Geophys. Res.* 90:11783–11798.

Klinck, J. M., 1989. Geostrophic adjustment over submarine canyons. *J. Geophys. Res.* 94(C5): 6133–6144.

Kolpack, R. L., 1971. Biological and oceanographical survey of the Santa Barbara channel oil spill 1969–1970. *Vol. II, Physical, Chemical and Geological Studies*. Allan Hancock Foundation, Univ. of California, Los Angeles, 477pp.

Kundu, P., and S. Allen, 1976. Some three-dimensional characteristics of low-frequency current fluctuations near the Oregon coast. *J. Phys. Oceanogr.* 6:181–199.

LaFond, E. C., 1962. Internal waves—The sea: Ideas and observations on progress in the study of the seas. *Phys. Oceanogr.* 1:731–763.

Ledwell, J. R., and A. J. Watson, 1991. The Santa Monica Basin traces experiment: a study of diapycnal and isopycnal mixing. *J. Geophys. Res.* 96:8695–8718.

Ledwell, J. R., and B. Hickey, 1993. Evidence for enhanced boundary mixing in Santa Monica Basin. *J. Geophys. Res.* submitted.

Lynn, R. J., K. A. Bliss, and L. E. Eber, 1982. Vertical and horizontal distributions of seasonal mean temperature, salinity, sigma-*t*, stability, dynamic height, oxygen, and oxygen saturation in the California Current, 1950–1978. *Calif. Coop. Oceanic Fish. Invest. Atlas No. 30.* 513 charts + 12pp.

Lynn, R. S., and J. J. Simpson, 1987. California Current system—The seasonal variability of its physical characteristics. *J. Geophys. Res.* 92(C12):12947–12966.

Pares-Sierra, A., and J. J. O'Brien, 1989. The seasonal and interannual variability of the California Current system: A numerical model. *J. Geophys. Res.* 94(C3):3150–3180.

Pawka, S. S., D. L. Inman, and R. T. Guza, 1984. Island sheltering of surface gravity waves: Model and experiment. *Cont. Shelf Res.* 3(1): 35–53.

Pickard, G. L., 1964. *Descriptive Physical Oceanography.* Pergamon Press, Oxford, England. 214pp.

Reid, J. L., Jr., G. I. Roden, and J. G. Wyllie, 1958. Studies of the California Current system. *Calif. Coop. Ocean. Fish. Invest. Prog. Rep.* 5:27–57.

Seymour, R. J., and M. H. Sessions, 1976. A regional network for coastal engineering data. In: *Proceedings of the 15th International Conference on Coastal Engineering.* Am Soc. Civil Engineers, Honolulu, HI. pp. 60–71.

Sholkovitz, E. R., 1972. The chemical and physical oceanography and the interstitial chemistry of the Santa Barbara Basin. Ph.D. Dissertation, Univ. of California, San Diego. 183pp.

Sholkovitz, E. R., and J. M. Gieskes, 1971. A physical–chemical study of the flushing of the Santa Barbara Basin. *Limnol. Oceanogr.* 16(3): 479–489.

Simpson, J. J., T. D. Dickey, and C. J. Koblinsky, 1984. An offshore eddy in the California Current system—I. Interior dynamics. *Prog. Oceanogr.* 13:5–49.

Sverdrup, H. U., and R. H. Fleming, 1941. The waters off the coast of southern California, March to July 1967. *Bull. Scripps Inst. Oceanogr. Univ. Calif.* 4(10):261–387.

Thornton, S. E., 1981. Suspended sediment transport in surface waters of the California Current off southern California; 1977–78 floods. *Geo-Mar. Letters.* 1(1):23–28.

Tibby, R. B., 1941. The water masses off the west coast of North America. *J. Mar. Res.* 4(2):113–121.

Torgrimson, G., and B. Hickey, 1979. Barotropic and baroclinic tides over the continental slope and shelf off Oregon. *J. Phys. Oceanogr.* 9:945–961.

Tsuchiya, M., 1980. Inshore circulation in the Southern California Bight, 1974–1977. *Deep-Sea Res.* 27(2A):99–118.

Vesecky, J. F., S. V. Hsiao, C. C. Teague, O. H. Shemdin, and S. S. Pawka, 1980. Radar observations of wave transformations in the vicinity of islands. *J. Geophys. Res.* 85(C9):4977–4986.

Winant, C. D., 1983. Longshore coherence of currents on the southern California shelf during the summer. *J. Phys. Oceanogr.* 13:54–64.

Winant, C. D., and A. W. Bratkovich, 1981. Temperature and currents on the southern California shelf: A description of the variability. *J. Phys. Oceanogr.* 11(1):71–86.

Winant, C. D., and J. R. Olson, 1976. The vertical structure of coastal currents. *Deep-Sea Res.* 23: 925–936.

Winant, C. D., R. C. Beardsley, and R. E. Davis, 1987. Moored wind, temperature, and current observations made during the coastal ocean dynamics experiments 1 and 2 over the northern California continental shelf and upper slope. *J. Geophys. Res.* 92(C2):1569–1604.

Wooster, W. S., and J. H. Jones, 1970. The California undercurrent off northern Baja California. *J. Mar. Res.* 28(2):235–250.

Chapter 3

Chemical Oceanography and Geochemistry

Robert P. Eganhouse and M. Indira Venkatesan

Introduction

The Southern California Bight (SCB) comprises a network of deep sea basins close to shore that trap coastal sediments. Water circulation is constrained below the basin sills, some of which are at depths intersecting the oceanic oxygen minimum zone. Consequently, nearshore (inner) basin sediments are anoxic and preserve sedimentation records. In addition, upper water column motions are restricted by the land masses and by diminished local wind effects caused by coastal mountain topography. Because the California Current is an eastern rather than a western boundary current, the general circulation within the bight is dynamically restricted (Jackson et al. 1989). These complex circulation patterns influence the biological as well as geochemical environments in the region. The geochemistry of dissolved and particulate phases as discussed in this chapter pertains to the water column from the sea–air interface to water depths of about 2000 m. The sedimentary record under discussion is confined to a few hundred years. Chemical components are identified as autochthonous (formed *in situ* in the SCB, of marine origin) or allochthonous (not originally formed in the SCB, originating outside of the SCB, and mainly land derived).

The cycling and fate of chemical components and elements in the SCB are determined by a complex interplay of various biological, chemical, and physical processes (fig. 3.1). The elements most affected by biological activity are those used by organisms for their cellular, structural, and energetic needs. Among these are C, N, P, and Si. These elements are classified as "macronutrients" because they are assimilated in relatively large amounts (although sometimes present in low concentrations in the surrounding environment). One or more of these nutrients may become limiting, depending on the rates of supply and utilization. In the SCB the N:P ratio in surface water is about 6, whereas the (Redfield) ratio in living phytoplankton is 16. This suggests that nitrogen is limiting in bight waters (Eppley and Holm-Hansen 1986). Other elements whose behavior could potentially be affected by biological cycles are among the list of "micronutrients" (Fe, Mn, Cu, Zn, Mo, V, and Co). These elements are required by organisms only in trace amounts and typically occur at subnanomolar to micromolar concentrations. Recent studies have indicated that the cycling of other chemical constituents believed to be nonessential (for example, Cd and Ni) may also be directly affected by the activities of living organisms (Bruland et al. 1978; Bruland 1980). Finally, biological activity indirectly affects the cycling of a variety of other nonessential elements and compounds by numerous mechanisms (such as scavenging, complexation by biogenic ligands, and bioturbation). These processes, which together control the spatial and temporal distribution of chemical substances in the SCB, are discussed in this chapter. The distribution of trace metals, stable isotopes (^{13}C and ^{15}N), and anthropogenic elements (such as Pb, Cr, and Cu) are reviewed, along with sources and fluxes of specific organic compounds such as the anthropogenic pollutants DDT, PCBs, and PAHs. The coupling among biological, physical, and chemical components of the system is emphasized to embody the dynamics of the ecosystem.

The dissolved, suspended, and sinking organic and inorganic matter and the deposited sediments are dynamically interrelated in the marine ecosystem. The dissolved phase is arbitrarily defined as those materials passing through filters having nominal pore sizes of 0.5–1.2 μm, whereas the suspended particles are those retained by the same filters (Williams 1986a). These particles are too small to sink rapidly through the water column (sinking flux is $<1 \text{ m d}^{-1}$) (McCave 1975) and thus can be transported by currents to adjacent areas. The suspended particles are made up of organic detritus and clay minerals in addition to significant bacterial populations (Williams 1986a). Because of the ease of lateral advection, suspended particles may not reflect the true characteristics of the overlying surface waters.

Rapidly sinking materials are collected by sediment or particle interceptor traps deployed in the water column over various time intervals ranging from a few days to a few months. Large particles such as fecal pellets and "marine snow" are aggregates of living and detrital organic material (Silver and Alldredge 1981). These aggregates account for most of the vertical settling flux of organic and inorganic matter because of their high sinking rates (Deuser and Ross 1980; Honjo and Roman 1978). They constitute only a small part of the total particulate matter pool in seawater. Settling (sinking) particles can play a major role in removal of dissolved components (radionuclides, metals, organics, and other materials) from the euphotic zone into subsurface waters (Goldberg 1961; Lal 1977). Studies by Williams and Zirino (1964), Suess (1970), and Meyers and Quinn (1973) suggest that metal oxides scavenge some of the dissolved organic matter, whereas amino acids and lipids are probably adsorbed onto clay minerals and carbonates. Moreover, particles repackaged (metabolically altered) by bacteria and other micro-

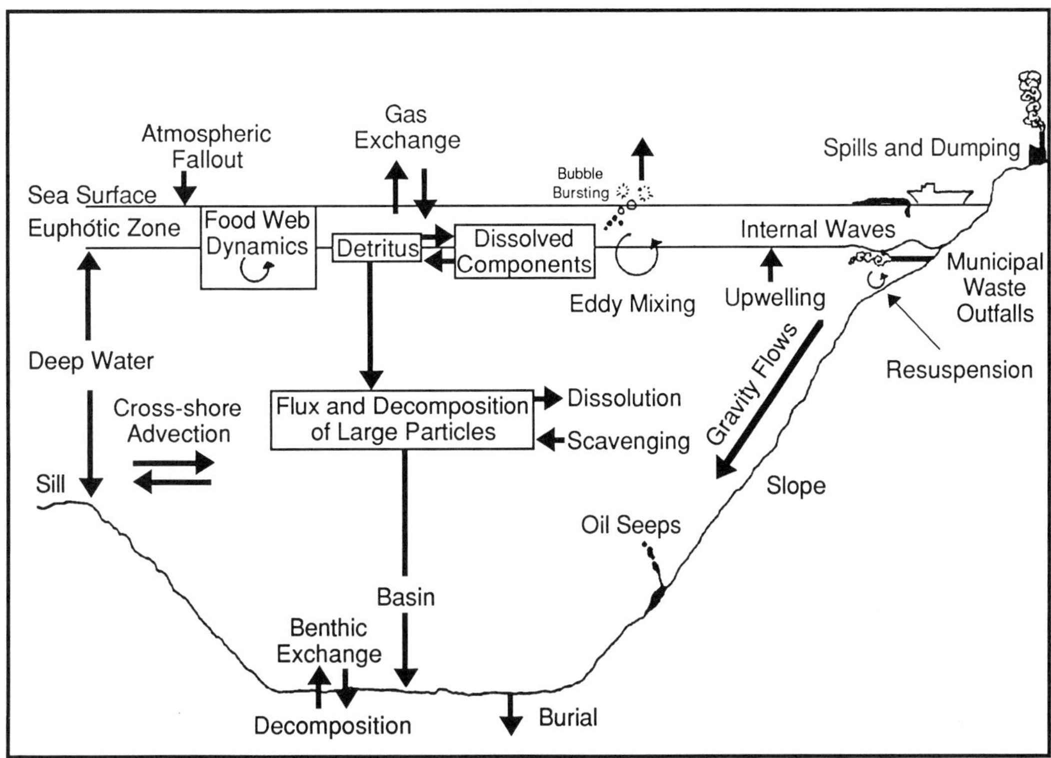

Figure 3.1. Inputs and biogeochemical processes in the SCB. Arrows between boxes suggest flows of matter and energy. (Modified from Carlucci et al. 1986; Williams 1986a.)

heterotrophs from dissolved and suspended organic material are believed to provide organic matter to zooplankton at depth (Fuhrman et al. 1980) (see fig. 3.1). Upwelling processes (from March to June in the SCB) can also transport fine particles to the surface, where they coalesce to form larger particles. These, in turn, can be ingested by zooplankton and expelled as sinking fecal pellets. Thus, some chemical components are found at enhanced concentrations farther offshore as a result of the remobilization and recycling of fine particles (Williams 1986a). Both dissolved material and suspended particles are also scavenged by marine snow (Silver and Alldredge 1981).

Although a number of data sets are available on the vertical distribution of soluble and suspended particulate inorganic and organic components in the SCB, especially from the nearshore environment (see references in Eppley 1986; Johnson et al. 1988; Williams and Druffel 1987), the chemistry of these phases is complex and only poorly understood. Information regarding sinking particulate matter is just beginning to accumulate (Crisp et al. 1979; Nelson et al. 1987; Small et al. 1989; Venkatesan and Kaplan 1992; Williams et al. 1992).

The fate of chemical (inorganic and organic) components in the SCB sediments has been both studied and reviewed extensively by a limited number of groups such as Southern California Coastal Water Research Project (SCCWRP), the University of California at Los Angeles and at San Diego, the Los Angeles County Sanitation Districts (Joint Water Pollution Control Plant, JWPCP), and the Department of Energy California Basin Study (CaBS) Program (Eganhouse and Kaplan 1988; Eppley 1986; Katz and Kaplan 1981; Finney and Huh 1989a; Jackson et al.

1989; Mankiewicz et al. 1978; Stull et al. 1986; Thompson et al. 1986; Venkatesan et al. 1980). Dissolved inorganic nutrients and other hydrographic properties have also been measured in coastal waters of the SCB since the early part of this century (Williams 1986a). Major research programs in progress or recently completed in the SCB are listed in table 3.1.

This chapter is presented in four sections. The first section discusses sources of inorganic and organic components to the SCB and attempts to provide budgets of the various inputs (organic carbon, water, suspended, and sinking particles). The dynamics of the water column are treated in the second section which concludes with an overview of the transport and fate of anthropogenic contaminants in the water column. The third section focuses on the distribution and post-depositional fate of organic and trace inorganic substances in sediments of the SCB. Major findings are summarized in the last section, where gaps in current knowledge are identified and recommendations are made concerning areas for future research.

Sources of Organic Matter and Trace Elements

The organic matter and trace elements in marine ecosystems are contributed by autochthonous (marine) as well as allochthonous (terrestrial) sources. The marine component in the SCB derives from primary production and submarine oil seepage. The major inputs of terrestrial origin include domestic and industrial waste discharges, surface runoff from rivers and urban storm drains, dry and wet atmospheric fallout, ocean dumping, and shales eroded from coastal areas. The nature and magnitude of waste discharges and runoff are better characterized than diffuse sources such as atmospheric fallout and ocean dumping. Data on erosion of the Monterey Shale along the coast of California are scant. Oil seepage is both episodic and chronic, whereas inputs from sewage are essentially constant (see table 12.1 in chap. 12). In contrast, storm runoff is episodic and mostly active during winter months (December to February, but occasionally extending from September to March). Strong northeasterly Santa Ana winds during spring and summer (March to July) blow seaward down slopes and valleys from the deserts and influence aerial fallout in the SCB. An attempt is made here to evaluate and estimate the input of organic and trace metal constituents from these various recognizable sources to the SCB. (For a more detailed treatment of spatial and temporal distribution of selected anthropogenic inputs, see chap. 12.)

Autochthonous Sources

The mean primary production (P) of the SCB is calculated to be 390 mg C m^{-2} d^{-1} based on the empirical algorithm developed by Eppley and Holm-Hansen (1986):

$$P = \exp(-3.78 - 0.372T + 0.227D)$$

where D is the day length set at 12 hours and T is the temperature anomaly assumed to be zero. Using this expression, one arrives at a total production over the approximate area of the SCB (78,000 km^2) (after Emery 1960) that ranges from 1.1 to 1.6 $\times$ 10^{7}t C yr^{-1} (table 3.2). This compares with the average global oceanic production of approximately 3 $\times$ 10^{10}t C yr^{-1}. Primary production in the SCB (approximately 400 mg C m^{-2} d^{-1}) is nearly twice the average oceanic value, and it falls between ranges estimated for the Peru upwelling region (approximately 1000 mg C m^{-2} d^{-1}) and the Scotia Sea of the Antarctic Ocean or the central subtropical gyre of the North Pacific (approximately 200 mg C m^{-2} d^{-1}) (Eppley and Holm-Hansen 1986). The study by Eppley and Holm-Hansen (1986) covered a region encompassing the San Diego Trough, the Santa Monica and San Pedro basins, and the narrow coastal strip of continental shelf between Los Angeles and San Diego (fig. 1.4 in Carlucci et al. 1986). It

Table 3.1. *Major Studies in Oceanography and Geochemistry in the SCB*

Type of Study	Program (Agency)	Time Frame	Source of Data
Nutrients, hydrography, elemental cycling, and food chain relationships	Marine Life Research Group and Food-Chain Research Group (FCRG) (Department of Energy)	1945–1984	Papers, reports, Eppley 1986
Hydrographic survey program	California Cooperative Oceanic Fisheries Investigations (CalCOFI) (State of California)	1949–present	Annual reports of CalCOFI conferences, atlases, Eppley 1986
Upwelling studies	Organization of Persistent Upwelling Structures (OPUS) (National Science Foundation)	1981–present	Reports, papers
Chemical and biological studies	Southern California Baseline Study (Bureau of Land Management)	1976–1979	Reports, papers
Coastal pollution	Southern California Coastal Water Research Project (SCCWRP) (Five local sanitation districts)	1969–present	Annual, biennial reports, papers
Biogeochemical cycling	California Basin Studies (CaBS) (Department of Energy)	1985–1991	Reports, papers

did not extend offshore beyond 107 km or north toward Santa Barbara Basin and Point Conception where primary productivity is generally higher (Owen and Sanchez 1974). The total bight-wide production estimate is, therefore, probably at the lower limit.

In addition to phytoplankton, microzooplankton contribute particular organic matter to the ocean. On average, about 132 mg C m^{-2} d^{-1} of particulate organic carbon were produced by microzooplankton in 1967 (Beers and Stewart 1970). Carbon from microzooplankton composed about 20% of total zooplankton carbon in the upper 100 m. Extrapolating from this value, the carbon contribution from total zooplankton of the SCB is estimated to be about 1.9×10^7 t yr^{-1} (table 3.2), assuming again the area of the SCB to be 78,000 km^2. Zooplankton and phytoplankton apparently contribute equally to the organic carbon and particulate matter pools in the ocean. Only a small fraction of these inputs is eventually deposited in the sediments. While about two-thirds of the primary production may be recycled in the euphotic zone, the remaining third sinks into deeper waters. The sinking flux of particles appears to be correlated with surface primary production (Deuser and Ross 1980; Honjo 1982).

The carbon derived from primary production is augmented by inputs from local natural oil seeps in the SCB that have been active throughout Holocene time (Wilson et al. 1974). The circum-Pacific belt, including the SCB, is estimated to contribute approximately 48% of the total global marine petroleum seepage (0.27×10^6 of 0.6×10^6 t) (Wilson et al. 1974). Of the 190 seepages cited by Wilson et al. (1974), about 60 zones were located by Wilkinson (1972) within the 2600-km^2 offshore California area from Point Conception to Long Beach. Some zones are continu-

Table 3.2. *Particulate, Water, and Total Organic Carbon Budgets for the SCB*

	Input Rates		
	Particles[a] ($t\ yr^{-1}$)	Water ($l\ yr^{-1}$)	Total Organic Carbon ($t\ yr^{-1}$)
Autochthonous inputs			
Primary production	4.1×10^{7} [b]	—	$1.11–1.6 \times 10^{7}$ [c]
Zooplankton	3.8×10^{7} [d]	—	1.9×10^{7} [d]
Oil seeps	$2 \times 10^{3}–5.6 \times 10^{4}$	—	$1.7 \times 10^{3}–4.8 \times 10^{4}$
Advective flow of California Countercurrent and Undercurrent	—	$2.5–5.7 \times 10^{16}$ [e]	$2.3–5.1 \times 10^{7}$ [e]
Allochthonous inputs			
Municipal waste discharges	8.3×10^{4} [f]	1.6×10^{12} [f]	2.9×10^{4} [f]
Industrial waste discharges:[g]			
Petroleum related	3.4×10^{3}	1.1×10^{11}	—
Metallic, fish cannery, miscellaneous chemical	2.1×10^{3}	3.5×10^{10}	—
Thermal discharges	—	7.7×10^{12}	—
Surface runoff			
Sediments	9.0×10^{6} [h]		—
Suspended solids	$1.5–6.0 \times 10^{5}$ [g,i]	2.4×10^{11} [g]	$1.9–7.4 \times 10^{4}$ [g,i]
Shale erosion	6.4×10^{4}	—	$6.4 \times 10^{2}–1.2 \times 10^{4}$
Dumping (dredge material)	$5.8 \times 10^{5}–1.4 \times 10^{6}$	—	$9.5 \times 10^{3}–2.4 \times 10^{4}$ [j]
Atmospheric deposition:			
Dry fallout			
fine particles	$8.9 \times 10^{3}–8.9 \times 10^{4}$	—	$2.1 \times 10^{3}–2.1 \times 10^{4}$
fine elemental carbon	—	—	$3.4 \times 10^{2}–3.4 \times 10^{3}$
total suspended particles	$2.9 \times 10^{4}–2.9 \times 10^{5}$	—	—
Rain washout	?	?	?
Particle interceptor trap measurements	$1.9–7.1 \times 10^{7}$ [k]	—	$1.3–3.6 \times 10^{6}$ [k]

[a] Particles, sediments, or suspended solids except for oil seeps, which is liquid.

[b] Based on phytoplankton production (500 g m^{-2} yr^{-1}) and attached marine algae and sea grass (20 g m^{-2} yr^{-1}) data of Emery (1960).

[c] Based on primary productivity = 0.39 g C m^{-2} d^{-1} (Eppley and Holm-Hansen 1986) and satellite chlorophyll data = 0.46–0.56 g C m^{-2} d^{-1} (Smith and Baker 1982).

[d] Average POC of microzooplankton = 0.132 g C m^{-2} d^{-1} (Beers and Stewart 1970). This constitutes about 20% of total zooplankton carbon in the upper 100 m. Extrapolating to total zooplankton, POC = 0.66 g C m^{-2} d^{-1}. Assuming TOC is 50% of the particulate flux from zooplankton (macro- and microzooplankton), total particulate flux from the zooplankton can be estimated.

[e] Hickey (chap. 2); organic carbon based on depth-integrated concentration of dissolved organic carbon = 75 μM (after Hansell et al. 1990).

[f] 1989 data (SCCWRP 1991) computed for particles and water. TOC calculated from % organic carbon in effluent particles (35%). (From Hendricks and Eganhouse 1992; Myers 1974.)

[g] SCCWRP (1973).

[h] Schwalbach and Gorsline (1985).

[i] Based on Eganhouse and Kaplan (1981) and Eganhouse (1982).

[j] Inputs calculated from a single dumpsite based on the range of amount allowable for disposal.

[k] Recent estimate for the SCB of average flux of total mass and organic carbon from February–May and May–October 1986 trap deployments off Santa Monica Basin from 850 and 350 m water depths (Williams 1986b; Jackson et al. 1989).

ously active while others are only sporadically active. Estimates of seepage rates range from 16 m^3 d^{-1} to more than 160 m^3 d^{-1} (Allen et al. 1970; Wilson 1973). Surprisingly little comprehensive compositional information on these natural oil seeps exists except for the reports of Delaney (1972), Sivadier and Mikolaj (1973), and Reed and Kaplan (1977). Since direct information pertaining to seepage rates from many areas is limited and of uncertain quality (Kvenvolden and Harbaugh 1983), the input of organic matter from seeps can only be crudely estimated. The active areas in the SCB from Point Conception to Long Beach are believed to have contributed seep material in the 1970s at rates of 2–56 × 10^3t yr^{-1}, which is based on multiplying the volume of seep (2.3–58 × 10^3 m^3 yr^{-1}) (Wilson et al. 1974; Fischer 1978) by the average specific gravity (0.97 g cm^{-3}) of California oils (Tissot and Welte 1984). Assuming that seep oil contains approximately 85% organic carbon, the total organic carbon (TOC) input to the SCB is approximated to be 1.7 × 10^3–4.8 × 10^4t yr^{-1} (table 3.2). This represents an upper limit estimate of the TOC contribution from seeps because a significant portion of nearshore seepage oil is probably gradually incorporated into the local food web (Spies and DesMarais 1983).

Advection via the California Current could play a significant role in mobilizing autochthonous as well as allochthonous materials in different regions of the SCB (Drake et al. 1985). However, limitations of the available data currently make it difficult to calculate the input of various constituents to the bight from advection. Mass transport rates of some trace components were estimated by SCCWRP (1973). Assuming that the trace metal and chlorinated hydrocarbon concentrations of open ocean waters are representative of California Current waters and that the advective transport of the California Current is about 2 × 10^{13} m^3 yr^{-1}, SCCWRP (1973) estimated that the mass transport rates of trace metals, DDT, and PCB by advection far exceeded the mass emission rates from all other sources. Measurements by Hickey and Kachel (pers. comm. 1989) indicate that the actual mass transport rates could be even higher since advective transport of the California Current is greater by a factor of about 10 (1.8–2.5 × 10^{14} m^3 yr^{-1}). On the other hand, estimates of mass transport rates such as these probably represent upper limits because the California Current generally flows outside of the Santa Rosa–Cortes Ridge and thus largely bypasses the SCB. The mean circulation in the SCB is dominated by the poleward surface flow (the Southern California Countercurrent) and the subsurface flow (the California Undercurrent) (see chap. 2). Considering the magnitude of this poleward transport (which varies from 2.5 to 5.7 × 10^{13} m^3 yr^{-1}), we have recalculated mass transport rates of several trace metals, DDT, and PCB here using the concentrations reported by SCCWRP (1973). These estimates are presented in table 3.3. An estimate of the organic carbon introduced and/or transported by current advection is included in table 3.2. Note again that the advective flow of the Southern California Countercurrent could transport enormous quantities of land-derived components to the entire SCB.

Allochthonous Sources

The four largest municipal waste dischargers (Hyperion, JWPCP, County Sanitation Districts of Orange County, and Point Loma) currently release 1.6 × 10^{12} l of water and 8.3 × 10^4 t of suspended solids into the coastal waters of the SCB annually (SCCWRP 1991). In addition, petroleum, metals, fish cannery, and other industries when combined with power plants discharge wastes and cooling waters on the order of 8 × 10^{12} l yr^{-1}. The volume of surface runoff from storm drains and rivers that enters the coastal waters is comparable to that from municipal wastes (2.4 × 10^{11} l yr^{-1}), but during episodic flooding, flows can be one or two orders of magnitude greater (Schwalbach and Gorsline 1985). A flow of about 1000 times as great as the total combined flows just listed is also estimated to

Table 3.3. *Estimated Mass Emission or Transport Rates* ($t\ yr^{-1}$) *of Selected Organic Compounds and Trace Metals to the SCB*

Constituent	Municipal Wastewater	Surface Runoff	Oil Seeps	Shale Erosion	Vessel[g] Coating	Ocean[g] Dumping	Rainfall Washout	Dry Fallout	Advective Transport[m]	Totals
Total organic carbon[a]	2.9×10^4	$1.9–7.4 \times 10^4$	$1.7–4.8 \times 10^3$	6.4×10^2–1.2×10^4	?	$0.95–2.4 \times 10^4$	?	$0.21–2.1 \times 10^4$	$2.3–5.1 \times 10^7$	$6.1–15 \times 10^4$
Hydrocarbon	1.7×10^4[b]	$<9.7 \times 10^3$[c]	$1.4–44 \times 10^3$[d]	?	?	?	?	?	?	$>2.7 \times 10^4$
Total DDT	0.02[e]	0.03[f]	—	—	<1	<1	?	0.5–2[g,j]	250–570	≤3
Total PCB	0.00[e]	0.05–0.10[f]	—	—	<1	<1	?	2[i]	250–570	≤4
Coprostanol	260[l]	?	—	—	—	?	—	—	—	260
Chromium	22[e]	18–31[f]	?	?	1	28	78[h]	51[k]	$0.5–1.1 \times 10^4$	198–211
Lead	27[e]	51–108[f]	?	?	10	28	1014[h]	1872[k]	$0.75–1.7 \times 10^3$	3000–3060
Copper	69[e]	33–62[f]	?	?	386	28	390[h]	242[k]	$0.75–1.7 \times 10^4$	1150–1180

[a] see table 3.2.

[b] Based on samples collected in 1979 (Eganhouse and Kaplan 1982a).

[c] Based on samples collected in 1978 (Eganhouse and Kaplan 1981). Assumes Los Angeles River contributes 30% of total flow to SCB.

[d] Based on seep hydrocarbon content of 71–79% (table 4; Reed and Kaplan 1977).

[e] 1989 data from SCCWRP (1991).

[f] Data of 1986–87 and 1987–88 water years for S. Clara River, Calleguas Cr., Ballona Cr., Los Angeles R., S. Gabriel R., S. Ana R., S. Diego R., Tijuana R. (SCCWRP 1992).

[g] SCCWRP (1973).

[h] Calculated from Lazrus et al. (1970) and Bruland et al. (1974).

[i] Bascom et al. (1979).

[j] Young et al. (1976).

[k] Young and Jan (1977). The estimated input in a 100-by-100 km zone off Los Angeles and Orange counties in 1975 is extrapolated here to the total area of the SCB.

[l] Venkatesan and Kaplan (1990).

[m] Advective flow rate and organic carbon, see table 3.2; concentration in water of other constituents from SCCWRP (1973).

be contributed by advected ocean water (2.5–5.7 × 10^{16} l yr^{-1}) to the SCB (chap. 2). It is therefore pertinent to estimate the influences of these important external sources on chemical balances in the SCB.

About three quarters of the municipal effluent produced in the coastal counties of southern California is directly discharged into the coastal waters of the SCB (SCCWRP, unpublished data). Part of the one-third not directly discharged to the ocean is reused. The remainder is released in drainage channels at different inland locations and transported to the bight via "dry weather" surface runoff. Consequently, significant amounts of suspended solids (table 3.2) from the effluents containing a wide spectrum of organic and inorganic constituents are probably deposited along the coastline or find their way to deeper parts of the offshore basins.

Municipal waste is a major contributor of anthropogenic heavy metals and chlorinated hydrocarbons to shelf sediments (e.g., Brown et al. 1986; Eganhouse and Kaplan 1988; Kettenring 1981; Stull et al. 1988; Young et al. 1977a). Quantitative estimates of inputs of these various components from waste discharges have been computed since 1971, much of this work emanating from the systematic studies at SCCWRP (Biennial and Annual Reports). Since 1971, the flow of water has increased by approximately 30%, while suspended solids have been reduced by approximately 68% (see fig. 12.2 in chap. 12) (SCCWRP 1991). Sharp decreases in emissions of trace contaminants from the outfalls have also been observed since monitoring began in 1971 (SCCWRP 1991). For example, the total input of DDT appears to have reached a plateau at around 20 kg yr^{-1} in 1989 from 21,600 kg yr^{-1} in 1971, and PCBs have declined steadily from 6200 kg yr^{-1} to essentially zero over this time period (figs. 3.2 and 12.3) (SCCWRP 1991). With the exception of Cd and Se, eight of the ten measured metals (As, Ag, Cr, Cu, Hg, Ni, Pb, and Zn) are being discharged at lowest reported rates (fig. 3.3).

Municipal wastewater particles are composed of as much as 30–35% organic carbon (Hendricks and Eganhouse 1992; Myers 1974). Considering that the average annual mass emission of total suspended solids from major discharges to the SCB during 1989 was approximately 8.4 × 10^4 t (SCCWRP 1991), municipal outfalls should contribute at least 2.9 × 10^4 t yr^{-1} of particulate organic carbon (POC) to the ocean. This estimate falls within the range of organic carbon possibly contributed to the SCB from oil seeps but is three orders of magnitude less than that from primary productivity. It is interesting to compare the input of particulate organic carbon from wastewater discharge with the natural flux of particles from planktonic and pelagic food web debris. While the emission rate of sewage-derived carbon is presently 2.9 × 10^4 t yr^{-1}, the natural flux of POC sinking out of the euphotic zone was estimated to be 25.6 g C m^{-2} yr^{-1} (Williams 1986b). Thus, the POC input from wastewater discharge is approximately equivalent to the natural planktonic flux calculated for an area equivalent to 1130 km^2. However, the input of municipal waste is localized. Recently improved wastewater treatment technologies and source controls to be implemented in the near future should further reduce sewage-derived contaminant emissions into the SCB (Schafer 1989).

Most industrial wastes originate from petroleum-related industries of onshore and offshore oil production, shipping, and other tanker activities. Petroleum waste discharges occur in the vicinity of Santa Barbara Channel and Santa Monica Bay and off Orange County coastal areas. The discharge of particles and water from other industries, such as metal-working plants and fish canneries (table 3.2), is approximately of the same order of magnitude as that from petroleum industries. The effluent flow from industrial wastes comprises about 9.1% of the total municipal wastewater discharge in the SCB, and the estimated mass emission rate of suspended solids from the former sources is

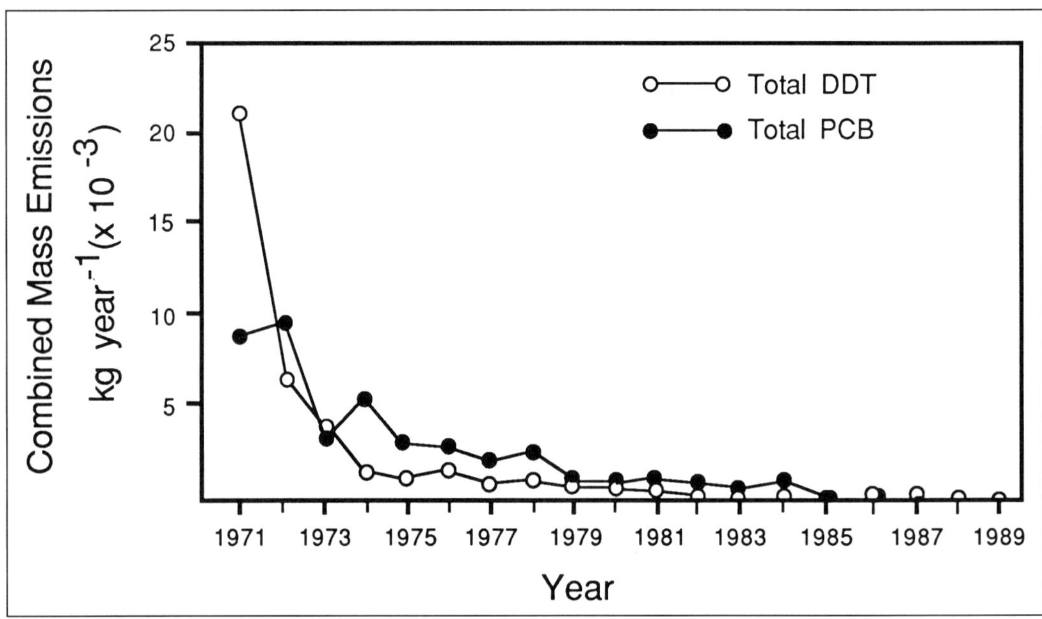

Figure 3.2. Mass emission rates (t yr^{-1}) of DDT and PCBs to the SCB. (From SCCWRP 1991.)

6.6% of that released from the latter dischargers (as of 1989). Practically no data exist in the open literature regarding the contribution of organic and inorganic constituents from industrial wastes to the California coastal waters.

Fourteen major thermal power generating stations and one nuclear power generating station (San Onofre) together discharge approximately 8×10^{12} l yr^{-1} of cooling water into the southern California coastal regime (including bays and estuaries) (table 3.2). Cooling water is indeed the largest type of discharge to the SCB. Young et al. (1977b) investigated the input of six trace metals (Cd, Cr, Cu, Ni, Pb, and Zn) in both seawater influent and effluent of eight major generating stations located along the Ventura, Los Angeles, and Orange county coastlines. The annual inputs of these trace metals from cooling waters were estimated to range from 0.3 to 2.1 t yr^{-1}. At that time, this constituted less than 1% of the combined input from storm runoff, aerial fallout, municipal wastewater, and thermal discharge. Retention basin waters (consisting of acid-cleaning wastes, fireside boiler wash water, and floor drainings) were not discharged during the sampling period. The release of basin waters would be expected to contribute additional trace metals (as much as four orders of magnitude) (Young et al. 1977b) as well as organic contaminants to the effluent cooling water. However, data pertaining to organic constituents are not readily available.

Approximately 300 rivers, streams, and storm drains discharge into the SCB, and 6 of these—all major rivers with tributaries (Ventura, Santa Clara, Calleguas, Los Angeles, San Gabriel, and Santa Ana)—make large contributions to coastal sediments. Using the data of Brownlie and Taylor (1981) and Taylor (1981), Schwalbach and Gorsline (1985) estimated the "actual" input of sediments from Point Conception to Baja California into the SCB to be about 9×10^6 t yr^{-1} (table 3.2). Taking into account the reduction in sediment yields by flood control and damming, Brownlie and Taylor (1981) projected "natural" discharge rates (which could be systematically higher in drainages accompanying urban developments) to be around 13 ×

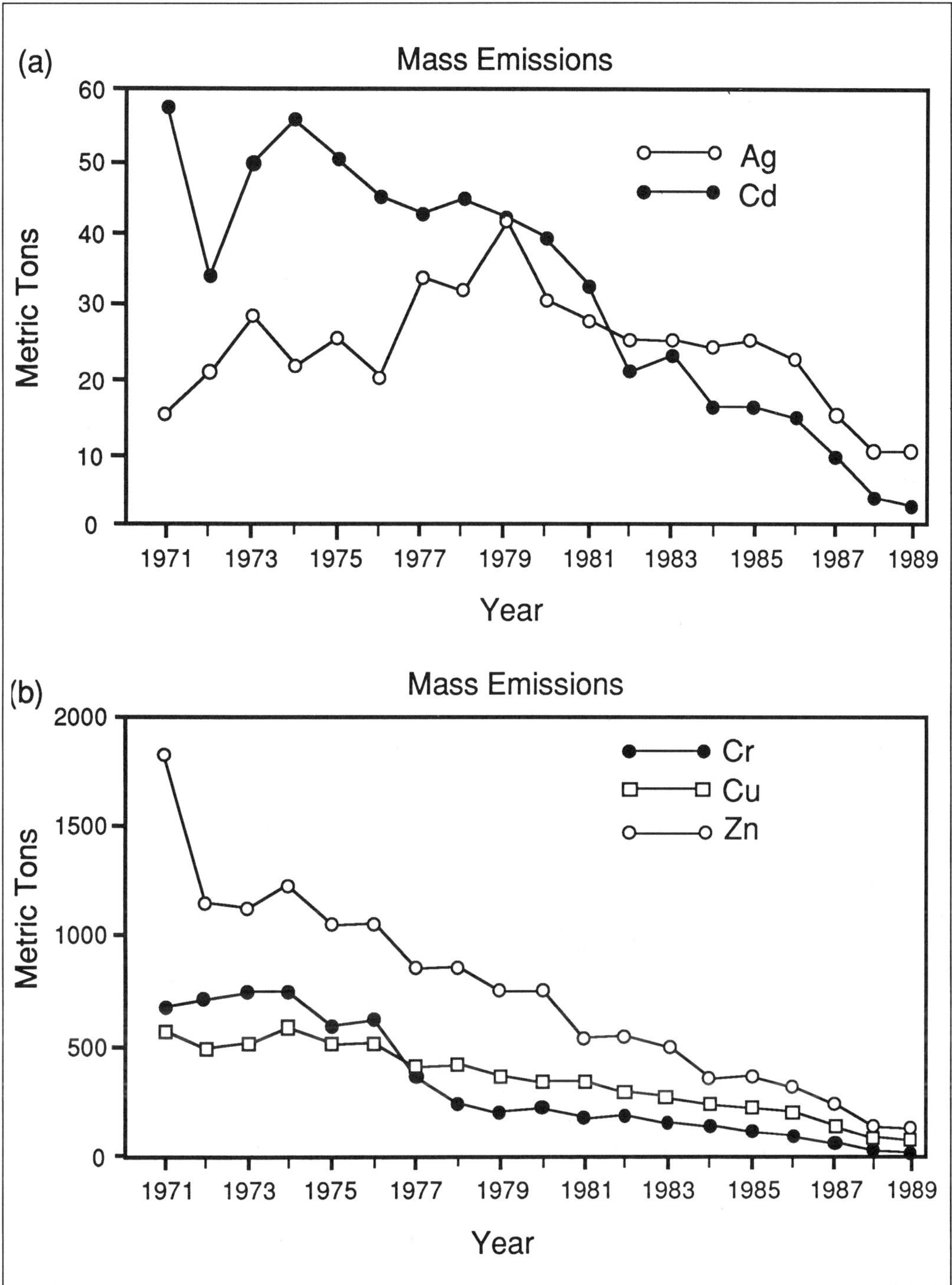

Figure 3.3. Emission rates (t yr^{-1}) of (a) silver and cadmium and (b) chromium, copper, and zinc to the SCB from four largest municipal waste dischargers from 1971 to 1989. (From SCCWRP 1991.)

10^6 t yr^{-1} during the Holocene. This is probably an underestimate of modern sediment inputs because agricultural land use and progressive deforestation tend to enhance natural sediment yields (Meade 1969). The discharge from streams could range up to one or two orders of magnitude greater than that observed during normal years (e.g., during the exceptional flood events of 1941 and 1969). These two major events alone contributed more than half of the influx of sediment from small streams during the past 50 years (Brownlie and Taylor 1981). Yet, very limited data exist to correlate storm flow characteristics with the mass emission rates of numerous constituents that are washed into the SCB during storm runoff and in dry weather flows (Eganhouse and Kaplan 1981; SCCWRP 1987).

As much as 70% of the total contemporary surface runoff could derive from storm flows, and a major portion of the suspended silt (94%) and total organic carbon (86%) are discharged into the SCB during the winter storm season lasting from November through April (SCCWRP 1973; Water Resources Data for California 1986). Mass emissions of oil and grease are apparently of equal magnitude in storm and dry weather flows (SCCWRP 1973).

The emission of suspended silts in storm runoff ranges from 1.5 to 6.0×10^5 t yr^{-1} (Eganhouse and Kaplan 1981; SCCWRP 1973; SCCWRP 1992) and appears to be fairly constant over a decade (table 3.2). The lower end of this range is nearly double the present-day suspended solids discharge rates from municipal wastewater outfalls, but the chemical composition of the two types of solids is quite different (Eganhouse and Kaplan 1982b; Eganhouse et al. 1981). Over time scales of decades to centuries, the Santa Clara River is the single most important source of sediment in southern California, although it discharges only about 10% of the total surface runoff flows from the Los Angeles Basin (SCCWRP 1973). During water years 1986–1987 and 1987–1988, the Los Angeles and San Gabriel rivers carried the greatest amounts of suspended solids to the northern part of the bight. Ballona Creek and the Los Angeles River contained silt concentrations in the range of 300–400 mg l^{-1} during the storm in 1970–1971. This is comparable to the average concentration of suspended solids (715 mg l^{-1}, ranging from 30 to >1000 mg l^{-1}) determined for the Los Angeles River after a storm event in 1978 (Eganhouse and Kaplan 1981). Apparently the stringent flood control measures on rivers such as the Los Angeles and Santa Ana have greatly reduced the mass emission rates of silt, although the Los Angeles River has averaged about 77% of the total storm runoff carried from the Los Angeles Basin over the decade (Young et al. 1980).

The estimate made by Schwalbach and Gorsline (1985) of the suspension (natural) discharge (12×10^6 t yr^{-1}) from rivers in southern California is an order of magnitude greater than recent storm runoff estimates (Eganhouse and Kaplan 1981; SCCWRP 1973, 1992). This is probably because Schwalbach and Gorsline (1985) have integrated the data over the entire Holocene. A large range in values is conceivable because terrigenous silt could enter in large pulses during exceptional events, as previously described. Furthermore, note that the data of Eganhouse and Kaplan (1981) are based on only one storm event.

The particulate organic carbon emission rate from surface runoff to the SCB in modern times is of the same order of magnitude as that from municipal waste outfalls (table 3.2). This emphasizes the potential importance of surface runoff to the ecosystem of the SCB. The limited data available on chlorinated hydrocarbons (SCCWRP 1973) suggest that the mass emission rate of these compounds from runoff during the early 1970s was roughly 1% of that observed from municipal wastewaters at that time, although the rates are similar a decade later (table 3.3) (Young et al. 1980). A comparative study of storm runoff in the Los Angeles River by

Young et al. (1980) showed that most of the trace metals investigated (Ag, Cd, Cr, Cu, Hg, Ni, Zn, Fe, and Mn) exhibited only a minor decline in flow-weighted mean concentrations over the decade. However, total PCBs decreased by about a factor of eight (Los Angeles River storm runoff in 1971–1972 was 2.6 mg l^{-1}, while in 1979–1980 it was 0.31 mg l^{-1}). This trend is to be expected because the use of PCBs was banned in the mid-1970s. Similarly, the concentration of lead decreased by a factor of six (Los Angeles River storm runoff in 1971–1972 was 940 mg l^{-1}, while in 1979–1980 it was 160 mg l^{-1}). This probably reflects federal regulation (initiated in 1975) of leaded gasoline and other fuel additives (National Research Council 1980). Comparison of trace metal concentrations in runoff and wastewaters during 1979 and the entire decade (1970–1979) shows that the annual mass emissions of most constituents via wastewater discharge exceed (by an order of magnitude) runoff emissions from Los Angeles Basin storm channels (Young et al. 1980). Lead and zinc are exceptions because of their automotive uses.

Harmful wastes have been barged farther out into the sea and dumped, either without packaging or after encapsulation in metallic drums. Few data are available in the open literature on the contaminants contained in these wastes, but they include radioactive and industrial wastes, oil drilling muds and cuttings, garbage, and military explosives. Drums could probably corrode in a decade, gradually releasing the enclosed materials. Only oil refinery and chemical wastes are believed to contribute significant amounts of pollutants in the SCB (SCCWRP 1973). Practically all dumping ceased as of 1972–1975 (Chartrand et al. 1985), with the exception of sediments dredged from the Los Angeles, Long Beach, and San Diego ports and from the naval station at San Diego. Currently, site LA-5 (at the outer edge of the continental shelf at a depth range of 130–190 m and 14 km from the entrance to San Diego Bay) is an active interim site for disposing of dredged material. Sites LA-2 (off Point Fermin, 9 km from the breakwater at San Pedro) and LA-3 (8 km to the south, southwest of Newport Beach Harbor) are inactive because their site designations have lapsed. The latter two sites will probably be designated as permanent sites in the future (Patrick Cotter, EPA, pers. comm. 1989).

The average amount of dredge spoils allowable for disposal at these three sites is 200,000–500,000 yd^3 (2.62–6.54 $\times$ 10^5 m^3) per site per year. Assuming that the dredged sediments from the southern California harbors have an average total organic carbon content of 1.65% (based on three replicate samples from 12 harbor locations extending from San Diego Bay to Los Angeles Harbor) (Anderson and Gossett 1987), we can estimate the potential organic carbon input from the allowable volume of dredged material dumped into a single dumpsite. An average specific gravity of 2.2 is also assumed in the calculations. About 9.5 $\times$ 10^3–2.4 $\times$ 10^4 t yr^{-1} of organic carbon is estimated to be contributed to the SCB environment from a single dumpsite (table 3.2).

Sedimentary rocks such as those of the western Transverse Ranges generally disintegrate to produce fine-grained debris. Surface runoff should therefore receive sizeable contributions from eroded shales such as the Monterey Shale (Miocene) along the coast of California. Unfortunately, data on the input of eroded shale to the SCB do not exist. However, the sediment budgets discussed by Schwalbach and Gorsline (1985) from southern California drainages obviously encompass contributions from shale erosion. Taylor (1983) estimated that an average of approximately 12 $\times$ 10^6 m^3 of sedimentary debris is eroded annually from the southern California coastal drainage systems. Fifty percent of this volume is composed of fine silt and clay, and a sizeable fraction is most likely derived from the eroding coastal shales. Thus, a gross estimate of mass emissions from shale erosion can be attempted here.

Of the mainland shoreline distance of

327 km, 87% is estimated to be erosional (Emery 1960), and exposures of the Monterey Shale are found almost continuously along this shoreline (Obradovich and Naeser 1981). Assuming that up to a 1-m depth of this shale is eroded at a rate of approximately 10 cm yr^{-1}, the volume of sediment input to the SCB from shale erosion amounts to 2.9×10^{10} cm^3 yr^{-1}. Using an average specific gravity of 2.2 (for diatomite, which is common in Monterey Shale), we estimate the sediment input from shale erosion to be about 6.4×10^4 t yr^{-1}. This constitutes roughly 1% of the contribution of sediment load from surface runoff, or 77% from waste discharge. The total organic carbon (TOC) input from the shale is approximately 6.4×10^2–1.2×10^4 t yr^{-1} (table 3.2) when estimates are based on the TOC percentage for shale (1–18%) (Curiale et al. 1985). The upper limit of shale-derived TOC leads to an estimated input that roughly equals the contribution from surface runoff or waste discharge.

Thousands of different organic compounds are emitted to the atmosphere from natural and pollution sources (e.g., Finlayson-Pitts and Pitts 1986; Kawamura and Kaplan 1987; Gagosian et al. 1987; Mazurek and Simoneit 1984). It is estimated by the South Coast Air Quality Management District (SCAQMD) that, in 1985, 1637 t d^{-1} of particles were emitted into the atmosphere in the south coast air basin (SCAQMD 1988). These particles undergo photochemical reactions that generate numerous compounds and adsorb sulfate, nitrate, and other reactive species (Arey et al. 1989; Finlayson-Pitts and Pitts 1986).

There have been suggestions that some of these species, especially PAHs, may accumulate in lipid-containing films on the ocean surface. These could become toxic to eggs and larvae (Hardy and Gucinski 1988). However, studies of atmospheric nitrogen budgets in the Pacific Ocean suggest that atmospheric nitrate deposition may add to the nutrient content of this oceanic regime (Logan 1983; Uematsu et al. 1985; Prospero and Savoie 1989) and affect productivity. Each of the airborne compounds has its own characteristic chemical and physical properties and associated atmospheric sources, residence times, and sinks. Unfortunately, data on aerial fallout rates of trace components into the SCB are very limited. In view of the restricted rainfall in southern California (approximately 20–40 cm yr^{-1}) (National Climate Data Center 1988), dry aerial fallout rather than wet deposition is expected to be the dominant airborne mechanism in the SCB. Strong northeasterly Santa Ana winds transport fine material from the nearby arid terrain into the SCB from March through July. As evidence of the importance of dry deposition, ships offshore have been found coated with fine dust during these episodes (Emery 1960).

Gray et al. (1986) have described the spatial and temporal distribution of aerosol carbon concentrations over an entire annual cycle in the Los Angeles area using a reference site on San Nicolas Island. Their island data are used here to compute the flux of total suspended particles, total fine organic carbon, and elemental carbon in the SCB. The fine carbonaceous particles emitted from most combustion processes (Cass et al. 1982; Muhlbaier and Williams 1982; Siegla and Smith 1981) contain organic compounds as well as black nonvolatile soot that has a chemical structure akin to impure graphite (Rosen et al. 1978). The black soot is referred to as elemental carbon, light absorption by which plays a significant role in the earth's radiation budget (Cess 1983; Patterson et al. 1982).

Assuming that the concentration of components measured by Gray et al. (1986) for San Nicolas Island is representative of the entire SCB area of 78,000 km^2, a gross estimate of the fluxes is computed as follows (after the method of Duce and Gagosian 1982). Dry deposition of atmospheric particulate matter is estimated, using the deposition velocity vd, as

$$vd = F/M$$

where M is the mass of aerosol in the atmosphere in micrograms per cubic centimeter, F is the flux of atmospheric particles to the surface (g cm^{-2} s^{-1}) and *vd* is the deposition velocity (cm s^{-1}). For particles collected by Gray et al. (1986) that are less than 2.1 μm, a deposition velocity (*vd*) range of 0.05–0.5 cm s^{-1} at wind speeds of 5–10 m s^{-1} should be applicable (Slinn and Slinn 1980). With this velocity range and the known concentration of particles (Gray et al. 1986), *F* can be calculated. The estimated annual dry deposition for the SCB is presented in table 3.2. Data are available only for fine particles, which compose approximately 31% of the total suspended load. If these parameters are equally valid for the total suspended particles by analogy with the fine particles, the flux of the former can be estimated.

This computation shows that the total suspended particulates and the total fine organic carbon fluxes to the SCB from dry fallout are one or two orders of magnitude less than that of wastewater discharge or surface runoff. Similarly, by the extrapolation of La Jolla fallout, Chow and Earl (1970) found that lead from combustion of leaded gasoline could account for as much as 2400 t of fallout over the SCB. The annual dry fallout of chlorinated hydrocarbons in February 1972, however, has been estimated to be approximately 2–4 t, which at that time was lower than from municipal wastewaters (SCCWRP 1973; Young et al. 1976).

Rabinowitz (1972) noted that although the concentrations of lead in wild oats (7–22 μg g^{-1}) from the islands in the SCB were lower by an order of magnitude than those found on the mainland, the island concentrations were above natural levels (2–3 μg g^{-1}). This indicates a significant flux in the past of atmospheric lead, other trace metals, and probably also organic constituents to waters of the SCB from urban centers. However, emissions of certain inorganic and organic components into the atmosphere have been curtailed by source controls such as the regulation of leaded gasoline from 1975 (National Research Council 1980) and the cessation of DDT production after 1982 by Montrose Chemical Corporation of Los Angeles County.

The particle flux from dry fallout just computed may be a realistic estimate. Fleischer (1970) attributes the increasing fine silt- and clay-sized quartz offshore to eolian deposition superimposed on the fluvial gradient offshore. The range in the particle input rate quoted in table 3.2 is consistent with the estimate of Fleischer (1970), who reported the eolian contribution to be less than 10^6 t yr^{-1}.

Lazrus et al. (1970) determined the concentration of different metals in rainfall over Santa Catalina Island from September 1966 to January 1967. Assuming an average annual rainfall rate of 40 cm over the SCB (National Climate Data Center 1988), their rough estimate indicates that the mass deposition rates of mercury, lead, and manganese exceed the flux from wastewater and runoff (for example, mass deposition rate of lead was approximately 400 t yr^{-1} from rainwater versus 243 t yr^{-1} from municipal discharge in 1971) (after SCCWRP 1987). These results are consistent with the findings of Bruland et al. (1974), who noticed enhanced fluxes of lead, zinc, and copper in more recently deposited sediments as opposed to iron, nickel, and manganese whose concentrations are uniform with depth. A close similarity in anthropogenic sedimentary fluxes and rainfall washout fluxes for many elements was also obvious. This implies a significant atmospheric transport mechanism for these metals (Bertine and Goldberg 1977). Further measurements of anthropogenic metal fluxes in the outer basins would help confirm this observation.

Although numerous organic compounds (over 300) have been measured in air, rainwater, fog, and mist in and around Los Angeles (Grosjean 1982; Kawamura and Kaplan 1983; Kawamura and Kaplan 1986; Steinberg and Kaplan 1984), no similar studies have been conducted to date from the islands off coastal southern California. Hence, the task of computing fluxes of various organic constituents from rainfall into the SCB is currently impossible. Reductions in ocean discharges from municipal outfalls over the last decade (Schafer 1989) and consequent combustion of sludge for energy recovery undoubtedly warrant a more important role of atmospheric deposition than ever before in contaminant inputs to the SCB. Yet, research efforts involving wet and dry deposition over the SCB have not been extensive to date.

Table 3.3 presents a summary of mass emission rates of organic carbon and trace constituents from various sources to the SCB. All data under advective transport, except organic carbon, were calculated from concentration data compiled by SCCWRP (1973). Mass transport rates of organic carbon were calculated from the depth-integrated concentration of dissolved organic carbon in bight waters (approximately 75 μM) (Hansell et al. 1990) for Santa Monica Basin on the assumption that these values are representative of Southern California Countercurrent waters. The last column in table 3.3 represents the total constituent mass emission rates from municipal wastewater, surface runoff, oil seeps, shale erosion, vessel coating, ocean dumping, rainfall washout, and dry fallout.

For organic carbon and all listed trace components, the contribution from advective transport is far greater than the sum from all other sources. Of the nonadvective sources, municipal wastewaters, surface runoff, and atmospheric deposition are the primary inputs. A few other trace metals such as cobalt, iron, and manganese (not listed in table 3.3) from surface runoff exceed the contribution from municipal wastewaters (SCCWRP 1973). It is clear from these data that vessel coating has contributed major amounts of copper in the past, while ocean dumping may be a significant source of DDT and PCB to the SCB. Unfortunately, these latter estimates have great uncertainties at the present time. Dry and wet deposition probably contributed significant amounts of lead to the SCB until the late 1970s. This compilation (table 3.3) also indicates that data on contributions of such constituents as hydrocarbons and total organic carbon from other sources are lacking. Further study is required to determine the net advective transport of various constituents and to understand the influence of advective transport on the spatial distribution of these compounds.

The environmental fate of organic carbon is governed by the physical and chemical characteristics of individual organic constituents. These characteristics ultimately control their reactivity, mobility, and stability within the water and sedimentary columns. Although the relative magnitudes of contributions of organic carbon from different sources are compared here, it should be noted that organic carbon from various inputs is not equally reactive.

Water Column Processes

In this section we discuss processes affecting the cycling of organic and inorganic chemical species in coastal waters of the SCB. The approach we have taken is to first describe the natural system, whereby the water column is divided into (1) near-surface waters (the upper 200 m), (2) intermediate waters (>200 m to basin sill depths), and (3) deep basin waters. The final subsection provides a brief discussion of anthropogenic effects. Specific examples of perturbations of the system resulting from inputs of trace organics and metals are presented.

For present purposes we have adopted the simplified definition scheme of Williams (1986a), wherein the distribution of chemical species between *particulate* and *dissolved* phases is set at the 0.4–0.5 μm cutoff. Particulate matter is further subdivided into *suspended* and *sinking* particles. The former consist of mineral grains, bacteria, phytoplankton cells, protozoans, and detrital aggregates generally having diameters less than approximately 50–100 μm (Beers 1986; Shiller 1982). These particles are easily maintained in suspension by the action of currents, and consequently, they dominate the mass of particulate matter collected by traditional water bottle samplers (McCave 1975). By contrast, the large, rapidly sinking particles, although relatively rare in the water column, are believed to be responsible for most of the vertical flux of particulate matter to deeper waters and sediments (McCave 1975; see references in Fowler and Knauer 1986). They range in diameter from 10^2 to 10^4 μm, have settling speeds of $10–10^3$m d^{-1}, and consist largely of macroscopic aggregates of biogenic and lithogenic materials in the form of intact organisms, skeletal parts, and fecal pellets (marine snow) (Asper 1987). Because traditional water bottle sampling is inefficient at capturing the large sinking particles, sediment traps (or particle interceptor traps) and large volume *in situ* filtration systems (Bishop and Edmond 1976) have been developed.

The distinction between the dissolved and particulate phases is somewhat obscured by the existence of colloids that represent a size transition between dissolved monomolecular species and the majority of the mass of suspended particles in seawater (1–70 μm) (Sharp 1973; Stumm and Morgan 1981). If the 0.4–0.5 μm cutoff is used as an operational definition, aggregates of large organic compounds (such as high molecular weight humic substances), clays, and iron oxide phases can be classified as particles based on their separation by filtration. Meanwhile, smaller aggregates in the colloidal size range, small bacterial cells, and members of the picoplankton (0.2–2 μm) may pass the same filters (Small et al. 1989). The recent introduction of ultrafiltration techniques has extended the particle size cutoff to the subnanometer range, greatly enhancing the ability to distinguish between suspended, colloidal, and dissolved species. However, difficulties can arise with this technique because of adsorption of dissolved organic matter on the filters and clogging of the pores with consequent retention of particles smaller than the nominal pore size (Carlson et al. 1985).

Natural System Dynamics

Here we discuss biogeochemical processes occurring in three portions of the water column: (1) near-surface waters (0–200 m), (2) intermediate waters (200 m to basin sill depth), and (3) deep basin waters. These divisions reflect significantly different environments with regard to their biology, chemistry, and physics. In broadest terms, the chemical composition of a parcel of coastal seawater is determined by the properties of the source waters, the mixing rates of adjacent parcels (vertically and horizontally), and the rates of processes that result in consumption or production of a given chemical species. The spatial distribution of these species in the water column is thus governed by an interplay of the physical, chemical, and biological factors that affect these rates.

As discussed earlier (chap. 2), coastal waters in the SCB are density stratified. During summer months, solar heating of surface waters results in maximum temperatures that approach 20°C, and the mixed layer depth ranges from 5 to 10 m depending on location. Advection dominates in the intervening period, during which minimum surface temperatures are approximately 14°C, with the mixed layer ranging from 10 to 30 m in depth. Stratification strongly inhibits vertical mixing between layers of different density. Thus, vertical concentration gradients in the water column (particularly in the upper 200 m) are much stronger than horizontal gradients. Measurable variations in water chemistry along isopycnals result mainly from mixing of adjacent water parcels and temporal variations in source strength (Jackson 1986).

Although vertical gradients in temperature and salinity in the surface waters of the SCB result from seasonal solar heating superimposed on advection, gradients of nonconservative chemical constituents in the upper 200 m (for example, nutrients, particulate organic matter, and trace metals), while largely established outside of the SCB, are sustained and locally modified by biological activity. The intermediate waters (200 m to basin sill depths) are subject to less intense biological activity and show weak or no vertical concentration gradients. Thus, the spatial distribution of chemical species in this portion of the water column is dominated by advection. Deep basin waters show weak density stratification, but horizontal advection is greatly restricted by the presence of basin slopes. Exchange occurs mainly through eddy diffusional mixing between deep and upper (above sill depth) basin waters and variable flows over the sills. Rapid exchange of deep basin waters, presumably brought on by strong coastal upwelling, has been documented for at least one inner basin: Santa Barbara (Liu 1979; Reimers et al. 1990; Sholkovitz and Gieskes 1971). Recent studies of San Pedro and San Nicolas basins by Berelson (1991) indicate that basin flushing over longer time scales and controlled by regional oceanographic events is operating in other parts of the bight. Because the sills of inner basins (Santa Barbara, Santa Monica, and San Pedro) intersect the oxygen minimum zone, the oxygen content of bottom waters is generally low ($<5\ \mu M\ O_2$) (Emery 1960). This leads to suboxic metabolism (e.g., denitrification) in basin waters and at the benthic boundary layer. Such reactions and chemical exchange across the sediment–water interface affect the distribution of biologically active substances in the deep basin environment.

NEAR-SURFACE WATERS

Air–Sea Interactions. The sea surface is of great importance to the biogeochemical cycles of numerous elements. It acts as both a source and sink for natural and anthropogenic substances and supports an active and unique biological community (the neuston). Materials deposited at the surface of the ocean from the atmosphere may become associated with particulate matter, some of which probably originates by particle aggregation during wave-induced bubble transport (Wallace and Duce 1978). The aggregation of particulate matter also facilitates removal of materials to deeper waters. Once formed, these particles may undergo dissolution, complexation, and decomposition reactions. At the same time, sea salts and other dissolved and particulate inorganic species (NO_3^- and PO_4^{-3}) and organic species as well as gases are injected into the atmosphere via bubble bursting or diffusive processes (Duce and Hoffman 1976; Lion and Leckie 1981). Unfortunately, knowledge of the exchange of chemicals across the air–sea interface in the SCB is extremely limited. However, information concerning the chemistry of sea surface films has recently begun to accumulate.

Williams (1967) first documented the enrichment of dissolved and particulate organic matter (carbon, nitrogen, and phosphorus) and nitrate in sea surface film samples relative to subsurface (15–20 cm) seawater taken off San Diego. Particulate organic matter (POM) is enriched in nitrogen and phosphorus relative to the dissolved organic matter (DOM) in the surface film. The enrichment of nitrogen and phosphorus in surface film particles suggests a recent origin for these materials, whereas the dissolved material may have undergone more biological reworking.

Henrichs and Williams (1985), Williams et al. (1986), and Williams (1986b) performed extensive physical, chemical, and microbiological analyses of sea surface films and subsurface water samples from eutrophic and oligotrophic waters off Baja California and within the SCB. These sites were free of contamination sources, and the films were believed to be representative of natural (unperturbed) coastal waters. They found no significant differences in chemical composition among films collected in these environments. However, film and subsurface waters showed qualitative and quantitative compositional differences. When the amounts of proteinaceous (P), carbohydrate (CA), and lipoidal fractions (L) of the films were combined, only approximately 50% of the particulate organic carbon (POC) and approximately 26% of the dissolved organic carbon (DOC) could be accounted for (table 3.4). The corresponding percentages of POC and DOC composed of the combined P, CA, and L fractions for subsurface waters are 52

Table 3.4. *Distribution of Dissolved and Particulate Organic Carbon among Various Compound Classes in Surface Film and Subsurface Seawater Samples*

Sample[a]	POC[b] (μM)	P (% of POC)[c]	CA (% of POC)[c]	L (% of POC)[c]	(P + CA + L)	(MP + B)
Cruise SF-2						
S. film	20	13	22	31	66	20
10 cm	12	16	23	19	54	23
SF/10 cm	1.7	0.8	1.0	1.6	1.2	0.9
Cruise SF-3						
S. film	37	16	12	4	33	12
10 cm	17	15	16	17	48	15
SF/10 cm	2.2	1.1	0.8	0.2	0.7	0.8
Cruise SF-4						
S. film	1.7	—	—	—	—	—
10 cm	6.6	—	—	—	—	—
SF/10 cm	2.6	—	—	—	—	—

Sample[a]	DOC[b] (μM)	P (% of DOC)[c]	CA (% of DOC)[c]	L (% of DOC)[c]	(P + CA + L)
Cruise SF-2					
S. film	100	6	17	3.4	27
10 cm	76	4	17	2.8	23
SF/10 cm	1.3	1.5	1.0	1.2	1.2
Cruise SF-3					
S. film	144	12	21	1.1	34
10 cm	96	6	17	2.3	25
SF/10 cm	1.5	2.0	1.2	0.5	1.4
Cruise SF-4					
S. film	86	2.1	13	3.8	19
10 cm	83	1.6	11	2.1	15
SF/10 cm	1.0	1.3	1.2	1.8	1.4

[a]S. film—surface film; 10 cm—subsurface water from 10 cm below air-sea interface; SF/10 cm—surface film/subsurface water (mean concentration ratio). Cruises SF-2, SF-3 were off Baja, California, whereas SF-4 was in the vicinity of San Clemente and Santa Catalina islands (see Williams 1986b).

[b]POC, DOC are particulate and dissolved organic carbon concentration means, respectively ($n = 5$ for Cruise SF-2, $n = 3$ for Cruise SF-3, $n = 4$ for Cruise SF-4).

[c]Mean percent total POC or DOC represented by different fractions of the carbon pools. P, CA—protein and carbohydrate carbon, respectively, passing 35 μm Nitex netting, but retained by 1.0 μm glass fiber filter; L—50% of lipid carbon passing the 35 μm Nitex netting; MP, B—microplankton and bacterial carbon, respectively, passing the 35 μm netting, but retained by a 0.2 μm Nuclepore filter.

Modified from Williams et al. (1986); Williams (1986b).

and 21%, respectively. It should be noted that the DOC measurements made in these studies were based on the classic persulfate oxidation method of Menzel and Vaccaro (1964). Recent work by Sugimura and Suzuki (1988) indicates that this method may underestimate (by as much as a factor of two) the actual DOC concentration. If so, this would further reduce the percentage of DOC represented by the P, CA, and L fractions.

Hydrolyzable amino acids are enriched (relative to carbohydrates) in the surface film when compared to subsurface waters (Henrichs and Williams 1985). Similarly, nonpolar hydrolyzable amino acids are more abundant than other amino acids in the surface film samples. On average, lipids contribute approximately 18% of the POC and $\leq$ 3% of the DOC pools. The identities of the remaining uncharacterized fractions (that is, the majority of the carbon) remain unknown, although measurements of the hydrophobic humic substances suggest that this fraction may account for as much as an additional 30% of the DOC (Williams et al. 1986). These findings are generally consistent with observations by others (see references in Hunter and Liss 1981) that the majority of the DOC and POC in sea surface films is present as complex macromolecular substances having properties similar to proteoglycans and glycoproteins. However, the lack of hydrolyzable amino acids in humic substances isolated with XAD–2 resin and their abundance in dissolved organic matter fractions retained by Sep-Pak C_{18} led Williams (1986b) to suggest that the amino acids were associated with nonpolar moieties such as lipoproteins or glycolipids.

Approximately 16–19% of the POC appears to come from living microplankton (predominantly dinoflagellates >1 μm) and bacterial (>0.2 μm) carbon. These organisms comprise approximately half of the protein plus carbohydrate carbon. Surprisingly, the surface film does not appear to be consistently enriched with respect to numbers of total or metabolizing bacteria relative to subsurface waters, but the amino acid utilization rate (on a per cell basis) of the metabolically active bacteria, as measured by [^{3}H]glutamic acid and [^{3}H]leucine uptake, is higher in films (Carlucci et al. 1985). This may reflect the greater availability of amino acids in surface films.

In the vicinity of urban centers, sea surface films receive significant contributions of anthropogenic substances. This leads to enrichment factors (i.e., $conc_{\mu layer}/conc_{subs\ water}$) for hydrophobic organic substances and heavy metals ranging from 10^1 to 10^5 (see references in Hardy 1982; Lion and Leckie 1981). Cross et al. (1987) conducted a survey of nearshore sites in the vicinity of the San Pedro Channel, Los Angeles and Long Beach harbors, San Pedro Bay, and Santa Monica Bay. They found that the concentrations of Ag, Cr, Cu, Fe, Mn, Ni, Pb, Zn, chlorinated hydrocarbons (DDT and PCB), and polycyclic aromatic hydrocarbons (PAHs) in microlayer samples collected at offshore stations were approximately two or three orders of magnitude lower than those in samples obtained from harbor stations. In general, the more contaminated sites contained higher particulate (as opposed to dissolved) metal concentrations. The PAH compositions of the harbor samples were dominated by lower molecular weight species (for example, naphthalene, phenanthrene, and their alkylated homologs), suggesting an uncombusted fossil fuel origin. Offshore samples contained larger relative amounts of the higher molecular weight PAH, indicating possible inputs from the atmosphere. Higher abundances of *o,p'*- and *p,p'*-DDT isomers (as opposed to the metabolite DDE) were also observed in the Los Angeles Harbor sample. This probably reflects the continued input of DDT wastes via runoff from the dominant source of these compounds in southern California—the Montrose Chemical Corporation (see section on Distribution and Fate of Chemical Constituents in Sediments). In contrast, the DDT composition at the Santa Monica Bay site was dominated by *p,p'*-DDE, most likely derived from older weathered sources such as the Palos Verdes shelf (MacGregor 1976). A recent study of microlayer chemistry in Los Angeles Harbor by Eganhouse et al. (1990) has shown similar results to that of Cross et al. (1987), including evidence that the PCB contamination originates, at least in part, from inadvertent introduction of debris generated locally by the shredding of scrap metal.

Nutrients: Distribution and Cycling. Waters within the SCB originate from at least five identified sources: (1) the westwind drift, (2) the subarctic Pacific via the California Current, (3) the equatorial Pacific, (4) the central North Pacific (which mixes with subarctic Pacific waters at the surface of the western margin of the California Current), and (5) a mixture of deep subarctic and equatorial waters (Jackson 1986). Near-surface waters in the SCB appear to be influenced strongly by contributions from the northern source from the California Current (Sverdrup and Fleming 1941), which is characterized by low salinity, low temperature, and relatively high oxygen and phosphate content (Emery 1960), as well as from the southern water, which is more saline, less oxygenated, and warmer. Transport in the upper water column is to the northwest from the large counterclockwise eddy known as the Southern California Countercurrent. Figure 3.4 depicts profiles of oxygen, phosphate, nitrate, nitrite, ammonium, and silicate for stations in San Pedro–Santa Monica Basin (Williams 1986a) and for many of the same constituents at station 90.70 (32°5.1′ N, 120° 38.5′ W, located in the California Current) from CalCOFI Cruise 8805. The qualitative similarity of many of the profiles is evident.

Referring to figure 3.4a, typical concentrations of dissolved nutrients in surface waters of the SCB (0–20 m) are as follows: nitrate 5–200 nM, nitrite 0–0.1 μM, phosphate 0.1–0.5 μM, silicate <5 μM, and ammonium 0.3 μM. With increasing water depth, concentrations of nitrate, phosphate, and silicate rise dramatically until, at a depth of approximately 100 m, the rate of change declines. Dissolved oxygen shows a trend that is inverse to nutrients, with surface water concentrations at or near saturation levels (approximately 0.27 mM), decreasing rapidly with increasing depth below about 50–100 m. In deep waters of the inner basins, concentrations of oxygen are extremely low (less than approximately 5 μM). Concentrations of nitrate typically decline below sill depths because of denitrification in the suboxic basin environment (Johnson et al. 1988; Liu 1979; Sholkovitz 1972). Thus, nitrate may reach a maximum concentration at water depths just above the sill depth of the inner basins. Nitrite and ammonium show subsurface maxima in the upper water column in response to activities of nitrifying bacteria. These nitrogen species are essentially confined to the upper 100 m due to their rapid assimilation.

Seasonal variations in the temperature and density structure of the water column are similar within the SCB. The main differences appear in the rate at which isotherms descend or rise seasonally for nearshore (shelf) and offshore areas. The euphotic zone, defined as the depth below which respiration exceeds photosynthesis (or below which light illumination is $<1\%$ that at the surface), ranges from approximately 50 to 100 m. Within the euphotic zone, light intensity and nutrient concentrations vary inversely such that phytoplankton at each depth experience unique conditions for growth (Eppley and Holm-Hansen 1986). An example of the distribution of light intensity, nitrate concentration, and other hydrographic parameters in the upper water column for an offshore region of the SCB is shown in figure 3.5. In this figure, the standing stock of phytoplankton (represented by chlorophyll *a*) is low in the surface waters where the photosynthetic rate and light levels are high. Nitrate concentration is essentially nil. Below the mixed layer (approximately 30 m), photosynthesis is light limited, but the standing stock increases dramatically, giving rise to a subsurface chlorophyll *a* maximum. This is possible because of the increased avail-

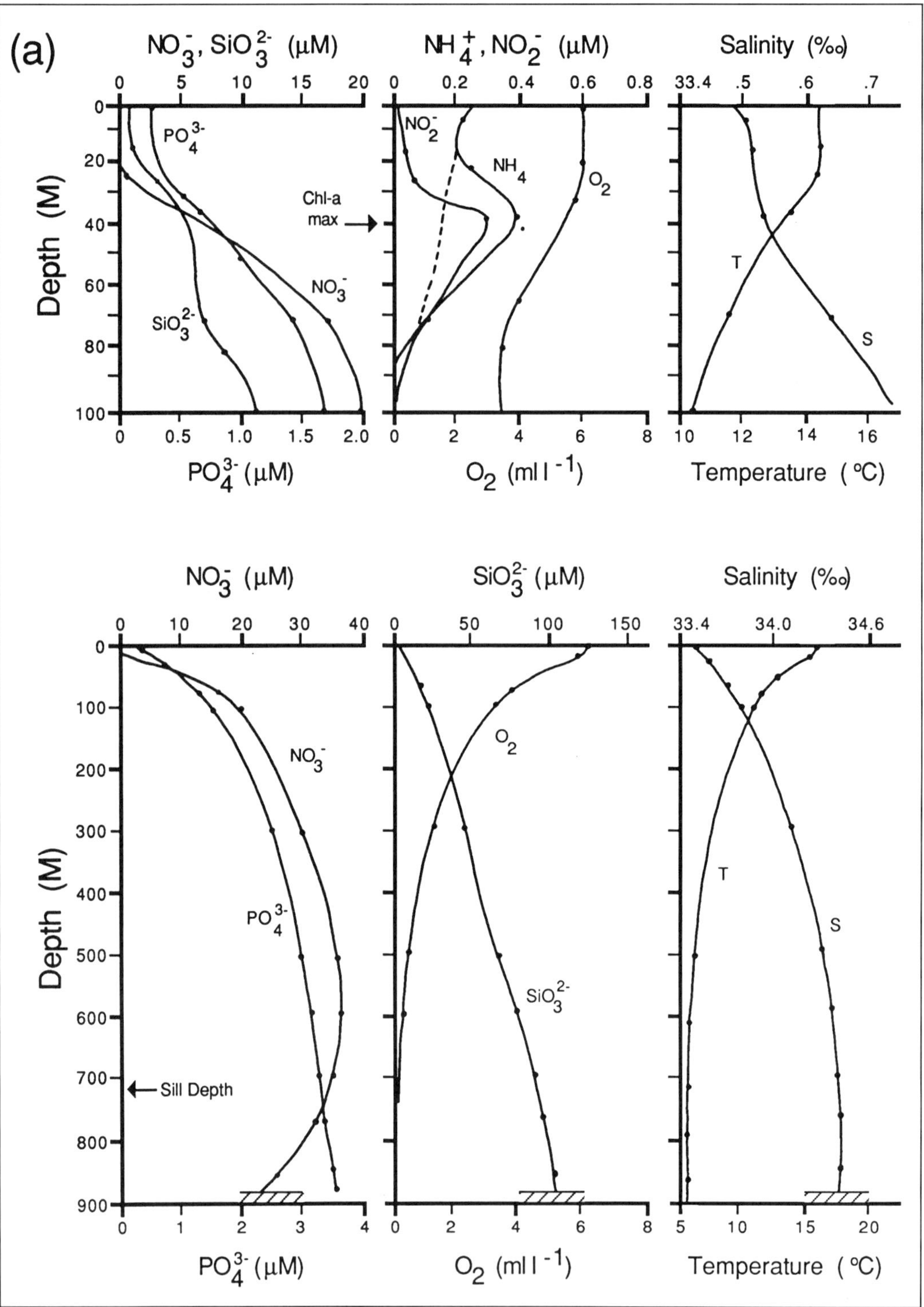

Figure 3.4. Vertical profiles of nutrients, dissolved oxygen, temperature, and salinity at (a) a location in San Pedro–Santa Monica basin, March 23–29, 1982 (after Williams 1986a) and (b) at CalCOFI station 90.70 in the California Current (CalCOFI cruise 8805, April 5, 1988).

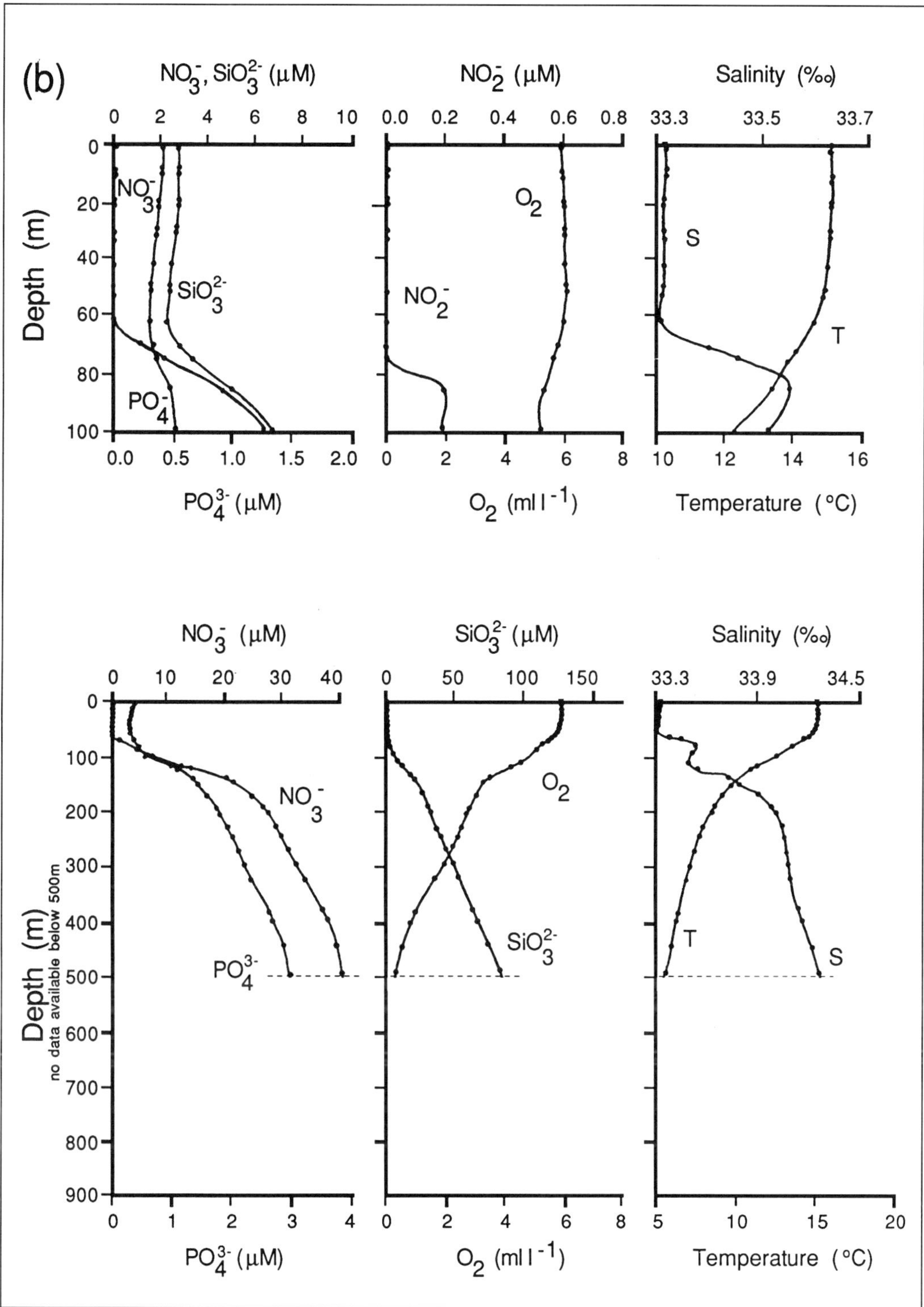

Figure 3.4. (continued)

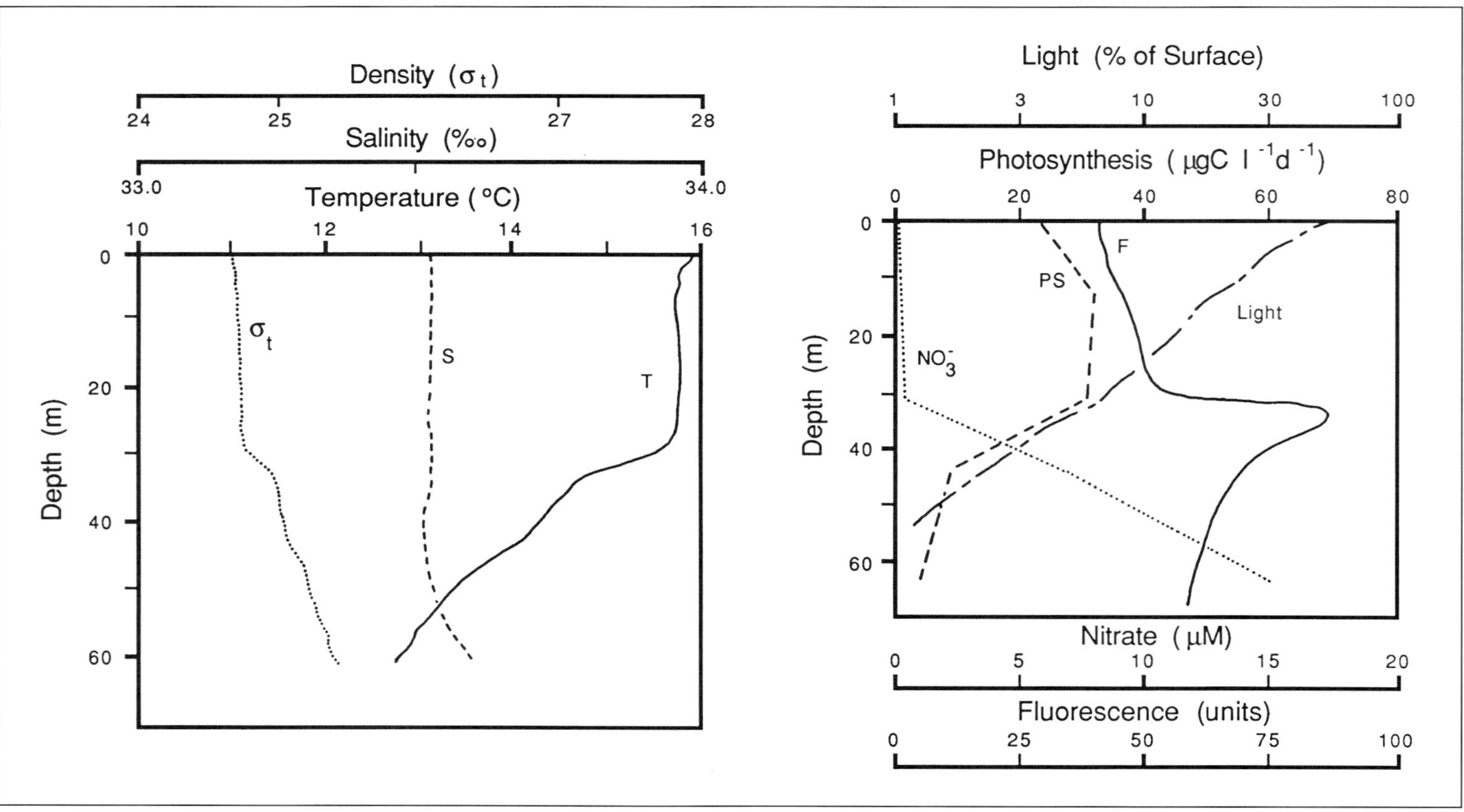

Figure 3.5. Distribution of light intensity, photosynthetic rate, and hydrographic properties in the euphotic zone at a site 52 km offshore along CalCOFI line 90. (After Eppley and Holm-Hansen 1986.)

ability of nitrate and lack of vertical mixing of the phytoplankton in deeper waters.

Eppley et al. (1979a) have demonstrated an inverse relationship between the depth of the nitracline and primary production, presumably reflecting a balance between the vertical mixing rate of nitrate and its uptake by phytoplankton. When integrated over the upper 50 m of the water column along a cross-shelf transect extending 125 km offshore, the concentration of nutrients increases at inshore locations (Mullin 1986). This gradient results from inshore shoaling of the nutricline (Eppley et al. 1978; Eppley 1992). However, when concentrations are integrated over depths corresponding to the euphotic zone, no difference between inshore and offshore stations is apparent. Figure 3.6 illustrates the distribution of nitrate as of August 1988 along two cross-shore CalCOFI transects originating near the coastline and extending into the California Current. Although inshore shoaling of the nitracline is evidenced along both transects, the cross-shore variations are irregular. In fact, there appears to be a regional bulge in the nitracline between stations 50 and 60 (at the outer margin of the SCB). As expected from the findings of Eppley et al. (1978), these cross-shore variations in depth of the nitracline compare favorably with the distribution of phytoplankton biomass revealed by satellite imagery of chlorophyll concentration (Eppley 1986).

There are several mechanisms by which nutrient-rich deep waters can be mixed into the euphotic zone. One of these, coastal upwelling, results in the offshore displacement of warm, low-density, nutrient-depleted surface waters. These waters are replaced through onshore advection of deeper waters with consequent shoaling of the thermocline over the shelf. Coastal upwelling on the west coast of North America is most intense in areas offshore of Oregon and northern California. However, upwelling also occurs in the SCB, most significantly in the area off Point Conception, during the early spring and summer months (March to July) with the onset of north–northwesterly winds. Figure 3.7 shows an 18-month time record for nutrient and oxygen concentrations at a station 10 km off Corona del Mar (Barcelona et al. 1982). These data illustrate the effect of coastal upwelling events on temperature, salinity, and nutrient distributions. The base of the thermocline is compressed from 100 m to 40 m as "plumes" of colder, nutrient-rich deep waters intrude the upper 200 m of the water column. The effects at this site are greatest below 100 m, and there is little or no observable change in nutrient concentration in the euphotic zone (approximately upper 50 m) at this location (water depth is 500 m). One would expect greater perturbations of nutrient distributions in the upper water column inshore of these depths with consequent impacts on primary production. An example was provided by Tont (1981), who demonstrated a relationship between records of phytoplankton abundance (namely, diatom blooms) off the Scripps Institution of Oceanography pier over a 20-year period and the occurrence of upwelling episodes.

Another mechanism by which nutrients can be transported vertically from deeper water to the euphotic zone in nearshore shelf regions are the meter-range oscillations in isopycnals generated by semidiurnal tidal motions and internal waves. Winant and Bratkovich (1981) have shown that motions occurring at these frequencies make a significant contribution to the power spectrum of longshore and cross-shelf currents in the SCB. Semidiurnal tidal motions and internal waves modulated at frequencies on the order of 5–15 minutes can, thus, pump nutrients into the surface mixed layer of the ocean. Armstrong and LaFond (1966) observed short-term variations in nutrient concentrations at fixed depths near the nutricline that were correlated with temperature fluctuations oscillating at frequencies characteristic of internal waves. Cullen et al. (1983) extended these observations over several days and showed that internal tides probably bring about mixing between layers by dissipation of shear instabilities.

As previously noted, density stratification

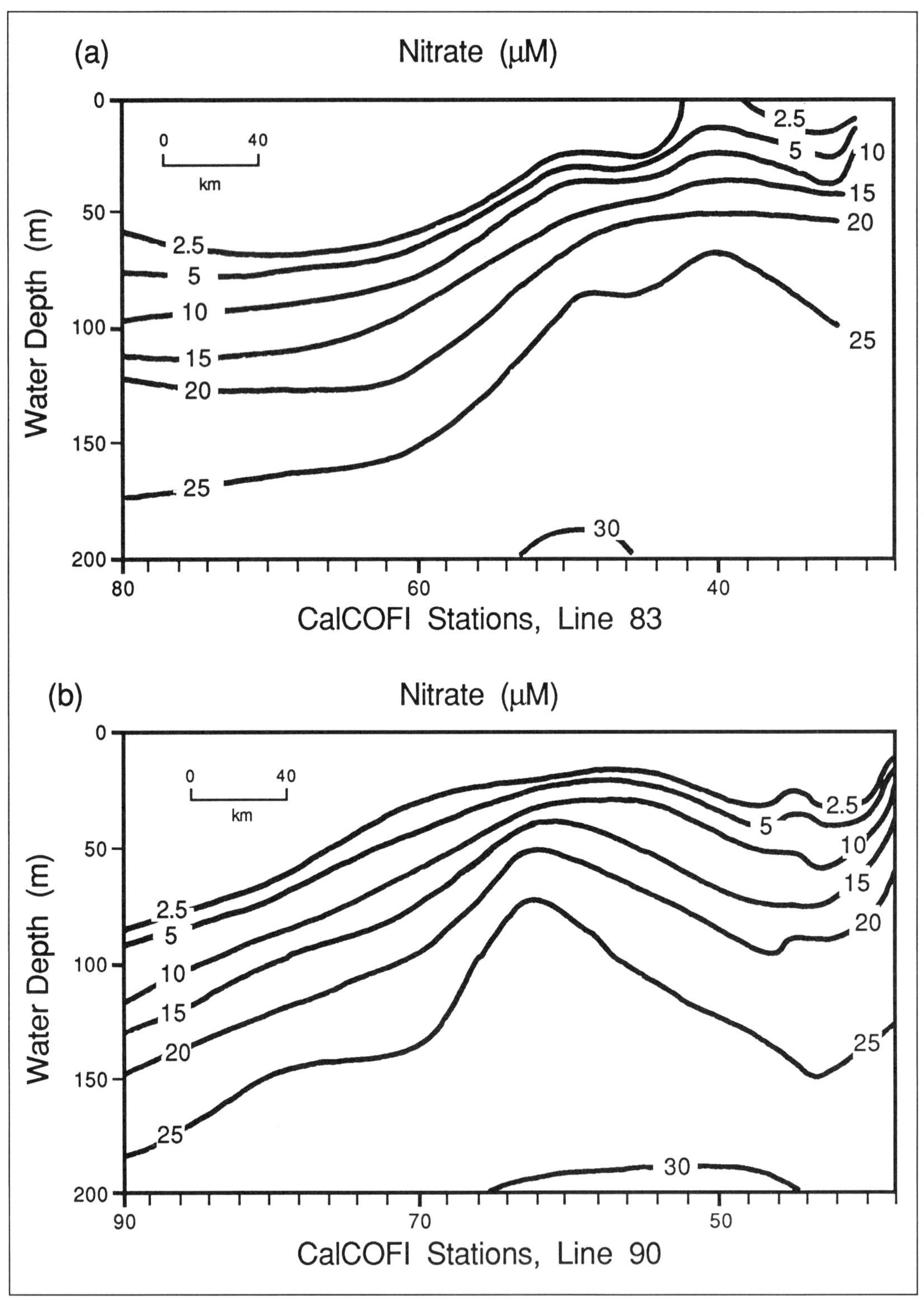

Figure 3.6. Distribution of nitrate along two cross-shore transects: (a) line 83 and (b) line 90. (After CalCOFI data, cruise 8808, August 1988.)

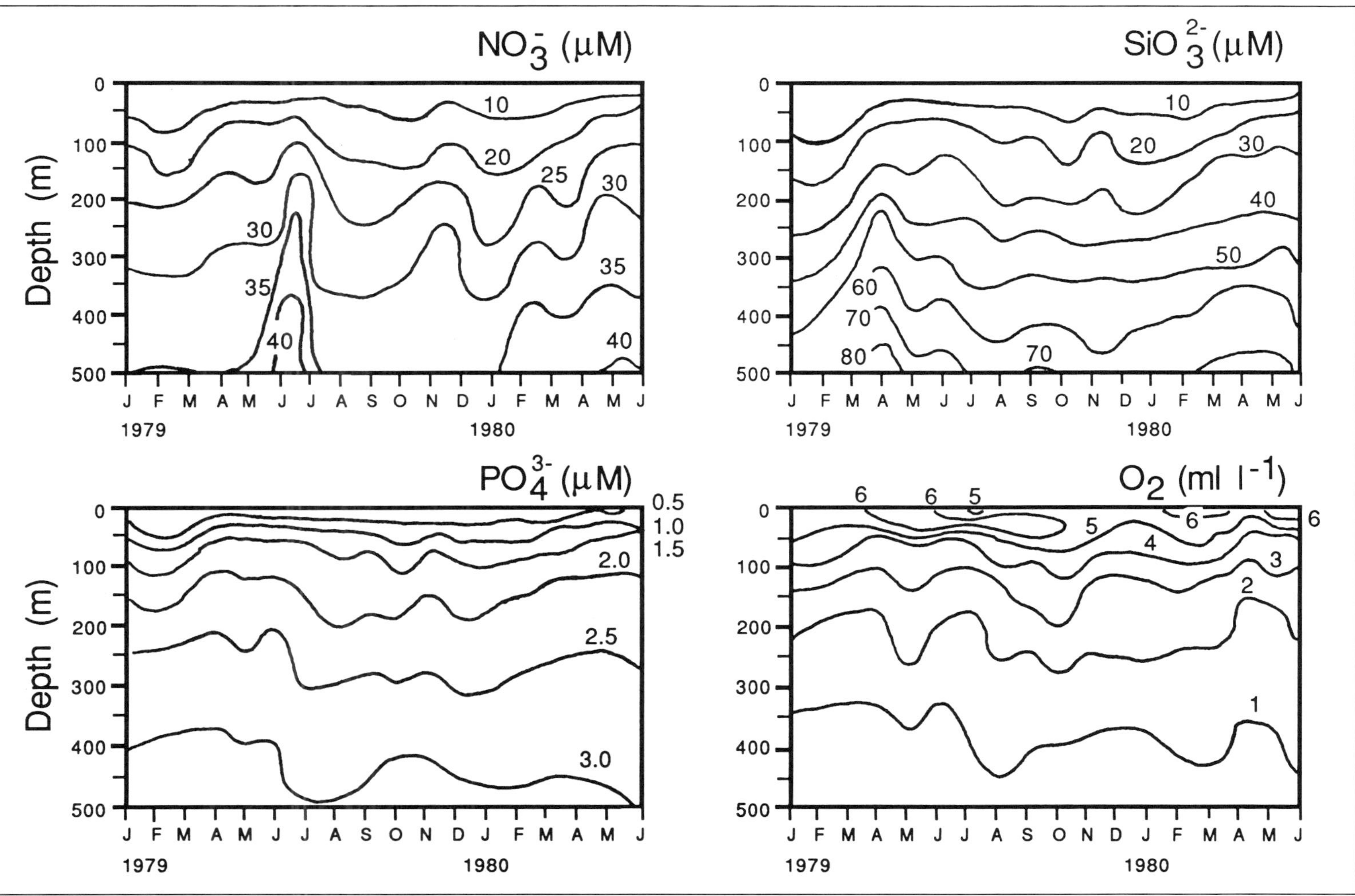

Figure 3.7. Temporal variations in nutrient, temperature, and oxygen distributions over an 18-month period at a station 10 km offshore Corona del Mar. (After Barcelona et al. 1982.)

of the water column results in the vertical zonation of certain biological communities. Distributions of metabolites generated by these communities reflect the net result of production and utilization processes. Such zonation was first noted in the SCB by Holm-Hansen et al. (1966) for waters overlying Santa Catalina Basin. One of these metabolites, ammonium, is produced during the heterotrophic degradation of organic matter (ammonification). Ammonium is rapidly taken up by phytoplankton and in southern California coastal waters has been estimated to account for 30–40% of the nitrogen assimilated during primary production (Eppley et al. 1979a). Ammonium can also be oxidized to nitrite by two genera of nitrifying bacteria, *Nitrosococcus* and *Nitrosomonas*. Nitrite, in turn, is oxidized to the most common form of combined nitrogen, nitrate, by bacteria of the genus *Nitrobacter*.

As mentioned earlier, ammonium and nitrite concentrations in the SCB in the submicromolar range are essentially restricted to the upper water column (<100m) (see fig. 3.4a). A possible exception is the occurrence of measurable ammonium in suboxic deep waters of inner basins such as the Santa Barbara Basin (Liu 1979), where nitrification rates are reduced and degradation of organic matter by denitrification and sulfate reduction (in sediments) can occur. Subsurface maxima in ammonium concentrations have been observed by Eppley et al. (1979b). However, the vertical distribution of ammonium is highly variable both spatially and temporally. Local sources of ammonium may be important, as evidenced by the elevated concentrations in waters of Santa Monica Bay that are believed to be due to the discharge of municipal wastes (Eppley et al. 1979b; Thomas and Carsola 1980).

Nitrite profiles typically exhibit a subsurface maximum (primary nitrite maximum) positioned at or near the base of the euphotic zone in the upper part of the nitracline (Mullins et al. 1985; Olson 1981a). Olson (1981a, b) demonstrated by ^{15}N tracer studies that the primary nitrite maximum in southern California arises from oxidation of ammonium, presumably by nitrifying bacteria. He presented a model (fig. 3.8) for the formation of a nitrite maximum, the position of which was determined by photoinhibition of ammonium oxidation in surface waters coupled with reduced uptake of nitrite by phytoplankton. (Ward [1985] and Ward et al. [1982] later confirmed the photoinhibition of ammonium-oxidizing bacteria off southern California and the Washington coast.) The reduced uptake of nitrite was postulated as being the result of differential rates of uptake for nitrite and nitrate within the nitracline. The low abundance of nitrate in the upper part of the nitrite maximum was subsequently attributed to greater photoinhibition of the nitrite-to-nitrate conversion (Olson 1981b). Olson (1981a) further proposed that the declining nitrite concentrations in deeper waters was attributable to increased conversion of nitrite to nitrate.

Particles: Formation, Distribution, and Fluxes. The dominant source of particulate matter in surface waters (especially the euphotic zone) of the SCB is autochthonous biological production. These particles include free-living primary producers and associated members of the planktonic food web, suspended detritus, fecal pellets, and macroscopic aggregates (marine snow). The latter range in size from < 1 mm to several centimeters in diameter and are believed to sink at rates of tens to hundreds of meters per day (Alldredge 1979; Asper 1987; Shanks and Trent 1980). Consequently, they, along with zooplankton fecal pellets, are believed to be important sources of the vertical flux of particulate matter to the benthos (Dunbar and Berger 1981). Although patchy in distribution and compositionally variable, marine snow is present throughout the water column and consists of complex associations of planktonic organisms, fecal pellets, exoskeletons, lithogenic debris, and bacteria (Hebel et al. 1986; Silver et al. 1978; Silver and Alldredge 1981). The

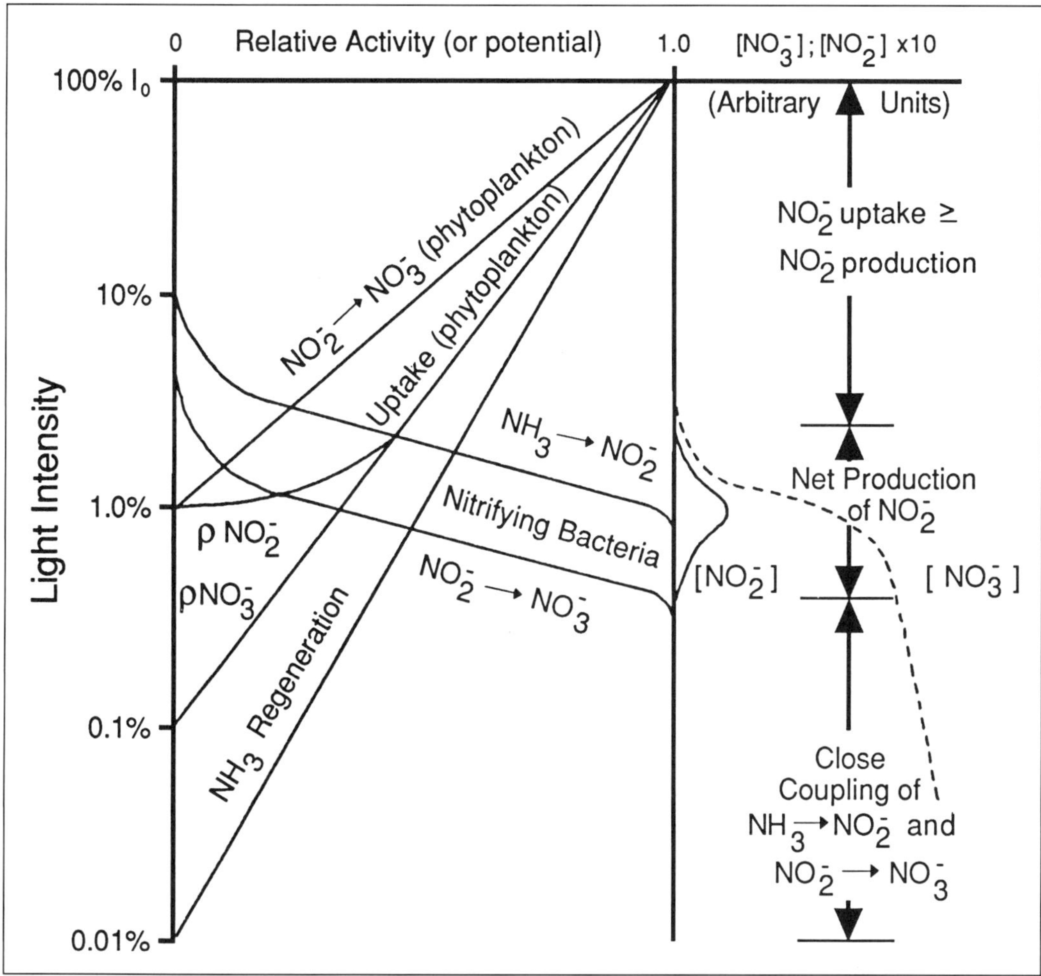

Figure 3.8. Conceptual model of processes contributing to formation of a primary nitrite maximum. PNO_3^- and PNO_2^- are nitrate and nitrite uptake rates, respectively. (After Olson 1981a.)

density of macroscopic aggregates in surface waters is highly variable (Fowler and Knauer 1986), but ^{14}C production rates within them can exceed that of the surrounding aggregate-free seawater by as much as 50 times (Alldredge and Cox 1982). Under upwelling conditions, as much as 20–60% of all primary production can occur on them (Knauer et al. 1982; Prézelin and Alldredge 1983). Alldredge (1979) estimated that, on average, 34% of the POC and 26% of the particulate organic nitrogen (PON) in surface waters of the Santa Barbara Channel was present in the form of macroscopic aggregates.

Knowledge of the chemical composition of particulate organic matter (POM) in the upper water column of the SCB is limited. Siezen and Mague (1978) found that nearly all of the PON and approximately 16–52% of the POC in near surface waters (<55 m) could be accounted for by protein amino acids. Similarities in the proportions of protein organic carbon reported for cultured phytoplankton led Eppley et al. (1977) to conclude that the vast majority of the POC was of recent biogenic origin. If this is the case, the remaining 50% of the POC would be expected to consist of carbohydrates and lipids.

Similarly, Holm-Hansen et al. (1966) demonstrated that the POM in the upper 75 m of the water column over Santa Catalina Basin had a C:N:P ratio of 100:18:3 (similar to living phytoplankton grown in nitrogen-deficient waters), whereas suspended POM from deeper sections yielded ratios depleted in nitrogen and phosphorus. These observations support the idea that the vast majority of the particulate organic matter in the surface waters of the SCB is of recent biological origin and that the fraction represented by relatively older, refractory detritus is minor. Whereas the living plankton biomass generally accounts for less than one-half of the total POC in surface waters, blooms can lead to abundances exceeding 70% (Eppley et al. 1983). Data of Small et al. (1989) for Santa Monica Bay suggest that as much as 25–59% of the standing stock of POC within the euphotic zone can be composed of bacterial carbon.

Figure 3.9 illustrates the distribution of POC and PON along a cross-shore transect corresponding to CalCOFI line 87 and vertical profiles at a station along this transect within Santa Monica Basin. Surface water POC concentrations at nearshore stations generally reach 100–300 $\mu g\ C\ l^{-1}$ and decline by approximately an order of magnitude within a distance of 100 km offshore. Similar patterns are seen for PON. The high particulate organic matter concentrations observed at the inshore stations along transect 87 in figure 3.9a are believed to reflect elevated plankton stocks caused by nutrient enrichment associated with municipal wastewater discharges in Santa Monica Bay. There is often a subsurface POC (and PON) maximum corresponding to the chlorophyll *a* maximum (Small et al. 1989). With increasing water depth, however, POC concentrations decrease rapidly until, at depths below 100–300 m, a concentration of approximately 1 $\mu g\ l^{-1}$ is reached. If one assumes that the POC observed at greater depths largely represents the background refractory detrital component throughout the water column, it is seen that this fraction makes up less than 1% of the total POC pool in surface waters. This is consistent with the hypothesis that the vast majority of the POC in the euphotic zone is of recent biogenic origin.

Particulate organic carbon near the surface of the ocean is either recycled or lost to deeper waters by sinking. Aside from its importance to nutrient generation at depth and its vital role in supplying energy to the benthos, sinking particulate matter provides an effective vehicle for transporting trace inorganic and organic pollutants to the deep ocean (e.g., Bruland and Franks 1979; Bruland et al. 1981; Crisp et al. 1979; Venkatesan and Kaplan 1988). Consequently, it is important to characterize the factors affecting the residence time of particulate matter in surface waters and its vertical flux out of them.

Eppley et al. (1983) defined the residence time of POC in surface waters of the SCB as the standing stock of POC divided by "new production" (estimated either by nitrate assimilation rates and the POC:PON ratio or by fluxes measured in sediment traps positioned below the euphotic zone). They found that the residence time of POC was inversely related to the total ^{14}C production. Furthermore, residence times estimated for offshore sites during the period 1974–1979 ranged from less than 10 days (high ^{14}C production) to several hundreds of days (during oligotrophic conditions), the wide differences being attributable to changes in ocean climate (for example, El Niño) and upwelling intensity. The long residence times observed during certain periods (1976 and 1977), rather than reflecting resistance of the POC to degradation, imply a high rate of recycling, with turnover times on the order of 11 days. Rapid turnover means that the particulate organic matter is recycled numerous times before sinking to deeper waters. Eppley et al. (1983) calculated the number of times a POC atom would be recycled before escaping the euphotic zone at between 1 (during upwelling) and nearly 20 (during oligotrophic periods). This suggests that during periods of high

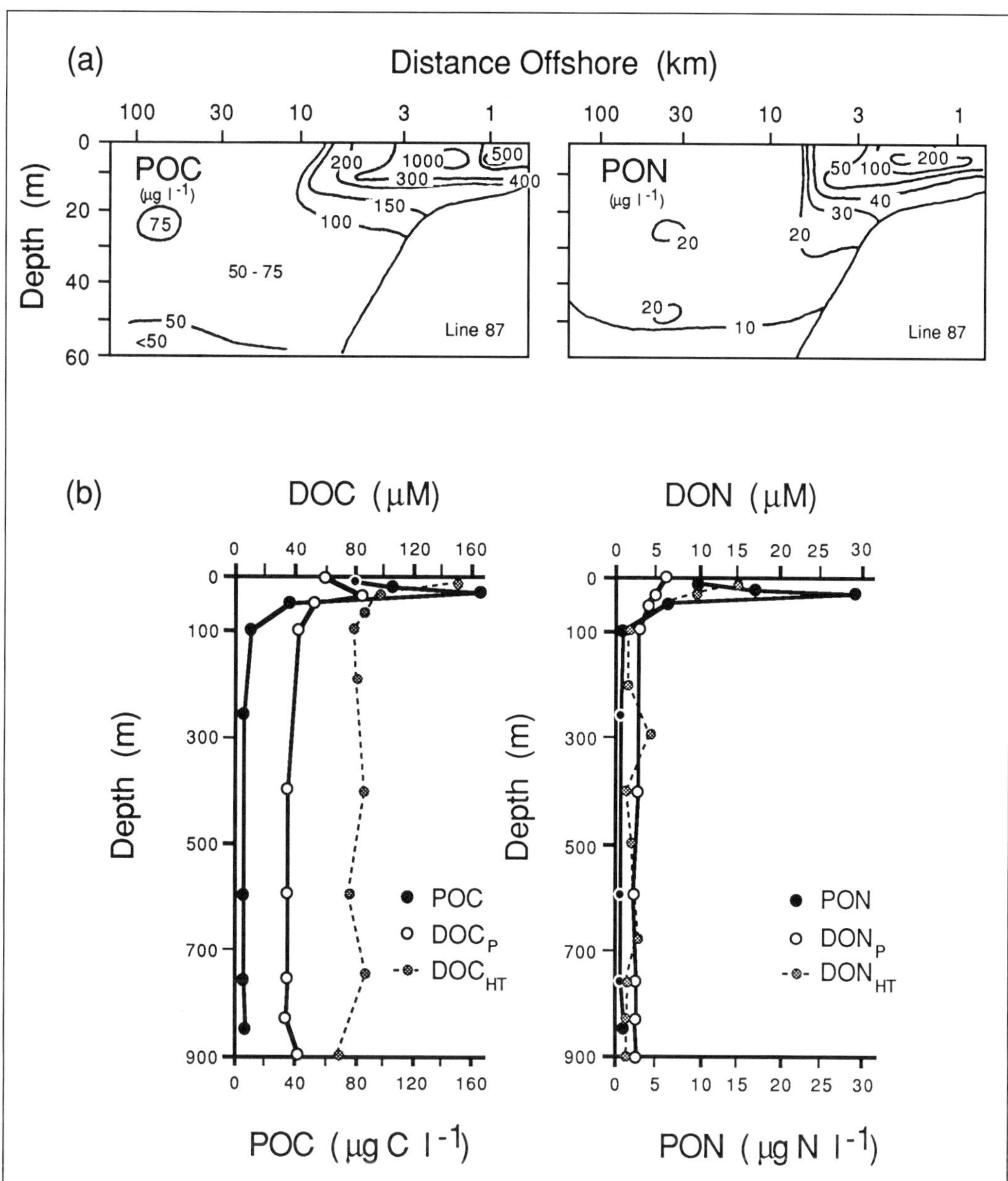

Figure 3.9. Distribution of particulate organic carbon (POC) and particulate organic nitrogen (PON) (a) along a cross-shore transect corresponding to CalCOFI line 87 (after Eppley et al. 1977) and (b) POC, PON, DOC_P, and DON_P profiles from station 305, cruise SCBS, May 18, 1981 (after Williams 1986a), are based on persulfate oxidation, whereas DOC_{HT} and DON_{HT} profiles from the same station in July 1990 (after Hansell et al. 1990) are based on the high-temperature catalytic combustion method.

production, POC generated in surface waters would likely settle to the sea floor, whereas oligotrophic conditions would favor transport of suspended particles generated in the euphotic zone out of the SCB. This is consistent with the paradigm that high productivity enhances operation of the grazing food chain and therefore promotes the generation of large particles such as zooplankton fecal pellets and macroscopic aggregates (Prézelin and Alldredge 1983) capable of sinking from the euphotic zone. Oligotrophic conditions would be expected to favor functioning of the microbial loop (Azam 1986), whereby recycling of POM is carried out efficiently and production of large particles with rapid sinking rates is reduced. However, recent modeling efforts by Jackson and Eldridge (1992) using CaBS data indicate that food web dynamics in the SCB may be considerably more complex than either of these simple models would suggest. This may help explain why no apparent seasonal trend in POC and PON residence times was found within Santa Monica Basin over a 2.5-year period (Small et al. 1989).

Deuser and Ross (1980) first noted the seasonal coupling of primary productivity and the vertical flux of particulate matter to the deep sea. It was evident that this coupling could only be achieved by transport of particulate matter at rates on the order of 10^1–10^3 m d^{-1}, a requirement that established (along with earlier theoretical considerations) (McCave 1975) the importance of the relatively rare large particles in dominating the vertical flux of particulate matter in the ocean. These relationships were formalized in a quantitative empirical model by Suess (1980), who used the results of early sediment trap studies. Eppley and Peterson (1979) showed that the proportion of "new production" arising from input of nutrients to the euphotic zone (through upwelling and diffusion from deeper water) was directly related to total production in cases where total production was below 200 g C m^{-2} yr^{-1}. The proposed relationship was developed using data from sites in the SCB, among others. As a percentage of total production, new production within the SCB was estimated at 15–60% or greater, depending on location and time (see also Small et al. 1989). Under steady-state conditions, new production should equal the downward vertical flux of particulate matter from the euphotic zone. If so, simultaneous measurement of vertical fluxes (using sediment traps) and total production in the euphotic zone should afford an opportunity to estimate new production and test its relationship to total production.

Early work by Knauer and Martin (1981) for a coastal site 80 km off the Monterey Peninsula indicated that the flux of particulate carbon and nitrogen at the base of the euphotic zone represented 22% of the primary productivity (table 3.5). These studies were performed over a 6-day period and involved contemporaneous measurements of primary productivity. Previous investigations (Knauer et al. 1979) conducted under upwelling and nonupwelling conditions at a site 30 km offshore had suggested higher percentages (upwelling: 53% carbon, 39% nitrogen; nonupwelling: 34% carbon, 23% nitrogen). However, primary productivity was not measured in these studies. In both cases, the estimated vertical flux of particulate organic matter decreased with increasing water depth at the site, the greatest changes occurring in the upper 500 m of the water column. This identifies the upper water column as an active site of recycling or remineralization, as reflected in vertical profiles of nitrate, silicate, and phosphate (fig. 3.4).

Nelson et al. (1987) reported results of sediment trap studies in the Santa Monica–San Pedro Basin in which free-drifting sediment traps (FST) deployed at 100 m for 2.5 days were used to estimate particle fluxes from the euphotic zone (approximately 53 m). Comparison of the fluxes of POC with primary productivity measured in the overlying waters during the trap deployments suggested that only 5% of the carbon being fixed in the euphotic zone was reaching deeper waters.

Table 3.5. *Summary of Sediment Trap Studies from the SCB and Adjacent Areas*

Study	Location[a]	Deployment Depth[b] (m)	Flux (g m^{-2} d^{-1})			Other Parameters[c]
			Mass	OC	N	
Soutar et al. (1977)	SBB	150	0.55	—	—	
	SBB	30/A	0.82	—	—	
	SBB	10/A	4.65	—	—	
	SBB	100	2.19	—	—	
	SBB	10/A	2.46	—	—	
	SBB	150	0.82	—	—	
	SBB	10/A	1.6	—	—	
	SPB	150	0.3	—	—	
	SPB	10/A	0.3	—	—	
	SB	150	1.4	—	—	
	SB	180	0.5	—	—	
	SB	40/A	0.8	—	—	
	SB	10/A	0.8	—	—	
	SB	150	0.3	—	—	
	SB	10/A	8.0	—	—	
Knauer et al. (1979)	30 km off Monterey	50	—	0.43	0.057	^{210}Pb, FP, P
		250	—	0.25	0.025	
		700	—	0.12	0.013	
		50	—	0.091	0.011	
		250	—	0.052	0.006	
		700	—	0.049	0.004	
Knauer and Martin (1981)	80 km off Monterey	35	—	0.31	0.061	PP
		65	—	0.15	0.026	
		150	—	0.080	0.012	
		500	—	0.025	0.008	
		750	—	0.023	0.002	
		1500	—	0.071	0.009	
Dymond et al. (1981)	SBB	341	1.83	0.064	—	Al, Si, Mg, K, Ca, Fe, Mn, Ni, Cu, Zn, Ba, $CaCO_3$
	SBB	381	1.96	0.073	—	
	SBB	213	2.12	0.074	—	
	SBB	328	1.27	0.047	—	
	SBB	162	1.01	0.032	—	
Crisp et al. (1979)	SBB	30/A	1.12	0.040	—	Markers, AHC, SHC, UCM, $\delta^{13}C$, $\delta^{34}S$, $\delta^{15}N$, ESR
	SNB	30/A	0.27	0.009	—	
	SMB	30/A	0.49	0.76	—	
	SPB	30/A	0.76	0.034	—	
Bruland et al. (1981)	SBB	30/A	2.01	—	—	Fe, Cu, Zn, Cd, Pb, ^{210}Pb, ^{228}Th, ^{232}Th
	SBB	30/A	1.96	—	—	
	SNB	30/A	0.40	—	—	
	SMB	30/A	0.46	—	—	
	SPB	30/A	1.00	—	—	
Karl and Knauer (1984)	160 km off Big Sur	100	—	0.064	0.010	ATP, RNA, nucleic acid synthesis, PP, O_2, FP, LH, Al
		300	—	0.031	0.0056	
		500	—	0.023	0.0042	
		600	—	0.022	0.0034	
		700	—	0.015	0.0020	

Table 3.5. *Summary of Sediment Trap Studies from the SCB and Adjacent Areas* (continued)

Study	Location[a]	Deployment Depth[b] (m)	Flux ($g\ m^{-2}\ d^{-1}$) Mass	OC	N	Other Parameters[c]
		900	—	0.031	0.0053	
		1100	—	0.021	0.0021	
		1400	—	0.015	0.0023	
		1700	—	0.015	0.0023	
Noriki and Tsunogai (1986)	California Current (EP5)	510	0.081	0.030	—	Percent clay, TVS opal, $CaCO_3$
		720	0.082	0.031	—	
		1250	0.070	0.026	—	
		3370	0.054	0.018	—	
		4220	0.046	0.013	—	
	California Current (EP7)	500	0.019	0.008	—	
		720	0.015	0.005	—	
		1250	0.015	0.003	—	
		3380	0.017	0.004	—	
		4220	0.016	0.003	—	
Matsueda and Handa (1986)	ENP (Stn. 5)	740	0.099	0.023	0.0032	HC, lipids, markers
		940	0.104	0.018	0.0023	
		1440	0.044	0.007	0.0010	
		3440	0.047	0.007	0.0009	
		4240	0.069	0.009	0.0011	
	ENP (Stn. 7)	720	0.020	0.004	0.0006	
		920	0.017	0.003	0.0004	
		1420	0.017	0.002	0.0003	
		3420	0.016	0.001	0.0001	
		3820	0.016	0.001	0.0001	
	ENP (Stn. 11)	670	0.039	0.004	0.0006	
		870	0.002	0.003	0.0005	
		1370	0.037	0.002	0.0003	
		3370	0.030	0.002	0.0002	
		3670	0.021	0.002	0.0002	
Nelson et al. (1987)	(see table 3.6)					CO_3-C, chl-*a*, phaeo-*a*, transmis. PP, SI, NO_3^-, $\delta^{15}N$
CaBS study[d]	SMB-SPB	100	0.57	0.068	0.012	^{210}Pb, ^{232}Th, ^{228}Th, ^{230}Th, ^{239}Pu, ^{240}Pu, ^{234}Th, ^{210}Po, $\delta^{13}C_{OC}$, $\delta^{15}N$
		500	0.63	0.049	0.006	
	2/86–5/86	700	0.75	0.052	0.006	
		850	0.82	0.055	0.006	
		100	0.24	0.045	0.008	lipids, markers, AHC
	5/86–10/86	500	0.41	0.041	0.006	
		700	0.54	0.042	0.006	
		850	0.58	0.039	0.005	

Table 3.5. *Summary of Sediment Trap Studies from the SCB and Adjacent Areas* (continued)

Study	Location[a]	Deployment Depth[b] (m)	Flux ($g\ m^{-2}\ d^{-1}$) Mass	OC	N	Other Parameters[c]
		200	0.39	0.51	0.021	
	4/87–10/87	500	0.57	0.041	0.008	
		700	0.68	0.037	0.006	
		850	0.54	0.030	0.004	
		200	0.44	0.089	0.016	
	2/88–10/88	500	0.46	0.045	0.007	
		700	0.49	0.039	0.005	
		850	0.55	0.046	0.006	

[a] Location designations as follows: SBB—Santa Barbara Basin; SMB—Santa Monica Basin; SPB—San Pedro Basin; SB—Soledad Basin; ENP—Eastern North Pacific; SNB—San Nicholas Basin.

[b] Deployment depths given in form *n*/A represent traps positions *n* meters above sea floor.

[c] Abbreviations: TVS—total volatile solids; P—phosphorus; FP—fecal pellets; PP—primary productivity; AHC—aromatic hydrocarbons; SHC—saturated hydrocarbons; marker—molecular markers; UCM—unresolved complex mixture (hydrocarbons); ESR—electron spin resonance; LH—larvacean houses; HC—total hydrocarbons; chl-*a*—chlorophyll *a*; phaeo-*a*—phaeophorbide *a*; transmis.—transmissivity; SI—surface irradiance.

[d] See Huh et al. (1990), Jackson et al. (1989), Landry et al. (1992), Venkatesan and Kaplan (1992). Deep basin fluxes using Soutar cone trap.

This flux is approximately an order of magnitude lower than predictions of new production based on previous productivity measurements in this area (that is, new production was about 50% of total production) (Eppley and Peterson 1979), raising questions about the general applicability of short-term sediment trap fluxes for the estimation of new production. Sediment traps are areal integrators, incorporating fluxes of small and large particles settling from different points of origin as determined by current advection. Difficulties can thus arise when short-term deployments are used because of weak temporal and spatial coupling of biological processes occurring in the euphotic zone and fluxes measured in the underlying water column.

Recent efforts by the CaBS project have resulted in further *direct* comparisons between estimates of new production and the flux of particulate carbon and nitrogen below the euphotic zone (at 100 m) at sites within the Santa Monica–San Pedro Basin as measured by free-drifting sediment traps (Jackson et al. 1989; Small et al. 1989). During five cruises from 1985 to 1987, they found trap fluxes of PON ranging from 14.8 to 49.4 mg N m^{-2} d^{-1} with a mean of 28.6 mg N m^{-2} d^{-1}. By comparison, new production estimates ranged from 18.2 to 33.0 mg N m^{-2} d^{-1} with a mean of 23.2 mg N m^{-2} d^{-1}. The ratios of trap flux to new production varied from 0.81 to 2.4 with an average of 1.3, suggesting that the estimates are in reasonable agreement. In general, the comparisons made for POC showed that trap fluxes exceeded new production estimates. This signals efficient retention of nitrogen during recycling and selective release of carbon, presumably during zooplankton grazing, to deeper waters by fecal pellet production, release of larvacean houses, or formation of macroscopic aggregates enriched in carbon (Alldredge 1979). Based on these measurements, the flux of particulate organic carbon from the euphotic zone would appear to represent approximately 45% of the primary production, in agreement with earlier work (Eppley and Peterson 1979). The remaining POC is apparently recycled rapidly in the upper 100 m of the water column (Small et al. 1989).

Dissolved Organic Matter. The vertical distribution of dissolved organic carbon (DOC) and nitrogen (DON) are shown in figure 3.9b for a station in the center of Santa Monica Basin. The data of Williams (1986a), indicated as DOC_P and DON_P, were obtained using the classic persulfate oxidation infrared detection method (Menzel and Vaccaro 1964) with preliminary ultraviolet (UV) irradiation for DON. Recent work with high temperature catalytic combustion procedures indicates that the classic wet oxidation methods may underestimate marine DOC and DON concentrations by as much as 50–500%, the lost fraction apparently consisting of biochemically labile components of the DOM pool (Sugimura and Suzuki 1988; Suzuki and Itoh 1985; Williams and Druffel 1988). Although controversy surrounding the determination of DOC and DON has not ceased, data recently obtained by Hansell et al. (1990) using the method of Sugimura and Suzuki (1988), indicated as DOC_{HT} and DON_{HT} on figure 3.9b, are also shown for purposes of comparison. Based on these limited data for samples taken at different times, the high-temperature combustion method appears to yield DOC concentrations that are roughly twice those obtained by the wet oxidation procedure. In other respects, the vertical profiles appear quite similar. In the case of DON, however, concentrations measured by the high-temperature combustion method are only higher in surface waters (<100 m). At greater depths, the DON concentrations obtained by these two methods are essentially indistinguishable.

Concentrations of DOC and DON within the euphotic zone (upper 100 m) are in the range of 50–80 μM, and 3–14 μM, respectively. With increasing depth below the euphotic zone, concentrations reach background levels of approximately 33–50 μM (DOC) and 1–4 μM (DON). The biologically labile fraction of dissolved organic matter generated in the euphotic zone (by exudation, sloppy feeding, excretion, or autolysis) is rapidly recycled by bacteria and remineralized or transferred to higher trophic levels or the detrital pool via the "microbial loop" (see references in Azam 1986) or other pathways (Jackson and Eldridge 1992). Alternatively, a portion may be directly utilized by phytoplankton (urea and amino acids). It has been noted that the concentrations of DON and dissolved organic phosphorus (DOP) vary inversely with nitrate and phosphate, respectively, in the euphotic zone, and that when $[DON + NO_3^-]$ is plotted against $[DOP + PO_4^{-3}]$, the slope approximates the Redfield ratio of 16 (Jackson and Williams 1985). This supports the view that at least a portion of the DON and DOP is available for direct utilization by autotrophic organisms and may therefore be an important source of nutrients. C:N:P ratios of DOM are in the range of 100:9.4:0.4 to 100:7.0:0.16 and do not show systematic variations with water depth (Holm-Hansen et al. 1966; Williams 1986a). The reduced nitrogen and phosphorus contents of the DOM compared with the Redfield ratio, its *relatively* invariant concentration with depth below the euphotic zone, and the constancy of the stable carbon isotope signature (approximately −21.2 to −24.4‰; Williams and Gordon 1970; −22.2 to −20.9‰; Williams et al. 1992) suggest that the majority of this material is refractory.

Williams et al. (1992) have shown that the ^{14}C activity of the UV-oxidizable fraction of DOC in Santa Monica Basin ranges from −197‰ to −462‰, decreasing systematically with depth in the water column. The vertical profile parallels that of dissolved inorganic carbon ($DI^{14}C$) with an offset of approximately 300‰ (DIC showing more activity) due to introduction of bomb radiocarbon. Similar ^{14}C activity relationships between DIC and DOC pools have been reported for the north-central Pacific (Druffel and Williams 1990; Williams and Druffel 1987). At the same time, sinking and suspended particulate organic carbon activities more closely resemble $DI^{14}C$, particularly in surface waters of Santa Monica Basin (<100 m). Based on these observations, it seems that the UV-oxidizable fraction of the DOC is largely inherited from waters advecting into the bight, whereas the remaining labile component of the DOC pool (as revealed by high-temperature combustion) may be locally derived.

Williams (1986a) has reviewed available literature on the composition of DOM off southern California. If all of the organic compounds identified in surface waters are summed (in terms of carbon molar equivalents), only about 25% of the DOC can be accounted for. The tally is as follows: total carbohydrates 12 μM (14.4%); dissolved free amino acids 0.13 μM (<0.2%); total hydrolyzable amino acids 2.2 μM (2.6%); urea 5.0 μM (6%); dimethylsulfoxide 0.1 μM (<0.2%); and oxalic, glyoxalic, and pyruvic acids 0.38 μM (0.4%). The remainder apparently represents a mixture of humic substances, lipids (minor), and other as yet unidentified compounds. These estimates, again, are based on DOC determined by persulfate oxidation. If it is established that the concentrations of DOC obtained by high-temperature combustion supersede those from chemical oxidation, then the percentages just quoted would be reduced, perhaps by a factor of two.

INTERMEDIATE WATERS

Overview. Waters confined between 200 m and basin sill depths (which vary between 500 and 1900 m) (Emery 1960) are believed to be primarily of southern origin (Liu and Kaplan 1989; Sverdrup and Fleming 1941). This means they are more saline, have lower dissolved oxygen contents, and have generally higher nitrate concentrations than the northern source that dominates surface waters in the SCB. Water motion is poleward (northwestward) and directed along isobaths by the constraints of local bottom topography. The mixing of southern and northern (and other) waters is not complete at all depths. As a result, high-resolution vertical profiles of chemical constituents reveal the interdigitation of water lenses with distinct biochemical signatures (Holm-Hansen et al. 1966; Strickland 1968). Work by Liu and Kaplan (1989) shows that nitrate in intermediate waters of the SCB is enriched in ^{15}N (approximately 7–12‰). Because denitrification is restricted to inner basin bottom waters, the most likely source of this nitrate is Eastern Tropical North Pacific water. This apparently is carried into the SCB at depth and transported north by the California Undercurrent as a band of phosphate-rich, high-salinity water.

Waters below the euphotic zone do not support the growth of phytoplankton, hence heterotrophic processes predominate. (The only known exception to this generalization is the possibility of carbon production within the oceanic oxygen minimum zone where chemolithotrophic fixation of CO_2 may be occurring [Karl and Knauer 1984].) Thus, the products of *in situ* decomposition of sinking particles (that is, NO_3^-, PO_4^{-3}, and CO_2) are superimposed on chemical distributions inherited from the source waters. Emery (1960) noted that computation of the predicted dissolved oxygen content of intermediate waters based on T-S relationships for southern and northern sources resulted in a deficit. Liu and Kaplan (1989) likewise suggest that the apparent oxygen utilization changes within intermediate waters. These observations demonstrate that *in situ* decomposition processes occurring in this part of the water column are not only detectable but significant.

Vertical Flux and Remineralization of Particulate Matter. Until recently there have been few studies of the vertical flux and decomposition of sinking particulate organic matter in the SCB. A series of investigations carried out in the 1970s focused on the estimation of fluxes of particles, fecal pellets, organic carbon, nitrogen, and trace inorganic and organic constituents to basin sediments (Bruland and Franks 1979; Crisp et al. 1979; Dunbar and Berger 1981). Other programs were primarily concerned with questions of trap calibration and design (Bruland et al. 1981; Dymond et al. 1981; Soutar et al. 1977). Information relevant to these and other studies undertaken within the California Current or eastern Pacific is summarized in table 3.5. Only within the last decade have multilevel and time series trap arrays been used for purposes of understanding particle transport and rates of remineralization during sedimentation.

Nelson et al. (1987) reported results of two short-term free-drifting arrays with sediment traps positioned at water depths of 100, 300, and 500 m in the Santa Monica–San Pedro Basin and a long-term (six-month) trap deployment at 500 and 824 m in San Pedro Basin. The data summarized in table 3.6 show that with increasing depth, the fluxes of organic carbon and nitrogen to the short-term traps deployed at 500 m were ≤50% of those at 100 m. These differences reflect processes of disaggregation, dissolution, and decomposition of sinking particles. Vertical sampling resolution was insufficient to examine the possibility of lateral advection at these water depths. Only a small fraction (<3%) of the particulate organic carbon flux at all depths during the first short-term deployment could be accounted for by intact microorganisms. However, phytoplankton carbon, as measured by chlorophyll *a* and phaeophorbide *a*, comprised a significant proportion of the total organic carbon flux (10–70%), with increasingly lower percentages found in traps at greater depth in the water column. This is consistent with stable carbon and nitrogen isotope measurements of particles trapped in the central parts of San Pedro Basin by Crisp et al. (1979) and Santa Monica Basin by Williams et al. (1992), which show that the sinking POM is primarily of marine origin. The majority of the phytoplankton carbon is apparently associated with fecal material, much of which was derived from the pelagic crab *Pleuronocodes planipes*. Other investigators have found that fecal pellets can dominate the vertical flux of particles in waters of the SCB (Dunbar and Berger 1981; Soutar et al. 1977). However, it is likely that fecal pellets, particularly those produced by smaller zooplankton, are transported as aggregates that are not effectively preserved in sediment traps (Asper 1987; Small et al. 1989).

In the case of the long-term experiment, the deeper trap (positioned 50 m above the basin floor) collected nearly double the amount of solids, organic carbon, and nitrogen as the trap positioned at 500 m. This reflects the introduction of resuspended sediments, presumably originating from shelf and slope environments, into basin waters. Similar results have been reported by Dymond et al. (1981) and Soutar et al. (1977) for Santa Barbara Basin and by Venkatesan and Kaplan (1992) and Williams et al. (1992) for Santa Monica Basin. Moreover, Shiller (1982) provided evidence that resuspension of slope sediments contributed to the enrichment of suspended particulate aluminum at mid-water depths (below the upper 100 m) in Santa Barbara Basin. The episodic nature of resuspension is exemplified by the variable particulate aluminum profiles found on four separate occasions in this basin (Shiller 1982).

Difficulties arise in comparing fluxes obtained using different sediment traps because of variations in trapping efficiency related to trap geometry and its interaction with differing hydrodynamic conditions (U.S. Global Ocean Flux Study 1989). However, examination of table 3.5 reveals that the intrabasin variations in solids fluxes are as large as interbasin fluxes. The nature and cause of intrabasin variations are unknown, although seasonal upwelling, lateral advection of resuspended material, and turbidity currents have been invoked as mechanisms. Studies currently ongoing in San Pedro Basin examine the long-term variability of short-term particle fluxes to this basin (L. Sautter and C. Pilskaln pers. comm. 1989). Figure 3.10 provides data on a series of 1-week trap collections at a water depth of 500 m over a period of 7 months. Fluxes of total particulate matter vary by more than an order of magnitude, and organic carbon fluxes vary by a factor of six. Moreover, the week-to-week variability is quite high. In general, the fluxes are in line with those reported for San Pedro Basin in earlier studies (Crisp et al. 1979; Nelson et al. 1987). However, the unusually high C:N ratios occurring in the months of Janu-

Table 3.6. *Summary of Short-Term and Long-Term Sediment Trap Studies Conducted in Santa Monica–San Pedro Basin, 1983*

Trap Depth	Trap Preservatives[a]	Flux ($mg\ m^{-2}\ d^{-1}$) Mass	OC	N	% Phytoplankton[b]	% Microorganisms[c]
Short-term deployment No. 1[d]						
100	L	216	34.5	5.4	22.0	2.0
100	F	242	35.0	5.3	—	—
300	L	131	18.6	3.1	9.5	2.6
300	F	138	13.6	2.2	—	—
500	L	193	14.6	1.8	—	2.9
500	F	256	17.1	2.2	—	—
Short-term deployment No. 2						
100	L	327	30.0	4.2	69.5	—
100	F	220	26.3	3.2	—	—
300	L	391	37.0	5.1	26.7	—
300	F	393	36.6	4.7	—	—
500	L	240	18.4	2.7	22.7	—
500	F	419	24.9	3.2	—	—
Long-term deployment[d]						
500	F	167	13.6	1.6	—	—
500	M	239	16.6	2.0	—	—
824	F	440	22.1	2.5	—	—
824	M	401	24.7	3.0	—	—

[a] Trap preservatives used: F—formalin; L—no preservative used; M—$HgCl_2$.

[b] Percent of total organic carbon flux represented by phytoplankton (based on pigment fluxes and 40:1 carbon:chlorophyll ratio).

[c] Percent of total organic carbon flux represented by *intact* microorganisms based on microscopic analysis and cell volume-to-carbon biomass conversion.

[d] Deployments as follows: short-term No.1—59 hours, 5/17–23/83; short-term No. 2—59 hours, 5/22–25/83; long-term—190 days, 5/23–9/29/83.

After Nelson et al. (1987).

ary and February are anomalous. These indicate introduction of terrigenous material and/or resuspension and lateral transport of sediments from the shelf and slope. The occurrence of a spring bloom during the months of March to May is indicated by maximum fluxes of carbonate and lower C:N ratios approximating the Redfield ratio. Nelson et al. (1987) found C:N ratios ranging from 6.0 to 8.2 (means of 6.9 and 7.5 for first and second deployments, respectively) for the short-term deployments in May, whereas the ratios in the long-term deep traps were 8.2 to 8.8 (mean of 8.4).

The largest database on the origin and fate of sinking particulate matter has been developed in the context of the multidisciplinary CaBS study conducted from 1985 to 1988. Some of the results of free-drifting sediment traps deployed for short periods of time (24 hours) have previously been discussed in the context of particulate matter fluxes from the euphotic zone. We now shift attention to studies involving traps moored for longer periods (100–150 days) at depths of 200–850 m in Santa Monica Basin. Although most of these data have yet to be published, we present a brief summary based on early re-

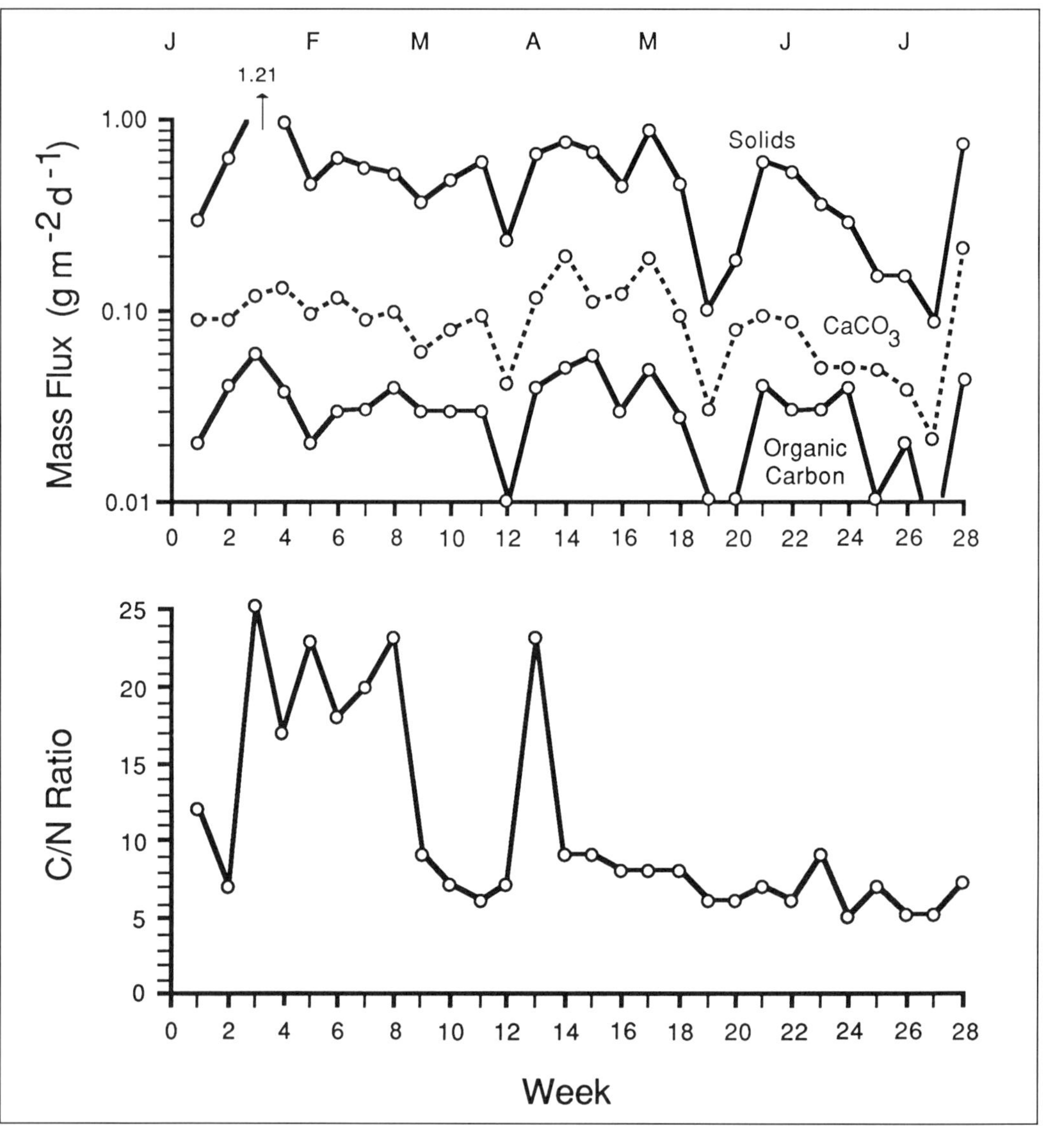

Figure 3.10. Short-term (1-week deployments) variations in the vertical flux of particulate matter, total organic carbon, and calcium carbonate ($CaCO_3$) to sediment traps positioned at 500 m in San Pedro Basin, 1988. (After Sautter and Pilskaln pers. comm. 1989.)

ports by Jackson et al. (1989), Williams (1988), Williams et al. (1992), and Venkatesan and Kaplan (1988, 1992).

Sediment traps of the Soutar cone design (Bruland et al. 1981) were deployed over the eastern slope of Santa Monica Basin (bottom depth approximately 395 m) at depths of 100 and 350 m during 1986 (February and May), 1987 (April), and 1988 (February) for periods ranging from 99 to 233 days (table 3.5). In addition, multiple traps were deployed over similar time periods at a location near the center of the basin (bottom depth approximately 910 m) at depths of 100 (or 200), 500, 700, and 850 m. Particle fluxes within the central basin appear to be on the order of 240–

570 mg m^{-2} d^{-1} at shallower depths (100 m), increasing to an average of 620 mg m^{-2} d^{-1} at 850 m water depth. Comparison of samples collected at this depth during May 1986 and February 1988 indicates higher fluxes during the spring deployment. This may reflect a seasonal linkage between benthic flux and primary production in overlying waters. However, the data are too sparse to make definitive statements. The average flux agrees well with sediment accumulation rates estimated from ^{210}Pb profiles (630 mg m^{-2} d^{-1}) (Huh et al. 1987; Jackson et al. 1989). For comparison, Bruland et al. (1981) and Crisp et al. (1979) reported fluxes of 463 and 493 mg m^{-2} d^{-1}, respectively, for studies conducted in the same basin in 1977.

Organic carbon shows a flux of approximately 55 mg m^{-2} d^{-1} at 100 m and decreases to 47 mg m^{-2} d^{-1} at 850 m. Below a depth of 500 m there is essentially no change in the flux of organic carbon, and the accumulation measured in the bottom trap agrees well with that estimated by summing burial and diagenetic losses (44 mg m^{-2} d^{-1}). The stable isotopic composition of the sinking POM reflects a largely marine autochthonous origin ($\delta^{15}N$ = +5.5 to + 10.2‰; $\delta^{13}C$ = −20.6 to −22.4‰) (Williams et al. 1992). The variable and occasionally low nitrogen isotope ratios indicate a contribution from terrestrial sources (Sweeney and Kaplan 1980a). Similar depletion of ^{15}N was found for the slope particles, where one would expect to encounter terrigenous debris. In general, these compositions compare well with the range (+7.09 to +8.05‰) reported by Nelson et al. (1987) for total organic matter and are slightly lower than Crisp et al. (1979) noted for humic acid and kerogen isolates (+8.3 to +9.4 and +10.4 to +13.5‰).

Venkatesan and Kaplan (1988, 1992) reported results of molecular analyses for the trap deployments. Above 500 m, fluxes of all lipid classes (hydrocarbons, alkanols, fatty acids, and sterols) decreased with increasing water depth at rates higher than those observed for organic carbon. This suggests that the lipids associated with sinking particles represent one of the more labile fractions of the particulate organic matter in waters overlying Santa Monica Basin. In deep basin waters (>500 m), fluxes of all constituents increase or remain constant. This probably reflects the lateral transport of slope sediment into the central basin (Williams et al. 1992). The composition of the fatty acids and alkanols indicated contributions from both marine and terrestrial sources, with the former predominating at all depths. However, the terrestrially derived homologs ($>C_{24}$) increased in relative abundance in the deeper traps. This may reflect a greater resistance to degradation, possibly conferred by association of these compounds with refractory plant parts such as cuticles. This observation is consistent with the fact that these higher molecular weight homologs are found even in sediments deposited on the outer continental shelf (see the section on Distribution and Fate of Chemical Constituents in Sediments). Sediment traps deployed near the slope generally collected greater amounts of lipids than those situated in the basin interior at comparable depths. This is expected if a significant fraction of the trapped particles is derived by resuspension and transport from shallower slope or shelf areas. Comparison of fluxes for the fecal sterol coprostanol at basin and slope sites confirms the dilution of anthropogenic (that is, sewage) inputs with distance from land (Venkatesan and Kaplan 1990).

DEEP BASIN WATERS

Offshore basins of southern California contain waters originating from currents moving over the sills under the influence of gravity. As described by Emery (1960), the temperature and oxygen content of the waters in each basin are largely determined by the quality of waters intersecting the sill depth. Because of the strong vertical gradients exhibited by these properties and differences in the sill depths of the offshore basins, each basin is characterized by unique temperature and dissolved oxygen levels. Under normal conditions temperature and salinity within a given basin are nearly uniform from the sill to the basin floor (fig. 3.4). The same is not true of nutrients and dissolved oxygen. Unfortunately, information on the chemical composition of deep basin waters is limited, consisting largely of data generated in specific investigations.

Because the sills of the inner basins are located at the depths within the oxygen minimum zone (approximately 500–700 m), the dissolved oxygen concentration of deep waters in these basins is low and bottom sediments may be anoxic and essentially free of benthic fauna. In the Santa Barbara Basin (Liu 1979; Sholkovitz and Gieskes 1971) and Santa Monica Basin (Johnson et al. 1988; Williams 1986a), oxygen concentrations in waters below sill depth are sufficiently depressed to permit active denitrification in bottom waters (and surface sediments). This leads to declining water column nitrate concentrations with approach to the sediment–water interface (fig. 3.4a). Santa Barbara Basin is the shallowest and best studied of the inner basins. We use it here for discussion, recognizing that its behavior is probably different from that of other basins in the SCB (Berelson 1991).

Detailed descriptions of the physics and chemistry of Santa Barbara Basin waters are given in Emery (1960), Liu (1979), Shiller (1982), and Sholkovitz (1972). Santa Barbara Basin reaches a maximum depth of 590 m with a sill depth of 475 m. Exchange of water apparently occurs by spillover of waters at sill depth, most likely from the west. Frequent small-scale spillovers probably involve movement of cooler, more saline, nitrate-rich and oxygenated waters over the sill and along the slope to the basin bottom. Such events sometimes result in increased turbidity (Sholkovitz and Soutar 1975) and observable reversals of the nitrate profile near the sediment–water interface (Liu 1979; Shiller 1982). Nitrate is perhaps the best tracer of waters introduced from outside the basin. Under these conditions basin waters are only partially replaced (perhaps 10%) (Shiller 1982), but the events lead to small but detectable, perturbations of the "normal" condition, manifestations of which are discussed in the next paragraph. Although the frequency of these small spillovers is unknown, Liu (1979) reported that among thirteen observations that had been taken during normal conditions up to August 1978, signs of bottom water disturbance were noted for six of them. If one adds to this list the four sampling periods reported by Shiller (1982), three of which showed evidence of spillover, such perturbations would appear to be the rule (53%) rather than the exception.

Normal conditions in the bottom waters of Santa Barbara Basin include active denitrification within a zone estimated to extend approximately 20–40 m above the basin floor (fig. 3.11) (Liu 1979). Denitrification in bottom waters and surface sediments results in production of a subsurface nitrate concentration maximum, typically positioned just above sill depth. Nitrate concentrations below the sill decrease by as much as 50% (from about 30–35 μM above the sill to approximately 16 μM near the bottom). Moreover, because of isotopic fractionation during denitrification (Sweeney et al. 1978), this nitrate is enriched in ^{15}N relative to overlying waters by 1-13‰, depending upon the extent to which nitrate has been consumed. For example, Liu (1979) found a correlation between $\ln[NO_3^-]$ and its $\delta^{15}N$ value in basin waters with a slope of -7.8 ($r^2 = 0.85$). Although present in small concentrations throughout much of the upper water column (surface to sill, approximately 0–0.4 μM), nitrite exhibits a secondary maximum (approximately 0.3–1.1 μM) below sill depth. This maximum is attributable to denitrification, but the concentrations of nitrite are, nonetheless, small, and in no case has nitrite been found to exceed 5% of the combined nitrate + nitrite pool. Nitrous oxide (N_2O) shows a profile characterized by saturation concentrations (approximately 10 nM) at the surface, increasing concentrations with water depth to a maximum near the sill of approximately 40 n*M*, followed by a precipitous decline below the sill to approximately 1 n*M*. Nitrous oxide is associated with nitrification, but is consumed preferentially over nitrate during denitrification (Liu and Kaplan 1982). Hence, its distribution in Santa Barbara Basin is controlled by production in the upper water column, loss to the atmosphere at the surface, removal during denitrification below the sill, and exchange between basin and overlying water masses.

Ammonium, present at low concentrations throughout the upper water column (approximately 0.3 μM), increases in concentration near the basin floor (to approximately 0.8 μM), presumably as a result of suboxic metabolism (that is, denitrification in water and sediments) or anaerobic metabolism (sulfate reduction in sediments). Oxidation of ammonium by nitrifying bacteria is effectively prevented by the low oxygen concentrations. Phosphate is present within surface waters at concentrations of approximately 0.5 μM. Concentrations increase rapidly with depth down to approximately 100–150 m and more gradually thereafter to approximately 3.5 μM at the sill. Throughout the upper portion of the water column, nitrate and phosphate concentrations are correlated, with a slope of 14.4 (Liu 1979), slightly lower than the Redfield ratio (16). Below the sill, a negative correlation is observed between nitrate and phosphate concentrations (slope = -16.8). Since denitrification yields a $\Delta N{:}\Delta P$ ratio of -84.8 (Richards 1965), Liu (1979) reasoned that phosphate must be generated by processes other than denitrification. Some possibilities include oxygen respiration in the water column, sulfate reduction in sediments, and solubilization of iron phosphate. The first two processes are known to occur. The third of these, solubilization of iron phosphate, is consistent with observations made by Shiller (1982) and Shiller et al. (1985), who examined the chemistry of particulate matter in deep basin waters. They found that particulate phosphorus and iron were positively correlated at all times (that is, during both normal and spillover conditions), but that aluminum, a terrigenous resuspension indicator, tended to follow nitrate. Both iron and phosphorus were low during spillover periods when aluminum and nitrate were high. Shiller (1982) therefore proposed that particulate phosphorus enters the deep basin environment from the east via resuspension of sediments enriched in ferric-hydroxo-phosphate precipitates from the Ventura shelf. The latter were thought to originate from the salt marsh off the Santa Clara River.

Major episodic "flushing" events have also been described for the Santa Barbara Basin (Liu 1979; Sholkovitz and Gieskes 1971). The frequency of basin flushing is uncertain because of the small number of observations and the poor temporal resolution of the hydrographic surveys. However, Sholkovitz and Gieskes (1971) suggested that flushing in Santa Barbara Basin might occur seasonally, with complete exchange of basin water requiring as little as 1 or 2 months. Major flushing events are apparently triggered by offshore transport of surface waters during intense upwelling and replacement of deep basin water by denser upwelled water entering from outside the basins at sill depth (Sholkovitz and Gieskes 1971). After examining quarterly CalCOFI hydrographic data taken between 1986 and 1989, Reimers et al. (1990) concluded that flushing events are indeed coupled to seasonal upwelling and that the characteristic varving in basin floor sediments is related to the periodic alternation of basin water chemistry and its effects on microbial communities inhabiting the benthic boundary layer.

Figure 3.11 illustrates differences in the vertical distribution of chemical species between normal conditions within Santa Barbara Basin and after complete flushing has occurred. Upon completion of flushing, oxygen and nitrate concentrations in bottom waters increase measurably and exhibit relatively uniform profiles from the sill down. Tem-

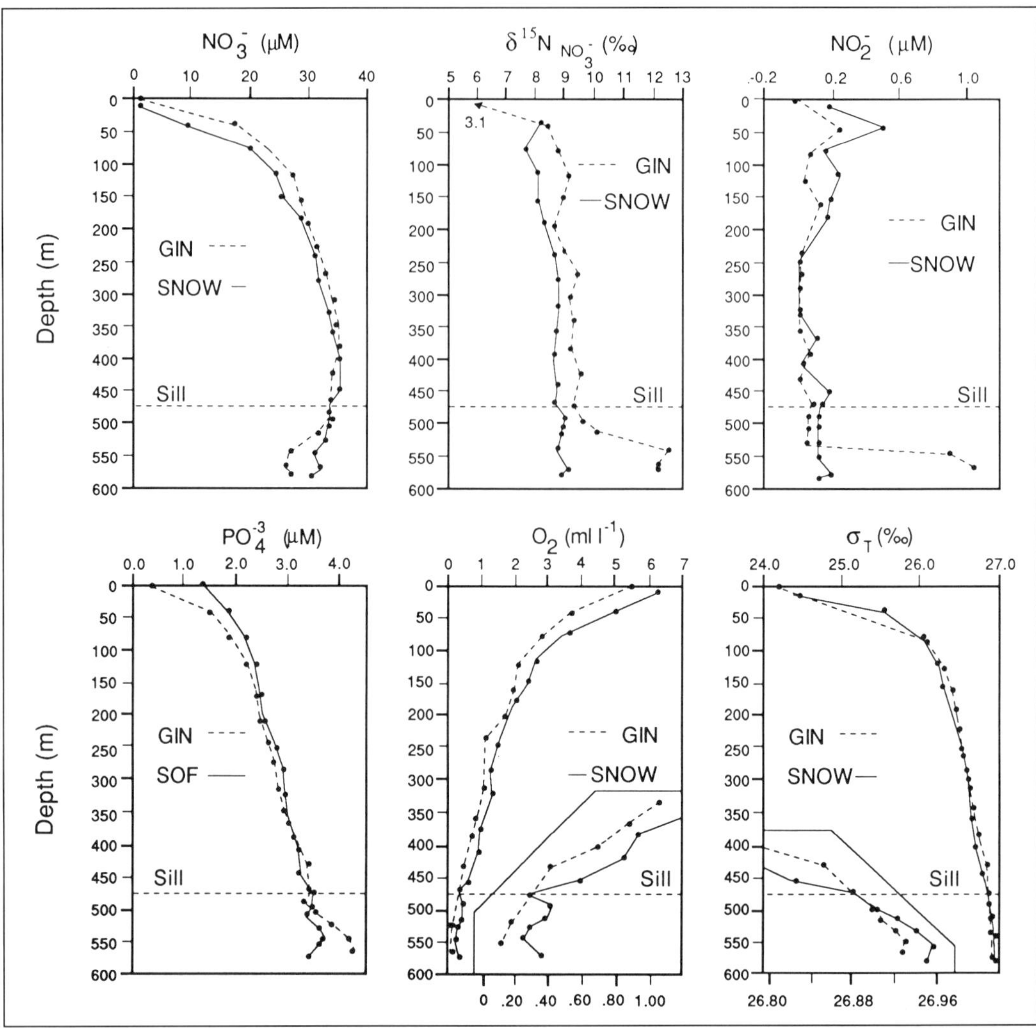

Figure 3.11. Vertical profiles of nitrate, nitrite, phosphate, oxygen, σ_T, and δ^{15}N-nitrate during "normal" (GIN cruise, August 22–25, 1978) and "flushed" conditions (SNOW and SOF, August 23–25, 1977, and April 18–22, 1987, respectively). (After Liu 1979.)

perature, δ^{15}N-nitrate, and nitrite and phosphate concentrations decrease, reflecting the replacement of basin waters with colder upwelled waters depleted in phosphate. The lowering of nitrite concentrations and δ^{15}N-nitrate values indicates that denitrification has not yet been established. Profiles taken during the course of flushing events (prior to completion) and after flushing show trends that are intermediate between those found during the normal and flushed conditions depicted here.

Anthropogenic Effects

Approximately 21 million human beings inhabit the coastal counties that adjoin the SCB. One might reasonably ask, What effects, if any, do human activities have on the marine ecosystem off southern California? Since this question is addressed at length in chapter 12, the purpose of this section is to present selected examples of anthropogenic effects in the water column. From information supplied in foregoing sections of this chapter, it should be evident that, except for the possible localized enhancement of primary productivity associated with municipal waste discharge (via introduction of nutrients) in the northern part of the SCB, the dynamics of dissolved and particulate matter are controlled by natural physical, chemical, and biological processes—especially the latter. These processes, in turn, are largely influenced by variable climatic conditions exemplified by the occurrence of El Niño episodes, the intensity of seasonal upwelling, and episodic storms that affect nearshore waters for short time periods. The principal goal of environmental science is to understand how the natural system functions and interacts with the products of human activity.

Unfortunately, information on the contamination of dissolved and suspended particulate phases of the water column within the SCB is sparse. This is partly caused by the difficulty of measuring trace inorganic and organic constituents at or near their natural abundance levels in seawater (e.g., Bruland et al. 1979; de Lappe et al. 1983; Patterson and Settle 1976). In addition, because many of these substances are particle reactive, it has proven more useful to examine depositional records preserved in anoxic basin sediments. However, with the advent of sediment traps, it is now possible to estimate the vertical fluxes of these contaminants on environmentally relevant time scales. In contrast, inner basin sediments can, at best, be examined with a temporal resolution of approximately one year. Furthermore, sediment traps can be deployed at sites where the sediment record is disrupted or destroyed by bioturbation or physical mixing processes (e.g., central or outer basins; Bruland et al. 1981). For these reasons, we turn attention to the role of sinking particles in transporting trace organic and inorganic contaminants to the benthos.

TRACE ORGANICS

The existing database for trace organic contaminants in the water column of the SCB is extremely limited. The most extensive results come from the BLM studies on hydrocarbon distributions in surface and deep (10 m off the bottom) water samples collected at 41 stations within the SCB (Risebrough and Burlingame 1978). Other investigations include those concerned with hydrocarbon contamination of surface waters in the Coal Oil Point area (Stuermer et al. 1982); various offshore sites in southern California (de Lappe et al. 1980, 1983); the sea surface microlayer (Cross et al. 1987); DDT residues in the California Current (Cox 1971); and the results of sediment trap studies (Crisp et al. 1979; Jackson et al. 1989; Venkatesan and Kaplan 1992).

Because coverage of this subject is so patchy and because sediment trap studies have had a tremendous impact on our understanding of contaminant transport processes, we focus on the results presented by Crisp et al. (1979). They investigated particulate matter collected in sediment traps positioned 30–60 m above the bottom of four basins: Santa Barbara, San Nicolas, Santa Monica, and San Pedro. Table 3.7 summarizes some of the data reported in this study.

The highest fluxes of particulate matter were found in Santa Barbara Basin, followed in decreasing order by San Pedro, Santa Monica, and San Nicolas basins. Hydrocarbon concentrations in the trapped particles, however, were in the order San Pedro Basin>Santa Monica Basin>Santa Barbara Basin>San Nicolas Basin. Consequently, the ranking of the basins by their rates of hydrocarbon deposition approximated that found for the concentrations. Hydrocarbon fluxes in San Pedro Basin were approximately 25 times those observed in San Nicolas Basin, owing to the proximity of the former to the dominant anthropogenic hydrocarbon sources (Los Angeles County waste outfall and Los Angeles–Long Beach Harbor) (Eganhouse and

Table 3.7. *Summary of Sediment Trap Studies Carried Out in Santa Barbara, San Nicolas, Santa Monica, and San Pedro Basins, 1977*

Basin	HC Concentration[a] ($\mu g\ g^{-1}$)	Flux ($g\ m^{-2}\ yr^{-1}$)[a] Mass	THC	Ratios[b] CPI	Pr/Ph	Hop/C_{28}
San Nicolas	87	100	0.009	3.3	4.3	4.8
Santa Barbara	177	410	0.073	3.3	7.0	2.2
Santa Monica	394	180	0.071	1.9	3.2	5.3
San Pedro	890	280	0.249	1.5	2.1	4.8

[a]Hydrocarbon concentration for Santa Barbara Basin and hydrocarbon flux for San Pedro Basin have been corrected from that given in Crisp et al. (1979).

[b]CPI—carbon preference index for C_{20} to C_{32}; Pr/Ph—pristane/phytane ratio; hop/C_{28}—hopane/17α(H),18α(H),21β(H)-28,30-bisnorhopane ratio.

After Crisp et al. (1979).

Kaplan 1981; Eganhouse and Kaplan 1982a). The vast majority of both saturated and aromatic hydrocarbon fractions in all samples were composed of chromatographically unresolved mixtures (>80% in all but one case), signaling the dominance of petroleum inputs. Biogenic hydrocarbons present in these sinking particles included a series of *n*-alkanes from *n*-C_{15-17} (marine) and *n*-C_{29-31} (terrestrial), a collection of alkenes having 16–30 carbons with 1–6 double bonds, and trace amounts of $C_{27,29,30}$ pentacyclic triterpanes of the 17β(H),21β(H) stereochemistry. Normal alkanes from the more remote basins (Santa Barbara and San Nicolas) showed less evidence of petroleum contamination than did the inner basins near Los Angeles (San Pedro and Santa Monica), as indicated by the higher relative abundance of odd chain length homologs. In addition, pristane:phytane ratios of the former basins were significantly higher.

Petroleum arising from natural seepage in the SCB is characterized by the abundance of an unusual pentacyclic triterpane (see fig. 3.12), structure I, 17α(H),18α(H),21β(H)-28,30-bisnorhopane (Simoneit and Kaplan 1980). This compound is not present in significant amounts in either municipal waste effluents of southern California (Eganhouse and Kaplan 1982b) or in stormwater runoff (Eganhouse et al. 1981). It is only rarely found in other oils. Although it was observed by Crisp et al. (1979) in all trap samples, in no case did it predominate (among the triterpanes), as it sometimes does in seepage oil. It was most abundant (relative to hopane) (see fig. 3.12, structure II) in the Santa Barbara Basin sample, reflecting the importance of seepage in this area.

A variety of aromatic and halogenated hydrocarbons were also detected. DDT metabolites and linear alkylbenzenes were found in samples from Santa Monica and San Pedro basins, but not San Nicolas and Santa Barbara basins. The DDT compounds were probably derived from historical deposits on the Palos Verdes Shelf that originated primarily from the Los Angeles County waste outfalls in earlier times (MacGregor 1976). The alkylbenzenes are molecular markers of municipal waste and probably came from contemporary discharges or resuspension of historical deposits on the shelf and slope (Eganhouse and Kaplan 1982b; Eganhouse et al. 1983). In addition, all four samples contained polycyclic aromatic hydrocarbons derived from a combination of fossil fuel and combustion sources. The occurrence of these compounds at all sites suggests either local sources within each basin (seepage) or long range transport via currents and the atmosphere. Examina-

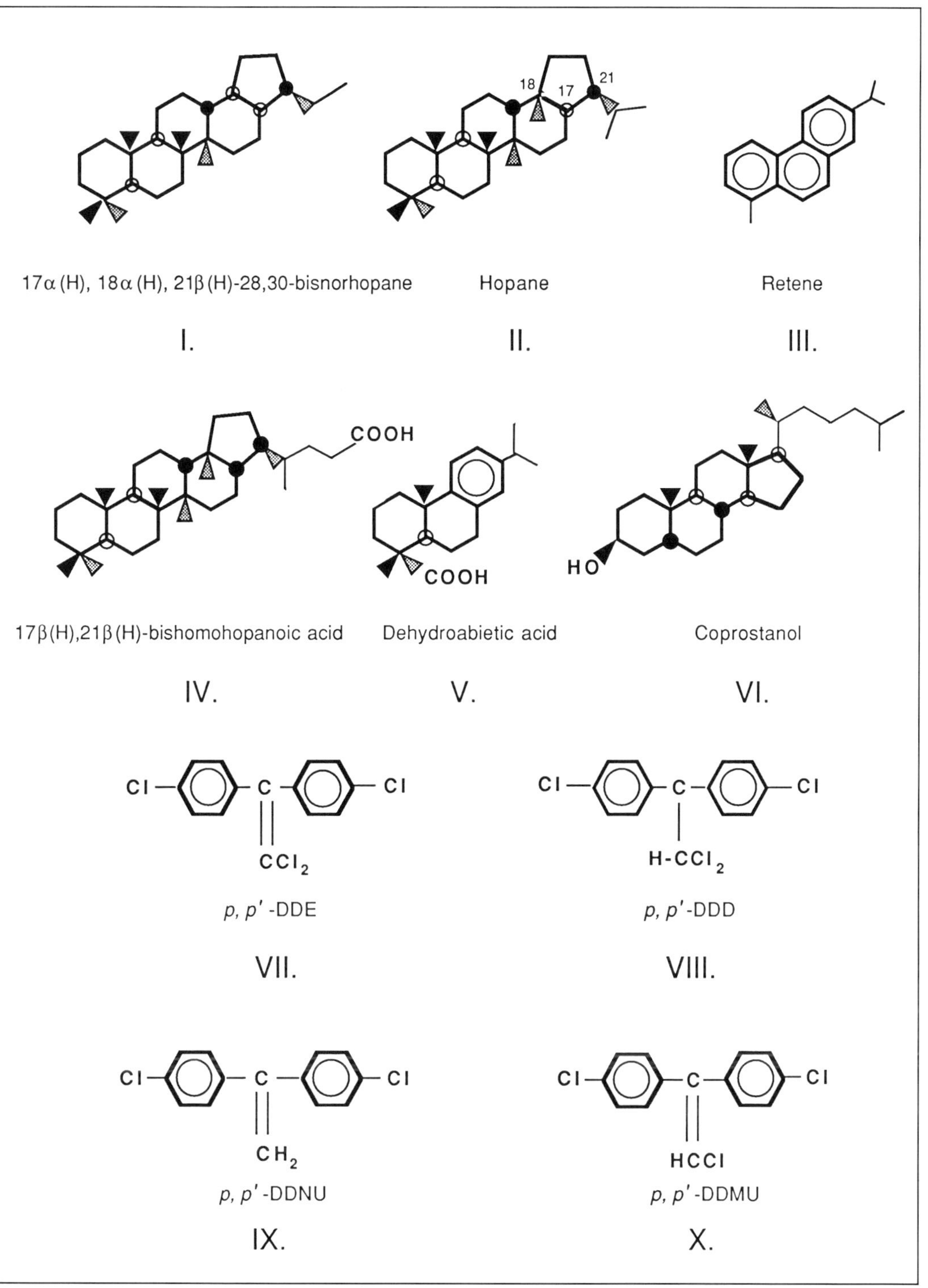

Figure 3.12. Structures of selected compounds found in the SCB.

tion of humic acids and kerogens isolated from these samples for stable isotope composition indicated that they were predominantly of marine origin ($\delta^{15}N$ = +8.3 to +13.5‰; $\delta^{13}C$ = −19.7 to −22.6‰). Elemental analysis (CHN), however, suggested some terrestrial input.

It is unfortunate that no comparable data exist for other central or outer basins within the SCB. The closest approximations are results obtained by Matsueda and Handa (1986) for the eastern north Pacific within the California Current (table 3.5). They reported total hydrocarbon fluxes in the range of 0.26–2.0 mg m^{-2} yr^{-1} (0.0007–0.0056 mg m^{-2} d^{-1}) for traps positioned at approximately 700 m at three sites. The composition of these hydrocarbons indicated a dominant biogenic source. The presence of a series of unsaturated hydrocarbons presumed to be derived from planktonic organisms suggested that the hydrocarbons had probably not been reworked extensively. If one assumes that the rates at which hydrocarbons degrade during particle sinking within the SCB are similar to those in the California Current, then the difference in hydrocarbon fluxes (at a fixed depth in the water column) between SCB basins and the California Current reflect differences in the magnitude of petroleum inputs plus inputs from primary production. Primary production in nearshore regions of the SCB is less than or equal to five times that within the California Current (Eppley 1986). If one further assumes that the hydrocarbon flux at some depth in the ocean is proportional to primary production in the euphotic zone, the biogenic component of the total hydrocarbon flux in the San Nicolas Basin site represents from 10 to 100%. The lower percentage is more consistent with the molecular evidence previously discussed.

Accordingly, the proportion of the total hydrocarbon flux composed of biogenic hydrocarbons in the inner basins (Santa Monica, Santa Barbara, and San Pedro) would be approximately 3% or less. If one further assumes that the average hopane:bisnorhopane ratio for oil seepage in the SCB is 1 (table 3.11) and that the nonseepage petroleum hydrocarbon ratio is approximately 11 (Eganhouse and Kaplan 1988), one can estimate the percentage of seep-derived hydrocarbons being transported to the basins as follows: Santa Barbara, 39%; San Nicolas, 15%; Santa Monica, 12%; and San Pedro, 12%. The corresponding seepage fluxes would be Santa Barbara, 28; San Nicolas, 1.4; Santa Monica, 8.5; and San Pedro, 30 mg m^{-2} yr^{-1}. Assuming the background biogenic flux is approximately 1.3 mg m^{-2} yr^{-1} (5 × 0.26 mg m^{-2} yr^{-1}; see above), anthropogenic hydrocarbon fluxes are computed (by difference) to be Santa Barbara, 44; San Nicolas, 6.3; Santa Monica, 61; and San Pedro, 218 mg m^{-2} yr^{-1}.

Although these estimates are approximate and based on very few data, they reveal some interesting trends. The flux of seep-derived hydrocarbons in Santa Barbara and San Pedro basins are approximately equal and are about 100 times greater than biogenic inputs. The flux of seep hydrocarbons in San Nicolas Basin is minor. This is consistent with its isolation from the primary sources, offshore Santa Barbara and Santa Monica Bay (Hartman and Hammond 1981). The fact that Santa Monica Basin has significantly smaller fluxes of seep oil than San Pedro Basin may indicate transport downcoast by longshore currents (see chap. 2). With the exception of Santa Barbara Basin, anthropogenic fluxes in the basins exceed natural seepage plus biogenic fluxes by more than a factor of four. Moreover, they exceed the seep-related fluxes in San Nicolas, Santa Monica, and San Pedro basins by factors of more than five, seven, and seven, respectively. The relative isolation of Santa Barbara Basin from the heavily populated Los Angeles region and the pervasive oil seepage (and perhaps production) in this area explain the more favorable anthropogenic: seepage flux ratio. Nevertheless, the anthropogenic hydrocarbon flux in Santa Barbara Basin would appear to be only slightly less (approximately 30%) than that in Santa Monica Basin. In contrast, the efficient trapping of particle-associated hydrocarbons in the inner basins is illustrated by the greatly reduced flux of anthropogenic hydrocarbons in San Nicolas Basin. Clearly, the inner basins adjoining Los Angeles (Santa Monica and San Pedro) are being heavily contaminated by petroleum hydrocarbons, the vast majority of which are anthropogenic.

TRACE METALS

A considerably larger body of information exists on the trace metal concentrations in dissolved, suspended particulate, and sinking particulate phases in the SCB. However, many of the data generated prior to the mid-1970s are of limited use because of inadequacies in the sample collection and handling procedures used at that time (Bruland et al. 1979; Patterson and Settle 1976). The only systematic, regional survey in which acceptable procedures were used occurred in the context of the BLM baseline studies (Bruland and Franks 1978, 1979), In recent years, the trace metal investigations have become more disparate, generally focusing on the geochemistry of one or a few elements (e.g., Johnson et al. 1988; Murnane et al. 1989; Shiller et al. 1985). The literature for this area has been reviewed up to about 1980 by Katz and Kaplan (1981) and more recently by Williams (1986a).

Bruland et al. (1981) conducted an experiment designed to compare particle fluxes in four offshore basins (Santa Barbara, Santa Monica, San Nicolas, and San Pedro) with sediment accumulation rates. The latter were based on ^{210}Pb measurements made on sediments collected in each of the basins using box cores. In addition to the radioisotopic and particle mass flux measurements, the concentrations of Pb, Cr, Zn, Cu, Cd, Fe, and Ni were determined in the trapped particles. Several years earlier, Bruland et al. (1974) had reported trace metal measurements of ^{210}Pb-dated cores from three of these basins, the results of which are discussed in a later section. Bruland et al. (1981) demonstrated that the mass fluxes derived from the particle interceptor traps (PIT) fell within 25% of the sediment accumulation rates. Moreover, comparison of the total and 1*N*-HNO_3-leachable concentrations in the PIT samples with those in the surface sediments showed that the trace metal chemistry of trapped particles accurately represented that of the surface sediments. Because San Nicolas Basin does not have anoxic varved sediments, accurate sedimentation rates could not be determined for this basin using radiometric dating. Thus, the observation of anomalously high concentrations of Pb in the San Nicolas Basin PIT samples (whereas concentrations of other metals correlated well between PIT and sediment samples) signaled dilution of modern stable lead inputs by less contaminated sediments deposited at greater depth, probably by bioturbation. As previously noted, similar discrepancies were not found for other metals, some of which (Cd and Zn) originate largely from anthropogenic inputs (such as municipal wastes and surface runoff) that enter the marine environment directly near the shore. In other words, trace metals introduced to the nearshore water column are effectively screened from entering central and outer basins, whereas contaminants transported through the atmosphere represent a larger fraction of the metal burden of sinking particles and sediments further offshore (see Bertine and Goldberg 1977). Because the acid-leachable fraction accounted for 93% of the total Pb in the San Nicolas Basin PIT sample, an anthropogenic origin was clearly implicated (Ng and Patterson 1982; Shokes and Mankiewicz 1979).

Table 3.8 provides estimates of the fluxes of trace metals to the sediment traps positioned in the inner basins in 1977, as reported in Bruland and Franks (1979) and Bruland et al. (1981), along with trace metal accumulation rates derived from analyses of sediment cores performed in 1972 (Bruland et al. 1974). Despite a 5-year hiatus, the fluxes and sediment accumulation rates generally fall within a factor of three. Based on the

Table 3.8. *Estimated Fluxes and Sediment Accumulation Rates of Particulate Matter, Pb, Cr, Zn, Cu, and Cd in Three Inner Basins*[a]

	Trap Flux ($\mu g\ cm^{-2}\ yr^{-1}$)					
Basin	TPM[b]	Pb	Cr	Zn	Cu	Cd
Santa Barbara Basin	72.4	1.67	—[c]	4.92	1.45	0.080
San Pedro Basin	36.4	2.66	—[c]	4.37	1.71	0.131
Santa Monica Basin	16.9	0.42	1.86	1.86	0.54	0.042
	Sediment Accumulation Rate ($\mu g\ cm^{-2}\ yr^{-1}$)					
Basin	TPM[b]	Pb	Cr	Zn	Cu	Cd
Santa Barbara Basin	90.	3.1	13.6	11.9	4.0	0.21
San Pedro Basin	28.	1.96	5.9	5.0	2.6	—[d]
Santa Monica Basin	24.	1.14	4.7	4.9	2.1	—[d]
	Trap Flux/Sediment Accumulation Rate					
Basin	TPM[b]	Pb	Cr	Zn	Cu	Cd
Santa Barbara Basin	0.804	0.537	—	0.413	0.362	0.379
San Pedro Basin	0.603	1.356	—	0.874	0.658	—
Santa Monica Basin	0.612	0.371	0.386	0.379	0.258	—

[a] Based on sediment trap results of Bruland and Franks (1979) and Bruland et al. (1981) and ^{210}Pb dating of sediments by Bruland et al. (1974).

[b] TPM—total particulate matter ($mg\ cm^{-2}\ yr^{-1}$).

[c] Data believed inaccurate because of contamination (Bruland and Franks 1979), not reported here.

[d] Data not reported (Bruland et al. 1974).

finding by Bruland et al. (1981) that PIT and surface sediment samples collected at the same time have similar concentrations, one might assume that the differences reflect changes in the fluxes of these metals to basin sediments over time. If so, a measurable and consistent reduction seems to have occurred in the flux of Pb, Cr, Zn, Cu, and Cd between 1972 and 1977 in all cases except one: Pb in San Pedro Basin. These reductions would be predicted on the basis of declining wastewater discharges (fig. 3.3) (SCCWRP 1989) and dramatic reductions in the usage of leaded gasoline since the early 1970s (Ng and Patterson 1982).

The anomalous behavior of Pb in San Pedro Basin may reflect the fact that this basin is situated immediately adjacent to the Los Angeles County wastewater discharge on the Palos Verdes Shelf. During the 1971–1977 period, this plant discharged between 53 and 62% of all lead entering the SCB from municipal wastewater treatment plants. Whereas the combined emissions decreased by 37%, Pb emissions from this plant decreased by only 22%. Ng and Patterson (1982) estimated that all, one-third, and one-fourth of the industrial lead accumulating in sediments of Santa Barbara Basin, Santa Monica Basin, and San Pedro Basin, respectively, originate from nonsewage sources (that is, eolian and fluvial). Thus, reduced inputs from nonsewage sources would be expected to have the least effect on San Pedro Basin, followed by Santa Monica and Santa Barbara basins. However, declining inputs from wastewater discharge would most affect Santa Monica and San Pedro basins. The results shown in table 3.8 suggest that atmospheric fluxes probably declined at a greater rate than those from sewage discharge. Further changes in the sediment accumulation rates of Pb and other metals in the post–1977 era are discussed in the next section.

Distribution and Fate of Chemical Constituents in Sediments

This section discusses the distribution and postdepositional fate of organic and trace inorganic substances in sediments of the SCB. The first half is largely descriptive and concerns two topics: (1) interrelationships between the spatial distribution of chemical substances in sediments of the SCB, their sources, and physical transport processes; and (2) the accumulation of these materials in basin and shelf environments. The second part of this section addresses postdepositional processes affecting the fate of sediments.

The bight-wide spatial distribution of chemical substances largely reflects where the source materials are introduced and processes affecting particulate matter during transport and sedimentation. As discussed earlier, particles entering or formed in the ocean can undergo decomposition and dissolution or can be altered by interactions with dissolved materials through adsorption and desorption reactions. Aggregation or biological repackaging (see fig. 3.1) changes the particle size distribution and chemical characteristics of these materials. Upon deposition on the sea floor, particulate matter undergoes further transformations as a result of numerous physical, biological, and chemical processes occurring at or near the sediment surface. Hence, the chemical composition of sediments reflects not only the original source materials, but also the processes acting on them during transit in the water column and incorporation into the seabed.

One of the principle aims of geochemistry is to understand the factors controlling sediment chemistry. This is important for a variety of reasons. The surface layers of bottom sediments and associated near-bottom waters, sometimes referred to as the benthic boundary layer, represent a zone that is very active biologically and geochemically (Craven et al. 1986; Craven and Jahnke 1992; Emery and Rittenberg 1952; Jahnke 1990; Reimers and Smith 1986; Reimers et al. 1990; Smith et al. 1987). Biologically mediated decomposition of labile organic matter and dissolution of mineral phases in this zone leads to the buildup of metabolic products and other soluble species (such as CO_2, NH_4^+, CH_4, HS^-, PO_4^{-3}, and SiO_3^{-2}) in pore waters. At the same time, depletion of oxidants (O_2, NO_3^-, Fe^{+3}, Mn^{+4}, SO_4^{-2}, and HCO_3^-) by indigenous heterotrophic organisms affects the redox potential and alkalinity of interstitial water. This may lead to chemical precipitation reactions (e.g., $Fe^{+2} + S^{-2} \rightarrow FeS^S \xrightarrow{s^\circ} FeS_2$) that bring about transfer of materials from pore waters to the solid phase (Leslie et al. 1990). These processes are collectively referred to as organic diagenesis.

The onset of diagenesis may result in establishment of concentration gradients of soluble constituents between pore waters and the overlying seawater. If advective fluxes of pore waters (as a result of compaction) are small and bioturbation is absent, these concentration gradients will drive chemical exchange across the sediment–water interface via molecular diffusion. To the extent that physical resuspension and bioturbation occur, the rate of exchange may be further enhanced (Aller 1982). Thus, processes occurring at the seabed directly link sediment chemistry with that of the overlying waters and provide a means of coupling benthic respiration with primary production in the euphotic zone.

A second concern about sediment chemistry relates to interactions between the biological food web and toxic chemicals deposited in sediments. The sea floor can act as a source of many organic and inorganic contaminants to benthic and pelagic populations (Gossett et al. 1983; Young et al. 1988). Remobilization occurs by desorption from sediment particles and diffusion into overlying or interstitial waters (Brownawell and Farrington 1986). Once mobilized, these substances may be taken up directly by organisms via absorption across membranes. Deposit-feeding benthic organisms are also exposed to contaminants through ingestion of sediment particles. Compounds having the required stability and lipophilicity can become concentrated in higher trophic levels through food web transfer, a process known as biomagnification. If concentrations reach sufficiently high levels, adverse biological effects may occur, as in the case of the population decline of the California brown pelican caused by DDT-induced eggshell thinning (Keith et al. 1970). Because humans occupy a position at the highest trophic level, ingestion of contaminated seafood by humans may also present significant health risks. Thus, toxic chemicals originally deposited in marine sediments can be transferred and exert effects throughout the biosphere.

Sedimentary Organic Matter

The concentration of organic matter in sediments depends on its rate of supply, its preservation before and after deposition, and the overall sedimentation rate (Tissot and Welte 1984). These factors in combination with the physical oceanographic setting establish the depositional environment.

The relative importance of environmental factors for the preservation of sedimentary organic matter is presently a subject of controversy (Calvert and Pederson 1992). One view holds that preservation is enhanced when the oxygen content of the overlying water column is reduced (Emerson 1985), such as in the case of the SCB's inner basins, whose sill depths intersect the oceanic oxygen minimum zone of Pacific Intermediate Water (Emery 1960; Reid 1965). Accordingly, higher sedimentation rates would be expected to favor preservation since freshly deposited sediments are more rapidly buried to depths below which oxygen is no longer available. Likewise, high productivity should aid the preservation of deposited organic matter by enhancing the vertical flux of large particles in the form of fecal pellets and macroscopic aggregates (thus, increasing the overall sedimentation rate), as well as by creating a greater demand on oxygen available to heterotrophic organisms in the water column and at the sea floor. If high productivity is coupled with rapid sedimentation, as in the case of Santa Barbara Basin, the oxygen demand may be sufficient to establish anoxia at or near the sediment–water interface, thereby preventing larger life forms from inhabiting (and disturbing) the sediments. These assertions are based on the premise that anaerobic decomposition of labile organic matter proceeds more slowly (or less efficiently) than aerobic degradation, an assumption that has recently been questioned (Henrichs and Reeburgh 1987). Although our understanding of the relationship between depositional environment and organic carbon burial are incomplete, it is clear that a complex interplay of factors is involved.

AREAL DISTRIBUTION OF ORGANIC MATTER IN SEDIMENTS

Emery (1960) summarized results of early studies on the concentration of total organic matter in sediments of the SCB. In general, these investigations showed that organic matter comprises less than 1% (dry wt) on the mainland shelf, island shelves, and bank tops. In contrast, concentrations of organic matter in basin slopes, sills, and floors range from 5 to 10%.

Several investigators (see references in Emery 1960) have noted a trend of increasing organic matter concentrations progressing from inner to outer basins, with a precipitous decline in organic content when the continental slope is reached. The maximum in organic content in the outer basin sediments has been attributed to a balance between dilution by detrital (lithogenous) sediments under conditions of high sedimentation (farther inshore) and poor preservation under conditions of low sedimentation rates (farther offshore). Emery (1960) and others (Anderhalt and Reed 1978; Choi and Chen 1976; Gorsline 1992; Thompson et al. 1987) have also noted an inverse correlation between organic content and grain size. Analyses of size-fractionated sediments reveal higher concentrations of organic nitrogen in the finer grain size fractions, with the result that up to 65% of the organic matter may be associated with sediment grains ranging from 1 to 16 μm in diameter (Emery 1960). The association of organic matter with finer sediment fractions is probably attributable to either the nearly equivalent settling velocities of organic particles and fine-grained sediments or the adsorption of organic matter to fine-sized minerals, especially clays (Meyers and Quinn 1973). Because the grain size of basin sediments generally decreases with distance offshore (Emery 1960), it is not surprising that sediments of the offshore basins should be richer in organic matter with increasing distance from land. Conversely, sediments from bank tops and shelf regions that are subject to greater reworking by waves and currents have correspondingly lower organic matter contents.

VERTICAL DISTRIBUTION OF ORGANIC MATTER IN SEDIMENTS

Knowledge of the vertical distribution of organic matter in sediments of the SCB is limited by a shortage of data and the disparate nature of the studies that have been undertaken. The analysis of sediment cores has been largely restricted to the basin and mainland shelf environments. Consequently, the following generalizations are based on rather poor coverage of the SCB as a whole.

Numerous investigators (Anderhalt and Reed 1978; Bruland et al. 1974; Doose 1980; Emery 1960; Gorsline et al. 1968; Kalil 1976; Shaw 1988) have documented a decline in the concentration of organic carbon with increasing depth (down to approximately 6–8 m) in consolidated basin sediments. Most often there is a rapid decrease in near subsurface sediments (no more than 50% within the top 1 m), with smaller or no changes at greater depth. Patterns within a given basin appear somewhat variable, reflecting either differences in the rates and modes of sediment supply or possibly the character of the organic matter. These downcore decreases have usually been attributed to diagenetic losses (Rittenberg et al. 1955; Sholkovitz 1973). However, it is unclear whether the rates of sulfate reduction and methanogenesis in Santa Barbara and Tanner basins (Doose 1980; Kaplan et al. 1963; Kalil 1976) are sufficient to account for all of the carbon presumed to have been lost. Recent work by Reimers (1987) and Reimers and Smith (1986) has shown that in San Clemente and Santa Catalina basins and deeper waters of the Patton Escarpment, sedimentary organic carbon concentrations decline rapidly within the upper few millimeters of the sediment column. This apparently results from aerobic respiration of readily degradable organic matter mediated by the indigenous bacterial, meiofaunal, and macrofaunal assemblages. Similarly, Jahnke (1990) has found that roughly three-fourths of the organic matter remineralization in shallow sediments of the Santa Monica Basin occurs within the upper 2 cm.

In shallow cores (approximately 30 cm) from Santa Monica Basin, Finney and Huh (1989a, b) identified a subsurface maximum in organic carbon content that corresponds with a radiometric ^{210}Pb date of 1971 (fig. 3.13a). Similar subsurface maxima have been reported for the San Pedro Shelf (Eganhouse and Kaplan 1988; Stull et al. 1986) in the vicinity of the Los Angeles County Sanitation District's (LACSD) municipal wastewater discharge in 60 m of water off White Point (fig. 3.13b). In the year 1971, emissions of organic-rich solids from the LACSD treatment plant reached a peak (Eganhouse et al.

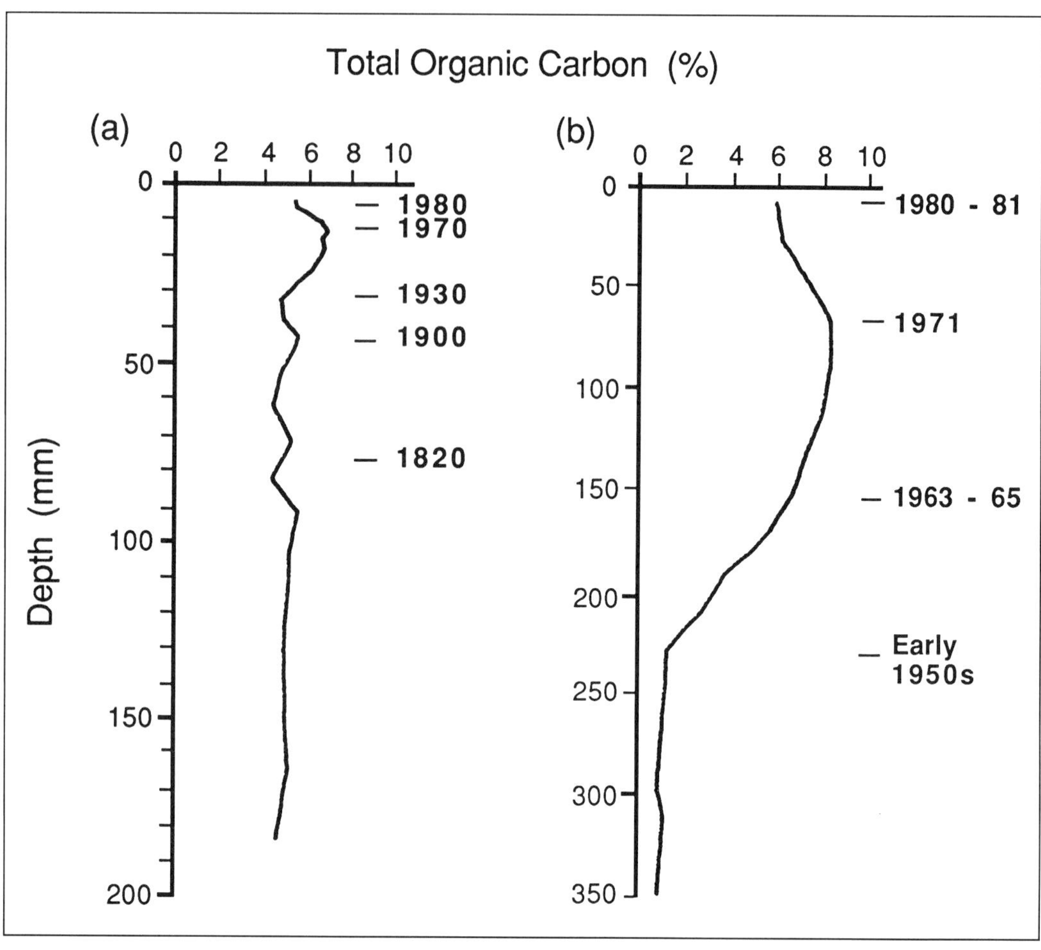

Figure 3.13. Vertical concentration profiles of total organic carbon in shallow age-dated cores taken in (a) Santa Monica Basin (Finney and Huh 1989a) and (b) San Pedro Shelf (Eganhouse and Kaplan 1988).

1983). Because of the generally poleward flow of water over basinal slopes (see chap. 2), particulate matter originating from the outfall is carried to the north-northwest, as evidenced by the pattern of accumulation of waste-derived contaminants in surface sediments near the outfalls (see chap. 12). Thus, the near-surface sediments of the Santa Monica Basin floor and the San Pedro Shelf appear to record the deposition of anthropogenic carbon. The maximum degree of organic carbon enrichment (over background levels) believed to be due to anthropogenic inputs has been estimated at approximately 800% and 29–58%, respectively, for shelf sediments (San Pedro Shelf) (Eganhouse and Kaplan 1988) and inner basin sediments (Santa Monica Basin) (Finney and Huh 1989a). Independent estimates for the contemporary accumulation flux of sewage-derived organic carbon to Santa Monica Basin based on sterol analyses suggest a smaller contribution (about 2%) (Venkatesan and Kaplan 1992).

ACCUMULATION RATES OF ORGANIC MATTER

Emery (1960) and Emery and Bray (1962) presented data (based on ^{14}C measurements of sediments collected by piston cores 3–5 m in length) on the average accumulation rates of organic matter in 13 offshore basins in the SCB. For comparison, accumulation rates were also given for onshore basins, the continental slope, abyssal sediments, and basin slope environments. These data and others from the literature are provided in table 3.9.

Accumulation rates of organic matter vary by a factor of 50 (from 0.05 to 2.7 mg C cm^{-2} yr^{-1}) with a continuous decline in the offshore direction. The higher accumulation rates in inner basins (e.g., Santa Barbara, Santa Monica, and San Pedro) reflect the combined effects of higher productivity, shallower depths to basin floors, low-oxygen tension of basin waters, and higher overall sediment accumulation rates (Schwalbach and Gorsline 1985). In contrast, the outer basins (e.g., Tanner and West Cortes) are removed from sources of terrigenous detrital sediments (hence, lower sediment accumulation rates), have surface waters exhibiting lower primary production (see chap. 4) (Mullin 1986), and have basin floors at greater water depths. In addition, basin sills generally lie below the depth of the oxygen minimum zone (with O_2 concentrations at >0.02 mM), possibly permitting more efficient oxidation of particles moving through the water column.

Schwalbach and Gorsline (1985) recently updated Emery's database for the six northern (inner and central) basins, thereby providing considerably more detail of the interbasin variations. They determined the average (Holocene) sedimentary accumulation rates of organic carbon within the Santa Barbara, Santa Monica, San Pedro, Santa Cruz, San Nicolas, and Santa Catalina basins, using a combination of high-resolution acoustic reflection profiling, ^{14}C and ^{210}Pb radiometric dating, and carbonate analysis. Accumulation rates of organic carbon ranged from 0.5 to slightly over 4 mg C cm^{-2} yr^{-1}, in reasonable agreement with, but slightly higher than, Emery's (1960) estimates of 0.9 to 2.7 mg C cm^{-2} yr^{-1}. Within individual basins, the highest rates are generally found in the central floors, possibly reflecting the concentration of low-density organic-rich material by deep water gyres (Gorsline 1992; Malouta et al. 1981). Dilution by direct terrigenous inputs is evidenced near the heads of canyons intersecting shelves of the inner basins.

SOURCES OF ORGANIC MATTER: USE OF STABLE ISOTOPE RATIOS

The light elements S, O, C, H, and N comprise the building blocks of all forms of life on earth. Each of these elements exists as two or more stable isotopes whose physical properties differ from each other slightly because of differences in their atomic mass (Bigeleisen and Mayer 1947). These differences lead to measurable variations in the relative abundance of the isotopes which, in turn, reflect the origin(s) and processes occurring during cycling of the elements. Two general types of isotope effects are observed: equilibrium and kinetic. These lead to isotopic fractionation (i.e., changes in isotope ratios).

Isotopic composition is usually defined in terms of isotope ratios rather than absolute abundances. By convention, the measured isotope ratio is referred to that of an accepted standard (see Hoefs 1987), the results being presented in the familiar "del" notation whereby:

$$\delta = [R_{sample} - R_{\mathrm{std}}/R_{\mathrm{std}}] \times 1000$$

where R is the isotopic ratio (heavy to light, for example, $^{13}C/^{12}C$).

Emery (1960) presented data on the stable isotopic composition of organic carbon in SCB basin sediments along with results for the continental slope and the deep sea floor. Most ratios fell within a range of approximately -20 to -23‰ (grand mean of -21.7‰), indicating that the bulk of the organic matter in these sediments is derived from marine, not terrestrial, primary production. Emery (1960) also noted a consistent depletion of ^{13}C with depth in the sediments, which was attributed to selective decomposition of proteins and carbohydrates, although no evidence based on molecular analysis was provided.

Table 3.9. *Estimated Accumulation Rates of Organic Carbon (mg C cm^{-2} yr^{-1}) in Sediments of the SCB*

Basin	Emery (1960)[a]	Others	
		^{210}Pb	Sediment Traps
Santa Barbara	2.7	3.8[b]	2.6[c], 1.5[d], 2.6[e], 2.2[f], 2.7[g]
Santa Monica	2.7	1.0[b], 1–2[h] 1.0[i], 0.6[j]	1.0[c], 1.1[d], 1.7[j]
San Pedro	0.94	1.4[b]	1.6[c], 1.3[d]
San Diego	0.5		
Santa Cruz	1.2		
Santa Catalina	1.3		
San Clemente	0.78		
San Nicolas	0.61	0.63[b]	0.63[c], 0.43[d]
East Cortes	0.56		
No Name	0.22		
Tanner	0.56		
West Cortes	0.33		
Long	0.5		
Continental slope	0.44		
Deep-sea floor	0.06		
Top of basin slope	0.06		
Ventura Basin	2.8		
Los Angeles Basin	1.8		

[a] Accumulation rates have been converted to organic carbon basis by dividing values given in Emery (1960) by 1.8.

[b] Based on sediment accumulation rate data of Bruland et al. (1981) and organic carbon data of Anderhalt and Reed (1978).

[c] Based on vertical flux measurements (Bruland et al. 1981) and organic carbon data of Crisp et al. (1979).

[d] Based on vertical flux measurements and organic carbon data reported by Crisp et al. (1979).

[e] Based on data given by Dymond et al. (1981) for Soutar cone.

[f] Dunbar and Berger (1981).

[g] Based on data given in Dymond et al. (1981) for Gardner trap.

[h] Malouta et al. (1981).

[i] Huh et al. (1987) and Finney and Huh (1989a).

[j] Jackson et al. (1989).

Later Nissenbaum and Kaplan (1972), Nissenbaum (1973), and Choi and Chen (1976) showed that humic substances (not including humin) may comprise from 10 to 35% of the organic carbon in sediments from San Diego Trough, Santa Barbara Basin, Santa Monica Basin, and San Pedro Channel and Harbor. Isotopic analysis of humic substances isolated from soils and marine sediments revealed different compositions: $\delta^{13}C$ values for terrestrial humic acids in the range of −25 to −26‰ and marine humic acids generally about −20 to −22‰ (Nissenbaum and Kaplan 1972; Stuermer et al. 1978; Simoneit et al. 1979b). An exception to this pattern was found for Santa Monica Basin (−27.4%), presumably because of the influx of terrigenous debris. Nissenbaum and Kaplan (1972) suggested that marine humic sedimentary acids could be formed by condensation reactions occurring in interstitial waters. According to this scenario, the reactants would be degradation products derived originally from marine phytoplankton. Thus, marine sedimentary humic substances would inherit the isotopic composition of carbon fixed in the cells of phytoplankton.

In an extension of these concepts, Peters et al. (1978) investigated the stable carbon and nitrogen isotopic composition of sedimentary organic matter from coastal sites along the northeastern Pacific. Samples were collected from Santa Catalina Basin, Santa Barbara Basin, Tanner Basin, and San Pedro Shelf. Figure 3.14 illustrates the correlation between $\delta^{13}C$ and $\delta^{15}N$ for total organic matter and kerogen in these samples. The regression line is nearly parallel to that connecting the isotopic compositions of the presumed inorganic precursors (that is, marine, HCO_3^- and NO_3^-; terrestrial, CO_2 and N_2), indicating that coastal sediments largely reflect the differences in isotopic composition of the corresponding source materials. This finding made it possible to use the isotopic composition of two of the most abundant elements in organic matter for purposes of source differentiation.

In a series of papers, Sweeney and coworkers (Sweeney and Kaplan 1980a, b; Sweeney et al. 1978; Sweeney et al. 1980) further described the application of nitrogen isotopes in measuring relative contributions of terrestrial and marine nitrogen to nearshore basin and shelf sediments. Sweeney and Kaplan (1980a) found the isotopic composition of pore water ammonium in Santa Barbara Basin sediments (mean +10.2‰) to be similar to that of marine phytoplankton (mean +8.9‰), suggesting that most of the diagenetic remineralization of nitrogen in these sediments involves marine, not terrestrial, organic matter (see also Doose 1980). The sediments exhibited isotope ratios of +2.8 to +9.4‰ with most values falling above +6‰ (fig. 3.15a). These were compared to ratios for marine phytoplankton (+8.9 ± 1.8‰) and a terrigenous end-member of +2‰ in a simple two-source mixing model. Model calculations indicated that most of the nitrogen in Santa Barbara Basin sediments was of marine origin (fig. 3.15a). Thus, downcore variations in the nitrogen isotope ratios of the sediments were interpreted in terms of changing relative inputs of marine and terrestrial (organic) nitrogen as a function of time (fig. 3.15b). Sweeney and Kaplan (1980a) also noted a sudden and progressive decline in the fraction of marine nitrogen below a depth of 3 m. Because little additional ammonium was released below this sediment horizon, the change in isotopic composition was interpreted as indicating a shift toward greater sedimentation of terrigenous organic matter in earlier times. Such a shift may have had its origins in climatic (hence, sea level) fluctuations during the Holocene, as suggested by Kalil (1976) and Fleischer (1972).

A puzzling situation exists for the nitrogenous component of Santa Barbara Basin sediments. As discussed, Sweeney and Kaplan (1980a) found that a significant fraction (approximately 25–40%) of the nitrogen in these sediments was derived from terrestrial sources. Nevertheless, there is little evidence of terrigenous inputs from the carbon isotope record: $\delta^{13}C$ = −20.4 to −21.6‰ in Emery (1960) and $\delta^{13}C$ = −20.6 to −22.6‰ in Doose (1980). This discrepancy can be explained in part by the greater sensitivity of the nitrogen isotope ratio to small contributions of one source material because of the larger difference in end-member isotope compositions (+2 to +10‰ for nitrogen versus −21 to −26‰ for carbon). However, it is well known that the refractory fraction of terrestrial organic matter is nitrogen poor relative to that derived from algal sources (Nissenbaum and Kaplan 1972; Stuermer et al. 1978). Therefore, if the terrigenous nitrogen in Santa Barbara Basin sediments represents approximately 33% of the total, as suggested by Sweeney and Kaplan (1980a), the fraction of

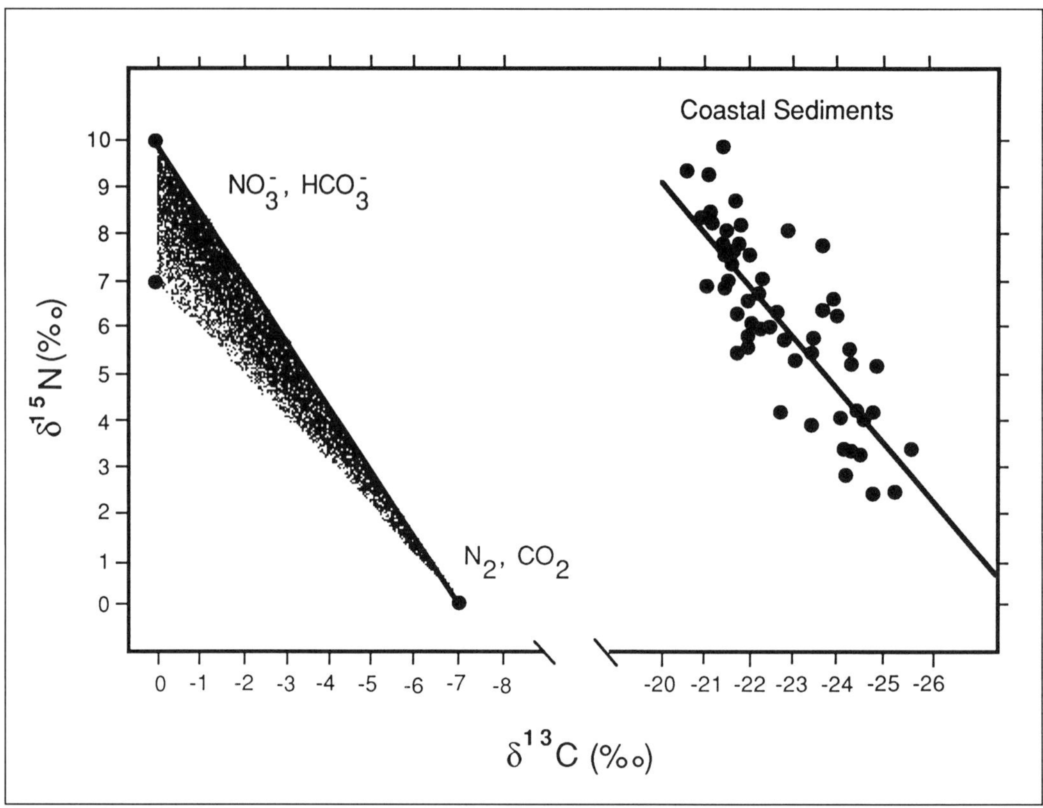

Figure 3.14. Distribution of stable nitrogen and carbon isotope ratios for marine and terrestrial inorganic subtrates (NO_3^-, HCO_3^-, N_2, and CO_2) and sedimentary organic matter from the northeastern Pacific. (After Peters et al. 1978.)

terrigenous carbon would have to be greater than 33%, all other things being equal. This quantity of terrigenous carbon should provide a detectable isotope signal. In addition to the high productivity of the region, Santa Barbara Basin receives significant inputs of terrigenous debris (Fleischer 1972). It is therefore surprising that the carbon isotopic composition of the sediments does not reflect the presence of more terrestrially derived organic carbon.

The most recent work on the isotopic composition of basin sediments comes from Williams (1988) and Williams et al. (1992) in the context of the CaBS project. In this investigation, the organic carbon and nitrogen isotopic composition of suspended and sinking (sediment trap) particles and sediments from Santa Monica Basin were measured and compared. First-order calculations indicated that sinking particles collected on the basin slope were composed of 21–93% terrestrial material, whereas at the basin center the proportions ranged from 16 to 58% (based on $\delta^{15}N$ measurements). In contrast, the corresponding slope and basin estimates, using stable carbon isotope ratios, were 5–14% and 0–27%, respectively. The difference was interpreted as the result of incorporation of carbon-rich marine sedimentary material into traps after downslope transport.

Williams (1988) and Williams et al. (1992) found that downcore variations in $\delta^{13}C$ and $\delta^{15}N$ values of total sedimentary organic matter in Santa Monica Basin were not strongly correlated. However, the distribution of

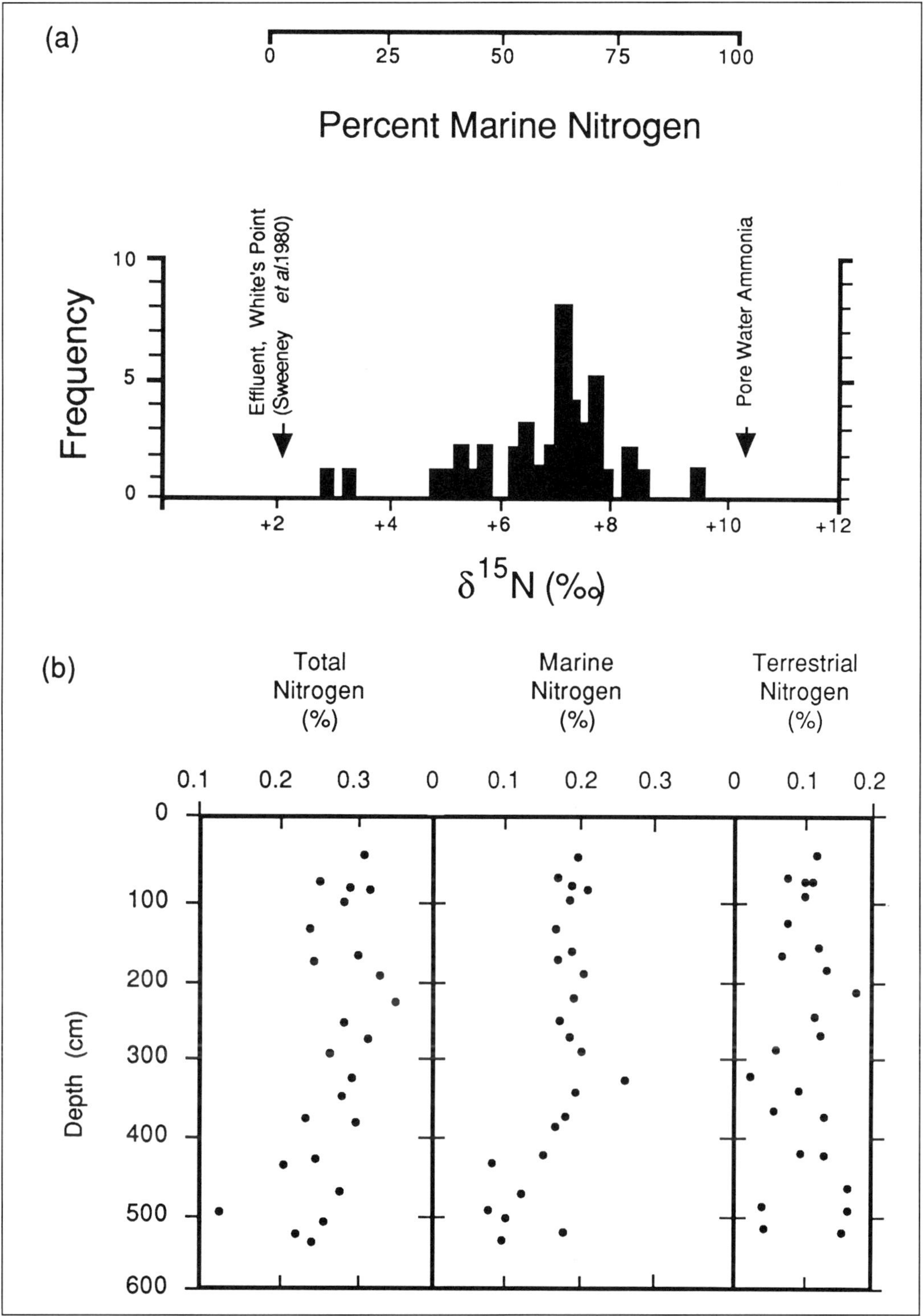

Figure 3.15. Distribution of nitrogen in Santa Barbara Basin sediments: (a) frequency distribution of $\delta^{15}N$ of total nitrogen and (b) proportion of marine and terrestrial nitrogen based on stable isotope ratios. (After Sweeney and Kaplan 1980a.)

$\delta^{13}C$ and $\delta^{15}N$ values falls close to the regression line presented by Peters et al. (1978) for northeast Pacific coastal sediments. Generally speaking, $\delta^{15}N$ values varied more than those for $\delta^{13}C$, perhaps reflecting greater natural variation in the isotopic composition of seawater NO_3^--nitrogen, particularly where denitrification is an important phenomenon (Cline and Kaplan 1975; Liu 1979). More important, the sediments differed markedly from the suspended and sinking particles, the primary difference being that suspended particles typically are more depleted in ^{13}C than the sediments, whereas the sinking particles are similar to or slightly enriched in ^{13}C relative to the sediments. Williams et al. (1992) concluded that particles maintained in suspension may have experienced more reworking than those rapidly transported through the water column and incorporated into sediments.

A series of geochemical studies involving stable isotopes were carried out on the sediments of the San Pedro Shelf, site of the LACSD municipal waste outfall system (Eganhouse and Kaplan 1988; Myers 1974; Sweeney et al. 1980; Sweeney and Kaplan 1980b). Myers (1974) measured the stable carbon isotopic composition of LACSD wastewater effluent and sediments from the vicinity of the discharge. He concluded that while the sedimentary organic carbon was dominated by inputs from the wastewater discharge, only 1% of particulate organic carbon released from the outfalls was actually retained in shelf sediments. Sweeney and Kaplan (1980b) and Sweeney et al. (1980) extended this work by examining the nitrogen isotopic composition of flocculent material present in the rocky intertidal zone of Palos Verdes and subtidal sediments taken from the vicinity of the outfalls. They were able to demonstrate that approximately 50% of the flocculent samples tested contained recognizable amounts of waste-derived nitrogen. They applied a two-source mixing model for estimation of the fraction of sewage and marine components of the sedimentary organic nitrogen. Their quantitative results were similar to those obtained by Myers (1974), confirming the extensive impact of the sewage discharge on shelf sediments.

Composition of Sedimentary Organic Matter

Up to this point, discussion has focused on the distribution of total sedimentary organic matter and certain bulk properties of this material. Although the elemental and isotopic composition of sediments offer important information, detailed chemical analysis provides greater insight into the origins and history of this material. For this reason, we now turn to studies involving the examination of specific compound classes.

The organic matter in marine sediments comes from the products of photosynthesis in the ocean and on land. Although the basic molecular building blocks of biopolymers are universal, there are important differences between the chemical composition and specific molecular structures and distributions in phytoplankton, higher plants, and heterotrophic microorganisms. To the extent that these cellular constituents survive, they provide clues to the original sources of the organic matter. Because the various components of cells differ in their susceptibility to biodegradation and reactivity, sediment chemistry reflects the net result of input and biological recycling in the water column and sediments.

Figure 3.16 presents a schematic conceptualization of the transformations involved in early (organic) diagenesis. It is clear from this diagram that the chemical substances incorporated into sediments comprise an exceedingly complex mixture of natural (and possibly anthropogenic) compounds as well as transformation products generated during early diagenesis. These substances are usually isolated for analysis on the basis of their physicochemical properties (solubility, adsorption, and partitioning). Those that partition into an organic solvent such as dichlorometh-

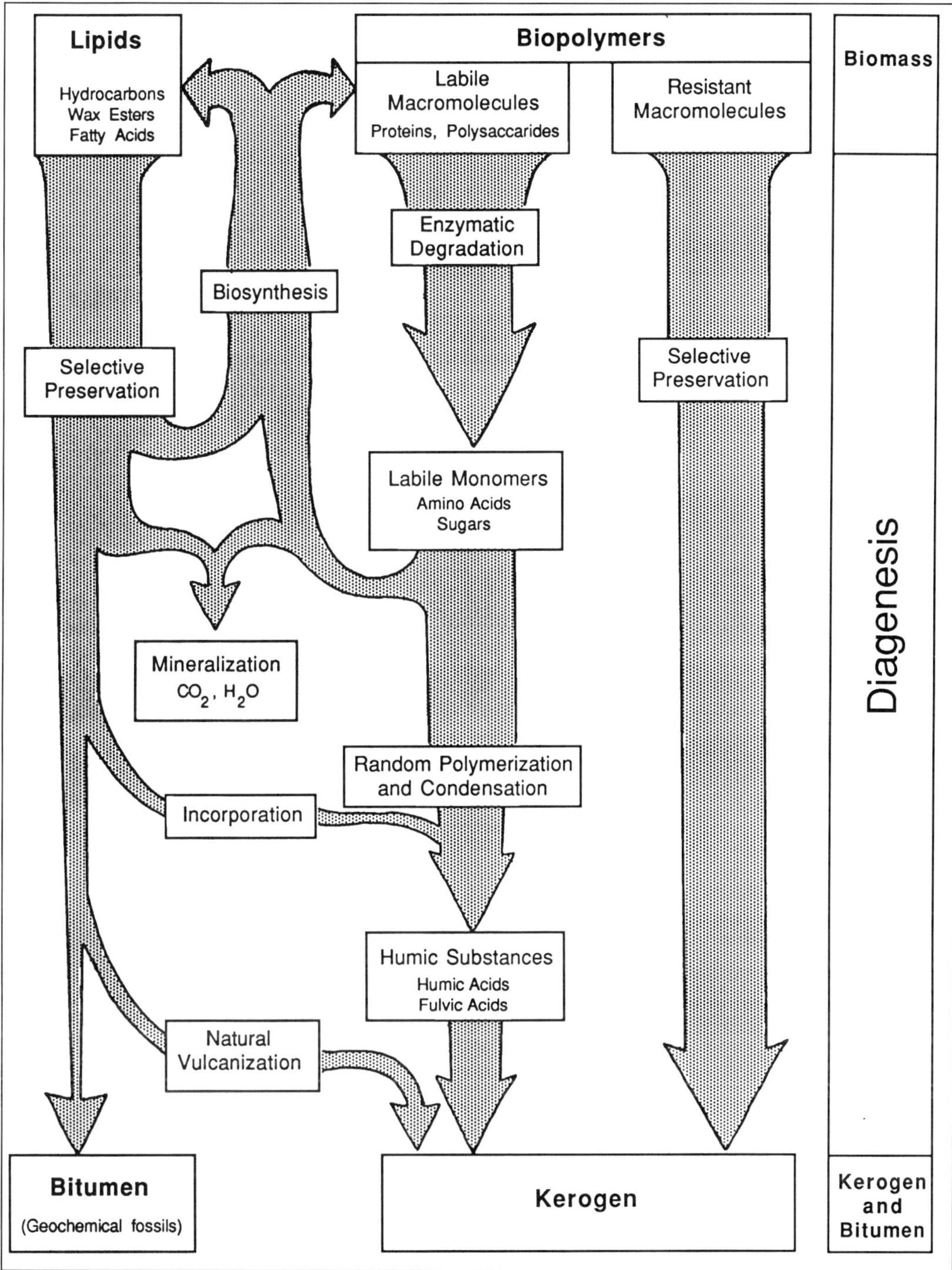

Figure 3.16. Conceptual model of transformations affecting organic matter during diagenesis. (After Tegelaar et al. 1989; Tissot and Welte 1984.)

ane or methanol are commonly referred to as lipids. In fact, this fraction may consist of a large number of biogenic and synthetic substances. Other fractions, representing the more labile constituents of biogenic organic matter, include the amino acids (combined and free) and carbohydrates (e.g., sugars, cellulose, and hemicellulose). Once broken into soluble constituents, these biomolecules are readily utilized by microorganisms for their energetic and biosynthetic needs. Consequently, they may be recycled, released from the sediments as DOM, or incorporated into more complex refractory materials. The refractory organic matter found in sediments includes humic substances (humic acids, fulvic acids and humin) and kerogen. These are widely believed to represent high molecular weight condensation products of reactive biological precursors, although there is some dispute as to the timing of their formation (Brown et al. 1972; Hatcher et al. 1983). Some investigations have revealed the existence in some living organisms of highly resistant biopolymers that may represent another previously overlooked source of humic substances and kerogen (Goth et al. 1988; Nip et al. 1986; and Philp and Calvin 1976, among others). The extent to which these resistant biopolymers contribute to the refractory pool of sedimentary organic carbon in the SCB is unknown. We turn now to what is known about the chemical composition of sedimentary organic matter in the SCB.

Emery (1960) summarized the only comprehensive analyses of organic matter in SCB sediments ever published. The data, shown in table 3.10, are for sediments from Santa Barbara Basin. As much as 50% of the organic matter in these recent sediments appears to be sequestered in highly resistant fractions such as the lignin-humus complexes (humic substances and kerogen) and residual fractions of nitrogen-bearing organics. Lipoidal and simple biogenic components (e.g., amino acids and sugars) comprise less than 10% each of the total material. Similar compositions have been reported for reducing sediments from other regions (Brown et al. 1972; Spiker and Hatcher 1984). In subsequent studies, constituents of these latter compound groups, especially the lipids, were studied in greater detail principally because of their relative simplicity and ease of analysis.

HYDROCARBONS

Emery (1960) reported early work on the hydrocarbon geochemistry of the SCB. It was not until the late 1970s, however, with the initiation of the Southern California Baseline Study sponsored by the Bureau of Land Management (BLM) that an extensive database on hydrocarbon composition and abundance was established for large portions of the SCB. Some of this work appears in the primary literature (Crisp et al. 1979; Reed et al. 1977; Simoneit and Kaplan 1980; Venkatesan et al. 1980). However, much of it can be found only in the extensive compilations of data presented in BLM final reports or student dissertations (Mankiewicz 1981). Reed et al. (1977) summarized early findings of the 1975–1976 BLM survey. In this study, 145 benthic box-core samples and 52 sandy intertidal samples were collected and analyzed for their hydrocarbon content and composition. Reed and his co-workers reported concentrations of total hydrocarbons in the range of <50 $\mu g\ g^{-1}$ for surface sediments from the outer banks and ridges and the mainland shelf areas to the south of Los Angeles. Central basin (Santa Catalina and San Nicolas) sediments were found to contain 50–100 $\mu g\ g^{-1}$, whereas the highest concentrations (200–1350 $\mu g\ g^{-1}$) were measured in Santa Monica and San Pedro basins or near Coal Oil Point (approximately 600 $\mu g\ g^{-1}$) in the vicinity of known oil seeps. The aromatic hydrocarbons represented 18–55% of the total hydrocarbons (mean = 38.1%, n = 11). These results are in reasonable agreement with data given earlier by Emery (1960) and others in later reports (Reed and Kaplan 1977; Simoneit and

Table 3.10. *Chemical Composition of Recent Marine Sediments from Santa Barbara Basin*

Organic Fraction	% Total Organic Matter
Ether extractables	1
Alcohol extractables	5
Hemicellulose	2
Cellulose	1
Nitrogenous compounds	40
Amino acids	1
Complex proteins	19
Resistant compounds	17
Water soluble, N-free	3
Acid soluble, N-free	7
Lignin–humus complexes	31

From Emery (1960).

Kaplan 1980; Simoneit et al. 1979a,b; Venkatesan et al. 1980).

Figure 3.17 shows the distribution of total hydrocarbons in surface sediments determined in the second year (1976–1977) of the BLM study, along with data from other sources. Here, concentrations are normalized to total organic carbon as a means of compensating for the effects of varying grain size. The strong gradient in hydrocarbon concentrations with radial distance from the Los Angeles Basin is evident. Highest concentrations (approximately 400 mg g^{-1}OC) have been found in sediments of the San Pedro Shelf (Eganhouse 1978). San Pedro and Santa Monica basin sediments exhibit concentrations in the range of 14–62 mg g^{-1} OC (Eganhouse 1978; Venkatesan et al. 1980). These shelf and inner basin sediment concentrations exceed (by one to more than three orders of magnitude) those found in surface sediments of the outer basin and bank environments (0.2–3.3 mg g^{-1} OC) (Mankiewicz 1981; Rapp and Kvenvolden 1982), sediments from the continental margin and rise (2.6–7.5 mg g^{-1} OC) (Simoneit et al. 1979a), and deeper subsurface basin sediment horizons (1.8–13.4 $\mu g\ g^{-1}$ OC) (Simoneit and Kaplan 1980; Venkatesan et al. 1980). As will be shown, the enrichment of hydrocarbons in nearshore surface sediments of the northern SCB is related to the proximity of anthropogenic inputs (municipal wastewaters, storm runoff, and atmospheric input). Concentrations in Santa Barbara Basin sediments are relatively elevated (approximately 6 mg g^{-1} OC) but are still a factor of three less than the highest found in nearby shelf areas (Mankiewicz 1981). Reed and Kaplan (1977) suggest that the Santa Barbara Basin sediments are undoubtedly affected by local seepage of oil so prevalent in this area.

Figure 3.18 illustrates the differences in composition of the saturated hydrocarbons isolated from sediments along a transect originating in Santa Monica Bay and extending to the outer continental shelf west of Tanner Bank. The significant changes with distance offshore include a decrease in the proportion of unresolved versus resolved components, the increased relative abundance of normal versus acyclic isoprenoid alkanes, and increases in the proportions of high molecular weight normal (i.e., straight chain) hydrocarbons. Reed et al. (1977) and others have demonstrated that most sediments from the SCB contain a chromatographically unresolved complex mixture of branched and cyclic hydrocarbons whose presence is characterized by a "hump" in the gas chromatogram (see fig. 3.18). This feature, representing from 70 to >90% of the total hydrocarbons detected by gas chromatography, is typical of weathered petroleum and signals the widespread distribution of anthropogenic and natural seepage hydrocarbons in the SCB. Other indicators of petroleum commonly found in sediments include multiple homologous series of acyclic isoprenoids; normal alkanes distributed over a wide boiling range (exhibiting little or no odd–even carbon chain length preference); elevated phytane:pristane ratios; and tricyclic diterpanes, pentacyclic triterpanes, and steranes having stereochemistry typical of thermally mature organic matter (see references in Simoneit and Kaplan 1980). In general, the prevalence of these indi-

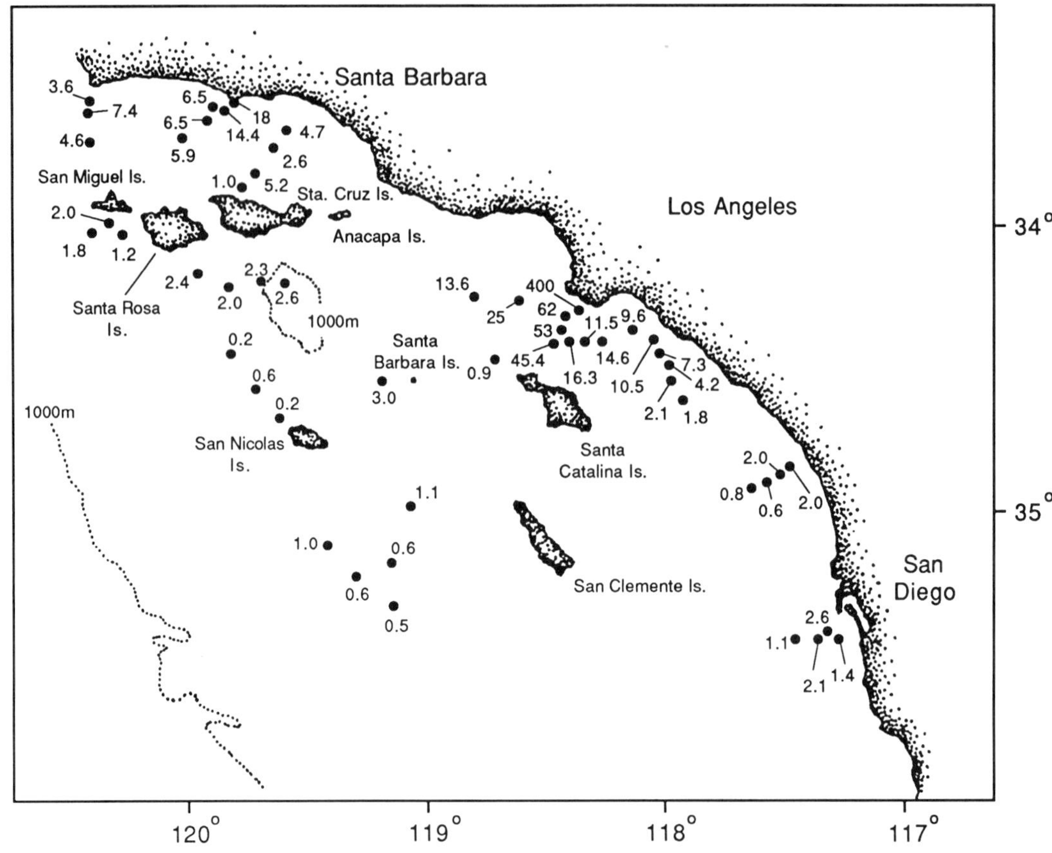

Figure 3.17. Distribution of hydrocarbons (mg THC g^{-1} OC) in surface sediments of the SCB. (After Eganhouse 1978 and Mankiewicz 1981.)

cators of petroleum contamination is greatest in mainland shelf and inner basin sediments in the vicinity of Los Angeles, again reflecting the importance of various anthropogenic activities associated with waste disposal, shipping, and oil production (Mankiewicz 1981; Venkatesan et al. 1980; Venkatesan and Kaplan 1992). Farther offshore and especially to the south, saturated hydrocarbon compositions show increasing relative contributions of biogenic material from both autochthonous and allochthonous sources with lesser amounts of petroleum. However, the hydrocarbon mixtures are complex and their areal distribution is quite heterogeneous.

The presence and distributions of unique biomarker compounds give further evidence of inputs from specific biogenic and petroleum sources. Figure 3.19 shows distributions of normal alkanes for selected sediments from the BLM studies reported by Simoneit and Kaplan (1980). These samples depict variations in the major features seen in sediments from different parts of the SCB. The *n*-alkane distributions show contributions of petroleum (n-C_{15}–C_{36}, with little or no odd–even predominance; CPI [carbon preference index] approximately 1.0), bacteria (n-C_{19}), bacterial resynthesis products (n-C_{18}–C_{25}, maximum at n-C_{22} or n-C_{23}), algae (n-C_{15}–C_{19}, CPI $>>$ 1.0), and higher plant epicuticular waxes ($>n$-C_{26}, maximum or at n-C_{29}, n-C_{31}, CPI $>>$ 1.0). In general, the higher plant wax *n*-alkanes are found throughout the SCB, indicating long-range transport via the atmosphere or as suspended load. In the near-

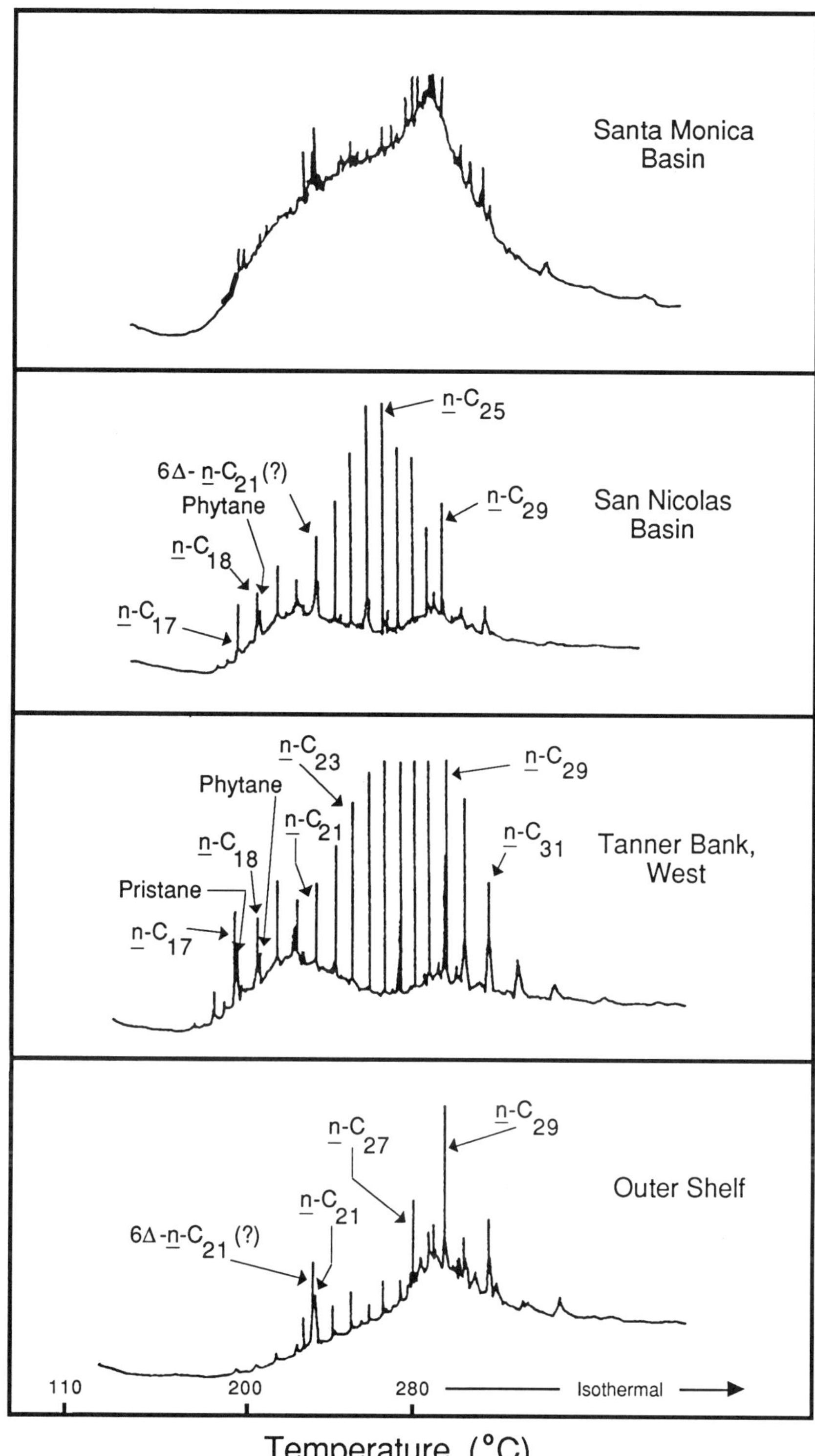

Figure 3.18. Gas chromatograms of saturated hydrocarbon fractions isolated from surface sediments collected along a cross-shelf transect in the SCB. (After Reed et al. 1977.)

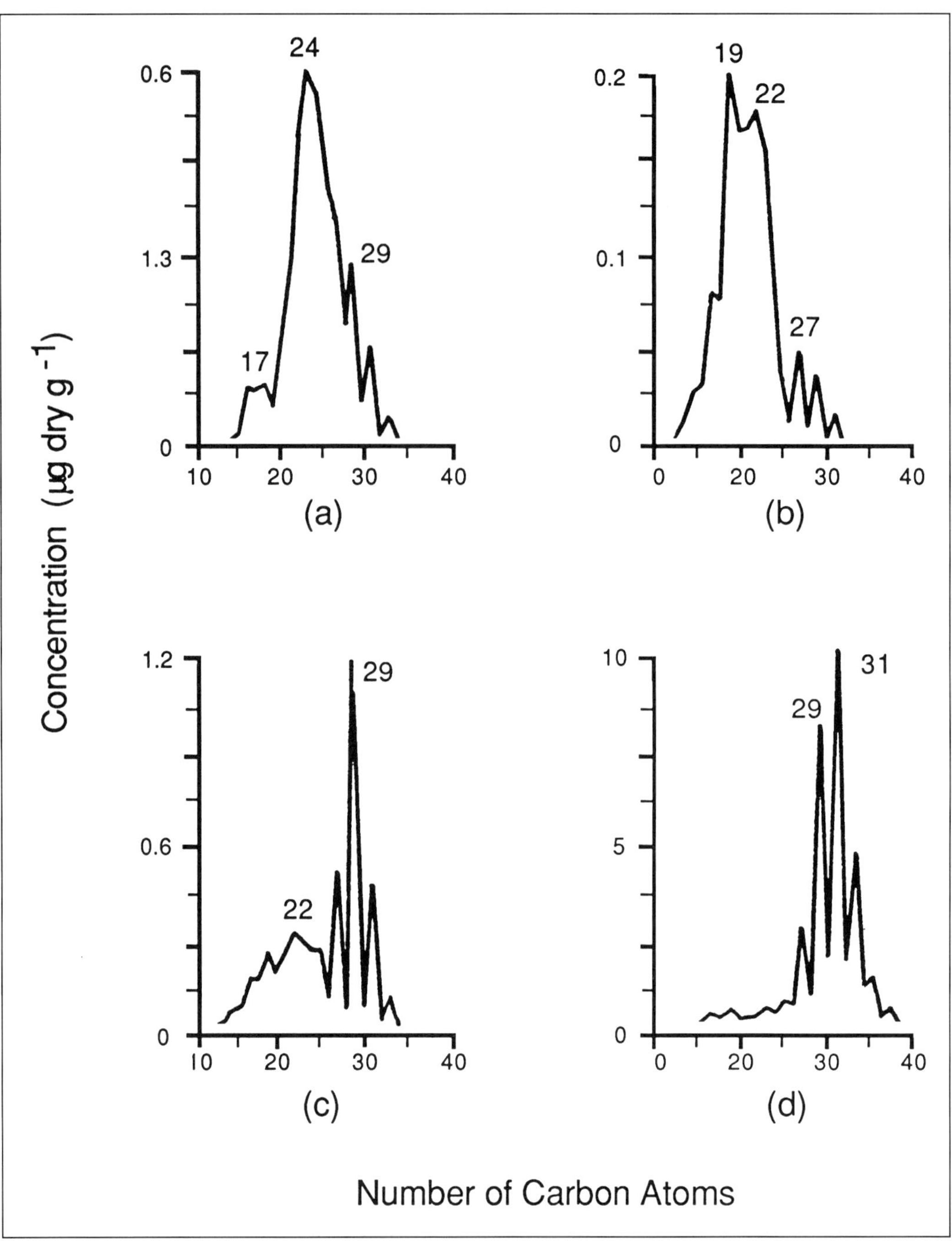

Figure 3.19. Normal alkane distribution plots for sediments from the SCB: (a) San Nicolas Basin (surface); (b) southwest of Santa Cruz Island (20–26 cm); (c) Santa Monica Basin (20–26 cm); and (d) Santa Monica Basin (surface). (After Simoneit and Kaplan 1980.)

shore region, *n*-alkane patterns are sometimes dominated by petroleum or bacterial degradation products or, alternately, by terrigenous debris, depending upon the proximity of the sediments to sources of these materials.

A group of compounds commonly observed in upper sediment layers and presumed to be of marine biogenic origin are a series of C_{25} cyclic and acyclic olefins eluting between n-C_{20} and n-C_{22} (Venkatesan et al. 1980; Reed et al. 1977; Simoneit and Kaplan 1980). These compounds appear to be distributed differently from each other in the SCB, possibly in response to variations in grain size or oxygen levels (Mankiewicz 1981). Equally ubiquitous are the polycyclic diterpanes, triterpanes, and steranes. Of particular note is the C_{28} pentacyclic triterpane, 17α(H),18α(H),21β(H)-28,30-bisnorhopane (see fig. 3.12, structure I). This compound is abundant in the Monterey Shale and California oils (Seifert et al. 1978), but is absent or present in exceedingly low concentrations in most non-California oils and in municipal wastes, urban runoff, and atmospheric aerosols from southern California (Eganhouse and Kaplan 1982b; Eganhouse et al. 1981; Simoneit and Mazurek 1982). Therefore, it has been proposed as a unique molecular marker of oil seepage in the SCB (Simoneit and Kaplan 1980).

The abundance ratio of a related but ubiquitous (nonspecific) C_{30} pentacyclic triterpane, hopane (see fig. 3.12, structure II) to the C_{28} triterpane has been used to differentiate contributions of natural seepage and anthropogenic inputs (Eganhouse and Kaplan 1988; Shokes and Mankiewicz 1979) in shelf and basin sediments. Table 3.11 lists this ratio for sediments from Santa Barbara, Santa Monica, San Nicolas, and Tanner basins, the outer shelf break, the abyssal sea floor, the San Pedro Shelf, and particles collected in sediment traps from some of these areas. The background ratio (given by deeper sediments deposited prior to recent historical times) ranges from 0.4 to 1.0, whereas surface sediments and sinking particles collected in the vicinity of the inner basins exhibit ratios from approximately 1.3 to 5.2. An exception to this pattern would seem to be the subsurface sediments collected (25–31 cm) from Tanner Basin, which have a ratio of 2.5. From these data, it is clear that sediments from San Pedro and Santa Monica basins are most heavily contaminated by anthropogenic hydrocarbons as evidenced by high ratios (2–5), presumably originating from the city of Los Angeles (Hyperion) and LACSD outfall systems as well as shipping and industrial activities in and around San Pedro Harbor. In the case of the San Pedro Shelf, Eganhouse and Kaplan (1988) estimated that sewage effluent and local seepage contribute greater than 95% of the sedimentary hydrocarbons, with the former dominating in the upper 34 cm at one site.

Surface sediments in Santa Barbara Basin and those presently being deposited in San Nicolas Basin (note PIT sample from San Nicolas Basin in table 3.11) would also appear to contain detectable quantities of nonseepage (i.e., anthropogenic) hydrocarbons. Results for sediments from the Patton Escarpment and continental rise (Simoneit et al. 1979a) suggest that little if any anthropogenic petroleum reaches the continental slope. Thus, hydrocarbons introduced to coastal waters by humans heavily influence nearshore shelf and basin sediment chemistry but are not widely dispersed offshore.

Studies of the vertical distribution and accumulation rate of hydrocarbons in southern California sediments are limited (Eganhouse and Kaplan 1988; Emery 1960; Shokes and Mankiewicz 1979; Venkatesan et al. 1980; Venkatesan and Kaplan 1988, 1992). Venkatesan et al. (1980) presented geochemical data for age-dated box cores taken from San Pedro and San Nicolas basins (fig. 3.20). This was part of a larger study (Shokes and Mankiewicz 1979) that also included Santa Barbara and Santa Monica basin sediments. In the San Pedro Basin core (fig. 3.20a), the concentrations of total, aliphatic, and aromatic hy-

Table 3.11. *Ratio of Hopane (C_{30}) to 17α(H),18α(H),21β(H)-28,30-Bisnorhopane (C_{28}) in Sediments of the SCB*

Location	Sample Type[a]	Station/ Depth	C_{30}/C_{28} Ratio	Reference[b]
Santa Barbara Basin	SED	193/(surface)	2.2	1
		193/(25–31 cm)	0.8	1
	PIT		2.2	4
Santa Monica Basin	SED	380/(surface)	4.1	1
		380/(20–26 cm)	0.8	1
		BC113/(0–2 cm)	3.3	7
		BC102/(0–2 cm)	3.6	7
	PIT		5.2	4
			3.8	7
			3.2	7
San Pedro Basin	SED	823/(0–5 mm)	5.1	2
		823/(20–25 mm)	3.7	2
		823/(45–50 mm)	2.9	2
		823/(75–80 mm)	1.3	2
		823/(12–13 cm)	0.6	2
		372/(28–30 cm)	1.5	1
	PIT		5.2	4
San Pedro Shelf	SED	3C1/(0–2 cm)	5.3	3
		3C1/(2–4 cm)	5.0	3
		3C1/(4–6 cm)	5.3	3
		3C1/(6–8 cm)	5.3	3
		3C1/(8–10 cm)	6.7	3
		3C1/(10–12 cm)	5.0	3
		3C1/(12–14 cm)	5.3	3
		3C1/(14–16 cm)	6.7	3
		3C1/(16–18 cm)	5.0	3
		3C1/(18–20 cm)	4.8	3
		3C1/(20–22 cm)	3.2	3
		3C1/(22–24 cm)	2.4	3
		3C1/(24–26 cm)	1.9	3
		3C1/(26–28 cm)	2.4	3
		3C1/(28–30 cm)	1.6	3
		3C1/(30–32 cm)	1.6	3
		3C1/(32–34 cm)	2.0	3
		3C1/(34–36 cm)	0.4	3
San Nicolas Basin	SED	813/(15–20 mm)	1.8	2
		813/(75–80 mm)	0.5	2
		748/(28–32 cm)	1.4	1
	PIT		4.7	4

Table 3.11. *Ratio of Hopane (C_{30}) to 17α(H),18α(H),21β(H)-28,30-Bisnorhopane (C_{28}) in Sediments of the SCB* (continued)

Location	Sample Type[a]	Station/ Depth	C_{30}/C_{28} Ratio	Reference[b]
Tanner Basin	SED	575/(25–31 cm)	2.5	1
		YELS 5/(11–20 cm)	1.2	6
		YELS 5/(291–300 cm)	0.6	6
Outer Bank	SED	12G/(11–29 cm)	0.6	5
Continental Margin Rise	SED	4G/(0–10 cm)	0.9	5
		4G/(136–146 cm)	0.9	5
		3G/(118–128 cm)	1.0	5
		2G/(0–10 cm)	1.0	5
		PE/(5–20 cm)	1.0	5
Weathered California Oil	OIL		0.6	5
Redondo Seep	OIL		1.2	3

[a] Sample types: SED–sediment; PIT—particle interceptor trap; OIL—petroleum.

[b] References: 1—Simoneit and Kaplan 1980; 2—Venkatesan et al. 1980; 3—Eganhouse and Kaplan 1988; 4—Crisp et al. 1979; 5—Simoneit et al. 1979a; 6—Simoneit et al. 1979b; 7—Venkatesan and Kaplan 1992.

drocarbons exhibit a subsurface maximum (approximately 25 mm) and decrease continuously at greater depths, reaching apparent background levels at 12–13 cm, a depth estimated to correspond to the mid-1800s. Over this same depth interval, the percent unresolved complex mixture, total hydrocarbons: organic carbon ratio, hopane:bisnorhopane ratio, and concentration of total DDT also decrease dramatically, with the highest values confined to the upper 50 mm. The most rapid change in these parameters is observed at a depth dated at about 1945; the subsurface maximum in total hydrocarbon concentration occurs in the 1966–1968 period. Although temporal resolution is poor because of slower sedimentation in the basin, this profile clearly mimics the stratigraphic record established on the San Pedro Shelf (Eganhouse and Kaplan 1988) (see fig. 3.20b). In the latter case, downcore variations in the concentration of a number of constituents (including total hydrocarbon concentrations) have been correlated to historical changes in the quality and quantity of wastes discharged from the LACSD wastewater outfall system. These data indicate that human activities, especially the discharge of municipal wastes, have led to significant impacts on nearshore shelf and basin sedimentary environment over the last 40–50 years, resulting in enrichment of sedimentary hydrocarbon burdens over "background" levels by as much as two orders of magnitude.

Examination of core data from Santa Monica Basin (fig. 3.20c) indicates a record similar to that found in San Pedro Basin. San Nicolas Basin sediments (fig. 3.20d), in contrast, contain much lower concentrations of total hydrocarbons (HC) (approximately 22–100 $\mu g\ g^{-1}$; 0.5–1.7 mg HC g^{-1} OC) which exhibit a general decline with increasing depth. Variations in other measured parameters show near-constant concentrations in surface layers, suggesting the importance of bioturbation in these well-oxygenated sediments (Berelson et al. 1982; Berelson et al. 1987). Trace amounts of *o,p'*- and *p,p'*-DDE are restricted to surface layers, and hopane:bisnorhopane ratios are generally lower than those

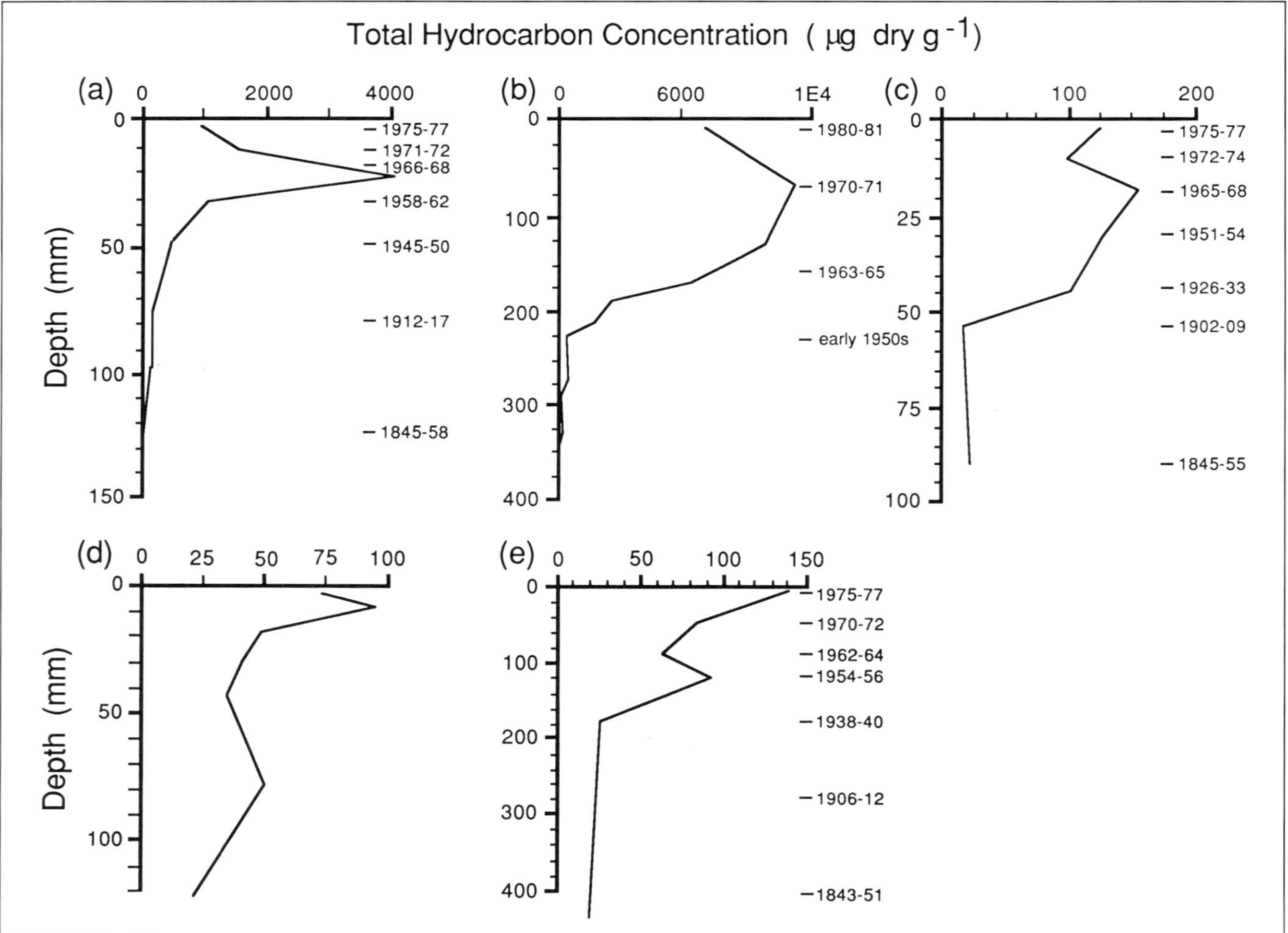

Figure 3.20. Vertical concentration profiles of total hydrocarbons in age-dated cores from southern California: (a) San Pedro Basin; (b) San Pedro Shelf; (c) Santa Monica Basin; (d) San Nicolas Basin; and (e) Santa Barbara Basin. (a, c, d, and e after Shokes and Mankiewicz 1979; b after Eganhouse and Kaplan 1988.)

found in San Pedro Basin sediments and nearer ratios characteristic of background seepage (table 3.11). In combination, these results suggest a lower degree of contamination by anthropogenic hydrocarbons, presumably the result of the greater distance of San Nicolas Basin from the primary sources.

Santa Barbara Basin sediments (fig. 3.20e) exhibit a nearly exponential decrease in total hydrocarbon concentration with depth (with the exception of the horizon corresponding to the 1954–1956 period). Although it is clear that a significant fraction of the hydrocarbons may be coming from anthropogenic sources (based on the hopane:bisnorhopane ratio) (table 3.11), the hydrocarbon profile does not appear to reflect the rapid changes in post-1971 anthropogenic hydrocarbon fluxes (such as the absence of a subsurface maximum) seen in other inner basin sediments.

The major sources of polynuclear aromatic hydrocarbons (PAHs) to the SCB are believed to be combustion of fossil fuels and petroleum or oil shales. The BLM and Department of Energy (DOE) studies revealed a complex assemblage of sedimentary aromatic hydrocarbons that are generally present in lower abundance than the saturated hydrocarbons, except in the northern Channel Islands region (Amit et al. 1980; Reed et al. 1977; Shaltiel et al. 1979; Venkatesan et al. 1980). Again, an unresolved complex mixture typically dominates the aromatic fraction in surface sediments, particularly in the nearshore zone. Examination of sediment cores from San Pedro and San Nicolas basins (Shokes and Mankiewicz 1979; Venkatesan et al. 1980) indicates that PAHs from both sources are present and that the more recently deposited sediments (that is, post-1900) have been influenced to a greater extent than deeper sections by contributions of specific compounds of pyrolytic origin, including pyrene, benzopyrene, and benzofluoranthenes.

Samples taken farther offshore from the Patton Escarpment and continental rise (Shaltiel et al. 1979) are dominated by phenanthrane, pyrene, and fluoranthene. Retene, an indicator of resinous higher plants (Simoneit and Kaplan 1980) (fig. 3.12, structure III), has also been found. Lower molecular weight species (naphthalene and phenanthrene) are accompanied by significant amounts of higher alkylated homologs (e.g., C_1-, C_2-, and C_3-). However, distributions of PAHs having four or more fused rings are dominated by the parent compounds, the latter again suggesting a pyrolytic origin. Together, these results demonstrate the effective offshore transport of terrigenous biogenic and pyrolytic PAHs either through the atmosphere or by currents. Superimposed on this signature is input of a chronic nature from oil seepage or shale erosion, which is probably largely responsible for the pervasive occurrence of the more labile lower molecular weight species.

Recent work by Anderson and Gossett (1987) in nearshore (mainland shelf, harbor, and river) sediments has shown that concentrations of total PAHs (43 individual compounds) range from approximately 0.15 to 16 $\mu g\ g^{-1}$ (13–570 $\mu g\ g^{-1}$ OC). Highest concentrations were found in the vicinity of municipal waste discharges and harbors and at the mouths of urban drainage channels. Cluster analysis of the stations on the basis of their PAH composition revealed three major groups representing different mixtures of PAH types. PAHs in sediments from the San Pedro Shelf are dominated by petroleum residues introduced by the discharge of municipal wastes (Eganhouse and Gossett 1991). Estimated accumulation rates of individual PAHs on this part of the shelf exceed those expected from atmospheric deposition by more than two orders of magnitude.

ACIDS AND ALCOHOLS

Analyses for fatty acids and other polar lipids were not carried out as a part of the BLM baseline survey. However, several investigators have reported results based on analyses of samples collected in the BLM and other programs (Eganhouse 1978; Simoneit and Kaplan 1980; Simoneit and Mazurek 1981;

Simoneit et al. 1979a, b; Venkatesan and Kaplan 1988, 1992). Generally, sediments in the SCB contain *n*-fatty acids at concentrations of 10–300 μg g^{-1} (1–13 mg g^{-1} OC). This range corresponds to surface and subbottom sediments collected from the edge of the outer continental shelf (Simoneit et al. 1979a) and subbottom sediments from the mainland shelf and offshore basins (Simoneit and Kaplan 1980). Concentrations of total fatty acids as high as 3300 μg g^{-1} have been reported for the San Pedro Shelf in the immediate vicinity of the LACSD outfall system (Eganhouse 1978).

The homolog distribution of *n*-fatty acids is always bimodal. Homologs $<C_{20}$ exhibiting a maximum at C_{16} originate from marine autochthonous sources, and those having carbon chains $>C_{20}$ with a maximum at C_{24} or C_{26} are derived from higher plant waxes. Monomethyl-branched fatty acids (*iso* and *anteiso*) having odd numbers of carbons from C_{13} to C_{19} are ubiquitous and reflect bacterial contributions. Isoprenoid acids with 15, 16, 17, 19, and 20 carbons have also been reported in Tanner Basin sediments (Hoering 1967a). These are most likely breakdown products of the phytol side chain of chlorophyll. Minor amounts of triterpenoidal acids (e.g., 17β(H),21β(H)-bishomohopanoic acid) (fig. 3.12, structure IV) presumably derived from bacterial cell membranes are typically observed, as are terpenoid acids indicative of terrestrial flora (e.g., dehydroabietic acid) (fig. 3.12, structure V). Because of their lability, only small amounts of unsaturated fatty acids are seen in surface sediments, the exception being sediments impacted by municipal wastes on the San Pedro Shelf (Eganhouse 1978).

Hoering (1967b) and Sever and Parker (1969) identified long-chain fatty alcohols in Holocene sediments of Tanner and San Nicolas basins, respectively. These compounds range in chain length from C_{12} to C_{28} with a pronounced even–odd carbon chain length preference. In both of these studies, bimodal distributions with maxima at C_{16} and C_{22} were found. The primary sources of these compounds are probably the lipids of zooplankton, especially calanoid copepods (C_{12}–C_{22}) (Boon and de Leeuw 1979; Sargent et al. 1976) and higher plant wax esters (Fukushima and Ishiwatari 1984; Tulloch 1976). Ikan et al. (1975b) reported both saturated and monounsaturated C_{22} and C_{24} *n*-alkanols in Tanner Basin sediments. They also found isoprenoid alcohols, phytol, and dihydrophytol, presumed to arise from hydrolysis of phaeophytin and dihydrophaeophytin in heat-treated sediments, although these same compounds were not found in the unaltered sediments. Wax esters have also been observed in Tanner Basin sediments (Fukushima and Ishiwatari 1984). The distributions of constituent saturated alcohol and acid moieties appear to be C_{12}–C_{30} and C_{14}–C_{30}, respectively, with maxima observed at C_{16} and C_{24} for each. The alcohol compositions are characterized by strong even–odd predominance and a greater abundance of the higher molecular weight ($>C_{20}$) homologs. This pattern most likely reflects the enhanced stability of wax esters associated with terrigenous debris when compared with those originating from autochthonous sources.

Until recently only limited studies have been made of cyclic alcohols in the SCB. Ikan et al. (1975b) detected a number of sterols, including 22-dehydrocholesterol, cholesterol, brassicasterol, Δ–7-ergosterol, campesterol, and β-sitosterol in sediments of Tanner Basin. Cholesterol, an ubiquitous constituent of eucaryotic organisms, was the dominant steroidal alcohol detected. Most of these steroids are probably of marine origin, although β-sitosterol has been used variously as a marker of terrestrial plant lipids despite its occurrence in some marine species (see Huang and Meinschein 1979; Smith et al. 1984).

Eganhouse (1979) reported the occurrence of a variety of stenols and stanols in LACSD effluent particles and sediments from the San Pedro Shelf. The dominant species was 5β(H)-cholestan–3β-ol, coprostanol, a specific product of the biohydrogenation of cholesterol in the intestinal tracts of mammals (fig. 3.12, structure VI). More recently, Venkatesan and Kaplan (1990, 1992) have used coprostanol to estimate the contribution of sewage-derived carbon to Santa Monica Basin sediments. Using both sediment trap data and analyses of sewage effluent and core sediments, they determined that the total waste-derived carbon in deeper basin sediments is approximately 2% of the total sedimentary organic carbon. Venkatesan and Kaplan (1990) demonstrated that the percentage of coprostanols ([coprostanol + epicoprostanol]/total sterols × 100) decreases with distance along a transect originating in Santa Monica Bay and terminating in Santa Monica Basin. This suggests dilution of waste-derived particles during offshore transport. Examination of sections from one age-dated Santa Monica Basin core revealed the existence of a subsurface maximum in total coprostanol content (as % total sterols) and rapidly diminishing concentrations at greater subbottom depths. The concentration maximum corresponded to a sediment horizon dated at about 1962, suggesting a loose relationship to the pattern of solids emissions for the LACSD outfall system off Palos Verdes (see chap. 12).

More recently, Venkatesan (1989) and Venkatesan et al. (1990) reported the occurrence of several pentacyclic triterpenols in sediments and sinking particles from Santa Barbara and Santa Monica basins. A series of 17β(H),21β(H)-hopanols (having 30, 31, and 32 carbons) are presumed to have originated diagenetically from bacteriohopanetetrol or to represent biosynthetic products of bacteria. Tetrahymanol appears to be of microbiological origin and most likely represents the biological precursor of gammacerane, a widespread constituent of oils and ancient sediments (Ten Haven et al. 1989; Venkatesan 1989).

CHLORINATED HYDROCARBONS

Polychlorinated biphenyls (PCBs) are a group of halogenated hydrocarbons comprising up to 209 individual compounds (congeners). Individual PCB congeners differ in the numbers and positions of chlorines attached to a biphenyl nucleus (Ballschmiter and Zell 1980). Consequently, they have widely varying physicochemical properties and toxicities. Owing to their thermal and chemical stability and favorable electrical properties, PCBs were synthesized from 1929 to 1977, mainly by Monsanto Chemical Company, in the form of several complex commercial mixtures (under the trade name Aroclor in the United States) for a wide variety of domestic and industrial uses. Inadequate safeguards in the manufacture, use, and disposal of PCBs led to their global distribution. In the late 1960s several severe poisoning incidents associated with PCB contamination led to the eventual ban on their production and usage in the mid–1970s (see references in Alford-Stevens 1986; Cairns and Siegmund 1981).

Although use of PCBs dates back to 1929, the presence of these compounds in the SCB was not appreciated until 1973–1974 when Young et al. (1975) developed an inventory of inputs from a variety of sources. The most important of these were municipal wastes (5400 kg yr^{-1}) and aerial fallout (1800 kg yr^{-1}). Within the space of a few years, it became evident that very high levels of PCB contamination existed in sediments deposited near submarine waste discharge systems (Young et al. 1975; Young and Heesen 1978; Word and Mearns 1979), urbanized bays, and harbors (see references in Chen and Lu 1974; Choi and Chen 1976; Eganhouse et al. 1990; Soule and Oguri 1980; Mearns et al. 1991). Subsequently, Word and Mearns (1979) reported results of a 1977 survey of surface (0–2 cm) sediments along the coastal shelf at the 60-m isobath. From these studies it was determined that most of the PCB contamination on the mainland shelf was concentrated in the vicinity of Los Angeles and San Diego (fig. 3.21a), with total PCB concentrations in San Pedro Shelf and Santa Monica Bay sediments reaching 10.9 and 0.5 $\mu g\ g^{-1}$, respectively. A 1985 study, in which many of the same sites were

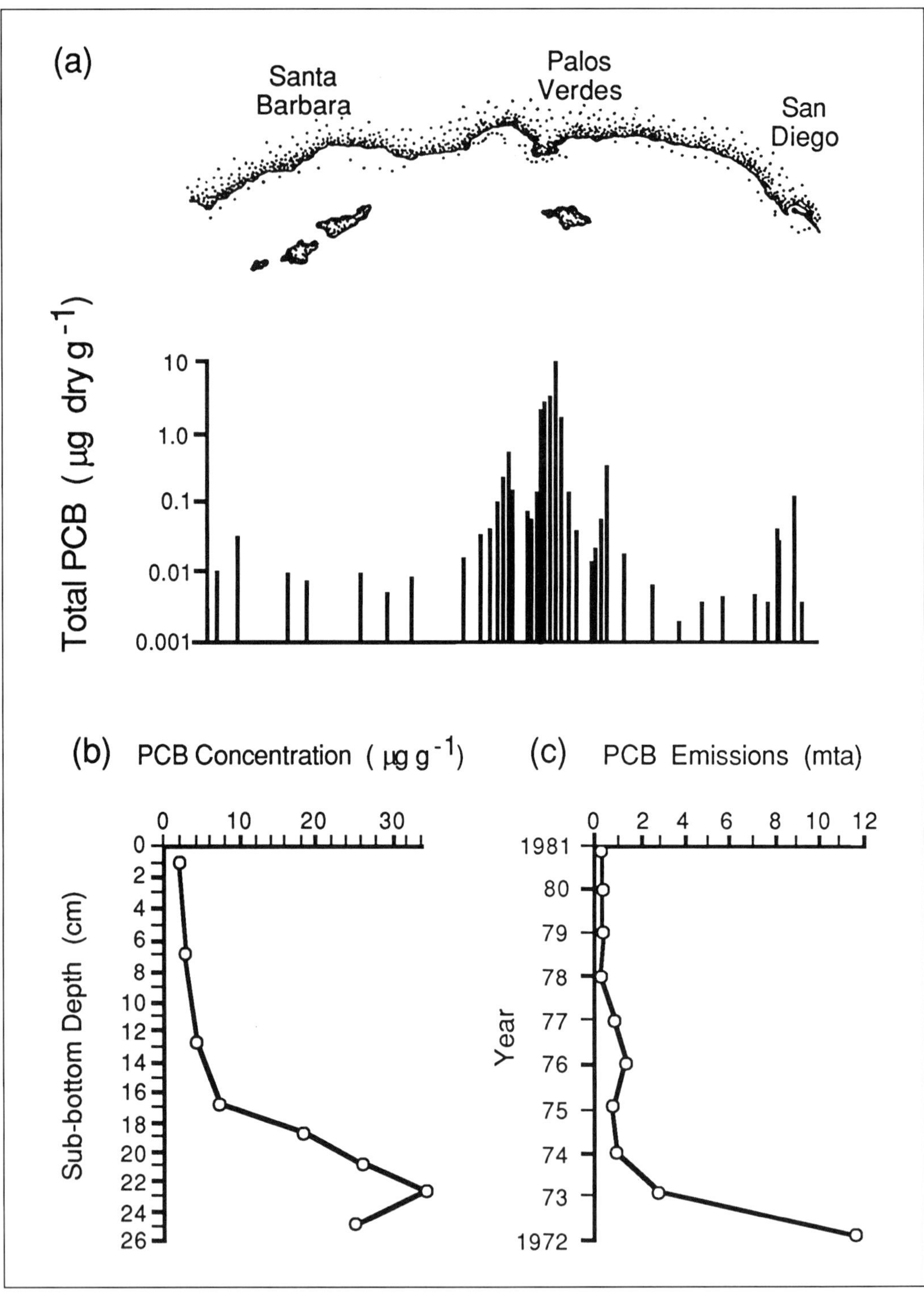

Figure 3.21. Polychlorinated biphenyls (PCBs): (a) concentrations (µg dry g^{-1}) in surface sediments along the 60-m isobath (Word and Mearns 1979); (b) vertical concentration profile in San Pedro Shelf sediments; and (c) mass emissions (t yr^{-1}) from LACSD. (b and c after Stull et al. 1988.)

resampled (Thompson et al. 1987), confirmed the persistence of these hot spots.

Young et al. (1975) showed that concentrations of total PCBs in sediments taken in 1971 from the San Pedro Shelf were highest within 2 cm of the surface (approximately 12 $\mu g\ g^{-1}$) and declined precipitously with increasing depth in the core. At that time, the inventory of total PCBs in the San Pedro Shelf sediments was estimated at 6 t. In a later study using cores collected in 1981, Stull et al. (1986, 1988) presented evidence that steady declines in PCB emissions from the LACSD waste treatment plant paralleled vertical PCB concentration profiles in nearby shelf sediments (fig. 3.21b). These changes have apparently resulted in the burial of the highly contaminated sediments (deposited in the early 1970s) at greater subbottom depths, such that there now exists a subsurface maximum in total PCB (and other trace contaminants) concentration approaching 40 $\mu g\ g^{-1}$ near the outfalls.

The only attempt to measure PCBs in offshore basin sediments was made by Hom et al. (1974), who reported finding concentrations of total PCBs of approximately 100 ng g^{-1} in surface sediments of Santa Barbara Basin dated about 1967. Age dating of the core by ^{210}Pb allowed Hom et al. (1974) to establish that PCBs began to accumulate in sediments around 1945 and continued to increase steadily with time to 1967. This agrees with national trends in the usage of PCBs (Summers et al. 1987) suggesting that peak accumulation rates probably occurred around 1970. Subsequently, Reed et al. (1977) and Venkatesan et al. (1980) have noted anecdotally the presence of PCBs in sediments of San Pedro and San Nicolas basins. This indicates that PCBs are transported offshore via the atmosphere and ocean currents. Young et al. (1975) reported atmospheric depositional fluxes of Aroclor 1254 ranging from approximately 7 to >35 ng $cm^{-2}\ yr^{-1}$ in the Los Angeles Basin during 1974. This range compares favorably with the estimated flux to Santa Barbara Basin sediments of approximately 12 ng $cm^{-2}\ yr^{-1}$ as of 1967 (Hom et al. 1974), indicating the eolian transport could play a significant role in the contamination of sediments deposited in offshore basins. Unfortunately, no bight-wide surveys have been carried out to determine the extent or magnitude of sediment contamination in regions beyond the mainland shelf.

Dichlorodiphenyltrichloroethane, or DDT, became the most widely used insecticide following World War II, when annual U.S. production reached a peak of approximately 10^8 kg in 1958–1960 (Eisenreich et al. 1989). Given the widespread application of this pesticide to soils and its inherent stability, DDT soon became globally distributed, and great concern arose about its environmental effects. In addition to the major isomer produced in the industrial syntheses, *p,p'*-DDT, smaller amounts of *o,p'*-DDT, and a series of metabolic degradation products (DDE, DDD, DDNU, and DDMU) (see fig. 3.12, structures VII–X) are commonly observed in contaminated environmental samples, the most abundant of which is usually *p,p'*-DDE. These compounds are also toxic and relatively persistent.

Southern California is unique as the site of the world's largest DDT manufacturing plant (Montrose Chemical Corporation), which operated for approximately 35 years. Wastes from this plant entered the influent stream of the LACSD from the early 1950s until 1971. There are no reliable estimates of the amount of DDT residues that may have been introduced to the treatment plant during this period because routine monitoring for DDT in influent and effluent streams did not begin until December 1970 (MacGregor 1974). Consequently, the amounts of DDT that entered the coastal waters off White Point, Palos Verdes, during the period of greatest inputs to the treatment system are unknown. Chartrand et al. (1985) has estimated that as much as 1800 t of DDT may have passed through the treatment plant. Mass emissions of total DDT from the LACSD treatment plant in 1971 were computed to be approximately

21.6 t, but with elimination of the source of DDT to the plant, major reductions in effluent emissions were accomplished in succeeding years (see chap. 12) (Smokler et al. 1979; Stull et al. 1988).

The discharge of these wastes led to the buildup of DDT residues in the sediments of San Pedro Shelf to such an extent that approximately 200–300 t now reside in a thin layer (10–100 cm) of sediments overlying relatively clean deposits laid down prior to the discharge of wastes at this site (MacGregor 1976; McDermott et al. 1974). Synoptic bight-wide surveys for DDT residues in surface sediments of the shelf (Thompson et al. 1987; Word and Mearns 1979; Mearns et al. 1991) consistently reveal that DDT contamination is widespread and that the concentration gradient is stronger to the south (fig. 3.22a). This indicates that the dominant dispersal mechanisms for effluent-derived DDT residues is advection by northward-flowing currents. Moreover, the distribution of contaminant isopleths in sediments on San Pedro Shelf and slope indicates that the LACSD plant was the major source of DDT (compare fig. 3.22b with either Word and Mearns 1979 or Mearns et al. 1991), although leakage of barged wastes in San Pedro Channel may be a possible second source (Chartrand 1988). In addition, it is clear that other municipal waste discharges have led to localized accumulations of DDT residues in sediments around the Los Angeles (Santa Monica Bay), Orange County (Newport–Huntington Beach), and San Diego (Point Loma) submarine outfall systems (McDermott et al. 1974).

Sediments from the outer basins have not been examined for the presence of DDT metabolites. However, *o,p'*-DDE and *p,p'*-DDE were observed in surface sediments of the Santa Barbara, Santa Cruz, and San Nicolas basins (Hom et al. 1974; Shokes and Mankiewicz 1979; Venkatesan et al. 1980) and shallower areas throughout most of the northern SCB (Mankiewicz 1981), reflecting its widespread dispersal, presumably by the Southern California Countercurrent (see chap. 2).

At least part of the DDT residues observed in seawater from the California Current could come from southern California metropolitan areas via the atmosphere (Cox 1971). If so, this would implicate southern California as a source of DDT for offshore regions to the south of the SCB. Young et al. (1976) demonstrated that aerial fallout of DDT could in fact account for an input of as much as 0.5–2.2 t yr^{-1} of DDT. This greatly exceeds the current combined emissions from all municipal waste discharges in the SCB, which are estimated at approximately 50 kg yr^{-1} (see chap. 12) (SCCWRP 1989). The study of Young et al. (1976) reported fluxes of total DDT in the vicinity of Santa Barbara Basin (that is, the northern Channel Islands) during 1973–1974 that were on the order of 1.4–5.0 ng cm^{-2} yr^{-1}. By comparison, Hom et al. (1974) estimated a flux of DDE to Santa Barbara Basin sediments of approximately 19 ng cm^{-2} yr^{-1} in 1967. Although these studies straddle the time when DDT inputs to the LACSD plant were terminated (1971), the comparison suggests that in the 1960s, transport of DDT to locations distal from the point source(s) was dominated by advection. Whether this remains true today is unknown.

Information on the accumulation rates of DDT in basin sediments is somewhat limited. As previously noted, Hom et al. (1974) estimated the accumulation rate of DDE in Santa Barbara Basin sediments to be approximately 19 ng cm^{-2} yr^{-1} in 1967. DDT first appears in the sediment horizons dated by ^{210}Pb at about 1952. Although Montrose Chemical Corporation began manufacturing DDT in 1947, it was not until 1953 that it received a permit to discharge its wastes to the LACSD treatment system. The coincidence of these dates further suggests that DDT fluxes to Santa Barbara Basin may have been dominated by advection of waste-derived particles.

MacGregor (1976), Venkatesan et al. (1980), and Eganhouse and Kaplan (1988) have also documented the contamination of San Pedro Basin sediments, where concentrations of total DDT reach 2–4 μg g^{-1} (approxi-

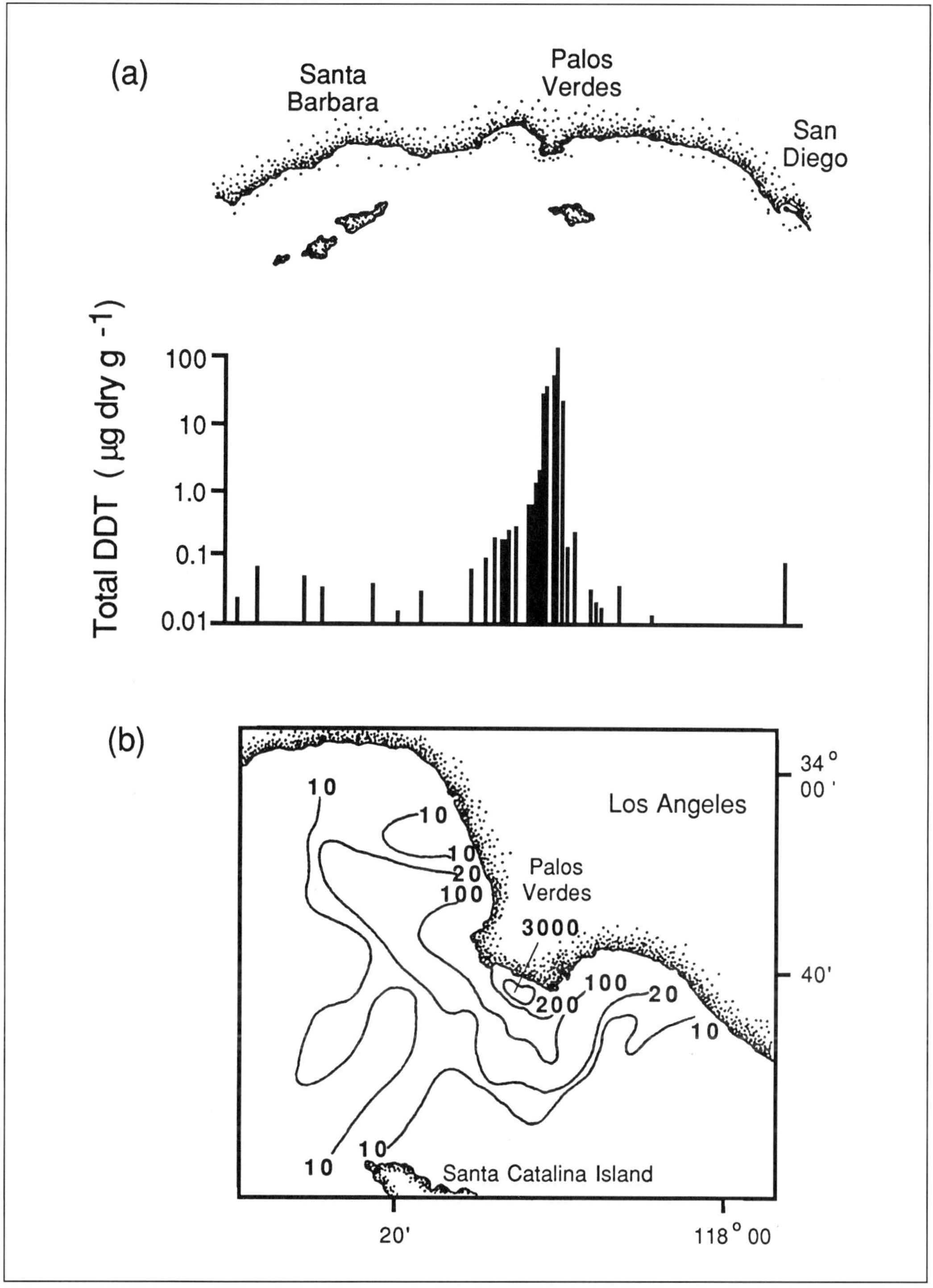

Figure 3.22. Total DDT in sediments of the SCB: (a) concentrations (μg dry g^{-1}) total DDT along the 60-m isobath (Word and Mearns 1979); (b) inventory total DDT (mg m^{-2}) in sediments off Santa Monica Bay–San Pedro Channel. (From MacGregor 1976.)

mately 70 μg g^{-1} OC) in surface layers. It was determined that DDT production wastes (comprising approximately 350–700 t of total DDT) were dumped into San Pedro Channel from 1947 to 1961 (Chartrand et al. 1985). Thus, emissions from the LACSD outfalls may not have been the only source of DDT residues to sediments of the San Pedro and Santa Monica basins. This may account for the observation of a second subsurface maximum in total DDT concentration in San Pedro Basin sediments dated at approximately 1958–1962 by Venkatesan et al. (1980). Moreover, because the acid wastes barged offshore contained significant quantities of undegraded DDT (as opposed to metabolites such as DDE), the higher *p,p'*-DDT/*p,p'*-DDE ratios reported by MacGregor (1976) for San Pedro Basin sediments when compared with those found on the nearby shelf may reflect contributions from dumped wastes as opposed to outfall-discharged wastes. This hypothesis has been forwarded by Risebrough et al. (1992), who recently presented evidence of high DDT concentrations in tar balls and tar cakes from historical dumpsites in San Pedro Basin. In these samples *o,p'*-DDT and *p,p'*-DDT isomers comprise 44–98% of the total DDT (total DDT = total DDTs, DDEs, DDDs, and *p,p'*-DDMU). Formerly, the abundance of undegraded DDT residues was attributed solely to differences in the availability of oxygen in shelf and basin environments.

Amino Acids and Lignin. Since Emery's (1960) anecdotal reference to amino acid analyses in sediments of Santa Cruz Basin, very little research on amino acids has been carried out in the SCB. Degens et al. (1964) reported the occurrence of 17 essential amino acids in sediments near Guadalupe Island, Mexico. Although these waters are not a part of the SCB, the results are probably applicable to this area. Degens et al. (1964) found approximately 350 μg g^{-1} of total amino acids at the water–sediment interface with high abundances of arginine, lysine, β-alanine, α-alanine, glycine, and proline. Acidic amino acids (glutamic and aspartic) were found in low abundance. Pollock and Kvenvolden (1978) reported stereoisomeric compositions of eight amino acids isolated from sediments deposited between Cortes and Tanner Banks. Their results showed that aspartic acid and alanine had elevated levels of the D-isomer (versus the L isomer) when compared with other amino acids and that their compositions were more like those found in soils. The authors suggested that this reflected bacterial contributions of these specific amino acids.

Virtually nothing is known about the carbohydrate composition of sediments in the SCB despite the fact that these compounds may comprise as much as 10–15% of the total sedimentary organic matter (Emery 1960).

Lignin is a complex cross-linked polymeric material biosynthesized by condensation of aromatic alcohols. It functions as supporting tissue for higher plants and is not found in marine phytoplankton. Relative to other biopolymers such as cellulose or proteins, it is more stable to microbial degradation. Consequently, it has proven useful as a molecular marker of terrestrial higher plants in coastal marine sediments (Hedges and Parker 1976; Hedges and Mann 1979). Chemical oxidation of lignin produces a series of aldehydes, alcohols, and acids, the distribution of which yields information about specific sources of lignin and its state of degradation. Studies using lignin oxidation products in the SCB are a recent phenomenon. Steinberg et al. (1984) presented data on the concentrations of lignin oxidation products in Tanner Basin kerogen. Only small amounts of lignin oxidation products were detected (approximately 0.001–0.3 mg g^{-1} OC). This is consistent with the hemipelagic nature of these sediments (Ikan et al. 1975a; Stuermer et al. 1978). More recent studies by Venkatesan and Kaplan (1992) have shown that approximately 4% of the organic carbon in Santa Monica Basin sediments can be attributed to lignin phenols.

Humic Substances and Kerogen. Humic substances include humic acids, fulvic acids, and humin. Humic acids are high molecular weight (2000 to >10^6) organic substances that form in both marine and terrestrial environments and are soluble in basic solution but are insoluble at pHs of <2. Fulvic acids are lower molecular weight (700 to 10^4), more highly functionalized compounds that are soluble in both acidic and basic solutions. Humin, sometimes referred to as protokerogen, is sedimentary organic matter that is insoluble in both basic and acidic media but has moieties that are hydrolyzable. It is considered to be an intermediate product of early diagenesis and an immature precursor to kerogen, a more highly aromatized, defunctionalized, refractory, carbon-rich material. Kerogen is believed to be the main source of petroleum that is released upon further burial (and heating) of sediments (see Tissot and Welte 1984).

According to the nomenclature of Emery (1960), compounds represented by humic substances + kerogen could comprise as much as 50% of the total organic matter of recent sediments from southern California. This could explain the high burial efficiency of organic carbon (27–70%) reported for five offshore basins by Jahnke (1990). Unfortunately, there have been no systematic studies of the abundance or chemical composition of humic substances and kerogen in southern California. The available data are distributed through reports from a number of research investigations, most of which are concerned with the thermal alteration products of these materials. Table 3.12 summarizes these data.

In spite of the limited nature of the data, it is clear that humic acids generally comprise a minor component of the sedimentary organic carbon pool in basin sediments (approximately 3–7%) when compared with the organic matter remaining after removal of lipids and humic-fulvic acids (i.e., kerogen, humin, or some mixture of these). The latter can represent as much as 36–72% of the total organic carbon. In these cases, the sediments were near-surface samples. This suggests that the kerogen was either a natural constituent of the biota (Goth et al. 1988; Philp and Calvin 1976) or it was formed rapidly during the early stages of diagenesis. As seen in table 3.12, the humic acids are more oxygen- and nitrogen-rich than the corresponding kerogens. However, the carbon and nitrogen isotope ratios of the humic acids are approximately equivalent to kerogens isolated from the same samples. This indicates that the kerogens may represent defunctionalized, more highly condensed coevals of the humic acids. It is possible that some of the "kerogen" may be relict material derived from erosion of older deposits (Miocene shale, for example). This may help to explain the unusually "old" ^{14}C dates found in surface sediments (Emery 1960; Williams et al. 1992). However, the size of the kerogen pool and its isotopic similarity to the humic acids argue for either an authigenic or direct biogenic (preformed) origin.

As expected, the stable isotopic compositions of the humic acids and kerogens reflect a dominantly marine origin. Kerogens from the San Pedro Shelf would appear to contain a significant terrigenous component, as indicated by their slightly heavier carbon and lighter nitrogen isotope ratios. These compositions are consistent with molecular evidence (Eganhouse and Kaplan 1988) of anthropogenic impacts in this area. Comparison of stable nitrogen isotope ratios for sinking particles and basin sediments (see table 3.12 and Crisp et al. 1979; Williams et al. 1992) indicates that the sediments are generally depleted in ^{15}N relative to the particles, whereas the carbon isotope ratios show better agreement.

Trace Metals

The distribution of heavy metals in sediments of the SCB has been studied extensively. Whereas some metals are known to be highly toxic to marine life (e.g., Hg, Pb, Cu, and Cd), others are viewed primarily as indicators of anthropogenic contamination (e.g., Cr, Se, and Zn), and their biological importance is either minimal or poorly understood. The existing database for Cd, Cr, Ni, Zn, Pb, and Cu is quite large, but much less

Table 3.12. *Elemental and Stable Isotopic Composition of Humic Acids and Kerogens Isolated from Sediments of the SCB*

	Elemental Composition						Isotope Ratios		
Location	%OC[a]	%C	%N	%H	H/C	N/C	δ¹³C	δ¹⁵N	References[b]
Humic Acids									
Santa Barbara Basin		54	4.5	6.5	1.6	0.14	–22.5	—	1
Santa Monica Basin	—	59	6.2	5.9	1.4	0.17	–27.4	—	1
San Pedro Basin	—	54	3.9	6.4	1.6	0.12	–22.7	—	1
	—	—	—	—	—		–22.4	8.1–10.2	2
Santa Cruz Basin	—	52	5.2	6.6	1.7	0.16	–21.8	—	1
San Nicolas Basin	—	—	—	—	—	—	–21.1	8.6–9.2	2
Tanner Basin	3	38	8.1	4.9	1.6	0.18	–20.6	9.1	3
	—	52	4.7	6.1	1.4	0.08	–22.0	—	1
	7	51	7.1	5.2	1.2	0.12	–21.7	—	4
Kerogen									
Santa Barbara Basin	—	—	—	—	—	—	–22.1, –22.3		6
	45	52	4.0	5.9	1.4	0.07	–22.4	—	5
San Pedro Basin	—	—	—	—	—	—	–22.2	8.5–9.6	2
San Pedro Shelf	—	—	—	—	—	—	–23.3, –22.2	3.8–6.8	6
San Nicolas Basin	—	—	—	—	—	—	–21.5	8.9–10	2
Tanner Basin	36	48	5.1	5.3	1.3	0.09	–21.4	7.2	3
	72	57	6.0	6.5	1.4	0.09	–22.4	—	4,5
	—	47–52	5.2–5.5	—	—	0.08–0.10	–21.5––21.6	7.2–7.3	7
	—	—	—	—	—	—	–21.4––22.3	5.7–9.9	6
	—	60	6.3	6.3	1.2	0.09	–21.5	8.1	8

[a] %OC—percent of organic carbon made up by specific organic fraction (e.g., kerogen or humic acids).

[b] References: 1—Nissenbaum and Kaplan 1972; 2—Venkatesan et al. 1980; 3—Stuermer et al. 1978; 4—Ishiwatari et al. 1977; 5—Ishiwatari et al. 1978; 6—Peters et al. 1978; 7—Minagawa et al. 1984; 8—Halpern 1981.

is known about As, Hg, Sn, Sb, Se, and Co (Katz and Kaplan 1981; Mearns et al. 1991). In some cases, the lack of information about the latter group stems from analytical difficulties.

By far most of the measurements have been made on sediments from the mainland shelf, particularly in the vicinity of major wastewater outfall systems and in harbors and bays. Many of these data are reported in the annual and biennial reports of SCCWRP as well as related publications (Eganhouse et al. 1976; Galloway 1979; Hershelman et al. 1981; Stull and Baird 1985; Thompson et al. 1987; Word and Mearns 1979). There have also been a number of investigations of basin sediments (Bruland et al. 1974; Chow et al. 1973; Murnane et al. 1989; Ng and Patterson 1982; Shaw 1988; Shokes and Mankiewicz 1979; Young et al. 1973), many of which were conducted under the auspices of the BLM baseline survey in the mid–1970s and more recently by the DOE-sponsored CaBS project (Finney and Huh 1989a,b; Huh et al. 1992). The only attempt to examine the bight-wide distribution of metals occurred during the BLM study, results of which are found in the final technical report (Chow and Earl 1979).

The reader should also be aware of two

major reviews of this subject. Katz and Kaplan (1981) discussed the literature developed during the 1970s and attempted to synthesize information on sediment distributions and correlate them with the sources of various metals. More recently, Mearns et al. (1991) reviewed available data up to 1990 for the mainland shelf and examined temporal trends for selected metals measured in various coastal surveys. In view of the enormous amount of information developed over the last two decades on heavy metal distributions in sediments, we present results from only a small number of the investigations.

Establishing the "background" concentration of a given metal in sediments of the SCB is complicated by variations in the sediment grain size distribution and the abundance of organic matter because these characteristics largely control metal content (Chow and Earl 1979; Thompson et al. 1987). Nevertheless, Katz and Kaplan (1981) attempted to estimate the background concentrations of heavy metals on the basis of measurements made on surface sediments from "clean" sites (Word and Mearns 1979; Chow and Earl 1979) and from deep sections of basin and mainland shelf cores whose deposition was known to predate anthropogenic influences (Chow et al. 1973; Bruland et al. 1974; Galloway 1979; Kettenring 1981; Shokes and Mankiewicz 1979). Table 3.13 summarizes these data and includes for comparison recent results presented by Thompson et al. (1987) for the mainland shelf.

The data of Word and Mearns (1979), Thompson et al. (1987), and Chow and Earl (1979) compare well despite the fact that vastly different sample digestions were used in the first two studies. Sediments analyzed in the Word and Mearns (1979) study came from a survey along the 60-m isobath, whereas the data of Chow and Earl (1979) represent sediments from the mainland shelf, inner and central basin, and outer bank and ridge environments. Data from Thompson et al. (1987) are for sediments taken at 13 of the same sites as Word and Mearns (1979) as well as 25 additional stations over shallower and deeper parts of the shelf (30 m and 150 m). It is noteworthy that the background concentrations developed from historical core sections are in all cases higher than those found in clean surface sediments. This probably reflects the influence of finer grain size and higher organic matter content of the basin sediments (Chow and Earl 1979).

Numerous studies have documented the accumulation of heavy metals in sediments deposited near the major wastewater outfall systems (Eganhouse et al. 1976; Galloway 1979; Hershelman et al. 1981; Stull and Baird 1985). These investigations have shown that concentrations of many heavy metals in waste-impacted shelf sediments exceed "background" concentrations by one to two orders of magnitude. Nevertheless, only a small fraction of the metals discharged to the mainland shelf from outfall systems appear to be stored in the sediments. Galloway (1979), Hendricks and Young (1974), and Huh et al. (1992) have independently estimated that more than 85% of waste-associated metals escape deposition on the Palos Verdes Shelf, either being released from particles into solution (Sweeney et al. 1980) or transported offshore in association with fine suspended matter.

The first evidence that inner basin sediments were also receiving significant amounts of trace metal contamination due to human activities came in 1973. Young et al. (1973) and Chow et al. (1973) reported increased rates of mercury and lead accumulation in age-dated sediments of the Santa Barbara, Santa Monica, and San Pedro basins. These authors noted a significant change in the flux of mercury at approximately the turn of the century; more rapid increases occurred in the 1940s. Similar results were reported by Chow et al. (1973), who estimated that as of 1970 anthropogenic fluxes of Pb exceeded natural fluxes by a factor of two. This shift was ascribed to contributions made by atmospheric deposition of lead derived from combustion of leaded gasoline as well as inputs

Table 3.13. *"Background" Concentrations of Various Heavy Metals in Sediments of the SCB*

Metal	Surface Sediments (μg dry g^{-1})			Core Base[b]
	Chow and Earl (1979)[a]	Word and Mearns (1979)[a]	Thompson et al. (1987)	
Ag	—	—	0.03	1.31
Cd	0.40	0.43	0.14	0.87
Cr	43.5	25.5	25.4	79.9
Cu	12.5	9.0	10.4	34.3
Ni	15.2	15.4	12.9	49.4
Pb	16.9	10.5	4.8	9.54
Zn	53.7	44.4	48.0	108.
Co	—	—	—	9.19

[a] Data from Katz and Kaplan (1981).

[b] Core base data represent means for bottoms of cores collected in offshore basin and marginal shelf environments (see Katz and Kaplan 1981).

from surface runoff and municipal waste discharges.

Subsequent studies (Bruland et al. 1974; Ng and Patterson 1982; Finney and Huh 1989a, b) substantiated these early observations and extended the chronology and number of trace metals determined. Figure 3.23 illustrates a series of age-dated box core profiles of stable Pb from the Santa Monica Basin corresponding to collections made in 1971, 1977, and 1986. These samples were taken in approximately the same area within the zone where anoxic conditions are found at the sediment-water interface, allowing for preservation of undisturbed varved sediment layers. The progressive development of a well-defined subsurface maximum corresponding to a date of approximately 1970 is evident. This subsurface maximum has been attributed to peak emissions of Pb from the LACSD treatment plant around 1971 and the usage of tetraethyl lead as an additive to gasoline during the late 1960s (Finney and Huh 1989a,b). Similar trends have been noted for other metals (Cr and Zn) and for organic carbon in the Santa Monica Basin as well as for sediments from the Palos Verdes Shelf (Stull et al. 1988; Olmez et al. 1991) and San Pedro Basin (Finney and Huh 1989b). These trends suggest that waste emissions from the LACSD and other treatment plants are having a significant impact on the heavy metal geochemistry of these inner basin sediments.

Shokes and Mankiewicz (1979) performed mild acid leaching as well as rigorous acid dissolution techniques for releasing metals from sediments. They found strong correlations between total and acid-leached metal concentrations, and they suggest that the mild acid leaching treatment yields primarily anthropogenic metals, the remainder being largely incorporated into the crystal lattice of detrital minerals. Ng and Patterson (1982) demonstrated that stable Pb in acid leachates and leached residues of recent sediments from three basins (Santa Monica, Santa Barbara, and San Pedro) are isotopically distinct. The leachates have $^{206}Pb/^{207}Pb$ ratios resembling those of atmospheric particles and tetraethyl lead, whereas the Pb isotopic ratios of leached residues are typical of Pb derived from marginal provenances in southern California. Ng and Patterson (1982) further demonstrated downcore variations in the isotopic composition of leachate lead that paralleled those of tetraethyl lead used in the United States. Inspection of fig. 3.23b clearly shows that the rise in Pb concentrations in Santa Monica

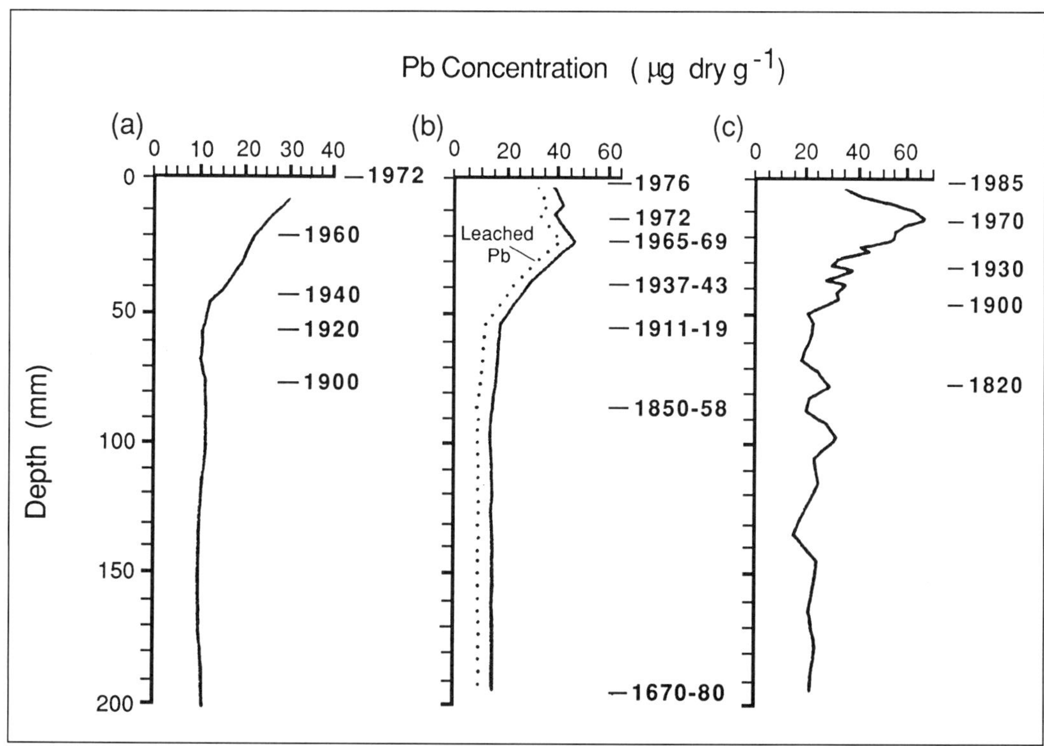

Figure 3.23. Comparison of lead (Pb) distributions in age-dated sediment cores collected from Santa Monica Basin: (a) 1971 (Bruland et al. 1974); (b) 1977 (Ng and Patterson 1982); and (c) 1986 (Finney and Huh 1989a).

Basin sediments through time is attributable to additions made to the leachable Pb pool. On the basis of differences between the historical lead flux rates and the assumption that atmospheric lead is removed to sediments via adsorption only to clay particles, Ng and Patterson (1982) calculated that the proportions of sewage-derived industrial lead being deposited in Santa Barbara, Santa Monica, and San Pedro basin sediments were 0, 66, and 75%, respectively.

Table 3.14 provides estimates of the sedimentary accumulation rates of anthropogenic and natural Pb, Cr, Cd, Cu, Ag, and Zn to Santa Barbara, Santa Monica, and San Pedro basins. Natural fluxes of all elements to Santa Barbara Basin exceed those in other basins because of the higher sedimentation rate. Estimates for the natural fluxes of Cr, Zn, and Pb made by the various investigators are in reasonable agreement, and the anthropogenic fluxes to Santa Monica Basin in 1970–1971 as determined by Bruland et al. (1974) are within a factor of two of peak fluxes (occurring about 1965–1970) reported by Finney and Huh (1989a). Data given by Shokes and Mankiewicz (1979) for pre- and post–1970 fluxes of Pb also fall close to those reported by Bruland et al. (1974). According to these data, it is clear that prior to 1970 the fluxes of Pb to the basins generally increased with time, enrichments due to anthropogenic additions being in the order San Pedro > Santa Monica > Santa Barbara. Following 1970, anthropogenic fluxes to Santa Monica Basin have declined from approximately 0.5–0.9 $\mu g\ cm^{-2}\ yr^{-1}$ (in 1970–1971) to 0.21 $\mu g\ cm^{-2}\ yr^{-1}$ (in 1986). This decline would, again, seem to reflect the decreasing inputs of wastewater lead and atmospheric lead known to have occurred during this same time.

The decline in mass flux of Cr to sediments

Table 3.14. *Estimated Fluxes of Selected Metals to Sediments in Inner Basins of the SCB*

Study	Flux Category[a]	Basins ($\mu g\ cm^{-2}\ yr^{-1}$)		
		Santa Barbara	Santa Monica	San Pedro
Chromium				
Bruland et al. (1974)	A—1970	2.9	2.6	3.1
	N	10.7	2.1	2.8
Finney and Huh (1989a)	A—1986		0.53	
	A—(1970–1971)		1.5	
	N		1.5	
Finney and Huh (1989b)	A—(1983–1987)			0.32
Zinc				
Bruland et al. (1974)	A—1970	2.2	2.1	1.9
	N	9.7	2.8	3.1
Finney and Huh (1989a)	A—(1983–1986)		0.002	
	A—(1970–1971)		0.84	
	N		1.9	
Cadmium				
Bruland et al. (1974)	A—1970	0.07		
	N	0.14		
Copper				
Bruland et al. (1974)	A—1970	1.4	1.1	1.4
	N	2.6	1.0	1.2
Silver				
Bruland et al. (1974)	A—1970	0.10	0.09	0.09
	N	0.11	0.03	0.05
Lead				
Bruland et al. (1974)	A—1970	2.1	0.9	1.7
	N	1.0	0.24	0.26
Shokes and Mankiewicz (1979)	A—pre-1970	1.01	0.61	1.2
	A—post-1970	0.19	0.88	1.55
	N	1.66	0.19	0.3
Finney and Huh (1989a)	A—(1983–1986)		0.21	
	A—(1970–1971)		0.53	
	N		0.18	
Finney and Huh (1989b)	A—(1983–1987)		0.10	

[a]Flux category: A—anthropogenic, N—natural background.

of Santa Monica Basin from 1970–1971 to 1986 (1.5–0.53 $\mu g\ cm^{-2}\ yr^{-1}$) is similar to that observed for lead. However, reductions in the flux of Zn are far more dramatic (0.8–0.002 $\mu g\ cm^{-2}\ yr^{-1}$, respectively). Again, these trends may reflect the effects of reduced emissions of municipal wastes, particularly from the LACSD, during this same time period. In support of this hypothesis is the reduction in metal concentrations noted for

Hg, Pb, Cr, Zn, Cd, Cu, and Se in surface sediments from the San Pedro Shelf since the early 1970s (Eganhouse et al. 1978; Mearns et al. 1991; Stull and Baird 1985).

Efforts to determine the origin and fate of trace metals to sediments in southern California are complicated by many factors. Katz and Kaplan (1981) reviewed the available data and discussed problems with application of simple two-source mixing models to metal deposition in shelf and basin sediments. Although the trace metal composition of surface sediments deposited in the immediate vicinity of the outfalls corresponds reasonably well with that of contemporary effluent particles from the LACSD (Bruland et al. 1974; Galloway 1979), the trace metal composition of the effluent has changed with time. Because sediments that are bioturbated or otherwise mixed by resuspension events integrate sedimentation events and changing redox conditions on time scales of years to decades, these temporal changes may lead to apparent differences between (surface) sedimentary and effluent compositions. Moreover, it has been observed that sedimentary compositions become increasingly dissimilar to the effluent pattern with greater distance from the outfall systems (Finney and Huh 1989a; Katz and Kaplan 1981). This has been attributed to (1) differences in the distribution of elements among different particle sizes (that is, fractionation by settling of different size particles) (Chen and Hendricks 1974; Katz and Kaplan 1981), (2) differences in the rates and magnitude of mobilization of the elements from suspended particles (Chen and Hendricks 1974; Faisst 1976; Finney and Huh 1989a; Kettenring 1981; Lu and Chen 1977; Sweeney et al. 1980), and (3) mixing of effluent particles from different treatment plants.

Recent work by Heggie and Lewis (1984), Shaw (1988) and Shaw et al. (1990) has also shown that the postdepositional distribution and fate of several metals (Ni, Co, Cu, Mo, V, and Cr) in offshore basins are determined by the redox-dependent cycling of manganese oxides. When bottom water oxygen concentrations are $\leq 10\ \mu M$, Mn is mobilized within sediments under reducing conditions and reprecipitated in oxidizing sediment horizons near the sediment–water interface (Froehlich et al. 1979). Hence, under the strongly reducing conditions found in bottom waters of the inner basins (e.g., Santa Monica Basin), Mn does not appear to accumulate appreciably and is probably transported offshore. Ni and Co are scavenged by manganese oxides, leading to enrichment of these metals in the oxic zone of sediments and enhanced accumulation. Consequently, inner basin sediments that do not preserve manganese oxides are more likely to release these metals into the overlying waters (Johnson et al. 1988). Cr, V, and Mo appear to reach the sediments in their reduced (insoluble) forms, perhaps in association with biogenic material. Of the three, only Cr is depleted in bottom waters, and this phenomenon is an indication that scavenging of dissolved Cr by sinking particles may be occurring. The accumulation of these metals is thus evidenced under reducing conditions, and if bottom waters approach anoxia, stripping from overlying seawater may result. Cu accumulation is apparently less sensitive to sedimentary redox conditions and more closely reflects the flux of biogenic particles. Finally, Shaw (1988) found significant seasonal variations in the redox conditions in offshore basin sediments (presumably in response to variation in surface water productivity). Thus, biological processes occurring in the water column may exert controls on the exchange of metals across the sea floor off southern California.

Early Diagenetic Processes

The inner basins in the northern portion of the SCB provide a nearly ideal setting for investigating early diagenetic transformations occurring in sediments. Suboxic conditions in deep bottom waters and the corresponding lack of significant bioturbation have led to conditions that approach steady-state diagenesis (Koide et al. 1972). At the same time, each basin environment is unique with regard to the types and amounts of inorganic and organic source materials being deposited in its sediments. Variations in physiography and oceanographic conditions also contribute to differences in biological activity within the sediments.

Following burial, sedimentary organic matter undergoes decomposition as a result of respiration by benthic fauna and heterotrophic microorganisms, primarily bacteria. When oxygen is available, aerobic respiration predominates and benthic macrofauna may be abundant. When oxygen is absent or available in limited amounts (approximately <20 μm), as in the case of water at subsill depths in Santa Barbara, Santa Monica, and San Pedro basins (Emery 1960; Shaw 1988), bacterial degradation predominates and metazoans are either not found or are present in small numbers.

Studies of the distribution of sediment bacteria and pore water profiles in cores from such environments have revealed vertical zonation of sediments which reflect specific diagenetic reactions (fig. 3.24). Such zonation represents the activities of a succession of bacterial communities utilizing different terminal electron acceptors over various depths in the sediments as the supply of each electron acceptor is exhausted. According to this scheme, the sequence of biologically mediated reactions proceeds in the direction of increasingly smaller free energy yield (Aller 1982; Claypool and Kaplan 1974; Shaw 1988) and leads to the buildup of specific metabolites in pore waters of different sediment layers (cf. Bender et al. 1989, among others). Changes in redox potential facilitates mobilization of chemical species which may have reached the sediments in oxidized form but are soluble under reducing conditions (e.g., Fe^{+2} and Mn^{+2}). The cycling of other species (e.g., Co, Ni, and Ge) is controlled by association with such oxide phases prior to or following sedimentation. The biogeochemical zones are not always resolved or fully developed as the dominance of one reaction at a given sediment horizon is determined by the overall sedimentation rate, biological and physical mixing rates, the porosity of the sediments, and the chemical composition of the overlying seawater (Leslie et al. 1990; Shaw et al. 1990). However, the simple model illustrated in figure 3.24 is a useful conceptual framework within which pore water profiles can be interpreted

As previously noted, the production of soluble metabolites via decomposition of labile organic matter leads to measurable changes in the chemical composition of interstitial waters (e.g., Brooks et al. 1968; Shaw 1988; Sholkovitz 1973). If the sedimentation rate is constant over time scales longer than that required to establish steady-state diagenesis, the flux of any chemical species across the sediment–water interface can be determined from measured concentration gradients in pore waters using suitable advection–diffusion–reaction models (e.g., Berner 1980). Knowledge of these fluxes is important to understanding the biological cycling of the elements and environmental controls of benthic respiration processes. Calculation of diffusive fluxes from pore water profiles is only one of several means by which fluxes of important chemical species (such as nutrients) to (or from) overlying waters can be determined. Other approaches include direct measurements using *in situ* benthic sampling chambers (Bender et al. 1989; Berelson and Hammond 1986; Jahnke 1990), incubation of recovered sediments at *in situ* temperatures (Jahnke 1990), modeling of water column properties (Berelson et al. 1987), and under appropriate conditions, changes in solid phase chemistry (percent organic carbon) (Martens and Klump 1984; Alperin et al. 1992).

Information on chemical fluxes across the sediment–water interface has been obtained only within the last decade, and the development of *in situ* sampling techniques is even more recent. In view of the large number of studies that have been performed in Santa Barbara Basin and the relatively incomplete database for other basin environments, we restrict ourselves to Santa Barbara Basin in the following discussion of early diagenesis.

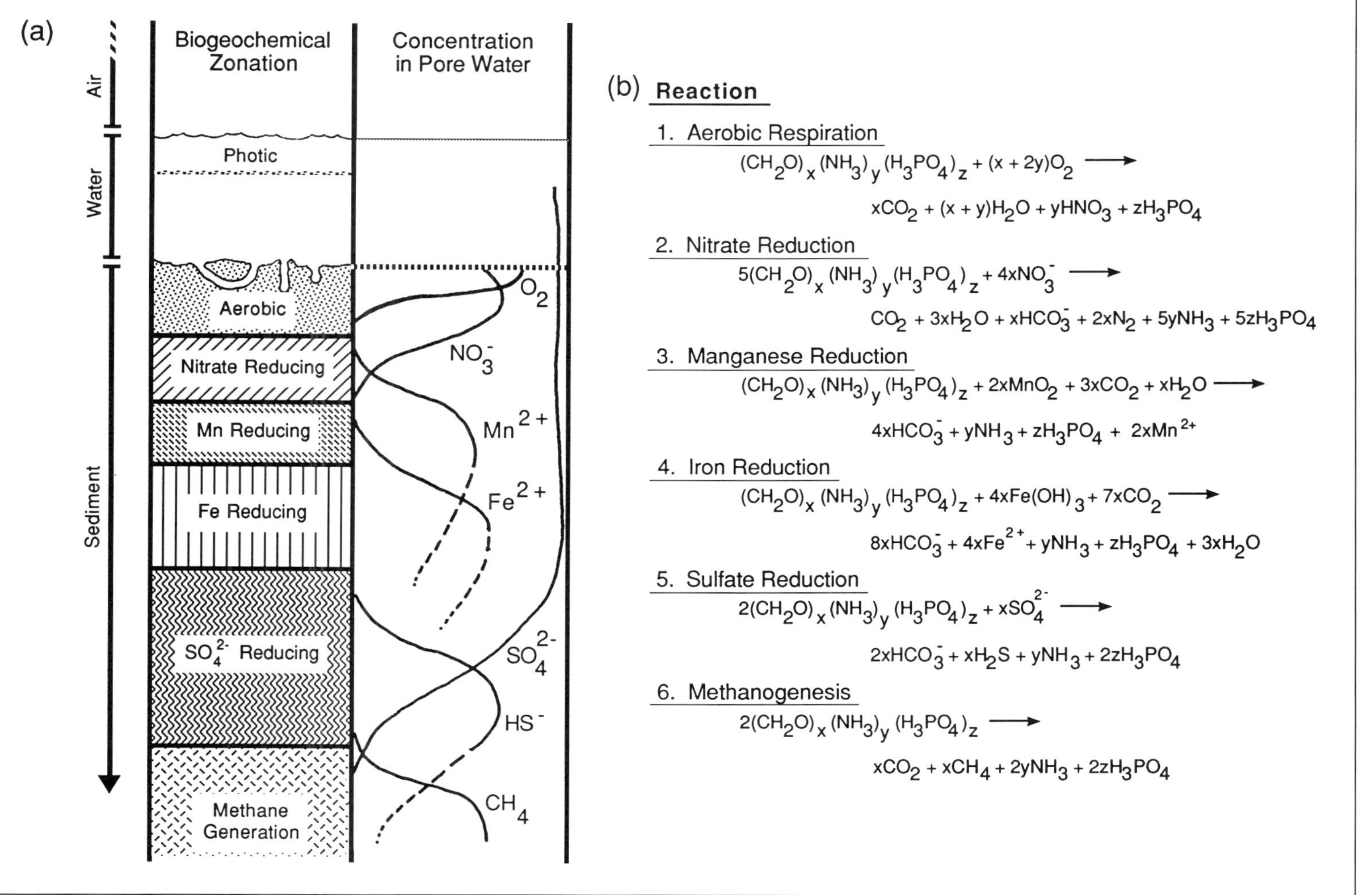

Figure 3.24. Conceptual model of (a) biogeochemical zonation of marine sediments and (b) idealized reactions associated with known metabolic pathways in marine sedimentary environments. (Modified from Aller 1982; Claypool and Kaplan 1974; Shaw et al. 1990.)

A subsequent section on chemical exchange across the sediment–water interface considers interbasin differences.

SANTA BARBARA BASIN: A CASE STUDY OF DIAGENETIC PROCESSES

Santa Barbara Basin has been the site of numerous investigations of pore water chemistry and early diagenesis (table 3.15). Early work demonstrated that pore waters from the deeper basin floor sediments were highly reducing as evidenced by negative values of Eh (approximately −90 to −150 mV) (Brooks et al. 1968; Kaplan et al. 1963; Rittenberg et al. 1955) found even in surficial layers of the sediment. Bottom waters in the basin are nearly anoxic (2.2–4.4 μM O_2) and active denitrification is known to occur (Liu 1979; Sholkovitz and Gieskes 1971). The surface of the sediments in the anoxic portion of the basin floor are commonly observed to be covered by a mat of colorless sulfur bacteria, *Beggiatoa* spp., and oxygen is absent from all but the upper few millimeters of the sediment (Reimers et al. 1990).

Nitrate and nitrite typically present at concentrations of approximately 15–40 and 0.1–0.4 μM, respectively, in deep basin waters (Barnes et al. 1975; Liu 1979; Rittenberg et al. 1955; Sholkovitz and Gieskes 1971; Sholkovitz 1973) were not found in sediments by Rittenberg et al. (1955). At the same time, pore water ammonium concentrations showed a consistent increase with depth (0.61–11.4 mM) down to 200 cm, whereas sedimentary nitrogen decreased over the same interval. These observations led Rittenberg et al. (1955) to conclude that denitrification within the sediments was not occurring and that deficits between the organic nitrogen lost diagenetically and ammonium accumulated in the pore waters could be accounted for by diffusion of ammonium from the sediments into the overlying waters. In a later study, Sholkovitz (1973) reported finding nitrate in the upper 1 cm of sediments. This is consistent with recent work (in which box-cored sediments were extruded at *in situ* temperatures under an inert atmosphere) demonstrating that nitrate is, indeed, detectable within the upper 2 cm of the Santa Barbara Basin sediments, thus signaling the importance of denitrification as a pathway for oxidation of recently deposited organic carbon (C. Reimers pers. comm. 1989). The discrepancy between these studies can be explained by the fact that the early investigations employed gravity coring devices that greatly disturbed the uppermost layers of the sediments where denitrification is most intense. Reimers employed a box corer capable of recovering relatively undisturbed surface sediments. Barnes et al. (1975) also suggested that nitrate and nitrite (as well as excess N_2) were present throughout the upper 10 cm of the Santa Barbara Basin sediments and that active denitrification accompanied by diffusion of nitrate into the sediments and N_2 out of the sediments was occurring. Based on the data of Barnes et al. (1975), Liu and Kaplan (1984) calculated a first-order rate constant for denitrification of 12 yr^{-1} (or 130 $\mu mol\ cm^{-2}\ yr^{-1}$) at the sediment–water interface of Santa Barbara Basin. This exceeds denitrification rate constants for hemipelagic sediments from other areas by two or three orders of magnitude, presumably driven by the high flux of metabolizable organic matter in Santa Barbara Basin sediments.

Sulfate reduction is a major respiratory pathway for the decomposition of metabolizable organic matter below the sediment–water interface in Santa Barbara Basin. Kaplan et al. (1963) suggested that sulfate reduction could account for nearly all of the organic matter decomposition over the upper 4 m of the sediment column (a reduction in organic carbon of 16% from that found in surface sediments). Sulfate concentrations decline in pore waters from 28 mM (Sholkovitz 1973) to effectively zero at depths below 200 cm (fig. 3.25). Early studies (Goldhaber and Kaplan 1973; Kaplan et al. 1963) indicated the presence of measurable amounts (albeit lower than seawater) of sulfate at depth (that is,

Table 3.15. *Investigations of Interstitial Water Chemistry in Southern California Basins*

Investigation	Basins[a]	Pore Water Constituents Analyzed[b]
Rittenberg et al. (1955)	SBB, SMB, SCtB	pH, Eh, NH_4^+, NO_2^-, NO_3^-, PO_4^{-3}, Si
Emery and Hoggan (1958)	SBB, SMB, SCtB	NH_4^+, N_2, Ar, CO_2, CH_4, VHC
Kaplan et al. (1963)	SBB, SMB, SCtB, SDT	Eh, pH, SO_4^{-2}, H_2S, ΣS^{-2}
Brooks et al. (1968)	SBB, SCtB, TB SCzB	pH, Eh, PO_4^{-3}, Na^+, K^+, Mg^{+2}, Ca^{+2}, Sr^{+2}, Cd, Co, Cu, Fe, Ni, Zn
Presley and Kaplan (1968)	SPB, SCtB	pH, Eh, Ca^{+2}, SO_4^{-2}, ΣCO_2, $\delta^{13}C_{CO_2}$, Fe, Zn, Co, Ni
Sholkovitz (1973)	SBB	pH, NH_4^+, NO_2^-, NO_3^-, PO_4^{-3}, Si, Cl^-, HCO_3^-, TAlk, ΣS^{-2}, SO_4^{-2}, K^+, Ca^{+2}, Mg^{+2}
Barnes et al. (1975)	SBB	NO_2^-, NO_3^-, N_2, Ar, CH_4
Bischoff et al. (1975)	SBB	Mg^{+2}
Kalil (1976)	SBB, SCtB, TB, ECB, SPS	pH, Eh, NH_4^+, PO_4^{-3}, TAlk, SO_4^{-2}, Ca^{+2}, Mg^{+2}
Warford (1977)	SBB, SPS, SMBy	pH, Eh, Es, SO_4^{-2}, H_2S, CH_4, CH_3COO^-, HCO_3^-
Doose (1980)	SBB, SPS	pH, Eh, NH_4^+, $\delta^{15}N_{NH_3}$, PO_4^{-3}, HCO_3^-, ΣCO_2, $\delta^{13}C_{CO_2}$, CH_4, $\delta^{13}C_{CH_4}$, H_2S, SO_4^{-2}, Ca^{+2}, Mg^{+2}, VOA
Sweeney and Kaplan (1980a)	SBB	NH_4^+, $\delta^{15}N_{NH_{4+}}$, SO_4^{-2}
Heggie and Lewis (1984)	SClB	Co, NO_3^-, Mn
Berelson et al. (1987)	SPB, SNB	pH, Eh, NH_4^+, PO_4^{-3}, ΣCO_2, Si, H_2S, Ca^{+2}, SO_4^{-2}
Smith et al. (1987)	SCtB	pH, TAlk, Si
Jahnke (1986, 1987, 1988, 1990)	SMB	pH, NH_4^+, PO_4^{-3}, Si, ΣCO_2, SO_4^{-2}, TAlk, Fe, Mn, F
Johnson et al. (1988)	SMB	Co, Cu
Shaw (1988); Shaw et al. (1990)	SMB, SNB, SCzB SClB, PE	Fe, Mn, Co, Cu, Cr, V, Mo, Si, Ni, NO_3^-, NO_2^-, O_2
Bender et al. (1989)	SClB	Mn, Si, NO_3^-, NH_4^+, PO_4^{-3}, O_2
Murnane et al. (1989)	SPB, SNB	ΣCO_2, Ge, SO_4^{-2}, PO_4^{-3}, Si, NH_4^+, Fe
Leslie et al. (1990)	SPB, SCtB, SNB	ΣCO_2, NH_4^+, H_2S, Fe^{+2}, PO_4^{-3}, SO_4^{-2}, Mg^{+2}, Eh, pH

[a] SBB—Santa Barbara Basin; SMB—Santa Monica Basin; SPB—San Pedro Basin; SCzB—Santa Cruz Basin; SCtB—Santa Catalina Basin; SClB—San Clemente Basin; TB—Tanner Basin; SNB—San Nicolas Basin; SDT—San Diego Trough; ECB—East Cortes Basin; SPS—San Pedro Shelf; SMBy—Santa Monica Bay; PE—Patton Escarpment.

[b] TAlk—titration alkalinity; VHC—volatile hydrocarbons; VOA—volatile organic acids.

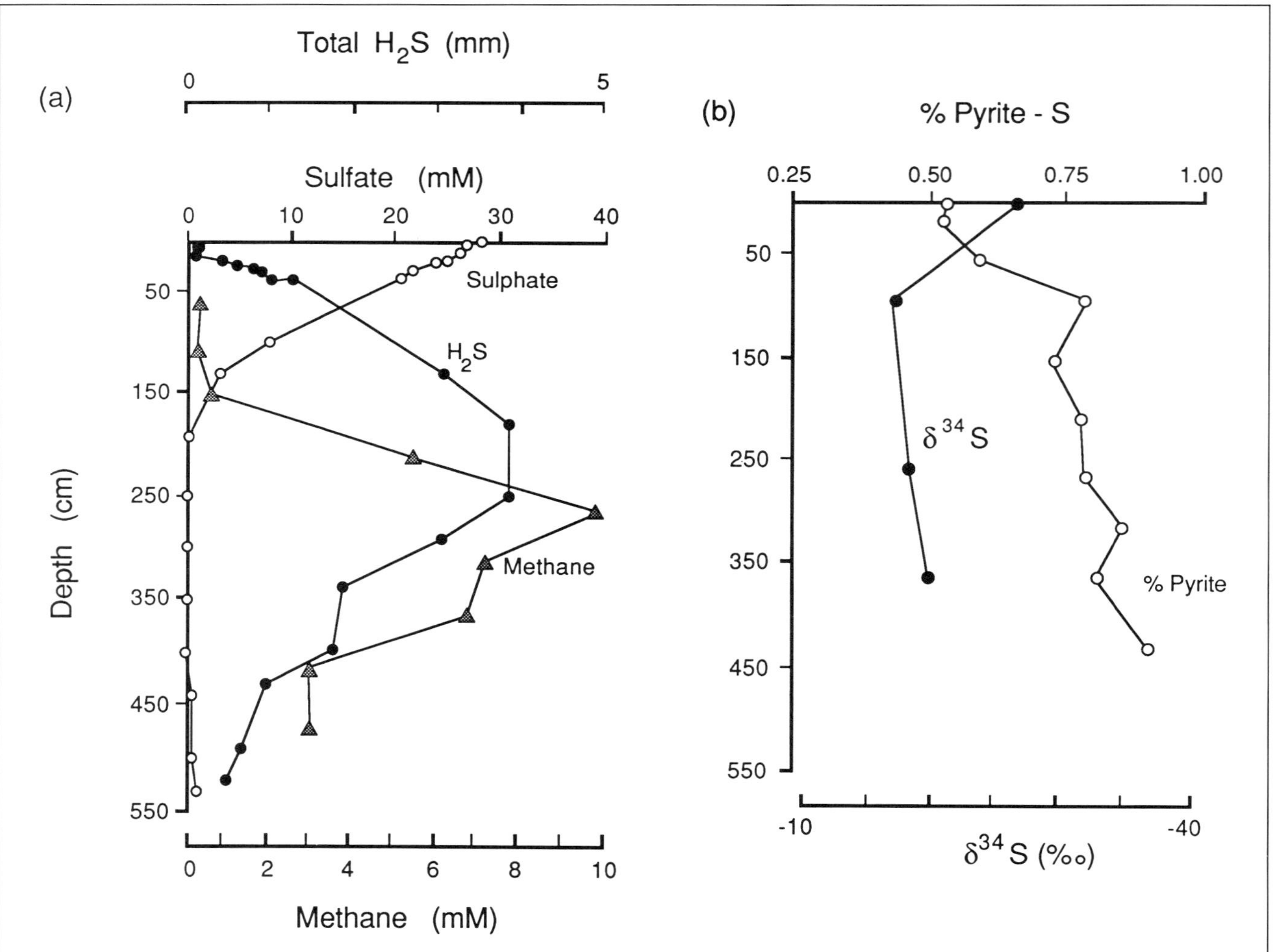

Figure 3.25. Distribution of pore water and solid phase constituents in Santa Barbara Basin sediments: (a) concentrations of dissolved sulfate, sulfide, and methane in pore waters (after Doose 1980) and (b) percent pyrite and $\delta^{34}S$-pyrite (after Kaplan et al. 1963).

>2 m); however, these results were probably artifacts attributable to the oxidation of hydrogen sulfide or pyrite during sample handling. Exhaustion of sulfate at depth indicates that the rate of sulfate removal exceeds the rate of supply. It does not necessarily signal that the system is closed with respect to diagenesis (Sholkovitz 1972).

Dissolved sulfide, the product of sulfate reduction (see reaction 5 in fig. 3.24b) (fig. 3.25), is found even in near-surface sediments (Doose 1980), with concentrations increasing with depth in the sediment column and reaching a maximum of approximately 4 m*M* at the 200-cm horizon (Doose 1980; Warford et al. 1979). (The absence of sulfide in surface sections noted by Goldhaber and Kaplan [1973] may again be related to oxidation effects during sample handling rather than *in situ* reaction with iron oxyhydroxide phases.) Estimation of sulfate reduction rates by modeling of sulfate pore water profiles under the assumption of closed system diagenesis yields a value of 5.9×10^{-4} mol l^{-1} yr^{-1}, which is intermediate in the range of estimates reported for a variety of marine environments (Goldhaber and Kaplan 1975). Sholkovitz (1973) constructed models to explain the downcore increases in ammonium and phosphate based on release of these nutrients via sulfate reduction. He found reasonable agreement between predicted and measured concentration profiles when sedimentary (not planktonic) C:N:P ratios were used, suggesting that sulfate reduction could account for the pore water inventory of these metabolites.

Kaplan et al. (1963) pointed out that the inventory of pyrite (FeS_2) exceeds that available as pore water SO_4^{-2} in the absence of additional diffusion of seawater sulfate into the sediments. Since pyrite is found in high concentrations (approximately 0.5–0.6%) even at the surface of the basin sediments, much of the pyrite formation must occur at or near the sediment–water interface, possibly in microniches. Sholkovitz (1973) found this to be true also for surficial sediments from the Santa Barbara Basin slope whose overlying waters contain greater amounts of oxygen (13–18 μM). The hypothesis that sulfate reduction occurs throughout the upper sediment column (down to approximately 2 m) was confirmed by stable isotopic analysis of the various sulfur species. For example, pyrite shows strong depletion of ^{32}S ($\delta^{34}S = -26.6$‰), whereas pore water SO_4^{-2} is nearer to, but slightly heavier than, seawater SO_4^{-2} (approximately +26.9 versus +20.4‰) as expected for the fractionation associated with sulfate reduction when a large pool of SO_4^{-2} is available (fig. 3.25b) (Goldhaber and Kaplan 1973). The isotopic composition of total sedimentary sulfur (−10 to −15‰) is also significantly depleted in ^{32}S relative to seawater sulfate, further suggesting the importance of sulfate reduction at the sediment–water interface where large quantities of sulfate can be made available. Increasing concentrations of pyrite with depth indicate that continued pyrite formation occurs following burial (fig. 3.25b). Similar trends have been found in other offshore basins (Leslie et al. 1990). At the same time, dissolved sulfide concentrations appear to reach a peak at approximately 250 cm below the sediment–water interface (fig. 3.25a). The decline, albeit irregular, in dissolved sulfide concentration with increasing depth signals further reaction of hydrogen sulfide with iron in deeper sections where a supply of sulfate no longer exists. The correspondence between declining sulfate concentrations and increasing dissolved sulfide and pyrite concentrations is indicated by the measurable enrichment of ^{34}S in the reduced sulfur species with increasing depth in the sediment column. For example, the isotope ratios of the dissolved sulfide and pyrite increase from −26.5 to +17.4‰ and −26.6 to −19.9‰, respectively, reflecting contributions from the isotopically heavier residual sulfate pool (fig. 3.25b).

Because the rate of sulfate reduction exceeds the supply of sulfate from overlying waters, dissolved SO_4^{-2} is exhausted within 2 m of the sediment–water interface (i.e., sulfate reduction is essentially complete at this depth). Methanogenesis (see fig. 3.24b, reaction 6) then becomes the thermodynamically favored metabolic process, whereby methane is produced via reduction of bicarbonate, disproportionation of acetate, or both. Emery and Hoggan (1958) were first to report the occurrence of biogenic methane in subsurface sediments of Santa Barbara Basin. Subsequent studies (Barnes and Goldberg 1976; Doose 1980; Kosiur and Warford 1979; Warford et al. 1979; Warford 1977) demonstrated that methane is present in small quantities (0.05–0.3 mM) in pore waters of sediments from the upper 150–200 cm where active sulfate reduction occurs (fig. 3.25a). With increasing depth below this zone, methane concentrations increase rapidly, reaching 10–12 mM at a depth of approximately 250 cm. At greater subbottom depths, methane concentrations decline in an irregular fashion. Thus, the sediments of Santa Barbara Basin appear to exhibit a distinct zonation with respect to methanogenesis and sulfate reduction.

Several hypotheses have been forwarded in attempts to explain the sharp zonation of freshwater and marine sediments into what appears to be mutually exclusive sulfate reducing and methanogenic zones. Among these are (1) inhibition of methanogenesis by sulfate reduction via production of toxic products (e.g., hydrogen sulfide), (2) competition for and availability of suitable substrates (such as H_2 and acetate) in the sulfate reduction zone, and (3) oxidation of methane by sulfate-reducing bacteria. The first of these, inhibition by production of toxic metabolites, seems to be problematic in the case of Santa Barbara Basin in view of the persistence of dissolved sulfide at high concentrations in the zone where methanogenesis is important (fig. 3.25a). Moreover, Warford et al. (1979) demonstrated that methanogenesis is stimulated, not inhibited, when sulfate is added to Santa Barbara Basin sediments collected from the base of the sulfate reduction zone. This strongly suggests that the products of sulfate reduction (such as HS^-), rather than inhibiting methanogens, may provide necessary substrates (HCO_3^-) for them.

Claypool and Kaplan (1974) proposed that the dominant substrate for methanogenesis in Santa Barbara Basin sediments is bicarbonate. This would appear to be the case according to experiments of Warford et al. (1979), who incubated basin sediments with 2-^{14}C-acetate and $H^{14}CO_3^-$. Additional field and laboratory evidence for a bicarbonate source was provided by Doose (1980), who observed steadily increasing $\delta^{13}C$ values for methane and HCO_3^- with depth throughout the zone of methanogenesis (150–500 cm) (fig. 3.26). Isotopic fractionation results when methanogenic bacteria preferentially utilize $H^{12}CO_3^-$, causing residual bicarbonate (and later-formed methane) to become increasingly heavy. Under these circumstances, the resulting isotope profiles exhibit the classic Rayleigh relationship (Hoefs 1987). The instantaneous fractionation factor ($\alpha_{methane\text{–}bicarbonate}$) determined from the Santa Barbara Basin profiles (1.07) was found to agree with measurements made using laboratory cultures of methanogenic bacteria from Santa Barbara Basin. Together, these facts indicate that substrate (that is HCO_3^-) limitation, rather than competition, is a more likely explanation for the reduced methane production rates observed by Warford et al. (1979) with increasing depth in the sulfate reducing zone.

Based on modeling of methane pore water profiles, Barnes and Goldberg (1976) concluded that consumption of methane in the sulfate reduction zone was occurring. This was later confirmed by Warford et al. (1979), who performed incubations of Santa Barbara Basin sediments with and without addition of ^{14}C-labeled substrates. They showed that methane production occurs at all depths in the sediments, whereas the rate of loss of methane (due to methane oxidation) is reduced in sediments below the sulfate reduction zone where high concentrations of methane occur naturally (Kosiur and Warford 1979). Methane production rates were found to range

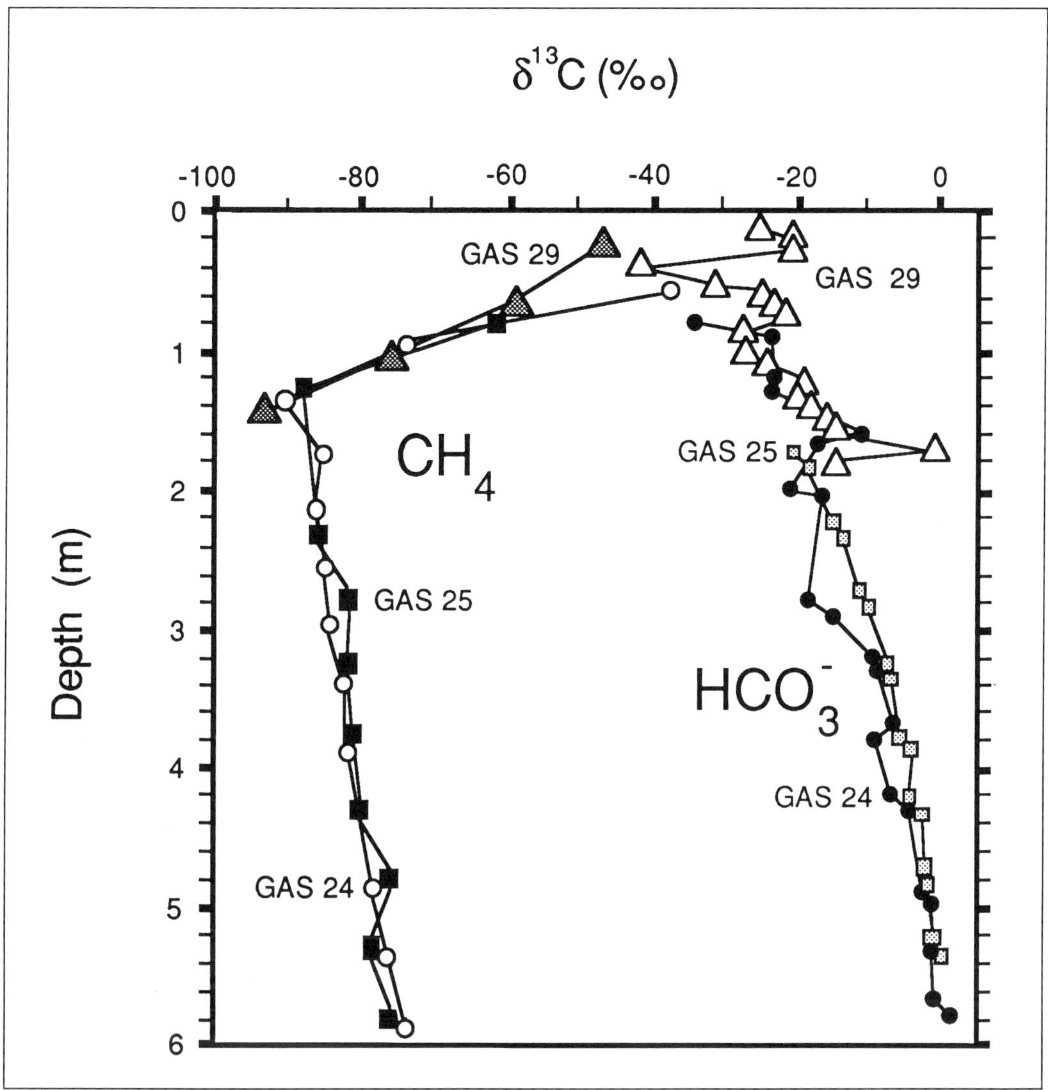

Figure 3.26. Vertical profiles of stable carbon isotope ratios ($\delta^{13}C$) for methane (CH_4) and bicarbonate (HCO_3^-) in pore waters of sediments from Santa Barbara Basin. (After Doose 1980.)

from 5 to >40 μmol yr^{-1} with oxidation rates reaching a maximum of 360 μmol yr^{-1} in the sulfate reducing zone (30–35 cm). Doose (1980) provided direct evidence of methane oxidation from field measurements of the stable carbon isotopic composition of methane and bicarbonate in interstitial waters. In the sulfate reduction zone, $\delta^{13}C$ values of methane and dissolved bicarbonate become heavier (from −93 to −37‰) and lighter (from 0 to <−22‰), respectively, than in deeper sediments (fig. 3.26). This is ascribed to methane oxidation whereby anaerobic consumption of methane by sulfate reducing bacteria leads to preferential utilization of $^{12}CH_4$ and the enrichment of ^{13}C in the residual methane pool. Meanwhile, bicarbonate becomes increasingly enriched in ^{12}C such that near the sediment–water interface, its isotopic composition becomes lighter than the sedimentary organic carbon (approximately −23‰).

Sholkovitz (1972, 1973) and Brooks et al. (1968) carried out studies of the inorganic geochemistry of Santa Barbara Basin pore waters. They reported decreases in the concentration of Ca^{+2} and Mg^{+2} and increases in alkalinity and in SiO_3^{-2}, PO_4^{-3}, and NH_4^+ concentrations with increasing depth (fig. 3.27). Sholkovitz (1973) developed diagenetic models for the upper 70 cm of Santa Barbara Basin sediments based on the assumption that sulfate reduction dominates the decomposition of labile sedimentary organic matter and that removal of calcium (as authigenic carbonate) and production of ammonium ion contribute to alkalinity. Comparison of model predictions for the downcore changes in alkalinity with the measured values showed good agreement. Mass balance calculations suggested the addition of approximately 0.5–1% (by weight) of authigenic carbonate to the solid carbonate pool. However, the depletion of Mg^{+2} in pore waters could not be attributed to diagenetic sulfate reduction or to other reactions strongly affecting alkalinity. Sholkovitz (1973) proposed that Mg^{+2} was removed by cation exchange at the surfaces of clay minerals whose exchange sites were exposed by dissolution of iron oxide–hydroxide coatings under the reducing conditions in Santa Barbara Basin sediments (see also Leslie et al. 1990). This hypothesis was later tested by Bischoff et al. (1975), who performed cation exchange capacity measurements on sediments from several basins in southern California. These investigators found that the Santa Barbara Basin sediments had higher Mg^{+2} cation exchange capacity values than sediments from the Santa Cruz, Santa Catalina, and San Nicolas basins. They postulated that the production of H_2S in Santa Barbara Basin sediments provided the necessary sink for Fe^{+2} mobilized under highly reducing conditions. A flux of Mg^{+2} into sediments of 10 μmol cm^{-2} yr^{-1} was proposed on the basis that changes in the exchangeable Mg^{+2} concentrations on the sediments greatly exceed the pore water supply.

SEDIMENT–WATER EXCHANGE

Table 3.16 presents published data for the exchange of nutrients and other chemical constituents between sediments and overlying waters for several offshore basins and the Patton Escarpment. The estimates come from studies that employed *in situ* benthic chambers, modeling of pore water concentration profiles, modeling of water column property gradients, sediment incubations, and basin water mass balances. Only a small number of offshore basins have been studied, and estimates of benthic fluxes derived by more than one method exist only for Santa Monica, San Pedro, Santa Catalina, San Clemente, and San Nicolas basins.

From the limited data available, it is clear that there is some variation (generally within a factor of two or three) in the oxygen fluxes estimated for a given basin. This variability reflects uncertainties and differences in the analytical methods used by various investigators, possible difficulties with the assumptions and mixing coefficients employed in calculations for modeling of pore water and water column profiles, and temporal and spatial variations of measured fluxes. Differentiating among the sources of variation is difficult.

Work by Berelson et al. (1987) in San Nicolas Basin indicates that spatial and temporal variations in oxygen fluxes within a basin may be relatively small. They found good agreement between directly measured benthic fluxes (*in situ* chamber) and those estimated from water column properties (table 3.16). Because the water column standing crop integrates over longer time periods and larger areas than waters enclosed in a benthic chamber (3–5 days; 730 cm^2), the similarity of the flux estimates determined by these two methods suggests that benthic oxygen exchange rates are not likely to have experienced large excursions with time. Bender et

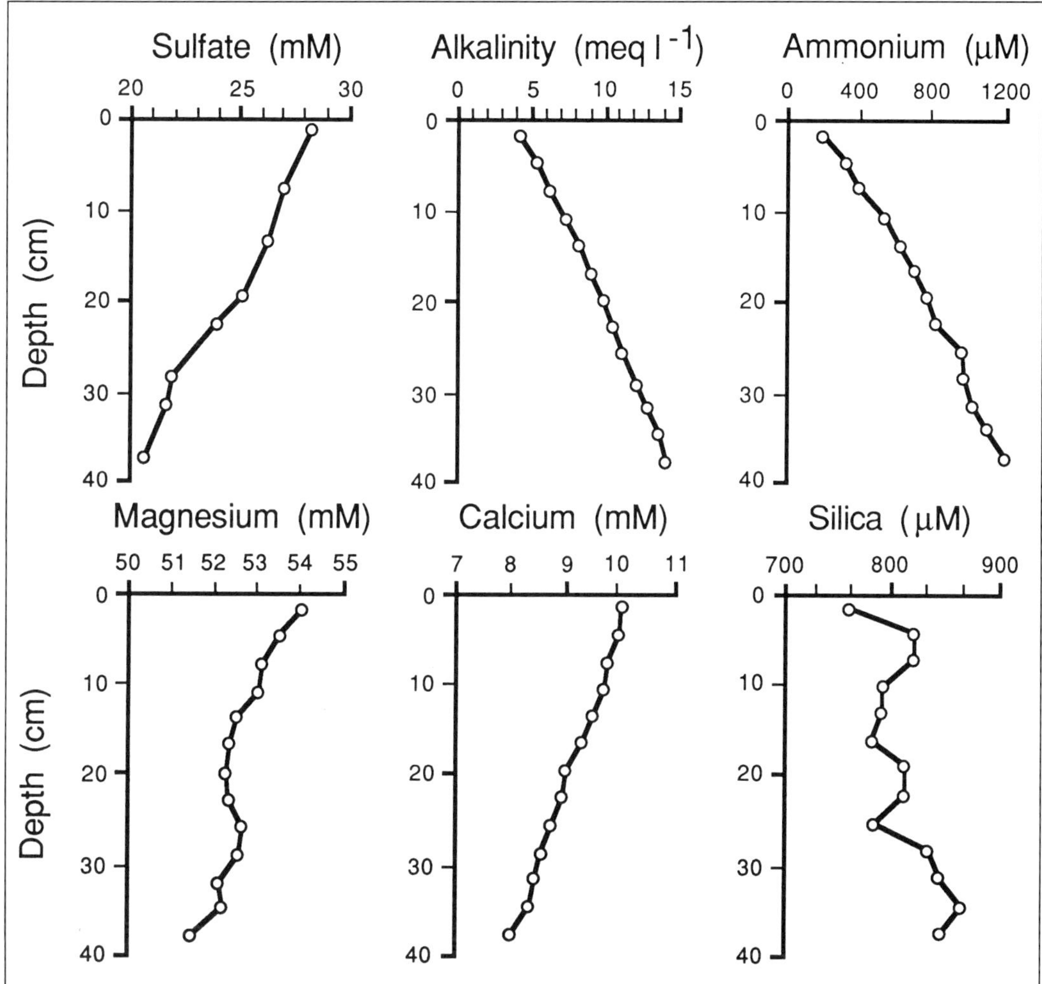

Figure 3.27. Distribution of inorganic pore water constituents in Santa Barbara Basin sediments. (After Sholkovitz 1973.)

al. (1989) found little difference between oxygen fluxes determined from benthic chamber experiments conducted in San Clemente Basin in January and April 1986. In contrast, Smith and Baldwin (1984) report variations in benthic oxygen uptake for sediments from the base of the Patton Escarpment ranging up to a factor of four during different seasons. Reimers and Smith (1986) suggest that seasonal variations in the rate of oxygen consumption in sediments of the Patton Escarpment may be related to productivity of the overlying waters. (It is important to note that this site is not within a basin and may be subject to greater variation with respect to oxygen flux due to spatial heterogeneities [Reimers 1987].) In the course of more extensive investigations at five stations along a transect from the central Pacific to Santa Catalina Basin, Smith (1987) documented statistically significant temporal variations in benthic oxygen consumption. At the latter (basin) site, however, the maximum range of oxygen consumption rates was only 2.6–3.4 mmol O_2 m^{-2} d^{-1}. Discrepancies between the depth of pore water manganese removal and solid

Table 3.16. *Fluxes of Chemical Constituents Across the Sediment–Water Interface in Basins of the SCB*

Basin[b]	Chemical Fluxes (mmol m^{-2} d^{-1})[a]								
	O_2	NO_3^-	SO_4^{-2}	NH_4^+	ΣCO_2	PO_4^{-3}	SiO_3^{-2}	Alkalinity[c]	Radon Flux[d]
SMB[e]	(–0.36 ± 0.18)	(–1.1 ± 0.31)	–0.4	—	(2.7 ± 1.33)	(0.097 ± 0.046)	(1.47 ± 0.31)	(2.15 ± 1.04)	—
SMB[f]	–0.6	—	(–0.4 ± 0.08)	—	—	—	(1.9 ± 0.5)	—	—
SPB[g]	–0.5	–0.7	–0.22	—	2.0	—	0.7	1.8	63
SPB[h]	—	–0.8	—	<0.03	1.9	15	0.7	1.7	86
SPB[i]	—	—	—	0.09	—	110	0.6	—	63
SPB[j]	–1.8	–1.5	—	—	—	—	—	—	—
SCtB[k]	(–2.6––3.4)	–0.2	—	0.01	—	—	—	—	—
SCtB[l]	(–0.5 ± 0.2)	—	—	—	—	—	—	—	—
SNB[g]	–1.0	–0.8	–0.13	—	2.9	—	1.1	1.6	490
SNB[h]	–0.7	–0.4	—	<0.03	1.7	17	0.7	1.7	490
SNB[m]	–0.25	—	—	—	—	—	—	—	—
SDT[n]	–2.4	0.3	—	0.53	—	0.01	—	—	—
SClB[o]	(–0.5––1.3)	(–0.06––0.17)	—	—	—	—	(0.3–0.8)	—	—
SClB[p]	(–1.4 ± 0.4)	(–0.02––0.13)	—	—	—	—	(0.3–1.0)	—	—

[a] Fluxes into sediments indicated as negative values.

[b] Basins: SMB—Santa Monica Basin; SPB—San Pedro Basin; SCtB—Santa Catalina Basin; SNB—San Nicolas Basin; SClB—San Clemente Basin; SDT—San Diego Trough.

[c] Alkalinity in mequivalents m^{-2} d^{-1}.

[d] Radon flux in atoms m^{-2} s^{-1}.

[e] Benthic chamber, shipboard incubation of sediments (Jahnke 1990).

[f] Pore water modeling (Jahnke 1990).

[g] Water column modeling (Berelson et al. 1987).

[h] Benthic chamber (Berelson et al. 1987).

[i] Pore water modeling (Berelson et al. 1987).

[j] Hydrographic mass balance (Jackson 1982).

[k] Benthic chamber (Smith et al. 1983; Smith 1987).

[l] Modeling of *in situ* pore water O_2 microprofiles (Reimers 1987).

[m] Benthic chamber (Berelson and Hammond 1986).

[n] Benthic chamber (Smith et al. 1979).

[o] Benthic chamber (Bender et al. 1989).

[p] Pore water modeling (Bender et al. 1989); modeling of *in situ* pore water O_2 microprofiles (Reimers 1987).

phase manganese enrichment in sediments of several basins led Shaw (1988) to conclude that changes in the redox conditions do occur, possibly within a seasonal time frame. Such changes would be expected to result in temporal variations in the flux of oxygen into sediments. To address the question of spatial variability, Smith et al. (1983) measured oxygen consumption rates at three different stations in Santa Catalina Basin within a period of 60 days during the months of January and February 1979. The rates could not be distinguished statistically (2.5–2.9 mmol O_2 m^{-2} d^{-1}). Reimers et al. (1986) suggested that intrasite variations in the flux of oxygen into sediments of Santa Catalina Basin and the base of the Patton Escarpment can arise from sediment heterogeneities on the scale of millimeters to centimeters, further highlighting the difficulty of comparing benthic flux estimates at a given site.

In spite of the limited size of the database and the difficulties just elaborated, some trends are evident. Estimates for the flux of oxygen into sediments of different basins range from 0.25 to 3.4 mmol m^{-2} d^{-1}. The highest fluxes are approximately one order of magnitude greater than oxygen consumption rates measured at the sea floor in the oligotrophic central North Pacific (approximately 0.1–0.4 mmol m^{-2} d^{-1}) (Reimers 1987; Smith et al. 1983) or those estimated for pelagic Atlantic and Pacific sites (<approximately 0.3 mmol m^{-2} d^{-1}) (Bender and Heggie 1984). This undoubtedly reflects the higher productivity of SCB waters. Berelson et al. (1987) suggest that the magnitude of benthic oxygen fluxes varies directly with basin water oxygen content (tables 3.16 and 3.17). A different conclusion was reached by Smith et al. (1983) for a transect originating in the eastern Pacific and terminating in Santa Catalina Basin. Smith et al. (1983) developed an equation to describe benthic oxygen demand in which the dependent variables were water depth and macrofaunal abundance. It is most likely that benthic oxygen flux is a complex function of organic matter supply and character (degradability), oxygen availability, sedimentation rate, and benthic community composition. The relative importance of these factors in controlling organic carbon remineralization, and hence oxygen utilization, are a continuing subject of controversy (Emerson 1985; Jahnke 1990; Reimers 1989).

Smith et al. (1987) attempted to estimate the contributions made by different groups of organisms to oxygen consumption in the benthic boundary layer (a region they defined as encompassing the overlying 100 m of seawater and basin sediments) of Santa Catalina Basin. They were able to demonstrate that approximately 90% of the oxygen demand could be accounted for by microorganisms inhabiting the sediments, particularly the near-surface layers. However, the flux of particulate organic carbon to deep basin waters as measured by sediment traps accounted for only 17–43% of that estimated to be remineralized by the benthic boundary layer biological communities. Later work (Smith 1987) confirmed this deficit for five stations with periodic sampling over a time span of several years. The discrepancies suggest difficulties with the methods used to measure vertical fluxes and/or benthic respiration or the existence of other, as yet unknown, sources of organic matter to the benthos.

Ultimately, the consumption of organic matter at the sea floor is reflected in the demand for all available electron acceptors under the conditions that exist at and below the sediment–water interface. Thus, the inner basins, whose waters are relatively oxygen poor, sustain benthic metabolism to a significant extent under suboxic or anoxic conditions. Central and outer basin waters are deeper and more oxygenated and their sediments are subject to bioturbation (Berelson et al. 1987; Reimers 1987; Smith and Hamilton 1983; Smith et al. 1987) with the result that organic remineralization in surface sediments is largely dominated by aerobic processes. Thus, the importance of nitrate reduction and sulfate reduction as metabolic pathways versus aerobic destruction (O_2 as electron

Table 3.17. *Organic Carbon Remineralization Rates (mmol C m^{-2} d^{-1}) for Different Metabolic Pathways in Sediments of Santa Monica, San Pedro, San Nicolas, and San Clemente Basins*

Basin[a]	O_2 Concentration Basin H_2O[b]	OC Mineralization Rate[c]				Vertical Flux Organic Carbon[g]
		O_2	NO_3^-	SO_4^{-2}	Total	
SMB[d]	13	–0.36 (14)	–1.4 (55)	–0.8 (31)	–2.56	2.3–3.9
SPB[e]	9	–0.50 (27)	–0.88 (48)	–0.44 (24)	–1.82	3.0–3.6
SNB[e]	22	–1.0 (44)	–1.0 (44)	–0.26 (12)	–2.26	1.0–1.4
SClB[f]	56	–0.85 (70)	–0.2 (17)	–0.16 (13)	–1.21	ND[h]

[a] SMB—Santa Monica Basin; SPB—San Pedro Basin; SNB—San Nicolas Basin; SClB—San Clemente Basin.

[b] Concentration of oxygen (mM^{-1}) in deep basin (subsill depths) waters. Sources: Emery (1960); Bender et al. (1989).

[c] Organic carbon mineralization rates (mmol C m^{-2} d^{-1}) for specific metabolic pathways; negative values indicate consumption of organic carbon. Stoichiometry given by reactions in figure 3.24. Numbers in parentheses represent percentages of combined mineralization rates represented by individual pathways.

[d] Jahnke (1990).

[e] Berelson et al. (1987).

[f] Bender et al. (1989).

[g] Based on sediment trap data presented in table 3.9.

[h] Not determined.

acceptor) of sedimentary organic matter varies from basin to basin. (Methanogenesis has purposely been excluded from this discussion because, except in Santa Barbara Basin, early sediment diagenesis in the basins appears to be dominated by reduction of O_2, NO_3^-, Mn^{+4}, Fe^{+3}, and SO_4^{-2}.) Cycling of manganese is difficult to address quantitatively insofar as the identities of oxidant(s) involved in reactions with Mn^{+2} are not known (Shaw 1988). Except for Santa Monica and Santa Cruz basins, however, basin sediments would appear to be closed with respect to manganese cycling. Computation of the excess inventory of dissolved and solid phase manganese in the manganese cycling zone of basin sediments (Shaw 1988) with the assumption of 2:1 (Mn:C) stoichiometry for oxidation of organic matter via manganese reduction (reaction 3, fig. 3.24) leads to the conclusion that under steady-state conditions, Mn^{+4} reduction plays a quantitatively insignificant role in benthic respiration in basin sediments. The same applies for iron, except that cycling is closed even in anoxic inner basin sediments (Leslie et al. 1990). Similar conclusions were reached by Bender et al. (1989) and Bender and Heggie (1984), who estimated organic carbon oxidation rates for reduction of O_2, NO_3^-, Mn^{+4}, Fe^{+3}, and SO_4^{-2} in deep ocean sediments (including San Clemente Basin) by modeling pore water profiles.

For all basin sediments examined thus far (except the San Diego Trough and one site in Santa Catalina Basin) (Smith et al. 1983), there appears to be a measurable flux of nitrate and sulfate from overlying waters into the sediments in support of active nitrate and sulfate reduction (table 3.16). In general, the magnitude of the fluxes of these two electron acceptors varies inversely with the oxygen concentrations in basin waters (compare tables 3.16 and 3.17). Although one indication of the importance of nitrate reduction is the flux of nitrate into the sediments, the significance of measured (or estimated) nitrate fluxes is complicated by cycling of nitrogen within the near-surface sediments wherein ammonium generated by oxidation of organic matter in the O_2 reduction zone is oxidized to nitrate (nitrification). By this mechanism, a second source of nitrate becomes available for denitrification in microniches or proximal suboxic layers (Bender et al. 1989; Berelson et al. 1987). In addition, nitrate is consumed by upward-diffusing ammonium (generated by sulfate reduction) as demonstrated by the increasing ammonium concentrations at depths near or below that where nitrate concentrations approach zero (Bender et al. 1989). In outer basins and at the shelf break, the expansion of nitrification and denitrification zones is evidenced by the occurrence of a subsurface maximum in the NO_3^- profile, whereas these zones are compressed in inner and central basin sediments and nitrate concentrations decline exponentially with increasing subbottom depth (Shaw 1988; Shaw et al. 1990). For these reasons, sulfate profiles may provide the most reliable indicator of the intensity of anaerobic metabolism of organic matter in various basin environments.

The relative importance of sulfate reduction in different basins is indicated by the subbottom depths at which pore water sulfate concentrations begin to decline and by the magnitude of the observed concentration gradients (fig. 3.28). These depths increase and the magnitude of the gradients decreases in basins along north–south and east–west transects. Similar observations have been made by Shaw (1988) and Shaw et al. (1990) based on oxygen, nitrate, iron, and manganese pore water profiles. When the fluxes of oxygen, nitrate, and sulfate are converted to a molar carbon basis (stoichiometries given by equations 1, 2, and 5 in figure 3.24b), remineralization rates of organic carbon are obtained as shown in table 3.17. (These rates do not include manganese, iron, or carbonate reduction and thus must be regarded as first approximation lower limit estimates for the total flux of electrons. In addition, consumption of oxygen during nitrification and nitrate during reaction with ammonium cannot be explicitly separated.) The range in the *total* rate of sedimentary organic carbon remineralization among these basins is within a factor of two (1.2–2.6 mmol C m^{-2} d^{-1}). Since the flux of particulate organic carbon to basin sediments generally decreases with distance offshore (table 3.9), the similarity of the remineralization rates for Santa Monica and San Clemente basins would lead one to predict that a higher percentage of the organic carbon being deposited in outer and central basin sediments is recycled when compared with the inner basins (all else being equal). If so, this could be related to the higher oxygen concentrations of offshore basin waters, the lower sedimentation rates, and the greater lability of autochthonous organic matter, the relative abundance of which would be expected to increase with distance from land. If the vertical flux of organic carbon as determined by sediment trap data is compared with the combined remineralization rates (table 3.17), however, deficits in the carbon budget sometimes occur (that is, vertical $flux_{sed\ traps}$ < combined carbon mineralization rate). Such comparisons may suffer from the same difficulties previously described for estimating sediment–water exchange rates.

A more consistent *qualitative* pattern is observed when the carbon remineralization rates are partitioned between pathways associated with O_2, NO_3^-, and SO_4^{-2} fluxes (table 3.17). Here, differences in the importance of the metabolic pathways in the basins are revealed. As expected, oxygen and nitrate reduction account for increasingly greater proportions of benthic carbon recycling with distance offshore (and to the south) at the expense of sulfate reduction, an anaerobic process. Although the data are limited, this trend generally appears to correlate with increasing water depth, bottom water oxygen content, and particulate carbon flux, making it difficult to ascertain the controlling factors (Reimers 1989).

There have been a few attempts to construct carbon budgets for basins of the SCB. As noted earlier, however, Smith (1987) and Smith et al. (1983) were unable to account for the oxygen consumption in Santa Catalina Basin sediments from estimates of the vertical flux of particulate organic carbon based on the use of sediment traps. Berelson et al.

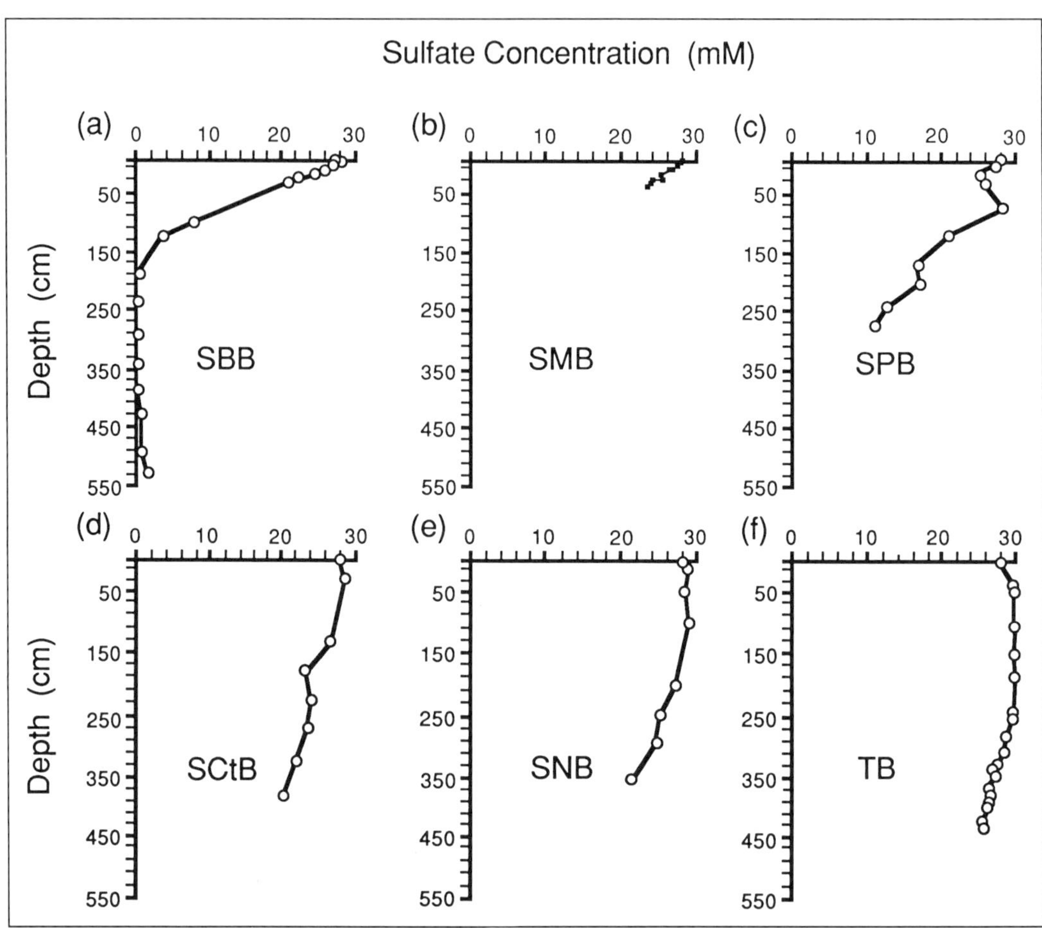

Figure 3.28. Vertical concentration profiles of dissolved sulfate in pore waters of offshore basin sediments: (a) Santa Barbara Basin (Doose 1980; Sholkovitz 1973); (b) Santa Monica Basin (Jahnke 1988); (c) San Pedro Basin (Berelson et al. 1987); (d) Santa Catalina Basin (Kalil 1976); (e) San Nicolas Basin (Berelson et al. 1987); and (f) Tanner Basin (Kalil 1976).

(1987) determined carbon and silica budgets for San Nicolas and San Pedro basins. They found that approximately 90% of the silica deposited on the basin floors was recycled and released to the overlying waters. Based on pore water profiles that show a shallow subsurface maximum in pore water silica content (approximately 5 cm) (Berelson et al. 1987; Smith et al. 1987), much of the silica recycling appears to occur rapidly after sediment deposition. Comparison of sediment organic carbon fluxes (burial + sediment–water exchange) with primary productivity measured by Smith and Eppley (1982) indicates that approximately 10% of the POC formed in surface waters of San Nicolas and San Pedro basins actually reaches the sediment surface and that about 33–38% of this carbon is subsequently recycled. Bender et al. (1989) reported that approximately 64% of the particulate organic carbon reaching San Clemente Basin sediments was recycled at the sea floor; they attributed the greater efficiency of POC degradation in San Clemente Basin (when compared to San Nicolas and San Pedro basins) to the higher oxygen content of its bottom waters.

Recent results developed by the CaBS

group for Santa Monica Basin provide the most comprehensive data set for an inner basin (Jackson et al. 1989; Jahnke 1990). Comparison of sediment trap fluxes, primary productivity measurements, sediment–water exchange, and burial flux estimates (fig. 3.29) indicate an approximate balance in the organic carbon budget for this basin. The long-term average flux of organic carbon to the sediments was found to be 1.72 mg C cm^{-2} yr^{-1} (47 mg C m^{-2} d^{-1}) based on sediment trap data (table 3.9). This represents approximately 4–11% of the primary production and approximately 9–26% of the organic carbon exported from the euphotic zone. In comparison, the rate of sediment organic matter consumption estimated from pore water profiles and benthic flux measurements was 29 mg C m^{-2} d^{-1}, whereas the long-term sedimentary burial rate of organic carbon is 17 mg C m^{-2} d^{-1}. Thus, approximately 10% of the photosynthetically fixed carbon reaches the floor of Santa Monica Basin, and of this, about two-thirds is recycled within the upper 40 cm of the sediment column. The turnover of organic carbon in the sediments appears to be roughly 10 times that accounted for by bacterial remineralization in the lower basin waters, again indicating the importance of diagenetic processes in bottom sediments.

Summary and Prospectus for Future Research

The distribution of chemical substances in the SCB is determined by the fluxes of materials into the bight, processes occurring in the water column and sediments, and losses attributed to air–sea exchange and advection out of the bight. Because of the size of this coastal region and the complexity of these processes, our understanding of bight-wide dynamics is still rather poor. However, the development of more highly sophisticated monitoring techniques (for example, remote sensing) and the initiation of large multidisciplinary studies during the last decade offer new possibilities for understanding intrabight dynamics and interactions between bight waters and larger scale oceanic processes, such as El Niño.

When calculated on a mass flux basis, advection dominates the input of most constituents to the SCB. However, other inputs such as those resulting from human activities (for example, the discharge of municipal wastes) are locally important and have led to the enrichment of many heavy metals, petroleum hydrocarbons, and synthetic organic substances in sediments of the continental marginal shelf and inner basins. While impressive databases exist for historical emissions from waste treatment plants over the past two decades, information on inputs of contaminants from the atmosphere and surface runoff is scarce. In other cases (oil seepage and shale erosion, for example), estimates are based on extremely sketchy information. This makes the establishment of reliable budgets impossible at the present time.

The chemistry of the water column has been studied primarily in the context of ongoing efforts to understand controls on primary and secondary production within the SCB. Vertical profiles of most constituents of the dissolved and suspended particulate phases are largely inherited from the California Current with contributions from the Central Pacific and Equatorial Pacific. In addition, the water column is density stratified. As a consequence, vertical gradients in all properties greatly exceed horizontal variations. To the extent that the source waters experience changing relative contributions with time, the composition of SCB waters is variable both spatially and temporally. Within the bight, nutrients, particularly nitrate, are maintained at very low concentrations in surface waters. Primary productivity is high in the SCB relative to open ocean waters aided by mild seasonal upwelling. Most of this carbon is recycled in the upper 100 m of the water column. However, larger particles transport approximately 10–30% of photosynthetic organic carbon to deeper waters, where further

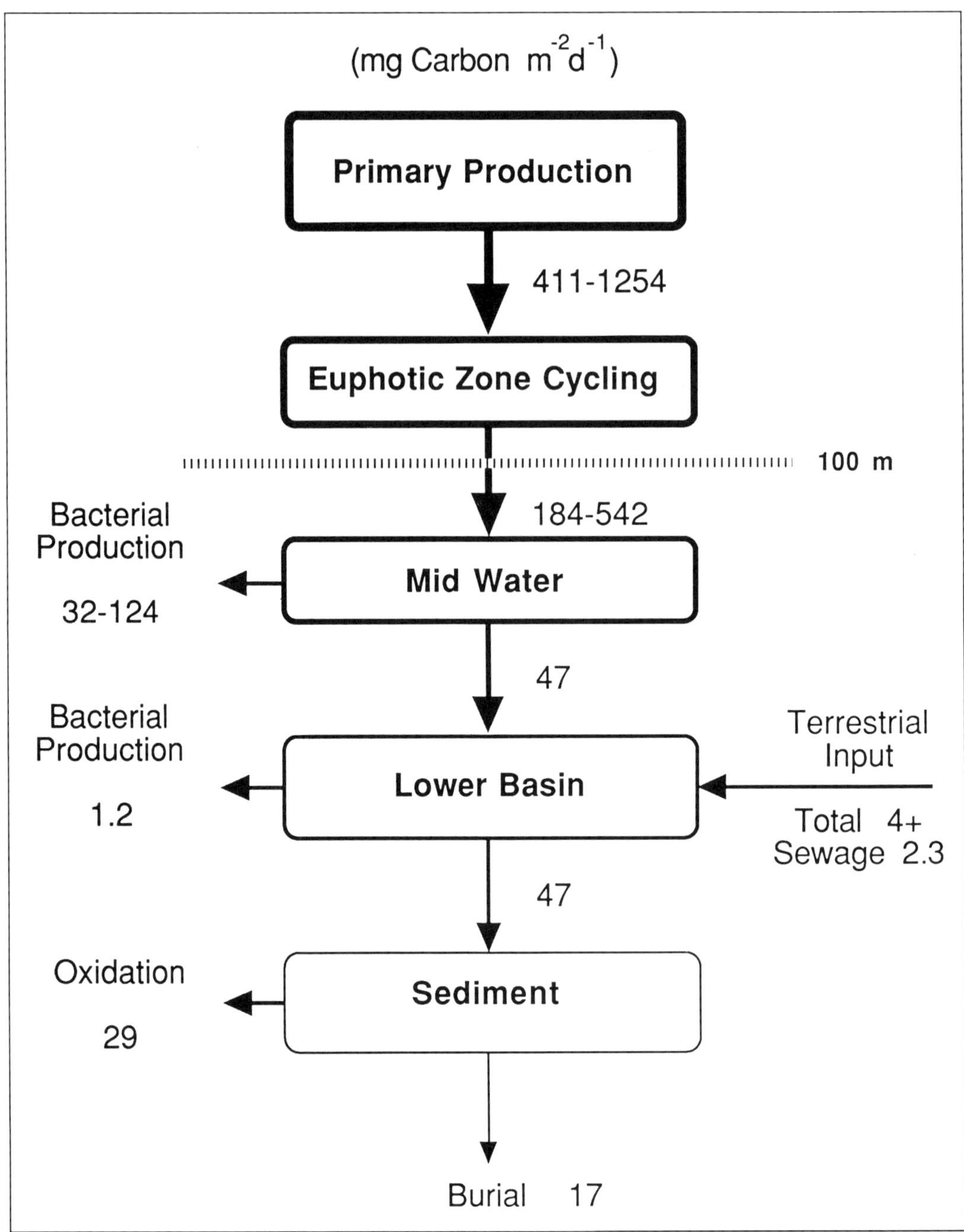

Figure 3.29. Carbon budget (mg C m^{-2} d^{-1}) for Santa Monica Basin. (After Jackson et al. 1989.)

bacterial decomposition leads to release of nitrogen and phosphorus. These nutrient-rich deeper waters represent the primary source of nitrogen (believed to be limiting) for "new production," terrestrial inputs contributing only minor amounts of nitrogen because of the arid climate of the region. We are just beginning to understand the dynamics of carbon cycling within the inner basins.

In contrast, very little effort has been made to characterize the distribution and fate of synthetic organics and trace metals in waters of the SCB. Most of what is known about these substances is based on numerous sediment surveys conducted along the inner shelf, slope, and basin environments, particularly near urban centers. Here, evidence of historical contamination is recorded in basin sediments where benthic life forms are absent. Much of this contamination is attributable to the discharge of municipal wastes through deep water outfalls, a practice that has been established since the middle part of this century. Although extensive surveys of the outer continental shelf were initiated in the mid-1970s, the areal coverage was largely restricted to the northern portion of the SCB. Few measurements have been made on the outer edge of the continental shelf or to the south. The available evidence suggests that most of these anthropogenic materials are associated with particulate matter. Consequently, they are trapped on the inner shelves, slopes, and basins aided by entry into the biological cycle in association with fecal pellets and other rapidly settling particles. Evidence of offshore transport of contaminants can be found, but dispersion and effective trapping by the inner and central basins prevent the buildup of anything but traces of these substances in sediments of the outer continental shelf. The relative importance of eolian and subaqueous transport mechanisms is unclear.

The topography of the SCB is highly irregular, having been formed by convergence of the Pacific and North American plates. A series of northwest–southeast trending basins, ridges, and islands form topographic features that not only influence water motions within the SCB but also provide a range of environments for benthic organisms and sedimentation. The inner basins are shallower and have sills that intersect the oceanic oxygen minimum zone. Consequently, deep basin waters are suboxic and sediments are anoxic, exhibiting varving due to the absence of burrowing benthic organisms. These sediments record depositional events and represent anoxic diagenetic environments where nitrate, manganese, iron, and sulfate reduction occur in upper layers of the sediment. The preservation of organic matter in basin sediments reflects a balance between productivity of surface waters, oxygen concentrations of basinal waters, water column depth, and dilution by lithogenous material, most of which originates in the northern part of the SCB. With distance from land, basin sediments become increasingly oxidizing due to deeper sill depths and lower productivity (hence sedimentation) in overlying waters. The enhanced preservation of organic matter in inner basin sediments accounts for their status as modern-day analogs of petroleum-bearing onshore basins. Recent studies suggest that early diagenesis may result in recycling of as much as 50–60% of the particulate organic carbon reaching the bottom. Ultimately, the (buried) sediments retain <5% of the particulate organic carbon produced in the euphotic zone.

Acknowledgments

The authors wish to thank the following individuals for reviewing all or portions of this chapter and providing valuable constructive comments: W. Berelson, S. C. Bight, R. Carpenter, J. Gieskes, D. Gorsline, I. R. Kaplan, C. Reimers, T. Shaw, E. Sholkovitz, P. M. Williams, and members of the Quality Review Board. We are also indebted to D. Hansell, C. Reimers, L. Sautter, and C. Pilskaln for kindly providing unpublished data for inclusion in this volume. Partial financial support was provided by the Department of Energy, Office of Health and Ecological Research, under grant #DE-FG03-85-ER60338.

Literature Cited

Alford-Stevens, A. L., 1986. Analyzing PCBs. *Environ. Sci. & Technol.* 20:1194–1199.

Alldredge, A. L., 1979. The chemical composition of macroscopic aggregates in two neritic seas. *Limnol. Oceanogr.* 24:855–866.

Alldredge, A. L., and J. L. Cox, 1982. Primary productivity and chemical composition of marine snow in surface waters of the Southern California Bight. *J. Mar. Res.* 40:517–527.

Allen, A. A., R. S. Schlueter, and P. J. Mikolaj, 1970. Natural oil seepage at Coal Oil Point, Santa Barbara, California. *Science.* 170:974–977.

Aller, R. C. 1982. The effects of macrobenthos on chemical properties of marine sediment and overlying water. In: P. L. McCall and M. J. S. Tevesz, eds. *Animal–Sediment Relations.* Plenum Press, New York, pp. 53–102.

Alperin, M. J., W. S. Reeburgh, and A. H. Devol, 1992. Organic carbon remineralization and preservation in sediments of Skan Bay, Alaska. In J. K. Whelan and J. W. Farrington, eds., *Organic Matter: Productivity, Accumulation and Preservation in Recent and Ancient Sediments.* Columbia University Press, New York. pp. 99–122.

Amit, O., M. I. Venkatesan, N. Shaltiel, E. Ruth, and I. R. Kaplan, 1980. The source of polycyclic aromatic hydrocarbons as implied from their depth profiles. *Annual Rep. to U. S. Dept. of Energy.* Contract No. EY–76–3–03–0034, 21pp.

Anderhalt, R., and W. E. Reed, 1978. Sediment texture and organic carbon characterization, benthic sediments, Southern California. *Annual Rep. to Bureau of Land Management, Southern California Baseline Study, Benthic, Year 2.* Vol. II, Sec. 15. BLM AA550-CT6–40, 48pp.

Anderson, J. W., and R. W. Gossett, 1987. Polynuclear aromatic hydrocarbon contamination in sediments from coastal waters of southern California. *Final Rep. to Calif. State Water Resources Control Board.* C–212, Contract No. 5–174–250–0. Prep. by South. Calif. Coastal Water Res. Proj., Long Beach, CA. 51pp.

Arey, J., R. Atkinson, B. Zielinska, and P. A. McElroy, 1989. Diurnal concentrations of volatile polycyclic aromatic hydrocarbons and nitroarenes during a photochemical air pollution episode in Glendora, California. *Environ. Sci. & Technol.* 23:321–327.

Armstrong, F. A. J., and E. C. LaFond, 1966. Chemical nutrient concentrations and their relationship to internal waves and turbidity off southern California. *Limnol. Oceanogr.* 11:538–547.

Asper, V. L., 1987. Measuring the flux and sinking speed of marine snow aggregates. *Deep-Sea Res.* 34:1–17.

Azam, F., 1986. Nutrient cycling and food web dynamics in the Southern California Bight: The microbial food web. In: R. W. Eppley, ed. Lecture notes on coastal and estuarine studies. Vol. 15. *Plankton Dynamics of the Southern California Bight.* Springer-Verlag, Berlin. pp. 274–288.

Ballschmiter, K., and M. Zell, 1980. Analysis of polychlorinated biphenyls (PCB) by glass capillary gas chromatography. *Fresenius Z. Anal. Chem.* 302:20–31.

Barcelona, M. J., L. C. Cummings, S. H. Lieberman, H. S. Fastenau, and W. J. North, 1982. Marine farming the coastal zone: Chemical and hydrographic considerations. *Calif. Coop. Oceanic Fish. Invest. Rep.* 23:180–187.

Barnes, R. O., and E. D. Goldberg, 1976. Methane production and consumption in anoxic marine sediments. *Geology.* 4:297–300.

Barnes, R. O., K. K. Bertine, and E. D. Goldberg, 1975. N_2:Ar, nitrification and denitrification in southern California borderland basin sediments. *Limnol. Oceanogr.* 20:962–970.

Bascom, W., N. Brooks, R. Eppley, T. Hendricks, G. Knauer, D. Pritchard, M. Sherwood, and J. Word, 1979. Southern California Bight. In: E. D. Goldberg, ed. *Assimilation Capacity of U. S. Coastal Waters for Pollutants.* NOAA (Natl. Oceanic Atmos. Adm.), U. S. Dep. Commerce, Boulder, CO. pp. 179–224.

Beers, J. R., 1986. Organisms and the food web. In: R. W. Eppley, ed. Lecture notes on coastal and estuarine studies. Vol. 15. *Plankton Dynamics of the Southern California Bight.* Springer-Verlag, Berlin. pp. 84–175.

Beers, J. R., and G. L. Stewart, 1970. Part VI. Numerical abundance and estimated biomass of microzooplankton. In: J. D. H. Strickland, ed. The ecology of plankton off La Jolla, California, in the period April through September

1967. *Bull. Scripps Inst. Oceanogr. Univ. Calif.* 17:67–87.

Bender, M. L., and D. T. Heggie, 1984. Fate of organic carbon reaching the deep sea floor: A status report. *Geochim. Cosmochim. Acta.* 48: 977–986.

Bender, M. L., R. A. Jahnke, R. Weiss, W. Martin, D. T. Heggie, J. Orchardo, and T. Sowers, 1989. Organic carbon oxidation and benthic nitrogen and silica dynamics in San Clemente Basin, a continental borderland site. *Geochim. Cosmochim. Acta.* 53:685–697.

Berelson, W. M., 1991. The flushing of two deep sea basins, southern California borderland. *Limnol. Oceanogr.* 36:1150–1166.

Berelson, W. M., and D. E. Hammond, 1986. The calibration of a new free-vehicle benthic flux chamber for use in the deep sea. *Deep-Sea Res.* 33:1439–1454.

Berelson, W. M., D. E. Hammond, and C. Fuller, 1982. Radon–222 as a tracer for mixing in the water column and benthic exchange in the southern California borderland. *Earth Planet. Sci. Lett.* 61:41–54.

Berelson, W. M., D. E. Hammond, and K. S. Johnson, 1987. Benthic fluxes and the cycling of biogenic silica and carbon in two southern California borderland basins. *Geochim. Cosmochim. Acta.* 51:1345–1363.

Berner, R. A., 1980. *Early Diagenesis.* Princeton University Press, Princeton, NJ. 241pp.

Bertine, K. K., and E. D. Goldberg, 1977. History of metal pollution in southern California coastal zone—Reprise. *Environ. Sci. & Technol.*, 11: 297–299.

Bigeleisen, J., and M. G. Mayer, 1947. Calculation of equilibrium constants for isotopic exchange reactions. *J. Chem. Phys.* 15:261–271.

Bischoff, J. L., J. J. Clancy, and J. S. Booth, 1975. Magnesium removal in reducing marine sediments by cation exchange. *Geochim. Cosmochim. Acta.* 39:559–568.

Bishop, J. K. B., and J. M. Edmond, 1976. A new large volume filtration system for the sampling of oceanic particulate matter. *J. Mar. Res.* 34: 181–198.

Boon, J. J., and J. W. de Leeuw, 1979. The analysis of wax esters, very long mid-chain ketones and sterol ethers isolated from Walvis Bay diatomaceous ooze. *Mar. Chem.* 7:117–132.

Brooks, R. R., B. J. Presley, and I. R. Kaplan, 1968. Trace elements in the interstitial waters of marine sediments. *Geochim. Cosmochim. Acta.* 32:397–414.

Brown, D. A., R. W. Gossett, G. P. Hershelman, C. F. Ward, A. M. Westcott, and J. N. Cross, 1986. Municipal wastewater contamination in the Southern California Bight: Part 1. Metal and organic contaminants in sediments and organisms. *Mar. Environ. Res.* 18:291–310.

Brown, F. S., M. J. Baedecker, A. Nissenbaum, and I. R. Kaplan, 1972. Early diagenesis in a reducing fjord, Saanich Inlet, British Columbia—III. Changes in organic constituents of sediment. *Geochim. Cosmochim. Acta.* 36:1185–1203.

Brownawell, B. J., and J. W. Farrington, 1986. Biogeochemistry of PCBs in interstitial waters of a coastal marine sediment. *Geochim. Cosmochim. Acta.* 50:157–169.

Brownlie, W. R., and B. D. Taylor, 1981. Sediment management for southern California mountains, coastal plains and shorelines: Part C, Coastal sediment delivery by major rivers in southern California. Environmental Quality Laboratory Report 17C. Calif. Inst. Tech., Pasadena, CA. 314pp.

Bruland, K. W., 1980. Oceanographic distributions of cadmium, zinc, nickel, and copper in the north Pacific. *Earth Planet. Sci. Lett.* 47:176–198.

Bruland, K. W., and R. P. Franks, 1978. Water column trace metal chemistry (suspended particulates). In: *Southern California Baseline Study and Analysis, 1975–1976.* Vol. II, Integrated summary report to Bureau of Land Management. pp. II–361–II–375.

Bruland, K. W., and R. P. Franks, 1979. Trace metal characterization of water column particulates. *Annual Rep. to Bureau of Land Management, Southern California Baseline Study, Benthic, Year 2.* Vol. II, Report 7.0, May 1979.

Bruland, K. W., K. Bertine, M. Koide, and E. D. Goldberg, 1974. History of metal pollution in southern California coastal zone. *Environ. Sci. & Technol.* 8:425–432.

Bruland, K. W., G. A. Knauer, and J. H. Martin, 1978. Cadmium in northeast Pacific waters. *Limnol. Oceanogr.* 23:618–625.

Bruland, K.W., R. P. Franks, G. A. Knauer, and J. H. Martin, 1979. Sampling and analytical methods for the determination of copper, cadmium, zinc and nickel at the nanogram per liter level in seawater. *Anal. Chim. Acta.* 105:233–245.

Bruland, K. W., R. P. Franks, W. M. Landing, and A. Soutar, 1981. Southern California inner basin sediment trap calibration. *Earth Planet. Sci. Lett.* 53:400–408.

Cairns, T., and E. G. Siegmund, 1981. PCBs: Regulatory history and analytical problems. *Anal. Chem.* 53:1183A–1193A.

Calvert, S. E., and T. F. Pederson, 1992. Organic carbon accumulation and preservation in marine sediments: How important is anoxia? In: J. K. Whelan and J. W. Farrington, eds. *Organic Matter: Productivity, Accumulation and Preservation in Recent and Ancient Sediments.* Columbia University Press, New York. pp. 231–263.

Carlson, D. J., M. L. Brann, T. H. Mague, and L. M. Mayer, 1985. Molecular weight distribution of dissolved organic materials in seawater determined by ultrafiltration: A re-examination. *Mar. Chem.* 16:155–173.

Carlucci, A. F., D. B. Craven, and S. M. Henrichs, 1985. Surface-film microheterotrophs: Amino acid metabolism and solar radiation effects on their activities. *Mar. Biol. (Berl.).* 85: 13–22.

Carlucci, A. F., R. W. Eppley, and J. R. Beers, 1986. Introduction to the Southern California Bight. In: R. W. Eppley, ed. Lecture notes on coastal and estuarine studies. Vol. 15. *Plankton Dynamics of the Southern California Bight.* Springer-Verlag, Berlin. pp. 1–12.

Cass, G. R., P. M. Boone, and E. S. Macias, 1982. Emissions and air quality relationships for atmospheric carbon particles in Los Angeles. In: G. T. Wolff and R. L. Klimisch, eds. *Particulate Carbon: Atmospheric Life Cycle.* Plenum Press. New York. pp. 207–243.

Cess, R. D., 1983. Arctic aerosols: Model estimates of interactive influences upon the surface–atmosphere clear-sky radiation budget. *Atmos. Environ.* 17:2555–2564.

Chartrand, A. B., 1988. Montrose Chemical Corporation: Strategies for managing a widespread point source contaminant. In: *Managing Inflows to California's Bays and Estuaries.* The Bay Institute of San Francisco, Sausalito, CA. pp. 50–56.

Chartrand, A. B., S. Moy, A. N. Safford, T. Yoshimura, and L. A. Schinazi, 1985. Ocean dumping under Los Angeles Regional Water Quality Control Board Permit: A review of past practices, potential adverse impacts, and recommendations for future action. Unpublished Report, California Regional Water Quality Control Board. 44pp.

Chen, K. Y., and T. Hendricks, 1974. Trace metals on suspended particulates. In: *Annual Report. South. Calif. Coastal Water Res. Proj.*, El Segundo, CA. pp. 147–152.

Chen, K. Y., and J. C. S. Lu, 1974. Sediment compositions in Los Angeles–Long Beach Harbors and San Pedro Basin. In: D. F. Soule and M. Oguri, eds., *Marine Studies of San Pedro Bay, California, Part VII. Sediment Investigations.* Allan Hancock Foundation, Univ. of Southern California, Los Angeles. 177pp.

Choi, W.-W., and K. Y. Chen, 1976. Associations of chlorinated hydrocarbons with fine particles and humic substances in nearshore surficial sediments. *Environ. Sci. & Technol.* 10:782–786.

Chow, T. J., and J. H. Earl, 1970. Lead aerosols in the atmosphere: Increasing concentrations. *Science.* 169:577–580.

Chow, T. J., and J. H. Earl, 1979. Trace metal distributions in southern California OCS benthic sediments. *Final Rep. to Bureau of Land Management*, Vol. II, Rep. no. 8.0, 45pp.

Chow, T. J., K. W. Bruland, K. K. Bertine, A. Soutar, M. Koide, and E. D. Goldberg, 1973. Lead pollution: Records in southern California coastal sediments. *Science.* 181:551–552.

Claypool, G. E., and I. R. Kaplan, 1974. The origin and distribution of methane in marine sediments. In: I. R. Kaplan, ed. *Natural Gases in Marine Sediments.* Plenum Press, New York. pp. 99–139.

Cline, J. D., and I. R. Kaplan, 1975. Isotopic fractionation of dissolved nitrate during denitrification in the eastern tropical North Pacific Ocean. *Mar. Chem.* 3:271–299.

Cox, J. L., 1971. DDT residues in seawater and particulate matter in the California Current system. *U.S. Natl. Mar. Fish. Serv. Fish. Bull.* 69: 443–450.

Craven, D. B., and R. A. Jahnke, 1992. Microbial utilization and turnover of organic carbon in Santa Monica Basin sediments. *Progr. Oceanogr.* 30:313–333.

Craven, D. B., R. A. Jahnke, and A. F. Carlucci, 1986. Fine-scale vertical distributions of microbial biomass and activity in California borderland sediments. *Deep-Sea Res.* 33:379–390.

Crisp, P. T., S. Brenner, and M. I. Venkatesan, E. Ruth, and I. R. Kaplan, 1979. Organic chemical characterization of sediment-trap particulates from San Nicolas, Santa Barbara, Santa Monica and San Pedro basins, California. *Geochim. Cosmochim. Acta.* 43:1791–1801.

Cross, J. N., J. T. Hardy, J. E. Hose, G. P. Hershelman, L. D. Antrim, R. W. Gossett, and E. A. Crecelius, 1987. Contaminant concentrations and toxicity of sea-surface microlayer near Los Angeles, California. *Mar. Environ. Res.* 23:307–323.

Cullen, J. J., E. Stewart, E. Renger, R. W. Eppley, and C. D. Winant, 1983. Vertical motion of the thermocline, nitracline and chlorophyll maximum layers in relation to currents on the Southern California Shelf. *J. Mar. Res.* 41:239–262.

Curiale, J. A., D. Cameron, and D. V. Davis, 1985. Biological marker distribution and significance in oils and rocks of the Monterey Formation, California. *Geochim. Cosmochim. Acta.* 49:271–288.

Degens, E. T., J. H. Reuter, and K. N. F. Shaw, 1964. Biochemical compounds in offshore California sediments and sea waters. *Geochim. Cosmochim. Acta.* 28:45–66.

Deines, P., 1980. The isotopic composition of reduced organic carbon. In: P. Fritz and J. C. Fontes, eds. *Handbook of Environmental Isotope Geochemistry. Vol. I.* Elsevier Press, Amsterdam, pp. 329–406.

Delaney, R. C., 1972. Trace metal weathering effect and characterization of oily pollutants from the Santa Barbara Channel. Master's Thesis, Univ. of California, Santa Barbara, CA. 131pp.

de Lappe, B. W., R. W. Risebrough, A. M. Springer, T. T. Schmidt, J. C. Shropshire, E. F. Letterman, and J. R. Payne, 1980. The sampling and measurement of hydrocarbons in natural waters. In: B. K. Afghan and D. McKay, eds. *Hydrocarbons and Halogenated Hydrocarbons in the Aquatic Environment.* Plenum Press, New York, pp.29–68.

de Lappe, B. W., R. W. Risebrough, and W. Walker II, 1983. A large-volume sampling assembly for the determination of synthetic organic and petroleum compounds in the dissolved and particulate phases of seawater. *Can. J. Fish. Aquat. Sci.* 40:322–336.

Deuser, W. G., and E. H. Ross, 1980. Seasonal change in the flux of organic carbon to the deep Sargasso Sea. *Nature.* 283:364–365.

Doose, P. R., 1980. The bacterial production of methane in marine sediments. Ph.D. Dissertation, Univ. of California, Los Angeles, 240pp.

Drake, D. E., D. A. Cacchione, and H. A. Karl, 1985. Bottom currents and sediment transport on San Pedro Shelf, California. *J. Sediment. Petrol.* 55:15–28.

Druffel, E. R. M., and P. M. Williams, 1990. Identification of a deep marine source of particulate organic carbon using bomb ^{14}C. *Nature.* 347: 172–174.

Duce, R. A., and R. B. Gagosian, 1982. The input of atmospheric n-C_{10} to n-C_{30} alkanes to the ocean. *J. Geophys. Res.* 87:7192–7200.

Duce, R. A., and E. J. Hoffman, 1976. Chemical fractionation at the air/sea interface. *Annu. Rev. Earth Planet. Sci.* 4:187–227.

Dunbar, R. B., and W. H. Berger, 1981. Fecal pellet flux to modern bottom sediment of Santa Barbara Basin (California) based on sediment trapping. *Geol. Soc. Am. Bull.* 92:212–218.

Dymond, J., K. Fischer, M. Clauson, R. Cobler, W. Gardner, M. J. Richardson, W. Berger, A. Soutar, and R. Dunbar, 1981. A sediment trap intercomparison study in the Santa Barbara basin. *Earth Planet. Sci. Lett.* 53:409–418.

Eganhouse, R. P., 1978. Hydrocarbons, free fatty acids and stable nitrogen isotopes in sewage effluent, outfall sediments and flocculents. *Annual Rep. to U. S. Dept. of Energy, Bureau of Land Management.* Contract no. EY–76–3–03–0034, 39pp.

Eganhouse, R. P., 1979. Dynamics of lipid transport to the ocean via urban stormwater runoff and the search for a sewage-specific lipid tracer. *Annual Report to U.S. Dept. of Energy, Bureau of Land Management.* Contract no. EY–76–3–03–0034, 20pp.

Eganhouse, R. P., 1982. Organic matter in municipal wastes and storm runoff: Characterization and budget to the coastal waters of southern California. Ph.D. Dissertation, Univ. of California, Los Angeles. 230pp.

Eganhouse, R. P., and R. W. Gossett, 1991. Historical deposition and biogeochemical fate of polycyclic aromatic hydrocarbons in sediments near a major submarine wastewater outfall in southern California. In: R. A. Baker, ed., *Organic Substances and Sediments in Water, Vol. 2: Processes and Analytical*, Lewis Publishers, Chelsea, MI. pp.191–220.

Eganhouse, R. P., and I. R. Kaplan, 1981. Extractable organic matter in urban stormwater runoff. 1. Transport dynamics and mass emission rates. *Environ. Sci. & Technol.* 15:310–315.

Eganhouse, R. P., and I. R. Kaplan, 1982a. Extractable organic matter in municipal wastewaters. 1. Petroleum hydrocarbons: Temporal variations and mass emission rates to the ocean. *Environ. Sci. & Technol.* 16:180–186.

Eganhouse, R. P., and I. R. Kaplan, 1982b. Extractable organic matter in municipal wastewaters. 2. Hydrocarbons: Molecular characterization. *Environ. Sci. & Technol.* 16:541–551.

Eganhouse, R. P., and I. R. Kaplan, 1988. Depositional history of Recent sediments from San Pedro shelf, California: Reconstruction using elemental abundance, isotopic composition and molecular markers. *Mar. Chem.* 24:163–191.

Eganhouse, R. P., J. N. Johnson, D. R. Young, and D. J. McDermott, 1976. Mercury in southern California waters: Inputs distribution and fate. *Tech. Mem. 227*, South. Calif. Coastal Water Res. Proj., El Segundo, CA. 57pp.

Eganhouse, R. P., D. R. Young, and J. N. Johnson, 1978. Geochemistry of mercury in Palos Verdes sediments. *Environ. Sci. & Technol.* 12: 1151–1157.

Eganhouse, R. P., B. R. T. Simoneit, and I. R. Kaplan, 1981. Extractable organic matter in urban stormwater runoff. 2. Molecular characterization. *Environ. Sci. & Technol.* 15:315–326.

Eganhouse, R. P., D. L. Blumfield, and I. R. Kaplan, 1983. Long-chain alkylbenzenes as molecular tracers of domestic wastes in the marine environment. *Environ. Sci. & Technol.* 17:523–530.

Eganhouse, R. P., R. W. Gossett, and G. P. Hershelman, 1990. Congener-specific characterization and source identification of PCB input to Los Angeles Harbor. *Report to the Los Angeles Regional Water Quality Control Board*, Sept. 20, 1990, 34pp.

Eisenreich, S. J., P. D. Capel, J. A. Robbins, and R. Bourbonniere, 1989. Accumulation and diagenesis of chlorinated hydrocarbons in lacustrine sediments. *Environ. Sci. & Technol.* 23: 1116–1126.

Eisenreich, S. J., N. A. Metzer, N. R. Urban, and J. A. Robbins, 1986. Response of atmospheric lead to decreased use of lead in gasoline. *Environ. Sci. & Technol.* 20:171–174.

Emerson, S., 1985. Organic carbon preservation in marine sediments. In: O. Sundquist and W. S. Broecker, eds. The carbon cycle and atmospheric carbon dioxide: Natural variations Archean to present. *Am. Geophys. Union Monograph 32.* American Geophysical Union, Washington, D.C. pp.78–87.

Emery, K. O., 1960. The sea off Southern California. John Wiley & Sons, New York. 366pp.

Emery, K. O., and E. E. Bray, 1962. Radiocarbon dating of California basin sediments. *Am. Assoc. Pet. Geol. Bull.* 46:1839–1856.

Emery, K. O., and D. Hoggan, 1958. Gases in marine sediments. *Am. Assoc. Pet. Geol. Bull.* 42:2174–2188.

Emery, K. O., and S. C. Rittenberg, 1952. Early diagenesis of California Basin sediments in relation to the origin of oil. *Am. Assoc. Pet. Geol. Bull.* 36:735–806.

Eppley, R. W., ed., 1986. Lecture notes on coastal and estuarine studies, Vol. 15. *Plankton Dynamics of the Southern California Bight.* Springer-Verlag, Berlin. 373pp.

Eppley, R. W., 1992. Chlorophyll, photosynthesis, and new production in the Southern California Bight. *Progr. Oceanogr.* In press.

Eppley, R. W., and O. Holm-Hansen, 1986. Primary production in the Southern California Bight. In: R. W. Eppley, ed. Lecture notes on coastal and estuarine studies. Vol. 15. *Plankton Dynamics of the Southern California Bight.* Springer-Verlag, Berlin. pp.176–215.

Eppley, R. W., and B. J. Peterson, 1979. Particulate organic matter flux and planktonic new production in the deep ocean. *Nature.* 282:677–680.

Eppley, R. W., W. G. Harrison, S. W. Chisholm, and E. Stewart, 1977. Particulate organic matter in surface waters off southern California and its relationship to phytoplankton. *J. Mar. Res.* 35:672–696.

Eppley, R. W., C. Sapienza, and E. H. Renger, 1978. Gradients in phytoplankton stocks and nutrients off southern California in 1974–76. *Estuarine Coastal Mar. Sci.* 7:291–301.

Eppley, R. W., E. H. Renger, and W. G. Harrison, 1979a. Nitrate and phytoplankton production in southern California coastal waters. *Limnol. Oceanogr.* 24:483–494.

Eppley, R. W., E. H. Renger, W. G. Harrison, and J. J. Cullen, 1979b. Ammonium distribution in southern California coastal waters and its role in the growth of phytoplankton. *Limnol. Oceanogr.* 24:495–509.

Eppley, R. W., E. H. Renger, and P. R. Betzer, 1983. The residence time of particulate organic carbon in the surface layer of the ocean. *Deep-Sea Res. Part A Oceanogr. Res. Pap.* 30:311–323.

Faisst, W. K., 1976. Digested sludge: Delineation and modeling for ocean disposal. Ph.D. Dissertation, Calif. Inst. Technol., Pasadena, CA. 193pp.

Finlayson-Pitts, B. J., and J. N. Pitts, Jr., 1986. *Atmospheric Chemistry*. Wiley Interscience, New York. 1098pp.

Finney, B. P., and C.-A. Huh, 1989a. History of metal pollution in the Southern California Bight: An update. *Environ. Sci. & Technol.* 23:294–303.

Finney, B. P., and C.-A. Huh, 1989b. High resolution sedimentary records of heavy metals from the Santa Monica and San Pedro Basins, California. *Mar. Pollut. Bull.* 20:181–187.

Fischer, P. J., 1978. Natural gas and oil seeps, Santa Barbara Basin. In: *The State Lands Commission 1977, California Gas, Oil, and Tar Seeps*. Sacramento. pp.1–62.

Fleischer, P., 1970. Mineralogy of hemipelagic basin sediments, California continental borderland. Ph.D. Dissertation, Univ. of Southern California, Los Angeles. 208pp.

Fleischer, P., 1972. Mineralogy and sedimentation history, Santa Barbara Basin, California. *J. Sediment. Petrol.* 42:49–58.

Fowler, S. W., and G. A. Knauer, 1986. Role of large particles in the transport of elements and organic compounds through the oceanic water column. *Prog. Oceanogr.* 16:147–194.

Froehlich, P. N., G. P. Klinkhammer, M. L. Bender, N. A. Luedtke, G. R. Heath, D. Cullen, P. Dauphin, D. Hammond, B. Hartman, and V. Maynard, 1979. Early oxidation of organic matter in pelagic sediments of the eastern equatorial Atlantic: Suboxic diagenesis. *Geochim. Cosmochim. Acta.* 43:1075–1090.

Fuhrman, J. A., J. W. Ammerman, and F. Azam, 1980. Bacterioplankton in the coastal euphotic zone: Distribution, activity and possible relationships with phytoplankton. *Mar. Biol. (Berl.)*. 60:201–207.

Fukushima, K., and R. Ishiwatari, 1984. Acid and alcohol compositions of wax esters in sediments from different environments. *Chem. Geol.* 47: 41–56.

Gagosian, R. B., E. T. Peltzer, and J. T. Merill, 1987. Long-range transport of terrestrially derived lipids in aerosols from the South Pacific. *Nature*. 325:800–803.

Galloway, J. N., 1979. Alteration of trace metal geochemical cycles due to the marine discharge of wastewater. *Geochim. Cosmochim. Acta.* 43:207–218.

Garlick, G. D., 1974. The stable isotopes of oxygen, carbon, and hydrogen in the marine environment. In: E. D. Goldberg, ed. *The Sea, Vol. 5*. John Wiley & Sons, New York. pp. 393–425.

Goldberg, E. D., 1961. Chemistry in the oceans. In: M. Sears, ed. *Oceanography*. American Association for the Advancement of Science, Washington, D.C. pp. 583–597.

Goldhaber, M. B., and I. R. Kaplan, 1973. The sulfur cycle. In: E. D. Goldberg, ed. *The Sea*. John Wiley & Sons, New York. pp. 569–655.

Goldhaber, M. B., and I. R. Kaplan, 1975. Controls and consequences of sulfate reduction in Recent marine sediments. *Soil Sci.* 119:42–55.

Gorsline, D. S. 1992, The geologic setting of Santa Monica and San Pedro basins, California continental borderland. *Progr. Oceanogr.* In press.

Gorsline, D. S., D. E. Drake, and P. W. Barnes, 1968. Holocene sedimentation in Tanner Basin, California continental borderland. *Geol. Soc. Am. Bull.* 79:659–674.

Gossett, R. W., D. A. Brown, and D. R. Young, 1983. Predicting the bioaccumulation of organic compounds in marine organisms using octanol/water partition coefficients. *Mar. Pollut. Bull.* 14:387–392.

Goth, K., J. W. de Leeuw, W. Puttmann, and E. W. Tegelaar, 1988. Origin of Messel oil shale kerogen. *Nature*. 336:759–761.

Gray, H. A., G. R. Cass, J. J. Huntzicker, E. K. Heyerdahl, and J. A. Rau, 1986. Characteristics of atmospheric organic and elemental carbon particle concentrations in Los Angeles. *Environ. Sci. & Technol.* 20:580–589.

Grosjean, D., 1982. Formaldehyde and other carbonyls in Los Angeles ambient air. *Environ. Sci. & Technol.* 16:254–262.

Halpern, H. I., 1981. An investigation of mineral-kerogen interactions and their relation to petroleum genesis. Ph.D. Dissertation, Univ. of California, Los Angeles. 216pp.

Hansell, D. A., P. M. Williams, and B. B. Ward, 1990. Profiles of DOC and DON in the Southern California Bight. *EOS, Trans., Am. Geophys. Union*. 71:1414.

Hardy, J. S., and H. Gucinski, 1988. The biological structure subjacent to the sea surface microlayer: Effects of anthropogenic contamination. *EOS, Trans., Am. Geophys. Union*. 69:1095.

Hardy, J. T., 1982. The sea surface microlayer: Biology, chemistry and anthropogenic enrichment. *Prog. Oceanogr.* 11:307–328.

Hartman, B., and D. E. Hammond, 1981. The use of carbon and sulfur isotopes as correlation parameters for the source identification of beach tar in the southern California borderland. *Geochim. Cosmochim. Acta.* 45:309–319.

Hatcher, P. G., E. C. Spiker, N. M. Szeverenyi, and G. E. Maciel, 1983. Selective preservation and origin of petroleum-forming aquatic kerogen. *Nature.* 305:498–501.

Hebel, D., G. A. Knauer, and J. H. Martin, 1986. Trace metals in large agglomerates (marine snow). *J. Plankton Res.* 8:819–824.

Hedges, J. I., and D. C. Mann, 1979. The lignin geochemistry of marine sediments from the southern Washington coast. *Geochim. Cosmochim. Acta.* 43:1809–1818.

Hedges, J. I., and P. L. Parker, 1976. Land-derived organic matter in surface sediments from the Gulf of Mexico. *Geochim. Cosmochim. Acta.* 40:1019–1029.

Heggie, D., and T. Lewis, 1984. Cobalt in pore waters of marine sediments. *Nature.* 311:453–455.

Hendricks, T. J., and R. P. Eganhouse, 1992. Modification and verification of sediment deposition models. *Final Report to California State Water Resources Control Board.* In preparation.

Hendricks, T. J., and D. R. Young, 1974. Modeling the fates of metals in ocean-discharged wastewaters. *Tech. Mem. 208,* South. Calif. Coastal Water Res. Proj., El Segundo, CA. 32pp.

Henrichs, S. M., and W. S. Reeburgh, 1987. Anaerobic mineralization of marine sediment organic matter: Rates and the role of anaerobic processes in the oceanic organic carbon economy. *Geomicrobiol. J.* 5:191–237.

Henrichs, S. M. and P. M. Williams, 1985. Dissolved and particulate amino acid and carbohydrates in the surface microlayer. *Mar. Chem.* 17:141–163.

Hershelman, G. P., H. A. Schafer, T. K. Jan, and D. R. Young, 1981. Metals in marine sediments near a large California municipal outfall. *Mar. Pollut. Bull.* 12:131–134.

Hoefs, J., 1987. *Stable Isotope Geochemistry.* 3rd Edition. Springer-Verlag, Berlin. 241pp.

Hoering, T. C., 1967a. Branched-chain fatty acids in Recent sediments. *Carnegie Inst. Wash. Year Book.* 67:201–202.

Hoering, T. C., 1967b. Fatty alcohols in sedimentary rocks. *Carnegie Inst. Wash. Year Book.* 67:202–203.

Holm-Hansen, O., J. D. H. Strickland, and P. M. Williams, 1966. A detailed analysis of biologically important substances in a profile off southern California. *Limnol. Oceanogr.* 11:548–561.

Hom, W., R. W. Risebrough, A. Soutar, and D. R. Young, 1974. Deposition of DDE and polychlorinated biphenyls in dated sediments of the Santa Barbara Basin. *Science.* 184:1197–1199.

Honjo, S., 1982. Seasonality and interaction of biogenic and lithogenic particulate flux at the Panama Basin. *Science.* 218:883–884.

Honjo, S., and M. R. Roman, 1978. Marine copepod fecal pellets: Production, preservation and sedimentation. *J. Mar. Res.* 36:45-56.

Huang, W.-Y., and W. G. Meinschein, 1979. Sterols as ecological indicators. *Geochim. Cosmochim. Acta.* 43:739–745.

Huh, C.-A., D. L. Zahnle, L. F. Small, and V. E. Noshkin, 1987. Budgets and behaviors of uranium and thorium series isotopes in Santa Monica Basin sediments. *Geochim. Cosmochim. Acta.* 51:1743–1754.

Huh, C.-A., L. F. Small, S. Niemnil, B. P. Finney, B. M. Hickey, N. B. Kachel, D. S. Gorsline, and P. M. Williams, 1990. Sedimentation dynamics in the Santa Monica-San Pedro Basin off Los Angeles: Radiochemical, sediment trap, and transmissometer studies. *Cont. Shelf Res.* 10:137–164.

Huh, C.-A., B. P. Finney, and J. K. Stull, 1992. Anthropogenic inputs of several heavy metals to nearshore basins off Los Angeles. *Progr. Oceanogr.* In press.

Hunter, K. A., and P. S. Liss, 1981. Organic surface films. In: E. K. Duursma and R. Dawson, eds. *Marine Organic Chemistry.* Elsevier, Amsterdam, The Netherlands. pp. 259–298.

Ikan, R., M. J. Baedecker, and I. R. Kaplan, 1975a. Thermal alteration experiments on organic matter in Recent marine sediment. II. Isoprenoids. *Geochim. Cosmochim. Acta.* 39:187–194.

Ikan, R., M. J. Baedecker, and I. R. Kaplan, 1975b. Thermal alteration experiments on organic matter in Recent marine sediment. III. Aliphatic and steroidal alcohols. *Geochim. Cosmochim. Acta.* 39:195–203.

Ishiwatari, R., M. Ishiwatari, B. G. Rohrback, and I. R. Kaplan, 1977. Thermal alteration experiments on organic matter from Recent marine sediments in relation to petroleum genesis. *Geochim. Cosmochim. Acta.* 41:815–828.

Ishiwatari, R., B. G. Rohrback, and I. R. Kaplan, 1978. Hydrocarbon generation by thermal alteration of kerogen from different sediments. *Am. Assoc. Pet. Geol. Bull.* 62:687–692.

Jackson, G. A., 1982. Sludge disposal in southern California basins. *Environ. Sci. & Technol.* 16:746–757.

Jackson, G. A., 1986. Physical oceanography of the Southern California Bight. In: R. W. Eppley, ed. Lecture notes on coastal and estuarine studies. Vol. 15. *Plankton Dynamics of the Southern California Bight.* Springer-Verlag, Berlin. pp. 13–52.

Jackson, G. A., and P. M. Eldridge, 1992. Food web analysis of a planktonic system off southern California. *Progr. Oceanogr.* In press.

Jackson, G. A., and P. M. Williams, 1985. Importance of dissolved organic nitrogen and phosphorus to biological nutrient cycling. *Deep-Sea Res. Part A Oceanogr. Res. Pap.* 32:223–235.

Jackson, G. A., F. Azam, A. F. Carlucci, R. W. Eppley, B. Finney, D. S. Gorsline, B. Hickey, C.-A. Huh, R. A. Jahnke, I. R. Kaplan, M. R. Landry, L. F. Small, M. I. Venkatesan, P. M. Williams, and K. M. Wong, 1989. Elemental cycling and fluxes off the coast of southern California. *EOS, Trans., Am. Geophys. Union.* 70:146–155.

Jahnke, R., 1986. The flux and recycling of bioactive substances in the surface sediments of the deep basins off southern California. *Prog. Rep. II.* West Coast Environmental Studies, California Basin Study (CaBS), 15pp.

Jahnke, R. A., 1987. The flux and recycling of bioactive substances in the surface sediments of the deep basins off southern California. *Prog. Rep. IV.* West Coast Environmental Studies, California Basin Study (CaBS), 12pp.

Jahnke, R. A., 1988. The flux and recycling of bioactive substances in the surface sediments of the deep basins off southern California. *Prog. Rep. V.* West Coast Environmental Studies, California Basin Study (CaBS), 9pp.

Jahnke, R. A., 1990. Early diagenesis and recycling of biogenic debris at the sea floor. Santa Monica Basin, California. *J. Mar. Res.*, 48:413–436.

Johnson, K. S., P. M. Stout, W. M. Berelson, and C. M. Sakamoto-Arnold, 1988. Cobalt and copper distributions in the waters of Santa Monica Basin California. *Nature.* 332:527–530.

Kalil, E. K., 1976. The distribution and geochemistry of uranium in Recent and Pleistocene marine sediments. Ph.D. Dissertation, Univ. of California, Los Angeles, 283pp.

Kaplan, I. R., 1975. Stable isotopes as a guide to biogeochemical processes. *Proc. R. Soc. Lond. B.* 189:183–211.

Kaplan, I. R., K. O. Emery, and S. C. Rittenberg, 1963. The distribution and isotopic abundance of sulphur in Recent marine sediments off southern California. *Geochim. Cosmochim. Acta.* 27:297–331.

Karl, D. M., and G. A. Knauer, 1984. Detritus-microbe interactions in the marine pelagic environment: Selected results from the VERTEX experiment. *Bull. Mar. Sci.* 35:550–565.

Katz, A., and I. R. Kaplan, 1981. Heavy metals behavior in coastal sediments of southern California: A critical review and synthesis. *Mar. Chem.* 10:261–299.

Kawamura, K., and I. R. Kaplan, 1983. Organic compounds in the rainwater of Los Angeles. *Environ. Sci. & Technol.*17:497–501.

Kawamura, K., and I. R. Kaplan, 1986. Compositional changes of organic matter in rainwater during precipitation events. *Atmos. Environ.* 20:527–535.

Kawamura, K., and I. R. Kaplan, 1987. Motor exhaust emissions as a primary source for dicarboxylic acids in Los Angeles ambient air. *Environ. Sci. & Technol.* 21:105–110.

Keith, J. O., L. A. Woods, Jr., and E. G. Hunt, 1970. Reproductive failure in brown pelicans on the Pacific coast. *Trans. 31st North. Am. Wildl. Nat. Resour. Conf.* pp. 150–177.

Kettenring, K. N., Jr., 1981. The trace metal stratigraphy and Recent sedimentary history of anthrogenous particulates on the San Pedro Shelf, California. Ph.D. Dissertation, Univ. of California, Los Angeles. 156pp.

Knauer, G. A., and J. H. Martin, 1981. Primary production and carbon-nitrogen fluxes in the upper 1,500 m of the northeast Pacific. *Limnol. Oceanogr.* 26:181–186.

Knauer, G. A., J. H. Martin, and K. W. Bruland, 1979. Fluxes of particulate carbon, nitrogen and phosphorus in the upper water column of the northeast Pacific. *Deep-Sea Res.* 26:97–108.

Knauer, G. A., D. Hebel, and F. Cipriano, 1982. Marine snow: Major site of primary production in coastal waters. *Nature.* 300:600–631.

Koide, M., A. Soutar, and E. D. Goldberg, 1972. Marine geochronology with ^{210}Pb. *Earth Planet. Sci. Lett.* 14:442–446.

Kosiur, D. R., and A. L. Warford, 1979. Methane production and oxidation in Santa Barbara Basin sediments. *Estuarine Coastal Mar. Sci.* 8:379–385.

Kvenvolden, K. A., and J. W. Harbaugh, 1983. Reassessment of the rates at which oil from natural sources enters the marine environment. *Mar. Environ. Res.* 10:223–243.

Lal, D., 1977. The oceanic microcosm of particles. *Science.* 198:997–1009.

Landry, M. R., W. K. Peterson, and C. C. Andrews, 1992. Flux of organic particulates in waters overlying Santa Monica Basin. *Progr. Oceanogr.* In press.

Lazrus, A. L., E. Lorange, and J. P. Lodge, Jr., 1970. Lead and other metal ions in United States precipitation. *Environ. Sci. & Technol.* 4:55–58.

Leslie, B. W., D. E. Hammond, W. M. Berelson, and S. P. Lund, 1990. Diagenesis in anoxic sediments from the California continental borderland and its influence on iron, sulfur, and magnetite behavior. *J. Geophys. Res.* 95:4453–4470.

Lion, L. W., and J. O. Leckie, 1981. The biogeochemistry of the air–sea interface. *Annu. Rev. Earth Planet. Sci.* 9:449–486.

Liu, K.-K., 1979. Geochemistry of inorganic nitrogen compounds in two marine environments: The Santa Barbara Basin and the ocean off Peru. Ph.D. Dissertation, Univ. of California, Los Angeles, 354pp.

Liu, K.-K., and I. R. Kaplan, 1982. Nitrous oxide in the sea off southern California. In: W. G. Ernst and J. G. Morin, eds. *The Environment of the Deep Sea.* Prentice-Hall, New York. pp. 73–92.

Liu, K.-K., and I. R. Kaplan, 1984. Denitrification rates and availability of organic matter in marine environments. *Earth Planet. Sci. Lett.* 68: 88–100.

Liu, K.-K., and I. R. Kaplan, 1989. The eastern tropical Pacific as a source of nitrogen–15 enriched nitrate in seawater off southern California. *Limnol. Oceanogr.* 34:820–830.

Logan, J. A., 1983. Nitrogen oxides in the troposphere: Global and regional budgets, *J. Geophys. Res.* 88:10785–10807.

Lu, J. C. S., and K. Y. Chen, 1977. Migration of trace metals in interfaces of seawater and polluted surficial sediments. *Environ. Sci. & Technol.* 11:174–182.

McCave, I. N., 1975. Vertical flux of particles in the ocean. *Deep-Sea Res.* 22:491–502.

McDermott, D. J., T. C. Heesen, and D. R. Young, 1974. DDT in bottom sediments around five southern California outfall systems. *Tech. Mem. 217.* South. Calif. Coastal Water Res. Proj. El Segundo, CA. 54pp.

MacGregor, J. S., 1974. Changes in the amount and proportions of DDT and its metabolites, DDE and DDD, in the marine environment off southern California, 1949–72. *U.S. Natl. Mar. Fish. Serv. Fish. Bull.* 72:275–293.

MacGregor, J. S., 1976. DDT and its metabolites in the sediments off southern California. *U.S. Natl. Mar. Fish. Serv. Fish. Bull.* 74:27–35.

Malouta, D. N., D. S., Gorsline, and S. E. Thornton, 1981. Processes and rates of Recent (Holocene) basin filling in an active transform margin: Santa Monica Basin, California continental borderland. *J. Sediment. Petrol.* 51:1077–1095

Mankiewicz, P. J., 1981. Hydrocarbon composition of sediments, water and fauna in selected areas of the Gulf of Mexico and southern California marine environment. Ph.D. Dissertation, Univ. of California, Los Angeles. 324pp.

Mankiewicz, P., J. Nemmers, R. Jordan, J. Payne, I. R. Kaplan, M. I. Venkatesan, S. Brenner, and E. Ruth, 1978. Source and depositional controls on the high molecular weight hydrocarbon content of surficial sediments from the southern California outer continental shelf. *Annual Rep. to Bureau of Land Management, Southern California Baseline Study, Benthic, Year 2.* Vol. II. Report 5.1. Prep. by Science Applications, Inc. Contract no. AA550-CT6–40.

Martens, C. S., and J. V. Klump, 1984. Biogeochemical cycling in an organic-rich coastal marine basin—4. An organic carbon budget for sediment dominated by sulfate reduction and methanogenesis. *Geochim. Cosmochim. Acta.* 48: 1987–2004.

Matsueda, H., and N. Handa, 1986. Vertical flux of hydrocarbons as measured in sediment traps in the eastern north Pacific Ocean. *Mar. Chem.* 20:179–195.

Mazurek, M. A., and B. R. T. Simoneit, 1984. Characterization of biogenic and petroleum-derived organic matter in aerosols over remote, rural and urban areas. In: L. H. Keith,ed. *Identification and Analysis of Organic Pollutants in Air.* Ann Arbor Science/Butterworth Publishers, Boston. pp. 353–370.

Meade, R. H., 1969. Errors in using modern stream-load data to estimate natural rates of denudation. *Geol. Soc. Am. Bull.* 80:1265–1274.

Mearns, A. J., M. Matta, G. Shigenaka, D. MacDonald, M. Buchman, H. Harris, J. Golas, and G. Lauenstein, 1991. Contaminant trends in the Southern California Bight: Inventory and assessment. *NOAA Tech. Mem.* NOSORCA 62. Seattle, WA.

Menzel, D. W., and R. F. Vaccaro, 1964. The

measurement of dissolved and particulate organic carbon in seawater. *Limnol. Oceanogr.* 9:138–142.

Meyers, P. A., and J. G. Quinn, 1973. Factors affecting the association of fatty acids with mineral particles in sea water. *Geochim. Cosmochim. Acta.* 37:1745–1759.

Middleton, G. V., and J. B. Southard, 1978. *Mechanics of Sediment Movement.* Soc. Econ. Paleontol. Mineral. Short Course No. 3.

Minagawa, M., D. A. Winter, and I. R. Kaplan, 1984. Comparison of Kjeldahl and combustion methods for measurement of nitrogen isotope ratios in organic matter. *Anal. Chem.* 56:1859–1861.

Muhlbaier, J. L., and R. L. Williams, 1982. Fireplaces, furnaces and vehicles as emission sources of particulate carbon. In: G. T. Wolff and R. L. Klimisch, eds. *Particulate Carbon: Atmospheric Life Cycle.* Plenum Press, New York. pp. 185–205.

Mullin, M. M., 1986. Spatial and temporal scales and patterns. In: R. W. Eppley, ed. Lecture notes on coastal and estuarine studies. Vol. 15. *Plankton Dynamics of the Southern California Bight.* Springer-Verlag, Berlin, pp. 216–273.

Mullins, H. T., J. B. Thompson, K. McDougall, and T. L. Vercoutere, 1985. Oxygen-minimum zone edge effects: Evidence from the central California coastal upwelling system. *Geology.* 13:491–494.

Murnane, R. J., B. Leslie, D. E. Hammond, and R. F. Stallard, 1989. Germanium geochemistry in the southern California borderlands. *Geochim. Cosmochim. Acta.* 53:2873–2882.

Myers, E. P., 1974. The concentration and isotopic composition of carbon in marine sediments affected by a sewage discharge. Ph.D. Dissertation, Calif. Inst. of Technol. Pasadena, CA. 179pp.

National Climate Data Center, 1988. *Local Climatic data, Summary with Comparative Data, Los Angeles, California, Civic Center 1988.* Asheville, NC. 6pp.

National Research Council, 1980. *Lead in the Human Environment.* National Academy of Sciences, Washington, D.C. 525pp.

Nelson, J. R., J. R. Beers, R. W. Eppley, G. A. Jackson, J. J. McCarthy, and A. Soutar, 1987. A particle flux study in the Santa Monica–San Pedro Basin off Los Angeles: Particle flux, primary production, and transmissometer survey. *Cont. Shelf Res.* 7:307–328.

Ng, A., and C. C. Patterson, 1982. Changes of lead and barium with time in California offshore basin sediments. *Geochim. Cosmochim. Acta.* 46:2307–2321.

Nip, M., E. W. Tegelaar, J. W. de Leeuw, P. A. Schenck, and P. J. Holloway, 1986. A new non-saponifiable highly aliphatic and resistant biopolymer in plant cuticles: Evidence from pyrolysis and carbon–13—NMR analysis of present day and fossil plants. *Naturwissenschaften.* 73:579–585.

Nissenbaum, A., 1973. The organic geochemistry of marine and terrestrial humic substances: Implications of carbon and hydrogen isotope studies. In: B. Tissot and F. Bienner, eds. *Advances in Organic Geochemistry.* Editions Technip, Paris. pp. 39–52.

Nissenbaum, A., and I. R. Kaplan, 1972. Chemical and isotopic evidence for the *in situ* origin of marine humic substances. *Limnol. Oceangr.* 17:570–582.

Noriki, S., and S. Tsunogai, 1986. Particulate fluxes and major components of settling particles from sediment trap experiments in the Pacific Ocean. *Deep-Sea Res. Part A Oceanogr. Res. Pap.* 33:903–912.

Obradovich, J. D., and C. W. Naeser, 1981. Geochronology bearing on the age of the Monterey Formation and siliceous rocks in California. In: R. E. Garrison, R. G. Douglas, K. E. Pisciotto, C. S. Isaacs, and J. S. Ingle, eds. Monterey Formation and related siliceous rocks of California. *Soc. Econ. Paleontol. Mineral.* Publ. 87–95.

Olmez, I., E. R. Sholkovitz, D. Hermann, and R. P. Eganhouse, 1991. Rare earth elements in sediments off southern California: A new anthropogenic indicator. *Environ. Sci. & Technol.* 25:310–316.

Olson, R. J., 1981a. Nitrogen–15 tracer studies of the primary nitrite maximum. *J. Mar. Res.* 39:203–226.

Olson, R. J., 1981b. Differential photoinhibition of marine nitrifying bacteria: A possible mechanism for the formation of the primary nitrite maximum. *J.Mar. Res.* 39:227–238.

Owen, R. W., Jr., and C. K. Sanchez, 1974. Phytoplankton pigment and production measurements in the California Current region, 1969–72. *NOAA Data Rep.* no. 91. 185pp.

Patterson, C., and D. Settle, 1976. The reduction of orders of magnitude errors in lead analyses of biological materials and natural waters by evaluating and controlling the extent and sources of industrial lead contamination introduced during sample collecting and analysis. In: *Accuracy in Trace Analysis: Sampling, Sample Handling, Analysis.* Natl. Bureau of Standards Spec. Publ. 422. pp. 321–351.

Patterson, E. M., B. T. Marshall, and K. A. Rahn, 1982. Radiative properties of the Arctic aerosol. *Atmos. Environ.* 16:2967–2977.

Peters, K. E., R. E. Sweeney, and I. R. Kaplan, 1978. Correlation of carbon and nitrogen stable isotope ratios in sedimentary organic matter. *Limnol. Oceanogr.* 23:598–604.

Philp, R. P., and M. Calvin, 1976. Possible origin for insoluble organic (kerogen) debris in sediments from insoluble cell-wall materials of algae and bacteria. *Nature.* 262:134–136.

Pollock, G. E., and K. A. Kvenvolden, 1978. Stereochemistry of amino acids in surface samples of a marine sediment. *Geochim. Cosmochim. Acta.* 42:1903–1905.

Presley, B. J., and I. R. Kaplan, 1968. Changes in dissolved sulfate, calcium and carbonate from interstitial water of near-shore sediments. *Geochim. Cosmochim. Acta.* 32:1037–1048.

Prézelin, B. B., and A. L. Alldredge, 1983. Primary production of marine snow during and after an upwelling event. *Limnol. Oceanogr.* 28: 1156–1167.

Prospero, J. M., and D. L. Savoie, 1989. Effect of continental sources on nitrate concentrations over the Pacific Ocean. *Nature.* 339:687–689.

Rabinowitz, M., 1972. Plant uptake of soil and atmospheric lead in southern California. *Chemosphere.* 4:175–180.

Rapp, J. B., and K. A. Kvenvolden, 1982. Organic geochemistry of sediments on the flanks of Tanner and Cortes banks offshore from southern California. *U.S. Geol. Surv. Open-File Rep. 82–0829.* U. S. Geological Survey, Denver. 12pp.

Reed, W. E., and I. R. Kaplan, 1977. The chemistry of marine petroleum seeps. *J. Geochem. Explor.* 7:255–293.

Reed, W. E., I. R. Kaplan, M. Sandstrom, and P. Mankiewicz, 1977. Petroleum and anthropogenic influence on the composition of sediments from the Southern California Bight. In: *Proceedings 1977 Oil Spill Conference (Prevention, Behavior, Control, Cleanup).* New Orleans, 8–10 March 1977. pp. 183–188.

Reid, J. L., 1965. Intermediate waters of the Pacific Ocean. *Johns Hopkins Oceanographic Studies,* no. 2. 85pp.

Reimers, C. E., 1987. An *in situ* microprofiling instrument for measuring interfacial pore water gradients: Methods and oxygen profiles from the North Pacific Ocean. *Deep-Sea Res.* 34: 2019–2035.

Reimers, C. E., 1989. Control of benthic fluxes by particulate supply. In: W. H. Berger, V. S. Smetacek, and G. Wefer, eds. *Productivity of the Ocean: Present and Past.* John Wiley & Sons, New York. pp. 217–233.

Reimers, C. E., and K. L. Smith, 1986. Reconciling measured and predicted fluxes of oxygen across the deep sea sediment–water interface. *Limnol. Oceanogr.* 31:305–318.

Reimers, C. E., D. M. Fischer, R. Merewether, K. L. Smith, and R. A. Jahnke, 1986. Oxygen microprofiles measured *in situ* in deep ocean sediments. *Nature.* 320:741–744.

Reimers, C. E., C. B. Lange, M. Tabak, and J. M. Bernhard, 1990. Seasonal spillover and varve formation in the Santa Barbara Basin, California. *Limnol. Oceanogr.* 35:1577–1585.

Richards, F. A., 1965. Anoxic basins and fjords. In: J. P. Riley, and G. Skirrow, eds. *Chemical Oceanography, Vol. 1.* Academic Press, London, pp. 611–635.

Risebrough, R. W., and A. L. Burlingame, 1978. Water column hydrocarbon chemistry (dissolved and particulate). In: *Southern California Baseline Study and Analysis, 1975–1976, Vol. II. Integrated Summary Report.* Science Applications, Inc., La Jolla, CA. pp. II:375-II:389.

Risebrough, R. W., W. M. Jarman, B. R. T. Simoneit, and A. B. Chartrand, 1992. Persistence of DDT and refinery wastes at offshore dump sites in the Southern California Bight. Submitted to *Mar. Environ. Res.*

Rittenberg, S. C., K. O. Emery, and W. L. Orr, 1955. Regeneration of nutrients in sediments of marine basins. *Deep-Sea Res.* 3:23–45.

Rosen, H., A. D. A. Hansen, L. Gundel, and T. Novakov, 1978. Identification of the optically absorbing component in urban aerosols. *Appl. Opt.* 17:3859–3861.

Sargent, J. R., R. F. Lee, and J. C. Nevenzel, 1976. Marine waxes. In: P. E. Kolattukudy, ed. *Chemistry and Biochemistry of Natural Waxes.* Elsevier Scientific Publishing Company, Amsterdam. pp. 49–91.

Schafer, H. A., 1989. Historical trends in municipal wastewater emissions. *J. Water Pollut. Control Fed.* 61:1395–1401.

Schwalbach, J. R., and D. S. Gorsline, 1985. Holocene sediment budgets for the basins of the

California continental borderland. *J. Sediment. Petrol.* 55:829–842.
Seifert, W. K., J. M. Moldowan, G. W. Smith, and E. V. Whitehead, 1978. First proof of C_{28}-pentacyclic triterpane in petroleum. *Nature.* 271:436–437.
Sever, J., and P. L. Parker, 1969. Fatty alcohols (normal and isoprenoid) in sediments. *Science.* 164:1052–1054.
Shaltiel, N., B. R. T. Simoneit, and E. Ruth, 1979. Aromatic hydrocarbons in sediments from the continental rise and bight of southern California. *Annual Report to the U.S. Dept. of Energy, Bureau of Land Management.* Contract no. EY–76–3–03–0034, 19pp.
Shanks, A. L., and J. D. Trent, 1980. Marine snow: Sinking rates and potential role in vertical flux. *Deep-Sea Res.* 27:137–143.
Sharp, J. H., 1973. Size classes of organic carbon in seawater. *Limnol. Oceanogr.* 18:441–447.
Shaw, T. J., 1988. The early diagenesis of transition metals in nearshore sediments. Ph.D. Dissertation, Univ. of California, San Diego. 164 pp.
Shaw, T. J., J. M. Gieskes, and R. A. Jahnke, 1990. Early diagenesis in differing depositional environments: The response of transition metals in pore water. *Geochim. Cosmochim. Acta.* 54: 1233–1246.
Shiller, A. M., 1982. The geochemistry of particulate major elements in the Santa Barbara Basin and observations on the calcium carbonate–carbon dioxide system in the ocean. Ph.D. Dissertation, Univ. of California, San Diego. 197pp.
Shiller, A. M., J. M. Gieskes, and N. B. Price, 1985. Particulate iron and manganese in the Santa Barbara Basin, California. *Geochim. Cosmochim. Acta.* 49:1239–1249.
Shokes, R. F., and P. J. Mankiewicz, 1979. Natural and anthropogenic fluxes of chemicals into the Southern California Bight as related to the potential impacts of offshore drilling. *Rep. to Bureau of Land Management, Benthic, Year 2.* Vol. II, Rep. 24.0, 96pp.
Sholkovitz, E. R., 1972. The chemical and physical oceanography and the interstitial chemistry of the Santa Barbara Basin. Ph.D. Dissertation, Univ. of California, San Diego. 178pp.
Sholkovitz, E. R., 1973. Interstitial water chemistry of the Santa Barbara Basin sediments. *Geochim. Cosmochim. Acta.* 37:2043–2073.
Sholkovitz, E. R., and J. M. Gieskes, 1971. A physical–chemical study of the flushing of the Santa Barbara Basin. *Limnol. Oceanogr.* 16:479–489.
Sholkovitz, E. R., and A. Soutar, 1975. Changes in the composition of the bottom water of the Santa Barbara Basin: Effect of turbidity currents. *Deep-Sea Res.* 22:13–21.
Siegla, D. C., and G. W. Smith, eds., 1981. *Particulate Carbon: Formation During Combustion.* Plenum Press, New York. 505pp.
Siezen, R. J., and T. H. Mague, 1978. Amino acids in suspended particulate matter from oceanic and coastal waters of the Pacific. *Mar. Chem.* 6:215–231.
Silver, M. W., and A. L. Alldredge, 1981. Bathypelagic marine snow: Deep-sea algal and detrital community. *J. Mar. Res.* 39:501–530.
Silver, M. W., A. L. Shanks, and J. D. Trent, 1978. Marine snow: Microplankton habitat and source of small-scale patchiness in pelagic populations. *Science.* 201:371–373.
Simoneit, B. R. T., and I. R. Kaplan, 1980. Triterpenoids as molecular indicators of paleoseepage in Recent sediments of the Southern California Bight. *Mar. Environ. Res.* 3:113–128.
Simoneit, B. R. T., and M. A. Mazurek, 1981. Organic geochemistry of sediments from the southern California borderland, DSDP/IPOD Leg 63. *Annual Rep. to U.S. Dept. of Energy.* Contract no. EY–76–3–03–0034, 24pp.
Simoneit, B. R. T., and M. A. Mazurek, 1982. Organic matter of the troposphere—II. Natural background of biogenic lipid matter in aerosols over the rural western United States. *Atmos. Environ.* 16:2139–2159.
Simoneit, B. R. T., M. A. Mazurek, and N. Shaltiel, 1979a. Lipid geochemistry of Recent sediments from the Patton Escarpment and southern California continental margin and rise. *Annual Rep. to U.S. Dept. Energy.* Contract no. EY–76–3–03–0034, 14pp.
Simoneit, B. R. T., M. A. Mazurek, and E. K. Kalil, Jr., 1979b. Organic geochemistry of Holocene sediments from Tanner Basin, Southern California Bight. *Annual Rep. to U.S. Dept. of Energy, Bureau of Land Management.* Contract no. EY–76–3–03–0034, 12pp.
Sivadier, H. O., and P. J. Mikolaj, 1973. Measurement of evaporation rates from oil slicks on the open sea. In: *Proceedings, Joint Conference on Prevention and Control of Oil Spills.* American

Petroleum Institute, Washington, D.C. pp. 475–479.

Slinn, S. A., and W. G. N. Slinn, 1980. Predictions for particle deposition on natural waters. *Atmos. Environ.*14:1013–1016.

Small, L. F., M. R. Landry, R. W. Eppley, F. Azam, and A. F. Carlucci, 1989. Role of plankton in the carbon and nitrogen budgets of Santa Monica Basin, California. *Mar. Ecol. Progr. Series.* 56:57–74.

Smith, C. R., and S. C. Hamilton, 1983. Epibenthic megafauna of a bathyal basin off southern California: Patterns of abundance, biomass and dispersion. *Deep-Sea Res. Part A Oceanogr. Res. Pap.* 30:907–928.

Smith, D. J., Eglinton, R. J. Morris, and E. L. Poutanen, 1984. Aspects of the steroid geochemistry of an interfacial sediment from the Peruvian upwelling. *Oceanol. Acta.* 6:211–219.

Smith, K. L., Jr., 1987. Food energy supply and demand: A discrepancy between particulate organic carbon flux and sediment community oxygen consumption in the deep ocean. *Limnol. Oceanogr.* 32:201–220.

Smith, K. L., Jr., and R. J. Baldwin, 1984. Seasonal fluctuations in deep-sea sediment community oxygen consumption: Central and eastern North Pacific. *Nature.* 307:624–626.

Smith, K. L., Jr., G. A. White, and M. B. Laver, 1979. Oxygen uptake and nutrient exchange of sediments measured *in situ* using a free vehicle grab respirometer. *Deep-Sea Res.* 26:337–346.

Smith, K. L., Jr., M. B. Laver, and N. O. Brown, 1983. Sediment community oxygen consumption and nutrient exchange in the central and eastern North Pacific. *Limnol. Oceanogr.* 28: 882–898.

Smith, K. L., Jr., A. F. Carlucci, R. A. Jahnke, and D. B. Craven, 1987. Organic carbon mineralization in the Santa Catalina Basin, California: Benthic boundary layer metabolism. *Deep-Sea Res. Part A Oceanogr. Res. Pap.* 34: 185–211.

Smith, P. E., and R. W. Eppley, 1982. Primary production and the anchovy population in the Southern California Bight: Comparison of time-series. *Limnol. Oceanogr.* 27:1–17.

Smith, R. C., and K. S. Baker, 1982. Oceanic chlorophyll concentrations as determined by satellite (Nimbus–7 coastal zone color scanner). *Mar. Biol. (Berl.).* 66:269–279.

Smokler, P. E., D. R. Young, and K. L. Gard, 1979. DDTs in marine fishes following termination of dominant California input: 1970–1977. *Mar. Pollut. Bull.* 10:331–334.

Soule, D. F., and M. Oguri, eds., 1980. The marine environment in Los Angeles and Long Beach harbors during 1978. *Marine Studies of San Pedro Bay, Part 17, Tech. Rep.* Univ. of Southern California Sea Grant Prog. Publ. USC SG LA 1980. 667pp.

Soutar, A., S. A. Kling, P. A. Crill, E. Duffrin, and K. W. Bruland, 1977. Monitoring the marine environment through sedimentation. *Nature.* 266:136–139.

South Coast Air Quality Management District, 1988. *Air Quality Management Plan.* 1988 revision. El Monte, CA.

Southern California Coastal Water Research Project, 1973. *Annual Report.* El Segundo, CA.

Southern California Coastal Water Research Project, 1974. *Annual Report.* El Segundo, CA.

Southern California Coastal Water Research Project, 1975. *Annual Report.* El Segundo, CA.

Southern California Coastal Water Research Project, 1976. *Annual Report.* El Segundo, CA.

Southern California Coastal Water Research Project, 1978. *Annual Report.* El Segundo, CA.

Southern California Coastal Water Research Project, 1980. *Biennial Report, 1979–80.* El Segundo, CA.

Southern California Coastal Water Research Project, 1982. *Biennial Report, 1981–82.* Long Beach, CA.

Southern California Coastal Water Research Project, 1987. *Annual Report.* Long Beach, CA.

Southern California Coastal Water Research Project, 1989. *Annual Report, 1988–89.* Long Beach, CA.

Southern California Coastal Water Research Project, 1991. *Annual Report, 1989–90.* Long Beach, CA.

Southern California Coastal Water Research Project, 1992. *Annual Report, 1990–91 and 1991–92.* Long Beach, CA. In press.

Spies, R. B., and D. J. DesMarais, 1983. Natural isotope study of trophic enrichment of marine benthic communities by petroleum seepage. *Mar. Biol.* 73:67–71.

Spiker, E. C., and P. G. Hatcher, 1984. Carbon isotope fractionation of sapropelic organic matter during diagenesis. *Org. Geochem.* 5:283–290.

Steinberg, S., and I. R. Kaplan, 1984. The determination of low molecular weight aldehydes in rain, fog and mist by reverse phase liquid chromatography of the 2,4-dinitrophenolhydroazone derivatives. *Int. J. Environ. Anal. Chem.* 18:253–266.

Steinberg, S., M. I. Venkatesan, and I. R. Kaplan, 1984. Analysis of the products of the oxidation of lignin by CuO in biological samples by reversed-phase high-performance liquid chromatography. *J. Chromatogr.* 298:427–434.

Strickland, J. D. H., 1968. A comparison of profiles of nutrient and chlorophyll concentrations taken from discrete depths and by continuous recording. *Limnol. Oceanogr.* 13:388–391.

Stuermer, D. H., K. E. Peters, and I. R. Kaplan, 1978. Source indicators of humic substances and proto-kerogen. Stable isotope ratios, elemental compositions and electron spin resonance spectra. *Geochim. Cosmochim. Acta.* 42: 989–997.

Stuermer, D. H., R. B. Spies, P. H. Davis, D. J. Ng, C. J. Morris, and S. Neal, 1982. The hydrocarbons in the Isla Vista marine seep environment. *Mar. Chem.* 11:413–426.

Stull, J. K., and R. B. Baird, 1985. Trace metals in marine surface sediments of the Palos Verdes Shelf, 1974 to 1980. *J. Water Pollut. Control Fed.* 57:833–840.

Stull, J. K., R. B. Baird, and T. C. Heesen, 1986. Marine sediment core profiles of trace constituents offshore of a deep wastewater outfall. *J. Water Pollut. Control Fed.* 58:985–991.

Stull, J. K., R. B. Baird, and T. C. Heesen, 1988. Relationship between declining discharges of municipal wastewater contaminants and marine sediment core profiles. In: T. P. O'Connor, and D. A. Wolfe, eds. *Urban Wastes in Coastal Marine Environments.* Krieger Publishing Co., Malabar, FL. pp. 23–32.

Stumm, W., and J. J. Morgan, 1981. *Aquatic Chemistry,* 2d ed. John Wiley & Sons, New York. 780pp.

Suess, E., 1970. Interaction of organic compounds with calcium carbonate. I. Association phenomena and geochemical implications. *Geochim. Cosmochim. Acta.* 34:157–168.

Suess, E., 1980. Particulate organic carbon flux in the oceans: Surface productivity and oxygen utilization. *Nature.* 288:260–263.

Sugimura, Y., and Y. Suzuki, 1988. A high temperature catalytic oxidation method for the determination of non-volatile dissolved organic carbon in seawater by direct injection of a liquid sample. *Mar. Chem.* 24:105–131.

Summers, J. K., C. F. Stroup, and V. A. Dickens, 1987. A retrospective pollution history of southern California. In: *Proceedings, Oceans '87. Vol. 5: Coastal and Estuarine Pollution.* pp. 1641–1645.

Suzuki, Y. T., and T. Itoh, 1985. A catalytic oxidation method for the determination of total nitrogen dissolved in seawater. *Mar. Chem.* 16:83–97.

Sverdrup, H. U., and R. H. Fleming, 1941. The waters off the coast of southern California, March to July 1937. *Scripps Inst. of Oceanogr. Bull.* 4:261–378.

Sweeney, R. E., and I. R. Kaplan, 1980a. Natural abundances of nitrogen–15 as a source indicator for near-shore marine sedimentary and dissolved nitrogen. *Mar. Chem.* 9:81–94.

Sweeney, R. E., and I. R. Kaplan, 1980b. Tracing flocculent industrial and domestic sewage transport on San Pedro shelf, southern California, by nitrogen and sulfur isotope ratios. *Mar. Environ. Res.* 3:215–224.

Sweeney, R. E., and K.-K. Liu, and I. R. Kaplan, 1978. Oceanic nitrogen isotopes and their uses in determining the source of sedimentary nitrogen. *N.Z. Dep. Sci. Ind. Res. Bull.* 220:9–26.

Sweeney, R. E., E. K. Kalil, and I. R. Kaplan, 1980. Characterization of domestic and industrial sewage in southern California coastal sediments using nitrogen, carbon, sulfur, and uranium tracers. *Mar. Environ. Res.* 3:225–243.

Taylor, B. D., 1981. Sediment management for southern California mountains, coastal plains and shoreline. Part B. Inland sediment movements by natural processes. In: *Calif. Inst. Technol. EQL Rep. No. 17–5B.* pp. B63-B78.

Taylor, B. D., 1983. Sediment yields in coastal southern California. *J. Hydraulic Engineering.* 109:71–85.

Tegelaar, W. W., J. W. de Leeuw, S. Derenne, and C. Largeau, 1989. A reappraisal of kerogen formation. *Geochim. Cosmochim. Acta.* 11:3103–3106.

Ten Haven, H. L., M. Rohmer, J. Rullkotter, and P. Bisseret, 1989. Tetrahymanol, the most likely precursor of gammacerane, occurs ubiquitously in marine sediments. *Geochim. Cosmochim. Acta.* 53:3073–3079.

Thomas, W. H., and A. J. Carsola, 1980. Ammonium input to the sea via large sewage outfalls, Part 1: Tracing sewage in southern California waters. *Mar. Environ. Res.* 3:277–289.

Thompson, B. E., G. P. Hershelman, and R. Gossett, 1986. Contaminants in sediments of two

nearshore basin slopes off southern California. *Mar. Pollut. Bull.* 17:404–409.
Thompson, B. E., J. D. Laughlin, and D. T. Tsukada, 1987. 1985 reference site survey. *Tech. Rep. C–221*, South. Calif. Coastal Water Res. Proj., Long Beach, CA. 50pp.
Tissot, B. P., and D. H. Welte, 1984. *Petroleum Formation and Occurrence*, 2d ed. Springer-Verlag, Berlin. 699pp.
Tont, S. A., 1981. Temporal variations in diatom abundance off southern California in relation to surface temperature, air temperature and sea level. *J. Mar. Res.* 39:191–201.
Tulloch, A. P., 1976. Chemistry of waxes of higher plants. In: P. E. Kolattukudy, ed. *Chemistry and Biochemistry of Natural Waxes*. Elsevier Scientific Publishing Company, Amsterdam. pp. 235–287.
Uematsu, M., R. A. Duce, and J. M. Prospero, 1985. Deposition of atmospheric mineral particles in the North Pacific Ocean. *J. Atmos. Chem.* 3:123–138.
U.S. Environmental Protection Agency, 1983. *National Air Quality and Emissions Trends Report, 1981*. EPA–450/4–83–011. Washington, D. C. pp. 42–45.
U.S. Global Ocean Flux Study, 1989. Sediment trap technology and sampling. U.S. GOFS Planning Report No. 10, August 1989, 94pp.
Venkatesan, M. I., 1989. Tetrahymanol: Its widespread occurrence and geochemical significance. *Geochim. Cosmochim. Acta.* 53:3095–3101.
Venkatesan, M. I., and I. R. Kaplan 1988. Organic geochemistry of particulates and sediments. Prog. Rep. to U.S. Dept. of Energy, 16pp.
Venkatesan, M. I., and I. R. Kaplan, 1990. Sedimentary coprostanol as an index of sewage addition in Santa Monica Basin, southern California. *Environ. Sci. & Technol.* 24:208–214.
Venkatesan, M. I., and I. R. Kaplan, 1992. Vertical and lateral transport of organic carbon and the carbon budget in Santa Monica Basin, California. *Progr. Oceanogr.* In press.
Venkatesan, M. I., S. Brenner, E. Ruth, J. Bonilla, and I. R. Kaplan, 1980. Hydrocarbons in age-dated sediment cores from two basins in the Southern California Bight. *Geochim. Cosmochim. Acta.* 44:789–802.
Venkatesan, M. I., E. Ruth, and I. R. Kaplan 1990. Triterpenols from sediments of Santa Monica Basin, Southern California Bight. *Org. Geochemistry.* 16:1015–1024.
Wallace, G. T., and R. A. Duce, 1978. Transport of particulate organic matter by bubbles in marine waters. *Limnol. Oceanogr.* 23:1155–1167.
Ward, B. B., 1985. Light and substrate concentration relationships with marine ammonium assimilation and oxidation rates. *Mar. Chem.* 16: 301–316.
Ward, B. B., R. J. Olson, and M. J. Perry, 1982. Microbial nitrification rates in the primary nitrite maximum off southern California. *Deep-Sea Res. Part A Oceanogr. Res.* Pap. 29:247–255.
Warford, A. L., 1977. Methanogenesis in polluted and naturally anoxic marine sediments. Ph.D. Dissertation, Univ. of California, Los Angeles, 199pp.
Warford, A. L., D. R. Kosiur, and P. R. Doose, 1979. Methane production in Santa Barbara basin sediments. *Geomicrobiol. J.* 1:117–137.
Water Resources Data for California, Water Year 1986. Vol. 1. Southern Great Basin from Mexican border to Mono Lake basin, and Pacific slope basins from Tijuana River to Santa Maria River. *U.S. Geological Survey Water-Data Rep.* CA–86–1, 301pp.
Wilkinson, E. D., 1972. *California Offshore Oil and Gas Seeps*. Publ. #TR09, Calif. Div. Oil Gas. 11pp.
Williams, P. M., 1967. Sea surface chemistry: Organic carbon and organic and inorganic nitrogen and phosphorus in surface films and subsurface waters. *Deep-Sea Res.* 14:791–800.
Williams, P. M., 1986a. Chemistry of the dissolved and particulate phases in the water column. In: R. W. Eppley, ed. Lecture notes on coastal and estuarine studies. Vol. 15. *Plankton Dynamics of the Southern California Bight*. Springer-Verlag, Berlin. pp. 53–83.
Williams, P. M., 1986b. The chemical composition of the sea-surface microlayer and its relation to the occurrence and formation of natural sea-surface films. In: F. L. Herr, and J. Williams, eds. *ORNL Workshop Proceedings: Role of Surfactant Films on the Interfacial Properties of the Sea Surface*. November 21, 1986, ORNL Report C–11–86, 32pp.
Williams, P. M., 1986c. Long-term particle interceptor trap (PIT) studies. *Prog. Rep. III, West Coast Environmental Program, Calif. Basin Study (CaBS)*. U.S. Dept. of Energy, Washington, D.C.
Williams, P. M., 1988. Compositional and isotopic studies of organic matter in Santa Monica Basin. *Prog. Rep. V, West Coast Environmental*

Program, Calif. Basin Study (CaBS). U.S. Dept. of Energy, Washington, D.C. 12pp.
Williams, P. M., and E. R. M. Druffel, 1987. Radiocarbon in dissolved organic matter in the central North Pacific Ocean. *Nature.* 330:246–248.
Williams, P. M., and E. R. M. Druffel, 1988. Dissolved organic matter in the ocean: Comments on a controversy. *Oceanography.* 1:14–17.
Williams, P. M., and L. I. Gordon, 1970. Carbon–13:carbon–12 ratios in dissolved and particulate organic matter in the sea. *Deep-Sea Res.* 17:19–27.
Williams, P. M., and A. Zirino, 1964. Scavenging of 'dissolved' organic matter from seawater with hydrated metal oxides. *Nature.* 204:462–464.
Williams, P. M., A. F. Carlucci, S. M. Henrichs, E. S. Van Vleet, S. G. Horrigan, F. M. H. Reid, and K. J. Robertson, 1986. Chemical and microbiological studies of sea-surface films in the southern Gulf of California and off the west coast of Baja California Mexico. *Mar. Chem.* 19:17–98.
Williams, P. M., K. J. Robertson, A. Soutar, S. M. Griffin, and E. R. M. Druffel, 1992. Isotopic signatures (^{14}C, ^{13}C, ^{15}N) as tracers of sources and cycling of soluble and particulate organic matter in the Santa Monica Basin, California. *Prog. Oceanogr.* In press.
Wilson, R. D., 1973. Estimate of annual input of petroleum to the marine environment from natural marine seepage. In: *Background Papers for a Workshop on Inputs, Fates and Effects of Petroleum in the Marine Environment. Vol. 1.* National Technical Information Service (NTIS), Springfield, VA. pp. 59–96.
Wilson, R. D., P. H. Monaghan, A. Osanik, L. C. Price, and M. A. Rogers, 1974. Natural marine oil seepage. *Science.* 184:857–865.
Winant, C. D., and A. W. Bratkovich, 1981. Temperature and currents on the southern California shelf: A description of the variability. *J. Phys. Oceanogr.* 11:71–86.
Word, J. Q., and A. J. Mearns, 1979. 60-meter control survey off southern California. *Tech. Mem. C229-TR.* South. Calif. Coastal Water Res. Proj., El Segundo, CA. 58pp.
Young, D. R., and T. C. Heesen, 1978. DDT, PCB, and chlorinated benzenes in the marine ecosystem off southern California. In: R. L. Jolley, W. H. Gorchev, and D. H. Hamilton, Jr., eds. *Water Chlorination: Environmental Impact and Health Effects.* Ann Arbor Science Publishers, Ann Arbor, MI. pp. 267–290.
Young, D. R., and T.-K. Jan, 1977. *Trace Metals in Coastal Power Plant Effluents.* Rep. to Southern California Edison, Rosemead, CA. Contract U0317016. South. Calif. Coastal Water Res. Proj., El Segundo, CA.
Young, D. R., J. N. Johnson, A. Soutar, and J. D. Isaacs, 1973. Mercury concentrations in dated varved marine sediments collected off southern California. *Nature.* 244:273–274.
Young, D. R., D. J. McDermott, and T. C. Heesen, 1975. Polychlorinated biphenyl inputs to the Southern California Bight. *Tech. Mem. 224,* South. Calif. Coastal Water Res. Proj., El Segundo, CA. 50pp.
Young, D. R., D. J. McDermott, and T. C. Heesen, 1976. Aerial fallout of DDT in southern California. *Bull. Environ. Cont. Toxicol.* 16:604–611.
Young, D. R., D. McDermott-Ehrlich, and T. C. Heesen, 1977a. Sediments as sources of DDT and PCB. *Mar. Pollut. Bull.* 8:254–257.
Young, D. R., T.-K. Jan, and M. D. Moore, 1977b. Metals in power plant cooling water discharges. In: *Annual Report.* South. Calif. Coastal Water Res. Proj., El Segundo, CA, pp. 25–31.
Young, D. R., T.-K. Jan, R. W. Gossett, and G. P. Hershelman, 1980. Trace pollutants in surface runoff. In: *Biennial Report, 1979–1980.* South. Calif. Coastal Water Res. Proj., Long Beach, CA. pp. 163–169.
Young, D. R., R. W. Gossett, and T. C. Heesen, 1988. Persistence of chlorinated hydrocarbon contamination in California marine ecosystem. In: T. P. O'Connor, and D. A. Wolfe, eds. *Urban Wastes in Coastal Marine Environments.* Krieger Publishing Co., Malabar, FL. pp. 33–41.

Chapter 4

Microbiology

Gill G. Geesey

Introduction

Microorganisms play an essential role in the maintenance of the waters of the Southern California Bight (SCB). They degrade not only the decaying plant and animal remains produced in these waters but also the anthropogenic matter introduced as sewage and industrial waste. Their degradative activities also result in the production of microbial biomass, which is now becoming recognized as a significant food source for other organisms in the water column and sediments. Chroococcoid cyanobacteria and nitrifying bacteria contribute to autotrophic processes in waters off southern California.

Some microorganisms in SCB marine waters pose a risk to public health. Humans as well as marine species are susceptible to disease-causing microorganisms that are either indigenous to these waters or introduced from other environments. In this chapter, a summary of the literature on the microbiology of the SCB is presented with the goal of identifying microbiological features and trends that may promote a more accurate understanding of ecological and geochemical processes within its boundaries. The result of this effort should help to identify specific areas of understanding and uncertainty and where there is no information.

Microbial Biomass

Water Column

ZoBell and Feltham (1934) conducted some of the first detailed studies of bacterial concentrations in marine waters off southern California. An enriched medium containing peptone, beef extract, and seawater yielded low numbers of colony-forming units (cfu) from water collected off the pier at Scripps Institution of Oceanography (SIO) in La Jolla. No great difference in bacterial density was found in samples collected off the pier at a depth of 5 m and at the same depth 16.1 km offshore. The ZoBell and Feltham (1934) investigation included studies on the effect of tidal, diurnal, depth, and seasonal variations. After a comprehensive study of the cultural requirements of bacteria isolated from waters of the SCB and other locations, ZoBell developed the 2216 Medium, which is now commercially marketed as Marine Broth or Marine Agar (ZoBell 1941a). Culture methods were the primary approach for bacterial enumeration until the mid-1970s.

NONCULTURE METHODS OF ESTIMATING MICROBIAL BIOMASS

Holm-Hansen and Booth (1966) described an adenosine triphosphate (ATP) assay for the determination of total biomass in seawater. The method has subsequently been adapted for evaluation of microbial biomass in the sea (Hamilton and Holm-Hansen 1967). ATP-based determinations of microbial biomass were presented by Sullivan et al. (1978). In their study, the term *microbial biomass* was operationally defined as ATP-derived carbon associated with material passing through a 203-μm pore size filter but retained by a 0.2-μm pore size filter. This fraction contained a variety of living material including bacteria and phytoplankton.

Since the mid-1970s, microbial biomass has been determined by a direct enumeration approach using epifluorescence microscopy. The fluorochrome acridine orange, 4′6-diamidino-2-phenylindole (DAPI), or a Hoechst dye is used to stain cells trapped on a black polycarbonate membrane. Since acridine orange has become the fluorochrome of favor, the technique is now known as "acridine orange direct counts" (AODC). Electron microscopy has also been used to a limited extent to determine microbial sizes for biomass estimates. The direct microscopic methods have yielded biomass values that far surpass earlier estimates based on culture methods.

HARBORS

ZoBell and Feltham (1934) compared bacterial densities in Mission Bay and San Diego

Bay with those obtained from oceanic samples. Bacterial densities in these harbors varied by over two orders of magnitude (table 4.1). In Mission Bay, the lowest counts were recovered at the mouth of the bay when the tide was flooding and the highest in parts of the bay most influenced by land runoff (ZoBell 1941a). In a subsequent survey, ZoBell and Feltham (1942) found that the average bacterial densities from 34 sampling days in the water collected during flood tide were significantly less than samples obtained during an ebbing tide (table 4.1).

A number of stations were sampled by Sullivan et al. (1978) in Los Angeles–Long Beach Harbor. Microbial biomass was highest near a cannery effluent outfall in the harbor (table 4.2). Biomass was higher inside the harbor than outside the breakwater, with bacterial densities one to two orders of magnitude higher in the harbor than offshore. Biomass was higher in surface waters than near the bottom in the harbor, whereas outside the breakwater, no significant difference in biomass was observed between surface and bottom waters.

Plate counts performed in conjunction with the AODC bacterial enumeration method demonstrated that only 1–10% of the bacteria were culturable on Marine Agar. Highest plating efficiency was obtained from sites receiving the highest organic loading. These results were similar to those obtained by Juge and Griest (1973) by standard plate counts of aerobic heterotrophic microorganisms in the harbor from 1973 to 1975. A number of bacterial isolates (162) were recovered during the study, 17 of which were characterized. The low and variable recoveries of bacteria from these areas typify the inadequacy of culture methods for enumerating bacteria in natural waters. Nevertheless, much of the bacterial density data in the SCB to date is based on culture methods.

More recent determinations by AODC of microbial biomass in harbors yield higher estimates of bacterial densities than earlier culture methods. A seasonal peak in bacterial densities within the Los Angeles–Long Beach Harbor occurred in October (5×10^9 cells l^{-1}) one year (1978) and in August (4×10^9 cells l^{-1}) the following year (Krempin et al. 1981). McGrath and Sullivan (1981) reported the same peak bacterial densities in the harbor in October 1978. The bacterioplankton peaks occurred during or immediately after peaks in chlorophyll *a* concentration. During winter, bacterial densities ranged between 1 and 2×10^9 cells l^{-1}. The mean phytoplankton biomass was 24 times the mean bacterioplankton biomass. In spite of the larger biomass contributed by the phytoplankton, the mean surface area of the bacterioplankton was 168 times that of the phytoplankton cell surface area. The large catalytic surface area maintained by the bacterioplankton helps to explain how these microorganisms are able to have such a large impact in spite of their relatively small contribution to the total biomass.

Sullivan and co-workers determined the densities and biomass of bacterioplankton and autofluorescent chroococcoid cyanobacteria in Los Angeles Harbor and outside the harbor in the San Pedro Channel in 1978 (Soule and Oguri 1979). The annual bacterioplankton densities in the harbor were generally higher than those outside the harbor (table 4.1). Monthly mean densities in the harbor were 2.5 times higher than those outside the harbor. Two blooms of bacterioplankton were observed: one in late spring and another in early fall. The cells in the harbor were generally larger than those found in coastal waters. A subsurface maximum (3-m depth) in bacterioplankton standing stock (14.8 μg C l^{-1}) was observed. It was estimated that bacterioplankton biomass contributed from 110 to 2610 kg C yr^{-1} throughout the area of the harbor that was included in the sampling program.

Autofluorescent cyanobacteria were observed throughout the year in the harbor (Soule and Oguri 1979). A total of 82% of these cells fell within the size range of 0.6–1.0 μm. Their standing stock was more than one order of magnitude less than the

Table 4.1. *Bacterioplankton Standing Stocks in the SCB*

Location	Enumeration Method	Bacterioplankton (cells per liter)	Autofluorescent Cyanobacteria (cells per liter)	References
San Diego and Mission bays	Culture	$<5 \times 10^5$ to 4.8×10^8	—	ZoBell and Feltham 1934
Mission Bay				
Flood tide	Culture	1.47×10^6	—	ZoBell and Feltham 1934
Ebb tide	Culture	2.95×10^8		
Los Angeles–Long Beach harbors (mean)	AODC[a]	1.6×10^8 to 5.5×10^8	0.2×10^7 to 6.3×10^7	Soule and Oguri 1979
Los Angeles Harbor near cannery effluent	AODC	1.0×10^{10} to 10×10^{10}	—	Sullivan et al. 1978
SIO Pier, La Jolla	AODC	0.66×10^9 to 2.9×10^9	—	Fuhrman and Azam 1980
SIO Pier, La Jolla	AODC	3.26×10^9	—	Fuhrman 1981
Del Mar nearshore	AODC	3×10^8	—	Fuhrman et al. 1980
1.5 km outside Los Angeles Harbor breakwater	AODC		—	
Winter 1978		0.4×10^9	0.1×10^7	Krempin et al. 1981
October 1978		2×10^9	2×10^7	
August 1979		4×10^9	7×10^7	Krempin and Sullivan 1981
15 km offshore in San Pedro Channel	AODC			
Surface		1.5×10^9	1.5×10^7	Krempin et al. 1981
30-m depth		1.7×10^9	3.0×10^7	
40-m depth		1.2×10^9	1.5×10^7	Krempin and Sullivan 1981
Santa Monica Basin, 1981	AODC			
Surface		4×10^8	—	Carlucci et al. 1986
835 m depth		$<1 \times 10^8$		
Santa Monica Basin, 1982	AODC			
Surface		10×10^8	—	Carlucci et al. 1986
800 m depth		1.5×10^8		
San Pedro Basin, 1983	AODC			
Surface		3×10^8	—	Carlucci et al. 1986
800 m depth		$<1 \times 10^8$		
San Pedro Sill, 1982	AODC			
Surface		9×10^8	—	Carlucci et al. 1986
700 m depth		1×10^8		
Offshore San Diego	Culture	1×10^6	—	ZoBell and Feltham 1934
6 km offshore from Dana Point	AODC			
May		3.7×10^8	—	Carlucci et al. 1984
October		4.5×10^8		

Table 4.1. *Bacterioplankton Standing Stocks in the SCB* (continued)

Location	Enumeration Method	Bacterioplankton (cells per liter)	Autofluorescent Cyanobacteria (cells per liter)	References
0.5 km off Santa Catalina Island	AODC			
Station 1				
Surface		4×10^8	4×10^7	Krempin and Sullivan 1981
40 m depth		3×10^8	$<1 \times 10^7$	
Station 2				
Surface		1.0×10^9	—	
30 m depth		1.5×10^9		

[a] AODC = acridine orange direct counts.

Table 4.2. *Planktonic Microbial Biomass*

	Biomass (μg C l^{-1})		
Location	Bacterioplankton	Cyanobacteria	References
Los Angeles–Long Beach Harbor near cannery effluent			Sullivan et al. 1978
Surface	406	—	
Bottom	273		
Other harbor locations			
Surface	178	—	
Bottom	79		
Average of several harbor locations		0.09–2.87	Soule and Oguri 1979
Outside Los Angeles Harbor breakwater			Sullivan et al. 1978
Surface	17.4	—	
Bottom	16.5		
San Pedro Channel 15 km offshore			Krempin and Sullivan 1981
Surface	12	0.12	
30-m depth	13	0.23	
40-m depth	9.4	0.12	
Seasonal average	9	1.5	
0.5 km off Santa Catalina Island			Krempin and Sullivan 1981
Site 1			
Surface	3.5	0.3	
40-m depth	2.3	<0.07	
Site 2			Krempin and Sullivan 1981
Surface	7.8	0.34	
30-m depth	12	0.15	

total bacterioplankton, contributing 0.09–2.87 μg C l^{-1} (tables 4.1 and 4.2), or 7% of the total bacterial biomass in the harbor.

Discontinuation of cannery effluent discharge into the harbor led to a decrease in bacterial concentrations in the water near the outfall from 237–1648 $\times 10^8$ to 35–43 $\times 10^8$ cells l^{-1} over a 9-month period in 1977–1978 (Soule and Oguri 1979). A 30-fold drop in total microheterotrophs occurred with the transition from primary to secondary waste effluents. This also resulted in a reduction in the difference in bacterial densities inside and outside the harbor. A 10-fold increase in marine bacterial standing stock in the harbor during the summer of 1978 coincided with a malfunction in the Terminal Island Water Treatment facility. It was not clear, however, whether the two events were related, since high bacterial counts also occurred outside the harbor up current from the harbor plume. A treatment plant upset between April and June 1979 led to an increase in the biological oxygen demand (BOD) in the harbor water, which was subsequently followed by a peak in bacterial density in August (McGrath and Sullivan 1981). These data demonstrate the positive correlation between the levels of utilizable organic matter in the water and the densities of bacteria. The organic matter appears to be derived from sewage and phytoplankton sources in harbors and bays of the SCB.

COASTAL WATERS

Samples collected from the pier at SIO at six different times yielded AODC-based bacterioplankton densities that varied from 0.66 to 2.9 $\times 10^9$ cells l^{-1} (Fuhrman and Azam 1980). Estimates of bacterial cell carbon, based on size measurements obtained by scanning electron microscopy (SEM), ranged from 4.6 to 39 fg (femtograms) C $cell^{-1}$. The bacterioplankton biomass at this nearshore site thus ranged from 3 to 113 μg C l^{-1}. A single biomass estimate based on size measurements made by epifluorescence microscopy fell within this range.

In another study of water samples collected off the SIO pier, Fuhrman (1981) determined a mean AODC-based bacterial density of 3.26 $\times 10^9$ cells l^{-1} and a mean cell volume of 0.145 μm^3 based on size determinations made by epifluorescence microscopy. Using the carbon-to-volume ratio of 0.2 pg C μm^{-3} recommended by Kogure and Koike (1987) for natural populations of bacteria in nearshore waters, we calculate a bacterioplankton biomass estimate of 95 μg C l^{-1} of seawater at this site.

Fuhrman et al. (1980) determined mean bacterioplankton densities throughout the water column at a nearshore station off Del Mar (table 4.1). Bacterioplankton biomass was estimated from volume measurements (mean of 0.046 μm^3) based on size determinations made by SEM. Biomass increased from 4 μg C l^{-1} at the surface to 12 μg C l^{-1} at 15-m depth and then declined to 2 μg C l^{-1} at 30 m. Bacterioplankton biomass contributed from 3.6 to 8.8% of the total plankton biomass at this nearshore station.

Bacterioplankton densities in water 6 km off Dana Point were similar to those obtained off the SIO pier in La Jolla (Carlucci et al. 1984). Samples collected at different times of the year showed no significant differences in bacterioplankton density (table 4.1).

SAN PEDRO CHANNEL

Surface waters of the San Pedro Channel exhibited trends in bacterioplankton densities that were similar to those observed in the waters of the Los Angeles–Long Beach Harbor (Krempin et al. 1981). A seasonal peak in bacterioplankton densities was observed during October 1978 and in August 1979 (table 4.1). The lowest densities occurred during the winter and spring months. The bacterioplankton averaged 5% of the total carbon biomass within the size range of 0.2–203 μm.

The density (table 4.1) and size range (0.6–1.0 m) of autofluorescent chroococcoid cyanobacteria in the San Pedro Channel 1.5 km outside the Los Angeles Harbor, up current of the harbor excurrent plume, was similar to the density and size range of these microorganisms in the harbor (Soule and Oguri 1979; Krempin and Sullivan 1981). Although the density of the autofluorescent cells was nearly two orders of magnitude lower than that of the total bacterioplankton, the seasonal trends of the two populations were similar: low in winter, increasing through the spring and summer, and peaking in early fall (table 4.1). Bacterioplankton densities were determined at different depth intervals at one station 15 km offshore in the San Pedro Channel in 1979 (Krempin and Sullivan 1981). A subsurface maximum was observed at a depth of 30 m (table 4.1).

Bacterioplankton densities in Santa Monica and San Pedro basins water were determined during 1981–1983 using the AODC technique (Carlucci et al. 1986). San Pedro sill water contained higher densities in 1982 than water in either basin in 1981 or 1983 (table 4.1). Yearly variations were observed in bottom waters of the Santa Monica Basin. The bacterioplankton population density in the basin water reflected no apparent effects of the 1983 El Niño event. Distribution of bacteria and total free amino acid concentrations were well correlated.

OFFSHORE

The mean bacterioplankton biomass estimated on a cruise in which a total of 6 stations (including inshore and offshore stations) were sampled was 4.6% (range of 1.7–16.7%) of the total plankton biomass (Fuhrman et al. 1980). On another cruise, which sampled 11 stations throughout the SCB, the mean bacterial biomass was 3.3% (range of 0.4–12.8%) of the total plankton biomass. Bacterial biomass tended to decrease with distance from shore. A subsurface maximum was common.

Bacterioplankton densities in the channel near Santa Catalina Island were lower than in Los Angeles–Long Beach harbors (Krempin and Sullivan 1981; Krempin et al. 1981). At one station 0.5 km off the island, the density decreased slightly with depth within the top 40 m (table 4.1).

Carlucci et al. (1985) compared bacterioplankton densities in surface films (neuston layer) with those in underlying water at several stations off San Clemente Island and Santa Catalina Island. Surface film densities averaged 4.4×10^8 cells l^{-1}, which were more than two times higher than the corresponding bacterioplankton densities in water collected at a depth of 10 cm.

In summary, bacterioplankton represent a very small fraction of the total plankton biomass. The standing stock is controlled by the availability of usable dissolved organic carbon. Higher bacterioplankton standing stocks occur in harbors, bays, and nearshore waters than in offshore waters. Maximum bacterioplankton biomass occurs in the surface film, and a submaximum occurs at subsurface depths in the water column.

PARTICLE-ASSOCIATED MICROBIAL BIOMASS

Suspended particles in the water column harbor populations of microorganisms. Macroscopic particles composed primarily of discarded appendicularian "houses" and found in samples collected at depths of 7–10 m in the Santa Barbara Channel, 5 km from shore at the leading edge of an upwelling plume, contained bacterial densities that were three orders of magnitude higher than those in the surrounding water (Prézelin and Alldredge 1983). A total of 78% of the bacterial population resided on the discarded particles. During nonupwelling conditions, aggregates of a different origin were found. These particles contained 7% of the total bacterial population.

The densities of bacteria on macroscopic aggregates collected 2 km south of Santa Cruz Island during July ranged from 12.5 to 16.9×10^5 bacteria per aggregate (Alldredge et al. 1986). These densities were lower than those observed on particles collected in the Atlantic Ocean. The authors of this 1986 study postulated that the El Niño event may have contributed to the low observed bacterial densities. Less than 3% of the total number of bacteria in the water column occurred in association with aggregates in samples collected from the SCB. Bacteria on macroscopic aggregates were significantly larger than free-living planktonic bacteria. The average volume of bacteria on aggregates was 1.02 μm^3, five times greater than that of free-living bacteria in the surrounding water. The contribution of particle-associated bacteria to the total water column bacterial standing stock thus appears to depend on local conditions within the water mass.

MICROZOOPLANKTON

Taylor et. al. (1985) determined the abundance of microzooplankton in waters of the SCB. Microzooplankton in Los Angeles Harbor was dominated by the ciliates *Strombidium* sp., *Euplotes* sp., and *Uronema* sp., which had a combined biomass of 1.0 $\mu g\ C\ l^{-1}$. In the San Pedro Channel, 19 km offshore, *Strombidium* sp. and six species of tintinnids were observed contributing a biomass of 1.1 $\mu g\ C\ l^{-1}$. More studies are needed to relate the standing stocks of these microorganisms to the bacterioplankton standing stocks in waters of the SCB.

Sediments

The earliest enumeration studies were carried out using culture methods. ZoBell and Feltham (1934) analyzed bottom deposits from 28 stations and found from 1.2×10^3 to 3.8×10^7 viable aerobes per gram wet mud; they obtained the highest counts from shallow bottoms. Anaerobic bacterial counts were only 5% of the aerobes in these sediments, but the proportion of anaerobes increased with depth in the sediment. At three stations in coastal waters off San Diego, ZoBell and Anderson (1936) found between 10^5 and 10^7 bacteria per gram sediment. Sediments off the Channel Islands at depths of 1000 to 2000 m yielded the highest bacterial densities ($8–35 \times 10^6$ bacteria g^{-1}) in the top 5 cm of sediment, with densities decreasing to 1–41 x 10^3 bacteria g^{-1} at depths below 40 cm. Aerobes predominated throughout these depth intervals. The ratio of anaerobes to aerobes increased from 1:64 in surface sediments to 1:2 at 66–68 cm depths.

ZoBell (1942) presented data showing a decrease in bacterial densities with depth in bottom deposits collected off southern California. Although the culture methods previously described grossly underestimate the total microbial densities in the sediment, they provide data that describe trends in microbial biomass and that have been mostly confirmed by more modern techniques. Culture techniques are still used to a great extent to evaluate the abundance of specific physiological groups of microbes present in aquatic habitats. In many instances, the only bacterial density data available for some areas of the SCB are derived from culture methods.

HARBORS AND ESTUARIES

The most extensive studies published on the microbial biomass of sediments in protected bodies of water within the SCB are those of ZoBell and Feltham (1942). Large variations in colony-forming units (cfu) were obtained from different cores collected within a 1-m^2 area in Mission Bay. Densities in Mission Bay sediments varied from $1.72–4.60 \times 10^8$ cfu g^{-1} dry wt sediment at the surface to $8.2–62 \times 10^3$ cfu g^{-1} dry wt at a depth interval of 85–90 cm below the sediment surface. Most of the bacteria were in the top 25 cm of sediment. Mud flats contained more bacteria than oceanic bottom mud.

Bacterial densities in the top 1 cm of bottom deposits in harbors and estuaries (as based on AODC) range from 1×10^9 to 5×10^{10} cells g^{-1} dry wt sediment. Using the value 1×10^{10} as the mean bacterial cell density, the value 0.29 μm^3 as the mean cell volume of sediment bacteria (Newell and Fallon 1982), and the value 0.2 pg C μm^{-3}

for the carbon-to-volume ratio (Kogure and Koike 1987), bacterial biomass in harbor deposits is estimated to be 580 μg C g^{-1} dry wt sediment.

COASTAL WATERS AND OFFSHORE WATERS

Sediment bacterial densities were determined at different distances from sewage outfalls along a 60-m depth survey in Santa Monica Bay and off the Palos Verdes Peninsula (Laughlin 1982). Bacterial densities, based on the AODC method, increased from 0.25×10^9 to 30×10^9 cells g^{-1} sediment as the infaunal index (degree of pollution based on macrofaunal populations) decreased from 77 (near ambient) to 6.8 (highly organically enriched). These data, like those for water column bacterial populations, suggest that sediment bacterial biomass is controlled by the availability of usable organic matter.

Rittenberg (1940) sampled green clay and silt bottom deposits between the coastline of north San Diego County and San Clemente Island at water depths ranging from 450 to 1100 m. The aerobic microbial densities, determined by a culture method, decreased rapidly with depth over the top 20 cm (table 4.3). At all levels, the aerobes outnumbered the anaerobes. Within some cores, the vertical distribution of bacteria departed from the general decrease in numbers with depth. These differences were thought to be due to extreme lateral and vertical variability in the bottom material. Depths as great as 355 cm still contained viable bacteria. It remains to be determined whether bacteria recovered from these depths in the sediment are active *in situ*.

Sediment bacterial densities were determined off Orange County on the slope where overlying water ranged from 300 to 600 m in depth (Thompson et al. 1984a). Bacterial densities based on direct enumeration epifluorescence microscopy (using the stain DAPI) ranged from 7×10^8 to 12×10^8 cells cm^{-3} sediment (table 4.3). No trend was observed between bacterial densities in the sediment and water column depth or between bacterial densities and sediment grain size. Using the factors previously described to convert sediment bacterial densities to sediment bacterial biomass, we obtain a value of 58 μg C cm^{-3} sediment. It was estimated that 10–36% of the total microbial carbon was contributed by bacteria and the remainder by protozoans and meiofaunal metazoans.

Sediment bacterial densities were determined on the upper and lower slope off Newport Beach (Thompson et al. 1984b). No significant differences were detected from a sample size of four. Upper and lower slope sediments also contained similar organic carbon content. Bacterial densities based on direct enumeration epifluorescence microscopy (using DAPI) ranged from 7×10^8 to 9×10^8 cells cm^{-3} in sediments that contained 2–4% organic matter. With our conversion factors, this equates to 46 μg bacterial C cm^{-3} sediment.

Carlucci et al. (1976) studied the forms of bacteria that exist in sediments of the San Diego Trough. Most bacteria were found to be rod-shaped and less than 0.4 μm in diameter, and most occurred in aggregates or were attached to particles. No bacterial biomass estimates were presented in this study.

Burnett (1979, 1981) determined the biomass in sediments contributed by nanobiota (defined as organisms in the size range between bacteria and meiofauna, 2–50 μm) at a station sampled approximately 24 km west of San Diego in the San Diego Trough at a depth of 1200 m. Density of organisms and biovolume were determined for six depth intervals ranging from 0–5 mm to 89–98 mm below the sediment–water interface. There appeared to be no decrease in biovolume with increasing depth. A subsurface maximum of nanobiota volume occurred at a depth of 5–10 mm. Yeastlike cells outweighed both protozoans and procaryotes, typically constituting over 70% of the nanobiota volume. Burnett's work suggested that the yeastlike cells formed a major part of the food web in these sediments. Among the protozoans, flagellates predominated in the majority of samples. However, amoebae constituted a

Table 4.3. *Sediment Bacterial Standing Stocks*

Location	Method	Density	References
Offshore San Diego to San Clemente Island	Culture		Rittenberg 1940
Surface			
Aerobes		1×10^6 to 8×10^6 cells g^{-1} wet wt sediment	
Anaerobes		1.5×10^3 cells g^{-1} wet wt sediment	
20-cm depth			
Aerobes		1×10^4 cells g^{-1} wet wt sediment	
Anaerobes		<500 cells g^{-1} wet wt sediment	
Upper slope off Newport Beach	(DAPI)DC		
300-m depth		12×10^8 cells cm^{-3}	Thompson et al. 1984a
380-m depth		7×10^8 cells cm^{-3}	
480-m depth		9×10^8 cells cm^{-3}	
620-m depth		7×10^8 cells cm^{-3}	

larger fraction of the protozoans in deeper parts of the sediment (23% at the top versus 43% in the 10–16 mm interval). Testate cells of foraminiferids and testacids fluctuated around 10%. Nanobiota contributed from 2.5 to 5.1 g m^{-2} biomass in the sediment. Studies such as these by Burnett need to be extended to other areas of the SCB.

Bacterial Population Structure

Virtually nothing is known of community structure at the microbial level in the SCB. Relatively few of the total number of species present have been identified. Differences between bacterial densities estimated by culture methods and those obtained by direct microscopic techniques suggest that the vast majority of the species in marine waters have not been cultured. Characterization of the different types of microorganisms present in various water masses will be an exciting challenge to future marine microbiologists. This information is essential for understanding the role of microorganisms in the sea. Where studies have focused on specific physiological types of microorganisms, the results generally reveal that the bacteria serve a function that is important to other trophic levels and to the successful operation of the ecosystem as a whole.

Distribution of Aerobic and Anaerobic Bacteria in Sediments

One of the earliest studies to address the question of microbial population structure was on the distribution of aerobic and anaerobic bacteria in sediments of San Diego Bay (ZoBell and Anderson 1936). Aerobes were more abundant than anaerobes in the top 3 cm of sediment, but the ratio of aerobes to anaerobes decreased from 74 colony-forming units (cfu) at the surface to 2 cfu at a depth of 66 cm. The ratio of aerobes to anaerobes in surface sediments near the Channel Islands was lower (10–14 cfu) than that in San Diego Bay. Sediments 50 km offshore from San Diego contained surface aerobic bacterial densities ranging from 1.8 to 7.5 $\times$ 10^6 cfu g^{-1} and anaerobic densities of only 1.5–7.5 $\times$ 10^3 cfu g^{-1} (Rittenberg 1940) (table 4.3). The sediments were green clays or silts derived from terrigenous sources. ZoBell and Anderson (1936) concluded that the distribution of aerobic and anaerobic bacteria was governed primarily by the type and quantity of organic matter rather than by distance from shore, depth of overlying water, or temperature.

Luminous Bacteria

Marine luminous bacteria comprise four species in two genera. They exist in a variety of niches and can be found as commensal, parasitic, saprophytic, and free-living forms. Water samples from the entrance of Mission Bay were collected at 3- to 4-week intervals over a 23-month period to evaluate seasonal variations in free-living luminous bacteria (Ruby and Nealson 1978). The total concentration of luminous bacteria varied from 1.5 to 7.5 ml^{-1} with the highest concentrations in late spring and the lowest in winter. *Beneckea harveyi* and *Photobacterium fischeri* together contributed 99% of the strains isolated in the study. *Photobacterium phosphoreum* never contributed more than 14% of the total luminous bacteria. *Photobacterium leiognathi* was not detected by the culture procedures employed in the study. In the summer, *B. harveyi* contributed 60–70% of the total population, while in the winter, it disappeared. *P. fischeri* varied inversely with *B. harveyi*. The abundance of *B. harveyi* was highly correlated with ambient surface water temperature.

Three taxa of luminous bacteria (*P. fischeri, P. phosphoreum,* and *Beneckea* spp.) were found to contribute to the enteric microbial populations of 22 species of surface- and mid-water-dwelling fish inhabiting waters of La Jolla kelp beds, subtidal areas of Santa Catalina Island, and Santa Barbara Basin (Ruby and Morin 1979). High densities of luminous bacteria (10^5–10^7 cfu ml^{-1} gut contents) were associated with many of the fish. The luminous bacteria enter the fish during ingestion of seawater or particles, traverse the alimentary tract, and survive the digestion process. They proliferate in the fecal material and are then redistributed in the surrounding seawater. The study suggested that the enteric habitat of fish serves these bacteria as a site of enrichment.

Bacterial Production

Production of microbial biomass is one of several pathways through which energy may flow in waters and sediments of the SCB. Estimates of the flux of energy through the microbial component of pelagic and benthic food webs must account for the amount that is used for the production of microbial cells. Knowledge of microbial production is of particular significance in instances where microbial biomass serves as an important food resource for other trophic levels.

ZoBell (1939) made the earliest bacterial production estimates in the SCB. On the basis of biochemical oxygen demand, sediment bacterial production was suggested to be as high as "several" milligrams of carbon per square meter per day in the top 5 cm. Unfortunately, no follow-up studies have been performed to verify or refine the results of this early work. Consequently, understanding of benthic bacterial production in the SCB is very limited.

In recent years, bacterioplankton production has been based on rates of thymidine incorporation into DNA. Thymidine incorporation is determined as the amount of all cold trichloroacetic acid–insoluble material produced from the addition of tritiated thymidine to water samples containing populations of planktonic microorganisms. It was shown that virtually all bacteria that utilize glucose and amino acids (active bacteria) could utilize exogenous thymidine (Fuhrman and Azam 1982). For nearshore waters (SIO pier), 45–55% of the bacteria were active.

The success of the thymidine incorporation approach depends on the validity of several assumptions, some of which have not yet been tested under conditions characteristic of natural waters. Recognizing these limitations, Fuhrman and Azam (1980) converted moles of thymidine incorporated per liter per day to cells per liter per day by multiplying by 0.2×10^{18} and 1.3×10^{18} to provide a range that reflects the uncertainty in the amount of DNA per cell. After more detailed studies, Fuhrman and Azam (1982) determined that the most realistic factors for conversion of moles of thymidine incorporated to cells produced are 1.7×10^{18} for nearshore populations and 2.4×10^{18} for offshore (>10 km) bacterioplankton populations. Results from thymidine uptake studies provide some of the best estimates currently available for bacterioplankton production in waters of the SCB.

Nearshore

Studies with water samples collected from the SIO pier demonstrated that the majority (93%) of total thymidine uptake occurred in particles that passed through a 1-μm pore size filter (Fuhrman and Azam 1980). This indicated that thymidine incorporation was primarily a measurement of free-living bacterioplankton production in the sea. Most of the bacterial growth occurred in the absence of particles >3 μm. In view of these results, it was suggested that most of the bacterial growth occurred at the expense of dissolved organic carbon (DOC) rather than particulate organic carbon (POC).

Natural assemblages of marine bacteria grew as a continuous culture when maintained in nearshore water (SIO pier) that was filtered through a 0.22-μm pore size membrane (Hagström et al. 1984). Bacteria grew at generation times of 6–39 hours. These data suggested that at least some bacteria in the sea can grow rapidly at the expense of ambient levels of dissolved organic matter (DOM).

Production rates of bacterioplankton in water collected from the SIO pier at six different sampling times varied from $0.13–8.8 \times 10^8$ cells l^{-1} d^{-1} on one day to $0.46–3.0 \times 10^9$ cells l^{-1} d^{-1} on another day (Fuhrman and Azam 1980). This equated to a microbial biomass production rate of 0.26–14 μg C l^{-1} d^{-1} when based on an average carbon content of 16 fg $cell^{-1}$ (estimated by epifluorescence photomicroscopy) and 2.1–13 μg C l^{-1} d^{-1} when based on an average carbon content of 4.6 fg $cell^{-1}$ (estimated by SEM) from the two different sampling periods, respectively (table 4.4).

On the basis of thymidine incorporation rates that yielded a realistic bacterial production rate of 8.91×10^8 to 2.51×10^9 cells l^{-1} d^{-1}, Fuhrman and Azam (1982) estimated bacteria doubling times of 1–4 days in nearshore waters. The ratio of bacterioplankton secondary production to total primary production in nearshore waters of the SCB was comparable to the ratios obtained for coastal waters of British Columbia and Antarctica (15–25%).

Offshore

Both bacterial abundance and thymidine incorporation in the SCB declined progressively with distance from shore to 100 km (Fuhrman et al. 1980). This trend paralleled that for phytoplankton. In offshore waters of the SCB, doubling times were estimated to be as long as a week or more. At one station in oligotrophic waters 50 km offshore, however, doubling times as short as 1 day were obtained (Fuhrman and Azam 1982). The fastest growth rates occurred at depths just below the depths of greatest phytoplankton abundance (Fuhrman and Azam 1982). Bacterial production in offshore waters is approximately 5–25% of primary production.

Thymidine incorporation rates were determined in samples collected at various stations in the SCB during March 1979 (Fuhrman et al. 1980). Using the conversion factors presented by Fuhrman and Azam (1980) and an average cell volume of 0.046 μm^3, we estimate bacterioplankton production to be 0.1–0.8 μg C l^{-1} d^{-1} (table 4.4). Similar val-

Table 4.4. *Bacterioplankton Production at Various Locations in the SCB*

Location	Moles Thymidine l^{-1} d^{-1}	Cells Produced/ Moles Thymidine Incorporated	Cell Volume (μm^3)	fg C μm^{-3}	fg C $cell^{-1}$	Cells l^{-1} d^{-1}	μg C l^{-1} d^{-1}	References
SIO Pier								
May 18, 1979	6.5×10^{-10} or 6.8×10^{-10}	2.0×10^{17} to 1.3×10^{18}	—	—	16	0.13×10^{8} to 8.8×10^{8}	0.26–14	Fuhrman and Azam 1980
May 16, 1979	2.3×10^{-9}	2.0×10^{17} to 1.3×10^{18}	—	—	4.6	0.46×10^{9} to 3.0×10^{9}	2.1–13	Fuhrman and Azam 1980
May 16, 1979	2.3×10^{-9}	1.7×10^{18}	—	—	4.6	3.9×10^{9}	18	Fuhrman and Azam 1980; Fuhrman and Azam 1982
Southern California Bight								
March 1979	10.8×10^{-11}	1.7×10^{18}	0.046	121	—	1.8×10^{8}	1	Fuhrman et al. 1980; Fuhrman and Azam 1982
August 1979	10.35×10^{-11}	1.7×10^{18}	0.046	121	—	1.8×10^{8}	1	Fuhrman et al. 1980; Fuhrman and Azam 1982
Nearshore summer	—	—	—	—	—	—	6–69	Fuhrman and Azam 1980; Azam and Fuhrman 1981
2 km south of Santa Cruz Island								
Particle-associated bacteria	—	1.4×10^{18}	1.02	120	122	0.48×10^{6} to 6.2×10^{6}	0.48–0.74	Alldredge et al. 1986
Free-living bacteria	—	1.4×10^{18}	0.18	120	0.02	2.1×10^{7} to 2.5×10^{8}	0.46–5.4	

ues were obtained from samples collected in August 1979.

Relationship of Bacterioplankton Production to Other Factors

Bacterial abundance, thymidine incorporation, and thymidine and glucose turnover rates in the SCB were all significantly correlated (Fuhrman et al. 1980). Thymidine incorporation per cell, which is an indicator of specific growth rate, did not correlate with bacterial abundance. Bacterioplankton growth rate was influenced more by the standing stock of the phytoplankton than by primary production of the phytoplankton. It was suggested that bacterial growth results more from DOM released by phytoplankton that has been disrupted and incompletely digested during predation by zooplankton and nekton than from leakage by healthy phytoplankton.

Fuhrman and Azam (1982) estimated that bacterioplankton consume 10–50% of the total fixed carbon. This relationship appears to be valid for inshore mesotrophic waters as well as oligotrophic offshore waters of the SCB. The conservative estimates of bacterial secondary production previously presented demonstrate that bacterial utilization of organic matter is an important pathway for energy passage through pelagic food webs. Based on thymidine incorporation results obtained for the SCB, Azam and Fuhrman (1984) suggested that this pathway is comparable in magnitude to the energy transfer between phytoplankton and macrozooplankton.

Particle-Associated Bacterial Production

Thymidine incorporation experiments were performed with water samples containing macroscopic aggregates of particulate matter collected at a station 2 km south of Santa Cruz Island at depths of 7–15 m (Alldredge et al. 1986). Thymidine incorporation by microorganisms on aggregates varied by more than an order of magnitude over a 6-day period. Rates of incorporation into the particles were 3–95 times higher than in surrounding seawater. Despite this high rate, less than 3% of the total thymidine incorporation in surface waters was on aggregates. Thymidine incorporation per bacterium was 1.1–30 times lower on aggregates than in the surrounding seawater. Carbon production on particles, based on the thymidine incorporation rate, ranged from 7.2×10^{-3} to 7.44×10^{-1} $\mu g\ C\ l^{-1}\ d^{-1}$ (table 4.4). Of total bacterial production 1–13% occurred on the particles collected from the waters off southern California as compared to 3.5–26% in waters of the Atlantic Ocean.

Bacteria as Food for Higher Organisms

Microbial food webs have been shown to act both as a sink for organic carbon by serving as the terminal element in the detrital food chain (Ducklow et al. 1986) and as a link in the food web through the conversion of DOC to nutritional particles that serve as food for other trophic levels (Fuhrman and Azam 1980). The functional significance of the microbial component in marine waters therefore rests on the relative importance of these two pathways.

Bacteria are a rich source of proteins, phosphates (nucleic acids and polyphosphates), carbohydrates (polysaccharides), fatty acids (lipids), and growth factors (vitamins or ectocrine compounds). Consequently, bacteria represent an important source of energy to animals in the sea.

ZoBell and Feltham (1937) were among the first to show that marine bacteria were consumed by the mussel *Mytilus californianus*. Digestive enzymes that lyse bacteria were recovered from the viscera of mussels. Mussels fed a bacterial diet demonstrated a 10–12% weight gain over a 9-month period. The same investigators showed that bacteria were ingested by the sand crab *Emerita analoga*, a detritus-feeding crustacean. The ingested bacteria were readily digested, but it was difficult to determine whether they provided nutritional value to the crab. Evidence was also presented for bacterial consumption by the sipunculid worm *Dendrostroma zostericola*. ZoBell and Feltham (1937) concluded that at least some bacteria are assimilated as food, but it seems doubtful that bacteria are sufficiently abundant in seawater to constitute an appreciable component of the diet of these marine animals. It is also doubtful whether macrofauna play a significant role in controlling bacterioplankton populations on a large scale.

Bacteria appear to play a major role in the diet of polychaetes and amphipods living in sediments of the SCB (Laughlin 1982). *Capitella capitata* removed large amounts of bacteria from the sediments they ingested. Concentrations of bacteria, measured by the AODC method, were 10,000 times higher in the gut of the worms than in the surface sediments where they fed. Concentrations of bacteria in the gut of the polychaete *Cistena californiensis* were two orders of magnitude higher than their densities in the sediment. As the sediments became more contaminated with sewage effluent, the densities of bacteria in the gut decreased even though the bacterial concentrations in the sediment increased.

Filamentous sulfur-oxidizing bacteria that develop around subtidal sulfur springs near White Point off the Palos Verdes Peninsula are ingested by the black abalone *Haliotis cracherodii* (Stein 1984). Foraging vent abalones actively consume the bacteria and restrict their nightly feeding activities to the bacterial mats surrounding the springs. Microbial biomass produced through sulfur oxidation in morphologically similar bacteria around deep sea hydrothermal vents is a form of chemosynthetic production. Thus, the nutrition of invertebrates foraging around the shallow vents in the SCB that is derived from the sulfur bacteria may reflect a source of energy based on chemosynthesis rather than photosynthesis.

Dense populations of oligochaetes and the maldanid polychaete *Paraxillella affinis pacifica* in an area of the Santa Barbara Channel that contains a natural oil seep were thought to be attributable to an abundant supply of food in the form of H_2S-oxidizing *Beggiatoa* mats and hydrocarbon-degrading bacteria that occurred in the sediments around the seeps (Spies and Davis 1979). A consistently larger but fluctuating density of invertebrates was noted in areas containing seeps than in otherwise similar areas without seeps. However, infaunal benthic invertebrate diversity in seep areas appeared similar to that in areas without seeps (Davis and Spies 1980). Sulfur isotope data were consistent with a pathway beginning with petroleum degradation via sulfate-reducing bacteria with the production of H_2S. The sulfide is subsequently oxidized to elemental sulfur via chemosynthetic *Beggiatoa*. The *Beggiatoa* biomass produced is then consumed by nematodes and other infauna (Spies and DesMarais 1983). Approximately 14% more carbon is available through this chemosynthetic pathway, one that appears to be energized initially by hydrocarbon degradation.

Two species of clams, *Lucinoma annalata* and *Calyptogena elongata*, which inhabit sulfide-rich sediments in the Santa Barbara Basin, harbor symbiotic sulfur-oxidizing bacteria in their gills (Vetter 1985). The bacteria appear to provide the mollusc with chemosynthetically fixed carbon via aerobic oxidation of sulfide in return for a stable environment in the gills.

These examples suggest that a portion of the sessile bacterial production in sulfide-rich bottom deposits of the SCB is due to chemosynthetic primary production rather than photosynthetic primary production and secondary production. Bacterial production in nearly all other marine waters considered to date (with the exception of regions of the deep sea that contain hydrothermal vents) is based solely on secondary production and photosynthetic primary production (cyanobacteria) estimates. Unfortunately, no estimates of the extent of bacterial chemosynthetic production in the SCB have been made, but recent evidence suggests that it may be widespread in bottom deposits of the Santa Barbara Basin (R. Francis pers. comm.). It is also not known to what extent the grazing activities of benthic organisms control production of the sessile bacterial populations in these habitats.

Thymidine incorporation studies suggest that secondary production of bacterioplankton is utilized by microflagellates (Fuhrman and Azam 1980). A "microbial loop" that returns energy released as DOM and POC from phytoplankton to the main food chain has been proposed (fig. 4.1).

Bactivory (ingestion of bacteria) by phagocytic protozoa (microflagellates) appears to have a significant impact on standing stocks and metabolic activities of bacterioplankton in many marine habitats. Heterotrophic microflagellates are now recognized as the main grazers of bacterioplankton (Fenchel 1986). Microflagellates in the size range of 3–10 μm are capable of filtering 10–70% of the water column per day (Fenchel 1986). Bacterial densities that support flagellate feeding activity are in the range of $0.5–2 \times 10^9$ cells l^{-1}. It has been estimated that the ratio of microflagellates to bacteria is approximately 1:1000. Oscillations have been observed, however, in the densities of the two populations. There is often a 3- to 4-day lag between the peaks that occur in the densities of the bacterioplankton and microflagellates (Fenchel 1986). This indicates that grazing is important in eutrophic waters such as in harbors and estuaries. It remains to be determined whether microflagellate grazing of bacterioplankton is significant in coastal and offshore waters of the SCB where bacterial densities are on the order of 1×10^6 cells l^{-1}.

Bactivory by microzooplankton and nanoplankton may regulate standing stocks, species composition, and metabolic activity of bacterioplankton. Taylor et al. (1985) determined the importance of bacterioplankton as food for microzooplankton in water samples collected from Mugu Lagoon, Los Angeles Harbor, and San Pedro Channel. Rates of bacteria consumption were 7.05×10^7, 0.14×10^7, and 0.003×10^7 bacteria l^{-1} h^{-1}, respectively. The microzooplankton population in the harbor was dominated by the ciliates *Strombidium* sp., *Euplotes* sp., and *Uronema* sp. *Strombidium* sp. and tintinnids were observed in channel samples.

Bactivory by microzooplankton enhanced release of DOM by 487 ± 354, 2 ± 0, and 1 ± 0 ng C l^{-1} h^{-1} in lagoon, harbor, and channel samples, respectively (Taylor et al. 1985). The presence of bactivores appeared to alter the DOM pool quantitatively and qualitatively. Grazing activities contributed organic compounds of molecular weights less than 500 and between 1000 and 10,000 daltons. It was suggested that bactivore grazing stimulates production of biologically available DOM, which in turn enhances heterotrophic activities and bacterial production. Bacterial ingestion by bactivorous plankton may be at steady state with bacterial production (Soule and Oguri 1979). A significant amount of the ingested bacterial biomass is used in energy metabolism.

Grazing on bacteria by bactivorous microzooplankton was suggested as the major pathway for the loss of adenylates from bacterioplankton (Taylor and Sullivan 1979). The turnover time of the bacterioplankton population by bactivory was estimated to be 12–29 hours in the Los Angeles Harbor. Grazing of bacterioplankton was also suggested to be a mechanism of phosphate transfer to other consumers within the food web. These data suggest that secondary production of microbial biomass in harbors and nearshore waters of the SCB is an important source of energy for higher trophic levels.

Decomposition of Organic Matter

One of the most important functions of bacteria in the sea is to recycle nutrients that are

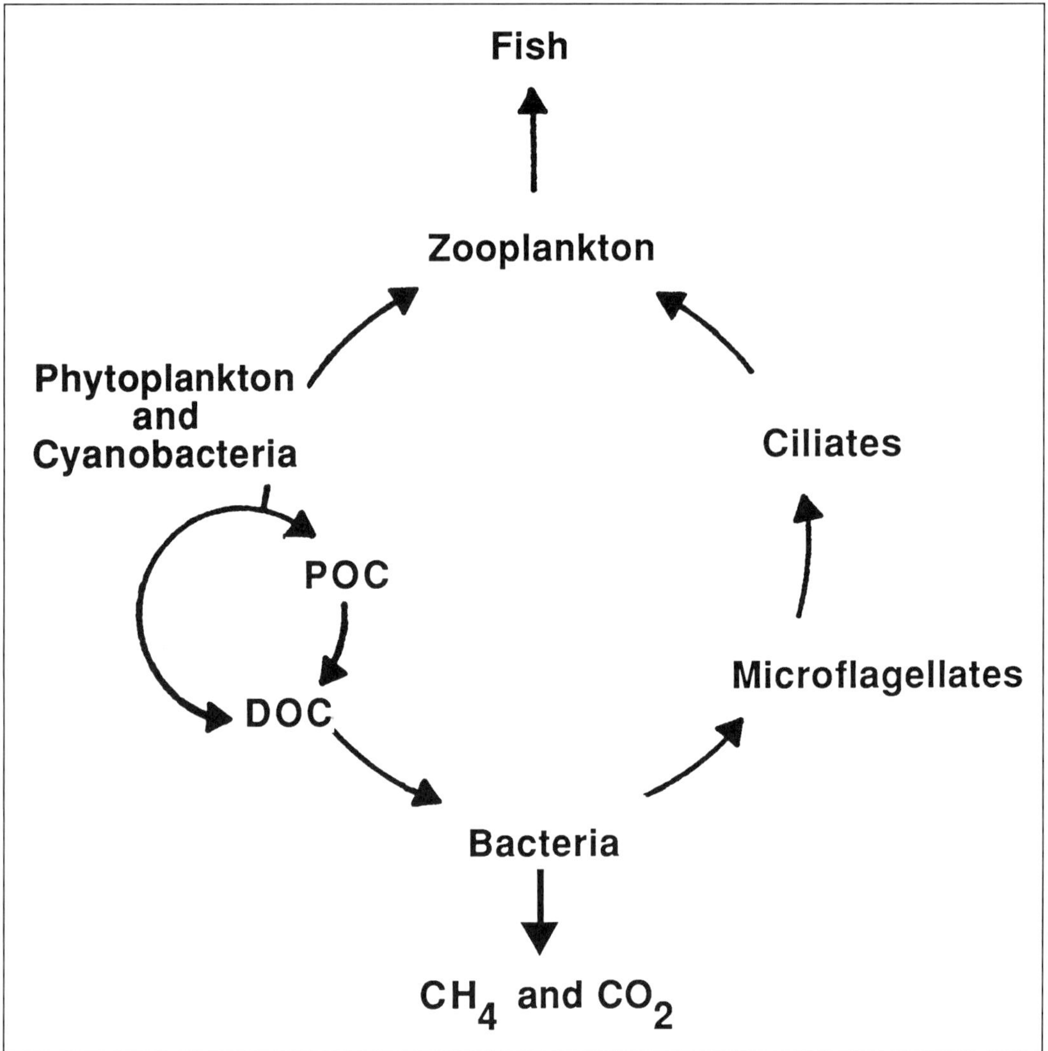

Figure 4.1. Microbial loop showing the utilization of bacteria as a food source for microflagellates. Bacterial biomass is derived from dissolved organic carbon (DOC) and particulate organic carbon (POC) from primary producers.

retained in dead plant and animal biomass. Essential elements such as carbon, nitrogen, sulfur, and phosphorus that are complexed in organic molecules are metabolized by heterotrophic microorganisms to the inorganic compounds required by phytoplankton and macroalgae for primary production. Marine microorganisms contribute to the carbon cycle by converting organic matter that exists in particulate and water-insoluble forms to dissolved compounds of a suitable size range that can be transported into the cell, where biochemical pathways convert the organic compounds to inorganic forms (fig. 4.1).

In offshore waters, high and low molecular weight organic compounds are derived from particulate matter, primarily in the form of phytoplankton. The compounds may be released directly from algal cells as DOC as a result of leakage and cell disruption or indirectly through the hydrolytic activities of microorganisms attached to the senescent algal cells. The organic matter in nearshore waters of the SCB is derived not only from phytoplankton but from terrestrial sources as well.

Most of the studies on the degradative activities of microorganisms in the SCB have been concerned with the dissolved, labile, low molecular weight products derived from phytoplankton. Little information exists on the relationship of the degradative activities of microorganisms in the SCB with the other forms of organic carbon that are present. Sewage is a major contributor of particulate and dissolved organic carbon derived from land-based activities for which degradation rates in the SCB are largely unknown. Hydrocarbons in the form of crude oil and refinery products represent another significant source of organic carbon in the SCB. Hydrocarbons are sparingly soluble in seawater. Consequently, microorganisms must first dissolve the hydrocarbon compounds before they can be degraded. Oil from natural seeps in the Santa Barbara Basin and from shipping and land-based industrial activities are the main sources of this form of organic carbon. Our understanding of pathways and rates of microbial degradation of hydrocarbons in marine environments, however, comes from studies outside the SCB. Some of the organic matter introduced to the SCB from terrestrial sources, such as plastics and pesticides, are exceedingly resistant to microbial degradation and have a high likelihood of accumulating in the sediments or food chain. At present there are no good estimates of microbial degradation of these compounds in the SCB.

Characterization of the mechanisms by which different forms of organic matter are degraded by microorganisms in the sea is also an area that deserves greater attention than given in the past. Most of what we know is based on the use of sugars and amino acids by enteric bacteria from humans. More effort needs to be directed toward understanding the mechanisms by which POC, hydrocarbons, and pesticides are degraded by marine microbial populations.

Hydrocarbons

The SCB contains as many as 60 zones of natural oil seepage. Seepage rate estimates range from <16 to $>160\ m^{-3}\ d^{-1}$ (Reed et al. 1977). In addition, anthropogenic input of hydrocarbons to the SCB is estimated to be at least as important as that introduced through natural seeps.

ZoBell et al. (1943) first demonstrated hydrocarbon-utilizing bacteria in seawater collected in the La Jolla region. All 60-ml samples that were examined contained microorganisms that utilized kerosene, paraffin oil, crude oil, and petroleum ether. Hydrocarbon-utilizing bacteria were most abundant at the mud–water interface, where 10^2–$10^5\ g^{-1}$ wet wt sediment were detected by culture methods. Their density decreased exponentially with depth in the sediment. However, one sample collected from an area near the Channel Islands contained hydrocarbon-utilizing bacteria at the bottom of a 5-m core.

Oil-oxidizing bacteria are most abundant in coastal waters and mud that are chronically exposed to oil. Of the culturable bacteria, 5–50% is able to make use of one or more hydrocarbons in such areas (ZoBell 1969). Along the coast of southern California, oil-oxidizing bacteria range in concentrations from <1 cfu 10 ml^{-1} to $>10^8$ cfu ml^{-1} mud (ZoBell 1969). The highest densities occurred in Los Angeles and Long Beach harbors. Samples of beach sand and coastal seawater rarely contained amounts of oxidizing bacteria greater than 10^4 cfu ml^{-1}. The populations of oil-oxidizing bacteria fluctuated dramatically from month to month at a single station. Densities varied by four to five orders of magnitude.

Hydrocarbon oxidation in marine systems requires the presence of free oxygen (ZoBell et al. 1943). Hydrocarbon oxidation is inhibited where oxygen concentrations are reduced to a level that permits sulfide production by sulfate-reducing bacteria. Surfaces in the form of combusted sediments promote hydrocarbon oxidation by marine bacteria (ZoBell et al. 1943). It was proposed that the sediment aids in dispersion of the hydrocarbons, which increases the surface area of the hydrocarbons and their accessibility to bacteria. Under well-oxygenated conditions, hydrocarbons may be oxidized at rates ranging from 0.02 to 2 g m^{-2} sediment per day at 20–30°C (ZoBell 1969). The end-products of hydrocarbon (paraffin oils) oxidation are carbon dioxide and water, although intermediate products may also accumulate.

Chitin

One of the most abundant forms of polymeric organic matter in the sea is chitin, which is produced by many marine species and incorporated into their skeletal structure. It is estimated that several billion tons of chitin are produced annually (ZoBell and Rittenberg 1938). If chitin were not decomposed by microorganisms, there would be a serious loss of usable carbon and nitrogen in the sea. Because of chitin's water-insoluble nature, microorganisms must rely on exoenzymes to break down the molecules into usable subunits.

Samples of bottom sediments (5- to 6-g portions) from beaches and shallow water along the southern California coast and in deep water nearly 320 km off the coast all demonstrated the presence of chitin-degrading bacteria (ZoBell and Rittenberg 1938). The distribution of chitinoclastic bacteria in sediments varied greatly within any given area. Chitinoclastic bacteria were most abundant in the topmost layers of the mud, where as many as 1000 per gram were recovered. Between 0.1 and 1.0% of the bacteria recovered from seawater by culture methods are chitinoclastic to some degree. From 100 to >1000 chitinoclastic bacteria were found per milliliter of stomach contents of squid and other cephalopods that digest chitinous food. This suggests that chitin-digesting bacteria may exist as endosymbionts, which aid in the digestion of chitin in these organisms.

Many chitinoclastic bacteria derive their complete carbon (or energy) and nitrogen requirements from chitin (ZoBell and Rittenberg 1938). They also liberate ammonia and acetic acid from chitin. Reducing sugars have been detected as products of chitin digestion in some cultures of isolated bacteria. Although there is ample evidence to suggest that chitin degradation via microorganisms occurs in the SCB, no rates are available at this time.

Protein

Protein is one of the most labile polymeric forms of organic carbon in the sea. Protease enzymes are produced by microorganisms that hydrolyze proteins into free amino acids. Hollibaugh and Azam (1983) determined protein degradation rates by natural populations of bacteria in the SCB. Protein as hemoglobin was degraded at a constant rate of 4% per hour by natural microbial populations in water samples collected near Santa Catalina Island. The protein was respired as carbon dioxide at a conservative rate of 0.2% per hour. The rate of degradation increased with increased protein concentration. A biphasic Wolff plot of the reaction suggested that more than one degradative enzyme was active over the added protein concentration range of 1 to 1000 $\mu g\ l^{-1}$. The fraction of degraded protein that was respired ranged from 8 to 10%.

The rate of protein (as bovine serum albumin) degradation by nearshore microbial populations at the SIO pier increased from 0.06% to a maximum of 0.8% per hour as the temperature of incubation increased from 0° to 25°C (Hollibaugh and Azam 1983). Essentially all of the protein degradation activity was associated with particles >0.2 μm in size. In nearshore waters, only about 40% of the degradative activity was associated with particles ≤1.0 μm in size (primarily bacteria and small algae), whereas in offshore waters near Santa Barbara, approximately 80% of the activity was associated with particles in this size range.

Accumulation of protein degradation products in the microbial population was not greatly inhibited by exogenously supplied free amino acids (Hollibaugh and Azam 1983). It was concluded that the amino acids from proteins are utilized preferentially over free amino acids present in surrounding seawater. The advantage of this strategy to the microorganisms is not yet known, since one would predict that it requires more energy for the microbes to hydrolyze the protein to gain access to the amino acids than to make use of the amino acids that already exist in a free state. It was also concluded that dissolved enzymes are not important in protein degradation in natural seawater. Statistical analysis of data collected from water taken nearshore showed that hemoglobin turnover rates correlated significantly with bacterial abundance and biomass, L-leucine turnover rates, chlorophyll concentration, and primary productivity.

Phytoplankton Exudates

Iturriaga (1981) determined that 18–19% of the carbon assimilated by phytoplankton in coastal waters of southern California is released into the surrounding seawater. It was estimated that this exudate is turned over at a rate of 2% per hour. More than 49% of the exudate was composed of compounds with molecular weights of less than 1000 daltons. The low molecular weight fraction was taken up and respired faster than larger sized fractions of phytoplankton extracellular products.

Amino Acids

Degradation of amino acids leads to the liberation of carbon dioxide and ammonia, which are required for phytoplankton growth. Amino acids are also metabolized into proteins and other cellular components that contribute to microbial biomass.

Sullivan et al. (1978) determined dissolved free amino acid (DFAA) usage by microbial populations in Los Angeles Harbor. A mixture of [^{14}C] amino acids was rapidly taken up by the microbial community in bottom waters of the harbor near a cannery effluent outfall. As much as 33% of the carbon taken up in the form of amino acids was respired as carbon dioxide. Uptake of the amino acid mixture occurred at a greater rate in bottom water microbial populations than in surface water populations even though total microbial biomass and bacterial densities were two- to threefold higher in surface waters. Most of the amino acid uptake (72–94%) was mediated by microorganisms in the size range of 0.2–1.0 μm. The size fraction versus amino acid uptake rate was similar to that reported by Azam and Hodson (1977a).

Turnover times of amino acids by microorganisms in harbor water (<25 hours) were among the shortest reported in marine systems (Sullivan et al. 1978). These results suggest that the heterotrophic microbial population is adapted to degrade the protein-rich cannery wastes discharged into the harbor at this location. Outside the harbor in the San Pedro Channel, turnover times ranged from 15 to 100 hours. Turnover times of amino acids by the size fraction 0.2–1.0 μm were found to be as much as 75 times shorter than the size fraction 5–203 μm.

Williams et al. (1976) determined amino acid concentrations and heterotrophic turnover of amino acids at a station 6 miles off the California coast. The turnover rate ranged from undetectable to 1.2 $\mu g\ l^{-1}\ d^{-1}$, with serine, aspartate, alanine, and glutamate having the highest rates. The highest turnover rates were obtained from samples collected at a depth of 25 m. Rates at this depth were 15–20 times higher than at a depth of 100 m. It was estimated that the flux of the amino acids studied amounted to a rate that was 1–10% of the photosynthetic carbon dioxide fixation rate.

Size fractionation studies demonstrated that the majority of leucine uptake in surface waters collected from a variety of stations in the SCB is mediated by particles <1 μm in size, a fraction considered to consist of free-living bacteria (Fuhrman and Azam 1983). Only 4.9% of the total uptake was contributed by >1-μm size particles.

Bacterial activities in subsurface waters of the SCB appear to be dominated by locally adapted free-living microbial populations. Bacteria attached to sinking particles with temperature optima characteristic of surface populations contributed only a small part of the leucine uptake at depths of 100–200 m (Fuhrman and Azam 1983). In subsurface waters, the >1-μm size fraction averaged 4.1% of the total leucine uptake. It is possible, therefore, that decomposition of dissolved organic matter in subsurface waters of the SCB does not depend on bacteria descending from the surface.

Carlucci et al. (1984) determined the uptake of dissolved free amino acids (DFAA) by natural microbial populations in the waters 6 km off the coast of Newport Beach. Their diel studies revealed that maximum rates of DFAA uptake occurred between mid-morning to mid-afternoon and that minimum uptake rates occurred during late evening. The turnover time for DFAA pools ranged from 6 to 48 hours. Turnover times for the DFAA pools were longest at night and shortest in the early morning. Seasonal differences were observed in the daytime and nighttime turnover times.

Glutamic acid utilization was threefold faster in May than in October (Carlucci et al. 1984). The DFAA utilized by the microbial population was believed to have been derived from the DOC exudate from phytoplankton. The study suggested that there is a coupling between primary production and amino acid flux through the microbial population. Microbial utilization, or production of total protein amino acids, was estimated to be 3.6 μg C 1^{-1} d^{-1} in spring and about one-half that rate in fall. Assimilation efficiency for DFAA averaged 65% for marine microbial populations.

Rates of heterotrophic utilization of amino acids were determined in seawater surface films (neuston layer) off the east coast of San Clemente Island and the southwest tip of Santa Catalina Island during November 1982 (Carlucci et al. 1985). Glutamic acid utilization rates at all stations ranged from 0.03 to 0.13 nmol l^{-1} h^{-1}. With one exception, there was no evidence of higher utilization rates of this amino acid in surface films than in water at a depth of 10 cm. The percentage of glutamate taken up that was respired to carbon dioxide varied from 8 to 41%. Respiration rates in the surface film were highest at night. Turnover times for glutamic acid varied from 91 to 174 hours in the neuston layer; these rates were generally equal to or longer than values obtained at the 10-cm depth. Surface film microheterotrophs had an average of 63% amino acid carbon assimilation efficiency, which was similar to euphotic zone heterotrophic populations.

Amino acid utilization by microheterotrophs was studied in waters of Santa Monica and San Pedro basins (Carlucci et al. 1986). The effect of different concentrations of glutamic acid on its use by microheterotrophs in euphotic (50 m) water with high concentrations of total free amino acids (TFAA) and in deep (500 m) water containing low TFAA concentrations was determined. The populations in the upper water layer did not respond to increases in amino acid concentrations over the range of 0.15–15 m*M*. Populations in the deeper water layer did respond to enrichments of the amino acid at concentrations between 7 and 15 n*M*. It was determined that DFAA additions must be kept below 2 n*M* to simulate "natural" amino acid uptake rates.

DFAA utilization rates decreased from 1 nmol l^{-1} h^{-1} at the surface to 0.001 nmol l^{-1} h^{-1} at a depth of 800 m (Carlucci et al. 1986). The absorption rate per cell was nearly an order of magnitude higher in euphotic zone water than in deep water. Bacterial cells in mid-waters absorbed the lower levels of DOC more efficiently than the euphotic zone bacterial cells. Both shallow and deep water populations consumed glutamic acid at a slower rate in 1982 and 1983 than in 1981. During an El Niño event that occurred in 1983 and was characterized by warm water and low productivity, substantial differences in glutamate consumption were noted between the different basins. DFAA uptake was similar in San Pedro Basin and sill waters. The fraction of the amino acids taken up that was respired as carbon dioxide ranged from 10 to 38% in shallow (0–30 m) water (Carlucci et al. 1986).

Turnover times for amino acids were shortest in subsurface euphotic zone waters, while the longest turnover times were recorded at mid-water depths (Carlucci et al. 1986). These turnover times were within the range of turnover times for euphotic zone DFAA pools determined by other researchers. TFAA utilization represented about 2–11% of primary production in the study areas. The higher values were measured during periods of relatively low primary productivity.

Glucose

Complete degradation of glucose results in the formation of carbon dioxide. Like amino acids, sugars such as glucose are also metabolized to other compounds that contribute to microbial biomass. Glucose is one of the most common sugars in coastal seawater since it is the structural subunit of plant cellulose.

Glucose consumption by the microbial population in surface waters of Los Angeles Harbor was studied by Sullivan et al. (1978). The glucose uptake rate near a cannery waste outfall was estimated to be 42 μg glucose l^{-1} d^{-1}. Approximately 25% of the glucose taken up by the microorganisms was oxidized to carbon dioxide. The bulk of the glucose taken up (88%) occurred in a size fraction (<1.0 and >0.2 μm) that contained mainly bacteria.

NEARSHORE AND OFFSHORE WATERS

Azam and Holm-Hansen (1973) indicated that usable organic compounds in the sea are generally on the order of 10^{-9} to 10^{-8} *M*. To evaluate heterotrophic activity in marine waters off southern California, they used [^{3}H]glucose added at concentrations that did not significantly alter the ambient substrate concentration. At one station in offshore waters, most of the heterotrophic activity (86%) at a depth of 10 m appeared to be due to cells that passed through a 3-μm pore size membrane. At a depth of 50 m, the portion of the total heterotrophic activity associated with <3 μm particles decreased to approximately 66%. The total heterotrophic activity at this depth was approximately twice that at 10 m. Approximately one-third of the heterotrophic activity was associated with particles greater than 3 μm in size.

Another study on the distribution of [^{3}H]glucose assimilation by various plankton size fractions was performed by Azam and Hodson (1977a). Depth profiles in the SCB 100 km offshore showed very little variation in the particle size fraction responsible for [^{3}H]glucose assimilation from the surface to 45 m (1% surface irradiance depth) at one station and from the surface to 150 m depth at another station. The <1-μm size fraction contained most of the glucose assimilation activity. Of the particulate chlorophyll, 10–15% passed through the 1-μm filter in nearshore samples collected from the SIO pier, while 20–40% passed through the filters in samples taken offshore in the SCB. Microflagellates were suggested as the possible source of the particulate chlorophyll in the 1-μm filtrates of the nearshore samples. The majority (70%) of the assimilation was due to particles in the size range of 0.4–0.6 μm. No significant amount of chlorophyll was detected in this size fraction. ATP determinations suggested that this size fraction contained 33 μg of living bacterial biomass, which is equivalent to about 1×10^8 cells l^{-1}, based on an ATP:C ratio of 1:250 obtained from laboratory cultures. It was concluded that roughly 90% of the microbial heterotrophic activity is due to free-living organisms, presumably bacteria.

Bacteria associated with particles exhibited D-glucose utilization rates that were one order of magnitude less than that of the free-living bacteria (Azam and Hodson 1977a). Variations in D-glucose concentration from 5×10^{-9} to 1×10^{-3} *M* did not appreciably change the fraction of heterotrophic activity ascribed to free bacteria. The results support the contention that bacteria are responsible for most of the microbial heterotrophic activity in the SCB, even though this size fraction comprises only 5–15% of the total biomass. The smallest bacteria are most active in utilization of D-glucose.

According to Fuhrman (1981), the uptake of radiolabeled organic compounds more closely follows bacterial numbers than bacterial volume. The various sizes of bacteria have roughly the same activity on a per cell basis. Their evidence suggests that the smaller marine bacterioplankton are as active as larger bacteria.

Glucose uptake kinetics of natural marine microbial populations in the waters of the SCB were studied by Azam and Hodson (1981). Wright-Hobbie plots of t/f versus A, where t/f is the reciprocal of the tracer uptake rate and A is the amount of tracer added to the sample, were nonlinear for a large number of seawater samples. This was interpreted as the existence of multiple transport systems with different transport constants (K_t) within the population. This is reasonable considering the variety of microbial species likely to be present. As the ambient glucose concentration increased, systems with lower affinity for glucose were employed in place of high-affinity systems. Multiphasic kinetics were observed for size fractions greater than 0.6 μm and less than 0.6 μm. Only 5–9% of the total glucose uptake was contributed by organisms in the larger size fraction. The size fraction containing small free-living bacteria accounts for the vast majority of the glucose uptake at glucose concentrations ranging from 2×10^{-9} to 6×10^{-4} M. Multiphasic kinetics were observed in nearshore waters as well as in waters 10 km offshore (Azam and Hodson 1981). Although actual transport constants were not determined because of difficulties in determining the natural substrate concentration, linear glucose uptake kinetics over the glucose concentration range of 9.8×10^{-11} to 1×10^{-8} M suggested that subnanomolar K_t glucose uptake systems were not present and that glucose uptake was unlikely in the picomolar range.

The maximum rate of glucose uptake increased as the exogenous glucose concentration was increased (Azam and Hodson 1981). Even with the greater capacity to take up this compound, turnover times increased from 9.5 hours after enrichments of 2.5×10^{-9} M glucose to over 6000 hours after enrichments of 10^{-3} M glucose. Respiration of glucose was low when exogenous concentrations of this substrate were in the nanomolar range (17–18% of the total glucose taken up) and increased at higher exogenous glucose concentrations until a maximum (25–28%) was obtained at 5×10^{-8} M glucose concentrations.

These results demonstrate that the ambient glucose concentration has a significant impact on the regeneration of glucose carbon in waters off southern California. The glucose turnover rate, which is the percentage of glucose in the water sample that is assimilated each hour, was found to average 1.23 per hour over 11 stations that included nearshore and offshore sites (Fuhrman et al. 1980).

Sediment Microbial Activities

Sediment particles offer an abundance of surface areas for sessile microbial colonization and growth. These surfaces provide the opportunity for the development of structured microbial communities. Immobilized sediment bacteria have a greater chance to interact and cooperate in the metabolism of various compounds than do free-living pelagic microbial populations. Consequently, microbial consortia can bring about biochemical transformations that are not normally favored in the water column.

Although sediment particles provide a large surface area for microbe-catalyzed reactions, their spacing impedes the diffusion of substances to and from the surface. Diffusion processes, therefore, often limit microbial activities and growth in the sediment environment. Oxygen depletion and hydrogen accumulation commonly occur during microbial decomposition of reduced organic matter in sediments, frequently leading to the establishment of highly reduced anaerobic conditions.

OXYGEN

During degradation of organic matter in sediments, oxygen is consumed by aerobic microorganisms. Oxygen consumption studies in sediments on the continental slope off Newport Beach yielded an average of 5.32 ml O_2 m^{-3} h^{-1} from 18 different measurements (Thompson et al. 1984b). A high coefficient of variation (61%) was obtained, which suggested heterogeneity due to differences in the organic carbon content, microbial densities, and infauna distribution. Values of 2.31 ml O_2 m^{-3} h^{-1} were reported from the floor of the San Diego Trough (Smith 1974).

HYDROGEN

Microbially mediated hydrogen evolution was demonstrated during organic enrichment of mud from various estuaries, bays, and coastal and offshore sites in marine waters off southern California (ZoBell 1947). The gas evolved was a mixture of hydrogen (1.5–3.5%), carbon dioxide (6–36%), and methane (60–90%). Unenriched mud containing 3.8% organic matter also contained active hydrogen-evolving bacteria. Up to 10^4 hydrogen-producing bacteria were detected per gram of sediment based on the minimum dilution method. Hydrogen producers decreased in abundance with depth in the sediment.

Hydrogen lyase is the microbial enzyme most commonly responsible for the liberation of hydrogen from organic matter (ZoBell 1947). Hydrogen production in sediments is most easily demonstrated by inhibiting hydrogen-oxidizing bacteria. These reactions are closely coupled in marine sediments. In one experiment, Mission Bay mud microorganisms consumed from 20–65 ml hydrogen per 20 g mud at 27°C over a period of 6 weeks. A variety of organic compounds were shown to promote hydrogen utilization in muds suspended in seawater medium. Muds obtained from various estuaries were shown to produce hydrogen sulfide at the expense of hydrogen. Part of the hydrogen appeared to be used for reduction reactions in addition to sulfate reduction.

METHANE PRODUCTION

Methane formation is considered to be the final step in anaerobic bacterial degradation of organic carbon. Most methanogenic bacteria grow optimally in pure cultures at an Eh less than −330 mV and within a pH range of 6 to 8. Claypool and Kaplan (1974) have proposed that bicarbonate is the primary precursor of methane in marine sediments. In sediments of the Santa Barbara Basin, methanogenesis is thought to be most prevalent at depths below the sulfate-reducing zone (150–200 cm depth) (Kaplan et al. 1963).

Warford et al. (1979) found that methane concentrations increase with depth from approximately 1 m*M* at 150 cm to 10 m*M* at 250 cm in Santa Barbara Basin sediments. Methane was produced from substrates that occurred naturally in the sediments. Production rates determined at different depth intervals demonstrated a submaximum at the surface (0–25 cm) of 0.442 nmol g^{-1} d^{-1}, a minimum of 0.016 nmol g^{-1} d^{-1} at a depth of 110–120 cm, and a maximum of 0.767 nmol g^{-1} d^{-1} at a depth of 300 cm. In the sulfate-reducing zone, the rate of methane production decreased with depth. The absence of methane production in the presence of carbon tetrachloride indicated that the methane formed was the result of microbial action rather than sediment outgassing. Rates of methane production did not appear to be limited by hydrogen. The Warford study proposed that the relatively low levels of methane in the sulfate-reducing zone were caused by rapid methane consumption rather than the absence of methanogenesis.

Radioisotope studies suggested that 70–85% of the methane produced in the sediments was derived from bicarbonate with only 2–11% derived from acetate (Warford et al. 1979). It was further demonstrated that some of the carbon dioxide evolved by sulfate-reducing bacteria during lactate oxidation was subsequently metabolized to methane by methanogens. Addition of sulfate appeared to stimulate the conversion of lactate to methane. These results suggest that some form of syntrophic relationship exists between the sulfate-reducing bacteria and the methanogens and that the sulfate-reducing bacteria are capable of providing substrates for methane-producing bacteria. Although it has been suggested that the activities of sulfate-reducing bacteria and methanogens are mutually exclusive, since they both compete for the same reductants and substrates, the results of Warford et al. (1979) suggest that substrate availability, not competition for substrates, controls methane production in the sulfate-reducing zone.

Methane recycling in sediments receiving sewage discharged from the Avalon outfall off Santa Catalina Island was determined by Warford and Kosiur (1979). The rate of methane production was 30.3 nmol g^{-1} d^{-1} during a period when raw sewage (1100 m^3 d^{-1}) was discharged. The density of methanogens estimated by culture method was 1×10^5 cfu ml^{-1} sediment. The rate of methanogenesis decreased by a factor of 30 and the density of methanogens in the sediment decreased by a factor of 20 after the sewage was treated with extended aeration and secondary clarification before discharge. Sulfate was reduced to sulfide during methane production, suggesting that the two reactions are not mutually exclusive. On the basis of this study, it was estimated that prior to wastewater treatment, only 0.13% of the sedimentary organic carbon was recycled annually. Following sewage treatment, less than 0.001% was recycled by methanogenesis.

METHANE OXIDATION

Kosiur and Warford (1979) proposed that methane oxidation occurred at the same sediment depths in the Santa Barbara Basin as methane production. The highest oxidation rate (357.6 μmol l^{-1} yr^{-1}) was observed near the sediment surface (30–35 cm) and decreased with depth, yielding values of approximately 8 μmol l^{-1} yr^{-1} at depths below 130 cm.

Methane oxidation occurred in zones containing high- and low-sulfate pore water concentrations (Kosiur and Warford 1979). It was estimated that the energy yield from methane oxidation was not sufficient to support bacterial growth. It was proposed that other substrates are put to use as energy sources during methane oxidation, using sulfate as the oxidant.

Sulfur Cycling

Marine microorganisms are important in the cycling of sulfur. Although sulfur is rarely considered a limiting element in the marine environment because of the abundance of magnesium sulfate in seawater, microbes are needed to oxidize reduced organic forms of the element—as it exists in living and dead material—to the sulfate form required by primary producers and many microorganisms (fig. 4.2). Sulfur-oxidizing bacteria also catalyze the oxidation of mineral sulfides that are common in reduced marine sediments. Sulfate-reducing bacteria impact the sulfur cycle because they convert biologically available sulfur from sulfate to nonmetabolizable toxic hydrogen sulfide.

Sulfate Reduction

In recent marine sediments rich in organic matter, the major process affecting sulfur geochemistry is bacterial reduction of sulfate to sulfide. Kaplan et al. (1963) presented evidence that bacterial reduction of sulfate is the single most important process in the sulfur

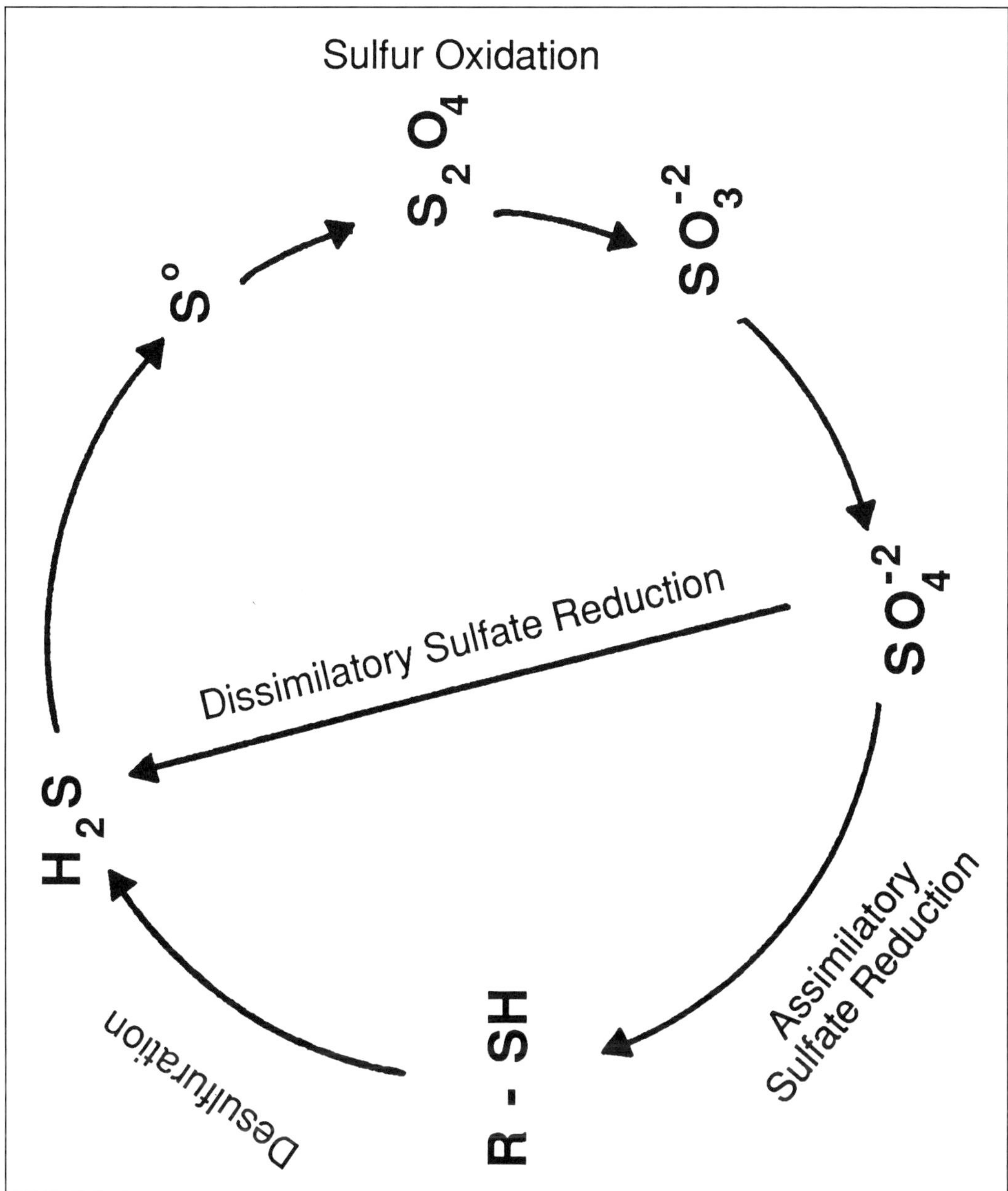

Figure 4.2. Involvement of marine bacteria in the cycling of sulfur. Microorganisms catalyze the reactions in each of the steps shown. H_2S is hydrogen sulfide, S^0 is elemental sulfur, S_2O_4 is thiosulfate, SO_3^{-2} is sulfite, SO_4^{-2} is sulfate, and R-SH is organic sulfide.

cycle. Evaluation of cores from the Santa Barbara Basin, Santa Monica Basin, Santa Catalina Basin, San Diego Trough, and a Newport Bay marsh showed that sulfate concentration decreased markedly with depth in the sediments and was significantly less than that in seawater. This finding indicated that sulfate was reduced in the sediments. Sulfate reduction occurred mainly in the upper 1–1.5 m of sediment. The rate of sulfate reduction in sediments of the Santa Barbara Basin was calculated to be 7×10^{-6} mg S^{2-} cm^{-3} h^{-1}.

ZoBell (1938) found 10^3 to 10^4 sulfate-reducing bacteria (SRB) per gram of sediment in reducing areas. However, Kaplan et al. (1963) estimated densities of 3.5×10^7 SRB cm^{-3} using sulfur isotope values and rates of sulfide production in the laboratory that yielded isotope values found in sulfides in surface sediments. Sulfate reduction was suggested to be a major pathway for organic carbon oxidation in oxygen-depleted sediments such as the Santa Barbara Basin. It was estimated that approximately 16% of the initial organic carbon in the surface sediment is oxidized via the sulfate reduction pathway.

Sulfide levels in Los Angeles Harbor water were high due to the large amounts of organic matter discharged from canneries (Soule and Soule 1981). Treatment of the waste before discharge resulted in reduced BOD levels and a reduction in sulfide levels by as much as an order of magnitude between the years 1974 and 1978. These observations suggest that in anaerobic marine environments, where sulfate ions are not limiting, sulfide production is controlled by the availability of organic carbon, which serves as an electron donor for sulfate reduction to sulfide.

Sulfur Oxidation

Filamentous sulfur-oxidizing bacteria have been shown to exist around intertidal and subtidal sulfur springs near White Point off the Palos Verdes Peninsula and around oil seeps in the Santa Barbara Channel (Spies and Davis 1979; Stein 1984). Stein (1984) indicated that these bacteria grow as a sessile mat around the point of discharge. Evidence suggests that the bacteria oxidize hydrogen sulfide to elemental sulfur (Jacq et al. 1989). The energy derived from this process leads to very high densities of bacterial biomass in localized areas on the bottom. The large biomass of filamentous bacteria around oil seeps in the Santa Barbara Channel was also thought to lead to the high benthic invertebrate biomass at these locations (Spies and Davis 1979).

Nitrogen Cycling

Plants, animals, and most microorganisms require combined forms of nitrogen for incorporation into cellular biomass; thus, the availability of combined nitrogen limits primary production in many marine ecosystems. Nitrogen in living and dead biomass occurs predominantly in the reduced amino form. During microbially mediated decomposition, ammonium is liberated via a process known as ammonification (fig. 4.3). Oxidation of ammonium to nitrite and nitrate is then carried out by a limited number of autotrophic nitrifying bacteria by the process of nitrification. Nitrification occurs exclusively in aerobic environments.

Ammonium, nitrite, and nitrate salts serve as small reservoirs of rapidly cycled nitrogen. Nitrate is reduced via assimilatory and dissimilatory reactions. Assimilatory nitrate reduction results in the incorporation of nitrogen into organic compounds needed for the production of cell biomass. The dissimilatory reactions can lead to the formation of free ammonia (nitrate ammonification) or, through a more complete reduction pathway, to the conversion of nitrate through nitrite to nitric oxide (NO) and nitrous oxide (N_2O) to molecular nitrogen. Denitrification occurs under strictly anaerobic conditions or conditions of reduced oxygen concentration. Atmospheric (molecular) nitrogen is the largest and most slowly cycled reservoir of nitrogen. Biological fixation of atmospheric nitrogen is mediated by a diverse number of procaryotes. Microbially mediated nitrogen fixation in the world's oceans is estimated to be 40 t yr^{-1}, or about 23% of the total annual global fixation (Atlas and Bartha 1987). The biogeochemical cycling of nitrogen is thus highly dependent on the activities of microorganisms.

The participation of microorganisms in ammonification and nitrification has been dem-

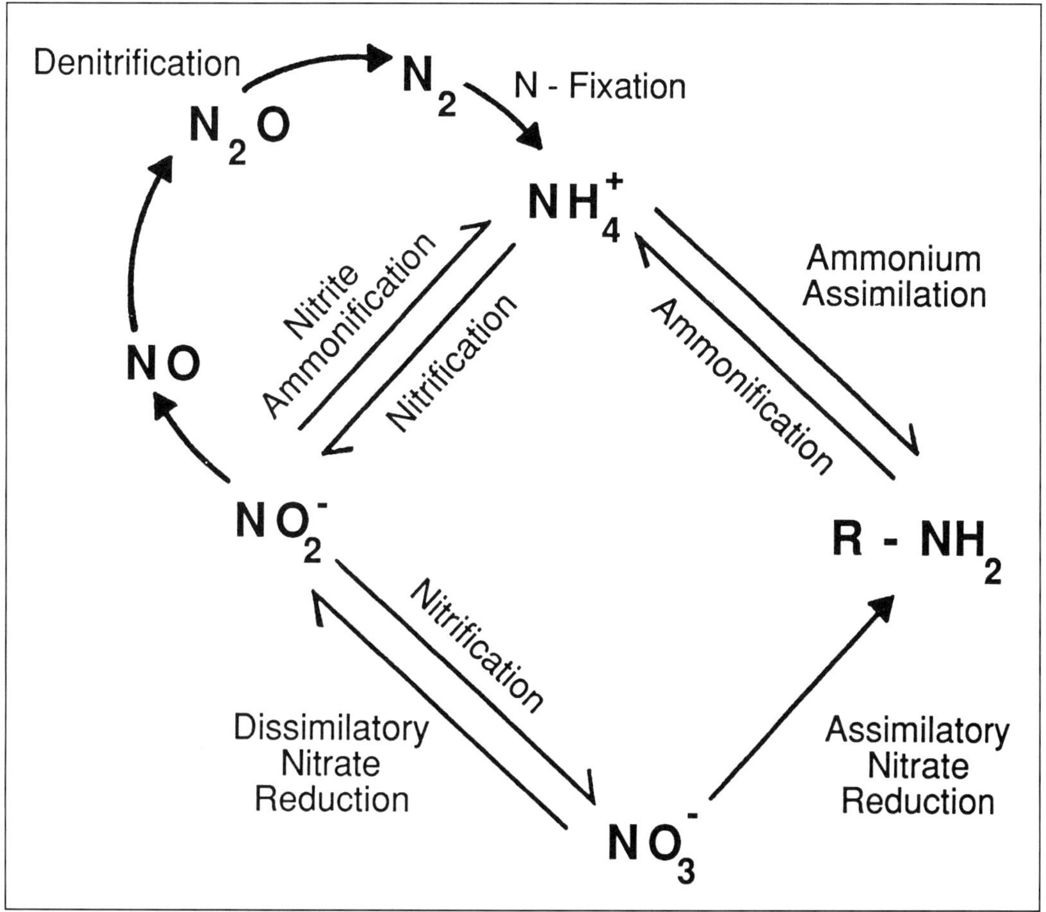

Figure 4.3. Involvement of marine bacteria in the cycling of nitrogen. Microorganisms catalyze the reactions in each step of the nitrogen cycle shown. N_2 is gaseous nitrogen, NH_4^+ is ammonium, R-NH_2 is organic nitrogen, NO_2^- is nitrite, NO is nitric oxide, N_2O is nitrous oxide, and NO_3^- is nitrate.

onstrated in the SCB. Other segments of the nitrogen cycle, such as denitrification and nitrogen fixation, need to be assessed in future studies of the area.

Ammonification

Nitrogen isotope studies suggest that the dissolved ammonium in pore water of sediments from the Santa Barbara Basin is produced from bacterial degradation of marine organic matter (Sweeney and Kaplan 1980). Bacterial degradation of marine planktonic organic matter was determined to be the main source of ammonia in sediment pore water. Only organic matter from marine sources is degraded to any extent by bacteria in the basin sediment. The zone of most rapid ammonia generation is the surface 2 m of sediment. Ammonification is also likely to occur in the water column. However, no information is available on the microbial participation in this process in waters of the SCB.

Nitrification

Primary production in the SCB appears to be driven by the rate of nitrogen input to the euphotic zone. Although there are numerous sources of nitrogen, the most significant is the ocean itself. The process of nitrification is critical to the replenishment of nitrate-N in euphotic waters of the SCB.

Stations sampled from Santa Monica Bay to 98 km offshore near the Channel Islands showed a typical nitrite maximum near the 1% light level which coincided with the upper part of the nitrate gradient (Olson 1981). Seasonal variations in the depth of the nitrite maximum were observed at a station in the San Pedro Channel. The maximum decreased from a depth of 35 m in March to 60 m in October. Highest nitrite concentrations were measured in June.

Olson (1981) used an ^{15}N tracer technique to determine microbial ammonia oxidation rates in the SCB. The microorganisms responsible for ammonia oxidation at various stations in the SCB exhibited a very high affinity for ammonia (K_s of 0.1 μM or less). No consistent increase in the rate of ammonia oxidation was observed over the ammonia concentration range of 0.1–20 μM. Ammonia appeared to be the main source of nitrite in the nitrite maximum. At depths above the maximum, nitrate reduction becomes a more important source of nitrite.

Carlucci et al. (1970) determined that the high nitrite concentrations in the secondary nitrite maximum (generally below 150 m) in oxygen-depleted waters of the eastern tropical Pacific Ocean arise from nitrate reduction. The energy for bacterial nitrate reduction originates from the biochemical oxidation of organic matter. Some nitrite in the secondary nitrite maximum may arise from oxidation of ammonia by nitrifying bacteria. Concentrations of ammonia found in seawater indicate that nitrite production by nitrifying bacteria is very slow.

The mean turnover time for nitrite in the nitrite maximum layer was 12.5 days in southern California coastal waters as compared to 25 days in the central North Pacific gyre (Olson 1981). Size fractionation experiments indicated that nitrifying bacteria mediated the observed ammonia-oxidizing activity. A large portion of the activity passed through a 0.6-μm pore size filter but was retained by a 0.2-μm pore size filter. Chemical inhibitors of nitrifying bacterial activity also inhibited ammonia oxidation when added to samples.

In these waters, ammonium oxidation rates were comparable to ammonium uptake rates. At the nitrite maximum, ammonium uptake by bacteria was comparable to that by phytoplankton. Studies in water collected from SIO pier and from the San Clemente Basin showed that nitrate reduction increases with light intensity while ammonia oxidation decreases (Olson 1981). The threshold for inhibition of ammonia oxidation is between 0.2% and 2% of sunlight.

Nitrite and nitrate uptake appear to be competitive processes (Olson 1981). Unlike ammonia oxidation, nitrite oxidation shows a marked dependence on nitrite concentration. A Lineweaver–Burk plot of the data obtained from water samples taken 3 miles off SIO pier at a depth of 60 m resulted in a half-saturation constant value of 0.07 μM nitrite. This value is similar to many natural nitrite concentrations in the marine environment. The highest nitrite oxidation potential was found below the nitrite maximum layer, while the ammonia oxidation activity was highest within the nitrite layer. The relatively low nitrite oxidation rate within the nitrite maximum is likely due to inhibition by light similar to that demonstrated for ammonia oxidation. Nitrite oxidation exceeded nitrite production at all depths except at the nitrite maximum. It was concluded that the main source of nitrite in oxygenated water is ammonia oxidation and that there is a gradual switch from nitrate reduction to ammonia oxidation as light intensity decreases.

The abundance of *Nitrosomonas* and *Nitrosococcus* was measured at five stations in the SCB, using an immunofluorescent assay (Ward 1982). Cell concentrations were fairly constant with depth. For *Nitrosococcus,* densities of 4 × 10^3 cells l^{-1} in November and

3×10^4 cells l^{-1} in July were obtained. *Nitrosomonas* densities were 4.6×10^4 cells l^{-1} in July. Nitrifiers were more abundant in bottom water samples than in surface samples, suggesting a sediment origin for part of the water column population.

Nitrifiers were found to constitute a very small portion of the total bacterial population in seawater (Ward and Carlucci 1985). Antisera developed against ammonium-oxidizing bacteria did not cross-react with nitrite-oxidizing bacteria and vice versa. Not all of the ammonium- and nitrite-oxidizing isolates reacted with antisera produced against genera of their respective physiological group. There was no significant difference in the distribution pattern of total ammonium-oxidizing and total nitrite-oxidizing bacteria in the water column of a station off the eastern tip of Santa Catalina Island. Most cells occurred singly. The total abundance of ammonium-oxidizing cells (*Nitrosococcus* plus *Nitrosomonas*) averaged 3.5×10^5 cells l^{-1}. Total cell concentration was highest in surface water. Total nitrite-oxidizing cells (*Nitrobacter* plus *Nitrococcus*) averaged 2.8×10^5 cells l^{-1}. Maximum concentration occurred at 9 m depth. Abundance at 700 m was greater than at 150 m. Nitrifying bacterial densities at this station were higher than those reported for another station in the SCB (Ward 1982).

Nitrifying bacteria densities were also determined in the water column 5 km off the SIO pier and 5 km off Del Mar (Ward 1982). The mean bacterial counts (over the upper 100 m of the water column), based on immunofluorescent assay for two species of ammonium-oxidizing bacteria, yielded 5.2×10^4 cells l^{-1} in July and 4.7×10^4 cells l^{-1} in November. *Nitrosomonas marinus* was always more abundant than *Nitrosococcus oceanus*.

Although ammonium concentrations were quite variable with depth and a well-defined nitrite maximum was present at the bottom of the thermocline, no significant difference in density was observed with depth in the upper 100 m for either species (Ward et al. 1982). Studies using ^{15}N tracers were performed to relate ammonia oxidation rates to nitrifying bacterial densities. Rates of ammonia oxidation were low at the surface and generally increased to a maximum in the nitrite maximum layer. Ammonia oxidation rates on a per cell basis increased two orders of magnitude from the surface to the nitrite maximum layer and below. The mean nitrite production rate was estimated to be 1.9×10^{-8} mol cell^{-1} d^{-1} over the upper 100 m of the water column in July and 1.6×10^{-8} mol cell^{-1} d^{-1} in November. The data suggest that ammonia concentrations do not control ammonia oxidation rates. The results report for the first time the presence of nitrifying bacterial populations of sufficient size to account for the high *in situ* rates of ammonia oxidation.

Virtually all the ammonia in water samples collected at depths of 7–10 m in the Santa Barbara Channel, 5 km from shore, and at the leading edge of an upwelling plume was associated with suspended macroscopic particles, which were composed primarily of discarded appendicularian houses (Prézelin and Alldredge 1983). Nitrate and nitrite were undetectable in the aggregates. Since ribulose 1,5-biphosphate carboxylase activity was high in photosynthetically suppressed aggregates, it was suggested that the enzyme activity is due to chemoautotrophic nitrifying bacteria. It was further suggested that the surfaces of aggregates are favorable environments for nitrifying bacteria. Such observations help to explain the consistent but low densities of these bacteria in pelagic waters.

Phosphorus Cycling

Phosphorus exists mainly as orthophosphate in the marine environment. Phosphate is cycled between particulate organic forms to soluble inorganic forms, the latter being the preferred source of phosphorus by phytoplankton (fig. 4.4).

Marine microorganisms have been shown to utilize dissolved phosphorylated organic

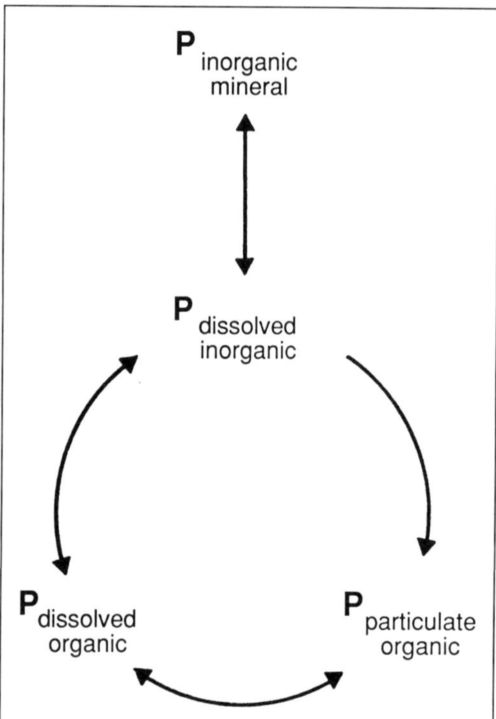

Figure 4.4. Involvement of marine bacteria in the cycling of phosphorus. Microorganisms catalyze the reactions that convert particulate inorganic phosphorus ($P_{inorganic\ material}$) to dissolved inorganic phosphorus ($P_{dissolved\ inorganic}$) and vice versa. They also catalyze the conversion of dissolved inorganic phosphorus to particulate dissolved organic phosphorus ($P_{dissolved\ organic}$) and particulate organic phosphorus ($P_{particulate\ organic}$) and catalyze reactions that lead to interconversion between the two latter forms of phosphorus.

compounds in seawater. Azam and Hodson (1977b) determined that dissolved adenosine triphosphate (DATP) in surface waters of the SCB varied from an average high value of 218 ng l^{-1} off the SIO pier to an average low value of 65 ng l^{-1} 4 miles offshore from Point Loma, San Diego. Turnover rates of DATP, defined as the fraction of the total DATP in the sample that is assimilated per hour, were determined from measurements of assimilation of [2,8-^{3}H]ATP by natural microbial populations in water samples. The turnover rates were indicated to be quite high (>0.59% per hour), although actual rates were not presented. With [^{14}C]DATP used as a tracer, it was determined that greater than 98% of the ^{14}C taken up was assimilated by the microbial population, with insignificant amounts being respired. Bacteria accounted for approximately 80% of the DATP assimilated.

Natural populations of microorganisms in nearshore and offshore waters of the SCB readily take up cyclic adenosine monophosphate (cAMP) (Ammerman and Azam 1981, 1982). Uptake occurred predominately by microorganisms (bacteria and some cyanobacteria) that passed through a 0.6-μm pore size filter. Rates of dissolved cAMP (DcAMP) turnover were approximately 1% per hour within 6 km of shore and approximately 0.1% per hour in water 100 km offshore.

It was proposed that DcAMP was taken up by specific, energy-dependent, high-affinity transport systems with measured transport constants (K_m) of 10–1000 p*M* (picomolar) (Ammerman and Azam 1981). The DcAMP concentrations in nearshore waters (off SIO pier) ranged from 2 to 30 p*M*, and DcAMP concentrations in the relatively oligotrophic waters 41 km offshore in the SCB varied from less than 1 to 3 p*M*. These concentrations appear too low for the transport systems to operate effectively. Only in the pore water of the sediments, where DcAMP concentrations were one to three orders of magnitude higher than water column concentrations, would these high-affinity bacterial transport systems be functional. Neither algal nor bacterial uptake of DcAMP could account for a sharp diurnal drop in the concentration of this compound in the water column between midnight and dawn. Since both [^{32}P]DcAMP and [2,8-^{3}H]DcAMP were taken up in the presence of excess 5′-AMP and inorganic phosphate (P_i), it was suggested that DcAMP was taken up intact.

McGrath and Sullivan (1981) determined the rate of total adenylates (ATP, ADP, and AMP) uptake in waters of Los Angeles Harbor and the San Pedro Channel. Four stations within the harbor and one station 1.5 km outside the breakwater were sampled on a monthly to quarterly basis. The rate of total adenylate uptake was approximately equal to that which was incorporated, indicating that soluble intracellular pools were small and turned over rapidly. The uptake rate of adenylates in harbor water equaled or was slightly greater than that in channel water. Uptake kinetic analysis suggested that the transport systems for adenylates were always undersaturated. There were no significant differences in the natural velocity of uptake or the turnover time between the bacterioplankton of harbor water and channel water for adenylate uptake. The plankton community in channel water had lower transport constants than in harbor water, indicating that the channel populations were adapted to lower total adenylate concentrations in the surrounding seawater. Uptake was due almost entirely to bacterial-sized rather than algal-sized organisms. These data thus demonstrate that bacteria in the SCB are important in converting dissolved organic phosphate to particulate organic phosphate.

Microorganisms in marine waters off the southern California coast release some of the phosphorus taken up in an organic form as free P_i. Picoplankton (organisms less than 1 μm in diameter) have a cell surface enzyme, 5′-nucleotidase, that rapidly hydrolyzes 5′-nucleotides and regenerates P_i (Ammerman and Azam 1985). Substrate specificity was not repressed by P_i concentrations (100 μM or less) that normally exist in the sea. Hydrolysis of organic phosphorus and uptake of the hydrolyzed P_i were coupled. High enzyme activity was detected in nearshore waters (off SIO pier) and those collected 3 km offshore. Activity was detected through the water column down to 50 m at a station 100 km offshore. Little activity was due to enzymes free in solution. The enzyme exhibited high affinity for 5′-nucleotides, with Michaelis constants of 3 and 67 n*M*, depending on the 5′-nucleotide concentration. In California coastal water, a typical rate of P_i consumption is 1–2 nmol l^{-1} h^{-1}. The rate of P_i release by the microbial enzyme was estimated to be 1 nmol l^{-1} h^{-1}. Thus, microbial 5′-nucleotidase accounts for 50–100% of the total P_i consumption rate in the sea. Of the P_i generated by the enzymatic activity, 10–15% was taken up by the bacteria. As P_i is depleted in the seawater, the amount of P_i incorporated by the microorganisms increases to about 50%.

Microorganisms also take up P_i from marine waters. A phytoplankton-enriched size fraction (1–203 μm) and a bacteria-enriched size fraction (0.2–1.0 μm) accounted for 47% and 53%, respectively, of the microbial P_i uptake in Los Angeles Harbor and San Pedro Channel water (Krempin et al. 1981). Although the mean phytoplankton biomass was 24 times the mean bacterioplankton biomass, the two populations took up P_i at comparable rates (7 nmol P l^{-1} h^{-1}) and turned over the P_i in the harbor water in similar time periods (140–150 hours). Rates of P_i uptake by microbial populations in channel water were lower than in harbor water. Turnover times of phosphorus by the bacteria fraction in the channel were similar to those in the harbor. Microbial biomass and P_i uptake rates were significantly correlated (r = 0.89) among samples collected at a station inside and outside the harbor.

Phosphate uptake at different depths in the harbor correlated with chlorophyll *a*, bacterioplankton, and *in situ* soluble reactive P_i (Krempin et al. 1981). The greatest uptake occurred at a depth of 3 m, where chlorophyll *a* and bacterial biomass peaked. In the channel, P_i uptake rates varied with microbial biomass above and below the thermocline.

Maximum velocity of P_i uptake by combined phytoplankton and bacterioplankton fractions was estimated to be 1.23 nmol P l^{-1} h^{-1}, while that of the bacterioplankton fraction alone was 5.17 nmol P l^{-1} h^{-1} in surface water 0.5 km north of Isthmus Cove, Santa Catalina Island (Krempin et al. 1981). Transport constants (K_t) for the combined plankton fractions and the bacterioplankton fraction alone were estimated to be 5.93 and 112 nmol P l^{-1}, respectively.

Krempin et al. (1981) suggested that if the flow of energy is more rapid through the detrital food web than through the phytoplankton, the bacterioplankton could be dominant in P_i uptake while phytoplankton dominate the biomass. Krempin's group concluded that although bacterioplankton may constitute a relatively small percentage (10–20%) of the total microbial biomass, they are a functionally significant part of the pelagic ecosystem and are responsible for the majority of orthophosphate assimilation in the outer Los Angeles Harbor.

Human Pathogens and Fecal Indicators

Southern California has many coastal beaches that are used year round by bathers. Shellfish are harvested in tidal areas and eaten by the population. Since sewage is also discharged into these coastal waters, it is necessary to monitor the water closely for infectious microbes. Because concentrations of pathogens are usually very low, indicator organisms, which reflect the presence of pathogens but are present in greater concentrations and thus easier to detect, are generally used as monitors of fecal pollution. Total coliforms (TC), fecal coliforms (FC), and fecal streptococci (FS) are usually employed to indicate the possible presence of enteropathogens such as *Salmonella* and *Shigella*. (In reality, there is no universal indicator organism for determining water quality.)

According to a California Department of Public Health report (1942), no epidemic disease traceable to use of Santa Monica Bay beaches for bathing was reported prior to the period of rapid population growth that occurred in the Los Angeles area after World War II, in spite of heavy pollution in which coliform counts exceeded 11,000 coliform MPN (most probable number) per 100 ml in surf near the Hyperion Wastewater Plant discharge area. More recently, however, there have been reports of increased incidence of sickness among lifeguards at beaches in Santa Monica Bay. Furthermore, beach closures are now common during and immediately following rainy periods, when coliform counts rise above acceptable levels as a result of the diversion of raw sewage to storm drains that empty into coastal waters of Los Angeles and Orange counties.

Occurrence of Indicator Microorganisms

After a survey involving 961 samples of seawater and 387 marine fish, ZoBell (1941b) concluded that coliform bacteria were not present in the ocean at places remote from sewage outfalls. *Escherichia coli* was found in the intestines of only a few fish that were taken near land. *Enterobacter* and *Citrobacter* were much more abundant than *E. coli* in the intestinal contents of fish. In bays, in estuaries, and in the vicinity of outfall sewers, as few as 1 coliform per liter often gave a presumptive test. The abundance of these organisms decreases rapidly with distance from the outfall.

ZoBell (1960) presented evidence that in the open ocean only 12 of 1468 1-ml samples of seawater yielded positive presumptive tests. The presence of *E. coli* was confirmed in none. Volumes of 1 ml of the intestinal contents of fish yielded 206 presumptive positive tests from a total of 630 tests. Only 9% were confirmed positive for *E. coli*. Approximately 75% of the gas producers were *Aeromonas* spp. In general, it is difficult to estimate bacterial pollution on the basis of these presumptive tests.

Rittenberg et al. (1958) determined the numbers of TC in sediments in the vicinity of sewage outfalls in Orange County (Santa Ana River mouth), White Point, and Santa Monica Bay. High numbers of presumptive positive tests were obtained within a radius of a few thousand feet of the outfalls off Orange County and in Santa Monica Bay. Sediments in an extensive area around the Orange and Los Angeles county outfalls, which discharged unchlorinated sewage, contained as many as 25,000 and 92,000 coliforms cm^{-2}, respectively. Samples collected around the Santa Monica Bay outfall, which discharged chlorinated sewage, contained much lower (250 coliforms cm^{-2}) levels of contamination. Like the pelagic population, the large majority of presumptive coliform positive samples from sediments failed to confirm on eosin methylene blue (EMB) agar. At least some of the false positive presumptive results were due to the presence of lactose-fermenting, anaerobic, spore-forming bacteria. Results with brilliant green bile broth suggested that this medium should not be used for the confirmed test when using marine sediments.

In 1972, the California State Water Quality Board established criteria for limits on total coliforms in waters along the shore and in areas where shellfish may be harvested for human consumption. Monthly monitoring of TC, standard plate counts, and BOD were instituted in 1972 in Los Angeles Harbor in response to discharges of cannery wastes (Soule and Oguri 1975). Mean standard plate counts showed considerable variation between yearly means in 1973 and 1974. Outfall areas were high, as was the Los Angeles River channel. Increases in TC generally occurred after storms. A three-year investigation of microbial activity in the harbor showed that deviation from normal marine bacterial flora reflects the nature of nonmarine or mixed waters introduced into the harbor. No one variable, with the possible exception of FC, could be used to estimate the presence or levels of microbial pollution.

Bacterial pollution did not appear to exhibit a seasonal pattern in San Diego Bay (Nusbaum and Garver 1955). Coliforms disappeared rapidly in the large, shallow south section of the bay, where the heaviest sources of pollution exist. At low tide, coliforms were detected as far as 6 miles from the sewage outfall, whereas at high tide, coliforms were detected no farther than 2 miles from the outfall.

Newport Bay has experienced increasing bacterial pollution since 1976 (Skinner 1984). San Diego Creek discharges approximately 16 million gallons of water each day into the bay with coliform counts as high as 240,000 per 100 ml. The percentage of water samples collected from the cleanest areas of the lower bay with coliform counts over 100 per 100 ml increased from 5% in 1976 to 53% in 1983.

Olson (1978) conducted a study of the effectiveness of the most probable number (MPN) method to enumerate coliforms in marine waters of Newport Beach. False negatives were found in approximately one-half of the 270 samples tested. More than 50% of the isolates giving false negatives were *E. coli*. Inclusion of the false negative tubes into the data increased the number of violations of the California ocean water contact sports standard for 1000 cfu 100 ml^{-1} at all sites (California State Water Resources Control Board 1988). More than 20% of the samples collected were in violation of this standard. It was suggested that the MPN method be modified to detect coliforms in the marine environment. No suitable modifications have yet been adopted in the standard methods for bacteriological examination of recreation waters.

Coliform concentrations were reported for waters in the Point Loma wastewater discharge area off San Diego (Bascom 1973). At all but one of the stations, concentrations of 1000 MPN per 100 ml or less were recovered 86% or more of the time from 1963 to 1970.

Since 1960, the city of Los Angeles has monitored coliform concentrations along the shoreline between Malibu Canyon and Palos Verdes Point and around a 5-mile outfall (Bascom 1973). Between 1960 and 1970, coliform concentrations averaged 0.6 MPN per 100 ml at the offshore stations. Coliform concentrations varied with season and tended to decrease in surface and subsurface waters with increasing distance from shore. Coliform concentrations greater than 1000 MPN per 100 ml occurred from nearly 0 to 15% of the time.

An intensive study was conducted in 1966 and 1967 at five major storm drain discharge areas (Bascom 1973). Coliform concentrations were significantly higher (23,000–230,000 MPN per 100 ml) in storm flows than in dry periods (62–6200 MPN per 100 ml). The north shoreline of Santa Monica Bay often had higher coliform counts after storms than did other shoreline areas.

A study of coliform concentrations in the surf near municipal discharges in Orange and San Diego counties revealed that the levels were within state standards for bathing beaches (Bascom 1973). Samples collected near storm drains had consistently higher coliform concentrations than samples from other areas, and the frequency of occurrence of samples containing greater than 1000 MPN per 100 ml was higher during the winter months than the summer months. The high values in the winter may have resulted from storm sewer discharge. In a study of waters near Orange County Sanitation District outfalls, Fay (1966) found that offshore subsurface (6-m depth) water contained higher coliform concentrations than corresponding offshore and surf waters. Subsurface concentrations of 10,000–50,000 MPN per 100 ml were often observed. Some surf stations appeared to be contaminated by the coliforms from subsurface offshore waters.

Coliforms were found in seawater surface films (neuston layer) in areas around sewage outfalls (Hyperion and Palos Verdes) at concentrations of approximately 300 cfu per 100 ml (Bascom 1973). Coliform concentrations in surface films in outer areas of Santa Monica Bay and around Santa Catalina Island were less than 2 cfu per 100 ml.

Coliforms were found on wastewater macroparticles at outfall sites (Bascom 1973). Particles in the vicinity of outfall sites contained coliform concentrations of 3500–21,000 cfu per particle. No coliforms were found on particles recovered from unimpacted seawater around Santa Catalina Island. These data suggest that coliforms are not a part of the normal particle-associated microfauna in waters of the SCB, but that particles do serve as a vehicle of entry into these waters for these microorganisms.

Kenis et al. (1972) evaluated bacteriological water quality in waters that received sewage effluent off San Clemente Island. The initial concentrations and dieoff rates of TC, FC, and FS were determined at 4°, 12°, and 25°C. The TC group was found to be more abundant than the FC and FS groups at time zero and at subsequent samplings. Dieoff of the TC group was slower at 4°C than at 25°C. TC abundance decreased rapidly with distance from the outfall.

The densities of TC were also determined on surfaces of kelp and rocks in the vicinity of the outfall off San Clemente Island (Kenis et al. 1972). Slime masses on submerged rocks around sources of raw sewage discharge concentrated TC by factors of >1000 relative to the surrounding seawater. Kelp concentrated TC by a factor of 40–60 relative to the surrounding water. Sewage dilution in the water column was generally in excess of 1000 times in water 30 m from the point of discharge. Seawater outside of a radius of several hundred feet from a discharge pipe releasing 95,000 l d^{-1} generally met bacteriological health standards for swimming beaches. It was concluded that it was necessary to treat the sewage from human wastes generated on the island.

The results of this study were also used to predict the impact of sewage discharge from a middle-class Navy ship having a similar number of inhabitants as well as from larger vessels, such as aircraft carriers, which carry approximately 10 times more people. It was concluded that even with large ships, the BOD of microorganisms degrading the sewage would not significantly alter the oxygen concentration more than 30 m from the point of discharge.

Dilution and Dieoff of Sewage Microorganisms

Dye and radioactive tracers were used in conjunction with MPN coliform enumeration by Rittenberg (1956) to estimate the dieoff and dilution of sewage microorganisms. The average time required for a 90% reduction in the MPN of coliform bacteria (T_{90}) from the Hyperion primary effluent in Santa Monica Bay was 3 hours. Nine hours after discharge, the T_{90} increased to 15 hours. This difference was attributed either to a greater resistance of the bacteria that survived the initial exposure to seawater or to disaggregation of bacterial clumps.

Indicator bacteria from sewage effluents discharged into coastal waters near Santa Barbara, Ventura, and Oxnard were monitored by Foxworthy and Kneeling (1969). Coliform counts decreased from 5050 per 100 ml at the outfall to 0 per 100 ml at a distance 1170 m from the outfall. The decrease was due to both dilution and dieoff. Little or no tendency for sedimentation of coliforms occurred except in the immediate vicinity of the outfall. Maximum coliform counts usually occurred at the surface.

An equation was described that predicted the concentration of coliform bacteria anywhere along the discharge plume:

$$c/c_o \text{ (dieoff)} = ae^{-k_1 t} + (100 - a)\, e^{-k_2 t}$$

where c and c_o are coliform concentrations at the source and at any point along the longitudinal axis of the plume, respectively, k_1 and k_2 have units of $1/t$ and are constants, a is a constant, and e is the natural log. The rate of dieoff was high near the source and decreased with time. After 1 hour of exposure to seawater, dieoff was nearly constant. On another day, however, the rate of dieoff was low at the source and increased with time, become nearly constant after 125 minutes. There was no evidence of growth of coliforms in the plume. However, within the first few minutes of contact with seawater, a significant increase in counts was observed in about 50% of the samples. In the majority of samples, maximum coliform density occurred within a few feet of the water surface all along the plume.

A parallel study by Foxworthy and Kneeling (1969) showed that fecal streptococci (FS) dieoff could also be adequately described by the equation. In one experiment, the initial FS density was about two orders of magnitude less than the coliform density. Although the experiments suggested that FS are better indicators of enteric bacterial pollution in the sea (because they exhibit the highest overall resistance to adverse effects of the marine environment), other studies suggest that this conclusion is erroneous. FS densities decreased initially after exposure to the seawater and then increased to a concentration greater than the initial maximum concentration. This contrasts with the coliform density that decreased continuously along the plume. It was concluded that the growth and mortality characteristics of FS are highly unpredictable.

The factors that contributed to coliform dieoff were evaluated. Since there was good correlation between the distribution of Rhodamine B dye and coliform densities in the plume, it was concluded that sedimentation was not a factor. Dieoff rates varied significantly with season. There was a greater tendency for dieoff in spring and summer and regrowth in winter. Although the influencing factors were not identified, this tendency did not appear to be due to consequences of phytoplankton growth. The rate of dieoff did not vary directly with temperature; rather, a complex relationship with temperature was observed. There was a tendency for higher plume regrowth at lower (<15°C temperatures.

Survival of Indicator Microorganisms in Seawater

Several studies have addressed the survival of coliforms in marine waters of the SCB. ZoBell (1936) determined viability in filtered and autoclaved water samples. The most significant destructive agents of coliform bacteria in seawater were suggested to be biological in nature, heat labile, and related to the normal population of organisms indigenous to the water masses. Inactivation of coliform bacteria is caused by adsorption and sedimentation processes, sunlight ultraviolet irradiation inactivation, lack of nutrients, toxic substances, bacteriophages, predation by protozoans and nematodes, and antagonism by other microorganisms such as *Bdellovibrio* and *Pseudomonas* sp. (Carlucci and Pramer 1959; Foxworthy and Kneeling 1969). Carlucci and Silbernagel (1965) concluded that seawater toxicity is not controlled by the physical and biological properties of a particular water mass but rather the same factors may be responsible for the rapid death of bacteria in all waters.

Nusbaum and Garver (1955) studied the effects of seawater from San Diego Bay on the survival of laboratory cultures of *E. coli* and coliforms in raw sewage. Viability was evaluated by the presumptive three-tube MPN method. The death rate of *E. coli* at an initial concentration of 10^7 cells ml^{-1} was about two times that of a population with an initial cell concentration of 5×10^3 cells ml^{-1}. The period of time over which the coliform population remained relatively stable before the effects of seawater inactivation occurred varied from less than 24 hours to 4 days. This variation appeared to be seasonal. In one experiment, the coliform density decreased from 4.9×10^5 to 330 per 100 ml over a 9-day period. It appears from the studies conducted to date that the factors controlling survival of sewage microorganisms are rather complex and not yet understood.

Microorganisms That Cause Disease in Marine Species

Microorganisms have been implicated in several diseases of marine species in the SCB. Two common sea stars, *Patiria miniata* and *Henricia leviuscula,* suffered heavy mortalities in 1978 and 1981–1983. The highest mortalities coincided with the warming trend associated with the 1982–1984 El Niño. The causative agent of the disease appears to be a previously undescribed marine bacterial species tentatively referred to as *Vibrio patiria* (Schroeter et al. 1991). *V. patiria* is *o*-nitrophenyl-β-D-galactopyranoside (ONPG) positive, and unlike *Vibrio vulnificus,* a human pathogen, and other reported ONPG-positive marine isolates, is sucrose positive. Since other bacteria were also isolated from the lesions, the possibility of a multiorganismal infection cannot be ruled out. The disease virtually eliminated populations of these common sea stars in the shallow subtidal waters along the southern California coastline. It appears, therefore, that the epizootics are a major perturbation to community structure since these sea stars are an important benthic predator.

An infectious fin rot disease in Dover sole (*Microstomus pacificus*) and other bottom fish occurred at a higher incidence around wastewater discharges on the Palos Verdes shelf than in other areas (Sherwood 1978). A total of 10% of all fishes collected from the shelf had fin erosion; 99% of the fish showing the symptoms were flatfish or rockfish. No systemic infectious agent has been identified. However, the total number and diversity of bacteria were greater on specimens from the shelf than in other areas. The presence of the disease in the environment has been correlated with high sediment bacterial densities (Mahoney et al. 1973). In the laboratory, fin erosion has been induced in fish exposed to bacteria (Oppenheimer 1958). Most recently, the etiological agent of the disease is suggested to be a unicellular protozoan parasite or X-cell that resembles a parasitic amoeba (Cross 1984). Amoebae that are morphologically similar to the X-cell are common in marine sediments and sewage sludge.

A new species of bacteria, *Vibrio damsela,* was identified as the causative agent of skin ulcers in the blacksmith *Chromis punctipinnis* (Love et al. 1981). Infected fish have been observed in King Harbor, off Redondo Beach, and off Santa Catalina Island during the summer and fall spawning season. Ulceration rates are generally about 10%, although some aggregations of fish exhibited incidences of infection as high as 70%. Scarification of the skin with 10^7–10^8 viable cells of the bacterium produced large ulcers within three days and death in all infected fish within four days. The disease was never observed in other species, although it could be artificially induced in the garibaldi (*Hypsypops rubicunda*). The disease could also be induced in blue and brown *Chromis* spp. found in the Caribbean (G. G. Geesey pers. comm.). Since *V. damsela* was also isolated from marine algae, it is likely indigenous to SCB waters.

In Florida, the Gulf coast, the Bahamas, and Hawaii, strains of *V. damsela* have been isolated from human wounds. In all but one case, wounds had been incurred in seawater or brackish water. Wounds were described as being erythematous, indurated, and with purulent drainage (Morris et al. 1982). In all cases, the wounds were self-limiting and the patients recovered fully. Infections have also developed through the consumption of raw shellfish. However, no human infections such as these have yet been reported in southern California.

A systematic search demonstrated the occurrence of *Vibrio cholerae* in water, sediment, and shellfish of the SCB as well as in other west coast regions (Kaysner et al. 1987a). Samples from San Diego Bay, Mission Bay, San Diego River, and Tijuana Slough all contained *V. cholerae*. The vast majority were non-01 strains. *V. cholerae* were isolated from shellfish beds that were considered safe by coliform standards. The bacterium occurred in lower incidence in west coast marine environments than in east coast and Gulf coast environments.

Virulent strains of *Vibrio vulnificus* were isolated from sediments of Mission Bay in 1984 (Kaysner et al. 1987b). Samples of water and shellfish from Mission Bay tested negative for this organism. Since all strains that were isolated hybridized with cytotoxin–hemolysin gene probe, they likely produced cytotoxin. *V. vulnificus* can cause fatal septicemia in patients with liver disease and may be responsible for necrotic wound infections in otherwise healthy individuals. There was a lower incidence of this bacterium in southern California waters than in northern California waters. In general, west coast waters exhibited a lower incidence of *V. vulnificus* than east coast or Gulf coast waters.

Summary and Prospectus for Future Research

Some of the earliest studies in marine microbiology were conducted in waters of the SCB. It was in these waters and underlying sediments that a fundamental understanding of microbial distribution and activities developed. In the past decade, detailed studies have provided an understanding of the utilization of DOC by bacterioplankton and bacterioplankton production in the SCB. There is now wide agreement that bacterioplankton biomass, production, and heterotrophic activities in coastal and offshore waters are closely coupled to phytoplankton. In harbors and estuaries, bacterioplankton activities are also influenced by terrestrial organic input. Bacterioplankton production and biomass in the Los Angeles–Long Beach Harbor appear to be sufficient to serve as an important energy source for microzooplankton. Whether this holds true for coastal and offshore waters remains to be determined.

Bacteria play an important role in the cycling of nitrogen, sulfur, and phosphorus in the SCB. Planktonic nitrifying bacteria exist in sufficient numbers to account for the high *in situ* rates of ammonia oxidation. Macroscopic particles appear to be active sites of nitrification. In recent marine sediments of the SCB, the major process affecting sulfur geochemistry is bacterial reduction of sulfate to sulfide. Sulfate reduction is likely to be the major pathway for organic carbon oxidation in oxygen-depleted sediments. Bacterioplankton are the main means for making phosphate that is present in very dilute concentrations of DOC available to higher organisms in the food web. Microorganisms also compete with the phytoplankton for inorganic phosphate and are responsible for most of the inorganic phosphate assimilation in parts of Los Angeles Harbor.

The impact of sewage and other forms of organic point source pollution on the microbiology of receiving waters in southern California coastal waters has been widely studied over the past several decades. Monitoring infectious microorganisms has become a routine practice for southern California coastal city and county agencies.

The role of microorganisms in maintaining the stability of marine ecosystems such as the SCB is only now beginning to be appreciated. As additional work is carried out in marine microbial ecology, we will discover new pathways in food webs that depend on microorganisms, and we will obtain reaction rates for microbial-driven processes that we can now assign to microorganisms only as "uncertain involvement."

One of the most critical areas for which we lack information in the SCB is sediment microbiology. We need comprehensive studies on microbial biomass, production, and degradation of particulate organic matter. Although vast quantities of anthropogenic solid waste are discharged into the SCB, little is known about the microbiological processing of these wastes in bottom deposits where they accumulate. This information is needed because microbial action plays a major role in the fate of this material. We also need to determine denitrification, sulfate reduction, sulfide oxidation, and mineral transformation rates in bottom deposits. Studies performed in other marine areas comparable to the SCB suggest that microorganisms play a major role in these reactions.

Another area that deserves attention is the autotrophic activity of microorganisms in the SCB. With the exception of the nitrifying bacteria, little effort has been directed toward the study of microorganisms that produce biomass from carbon dioxide. The activities of chroococcoid cyanobacteria are likely to dominate primary production in some areas of the SCB, but these prokaryotic organisms are generally overlooked during studies of primary production.

Finally, it should be emphasized that work must be initiated in the area of marine virology. Efforts should be increased to obtain a better understanding of the fate of pathogenic viral agents such as infectious hepatitis, polio, and viral gastroenteritis. Virus recovery, cultivation, and enumeration needs further scientific progress before widescale monitoring in marine waters can be undertaken.

The SCB is adjacent to a population center that is growing faster than any other metropolitan area in the United States. As in the past, we will continue to depend on the marine waters off southern California for many purposes. After all, it is the mild Mediterranean climate afforded by the SCB that makes its shoreline and inland areas so desirable a place to live. We must, therefore, continue to improve our analytical techniques to assess the microbiology of the SCB so that we can detect changes brought about by either natural or human causes and—we hope—prevent unnecessary deterioration of this valuable marine habitat.

Literature Cited

Alldredge, A. L., J. J. Cole, and D. A. Caron, 1986. Production of heterotrophic bacteria inhabiting macroscopic organic aggregates (marine snow) from surface waters. *Limnol. Oceanogr.* 31:68–78.

Ammerman, J. W., and F. Azam, 1981. Dissolved cyclic adenosine monophosphate (cAMP) in the sea and uptake of cAMP by marine bacteria. *Mar. Ecol. Prog. Ser.* 5:85–89.

Ammerman, J. W., and F. Azam, 1982. Uptake of cyclic AMP by natural populations of marine bacteria. *Appl. Environ. Microbiol.* 43:869–876.

Ammerman, J. W., and F. Azam, 1985. Bacterial 5′-nucleotidase in aquatic ecosystems: A novel mechanism of phosphorus regeneration. *Science.* 227:1338–1340.

Atlas, R. M., and R. Bartha, 1987. *Microbial Ecology.* Benjamin Cummings Publ., Menlo Park, CA. 533pp.

Azam, F., and J. A. Fuhrman, 1984. Measurement of bacterioplankton growth in the sea and its regulation by environmental conditions. In: J. E. Hobbie, and P. J. leB. Williams, eds. *Heterotrophic Activity in the Sea.* Plenum Press, New York. pp. 179–196.

Azam, F., and R. E. Hodson, 1977a. Size distribution and activity of marine microheterotrophs. *Limnol. Oceanogr.* 22:492–501.

Azam, F., and R. E. Hodson, 1977b. Dissolved ATP in the sea and its utilization by marine bacteria. *Nature*. 267:696–698.

Azam, F., and R. E. Hodson, 1981. Multiphasic kinetics for D-glucose uptake by assemblages of natural marine bacteria. *Mar. Ecol. Prog. Ser.* 6:213–222.

Azam, F., and O. Holm-Hansen, 1973. Use of tritiated substrates in the study of heterotrophy in seawater. *Mar. Biol.* 23:191–196.

Bascom, W., 1973. *The Ecology of the Southern California Bight: Implications for Water Quality Management.* TR 104. South. Calif. Coastal Water Res. Project, El Segundo, CA. 531pp.

Burnett, B. R., 1979. Quantitative sampling of microbiota of the deep-sea benthos II. Evaluation of technique and introduction to the biota of the San Diego Trough. *Trans. Am. Microsc. Soc.* 98(2):233–242.

Burnett, B. R., 1981. Quantitative sampling of the nanobiota (microbiota) of the deep sea benthos—III. The bathyal San Diego Trough. *Deep-Sea Res.* 28A(7):649–663.

California Department of Public Health, 1942. *Report on a Pollution Survey of Santa Monica Bay Beaches.* Berkeley.

California State Water Resources Control Board, 1988. *Water Quality Control Plan: Ocean Waters of California.* Sacramento. 13pp.

Carlucci, A. F., and D. Pramer, 1959. Factors affecting the survival of bacteria in seawater. *Appl. Microbiol.* 7:388–392.

Carlucci, A. F., and S. B. Silbernagel, 1965. Effect of different seawaters on the development of biochemically deficient mutants of *Serratia marinorubra. Appl. Microbiol.* 13:663–668.

Carlucci, A. F., E. O. Hartwig, and P. M. Bowes, 1970. Biological production of nitrite in seawater. *Mar. Biol.* 7:161–166.

Carlucci, A. F., S. L. Shimp, P. A. Jumars, and H. W. Paerl, 1976. *In situ* morphologies of the deep-sea and sediment bacteria. *Can. J. Microbiol.* 22:1667–1671.

Carlucci, A. F., D. B. Craven, and S. M. Henrichs, 1984. Diel production and microheterotrophic utilization of dissolved free amino acids in waters off southern California. *Appl. Environ. Microbiol.* 48:165–170.

Carlucci, A. F., D. B. Craven, and S. M. Henrichs, 1985. Surface-film microheterotrophs: Amino acid metabolism and solar radiation effects on their activities. *Mar. Biol.* 85:13–22.

Carlucci, A. F., D. B. Craven, K. J. Robertson, and S. M. Henrichs, 1986. Microheterotrophic utilization of dissolved free amino acids in depth profiles of southern California borderland basin waters. *Oceanol. Acta.* 9(1):89–96.

Claypool, G. E., and I. R. Kaplan, 1974. The origin and distribution of methane in a marine sediment. In: I. R. Kaplan, ed. *Natural Gas in Marine Sediments.* Plenum Press, New York. pp. 99–140.

Cross, J. N., 1984. Tumors in fish collected on the Palos Verdes shelf. In: W. Bascom, ed. *Biennial Report 1983–1984.* South. Calif. Coastal Water Res. Proj., Long Beach, CA. pp. 81–91.

Davis, P. H., and R. B. Spies, 1980. Infaunal benthos of a natural petroleum seep: Study of community structure. *Mar. Biol.* 59(1):31–41.

Ducklow, H. W., D. A. Purdie, P. J. Williams, and J. M. Davies, 1986. Bacterioplankton: A sink for carbon in a coastal marine plankton community. *Science.* 232:865–867.

Fay, R. C., 1966. Distribution of coliform bacteria about the submarine outfall of CSDOC during the night of June 20–21, 1966. Preliminary Report III to the County Sanitation Districts of Orange County, CA.

Fenchel, T., 1986. The ecology of heterotrophic microflagellates. *Adv. Microb. Ecol.* 9:57–97.

Foxworthy, J. E., and H. R. Kneeling, 1969. *Eddy Diffusion and Bacterial Reduction in Waste Fields in the Ocean.* Final Report to Federal Water Pollution Control Adm., Grant no. WP 00931 Allan Hancock Foundation, Univ. of Southern California, Los Angeles, CA. 176pp.

Fuhrman, J. A., 1981. Influence of method on the apparent size distribution of bacterioplankton cells: Epifluorescence microscopy compared to scanning electron microscopy. *Mar. Ecol. Prog. Ser.* 5:103–106.

Fuhrman, J. A., and F. Azam, 1980. Bacterioplankton secondary production estimates for coastal waters of British Columbia, Antarctica, and California. *Appl. Environ. Microbiol.* 39: 1085–1095.

Fuhrman, J. A., and F. Azam, 1982. Thymidine incorporation as a measure of heterotrophic bacterioplankton production in marine surface waters evaluation and field results. *Mar. Biol.* 66(2):109–120.

Fuhrman, J. A., and F. Azam, 1983. Adaption of bacteria to marine subsurface waters studied by temperature response. *Mar. Ecol. Prog. Ser.* 13:95–98.

Fuhrman, J. A., J. W. Ammerman, and F. Azam, 1980. Bacterioplankton in the coastal euphotic zone: Distribution, activity, and possible relationships with phytoplankton. *Mar. Biol.* 60(2–3):201–207.

Hagström, Å., J. W. Ammerman, S. Henrichs, and F. Azam, 1984. Bacterioplankton growth in seawater: II. Organic matter utilization during steady-state growth in seawater cultures. *Mar. Ecol. Prog. Ser.* 18:41–48.

Hamilton, R. D., and O. Holm-Hansen, 1967. Adenosine triphosphate content of marine bacteria. *Limnol. Oceanogr.* 12:319–324.

Hollibaugh, J. T., and F. Azam, 1983. Microbial degradation of dissolved proteins in seawater. *Limnol. Oceanogr.* 28:1104–1116.

Holm-Hansen, O., and C. R. Booth, 1966. The measurement of adenosine triphosphate in the ocean and its ecological significance. *Limnol. Oceanogr.* 11:510–519.

Iturriaga, R., 1981. Phytoplankton photoassimilated extracellular products; Heterotrophic utilization in marine environment. *Kiel. Meeresforsch. Sonderh.* 5:318–412.

Jacq, E., D. Prieur, P. Nichols, D. C. White, T. Porter, and G. G. Geesey, 1989. Microscopic examination and fatty acid characterization of filamentous bacteria colonizing substrata around subtidal hydrothermal vents. *Arch. Microbiol.* 152:64–71.

Juge, D. M., and G. Griest, 1973. The roles of microbial activity in the harbor ecosystem. In: D. F. Soule, and M. Oguri, eds. *Marine Studies of San Pedro Bay, California, Part 2. Biological Investigations.* Allan Hancock Foundation, Univ. of Southern California, Los Angeles, CA. pp. 29–63.

Kaplan, I. R., K. O. Emery, and S. C. Rittenberg, 1963. The distribution and isotopic abundance of sulfur in recent marine sediments off southern California. *Geochim. Cosmochim. Acta.* 27: 297–331.

Kaysner, C. A., C. Abeyta, Jr., M. M. Wekell, A. DePaola, Jr., R. F. Stott, and J. M. Leitch, 1987a. Incidence of *Vibrio cholera* from estuaries of the United States west coast. *Appl. Environ. Microbiol.* 53:1344–1348.

Kaysner, C. A., C. Abeyta, Jr., M. M. Wekell, A. DePaola, Jr., R. F. Stott, and J. M. Leitch, 1987b. Virulent strains of *Vibrio vulnificus* isolated from estuaries of the United States west coast. *Appl. Environ. Microbiol.* 53:1349–1351.

Kenis, P. R., M. H. Salazar, and J. A. Tritschler, 1972. *Environmental Study of the Sewage Outfall Area at San Clemente Island. An Open-Ocean Sewage Disposal Model.* Research Report, July 70–July 71. Naval Undersea Research and Development Center, San Diego, 41pp.

Kogure, K., and I. Koike, 1987. Particle counter determination of bacterial biomass in seawater. *Appl. Environ. Microbiol.* 53:274–277.

Kosiur, D. R., and A. L. Warford, 1979. Methane production and oxidation in Santa Barbara Basin sediments. *Estuarine Coastal Mar. Sci.* 8: 379–385.

Krempin, D. W., and C. W. Sullivan, 1981. The seasonal abundance, vertical distribution, and relative microbial biomass of chroococcoid cyanobacteria at a station in southern California coastal waters. *Can. J. Microbiol.* 27:1341–1344.

Krempin, D. W., S. M. McGrath, J. B. Soohoo, and C. W. Sullivan, 1981. Orthophosphate uptake by phytoplankton and bacterioplankton from the Los Angeles Harbor and southern California coastal waters. *Mar. Biol.* 64(1):23–33.

Laughlin, J. D., 1982. Gut analysis of polychaetes and amphipods. In: W. Bascom, ed. *Biennial Report 1981–1982.* South. Calif. Coastal Water Res. Proj., Long Beach, CA. pp. 59–66.

Love, M. S., D. Teebken-Fischer, J. Hose, J. J. Farmer, F. W. Hickman, and G. R. Fanning, 1981. *Vibrio damsela* sp. nov.: A marine bacterium causes skin ulcers on the damselfish *Chromis punctipinnis. Science.* 214:1139–1140.

McGrath, S. M., and C. W. Sullivan, 1981. Community metabolism of adenylates by microheterotrophs from the Los Angeles Harbor and southern California coastal waters. *Mar. Biol.* 62(2–3):217–226.

Mahoney, J. B., F. H. Midlige, and D. G. Devel, 1973. A fin rot disease of marine and euryhaline fishes in the New York Bight. *Trans. Am. Fish. Soc.* 102:596–605.

Morris, J. G., Jr., R. Wilson, D. G. Hollis, R. E. Weaver, H. G. Miller, C. O. Tacket, F. W. Hickman, and P. A. Blake, 1982. Illness caused by *Vibrio damsela* and *Vibrio hollisae. Lancet.* 1(8284):1294–1296.

Newell, S. Y., and R. D. Fallon, 1982. Bacterial productivity in the water column and sediments of the Georgia (USA) coastal zone: Estimates via direct counting and parallel measurements of thymidine incorporation. *Microb. Ecol.* 8:33–46.

Nusbaum, I., and R. M. Garver, 1955. Survival

of coliform organisms in Pacific Ocean coastal waters. *Sewage Ind. Wastes.* 27:1383–1390.

Olson, B. H., 1978. Enhanced accuracy of the coliform testing in seawater by a modification of the most probable number method. *Appl. Environ. Microbiol.* 36:438–444.

Olson, R. J., 1981. Nitrogen-15 tracer studies of the primary nitrite maximum. *J. Mar. Res.* 39(2):203–226.

Oppenheimer, C., 1958. A bacterium causing tail rot in the Norwegian codfish. *Publ. Inst. Mar. Sci. Univ. Texas.* 5:160–164.

Prézelin, B. B., and A. L. Alldredge, 1983. Primary production of marine snow during and after an upwelling event. *Limnol. Oceanogr.* 28(6):1156–1167.

Reed, W. E., I. R. Kaplan, M. Sandstrom, and P. Mankiewicz, 1977. Petroleum and anthropogenic influence on the composition of sediments from the Southern California Bight. In *Proceedings, 1977 Oil Spill Conference (Prevention, Behavior, Control, Cleanup).* New Orleans, LA, 8–10 March 1977. American Petroleum Institute, Washington, D.C. pp. 183–188.

Rittenberg, S. C., 1940. Bacteriological analysis of some long cores of marine sediments. *J. Mar. Res.* 3:191–201.

Rittenberg, S. C., 1956. *Studies on Coliform Bacteria Discharged from the Hyperion Outfall.* Final report submitted to Hyperion Engineers, Inc., 51pp.

Rittenberg, S. C., T. Mittwer, and D. Ivler, 1958. Coliform bacteria in sediments around three marine sewage outfalls. *Limnol. Oceanogr.* 3: 101–108.

Ruby, E. G., and J. G. Morin, 1979. Luminous enteric bacteria of marine fishes: A study of their distribution, densities, and dispersion. *Appl. Environ. Microbiol.* 38:406–411.

Ruby, E. G., and K. H. Nealson, 1978. Seasonal changes in the species composition of luminous bacteria in nearshore seawater. *Limnol. Oceanogr.* 23:530–533.

Schroeter, S. C., J. D. Dixon, J. E. Hose, and R. O. Smith, 1991. Mass mortalities of sea stars in southern California caused by a new marine *Vibrio.* (unpubl. ms.)

Sherwood, M. J., 1978. The fin erosion syndrome. In: W. Bascom, ed. *Annual Report.* South. Calif. Coastal Water Res. Proj., El Segundo, CA. pp. 203–221.

Skinner, J., 1984. Does swimming endanger your health? *The Newport News.* 1:4–5.

Smith, K. L., Jr., 1974. Oxygen demands of San Diego Trough sediments: An *in situ* study. *Limnol. Oceanogr.* 19:939–944.

Soule, D. F., and M. Oguri, 1975. Microbiological investigations in Los Angeles–Long Beach Harbors. In: *Environmental Investigations in Analysis for Los Angeles–Long Beach Harbors, Los Angeles, California.* Allan Hancock Foundation, Univ. of Southern California, Los Angeles, CA. p. 552.

Soule, D. F., and M. Oguri, 1979. Monthly standing stock measurements of the bacterioplankton and phytoplankton in Los Angeles Harbor and southern California coastal waters. In: *Ecological Changes in Outer Los Angeles–Long Beach Harbors Following Initiation of Secondary Waste Treatment and Cessation of Fish Cannery Waste Effluent—Marine Studies of San Pedro Bay, California.* Part 16. Allan Hancock Foundation and the Office of Sea Grant Programs, Los Angeles, CA. pp. 227–276.

Soule, D. F., and J. D. Soule, 1981. The importance of non-toxic urban wastes in estuarine detrital food webs. *Bull. Mar. Sci.* 31:786–800.

Spies, R. B., and P. H. Davis, 1979. The infaunal benthos of a natural oil seep in the Santa Barbara Channel. *Mar. Biol.* 50(3):227–237.

Spies, R. B., and D. J. DesMarais, 1983. Natural isotope study of trophic enrichment of marine benthic communities by petroleum seepage. *Mar. Biol.* 73(1):67–71.

Stein, J. L., 1984. Subtidal gastropods consume sulfur-oxidizing bacteria: Evidence from coastal hydrothermal vents. *Science.* 223:696–698.

Sullivan, C. W., A. Palmisano, S. McGrath, D. Krempin, and G. Taylor, 1978. Microbial standing stocks and metabolic activities of microheterotrophs in southern California coastal waters. In: D. F. Soule, and M. Oguri, eds. *Marine Studies of San Pedro Bay, California. Part 14. Biological Investigations.* Univ. of Southern California, Los Angeles, CA. pp. 1–24.

Sweeney, R. E., and I. R. Kaplan, 1980. Natural abundances of nitrogen-15 as a source indicator for nearshore marine sedimentary and dissolved nitrogen. *Mar. Chem.* 9:81–94.

Taylor, G. T., and C. W. Sullivan, 1979. The ingestion and utilization of labeled marine bacteria by bacterivorous plankton from Los Angeles Harbor and southern California coastal waters. Abstract of the Annual Meeting of the American Society for Limnology and Oceanography.

Taylor, G. T., R. Itturriaga, and C. W. Sullivan, 1985. Interactions of bactivorous grazers and heterotrophic bacteria with dissolved organic matter. *Mar. Ecol. Prog. Ser.* 23:129–141.

Thompson, B. E., J. N. Cross, J. D. Laughlin, G. P. Hershelman, R. W. Gossett, and D. T. Tsukada, 1984a. Sediment and biological conditions on coastal slopes. In: W. Bascom, ed. *Biennial Report 1983–1984*. South. Calif. Coastal Water Res. Proj., Long Beach, CA. pp. 37–67.

Thompson, B. E., J. D. Laughlin, and D. T. Tsukada, 1984b. Ingestion and oxygen consumption by slope echinoids. In: W. Bacsom, ed. *Biennial Report 1983–1984*. South. Calif. Coastal Water Res. Proj., Long Beach, CA. pp. 93–107.

Vetter, R. D., 1985. Elemental sulfur in the gills of three species of clams containing chemoautotrophic symbiotic bacteria: A possible inorganic energy storage compound. *Mar. Biol.* 88:33–42.

Ward, B. B., 1982. Oceanic distribution of ammonium-oxidizing bacteria determined by immunofluorescent assay. *J. Mar. Res.* 40:1155–1172.

Ward, B. B., and A. F. Carlucci, 1985. Marine ammonia and nitrite-oxidizing bacteria: Serological diversity determined by immunofluorescence in culture and in the environment. *Appl. Environ. Microbiol.* 50(2):194–201.

Ward, B. B., R. J. Olson, and M. J. Perry, 1982. Microbial nitrification rates in the primary nitrite maximum off southern California. *Deep-Sea Res. Part A Oceanogr. Res. Pap.* 29(2):247–255.

Warford, A. L., and D. R. Kosiur, 1979. The effect of wastewater treatment on methanogenesis in a marine outfall. *J. Water Pollut. Control Fed.* 51:37–42.

Warford, A. L., D. R. Kosiur, and P. R. Doose, 1979. Methane production in Santa Barbara Basin sediments. *Geomicrobiol. J.* 1:117–137.

Williams, P. J., T. Berman, and O. Holm-Hansen, 1976. Amino acid uptake and respiration by marine heterotrophs. *Mar. Biol.* 35:41–47.

ZoBell, C. E., 1936. Bactericidal actions of seawater. *Proc. Soc. Exper. Biol. Med.* 34:113–116.

ZoBell, C. E., 1938. Studies on the bacterial flora of marine bottom deposits. *J. Sed. Petrol.* 8:10–18.

ZoBell, C. E., 1939. Occurrence and activity of bacteria in marine sediments. In: P. D. Trask, ed. *Recent Marine Sediments*. Thomas Murby & Co., London. pp. 416–427.

ZoBell, C. E., 1941a. Studies on marine bacteria. 1. The cultural requirements of heterotrophic aerobes. *J. Mar. Res.* 4:42–75.

ZoBell, C. E., 1941b. The occurrence of coliform bacteria in ocean water. *J. Bacteriol.* 42:284.

ZoBell, C. E., 1942. Changes produced by microorganisms in sediments after deposition. *J. Sed. Petrol.* 12:127–136.

ZoBell, C. E., 1947. Microbial transformation of molecular hydrogen in marine sediments, with particular reference to petroleum. *Bull. Am. Assoc. Petrol. Geol.* 31:1709–1751.

ZoBell, C. E., 1960. Marine pollution problems in the southern California area. In: *Biological Problems in Water Pollution*. Trans. Second Seminar Biol. Problems in Water Pollution, Cincinnati, OH. pp. 177–183.

ZoBell, C. E., 1969. Microbial modification of crude oil in the sea. In: *Proceedings API/FWPCA Conference of Prevention and Control of Oil Spills*. Am. Petrol. Inst. Publ. No. 4040. pp. 317–326.

ZoBell, C. E., and D. Q. Anderson, 1936. Vertical distribution of bacteria in marine sediments. *Bull Am. Assoc. Petrol. Geol.* 20:258–269.

ZoBell, C. E., and C. B. Feltham, 1934. Preliminary studies on the distribution and characteristics of marine bacteria. *Bull. Scripps Inst. Oceanogr.* 3:279–296.

ZoBell, C. E., and C. B. Feltham, 1937. Bacteria as food for certain marine invertebrates. *J. Mar. Res.* 1:312–327.

ZoBell, C. E., and C. B. Feltham, 1942. The bacterial flora of a marine mud flat as an ecological factor. *Ecology*. 23:69–78.

ZoBell, C. E., and S. C. Rittenberg, 1938. The occurrence and characteristics of chitinoclastic bacteria in the sea. *J. Bacteriol.* 35:225–287.

ZoBell, C. E., C. W. Grant, and H. F. Haas, 1943. Marine microorganisms which oxidize petroleum hydrocarbons. *Bull. Am. Assoc. Petrol. Geol.* 27:1175–1193.

Chapter 5

Phytoplankton

John T. Hardy

Introduction

The Importance of Phytoplankton

The vast majority of life in the world's oceans depends either directly or indirectly on phytoplankton, tiny unicellular or colonial algae. These plants of the sea utilize carbon dioxide, present as dissolved bicarbonate in seawater, and light energy to convert the inorganic carbon to cellular material through photosynthesis. Phytoplankton form the base of the food web; they support grazing zooplankton, fish, and, through their decay, large quantities of marine bacteria. The success of zooplankton depends upon both the quantity and quality of their phytoplankton food supply. For example, in the Southern California Bight (SCB), the fecundity (egg production) of zooplankton depends upon the nutritive value (nitrogen content) of the phytoplankton on which they feed (Checkley 1980a, b).

Fish production, in turn, is highly dependent on the growth and productivity of phytoplankton and zooplankton (Ryther 1969). Empirical indices indicate that fishery yield increases exponentially with increasing primary production in a variety of marine and freshwater environments (Hanson and Leggett 1982; Nixon 1988). Furthermore, spatial and temporal patterns of phytoplankton occurrence are important to fisheries. The success of larval fish and their subsequent recruitment into the adult fish population often depend upon spatial and temporal co-occurrence of fish larvae with an abundance of their plankton food source (Lasker 1975; Cushing 1982; Mullin et al. 1985).

Systematics of Phytoplankton

The algae are an extremely diverse group of organisms. Their systematic classification is based on evolutionary relationships using criteria based on the structure of the flagella, the pigment composition, the structure of the chloroplast, and the relationship between the chloroplast and the nuclear envelope. The taxonomy and biology of algae have been discussed in a number of standard texts (Fritsch 1935; Prescott 1968; Bold and Wynne 1978). Most of the major algal groups have planktonic representatives, that is, forms that float freely in the water and are distributed primarily by water movements. An excellent general review of phytoplankton ecology is provided by Harris (1986).

Phytoplankton are single cell or colonial algal species that range in size over three orders of magnitude (Sieburth 1979). Some very small (bacteria size) species are classified as picoplankton (0.2–2 μm), most as nanoplankton (2–20 μm) or microplankton (20–200 μm), and a few large species as mesoplankton (0.2–20 mm). A strict definition of phytoplankton is probably not possible. Some of the species in the water column spend part of their life cycles as encysted forms in the bottom sediments; others are typically benthic and associated with bottom sediments, which may become suspended by turbulence. The blue-green algae are generally included, even though they are prokaryotes and are structurally similar to bacteria. Planktonic forms of the diatoms and chrysophytes, green algae, cryptophytes, and dinoflagellates are common, but there are very few planktonic red algae and no planktonic brown algae or charophytes (Bold and Wynne 1978; Sournia 1982).

In general, phytoplankton are photoautotrophic prokaryotes and eukaryotes, deriving their energy from photosynthesis, but some heterotrophic forms may be at least partially dependent on organic substrates (Droop 1974). Thus, factors that influence the growth and success of all photosynthetic organisms (that is, light, temperature, and supply of major nutrients) commonly affect phytoplankton. Growth rates range from a few doublings per day for the fastest growing species to one doubling every week or 10 days for the slowest growing species.

History of Phytoplankton Study in Southern California

Certain broad trends become evident when one reviews the historical development of knowledge concerning the phytoplankton of the SCB. First, the number of reports per year has followed a sigmoid growth curve. Although the reports reviewed here are not all inclusive, the bibliography reveals a trend. Only a few reports per year were published between the 1920s and the end of the 1950s, but the number accelerated rapidly to about 10 times the pre-1950 level during the decades of the 1960s and 1970s. The rate seems to have slowed slightly during the 1980s.

Second, as one would expect, the nature of the reports has changed over time. During the 1920s, descriptive studies focused on the taxonomy of phytoplankton to answer the fundamental question of "what was there." During the 1930s and 1940s, researchers began to examine the broad relationship between seasonal patterns and hydrographic (current) patterns, and Allen (1941) reported on long-term general patterns of the populations. During the 1960s, much of the work turned toward understanding biomass (chlorophyll and carbon) and productivity and their relationships to major nutrients. New studies examined the "red–tide" blooms of dinoflagellates. The decade of the 1970s saw an explosion in research activity. Numerous studies examined temporal and spatial patterns of distribution, trophic relationships with zooplankton, and the importance of trace elements and organics as growth factors. In the 1980s, researchers pointed to the importance of small (nano- and pico-) plankton in the ecosystem and quantified the residence times of carbon and nitrogen in the surface layer. Most recently, remote sensing has emerged as a valuable tool for understanding large–scale patterns in surface populations.

Sampling and Measurement Techniques

An understanding of the dynamics of phytoplankton communities requires the proper utilization of a large number of specialized field and laboratory techniques. These techniques have evolved and improved over many years. They include sampling design, use of sampling apparatus such as water bottles and nets, preservation and storage, concentrating and subsampling populations, taxonomic identifications, estimation of cell numbers, special microscopic techniques (including scanning electron microscopy), electronic cell counting, culturing, statistical methods of data analysis, and finally, interpretation.

A particularly useful and comprehensive manual on methodology for phytoplankton study was prepared by the United Nations Educational, Scientific and Cultural Organization (UNESCO) (Sournia 1978). It remains the best overall treatment of methodology, from initial sampling to final data interpretation. Specific methods for measuring biomass and primary productivity are found in Parsons et al. (1984) and Vollenweider (1974). Culture techniques for phytoplankton are described in Stein (1973). In addition, new innovations in methodology are reported regularly in the following journals: *Limnology and Oceanography, Phycologia, Journal of Phycology, Deep Sea Research* and *Journal of Plankton Research*.

Species Composition

Size Distribution

The size distribution of phytoplankton species populations determines several important characteristics of the community. Phytoplankton are typically divided into two size classes based on their retention in plankton nets. The larger phytoplankton, retained in nets of mesh size 20–90 μm, are commonly called *net plankton,* while those that pass through such nets are called *nanoplankton*. The ecological significance of cell size lies in the surface area to volume ratio, which affects the dynamics of phytoplankton productivity and energy flow through the food web.

Small cells generally have shorter generation times and higher growth rates in a given environment than do larger cells (Eppley et al. 1969). For nitrate and ammonium, uptake rates increase with increasing cell size, so larger net plankton are favored when nitrate concentrations are high, while nanoplankton are favored when nitrate concentrations are low (Eppley et al. 1969). Sinking rates generally increase as cell size increases. Larger cells have shorter residence times in the photic zone under stratified conditions and tend to be concentrated in regions of upward water flow.

Cell size also determines the distribution and abundance of herbivores that graze selectively on the basis of preferred food size. Plankton-feeding larvae and microzooplankton generally prefer small phytoplankton as food, whereas larger herbivorous copepods actively select larger planktonic species (Thorson 1950; Richman and Rogers 1969). Phytoplankton cell size can also affect the efficiency of energy transfer to large predators since nanoplankton-based food chains appear to require one or two more energy transfers to reach a given size consumer than do larger net plankton-based food chains (Ryther 1969; Parsons and LeBrasseur 1970). Typically, more than half of the phytoplankton biomass in the Santa Monica Basin passes through a 5-μm sieve and therefore is of a size that can be grazed by protozoans (Small et al. 1989).

Many species of phytoplankton inhabit the SCB (fig. 5.1). Their relative abundance in terms of numbers, biomass, and production varies greatly both spatially and temporally. The two most abundant and important components of the phytoplankton community are generally the diatoms (bacillariophytes) and the dinoflagellates (pyrrophytes). The size range of individual species of both groups is great—from a few micrometers to a few hundred micrometers.

The community of larger (>50 μm) net phytoplankton in the SCB includes a broad range of temperate water forms as well as forms that characteristically occur in either warmer or colder water. This diversity reflects the general transitional nature of the SCB's flora, which results from the physical oceanographic and mixing characteristics of the region. For example, incursions of exceptionally warm water currents in the area generally bring with them warm water species (Balech 1960).

The species composition and distribution of the smaller micro-, nano-, and picoplanktonic autotrophs is less well known, although data suggest that nanoplankton are generally extremely important contributors to primary productivity in both nearshore and offshore areas. For example, in Monterey Bay and in the California Current system, nanoplankton accounted for 60–99% of the observed productivity and standing crop (Malone 1971).

Seasonal and geographic variations in nanoplankton are remarkably stable, and variations in plankton productivity are due primarily to the net plankton. Malone (1971) found that (1) the nanoplankton fraction varied within narrow limits compared with fractions of the net plankton, (2) the nanoplankton assimilation ratios were consistently high and twice those of the net plankton, and (3) the net plankton productivity and standing crop increased relative to the nanoplankton during periods of upwelling. The most abundant components of the biomass, integrated over the euphotic zone, at sites from 1.4 to 12.1 km off La Jolla were often small dinoflagellates, monads, and flagellates (Reid et al. 1970). Studies that have included data on the nanoplanktonic and small picoplanktonic components of the SCB indicate that they commonly include the diatoms *Cylindrotheca closterium,* the dinoflagellates *Scrippsiella trochoidea* and *Prorocentrum vaginulum,* and the coccolithophorid *Emiliania* (= *Coccolithus*) *huxleyi* (Briand 1976; Reid et al. 1970).

As in other marine environments, recent data from southern California coastal waters suggest the importance of small photosynthetic prokaryotic chroococcoid cyanophytes (blue-green algae). For example, at a depth of 1 m in the San Pedro Channel, outside of Los Angeles Harbor, cyanobacteria abundances were relatively low in winter and increased throughout the spring, reaching a peak in late summer (Krempin and Sullivan 1981).

Species Assemblages

Phytoplankton assemblages in the SCB exist in great variety, richness, and complexity and

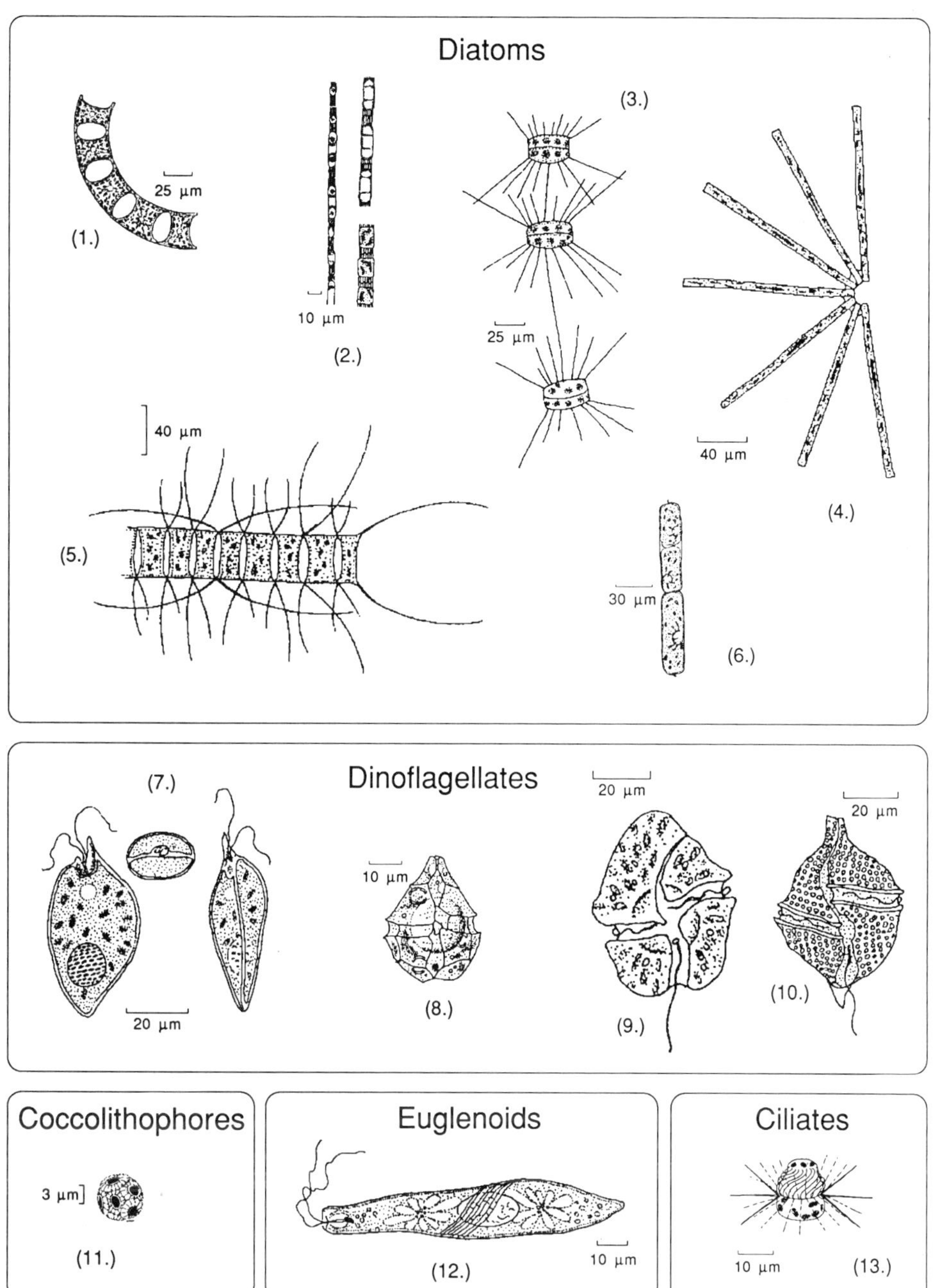

Figure 5.1. Some common phytoplankton of the SCB. Sizes can vary greatly among species and among individuals of the same species. DIATOMS: (1) *Eucampia zoodiacus*; (2) *Skeletonema costatum* (three views illustrate variation in morphology and size); (3) *Thalassiosira decipiens*; (4) *Thalassiothrix frauenfeldii* (colony); (5) *Chaetoceros decipiens*; (6) *Rhizosolenia fragilissima*. DINOFLAGELLATES: (7) *Prorocentrum* sp. (from left to right, lateral view, apical view, and median view); (8) *Scrippsiella* sp. (24–64 μm); (9) *Gymnodinium* sp. (10–150 μm); (10) *Gonyaulax* sp. (25–175 μm). COCCOLITHOPHORES: (11) *Emiliania* (= *Coccolithus*) *huxleyi* (acid-cleaned specimen showing details of $CaCO_3$ plates). EUGLENOIDS: (12) *Eutreptiella* sp. CILIATES: (13) *Mesodinium rubrum* (a ciliate protozoan containing photosynthetic dinoflagellate endosymbionts).

have not been adequately characterized. In general, phytoplankton community organization has been described in terms of two opposing hypotheses. The equilibrium hypothesis treats gradients of environmental conditions as essentially accessible to all species; the gradients simply define the differentiation of the resources available. The community composition at any given place and time is determined by the interplay of intrinsic biological properties of the species as modulated by local conditions. Coexistence of competitors is thought to depend on quantitative conditions and the extent of overlap of resource utilization (Allen 1977). The second, or nonequilibrium, hypothesis treats the separation of species and resources in time and space as virtual. Observed overlap in occurrence is interpreted merely as spillover of respective species centers of abundance—a coincidental phenomenon contingent on spatial and temporal transients. In this model, the persistence of a community depends very much on scale. This nonequilibrium view suggests that the community at any given place and time is greatly influenced by past environmental conditions and spatially distant events that cannot necessarily be reconstructed from information about the present local state of the environment.

Computerized statistical analyses of community species composition have shed new light on the temporal pattern of phytoplankton species assemblages in the SCB and its relationship to environmental variables. Goodman et al. (1984) demonstrated that the community off La Jolla can be separated into three species groups. They described the phytoplankton assemblages and changes in community composition over a 21-week period from mid-April to mid-September 1967 at three stations north of La Jolla. The stations were located 1.4, 4.6, and 12.1 km from shore at depths of 21, 175, and 175 m, respectively. The authors identified 147 phytoplankton taxa present in 63 samples with 114 of the taxa present in more than 10% of the samples; 29.8% were centric diatoms, 16.7% pennate diatoms, 43.9% dinoflagellates, 4.4% coccolithophorids, 3.5% flagellates, and 1.8% silicoflagellates. Canonical correlation was used to associate 25 taxa and 13 physical and chemical variables.

Goodman et al. (1984) also found that assemblages tended to persist at a given location for periods of 1–4 weeks. One assemblage (component I) was a typical warm water upwelling group dominated by *Skeletonema costatum* present at the inception of the spring bloom. A second component (II) was dominated by *Chaetoceros radicans* (also a temperate upwelling species), along with *Eucampia zoodicus* and two coccolithophorids, *Emiliania* (= *Coccolithus*) *huxleyi* and *Cyclococcolithus leptoporus*. Component III, which was dominated by *Gonyaulax polyedra,* did not show a marked agreement between stations except for a peak in week 20.

Phytoplankton component I showed very strong positive correlations with conditions characteristic of upwelling: a shallow pigment layer at which *in vivo* chlorophyll fluorescence dropped to near zero, low surface temperature, and high salinity and density at the pigment layer depth. Phytoplankton component II was highly correlated with a high concentration of nitrate relative to silicate at the pigment layer depth, again indicating an association with upwelling. Phytoplankton component III, dominated by *Gonyaulax polyedra,* was not an upwelling assemblage and was correlated with a shallow pigment layer, low silicate at the pigment layer depth, and high temperatures at both the surface and pigment layer depth (Goodman et al. 1984).

Reid et al. (1978) studied the vertical distribution of plankton assemblages in the nearshore part of the SCB during March 1976. By using three vessels, they were able to sample simultaneously at three locations: one north of the Palos Verdes Peninsula, another near Dana Point, and a third between Oceanside and La Jolla. Finding 20 species to be most abundant in the 58 samples (table 5.1), the authors of this study used a percent similarity index and principal component analysis

Table 5.1. *First 20 Taxa of Phytoplankton Species Listed in Order of Average Abundance in 58 Samples*

1.	*Exuviaella* sp.
2.	*Scrippsiella* sp. / *Peridinium trochoideum* (Stein) Lemm.
3.	*Skeletonema costatum* (Grev.) Cleve
4.	*Eucampia zodiacus* Ehrenberg
5.	*Prorocentrum gracile* Schutt
6.	*Calciosolenia murrayi* Schlauder
7.	*Thalassiothrix frauenfeldii* (Grun.) Grunow
8.	*Gymnodinium* sp. A
9.	*Gymnodinium splendens* Lebour
10.	*Mesodinium rubrum* Lohmann
11.	*Rhizosolenia fragilissima* Bergon (small form)
12.	*Eutrptiella gymnastica* Throndsen
13.	*Hemiaulus sinensis* Grev.
14.	*Ceratium kofoidii* Jorgensen
15.	*Torodinium robustum* Kofoid & Swezy
16.	*Gymnodinium* sp. S
17.	*Leptocylindrus danicus* Cleve
18.	*Cochlodinium catenatum* Okamura
19.	*Peridinium minutum* Kofoid
20.	*Gonyaulax polyedra* Stein

From Reid et al. 1978.

to investigate station-to-station differences among the 20 species. This method led to separation of the species into three groups (fig. 5.2). The most abundant species in the chlorophyll maximum layer (CML) were *Exuviaella* sp. and *Mesodinium rubrum*.

Reid et al. (1978) found only partial support for their hypothesis that the species assemblage is more similar in samples from the CML at nearby stations than between the CML and surface samples at the same station. A second hypothesis, that the species assemblage is more similar among stations of the CML than between the surface layer and CML, was not verified. A third hypothesis, that spatial differences in the phytoplankton assemblage are more abrupt in the offshore direction than in the longshore direction, was supported by the similarity analysis, but not by the principal component analysis. Distribution patterns were coherent over many kilometers in the longshore direction, while abrupt changes were evident in the offshore direction.

Winter (1985) reported finding four separate species assemblages of coccolithophores occurring during March in the SCB, each correlated with distinct water masses in (1) the California Current, (2) the Southern California Countercurrent, (3) the transitional zone, and (4) the nearshore. Coccolithophore assemblages were more uniform in June, indicating that the SCB was experiencing more stable conditions than were prevalent during March.

Species assemblages differ temporally and spatially. Near the thermocline, for example, an area of elevated chlorophyll concentration often occurs with a vertical species assemblage that is different from that of the surface layer. When the onshore–offshore gradient of abundance for individual species was examined, 20 of 44 taxa were found in significantly greater numbers horizontally onshore (notably *Nitzschia* sp., *Ceratium tripos,* and *Emiliania* [= *Coccolithus*] *huxleyi*) and 4 appeared in significantly greater abundance offshore (notably *Prorocentrum balticum*). Other data showed dominant species changes in onshore–offshore gradients (Reid et al. 1970). But when the same data were analyzed by multivariate techniques, temporal changes between stratified and upwelling conditions were found to be more significant than the persistent onshore–offshore floral changes (Goodman et al. 1984). Furthermore, offshore net caught diatoms of several species were often dominant (Sargent and Walker 1948).

Sampling of the mid-summer phytoplankton at five stations along each of six lines extending seaward from the coast at approximately 8-km intervals between Encinitas and San Onofre indicated significant differences in the spatial patterns of species. In terms of relative species composition, populations separated by vertical distances of a few tens of meters can be as different as, or more different from, populations separated by tens of kilometers in the horizontal dimension. Horizontal variation was generally greater at the surface than at the depth of chlorophyll maximum (Cullen et al. 1982a). The resi-

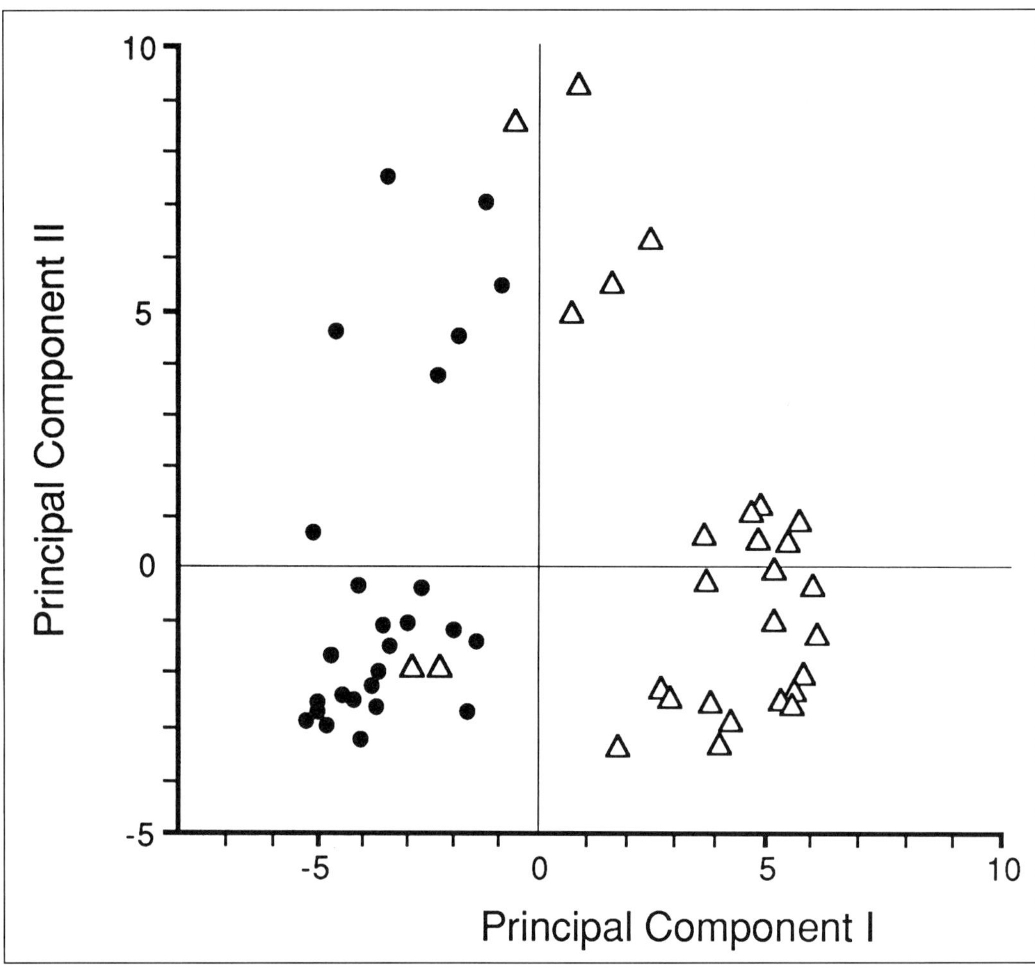

Figure 5.2. Principal component analysis of phytoplankton species assemblages collected in March 1976 between La Jolla and Santa Monica Bay. Triangles are samples from the chlorophyll maximum layer. Dots represent surface samples. The species assemblages fell into three clusters. The cluster at the lower left includes primarily surface samples taken between La Jolla and Dana Point. The cluster at the lower right includes most of the chlorophyll maximum samples from the same stations, although two of those appear in the first cluster. The upper looser cluster includes all the samples from Santa Monica Bay, both from the surface and the chlorophyll maximum layer. The most important species for component I were the dinoflagellates *Exuviaella* sp., *Gymnodinium* sp., and *Ceratium kofoidii*. Those most important for component II were the diatoms *Hemiaulus sinensis* and *Skeletonema costatum* and the protozoan *Mesodinium rubrum*. (From Reid et al. 1978.)

dence time of water near the coast was of great importance in determining the abundance and taxonomic composition of phytoplankton. Advection of offshore water toward the coast was a major determinant of the pattern (Cullen et al. 1982a).

Phytoneuston

The sea surface microlayer (upper 50 μm) is a unique habitat and a concentration point for both organisms and chemicals (Hardy 1987). A number of studies document enrichments in the concentration of microorganisms in the microlayer that are often several orders of magnitude greater than that occurring in subsurface waters (Hardy 1982).

Phytoneuston are the microalgae inhabiting the surface microlayer. Their existence has been recognized since early in this century (Naumann 1917), and their importance as a food source for surface-feeding zooplankton and fish larvae is well documented (Hardy 1982; Zaitsev 1971). Abundant populations of numerous species inhabit this layer. Research has not yet addressed the role of phytoneuston in the SCB, but studies in Puget Sound indicate that primary productivity within the microlayer is 2–53 times greater than in the subsurface water only a few centimeters below (Hardy and Apts 1989).

Standing Stock

Particulate Organic Carbon

Particulate organic carbon (POC) is a heterogeneous mixture of small particles in seawater. It includes living plankton, organic detritus such as fecal material, crustacean molts, and mucus material from gelatinous plankton and fish. Living plankton usually account for less than half of the total POC, but in blooms, they may comprise 70% or more of the measured POC.

Approximately one-third of the particulate organic matter produced by primary production sinks through the euphotic zone into deep water and settles on the sediments. The rate of sedimentation as measured in sediment traps is greatest during times of high surface production (Honjo 1982). Several studies have examined the residence time of particulate organic carbon (the quantity present in the water divided by its rate of input or removal). In general, the residence time of POC varies from 3 to more than 100 days between seasons and years (fig. 5.3) (Eppley et al. 1983). Eppley et al. (1977, 1983) considered POC to be the total mass of particulate carbon in the euphotic zone and assumed a steady state—that is, that the sinking flux out of the photic zone was equal to the production rate within the photic zone. They found that the residence time of POC decreased as a power function with increasing total carbon productivity. They suggested that this relationship provides a general guide to POC residence times in the surface layers of other oceanic regions of similar temperature and illumination.

A series of cruises spanning more than 5 years showed a marked temporal variability in both POC and primary production in the surface layer of the SCB. The area fluctuated between states of high production and short POC residence time and low production and long POC residence time, or between relatively eutrophic (nutrient-rich) and oligotrophic (nutrient-poor) states. These rates were correlated with anomalies in sea surface temperature, reflecting both low frequency changes in ocean climate and episodic upwelling.

The number of times an atom recycles before sinking is the rate of regenerated production divided by the rate of new production. This parameter varied in the SCB from a low of 1–3 during the seasonal maximum of upwelling in April and June 1975, a year of high production, to nearly 20 during the most oligotrophic periods in 1976 and 1977.

What proportion of the standing stock particulate organic matter is composed of phytoplankton? Eppley et al. (1977) found that POC was positively correlated with particulate organic nitrogen (PON), chlorophyll *a*, adenosine 5′-triphosphate (ATP), and total particulate volume, with 59–83% of the variation in one variable explainable by its correlation with the other. These correlations suggest that much of the POC is of biogenic origin. A total of 40% of the POC was associated with particles of less than 5 μm. Like POC, chlorophyll *a*, PON, ATP, and total particle volume in the chlorophyll maximum layers were higher onshore than offshore by a factor of two to three. During blooms of

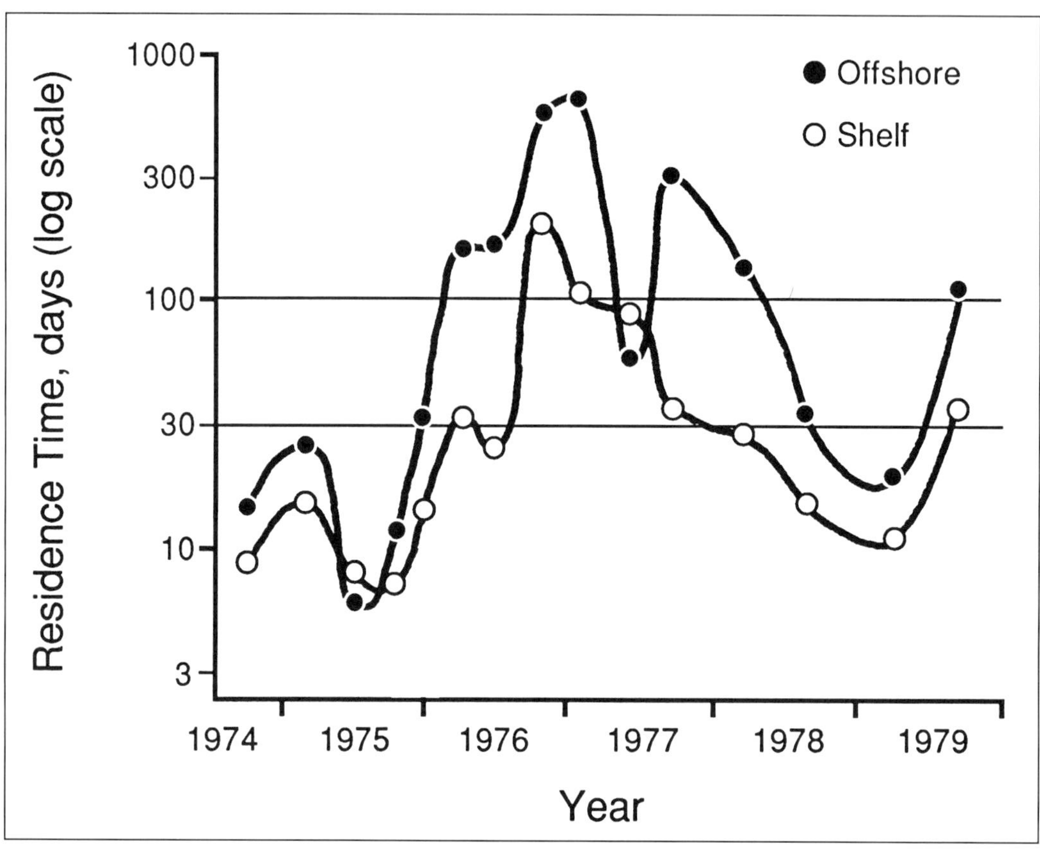

Figure 5.3. Residence time of particulate organic carbon (POC) in the euphotic zone from 1974 to 1979 in the SCB. Separate estimates are provided for shelf waters, <100 m depth, and offshore waters. The horizontal lines at 30 and 100 days correspond to minimum and mean estimates of the residence time of surface waters in the SCB from Hendricks (1979). When values lie above the lines, as in 1976, the surface circulation would carry organic particles out of the region before they settled out of the euphotic zone to deep water. (From Eppley et al. 1983.)

chain-forming diatoms, the particle size distribution was rather broad, whereas during blooms of dinoflagellates, specific particle size peaks were seen. For example, as much as 70% of the particulate carbon was associated with a single size distribution peak of *Prorocentrum micans*. Based on decomposition experiments conducted by others, Eppley et al. (1977) concluded that only about 10–20% of the measured POC was refractory. Thus, the observed distribution of particulate matter was largely a reflection of primary production and the standing stock of phytoplankton. Loss rates were density dependent, as would be expected if losses resulted from either grazing or sinking.

Small et al. (1989) examined the dynamics and fate of POC in the photic zone overlying the Santa Monica Basin during spring and fall seasons. The major analyses centered on five cruises (October 1985, 1986, 1987; May 1986; and April 1987). Particulate carbon fluxes (PCF) during the five periods ranged from 184 to 542 mg m^{-2} d^{-1}. When anomalies were ignored, carbon flux rates during spring were twice those of fall. Particulate carbon and nitrogen residence times in the photic zone averaged about 20 days and 39 days, respectively, with no seasonality. Rapidly sinking marine snow aggregates, invested with zooplankton fecal debris, probably represent a major part of the downward flux (Alldredge and Silver 1988).

Chlorophyll

Measurements of the biomass of phytoplankton have been facilitated by techniques of *in vivo* chlorophyll fluorescence (Lorenzen 1966), the measurement of ATP (Holm-Hansen and Booth 1966), and satellite remote sensing (Peláez and McGowan 1986). A total of 143 vertical profiles of *in vivo* fluorescence taken from 1974 to 1979 indicated that a subsurface chlorophyll maximum layer (CML) is generally present in the SCB. The CML is generally located well above the 1% light level in the vicinity of the nitracline, and it is also a biomass maxima. Although the CML occurs often where light and nutrient conditions are favorable for growth, the CML depth does not usually coincide with the depth of maximum productivity (Cullen and Eppley 1981).

Cullen et al. (1982a) sampled 30 stations in a 25-by-40-km grid close to the southern California coast and found that chlorophyll and phytoplankton biomass were higher nearshore and to the south where relatively large dinoflagellates dominated. At each station, the CML was near the nitracline and 10% light level. At some stations, the CML resulted from higher levels of phytoplankton biomass, and at other stations from increases in the chlorophyll content of small flagellated phytoplankton.

Remote Sensing

The coastal zone color scanner (CZCS) on the *Nimbus* 7 satellite has provided useful information on the distribution of phytoplankton by measuring chlorophyll over extensive areas of the SCB. Such data provide us with synoptic views of complex oceanographic regions which are impractical to obtain from ships alone (fig. 5.4). Satellite imagery has

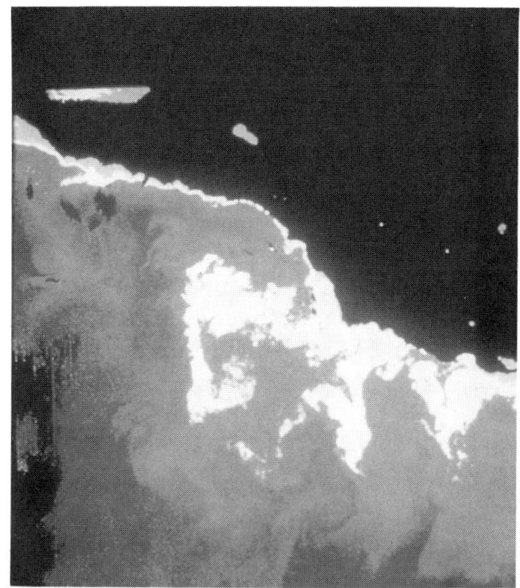

Figure 5.4. Phytoplankton pigment image off California derived from *Nimbus* 7 CZCS imagery. Red represents areas that were richest in plant pigment. A May–June cruise to the area had relatively few stations within these red areas, but those had 10-m chlorophyll values ranging from 2 to 8 mg m^{-3}. There were 28 stations within the yellow area with 10-m chlorophyll values ranging from >0.5 to <4 mg m^{-3}. The pale blue area had values of >0.06 mg m^{-3} but <0.5 mg m^{-3}. (From Peláez and McGowan 1986.)

also allowed the identification of persistent and striking biological features. Many of these recurring large-scale patterns were either unknown or only dimly perceived prior to the advent of satellite imagery. For example, *Nimbus* 7 CZCS imagery (fig. 5.4) revealed the occurrence far offshore of a large region of high phytoplankton pigment, a biological "hot spot" that loosely overlies a system of submarine ridges, banks, and basins that make up the southern California borderland. Also, shallow basins and enclosed shallow areas such as the Santa Barbara Channel consistently show high pigment content, with an approximately threefold change in phytoplankton pigment content over a distance of a few kilometers. These large-scale structures undergo significant monthly, seasonal, and annual changes, although the large-scale pigment patterns for a given season tend to reappear from one year to another (Peláez and McGowan 1986).

A longitudinally oriented meandering boundary 200–500 km offshore and several hundreds of kilometers long occurs. The inner side of this longitudinal boundary has high-pigment content interwoven with lower pigment structures located immediately offshore. A region of low-pigment content usually overlies the deep troughs and basins that characterize most of the SCB. This low-pigment area is a recurrent feature of the region. It is immediately offshore of a narrow coastal band of high-pigment content and is usually continuous with the low-pigment region south of the latitudinal boundary.

Surface chlorophyll, depth-integrated chlorophyll, and primary production may or may not be closely correlated, and there is considerable variance in the relationships. In some areas, such as the central North Pacific, surface chlorophyll is not correlated with either depth-integrated chlorophyll or integrated primary production, whereas good correlations have been found in the California Current. Therefore, the use of surface chlorophyll as an indicator of the total integrated water column biomass or production must be applied with caution (Hayward and Venrick 1982).

Peláez and McGowan (1986) used a 34-month CZCS data set to describe the major phytoplankton pigment patterns in the California Current over space and time. Measurements were complicated because of the variability in the depth of penetration of the signals received. During June 1971, they used empirical ground truthing and compared independent shipboard measurements of surface chlorophyll (estimated by a flow-through fluorometer) with measurements from the CZCS on a 200-km transect. They found a highly significant relationship (correlation coefficient of 0.92) between the satellite and shipboard measurements. Additional measurements in the California Current indicated that the surface chlorophyll concentrations correlated well with the integrated overall water column chlorophyll.

Remotely sensed near-surface chlorophyll data produce regular patterns in time and space, especially on the larger scales. However, until more is known about the relationship of surface chlorophyll to the state of the phytoplankton in the rest of the water column, caution must be used in interpreting such measurements in terms of population biology or production.

New technologies including airborne laser-induced chlorophyll fluorescence (Hoge and Swift 1986) and a new generation of satellite ocean color sensors (Hooker and Firestone 1992) should provide a wealth of new information about large-scale changes in phytoplankton biomass and productivity.

Growth and Productivity

Growth

Phytoplankton "grow" by cell division and thus increase in cell *number* rather than in *size*. As previously noted, the rate of cell division varies, according to both the natural intrinsic capability of the species and environmental conditions, from about one division per day to less than one division per week. One must be clear in distinguishing between changes in algal biomass and changes in the turnover rate of that biomass. The increase in algal biomass over a period of time may be referred to as *growth* or *new production*. Thus, growth is the increase in the pool size of biomass as represented, for example, by particulate organic carbon. *Specific growth*, or *regenerated production*, refers to the rate of flux of carbon through that pool. Short-term measurements of carbon or nitrogen uptake are really measurements of specific growth since it is the turnover rate that is being measured. It is possible for high specific growth rates to occur simultaneously with low growth rates if the phytoplankton particulate carbon pool turns over rapidly but does not get any larger because of grazing by zooplankton or other factors.

Maximum phytoplankton growth, or cell division, is frequently synchronized to occur (depending on the species) at a particular time of day. For example, some dinoflagellates show maximum cell division in the early morning hours (Weiler and Chisholm 1976). Division rates for dinoflagellates have been found to be about 0.16–0.5 d^{-1}, corresponding to doubling times of 1.4–4.4 days.

Phytoplankton Production

Numerous measurements of *primary production,* which is the photosynthetic conversion of inorganic carbon to organic cellular material by phytoplankton, have been conducted in the SCB. The efficiency of conversion of solar energy into organic matter in the SCB has been estimated to be well under 1% (Eppley and Holm-Hansen 1986). Most measurements have been made by the classic radioactive carbon incubation technique (Steemann-Nielsen 1952). Briefly, samples of seawater containing phytoplankton are collected from depth, inoculated with a known activity of ^{14}C, and then incubated either in a shipboard deck incubator or while resuspended *in situ*. After a period of time, the samples are collected, and the particulate ^{14}C is collected on a filter. The radioactivity is counted and is proportional to the amount of inorganic carbon fixed to organic carbon. Early measurements of primary production indicated values integrated over the water column of about 0.5 g C m^{-2} of sea surface per day (Holmes 1958). Data from more than 20 cruises, with more than 220 stations sampled since 1974, indicate that the average primary production in the inner and southern portions of the SCB is 0.39 g C m^{-2} d^{-1} over an area of approximately 30,000 km^2. The total production over this area is 64.3 $\times$ 10^6 t C yr^{-1}, or roughly 1.7 times more production per unit area than the average ocean value (Eppley et al. 1970; Owen 1974).

Small et al. (1989) summarized both historical data and the results of their own measurements of primary production for the Santa Monica Basin. Their summary indicates mean rates in spring (March–May) of 1.050 g C m^{-2} d^{-1} (range of 0.411–1.254 g C m^{-2} d^{-1}) and in fall (September–November) of 0.531 g C m^{-2} d^{-1} (range of 0.213–0.800 g C m^{-2} d^{-1}).

Photosynthesis and Chlorophyll

The efficiency of photosynthesis, often called the *photosynthetic assimilation number,* is the ratio of carbon fixed at light saturation to chlorophyll concentration in the population. In the SCB, as in many areas, the photosynthetic assimilation number increases during bloom formation or when ample light and nutrients become available during the spring, then decreases as factors limit production.

Phytoplankton show evidence of adaptation to available light quality and quantity, including a shift in pigment composition. The photosynthetic action spectrum, or photosynthetic activity at different specific wavelengths, has not been measured routinely at sea. However, measurements have been made of the action spectra of chlorophyll *a* fluorescence, which have been used to study the light adaptation of phytoplankton, including species in the SCB (Neori et al. 1984). Comparisons of absorption spectra with fluorescence excitation spectra indicate an apparent increase in efficiency of sensitization of chlorophyll *a* fluorescence in the blue and green spectral regions for low-light populations (Neori et al. 1984). In samples from deeper waters, both absorption of chlorophyll *a* and fluorescence excitation spectra showed enhancement in the blue to green portion of the spectrum from 700 to 560 nm relative to that at 440 nm. The data indicated that, in the natural population studied, shade adaptation caused increases in the concentration of photosynthetic accessory pigments relative to chlorophyll *a*. Changes in pigment concentration occurred over periods of less than one day.

Factors Affecting Production

CURRENTS AND UPWELLING

Primary production in the SCB is greatly influenced by basic physical processes, including wind, which affect both the stability of the water column and subsequent mixing and nutrient input to the euphotic zone. These processes change on a variety of temporal and spatial scales. An important controlling process is the advection of nutrient-rich, low-salinity water from the north in the offshore flow of the California Current. Time series data indicate that the strength of this cold current flow is positively correlated with increasing plankton biomass. For example, the anomalistic warm water years of 1957–1960 were correlated with low biological production, while the cold water years of 1953 and 1956 were associated with high production (Bernal and McGowan 1981).

Coastal upwelling is the vertical advection of water along the coast when previous surface waters are displaced offshore. Upwelling events are indicated by sudden declines of sea surface temperature, the shoaling of the nitracline, and the appearance of nitrate at the surface. Upwelling is usually driven by local winds blowing parallel to the shoreline and has been documented in the San Diego area (Dorman and Palmer 1981). At the Scripps Institution of Oceanography (SIO) pier, low-temperature anomalies corresponding to such upwelling episodes are accompanied by a rapid growth of phytoplanktonic diatoms (Tont 1981). Small-scale mixing processes are important to the nutrient input across the nitracline into the nutrient-depleted surface waters. Relatively short-term tidal and internal wave frequencies bring nitrate-rich water into the surface layer over short-term intervals (Cullen et al. 1983; Armstrong and LaFond 1966).

The upward transport of the denser subsurface water during upwelling brings nutrient-rich cold water to the surface. Satellite imagery, remote temperature sensing, and ship-based temperature sampling show that coastal upwelling in the SCB occurs within a few kilometers of shore, with exceptions at Point Conception, Palos Verdes Peninsula, and Point Loma. Water south of these points frequently has a clear, visible cold phase that extends for many kilometers. In general, upwelling is only a weak summer phenomenon in southern California, although May and June appear to be times of maximum upwelling in the vicinity of the SIO pier. In short, upwelling in the SCB is generally of short duration (a week or less) and generally occurs less frequently and less intensely than in other places along the coast north of Point Conception.

WATER TEMPERATURE

In the SCB, average seasonal temperatures range from a minimum of about 12°C to a maximum of about 24°C. As with biological processes in other organisms, temperature has a major influence on the growth and physiology of phytoplankton. Each species generally displays its own specific temperature optimum for growth (Smayda 1980) and for primary production (Eppley 1972).

The initial photochemical reactions of photosynthesis are not temperature dependent, but the subsequent enzymatic carbon transformation reactions are generally described by the van't Hoff equation (that is, an approximate doubling with a 10°C rise in temperature). Most marine phytoplankton have growth rates of about one or two doublings per day at 20°–25°C, whereas the fastest growing species approach three or four doublings per day at those temperatures. In the SCB, doublings are generally less than one per day averaged over the euphotic zone (Eppley 1972).

NUTRIENTS

Many of the micronutrients required by phytoplankton for growth, including Ca, K, S,

B, and others, are believed to be present in sufficient concentrations in seawater to meet or exceed the growth requirements of phytoplankton. Other micronutrients, such as N, P, and trace metals (for example, Fe, Mn, Co, Cu, and Zn), are in low concentrations and are potentially limiting to growth. Vitamins such as B_1, B_{12}, and biotin are also required by many species for growth. Data on the vitamin requirements of phytoplankton are extremely limited, but some studies indicate that populations in the SCB do require sufficient vitamin B_{12} to grow normally (Carlucci 1970).

Considerable stimulation in phytoplankton growth can be obtained in surface water samples from the SCB when they are experimentally enriched with N, P, and Si, indicating that these three elements are most likely to limit phytoplankton growth in the SCB (Eppley 1968). Evidence suggests that N is the most common limiting nutrient. Uptake rates of ^{15}N, ammonium, and urea have been measured at different concentration levels. In general, these uptake rates increase hyperbolically with increasing nutrient concentration. Such data suggest that increases in ambient N (for example, from coastal wastewater outfalls) will lead to increasing primary production (Eppley et al. 1979a, b).

Nitrate enters the euphotic zone from deep water, rain, and river and coastal wastewater discharges. The resulting production based on this nitrogen is termed *new production* in contrast to the *regenerated production* that results from utilization by the phytoplankton of ammonium, urea, and other forms of recycled N produced in the euphotic zone. The rates of utilization and regeneration of N in the SCB have been examined in some detail (McCarthy 1972; Eppley et al. 1979a, b; Harrison 1978; Redalje 1983). About two-thirds of the N demand is provided by recycled N (that is, ammonium and urea produced within the euphotic zone by the metabolism of animals and bacteria) and about one-third by nitrate via physical transport from deeper water. The physical processes that bring nitrate into the euphotic zone are, along with seasonal changes in light intensity and temperature, the driving forces for primary production. Nitrogen is recycled from ammonium to PON and back to ammonium over a time scale of approximately 2 weeks (Eppley et al. 1983). Thus, regenerated production is generally proportional to new production (Eppley et al. 1979a, 1983).

Sewage outfalls in southern California provide a source of nutrients, primarily N, which stimulate phytoplankton primary production. The combined flow of sewage outfalls in the SCB is large—about 4×10^9 l of primary and secondary effluent per day. Studies of the assimilation of ammonium, nitrate, and urea were conducted at outfall sites in the SCB (Institute of Marine Resources 1971; Thomas and Seibert 1974). Nutrient bioassays, that is, measurements of phytoplankton growth in response to different additions of nutrients, were performed on samples taken above major sewage outfalls. Assimilation of nitrogen fits the hyperbolic Michaelis–Menten kinetics such that the maximum growth rate increased with the concentration of nitrogen and the saturation constant was uniform over a range of dilution rates. In summary, these studies showed that N was the principal nutrient limiting phytoplankton growth and that N inputs from outfalls stimulated phytoplankton growth.

LIGHT

The quality (spectral distribution of wavelengths) and quantity (irradiance) of solar energy differ both spatially and temporally according to depth in the photic zone, cloud cover, day length, and seasonal sun angle. The portion of the solar spectrum that is utilized for photosynthesis, the so-called photosynthetically active radiation (PAR), is confined to the spectral region of 350–700 nm. The amount of energy penetrating the water column increases with decreasing angle to the solar zenith. Long wavelengths and very short wavelengths are absorbed relatively rapidly, with the least absorption occurring at approximately 490 nm in blue ocean water. In greener coastal water rich with phytoplankton, wavelengths of maximum penetration are shifted to about 550 nm.

General responses of phytoplankton photosynthetic rate to light intensity and quality have been discussed in detail elsewhere (Harris 1980; Raymont 1980). At low-light intensities, photosynthesis per unit chlorophyll generally increases linearly or hyperbolically with increasing light intensity until a maximum value of photosynthesis (P_{max}) is reached. At this point (the saturation light intensity), no further increase in photosynthesis occurs in response to light. At P_{max} the enzyme-controlled and temperature-sensitive "dark reactions" of photosynthesis will determine the overall rate of photosynthesis. The initial slope of the photosynthesis/irradiance (P/I) curve can differ according to species and to the past history of light exposure of the population.

Light saturation typically occurs at lower light levels for populations occurring deep in the euphotic zone and occurs at higher light intensities for populations near the surface (fig. 5.5). Such data indicate that phytoplankton are typically adapted to the incident levels of light where they occur; that is, so-called light or shade species occur, just as they do in terrestrial forest ecosystems. The compensation light intensity (that is, where photosynthesis and respiration balance each other in regard to oxygen exchange) generally occurs between 0.1 and 1% of surface irradiance (Platt and Jassby 1976; Raymont 1980).

Unlike high-latitude boreal areas where light levels can be limiting to production, evidence suggests that in the SCB, light intensity does not normally limit primary production. However, beyond saturation light levels, photoinhibition can occur and near-surface populations may be photoinhibited by high summer light intensities (Hardy and Apts 1989).

Both light-limited and light-saturated photosynthesis typically exhibit diel periodicity that is not uniform. An endogenous biological rhythm (the underlying basis of which has not been completely defined) may partially control these diel oscillations of photosynthesis. Studies in the Santa Barbara Channel, from north of Anacapa Island to south of Point Conception between May and August 1980, confirmed the presence of diel oscillations in the P/I relationships. The initial slope (alpha) and asymptote (P_{max}) of the P/I curve changed significantly over the day, and the oscillations appeared unrelated to changes in chlorophyll *a* concentrations. Amplitudes of the daily variations in photosynthesis ranged from approximately three to nine as measured by the maximum to minimum ratio for P_{max}. Diatom rich samples collected during an upwelling event and those dominated by dinoflagellates both have midday to early afternoon maxima in alpha and P_{max}. Samples from other locations had peak photosynthetic activity later in the afternoon. Results indicated that the P/I relationship is time dependent and, moreover, that changes in alpha and P_{max} are closely coupled for a variety of natural phytoplankton assemblages (Harding et al. 1982).

Ultraviolet-B (UV-B) radiation (280–320 nm) penetrates approximately 20 m through the water column in clear oceanic water. Evidence suggests that UV-B radiation inhibits photosynthesis in the upper water column (Smith et al. 1980; Behrenfeld et al. 1993). Many earlier measurements probably overestimated primary production to some extent because they were conducted in glass bottles or containers that filtered out the inhibiting UV-B component of the solar spectrum (Worrest 1983). Concerns are growing regarding global stratospheric ozone depletion, the resultant increases in UV-B radiation, and the potential detrimental effects on marine ecosystems (Hardy and Gucinski 1989).

GRAZING

Most zooplankton depend to a large degree on phytoplankton as a food source; therefore,

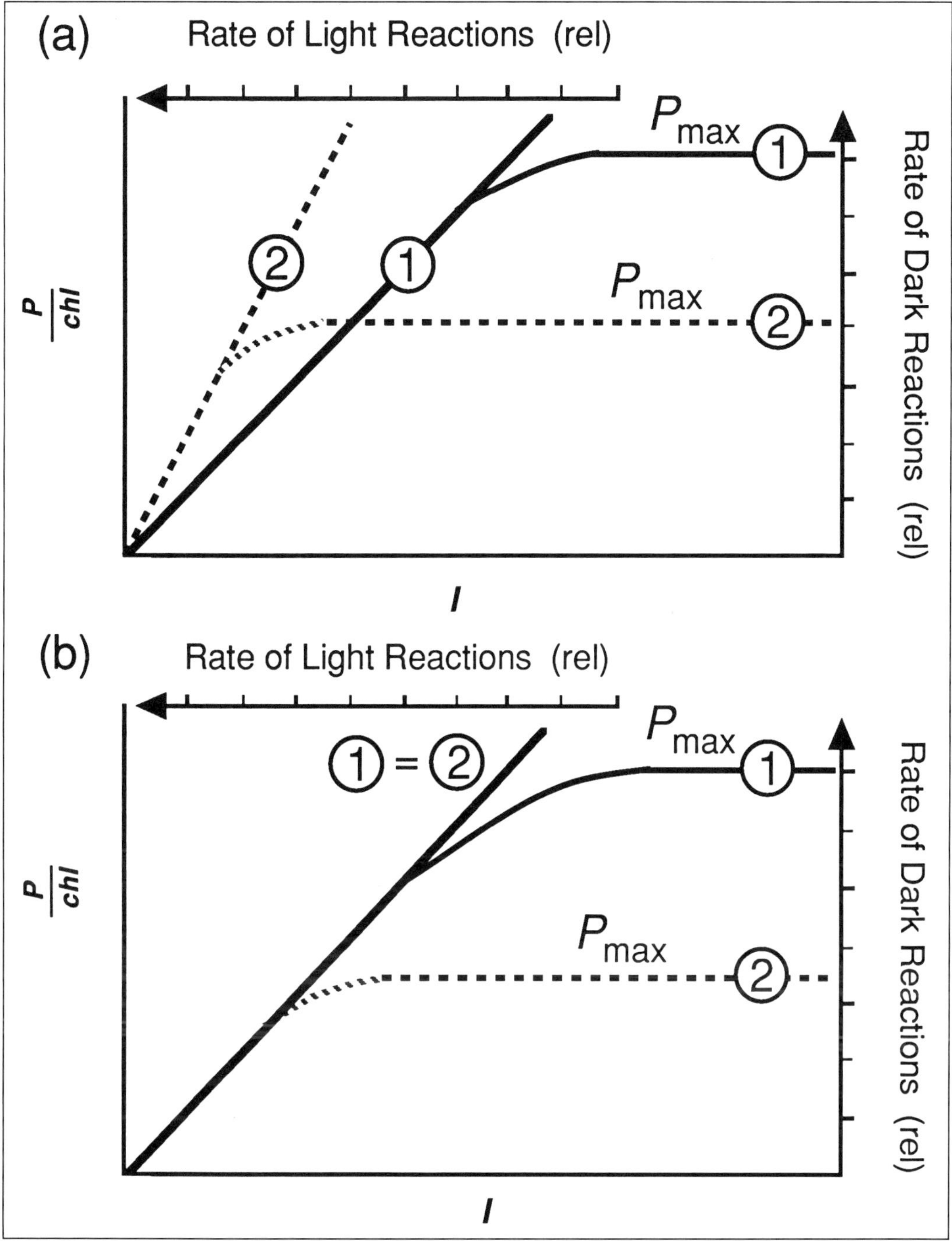

Figure 5.5. The effects of high- and low-light conditions on photosynthetic responses in phytoplankton. Population 1 is located at the top of the euphotic zone and population 2 at the bottom. (a) This situation reflects the response of increased chlorophyll content of population 2; that is, the light reaction rate is higher in population 2 than in population 1. (b) This situation reflects the response of the light reactions being equal. P = photosynthesis, chl = chlorophyll, and I = irradiance. (From Yentsch 1980.)

one might expect to find increases in zooplankton abundance correlated with increases in phytoplankton abundance. However, because of the rapid generation time of phytoplankton and the slower "response time" of zooplankton, one actually finds an alteration of positive and negative correlations in space and time (Harvey et al. 1935). The maximum concentration of phytoplankton tends to occur in spring, while the macrozooplankton maximum generally occurs in summer. The abundance of feeding juvenile stages of the copepod *Callinus pacificus* was positively correlated with the horizontal and vertical biomass of phytoplankton (Mullin and Brooks 1972). In one study using concurrent phytoplankton and zooplankton sampling, no correlation was found between phytoplankton biomass estimated microscopically and the biomass of protozoa, micrometazoa, or total microzooplankton (Beers et al. 1980). However, when estimated on a quarterly basis, macrozooplankton biomass was positively correlated with primary production over the past 30 years (Smith and Eppley 1982).

Modeling Phytoplankton Production

The Food Chain Research Group of SIO developed an empirical algorithm to predict primary production in the SCB from daylength and the temperature anomaly (departure from the long-term mean temperature) for sea surface temperature recorded at the SIO pier:

$$P = \exp(-3.78 - 0.372T + 0.227D)$$

The mean production, P, can be calculated from this equation by setting daylength (D = 12 hours) and temperature anomaly (T = 0) to give 0.392 g C m^{-2} d^{-1}. This algorithm explains, in a statistical sense, about 60% of the between-cruises variability in primary production measurements. It has been used to reconstruct the history of primary production back to 1920. Nitrate associated with the colder upwelled water is believed to drive production, and the temperature anomaly explained about 65% of the variability in nitracline depth on the shelf, but only about 25% at offshore stations.

Optical measurements, along with fluorometrically determined chlorophyll, have successfully been applied to characterize mesoscale productivity gradients in the SCB (Smith et al. 1987).

Satellite imagery has provided valuable information on the spatial and temporal distribution of chlorophyll biomass and its relationship to primary productivity. Using measurements of surface chlorophyll taken by the coastal zone color scanner aboard the *Nimbus* 7 satellite, along with measurements of photosynthetic primary production during simultaneous oceanographic cruises, the relationship between surface chlorophyll and productivity over large areas has begun to emerge. These measurements indicate that productivity in the SCB is intermediate when compared to other areas of the Pacific (for example, greater than the subtropical Pacific gyre), but is less than the Peru upwelling region. Data have been gathered concurrently by ships and satellites to calibrate satellite measurements against actual *in situ* measurements of chlorophyll. A general algorithm relating satellite chlorophyll measurements to shipboard measurements agreed within ±40%. Satellite images recorded between late February and mid-March 1979 revealed complex and changing chlorophyll patterns, including a significant decline in the mean chlorophyll concentration during the period of study. Fair agreement was obtained between shipboard production estimates (554 mg m^{-2} d^{-1}) and satellite estimates (403 mg m^{-2} d^{-1}) when averaged for the entire SCB during this period (Smith and Baker 1982; Smith et al. 1982).

Obtaining accurate measures of primary productivity from satellite imagery can be a difficult and complex task. First, only about the upper one-fifth of the euphotic zone chlorophyll is registered by satellite images, and the vertical distribution of chlorophyll and photosynthesis is not uniform. To estimate primary productivity from satellite imagery, one must derive an accurate proportionality estimate (f) relating the satellite-measured chlorophyll concentration to actual *in situ* chlorophyll concentration and to *in situ* primary productivity (in milligrams of carbon per square meter per day). This proportionality often has a large variance. In the richest waters, f approaches a minimum limit value of about 100 mg C m^{-2} d^{-1} mg^{-1} chlorophyll pigment m^{-3}. In the simplest systems, f is proportional to insolation (solar irradiance) (Eppley et al. 1985). However, as previously discussed, much of the variability in the relationship between time and space in the SCB is related to environmental variables other than solar irradiance.

Spatial and Temporal Patterns

General Pattern

Spatial and temporal patterns occur over a wide range of scales, complicating interpretations of the distribution and abundance of plankton. Environmental factors regulating growth lead to a complex spatial and temporal pattern of phytoplankton and productivity in the SCB. Every point in the water column is basically unique with regard to such variables as light intensity, nutrient mixture and concentration, and temperature. Small-scale biomass patchiness occurs even on scales of less than 1 m. For example, high chlorophyll concentrations are associated with mucilaginous aggregates of a few millimeters in size (Alldredge and Cox 1982). Physical factors of mixing and currents also determine the distribution of phytoplankton. Each species differs in its unique physiological requirements and optima for both light and nutrients. Topographic features of the SCB such as the complex of offshore islands and banks, which run from Santa Rosa and San Nicolas south to Tanner and Cortes banks, impose additional heterogeneity.

Santa Monica Bay in the central part of the SCB has unique geographic features, nutrients, and plankton populations that distinguish it from other nearshore sites (Reid et al. 1978). The bay generally has more dominant dinoflagellate populations, higher concentrations of ammonium, and closer proximity of the euphotic zone to nutrient regeneration at the bottom, as well as considerable anthropogenic inputs (Eppley et al. 1979a).

Onshore–Offshore Pattern

The terms *nearshore* and *offshore* are used rather loosely by different authors; for example, "nearshore" may mean an area extending out to 5 km in one study and as far out as 100 km in another study. In terms of the nutrient regime integrated over the entire euphotic zone in general, ammonia increases curvilinearly with distance from shore, silicate declines slightly, but phosphate and nitrate show no consistent gradient. This is because the euphotic depth increases offshore in approximately the same way as does the nitracline. Less is known about gradients of other dissolved materials. Dissolved organic carbon shows little pattern in surface waters within 10 km of shore. The concentrations of organic N, P, vitamin B_{12}, and thiamine generally decrease offshore (Strickland et al. 1970; Carlucci 1970).

As is typical of the west coast, plankton abundance and primary production in the SCB are generally higher nearshore than offshore. Since the continental shelf is only a few kilometers wide, internal waves from deep water typically move shoreward, injecting nutrient-rich water onto the shelf area (Cooper 1947). Also, episodic sediment disturbance and suspension are important mechanisms of nutrient regeneration in the shallow nearshore area (Fanning et al. 1982). Eppley et al. (1977) collected samples during quarterly cruises along three transects running approximately perpendicular offshore from (1) Santa Monica Bay, (2) near Dana Point, and (3) Del Mar. These samples were taken from six depths in the euphotic zone, which varied from 20 m nearshore to 50–70 m offshore. The three transects extended as far as 96–107 km offshore. They found that POC decreased along a gradient from onshore to offshore and decreased with increasing depth, although subsurface maxima were often present. The highest POC values were found in Santa Monica Bay (fig. 5.6).

The biomass of phytoplankton expressed as either per unit surface area or per unit volume decreases with increasing distance from shore within the first 10 km offshore (Eppley et al. 1970). This general pattern holds off Point Loma (Kiefer 1973) and off La Jolla (Barnett 1974). The CZCS satellite images confirm the decreasing biomass as one moves offshore, although offshore eddies and swirls exist that also contain high concentrations of phytoplankton (Smith and Baker 1982; Peláez and Guan 1982). Most data indicate that primary production also decreases in the offshore direction. Because of the greater supply of nutrients nearshore, a unit of chlorophyll is more productive nearshore than offshore (Cullen and Renger 1979).

Longshore Pattern

Statistical comparison of pairs of stations located at equal distances from shore revealed no significant longshore trends in nitrate, silicate, or phosphate in the upper 50 m (Mullin 1986). However, the lack of a longshore trend in nutrients may apply only to the portion of the SCB from Santa Monica Bay southward. Additional data that encompass the entire SCB do, in fact, show a significant trend of increasing nitrate from Santa Monica Bay to Point Conception. Evidence also points to small-scale longshore patchiness of phytoplankton biomass on scales between 0.1 and 10 km (Mullin 1979). Measurements based on *in vivo* fluorescence of chlorophyll at a single depth show predominant variability on scales exceeding 600 m (Star and Mullin 1979; Fiedler 1982a).

Significant differences in longshore abundances of phytoplankton species occurred between the north and south parts of the SCB. Out of 45 cases tested, 19 had greater abundances in the south—notably *Nitzschia sicula, Ceratium tripos,* and *Emiliania* (= *Coccolithus*) *huxleyi*. Only three species had greater abundances to the north (Cullen et al. 1982a). Surface phytoplankton occurring from Dana Point to San Diego were clearly divided into three distinct floral groups based on the 54 most abundant species. These "assemblage patches" had approximate longshore dimensions of 22–27 km (similar to the length of so-called coherent currents in the same area), suggesting that current patterns are important in determining the longshore distribution of these species assemblages (Eppley et al. 1984).

Phytoplankton species assemblages also differed in longshore transects between La Jolla and Santa Monica Bay (Reid et al. 1978). Comparison of the longshore- and offshore-directed transects indicates that, at both the chlorophyll maximum layer and at the surface, species composition varied more in the offshore direction than in the longshore direction.

Vertical Pattern and Vertical Migration

In addition to horizontal patterns, the abundance of individual species, total biomass, and productivity of phytoplankton generally show marked differences vertically through the water column. Differences in vertical stratification of biomass occur at the same depth from stations sampled within a few kilometers of each other and only a few days apart.

The use of *in vivo* fluorescence to measure biomass, by either pumping samples from depth to a sensing instrument on deck or by using submersible detectors, has led to a recognition of the importance of fine-scale vertical distribution in the SCB. Biomass, as measured by chlorophyll fluorescence, is generally maximum below the surface, often occurring at a depth of about 20 m. The maximum may become more intense and shal-

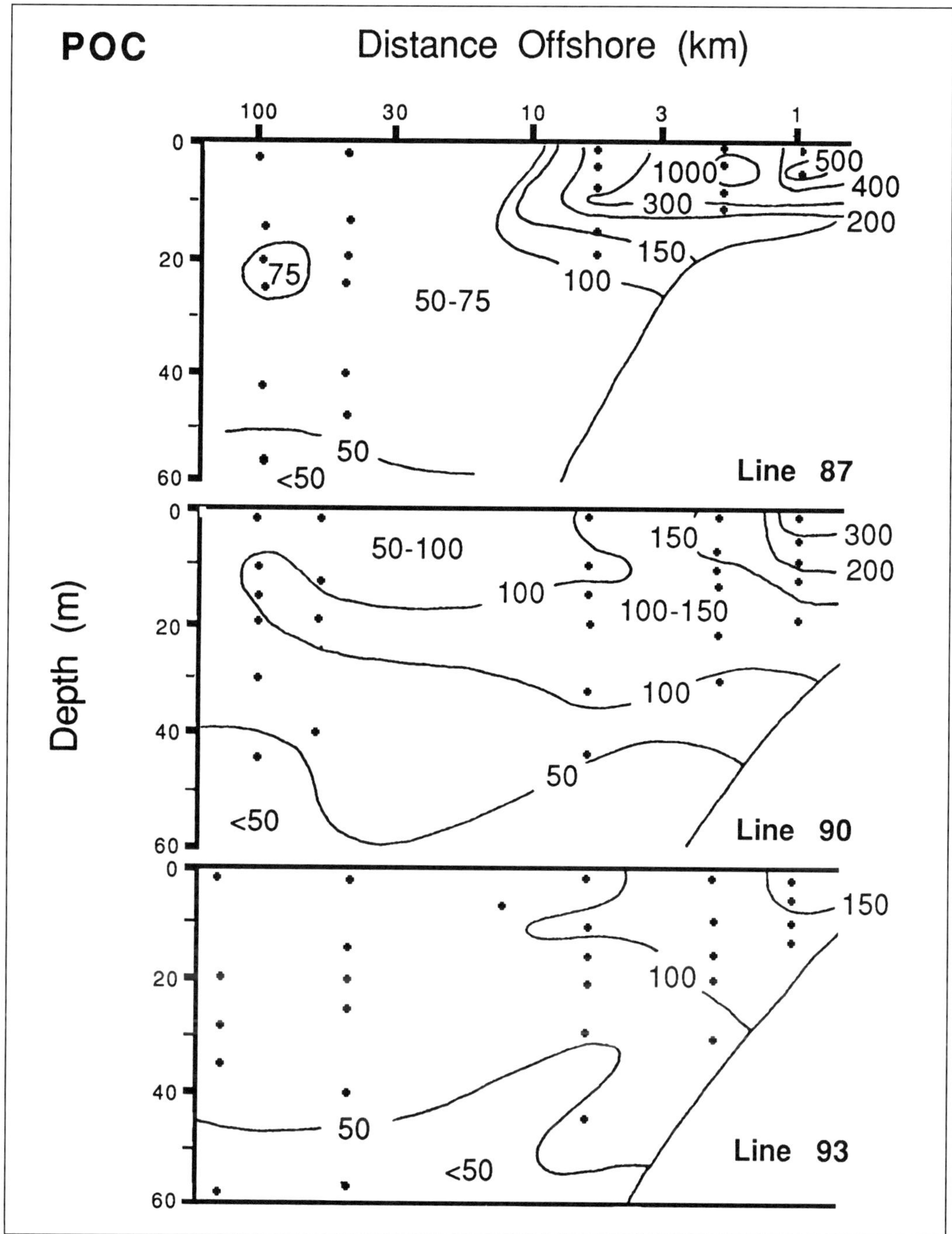

Figure 5.6. Vertical sections of particulate organic carbon (POC), expressed in milligrams per liter, during September 6–17, 1975. Distance offshore is shown on a logarithmic scale. Line 87 is from Santa Monica Bay, line 90 from Dana Point, and line 93 from Del Mar. (From Eppley et al. 1977.)

lower as one moves closer to shore, generally following the nutricline (Cullen and Eppley 1981; Cullen et al. 1982a). Other measurements, including microscopic phytoplankton abundance, ATP, and bacterial biomass, are often also positively correlated with the vertical distribution of chlorophyll in the euphotic zone (Fuhrman et al. 1980; Cullen and Eppley 1981).

The depth of maximum abundance usually differs between individual species. Differences in the vertical distribution of individual species result from differences in physiology with respect to light, nutrients, buoyancy, and motility. Large dinoflagellates are often plentiful near the surface, whereas the numerically dominant diatoms are most numerous at or below 20 m.

Primary production depends largely on the presence of light and chlorophyll and less directly on nutrients. It generally shows a subsurface maximum at both nearshore and offshore stations of the SCB and tends to be shallower than the chlorophyll maximum layer (Cullen and Eppley 1981). In Santa Monica Bay, maximum production is at the surface, as is the greatest concentration of chlorophyll. At nearshore stations, production per unit of chlorophyll is maximum at subsurface depths, probably because of the favorable nutrient supply there and perhaps also because of photoinhibition occurring at the surface. More recent data by Small et al. (1989) confirm subsurface maxima for both chlorophyll and primary production in the Santa Monica Basin (fig. 5.7).

In the SCB, as in many other coastal areas, dinoflagellates migrate vertically through the water column with a recurring diel pattern. For example, *Gonyaulax polyedra* was found to migrate downward in the evening and then re-form in surface patches in the morning (Holmes et al. 1967). A layer of dinoflagellates dominated by *Prorocentrum micans, Perirdinium depressum,* and *Ceratium furca* migrates downward in the evening (Eppley et al. 1968). Many studies have described particular patterns of depth distribution for different species (Allen 1928; Cullen et al. 1982a; Kimor 1981; Reid et al. 1978). Several studies have suggested that the relative species composition and abundances were more similar to each other in samples from the surface than in samples from the deep chlorophyll maximum. In these studies, the horizontal separation between stations was tens of kilometers and the vertical separation was only tens of meters (Reid et al. 1978; Cullen et al. 1982a).

Long-Term Temporal Pattern

Temporal patterns can be divided into short-term "events" on a scale of hours, days, or a few months and longer term seasonal or recurring annual trends. Long-term trends in species composition and population fluctuations have been observed at a few sites, including Santa Monica Bay from 1957 to 1972 (Kleppel 1980) and the SIO Pier from 1920 to 1939 (Allen 1936, 1941). In general, diatoms have several major peaks of abundance that are 5–6 weeks in duration, usually during the first half (but occasionally the latter half) of each summer (Tont 1976, 1981; Tont and Platt 1979). A high correlation in the occurrence of blooms was generally observed between San Diego and Port Hueneme, although the dominant species in the two locales were frequently different. The majority of these blooms occurred in conjunction with upwelling events. Sea surface temperature decreases of 2.5°C indicating upwelling were often associated with diatom standing stock increases of four orders of magnitude.

The biomass of the larger net-caught diatoms tends to be maximal in late winter or spring, although fall blooms also occur (Allen 1936). Large dinoflagellates tend to bloom in summer and slightly earlier at La Jolla than at Point Hueneme, but winter blooms are also known (Allen 1941). Unlike at La Jolla, phytoplankton densities at Point Hueneme show seasonal variations that exceed the variability on shorter time scales (Tont and Platt 1979). Dinoflagellate abundances fluctuate, with some peaks continuing to appear in the fall (Reid et

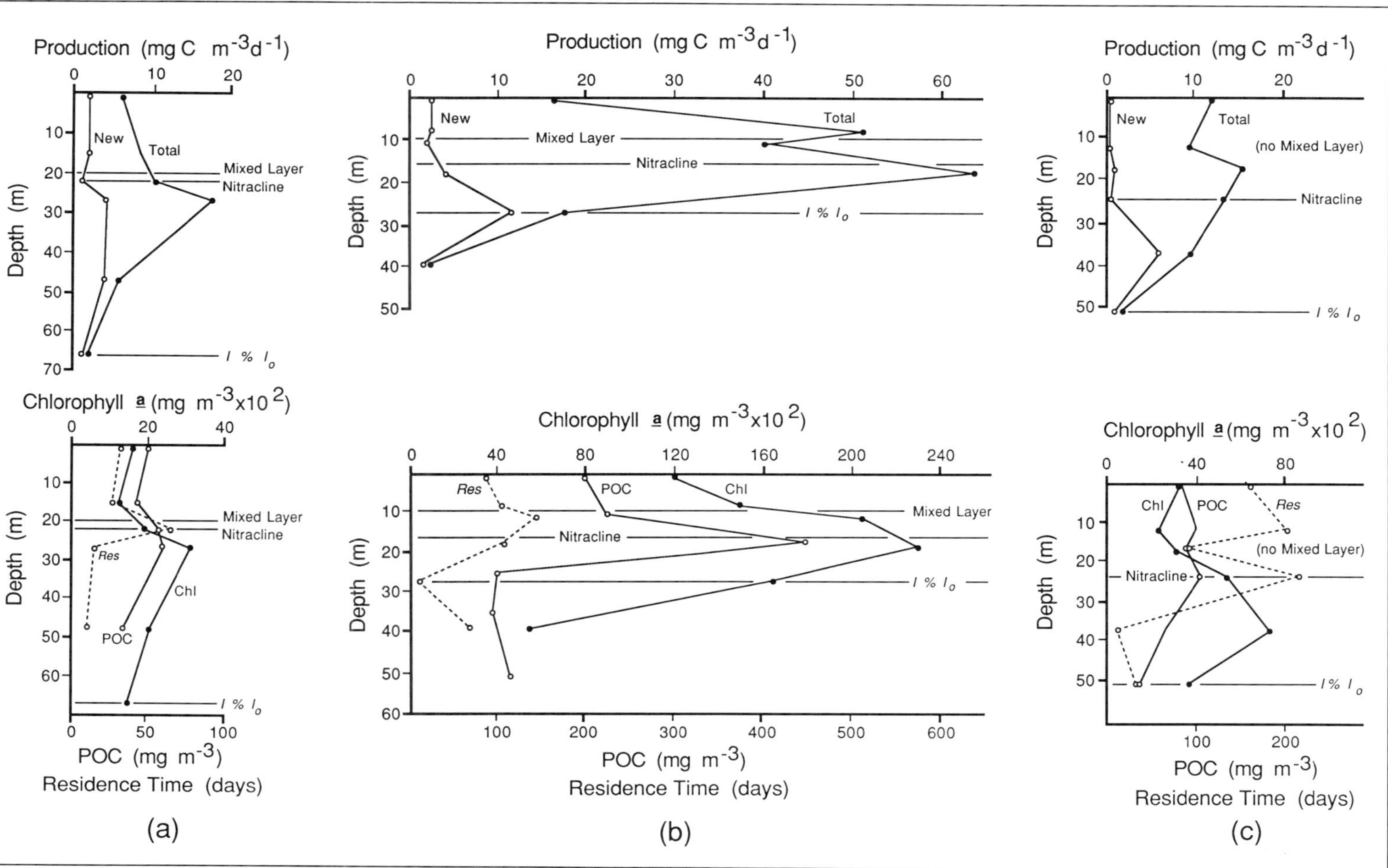

Figure 5.7. Vertical profiles of primary and new production, chlorophyll, POC, and particle residence time in the Santa Monica Basin. (a) October 1986; (b) May 1986; (c) April 1987. 1% I_0 = depth of 1% of surface irradiance. (From Small et al. 1989.)

al. 1970). Weekly samples at a depth of 2 m off the end of the Seal Beach Pier over a period of a year indicated abundant populations of diatoms or dinoflagellates (or both) blooming in the spring, followed by a second pulse of diatoms appearing in later summer. *Asterionella japonica* was an important component, succeeded later by *Skeletonema costatum*. The numbers of these major groups of phytoplankton remained low throughout the winter, while nanoplanktonic flagellates such as the chrysophytes and cryptophytes showed an increase in relative and absolute abundance.

Patterns of species abundance in SCB phytoplankton also differ from year to year (Esterly 1928; Sleggs 1927; Allen 1941; Goodman et al. 1984). The year-to-year differences in the size of coastal diatom blooms are believed to be related to the timing and intensity of local coastal upwelling and interannual changes in the flow of the California Current (Tont 1976, 1981).

Relationship Between Large- and Small-Scale Patterns

Persistent large-scale patterns such as onshore–offshore or gross vertical gradients of biomass do exist. Climatic changes on interannual and interdecadal scales also exist, as well as small-scale spatial and temporal variabilities that affect the patterns of phytoplankton distribution. However, the coupling between scales is not well understood. Also, for the most part, the time scale of phytoplankton patches and possible interannual changes in the supply of food for grazing animals remain unknown.

Enclosed Embayments

Studies of phytoplankton communities in harbors of southern California indicate general patterns of seasonal variation similar to those occurring nearshore. Specifically, low winter productivity occurs starting in December, followed by a spring bloom in April, a minor drop in productivity in May, and secondary summer and fall blooms that are localized and sporadic. In Marina del Rey, phytoplankton productivity and chlorophyll values decrease from the marina entrance inward. In contrast, in Long Beach Harbor, both productivity and chlorophyll values are consistently greater inside the harbor than outside, and productivity and chlorophyll are higher than in Marina del Rey by a factor of two or three (Soule and Oguri 1977).

The diatom flora of California lagoons have been classified into six categories (Carpelan 1978). Categories were defined on the basis of tolerance to many factors in addition to salinity. Stenotypic species were limited to environments with slow, small, and predictable changes. Species within other categories can tolerate rapid, large, and random changes in the environment.

Shallow-Water Benthic Microflora

Benthic microflora represent important contributors to primary productivity in shallow-water intertidal habitats from the tropics to the Arctic. The productivity of such benthic microflora differs greatly, both spatially and temporally. Shaffer and Onuf (1983) studied Mugu Lagoon to determine the factors that contributed to the heterogeneity of the benthic microflora there. Over a 1-year period, duplicate monthly determinations of benthic microflora production were made at different locations along six parallel transects located at different distances from the mouth of Mugu Lagoon so that the spatial, temporal, sediment, and tidal height relationships could be examined. Multiple regression analysis indicated that chlorophyll *a,* solar radiation, water temperature, community respiration, sediment composition, and initial dissolved oxygen accounted for little of the observed variation. But when samples were analyzed separately for each month, the individual categories explained much of the variance. In the monthly categorization, each of the six independent variables was most important during at least one month. Light or chlorophyll was most important in 5 of 14 months. Gross primary production did not vary according to sediment type.

Phytoplankton Blooms

Like other coastal areas, the SCB can experience *blooms,* that is, dense growths and accumulations of phytoplankton. Short-term blooms of diatoms and other phytoplankton associated with upwelling events often occur in winter or spring and last for a few days to a few weeks. A typical year has three such blooms each lasting 5 to 6 weeks (Tont 1976). The variance in abundance of phytoplankton between bloom and nonbloom periods can be almost as great as the annual variation in abundance (Tont and Platt 1979).

Under certain oceanographic conditions, blooms are dense enough to alter the color of the water to red, yellow, green, or brown (Oguri et al. 1975). Although these blooms can be caused by different groups of organisms, including diatoms, they are most commonly caused by dinoflagellates.

Red Tide

Although not related to the tidal cycle, blooms of red-pigmented dinoflagellates are called *red tide.* Red tides can occur in the SCB almost any month of the year and are generally most pronounced nearshore (Oguri et al. 1975). Spring red tide blooms are dominated by *Prorocentrum micans,* while the more intensive and frequent blooms during July through October are dominated by *Gonyaulax polyedra* (Sweeney 1975).

Dinoflagellate blooms in the La Jolla region between May 1964 and December 1966, as well as blooms in previous years, were dominated by *Gymnodinium polyedra.* In the same region, *Prorocentrum micans* was responsible for red tides in 1924, 1933, and 1939 (Allen 1941) and in April 1965 (Holmes et al. 1967). A bloom of the dinoflagellate *Gymnodinium splendins* (= *G. sanguineum*) occurred between March and July as a subsurface phenomenon extending in a band as far as 100 km along the southern California coast (Kiefer and Lasker 1975). *Mesodinium rubrum,* a ciliated protozoan that is autotrophic because of its algal endosymbionts, also produces red tides in the SCB (Taylor 1982).

Red tide outbreaks have been attributed to a variety of causes, including upwelling, heated effluents, reduced salinity, high levels of water runoff, sewage wastewater nutrient enrichment, general inshore eutrophication, and the presence of a specific unknown growth factor (Clendenning 1959; Oguri et al. 1975; Pra Kash 1975; Sweeney 1979). However, none of these conditions have been identified definitively as the sole causative factor of an outbreak of red tide. Initial introduction of bloom-forming dinoflagellates may result from transport of dormant resting-stage cysts. For example, cysts of *Gonyaulax polyedra* are found on the surface of sediments where they could be carried upward into the photic zone by upwelling. Species that do not produce cysts, such as *Prorocentrum micans,* may be introduced by lateral transport of the water mass. In either case, light, temperature, salinity, and nutrient supply must be at or near optimal conditions for the species to flourish (Taylor and Pollingher 1987). Such conditions generally occur in the late spring, after the spring diatom bloom, and through the summer months. Depletion of nutrients (especially nitrates) as a result of the diatom spring bloom, as well as a more stable water column, probably favor the buildup of dinoflagellates.

Other Colored Blooms

Other colored phytoplankton blooms occur less frequently. The most common is "yellow water," caused by the nonthecate dinoflagellate *Gymnodinium flavum* (Kofoid and Swezy 1921; Lackey and Clendenning 1963; Cullen et al. 1982b). Yellow tides caused by this organism occurred off La Jolla during the summers of 1914, 1961, and 1980, with organism densities reaching 6.2×10^6 cells l^{-1} (Cullen et al. 1982b). *G. flavum* was also responsible for an extensive yellow tide that was documented by satellite imagery in July 1980 over the Coronado shelf south of San Diego (Peláez-Hudlet 1984).

Green tides in the SCB have resulted from blooms of the chlorophyte algae *Staurastrum* and *Halosphaera* in Santa Monica Bay in September 1980 (Kleppel et al. 1982) and the euglenoid *Eutreptia* sp. in Los Angeles Harbor (Harbor Environmental Projects 1975). Brown tides are usually caused by diatom blooms in the spring.

Diel Migration of Blooms

The historical record for dinoflagellate blooms in the La Jolla region, as well as a summary of blooms occurring in other areas of the SCB, has been provided by Eppley and Harrison (1975). They noted that bloom-forming dinoflagellates, particularly *Gonyaulax polyedra,* undergo diel vertical migrations to depths at which nutrients are high. This may be the reason why dinoflagellates, rather than the less motile diatoms, are the primary bloom formers. A fall bloom of the dinoflagellate *Ceratium furca* occurred off Seal Beach from a surface band in the upper 2 m. The cells migrated to a depth of 5 m within 2 hours after sunset and had dispersed over the approximate depth interval of 5–16 m 4.5 hours after the onset of darkness.

Negative Effects of Blooms

Some species of dinoflagellates contain toxins and, in coastal areas elsewhere, are responsible for paralytic shellfish poisoning, making shellfish unfit for human consumption. In southern California, toxic species are rare, but depletion of oxygen from decay of the abundant biomass shortly after blooms in semienclosed basins can have negative impacts (Oguri et al. 1975). For example, oxygen depletion has caused extensive fish kills in Santa Monica Bay (Somner and Clark 1946), Alamitos Bay (Reish 1963), Huntington Harbor (D. J. Reish pers. comm. 1989), and Los Angeles Harbor (Oguri et al. 1975).

Bloom-forming dinoflagellate species are believed to be less useful as food for herbivores than other phytoplankton (Huntley 1982; Fiedler 1982b). However, red tide dinoflagellate blooms apparently provide an abundant food source for grazing crustaceans in Los Angeles–Long Beach Harbor (Morey-Gaines 1979) and for phagotrophic dinoflagellates in the La Jolla area (Holmes et al. 1967).

Bioluminescence

Many marine organisms, including dinoflagellates, display bioluminescence. Studies of the vertical distribution of bioluminescence in the California Current indicate a nighttime maximum at the surface and a daytime maximum at 30–40 m depth. Furthermore, day-to-night differences in the color spectrum at the depth of maximum bioluminescence suggested that the type of luminescent organisms differs from day to night (Greenblatt et al. 1984).

Summary and Prospectus for Future Research

Phytoplankton represent an important part of the SCB ecosystem. The conversion of solar energy to organic carbon through photosynthesis is the basic process upon which virtually all marine life depends. The phytoplankton community of the SCB consists of a great diversity of unicellular and colonial algal species covering a size range of several orders of magnitude. Work on phytoplankton of the SCB has accelerated and shifted focus over the past three decades from descriptive to quantitative and process-oriented research. New sampling and measurement techniques have been essential to the progress of our understanding.

The success of each species depends on water currents, zooplankton grazing, competitive interactions with other species, and optimum levels of available light and nutrients. Differences in these environmental variables lead to unique assemblages of species which differ both spatially and temporally. The standing stock of phytoplankton has been quantified on the basis of particulate organic carbon and chlorophyll as well as cell numbers (abundance). It generally has a subsurface maximum and decreases in the offshore direction. However, satellite remote sensing of chlorophyll has revealed that the actual pattern is extremely complex, with many changing eddies.

The growth and productivity of SCB phytoplankton is intermediate when compared to other areas of the world's oceans. It is greater than that in the central ocean gyres, but less than that in many of the estuarine or nutrient-rich upwelling regions. Nitrogen availability is probably a critical factor controlling the growth of many species. Recent models provide a rough way of predicting temporal changes in productivity.

The abundance of phytoplankton in the SCB varies. Populations are generally more abundant in the spring and, to a lesser extent, in the fall than at other times. Vertically, they are often most abundant near the nitracline. Differences in abundance are generally greatest in the vertical direction, intermediate in the offshore direction, and least in the longshore direction. In enclosed embayments, shallow-water benthic microflora are important to the system.

Very abundant growths, or blooms, occur frequently. They tend to be associated with stable water conditions and warm temperatures and are generally dominated by dinoflagellate species. Red tide blooms can have detrimental effects through subsequent oxygen depletion of the water.

Our current understanding of the ecology of phytoplankton of the SCB remains inadequate in many ways. We need models to predict the effects of changing conditions on phytoplankton and subsequent effects on higher levels of the food web. The predictive capability of such models will only be as good as the data upon which they are built. To reduce the uncertainty of our predictions, we need new information in several important areas:

1. The taxonomy and role of picoplankton in the food web. Some evidence suggests that these minute organisms, because of their high turnover rate, may be more important than previously believed.
2. Food web relationships that support higher organisms. What species of phytoplankton are most important in supporting zooplankton and fish populations?
3. Species-specific productivity. Techniques are now available to determine what fraction of the total productivity at any time is due to individual species in the community.
4. Total productivity by remote sensing. Methods are needed for determining not only surface chlorophyll but also total water column productivity from satellite observations.
5. The role of surface microlayer (phytoneuston) populations. Do such dense film communities occur in the SCB? How important are they in the near-surface food web, in binding surface-active materials, or in affecting the transfer of gases across the air–water interface?
6. The impacts of contaminants. What are the effects of sewage nutrient enrichment, toxic chemical effluents, and deposition of air toxics on phytoplankton growth and community diversity?
7. The effects of global climate change. What will be the effects of increased water temperatures from global "greenhouse" warming or increased ultraviolet radiation from stratospheric ozone depletion on the phytoplankton community?

Only when these questions can be answered with some degree of certainty will we have the ability to predict the results of natural or human-induced environmental changes on the ecology of the SCB.

Literature Cited

Alldredge, A. L., and J. L. Cox, 1982. Primary productivity and chemical composition of marine snow in surface waters of the Southern California Bight. *J. Mar. Res.* 40(2):517–528.

Alldredge, A. L., and M. W. Silver, 1988. Characteristics, dynamics and significance of marine snow. *Prog. Oceanogr.* 20:41–82.

Allen, T. H. F., 1977. Scale in microscopic algal ecology: A neglected dimension. *Phycologia.* 16:253–257.

Allen, W. E., 1928. Catches of marine diatoms and dinoflagellates taken by boat in southern California waters in 1926. *Bull. Scripps Inst. Oceanogr., Tech. Ser.* 1:201–246.

Allen, W. E., 1936. Occurrence of marine plankton diatoms in a ten-year series of daily catches in southern California. *Am. J. Bot.* 23:60–63.

Allen, W. E., 1941. Twenty years' statistical studies of marine phytoplankton dinoflagellates of southern California. *Am. Midl. Nat.* 26:603–635.

Armstrong, F. A. J., and E. C. LaFond, 1966. Chemical nutrient concentrations and their relationship to internal waves and turbidity off southern California. *Limnol. Oceanogr.* 11:538–547.

Balech, E., 1960. The changes in the phytoplankton off the California coast. *Calif. Coop. Oceanic Fish. Invest. Rep.* 7:127–132.

Barnett, A. M., 1974. The feeding ecology of an omnivorous neritic copepod, *Labidocera trispinosa* Esterly. Ph.D. Dissertation, Univ. of California, San Diego. 215pp.

Beers, J. R., F. M. H. Reid, and G. L. Stewart, 1980. Microplankton population structure in southern California nearshore waters in late spring. *Mar. Biol.* 60:209–226.

Behrenfeld, M., J. T. Hardy, H. Gucinski, A. Hanneman, H. Lee, and A. Wones. 1993. Effects of ultraviolet-B radiation on primary production along latitudinal transects in the South Pacific Ocean. *Mar. Environ. Res.* 35: 349–363.

Bernal, P. A., and J. A. McGowan, 1981. Advection and upwelling in the California Current. In: F. A. Richards, ed. *Coastal Upwelling.* Amer. Geophys. Union, Washington, D.C. pp. 381–399.

Bold, H. C., and M. J. Wynne, 1978. *Introduction to the Algae.* Prentice Hall, Englewood Cliffs, NJ. 706pp.

Briand, F. J.-P., 1976. Seasonal variations and associations of southern Californian nearshore phytoplankton. *J. Ecol.* 64:821–835.

Carlucci, A. F., 1970. Part II. Vitamin B_{12}, thiamine, and biotin. In: J. D. H. Strickland, ed. *The Ecology of the Plankton off La Jolla, California, in the Period April Through September, 1967.* Bull. Scripps Inst. Oceanogr. Univ. Calif. 17:23–31.

Carpelan, L. H., 1978. Revision of Kolbe's System der Halobien based on diatoms of California lagoons. *Oikos.* 31:112–122.

Checkley, D. M., Jr., 1980a. The egg production of a marine planktonic copepod in relation to its food supply: Laboratory studies. *Limnol. Oceanogr.* 25(3):430–446.

Checkley, D. M., Jr., 1980b. Food limitation of egg production by a marine planktonic copepod in the sea off southern California. *Limnol. Oceanogr.* 25(6):991–998.

Clendenning, K. A., 1959. Red tide in relation to organic pollution of coastal water. In: *Quarterly Progress Report, 10 October–31 December 1958.* Institute of Marine Resources, Univ. of California, La Jolla, CA. pp. 12–13.

Cooper, L. H. N., 1947. Internal waves and upwelling of oceanic water from mid-depths on to a continental shelf. *Nature (Lond.).* 159:579–580.

Cullen, J. J., and R. W. Eppley, 1981. Chlorophyll maximum layers of the Southern California Bight and possible mechanisms of their formation and maintenance. *Oceanol. Acta.* 4(1):23–32.

Cullen, J. J., and E. H. Renger, 1979. Continuous measurement of the DCMU-induced fluorescence response of natural phytoplankton populations. *Mar. Biol.* 53(1):13–20.

Cullen, J. J., F. M. H. Reid, and E. Stewart, 1982a. Phytoplankton in the surface and chlorophyll maximum off southern California in August 1978. *J. Plankton Res.* 4(3):665–694.

Cullen, J. J., S. G. Horrigan, M. E. Huntley, and F. M. H. Reid, 1982b. Yellow water in La Jolla Bay, California, July 1980. I. A. bloom of the dinoflagellate *Gymnodinium flavum* Kofoid and Swezy. *J. Exp. Mar. Biol. Ecol.* 63(1):67–80.

Cullen, J. J., E. Stewart, E. Renger, R. W. Eppley, and C. D. Winant, 1983. Vertical motion of the thermocline, nitracline and chlorophyll maximum layers in relation to currents on the southern California shelf. *J. Mar. Res.* 41:239–262.

Cushing, D. H., 1982. *Climate and Fisheries.* Academic Press, London. 373pp.

Dorman, C. E., and D. P. Palmer, 1981. Southern California summer coastal upwelling. In: F. A. Richards, ed. *Coastal Upwelling.* Am. Geophys. Union, Washington, D.C. pp. 44–56.

Droop, M. R., 1974. The nutrient status of algal cells in continuous culture. *J. Mar. Biol. Assn. U.K.* 54:825–855.

Eppley, R. W., 1968. An incubation method for estimating the carbon content of phytoplankton in natural samples. *Limnol. Oceanogr.* 13:574–582.

Eppley, R. W., 1972. Temperature and phytoplankton growth in the sea. *U.S. Fish. Wild. Serv. Fish. Bull.* 70:1063–1085.

Eppley, R. W., and W. G. Harrison, 1975. Physiological ecology of *Gonyaulax polyhedra,* a red water dinoflagellate of southern California. In: V. R. Lo Cicero, ed. *Proceedings of the First International Conference on Toxic Dinoflagellate Blooms.* Mass. Sci. Tech. Foundation, Wakefield, MA. pp. 11–22.

Eppley, R. W., and O. Holm-Hansen, 1986. Primary production in the Southern California Bight. In: R. W. Eppley, ed. *Lecture Notes on Coastal and Estuarine Studies, Vol. 15. Plankton Dynamics of the Southern California Bight.* Springer-Verlag, Berlin, pp. 176–215.

Eppley, R. W., O. Holm-Hansen, and J. D. H. Strickland, 1968. Some observations on the vertical migration of dinoflagellates. *J. Phycol.* 4:333–340.

Eppley, R. W., J. N. Rogers, and J. J. McCarthy, 1969. Half-saturation constants for uptake of nitrate and ammonium by marine phytoplankton. *Limnol. Oceanogr.* 14:912–920.

Eppley, R. W., F. M. H. Reid, and J. D. H. Strickland, 1970. Part III. Estimates of phytoplankton crop size, growth rate, and primary production. In: J. D. H. Strickland, ed. *The Ecology of the Plankton off La Jolla, California, in the Period April through September 1967.* Bull. Scripps Inst. Oceanogr. Univ. Calif. 17:33–42.

Eppley, R. W., W. G. Harrison, S. W. Chisholm, and E. Stewart, 1977. Particulate organic matter in surface waters off southern California and its relationship to phytoplankton. *J. Mar. Res.* 35(4):671–696.

Eppley, R. W., E. H. Renger, and W. G. Harrison, 1979a. Nitrate and phytoplankton production in southern California coastal waters. *Limnol. Oceanogr.* 24:484–494.

Eppley, R. W., E. H. Renger, W. G. Harrison, and J. J. Cullen, 1979b. Ammonium distribution in southern California coastal waters and its role in the growth of phytoplankton. *Limnol. Oceanogr.* 24(3):495–509.

Eppley, R. W., E. H. Renger, and P. R. Betzer, 1983. The residence time of particulate organic carbon in the surface layer of the ocean. *Deep-Sea Res. Part A Oceanogr. Res. Pap.* 30(3):311–324.

Eppley, R. W., F. M. H. Reid, and E. Stewart, 1984. Length of phytoplankton species patches on the southern California shelf. *Cont. Shelf Res.* 3:259–266.

Eppley, R. W., E. Stewart, M. R. Abbott, and U. Heyman, 1985. Estimating ocean primary production from satellite chlorophyll: Introduction to regional differences and statistics for the Southern California Bight. *J. Plankton Res.* 7(1):57–70.

Esterly, C. O., 1928. The periodic occurrence of Copepoda in the marine plankton of two successive years at La Jolla, California. *Bull. Scripps Inst. Oceanogr., Tech. Ser.* 1:247–345.

Fanning, K. A., K. L. Carder, and P. R. Betzer, 1982. Sediment resuspension by coastal waters: A potential mechanism for nutrient recycling on the ocean's margins. *Deep-Sea Res.* 29:953–965.

Fiedler, P. C., 1982a. Zooplankton avoidance and reduced grazing responses to *Gymnodinium splendens* (Dinophyceae). *Limnol. Oceanogr.* 27(5):961–965.

Fiedler, P. C., 1982b. Fine scale spatial pattern in the coastal epiplankton: Description and functional significance. Ph.D. Dissertation, Univ. of California, San Diego. 94pp.

Fritsch, F. E., 1935. *The Structure and Reproduction of the Algae. Vol. I and II.* Cambridge Univ. Press, Cambridge, UK.

Fuhrman, J. A., J. W. Ammerman, and F. Azam, 1980. Bacterioplankton in the coastal euphotic zone: Distribution, activity and possible relationships with phytoplankton. *Mar. Biol.* 60(2–3):201–207.

Goodman, D., R. W. Eppley, and F. M. H. Reid, 1984. Summer phytoplankton assemblages and their environmental correlates in the Southern California Bight. *J. Mar. Res.* 42(4):1019–1050.

Greenblatt, P. R., D. F. Feng, A. Zirino, and J. R. Losee, 1984. Observations of planktonic bioluminescence in the euphotic zone of the California Current. *Mar. Biol.* 84(1):75–82.

Hanson, J. M., and W. C. Leggett, 1982. Empirical prediction of fish biomass and yield. *Can. J. Fish. Aquat. Sci.* 39:257–263.

Harbor Environmental Projects, 1975. Environmental investigations and analyses for Los Angeles Harbor. Report to U.S. Army Corps of Engineers. Prep. by Harbor Environmental Projects, Allan Hancock Foundation, Univ. South. Calif., Los Angeles, CA. 582pp.

Harding, L. W., Jr., B. B. Prezelin, B. M. Sweeney, and J. L. Cox, 1982. Diel oscillations of the photosynthesis irradiance relationship in natural assemblages of phytoplankton. *Mar. Biol.* 67(2):167–178.

Hardy, J. T., 1982. The sea surface microlayer: Biology, chemistry and anthropogenic enrichment. *Prog. Oceanogr.* II:307–328.

Hardy, J. T., 1987. Anthropogenic alteration of the sea surface. *Mar. Environ. Res.* 23:223–225.

Hardy, J. T., and C. W. Apts, 1989. Photosynthetic carbon reduction: High rates in the sea-surface microlayer. *Mar. Biol.* 101:411–417.

Hardy, J. T., and H. Gucinski, 1989. Stratospheric ozone depletion: Implications for marine ecosystems. *Oceanography*. November: 18–21.

Harris, G. P., 1980. The measurement of photosynthesis in natural populations of phytoplankton. In: I. Morris, ed. *The Physiological Ecology of Phytoplankton*. Univ. of California Press, Berkeley, CA. pp. 129–187.

Harris, G. P., 1986. *Phytoplankton Ecology: Structure, Function, and Fluctuation*. Chapman and Hall Ltd., New York. 384pp.

Harrison, W. G., 1978. Experimental measurements of nitrogen remineralization in coastal waters. *Limnol. Oceanogr.* 23(4):684–694.

Harvey, H. W., L. H. N. Cooper, M. V. Lebour, and F. R. S. Russell, 1935. Plankton production and its control. *J. Mar. Biol. Assoc. U.K.* 20:407–441.

Hayward, T. L., and E. L. Venrick, 1982. Relation between surface chlorophyll, integrated chlorophyll and integrated primary production. *Mar. Biol.* 69(3):247–252.

Hendricks, T. J., 1979. A study to evaluate the dispersion characteristics of waste discharges to the open coastal waters of California. Univ. of California, Institute of Marine Resources (IMR Ref. 79–9), 78pp.

Hoge, F. E., and R. N. Swift. 1986. Chlorophyll pigment concentration using spectral curvature algorithms: an evaluation of present and proposed satellite ocean color sensor bands. *Applied Optics*. 25:3677–3682.

Holmes, R. W., 1958. Surface chlorophyll *a*, surface primary production and zooplankton volumes in the eastern Pacific Ocean. *Rapp. P.-V. Réun. Cons. Int. Explor. Mer.* 144:109–116.

Holmes, R. W., P. M. Williams, and R. W. Eppley, 1967. Red water in La Jolla Bay, 1964–1966. *Limnol. Oceanogr.* 12:503–512.

Holm-Hansen, O., and C. R. Booth, 1966. The measurement of adenosine triphosphate in the ocean and its ecological significance. *Limnol. Oceanogr.* 11:510–519.

Honjo, S., 1982. Seasonality and interaction of biogenic and lithogenic particulate flux at the Panama Basin. *Science*. 218:883–884.

Hooker, S. B., and E. R. Firestone. 1992. An overview of Sea WiFS and ocean color. NASA Technical Memorandum 104566. Goddard Space Flight Center. Greenbelt, MD.

Huntley, M. E., 1982. Yellow water in La Jolla Bay, California, July 1980. II. Suppression of zooplankton grazing. *J. Exp. Mar. Biol. Ecol.* 63:81–91.

Institute of Marine Resources, 1971. Eutrophication in coastal waters: Nitrogen as a controlling factor. Final Report to the U.S. Environmental Protection Agency, Project #16010 EHC. Washington, D.C.

Kiefer, D. A., 1973. Fluorescence properties of natural phytoplankton populations. *Mar. Biol.* 22:263–270.

Kiefer, D. A., and R. Lasker, 1975. Two blooms of *Gymnodinium splendens*, an unarmored dinoflagellate. *U.S. Fish Wildl. Serv. Fish. Bull.* 73:675–678.

Kimor, B., 1981. Seasonal and bathymetric distribution of thecate and nonthecate dinoflagellates off La Jolla, California. *Calif. Coop. Oceanic Fish. Invest. Rep.* 22:126–134.

Kleppel, G. S., 1980. Nurients and phytoplankton community composition in southern California coastal waters. *Calif. Coop. Oceanic Fish. Invest. Rep.* 21:191–196.

Kleppel, G. S., E. Manzanilla, B. Teter, and

S. Petrich, 1982. Santa Monica Bay plankton distribution. In: W. Bascom, ed. *Biennial Report, 1981–1982.* South. Calif. Coastal Water Res. Proj., Long Beach, CA. pp. 71–84.

Kofoid, L. A., and O. Swezy, 1921. The free-living unarmored Dinoflagellata. *Univ. Calif. Publ. Zool. Mem. Vol. 5.* 562pp.

Krempin, D. W., and C. W. Sullivan, 1981. The seasonal abundance, vertical distribution, and relative microbial biomass of chroococcoid cyanobacteria at a station in southern California coastal waters. *Can. J. Microbiol.* 27:1341–1344.

Lackey, J. G., and K. A. Clendenning, 1963. A possible fish-killing yellow tide in California waters. *Quart. J. Fla. Acad. Sci.* 26:263–268.

Lasker, R., 1975. Field criteria for survival of anchovy larvae: The relation between inshore chlorophyll maximum layers and successful first feeding. *U.S. Natl. Mar. Fish. Serv. Fish. Bull.* 73:453–462.

Lorenzen, C. J., 1966. A method for the continuous measurement of *in vivo* chlorophyll concentration. *Deep-Sea Res.* 30:639–643.

Malone, T. C., 1971. The relative importance of nannoplankton and net plankton as primary producers in the California Current system. *U.S. Natl. Mar. Fish. Serv. Fish. Bull.* 69(4): 799–820.

McCarthy, J. J., 1972. The uptake of urea by natural populations of marine phytoplankton. *Limnol. Oceanogr.* 17(5):738–748.

Morey-Gaines, G., 1979. The ecological role of red tides in the Los Angeles–Long Beach Harbor food web. In: D. L. Taylor and H. Seliger, eds. *Toxic Dinoflagellate Blooms.* Elsevier North Holland, New York. pp. 315–320.

Mullin, M. M., 1979. Longshore variations in the distribution of plankton in the Southern California Bight. *Calif. Coop. Oceanic Fish. Invest. Rep.* 20:120–124.

Mullin, M. M., 1986. Spatial and temporal scales and patterns. In: R. W. Eppley, ed. *Lecture Notes on Coastal and Estuarine Studies. Vol. 15. Plankton Dynamics of the Southern California Bight.* Springer-Verlag, Berlin, pp. 216–273.

Mullin, M. M., and E. R. Brooks, 1972. The vertical distribution of juvenile *Calanus* (Copepoda) and phytoplankton within the upper 50 m of water off La Jolla, California. In: A. Y. Takenouti, eds. *Biological Oceanography of the Northern North Pacific Ocean.* Idemitsu Shoten, Tokyo. pp. 347–354.

Mullin, M. M., E. R. Brooks, F. M. H. Reid, J. Napp, and E. F. Stewart, 1985. Vertical structure of nearshore plankton off southern California. A storm and a larval fish food web. *U.S. Natl. Mar. Fish. Serv. Fish. Bull.* 83(2): 151–170.

Naumann, E., 1917. Beitrage zur kenntnis des teichnanno planktons. II. Uber das Neuston des Susswassers. *Biol. Zentralbl.* 37:98–106.

Neori, A., O. Holm-Hansen, B. G. Mitchell, and D. A. Kiefer, 1984. Photoadaptation in marine phytoplankton changes in spectral absorption and excitation of chlorophyll *a* fluorescence. *Plant Physiol.* 76(2):518–524.

Nixon, S. W., 1988. Physical energy inputs and the comparative ecology of lake and marine ecosystems. *Limnol. Oceanogr.* 33(4, pt. 2): 1005–1025.

Oguri, M., D. Soule, D. M. Joge, and B. C. Abbott, 1975. Red tides in Los Angeles–Long Beach harbors. In: V. R. Lo Cicero, ed. *Proceedings of the First International Conference on Toxic Dinoflagellate Blooms.* Mass. Sci. Tech. Foundation, Wakefield, MA. pp. 41–46.

Owen, R. W., Jr., 1974. Distribution of primary production, plant pigments and secchi depth in the California Current region, 1969. *Calif. Coop. Oceanic Fish. Invest. Atlas 20,* pp. xi–xiv + charts 98–117.

Parsons, T. R., and R. J. LeBrasseur, 1970. The availability of food to different trophic levels in the marine food chain. In: J. H. Steele, ed. *Marine Food Chains.* Univ. of California Press, Berkeley, CA. pp. 325–343.

Parsons, T. R., Y. Maita, and C. M. Lalli, 1984. *A manual of Chemical and Biological Methods for Seawater Analysis.* Pergamon Press, New York. 173pp.

Peláez, J., and F. Guan, 1982. California Current chlorophyll measurements from satellite data. *Calif. Coop. Oceanic Fish. Invest. Rep.* 23:212–225.

Peláez, J., and J. A. McGowan, 1986. Phytoplankton pigment patterns in the California Current as determined by satellite. *Limnol. Oceanogr.* 31(5):927–950.

Peláez-Hudlet, J., 1984. Phytoplankton pigment concentrations and patterns in the California Current as determined by satellite. Ph.D. Dissertation, Univ. of California, San Diego. 98pp.

Platt, T., and A. D. Jassby, 1976. The relationship

between photosynthesis and light for natural assemblages of coastal marine phytoplankton. *J. Phycol.* 12:421–430.

Pra Kash, A., 1975. Dinoflagellate blooms—An overview. In: V. R. Lo Cicero, ed. *Proceedings of the First International Conference on Toxic Dinoflagellate Blooms.* Mass. Sci. Tech. Found., Wakefield, MA. pp. 1–6.

Prescott, G. W., 1968. *The Algae: A Review.* Houghton Mifflin, Boston. 436pp.

Raymont, J. E. G., 1980. *Plankton and Productivity in the Oceans. Vol. 1 Phytoplankton.* Pergamon Press, New York. 489pp.

Redalje, D. G., 1983. Phytoplankton carbon biomass and specific growth rates determined with the labelled chlorophyll technique. *Mar. Ecol. Prog. Ser.* 11:217–225.

Reid, F. M. H., E. Fuglister, and J. B. Jordan, 1970. Part V. Phytoplankton taxonomy and standing crop. In: J. D. H. Strickland, ed. *The Ecology of the Plankton off La Jolla, California, in the Period April Through September 1967.* Bull. Scripps Inst. Oceanogr. Univ. Calif. 17:51–66.

Reid, F. M. H., E. Stewart, R. W. Eppley, and D. Goodman, 1978. Spatial distribution of phytoplankton species in chlorophyll maximum layers off southern California. *Limnol. Oceanogr.* 23:219–226.

Reish, D. J., 1963. Mass mortality of marine organisms attributed to the "red tide" in southern California. *Calif. Fish Game.* 49(4):265–270.

Richman, S., and J. N. Rogers, 1969. The feeding of *Calanus helgolandicus* on synchronously growing populations of the marine diatom *Ditylum brightwellii. Limnol. Oceanogr.* 14:701–709.

Ryther, J. H., 1969. Photosynthesis and fish production in the sea. *Science.* 166:72–76.

Sargent, M. C., and T. J. Walker, 1948. Diatom populations associated with eddies off southern California. *J. Mar. Res.* 7:490–505.

Shaffer, G. P., and C. P. Onuf, 1983. An analysis of factors influencing the primary production of the benthic microflora in a southern California lagoon. *Neth. J. Sea Res.* 17(1):126–144.

Sieburth, J. M., 1979. *Sea Microbes.* Oxford Univ. Press, New York. 491pp.

Sleggs, G. F., 1927. Marine phytoplankton in the region of La Jolla, California, during the summer of 1924. *Bull. Scripps Inst. Oceanogr., Tech. Ser.* 1:93–117.

Small, L. F., M. R. Landry, R. W. Eppley, F. Azam, and A. F. Carlucci, 1989. Role of plankton in the carbon and nitrogen budgets of Santa Monica Basin, California. *Mar. Ecol. Prog. Serv.* 56:57–74.

Smayda, T. J., 1980. Phytoplankton species succession. In: I. Morris, ed. *The Physiological Ecology of Phytoplankton.* Univ. of California Press, Berkeley, CA. pp. 493–570.

Smith, P. E., and R. W. Eppley, 1982. Primary production and the anchovy population in the Southern California Bight: Comparison of time series. *Limnol. Oceanogr.* 27(1):1–17.

Smith, R. C., and K. S. Baker, 1982. Oceanic chlorophyll concentrations as determined by satellite (*Nimbus* 7 coastal zone color scanner). *Mar. Biol. (Berl.).* 66(3):269–279.

Smith, R. C., K. S. Baker, O. Holm-Hansen, and R. Olson, 1980. Photoinhibition of photosynthesis in natural waters. *Photochem. Photobiol.* 31:585–592.

Smith, R. C., R. W. Eppley, and K. S. Baker, 1982. Correlation of primary production as measured aboard ship in southern California coastal waters and as estimated from satellite chlorophyll images. *Mar. Biol. (Berl.).* 66(3): 281–288.

Smith, R. C., R. R. Bidigare, B. B. Prézelin, K. S. Baker, and J. M. Brooks, 1987. Optical characterization of primary productivity across a coastal front. *Mar. Biol.* 96:575–591.

Somner, H., and F. N. Clark, 1946. Effect of red water on marine life in Santa Monica Bay, California. *Calif. Fish Game.* 32:100–101.

Soule, D. F., and M. Oguri, eds., 1977. Marine studies of San Pedro Bay, California. Part 13. The marine ecology of Marina del Rey harbor, California. A baseline survey for the County of Los Angeles Dept. Small Craft Harbors, 1976–1977. Available from NTIS, Springfield, VA; PB–278–323–1, 437pp.

Sournia, A., ed., 1978. *Phytoplankton Manual. Monographs on Oceanographic Methodology 6.* UNESCO, Paris. 337pp.

Sournia, A., 1982. Form and function in marine phytoplankton. *Biol. Rev.* 57:347–394.

Star, J. L., and M. M. Mullin, 1979. Horizontal undependability in the planktonic environment. *Mar. Sci. Comm.* 5(1):31–46.

Steemann-Nielsen, E., 1952. The use of radioactive carbon (C-14) for measuring organic production in the sea. *J. Cons. Cons. Int. Explor. Mer.* 18:117–140.

Stein, J. R., 1973. *Handbook of phycological methods. Culture methods and growth measurements*. Cambridge Univ. Press, Cambridge, U.K. 448pp.

Strickland, J. D. H., L. Sorozano, and R. W. Eppley, 1970. Part I. General introduction, hydrography, and chemistry. In: J. D. H. Strickland, ed. *The Ecology of the Plankton off La Jolla, California, in the Period April Through September 1967*. Bull. Scripps. Inst. Oceanogr. Univ. Calif. 17:1–22.

Sweeney, B. M., 1975. Red tides I have known. In: V. R. Lo Cicero, ed. *Proceedings of the First International Conference on Toxic Dinoflagellate Blooms*. Mass. Sci. Tech. Found., Wakefield, MA. pp. 225–234.

Sweeney, B. M., 1979. The organisms, opening remarks. In: D. L. Taylor and H. H. Seliger, eds. *Proceedings of the Second International Conference of Toxic Dinoflagellate Blooms*. Elsevier North Holland, New York. pp. 37–46.

Taylor, F. J. R., 1982. Symbioses in marine microplankton. *Ann. Inst. Oceanogr. (Paris)*. 58(S): 61–90.

Taylor, F. J. R., and U. Pollingher, 1987. Ecology of the dinoflagellates. In: F. J. R. Taylor, ed. *The Biology of Dinoflagellates*. Blackwell Scientific, Oxford. pp. 398–502.

Thomas, W. H., and D. L. R. Seibert, 1974. Distribution of nitrate, nitrite, phosphate and silicate in the California Current region. *Calif. Coop. Oceanic Fish. Invest. Atlas. No. 20*, pp. vii–x + charts 2–97.

Thorson, G., 1950. Reproductive and larval ecology of marine bottom invertebrates. *Biol. Rev. (Cambridge)*. 25:1–45.

Tont, S. A., 1976. Short-period climatic fluctuations: Effects on diatom biomass. *Science*. 194:942–944.

Tont, S. A., 1981. Temporal variations in diatom abundance off southern California in relation to surface temperature, air temperature and sea level. *J. Mar. Res*. 39:191–201.

Tont, S. A., and T. Platt, 1979. Fluctuations in the abundance of phytoplankton on the California coast. In: E. Naylor and R. G. Hartnoll, eds. *Cyclic Phenomena in Marine Plants and Animals*. Pergamon Press, Oxford. pp. 11–18.

Vollenweider, R. A., 1974. *A Manual on Methods for Measuring Primary Productivity in Aquatic Environments. IPB Handbook No. 12*. Blackwell Scientific, Oxford. 225pp.

Weiler, C. S., and S. W. Chisholm, 1976. Phased cell division in natural populations of marine dinoflagellates from shipboard cultures. *J. Exp. Mar. Biol. Ecol*. 25:239–247.

Winter, A., 1985. Distribution of living coccolithophores in the California Current system, southern California borderland. *Mar. Micropaleontol*. 9(5):385–394.

Worrest, R. C., 1983. Impact of solar ultraviolet-B radiation (290–320 nm) upon marine microalgae. *Physiol. Plant*. 58:428–434.

Yentsch, C. S., 1980. Light attenuation and phytoplankton photosynthesis. In: I. Morris, ed. *The Physiological Ecology of Phytoplankton*. Univ. of California Press, Berkeley, CA. pp. 95–128.

Zaitsev, Y. P., 1971. *Marine Neustonology* (translated from Russian). NOAA, U.S. Natl. Mar. Fish. Serv., and Natl. Sci. Found. 207pp.

Chapter 6

Zooplankton

John K. Dawson and Richard E. Pieper

Introduction

The zooplankton of the Southern California Bight (SCB) comprise a large and diverse group of organisms. The focus of this review will be on the interrelationships between the distribution and abundance of these organisms and on the oceanography that influences these distributions.

The literature on zooplankton in the SCB is extensive. The California Cooperative Oceanic Fisheries Investigations (CalCOFI) database is probably one of the largest long-time series of data found anywhere. Numerous studies and reports have also been completed on harbors, enclosed bays, and nearshore zooplankton. While some of these studies have been reported in available literature, many are the result of environmental impact studies that remain unpublished.

Any understanding of zooplankton ecology of the SCB must be tied to an understanding of its oceanographic variability (see chapter 2). Excellent reviews on the zooplankton (as well as on the oceanography) of the SCB can be found in Sverdrup et al. (1942), Emery (1960), and Eppley (1986). The SCB also represents a variable boundary (and mixture) of assemblages of zooplankton and their corresponding oceanic ecosystems that are comparable to, and related to, water mass variability (McGowan 1974). Steele (1978) and Haury and Pieper (1988) discuss the importance of understanding spatial and temporal heterogeneity, variability, and patchiness. The importance of zooplankton variability in the SCB is also discussed by Roesler and Chelton (1987) who summarized CalCOFI zooplankton data (displacement volumes) over a 32-year period from 1951 to 1982. They noted that nonseasonal zooplankton variability was dominated by very low-frequency patterns with periods of 3–5 years that are associated with variations in large-scale equatorward transport of the California Current (fig. 6.1). Years when California Current flow was higher than normal were associated with larger zooplankton volumes. Episodic bursts of zooplankton biomass of 3–4 months' duration were also observed superimposed on the low-frequency pattern.

The interannual variability just discussed should be considered the baseline for understanding higher frequency events and processes, including biological interactions. These smaller scale, higher frequency processes include seasonal changes and localized events such as coastal upwelling, eddies, plumes, tidal oscillations, bottom processes, diel cycles, wind stress, and turbulence. The extent to which these physical events control or modify zooplankton ecology is a function of the particular organism, including its size, swimming ability, reproductive state, food needs, and other requirements.

A wide variety of sampling devices and preservation techniques are available for zooplankton population assessment. Each must be selected for its appropriateness for particular organisms and for the scale and data rate of the specific study. Sampling, fixation, and preservation techniques are reviewed in two UNESCO monographs on oceanographic methodology (UNESCO 1968, 1976). The macrozooplankton in the CalCOFI program, for example, are sampled with bongo nets (505-μm mesh) towed obliquely from a depth of 140 m. Covering a large spatial grid, the CalCOFI data set is long enough in duration (32 years) and large enough in number of samples to detect the interannual patterns of macrozooplankton biomass.

Three oceanographic zones of zooplankton will be discussed separately in this chapter: harbor and bay, nearshore (shelf and shelf break), and offshore (open ocean and basins). Emphasis is placed on field observations and results. Additional reviews of zooplankton trophic structure and spatial and temporal patterns in the SCB can be found in Beers (1986) and Mullin (1986), both in Eppley (1986).

Harbor and Bay Zooplankton

The SCB contains three major ports serving the import and export needs of the southwestern United States. These are the port of San Diego in San Diego Bay and the ports of Los Angeles and Long Beach, both located in San Pedro Bay. While there has been little zooplankton work done in the San Diego Bay, several studies have been conducted in San Pedro Bay. These studies were the result of environmental impact report requirements (Allan Hancock Foundation 1976), as well as the impact of pollution (Environmental Quality Analysts and Marine Biological Consultants 1976) and pollution abatement investigations (Soule and Oguri 1979).

Copepods of the genus *Acartia* generally dominate the zooplankton in the SCB harbors. We believe that much of the taxonomic work on this genus is actually a mixture of the two species *A. tonsa* and *A. californiensis*. Since we do not wish to change the words of cited references, where a reference refers to *A. tonsa,* but we believe it to be a mixture of *A. tonsa* and *A. californiensis,* we will append an asterisk (for example, *A.tonsa**).

Los Angeles–Long Beach Harbors

To help assess the environmental impacts of a proposed U.S. Army Corps of Engineers dredge and landfill of outer Los Angeles–Long Beach Harbors, the Harbors Environmental Projects of the University of Southern California conducted an extensive sampling program from 1972 through 1974 of the outer and inner Los Angeles–Long Beach Harbors (Allan Hancock Foundation 1976). The dominant zooplankton species in the harbor included the calanoid copepods *Acartia tonsa** (57.9%) and *Paracalanus parvus* (10%),

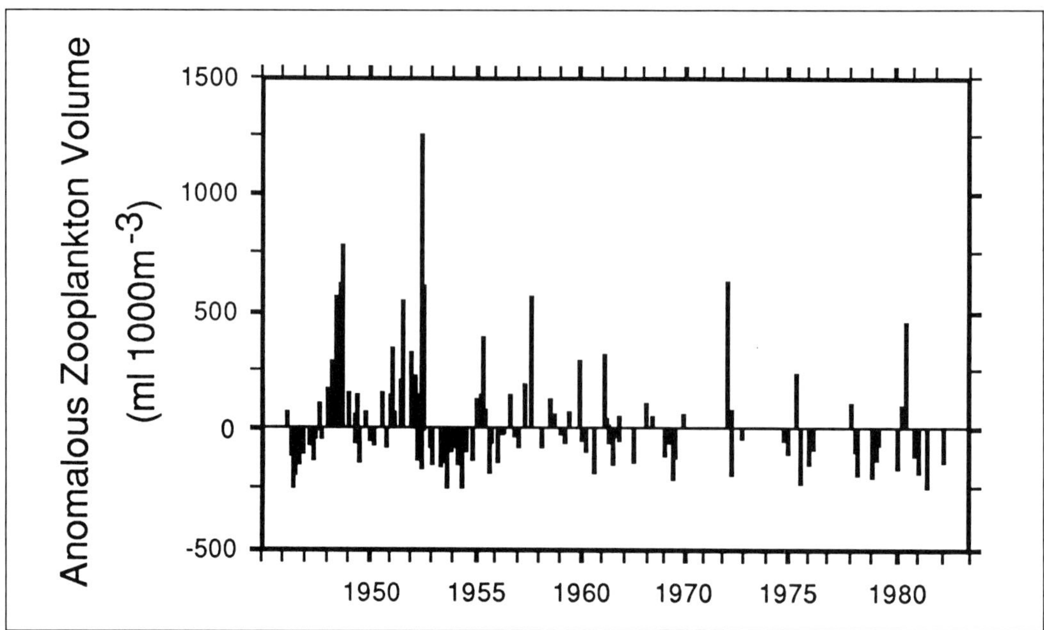

Figure 6.1. Seasonally corrected zooplankton biomass deviations from the mean over 32 years in the SCB. (From Roesler and Chelton 1987.)

the cladocerans *Podon polyphemoides* (11%) and *Evadne nordmanni* (4.7%), and the cyclopoid copepod *Corycaeus anglicus* (1.6%). Other abundant organisms, which may include one or more species, were larvaceans and cirripedia nauplii (approximately 5% each).

The spatial distribution of the dominant zooplankton reflected the environmental characteristics of the harbor's waters. The zooplankton of the inner harbor channels was characterized by high concentrations of the copepod *A. tonsa** and *Oithona oculata,* accompanied by a notable low abundance of other zooplanktonic species. The main decrease was in the three cladoceran species *Podon polyphemoides, Evadne nordmanni,* and *Penilia avirostris* which were one, two, and three orders of magnitude lower in concentration, respectively, in the inner harbor. The inner harbor channel was dominated by more than 75% *A. tonsa**.

Perhaps the most enlightening zooplankton distribution regarding the environmental conditions of the harbor was the distribution of the congeneric copepods *Acartia tonsa* and *A. californiensis*. It was believed that what had been called *Acartia tonsa* exclusively was actually a mixture of these two species. Dawson (1979) made a 1-month (November 1978) zooplankton study in the harbor aimed at resolving the distribution of these very closely related species of *Acartia*. Figure 6.2 shows the relative percentage distribution of these two species. *A. californiensis* dominated most of the inner harbor channels, and *A. tonsa* dominated the outer harbor. There appeared to be mixing of the two species, perhaps by tidal action, at stations A8, C1, and C2. The small baylet at station D10 was 85% *A. californiensis,* while the *A. californiensis* present at nearby D2 may have represented "spill out" from D10.

The lower concentrations of *A. californiensis* at the inner B stations (B5, B6, and B7) appeared anomalous when viewed with the rest of the inner harbor stations. This area, however, was found by Allan Hancock Foun-

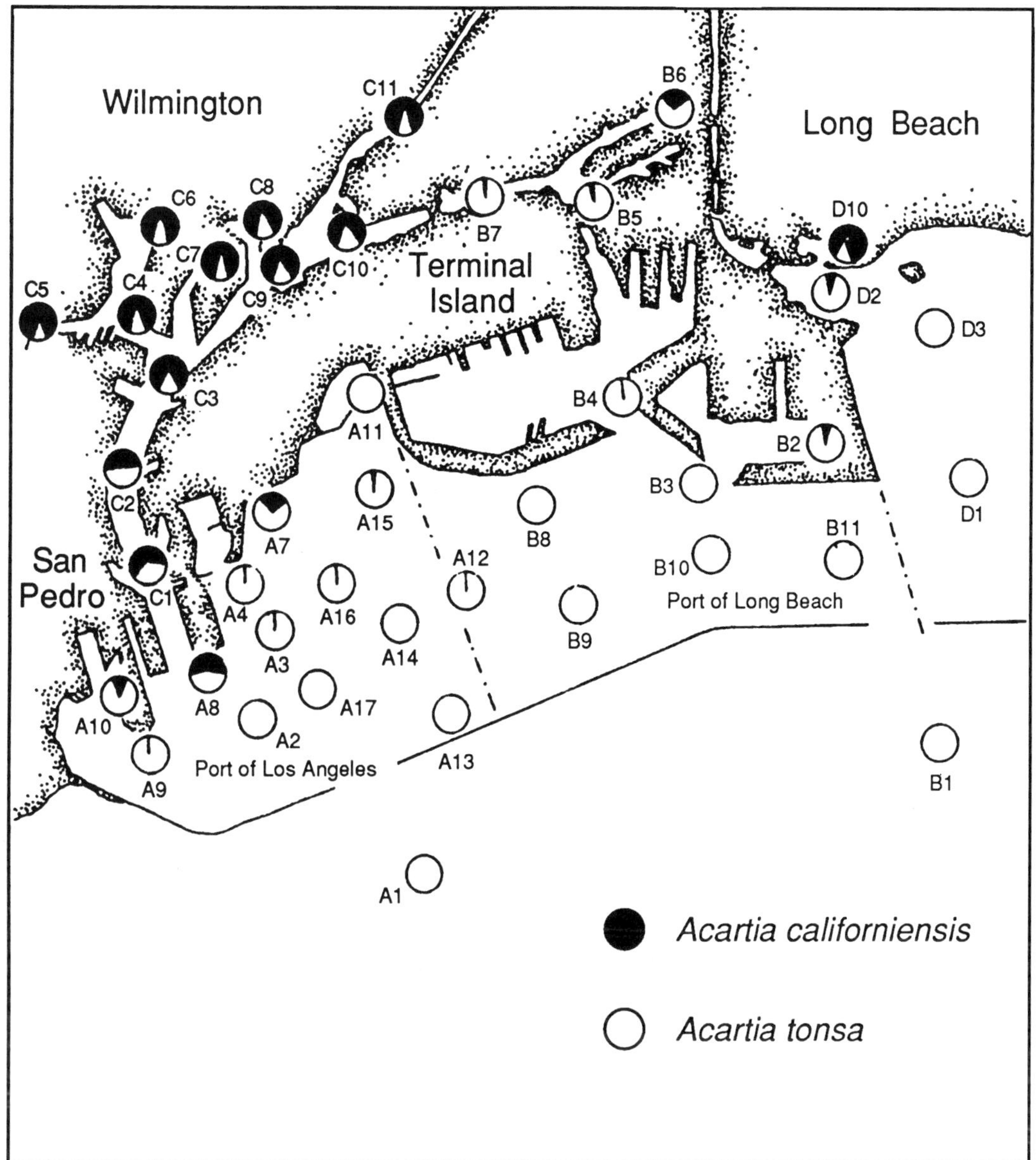

Figure 6.2. Distribution of *Acartia californiensis* and *A. tonsa* in Los Angeles–Long Beach Harbor, California, November 1978. (From Dawson 1979.)

dation (1976) to be one of overlapping characteristics of both inner- and outer-harbor plankton fauna.

The total zooplankton distributions within the inner and outer harbor were roughly similar (Allan Hancock Foundation 1976). It appeared that the near exclusion of *Cladocera* from the inner harbor and the increased abundance of *Acartia tonsa** were compensatory. Only the B stations were markedly higher in zooplankton, which may be credited to the overlapping of inner and outer zooplanktonic fauna in the area of stations B4 through B7.

Segregation of species between the inner

and outer harbor may result from environmental conditions imposed on each and from the competitive behavior between them. Allan Hancock Foundation (1976) has shown that the strongest and most consistent abiotic differences between the inner and outer harbor are pH and dissolved oxygen. The inner harbor pH (7.75) was lower than that of the outer harbor (8.02); the area also had a lower dissolved oxygen concentration (6.1 ppm) than the outer harbor (8.02 ppm). Whether or not such abiotic characteristics are responsible for the observed distribution of these species is unknown.

The seasonal distribution of the five dominant zooplankton species throughout the Los Angeles–Long Beach Harbor is shown in figure 6.3 (Allan Hancock Foundation 1976). The three dominant copepods—*Acartia tonsa**, *Paracalanus parvus,* and *Corycaeus anglicus*—show the same general seasonal pattern: increased abundance in the winter months and a decrease in the summer, with irregular spring peaks. These data conflict with that of Environmental Quality Analysts and Marine Biological Consultants (1978), which showed slightly dissimilar monthly abundance highs and lows among the three species in Long Beach Harbor. *A. tonsa** and *P. parvus* both showed peaks in May and July, with the latter species having additional peaks in December, February, and March. The low concentration of *A. tonsa** occurred in January–April, while *P. parvus* had low abundances in August–November. *Corycaeus anglicus* showed peak concentrations in the spring and fall, with a yearly low in summer.

Environmental Quality Analysts and Marine Biological Consultants (1978) also determined the vertical preferences of the major copepod species by daytime sampling at three depths (2, 6, and 12 m). *A. tonsa** showed a distinct preference for the upper 6 m of water, and from December through May, more than 50% of the population was located at 2 m. *P. parvus* concentrations increased with depth, the greatest concentration occurring at 12 m, and *C. anglicus* showed a distinct preference for 6 and 12 m.

The two dominant cladocerans, *Podon polyphemoides* and *Evadne nordmanni,* showed a seasonal pattern similar to the copepods (Allan Hancock Foundation 1976): low abundance in summer. Both species showed unusually high concentrations in the spring of 1972, but were never again sampled in such abundance.

In summary, copepods dominated the zooplankton in Los Angeles–Long Beach Harbor, and *Acartia* spp. were the dominant copepods. The distributions of *A. tonsa* and *A. californiensis* were important indicators of outer and inner harbor environments, respectively. The inability to separate *A. tonsa* and *A. californiensis* in the early studies represents a loss of valuable information regarding the harbors. This emphasizes the importance of accurate identification of closely related taxa in ecological studies.

Marina del Rey Harbor

Soule and Oguri (1977) reported on a 1-year study conducted in Marina del Rey, although sampling actually continued for an additional 2 years. The information on the latter 2 years was drawn from unpublished data, and results for all 3 years (July 1976 to June 1979) of these zooplankton investigations are summarized here.

The zooplankton of Marina del Rey were dominated by copepods, which comprised over 98.5% of the zooplankton; cladocerans contributed 0.84%. Other less significant groups included larvaceans (0.34%), brachyura zoea (0.14%), cirripedia nauplii (0.12%), and fish eggs and larvae (0.04%). Among the copepods, *Acartia* spp. dominated with 97.4% of the total zooplankton. This genus was composed mainly of *A. californiensis* (94.7%) and *A. tonsa* (2.7%). Other numerically less important copepods included *Paracalanus parvus* (1.0%) and *Corycaeus anglicus* (0.1%). Four species of cladocerans occurred; *Evadne nordmanni* was the most common (0.29%) followed by *Penilia avirostris* (0.22%), *Podon polyphemoides* (0.21%), and *Evadne spinifera* (0.12%).

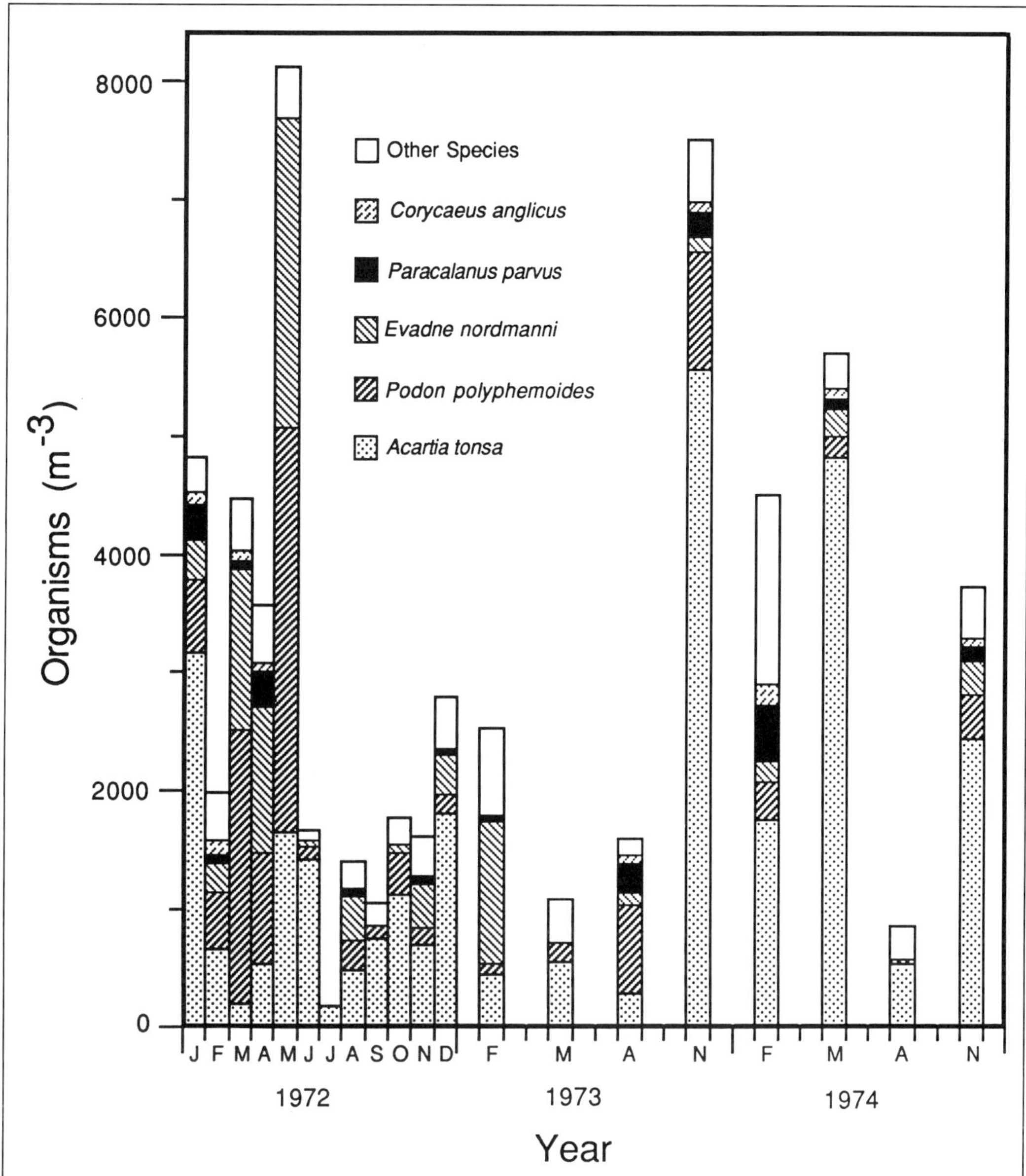

Figure 6.3. Mean temporal abundance of zooplankton and major component species in Los Angeles–Long Beach Harbor. (From Allan Hancock Foundation 1976.)

While zooplankton diversity was much less in Marina del Rey than that found in Los Angeles–Long Beach Harbors, the areal distribution of *A. tonsa* and *A. californiensis* followed a similar pattern. *A. californiensis* was most abundant in the inner main channel and slips, while *A. tonsa* was found primarily in the entrance channel and mouth of the harbor.

Newport Bay

Trinast (1976) described *Acartia californiensis* as a new species while studying the influence of tidal exchange on the population of *Acartia* spp. in upper Newport Bay (Trinast 1975). She found that *A. californiensis* was the numerically dominant zooplankton, and the loss of this copepod from the bay due to tidal exchange was minimized by its behavioral tendency to congregate in deeper water during ebb tide. The life cycle was studied in detail by Trujillo-Ortíz (1986), who considered it to be endemic to the northeastern Pacific and apparently restricted to estuaries and coastal lagoons. Additional reported SCB locations of *A. californiensis* are Laguna Peñasquitos (Trujillo-Ortíz 1986) and Mission Bay in San Diego (A. Fleminger pers. comm. 1968).

Nearshore Zooplankton

Seaward of the harbors and estuaries, oceanographers typically separate the coastal longshore regions of open water on the basis of bottom topography. The nearshore region encompasses those waters shoreward of the continental shelf slope break, or approximately the 200-m depth contour. This is a useful demarcation for study of zooplankton since the water over the continental shelf tends to be an area of high productivity. This augmented region of productivity (Ryther 1969) is usually associated with increased vertical mixing and, thus, greater nutrient recycling and upwelling, both of which are wind-forced phenomena.

The maintenance of a shelf zooplankton assemblage is largely dependent on the physical width of the shelf as well as on the frequency of offshore advection over the shelf. The presence of a wide continental shelf off the east coast of the United States has been shown to support distinct zooplankton assemblages (Grice and Hart 1962; Bowman 1971). Only occasional intrusions of Gulf Stream water disturbs this relatively stable zooplankton ecosystem. The relatively narrow shelf of the western United States has also been shown to support persistent zooplankton assemblages (Peterson and Miller 1975, 1977).

Microzooplankton

Because of their tendency to pass through typically sized plankton nets (250–300 μm), the microzooplankton were ignored by early oceanographers. The study of the microzooplankton as a food web component was an early objective of the Food Chain Research Group at Scripps Institution of Oceanography (SIO). Microzooplankton are now recognized as an important part of the consumer zooplankton, and this component of the plankton has been intensively studied in the SCB.

Microzooplankton are those animals feeding on particulate organic sources; they comprise protozoans as well as juvenile stages of metazoan zooplankters such as copepod nauplii and some early copepodites. Protozoans account for the greatest percentage of the microzooplankton numerically, while the micrometazoans dominate the biomass (Beers and Stewart 1967, 1969a, b, 1970). Because of their high reproductive capacities relative to the metazoans, protozoans have a markedly more important effect on the dynamics of the pelagic trophic web. Since protozooplankton can reproduce by simple asexual binary fission, they are able to respond rapidly to a changing environment. In addition, because generally higher physiological rates are found among smaller organisms, they are considered by Beers (1986) to be among the most important pelagic herbivores, a role generally reserved for copepods in the past. Beers and Stewart (1969b, 1970) have shown that the biomass of the microzooplankton is generally 20–25% of the total larger macrozooplankton, both inshore and offshore in the SCB.

In the SCB, Beers and Stewart (1970) made a weekly sampling study of the microzooplankton off La Jolla from April through September 1967. They sampled from the "pigment layer" at three stations (1.4 km, 4.6 km, and 12.1 km from shore, respectively) off La Jolla and from the bottom of the pigment layer to 100 m at the farthest offshore locations. The pigment layer was defined as that region from the surface to a depth of no significant chlorophyll fluorescence, averaging 50 m (Strickland et al. 1970). The average biomass, expressed as organic carbon, decreased from onshore to offshore in the pigment layer: 5.9, 3.0, and 2.4 mg C m^{-1} at stations 1, 2, and 3, respectively. Protozoans, dominated by ciliates, accounted for 32, 24, and 23% of the microzooplankton biomass at these stations. Copepod nauplii accounted for 59–65% of the total metazoan organic carbon at all stations. At stations 2 and 3, microzooplankton organic carbon accounted for 17 and 21%, respectively, of the total zooplankton organic carbon in the upper 100 m.

Variability in the microzooplankton over the 5-month period was generally attributed to periods of upwelling in the early part of the study and offshore advection later in the study. Periods delimited were approximately 1 month in length; the greatest temporal microzooplankton variation was between the initial period (first month of sampling) and the remainder of the study at station 1. At this station, the mean microzooplankton organic carbon per cubic meter in the first month was 375% of the rest of the values. Station 2 showed a 50% increase during the remaining periods. Station 3 showed a reversal of this trend; the remaining periods were 50% greater than the initial period.

While the Beers and Stewart (1970) study evaluated the cross-shelf and temporal variability of microzooplankton biomass, Beers et al. (1980) sampled the microzooplankton in small (5-m) increments to 50 m at several stations over the continental slope during a period of several days in late May and early June 1970. Their findings point out the extreme variability both vertically (up to 120-fold difference) and horizontally (greater than two orders of magnitude) in the areas off La Jolla. They suggest that this variability results from the effects of upwelling and advection from offshore.

A number of studies (for example, Brewer et al. 1981; Brewer and Smith 1982) have established the importance of the nearshore ecosystem to the early development of fish. The role of the microzooplankton in ichthyoplankton development was studied by Brewer and Kleppel (1986) in Santa Monica Bay. Gut analysis of fish larvae showed that they were feeding on a variety of microzooplankton. Copepod nauplii and copepodites were found in 34% of the northern anchovy and in 43% of the white croaker that contained food. Copepod eggs were also commonly found in the guts of these species of ichthyoplankton. Bivalve veligers and tintinnids were more frequently found in the white croaker than in the northern anchovy. While this may reflect selective feeding among the ichthyoplankton, it could also result from the integration of each species' microhabitat. Densities of microzooplankton in the water column were found to vary significantly over depth, with the highest concentration in samples near the bottom. The single highest fish larvae concentration was found in a bottom sample over the 22-m isobath. This was commensurate with the highest densities of copepod nauplii and copepodites, invertebrate eggs, tintinnids, bivalve veligers, and total microzooplankton.

Owen (1981) has shown that certain microzooplankters such as tintinnids, *Noctiluca*, and copepod nauplii are distributed both horizontally and vertically in microscale patches of less than 2 m. He concluded that since the microscale patches of food and nonfood items do not usually coincide, the strike success of fish larvae within food patches may be augmented.

The impact of microzooplankton on primary production in the nearshore environment of the SCB can be significant. Heinbokel and Beers (1979) estimated that generally less than 4%—but in some cases greater than 20%—of the total phytoplankton could be consumed by tintinnids. Since other ciliates are frequently more numerous than tintinnids and are mainly herbivorous, they concluded that ciliates can have a significant impact on phytoplankton production. Azam et al. (1983) also found that the protozooplankton are the chief consumers of the autotrophic pico- and nanoplankton, as well as of the bacteria. This makes them important contributors to the "microbial loop." A potential impact of microzooplankton on macrozooplankton populations may also exist. Kimor (1979) described predation by the dinoflagellate *Noctiluca miliaris* on the eggs of *Acartia tonsa* in samples off Del Mar. He concluded that this dinoflagellate may have a considerable influence on *Acartia* population abundance. Torrey (1902) also described *Noctiluca* to be a consumer of *Gonyaulax* sp. during a red tide bloom in the SCB.

Macrozooplankton

Just as the microzooplankton are characterized by their size, the macrozooplankton are similarly characterized, except that they are retained by the typical size (250–300 μm mesh) plankton nets. When free-swimming animals become large enough to avoid being collected by plankton nets, they are generally termed *micronekton* and may or may not be considered as part of the zooplankton. Examples of this category are the adult mysids and euphausiids. Although both groups are found in both the nearshore and offshore realms, mysids are found in higher numbers nearshore, while euphausiids are in higher concentrations in the offshore mesopelagic zone.

The macrozooplankton are a diverse group of animals composed of a number of major taxonomic categories. The medusae, ctenophores, and planktonic molluscs and tunicates are sometimes grouped into what is commonly termed *gelatinous zooplankton*. Planktonic tunicates include the thaliaceans, generally composed locally of salps and doliolids, and the larvaceans, which are exemplified by the very common genus *Oikopleura*. The chaetognaths (arrowworms) are important carnivorous zooplankters, but the majority of the zooplankton are made up of crustaceans. While cladocerans and ostracods are common, most of the crustaceans collected by net and pump are copepods. Planktonic copepods are primarily calanoids, but the cyclopoids and harpacticoids can sometimes be important contributors to the copepod fauna. Of the calanoid copepods, *Acartia, Paracalanus, Labidocera,* and *Calanus* are the most common genera collected nearshore in the SCB (Barnett and Jahn 1987).

ABUNDANCE AND DISTRIBUTION

Peterson et al. (1986) examined the nearshore nutrients and zooplankton biomass of transects sampled bimonthly from Ormond Beach, Playa del Rey, Seal Beach, and San Onofre during the years 1982–1984. Clear seasonality was shown in zooplankton volumes (fig. 6.4a), with a zooplankton maximum during the period April–June and a minimum during December–February. A plot of the mean cross-shelf zooplankton biomass for all transects (fig. 6.4b) shows that the station farther offshore (75-m isobath) always had a lower biomass than the inshore stations. During the February–June period of increasing zooplankton biomass, the maximum zooplankton biomass shifted from inshore to mid-transect depth (36 m), suggesting that the areas of highest zooplankton production and survivorship gradually shifted from nearshore to offshore as the season progressed.

The source of nutrients (nitrate) that could stimulate primary production and, thus, secondary production (zooplankton) was significantly correlated with zooplankton biomass at the 36-m and 75-m isobaths (fig. 6.4c). Processes such as tidal mixing, internal waves, and local upwelling were thought to be important in mixing outer shelf nutrient-rich waters with the shallow nearshore waters.

Barnett and Jahn (1987) described the cross-shelf and some longshore differences in zooplankton gradients and addressed the persistence of these gradients over time. Their sampling covers the years from 1976 to 1980 off the coast in the general vicinity of San Onofre, and they believe that it generally represents the over-shelf zooplankton dynamics of the southern half of the SCB.

The average cross-shelf abundance patterns of the nine most important zooplankton taxa found by Barnett and Jahn (1987) are shown in figure 6.5. The centers of distribution for three groups, *Acartia clausi, Oithona oculata,* and cypris larvae, show similar patterns. These zooplankton taxa have their highest concentrations near the bottom and closest to the shore, reflecting perhaps a near-

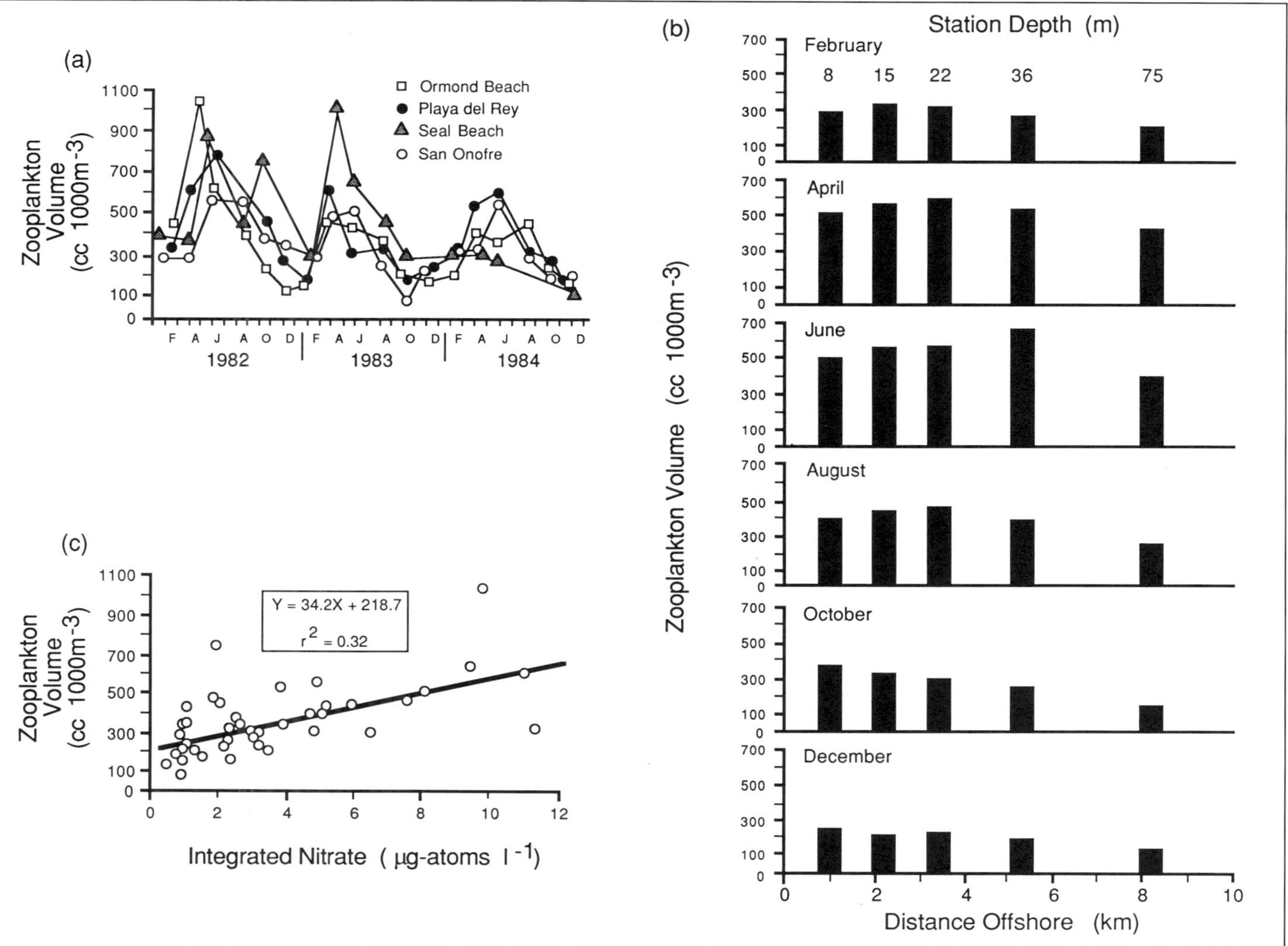

Figure 6.4. (a) Seasonal variations of zooplankton biomass from four major transects. (b) Cross-shelf distribution of zooplankton biomass; monthly average of data from the four transects, 1982–1984. (c) 75-m nitrate and zooplankton biomass regression. (From Peterson et al. 1986.)

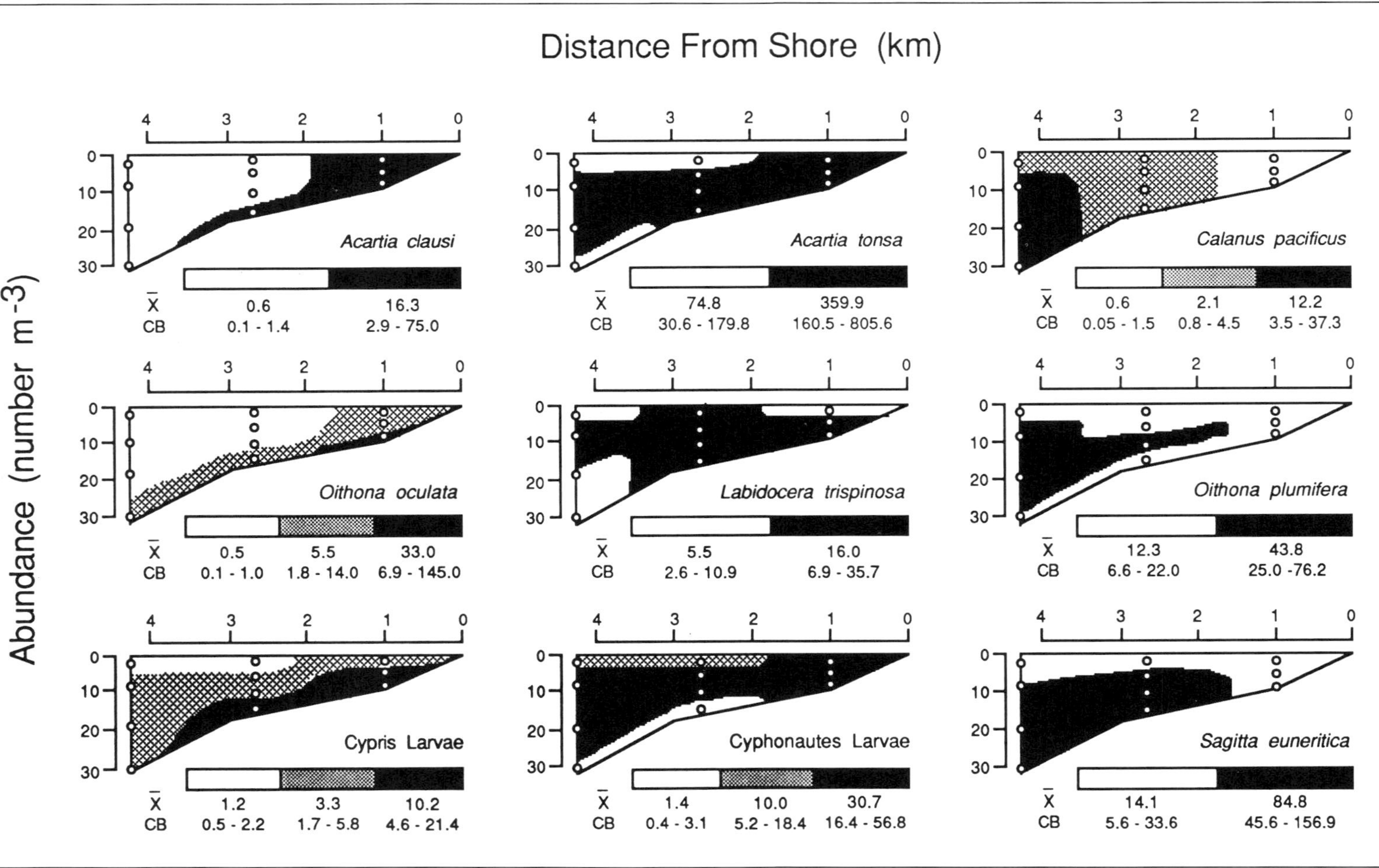

Figure 6.5. Mean cross-shelf abundance profiles of nine zooplankton taxa based on surveys taken between 1976 and 1980. Abundances are in number per cubic meter. (From Barnett and Jahn 1987.)

bottom rich layer of phytoplankton. They also reside in waters least likely to be dispersed offshore by intrusive advection because of reduced water velocity near the bottom. A second group of zooplankton species, *Calanus pacificus, Oithona plumifera,* and the chaetognath, *Sagitta euneritica,* are concentrated on the offshore end of the transect. The third set of taxa, *Acartia tonsa, Labidocera trispinosa,* and bryozoan larvae (cyphonautes), show a distribution covering both inshore and offshore areas. A similar transect 12 km to the south show no observable differences of cross-shelf species distributions; however, this distance may be too short to conclude an absence of SCB longshore differences in nearshore species.

Barnett and Jahn (1987) used clustering statistics to determine assemblages of zooplankton groups that occupied various cross-shelf positions at different times during the year of biweekly sampling (24 biweekly cruises). The 23 taxa used in this clustering analysis represented forms that were present during at least 17 of the 24 samplings. Five distinct groupings of zooplankton taxa were identified. Two groups (Groups I and II) were classified as the nearshore and offshore assemblages in the fall–winter period, and three groups (Groups III, IV, and V) made up the nearshore, transitional, and offshore assemblages in the spring–summer period. These group associations are shown in table 6.1.

Several species had single group associations: *Penilia avirostris* and *Evadne nordmanni* nearshore and *Calanus pacificus* and *Rhincalanus nasutus* offshore in the fall–winter period. Other taxa showed little temporal change (*Eucalanus californicus* offshore and *Podon polyphemoides* nearshore). A number of species (*Acartia tonsa, Corycaeus anglicus, Paracalanus parvus, Labidocera trispinosa, Oithona plumifera,* and *Sagitta euneritica*) showed a definite seaward shift in their distribution from winter to summer.

Zooplankton size was also clustered by shelf position and season. Results showed that smaller species dominated nearshore with the smallest crustaceans in the nearshore group of spring–summer and the largest species in the two offshore groups. Of the four taxa in which life stages were separated (*A. tonsa, L. trispinosa, P. parvus,* and barnacle larvae), the more immature stages showed a more nearshore distribution. Possible explanations for this include a shoreward transport of eggs and larvae, offshore movement of more mature stages, and a more fecund nearshore population (Barnett and Jahn 1987).

To further substantiate seasonal cross-shelf gradients and to reveal shorter term fluctuations or stability, an analysis independent of the cluster analysis was completed. Variations of each taxa's position relative to its yearly cross-shelf mean were considered. A general agreement among taxa was observed over two blocks of time. There was a typical onshore shift from February to early April, a period generally following a storm-generated mixing, and an offshore shift of most taxa from mid-April to July, a time of periodic upwelling.

Stability of the zooplankton over the shelf appeared to be quite high (Barnett and Jahn 1987). During the upwelling season of April, May, and July, zooplankton cross-shelf zonation was not diluted even though chemical–physical gradients were not apparent. Zonation was actually most intense from April to August. Even during intense wintertime mixing, there were no obvious changes in the weaker winter cross-shelf zonation. Peterson et al. (1979) also reported the maintenance of the nearshore population of copepods despite intense upwelling off Oregon. It has been suggested that the 2-km strip of water immediately adjacent to the Oregon coast was not directly affected by upwelling (Wroblewski 1980). This may also be true for the SCB.

Star (1980, as cited by Mullin 1986) also described the onshore–offshore position (within the 75-m isobath) of the more common zooplankters near San Onofre in 1979 by cluster analysis. The copepods *Calanus, Corycaeus,* and *Paracalanus* and the cladoceran *Evadne* were grouped into an offshore zooplank-

Table 6.1. *Mean Abundance of 23 Zooplankton Taxa in Five Clustered Zooplankton Groups*[a]

		Mean Abundance (no. m^{-3})				
		Fall–Winter Nearshore		Spring–Summer Transitional	Fall–Winter Offshore	
Taxa	Group	I	III	IV	V	II
Evadne spinifera	Cladocera	7.6				14.9
Penilia avirostris	Cladocera	197.7				
Evadne nordmanni	Cladocera	48.2				
Podon polyphemoides	Cladocera	3.5	3.5			
Acartia clausi	Copepod		25.9			
Oithona oculata	Copepod		38.9			
Cirriped nauplii	Barnacle	20.4	32.3	37.2		
Cirriped cypris larvae	Barnacle	3.8	3.6	93.5		
Unidentified fish eggs	—	2.3	5.5	5.5		
Acartia spp.[b]	Copepod		1918.9	1223.8		
Labidocera trispinosa[b]	Copepod		41.1	84.4		
Acartia tonsa	Copepod	352.1		664.1		
Corycaeus anglicus	Copepod	149.9		195.5	96.3	
Paracalanus parvus[b]	Copepod	151.8		256.9		
Paracalanus parvus Adult	Copepod	119.0		171.9	81.9	
Labidocera trispinosa Adult	Copepod	5.0		9.0	3.0	
Oithona plumifera	Copepod	59.1		85.8	148.4	81.6
Sagitta euneritica	Chaetognath	96.0		248.0	222.0	
Cyphonautes larvae	Bryozoa	42.4	82.1	71.5	52.0	
Medusae	—			11.1		
Calanus pacificus	Copepod				155.0	
Rhincalanus nasutus	Copepod				11.3	
Eucalanus californicus	Copepod				3.3	1.7

[a] Values given are peak abundances of the taxa, that is, concentrations of one-third or greater of the maximum concentration found for that taxon.

[b] Immature.

After Barnett and Jahn 1987.

ton assemblage, and the copepods *Acartia, Oithona,* and *Labidocera* were all associated in a nearshore cluster.

Hirota (1974) observed a general decrease in abundance of the ctenophore *Pleurobrachia bachei* from onshore to offshore within 10 km of the coast. This ctenophore is an important summer–fall component of the nearshore SCB plankton, with the highest concentration occurring in August. In contrast to the normal diel vertical migration of zooplankton, *Pleurobrachia* migrate downward at night from their daytime level of the upper 15 m (at or above the thermocline) to about 30–40 m. Hirota (1974) speculated that this reverse vertical migration pattern helps this species maintain high abundances close to shore. By living in surface waters during the day, they are moved shoreward by westerly–northwesterly sea breezes. At night, the weaker land breeze moves the surface waters offshore. Gut analyses have shown no major change in feeding intensity between day and night. Over the entire size range of *Pleurobrachia,* the single most important food item was the copepod *Acartia tonsa,* comprising about one-third of the ctenophore diet in terms of numbers and amount of carbon ingested. Other numerically important food items included the copepods *Euterpina acutifrons* (11.4%) and *Corycaeus anglicus* (7.6%), nauplii (5.9%), eggs (5.5%), the cladoceran *Evadne nordmanni* (5.2%), and the copepod *Paracalanus parvus* (4.3%). In terms of biomass, the copepod *Labidocera trispinosa* (27.4%) and the chaetognath *Sagitta euneritica* (10.8%) were the most important dietary components after *Acartia tonsa* (34.8%). This chaetognath was not consumed in proportion to its abundance, however, implying that through rapid escape responses, the prey may limit the ctenophore's capture effectiveness.

Clutter (1967) found a strong horizontal zonation among a number of hypopelagic (approximately 3–30 cm above the sea floor) mysids in the nearshore environment between the La Jolla and Scripps canyons. Despite the overlapping of mysid distributions, similar zonations were evident among the three transects studied. *Mysidopsis californica* was the most shoreward species, followed by the numerically dominant *Metamysidopsis elongata, Acanthomysis macropsis,* and most seaward, *Neomysis kadiakensis.* Bathymetric distribution changed between transects among these species, but their shore to seaward order remained constant. The environmental factor most responsible for *Metamysidopsis elongata* distribution appeared to be the pattern of food availability that was imposed by the nearshore circulation.

Mysids also exhibit a vertical distribution pattern within the giant kelp (*Macrocystis pyrifera*) beds of the SCB. Clarke (1971) studied the distribution of mysids within these kelp beds and found a three-tier distribution categorized by habitat as canopy, subcanopy, and benthic species. The most common canopy species, *Acanthomysis sculpta,* swim among the surface fronds in enormous numbers, with *Siriella pacifica* occasionally present. Subcanopy species are those between the surface canopy and the bottom and commonly include *Acanthomysis macropsis* and *Neomysis rayi.* Other species present are *Acanthomysis columbiae, Acanthomysis* sp., *Mysidopsis* sp., *Neomysis kadiakensis,* and possibly *Proneomysis wailesi.* Benthic species directly associated with the giant kelp include *Heteromysis odontops,* which occupies kelp holdfasts. Hypopelagic mysids living adjacent to kelp beds over sandy bottoms include *Neomysis kadiakensis, Acanthomysis macropsis, Metamysidopsis elongata,* and *Mysidopsis californica.* These hypopelagic species are at times so abundant that dense swarms extend over large areas. Canopy and subcanopy species of mysids are the most important mysids eaten by kelp fish.

The diel migration of demersal zooplankton into open water from within a kelp forest ecosystem off Santa Catalina Island was studied by Hammer (1980). The migration from the kelp holdfasts into open water was primarily a nocturnal phenomenon. Hammer further observed that 86% of the zooplankton consisted of the gammarid amphipods *Batea transversa* and *Parapleustes oculatus,* followed by (in order of numerical importance) ostracods (*Vargula tsujii*), isopods (*Paracercepis cordata* and *Gnathia* sp.), and the natantian shrimp *Lysmata californica.* Migration into open water during the day was numerically dominated by copepods: the harpacticoids *Tisbe* spp. and *Porcellidium* spp., the cyclopoids *Oithona* spp. and *Oncaea* spp., and the calanoid copepodites *Clausocalanus* spp.

Zooplankton migrations off the sand substrate within a kelp forest showed similar copepod diel migrations. Hammer (1981) found, however, that the nocturnal migrations were dominated by the caprellid amphipod *Caprella* spp. (67%), followed by the tanaid *Leptochelia* spp. (13%), and the gammarid amphipods (6%), *Photis* spp. and *Lysianassa* spp. Stretch (1985) made a similar study from a sand substrate outside the kelp forest, using a different collecting technique. He found that among the demersal gammarid amphipods, the nocturnal migration was dominated (86%) by *Rudilemboides stenopropodus,* and approximately 11% of the total benthic gammarid fauna migrated into the water column.

FEEDING AND REPRODUCTION

The impact of macrozooplankton grazing pressure on phytoplankton standing crop in the nearshore environment has been shown to be significant. Mullin and Brooks (1970) found that *Calanus helgolandicus* (= *pacificus*) showed a daily ingestion rate of 14–26% of the phytoplankton production off La Jolla during a 5-month study. Eighty percent of this ingestion was by late copepodite stages and was greatest in the late spring when primary production was relatively low. Beers and Stewart (1970) have estimated that the microzooplankton ingested about 27% of the net production at the same stations over the same period.

Kleppel et al. (1988) demonstrated the occurrence of carnivorous feeding among crustacean zooplankton. As much as 97% of the carbon in the guts of crustacean zooplankton was contributed by animals at times of low phytoplankton biomass and production.

While the macrozooplankton clearly impact the phytoplankton, phytoplankton concentrations also have an effect on zooplankton production. For instance, egg production of *Paracalanus parvus* was positively correlated with concentrations of chlorophyll *a* (Checkley 1980), resulting in increased food limitation on egg production with distance offshore. Only in the generally eutrophic Santa Monica Bay did copepod egg production appear not to be food limited. Reproduction is also affected by the abiotic habitats in which copepod eggs develop. Uye and Fleminger (1976) showed that the eggs from four species of nearshore and embayment *Acartia* have specific abiotic hatching requirements. These varying hatching requirements might be the underlying factors governing the abundance of different species of *Acartia* in different areas and times.

While microzooplankton serve as food for developing fish larvae, macrozooplankton may feed directly on ichthyoplankton. Brewer et al. (1984) found various taxa of zooplankton attached to, or partially ingesting, ichthyoplankton in preserved samples from Santa Monica Bay. These taxa included three copepods (*Corycaeus anglicus, Labidocera trispinosa,* and *Tortanus discaudatus*), euphausiid larvae (*Nyctiphanes simplex*), an amphipod (*Monoculides* sp.), a chaetognath (*Sagitta euneritica*), and an unidentified decapod larvae. *C. anglicus* was the species most commonly found clinging to fish larvae. A total of 4.6% of the white croaker larvae had *C. anglicus* attached, while only 1.6% of the northern anchovy larvae and 0.3% of unidentified fish larvae had *C. anglicus* attached. At least 36 of the 54 (67%) bongo samples contained fish larvae with clinging *C. anglicus.* While some of this "attachment" may represent net feeding, Brewer et al. (1984) believed that the selective attack on fish larvae reflects natural predation.

PATCHINESS

The heterogeneous spatial distribution of zooplankton has long been recognized as an important aspect of the planktonic environment (Steele 1978). Investigations of heterogeneous distributions have been difficult because of the ephemeral nature of zooplankton aggregations (active or passive) and the need to sample rapidly in time and space to conduct such investigations.

Richter (1985) used a bottom-moored, upward-looking, high-frequency acoustics device to survey zooplankton patches passing through the acoustic beams with the prevailing current. Sampling was conducted off San Diego Bay in 18 m of water. He found that small zooplankton occurred in patches on a scale from 10 to 100 m and that patch biomass was from 4 to 15 times the background level. Large zooplankton, consisting possibly of large copepods and small euphausiids and amphipods, were usually found amid small zooplankton aggregations. Horizontal back-scattering varied independently of both the temperature and the chlorophyll maximum.

Mullin (1979) reported on simultaneous sampling from two ships in March 1976 off Del Mar, California. He collected zooplankton in 5-m increments from 35 m to the surface and compared these profiles on horizontal distances from 100 m to 10 km. Bottom depth did not exceed 50 m throughout the study, and samples were taken during both day and night. The predominant horizontal scales of variability fell into three groups: (1) larvaceans were homogeneous in distribution at all spatial scales; (2) *Calanus pacificus, Sagitta,* and *Evadne* increased in variability with distance; and (3) *Corycaeus anglicus* was quite variable even in closely spaced samples. This implied that while larvacean patches were on a scale larger than 10 km, *Corycaeus* formed patches smaller than 100 m. When day and night samples were treated separately, different patterns emerged. Samples taken during the day were less similar as the distance apart increased, while samples taken at night showed no decrease in percent similarity with increasing separation. This indicated that patch characteristics can change diurnally.

A year later, Star and Mullin (1981) sampled from a depth of 36 m along a longshore transect in the same general location. Using data from Star and Mullin (1981), Mullin (1986) discussed the horizontal distribution of various taxa along the transect (fig. 6.6). Similarities in horizontal distributions can be seen in (1) *Calanus* developmental stages, (2) *Eucalanus* developmental stages, and (3) *Metridia* and *Pleuromamma* genera. He suggested that these patterns might arise from common locations of birth or similar responses to the environment.

Offshore Zooplankton

Transition Zone

While the three separate zones are natural divisions progressing seaward from the coast, only a few studies have assessed the transition zone between nearshore and offshore, that is, the zone over the continental slope. Some of these studies were discussed in the previous section (Beers and Stewart 1967, 1969b; Hirota 1974; Barnett and Jahn 1987).

Pieper et al. (1990) also studied the transition zone off Santa Monica Bay by use of a 21-frequency array of transducers for acquiring an acoustically derived biovolume of zooplankton. In addition, temperature, conductivity, chlorophyll *a*, and depth data were collected. Two transects were sampled during the daylight hours in October 1982. The first one extended from Point Dume over the continental shelf (40-m bottom depth) to a station 23.5 km offshore in the Santa Monica Basin (900-m bottom depth). This transect consisted of seven vertical casts over 5.5 hours. The second, more southern transect ran 35 km from the outermost station just mentioned to a station off Santa Monica in about 40 m of water, and it consisted of six vertical profiles taken over 4.5 hours. Both transects showed a thermocline at about 20 m over the shelf and an undulating isotherm with a high over the shelf break and a wavelength of 15–20 km. The cause of this wavelike isotherm is unknown but may include surface and subsurface currents deflecting isolines to internal waves or local eddies.

The chlorophyll *a* and zooplankton biovolumes for the northern transect are shown in figure 6.7. The zooplankton biovolume section shows a maxima from 15 to 30 m over the outer edge of the shelf. The maximum gradually deepens offshore. The center of the high zooplankton biomass near the edge of the shelf corresponds to a small increase in chlorophyll *a*. However, chlorophyll *a* is generally reduced in the region of this zooplankton maximum. It is possible that the zooplankton have reduced the chlorophyll *a* concentration in this area by grazing. Offshore from the edge of the shelf, zooplankton biovolumes are variable with depth, yet are generally higher in the region of the chlorophyll *a* maximum (25–55 m). A zooplankton biovolume and chlorophyll *a* maximum also occur at the offshore end of the transect

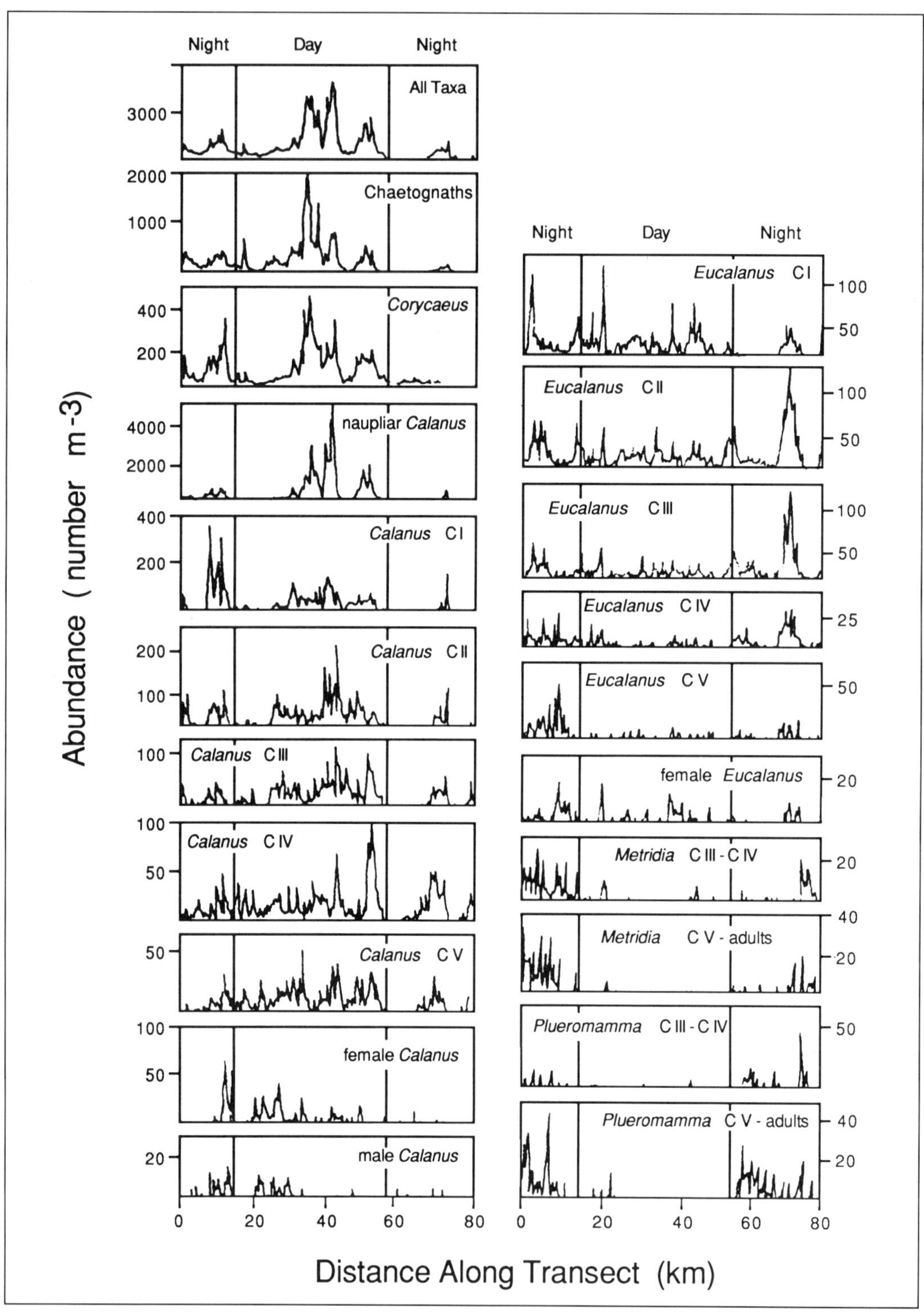

Figure 6.6. Horizontal distribution of zooplankton taxa at a depth of 34 m in a longshore transect in the SCB. (From Mullin 1986.)

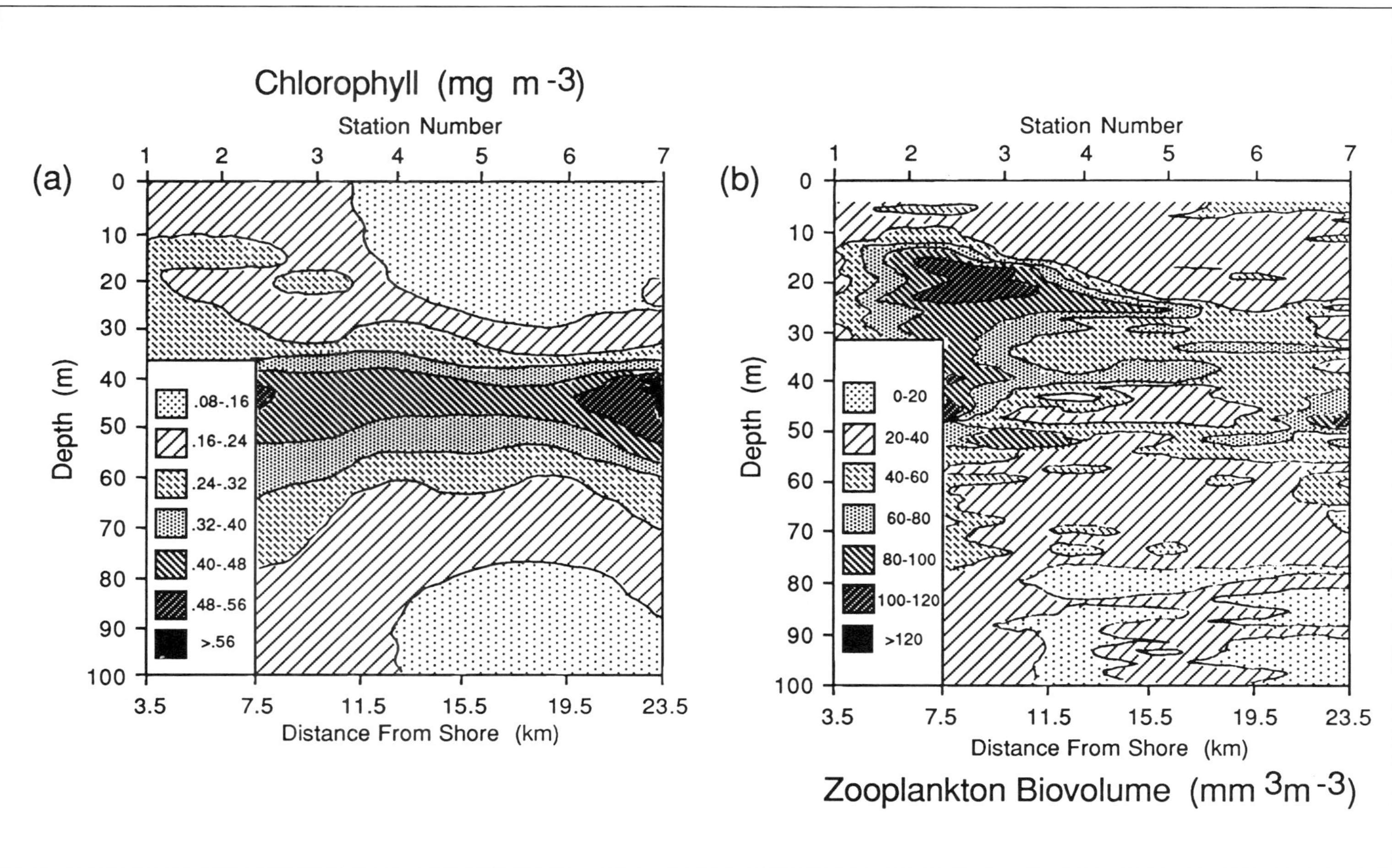

Figure 6.7. Horizontal sections of (a) chlorophyll *a* and (b) zooplankton biomass off Point Dume to 23.5 km offshore in the Santa Monica Basin, October 7, 1982.

between 40 and 50 m. The southern transect shows a similar coherence between chlorophyll *a* and zooplankton biomass, particularly over the shelf and slope (Pieper et al. 1990). These data indicate that the continental slope transition may be an area of increased productivity and may deserve greater attention of investigators in future work.

Long-Term Fluctuations

A number of investigators (Smith 1968; Ryther 1969; Cushing 1975; Walsh 1977; Eppley et al. 1979) have maintained that for eastern boundary currents, including the California Current, wind-driven coastal upwelling is the main source of new nutrients entering the euphotic zone. Other workers (Reid 1962; Bernal and McGowan 1981; Roesler and Chelton 1987) have found a correlation between zooplankton biomass, cold water temperature, and increased flow of the California Current. Chelton et al. (1982) analyzed 30 years of CalCOFI data to determine which of these factors plays the dominant role in California Current zooplankton biomass fluctuations. They compared the longshore component of wind stress with the mean monthly zooplankton volumes and concluded that, while wind-induced upwelling may play some role in zooplankton fluctuations, other factors are more important. They suggest that large fluctuations in zooplankton abundance appear to be unrelated to wind-driven upwelling, but are instead related to changes in the transport of the California Current—that is, greater transport is accompanied by greater productivity in the California Current and the SCB.

The causes of variable California Current flow have not been fully identified. The strength of the California Current may be correlated with El Niño occurrences in the eastern tropical Pacific. Positive El Niño events often correlate with low California Current transport, while negative El Niño events are associated with strong California Current flow. While not all California Current fluctuations can be associated with El Niño events, it appears that these events are important to California Current transport and thus to zooplankton fluctuations.

Microzooplankton

As already noted in the nearshore zooplankton section, Beers and Stewart (1969b) found a gradient of decreasing microzooplankton from onshore to offshore in the SCB. They also found an increasing concentration of microzooplankton relative to the concentration of chlorophyll *a* with distance offshore, and they suggested that the microzooplankton may play a more significant role in the offshore than in the nearshore realm.

Beers and Stewart (1969a) compared microzooplankton vertical structure in the nearshore and offshore zones. The ratio of microzooplankton in the euphotic zone to those found below the euphotic zone was 6.4:1 in the area over the continental shelf, whereas this ratio was much smaller (2:1) at the offshore station.

Temporal changes in the taxonomic composition of the radiolarian population in the upper 200 m of the SCB exhibited a distinct seasonality dependent upon the circulation pattern in the SCB (Casey 1966, as cited by Beers 1986). Radiolarians transported from equatorial waters dominated in the summer, and subarctic species were most abundant during the winter.

Macrozooplankton

Macrozooplankton of the offshore region often comprise many of the same species as found nearshore. In addition to these, more oceanic and deeper water species have been collected. Of the calanoid copepods, *Calanus, Pleuromamma,* and *Metridia* are common offshore genera in the SCB. *Calanus pacificus* (fig. 6.8) is the species most studied, and because it can be laboratory reared (Mullin and Brooks 1967; Paffenhöfer 1970), much has been learned of its trophic biology. The

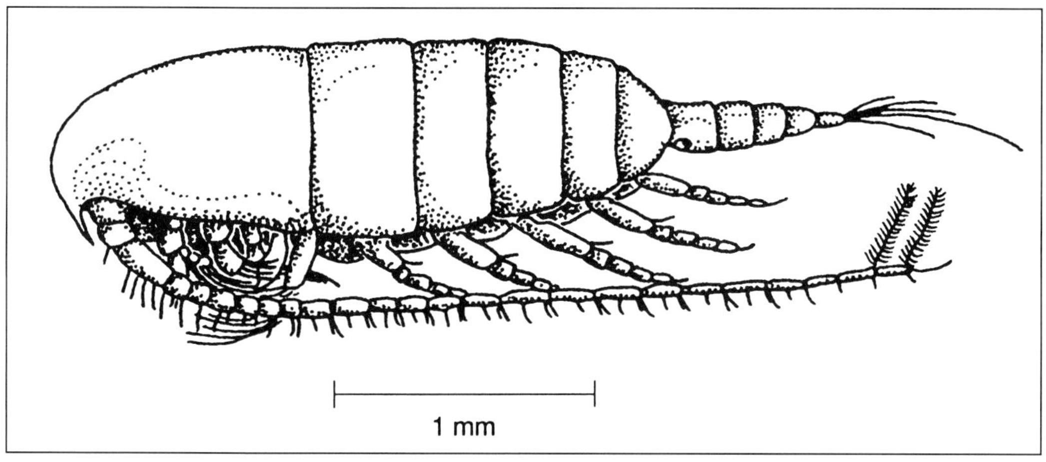

Figure 6.8. *Calanus pacificus,* female. (From Holliday and Pieper 1980.)

trophic knowledge thus gained, as well as that for other selected zooplankters, has been reviewed by Beers (1986). Micronekton (taxa capable of evading typical plankton nets) are also commonly found offshore. Euphausiids are typical members of the micronekton, and *Euphausia pacifica* and *Nematoscelis difficilis* (fig. 6.9) are ordinarily found in the offshore region of the SCB (Brinton 1976).

An occasional contributor to the micronekton of the SCB is the pelagic phase of the red crab *Pleuroncodes planipes.* Normally found off the Pacific coast of Baja California (Longhurst 1967), it occurs sporadically in the SCB in warm water intrusions from the south and, in particular, during El Niño events such as the winter–spring period of 1982–1984 (Beers 1986). Mass beach strandings have been reported as far north as Monterey (Glynn 1961). The red crab feeds on larger phytoplankton and microzooplankton (Longhurst et al. 1967) and represents a nutritional resource for various fish, marine mammals, and sea birds (Boyd 1963).

HORIZONTAL DISTRIBUTION

The causes and importance of interannual zooplankton variability in the SCB were summarized in the introduction to this chapter. Given that the SCB is in a transition region, an understanding of processes north, south, and west of the SCB is important in gaining insight about the zooplankton found within the area.

In the northern part of the CalCOFI grid and encompassing the northern part of the SCB, variability in zooplankton volumes appears to be associated with variations in California Current transport (Roesler and Chelton 1987). In this area, the zooplankton response to advection was rapid (on the order of months). Adult forms of zooplankton dominated collections when equatorward advection was strong (high-nutrient, nonfood limited), and larval forms dominated collections when advection was weak. In the southern part of the CalCOFI grid, which includes the southern part of the SCB, variability showed longer time scale responses (3–5 months) and the biomass was dominated by juvenile and larval stages of zooplankton. Roesler and Chelton (1987) concluded that zooplankton abundance in the southern area was controlled by local biomass response to changes in the advective environment. They suggested that biogeographic boundaries of subtropical species would move north and south in response to advective changes (see also McGowan and Miller 1980).

The majority of zooplankton samples collected during the years of CalCOFI sampling

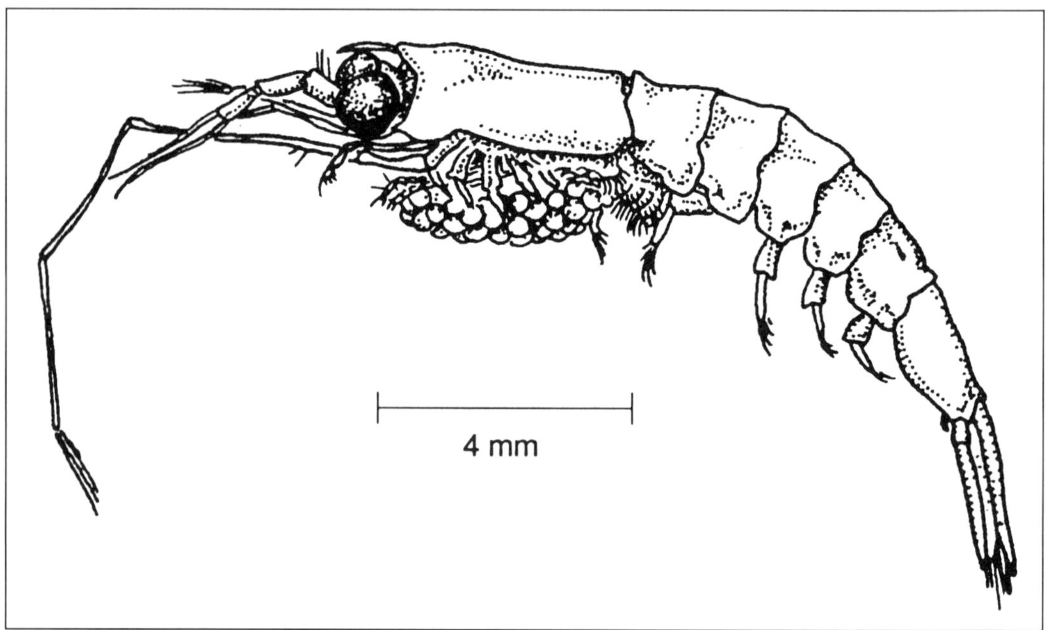

Figure 6.9. *Nematoscelis difficilis*. (From Brinton 1967.)

were not sorted by taxonomic group, but were instead analyzed by zooplankton displacement volume, a biomass analog. Bernal (1981) showed the mean zooplankton biomass of spring–summer and fall–winter over CalCOFI line 90 (onshore to offshore, starting off Los Angeles Harbor and passing southeast of Santa Catalina Island) for the years 1960–1969 (fig. 6.10a). The bimodal peak in the spring–summer biomass (fig. 6.10a) corresponded to variations in longshore transport. The peak at 180 km offshore occurred at the location of a net change in direction of flow from north to south in the upper 200 m. The farther offshore peak (270 km) occurred at the location of the maximum southward flow of the California Current.

Colebrook (1977) conducted one of the few studies of the distribution of zooplankton taxa from the CalCOFI data set. Five years of CalCOFI zooplankton samples (1955–1959) were selected for principal component analysis of the geographic distributions of 17 taxa. The first of the principal components was shown to be a north-to-south concentration gradient, and the second was an onshore-to-offshore gradient. These two components accounted for 61% of the variability found. Figure 6.11 shows a plot of the first vector (north–south) plotted against the second vector (onshore–offshore). From this figure, the relative contribution of each vector as well as the strength of that vector can be inferred by the relative position of the taxonomic category. For example, the radiolarians, amphipods, and ctenophores have a very strong north–south gradient, with the high concentration in the north, while the north–south concentration gradient of the mysids is highest in the south. The second vector influence is strongly seen in the cladocera, chaetognaths, and crustacean larvae concentrated onshore and in the ostracods, whose gradient is from offshore to onshore (high to low). The heteropods and pteropods are distributed slightly more to the south and offshore. The only group to show little clustering relative to either of these two vectors was the medusae.

Taxonomic groups, which may be com-

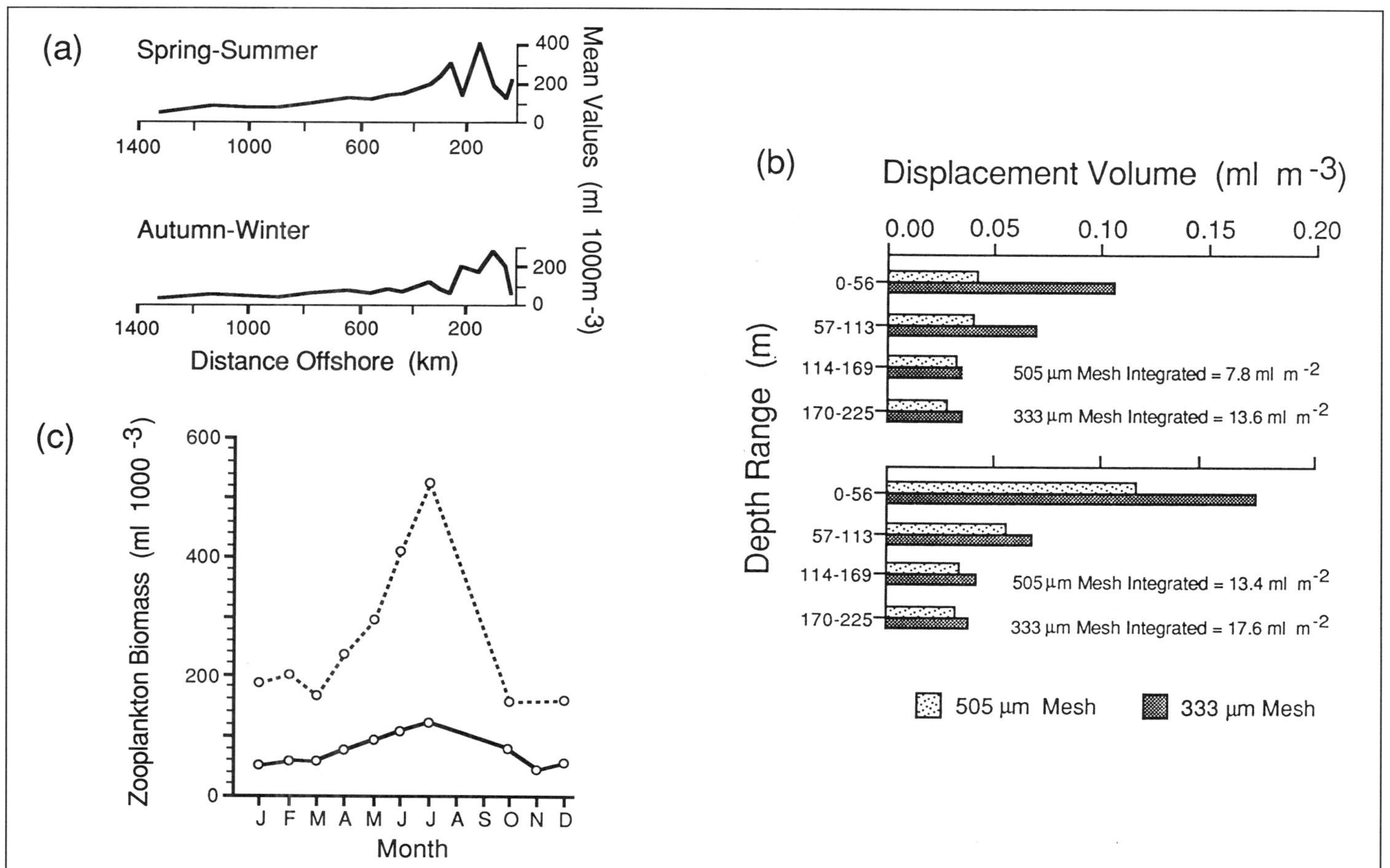

Figure 6.10. Zooplankton biomass (ml 1000 m^{-3}) in SCB. (a) Horizontal onshore to offshore mean values for 1960–1969 on CalCOFI line 90 for spring–summer and fall–winter periods. (From Bernal 1980.) (b) Vertical mean distribution for 12 daytime (upper) and 12 night-time (lower) stations in April 1965. (From Brooks and Mullin 1983.) (c) Monthly mean distribution from 1955 to 1956 (dashed line) during high zooplankton biomass period and from 1958 to 1959 (solid line) during period of low zooplankton biomass. (Data from Smith 1971.)

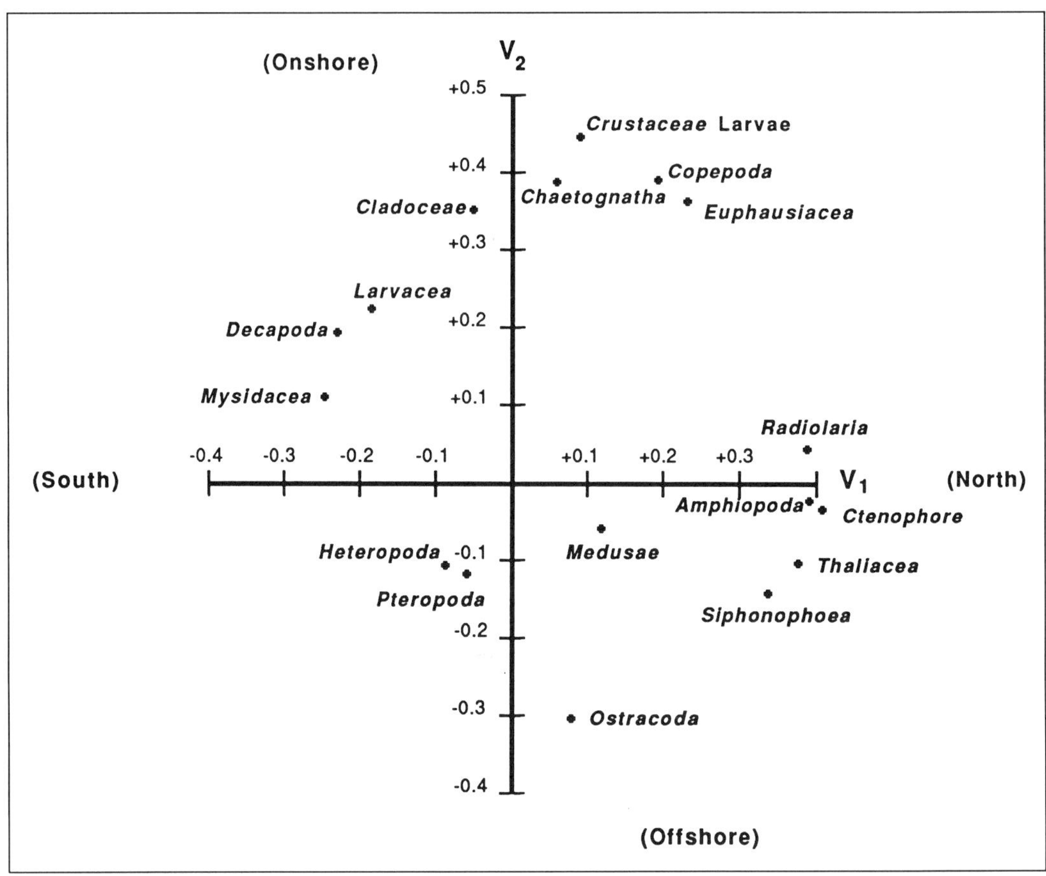

Figure 6.11. A plot of the first and second vectors from a principal components analysis of the geographic distributions of 17 zooplankton taxa. (After Colebrook 1977.)

posed of many independently distributed species, show an overall gradient longshore or onshore–offshore. Detailed distributions of these 17 taxa during this 5-year period can be found in Isaacs et al. (1969) for the spring and fall, Isaacs et al. (1971) for the winter, and Fleminger et al. (1974) for the summer. A further breakdown of Colebrook's (1977) taxonomic categories might reveal a distinct set of species representing north–south and onshore–offshore communities. Different species of euphausiids show such a pattern. Brinton (1962) reported that euphausiid distributions in the SCB show higher percentages of subarctic species to the north and higher percentages of equatorial species in the south. Transition species generally show patterns similar to subarctic species.

O'Connell (1971) studied zooplankton horizontal distributions in the SCB in a more limited manner, sampling in the autumn of 1961 and 1962 from a depth of 5 m with a towed pump system. Of the four taxonomic groups he considered, only chaetognaths showed a systematic decrease in abundance with an increase in distance from shore. Euphausiids and large and small copepods did not show this trend, but did sometimes show an increase in areas of low water temperature.

Fleminger (1967) grouped CalCOFI-collected copepods into species groups that fell into major biogeographic–habitat groups. The copepod fauna of the SCB represented combinations of transitional, subarctic, equatorial, and central species, with an absence of sharply zoned assemblages (Fleminger 1967).

Fenaux and Dallot (1980) have shown that for the southern California and Baja California coasts, the appendicularians can also be grouped by species corresponding approximately to the major water masses of the central and equatorial Pacific. Other taxa have also been reduced to species distributions for various years of the CalCOFI sampling; these include chaetognaths (fig. 6.12a) (Alvariño 1965), pelagic molluscs (McGowan 1967), thaliaceans (fig. 6.12b) (Berner 1967), and euphausiids (Brinton 1967, 1973; Brinton and Wyllie 1976).

VERTICAL DISTRIBUTION

The dynamics of vertical structure, or vertical migration, of zooplankton have interested biological oceanographers for decades. The pattern of zooplankton ascent at dusk and descent at dawn is well known. A persistent concern, however, is the amount of net avoidance among the more mobile macrozooplankton during the daylight hours near the surface. If this occurs, it could be perceived as a relative change in zooplankton concentrations between the surface and depth from day to night. Despite this possible source of ambiguity, a relative increase in surface concentration at night is generally considered as indicative of vertical migration.

The vertical distribution of zooplankton biomass was shown by Brooks and Mullin (1983) for 12 daytime and 12 nighttime samples in April 1965 in the SCB (fig. 6.10b). For collections from nets of two mesh sizes, the upper 56 m contained the most zooplankton. With the effects of vertical migration, the upper layers of the water column supported an even larger percentage of the zooplankton. Figure 6.13 shows the vertical distribution of selected species during the day and night. All species migrated into the upper 56 m from deeper depths, but a significant portion of *Eucalanus bungii* females remained below 169 m at night. For all species except *Euphausia pacifica,* no difference was seen between day and night in the number of animals in the water column. It is thought that *E. pacifica* probably migrated up from below 225 m, since Pieper (1979) has shown using acoustical data that *E. pacifica* has an upper depth distribution of 130–280 m during the day in the SCB.

Longhurst et al. (1966) reported on the daytime vertical distribution of a number of zooplankton taxa 10 miles offshore of La Jolla. *Acartia tonsa* was most abundant in the upper 50 m, *Calanus helgolandicus* (= *pacificus*) ranged between 50 and 120 m, and both *Eucalanus bungii* and *Pleuromamma borealis* were most abundant between 200 and 300 m. *Rhincalanus nasutus* was widely scattered throughout the water column, but was most abundant above 150 m.

The carnivorous copepods *Labidocera trispinosa* and *L. jollae* have vertical distributions very near the surface (Barnett 1974, as cited by Beers 1986). *L. jollae* copepodite stages through the adult stage were almost exclusively neustonic. The more common *L. trispinosa* showed some depth variation with developmental stages. The eggs were centered between 15 and 20 m and after hatching, the nonfeeding nauplius I and II descended to depths that were seldom below 30 m. Nauplius II to copepodite III were primarily neustonic, copepodite IV and V were centered between 5 and 15 m, and the adults had a center of abundance from 15 to 20 m. No strong diel vertical migration was found for this species.

The daytime distribution of the copepod *Metridia pacifica* was almost entirely below 150 m; the species migrated up to the surface in two stages (Enright 1977). With approaching dusk, they migrated into the depth range just below the thermocline (18–25 m). This depth was suggested to be a "staging area" for entry into the surface water as darkness falls. Vertical migration occurred at the remarkable rate of 30–90 m h^{-1} for durations of more than 1 hour.

A description of the vertical distributions and diel vertical migrations of the developmental stages of *Calanus pacificus* was assembled by Mullin (1986), using data of Mullin (1979) for nauplier and early copepodite stages and of Brooks and Mullin (1983) for later

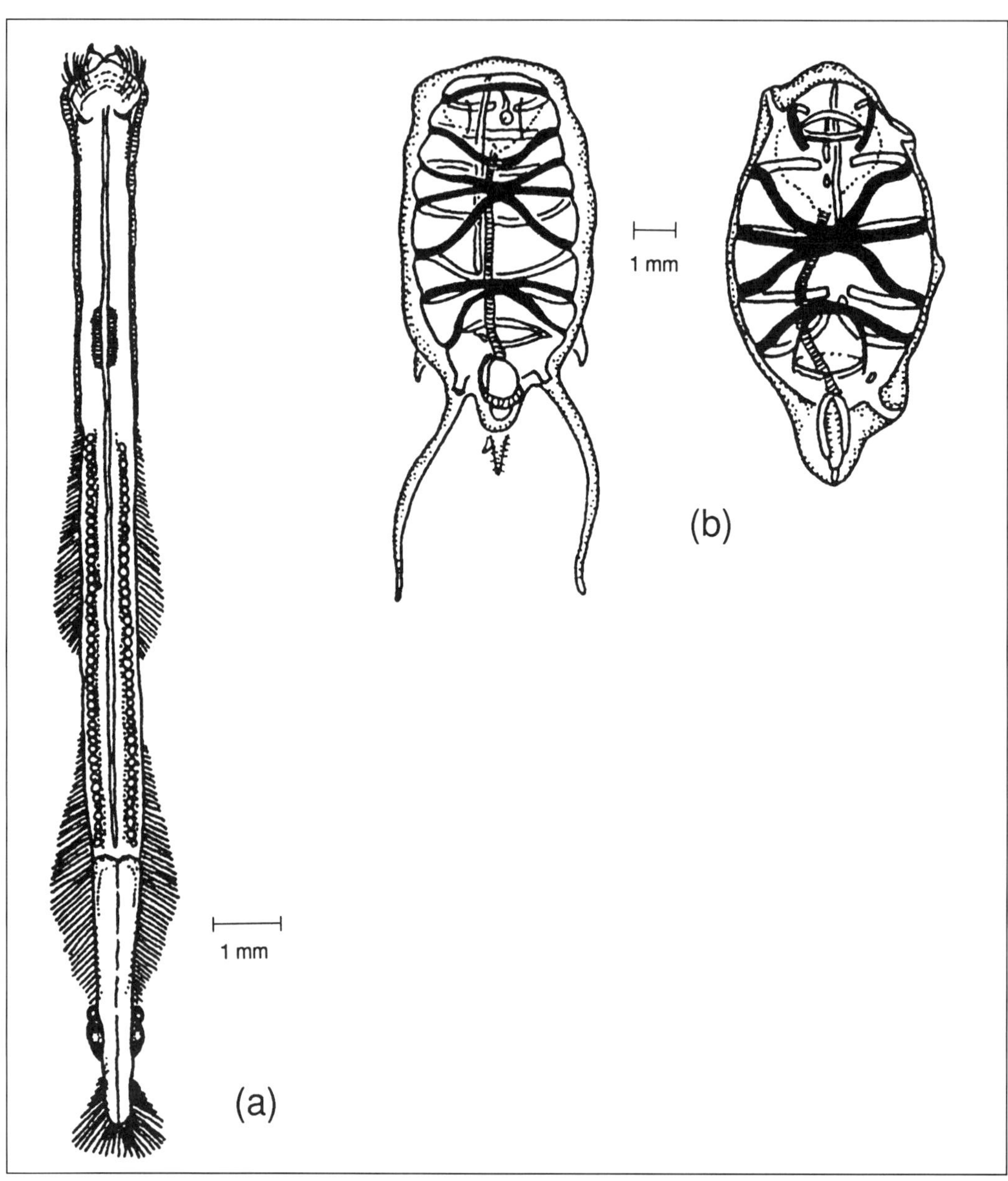

Figure 6.12. (a) The chaetognath *Sagitta euneritica*. (From Alvariño 1965.) (b) Two forms of *Thalia democratica*. (From Berner 1967.)

copepodite and adult stages (fig.6.14). Vertical migration behavior among the developmental stages changed ontogenetically. Similar day–night distributions were seen in nauplii through copepodite IV. Copepodite V, adult males and females showed clear vertical migration. Huntley and Brooks (1982) also demonstrated similar ontogenetic vertical migration patterns for *C. pacificus* reared in the SIO deep tank. We investigated the vertical distribution of *C. pacificus* developmental stages over a 24-hour period in the San Pedro Basin (March 1987) while following a 20-m window-shade drogue. We also observed that the vertical migration pattern changed ontogenetically, but with some major differences from the

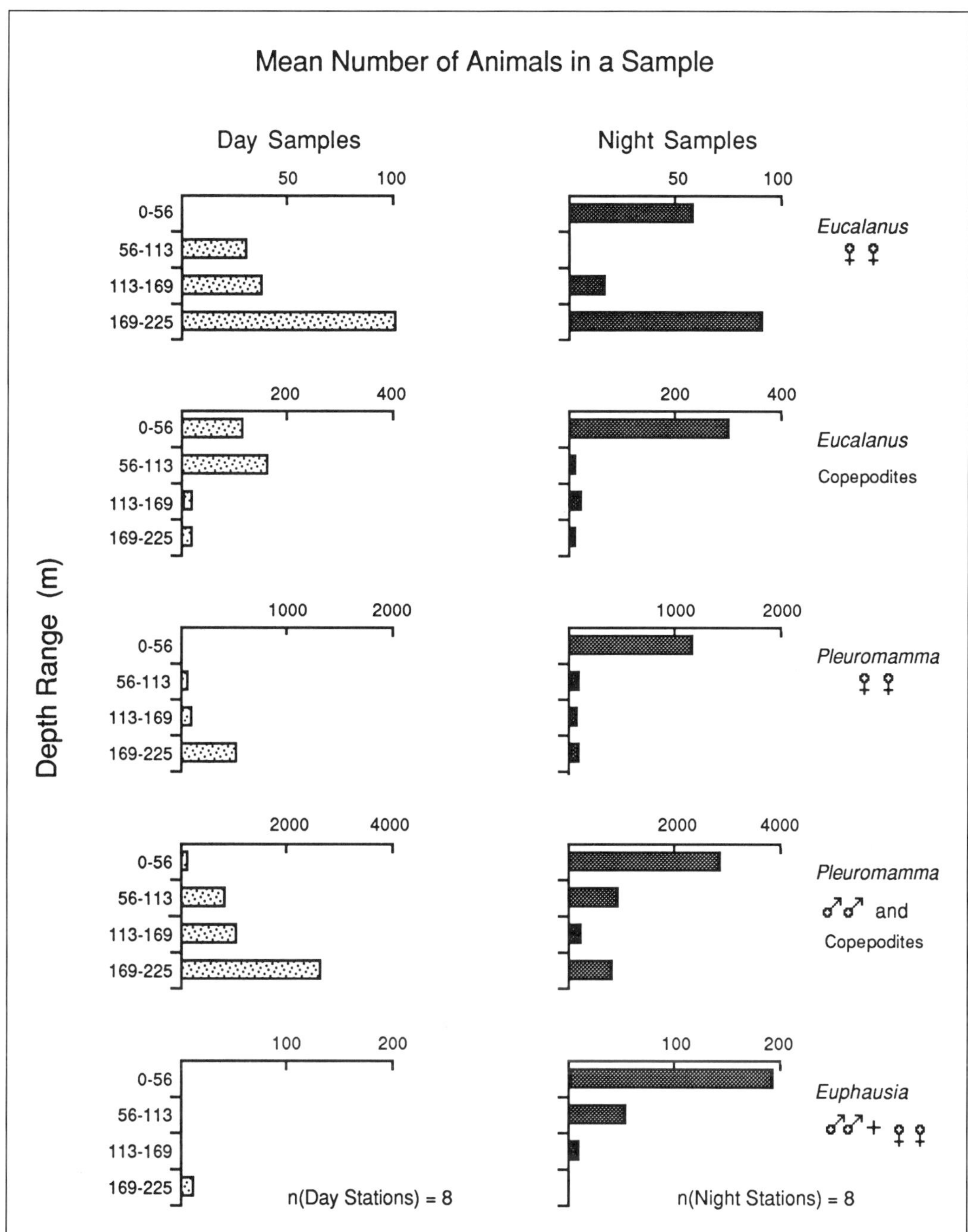

Figure 6.13. Day and night vertical distribution of selected species in April 1965. (From Brooks and Mullin 1983.)

previously mentioned studies. Figure 6.15 shows the developmental stages of *C. pacificus* and their respective day–night vertical distribution characteristics. As the copepods developed from nauplius through mature copepodite stages, the diel vertical migration pattern reversed itself. The nauplier and early copepodite stages showed higher surface concentra-

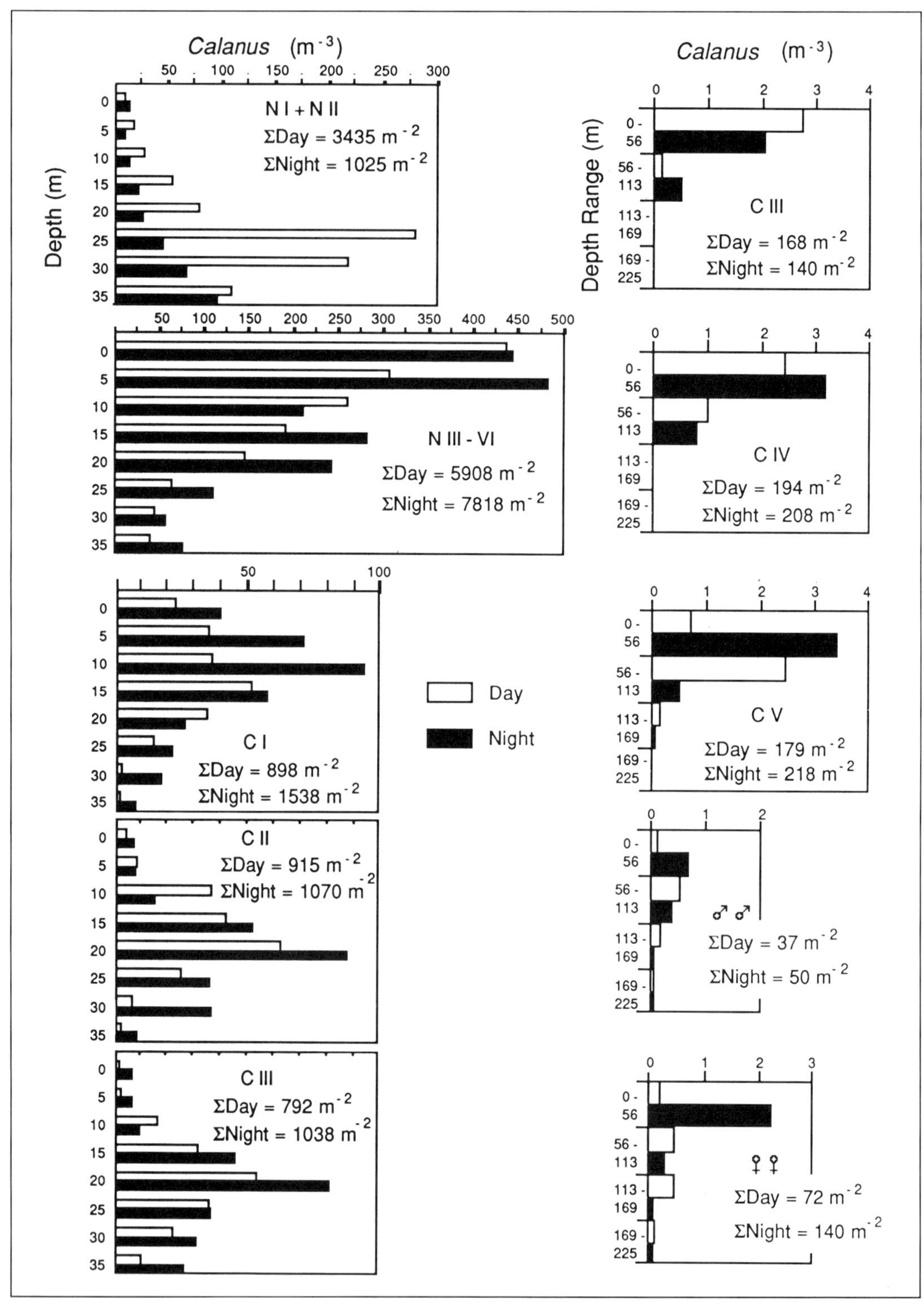

Figure 6.14. Day and night vertical distribution of developmental stages of *Calanus pacificus*. (From Mullin 1986.)

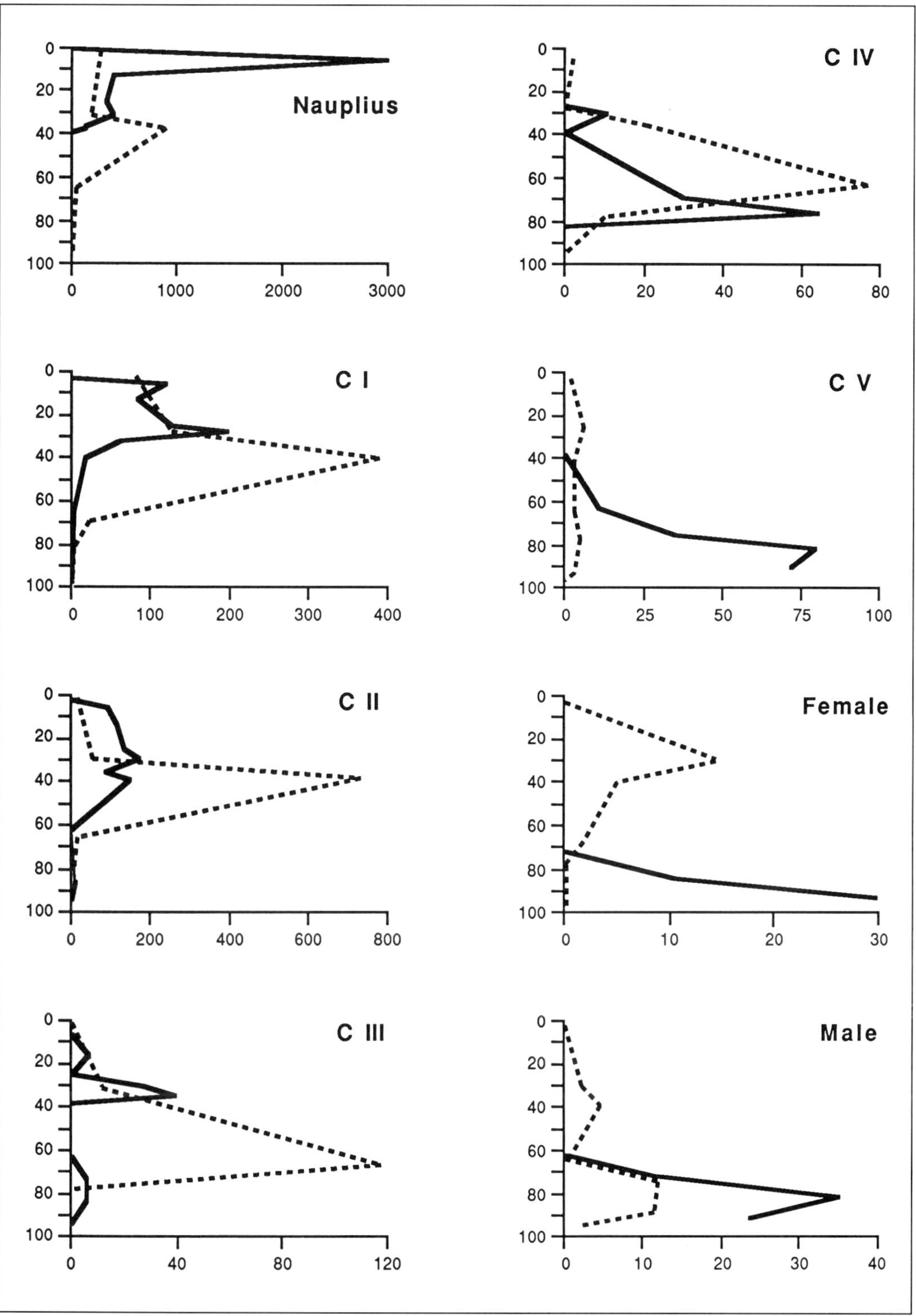

Figure 6.15. Vertical migration pattern of *Calanus pacificus* developmental stages, San Pedro Basin, March 1987. Concentration (number per cubic meter) versus depth. The solid lines are for daytime and the dashed lines for nighttime.

tions during the day and a maximum at depth at night. This changed between copepodite III and IV to a traditional diel pattern, that is, a deeper concentration during the day moving up to the surface at night for copepodite IV, copepodite V, and the adult males and females.

Our 1987 study may differ from those of Mullin (1979) and Brooks and Mullin (1983) for methodological reasons. If there is a reversal in vertical migration behavior between early and late developmental stages, as we found, these two studies might have missed it. Brooks and Mullin's (1983) sampling was too coarse to see the vertical migration patterns in the upper 56 m, and Mullin (1979) did not sample deep enough to detect the early stage migrations to below 35 m. Conversely, there is no apparent sampling problem in the deep tank data reported by Huntley and Brooks (1982). Also, sampling from our study was from depths above and below the drogue were advection could have transported in (or out) various "pulses" of organisms. Thus, the conflicting results between our work and that of Huntley and Brooks remain unresolved.

Further complicating the pattern, Pieper and Holliday (1984) reported that female *Calanus pacificus* from the San Pedro Basin had a trimodal vertical distribution in the upper 100 m during the daytime in collections from May 1978. This distribution appeared to be related to the reproductive phase of the female. At a near-surface peak (9 m), 73% of the females were gravid and their presence coincided with the peak distribution of male *C. pacificus.* At the mid-depth peak (36–40 m), 94% of the females were gravid and all of the nongravid females at this depth and above had attached spermatophores. In deeper water (80–100 m), 60% of the females were gravid and no spermatophores were attached to nongravid individuals. In addition, 90% of these females were relatively transparent, possibly indicating energetically spent individuals.

In addition to diel vertical migration, *C. pacificus* also displays a seasonal vertical distribution pattern that may be a survival strategy related to food availability. In the fall of 1982, Alldredge et al. (1984) observed copepodite V *C. pacificus* in a state of diapause in the Santa Barbara Basin. A 20-m-thick layer of these copepods was observed from a one-person submersible at a depth of 450 m. This nonmigrating layer occurred at an estimated concentration of up to 26×10^6 individuals per cubic meter just above the sill depth, below which conditions became anoxic. Numerous predators were observed feeding on the aggregation, including the deep sea smelt *Leuroglossus stilbius* and the physonectid siphonophore *Apolemia* sp. Alldredge et al. (1984) suggested that *C. pacificus* enters a diapause stage in response to the seasonal upwelling cycle along the California coast. *C. pacificus* might reside in the surface water during the seasonal upwelling period in winter and early spring. With the decline of food in the fall, they would enter this deep-water torpid phase, awaiting the following year's upwelling. In 1984, 2 years after Alldredge made her dives, B. H. Robison (pers. comm. 1984) made similar dives reconfirming the existence of this deep layer in the Santa Barbara Basin.

Investigations of deeper water zooplankton have largely centered around studies of sound-scattering layers in the SCB. Clarke (1966) studied the vertically migrating mesopelagic organisms in the Santa Barbara Channel in 1965. A 12-kHz fathometer was used for sound-scattering records, and a 6-ft Isaacs-Kidd mid-water trawl, outfitted with a photometer, thermistor, pressure sensor, and a flowmeter, was used to collect the organisms. Clarke reported on three migrating taxa that had their maximum concentration at a particular light intensity (isolume). The euphausiids were shallowest and showed day–night depth ranges of 150–200 m and 0–50 m, while sergestids had peak day–night densities at 225–325 m and 75–225 m. The bulk of the lantern fish, primarily *Lampanyctus* (= *Stenobrachius*) *leucopsarus,* were found below the sergestids, and they migrated with a deeper isolume.

Clarke (1966) found that the pasiphaeid shrimp showed no marked correlation with any particular light level; however, *Pasiphaea pacifica* appeared to migrate in a diurnal pattern. This species was most abundant in the day at depths greater than 400 m and at night from 225 to 275 m. *Pasiphaea emarginata* lives in such deep water that its vertical range was not sampled adequately.

Ebeling et al. (1970) investigated the pelagic communities of the shallow (0–150 m), upper mesopelagic (150–400 m), lower mesopelagic (400–600 m), and bathypelagic (> 600 m) depth zones over the Santa Barbara and Santa Cruz basins. They collected from 1964 to 1969, using—as did Clarke—an instrumented 6-ft Isaacs-Kidd mid-water trawl. Four resident community groups were identified from these areas: (I) mid-depth predators over the shallow (600 m) Santa Barbara Basin, (II) shallow invertebrate zooplankton equally abundant over both basins, (III) mesopelagic offshore mid-depth community, and (IV) bathypelagic predators over the deeper (2000 m) Santa Cruz Basin. These resident communities also had dispersed within them transient groups, usually seasonally abundant, that were advected from the north, south, and west of the SCB.

Group I was largely composed of fish in a community of mid-depth predators over the Santa Barbara Basin. This community was dominated by the deep-sea smelt *Leuroglossus stilbius,* the lantern fish *Lampanyctus leucopsarus,* and the pasiphaeid shrimp *Pasiphaea pacifica* and *P. emarginata.* Also common were the amphipods *Hyperia* spp. and *Paracallisoma coesus* and the medusae *Tiaropsidium kelseyi* and *Aegina citrea.* Ebeling et al. (1970) believed that the larger predators in this group prey on the shallower communities, especially at night when they merge near the surface following vertical migration.

Group II was made up of various zooplankters that apparently flourished during periods of thermal stratification in the summer and fall. This group was composed of the euphausiids *Nematoscelis difficilis* and *Euphausia pacifica*; the amphipods *Paraphronema crassipes, Primno macropa,* and *Phronema sedentaria*; the shrimp *Sergestes similis*; and various species of chaetognaths.

Group III, or the offshore mid-depth community, was more diverse than its nearshore counterpart (Group I). Two decapods dominated this community: *Pasiphaea chacei* and *Gennadas propinquus.* Other important members included the medusae *Atolla wyvelli* and *Colobonema sericeum,* the amphipod *Scina* spp., the mysid *Gnathophausia ingens,* and the two fish *Idiacanthus antrostomus* and *Cyclothone signata.*

Group IV, or the offshore deep community of the Santa Cruz Basin, was characterized by species that largely avoided the Santa Barbara Basin. This community appeared relatively stable throughout the year, and dominant species were one or two orders of magnitude less abundant than the overlying offshore mid-depth community (Group III). This group was dominated by the deep-sea fish *Cyclothone acclinidens, Melanostigma pammelas, Scopelogadus mizolepis bispinosus,* and *Holtbyrnia melanocephala*; the shrimp *Hymenodora frontalis*; and the mysids *Eucopia* spp. and *Boreomysis* spp.

Ebeling et al. (1970) observed that sound scattering (12 kHz) showed strong upper and lower layers in both the Santa Barbara and Santa Cruz basins. The shallow invertebrate community (Group II), along with some transitory groups of animals, appeared to be associated with the sound-scattering layers. The mid-depth communities, however, tended to mix with the shallow invertebrates through vertical migration.

The benthopelagic (very near the bottom) zooplankton of the Santa Catalina Basin was sampled by Gowing and Wishner (1986) at 1 and 50 m off the bottom (1300 m) using a multiple-sampling, opening–closing net system operated from a submersible. Their samples averaged 85% copepods, and a gut analysis of calanoid copepods showed them to be predominantly detritivores (63%) and carnivores (23%). Bacteria, some of which were metal precipitating, were also found in the gut. Considering the relative high ingestion rates (5.4 μg C mg dry wt^{-1} h^{-1}) of these zooplankters (Wishner and Gowing 1987), these benthopelagic copepods may play a significant role in the distribution of metals in the near bottom water by ingesting the bacteria and capsules and delivering the metal capsules to the bottom via fecal pellets (Gowing and Wishner 1986).

TEMPORAL STRUCTURE

The seasonal distribution (monthly means) of zooplankton biomass from samples taken along CalCOFI lines 80 (offshore from Point Conception) and 90 in the SCB are shown in figure 6.10c. To show the effects of differential transport of the California Current, data averaged from 1955 and 1956 are plotted as indicators of cold-water years (high California Current transport), and data from 1958 and 1959 are used as indicators of warm-water years. The warm- and cold-water years were chosen from the data presented by Chelton (1981), and the zooplankton volume data were gathered from oblique tows at depths of 0–140 m (Smith 1971). Although the data indicate a similar seasonal trend, with biomass peaking in July, a dramatic difference in biomass between these two periods is also shown.

There seems little doubt that change in sea surface temperature is related to relative reduction in biomass of the zooplankton. However, a marked similarity is also shown in the pattern of year-to-year fluctuations in upwelling, as indicated by principal component analysis for the first seven months of each year and by fluctuations in the biomass of the zooplankton. This implies some causal relationship (Colebrook 1977). Upwelling not only brings nutrients to the surface but also affects temperature stratification of the water column. Thus, the considerable interannual variability in zooplankton biomass can be associated with variations in the strength of the California Current (Roesler and Chelton 1987) as well as with the intensity of coastal upwelling (Colebrook 1977).

FEEDING

Calanoid copepods have been the focus of most feeding studies because they are both important in the pelagic ecosystem and are amenable to laboratory culture. *Calanus pacificus* is the local species of copepod that has received the most attention in regard to feeding dynamics. The laboratory culture of this species has been reviewed by Beers (1986) and it is considered to be primarily herbivorous. Landry (1980), however, has shown that *C. pacificus* has the ability to switch between herbivory and carnivory. Kleppel et al. (1988) suggested that *C. pacificus* tends toward herbivory when algal biomass and production are high and toward carnivory when algal biomass and production are low. Other "herbivorous" copepod species, such as *Rhincalanus nasutus*, were shown by Mullin and Brooks (1967) to feed actively in the laboratory on newly hatched *Artemia* nauplii. While the carnivorous predation rates of *C. pacificus* are not as high as that of the carnivorous *Labidocera trispinosa*, Landry (1978) believed that, given the generally higher standing stock of *C. pacificus* in the SCB, its impact on nauplier standing stock may be significant. Paffenhöfer and Strickland (1970) found no evidence that *C. pacificus* would feed on natural detritus from the SCB, but they did determine that it would feed on freshly prepared detritus from phytoplankton as well as *Calanus* fecal material.

Barnett (1974) studied the dietary aspects of the copepod *Labidocera trispinosa* (reviewed by Beers 1986). He found that there was a dietary change from primary herbivory in late nauplier to early copepodite stages, to omnivory by late copepodite stage, and to mainly carnivory in the adults. Early food preferences included the diatom *Rhizosolenia* spp., naked dinoflagellates of 10–40 μm (for example, *Gyrodinium* sp.), the armored dinoflagellate *Prorocentrum micans*, and various ciliates. Late copepodites showed a preference for the naked dinoflagellate *Gymnodinium* spp., the armored dinoflagellate *P. micans, Ceratium* spp., and *Gonyaulax polyedra,* as well as for the zooplankton copepod nauplii and copepodites of small species, copepod eggs, and small cladocerans. Adult *L. trispinosa* showed a preference for all stages of various species of copepods and other small zooplankton, with the only phytoplankton in the diet being *P. micans*. Over the complete generation of *L. trispinosa,* phytoplankton were calculated to comprise less than 10% of the total carbon intake. Landry (1978) has reported that the ability of *L. trispinosa* to capture copepod prey decreases abruptly after the copepod develops beyond the nauplier stages, but that the capture rates are greatest for the largest nauplii. He concluded that the greatest impact of *L. trispinosa* on the natural zooplankton assemblage would be on the later nauplier stages of larger prey species. Since the larger nauplii in plankton samples from the SCB are usually those of *Calanus pacificus* (J. Dawson and R. Pieper pers. comm. 1990), feeding by *L. trispinosa* may have a large impact on this population.

Among the gelatinous zooplankton, Silver (1975) studied the feeding habitats of three common California Current salps, *Salpa fusiformis, Thalia democratica,* and *Cyclosalpa bakeri*—all of which are species known to swarm. She found that the species composition of phytoplankton ingested by swarm-forming *S. fusiformis* differed from those ingested by *C. bakeri,* implying that these species are in biologically different waters when they swarm. Since swarms can persist up to several months in duration, she concluded that the phytoplankton consumed by these swarming salps could pose a considerable resource depletion for the other zooplankton.

While swarming salps may have a depleting effect on the phytoplankton, Dunbar and Berger (1981) have shown that they may be of considerable advantage to the benthic community. They showed that in the spring of 1978 in the Santa Barbara Basin, what appeared to be salp fecal pellets accounted for more than 70% of the total organic carbon input to the benthic environment. Although this observation is an isolated one, zooplankton fecal pellet flux is an energetic input to the benthic community and is an area of active research at this time.

Summary and Prospectus for Future Research

Over the past 100 years of zooplankton research in the SCB, much of the sampling has been carried out over short time periods (the length of a cruise) and in limited geographic areas. These data have been invaluable in increasing our knowledge of distributions, behavior, physiology, and other aspects of zooplankton ecology. Only the 32-year CalCOFI data set, however, is extensive enough to show the interrelationships between macrozooplankton biomass and eastern Pacific and global circulation patterns. These results have emphasized the large interannual fluctuations in biomass and the aperiodic nature of these variations. In fact, yearly fluctuations (see fig. 6.1) are much larger than the yearly mean (see fig. 6.10c).

Although it contains some unique species, the SCB is largely a transition zone between subarctic, central, and equatorial species. Thus, biomass fluctuations may also be accompanied by changes in species composition. The boundary (or clinal region) between cold, nutrient-rich California Current water (and its associated subarctic species) can vary in position relative to warmer, nutrient-poor water from the south (equatorial water) and west (central water).

Smaller scale processes are superimposed on these large-scale fluctuations. A specific process (such as coastal upwelling off the Palos Verdes Peninsula) will occur every year. However, its impact relative to zooplankton production and species composition would likely change in different years. Similarly, the impact of biological interactions, such as fish predation on zooplankton, would be expected to vary with variations in species composition, such as species of predators and their food preferences.

A number of marine zooplankton researchers met at Lake Arrowhead, California, in 1988 to discuss future directions in zooplankton research (Marine Zooplankton Colloquium 1, 1989). The following conclusions from this colloquium are particularly important to the understanding of zooplankton ecology in the SCB. Similarly, they should be emphasized in planning future research programs. The following statements and their order were selected by the present authors.

1. We need to study individual zooplankton relative to their ambit, rather than according to sampling convenience, to understand how they respond in their environment.
2. We need to better understand how environmental variance, rather than mean conditions, affects zooplankton ecology, physiology, and behavior.
3. When possible, we need to conduct these studies *in situ* and use noninvasive and nondisturbing approaches.
4. Most of our scientific approaches so far have been geared toward intermittent observations on individuals, populations, and communities.
5. We need to strive for either continuous observations or observations on time or space scales that are sufficiently frequent for the processes to be studied.
6. Without an understanding of physical, chemical, and numerous biological processes, an understanding of individual performances, population dynamics, and community function of zooplankton cannot be achieved.

Conversely, modelers would like to know average conditions and basic patterns for predictive systems models. Also, organizations interested in environmental impacts would like to determine baselines of many parameters to enable them to assess human impact on the system. The 32-year CalCOFI data set appears to fulfill the requirements of zooplankton ecologists, modelers, and environmental managers for zooplankton biomass on interannual scales. Much of the data discussed in this chapter, while extremely valuable in furthering our knowledge of zooplankton ecology, does not satisfy the present recommendations or needs of these groups. Zooplankton biologists suggest a need for studies more directly applicable to the ambit of a particular organism (smaller scale studies). Conversely, the oceanographic variability (including both currents and physical morphology) of the SCB makes it difficult to obtain a baseline to determine human impact when the variance of such a baseline is so high.

Our challenge for the future is to find ways to provide data for all potential interests. Zooplankton biologists have a primary interest in understanding population and community structure and function. The species is the basis of such understanding. Similarly, we suggest that baselines of a particular species or group of species might eventually provide the necessary background to determine potential environmental impacts.

The shelf break and slope region appears to be important biologically and is also understudied. Since this area is both a boundary region and a mixing zone, it will be most important to obtain concurrent data on as many parameters as possible (both physical and biological). Studies in this area might provide important data on food aggregations for fishes (chap. 9), marine birds (chap. 10), and marine mammals (chap. 11).

Literature Cited

Allan Hancock Foundation, 1976. Environmental investigations and analyses, Los Angeles–Long Beach harbors 1973–1976. Final Report to U. S. Army Corps of Engineers, L. A. District. Harbors Environmental Projects. Univ. of Southern California, Los Angeles, CA. 737pp.

Alldredge, A. L., B. H. Robison, A. Fleminger, J. J. Torres, J. M. King, and W. M. Hamner, 1984. Direct sampling and *in situ* observation of a persistent copepod aggregation in the mesopelagic zone of the Santa Barbara Basin. *Mar. Biol.* 80:75–81.

Alvariño, A., 1965. Distributional atlas of Chaetognatha in the California Current region. *Calif. Coop. Oceanic Fish. Invest. Atlas No.3,* 291pp.

Azam, F., T. Fenchel, J. G. Field, J. S. Gray, L. A. Meyer-Reil, and F. Thingstad, 1983. The ecological role of water column microbes in the sea. *Mar. Ecol. Prog. Ser.* 10:257–263.

Barnett, A. M., 1974. The feeding ecology of an omnivorous neritic copepod, *Labidocera trispinosa* (Esterly). Ph.D. Dissertation, Univ. of California, San Diego. 232pp.

Barnett, A. M., and A. E. Jahn, 1987. Pattern and persistence of a nearshore planktonic ecosystem off southern California. *Cont. Shelf Res.* 7:1–25.

Beers, J. R., 1986. Organisms and the food web. In: R. W. Eppley, ed. *Lecture Notes on Coastal and Estuarine Studies, Vol. 15. Plankton Dynamics of the Southern California Bight.* Springer-Verlag, Berlin. pp. 84–175.

Beers, J. R., and G. L. Stewart, 1967. Microzooplankton in the euphotic zone at five locations across the California Current. *J. Fish. Res. Board Can.* 24:2053–2068.

Beers, J. R., and G. L. Stewart, 1969a. The vertical distribution of microzooplankton and some ecological observations. *J. Cons. Int. Explor. Mer.* 33:30–44.

Beers, J. R., and G. L. Stewart, 1969b. Microzooplankton and its abundance relative to the larger zooplankton and other seston components. *Mar. Biol.* 4:182–189.

Beers, J. R., and G. L. Stewart, 1970. Part VI. Numerical abundance and estimated biomass of microzooplankton. In: J. D. H. Strickland, ed. *The Ecology of the Plankton off La Jolla, California, in the Period April Through September 1967.* Bull. Scripps Inst. Oceanogr., Univ. Calif. 17:67–87.

Beers, J. R., F. M. H. Reid, and G. L. Stewart, 1980. Microplankton population structure in southern California nearshore waters in late spring. *Mar. Biol.* 60:209–226.

Bernal, P. A., 1980. Large-scale biological events in the California Current: The low frequency response of the epipelagic ecosystem. Ph.D. Dissertation, Univ. of California, San Diego. 184pp.

Bernal, P. A., 1981. A review of the low-frequency response of the pelagic ecosystem in the California Current. *Calif. Coop. Oceanic Fish. Invest. Rep.* 22:49–62.

Bernal, P. A., and J. A. McGowan, 1981. Advection and upwelling in the California Current. In: F. A. Richards, ed. *Coastal Upwelling.* American Geophysical Union, Washington, D.C. pp. 381–399.

Berner, L. D., 1967. Distributional atlas of *Thaliacea* in the California Current region. *Calif. Coop. Oceanic Fish. Invest. Atlas No. 8,* 322pp.

Bowman, T. E., 1971. The distribution of calanoid copepods off the southeastern United States between Cape Hatteras and southern Florida. *Smithson. Contrib. Zool. No. 96,* 58pp.

Boyd, C. M., 1963. Distribution, trophic relationships, growth and respiration of a marine decapod crustacean *Pleuroncodes planipes* Stimpson (Galatheidae). Ph.D. Dissertation, Univ. of California, San Diego. 116pp.

Brewer, G. D., and G. S. Kleppel, 1986. Diel vertical distribution of fish larvae and their prey in nearshore waters of southern California. *Mar. Ecol. Prog. Ser.* 27:217–226.

Brewer, G. D., and P. E. Smith, 1982. Northern anchovy and Pacific sardine spawning off southern California during 1978–1980: Preliminary observations on the importance of the nearshore coastal region. *Calif. Coop. Oceanic Fish. Invest. Rep.* 23:160–171.

Brewer, G. D., G. S. Kleppel, and M. Dempsey, 1984. Apparent predation on ichthyoplankton by zooplankton and fishes in nearshore waters of southern California. *Mar. Biol.* 80:17–28.

Brewer, G. D., R. J. Lavenberg, and G. E. McGowen, 1981. Abundance and vertical distribution of fish eggs and larvae in the Southern California Bight: June and October 1978. *Rapp. P.-V. Réun. Cons. Int. Explor. Mer.* 178:165–167.

Brinton, E., 1962. The distribution of Pacific euphausiids. *Bull. Scripps Inst. Oceanogr., Univ. Calif.* 8:51–270.

Brinton, E., 1967. Distributional atlas of Euphausiacea (Crustacea) in the California Current region, Part I. *Calif. Coop. Oceanic Fish. Invest. Atlas No. 5.,* 275pp.

Brinton, E., 1973. Distributional atlas of Euphausiacea (Crustacea) in the California Current region, Part II. *Calif. Coop. Oceanic Fish. Invest. Atlas No. 18,* 336pp.

Brinton, E., 1976. Population biology of *Euphausia pacifica* off southern California. *U.S. Fish. Wildl. Serv. Fish. Bull.* 74:733–762.

Brinton, E., and J. G. Wyllie, 1976. Distributional atlas of euphausiid growth stages off southern California, 1953 through 1956. *Calif. Coop. Oceanic Fish. Invest. Atlas No. 24,* 289pp.

Brooks, E. R., and M. M. Mullin, 1983. Diel changes in the vertical distribution of biomass and species in the Southern California Bight. *Calif. Coop. Oceanic Fish. Invest. Rep.* 24:210–215.

Casey, R. E., 1966. A seasonal study on the distribution of polycystine radiolarians from waters overlying the Catalina Basin, southern California. Ph.D. Dissertation, Univ. of Southern California, Los Angeles.

Checkley, D. M., Jr., 1980. Food limitation of egg production by a marine planktonic copepod in the sea off southern California. *Limnol. Oceanogr.* 25(6):991–998.

Chelton, D. B., 1981. Interannual variability of the California Current—Physical factors. *Calif. Coop. Oceanic Fish. Invest. Rep.* 22:34–48.

Chelton, D. B., P. A. Bernal, and J. A. McGowan, 1982. Large-scale interannual physical and biological interaction in the California Current. *J. Mar. Res.* 40(4):1095–1125.

Clarke, W. D., 1966. Bathyphotometric studies of the light regime of organisms of the deep scattering layers. *General Motors Corp., Sea Op. Div. Tech. Rep. 66–02.* Santa Barbara, CA. 47pp.

Clarke, W. D., 1971. Mysids of the southern kelp region. In: W. J. North, ed. *The Biology of Giant Kelp Beds* (Macrocystis) *in California.* Beihefte, Nova Hedwigia, Heft. 32:369–380.

Clutter, R. I., 1967. Zonation of nearshore mysids. *Ecology.* 48:200–208.

Colebrook, J. M., 1977. Annual fluctuations in biomass of taxonomic groups of zooplankton in the California Current, 1955–59. *U.S. Fish. Wildl. Serv. Fish. Bull.* 75:357–368.

Cushing, D. H., 1975. *Marine Ecology and Fisheries.* Cambridge Univ. Press, Cambridge, U.K. 278pp.

Dawson, J. K., 1979. The recognition and occurrence of *Acartia californiensis* and *Acartia tonsa* in the Los Angeles–Long Beach harbors. *Harbors Environmental Projects, Univ. South. Calif. Spec. Publ.* May 28, 1979.

Dunbar, R. B., and W. H. Berger, 1981. Fecal pellet flux to modern bottom sediment of Santa Barbara Basin (California) based on sediment trapping. *Geol. Soc. Am. Bull.* 92:212–218.

Ebeling, A. W., G. M. Cailliet, R. M. Ibara, F. A. DeWitt, Jr., and D. W. Brown, 1970. Pelagic communities and sound scattering off Santa Barbara, California. In: G. B. Farquhar, ed. *Proc. Int. Symposium on Biological Sound Scattering in the Ocean.* Maury Center for Ocean Science, Washington, D.C. pp. 1–19.

Emery, K. O., 1960. *The Sea Off Southern California.* John Wiley & Sons, New York. 366pp.

Enright, J. T., 1977. Copepods in a hurry: Sustained high-speed upward migration. *Limnol. Oceanogr.* 22:118–125.

Environmental Quality Analysts and Marine Biological Consultants, 1976. Marine monitoring studies, Long Beach generating station—1975 annual report. Prep. for Southern California Edison Company. Costa Mesa, CA. 410pp.

Environmental Quality Analysts and Marine Biological Consultants, 1978. Final report 1974–1978. Marine monitoring studies, Long Beach generating station. Prep. for Southern California Edison Company. Pasadena, CA. 547pp.

Eppley, R. W., ed., 1986. *Lecture Notes on Coastal and Estuarine Studies, Vol. 15. Plankton Dynamics of the Southern California Bight.* Springer-Verlag, Berlin. 373pp.

Eppley, R. W., E. H. Renger, and W. G. Harrison, 1979. Nitrate and phytoplankton production in Southern California waters. *Limnol. Oceanogr.* 24(3):483–494.

Fenaux, R., and S. Dallot, 1980. Répartition des Appendiculaires au large des côtes de Californie. *J. Plankton Res.* 2:145–167.

Fleminger, A., 1967. Distributional atlas of calanoid copepods in the California Current region, Part II. *Calif. Coop. Oceanic Fish. Invest. Atlas No. 7,* 213pp.

Fleminger, A., J. D. Isaacs, and J. G. Wyllie, 1974. Zooplankton biomass measurements from CalCOFI cruises of July 1955 to 1959 and remarks on comparison with results from October, January and April cruises of 1955 to 1959. *Calif. Coop. Oceanic Fish. Invest. Atlas No. 21,* 118pp.

Glynn, P. W., 1961. The first recorded mass stranding of pelagic red crabs, *Pleuroncodes planipes,* at Monterey Bay, California, since 1859, with notes on their biology. *Calif. Fish Game.* 47:97–101.

Gowing, M. M., and K. F. Wishner, 1986. Trophic relationships of deep-sea calanoid copepods from the benthic boundary layer of the Santa Catalina Basin, California. *Deep-Sea Res.* 33:939–962.

Grice, G. D., and A. D. Hart, 1962. The abundance, seasonal occurrence and distribution of the epizooplankton between New York and Bermuda. *Ecological Monographs.* 32:287–309.

Hammer, R. M., 1980. Ecology of kelp forest demersal zooplankton. Ph.D. Dissertation, Univ. of Southern California, Los Angeles. 134pp.

Hammer, R. M., 1981. Day–night differences in the emergence of demersal zooplankton from a sand substrate in a kelp forest. *Mar. Biol.* 62(4):275–280.

Haury, L. R., and R. E. Pieper, 1988. Zooplankton: Scales of biological and physical events. In: D. F. Soule and G. S. Kleppel, eds. *Marine Organisms as Indicators.* Springer-Verlag, New York. pp. 35–72.

Heinbokel, J. F., and J. R. Beers, 1979. Studies on the functional role of tintinnids in the Southern California Bight. III. Grazing impact of natural assemblages. *Mar. Biol.* 52(1):23–32.

Hirota, J., 1974. Quantitative natural history of *Pleurobrachia bachei* in La Jolla Bight. *U.S. Fish Wildl. Serv. Fish. Bull.* 72:295–335.

Holliday, D. V., and R. E. Pieper, 1980. Volume scattering strengths and zooplankton distributions at acoustic frequencies between 0.5 and 3 MHz. *J. Acoust. Soc. Am.* 67:135–146.

Huntley, M. E., and E. R. Brooks, 1982. Effects of age and food availability on diel vertical migration of *Calanus pacificus. Mar. Biol.* 71:23–31.

Isaacs, J. D., A. Fleminger, and J. K. Miller, 1969. Distributional atlas of zooplankton biomass in the California Current region: Spring and fall 1955–1959. *Calif. Coop. Oceanic Fish. Invest. Atlas No. 10,* 252pp.

Isaacs, J. D., A. Fleminger, and J. K. Miller, 1971. Distributional atlas of zooplankton biomass in the California Current region: Winter 1955–1959. *Calif. Coop. Oceanic Fish. Invest. Atlas No. 14,* 122 pp.

Kimor, B., 1979. Predation by *Noctiluca miliaris* Souriray on *Acartia tonsa* Dana eggs in the inshore waters of southern California. *Limnol. Oceanogr.* 24:568–572.

Kleppel, G. S., D. Frazel, R. E. Pieper, and D. V. Holliday, 1988. Natural diets of zooplankton off southern California. *Mar. Ecol. Prog. Ser.* 49:231–241.

Landry, M. R., 1978. Predatory feeding behavior of a marine copepod, *Labidocera trispinosa. Limnol. Oceanogr.* 23:1103–1113.

Landry, M. R., 1980. Detection of prey by *Calanus pacificus:* Implications of the first antennae. *Limnol. Oceanogr.* 25:545–549.

Longhurst, A. R., 1967. The pelagic phase of *Pleuroncodes planipes* Stimpson (Crustacea, Galatheida) in the California Current. *Calif. Coop. Oceanic Fish. Invest. Rep.* 11:142–154.

Longhurst, A. R., C. J. Lorenzen, and W. H. Thomas, 1967. The role of pelagic crabs in the grazing of phytoplankton off Baja California. *Ecology.* 48:190–200.

Longhurst, A. R., A. D. Reith, R. E. Bower, and D. L. R. Seibert, 1966. A new system for the collection of multiple serial plankton samples. *Deep-Sea Res.* 13:213–222.

Marine Zooplankton Colloquium 1, 1989. Future marine zooplankton research—A perspective. *Mar. Ecol. Prog. Ser.* 55:197–206.

McGowan, J. A., 1967. Distributional atlas of pelagic molluscs in the California Current region. *Calif. Coop. Oceanic Fish. Invest. Atlas No. 6,* 218pp.

McGowan, J. A., 1974. The nature of oceanic ecosystems. In: C. B. Miller, ed. *The Biology of the Oceanic Pacific.* Oregon State Univ. Press, Corvallis, OR. pp. 9–28.

McGowan, J. A., and C. B. Miller, 1980. Larval fish and zooplankton community structure. *Calif. Coop. Oceanic Fish. Invest. Rep.* 21:29–36.

Mullin, M. M., 1979. Longshore variations in the distribution of plankton in the Southern California Bight. *Calif. Coop. Oceanic Fish. Invest. Rep.* 20:120–124.

Mullin, M. M., 1986. Spatial and temporal scales and patterns. In: R. W. Eppley, ed. *Lecture Notes on Coastal and Estuarine Studies, Vol. 15. Plankton Dynamics of the Southern California Bight.* Springer-Verlag, Berlin. pp. 216–273.

Mullin M. M., and E. R. Brooks, 1967. Laboratory culture, growth rate, and feeding behavior of a planktonic marine copepod. *Limnol. Oceanogr.* 12:657–666.

Mullin, M. M., and E. R. Brooks, 1970. Part VII. Production of the planktonic copepod, *Calanus helgolandicus.* In: J. D. H. Strickland, ed. *The Ecology of the Plankton off La Jolla, California, in the Period April Through September 1967.* Bull. Scripps Inst. Oceanogr., Univ. Calif. 17:89–103.

O'Connell, C. P., 1971. Variability of near-surface zooplankton off southern California, as shown by towed-pump sampling. *U.S. Fish. Wildl. Serv. Fish. Bull.* 69:681–698.

Owen, R. W., Jr., 1981. Microscale plankton patchiness in the larval anchovy environment. *Rapp. P.-V. Réun. Cons. Int. Explor. Mer.* 178:364–368.

Paffenhöfer, G.-A., 1970. Cultivation of *Calanus helgolandicus* under controlled conditions. *Helgol. Wiss. Meeresunters.* 20:346–359.

Paffenhöfer, G.-A., and J. D. H. Strickland, 1970. A note on the feeding of *Calanus helgolandicus* on detritus. *Mar. Biol.* 5:97–99.

Peterson, J. H., A. E. Jahn, R. J. Lavenberg, G. E. McGowen, and R. S. Grove, 1986. Physical-chemical characteristics and zooplankton biomass on the continental shelf off southern California. *Calif. Coop. Oceanic Fish. Invest. Rep.* 27:36–52.

Peterson, W. T., and C. B. Miller, 1975. Year to year variations in the planktology of the Oregon upwelling zone. *U.S. Natl. Mar. Fish. Serv. Fish. Bull.* 73:642–653.

Peterson, W. T., and C. B. Miller, 1977. Seasonal cycle of zooplankton abundance and species composition along the central Oregon coast. *U.S. Natl. Mar. Fish. Serv. Fish. Bull.* 75:717–724.

Peterson, W. T., C. B. Miller, and A. Hutchinson, 1979. Zonation and maintenance of copepod populations in the Oregon upwelling zone. *Deep-Sea Res.* 26:467–494.

Pieper, R. E., 1979. Euphausiid distribution and biomass determined acoustically at 102 kHz. *Deep-Sea Res.* 26:687–702.

Pieper, R. E., and D. V. Holliday, 1984. Acoustic measurements of zooplankton distributions in the sea. *J. Cons. Int. Explor. Mer.* 41:226–238.

Pieper, R. E., D. V. Holliday, and G. S. Kleppel, 1990. Quantitative zooplankton distributions from multifrequency acoustics. *J. Plankton Res.* 12:433–441.

Reid, J. L., Jr., 1962. On circulation, phosphate–phosphorus content and zooplankton volumes in the upper part of the Pacific Ocean. *Limnol. Oceanogr.* 7:287–306.

Richter, K. E., 1985. Acoustic determination of small-scale distributions of individual zooplankters and zooplankton aggregations. *Deep-Sea Res.* 32:163–182.

Roesler, C. S., and D. B. Chelton, 1987. Zooplankton variability in the California Current, 1951–1982. *Calif. Coop. Oceanic Fish. Invest. Rep.* 28:59–96.

Ryther, J. H., 1969. Photosynthesis and fish production in the sea. *Science.* 166:72–76.

Silver, M. W., 1975. The habitat of *Salpa fusiformis* in the California Current as defined by indicator assemblages. *Limnol. Oceanogr.* 20:230–237.

Smith, P. E., 1971. Distributional atlas of zooplankton volume in the California Current region, 1951 through 1966. *Calif. Coop. Oceanic Fish. Invest. Atlas No. 13,* 144pp.

Smith, R. L., 1968. Upwelling. *Oceanogr. Mar. Biol. Annu. Rev.* 6:11–46.

Soule, D. F., and M. Oguri, eds., 1977. Marine studies of San Pedro Bay, California. Part 13. The marine ecology of Marina del Rey Harbor, California. A baseline survey for the County of Los Angeles Dept. Small Craft Harbors 1976–1977. Harbors Environmental Projects. Univ. of Southern California, Los Angeles. 424pp.

Soule, D. F., and M. Oguri, eds., 1979. Marine studies of San Pedro Bay, California. Part 16. Ecological changes in outer Los Angeles–Long Beach harbors following initiation of secondary waste treatment and cessation of fish cannery waste effluent. Harbors Environmental Projects. Univ. of Southern California, Los Angeles. 597pp.

Star, J. L., 1980. Variation and covariation in the plankton of the euphotic zone in the north Pacific Ocean. Ph.D. Dissertation, Univ. of California, San Diego. 141pp.

Star, J. L., and M. M. Mullin, 1981. Zooplanktonic assemblages in three areas of the North Pacific as revealed by continuous horizontal transects. *Deep-Sea Res.* 28(11):1303–1322.

Steele, J. H., ed., 1978. *Spatial Pattern in Plankton Communities.* Plenum Press, New York. 470pp.

Stretch, J. J., 1985. Quantitative sampling of demersal zooplankton: Reentry and airlift dredge sample comparisons. *J. Exp. Mar. Biol. Ecol.* 91:125–136.

Strickland, J. D. H., L. Solorzano, and R. W. Eppley, 1970. Part I. General introduction, hydrography, and chemistry. In: J. D. H. Strickland, ed. *The Ecology of the Plankton off La Jolla, California, in the Period April through September 1967.* Bull. Scripps Inst. Oceanogr., Univ. of California. 17:1–22.

Sverdrup, H. U., M. W. Johnson, and R. H. Fleming, 1942. *The Oceans, Their Physics, Chemistry, and General Biology*. Prentice-Hall, New York. 1087pp.

Torrey, H. B., 1902. An unusual occurrence of dinoflagellates on the California coast. *Am. Nat.* 36:187–192.

Trinast, E. M., 1975. Tidal currents and *Acartia* distribution in Newport Bay, California. *Estuarine Coastal Mar. Sci.* 3:165–176.

Trinast, E. M., 1976. Preliminary note on *Acartia californiensis*, a new calanoid copepod from Newport Bay, California. *Crustaceana*. 31:54–58.

Trujillo-Ortíz, A., 1986. Life cycle of the marine calanoid copepod *Acartia californiensis* Trinast reared under laboratory conditions. *Calif. Coop. Oceanic Fish. Invest. Rep.* 27:188–204.

UNESCO, 1968. Zooplankton sampling. *Monographs on Oceanographic Methodology (2)*. The UNESCO Press, Paris. 174pp.

UNESCO, 1976. Zooplankton fixation and preservation. *Monographs on Oceanographic Methodology (4)*. The UNESCO Press, Paris. 350pp.

Uye, S., and A. Fleminger, 1976. Effects of various environmental factors on egg development of several species of *Acartia* in southern California. *Mar. Biol.* 38:253–262.

Walsh, J. J., 1977. A biological sketchbook for an eastern boundary current. In: E. D. Goldberg, I. N. McCave, J. J. O'Brien, and J. H. Steel, eds. *The Sea, Volume 6*. Interscience, New York. pp. 923–968.

Wishner, K. F., and M. M. Gowing, 1987. *In situ* filtering and ingestion rates of deep-sea benthic boundary-layer zooplankton in the Santa Catalina Basin. *Mar. Biol.* 94:357–366.

Wroblewski, J. S., 1980. A simulation of the distribution of *Acartia tonsa* during Oregon upwelling, August 1973. *J. Plankton Res.* 2:43–68.

Chapter 7

Benthic Macrophytes

Steven N. Murray and Richard N. Bray

Introduction

The Southern California Bight (SCB) is characterized by a rich flora of benthic macroalgae and seagrasses that occupies a diversity of coastal environments. Because of the predominantly east–west orientation of the northern mainland coast and the presence of the eight southern California Channel Islands (fig. 7.1), the SCB includes nearly 1000 km of coastline. The shallow coastal habitats of the SCB show considerable variation in wave action, ocean water masses, thermal regimes, and substrata. The large coastal area and the high degree of habitat heterogeneity contribute to the great diversity of macrophytes documented for the SCB (Abbott and Hollenberg 1976; Murray et al. 1980).

Knowledge concerning the macroalgae and seagrasses of the SCB has increased considerably in the past decade. Prior to the mid–1970s, relatively little published information was available (Murray 1974). However, due largely to the results of an extensive 3.8-year sampling program (1975–1979) sponsored by the U.S. Department of the Interior, quantitative descriptions of the seasonal distributions and abundances of intertidal macrophytes throughout the SCB are recorded in lengthy government reports (Littler 1977, 1978, 1979a) that have been summarized by Littler (1980a, b) and Littler et al. (1991). Most research on subtidal macrophytes in the SCB has been focused on kelp communities and the biology of *Macrocystis* (see reviews by Dayton and Tegner 1984a; Dayton 1985;

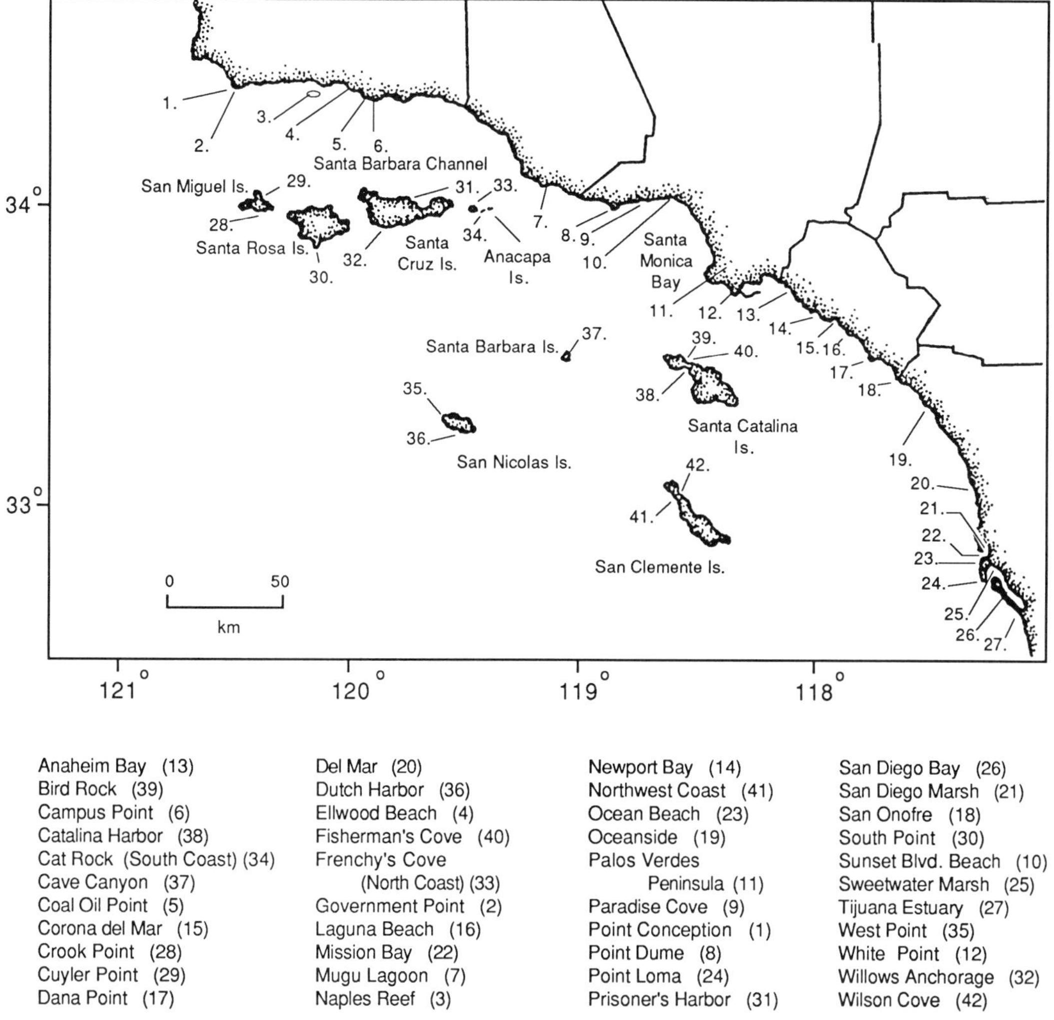

Figure 7.1. Locations of embayment, intertidal, and subtidal (kelp bed) sites discussed in this chapter.

Foster and Schiel 1985; Schiel and Foster 1986; North et al. 1986; Tegner and Dayton 1987). Knowledge of subtidal macrophyte communities other than those dominated by kelp is limited. Deep-water algae (>30m) are virtually unknown despite the availability of submersibles and the development of new video technology, both of which make deep-water habitats accessible for study.

This chapter summarizes and evaluates the current state of knowledge of the benthic macrophytes (macroalgae, seagrasses, and halophytes) inhabiting the SCB. The following areas are emphasized: (1) macroalgal and seagrass floristics and biogeography; (2) spatial patterns and temporal dynamics of macrophyte communities for embayment (salt marsh, estuarine, and lagoon), rocky intertidal, and kelp forest habitats; and (3) productivity of macrophyte populations and communities.

Floristics and Biogeography

Floristics

An account of phycological work along the Pacific North American coastline has been written by Papenfuss (1976), who outlined the sources of earlier collections and taxonomic works that serve as the foundation for our knowledge of the California seaweed flora. The work of Abbott and Hollenberg (1976), however, has provided a major advance by including taxonomic descriptions, distributional information, and keys for 669 California algae (72 Chlorophyta, 137 Phaeophyta, 1 Xanthophyta, and 459 Rhodophyta) in a single comprehensive volume. Of these species, 492 (73.5%) occur in the SCB, including 59 Chlorophyta, 86 Phaeophyta, and 347 Rhodophyta, making the composition of the SCB seaweed flora 70.5% red, 17.5% brown, and 12.0% green. Four species of seagrasses (Potamogetonaceae) occur in the SCB: *Phyllospadix scouleri*, *Phyllospadix torreyi*, *Ruppia maritima*, and *Zostera marina*.

Murray (1974) pointed out that prior to Abbott and Hollenberg's (1976) synthesis most research on the SCB seaweed flora consisted of studies confined to an isolated group and that studies where the principal aim was to produce a complete flora for a specific region were almost entirely lacking. During the past 15 years, knowledge of seaweed and seagrass distributions in the SCB has been augmented mostly by species lists for intertidal sites (Littler 1979a; Harris 1980; Apt et al. 1988; Murray and Littler 1989).

It appears that floral richness is greater in the SCB and that the composition of the dominant species differs from that in central and northern California. Abbott and Hollenberg (1976) commented that less biomass but greater diversity characterizes the seaweed flora south of Point Conception, where the flora is often dominated by shorter, more densely branched species of red algae instead of larger, fleshy forms. Greater floral diversity in the SCB has been quantitatively verified by Murray et al. (1980), who determined that maximal floral richness (446 species) for California was obtained between 33° and 34°N latitude and that floral diversity increased southward and was positively correlated with decreasing latitude (fig. 7.2). Murray et al. (1980) suggested that the high diversity of the SCB seaweed flora may be related to the greater amount of shoreline habitat found south of Point Conception and to the various exposures of island habitats to the warm and cold ocean currents prevalent in the SCB.

Studies of the southern California Channel Islands (Littler 1979a) suggest that the list of macrophytes occurring in the SCB will undergo future expansion. It is probable that many of the seaweeds now thought to terminate their distributions along the central California coast will eventually be collected from island sites exposed to cold water masses (e.g., San Miguel, San Nicolas, and Santa Rosa islands) or from deep subtidal habitats or offshore-seamounts. Stewart (1984) has observed that several species common in the low intertidal zone in central and northern California occur in deep subtidal sites in the SCB, suggesting that colder water species can survive at more southerly latitudes by occupying deep habitats below the thermocline. Also, Lewbel et al. (1981) have concluded that the flora and fauna of Cortes and Tanner banks, two shallow (approximately 20-m-deep) seamounts located 180 km off the San Diego coast, have closer affinities to central California biota than to SCB biota. Expanded work in shallow, warmer water habitats off San Diego County and on San Clemente and Santa Catalina islands, particularly during the late summer months when surface sea temperatures reach highest levels, should also result in additions to the SCB flora of species with subtropical distributional centers.

Biogeography

The geologic history and the present-day heterogeneity in exposure of the mainland and offshore islands and seamounts to the prevailing mixture of circulating water masses make the SCB a region of considerable biogeographic complexity (see chaps. 1 and 2). To date, biogeographic information on SCB macrophytes is confined almost entirely to the descriptive phase of floristic analysis dis-

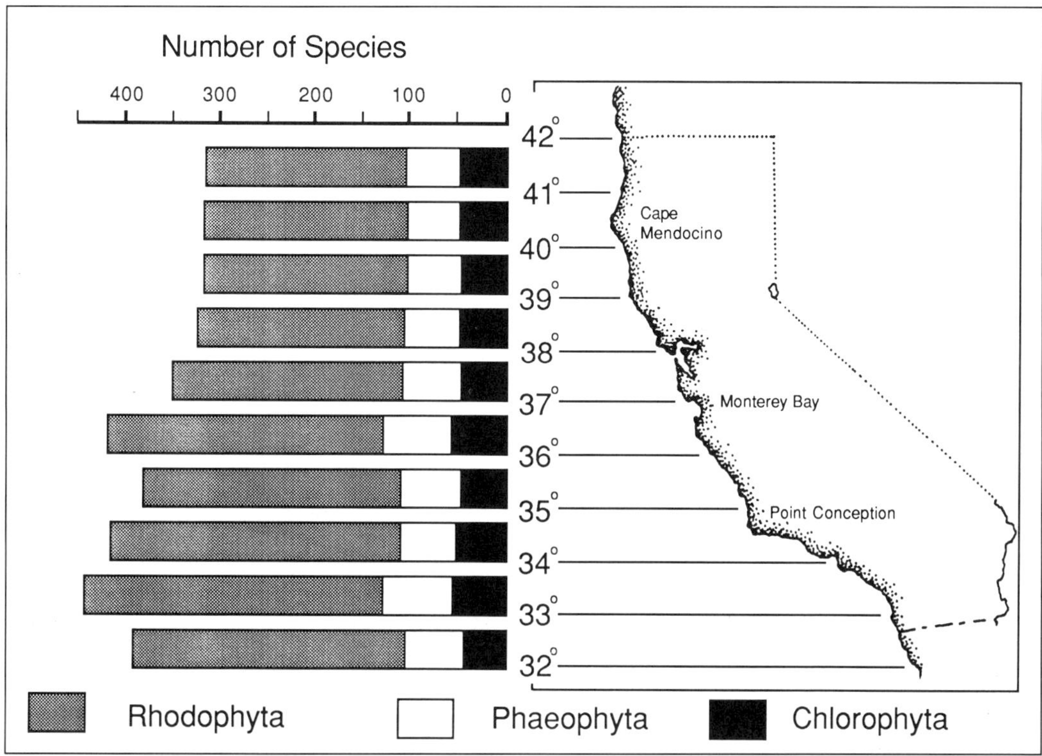

Figure 7.2. Numbers of species of marine Chlorophyta, Phaeophyta, and Rhodophyta as a function of latitude for California. Greatest floral richness occurs in the SCB. (From Murray et al. 1980.)

cussed by Garbary (1987). Such analyses performed during the past decade (Murray et al. 1980; Thom 1980; Murray and Littler 1981) have detailed the patterns of macrophyte distribution within the SCB. They form a partial foundation for the formulation and testing of hypotheses on the nature of factors responsible for limiting seaweed distributions and for the use of vicariance methodology (Garbary 1987) to evaluate the modern and evolutionary histories of the SCB seaweed flora.

Setchell (1893) and Saunders (1901) provided perhaps the first discussions of the importance of Point Conception (fig. 7.1), the northern boundary of the SCB, to macrophyte distributions in the eastern North Pacific. Saunders (1901) suggested that at least three distinct regions of geographic distribution could be recognized for Pacific North American seaweeds, including a "southern region" extending from Point Conception southward, "perhaps to the equator." Setchell and Gardner (1903) revised the description of Saunders' (1901) southern region by defining a "Subtropical Region" ranging from Point Conception southward to the area near Magdalena Bay, Mexico. According to Setchell and Gardner (1903), warmer water red algal species and brown algal genera such as *Egregia, Eisenia*, and *Pelagophycus* characterized the area south of Point Conception, but were absent to the north.

With few exceptions (e.g., Michanek 1979), phycologists have continued to recognize Point Conception as a significant distributional barrier since the turn of the century (Scagel 1963a, b) and have almost universally characterized the flora of the SCB as warm temperate. During the past decade, however, evidence has accumulated suggesting that the SCB is better characterized as a complex transitional region between cold- and warm-temperate biotas (Murray and Littler 1981).

An analysis of the distributions of marine algae (Murray et al. 1980) has revealed that about 32% of California seaweeds exhibited range terminations in the SCB between 33° and 35°N latitude (fig. 7.3). More terminations of species ranges occurred throughout the lower 75% of the SCB than in the latitudinal increment of coastline containing Point Conception, however, supporting the conclusion that Point Conception is not as sharp a boundary for species distributions as has been implied in the past.

Further evidence for the transitional nature of the SCB is provided by quantitative analyses of the compositions of insular and mainland intertidal floras, which appear to show affinities with both warm- and cold-temperate biotas (Murray et al. 1980; Thom 1980; Murray and Littler 1981). Murray and Littler (1981) defined five distinct biogeographic units in the SCB based on analyses of 21 sites by multivariate classification and ordination techniques (figs. 7.1 and 7.4). With the exception of a unique site on San Nicolas Island (West Point), island sites were segregated into those most exposed to the warm waters of the Southern California Countercurrent (San Clemente, Santa Catalina, and Anacapa islands), those bathed mostly by the colder waters of the California Current (San Nicolas, San Miguel, and Santa Rosa islands), and those receiving mixed exposure to the two current systems (Santa Barbara and Santa Cruz islands).

In agreement with earlier predictions, Murray and Littler (1981) also demonstrated (fig. 7.4) that the floras of the island sites proximal to the California Current had much greater affinities (92.3–98.4% overlap of taxa) with macrophytes north of Point Conception than did the floras of sites bathed principally by the Southern California Countercurrent (71.3–83.1%) or those located in mixed waters (80.4–91.8%).

A gradient in the composition of intertidal macrophyte communities also appears to occur along the SCB mainland (Thom 1980; Murray and Littler 1981). Of the seven mainland sites analyzed by Murray and Littler (1981), three groups were observed (fig. 7.4): (1) sites nearest Point Conception and the influence of the California Current (Government Point); (2) sites in the Santa Barbara Channel and the northern reaches of Santa Monica Bay (Coal Oil Point and Paradise Cove); and (3) sites located south of the Palos Verdes Peninsula and most proximal to the flow of the Southern California Countercurrent (White Point, Corona del Mar, Dana Point, and Ocean Beach). Members of these groups showed similar patterns of species overlap with the central California flora to those indicated for island sites (fig. 7.4), with greatest affinities recorded for the first group (96.4%), intermediate affinities for the second group (89.2–89.7%), and lowest affinities for the third group (74.6–87.7%). Using reciprocal averaging, Thom (1980) also identified a gradient in the composition of the intertidal algal floras of the SCB mainland. Thom observed that this gradient was not strictly latitudinal; he reported gradual and continuous compositional changes and noted substantial differences when sites at the northern and southern ends of the SCB were compared.

Historically, seawater temperature has been accorded a preeminent role among the characteristics of ocean water masses believed to influence the worldwide distributions of marine macrophytes (Setchell 1915; Michanek 1979; Druehl 1981). It is not surprising, then, that studies of seaweed distributions in the SCB implicate temperature as the principal environmental factor correlating with seaweed distributions (Murray and Littler 1981). On a geographic scale, however, patterns of sea surface temperatures are often associated with the generalized flow of surface water masses in the SCB, and the distributions of seaweeds appear to be correlated equally well with both.

As pointed out by Garbary (1987), only in recent years have studies of the current

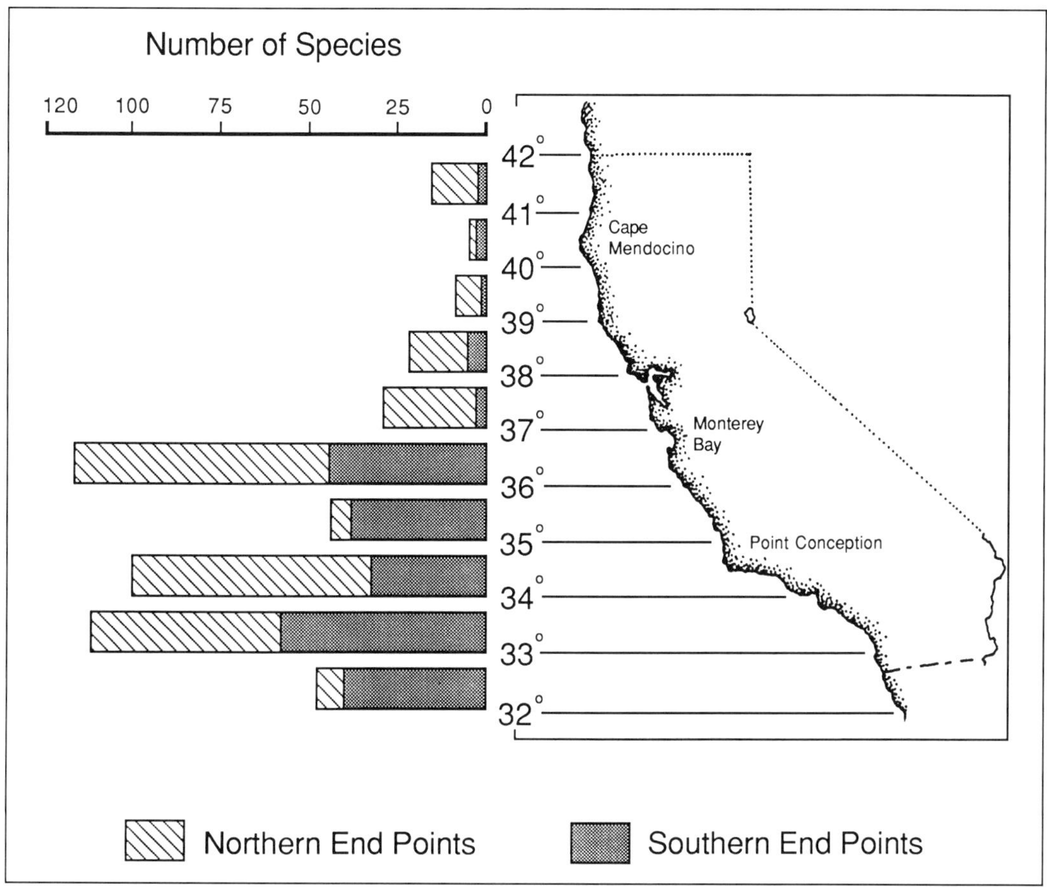

Figure 7.3. Numbers of northern and southern distributional limits of California marine macroalgae plotted by latitude. Areas with greatest numbers of species range terminations include Monterey Bay and the SCB. (From Murray et al. 1980.)

geographic distributions of macrophytes progressed from primarily descriptive to primarily experimental in their approaches. Experimental studies include those designed to test hypotheses of seaweed distributions by analyses of physiological tolerances and environmental regulation of life history events. For example, for the North Atlantic flora, experimental tests of hypotheses of the relationships between biogeographic boundaries and temperature extremes for growth or within which a species can complete its life history have been performed for numerous seaweeds (Hoek 1982a, b; Cambridge et al. 1984; Yarish et al. 1984, 1986). These studies provide interesting ecological insights and are clearly relevant to understanding the current geographic distributions of North Atlantic seaweeds (Garbary 1987).

For eastern North Pacific macrophytes, the relationship between geographic distribution and temperature requirements for growth (Stewart 1984), or temperature tolerances (Lüning and Freshwater 1988), has received little attention. Stewart (1984) measured the vegetative growth rates of five species of red algae in response to different water temperatures and found that all cultured thalli grew best under warmer (16° or 20°C) seawater temperatures. She was unable to segregate the seaweeds into groups of warm- or cold-water origin consistent with their SCB geo-

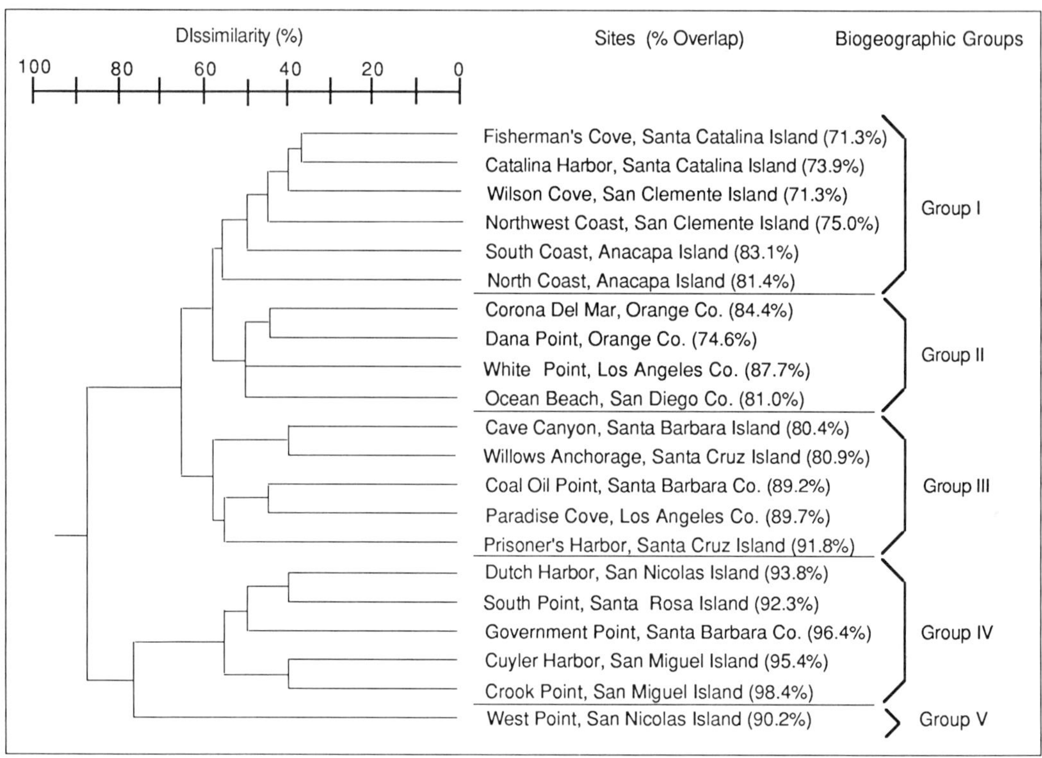

Figure 7.4. Classification of rocky intertidal sites in the SCB based on the presence or absence of macrophyte taxa. Dendrogram depicts the percentage dissimilarity between sites and the interpreted biogeographic units. Listed in parentheses next to site names are the percentages of overlap or the percentages of the recorded taxa that also occur in colder California waters north of Point Conception. See figure 7.1 for site locations. (From Murray and Littler 1981.)

graphic distributions. Lüning and Freshwater (1988), however, suggest that upper survival temperatures correspond with the geographic distributions of several subtidal cold-water Pacific macrophytes. These authors examined the temperature tolerances of 51 species of Washington state seaweeds, many of which occur in the SCB, by measuring net photosynthetic performance following a 1-week incubation period at temperatures ranging from $-1.5°$ to 30.0°C. Lüning and Freshwater (1988) found agreement between maximal incubation temperatures for which a positive net photosynthetic response could still be obtained and the average maximum summer seawater temperature occurring at the southern distributional limit for many subtidal species.

Coastal ocean currents play an as-yet unclear role in determining or maintaining seaweed distributions by transporting spores, propagules, vegetative fragments, and floating reproductive plants from donor to recipient sites (Hoek 1987).

Macrophytes recruit new individuals into suitable habitats in three ways. The first is vegetative reproduction, or the growth of new individuals from vegetative fragments. A second mechanism is the detachment and subsequent reestablishment of whole thalli. The third process involves the production and dispersal of motile or nonmotile reproductive cells, that is, zygotes or spores. Generally, macrophyte dispersal and recruitment by spores or zygotes is thought to be limited to relatively short distances. For example, the radius for recruitment from spores for most large brown seaweeds is probably less than

10 m (Dayton et al. 1984), although Reed et al. (1988) provide evidence for greater dispersal of kelp zoospores and indicate that small, filamentous brown algae recruit effectively at distances up to 500 m. Ocean currents can greatly increase dispersal distances by detaching and transporting reproductively active thalli.

For the SCB, nothing is known of the availability of planktonic reproductive cells in the circulating water masses or of the potential role of the transportation of detached plants, other than for *Macrocystis* (Dayton et al. 1984). Sampling of reproductive cells from surface water by filtration (Hruby and Norton 1979; Zechman and Mathieson 1985) and by settlement on slides (Amsler and Searles 1980) has yielded information for North Atlantic sites. Except for research on kelp (Reed et al. 1988), however, these studies have yet to be performed in SCB waters. Although drifting algae are evident throughout the SCB, their role in dispersal has yet to be investigated.

Patterns of Structure and Seasonal–Interannual Variation of Macrophyte Communities

The structures of macrophyte communities vary vertically and horizontally as a function of biotic and abiotic agents. Biotic interactions, such as interspecific competition between primary producers (macrophytes) or between macrophytes and sessile animals, and herbivory are known to be determinants of macrophyte community structure (Chapman 1986).

With the exception of kelp and associated organisms, research emphasis on SCB macrophytes has historically been mostly descriptive in nature and has not focused on elucidating biological mechanisms of community organization. The potential role of biotic agents in structuring SCB macrophyte communities in rocky intertidal habitats often must be extrapolated from work performed elsewhere. However, our understanding of the organization and dynamics of kelp forests is perhaps best known for communities found in the SCB. For rocky intertidal and embayment communities, the lack of experimental work in the SCB (with some notable exceptions, e.g., Sousa 1979a, b, 1980, 1984; Sousa et al. 1981; Kastendiek 1982; Taylor and Littler 1982; Taylor and Hay 1984) has resulted partially from difficulties in establishing and monitoring manipulative experiments on the readily accessible mainland sites where human interference is almost a certainty. Additionally, long-term (>10 yrs) quantitative records of species distributions and abundances for the SCB are extremely rare, a condition that is shared with most shallow water marine ecosystems (Lewis 1976; Jones et al. 1979; Dayton and Tegner 1984a). Wolfe et al. (1987) argue that this lack of long-term records exists because such data are expensive and time-consuming to obtain, and there are strong disincentives for long-term research programs built into the policies of both academic institutions and governmental research and regulatory agencies.

For the SCB, data depicting seasonal or interannual variations of benthic macrophytes are mostly confined to rocky intertidal and kelp forest populations. Many of these data have been obtained in response to concerns about environmental deterioration and have been used to establish so-called baselines from which to record change or to assess damage caused by anthropogenic inputs (e.g., sewage) or activities (e.g., petroleum exploration or kelp harvesting). The patterns of structure and seasonal and interannual variation are described here for the benthic macrophytes of protected embayment, rocky intertidal, and kelp forest communities of the SCB.

Embayments: Salt Marshes, Lagoons, and Modified Estuaries

Southern California has no large rivers that flow into the sea on a year-round basis. This lack of freshwater input is caused by the arid climate, low rainfall, recent geologic setting, and construction of flood-control dams. Because of the dense human population, the drainage patterns of naturally occurring smaller rivers and streams in the SCB have been heavily modified for flood control and the mouths of some of these rivers have also been extensively modified to create marinas or harbors (see chap. 12). Flow of freshwater into the sea in the SCB is highly seasonal and most often low in volume, which results in mean salinities (32.5‰) of the overlying waters of embayments that are highly comparable to the salinity of coastal seawater (Ho 1974). Therefore, circulation patterns and flushing of embayments are largely dependent upon tidal exchange. Because of the extensive human modification and the relatively low volume of freshwater drainage, expansive estuarine and salt marsh communities are absent from the SCB.

The benthic macrophyte communities of SCB embayments are mostly confined to mudflats of river mouths, harbors, marinas, and lagoons in areas covered periodically or permanently with shallow water. Marine algae and seagrasses occupy these shallow-water habitats, as do halophytes, or salt-tolerant flowering plants. The latter dominate the tidally influenced edges of highly protected sedimentary bays and lagoons that form salt marshes. Seagrasses are most prominent in the SCB in highly protected areas that are continually submersed by waters of ocean salinity, such as the permanently inundated lower reaches of marshlands or modified embayments. In the SCB, eelgrass (*Zostera marina*) is the most common seagrass species occurring in embayments. It is particularly widespread in relatively unaltered habitats such as Anaheim Bay and Upper Newport Bay (Ho 1974).

Marine macroalgae occur throughout salt marsh habitats, although smaller, largely microscopic, edaphic (sediment-associated) algae usually predominate on the sedimentary surfaces. For example, Ho (1974) has listed a range of only 13–20 species of benthic macroalgae for four SCB embayments. The green alga *Enteromorpha clathrata* var. *crinita* (as *E. crinita*) is believed to be the most common and abundant (Ho 1974) of about 27 species of macroalgae that characterize low wave energy, soft substrata in SCB embayments (fig. 7.5). The brown seaweeds appear to be poorly represented, whereas green algae tend to predominate in SCB embayment habitats.

About 40 salt marsh habitats occupying a combined area of less than 5000 ha can currently be found between Point Conception and Mexico (fig. 7.1 and table 8.1). These individual marshes are small in area and most have been heavily modified. Human usage of SCB embayments, in fact, has been so severe that Speth (1969a, b) has estimated that a nearly 75% reduction in the extent of coastal marsh habitat has occurred in southern California during the past 300 years. Salt marsh habitats in the SCB are mostly confined to the mainland, with only a few small embayment pockets occurring (Riznyk 1974) on five of the eight offshore islands. Perhaps the largest and best known (Straughan 1979) of the island embayment habitats is the small salt marsh located within Catalina Harbor, Santa Catalina Island (fig. 7.1).

MacDonald (1977a) divides California marshes into three geographic groups based on species composition: (1) a northern group extending from Point Arena (39°N) northward; (2) central California marshes, including those of San Francisco Bay and Monterey Bay; and (3) a southern group, which extends from Morro Bay southward. Within the southern group, MacDonald (1977b) refers to southern California salt marshes as belonging to the Dry Mediterranean Group and having the highest plant diversity of Pacific coast marshes. The species composition of the central California marshes is known to differ from that of southern California (Zedler 1982a). *Salicornia bigelovii, Salicornia subterminalis, Monanthochloe littoralis*, and *Batis maritima*, for example, are four halophytes that are important components of SCB marshes, but are unknown from north of Point Conception.

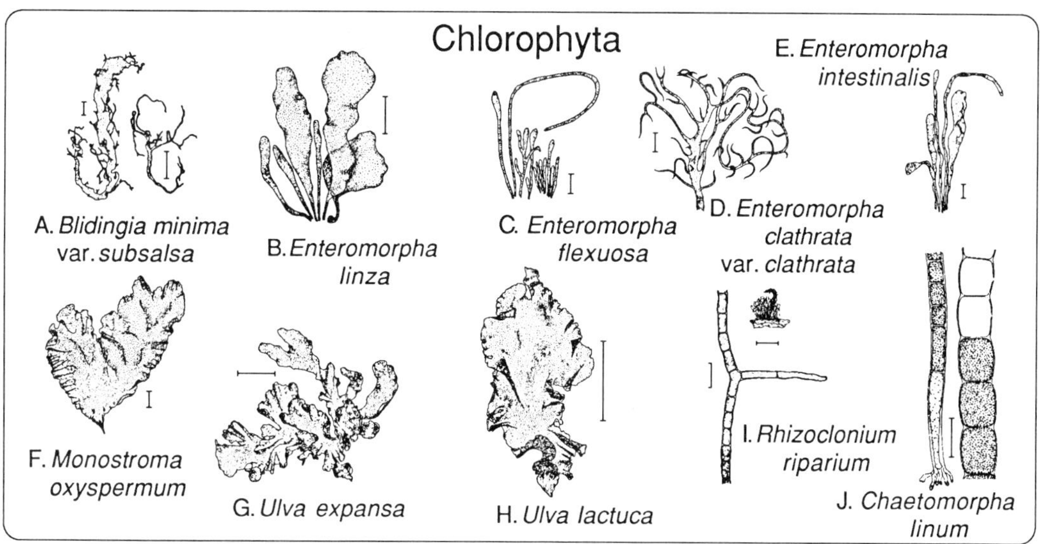

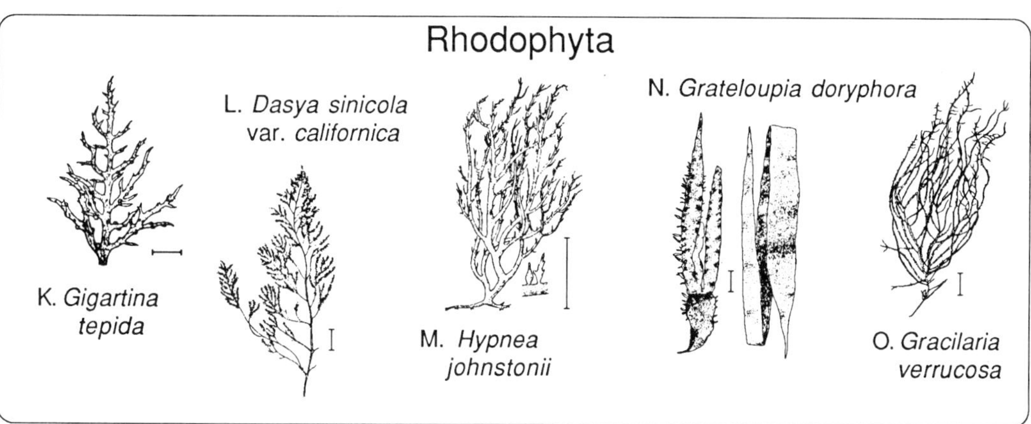

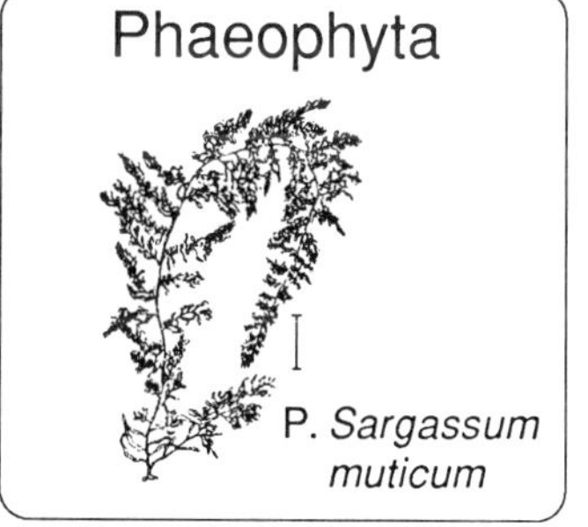

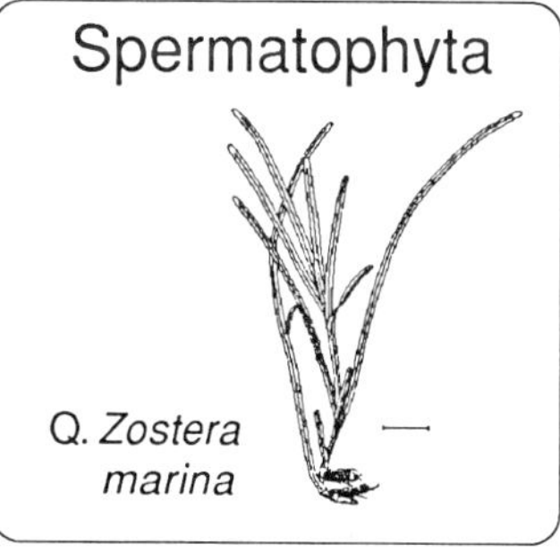

Figure 7.5. Marine macroalgae and seagrasses characteristic of low wave energy, soft substrata habitats of SCB embayments based on information from Abbott and Hollenberg (1976) and Zedler (1982a, b). Scales: 250 μm (J); 0.1 mm (I bottom); 5 mm (K); 1 cm (A, E, F, L); 2 cm (B, D); 3 cm (C, I top, M–O, Q); 5 cm (G, H, P). (Sources of illustrations: A–H, J–P from *Marine Algae of California* by I. A. Abbott and G. J. Hollenberg with permission of the publishers, Stanford University Press; I, Q reproduced from Scagel 1972 courtesy of the Royal British Columbia Museum, Victoria.)

MacDonald (1977a) recognizes two major types of coastal marshes for the SCB. The first of these occur in extensive, relatively deep water embayments that have a large enough tidal flow to maintain continuous contact with the sea. These include Mugu Lagoon, Anaheim Bay, Newport Bay, Mission Bay, San Diego Bay, and the Tijuana Estuary (fig. 7.1). These marshes have similar halophytes, and they exhibit more complex zonation than marshes in the San Francisco Bay region. The second group, which comprises the remainder of SCB marshes, includes marshes of smaller, shallow embayments that are closed periodically from the ocean. Environmental conditions are characteristically more variable and more extreme in this group than in the first major group of SCB marshes (MacDonald 1977a).

The vertical distribution of the dominant salt marsh plants of the Tijuana Estuary (fig. 7.6) is characteristic of most marshes of the southern California region (Zedler and Nordby 1986). However, the smaller, shallower marshes generally (1) lack *Spartina foliosa, Batis maritima*, and *Salicornia bigelovii* in low and middle zones; (2) have strong representation of high-shore marsh species; and (3) contain large numbers of opportunistic halophytes (MacDonald 1977a). The lower tidal levels of the Tijuana Estuary are dominated by *Spartina foliosa* (cordgrass), which, according to MacDonald (1977a), is the primary tidal flat colonist and most abundant constituent of low marsh vegetation throughout the SCB.

Marshes also provide habitat for algae, which often grow on mud surfaces as mats or attached to the leaves and bases of marsh halophytes. Zedler (1982a, b) points out that in the SCB, halophyte canopy coverage is more open, seasonal droughts occur, and interstitial soil water is often more saline than seawater. She suggests that the algal communities are likely different from those occurring in marshes of the Atlantic and Gulf of Mexico. Despite the importance of these algae as primary producers (Zedler 1980), researchers know relatively little about the composition of the matlike growths existing beneath the larger canopies of vascular halophytic plants in SCB marshes. However, algal growths occurring in the Tijuana Estuary are an exception. A total of 38 common species was identified by Zedler (1982a, b) from algal mat samples taken in the Tijuana Estuary, including 2 species of Chlorophyta, 4 of Cyanophyta, and 32 of Bacillariophyta. The larger algae, however, were limited to two species of Chlorophyta: *Rhizoclonium riparium* and *Enteromorpha clathrata* var. *crinita*.

The distribution of halophytes in marsh habitats is principally controlled by tidal features (McLusky 1971). Tides determine the intensity and frequency of sedimentation and erosion, the vertical extent of the marsh, and the frequency and duration of submergence; they also influence water quality (Ranwell 1972). For arid regions such as the SCB, tidal influx of seawater represents the major source of moisture for coastal marshes (Zedler 1977). Therefore, it follows that any significant alteration of tidal circulation will have major impact on the ecology of salt marsh communities. Modification of tidal flow and the natural landscape has been so extensive that no examples of pristine systems exist in the SCB today (see chap. 12).

Rocky Intertidal

The structures of rocky intertidal macrophyte communities of the SCB were largely unknown prior to the 1975–1979 studies by Littler (1977, 1978, 1979a). This was particularly true for the southern California Channel Islands, where research had been limited (Murray 1974) to only a few quantitative ecological papers (Hewatt 1946; Caplan and Boolootian 1967; Murray and Littler 1974; Littler and Murray 1974a, 1975). For the SCB mainland, knowledge of rocky intertidal macrophyte communities had been gained largely from surveys to evaluate effects of human perturbation (Dawson 1959, 1965; Widdowson 1971; Nicholson and Cimberg

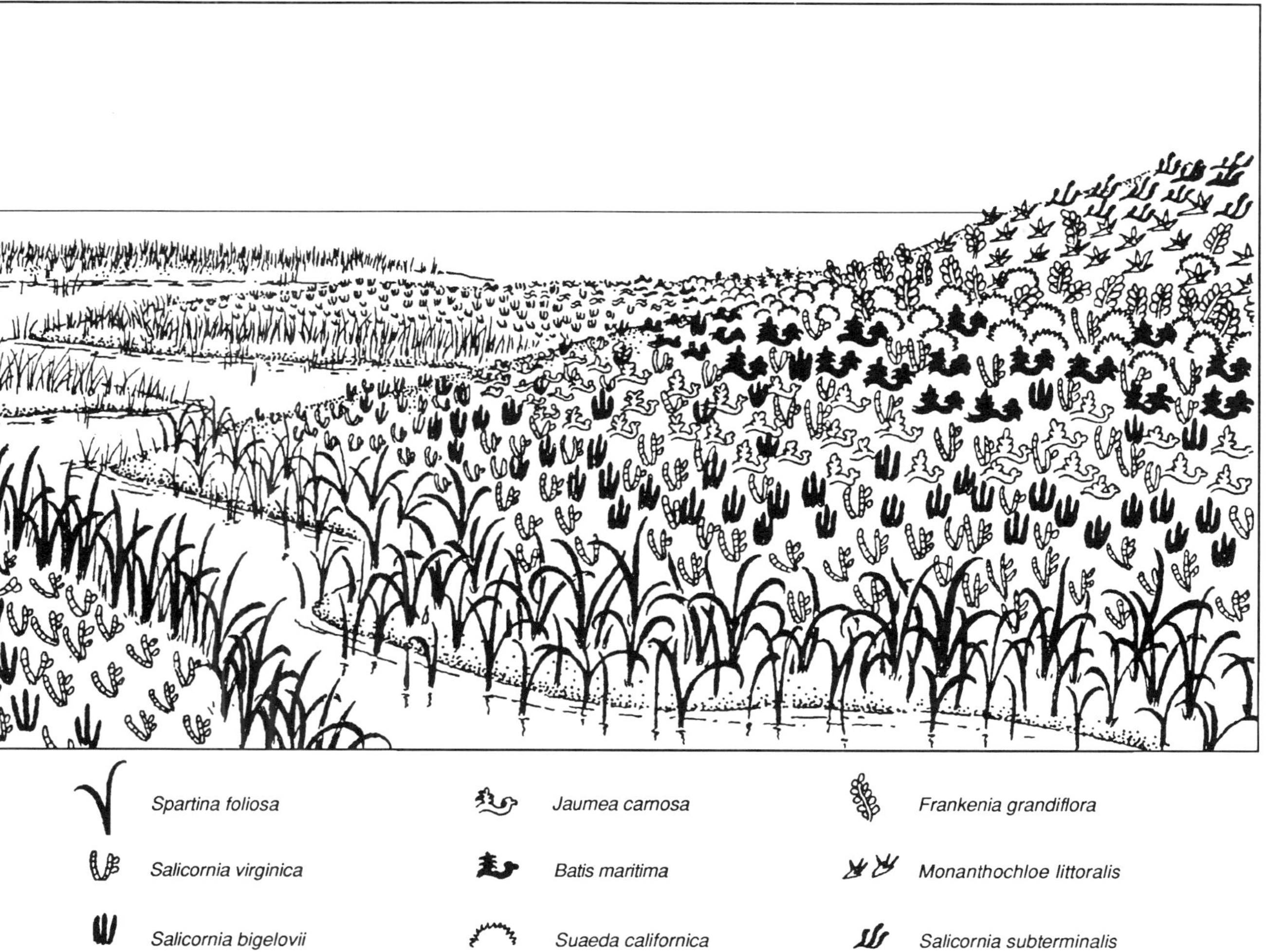

Figure 7.6. Schematic illustration of the patterns of vertical distribution of the common salt marsh halophytes occupying soft substrata in the Tijuana Estuary. The figured zonation patterns and the indicated species are characteristic of many SCB marshes. (After Zedler 1982a and Zedler and Nordby 1986.)

1971; Foster et al. 1971; Thom and Widdowson 1978) and studies of zonation (Stephenson and Stephenson 1972). During the past 10 years, ecological research on intertidal seaweeds and seagrasses in the SCB has increased. The data reviewed by Littler (1980a, b) and Littler et al. (1991) provide the foundation for most of the following discussion, which depicts the spatial patterns of structure of rocky intertidal macrophyte communities in the SCB. We present schematic profiles of rocky intertidal macrophyte communities for four representative sites in the SCB (fig. 7.7), as follows: (1) cold-water, sand-influenced sites at Dutch Harbor, San Nicolas Island (fig. 7.7a) and Government Point (fig. 7.7b); (2) a mixed cold- and warm-water, semiexposed site at Cave Canyon, Santa Barbara Island (Fig. 7.7c); and (3) a warm-water, low wave energy site at Fisherman Cove, Santa Catalina Island (fig. 7.7d).

For intertidal habitats, tidal characteristics represent the most important abiotic gradient at a given location. Tidal characteristics and the resultant periods of tidal emersion, however, vary little from site to site in the SCB. Variations in salinity also are slight and, except during times of heavy precipitation and runoff (Littler and Littler 1987), appear to be largely inconsequential to rocky intertidal and subtidal macrophyte populations. The abiotic environmental parameters of greatest geographic importance appear to be (1) substratum characteristics; (2) surface water masses and accompanying temperature regimes; and (3) wave action. In the following sections, we describe the spatial patterns of macrophyte community structure as a function of these parameters. We also examine the potential role of grazers in determining the distributions and abundances of intertidal macrophytes in the SCB.

ROLE OF SUBSTRATUM

Most macrophytes grow in shallow subtidal and intertidal habitats where the substratum is sufficiently stable to permit attachment. In the SCB, instability of the rock substrata presents a constraint on macrophyte distributions; large areas exist, particularly along the mainland, where substrata are composed entirely of unstable sand or loose cobbles and boulders. Sandy beaches and unstable boulder or gravel and cobble beds comprise more than 80% of the upper mainland shoreline, compared to less than 20% contributed by stable bedrock (Littler and Littler 1980a). This lack of stable rock surfaces is particularly evident in the upper intertidal zone along the mainland where sand and cobble deposits often occur, even on what are otherwise rocky shores (Littler et al. 1991). In contrast, the shorelines of the southern California Channel Islands consist mostly of stable rock (62.5%) and have a much smaller percentage of unstable sandy beaches (Littler and Littler 1979).

The dynamics of sand movement and the presence of unstable boulders or cobbles can significantly modify macrophyte communities occupying rocky intertidal sites. Onshore transportation and deposition of sand, often occurring during summer periods, result in the burial and abrasion of rock surfaces and attached organisms. During periods of enhanced wave energy, loose boulders and cobbles overturn and move in frequencies that are dependent upon their size (Sousa 1979a). Burial or scouring by sand (Taylor and Littler 1982; Littler et al. 1983, 1991) and the movement of unstable cobbles or boulders (Sousa 1979a, b; Littler and Littler 1984; Murray and Littler 1984) affect both macrophyte species composition and abundances.

Three categories of macrophytes are often found in abundance on SCB shores that are influenced by sand scour and sand deposition (fig. 7.8): (1) opportunistic macrophytes that readily colonize disturbed surfaces; (2) resistant macrophytes that tolerate sand abrasion and deposition; and (3) "sand-loving" or psammophytic macrophytes that, for unknown reasons, appear to be best represented in sand-perturbated communities. The first group of macrophytes functions as opportunists in sand-disturbed communities

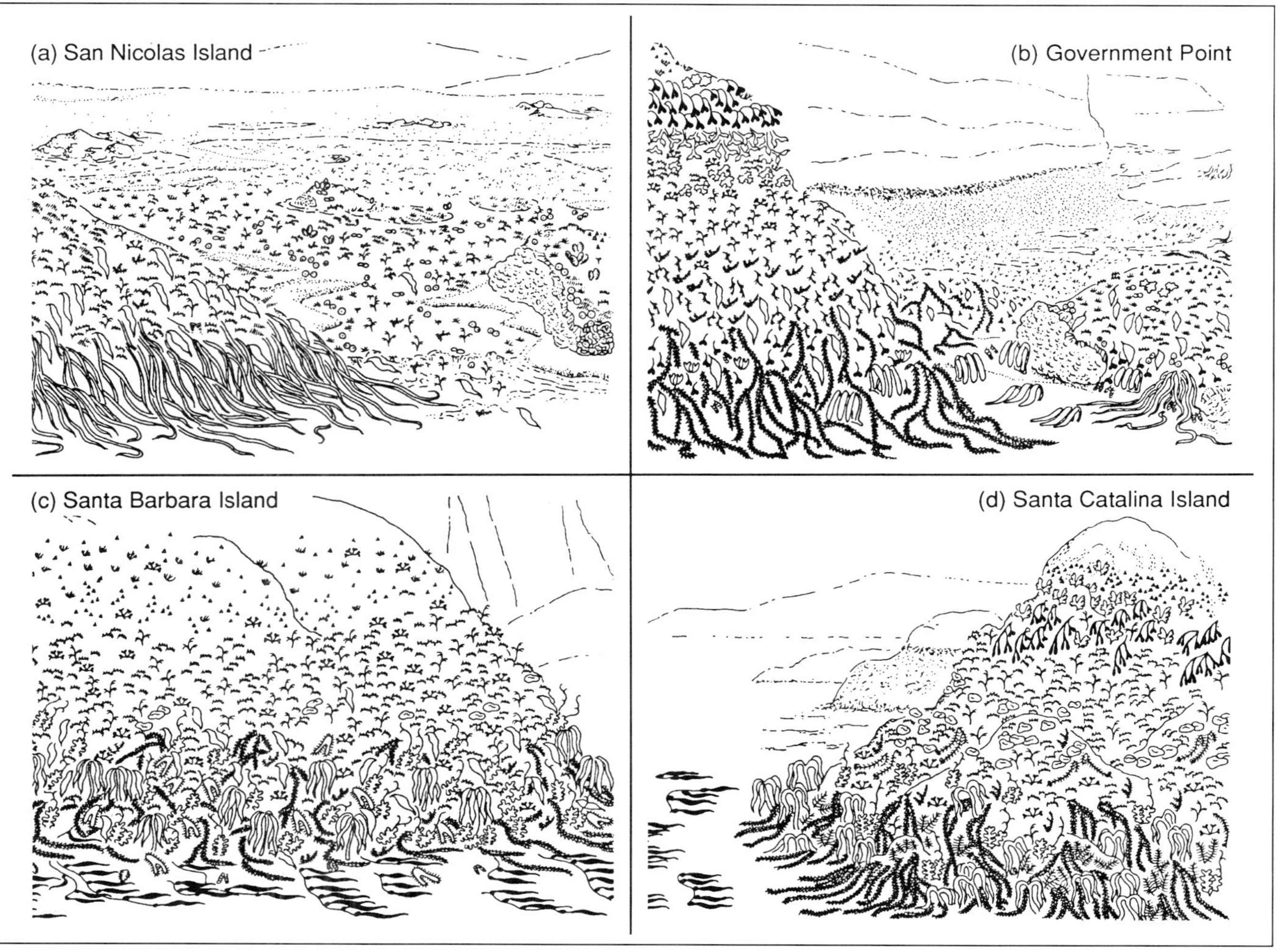

Figure 7.7. Schematic profiles of the vertical distributions of abundant macrophytes and selected sessile macroinvertebrates for four rocky intertidal sites in the SCB. (a) Dutch Harbor, San Nicolas Island. (b) Government Point, Santa Barbara County. (c) Cave Canyon, Santa Barbara Island. (d) Fisherman Cove, Santa Catalina Island. (After Littler 1978, 1979a.)

Key

Upper Shore Species

- *Chthamalus/Balanus* spp.
- *Porphyra* spp.
- *Endocladia muricata*
- *Hesperophycus harveyanus*
- *Pelvetia fastigiata*

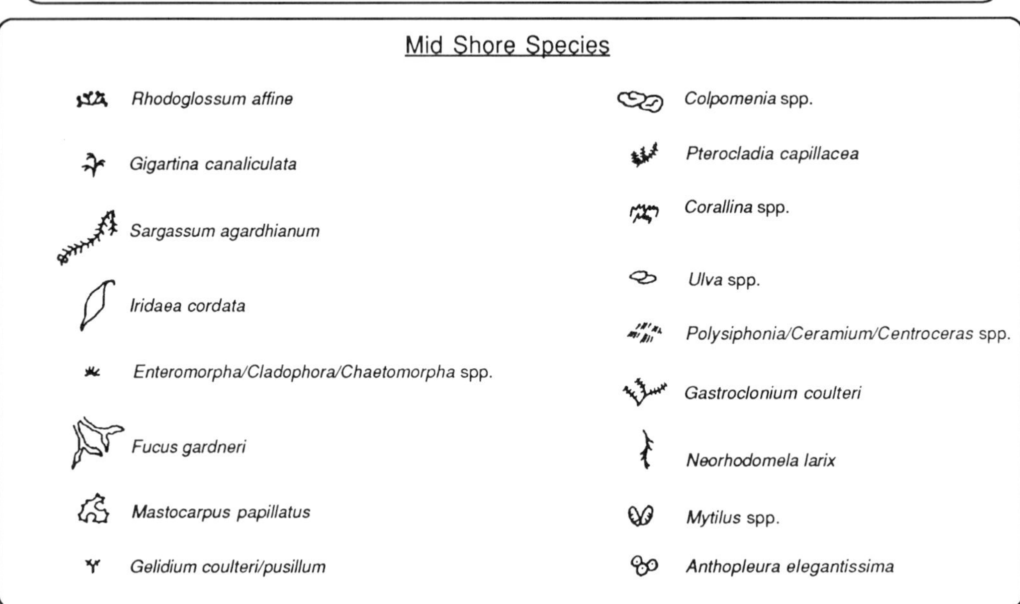

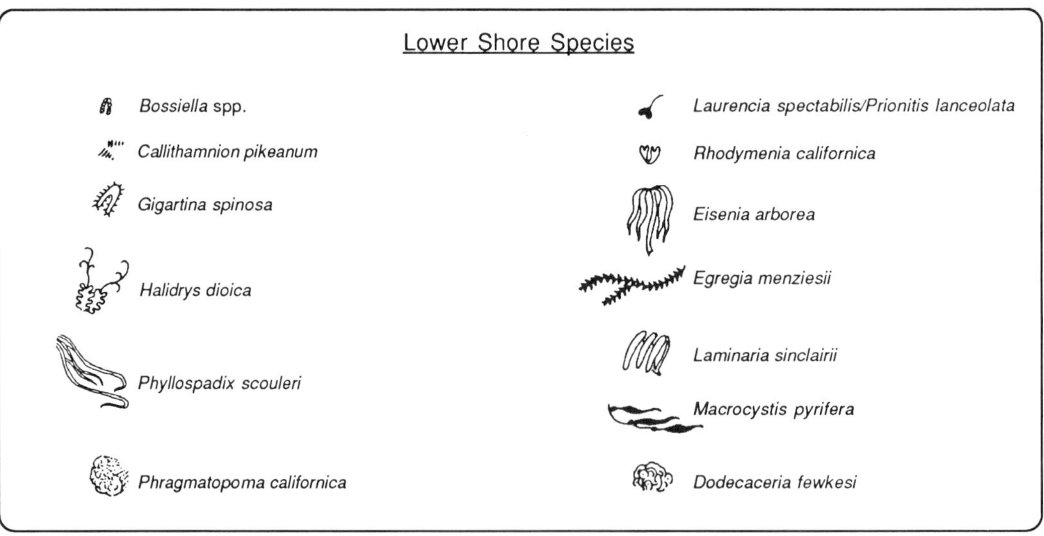

Figure 7.7. (continued)

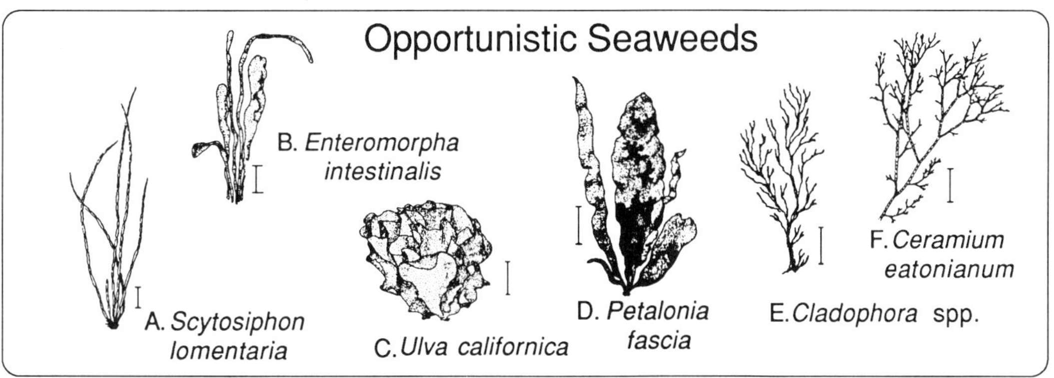

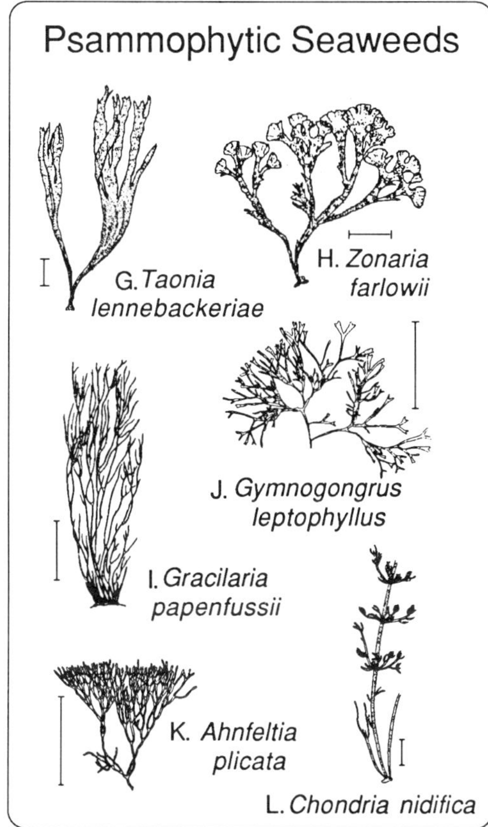

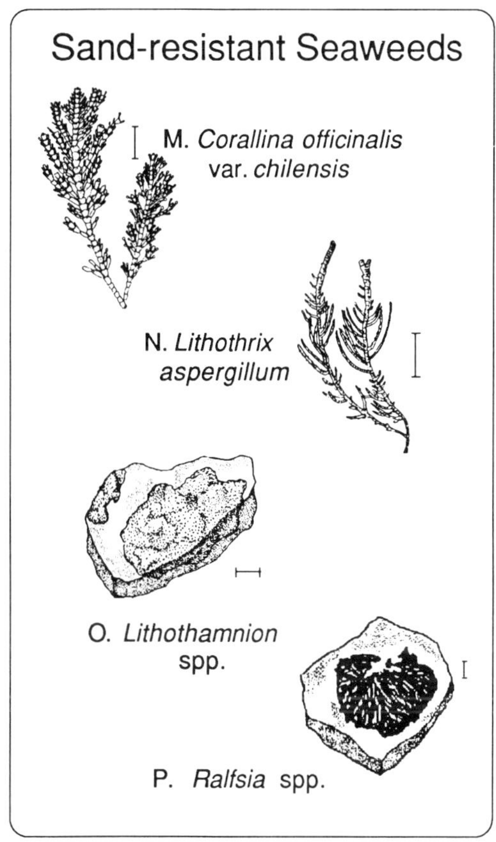

Figure 7.8. Categories of macrophytes found in abundance on SCB rocky shores subject to sand scour and deposition. Scales: 5 mm (F); 1 cm (B, C, L, M, O, P); 2 cm (A, D, G–I, N); 3 cm (E, J, K). (Sources of illustrations: E, G, H, N–P are original drawings; the remainder are reproduced from Abbott and Hollenberg 1976 with permission of the publishers.)

and includes morphologically simple sheet (*Ulva* spp., *Petalonia fascia*), tubular (*Enteromorpha* spp., *Scytosiphon lomentaria*), filamentous (*Cladophora sakaii, Cladophora columbiana, Chaetomorpha linum, Tiffaniella snyderiae*), and small, lightly corticated (*Ceramium* spp., *Centroceras clavulatum, Polysiphonia* spp., *Pterosiphonia* spp.) algae. The second group of sand-resistant seaweeds includes tough, crustose species (e.g., Ralfsiaceae) and articulated coralline algae that often form sand-trapping turfs (Stewart 1982, 1983) on SCB shores. The psammophytic macrophytes (in the sense of Daly and Mathieson 1977) include a variety of mostly brown and red algal species that "appear to favor" sand-influenced rocky habitats. Among these are the brown algae *Zonaria farlowii, Phaeostrophion irregulare*, and *Taonia lennebackeriae* and the red seaweeds *Ahnfeltia plicata, Chondria nidifica, Gigartina volans, Gracilaria* spp., *Gymnogongrus leptophyllus, Neorhodomela larix, Pterocladia media*, and *Sarcodiotheca gaudichaudii*. Also, lower intertidal seagrasses (*Phyllospadix scouleri*), which have rhizomatous root systems, appear to be favored by the accumulation of sand (Littler et al. 1983).

Littler et al. (1983) have offered hypotheses to explain the increased abundances of each of these three groups of macrophytes that tend to characterize sand-impacted rocky shores. They hypothesized that sand-impacted shores support abundant growths of opportunistic algae because these seaweeds are able to colonize space rapidly following the frequent mortality of organisms due to sand abrasion or burial. Great abundances of morphologically simple opportunistic algae also characterize intertidal habitats disturbed by sewage effluent (Littler and Murray 1975; Murray and Littler 1978), freshwater inundation (Littler and Littler 1987), episodes of severe aerial exposure (Seapy and Littler 1982), and movements of loose substratum (Sousa 1979a, b, 1980; Littler and Littler 1984).

For the sand-resistant macrophytes, Littler et al. (1983) have hypothesized that the development of morphological and physiological adaptations enables these seaweeds to resist injury and survive burial. Seaweeds with morphologically tough thalli, such as crustose algae or calcified articulated corallines, are known to be resistant to mechanical damage (Littler and Littler 1980b). Along with certain gelidiacean taxa (Stewart 1983), they appear readily able to withstand sand burial and abrasion.

Littler et al. (1983) have hypothesized that psammophytic taxa exhibit growth or reproductive characteristics that are enhanced by the presence of sand, and hence, they secure a competitive advantage in such habitats. An alternative hypothesis explaining the increased abundances of psammophytic algae is the effect of differential herbivory pressure on sand-impacted versus relatively sand-free rocky shores. Littler (1980a) observed reductions in the densities of macroinvertebrates at sand-impacted sites in southern California, suggesting the possibility that reduced grazing pressure may play an indirect or direct role in the success of certain psammophytic taxa. However, the causal mechanisms for increased abundances or, in certain cases, the almost exclusive relationship between the distributions of psammophytic taxa and the heavy influence of sand are unknown and in need of investigation.

Macrophyte populations inhabiting unstable intertidal boulder fields typically exhibit a mosaic of successional stages and often consist of a profusion of morphologically simple opportunistic algae similar to those flourishing in sand-impacted intertidal sites. This is credited to the frequently observed mixture of early and late successional populations that result from spatial and temporal variation in the frequency of substratum disturbance (Sousa 1979a; Littler and Littler 1984). When boulders are overturned, existing algae are removed or killed by grazing, anoxia, exposure to light levels below photosynthetic compensation, or damage from crushing or abrasion (Sousa 1980). Recolonization of freshly cleared space on boulders then occurs either by vegetative regrowth or by recruitment from spores, and opportunistic algae such as *Ulva* consistently dominate the most recently exposed boulders, whereas late successional species populate less frequently disturbed surfaces (Sousa 1979a, b; Littler and Littler 1984).

Sousa (1980) has experimentally demonstrated that the algal composition on the top surface of a boulder depends on the length of time since it was last rolled over by waves or swell. Consequently, smaller boulders, which experience the greatest amount of movement, are typically dominated by early successional forms, whereas larger boulders that experience less frequent disturbance support later successional associations. The greatest diversity appears on middle-sized boulders, which support mixed associations of early and late successional species (Sousa 1979a). Hence, the cover of opportunistic species that forms early successional associations can be utilized as an index of the frequency of habitat disturbance. Littler and Littler (1987) had observed high abundances of morphologically simple opportunistic algae dominating cover at a boulder beach site in Corona del Mar. Earlier, Littler and Littler (1984) had similarly compared macrophyte populations occupying stable bedrock and boulder field habitats in the Gulf of California and reported that early successional opportunists dominated cover in the latter habitat, whereas late successional predation-tolerant species were most abundant in the former.

A second characteristic of intertidal macrophyte communities occupying habitats subject to substratum instability or other disturbances (such as sand scour) is the high abundance of fleshy and calcified crustose algae and erect articulated corallines (Murray and Littler 1984). In contrast to opportunistic seaweeds, these forms are usually long-lived perennials that recover rapidly from basal thallus parts that remain after mechanical disturbance (Murray and Littler 1979). Although they are low primary producers (Littler 1980c; Littler and Arnold 1982), these seaweeds have tough thalli (Littler and Littler 1980b) and are particularly well suited for persistence in disturbed intertidal habitats.

Knowledge of boulder field habitats in the SCB is based mostly on studies at Ellwood Beach (Sousa 1979a, b, 1980, 1984) and Corona del Mar (Littler and Littler 1987; Littler et al. 1991) (fig. 7.1). Corona del Mar, which is subject to considerable human usage, is typical of many such sites in the SCB because it is dominated by opportunistic seaweeds and high abundances of crustose and articulated coralline algae (Littler 1980a). Patches of early successional associations of *Ulva* and small turf- or mat-forming red algae, such as *Ceramium* spp., *Gelidium* spp., and *Centroceras clavulatum* (Littler and Littler 1987), dominate Corona del Mar's macrophyte communities. In addition, abundant growths of morphologically tough, disturbance-resistant, crustose (Ralfsiaceae and Corallinaceae) and articulated coralline (*Corallina officinalis* var. *chilensis*) algae also occur at Corona del Mar. Murray and Littler (1984) found similar high concentrations of both crustose and articulated coralline algae, together with moderate growths of small opportunistic algae at a rocky intertidal site near White Point inshore from a major sewage outfall. The site is presently disturbed by intense human usage and the periodic movement of sand, gravel, and upper intertidal cobbles.

The geologic nature of the rocky substratum comprising the shore also varies throughout the SCB, ranging from irregular flow breccia to smooth sandstone or siltstone (Littler et al. 1991). However, despite indications of potential biological significance, little is known about the effects of substratum composition and topography on the distributions and abundances of intertidal macrophytes. For example, Littler (1980a) notes that the roughly textured volcanic rock surfaces occurring on certain island sites hold small pockets of moisture and have higher densities of macroinvertebrates compared to sites consisting of smooth sandstone or siltstone.

ROLE OF OCEAN WATER MASSES AND TEMPERATURE REGIMES

Analyses of the distributions of intertidal macrophytes have established a correlation between species distributions in the SCB and the generalized flow of surface current systems and their accompanying thermal characteristics (Murray and Littler 1981). Therefore, it follows that the structure of intertidal macrophyte communities also varies geographically and is strongly influenced by exposure to ocean water masses.

Although there is considerable overlap (71.3–98.4%) in species composition (fig. 7.4) between the intertidal macrophyte floras of SCB sites and the flora north of Point Conception, there are also pronounced differences in species composition and abundances (Murray and Littler 1981). Approximately 30 seaweed species appear to characterize intertidal habitats bathed by warm-water masses in the SCB (fig. 7.9), whereas 20 species appear to be indicators of cold-water sites (fig. 7.10). Many of these species are only occasional or inconspicuous components of any intertidal flora. However, the dominant species in terms of abundance often differ sharply between cold- and warm-water intertidal habitats (fig. 7.7).

Abbott and Hollenberg (1976) described the seaweed communities north of Point Conception as being characterized by laminarialean brown algae and large, fleshy red algae (e.g., *Iridaea*) and as having greater standing biomass than those in the SCB, where fucalean brown algae and shorter, more densely branched red algal species (e.g., *Laurencia* and *Pterocladia*) predominate. Perhaps the most obvious difference between warm- and cold-water macrophyte communities in the SCB occurs in the middle and lower intertidal zones (table 7.1 and fig. 7.7).

The middle intertidal zones of warmer water sites in the SCB consistently support high abundances of articulated coralline algae that often form extensive turfs (fig. 7.7d). Most of the coralline species that make up these turfs (table 7.1) also occur at sites exposed to cold-water masses in the SCB and north of Point Conception, where they are much less abundant and rarely form extensive turfs. *Colpomenia sinuosa* and various species of fleshy red algae are often present as epiphytes at the warm-water sites of the SCB (table 7.1 and fig. 7.7d). In addition, seasonally occurring patches of *Endarachne binghamiae* in the winter and *Laurencia pacifica*, *Laurencia snyderiae*, and *Sargassum agardhianum* in the summer appear frequently at warmer water sites in the SCB.

In contrast, the upper middle intertidal shorelines of most cold-water habitats in the SCB, and to the north as well, are characterized by abundant growths of *Endocladia muricata*, usually growing in association with the acorn barnacle *Balanus glandula*, and a zone or zones dominated by cover of larger, fleshy and saxicolous red algae such as *Mastocarpus papillatus* (table 7.1 and fig. 7.7a, b). Cold-water indicator species such as *Fucus gardneri* (= *F. distichus* ssp. *edentatus*) and *Iridaea cordata* var. *cordata* (= *I. flaccida*, see Foster 1982) appear to be absent from all but those intertidal habitats receiving continuous exposure to cold-water masses. These species provide good indicators of sites in the SCB that have species compositions similar to those along the central California coastline.

The lower intertidal zones of sites throughout the SCB are dominated by large brown algae. For warmer water sites, this brown algal zone consists mostly of *Halidrys dioica*, *Egregia menziesii*, and *Eisenia arborea* (table 7.1 and figs. 7.7d and 7.9). The fleshy red algae that occupy lower intertidal habitats of warm-water sites generally differ in species composition and have lower standing stocks than cold-water sites (table 7.1). Also colder water sites of the SCB often have lower intertidal zones dominated by *Laminaria* spp.; *Cystoseira osmundacea* appears to take the place of *Halidrys dioica* (Littler and Littler 1980a), and large, fleshy and saxicolous red algae are abundant along the lower shoreline (table 7.1 and fig. 7.7a, b).

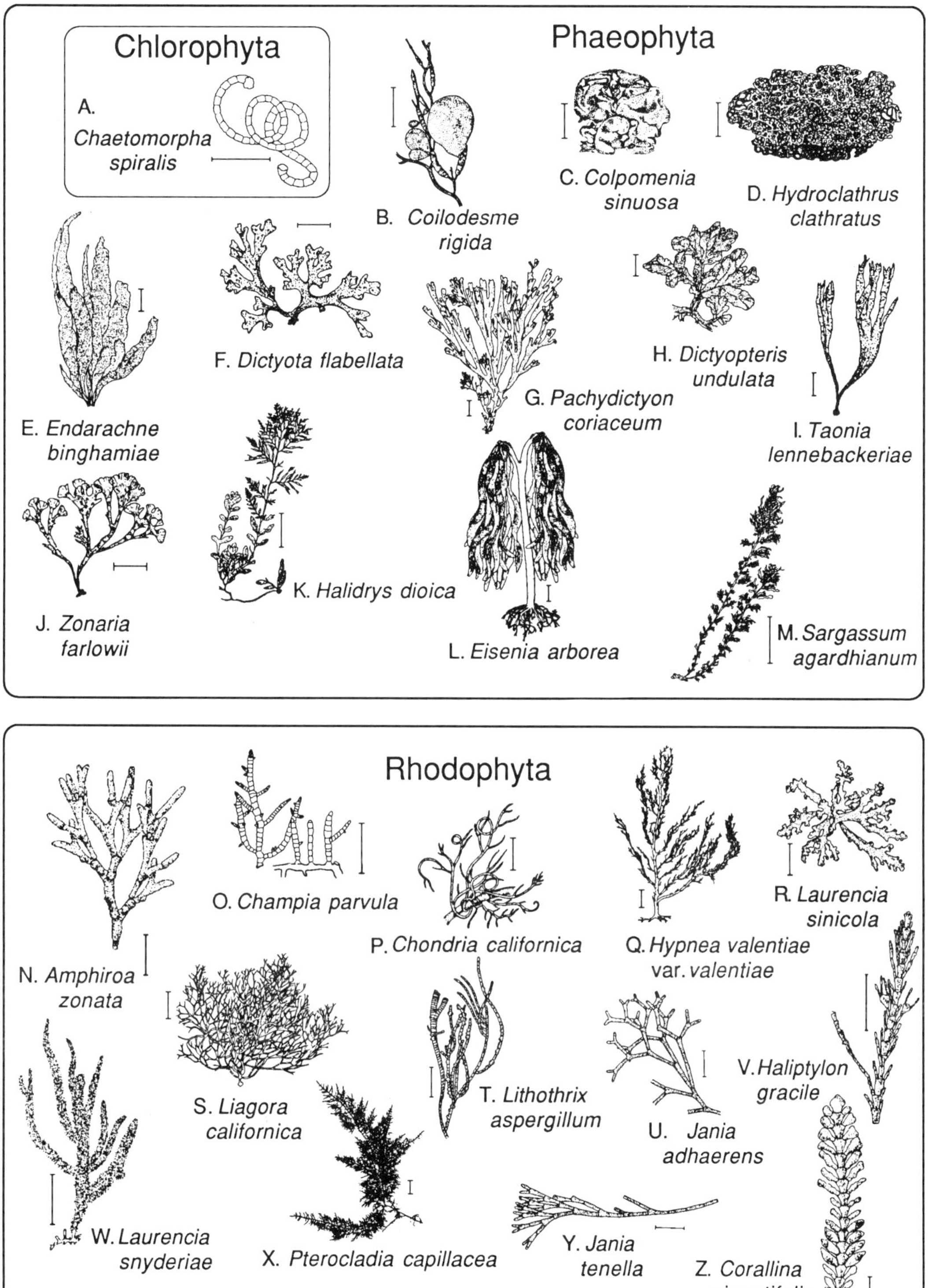

Chlorophyta
A. Chaetomorpha spiralis
Phaeophyta
B. Coilodesme rigida
C. Colpomenia sinuosa
D. Hydroclathrus clathratus
E. Endarachne binghamiae
F. Dictyota flabellata
G. Pachydictyon coriaceum
H. Dictyopteris undulata
I. Taonia lennebackeriae
J. Zonaria farlowii
K. Halidrys dioica
L. Eisenia arborea
M. Sargassum agardhianum
Rhodophyta
N. Amphiroa zonata
O. Champia parvula
P. Chondria californica
Q. Hypnea valentiae var. valentiae
R. Laurencia sinicola
S. Liagora californica
T. Lithothrix aspergillum
U. Jania adhaerens
V. Haliptylon gracile
W. Laurencia snyderiae
X. Pterocladia capillacea
Y. Jania tenella
Z. Corallina pinnatifolia

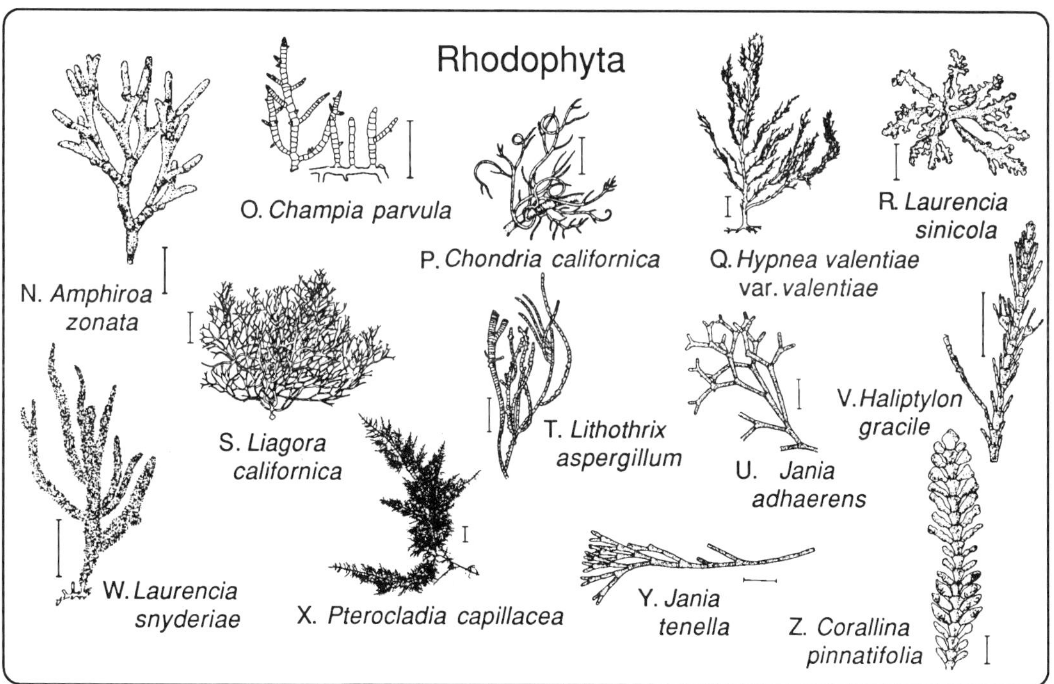

Figure 7.9. Macrophytes characteristic of rocky intertidal habitats bathed by warm-water masses in the SCB. Scales: 1 mm (Z); 2 mm (U); 3 mm (V, Y); 5 mm (A); 1 cm (O, P, X); 2 cm (B–J, N, S, T); 3 cm (Q, R, W); 5 cm (K, M); 10 cm (L). (Sources of illustrations: F, I, J are original drawings; the remainder are reproduced from Abbott and Hollenberg 1976 with permission of the publishers.)

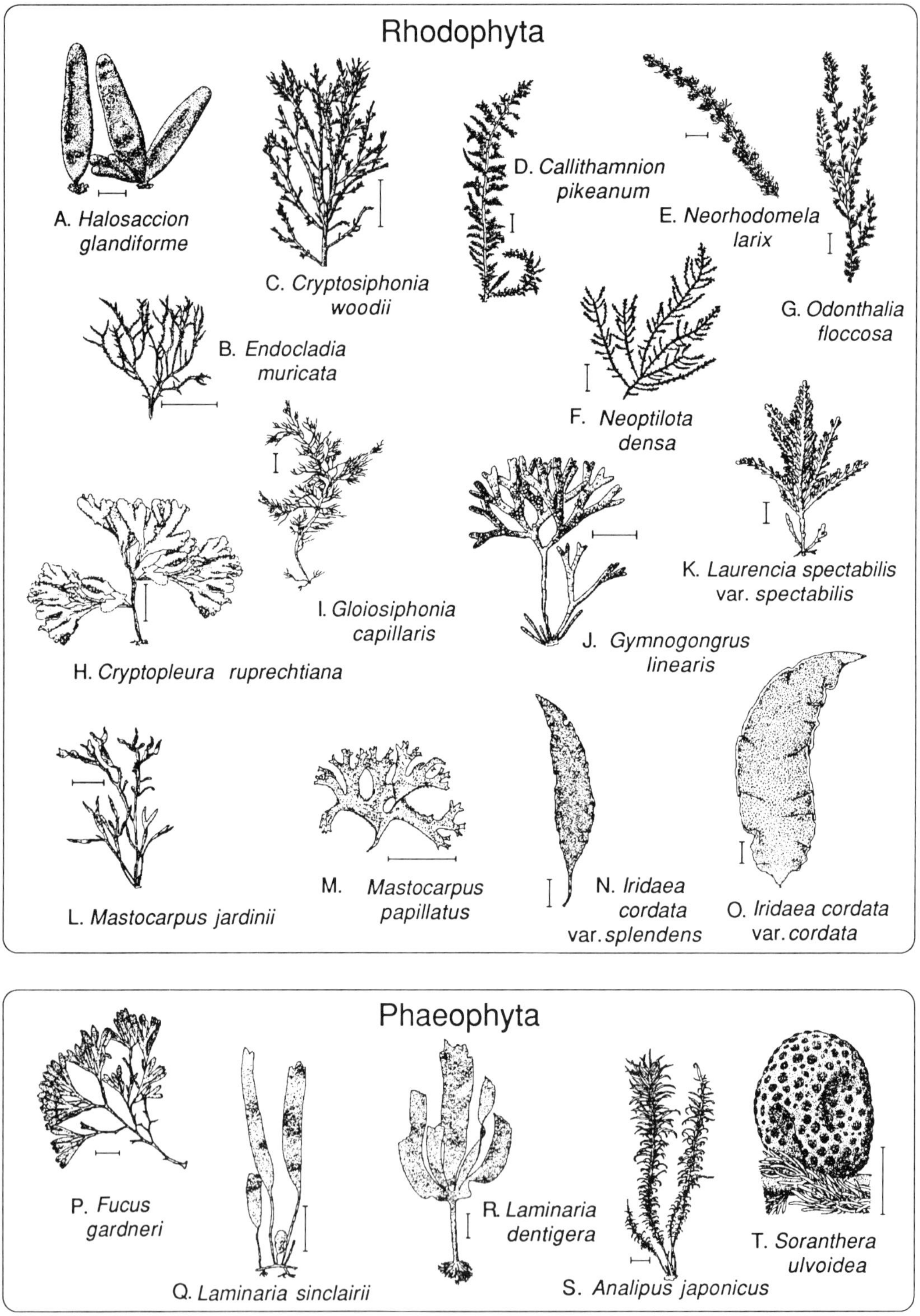

Rhodophyta

A. *Halosaccion glandiforme*

B. *Endocladia muricata*

C. *Cryptosiphonia woodii*

D. *Callithamnion pikeanum*

E. *Neorhodomela larix*

F. *Neoptilota densa*

G. *Odonthalia floccosa*

H. *Cryptopleura ruprechtiana*

I. *Gloiosiphonia capillaris*

J. *Gymnogongrus linearis*

K. *Laurencia spectabilis* var. *spectabilis*

L. *Mastocarpus jardinii*

M. *Mastocarpus papillatus*

N. *Iridaea cordata* var. *splendens*

O. *Iridaea cordata* var. *cordata*

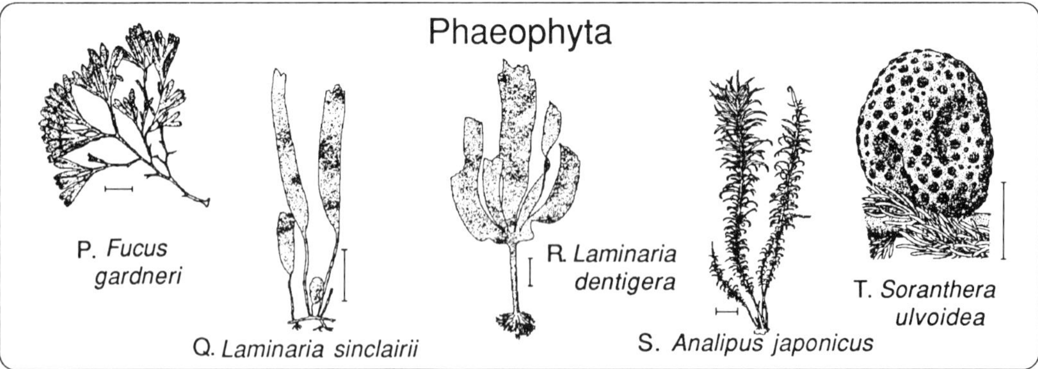

Figure 7.10. Macrophytes characteristic of rocky intertidal habitats bathed by cold-water masses in the SCB. Scales: 1 cm (B, D, E, S, T); 2 cm (A, C, F, G, I, K); 3 cm (J, L, N, O); 5 cm (H, M, P–R). (All drawings are reproduced from Abbott and Hollenberg 1976 with permission of the publishers.)

Table 7.1. *Characteristics of Rocky Intertidal Macrophyte Communities of Cold- and Warm-Water Habitats in the SCB*

Cold-Water Habitats	Warm-Water Habitats
High Intertidal Zone	
1. Lower portion of upper shores with some protection from wave action often dominated by *Pelvetia fastigiata* f. *fastigiata* at mainland sites and by either *P. fastigiata* f. *fastigiata* or broader-branched, olive-colored morphs of *P. fastigiata* f. *gracilis* on island sites.	1. Lower portion of upper shores with some protection from wave action usually characterized by *Pelvetia fastigiata* f. *fastigiata* on the mainland and by thinly branched, yellow-brown morphs of *P. fastigiata* f. *gracilis* on island sites.
2. Tendency for increased abundances of morphologically simple seaweeds (e.g., *Porphyra perforata, Ulva* spp. *Scytosiphon dotyi*) to occur in patches on upper shore.	2. Low wave energy (protected) island sites characterized by upper shore band of *Hesperophycus harveyanus*, a species that is relatively scarce on the mainland.
Middle Intertidal Zone	
1. General absence of extensive, compact turfs of articulated coralline algae. When present, articulated corallines usually occur as individual thalli and do not form a turf.	1. Presence of extensive turfs of numerous compact thalli of articulated coralline algae consisting of one or more of the following taxa: *Corallina officinalis* var. *chilensis, C. pinnatifolia, C. vancouveriensis, Lithothrix aspergillum,* and at more southerly sites, species of *Jania* and *Amphiroa*.
2. Predominance of larger, fleshy, saxicolous seaweeds, including abundant populations of *Endocladia muricata*, often in association with the barnacle *Balanus glandula, Mastocarpus papillatus, Iridaea cordata* var. *cordata*, and *Rhodoglossum affine*.	2. Generally a predominance of smaller, fleshy, turf-associated algae, often growing as epiphytes on turf-forming articulated coralline algae. In disturbed habitats, turf associates include small filamentouslike species such as *Ceramium* spp., *Polysiphonia* spp., *Gelidium pusillum*, and *Centroceras clavulatum*. In less disturbed habitats, larger seaweeds such as *Pterocladia capillacea, Gigartina canaliculata, Colpomenia* spp., and *Gelidium coulteri* are more prevalent.
3. Occasional occurrence of *Neorhodomela larix, Callithamnion pikeanum*, and *Fucus gardneri*, which become more abundant along with *Iridaea cordata* var. *cordata* farther north.	3. Seasonal occurrence of *Endarachne binghamiae* in winter months and *Laurencia pacifica* in summer and fall. On island sites, populations of *Laurencia snyderiae* and *Sargassum agardhianum* are common during the summer and early fall.
Low Intertidal Zone	
1. Lower brown algal zone dominated by populations of *Laminaria dentigera* and *Egregia menziesii*; increased abundances of *Cystoseira osmundacea* in place of *Halidrys dioica*. Generally, an absence of intertidal populations of *Eisenia arborea*.	1. Lower brown algal zone dominated by extensive populations of *Halidrys dioica, Egregia menziesii*, and *Eisenia arborea*, particularly during the summer and early fall when *Halidrys* is reproductive and maximal standing stocks of *Egregia* and *Eisenia* occur. Complete absence of intertidal populations of *Laminaria dentigera*.

Table 7.1. *Characteristics of Rocky Intertidal Macrophyte Communities of Cold- and Warm-Water Habitats in the SCB* (continued)

Cold-Water Habitats	Warm-Water Habitats
2. High abundances of fleshy, saxicolous red seaweeds, such as *Laurencia spectabilis* var. *spectabilis, Cryptopleura ruprechtiana, Iridaea cordata* var. *cordata, Prionitis lanceolata, Microcladia coulteri,* and *Phyllospadix* spp.; the last species bearing large, epiphytic thalli of *Smithora naiadum* during the summer and early fall.	2. Lower abundances of fleshy, saxicolous red algae. Generally prominent species include *Pterocladia capillacea, Gelidium purpurascens, Rhodoglossum affine,* and on occasion, *Gigartina spinosa.* Complete absence of *Iridaea* spp. *Smithora,* when present on *Phyllospadix* spp., usually much smaller.

From Stephenson and Stephenson 1972; Littler and Murray 1975; Littler 1977, 1978, 1979a, 1980a,b; Seapy and Littler 1978, 1982; Horn et al. 1983; Murray and Horn 1989; Littler et al. 1991; and personal observations of the senior author.

The persistence of coralline algal turfs on SCB shores may be related to their ability to retain space and recover rapidly following disturbance as well as to attributes of their growth form that increase resistance to desiccation stress and mortality caused by herbivory (Taylor and Hay 1984; Taylor 1985). Articulated corallines have a heterotrichous growth mode that allows for regeneration of upright axes from basal, crustose holdfasts. The significance of heterotrichy to the survival of *Corallina officinalis* var. *chilensis* (as *Corallina officinalis*) has been explored by Littler and Kauker (1984), who showed that *Corallina* can reestablish cover of erect axes four times more rapidly by regrowth from remnant basal crusts than by recolonizing sterilized surfaces by spores or vegetative fragments. Murray and Littler (1979) have also emphasized the rapidity with which articulated coralline algae are able to regenerate uprights from remnant pieces of basal crust.

In southern California, hot and dry Santa Ana winds predictably occur during the late fall and winter (see chap. 1) and coincide with the seasonal switch from night and early morning to afternoon minus tides, thus exposing intertidal algae to the stress of extreme heat and desiccation (Seapy and Littler 1982; Littler et al. 1991). Despite the extensive mortality of upright coralline algal fronds that correlates with this period, the basal crusts remain viable and can rapidly reestablish predisturbance standing stocks (Littler and Kauker 1984; Taylor and Hay 1984). Moreover, the compact turf growth form increases desiccation resistance (Taylor and Hay 1984; Taylor 1985), although at the cost of reduced productivity. It appears that once the basal crusts of articulated coralline algae are established, they can hold primary space and spread laterally by the marginal growth of apical cells (Littler and Kauker 1984). Stewart (1983) reported that the cells of these basal crusts can remain pigmented and presumably viable even after many months of sand burial. Consequently, turf-forming, articulated coralline algae appear well adapted to survive periods of enhanced desiccation stress and sand burial and scour, which are recurrent components of the disturbance regime characterizing SCB shores.

Littler (1980a) has suggested that this highly compact, epiphytized, coralline algal turf may in fact be characteristic of communities in stressed environments and that the composition of the turf epiphytes changes as a function of habitat disturbance. It appears (Littler and Murray 1975; Murray and Littler 1984) that the turf epiphytes at more disturbed SCB sites consist of smaller filamentous species (e.g., *Ceramium* spp., *Polysiphonia* spp., and *Centroceras clavulatum*), whereas larger fleshy species (e.g., *Pterocladia capillacea* and *Gigartina canaliculata*) appear in less disturbed areas (table 7.1 and fig. 7.7d). This view complements reports (Widdowson 1971; Thom and Widdowson 1978) that, since the 1950s, dominance has shifted from leafy, frondose algae toward turf-forming and articulated coralline species on the SCB mainland—a trend that is associated with an increase in anthropogenic disturbances.

Although macrophyte species assemblages in the SCB strongly correlate with ocean water masses (Murray and Littler 1981), it appears that other site-specific factors often modify and obscure the structures of intertidal macrophyte communities (Littler et al. 1991). For example, significant differences exist between mainland and island intertidal macrophyte communities, even between sites primarily exposed to the same water mass (Littler 1980a, b; Littler et al. 1991). Standing stocks of large brown algae (such as *Egregia, Eisenia*, and *Halidrys*) and surf grasses (*Phyllospadix* spp.) are generally higher, and middle intertidal coralline algae turfs are less compacted and bear larger algal epiphytes on less disturbed island shores than similar turfs occupying sites on the more heavily human-influenced mainland.

Quantitative classification analyses of whole intertidal communities, based on abundance data, produce patterns only partially consistent with those generated from presence–absence data, and these analyses reinforce the importance of unique site-specific conditions. Littler (1980a) believed that the influence of oceanic current systems and accompanying sea temperatures account for much of the observed pattern in community composition in the SCB; nonetheless, he strongly emphasized the confounding effects of the very high degree of site-specific autonomy. His classification analyses of 10 SCB intertidal communities, which were based on both macrophyte and macroinvertebrate wet and dry biomass and cover data, failed to produce patterns entirely consistent with those expected for a cold–warm water mass gradient. This led Littler (1980a) to conclude that various site-specific variables—such as the degree of wave action, coastal upwelling, substratum hardness and stability, frequency and degree of sand inundation, and levels of natural and human-induced disturbances—alter species composition and abundance profiles and dilute broad-scale patterns. A subsequent classification analysis of 22 intertidal sites also failed to entirely associate the community structures of intertidal sites in the SCB with the principal water mass to which they were exposed (Littler et al. 1991). Littler et al. (1991) identified two major groups, one containing sites subjected to seasonal sand inundation and scouring and a second whose members were more consistently free of sand influence. Both groups contained sites bathed in both warm- and cold-water masses. The high degree of environmental heterogeneity in the SCB manifests itself clearly in the complex relationships exhibited among the structures of the relatively unique intertidal communities (Littler et al. 1991).

ROLE OF WAVE ACTION

The degree of exposure to wave energy has long been recognized as a factor of major importance in structuring coastal communities (Lewis 1977; Stephenson and Stephenson 1972). Compared to California's central and northern coasts, most of which receive direct exposure to open ocean swell (Ricketts et al. 1985), nearly all of the SCB mainland coast and the leeward sides of the Channel Islands are protected from heavy wave action. Consequently, the effects of wave exposure on rocky shore communities of the eastern North Pacific are best known from research of habitats north of Point Conception (Stephenson and Stephenson 1972; Seapy and Littler 1978).

Intertidal communities appear to respond to increased exposure to wave shock by elevations in species distributions and changes in species composition (Lewis 1964; Stephenson and Stephenson 1972; Seapy and Littler 1978). Macrophyte morphologies also differ, with populations growing at wave-exposed sites usually being tougher, shorter, and narrower and having stronger attachment structures than those growing under lower wave energy conditions (Norton et al. 1981, 1982).

For wave-exposed shores near Pacific Grove, Stephenson and Stephenson (1972) depicted increased zonal elevations and shifts in species composition toward macrophytes such as *Lessoniopsis littoralis, Endocladia muricata*, and *Postelsia palmaeformis*; articulated coralline algae and "Lithothamnia" (calcareous crustose algae); and filter-feeding invertebrates such as *Pollicipes polymerus* and *Mytilus californianus*. A major repercussion of the increased abundances of macroinvertebrates and shifts in the standing stocks of most macrophytes on wave-exposed shores appears to be a change in trophic structure (Lewis 1964; Seapy and Littler 1978; McQuaid and Branch 1985). For example, on South African shores, the predominance of sessile, filter-feeding animals results in a shift toward the importation of energy from the water column through the activities of filter feeders, whereas on sheltered shores, an algal-mobile herbivore trophic structure predominates and emphasis is placed on energy export through the expanded role of primary producers (McQuaid and Branch 1985; McQuaid et al. 1985).

For rocky intertidal habitats in the SCB, most macrophytes considered to be indicators of wave-exposed sites (fig. 7.11) have northerly, cold-temperate distributions and appear in the SCB only at sites containing elements of a cold-water biota. This reflects the much greater understanding we have of the biota of wave-exposed habitats along the shoreline north of Point Conception where open coast, high wave energy beaches predominate. Previous research has not focused on variations caused by wave exposure in the structure and organization of rocky intertidal communities in the SCB. Also, of the rocky intertidal sites surveyed, none compare to the steep-sloped, wave-swept areas of central California.

Unfortunately, quantitative measures of the degree of wave action prevalent at the sites studied by Littler and his colleagues (Littler 1979a; Littler et al. 1991) are not available, thereby making it impossible to determine their relative positions on a common wave-exposure scale. The most exposed of these sites (Ocean Beach, San Diego County) contained fewer macroinvertebrate taxa and more macrophyte species and reduced standing stocks of both types of macroorganisms than did sites sheltered from wave action. This contrasts with trends described for wave-swept North Atlantic shores (Lewis 1964), South African shores (McQuaid and Branch 1984, 1985; McQuaid et al. 1985), and cold-temperate Pacific shores (Stephenson and Stephenson 1972; Seapy and Littler 1978), where enhancement of macroinvertebrate and reduction of macrophyte standing stocks have been consistently demonstrated. Littler et al. (1991) have suggested that reductions in the standing stocks of macroorganisms at Ocean Beach may be related to the soft, friable nature of the sandstone substratum, which has a tendency to erode when exposed to high wave energy.

ROLE OF GRAZING

Grazing animals play an extremely important role in determining macrophyte distributions and abundances in shallow-water marine systems (Gaines and Lubchenco 1982; Hawkins and Hartnoll 1983). In temperate intertidal habitats, such as the SCB, grazers consist primarily of prosobranch, trochiid, and mesogastropod molluscs, sea urchins and asteroids, isopods, amphipods, and crabs (Gaines and Lubchenco 1982; Hawkins and Hartnoll 1983) (see also chap. 8). Dipteran larvae also can be important grazers of upper zone ephemeral algae in eastern North Pacific intertidal communities (Robles 1982). The composition of grazers differs from that of subtropical and tropical locales where herbivorous fishes are believed to be much more numerous and important (Gaines and Lubchenco 1982; Hawkins and Hartnoll 1983).

The influence of grazers on the distributions and abundances of intertidal algae has

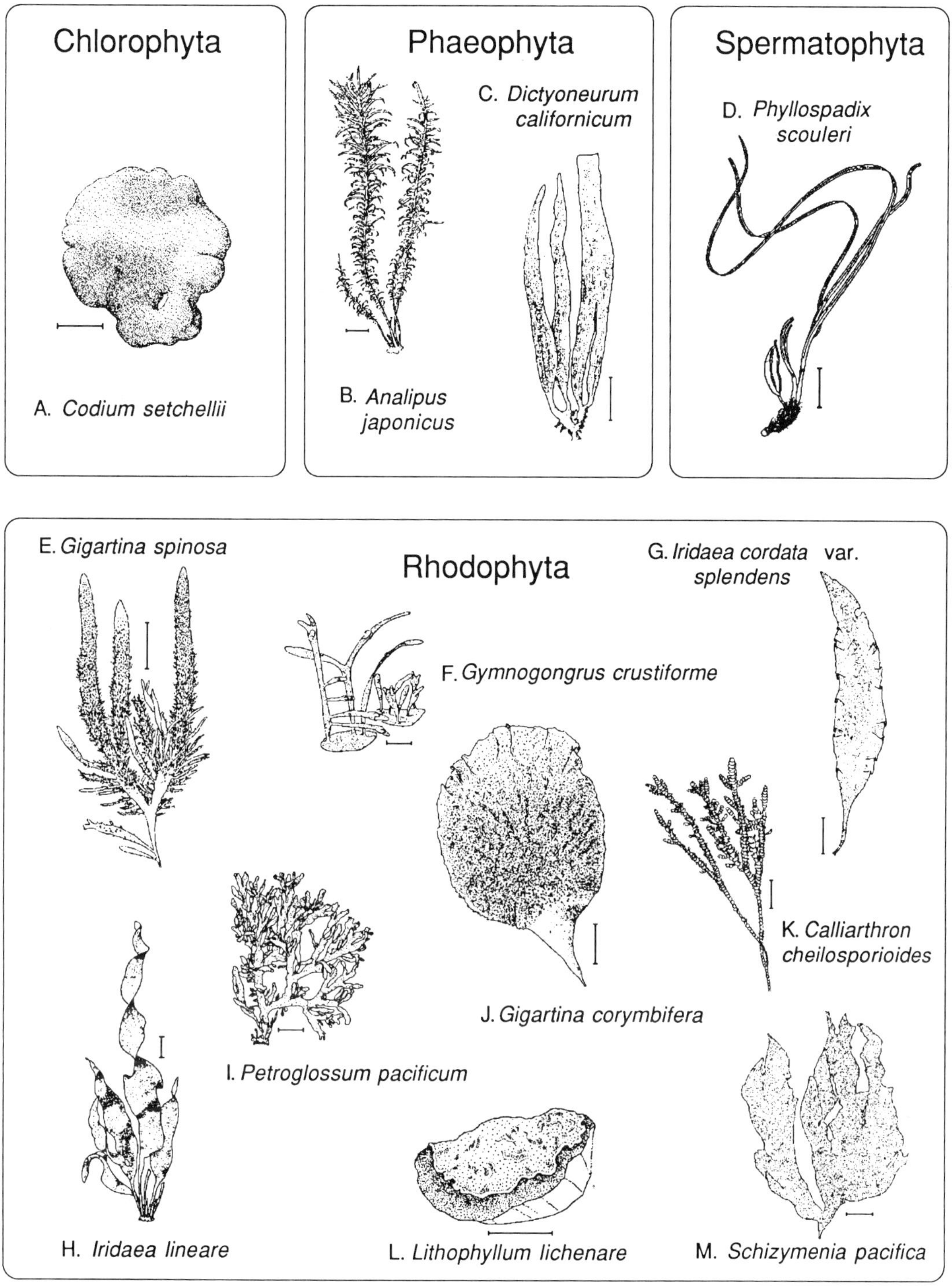

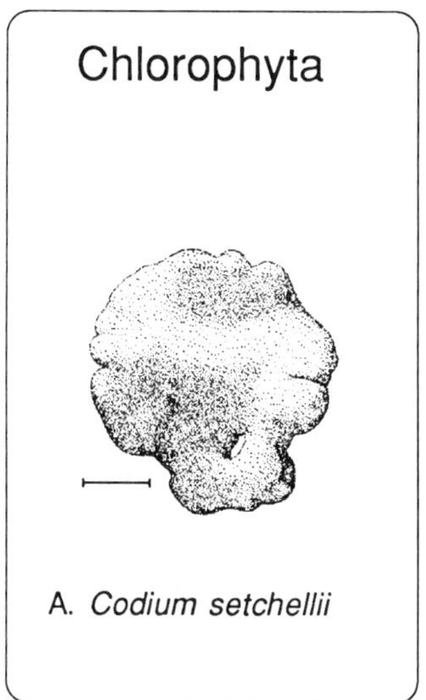

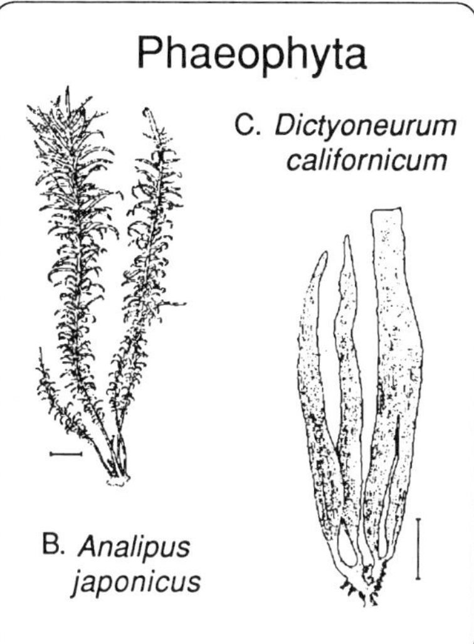

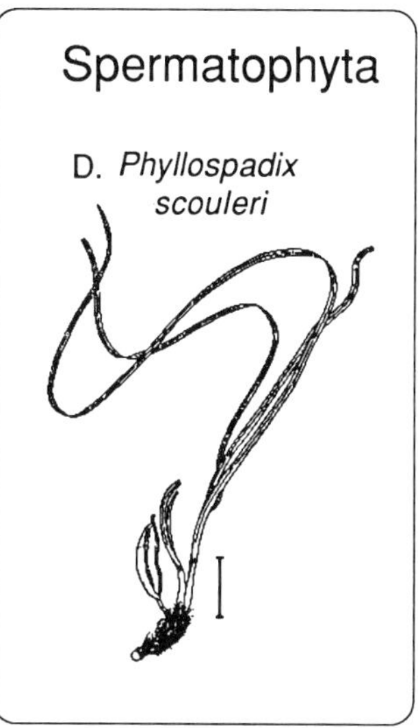

Chlorophyta
A. Codium setchellii
Phaeophyta
B. Analipus japonicus
C. Dictyoneurum californicum
Spermatophyta
D. Phyllospadix scouleri
Rhodophyta
E. Gigartina spinosa
F. Gymnogongrus crustiforme
G. Iridaea cordata var. splendens
H. Iridaea lineare
I. Petroglossum pacificum
J. Gigartina corymbifera
K. Calliarthron cheilosporioides
L. Lithophyllum lichenare
M. Schizymenia pacifica

Figure 7.11. Macrophytes characteristic of rocky intertidal habitats exposed to strong wave action in the SCB. Scales: 2 mm (F); 1 cm (B, I); 2 cm (A, C, K, L); 3 cm (D, G, H, M); 5 cm (E, J). (Sources of illustrations: D is reproduced from Scagel 1972 courtesy of the Royal British Columbia Museum, Victoria; L is an original drawing; remainder are reproduced from Abbott and Hollenberg 1976 with permission of the publishers.)

received much attention, particularly since the important limpet removal experiments performed in the North Atlantic over 40 years ago (Jones 1948; Southward 1964). Grazers can have pronounced effects on macrophyte communities. They can (1) determine the upper and lower limits of algal growth (Underwood 1980; Underwood and Jernakoff 1984); (2) lead to patchiness in local algal distributions (Lubchenco 1983; Jernakoff 1983; Gaines 1985); and (3) play a role in determining algal distributions and abundances in pools and on emergent surfaces (Lubchenco and Gaines 1981; Dethier 1982; Underwood and Jernakoff 1984; Duggins and Dethier 1985), and over wave-exposure gradients (Southward 1956). Unfortunately, experiments revealing the effects of grazers on intertidal macrophytes in the SCB have been few, and research emphasis has been on descriptive studies and on the role played by agents of physical disturbance in determining macrophyte distributions and abundances (Littler et al. 1991).

In addition to localized effects, latitudinal variations in the composition of the algal floras along the North and Central American coastlines have been linked to geographic differences in herbivory. Gaines and Lubchenco (1982) calculated that the proportion of the algal flora with calcified erect, calcified crustose, and fleshy branched thalli increased, whereas large algal thalli (20 cm to >1 m) and algae with bladelike morphologies decreased along the western North American coast from temperate to tropical latitudes. They hypothesized that these latitudinal patterns were influenced by changes in the intensity of herbivory, with tropical sites (i.e., Panama) being subject to greater grazing pressure than sites in the Pacific Northwest, due in part to the year-round foraging of highly mobile, herbivorous fishes. Many studies show that, in tropical and temperate seas alike, strong grazing pressure can lead to algal communities dominated by crustose species with tough thalli that are resistant to grazing and physical stress (Chapman 1986). Similar observations have been made for articulated coralline algae, whereas fleshy seaweeds have been shown to be more susceptible to grazer-induced disturbance (Littler and Littler 1980b).

With the exception of the work of Sousa et al. (1981), experimental tests of hypotheses, such as those of Gaines and Lubchenco (1982) concerning latitudinal variation in the composition of intertidal macrophyte communities, have not been performed in the SCB. As pointed out previously, shores of the SCB are often characterized by high abundances of turf-forming articulated coralline algae. In addition, larger brown algae generally are more abundant in the lower intertidal zone at latitudes north of the SCB, particularly along the mainland coast (Stephenson and Stephenson 1972; Sousa et al. 1981). Experiments by Sousa et al. (1981) suggest that reductions in the abundances of large brown algae in low intertidal habitats in the SCB result from sea urchin grazing, preemption of space by perennial red algae which readily proliferate vegetatively into freshly cleared spaces, and the frequency of space-clearing disturbances. These experiments (Sousa et al. 1981) emphasize the ability of many perennial red seaweeds in the SCB to grow vegetatively into cleared areas, in contrast with the growth patterns of most long-lived red algae of the Pacific Northwest (Dayton 1975). This tendency toward vegetative propagation also appears to be well developed in intensely grazed tropical turf communities (Hay 1981a, b; Sousa et al. 1981).

PATTERNS OF SEASONAL AND INTERANNUAL VARIATION

Although there are no long-term quantitative records for macrophyte populations at any intertidal site in the SCB, the available qualitative evidence suggests that considerable change has occurred in the composition and relative abundances of the macroalgal and seagrass populations during the past 90 years (Murray 1974; Harris 1980). Because of the absence of appropriate data, neither the specific changes that have occurred nor their causes can be clearly demonstrated. However, it has been argued that these changes are directly related to human activities (Dawson 1959, 1965; Widdowson 1971; Thom and Widdowson 1978).

The first observations of long-term and short-term variations in the compositions of the intertidal macrophyte floras of the SCB were provided by Dawson (1959, 1965), who described a decline in overall diversity and changes in the relative abundances of certain macrophyte forms, then attributed these changes to discharged human wastes. Basing his calculations on comparisons of his 1956–1959 survey data with the 1895–1912 herbarium records of William A. Setchell and Nathaniel L. Gardner, Dawson estimated the reduction in species richness to be as great as 50–70% at two sewage-polluted sites (White Point and Sunset Boulevard Beach) (fig. 7.1) since the turn of the century. Dawson consistently noted depauperate macrophyte floras and enhanced growths of articulated coralline algae—that is, *Bossiella chiloensis* (as *B. insularis*), *Bossiella orbigniana* ssp. *orbigniana* (as *B. cooperi*), and *Corallina vancouveriensis*—together with reduced abundances of fleshy red algae. He believed that sewage pollution was responsible for these changes. Short-term seasonal variations in intertidal macrophytes were also observed by Dawson (1965), who thought them to be caused by physical parameters such as shifting sands, high surf activity, heavy precipitation, or desiccating winds during low tide periods.

Subsequent studies have used Dawson's stations and observations as a baseline for detecting changes that have taken place during the past 30 years. These studies include resurveys performed to detect effects of the 1969–1970 Santa Barbara oil spills (Foster et al. 1971; Nicholson and Cimberg 1971) and to document long-term variations in the SCB macrophyte flora (Widdowson 1971; Thom and Widdowson 1978; Harris 1980, 1983). Unfortunately, Dawson's (1959, 1965) surveys were floristic accounts and did not generate data from which to answer ecological questions concerning changes in species distributions and abundances.

Views similar to those of Dawson (1959, 1965) are expressed by Widdowson (1971) and Thom and Widdowson (1978), who believe that more recent changes in the composition of intertidal macrophyte communities in the SCB are due to anthropogenic agents and activities. Approximately 15 years following Dawson's surveys, Widdowson (1971) recorded a general and widespread reduction in macrophyte diversity in the SCB accompanied by increased abundances of disturbance-resistant turf-forming species and decreases in leafy and massive forms. He implicated human usage, air pollution, and pollution from sewage outfalls, in that order, as the three most important factors associated with the observed changes. Approximately 5 years later, Thom and Widdowson (1978) noted a continued shift toward turf-forming algae and away from massive algal forms, but were unable to detect further declines in macrophyte diversity. Only Harris (1980, 1983) suggested that an increase in algal diversity has occurred in recent years, but she also attributed this increase to changes in human activities (in this case, improvements in the treatment and discharge of sewage effluent along the Palos Verdes Peninsula).

Seasonal variations in the distributions and abundances of marine macrophyte populations are well documented for temperate waters and have often been attributed to fluctuations in abiotic environmental parameters such as light, temperature, storm frequency, and tidal regimes (Emerson and Zedler 1978; McQuaid 1985; Murray and Horn 1989). Quantitative studies of macrophyte seasonality in the SCB are restricted to the research of Littler and his colleagues (Murray and Littler 1977; Littler 1979a, 1980a, b; Littler et al. 1991) who described seasonal and interannual variations in macrophyte communities at 12 sites over 3.8 years, a period insufficient to clearly delineate seasonal or annual cycles in highly variable coastal ecosystems.

Seasonal tendencies in intertidal macrophyte communities in the SCB appear to be difficult to detect, suggesting that local or site-specific conditions predominate over regional temporal cycles. Littler et al. (1991) emphasized the site-specific importance of stochastic, abiotic disturbances such as (1) heavy precipitation events with accompanying sediment loading and reduced salinities; (2) periods of extreme heat, low humidity, and high winds occurring during periods of tidal emersion; (3) large storm-generated waves; and (4) longshore sand transport. Perhaps the environmental variable of greatest importance is the seasonal pattern of daytime tidal emersion, a parameter frequently correlated with changes in the abundances of intertidal macrophytes in the SCB (Emerson and Zedler 1978; Gunnill 1980; Seapy and Littler 1982; Littler et al. 1991). Although the first two disturbances characteristically occur during the late fall and winter in southern California, the last two are temporally and spatially highly variable, perhaps contributing to the lack of clear-cut seasonal patterns.

Despite the high degree of site-to-site variation, Littler (1980a,b) and Littler et al. (1991) have identified two general patterns of macrophyte seasonality for southern California shores: (1) slight reductions in standing stocks and diversity following the fall–winter periods of daytime tidal emersion and (2) growth and accumulation of large seaweed standing stocks throughout the summer periods of long daylight hours and warmer sea temperatures. It appears that where patterns can be detected, seasonality of SCB macrophytes conforms with those for most mid-latitudinal locales, where standing stocks appear to be greatest during late spring through summer and least during fall or winter (Horn et al. 1983; Murray and Horn 1989). This pattern contrasts with that described for certain tropical intertidal communities where summer appears to be the season of least macrophyte abundance (Ngan and Price 1980; Hodgkiss 1984).

Subtidal Kelp Forests

Extensive stands of giant kelp (*Macrocystis*) are conspicuous and well-known features along much of the coastline in the SCB. These "kelp forests" or "kelp beds" extend from the sea floor to the surface to form a vertically structured habitat throughout the water column and exert a marked influence on physical and biological features of the coastline.

The surface area of kelp forests throughout the SCB has been quite variable through time. Aerial surveys in the mid–1970s indicated a kelp canopy area of approximately 88 km^2 (table 7.2). About half of the kelp occurs along the mainland and 28% and 20% occurs in the southern and northern island groups, respectively, with San Nicolas Island alone providing more than 14% of the total SCB kelp cover.

There is some controversy concerning the number of species of *Macrocystis* occurring in the SCB (Foster and Schiel 1985; Brostoff 1988). Most authors, including Abbott and Hollenberg (1976), refer exclusively to *Macrocystis pyrifera* in southern California. However, a second form, which has holdfast features resembling *M. angustifolia* from the Southern Hemisphere, commonly occurs from Santa Barbara to the northwestern end of the SCB and has been found as far south as San Clemente Island (Neushul 1971). The *M. angustifolia* form has a rhizomatous holdfast and reproduces vegetatively, and it has more stipes per plant and lives longer than *M. pyrifera* (Neushul 1971). Brostoff (1988) suggests that there is ecotypic variation in *Macrocystis* holdfast systems, probably related to substratum characteristics, and he argues for a single species, *M. pyrifera*, in southern California.

The early history of research on kelp forests has been summarized by North (1971), Neushul (1981), and Harger (1983). Most research on kelp forests is fairly recent, beginning in the 1950s with the advent of scuba diving. Nonetheless, the amount of research

Table 7.2. *Kelp Canopy Coverage (in km²) in the SCB, 1975–1977*[a]

Location	Average Area	Coefficient of Variation
Southern California mainland (Point Dume to Mexico border)	46.0	4.0
Southern California islands		
Southern island group		
San Clemente	9.0	24.9
Santa Catalina	0.7	22.6
Santa Barbara	2.3	23.2
San Nicolas	12.6	19.9
Southern island total	24.6	20.4
Northern island group		
Anacapa	0.9	14.3
Santa Cruz	4.8	21.7
Santa Rosa	8.0	12.4
San Miguel	3.6	12.7
Northern island total	17.4	7.6
Total kelp canopy coverage	88.0	5.3

[a]Average area and coefficients of variation were computed from eight aerial surveys taken approximately quarterly between September 1975 and June 1977 in 50 beds throughout the SCB. (Original data from Table III-1.4-1 in Hodder and Mel 1978.)

on *Macrocystis* easily exceeds that performed on all other species of algae in the SCB. Just in the past 7 years, there have been several major reviews of kelp forests in California (Dayton and Tegner 1984a; Dayton 1985; Foster and Schiel 1985; North et al. 1986), as well as more specialized treatments on the effects in the SCB of waste disposal (Bascom 1983) and El Niño events (Tegner and Dayton 1987). In addition, kelp forests in the SCB received major emphasis in general reviews of temperate subtidal algal stands (Dayton 1985; Schiel and Foster 1986) and of the ecological effects of Pacific sea otters (VanBlaricom and Estes 1988).

Examination of current knowledge about subtidal kelp forests reveals several problems. First, almost all research has focused on the large sporophyte phase of *Macrocystis*, no doubt because of its great commercial, recreational, and ecological importance. Remarkably little is known about the microscopic stages of *Macrocystis* until recently (Dean et al. 1984, 1988; Deysher and Dean 1986a, b; Reed et al. 1988; Reed 1990). Second, the knowledge of kelp ecology is probably biased in that subtidal studies are constrained by physical and logistical limitations of the researcher. Thus, many of the factors that affect various stages of subtidal macrophytes probably have not been studied at the appropriate scales (Schiel and Foster 1986). Regardless of these research deficiencies, the emerging picture shows that kelp forests are dynamic systems that exhibit a high degree of spatial and temporal variation. Factors affecting the distributions and abundances of *Macrocystis* and other macrophytes in the SCB differ in relative importance across these spatial and temporal scales.

Macrophytes within a kelp forest form horizontal strata much like the layering in a terrestrial rain forest (Dayton et al. 1984; Foster and Schiel 1985). Kelp forests in the

SCB may consist of one or more of the following strata: (1) a floating canopy of *Macrocystis*; (2) an erect understory of stipitate forms, including *Pterygophora californica*, *Egregia menziesii*, and *Eisenia arborea*; (3) a prostrate, low-growing canopy of species such as *Agarum fimbriatum*, *Laminaria farlowii*, and *Cystoseira osmundacea*; (4) bottom-dwelling understory species often composed of articulated corallines (e.g., *Calliarthron*) and many foliose and filamentous red and brown algae; and (5) a crustose layer generally consisting of calcified forms such as *Lithophyllum* and *Lithothamnion*.

Kelp forests differ in structure within and between beds throughout the SCB. For example, macrophytes in the Point Loma kelp forest are arranged in distinct patches dominated by members of only one or two strata (Dayton et al. 1984). Thus, patches consist mainly of either *Macrocystis* alone, a mix of *Laminaria* and *Cystoseira*, or a mix of *Pterygophora* and *Eisenia*. Dayton et al. (1984) identified and monitored eight patches within the Point Loma kelp forest between 1971 and 1981. The boundaries of these patches remained relatively constant during this period, and there was evidence that at least one *Pterygophora* patch had persisted since 1946. This patchiness is not universal throughout the SCB. In the San Onofre kelp forests, *Macrocystis* and *Pterygophora* exist in mixed stands rather than in distinct patches (Dean et al. 1984), and kelp patches at San Nicolas Island frequently switch to barren substrata (Harrold and Reed 1985).

Major physical differences exist among kelp forests in the nature and profile of the substratum, degree of wave exposure, and clarity and nutrient concentrations in the water column. Hodder and Mel (1978) have presented detailed descriptions of 50 existing or historical mainland and island forests in the SCB. Some of the more intensely studied forests (see fig. 7.1) are illustrated in fig. 7.12 to emphasize the variety of kelp communities found in the SCB.

Much of the substratum between Santa Barbara and Point Conception at water depths of 5–20 m consists of low-relief mudstone interspersed with sand flats and occasional low-relief rocky outcrops. This part of the SCB is usually well protected from large swells by Point Conception and by the southern California Channel Islands, allowing *Macrocystis pyrifera* ("*M. angustifolia* form") to become established on sand as well as on low-relief rock. The forest at Campus Point, Goleta, is typical of this section of the mainland coast (fig. 7.12a). In contrast, Naples Reef, 1.6 km offshore from Goleta, is a 2.2-ha high-relief rocky reef that is isolated from inshore kelp forests by extensive sand and cobble fields (fig. 7.12b). The giant kelp here is typical of *M. pyrifera* and is the form that dominates forests farther south.

The substratum of much of the mainland forests south of Santa Barbara ranges from sand and cobbles intermixed with low-relief siltstone (San Onofre, fig. 7.12c) to a wide mudstone–sandstone terrace interspersed with sand, cobbles, and rocks (Point Loma, fig. 7.12d). These forests are more exposed to storm-generated swells.

The physical settings for forests around the eight southern California Channel Islands are more variable among locations than those along the mainland (Hodder and Mel 1978). In some areas, such as Fisherman Cove at Santa Catalina Island (fig. 7.12e), the steepness of the bottom profile restricts forests to a thin ribbon adjacent to shore. Forests on other islands are wider because of the broad, shallow aprons of firm substratum. The islands also differ dramatically in exposure to oceanic swells. San Miguel Island, for example, is the most exposed both to winter swells from the northwest and to summer swells from the south. In contrast, coves along the leeward side of the Channel Islands (e.g., Fisherman Cove) are protected from oceanic swells. Compared to mainland water, island water is clear. Furthermore, at least near Fisherman Cove, water is more oligotrophic, so plants are often exposed to more irradiance but lower nutrient levels (Gerard 1982a; Zimmerman and Kremer 1984).

Kelp forests in the SCB are dynamic over

short-term and long-term time scales. Variation in the extent of kelp forests in the SCB has attracted considerable interest, initially because of the decline in forests off Palos Verdes and Point Loma and more recently because of the destruction of kelp forests throughout much of the SCB from unusually severe winter storms and the 1983 El Niño event. Long-term variation (years to decades) in kelp forests and the effects of El Niño events have received major coverage in recent reviews (Dayton and Tegner 1984a; Tegner and Dayton 1987).

The decline in kelp forests off the Palos Verdes Peninsula and Point Loma that began in the mid–1940s and continued through the mid–1970s (fig. 7.13) was attributed to a complex combination of factors involving human activities and natural events. Pollution increased from sewage outfalls, and grazing by sea urchins increased as fishing pressure intensified on their competitors (abalone) and predators (lobsters and California sheephead). Furthermore, the elevation of temperatures and depletion of nutrients experienced during the 1957–1959 El Niño placed additional physiological stress on *Macrocystis* (Wilson and North 1983). Not all kelp forests in the SCB suffered comparable declines, and at least several kelp forests between Santa Barbara and Point Conception have remained fairly constant over the past 50 years (Coon 1981).

Repeated aerial surveys of kelp forests throughout the SCB provide quantitative data on fluctuations in coverage. During Hodder and Mel's (1978) 21-month study, total kelp coverage in the SCB in eight quarterly estimates ranged from 81.6 to 94.8 km^2. Coefficients of variation in coverage differed among locations, ranging from 19% along mainland forests to 26% in the southern Channel Island group. Harger (1983) examined variations in the 32 forests along the SCB mainland and the forests surrounding Santa Catalina Island. The Santa Barbara region accounted for 65% of the kelp coverage, although it represented only 18% of the coastline studied. The Santa Barbara forests were larger and had greater canopy cover and exhibited a lower variation in cover from one year to the next (table 7.2). Forests around Los Angeles and San Diego were smaller, had less canopy cover, and showed greater yearly variation (Harger 1983).

There is much speculation about the causes of long-term variation in the extent of kelp forests in the SCB. Although no new information has emerged on the causes of these historical fluctuations in the areal coverage of kelp forests, recent laboratory and field studies have increased our understanding of the role of nine factors that affect kelp forest macrophytes in the SCB. The factors are discussed here separately, although these and other factors interact over all scales of space and time (Foster and Schiel 1985, 1988).

ROLE OF SUBSTRATUM

Most subtidal macrophytes need a firm anchor. *Macrocystis* sporophytes have gas-filled pneumatocysts at the base of each blade that make plants buoyant and cause fronds to rise through the water column. This tendency to float must be countered by attachment to the substratum through the holdfast system. If the substratum is not sufficiently massive or strong, whole plants, including the holdfasts and their substratum, can be lifted and moved away. This has been observed at San Onofre, where *Macrocystis* anchors to boulders. Storms move entire plants, with boulders still attached, to other areas where they may become reestablished (Foster and Schiel 1985; Dean et al. 1989). At Santa Catalina Island, *Macrocystis* plants that had settled on small, unstable substrata were swept as far away as 140 m when the buoyancy from developing pneumatocysts exceeded the force required to maintain contact with the substratum (Wells 1983).

Most established *Macrocystis pyrifera* forests in the SCB occur over hard substrata. Along much of the coastlines of Ventura and Santa Barbara counties, however, stands of the *M. angustifolia* form are often attached directly to the sandy bottom. The *M. angusti-*

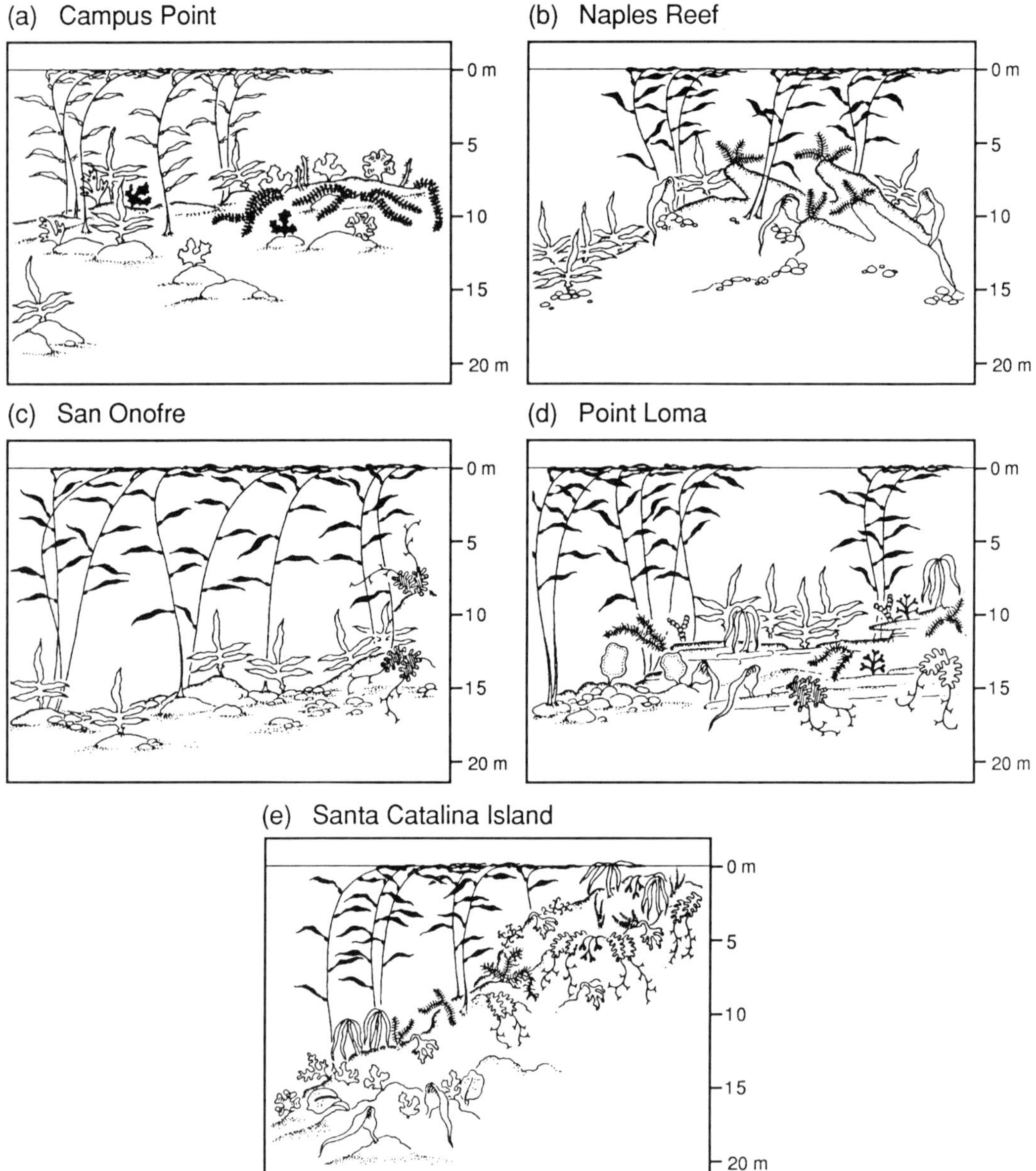

Figure 7.12a–e. Cross-sectional views of selected kelp bed communities in the SCB. (After Turner et al. 1968, Dayton et al. 1984, Dean et al. 1984, Foster and Schiel 1985, Ebeling and Laur 1985, Ebeling et al. 1985, Larson and DeMartini 1984, R. Bray pers. observ.)

folia type of holdfast is more extensive and produces more stipes than *M. pyrifera*, possibly as an adaptation for living on such unconsolidated substrata (Neushul 1971).

Because of difficulties of studying microscopic life history stages in the field, nothing is known about substratum preferences, if any, of the microscopic gametophytes or the earliest stages of the young sporophytes. However, Wells (1983) examined the substratum to which 80 young *Macrocystis* sporophytes were attached at Santa Catalina Island. Available substrata were categorized as other algae (that is, articulated and crustose coral-

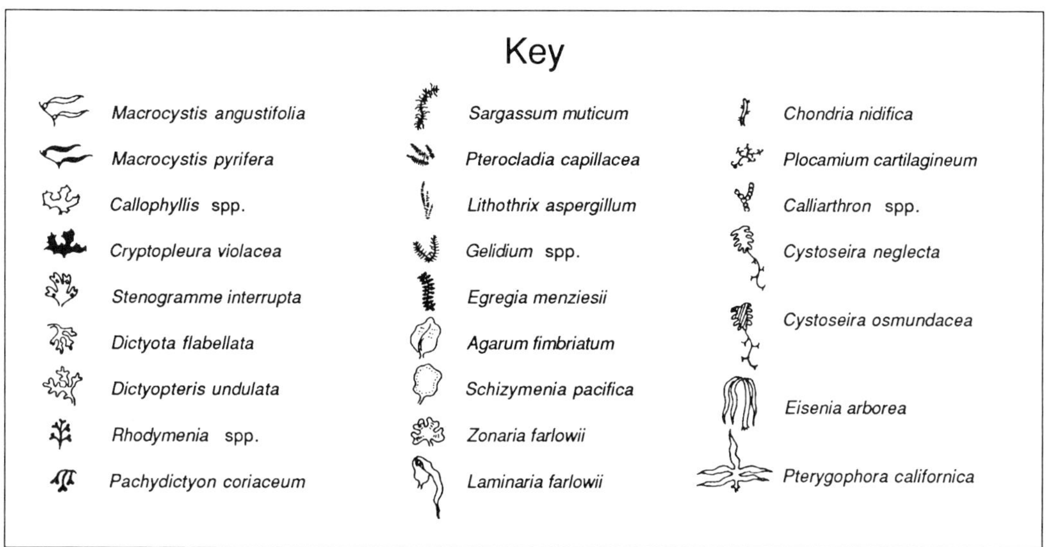

Figure 7.12. (continued)

lines) and bare rock. Wells found more kelp sporophytes on bare rock than were expected, based on substrata availability; numbers of sporophytes on the two algal substrata were proportional to the abundance of these substrata in the field.

Kelp sporophytes can also become established on newly exposed substrata. Harris et al. (1984) monitored the reestablishment of the macrophyte assemblage at Naples Reef after severe storms removed large numbers of macrophytes and grazers and created new surfaces by breaking off sections of the rocky substratum. Plants that appeared on old surfaces encrusted with corallines consisted of blades and filaments of red algae; these probably regenerated from basal fragments that survived the storms. Newly exposed surfaces, however, developed dense stands of filamentous brown algae. Within two months, small *Macrocystis* and *Pterygophora* sporophytes were observed over the reef but were most abundant on new surfaces (Harris et al. 1984).

ROLE OF OCEAN WATER MASSES AND TEMPERATURE REGIMES

Kelp abundances, growth, and recruitment are closely linked to the dynamic thermal environment in the SCB. The effects of the thermal environment can be either direct or indirect in controlling rates of chemical reactions because of the many other factors (such as nutrient concentrations) that are correlated with temperature. The general temperature regime within a kelp forest in the SCB follows a seasonal pattern. Summer and fall have a well-developed thermocline; winter and spring are nearly isothermal. Coldest sea surface temperatures generally occur between January and March, but lowest bottom temperatures occur around June (Quast 1971a, b).

Subtidal macrophyte assemblages differ throughout the SCB, presumably in relation to surface water currents and accompanying thermal regimes (Neushul et al. 1967; North 1971). Detailed information of geographic variation in the species composition and structure of kelp forests in the SCB is scant and is lacking for many areas among the southern California Channel Islands. Surveys from the low intertidal zone to depths of 38 m on Anacapa Island indicate that the subtidal flora contains a mixture of northern and southern elements (Clarke and Neushul 1967). *Pelagophycus porra*, common in deeper kelp forests farther south, was notably absent from Anacapa Island forests, although this

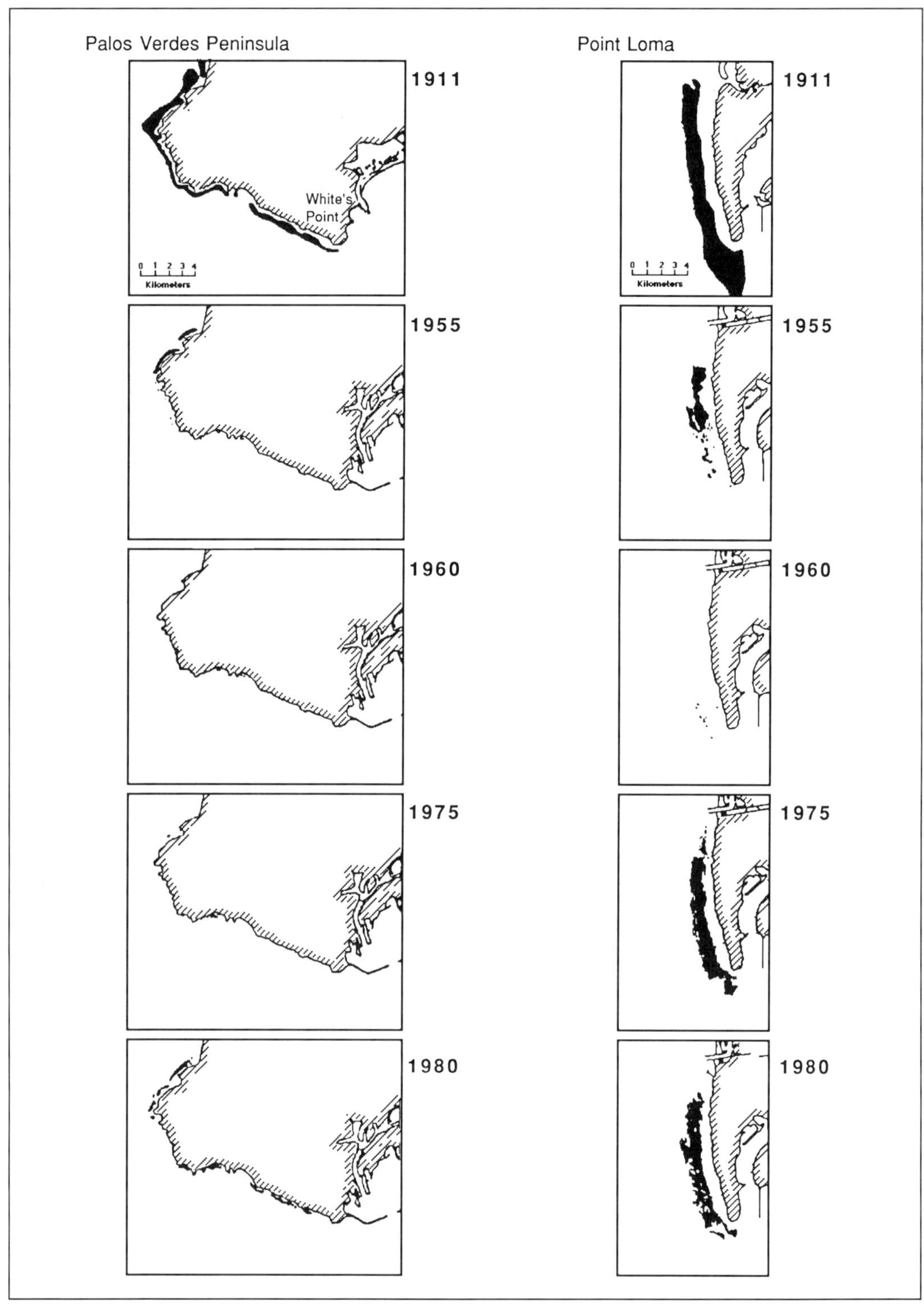

Figure 7.13. Historical patterns of surface coverage of kelp bed canopies at Palos Verdes Peninsula, Los Angeles County, and Point Loma, San Diego County. (After Wilson and North 1983.)

also could be due to the lack of appropriate rocky substratum in deep water. Neushul et al. (1967) commented that islands exposed to more northerly cold and southerly warm ocean currents than Anacapa have more cold- and warm-water representatives, respectively, but these authors provided no supporting data for their comments. If the distributional pattern is similar to that described earlier for intertidal algae, then subtidal assemblages of San Miguel, San Nicolas, and the western half of Santa Rosa Island should have cold-water representatives. In contrast, macrophyte assemblages off Santa Catalina and San Clemente islands should contain more warm-water species. However, such studies have not yet been undertaken.

Species lists compiled by Foster and Schiel (1985) provide some insight into the variation in kelp forests within and to the north of the SCB (table 7.3). Some species occurring in central California kelp forests, such as *Nereocystis luetkeana* and *Laminaria dentigera*, are unknown in most kelp forests in the SCB. Others, such as *Pelagophycus porra, Eisenia arborea, Agarum fimbriatum*, and *Laminaria farlowii*, are frequent kelp forest inhabitants in the SCB, particularly at warmer water sites (North 1971), but are not reported by Foster and Schiel (1985) to be conspicuous components of central California forests. It is important to realize that considerable site-to-site variation occurs in the composition of kelp forests since none of the aforementioned macrophytes was reported to occur at all sites in either central or southern California (table 7.3). Thus, as in rocky intertidal habitats, unique local factors may override the broad distributional effects of water currents and their correlates. A caveat is that Foster and Schiel (1985) obtained information from various sources suggesting that some of the site-specific differences may be due to variations in sampling methods and effort. Clearly, more information is needed on spatial and temporal distributions of subtidal macrophytes inhabiting kelp forests in the SCB.

ROLE OF NUTRIENTS

Physical processes that supply cold, nitrate-rich water to kelp forests in the SCB include wind-generated upwelling, horizontal advection, storm-associated mixing, coastal runoff, and the episodic ascent of the thermocline (North et al. 1982; Zimmerman and Kremer 1984). Biological processes may include local regeneration of ammonium from reef invertebrates and fishes, especially at night (Bray et al. 1986, 1988). Not all processes are influential in all locations. Winter runoff and spring upwelling cause major spikes in nitrate near San Onofre (North et al. 1982) but make only a minor contribution at Santa Catalina Island, where upward movement of the thermocline provides major, though episodic, inputs (Zimmerman and Kremer 1984).

The regular reduction of the *Macrocystis* canopy in southern California during the summer and early fall months has long been attributed to extended periods of water temperatures greater than 20°C (North 1971). Nutrient concentrations, however, are inversely related to water temperature, and evidence indicates that nutrient depletion might be a more significant factor (Jackson 1983; Zimmerman and Kremer 1984). *Macrocystis* continues photosynthesis at elevated temperatures under appropriate conditions. Photosynthesis peaks at 20°C but remains high above 25°C, and ratios of daily photosynthesis to respiration remain unchanged at water temperatures above 20°C (Zimmerman 1983; Arnold and Manley 1985). In addition, the temperature optimum for phosphate uptake by blades of adult *Macrocystis pyrifera* is 24°C (Manley 1985). North et al. (1986) pointed out that two basic physiological processes, photosynthesis and nutrient uptake, occur unimpeded above 20°C.

Field additions of nutrients provide some additional, but not always consistent, support for the hypothesis that nutrients limit summer growth of adult sporophytes. Application of nitrate in a natural forest at Santa

Table 7.3. *Conspicuous Components of Kelp Forests from Central and Southern California after Foster and Schiel (1985)*[a]

Species	Central California Kelp Forests					Southern California Kelp Forests				
	Greyhound Rock	Sandhill Bluff	Point Cabrillo	Stillwater Cove	Granite Creek	Campus Point	Anacapa Island	Catalina Island	Del Mar	Point Loma
Floating canopy species										
Macrocystis integrifolia Bory				X						
Macrocystis pyrifera (L.) C. Ag.	X	X	X	X	X		X	X	X	X
"*Macrocystis angustifolia*" type of *M. pyrifera*						X				
Nereocystis luetkeana (Mert.) Post. & Rupr.	X				X					
Pelagophycus porra (Lem.) Setch.								X		X
Stipitate, erect understory species										
Desmarestia ligulata var. *ligulata* (Lightf.) Lamour	X				X				X	
Egregia menziesii (Turn.) Aresch.			X	X	X	X				X
Eisenia arborea Aresch.							X	X		
Laminaria dentigera Kjellm.	X	X		X	X					
Pterygophora californica Rupr.	X	X		X	X	X	X		X	X
Prostrate, low-profile canopy species										
Agarum fimbriatum Harv.							X	X		
Costaria costata (C. Ag.) Saunders					X					
Cystoseira neglecta S. & G.								X		
Cystoseira osmundacea (Turn.) C. Ag.			X	X	X					X

Dictyoneuropsis reticulata (Saund.) Smith			X							
Dictyoneurum californicum Rupr.	X									
Laminaria farlowii Setch.							X	X	X	X
Phyllospadix spp.			X				X			X
Sargassum muticum (Yendo) Fensh.								X		
Bottom-dwelling understory species										
Acrosorium uncinatum (Turn.) Kyl								X		
Bossiella californica (Dec.) Silva				X						
Botryocladia pseudodichotoma (Farl.) Kyl.			X							
Calliarthron cheilosporioides Manza				X	X					
Calliarthron tuberculosum (Post. & Rupr.) Dawson				X						
Callophyllis spp.						X				
Chondria nidifica Harv.						X				
Corallina officinalis var. *chilensis* (Dec.) Kütz.						X				
Cryptopleura ruprechtiana (J. Ag.) Kyl.	X			X						
Cryptopleura violacea (J. Ag.) Kyl.						X				
Dictyopteris undulata Holmes								X		
Dictyota flabellata (Coll.) S. & G.								X		
Dictyota binghamiae J. Ag.				X	X					
Gelidium spp.								X		

Table 7.3. *Conspicuous Components of Kelp Forests from Central and Southern California after Foster and Schiel (1985)*[a] (continued)

Species	Central California Kelp Forests					Southern California Kelp Forests				
	Greyhound Rock	Sandhill Bluff	Point Cabrillo	Stillwater Cove	Granite Creek	Campus Point	Anacapa Island	Catalina Island	Del Mar	Point Loma
Gigartina corymbifera (Kütz.) J. Ag.			X							
Laurencia subopposita (J. Ag.) Setch.				X						
Lithothrix aspergillum Gray						X		X		
Pachydictyon coriaceum (Holmes) Okam.								X		
Phycodrys setchellii Skottsb.	X									
Plocamium cartilagineum (L.) Dix.				X	X			X		X
Polyneura latissima (Harv.) Kyl.	X									
Prionitis lanceolata (Harv.) Harv.			X							
Pterocladia capillacea (Gmel.) Born. & Thur.								X		
Pterosiphonia dendroidea (Mont.) Falk						X				
Rhodymenia pacifica Kyl.									X	
Rhodymenia spp.			X							X
Stenogramme interrupta (C. Ag.) Mont.						X				
Zonaria farlowii S. & G.						X		X		

[a] Species are categorized according to vegetation layers based on the system outlined in Dayton et al. (1984). Crustose algal species have been omitted.

Catalina Island resulted in increased frond elongation rates but not frond initiation rates (Zimmerman 1983). In another experiment, tissue nitrogen of fertilized plants remained constant or increased, while in unfertilized controls it dropped below 1%, a value believed to be below that required for growth (Gerard 1982a; North and Zimmerman 1984). Thus, experimental uncoupling of the temperature–nutrient relationship indicates that *Macrocystis* can tolerate temperatures normally found during summer months in the SCB. The seasonal variation in the extent of surface canopies in some areas of the SCB and parallel trends in nutrient availability have raised the possibility that nutrients potentially limit kelp productivity. Comparisons of studies conducted on mainland and island kelp forests indicate that the prevalence of nutrient limitation in the SCB varies along many different scales of time and space.

The 1982–1983 El Niño event allowed examination of responses of adult *Macrocystis* to both extreme nutrient and thermal stress. In the SCB, waters warmer than 15°C are nutrient depleted, with nitrate levels typically below 0.5 μM (Zimmerman and Kremer 1984). At a Laguna Beach kelp forest, a prolonged period of water temperatures greater than 20°C began in June 1983 and the canopy disappeared within 7 weeks (Gerard 1984). Gerard (1984) attributed the kelp deterioration to a complex combination of high temperatures and nutrient starvation as canopy blade tissue nitrogen dropped to the critical 1.1% level, with concurrent reductions in chlorophyll content and photosynthetic capacity.

At Santa Catalina Island, the 15°C isotherm was depressed to 50 m, twice its normal depth (Zimmerman and Robertson 1985), during the 1982–1983 El Niño. Thus, kelp forests were deprived of nutrients trapped below the thermocline and were not exposed to the normal periodic upward excursions of cold, nutrient-rich water (Zimmerman and Kremer 1984). Frond elongation and initiation rates were much lower during El Niño than during 1981, and the surface canopy was entirely eliminated (Zimmerman and Robertson 1985). Similar events were observed in the Point Loma kelp forest (Tegner and Dayton 1987). Bottom water temperatures, which rarely reach 16°C during normal years, were above 16°C between August and October 1983, reaching a maximum of 21°C. Tissue nitrogen in canopy blades dropped below 1.1%, and most plant tissue within the top 6–8 m deteriorated (Tegner and Dayton 1987).

The microscopic gametophytes and young sporophytes of *Macrocystis* tolerate a wide range of temperatures. Maximal photosynthesis is maintained between 15° and 20°C for gametophytes and between 10° and 20°C for embryonic sporophytes; net photosynthesis is positive for both stages between 10° and 25°C (Fain and Murray 1982). In addition, a strong relationship exists between the production of sporophytes and irradiance. This relationship is not altered with temperatures ranging between 10° and 20°C (Deysher and Dean 1986a), although in an earlier field experiment, sporophyte production dropped dramatically above 16.5°C (Deysher and Dean 1986b). However, this was likely due to nutrient stress rather than to elevated temperatures per se. Recruitment of *Macrocystis* sporophytes was significantly higher on nutrient-enriched substrata compared to unfertilized controls (Deysher and Dean 1986b).

ROLE OF WATER MOTION AND WAVE ACTION

Water movements affect benthic macrophytes in various ways (Neushul 1972; Lobban et al. 1985). Water movements are beneficial in that they transport nutrients and reproductive cells, but violent water movements can cause shearing, scouring, burial, and battering and may ultimately devastate subtidal macrophyte thalli.

It is important to distinguish the broad types of nearshore water movement. Currents are unidirectional and are generally large-scale processes caused by temperature differences, winds, tides, and buildup of water on the shore. Surge is a back and forth movement that occurs when surface waves enter shallow water. Water velocities in storm surge can far surpass those caused by currents. A 4-m-high swell can result in water velocities >1 m s^{-1}, a value resulting in forces comparable to 56 m s^{-1} winds (Foster and Schiel 1985).

Neushul (1972) pointed out four water motion regions within a kelp forest: (1) the current zone in midwater; (2) the surge zone closer to the bottom; (3) the turbulent–laminar zone; and (4) the still-water zone. The last two zones are in the boundary layer adjacent to the substratum. Several species of understory algae extend into the surge zone and are adapted morphologically to surge zone forces. For example, *Eisenia arborea* responds to surge by bending, which brings the plant closer to the bottom where water velocities are lower; frond clumping also helps this kelp to reduce form drag (Charters et al. 1969). Juvenile stages of other understory kelps bear morphological similarities to juvenile *Eisenia* in that they have a short stipe, a small holdfast, and a single blade. Neushul (1972) speculated that these structural similarities might represent adaptations to living in the surge zone.

Fronds of adult *Macrocystis* plants live in the current zone and stream out in the direction of current flow, as can be seen at the sea surface and in fathometer recordings (Neushul 1972). Clarke and Neushul (1967) reported that strong water currents at Anacapa Island submerged the *Macrocystis* canopy as deep as 10 m below the surface, even at depths down to 40 m. Circular sand patches around *Agarum fimbriatum* indicate that water movements cause a sweeping action in plants (Clarke and Neushul 1967). These periodic changes in the distribution of plant biomass may affect light penetration and primary productivity in a kelp forest (Gerard 1986). Despite the temporary effects that unidirectional currents can impose, there are no reports that such currents by themselves cause mortality in adult kelp forest organisms (Foster and Schiel 1985).

Wave surge has a major impact on population dynamics of kelp forests in the SCB. The immense blade and stipe biomass in the water column that contributes to the success of *Macrocystis* as a primary producer also produces drag forces during surge conditions that can result in plants being torn from the substratum. This can have a "snowballing" effect because detached kelp thalli become entangled with neighboring plants, increasing both the drag and subsequent damage in the forest (Dayton et al. 1984).

Knowledge of water flow in and about a kelp forest is important because currents supply nutrients and planktonic food and disperse reproductive cells. The most extensive current measurements were taken in the Point Loma kelp forest by Jackson and Winant (1983). Water movements were strongest along the edges and weakest toward the middle of this 7-km forest. Net water flow through the forest was longshore at approximately 1 cm s^{-1}, although semidiurnal tides moved water back and forth 0.5–1 km every 12 hours. Jackson and Winant (1983) estimated that it would take 1 week for water to pass entirely through the kelp forest.

Because the residence time for a parcel of water in the forest is long relative to the uptake rates for nutrients and the development time for many larvae, nutrients and plankton may be depleted toward the middle of a large forest (Jackson and Winant 1983). Several observations support this filtering effect. During the 1983 El Niño, summer mortality of *Macrocystis* plants was lower at the northern and southern ends of the Point Loma forest, located at the current front, than at other locations. Dayton and Tegner (1984a) suggested that the northern and southern plants scrubbed nutrients from the longshore currents, depriving centrally located plants. Recruitment of the red sea urchin (*Strongylocentrotus franciscanus*) was also consistently greater along the outside edge of the Point Loma forest (Tegner and Dayton 1981). At

Naples Reef, Bray (1981) repeatedly measured a decrease in zooplankton at the downcurrent ends; much of the decline was probably due to foraging by blacksmith (*Chromis punctipinnis*), a planktivorous reef fish that forms large daytime feeding aggregations at the incurrent kelp margin.

The interaction between an alga and the water immediately surrounding it has been discussed by Lobban et al. (1985). In still water, nutrients move to plants through molecular diffusion; in moving water, nutrient transport can increase by an order of magnitude (Neushul 1972). Laboratory studies by Wheeler (1980) indicate that carbon and nutrient uptake by *Macrocystis* saturates at current velocities of 3–5 cm s^{-1}. He suggested that velocities of <4 cm s^{-1} might limit productivity of kelp forests. Although water currents may often be <1 cm s^{-1} within a large forest, kelp movement from passing swells and surge combine to produce flow rates over kelp blades that are sufficiently high to saturate nutrient uptake, even in large forests during calm conditions (Gerard 1982b; Jackson and Winant 1983).

Water motion also affects settlement of motile reproductive cells. Sinking rates of microscopic spores and zygotes are variable but generally slow (Coon et al. 1972). Once these cells are near the bottom, attachment occurs mainly through turbulent deposition established near irregular surfaces (Lobban et al. 1985). The swimming movements of *Macrocystis* zoospores, which can reach 5 mm s^{-1}, probably enhance attachment (North 1972). Zoospores are generally adhesive, so they tend to stick to the substratum. For example, spores of *Gracilaria lemaneiformis* (as *Gracilariopsis sjoestedtii*) can withstand a dislodging force up to nearly 100 times their weight (Charters et al. 1972). Once attached, reproductive cells are small enough that they do not protrude far into the boundary layer. Nonetheless, they are susceptible to dislodgment in high currents, and thus they have several different strategies for remaining attached as they grow (Lobban et al. 1985).

ROLE OF STORMS

From the large piles of kelp strewn on beaches of the SCB, early workers recognized the devastating effects of strong surge. For 12 years, ZoBell (1971) regularly surveyed the amount of kelp that washed up on beaches in San Diego County. On the average, more seaweeds were found on beaches from November to February than during the remaining months combined. Sixty percent of the drift biomass was *Macrocystis*. Although portions of *Macrocystis* fronds slough off during periods of high water temperature and low nutrients, there was no relationship between temperature and the amount of kelp on the beach. Instead, drift kelp was correlated with wave height and wind velocity (ZoBell 1971).

These observations on accumulation of kelp on beaches are now well supported by long-term quantitative data on the disappearance of *Macrocystis* plants from subtidal communities. Storm surge and associated entanglement have been cited as major causes of adult *Macrocystis* mortality in every kelp forest in the SCB that has been intensely studied over a long period of time, including Point Loma (Dayton and Tegner 1984a, b), Del Mar (Rosenthal et al. 1974), Palos Verdes Peninsula (Wilson and Togstad 1983), and Naples Reef (Ebeling et al. 1985).

The best evidence for the impact of violent water motion on kelp forests of the SCB comes from a series of unusually severe storms that occurred in the winter of 1982–1983.

Dayton and Tegner (1984b) described the damage at the Point Loma kelp forest. Wave heights at the entrance to Mission Bay, several kilometers to the north, exceeded 3 m, and most wave periods reached lengths of more than 20 seconds (Dayton and Tegner 1984a, b). As a result, the surface canopy of *Macrocystis* declined from 600 to 40 ha. Off Palos Verdes, eight major storms occurred between January and April with deep-water wave heights at >3 m and periods ranging from 17 to 22 seconds. Aerial surveys showed that the surface canopy declined from 196 to 14 ha (Wilson and Togstad 1983). Davis (1986) reported that the 1982–1983 storms greatly reduced surface canopies of *Macrocystis* at 14 permanent study sites at San Miguel, Santa Rosa, Santa Cruz, Anacapa, and Santa Barbara islands. Wave heights in the Santa Barbara Channel reached at least 6 m, causing detachment of chunks of shale averaging 1.3 m in breadth at Naples Reef (Ebeling et al. 1985). The kelp forest on Naples Reef was destroyed by previous storms in 1980 and had not become reestablished prior to 1982–1983. However, sea urchins exposed on elevated parts of the nearly bare rock reef were decimated by the severe 1982–1983 storms, allowing kelp recovery (Ebeling et al. 1985). Zimmerman and Robertson (1985) reported, however, that Bird Rock, a protected area on leeward Santa Catalina Island, maintained a significant canopy throughout the 1982–1983 winter storms.

These studies revealed several important patterns that may occur from storm damage. For example, the extent of damage is not necessarily uniform throughout a kelp forest. *Macrocystis* loss was highest in the shallow, inshore margins of the Point Loma forest, and at the same depth, mortality was less in the middle of the forest than at the northern and southern ends (Dayton et al. 1984). Storms are particularly destructive when they coincide with low tides. The most severe storm damage at Palos Verdes in 1983 took place when heavy seas occurred during the most extreme tides of the year (Wilson and Togstad 1983). Understory algae experience far less storm damage than the surface canopy. In the Point Loma kelp forest, patches of *Pterygophora, Gelidium, Calliarthron*, and *Eisenia* and a mixture of *Macrocystis* and *Laminaria* persisted with little to no change through the storms, and the forest has been maintained since 1971 (Dayton and Tegner 1984a, b).

The effects of storms also depend on the structure of the existing community. Ebeling et al. (1985) described dramatic differences in the effects of winter storms at Naples Reef. The first storms, occurring during the winter of 1980, removed the entire surface canopy of *Macrocystis*, but spared most of the *Pterygophora* understory. The loss of kelp reduced the drift algae available to sea urchins, so the urchins consumed most living plants. The reef changed from one that supported a luxuriant kelp forest to one of almost entirely bare rock. The 1983 winter storms reversed the community structure by creating additional bare substratum and decimating the exposed sea urchins. Kelp recruitment was heavy. Without the grazing pressure by sea urchins, a luxuriant macrophyte assemblage developed that resembled the pre–1980 condition (Ebeling et al. 1985).

The pattern of kelp loss credited to storms is not consistent among "storm years," and some forests undergo marked cyclical fluctuations. For example, the forests between Del Mar and Oceanside appear to undergo a 4-year cycle in which there is a massive loss of adult kelp plants followed by a large recruitment and the eventual development of another canopy (North 1971). Other forests in the SCB are remarkably constant, especially those between Santa Barbara and Point Conception (table 7.2) (Coon 1981; Harger 1983). Differences in the variation between these two areas might be due in part to the ways in which kelp in these two areas reproduce. The cyclical variation in the southern forests resulted from sexual reproduction and the establishment of a large juvenile population. In contrast, northern forests were replenished through vigorous vegetative reproduction from established holdfasts, and juvenile plants were scarce (North 1971).

ROLE OF GRAZERS

The effects of grazing on macroalgae range from local loss of thallus tissue to a large-scale change and possible disappearance of macrophyte assemblages. Carnivores, such as sea otters and some fishes, may control the numbers of such herbivores. Schiel and Foster (1986) and VanBlaricom and Estes (1988) provide recent summaries of the effects of grazers and their predators on subtidal macrophytes in temperate zones. The role of invertebrate grazers within SCB kelp forests is discussed in chapter 8. In this section, we offer a broad overview of the effect of grazing on kelp forest macrophytes in the SCB.

Research in the SCB has focused on macroscopic grazers, although meio- and microfauna are also known to eat algae (Dayton 1985). The major macroherbivores in kelp forests are certain molluscs, crustaceans, sea urchins, and fishes. Foster and Schiel (1985) listed 40 species of common invertebrate grazers in California kelp forests, including 26 molluscs, 9 arthropods, and 5 echinoderms. Many fishes also consume algae. Quast (1971a) found algal fragments in 23 of 45 species of fish commonly seen in kelp forests. However, unlike tropical reefs where numerous fish species are strict herbivores (Choat 1982; Gaines and Lubchenco 1982), algae are generally an infrequent component in the diets of all but the kyphosid fishes *Girella nigricans, Medialuna californiensis*, and *Hermosilla azurea* (Quast 1971a); several embiotocids, especially *Brachyistius frenatus* and *Phanerodon furcatus* (Bray and Ebeling 1975); the labrid *Oxyjulis californica* (Bray and Ebeling 1975; Bernstein and Jung 1979); and the pomacentrid *Hypsypops rubicundus* (Foster 1972).

Almost all studies of grazing in subtidal macrophyte communities in the SCB have concentrated on the effects on the sporophytes of laminarians, principally *Macrocystis pyrifera*. With the exception of Dean et al. (1988), little work has been done on other algal groups or on the microscopic stages of kelps.

Sea urchins are generally considered the most important subtidal herbivores in the SCB in terms of frequency and severity of grazing. Most overgrazing by sea urchins is attributed to two species, *Strongylocentrotus franciscanus* and *S. purpuratus* (Ebeling et al. 1985), although *S. franciscanus* grazing had an insignificant effect on juvenile *Macrocystis* sporophytes off San Onofre (Dean et al. 1989). *Lytechinus anamesus* can attack early developmental stages of kelps in deeper portions of kelp forests (Dean et al. 1988, 1989); also, *Centrostephanus coronatus* consumes fleshy algae at Santa Catalina Island (Vance 1979). Periodic reduction and disappearance of macrophytes due to sea urchin overgrazing have often been observed in SCB kelp forests (North 1983; Dayton et al. 1984; Dean et al. 1984; Ebeling et al. 1985).

The interactions between macrophytes and sea urchins are dynamic, and subtidal rocky habitats in the SCB may periodically switch between macrophyte-dominated and sea urchin-dominated states (Ebeling et al. 1985; Harrold and Reed 1985). These reversals appear to be brought about by changes in the abundance, distribution, and feeding behavior of the sea urchins. In macrophyte-dominated situations, sea urchins are often found in crevices or in small aggregations where they forage principally on drift algae. During sea urchin-dominated states, the urchins are more exposed and form large moving fronts. The sea urchins in such a front remove essentially all attached macrophytes in their path, creating "barren grounds" or crustose coralline communities that persist because of grazing activities of other sea urchins behind the front. Leighton (1971) described one such front that completely destroyed all macrophytes in a 100 × 200 m kelp stand within 27 days. The dramatic change to active foraging on attached algae may be triggered by a decrease in sea urchin predators (Cowen 1983) or by a scarcity of algal drift (Ebeling et al. 1985). Sea urchin fronts also appear to be more common in areas of low-relief substratum, possibly because of the lack of shelters for their lobster and sheephead predators or the lack of trapped drift algae (Tegner and Dayton 1981; Dean et al. 1984).

Overgrazing by sea urchins is not inevitable, however, and sea urchins may coexist with large macrophyte populations in the SCB without any measurable effect. Foster and Schiel (1988) challenged the generalization that nearshore hard-bottom communities exist in two extreme states: one forested with few urchins, the other deforested with many sea urchins. Long-term coexistence between sea urchins and macrophytes has been reported for Point Loma (Dayton et al. 1984), San Onofre (Dean et al. 1984), San Clemente (Rosenthal et al. 1974), and Santa Catalina Island (Nelson and Vance 1979).

In addition to the wholesale removal of macrophytes, grazers can affect macrophyte populations in more subtle ways. Selective grazing can alter the species composition and abundance of kelps and other subtidal macrophytes (Schiel and Foster 1986). Laboratory studies have shown that sea urchins and other grazers prefer certain species of algae when presented equal opportunities to choose (Leighton 1966, 1971; Vadas 1977). Leighton (1971) found that 7 of the 11 invertebrate grazers common in the Point Loma kelp forest preferred *Macrocystis pyrifera* to *Egregia menziesii, Laminaria farlowii, Pterygophora californica, Cystoseira osmundacea*, and *Gigartina spinosa*. An important question with this type of laboratory study, however, is whether these preferences can explain macrophyte assemblages in the field (Schiel and Foster 1986). Field experiments by Dean et al. (1988) demonstrated that *Lytechinus* prefers microscopic stages of laminarians over newly recruited *Cystoseira osmundacea*; such differential grazing could cause a long-lasting change in algal community composition (Dean et al. 1988). There is also strong evidence that encrusting coralline algae in the SCB are avoided by most grazers (Dayton et al. 1984; Breitburg 1984).

The effect of grazers can also depend on the macrophyte assemblage. Field observations indicate that two fishes, *Medialuna californiensis* and *Girella nigricans*, graze heavily on small *Macrocystis* and *Pterygophora* sporophytes (Grant et al. 1982; Harris et al. 1984). Sporophytes developing within stands of filamentous brown algae (e.g., *Giffordia* and *Ectocarpus*), however, appear to be hidden and consequently experience less grazing pressure than those growing among a red algal turf (Harris et al. 1984).

Recently the U.S. Fish and Wildlife Service proposed to transplant sea otters to San Nicolas Island in an effort to establish a population in the SCB geographically separated from their center of distribution to the north (VanBlaricom and Estes 1988) (chap. 11). Sea otters are effective predators on sea urchins and other grazers (see chap. 11). At least initially, this may result in growth of algal populations (VanBlaricom and Estes 1988). There is some question, however, about the relative importance of sea otters and sea urchins compared to other factors that control macrophyte abundance. Foster and Schiel (1988) point out that sea urchins have deforested less than 10% of the sites examined outside the sea otter's present range. They argue that other factors may be at least as influential in kelp forest communities.

ROLE OF LIGHT

Light plays a key role in determining the distribution, abundance, and productivity of macrophytes (Lobban et al. 1985; North et al. 1986). The light regime within a kelp forest is quite different from that in adjacent open waters. Light characteristics in the open sea are largely controlled by the atmosphere, the sea surface, and the water column, whereas within a kelp forest below the surface, the amount of available light is largely dependent upon the overlying kelp canopy, which can reduce irradiance levels by 30–90% (Reed and Foster 1984). The greatest reduction in irradiance takes place within the top meter, where well over 50% of total kelp blade area occurs. Even on a sunny day, dense canopies reduce irradiance at depths of 1 m to values below photosynthetic saturation for many macrophytes (Gerard 1984). Near-bottom irradiances beneath kelp canopies are typically at or below 1% of surface illumination, approximately the lower limit for laminarian growth (Reed and Foster 1984). Kelp also makes subsurface irradiance and light penetration more variable (Gerard 1984). Water currents open and close kelp canopies, which results in periodic shading of underlying vegetation. The influence of canopy movements on the variation in total irradiance is large near the surface but negligible near the bottom (Gerard 1984). Subsurface light levels are further decreased by turbidity caused by terrestrial runoff, plankton, resuspended sediments, and particulate and dissolved organic material (Foster and Schiel 1985). Geographic patterns and temporal changes in turbidity levels have been of concern in the SCB because of their relationship to inputs of sewage and thermal effluents (Barilotti 1983; Dean and Deysher 1983). Consequently, much research has focused on determinations of the quantity of light required for photosynthetic compensation and saturation of adult (Arnold and Manley 1985; Gerard 1986) and embryonic sporophytes and gametophytes (Fain and Murray 1982) and gametogenesis (Deysher and Dean 1986a).

Adult *Macrocystis* sporophytes generally seem well adapted to the light regime within kelp forests. Although relatively few canopy blades sustain light-saturated photosynthesis, they are capable of maintaining high photosynthetic rates in the fluctuating irradiances characteristic of the canopy (Gerard 1986). Underlying tissues receive photosynthate by translocation (Lobban et al. 1985). However, the reduction of bottom irradiances due to shading by overlying canopies suggests that other macrophytes living along the sea floor are light limited. Competition for light among flora is a major biotic factor in determining subtidal community structure and dynamics. Experimental removal of canopies of several brown and red algal species has resulted in rapid establishment of young *Macrocystis* sporophytes and other understory algae (Dayton et al. 1984; Wells 1985). In addition, monitoring of tagged *Macrocystis* sporophytes over 2 years indicated that survival of juveniles was lower under an adult canopy than at positions receiving more light at the margins of or outside a kelp forest (Dean et al. 1989). Evidence also exists for intracohort density-dependent mortality in juvenile kelp sporophytes, possibly as a result of competition for light (Dean et al. 1989; Reed 1990).

ROLE OF SEDIMENTATION

Sedimentation can affect all life history stages of macrophytes. *Pterygophora* and *Eisenia* seem more resistant to burial than the basal stipes of *Macrocystis* (North et al. 1986). The stipes and blades of the *M. angustifolia* form of *M. pyrifera* along the northern SCB disintegrate after a few weeks of burial, although the extensive haptera apparently bury and anchor in fine sediments (North et al. 1986). Sedimentation and abrasive scour can kill *Macrocystis* spores in the laboratory and affect production of sporophytes from gametophytes outplanted on artificial substrata (Devinny and Volse 1978; Dean and Deysher 1983). Sediments from discharged wastes can interfere with plant growth, enhance growth of sea urchins, and increase concentrations of toxicants (Barilotti 1983).

ROLE OF RECRUITMENT

Dispersal of the microscopic stages of kelps appears to be limited. The effective radius of a spore released from large brown macrophytes is probably less than 10 m (Dayton et al. 1984). But *Macrocystis* is also clearly capable of long-distance dispersal. Young plants frequently appear on artificial reefs and other structures that are often at a considerable distance from adults (Grant et al. 1982). Drifting plants and reproductive fragments may enable recruitment over larger spatial scales. Attaching bags of *Macrocystis* sporophylls to the substratum can be an effective way to increase recruitment in an area (Dayton et al. 1984).

Reed et al. (1988) examined patterns of recruitment as a function of distance from adults in two kelps, *Macrocystis pyrifera* and *Pterygophora californica*, and in filamentous brown algae, mostly *Ectocarpus siliculosus*. There was significant temporal variation in recruitment of the macrophytes both near and away from the adults. The two kelps had clear dispersal shadows, evidenced by the rapid drop in recruitment densities with increasing distance from the source; the decrease became significant in just 3 m. This pattern was not consistent in time, however. Episodic recruitment events occurred during which gametophyte density on glass slides was unrelated to distance from the source. Even near the source, however, recruitment patterns of the filamentous browns were less variable over time and appeared unrelated to distance; recruitment densities were fairly consistent out to 500 m.

These differences in disperal patterns between the kelps and the filamentous brown algae were related in part to behavioral differences in zoospores. Spores of filamentous browns are positively phototactic, enabling them to remain suspended for a longer period of time, whereas those of *Macrocystis* and *Pterygophora* are not phototactic and thus settle quickly. Zoospore dispersal of the two kelps expands during winter storms when turbulent water suspends the spores in the water column (Reed et al. 1988).

There is a strong temporal component in reproduction and recruitment of subtidal macrophytes in the SCB. Many of the large macrophytes found in kelp forests (*Eisenia arborea, Laminaria farlowii*, and *Pterygophora californica*) produce sori year round, but show peaks mainly in the fall and early winter; an exception is *Agarum fimbriatum*, which fruits mainly from April through September (McPeak 1981).

Macrocystis is also reproductive throughout the year, although daily and seasonal zoospore release rates vary substantially (Anderson and North 1966). The production of sporophylls is variable and depends on vegetative biomass and location. Sporophyll production by *Macrocystis* at Naples Reef is lower than that in plants closer to shore, perhaps because of increased shading at the offshore site (Reed 1987).

We have made little progress in our knowledge about the survival of kelp gametophytes in nature. Gametophytes of *Macrocystis* have never been observed directly in the field (Neushul 1983). Survivorship is generally regarded as low (Reed 1990). Dayton (1985), however, speculated that some gametophytes that settle in cracks and other refuges might live for long periods of time and thus provide a long-term "seed bank." In the laboratory, gametophytes propagate vegetatively, and gametophyte cultures have been grown for at least 7 years (Neushul 1983).

Recruitment of kelp sporophytes is highly variable in time (Reed 1990). Dayton (1985) reported that recruitment occurred two or three times in 10 years. Heavy macrophyte recruitment often follows catastrophic events, such as the winter storms of 1980 and 1983 (Dayton and Tegner 1984a; Ebeling et al. 1985). Close examination of the dynamics of macrophyte patches has revealed recruitment processes on a smaller scale (Dayton et al. 1984). Experimental removal of patches of *Laminaria, Cystoseira*, and *Pterygophora* resulted in dense *Macrocystis* recruitment when adult *Macrocystis* were nearby; other clearing experiments indicated that *Pterygophora, Laminaria*, and *Cystoseira* also recruit in cleared areas (Dayton et al. 1984). Asymmetrical competition exists between microscopic stages of *Pterygophora* and *Macrocystis*. Settlement of *Pterygophora* spores inhibits recruitment of *Macrocystis* sporophytes, but *Macrocystis* spores have no effect on recruitment of *Pterygophora* sporophytes (Reed 1990). Shading by the introduced alga *Sargassum muticum* also inhibits *Macrocystis* recruitment (Ambrose and Nelson 1982). Other algae may increase the survivorship of young kelps. Turf algae provide *Macrocystis* and *Pterygophora* a refuge from fish grazing (Harris et al. 1984). Recruitment of *Macrocystis* sporophytes is correlated with temperature, irradiance, and seston flux. "Recruitment windows" occur when temperatures are below 16.3°C and irradiation levels are greater than 0.4 E m^{-2} d^{-1} (Deysher and Dean 1986b).

Thus, there may be two spatial patterns in recruitment. Local recruitment patterns for some macrophytes (such as *Macrocystis* and

Pterygophora) result from the spore shadow produced near adult stands. Within this region (<10 m), recruitment quickly drops off with increasing distance (Dayton et al. 1984; Reed et al. 1988). Recruitment rates farther from established adult stands are generally lower. Recruitment at this scale may be patchy if spores come from drifting adults or reproductive fragments, or it may be more uniform due in part to the behavior of the reproductive cells and mixing in the water column (Dayton et al. 1984; Reed et al. 1988).

Temporal patterns may depend on rare occurrences of favorable environmental conditions. Recruitment patterns may also depend on survivorship patterns of the reproductive cells. If these are short-lived, then recruitment will be closely linked to spore production. Alternatively, if some reproductive cells can lie dormant for extended periods, then recruitment events might be relatively independent of the release and dispersal of spores.

Macrophyte Productivity

Shallow coastal waters are known to be highly productive, particularly in areas of upwelling and in areas of high macrophyte abundances (Mann 1982). On rocky coasts, macroalgae tend to dominate, requiring only a hard surface for attachment, whereas seagrasses and halophytes are typically more abundant under conditions of sedimentation where their roots can penetrate the substratum. Seagrasses are often abundant in the low wave energy, protected waters of estuaries and salt marshes where they generally occur as the lowest intertidal component of salt marsh communities otherwise dominated by halophytic flowering plants (Chapman 1976). In addition, seagrasses occur independently of salt marsh communities, forming subtidal beds in certain low wave energy, shallow waters where the bottom is largely composed of sedimentary deposits. These habitats are rare in the SCB, and consequently, extensive seagrass beds do not occur as they do along many coasts. Only on unstable sandy stretches of the coastline are macrophyte standing stocks and productivity generally low.

Macrophytic communities appear to be the most important contributors to the productivity budgets of shallow coastal seas. Mann (1982) estimates that the production of coastal phytoplankton, in areas not subject to upwelling, ranges from 50 to 250 g C m^{-2} yr^{-1} compared to 300–1000 g C m^{-2} yr^{-1} for marine flowering plants (salt marsh halophytes, seagrass beds, and mangroves) and 400–1900 g C m^{-2} yr^{-1} for seaweed communities. In St. Margaret's Bay, Nova Scotia, for example, Mann (1972) determined that the macrophytic algae accounted for about 75% of the total primary production, averaging 603 g C m^{-2} yr^{-1} compared to about 200 g C m^{-2} yr^{-1} contributed by phytoplankton. Knowledge concerning the productivity of macrophyte populations and communities is critical to the understanding of the energetics of SCB coastal benthic systems. Understanding of the photosynthetic performances, and hence productivity potentials, of individual macrophytes in the SCB is perhaps developed as broadly as for any other region of the world.

Productivity of Macrophyte Populations

Measurements of photosynthetic performances have been published for more than 120 macrophyte populations belonging to over 60 taxa. Most of these measurements have been obtained *in situ* using comparable methodologies (Littler and Arnold 1985) that recognize potential sources of error (Littler 1979b) and variability (Littler and Arnold 1980) and have been obtained in the field at ambient sea temperatures and under natural daylight levels known to saturate but not inhibit photosynthesis (Arnold and Murray 1980).

The light-saturated net productivities (generally obtained by expressing dissolved oxygen flux during photosynthesis in terms of carbon fixation) of individual SCB macrophytes appears to range from <0.1 to 11.2 mg C g dry wt^{-1} h^{-1} (Littler and Arnold 1982). The magnitude of net productivity for species of temperate seaweeds has been found to correlate with thallus form (fig. 7.14), but not with the division to which a species belongs or with its geographic location (Littler 1980c, 1981; Littler and Arnold 1982).

For SCB macrophytes, the mean primary productivities of the functional form groups of Littler and Arnold (1982) are presented in figure 7.14. The "coarsely branched group" identified by Littler and Arnold (1982) has been subdivided into four categories based on the external appearance (i.e., breadth and depth) of the thallus. Coarsely branched group seaweeds are essentially fleshy and nonfilamentous, without thick, leathery, or thin sheetlike thalli. The four subcategories of Littler and Arnold's (1982) coarsely branched group identified here are (1) entire or multilobed blades or bladelike thalli (e.g., *Iridaea cordata* var. *cordata, Rhodymenia californica* var. *californica, Dictyopteris undulata*, and *Callophyllis flabellulata*); (2) saccate or cushionlike thalli (e.g., *Colpomenia sinuosa* and *Halosaccion glandiforme*); (3) spongy and cylindrical morphologies (e.g., *Codium fragile* and *Nemalion helminthoides*) or compressed morphologies (e.g., *Cylindrocarpus rugosus*); and (4) non-spongy, cylindrical, or compressed and branched forms (e.g., *Gigartina canaliculata, Endocladia muricata, Neorhodomela larix*, and *Sarcodiotheca gaudichaudii*). Mean net productivities of SCB macrophyte form groups (fig. 7.14) range from 0.2 mg C g^{-1} h^{-1} ("crustose group") to 5.0 mg C g^{-1} h^{-1} ("sheet group").

Productivity of Macrophyte Communities

Although considerable information has been obtained concerning the potential productivities of individual macrophyte populations, our knowledge of the production of SCB macrophyte communities is poorly developed. Few estimates are available for rocky intertidal or salt marsh macrophyte communities, and no measurements of the production of deep-water algal communities have been performed in the SCB. Most of the research on kelp forests has concentrated solely on *Macrocystis* and has neglected the role of other seaweeds comprising the kelp forest community.

EMBAYMENT COMMUNITIES: SALT MARSHES

As pointed out by Mann (1982), at least 90% of the research on system processes in salt marshes has been performed in North America; most of this work, however, has taken place in the extensive salt marsh habitats bordering the Atlantic Ocean and the Gulf of Mexico. Knowledge of the productivity of salt marsh macrophyte populations and communities in the SCB is much more poorly developed.

Assuming a carbon content of 19–35% for vascular marsh plants, Zedler (1982a) indicated aboveground net productivity values (based upon measurements of harvested biomass) ranging from 19–35 g C m^{-2} yr^{-1} for low marsh habitats in Mugu Lagoon (Onuf et al. 1978) to 209–385 g C m^{-2} yr^{-1} for hypersaline Sweetwater River marsh (Eilers 1981, as reanalyzed by Zedler 1982a). Greater annual net productivity is believed to occur in marsh systems where tidal flow has been eliminated during the growing season and soil salinities have been reduced by freshwater runoff (Zedler et al. 1980). The net productivity values of SCB salt marsh halophytes appear to compare favorably with those obtained in other marsh systems, with Mann (1982) reporting the net productivity of most marshes to range from 200 to 400 g C m^{-2} yr^{-1}.

Contributions of the macroalgal constituents to the productivity budgets of SCB salt marshes are largely unknown, except for the Tijuana Estuary. Here, Zedler (1982a) compared the relative productivities of salt marsh halophytes and mat-forming macroalgae. These comparisons indicate that halophytes produce 243–340 g C m^{-2} yr^{-1} com-

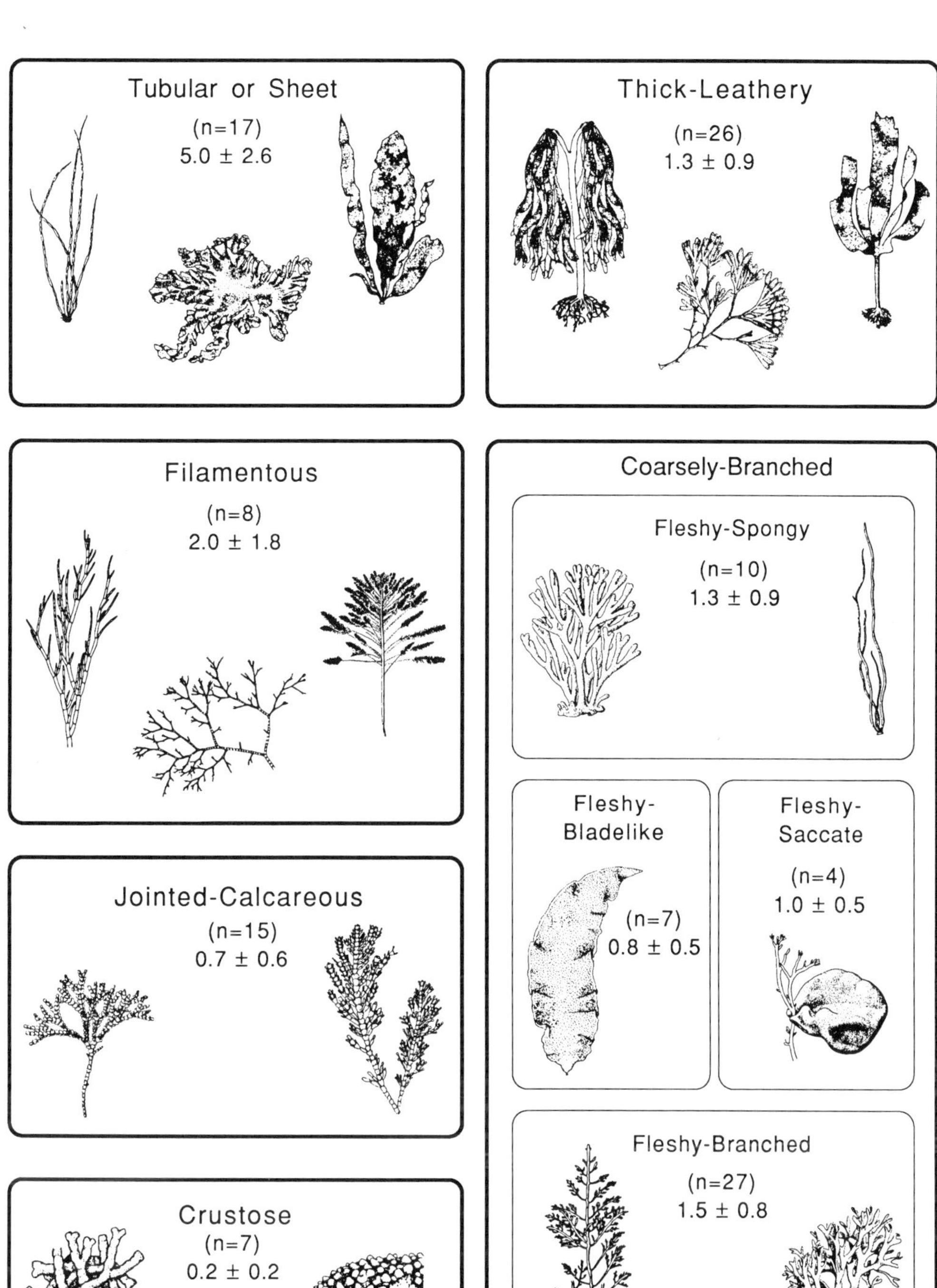

Figure 7.14. Mean productivities (±1 S.D.) in mg C g dry wt $^{-1}$ h^{-1} of macrophyte functional form groups designated by Littler and Arnold (1982). In this illustration, the "coarsely branched group" has been divided into four categories: fleshy-saccate, fleshy-spongy, fleshy-bladelike, and fleshy-branched. (Values for individual populations used to calculate group means were obtained or estimated from Littler and Murray 1974a, b, 1977, 1978; Littler 1980c, 1981; Littler and Arnold 1980, 1982; Littler et al. 1979; Arnold and Murray 1980; and Theis 1985; sources of illustrations are reproduced from Abbott and Hollenberg 1976 with permission of the publishers.)

pared to 185–341 g C m^{-2} yr^{-1} for the algal mat, resulting in net algal-halophyte production ratios of 0.8–1.4. Zedler (1982a) suggested that macroalgal mats produce more carbon in the Tijuana Estuary than in comparable systems on the eastern American coast where algal–halophyte production ratios range from about 0.25 to 0.33 (Teal 1962; Gallagher and Daiber 1974; Van Raalte et al. 1976).

ROCKY INTERTIDAL COMMUNITIES

For rocky intertidal habitats in the SCB, knowledge of macrophyte community productivity is confined to a wave-protected site near Wilson Cove, San Clemente Island, although comparable data are available for nearby regions (Seapy and Littler 1978; Littler and Littler 1981). For these communities, production has not been measured directly by monitoring changes in macrophyte biomass but has been calculated from short-term, light-saturated, net photosynthetic rates and data concerning the percentage of habitat covered by individual macrophyte populations (Littler and Murray 1974a).

Macrophytes from pristine Wilson Cove produced an estimated 150.5 mg C m^{-2} h^{-1} based on the productivity contributions of 18 of the most abundant populations (table 7.4). This was nearly the same rate as that calculated for an adjacent sewage-impacted habitat (152.5 mg C m^{-2} h^{-1}) containing fewer species and less macrophyte cover (Littler and Murray 1974b). The rates for Wilson Cove macrophytes are intermediate to those determined for geographically proximal communities where totals have been reported ranging from 203.6 mg C m^{-2} h^{-1} for Cayucos, central California (Seapy and Littler 1978) to 47.9 mg C m^{-2} h^{-1} for Punta Pelicano, Gulf of California (Littler and Littler 1981) (table 7.4).

Estimates of the seasonal variation in community productivity are available only for San Clemente Island macrophytes. These estimates are based on extrapolations of hourly rates to daily rates and seasonal measurements of macrophyte photosynthesis (Littler et al. 1979). As emphasized by Littler and Arnold (1985), the results of extrapolating such short-term rates to even daily rates are at best only crude estimates and need to be considered with caution. Known diurnal variations in net photosynthetic rates resulting, for example, from endogenous periodicities (Britz and Briggs 1976; Hoffman and Dawes 1980), photoinhibition (Ramus and Rosenberg 1980), or light limitation (Coutinho and Zingmark 1987) are not taken into account.

Using as a basis the contributions of the 13 most abundant intertidal macrophyte populations, Littler et al. (1979) calculated daily community production rates ranging from 1220 mg C m^{-2} d^{-1} for fall to 470 mg C m^{-2} d^{-1} for spring, closely paralleling seasonal fluctuations in sea temperature (table 7.4). Four taxa (encrusting blue-green algae, *Corallina officinalis* var. *chilensis, Pterocladia capillacea*, and *Egregia menziesii*) provided approximately 75% of the estimated yearly community productivity, which averaged 900 mg C m^{-2} d^{-1} for the four seasons. These values are slightly less, but they compare favorably with reports of daily production for seaweed communities from the Canary Islands (1500–3000 mg C m^{-2} d^{-1}) (Johnston 1969) and Nova Scotia (4800 mg C m^{-2} d^{-1}) (Mann 1972).

SUBTIDAL KELP FORESTS

It is difficult to obtain accurate estimates of the primary productivity of kelp forest communities because of their large size and the diversity of seaweeds that form them. Thus, previous research has concentrated on *Macrocystis*, the dominant kelp genus for much of the coast of Pacific North and South America, or on other kelps such as *Laminaria* in the North Atlantic and along the coasts of Japan and China and *Ecklonia* in Australia and South Africa.

Estimates of *Macrocystis* productivity for California populations range (Dayton 1985) from 350 to 1500 g C m^{-2} yr^{-1}, but most values are less than 1000 g C m^{-2} yr^{-1}. Ac-

Table 7.4. *Productivities of Rocky Intertidal Macrophyte Communities for Sites in the Proximity of the SCB*[a]

Community Description	Method	Period	Net Productivity	Source
Wilson Cove, San Clemente Island				
Undisturbed, wave-protected; large metamorphic rocks	Net photosynthesis; DO conversions to C	Hour	150.5	Littler and Murray 1974a, b
Seasonal values	Net photosynthesis; DO conversions to C	Day		
Fall			1220	Littler et al. 1979
Winter			790	Littler et al. 1979
Spring			470	Littler et al. 1979
Summer			1100	Littler et al. 1979
Sewage-perturbated, wave-protected; large metamorphic rocks	Net photosynthesis; DO conversions to C	Hour	152.5	Littler and Murray 1974b
Cayucos Point, central California				
Wave-protected; boulder beach	Net photosynthesis; DO conversions to C	Hour	203.6	Seapy and Littler 1978
Wave-exposed; sea stack	Net photosynthesis; DO conversions to C	Hour	139.8	Seapy and Littler 1978
Punta Bufeo, upper Gulf of California				
Seasonally disturbed; boulder beach	Net photosynthesis; DO conversions to C	Hour	56.2	Littler and Littler 1981
Punta Pelicano, upper Gulf of California				
Seasonally disturbed; boulder beach	Net photosynthesis; DO conversions to C	Hour	47.9	Littler and Littler 1981
Bahia Santa Rosalia, Pacific Baja California				
Low disturbance, seasonally constant; boulder beach	Net photosynthesis; DO conversions to C	Hour	81.7	Littler and Littler 1981

[a]Net productivities are reported in mg C m^{-2} per unit period for data obtained by conversion of net photosynthesis values; all DO conversions to C are reported assuming a photosynthetic quotient of 1.0.

cording to Mann (1982), a rough estimate of average *Macrocystis* productivity might be 800–1000 g C m^{-2} yr^{-1}. This range is slightly less than estimates for *Laminaria* populations (120–1900 g C m^{-2} yr^{-1}), but is quite comparable with productivity values reported for *Ecklonia* (620–1000 g C m^{-2} yr^{-1}) (Mann 1982; Dayton 1985). Researchers employed several different approaches to obtain these figures, and it is likely that the large range of estimates is in part due to the differences in measurement techniques. Estimates of *Macrocystis* productivity have been made from extrapolations of short-term photosynthetic rates, which according to Arnold and Manley (1985) range from 0.06 to 3.42 mg C g dry wt^{-1} h^{-1}, by field growth measurements (Coon 1981) and by harvest estimates (Foster and

Schiel 1985). Estimates of kelp forest production have also been made by measuring oxygen flux throughout the entire forest (Jackson 1977). Changes in oxygen concentration were used by Jackson (1977) as the basis for calculating the gross production of a Point Loma kelp forest at 9.5 g C m^{-2} d^{-1}. According to Jackson (1987), McFarland and Prescott (1959) obtained similar values for a kelp forest in Paradise Cove at the northwest end of Santa Monica Bay (12 g C m^{-2} d^{-1}), as did Towle and Pearse (1973) for a Monterey population of *Macrocystis* (6.8 g C m^{-2} d^{-1}).

Jackson (1987) developed a model that predicts annual biomass and productivity of *Macrocystis pyrifera* as a function of factors that affect light. This model incorporates laboratory and field data on parameters such as daily and seasonal patterns of light flux, water clarity, water depth, plant density, and photosynthetic performance. For a "typical" kelp forest in the SCB, the Jackson model predicts that daily net production would reach almost 3 g C m^{-2} d^{-1}; annual net productivity was estimated at 537 g C m^{-2} yr^{-1} and annual gross productivity at 1567 g C m^{-2} yr^{-1}. This model further predicts a strong seasonal cycle in *Macrocystis* production, with peaks in the mid-summer and minima in early winter, in agreement with previous empirical measurements.

Summary and Prospectus for Future Research

Current records suggest that the SCB seaweed flora consists of 492 species (347 Rhodophyta, 86 Phaeophyta, and 59 Chlorophyta). Four species of seagrasses and approximately 17 halophytes are also commonly encountered in SCB coastal waters and marshes. The SCB is a region of considerable biogeographic complexity, and, although generally referred to as supporting a warm temperate flora, is a transitional region between cold- and warm-water biotas. Studies of the physiological tolerances, environmental regulation of reproduction and life history events, dispersal and recruitment, and the recent and evolutionary affinities of SCB macrophytes are needed to better understand the nature of the SCB flora.

The patterns of structure of certain rocky intertidal and kelp forest communities are well known. However, few data exist for lagoon, modified estuary (harbors and marinas), salt marsh (except for the Tijuana Estuary), nonkelp subtidal, and deep-water macrophyte communities in the SCB. Much more experimental work is required to gain understanding of factors responsible for community organization in the SCB. Embayment (salt marsh, lagoon, and modified estuarine) habitats in the SCB are limited in their extent and historically have been subject to considerable anthropogenic disturbance. Salt marsh halophytes have received the greatest scientific attention; much less is known about the macroalgae that form matlike growths beneath the halophyte canopies. SCB halophyte canopy coverage is more open, seasonal droughts are more common, and macroalgal communities differ from more thoroughly studied Atlantic and Gulf Coast marshes.

The abiotic environmental features of greatest importance in determining the geographic distributions of rocky intertidal macrophytes in the SCB are (1) substratum characteristics; (2) ocean water masses and thermal regimes; and (3) wave action. The burial and abrasion of organisms and the movement of unstable cobbles or boulders result in increased abundances of opportunistic, disturbance-resistant, and "psammophytic" macrophytes in the SCB rocky intertidal habitats. Intertidal macrophyte communities exposed to warmer water masses are characterized by articulated coralline algal turfs and different species of upper and lower shore fucalean and laminarialean brown algae and various species of smaller red seaweeds. The factors responsible for these patterns, however, have yet to be demonstrated experimentally. Several studies performed in the SCB document shifts in algal types toward more disturbance-resistant seaweeds and reveal a general decline in species diversity. There are no long-term records of the distribution and abundances of macrophytes for the Channel Islands. Continued studies of sites established by Littler would make an important contribution to our knowledge of long-term fluctuations in the abundances of intertidal macrophytes in the SCB.

For intertidal macrophytes, consistent seasonal tendencies are often difficult to detect in the SCB, and local conditions predominate over regional cycles. Where seasonality has been demonstrated over only a 2–3 year period, it appears that (1) reductions occur in standing stocks and diversity following the fall and winter periods of extended daytime tidal emersion, and (2) increases in standing stocks take place throughout the summer months of long daylight hours and warmer sea temperatures.

The most intensely studied macrophyte associations in the SCB are the *Macrocystis*-dominated kelp forests that occupy large subtidal areas. However, the structure and dynamics of kelp-forest communities are best known for only a few sites. The nature of the substratum, including sedimentation, in concert with water motion has an important influence on kelp communities. Most kelp forests occur on hard, rocky substrata, although the "*Macrocystis angustifolia* form" of *M. pyrifera* often forms forests attached only to sand and cobbles. Bouts of wave-induced surge and storm waves can detach and entangle kelp thalli and are one of the major causes of damage to kelp forests. Water motion regimes also affect the settlement of macrophyte reproductive cells and invertebrate larvae, carbon and nutrient uptake by algae, and the availability of planktonic food for various heterotrophs associated with kelp fronds.

The role of herbivorous urchins in reducing biomass in subtidal habitats is well documented. Field experiments have elucidated the competitive interactions between canopy-forming and lower strata macrophytes, as well as the development and maintenance of various populations in SCB kelp forests.

Kelp forests are dynamic over short- and long-term (>80-year) scales. Oscillations in kelp coverage have been attributed to a complex set of factors related to human activities, urchin grazing, and storm and El Niño episodes. Kelp abundances are closely related to the thermal environment, and warm ocean temperatures have been implicated in recorded reductions in *Macrocystis* canopy coverage in the SCB. Recent evidence, however, suggests that nutrient availability, which is inversely related to water temperature, may be more important to this decline. Subtidal macrophytes can be dramatically affected by catastrophic storms. Understory algae in kelp forests generally experience less damage and can colonize and persist on substrata previously occupied by *Macrocystis* for long periods following storms. Knowledge of the deep-water macrophyte populations of the SCB is essentially nonexistent.

Estimates of primary productivity have been obtained for a large number of SCB macrophytes. Based on short-term measurements, light-saturated productivities ranging from 0.1 to 11.2 mg C g dry wt^{-1} h^{-1} have been recorded. Means of 5.0 mg C dry wt^{-1} h^{-1} for sheetlike algal forms to 0.2 mg C g dry wt^{-1} h^{-1} for crustose species represent the productivity extremes of macrophyte functional-form groups. Estimates of the production of intertidal macrophyte communities of southwestern North America range from 47.9 to 203.6 mg C m^{-2} h^{-1}. Seasonal productivity budgets have been calculated for only one SCB intertidal site and were found to be greatest (1220 mg C m^{-2} d^{-1}) during fall and least (470 mg C m^{-2} d^{-1}) during spring.

The net productivities of salt marsh halophytes in the SCB range from 19–35 to 209–385 g C m^{-2} yr^{-1}, with greater annual production occurring in marsh systems where tidal flow is eliminated during the growing season and soil salinities are reduced by freshwater runoff. In the Tijuana Estuary, algal mats produce 0.8–1.4 of the carbon fixed by halophytes. Empirical estimates of *Macrocystis* production range from 350 to 1000 g C m^{-2} yr^{-1} in California. These values are consistent with the 537 g C m^{-2} yr^{-1} predicted by the recent Jackson model.

Literature Cited

Abbott, I. A., and G. J. Hollenberg, 1976. *Marine Algae of California*. Stanford Univ. Press, Palo Alto, CA. 827pp.

Ambrose, R. F., and B. V. Nelson, 1982. Inhibition of giant kelp recruitment by an introduced brown algae. *Bot. Mar.* 25:265–267.

Amsler, C. D., and R. B. Searles, 1980. Vertical distribution of seaweed spores in a water column offshore of North Carolina. *J. Phycol.* 16:617–619.

Anderson, E. K., and W. J. North, 1966. *In situ* studies of spore production and dispersal in the giant kelp, *Macrocystis*. *Proc. Int. Seaweed Symp.* 15:73–86.

Apt, K., C. D'Antonio, J. Crisp, and J. Gauvain, 1988. Intertidal macrophytes of Santa Cruz Island, California. The Herbarium, Dept. of Biological Sciences, Univ. of California, Santa Barbara, Publ. No. 6. 87pp.

Arnold, K. E., and S. L. Manley, 1985. Carbon allocation in *Macrocystis pyrifera* (Phaeophyta): Intrinsic variability in photosynthesis and respiration. *J. Phycol.* 21:154–167.

Arnold, K. E., and S. N. Murray, 1980. Relationships between irradiance and photosynthesis for marine benthic green algae (Chlorophyta) of differing morphologies. *J. Exp. Mar. Biol. Ecol.* 43:183–192.

Barilotti, D. C., 1983. Measurements needed to determine the ecologically important effects of discharged wastes in kelp bed habitats. In: W. Bascom, ed. *The Effects of Waste Disposal on Kelp Communities*. South. Calif. Coastal Water Res. Proj., Long Beach, CA. pp. 163–180.

Bascom, W., ed., 1983. *The Effects of Waste Disposal on Kelp Communities*. South. Calif. Coastal Res. Proj., Long Beach, CA. 328pp.

Bernstein, B. B., and N. Jung, 1979. Selective pressures and coevolution in a kelp canopy community in southern California. *Ecol. Monogr.* 49:335–355.

Bray, R. N., 1981. Influence of water currents and zooplankton densities on daily foraging movements of blacksmith, *Chromis punctipinnis*, a planktivorous reef fish. *U.S. Natl. Mar. Fish. Serv. Fish. Bull.* 78:829–841.

Bray, R. N., and A. W. Ebeling, 1975. Food, activity, and habitat of three "picker type" microcarnivorous fishes in the kelp forests off Santa Barbara, California. *U.S. Natl. Mar. Fish. Serv. Fish. Bull.* 73:815–829.

Bray, R. N., L. J. Purcell, and A. C. Miller, 1986. Ammonium excretion in a temperate-reef community by a planktivorous fish, *Chromis punctipinnis* (Pomacentridae), and potential uptake by young giant kelp, *Macrocystis pyrifera* (Laminariales). *Mar. Biol.* 90:327–334.

Bray, R. N., A. C. Miller, S. Johnson, P. R. Krause, D. L. Robertson, and A. M. Westcott, 1988. Ammonium excretion by macroinvertebrates and fishes on a subtidal rocky reef in southern California. *Mar. Biol.* 100:21–30.

Breitburg, D. L., 1984. Residual effects of grazing: Inhibition of competitor recruitment by encrusting coralline algae. *Ecology*. 65:1136–1143.

Britz, S. J., and W. R. Briggs, 1976. Circadian rhythms of chloroplast orientation and photosynthesis capacity in *Ulva*. *Plant Physiol.* 58:22–27.

Brostoff, W. N., 1988. Taxonomic studies of *Macrocystis pyrifera* (L.) C. Agardh (Phaeophyta) in southern California: Holdfasts and basal stipes. *Aquat. Bot.* 31:289–305.

Cambridge, M., A. M. Breeman, R. van Oosterwijk, and C. van den Hoek, 1984. Temperature responses of some North Atlantic *Cladophora* species (Chlorophyceae) in relation to their geographic distribution. *Helgol. Meeresunters.* 38:349–363.

Caplan, R. I., and R. A. Boolootian, 1967. Intertidal ecology of San Nicolas Island. In: R. N. Philbrick, ed. *Proceedings of the Symposium on the Biology of the California Islands*. Santa Barbara Botanic Garden, Santa Barbara, CA. pp. 203–217.

Chapman, A. R. O., 1986. Population and community ecology of seaweeds. *Adv. Mar. Biol.* 23:1–161.

Chapman, V. J., 1976. *Coastal Vegetation*. 2d ed. Pergamon Press, Oxford. 292pp.

Charters, A. C., M. Neushul, and D. C. Barilotti, 1969. The functional morphology of *Eisenia arborea*. *Proc. Int. Seaweed Symp.* 6:89–105.

Charters, A. C., M. Neushul, and D. A. Coon, 1972. Effects of water motion on algal spore attachment. *Proc. Int. Seaweed Symp.* 7:243–247.

Choat, J. H., 1982. Fish feeding and the structure of benthic communities in temperate waters. *Annu. Rev. Ecol. Syst.* 13:423–449.

Clarke, W. D., and M. Neushul, 1967. Subtidal ecology of the southern California coast. In: T. A. Olson, and F. J. Burgess, eds. *Pollution and Marine Ecology*. John Wiley & Sons, New York. pp. 29–42.

Coon, D., 1981. Measurements of harvested and unharvested populations of the marine crop plant *Macrocystis*. *Proc. Int. Seaweed Symp.* 8: 678–687.

Coon, D., M. Neushul, and A. C. Charters, 1972. The settling behavior of marine algal spores. *Proc. Int. Seaweed Symp.* 7:237–242.

Coutinho, R., and R. Zingmark, 1987. Diurnal photosynthetic responses to light by macroalgae. *J. Phycol.* 23:336–343.

Cowen, R. K., 1983. The effect of sheephead (*Semicossyphus pulcher*) predation on red sea urchin (*Strongylocentrotus franciscanus*) populations: An experimental analysis. *Oecologia (Berl.)*. 58: 249–255.

Daly, M. A., and A. C. Mathieson, 1977. The effects of sand movement on intertidal seaweeds and selected invertebrates at Bound Rock, New Hampshire, U.S.A. *Mar. Biol.* 43:45–55.

Davis, G. E., 1986. Kelp forest dynamics in Channel Islands National Park, California 1982–85. In: *Channel Islands National Park and National Marine Sanctuary Natural Science Study Report*. CHIS-86-001. pp. 1–11.

Dawson, E. Y., 1959. A primary report on the benthic marine flora of southern California. In: *An Oceanographic and Biological Survey of the Continental Shelf Area of Southern California*. Publs. Calif. State Water Qual. Control Bd. Vol. 20. pp. 169–264.

Dawson, E. Y., 1965. Intertidal algae. In: *An Oceanographic and Biological Survey of the Southern California Mainland Shelf.* Publs. Calif. State Water Qual. Control Bd. Vol. 27. pp. 220–231, 351–438.

Dayton, P. K., 1975. Experimental evaluation of ecological dominance in a rocky intertidal algal community. *Ecol. Monogr.* 45:137–159.

Dayton, P. K., 1985. Ecology of kelp communities. *Annu. Rev. Ecol. Syst.* 16:215–245.

Dayton, P. K., and M. J. Tegner, 1984a. The importance of scale in community ecology: A kelp forest example with terrestrial analogs. In: P. W. Price, C. N. Slobodchikoff, and W. S. Gaud, eds. *A New Ecology: Novel Approaches to Interactive Systems*. John Wiley & Sons, New York, pp. 457–481.

Dayton, P. K., and M. J. Tegner, 1984b. Catastrophic storms, El Niño, and patch stability in a southern California kelp community. *Science* 224:283–285.

Dayton, P. K., V. Currie, T. Gerrodette, B. D. Keller, R. Rosenthal, and D. Ven Tresca, 1984. Patch dynamics and stability of some California kelp communities. *Ecol. Monogr.* 54:253–289.

Dean, T. A., and L. E. Deysher, 1983. The effects of suspended solids and thermal discharges on kelp. In: W. Bascom, ed. *The Effects of Waste Disposal on Kelp Communities*. South. Calif. Coastal Water Res. Proj., Long Beach, CA. pp. 114–135.

Dean, T. A., S. C. Schroeter, and J. D. Dixon, 1984. Effects of grazing by two species of sea urchins (*Strongylocentrotus franciscanus* and *Lytechinus anamesus*) on recruitment and survival of two species of kelp (*Macrocystis pyrifera* and *Pterygophora californica*). *Mar. Biol.* 78:301–313.

Dean, T. A., F. R. Jacobsen, K. Thies, and S. L. Lagos, 1988. Differential effects of grazing by white sea urchins on recruitment of brown algae. *Mar. Ecol. Prog. Ser.* 48:99–102.

Dean, T. A., K. Thies, and S. L. Lagos, 1989. Survival of juvenile kelp: The effects of demographic factors, competition, and grazers. *Ecology*. 70:483–495.

Dethier, M. N., 1982. Pattern and process in tidepool algae: Factors influencing seasonality and distribution. *Bot. Mar.* 25:55–66.

Devinny, J. S., and L. A. Volse, 1978. Effects of sediments on the development of *Macrocystis pyrifera* gametophytes. *Mar. Biol.* 48:343–348.

Deysher, L. E., and T. A. Dean, 1986a. Interactive effects of light and temperature on sporophyte production in the giant kelp *Macrocystis pyrifera*. *Mar. Biol.* 93:17–20.

Deysher, L. E., and T. A. Dean, 1986b. *In situ* recruitment of sporophytes of the giant kelp, *Macrocystis pyrifera* (L.) C. A. Agardh: Effects of physical factors. *J. Exp. Mar. Biol. Ecol.* 103:41–63.

Druehl, L. D., 1981. Geographical distribution. In: C. S. Lobban, and M. J. Wynne, eds. *The Biology of Seaweeds*. Univ. of California Press, Berkeley and Los Angeles. pp. 306–325.

Duggins, D. O., and M. N. Dethier, 1985. Experimental studies of herbivory and algal competition in a low intertidal habitat. *Oecologia (Berl.)*. 67:183–191.

Ebeling, A. W., and D. R. Laur, 1985. The influence of plant cover on surfperch abundance at an offshore temperate reef. *Environ. Biol. Fish.* 12:169–179.
Ebeling, A. W., D. R. Laur, and R. J. Rowley, 1985. Severe storm disturbances and reversal of community structure in a southern California kelp forest. *Mar. Biol.* 84:287–294.
Eilers, H. P., 1981. Production in coastal salt marshes of southern California. EPA–60013–81–023. Available from: NTIS, Springfield, VA; PB81–171845. Prep. by Dept. of Geography, California State Univ., Fullerton, CA. 100pp.
Emerson, S. E., and J. B. Zedler, 1978. Recolonization of intertidal algae: An experimental study. *Mar. Biol.* 44:315–324.
Fain, S. R., and S. N. Murray, 1982. Effects of light and temperature on net photosynthesis and dark respiration of gametophytes and embryonic sporophytes of *Macrocystis pyrifera*. *J. Phycol.* 18:92–98.
Foster, M. S., 1972. The algal turf community in the nest of the ocean goldfish, *Hypsypops rubicunda*. *Proc. Int. Seaweed Symp.* 7:55–60.
Foster, M. S., 1982. Factors controlling the intertidal zonation of *Iridaea flaccida* (Rhodophyta). *J. Phycol.* 18:285–294.
Foster, M. S., and D. R. Schiel, 1985. Ecology of giant kelp forests in California: A community profile. *U.S. Fish Wildl. Serv. Biol. Rep.* 85(7.2), 152pp.
Foster, M. S., and D. R. Schiel, 1988. Kelp communities and sea otters: Keystone species or just another brick in the wall? In: G. R. VanBlaricom, and J. A. Estes, eds. *Ecological Studies: Analysis and Synthesis, Vol. 65. The Community Ecology of Sea Otters.* Springer-Verlag, New York. pp. 92–115.
Foster, M. S., M. Neushul, and R. Zingmark, 1971. The Santa Barbara oil spill. Part 2: Initial effects on intertidal and kelp bed organisms. *Environ. Pollut.* 2:115–134.
Gaines, S. D., 1985. Herbivory and between-habitat diversity: The differential effectiveness of defenses in a marine plant. *Ecology.* 66:473–485.
Gaines, S. D., and J. Lubchenco, 1982. A unified approach to marine plant–herbivore interactions. II. Biogeography. *Annu. Rev. Ecol. Syst.* 13: 111–138.
Gallagher, J. L., and F. C. Daiber, 1974. Primary production of edaphic algal communities in a Delaware salt marsh. *Limnol. Oceanogr.* 19:390–395.
Garbary, D. J., 1987. A critique of traditional approaches to seaweed distribution in light of the development of vicariance biogeography. *Helgol. Meeresunters.* 41:235–244.
Gerard, V. A., 1982a. Growth and utilization of internal nitrogen reserves by the giant kelp *Macrocystis pyrifera* in a low-nitrogen environment. *Mar. Biol.* 66:27–35.
Gerard, V. A., 1982b. *In situ* water motion and nutrient uptake by the giant kelp *Macrocystis pyrifera*. *Mar. Biol.* 69:51–54.
Gerard, V. A., 1984. The light environment in a giant kelp forest: Influence of *Macrocystis pyrifera* on spatial and temporal variability. *Mar. Biol.* 84:189–195.
Gerard, V. A., 1986. Photosynthetic characteristics of giant kelp (*Macrocystis pyrifera*) determined *in situ*. *Mar. Biol.* 90:473–482.
Grant, J. J., K. C. Wilson, A. Grover, and H. A. Togstad, 1982. Early development of Pendleton artificial reef. *Mar. Fish. Rev.* 44:53–60.
Gunnill, F. C., 1980. Recruitment and standing stocks in populations of one green alga and five brown algae in the intertidal zone near La Jolla, California, during 1973–1977. *Mar. Ecol. Prog. Ser.* 3:231–243.
Harger, B., 1983. A historical overview of kelp in southern California. In: W. Bascom, ed. *The Effects of Waste Disposal on Kelp Communities.* South. Calif. Coastal Water Res. Proj., Long Beach, CA. pp. 70–83.
Harris, L. G., A. W. Ebeling, D. R. Laur, and R. J. Rowley, 1984. Community recovery after storm damage: A case of facilitation in primary succession. *Science.* 224:1336–1338.
Harris, L. H., 1980. Changes in intertidal algae at Palos Verdes. In: W. Bascom, ed. *Biennial Report for the Years 1979–1980.* South. Calif. Coastal Water Res. Proj., Long Beach, CA. pp. 35–75.
Harris, L. H., 1983. Changes in intertidal algae at Palos Verdes. In: W. Bascom, ed. *The Effects of Waste Disposal on Kelp Communities.* South. Calif. Coastal Water Res. Proj., Long Beach, CA. pp. 274–281.
Harrold, C., and D. C. Reed, 1985. Food availability, sea urchin grazing, and kelp forest community structure. *Ecology.* 66:1160–1169.
Hawkins, S. J., and R. G. Hartnoll, 1983. Grazing of intertidal algae by marine invertebrates. *Oceanogr. Mar. Biol. Annu. Rev.* 21:195–282.
Hay, M. E., 1981a. The functional morphology of

turf-forming seaweeds: Persistence in stressful marine habitats. *Ecology*. 62:739–750.

Hay, M. E., 1981b. Herbivory, algal distribution and the maintenance of between habitat diversity on a tropical fringing reef. *Am. Nat.* 118: 520–540.

Hewatt, W. G., 1946. Marine ecological studies on Santa Cruz Island, California. *Ecol. Monogr.* 16:185–210.

Ho, J. S., 1974. Marshes and bays. In: *A Summary of Knowledge of the Southern California Coastal Zone and Offshore Areas. Vol. II. Biological Environment.* U.S. Dept. of Interior, Bureau of Land Management, Washington, D.C. Prep. by South. Calif. Ocean Studies Consortium, Long Beach, CA. pp. 14.1–14.59.

Hodder, D., and M. Mel, 1978. Kelp survey of the Southern California Bight. Science Applications, Inc. Tech. Rep. Vol. III. Rep. 1.4 to Bureau of Land Management (Year II SCOSC Program), Contract No. AA550-CT6–40, La Jolla, CA. 105pp.

Hodgkiss, I. J., 1984. Seasonal patterns of intertidal algal distribution in Hong Kong. *Asian Mar. Biol.* 1:49–57.

Hoek, C. van den, 1982a. Phytogeographic distribution of groups of benthic marine algae in the North Atlantic Ocean. A review of experimental evidence from life history studies. *Helgol. Meeresunters.* 35:153–214.

Hoek, C. van den, 1982b. The distribution of benthic marine algae in relation to the temperature regulation of their life histories. *Biol. J. Linn. Soc.* 18:81–144.

Hoek, C. van den, 1987. The possible significance of long-range dispersal for the biogeography of seaweeds. *Helgol. Meeresunters.* 41:261–272.

Hoffman, W. E., and C. J. Dawes, 1980. Photosynthetic rates and primary production by two Florida benthic red algal species from a salt marsh and a mangrove community. *Bull. Mar. Sci.* 30:358–364.

Horn, M. H., S. N. Murray, and R. R. Seapy, 1983. Seasonal structure of a central California rocky intertidal community in relation to environmental variations. *Bull. South. Calif. Acad. Sci.* 82:79–94.

Hruby, T., and T. A. Norton, 1979. Algal colonization on rocky shores in the Firth of Clyde. *J. Ecol.* 67:65–77

Jackson, G. A., 1977. Nutrients and production of giant kelp, *Macrocystis pyrifera*, off southern California. *Limnol. Oceanogr.* 22:979–995.

Jackson, G. A., 1983. The physical and chemical environment of a kelp community. In: W. Bascom, ed. *The Effects of Waste Disposal on Kelp Communities.* South. Calif. Coastal Water Res. Proj., Long Beach, CA. pp. 11–37.

Jackson, G. A., 1987. Modelling the growth and harvest yield of the giant kelp *Macrocystis pyrifera*. *Mar. Biol.* 95:611–624.

Jackson, G. A., and C. D. Winant, 1983. Effect of a kelp forest on coastal currents. *Cont. Shelf Res.* 2:75–80.

Jernakoff, P., 1983. Factors affecting the recruitment of algae in a midshore region dominated by barnacles. *J. Exp. Mar. Biol. Ecol.* 67:17–31.

Johnston, C. S., 1969. The ecological distribution and primary production of macrophytic marine algae in the eastern Canaries. *Int. Rev. Gesamten Hydrobiol.* 54:473–490.

Jones, N. S., 1948. Observations and experiments on the biology of *Patella vulgata* at Port St. Mary, Isle of Man. *Trans. Liverpool Biol. Soc.* 56:60–77.

Jones, W. E., A Fletcher, S. Bennell, B. McConnell, and S. M. Smith, 1979. Changes in littoral populations as recorded by long-term shore surveillance. 1. Selected examples of cyclic changes. In: E. Naylor, and R. G. Hartnoll, eds., *Cyclic Phenomena in Marine Plants and Animals.* Proc. 13th Eur. Mar. Biol. Symp. Pergamon Press, Oxford. pp. 93–100.

Kastendiek, J. E., 1982. Competitor-mediated coexistence: Interactions among three species of benthic macroalgae. *J. Exp. Mar. Biol. Ecol.* 62:201–210.

Larson, R. J., and E. E. DeMartini, 1984. Abundance and vertical distribution of fishes in a cobble bottom kelp forest off San Onofre, California. *U.S. Natl. Mar. Fish. Serv. Fish. Bull.* 82:37–53.

Leighton, D. L., 1966. Studies of food preferences in algivorous invertebrates of southern California kelp beds. *Pac. Sci.* 20:104–113.

Leighton, D. L., 1971. Grazing activities of benthic invertebrates in southern California kelp beds. *Nova Hedwegia.* 32:421–453.

Lewbel, G. S., A. Wolfson, T. Gerrodette, W. H. Lippincott, J. L. Wilson, and M. M. Littler, 1981. Shallow-water benthic communities on California's outer continental shelf. *Mar. Ecol. Prog. Ser.* 4:159–168.

Lewis, J. R., 1964. *The Ecology of Rocky Shores.* The English Universities Press Ltd., London. 323pp.

Lewis, J. R., 1976. Long-term ecological surveillance: Practical realities in the rocky littoral. *Oceanogr. Mar. Biol. Annu. Rev.* 14:371–390.
Lewis, J. R., 1977. The role of physical and biological factors in the distribution and stability of rocky shore communities. In: B. F. Keegan, P. O'Ceidigh, and P. J. S. Boaden, eds. *Biology of Benthic Organisms.* Proc. 11th Eur. Mar. Biol. Symp. Pergamon Press, Oxford. pp. 417–424.
Littler, M. M., ed., 1977. Spatial and temporal variations in the distribution and abundance of rocky intertidal and tidepool biotas in the Southern California Bight. Southern California Baseline Study. Final Report, Vol. III, Rep. 2.1. U.S. Dept. of Interior, Bureau of Land Management, Washington, D.C.
Littler, M. M., ed., 1978. The annual and seasonal ecology of southern California rocky intertidal, subtidal and tidepool biotas. Southern California Baseline Study. Final Report, Vol. III, Rep. 1.1. U.S. Dept. of Interior, Bureau of Land Management, Washington, D.C.
Littler, M. M., ed., 1979a. The distribution, abundance, and community structure of rocky intertidal and tidepool biotas in the Southern California Bight. Southern California Baseline Study. Final Report, Vol. II, Rep. 1.0. U.S. Dept. of Interior, Bureau of Land Management, Washington, D.C.
Littler, M. M., 1979b. The effects of bottle volume, thallus weight, oxygen saturation levels, and water movement on apparent photosynthetic rates in marine algae. *Aquat. Bot.* 7:21–34.
Littler, M. M., 1980a. Overview of the rocky intertidal systems of southern California. In: D. M. Power, ed. *The California Islands: Proceedings of a Multidisciplinary Symposium.* Santa Barbara Museum of Natural History, Santa Barbara, CA. pp. 265–306.
Littler, M. M., 1980b. Southern California rocky intertidal ecosystems: Methods, community structure, and variability. In: J. H. Price, D. E. G. Irvine, and W.H. Farnham, eds. *The Shore Environment. Vol. 2: Ecosystems.* Academic Press, London. pp. 565–608.
Littler, M. M., 1980c. Morphological form and photosynthetic performances of marine macroalgae: Tests of a functional form hypothesis. *Bot. Mar.* 22:161–165.
Littler, M. M., 1981. The relationship between thallus form and the primary productivity of seaweeds. *Proc. Int. Seaweed Symp.* 8:398–403.
Littler, M. M., and K. E. Arnold, 1980. Sources of variability in macroalgal primary productivity: Sampling and interpretative problems. *Aquat. Bot.* 8:141–156.
Littler, M. M., and K. E. Arnold, 1982. Primary productivity of marine macroalgal functional-form groups from southwestern North America. *J. Phycol.* 18:307–311.
Littler, M. M., and K. E. Arnold, 1985. Electrodes and chemicals. In: M. M. Littler, and D. S. Littler, eds. *Handbook of Phycological Methods. Ecological Field Methods: Macroalgae.* Cambridge Univ. Press, London. pp. 349–375.
Littler, M. M., and B. J. Kauker, 1984. Heterotrichy and survival strategies in the red alga *Corallina officinalis. Bot. Mar.* 27:37–44.
Littler, M. M., and D. S. Littler, 1979. Rocky intertidal island survey. Southern California Intertidal Survey Year III. U.S. Dept. of Interior, Bureau of Land Management, Pacific OCS Region, Los Angeles, CA. Prep. by Science Applications, Inc., La Jolla, CA. 5pp.
Littler, M. M., and D. S. Littler, 1980a. Mainland rocky intertidal aerial survey from Point Arguello to Point Loma, California. U.S. Dept. of Interior, Bureau of Land Management, Pacific OCS Region, Los Angeles, CA, 12pp.
Littler, M. M., and D. S. Littler, 1980b. The evolution of thallus form and survival strategies in benthic marine macroalgae: Field and laboratory tests of a functional form model. *Am. Nat.* 116:25–44.
Littler, M. M., and D. S. Littler, 1981. Intertidal macrophyte communities from Pacific Baja California and the upper Gulf of California: Relatively constant vs. environmentally fluctuating systems. *Mar. Ecol. Prog. Ser.* 4:145–158.
Littler, M. M., and D. S. Littler, 1984. Relationships between macroalgal functional form groups and substrata stability in a subtropical rocky-intertidal system. *J. Exp. Mar. Biol. Ecol.* 74:13–34.
Littler, M. M., and D. S. Littler, 1987. Effects of stochastic processes on rocky-intertidal biotas: An unusual flash flood near Corona del Mar, California. *Bull. South. Calif. Acad. Sci.* 86:95–106.
Littler, M. M., and S. N. Murray, 1974a. The primary productivity of marine macrophytes from a rocky intertidal community. *Mar. Biol.* 27:131–135.
Littler, M. M., and S. N. Murray, 1974b. Primary

productivity of macrophytes. In: S. N. Murray, and M. M. Littler, eds. *Biological Features of Intertidal Communities Near the U.S. Navy Sewage Outfall, Wilson Cove, San Clemente Island, California*. Tech. Pap. 396, Naval Undersea Ctr., San Diego. pp. 67–79, 80–85.

Littler, M. M., and S. N. Murray, 1975. Impact of sewage on the distribution, abundance and community structure of rocky intertidal macroorganisms. *Mar. Biol.* 30:277–291.

Littler, M. M., and S. N. Murray, 1977. Influence of domestic wastes on the structure and energetics of intertidal communities near Wilson Cove, San Clemente Island. Calif. Water Res. Center, Univ. of California, Davis. Contribution No. 164. Available from: NTIS, Springfield, VA; PB–271–072–1, 88pp.

Littler, M. M., and S. N. Murray, 1978. Influence of domestic wastes on energetic pathways in rocky intertidal communities. *J. Appl. Ecol.* 15:583–595.

Littler, M. M., S. N. Murray, and K. E. Arnold, 1979. Seasonal variations in net photosynthetic performance and cover of intertidal macrophytes. *Aquat. Bot.* 7:35–46.

Littler, M. M., D. R. Martz, and D. S. Littler, 1983. Effects of recurrent sand deposition on rocky intertidal organisms: Importance of substrate heterogeneity in a fluctuating environment. *Mar. Ecol. Prog. Ser.* 11:129–140.

Littler, M. M., D. S. Littler, S. N. Murray, and R. R. Seapy, 1991. Southern California intertidal ecosystems. In: A. C. Mathieson, and P. Nienhuis, eds. *Ecosystems of the World. Vol. 24. Intertidal and Littoral Ecosystems*. Elsevier Scientific Publ., B.V., Amsterdam, The Netherlands pp. 273–296.

Lobban, C. S., P. J. Harrison, and M. J. Duncan, 1985. *The Physiological Ecology of Seaweeds*. Cambridge Univ. Press, Cambridge. 242pp.

Lubchenco, J., 1983. *Littorina* and *Fucus:* Effects of herbivores, substratum heterogeneity, and plant escapes during succession. *Ecology*. 64:1116–1123.

Lubchenco, J., and S. D. Gaines, 1981. A unified approach to marine plant-herbivore interactions. I. Populations and communities. *Annu. Rev. Ecol. Syst.* 12:405–437.

Lüning, K., and W. Freshwater, 1988. Temperature tolerance of northeast Pacific marine algae. *J. Phycol.* 24:310–315.

MacDonald, K. B., 1977a. Coastal salt marsh. In: M. G. Barbour, and J. Major, eds. *Terrestrial Vegetation of California*. John Wiley & Sons, New York. pp. 263–294.

MacDonald, K. B., 1977b. Plant and animal communities of Pacific North American salt marshes. In: V. J. Chapman, ed. *Ecosystems of the World. Vol. 1. Wet Coastal Ecosystems*. Elsevier Scientific Publ. B.V., Amsterdam, The Netherlands. pp. 167–191.

Manley, S. L., 1985. Phosphate uptake by blades of *Macrocystis pyrifera* (Phaeophyta). *Bot. Mar.* 28:237–244.

Mann, K. H., 1972. Ecological energetics of the seaweed zone in a marine bay on the Atlantic coast of Canada. II. Productivity of the seaweeds. *Mar. Biol.* 14:199–209.

Mann, K. H., 1982. *Ecology of Coastal Waters. A Systems Approach*. Univ. of California Press, Berkeley. 322pp.

McFarland, W. N., and J. Prescott, 1959. Standing crop, chlorophyll content, and *in situ* metabolism of a giant kelp community in southern California. *Publ. Inst. Mar. Sci.*, Univ. of Texas, Austin, TX. 6:109–132.

McLusky, D. S., 1971. *Ecology of Estuaries*. Heinemann Educational Books Ltd., London. 144pp.

McPeak, R. H., 1981. Fruiting in several species of Laminariales from southern California. *Proc. Int. Seaweed Symp.* 8:404–409.

McQuaid, C. D., 1985. Seasonal variation in biomass and zonation of nine intertidal algae in relation to changes in radiation, sea temperature and tidal regime. *Bot.Mar.* 28:539–544.

McQuaid, C. D., and G. M. Branch, 1984. Influence of sea temperature, substratum, and wave exposure on rocky intertidal communities: An analysis of faunal and floral biomass. *Mar. Ecol. Prog. Ser.* 19:145–151.

McQuaid, C. D., and G. M. Branch, 1985. Trophic structure of rocky intertidal communities: Response to wave action and implications for energy flow. *Mar. Ecol. Prog. Ser.* 22:153–161.

McQuaid, C. D., G. M. Branch, and A. A. Crowe, 1985. Biotic and abiotic influences on rocky intertidal biomass and richness in the southern Benguela region. *South. Afr. Tydskr. Dierk.* 20:115–122.

Michanek, G., 1979. Phytogeographic provinces and seaweed distribution. *Bot. Mar.* 22:375–391.

Murray, S. N., 1974. Benthic algae and grasses. In: M. D. Dailey, B. Hill, and N. Lansing, eds. *A Summary of Knowledge of the Southern Califor-*

nia Coastal Zone and Offshore Areas. Biological Environment. U.S. Dept. of Interior, Bureau of Land Management, Washington, D.C. II: 9.1–9.61.

Murray, S. N., and M. H. Horn, 1989. Seasonal dynamics of macrophyte populations from an eastern North Pacific rocky-intertidal habitat. *Bot. Mar.* 32:457–473.

Murray, S. N., and M. M. Littler, 1974. Analyses of standing stocks and community structure of macro-organisms. In: M. M. Littler and S. N. Murray, eds. *Biological Features of Intertidal Communities Near the U.S. Navy Sewage Outfall, Wilson Cove, San Clemente Island, California.* Tech. Pap. Naval Undersea Ctr., San Diego. Contrib. No. 396. pp. 23–51, 80–85.

Murray, S. N., and M. M. Littler, 1977. Seasonal analyses of standing stock and community structure of macro-organisms. In: M. M. Littler, and S. N. Murray, eds. *Influence of Domestic Wastes on the Structure and Energetics of Intertidal Communities Near Wilson Cove, San Clemente Island.* Calif. Water Res. Ctr., Univ. Calif., Davis. Contrib. No. 164. pp. 7–32.

Murray, S. N., and M. M. Littler, 1978. Patterns of algal succession in a perturbated marine intertidal community. *J. Phycol.* 14:506–512.

Murray, S. N., and M. M. Littler, 1979. Experimental studies of the recovery of populations of rocky intertidal macro-organisms following mechanical disturbance. In: M. M. Littler, ed. *The Distribution, Abundance and Community Structure of Rocky Intertidal and Tidepool Biotas in the Southern California Bight. Southern California Baseline Study.* Final Report, Vol. II, Rep. 2.0. U.S. Dept. of Interior, Bureau of Land Management, Washington, D.C.

Murray, S. N., and M. M. Littler, 1981. Biogeographical analysis of intertidal macrophyte floras of southern California. *J. Biogeogr.* 8:339–351.

Murray, S. N., and M. M. Littler, 1984. Analysis of seaweed communities in a disturbed rocky intertidal environment near Whites Point, Los Angeles, Calif., USA. *Hydrobiologia.* 116/117:374–382.

Murray, S. N., and M. M. Littler, 1989. Seaweeds and seagrasses of southern California: Distributional lists for twenty-one rocky intertidal sites. *Bull. South. Calif. Acad. Sci.* 88:61–79.

Murray, S. N., M. M. Littler, and I. A. Abbott, 1980. Biogeography of the California marine algae with emphasis on the southern California islands. In: D. M. Power, ed. *The California Islands: Proceedings of a Multidisciplinary Symposium.* Santa Barbara Museum of Natural History, Santa Barbara, CA. pp. 325–339.

Nelson, B. V., and R. R. Vance, 1979. Diel foraging patterns of the sea urchin *Centrostephanus coronatus* as a predator avoidance strategy. *Mar. Biol.* 51:251–258.

Neushul, M., 1971. The species of *Macrocystis* with particular reference to those of North and South America. *Nova Hedwegia.* 32:211–222.

Neushul, M., 1972. Functional interpretation of benthic marine algal morphology. In: *Contributions to the Systematics of Benthic Marine Algae of the North Pacific.* Japanese Society of Phycology, Kobe (JPN). pp. 47–73.

Neushul, M., 1981. The domestication and cultivation of California macroalgae. *Proc. Int. Seaweed Symp.* 10:71–96.

Neushul, M., 1983. An overview of basic research on kelp and the kelp forest ecosystem. In: W. Bascom, ed. *The Effects of Waste Disposal on Kelp Communities.* South. Calif. Coastal Water Res. Proj., Long Beach, CA. pp. 282–300.

Neushul, M., W. D. Clarke, and D. W. Brown, 1967. Subtidal plant and animal communities of the southern California islands. In: R. N. Philbrick, ed. *Proceedings of the Symposium on the Biology of the California Islands.* Santa Barbara Botanic Garden, Santa Barbara, CA. pp. 37–55.

Ngan, Y., and I. R. Price, 1980. Seasonal growth and reproduction of intertidal algae in the Townsville region (Queensland, Australia). *Aquat. Bot.* 9:117–134.

Nicholson, N. L., and R. L. Cimberg, 1971. The Santa Barbara oil spills of 1969: A post-spill survey of the rocky intertidal. In: D. S. Straughan, comp. *Biological and Oceanographical Survey of the Santa Barbara Channel Oil Spill 1969–1970.* Vol. I. Allan Hancock Foundation, Univ. of Southern California, Los Angeles, CA. pp. 325–399.

North, W. J., 1971. Introduction and background. In: W. J. North, ed. *The Biology of Giant Kelp Beds* (Macrocystis) *in California.* Beihefte Nova Hedwegia. 32:1–97.

North, W. J., 1972. Mass-cultured *Macrocystis* as a means of increasing kelp stands in nature. *Proc. Int. Seaweed Symp.* 7:394–399.

North, W. J., 1983. The sea urchin problem. In:

W. Bascom, ed. *The Effects of Waste Disposal on Kelp Communities*. South. Calif. Coastal Water Res. Proj., Long Beach, CA. pp. 147–162.

North, W. J., and R. C. Zimmerman, 1984. Influences of macronutrients and water temperatures on summertime survival of *Macrocystis* canopies. *Hydrobiologia*. 116/117:419–424.

North, W. J., V. A. Gerard, and J. S. Kuwabara, 1982. Farming *Macrocystis* at coastal and oceanic sites. In: L. M. Srivastava, ed. *Synthetic and Degradative Processes in Marine Macrophytes*. De Gruyter, New York. pp. 247–262.

North, W. J., G. A. Jackson, and S. L. Manley, 1986. *Macrocystis* and its environment, knowns and unknowns. *Aquat. Bot*. 26:9–26.

Norton, T. A., A. C. Mathieson, and M. Neushul, 1981. Morphology and environment. In: C. S. Lobban, and M. J. Wynne, eds. *The Biology of Seaweeds*. Univ. of California Press, Berkeley. pp. 421–451.

Norton, T. A., A. C. Mathieson, and M. Neushul, 1982. A review of some aspects of form and function in seaweeds. *Bot. Mar*. 25:501–510.

Onuf, C. P., M. L. Quammen, G. O. Shaffer, C. H. Peterson, J. W. Chapman, J. Cernak, and R. W. Holmes, 1978. An analysis of the value of Central and Southern California coastal wetlands. In: P. W. Greeson, J. K. Clark, and J. E. Clark, eds. *Wetlands functions and values: the state of our understanding*. American Water Resources Assoc., Minneapolis, MN. pp. 189–199.

Papenfuss, G. F., 1976. Landmarks in Pacific North American marine phycology. In: I. A. Abbott, and G. J. Hollenberg, eds. *Marine Algae of California*. Stanford Univ. Press, Palo Alto, CA. pp. 21–46.

Quast, J. C., 1971a. Observations on the food of kelp bed fishes. *Nova Hedwegia*. 32:541–579.

Quast, J. C., 1971b. Fish fauna of the rocky inshore zone. *Nova Hedwegia*. 32:481–507.

Ramus, J., and G. Rosenberg, 1980. Diurnal photosynthetic performance of seaweeds measured under natural conditions. *Mar. Biol*. 56: 21–28.

Ranwell, D. S., 1972. *Ecology of Salt Marshes and Sand Dunes*. Chapman and Hall, London. 258pp.

Reed, D. C., 1987. Factors affecting the production of sporophylls in the giant kelp *Macrocystis pyrifera* (L.) C.Ag. *J. Exp. Mar. Biol. Ecol*. 113:61–69.

Reed, D. C., 1990. The effects of variable settlement and early competition on patterns of kelp recruitment. *Ecology*. 71:776–787.

Reed, D. C., and M. S. Foster, 1984. The effects of canopy shading on algal recruitment and growth in a giant kelp forest. *Ecology*. 65:937–948.

Reed, D. C., D. R. Laur, and A. W. Ebeling, 1988. Variation in algal dispersal and recruitment: The importance of episodic events. *Ecol. Monogr*. 58:321–335.

Ricketts, E. F., J. Calvin, J. W. Hedgpeth, and D. W. Phillips, 1985. *Between Pacific Tides,* 5th edition. Stanford Univ. Press, Palo Alto, CA. 652pp.

Riznyk, R. Z., 1974. Wetland resources. In: M. D. Dailey, B. Hill, and N. Lansing, eds. *A Summary of Knowledge of the Southern California Coastal Zone and Offshore Areas. Vol. III. Social and Economic Elements*. U.S. Dept. of Interior, Bureau of Land Management, Washington, D.C. pp. 21.1–21.48.

Robles, C., 1982. Disturbance and predation in an assemblage of herbivorous Diptera and algae on rocky shores. *Oecologia (Berl.)*. 54:23–31.

Rosenthal, R. J., W. D. Clarke, and P. K. Dayton, 1974. Ecology and natural history of a stand of giant kelp, *Macrocystis pyrifera* (Linnaeus) Agardh off Del Mar, California. *U.S. Natl. Mar. Fish. Serv. Fish. Bull*. 72:670–684.

Saunders, D. A., 1901. Papers from the Harriman Alaska expedition. XXV. The algae. *Proc. Wash. Acad. Sci*. 3:391–486.

Scagel, R. F., 1963a. Distribution of attached marine algae in relation to oceanographic conditions in the northeast Pacific. In: M. J. Dunbar, ed. *Marine Distribution*. Spec. Publ. R. Soc. Can. No. 5. pp. 37–50.

Scagel, R. F., 1963b. Some problems in algal distribution in the North Pacific. *Proc. Int. Seaweed Symp*. 4:259–264.

Scagel, R. F., 1972. Guide to common seaweeds of British Columbia. Handbook No. 27. British Columbia Provincial Museum, Department of Recreation and Conservation, Victoria, Canada.

Schiel, D. R., and M. S. Foster, 1986. The structure of subtidal algal stands in temperate waters. *Oceanogr. Mar. Biol. Annu. Rev*. 24:265–307.

Seapy, R. R., and M. M. Littler, 1978. The distribution, abundance, community structure, and

primary productivity of macroorganisms from two central California rocky intertidal habitats. *Pac. Sci.* 32:293–314.
Seapy, R. R., and M. M. Littler, 1982. Population and species diversity fluctuations in a rocky intertidal community relative to severe aerial exposure and sediment burial. *Mar. Biol.* 71: 87–96.
Setchell, W. A., 1893. On the classification and geographical distribution of the Laminariaceae. *Trans. Conn. Acad.* 9:333–375.
Setchell, W. A., 1915. The law of temperature connected with the distribution of the marine algae. *Ann. Mo. Bot. Gard.* 2:287–305.
Setchell, W. A., and N. L. Gardner, 1903. Algae of northwestern America. *Univ. Calif. Publ. Bot.* 1:165–419.
Sousa, W. P., 1979a. Disturbance in marine intertidal boulder fields: The nonequilibrium maintenance of species diversity. *Ecology.* 60: 1225–1239.
Sousa, W. P., 1979b. Experimental investigations of disturbance and ecological succession in a rocky intertidal algal community. *Ecol. Monogr.* 49:227–254.
Sousa, W. P., 1980. The response of a community to disturbances: The importance of successional age and species' life histories. *Oecologia (Berl.).* 45:72–81.
Sousa, W. P., 1984. Intertidal mosaics: Patch size, propagule availability, and spatially variable patterns of succession. *Ecology.* 65:1918–1935.
Sousa, W. P., S. C. Schroeter, and S. D. Gaines, 1981. Latitudinal variation in intertidal algal community structure: The influence of grazing and vegetative propagation. *Oecologia (Berl.).* 48:297–307.
Southward, A. J., 1956. The population balance between limpets and seaweeds on wave-beaten rocky shores. In: Annual Report, Mar. Biol. Stn. Port Erin, Isle of Man, No. 68. pp. 20–29.
Southward, A. J., 1964. Limpet grazing and control of vegetation on rocky shores. In: D. J. Crisp, ed. *Grazing in Terrestrial and Marine Environments.* Blackwell Scientific Publ., Oxford. pp. 265–273.
Speth, J. W., 1969a. Status report on the coastal wetlands of southern California as of February 1, 1969. *Calif. Dept. Fish Game,* 29pp.
Speth, J. W., 1969b. The fuss over coastal wetlands. *Outdoor Calif.* 30:6–7.
Stephenson, T. A., and A. Stephenson, 1972. *Life Between Tidemarks on Rocky Shores.* W. H. Freeman and Co., San Francisco. 425pp.
Stewart, J. G., 1982. Anchor species and epiphytes in intertidal algal turf. *Pac. Sci.* 36:45–49.
Stewart, J. G., 1983. Fluctuations in the quantity of sediments trapped among algal thalli on intertidal rock platforms in southern California. *J. Exp. Mar. Biol. Ecol.* 73:205–211.
Stewart, J. G., 1984. Algal distributions and temperature: Test of an hypothesis based on vegetative growth rates. *Bull. South. Calif. Acad. Sci.* 76:57–68.
Straughan, D. S., 1979. Sandy beach and slough community analysis. Southern California Baseline Study. Final Rep., Vol. II, Rep. 8.0. U.S. Dept. of Interior, Bureau of Land Management, Washington, D.C.
Taylor, P. R., 1985. The influence of sea anemones on the morphology and productivity of two intertidal seaweeds. *J. Phycol.* 21:335–340.
Taylor, P. R., and M. E. Hay, 1984. Functional morphology of intertidal seaweeds: Adaptive significance of aggregate vs. solitary forms. *Mar. Ecol. Prog. Ser.* 18:295–302.
Taylor, P. R., and M. M. Littler, 1982. The roles of compensatory mortality, physical disturbance, and substrate retention in the development and organization of a sand-influenced, rocky-intertidal community. *Ecology.* 63:135–146.
Teal, J. M., 1962. Energy flow in the salt marsh ecosystem of Georgia. *Ecology.* 43:614–624.
Tegner, M. J., and P. K. Dayton, 1981. Population structure, recruitment, and mortality of two sea urchins (*Strongylocentrotus franciscanus* and *S. purpuratus*) in a kelp forest. *Mar. Ecol. Prog. Ser.* 5:255–268.
Tegner, M. J., and P. K. Dayton, 1987. El Niño effects on southern California kelp forest communities. In: A. MacFadyen, and E. D. Ford, eds. *Adv. Ecol. Res.* 17:243–289.
Theis, C. L., 1985. Photosynthetic compensation relative to depth in three species of the green alga *Codium* from Santa Catalina Island. Master's Thesis, California State Univ., Fullerton, CA. 76pp.
Thom, R. M., 1980. A gradient in benthic intertidal algal assemblages along the southern California coast. *J. Phycol.* 16:102–108.
Thom, R. M., and T. B. Widdowson, 1978. A resurvey of E. Yale Dawson's 42 intertidal algal transects on the southern California main-

land after 15 years. *Bull. South. Calif. Acad. Sci.* 77:1–13.

Towle, D. W., and J. S. Pearse, 1973. Production of the giant kelp, *Macrocystis*, estimated by *in situ* incorporation of ^{14}C in polyethylene bags. *Limnol. Oceanogr.* 18:155–159.

Turner, C. H., E. E. Ebert, and R. R. Given, 1968. The marine environment offshore from Point Loma, San Diego County. *Calif. Dept. Fish Game Fish Bull.* No. 140, 85pp.

Underwood, A. J., 1980. The effects of grazing by gastropods and physical factors on the upper limits of distribution of intertidal macroalgae. *Oecologia (Berl.).* 46:201–213.

Underwood, A. J., and P. Jernakoff, 1984. The effects of tidal height, wave exposure, seasonality, and rock-pools on grazing and the distribution of intertidal macroalgae in New South Wales. *J. Exp. Mar. Biol. Ecol.* 75:71–96.

Vadas, R. L., 1977. Preferential feeding: On optimization strategy in sea urchins. *Ecol. Monogr.* 47:337–371.

VanBlaricom, G. R., and J. A. Estes, eds., 1988. *Ecological Studies: Analysis and Synthesis. Vol. 65. The Community Ecology of Sea Otters.* Springer-Verlag, New York. 237pp.

Vance, R. R., 1979. Effects of grazing by the sea urchin, *Centrostephanus coronatus*, on prey community composition. *Ecology.* 60:537–546.

Van Raalte, C. D., I. Valiela, and J. M. Teal, 1976. Production of epibenthic salt marsh algae: Light and nutrient limitation. *Limnol. Oceanogr.* 21:862–872.

Wells, R. A., 1983. Disturbance mediated competition between *Macrocystis pyrifera* and articulated coralline algae. In: W. Bascom, ed. *The Effects of Waste Disposal on Kelp Communities.* South. Calif. Coastal Water Res. Proj., Long Beach, CA. pp. 181–198.

Wells, R. A., 1985. Disturbance and the dynamics of a shallow subtidal algal assemblage in Big Fisherman's Cove, Santa Catalina Island, California. Ph.D. Dissertation, Univ. of Southern California, Los Angeles. 188pp.

Wheeler, W. N., 1980. Pigment content and photosynthetic rate of the fronds of *Macrocystis pyrifera. Mar. Biol.* 56:97–102.

Widdowson, T. B., 1971. Changes in the intertidal algal flora of the Los Angeles area since the survey by E. Yale Dawson in 1956–1959. *Bull. South. Calif. Acad. Sci.* 70:2–16.

Wilson, K. C., and W. J. North, 1983. A review of kelp bed management in southern California. *J. World Maricult. Soc.* 14:347–359.

Wilson, K. C., and H. Togstad, 1983. Storm caused changes in the Palos Verdes kelp forests. In: W. Bascom, ed. *The Effects of Waste Disposal on Kelp Communities.* South. Calif. Coastal Water Res. Proj., Long Beach, CA. pp. 301–305.

Wolfe, D. A., M. A. Champ, D. A. Flemer, and A. J. Mearns, 1987. Long-term biological data sets: Their role in research, monitoring, and management of estuarine and coastal marine systems. *Estuaries.* 10:181–193.

Yarish, C., A. M. Breeman, and C. van den Hoek, 1984. Temperature, light, and photoperiod responses of some northeast American and west European endemic rhodophytes in relation to their geographic distribution. *Helgol. Meeresunters.* 38:273–304.

Yarish, C., A. M. Breemen, and C. van den Hoek, 1986. Survival strategies and temperature responses of seaweeds belonging to different biogeographic distribution groups. *Bot. Mar.* 29:215–230.

Zechman, F. W., and A. C. Mathieson, 1985. The distribution of seaweed propagules in estuarine, coastal, and offshore waters of New Hampshire, U.S.A. *Bot. Mar.* 28:283–294.

Zedler, J. B., 1977. Salt marsh community structure in the Tijuana Estuary, California. *Estuarine Coastal Mar. Sci.* 5:39–53.

Zedler, J. B., 1980. Algal mat productivity: Comparisons in a salt marsh. *Estuaries.* 3:122–131.

Zedler, J. B., 1982a. The ecology of southern California coastal salt marshes: A community profile. Report No. FWS/OBS–81/54. Prep. by U.S. Fish Wildl. Serv., Biol. Serv. Prog., Washington, D.C. 110pp.

Zedler, J. B., 1982b. Salt marsh algal mat composition: Spatial and temporal comparisons. *Bull. South. Calif. Acad. Sci.* 81:41–50.

Zedler, J. B., and C. S. Nordby, 1986. The ecology of Tijuana Estuary, California: An estuarine profile. U.S. Fish Wildl. Serv. Biol. Rep. 85(7.5). Available from: NTIS, Springfield, VA; PB87–177861. Prep. by San Diego State Univ., Dept. of Biology, San Diego, CA. 104pp.

Zedler, J. B., T. Winfield, and P. Williams, 1980. Salt marsh productivity with natural and altered tidal circulation. *Oecologia (Berl.).* 44:236–240.

Zimmerman, R., 1983. Modeling kelp growth. In:

W. Bascom, ed. *The Effects of Waste Disposal on Kelp Communities*. South. Calif. Coastal Water Res. Proj., Long Beach, CA. pp. 233–242.
Zimmerman, R. C., and J. N. Kremer, 1984. Episodic nutrient supply to a kelp forest ecosystem in southern California. *J. Mar. Res*. 42: 591–604.
Zimmerman, R. C., and D. L. Robertson, 1985. Effects of El Niño on local hydrography and growth of the giant kelp *Macrocystis pyrifera* at Santa Catalina Island, California. *Limnol. Oceanogr*. 30:1298–1302.
ZoBell, C. E., 1971. Drift seaweeds on San Diego County Beaches. *Nova Hedwegia*. 32:269–314.

Chapter 8

Benthic Invertebrates

Bruce Thompson, John Dixon, Stephen Schroeter, and Donald J. Reish

Introduction

The marine benthic invertebrates in the Southern California Bight (SCB) include representatives of nearly all phyla. No estimates of the total number of species in the region have been made, but there must be more than 5000 of them in the region (Smith and Carlton [Light's Manual] 1975; Straughan and Klink 1980).

Classification

Benthic invertebrates inhabit the sea floor during most of their lives, but they may have pelagic gametes or larvae. Benthos are often classified according to their habitat and size. The most obvious division of habitats is by substrate, either rock or unconsolidated (gravel, sand, or mud) sediments, hereafter referred to simply as soft sediments. Benthic species can be epifaunal (attached or motile species that inhabit rock or sediment surfaces) or infaunal (those that live in rock or soft sediments). Benthic invertebrates can also be classified into several overlapping size categories. The *microbenthos* (= nannobenthos) are organisms that are 2–50 μm (Burnett 1981) and include protozoans but not bacteria. *Meiobenthos* are usually considered to be animals between 50 and 300 μm, a category that includes nematodes, harpacticoid copepods, ostracods, and others (Rex 1981). *Macrobenthos* (usually 300–1000 μm) such as polychaetes and ophiuroids, among others, are the more common benthos. Rather than by size, *megabenthos* are generally operationally defined by whether they are visible in photographs or captured in trawls. This chapter will focus on macro- and megabenthos.

Most of the benthic habitats in the region have received at least some study. Obviously, much more is known about intertidal and shallow subtidal benthic species than those of deeper areas. Thus, this chapter concentrates on ecological interactions and life histories from the shallow areas (< 30 m). Information about deeper habitats is usually more descriptive; however, a considerable amount of experimental work has been recently conducted in the deep basins. We have divided the benthic assemblages into three general habitat types based on water depth and expertise of the authors: (1) embayments (D. J. Reish); (2) open coastal intertidal and subtidal habitats to 30 m (J. Dixon, S. Schroeter); and (3) deep benthic habitats (> 30 m) (B. Thompson). Each of these habitat types is further divided into rocky and soft substrates.

Other reviews of the benthic invertebrates of the region have been made by the Southern California Coastal Water Research Project (SCCWRP) (1973), the Southern California Ocean Studies Consortium (SCOSC) (1974), and Bakus (1989). Taxonomic listings have been compiled by Reish (1972) and Straughan and Klink (1980).

Physical Setting

An embayment is any recess from the open coastline. Embayments include the mouths of rivers, estuaries, lagoons, marinas, harbors, and wetlands that are under tidal influence. The embayments of the SCB can be categorized as wetlands, recreational or residential marina, and commercial or military harbors. Many embayments have more than one of these characteristics and most still retain some remnants of wetlands.

The length of coast is roughly the same on the Channel Islands (530 km) and on the mainland between Point Conception and the Mexican border (490 km). However, about 70% of all *rocky* shores occur on the islands and about 70% of all *sandy* shores occur on the mainland. Both sand and rock are present along many sections of shore. Commonly, the upper beach is sand and the lower beach is made up of rock outcrops or boulders. Rock outcrops are common in the subtidal areas of the narrow mainland shelf, becoming more abundant as one proceeds from the southeast to the northwest along the coast, and are especially characteristic around the Channel Islands (Shepard 1973). The types of rock bottoms are generally (1) submerged rock platforms, (2) deformed sedimentary rock, and (3) boulder and cobble fields. Most of the sublittoral bottom off southern California is made up of fine sand and coarse silt.

The shelves (>30 m depth) along the mainland are of variable width, ranging from 1 km off Palos Verdes Peninsula to 12 km off Santa Barbara and Imperial Beach. Sediments on the inner shelf are usually coarse sand with a large carbonate fraction; on the outer shelf they are silty clay, with localized intrusions of coarse sand. The deep benthic habitat commences at the shelf break (the rapid steepening of the shelf slope), which occurs between 100 and 150 m depths. A series of synclinal folds produces a basin-island-basin ridge system such that 16 submarine basins separate the mainland shelf, the Channel Islands, and the offshore ridges and banks. The basins increase in depth farther from shore, reaching depths of 2571 m in the Velero Basin. An important feature of each basin system is its sill depth, which controls its circulation (see chaps. 1 and 2).

Embayment Assemblages

The ecological environment of embayments in the SCB ranges from (1) nearly pristine in nature, such as Mugu Lagoon or Anaheim Bay; (2) modified for boating and recreational activities, such as Alamitos Bay and Mission Bay; (3) dredged from previous agriculture fields, such as Ventura Marina and Marina del Rey; to (4) highly industrialized, such as Los Angeles–Long Beach Harbors and San Diego Bay (table 8.1). Discussion here is directed toward the more natural embayments; those bodies of water that have been extensively altered and subjected to contamination are treated in chapter 12.

The general features of an embayment in the SCB are depicted in figure 8.1. Environmental conditions vary not only in relation to the tidal height, but also from the inner reaches to the entrance. The tidal indicator plants are salt grass (*Distichlis spicata*), present in the higher high-tidal level (1.8–2.4 m); pickleweed (*Salicornia* spp.), present in the mean high-tidal level (1.2–1.8 m); and cord grass (*Spartina foliosa*), present at mean sea level (0–1.2 m). The green algae *Enteromorpha* sp. and *Ulva* sp. are generally present in the spring and fall months within the pickleweed–cord grass zones as well as on the open mud flats. (See chap. 7 for an additional discussion of plants found in SCB embayments.)

The California horn snail (*Cerithidea californica*) is unquestionably the characteristic animal of the intertidal mud flat in the SCB, where population levels of hundreds per square meter are common. The rock louse (*Ligia occidentalis*) is present in the higher tidal levels, where it is generally found under rocks during daylight hours. The fiddler crab (*Uca crenulata*) burrows into the mud banks in the high-tidal zone. In some SCB embayments, the population of this crab has been decimated because of habitat modification caused by boating or flood control measures. The mud or yellow shore crab (*Hemigrapsus oregonesis*) is more common and inhabits a wider zone than the fiddler crab.

Common pelecypods include accidentally introduced Atlantic species; the horse mussel (*Ischadium demissum*), which often lives in abandoned fiddler crab burrows or attaches its byssal threads to shell fragments; the littleneck clam (*Protothaca staminea*); and the California jackknife clam (*Tagelus californianus*), which burrows into the sediment of the intertidal mud flat.

Gastropods, in addition to *Cerithidea californica*, include the cloudy bubble snail (*Bulla gouldiana*), the striped seahare (*Choelidoneura inermis*), and the smaller snails *Acteocina inculta* and *Assiminea californica*. Among the smaller invertebrates, the polychaetes, especially those species belonging to the family Spionidae (*Polydora nuchalis*, *Polydora ligni*, and *Streblospio benedicti*) dominate the intertidal mud flat. *Capitella capitata* is also a common polychaete, as is one or more species of oligochaetes (Reish 1975). Amphipods are especially abundant during the summer months; they include such species as *Corophium acherusicum*, *Corophium insidiosum*, and *Grandidierella japonica*, which burrow and construct tubes in the upper 1–2 cm of sediment. In addition, numerous species of harpacticoid copepods and ostracods are present, along with chironomids and other insect larvae. While the intertidal mud flats of the SCB are exceedingly rich and diversified, this habitat has not been adequately described either qualitatively or quantitatively, nor do we have any knowledge of the biology and life histories of the majority of its inhabitants.

As one approaches the entrance of a nearly pristine embayment (fig. 8.1b), the sediments become sandy and the flora and fauna change. The sediments no longer support the different species of marsh plants nor the fiddler crab and its associated fauna of the high-tidal zone. An intertidal transect through a fine sandy beach near the entrance will yield polychaetes, pelecypods, gastropods, ghost shrimp, and holothurians, among the larger species. The high-tidal horizon yields few animals. The midtidal level is characterized by the ghost shrimp (*Callianassa californiensis*) and the California jackknife clam (*Tagelus californianus*), both of which are present in the upper reaches of embayments. Polychaetes

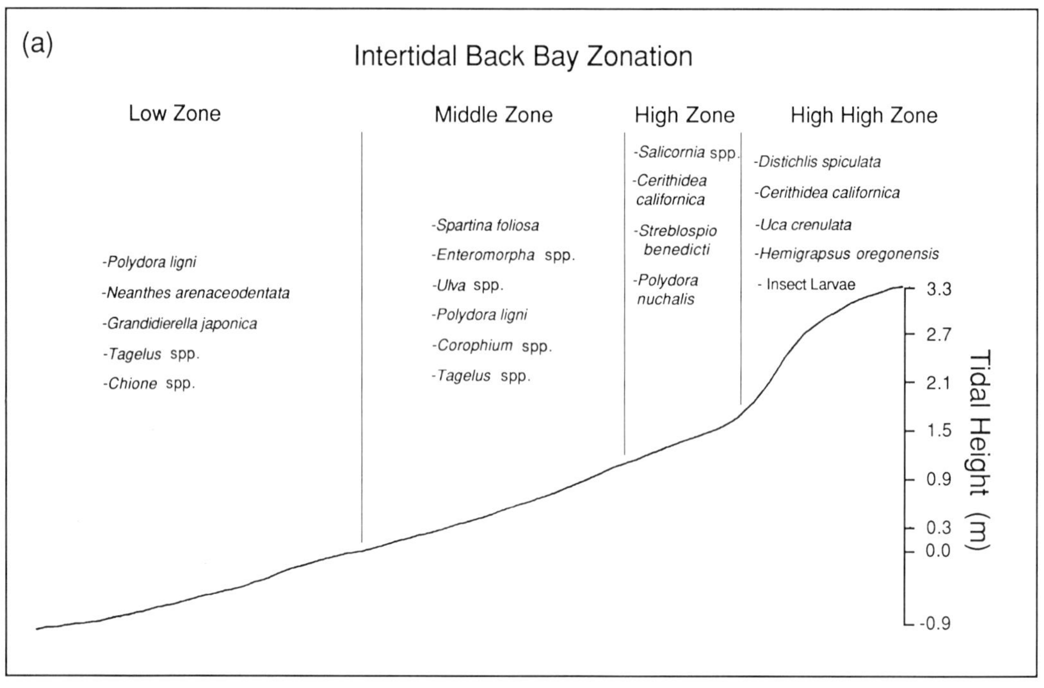

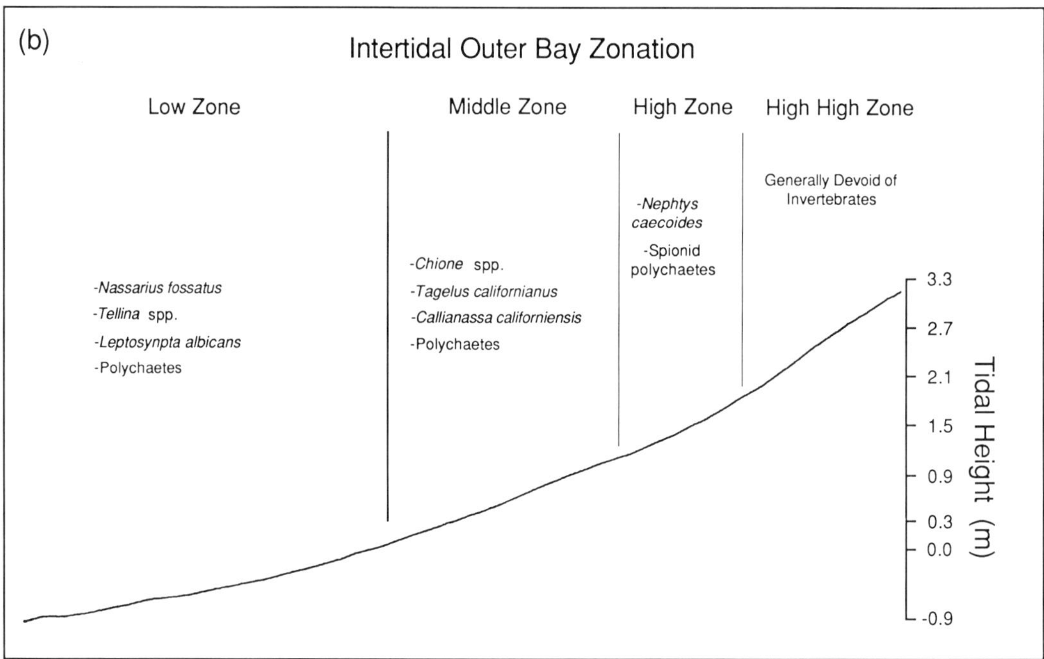

Figure 8.1. Cross-section of the intertidal environment of a southern California embayment. (a) Back bay. (b) Outer bay.

Table 8.1. *Embayments of the SCB*

Locality	General Characteristics				Reference
	Wetlands	Recreational and Residential	Marina	Commercial and Military Harbor	
Santa Barbara County					
Deveraux Lagoon	X				
Goleta Slough	X				Marine Biological Consultants 1974
Santa Barbara Harbor			X		
Carpinteria Slough	X				MacDonald 1976
Ventura County					
Ventura River	X				
Ventura Marina		X[a]	X[a]		Reish 1964a
Santa Clara River	X				
Channel Islands Harbor		X[a]	X[a]		Turner 1970; Reish 1977
Port Hueneme				X	
Mugu Lagoon	X			X	Onuf 1987
Los Angeles County					
Malibu Creek	X				
Marina del Rey		X	X[a]		Reish 1964a; Soule and Oguri 1980
Ballona Wetlands	X				Reish 1980; Schreiber 1981
Los Angeles–Long Beach Harbors			X	X	Reish 1959; Soule and Oguri 1980
Los Angeles River	X				James and Reish 1978
Shoreline Aquatic Park	X		X[a]		Reish 1975
Alamitos Bay	X	X			Reish 1963b
San Gabriel River	X				Reish 1956
Catalina Harbor	X		X		Reish 1964b
Orange County					
Anaheim Bay	X			X	Lane and Hill 1975
Huntington Harbor	X	X		X	Kauwling and Reish 1975

Table 8.1. *Embayments of the SCB* (continued)

Locality	General Characteristics				Reference
	Wetlands	Recreational and Residential	Marina	Commercial and Military Harbor	
Bolsa Chica Wetlands	X				Stein and Kreppert 1971
Santa Ana River	X				Tekmarine, Inc. 1986
Newport Bay	X	X	X		Barnard and Reish 1959; Seapy 1981
Doheny State Beach	X				Marine Biological Consultants 1972
San Diego County					
San Mateo Marsh	X				
Santa Margarita River	X				U.S. Fish and Wildlife Service 1981
Camp Pendleton Basin	X				
Oceanside Marina	X		X[a]		
San Luis Rey River	X	X			
Buena Vista Lagoon	X				Calif. State Coastal Conservancy 1982
Aqua Hedionda Lagoon	X				
Batiquitos Lagoon	X				Mudie et al. 1976
San Elijo Lagoon	X				Atlantis Scientific 1972
San Dieguito Lagoon	X				Mudie et al. 1976
Los Penasquitos Lagoon	X				
Mission Bay	X	X	X		Dexter 1983
San Diego River	X				
San Diego Bay	X	X	X	X	MacKenthum and Parrish 1968
Tijuana Estuary	X		X		Zedler and Nordby 1986

[a] Dredged from terrestrial environment.

are numerous and many of them extend into subtidal waters. They include *Diopatra splendidissima*, *Nephtys caecoides*, and *Lumbrineris* spp. as well as many different spionid genera whose tubes project above the sediment.

The animal life is more diversified in the lower intertidal zone. Many of the same species that occur in the midtidal zone occur here, but they are usually present in greater numbers. The carnivorous channel basket snail (*Nassarius fossatus*) is common and feeds upon the different pelecypods in the area. Pelecypods include the smaller *Tellina* spp. as well as the larger littleneck clam (*Protothaca staminea*), banded and smooth chiones (*Chione californiensis* and *Chione fluctifraga*), the purple clam (*Nuttallia nuttallii*), and the gaper clam (*Tresus nuttallii*). The holothurian *Leptosynapta albicans* burrows into the sediment. Other echinoderms that are occasional inhabitants include the sand dollar (*Dendraster excentricus*) and the southern sand star (*Astropecten armatus*).

The subtidal benthos of an embayment is rich and varied in species composition and number of specimens. For example, 290 different species were collected from Newport Bay. Of these, 34.8% were polychaetes, 20.6% were crustaceans, 14.1% were pelecypods, 8.2% were gastropods, and 23.3% were in the remaining groups (Marine Biological Consultants and Southern California Coastal Water Research Project 1980). Out of the 290 species taken, 17 comprised 80% of the total number of specimens collected (table 8.2). Polychaetes and oligochaetes were the most common species, with 3 species of amphipods and 1 pelecypod completing the list. Of particular interest is the observation that 8 of these 17 common species are considered to be introduced into the eastern Pacific Ocean embayments. These species include *Streblospio benedicti*, *Polydora ligni*, *Pseudopolydora paucibranchiata*, *Capitella capitata*, *Grandidierella japonica*, *Corophium acherusicum*, *Corophium insidiosum*, and *Theora lubrica*.

One subtidal community was distributed throughout much of Upper Newport Bay. It was comprised of 7 common species, namely *S. benedicti*, *P. paucibranchiata*, *P. ligni*, *C. capitata*, *Tubificoides pseudogaster*, *G. japonica*, and *C. insidiosum*. The other principal species formed temporary communities that were present for one sampling period (Marine Biological Consultants and Southern California Coastal Water Research Project 1980).

Although many embayments of the SCB have been characterized biologically to some extent, any comparison of species composition of the various invertebrate groups must be made with caution because of differences among studies in objectives and methods of collection. The polychaetes of SCB embayments have been the subject of more intensive studies. Comparisons of the dominant species of polychaetes in many of these embayments are shown in table 8.3. The data are summarized by locality as those species comprising at least 5% of the total polychaete population for that embayment. In addition, those dominant species that are present at other localities, but at populations less than 5%, are also included in table 8.3. Examination of this table indicates that none of the localities have a distinctive polychaete population. Dexter (1983) came to the same conclusion in her studies of SCB bays: no species was present at all localities studied and nearly all localities were characterized by one or two species not dominant elsewhere. *Mediomastus* spp. were the most commonly occurring dominant species. The dominant polychaete fauna in nearly pristine Anaheim Bay is similar to that in the residential marina complex of Huntington Harbour, an area dredged from mud flats in the 1960s. These two bodies of water share a common exit to the ocean, and the similarity of the polychaete composition can probably be explained by the settlement of larval-stage bay species in newly dredged Huntington Harbour.

The feeding habits of the principal species of polychaetes listed in table 8.3 were analyzed according to the feeding guild classification of Fauchald and Jumars (1979). The majority of these species are surface or subsurface deposit feeders that use tentacles in

Table 8.2. *Dominant Species in the Benthos of Newport Bay, November 1979 Through July 1980*[a]

Species	Number of Species	Rank	Percentage	Cumulative Percentage
Streblospio benedicti (P)[b]	4920	1	18.2	18.0
Tubificoides gabriellae (O)	2739	2	10.1	28.0
Mediomastus californiensis (P)	2254	3	8.3	36.0
Polydora ligni (P)	1855	4	6.9	43.0
Grandidierella japonica (A)	1770	5	6.5	50.0
Pseudopolydora paucibranchiata (P)	1762	6	6.5	56.0
Capitella capitata (P)	973	7	3.6	60.0
Corophium acherusicum (A)	945	8	3.5	63.0
Prionospio cirrifera (P)	825	9	3.1	66.0
Tubificoides pseudogaster (O)	713	10	2.6	69.0
Tubificoides apectinatus (O)	631	11	2.3	71.0
Euchone limnicola (P)	526	12	1.9	73.0
Theora lubrica (B)	437	13	1.6	75.0
Notomastus tenuis (P)	390	14	1.4	76.0
Mediomastus ambiseta (P)	330	15	1.2	77.0
Corophium insidiosum (A)	330	16	1.2	78.0
Sphaerosyllis californiensis (P)	328	17	1.2	80.0

[a] Seven stations, five sample periods.

[b] P = polychaete; O = oligochaete; A = amphipod; B = pelecypod.

From Marine Biological Consultants and Southern California Coastal Water Research Project 1980.

feeding. No differences were noted in the feeding mechanisms of the polychaetes inhabiting the different embayments of the SCB.

No pattern emerges that separates the highly industrialized Los Angeles–Long Beach Harbor from the other embayments. Of the 20 species listed in table 8.3, 6 are cosmopolitan in distribution, but these species are not confined to any one bay. These data do not indicate that a larger percentage of polychaetes inhabiting Los Angeles–Long Beach Harbor is cosmopolitan, as might be expected in a major world seaport where they are transported either as part of a fouling community on the hull of a ship or as "passengers" in ballast water (Carlton 1985). Earlier Barnard and Reish (1959) found no difference between Los Angeles–Long Beach Harbor and Newport Bay in the number of cosmopolitan species of amphipods and polychaetes. The similarity of the dominant polychaete fauna in SCB embayments may be attributed to the similarity in the physical and chemical characteristics of the water column, water depth, and particle size of the sediments as well as the movements of smaller ships and boats from one area to another.

Seasonal studies of benthic populations in SCB embayments are limited in number. However, seasonal differences were noted for the smaller invertebrates in Upper Newport Bay (Marine Biological Consultants and Southern California Coastal Water Research Project 1980), and Schreiber (1974) made monthly quantitative samples for polychaetes in Anaheim Bay. Some species, especially *Cossura candida*, *Prionospio cirrifera*, and *Rhynchospio arenicola*, showed peaks in population in late spring and early summer. Another group composed primarily of *Goniada littorea*, *Mediomastus californiensis*, and *Tharyx*

Table 8.3. *Comparisons of the Principal Species of Polychaetes Present in Some Embayments of the SCB*

Species	Marina del Rey 1985	Los Angeles–Long Beach 1959	Los Angeles–Long Beach 1973	Alamitos Bay 1954–1959	Anaheim Bay 1970–1971	Huntington Harbour 1970–1971	Newport Bay 1954	Newport Bay 1979	Mission Bay 1970–1971
Pseudopolydora paucibranchiata	xx[a]	xx	x	x	0	0	xx	x	0
Tharyx spp.	xx	xx	xx	x	x	x	xx	x	xx
Mediomastus spp.	xx	x	xx	xx	xx	xx	xx		x
Capitella capitata[b]	xx	xx	xx	x	x	x	x	xx	
Euchone limnicola	xx	x	x	x	x	xx			xx
Polydora ligni[b]	xx	0	x	0	x	x	x		
Cirriformia spirabrancha	xx	x	0	x	0	0			
Cossura candida	x	xx	xx	x	xx	x	xx	x	
Cirriformia luxuriosa	0	xx	x	x	xx	0		x	
Schistomeringos longicornis	x	x	x	xx	x	x	xx	x	x
Lumbrineris minima	0	x	0	xx	x	xx	xx	x	xx
Prionospio cirrifera[b]	x	x	x	xx	xx	xx	x	x	
Lumbrineris erecta	0	x	0	xx	x	x	x		x
Armandia bioculata	x	x	x	x	xx	xx	xx		x
Streblospio benedicti[b]	0	0	0		xx	xx		xx	
Hoploscoloplos elongatus	x	x	x	x	x	xx			xx
Fabricia limnicola	0	0	0						
Polydora nuchalis	x	0	0					xx	
Polydora socialis[b]	0	0	0					xx	
Pherusa neopapikata	0	0	0	0	0	0	0	0	xx

[a] xx denotes species comprising greater than 75%.

[b] Cosmopolitan in distribution.

Data from Soule and Oguri 1975, 1986; Reish 1959, 1975; Reish and Winter 1954; Barnard and Reish 1959; Seapy 1981; Dexter 1983.

parvus were abundant during the winter months. *Nerinides acuta, Prionospio pygmaeus,* and *Streblospio benedicti* exhibited early fall population peaks. Other commonly occurring species peaked two or three times during the 1971–1972 study. Peterson (1977), however, described a Mugu Lagoon community that consisted of larger invertebrates (namely, pelecypods, sand dollars, and ghost shrimp) and lacked seasonality. Apparently the populations of smaller size species (consisting of polychaetes, amphipods, and other smaller crustaceans), with their shorter life cycles, are more sensitive to varying environmental conditions than the larger forms. It is apparent that prior to undertaking a seasonal study of benthic animals in the SCB, the sampling procedures must be carefully considered to ensure adequate sampling of all sizes of invertebrates.

The interrelationships between species inhabiting embayments are not well known. Experimental studies by Peterson (1977), Peterson and Quammen (1982), and Levin (1981, 1984) on the mud flats of Mugu Lagoon and Mission Bay provide some insight on interspecific interactions, however. Caged littleneck clams, *Protothaca staminea,* for example, experienced a significantly greater growth rate than uncaged specimens that were subjected to siphon nipping by fish (Peterson and Quammen 1982). Examination of stomach contents revealed that 35% of the diet of the diamond turbot, *Hysopsetta guttulata*, consisted of the siphons of *Tagelus californianus*, and an estimated population of 3500 diamond turbots in the middle arm of Anaheim Bay ingested an estimated 1600 kg yr^{-1} of clam siphons (Lane 1975). The effects of siphon nipping, and subsequent regeneration of this part, and population dynamics have not been investigated.

Populations of the spionid polychaete *Pseudopolydora paucibranchiata* were uniformly distributed on an intertidal mud flat, while those of another spionid, *Streblospio benedicti*, were randomly distributed. Levin (1981) attributes this difference to the palpi fighting and biting behavior in *P. paucibranchiata* and the lack of such activity in *S. benedicti*.

Reproduction, Settlement, and Life Histories of Marine Invertebrates

For invertebrate inhabitants of SCB embayments, no thorough systematic study has been done of reproduction, larval ecology, settlement, or life histories that is comparable to the classic study pursued by Thorson (1946) for the Danish seas. Some studies do exist on the reproduction and development of individual species in embayments of the SCB (MacGinitie and MacGinitie 1949; Reish 1981), and laboratory culture of individual species has yielded valuable information. However, most data have been obtained by either suspension of test panels (Allan Hancock Foundation 1975; Soule 1976) or sediment bottle collectors (Reish 1961a; Soule 1976; Levin 1984).

Analyses of the principal invertebrates that settled in sediment bottle collectors over a 2-year period (1955–1957) in Los Angeles–Long Beach Harbor and of polychaetes settling on wooden test panels in the same area from March 1950 through March 1951 have provided data on the reproduction and settlement of many of these organisms (Reish 1961b, 1971). Because both the sediment bottle and test blocks are suspended in the water column, most animals present probably arrived as larvae, although the potential exists for benthic animals to be resuspended and subsequently trapped in the sediment bottle. These data furnished details on the seasonality of settlement, from which the reproductive period was inferred to have occurred previously.

The data for the seasonal settlement of the 14 more frequently occurring species within the jar are summarized in figure 8.2. Of the 14 species, 10 settled throughout the year. These species included the anthozoan *Diadumene leucolena*; the polychaetes *Eumida sanguinea, Podarke pugettensis, Platynereis bicanaliculata, Dorvillea articulata, Polydora pauci-*

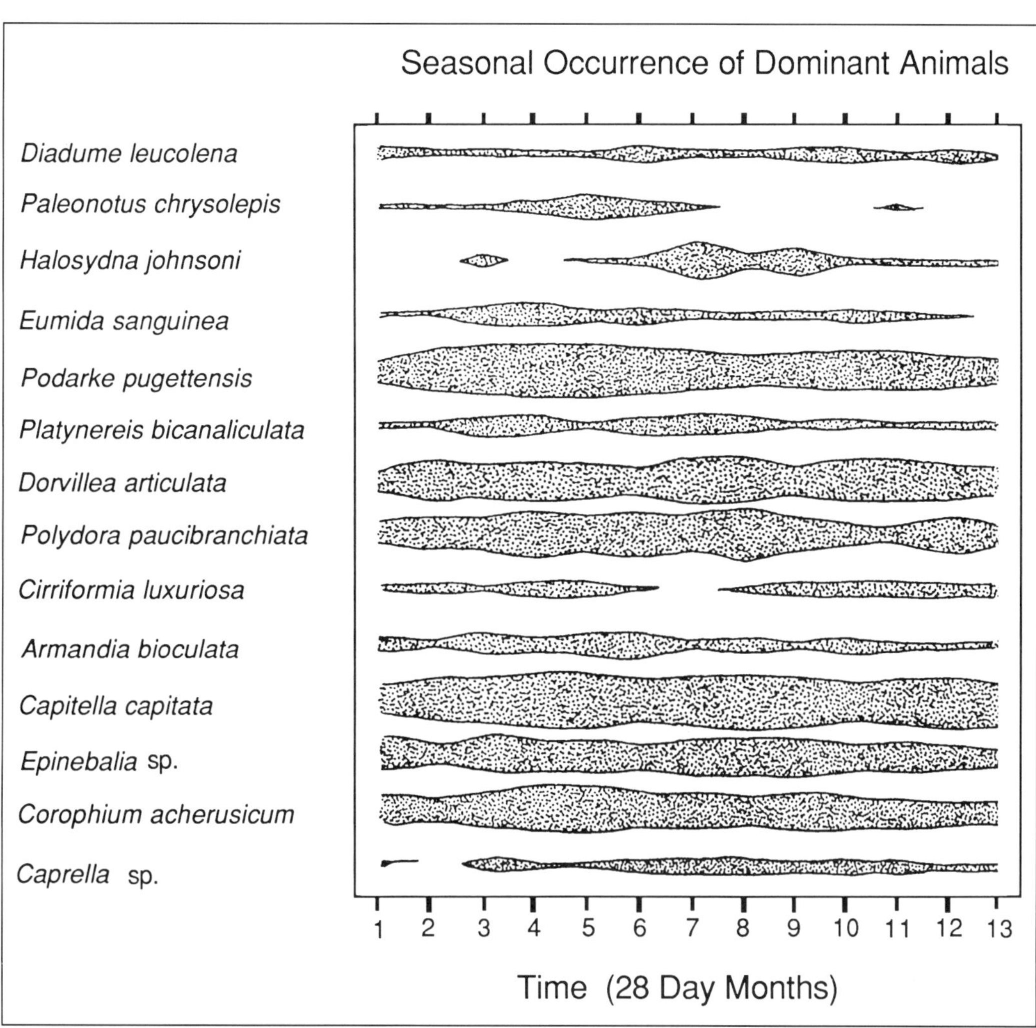

Figure 8.2. Seasonal settlement of the more frequently occurring invertebrates in sediment bottle collectors in Los Angeles–Long Beach harbors during 1956–1957. (After Reish 1961a.)

branchiata, *Armandia bioculata*, and *Capitella capitata*; and the crustaceans *Epinebalia* sp. and *Corophium acherusicum*. The polychaete *Paleonotus chrysolepis* reached its peak in the spring and did not occur during the summer months. Peak settlement occurred during the summer months for the polychaete *Halosydna johnsoni*. The polychaete *Cirriformia luxuriosa* and the amphipod *Caprella* sp. settled in all periods except one each.

Reish (1971) followed seasonal settlement of polychaetes on suspended wood blocks for 13 consecutive 28-day periods (over a year) in Los Angeles-Long Beach Harbor. Twenty-four species of polychaetes were encountered on test panels over the year. The number of species was greatest during the spring months and lowest during the winter period. In addition, a suppression in the number of species was noted whenever the dissolved oxygen content of the water was less than 4.0 mg l^{-1}. Species exhibiting spring peaks in settlement included *Paleonotus bellis*, *Brania limbata*, and *Platynereis bicanaliculata*. Spring through summer peaks were observed in *Polydora limnicola*. A summer peak was noted for *Armandia bioculata*, and a summer–fall period of settlement occurred for *Halosydna brevisetosa* and *Hydroides pacificus*. *Capitella capitata* and *Ophiodromus pugettensis* settled on panels throughout the year.

Comparisons of the polychaete settlement data in bottle collectors and test panels indicated a similar pattern. Larval settlement of many of the species occurred throughout the year, with some species possessing either a spring or spring–summer peak. Settlement was reduced during the colder months of the year, but there was an increase in the diversity of species present in the spring. Larval settlement was either by active (selective) or passive (nonselective) habitat selection (Butman 1987). Both hypotheses are apparently valid in these studies. Some species (e.g., *Capitella capitata*, *Platynereis bicanaliculata*, and *Halosydna brevisetosa*) settled both in the bottle collector and on test panels. These species could be considered passive habitat selectors; their apparent choice of a clean surface lacking intra- or interspecific competition may be the result of a selective process taking place. In contrast, species such as *Hydroides pacificus* required a solid surface for attachment, such as a test panel or the side of a bottle collector (Reish 1961b). It is also possible that *H. pacificus* may require the presence of lectins produced by bacteria to induce settlement, a possibility demonstrated in the laboratory with the settlement of another serpulid polychaete (Kirchman et al. 1982).

The limited number of experimental studies concerned with larval settlement indicate interspecific competition. For example, larval settlement of the polychaete *Pseudopolydora paucibranchiata* was unsuccessful when the population of the feather duster polychaete *Fabricia limnicola* was 1.0×10^5 to 2.0×10^5 m^{-2}, but settlement was successful when the concentration of *F. limnicola* was one order of magnitude less (Levin 1984). Removal of the ghost shrimp *Callianassa californiensis*, a deposit feeder, resulted in a high recruitment of the larvae of the purple clam *Nuttallia nuttallii*, but none of these clams appeared at the reference site where the ghost shrimp had not been removed (Peterson 1977). No similar experimental studies with subtidal infaunal invertebrate larvae have been conducted in SCB embayments, although interspecific competition undoubtedly occurs in this environment also.

The Food Web

Invertebrates play an important role in the SCB food web. Ware (1979) studied the food habits of the white croaker *Genyonemus lineatus* in Los Angeles Harbor. He examined the stomachs of 484 specimens collected at different localities in the outer harbor every 5 weeks over a 1-year period then compared the food habits at each specific locality with the benthic fauna at each site. While variations in the food habits of the white croaker existed, they appeared to be correlated with the benthic species present where the fish were caught. As the fish grew, the food choice changed from zooplanktonic to benthic species dominated by polychaetes, pericardean crustaceans, pelecypods, and nematodes. The benthic fauna in the harbor study area consisted of 191 taxa; 132 of these were present in the stomachs of the fish analyzed.

What role invertebrates play in the food web of shorebirds in an SCB embayment is not clear, largely because of the legal difficulties involved in obtaining bird stomachs for analysis. In studies of shorebird food habits, Quammen (1982, 1984), Barnard (1970), and Levin (1984) can only allude to the effect birds might have on SCB invertebrate populations.

Indications are that invertebrates are important food items for fish and shorebirds. Soule and Oguri (1980) determined the caloric value of 22 species of four invertebrate phyla inhabiting outer Los Angeles Harbor. Values ranged from 4069 to 5680 cal (ash-free) g^{-1}. The mean caloric values are summarized by phyla in table 8.4. These determinations show that the caloric value of polychaetes is significantly higher than that of crustaceans, molluscs, or echinoderms. No estimates of the invertebrate production in SCB embayments have been made, although

Table 8.4. *Mean Caloric Values of Four Invertebrate Phyla from Los Angeles Harbor (calories ash-free g^{-1})*

Animal Group	Number of Species	Number Analyzed	Mean	Standard Deviation
Polychaeta	10	23	4732.4	315.7
Crustacea	9	14	4416.9	384.9
Mollusca	5	16	4328.3	482.3
Echinodermata	1	3	4583.7	95.0

Data from Soule and Oguri (1980).

there are indications that large numbers of larvae are produced (Levin 1984). What percentage is consumed by larger organisms; how much organic material, such as mucus and gametes, is released into the environment; and the percentage of animals that die naturally must all be considered to ascertain the yearly production by invertebrates.

Open Coastal Intertidal and Subtidal Assemblages

Rocky Intertidal

To layman and specialist alike, rocky intertidal assemblages of plants and animals are the most well-known marine communities. The many popular field guides that have appeared since the early nineteenth century are testaments to the enduring public fascination with the striking zonation and variety of organisms on rocky shores. Many studies of physiological adaptations and of biological interactions have been stimulated by observations of changes in population characteristics along the intertidal exposure gradient. These studies have contributed substantially to a general understanding of community development and organization (Paine 1966, 1977; Connell 1972, 1975; Menge and Sutherland 1976, 1987; Underwood and Denley 1984).

VERTICAL ZONATION

As is the case throughout the world (Stephenson and Stephenson 1949, 1972), a marked vertical zonation occurs in the distribution of rocky intertidal organisms throughout the SCB. Despite differences in substrate types, currents, water temperatures, and wave energy in northern and southern California, the similarities in the vertical distribution of species on southern and northern shores are as striking as the differences (Stephenson and Stephenson 1972; Ricketts et al. 1968). Zonation is most apparent at locations with continuous rocky surfaces that span the intertidal. This is probably the exception rather than the rule in southern California. Most rocky shores in the SCB consist of scattered boulders or rock slabs or irregular horizontal benches. The upper beach is often sand. Under these conditions zonation is less obvious, although still present (Littler 1980).

At least 224 species of macroalgae and 315 species of macroinvertebrates were recorded by Littler and his colleagues in a study encompassing 22 sites scattered throughout the SCB (Littler 1979). Of these, only 13 algal taxa and 14 invertebrate taxa were common to all sites. Although any general description of zonation is inexact when applied to a specific site, a visitor to any rocky shore in the SCB will, nonetheless, recognize the pattern presented in figure 8.3.

The top of the shore from about mean high water to the highest area wet by splash, variously known as the uppermost horizon, the splash zone, or the supralittoral fringe, is characterized by the presence of lichens, blue-green algae, green algae such as *Enteromorpha* spp., patches of brown encrusting algae *Ralfsia* spp., and sparse populations of barnacles,

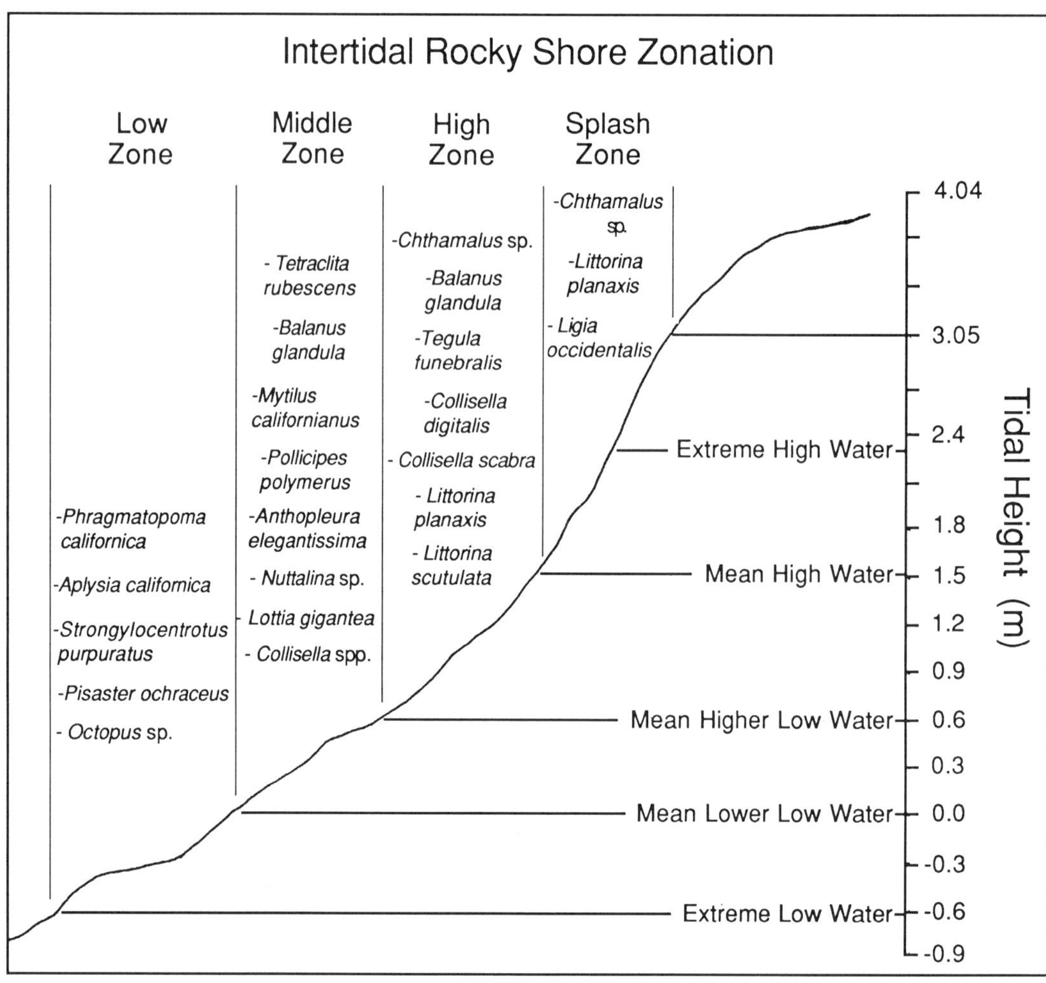

Figure 8.3. Intertidal zonation of a rocky shore in southern California.

Chthamalus spp. The nearly terrestrial isopod *Ligia occidentalis* is often abundant in the highest areas, especially among cobbles. At low tide, littorines and limpets occur in aggregations, particularly in cracks and depressions. *Littorina planaxis* and *Collisella digitalis* are the most abundant grazers on the high shore.

The shore from about mean high water to around mean higher low water is variously described as the high intertidal, upper midlittoral, or barnacle zone. *Chthamalus* spp. and *Balanus glandula* occur in dense populations and the red barnacle *Tetraclita rubescens* is often present. The abundance and diversity of macrophytes is greater here than it is higher on the shore. Grazers are extremely abundant as well. Periwinkles, numerous species of limpets, chitons, turban shells, and the crab *Pachygrapsus crassipes* impose severe constraints on algal populations. As a result of the intense grazing, a good deal of open space is usually present. That this is actually due to grazing is apparent when the herbivores are excluded experimentally (Cubit 1984) or suffer catastrophic mortality (Dixon 1978). Under these conditions, the high intertidal soon turns green.

The middle intertidal, lower midlittoral, or mussel zone is generally both submerged and exposed at least once a day. It extends from about mean higher low water to about mean lower low water (which is zero tidal datum, the arbitrary reference point for measuring tidal height). Algal cover is generally high. Red algae are well represented and green algae are seasonally abundant. Mussels are most abundant in this zone and often form mixed aggregations with the goose neck barnacle *Pollicipes polymerus*. On exposed rocky substrates, *Mytilus californianus* is the most numerous mussel, but in more protected localities, the bay mussel, *Mytilus edulis*, is also common. Dense aggregations of the cloning anemone *Anthopleura elegantissima* may blanket large areas of rock. Barnacles and the snails that prey upon them are also conspicuous members of the fauna. The predatory snails *Nucella emarginata* and *Acanthina spirata* are common. *Acanthina spirata*, which uses the spine on its shell as a battering ram to open barnacles (Perry 1983), is often very abundant. The snails *Pteropurpura* (= *Shaskyus*) *festivus* and *Roperia poulsoni* also eat barnacles but seem to prefer the clams *Penitella penita* and *Parapholas californica*, which live in holes that they bore in the rock substrate (Fotheringham 1974). *Octopus* spp. and the sea star *Pisaster ochraceous* are also important predators. Although they both take a great variety of prey, *P. ochraceous* relies mostly on mussels and *Octopus* spp. are responsible for much of the mortality of turban snails, limpets, and clams (Fotheringham 1974; Fawcett 1979, 1984; Wells 1980).

The low intertidal, infralittoral fringe, or laminarian zone extends from about mean lower low water to the lowest tide mark. Algae are generally the most conspicuous element of the biota in this low zone. Although *Egregia menziesii* and *Eisenia arborea* are often abundant, laminarian algae are not as distinctive a feature as they are on more northern Pacific shores (Stephenson and Stephenson 1972). The vascular plant *Phyllospadix* spp. (surfgrass) is common, as is the sand tube worm *Phragmatopoma californica*. Higher in the intertidal, *P. californica* occurs only as individuals or in small clumps, whereas on the low shore, these worms often form extensive reefs, cementing together and stabilizing boulders and preempting space. Large limpets are uncommon in the low intertidal, but sea hares, *Aplysia californica*, and purple sea urchins, *Strongylocentrotus purpuratus*, are formidable grazers.

The patterns of vertical distribution become more complicated in fields of boulders or raised slabs. Most boulders and slabs are less than a meter or so high, and greater overlap of populations generally occurs than on bigger rock surfaces. Species composition may also vary with boulder size, independent of tidal height (Dixon 1978). For example, the limpet *Collisella digitalis* is rare on small boulders ($<$20 cm high) at all tidal heights, and when transplanted onto such surfaces soon disappears relative to controls transplanted to larger boulders (J. Dixon pers. observ.; M. Fawcett pers. observ.). However, the limpet *Notoacmea fenestrata* is common on small boulders, even high on the shore, whereas they are seldom found at equivalent heights on large boulders located lower in the intertidal zone (Dixon 1978). These and other patterns related to boulder size are probably functions of boulder stability, exposure to waves and swash, intensity of sand scour, and frequency of burial.

It is obvious that species composition changes from upper to lower intertidal areas, and it is convenient to categorize shore levels by average environmental conditions and characteristic inhabitants. However, most of the many species actually exist in broadly overlapping populations (see Underwood 1978). The discrete "zones" that sometimes seem evident are usually defined by dense populations of one or two species (mussel zone) or by the relative scarcity of most species (barnacle zone). In southern California, the biomass of barnacles may actually be greater in the "mussel zone" than in the "barnacle zone" (e.g., see figures in Littler and Martz 1979; Seapy and Littler 1979), but they are less apparent lower on the shore because of the higher species diversity there.

In addition to changes in population density along the vertical exposure gradient, changes often occur in the size structure of populations. Vermeij (1972) suggested the generalization that the sizes of shells get larger with increased height on the shore for snails that inhabit the extreme upper shore, but that these shells decrease in size with intertidal height for species inhabiting the lower intertidal.

An increase in size with height on the shore has been reported for several high-level species that occur within the SCB: *Littorina planaxis* (North 1954), *Collisella digitalis* (Frank 1965; Haven 1971), and *Collisella scabra* (Sutherland 1970; Haven 1971). An increase in size in the direction of the low intertidal has been reported for *Collisella pelta* (Shotwell 1950), *Littorina scutulata* (Bock and Johnson 1967), and *Tegula funebralis* (Paine 1969; Doering and Phillips 1983). However, in the case of *Tegula*, the direction of size gradients varies from site to site (Fawcett 1984). Similar variability is reported for predatory snails (Butler 1979).

Vermeij (1972) suggested that size gradients result from differential mortality or movement of small individuals due to harsh physical factors on the high shore or predation in lower areas. The generalization does not appear to hold for *Tegula* on the low shore (Fawcett 1984). Where predators are abundant, *Tegula* increases rather than decreases in size with increased elevation as a result of the more effective escape response of large individuals. However, size does not appear to be a refuge in the low intertidal. The situation in the high intertidal may also be more complicated. For example, *L. planaxis* may increase in size with elevation, not because small individuals are unable to tolerate drying conditions on the high shore, but rather because large individuals cannot withstand the force of waves below the splash zone (North 1954).

HORIZONTAL PATCHINESS

The variability in the distribution and abundance of species from place to place within a given intertidal zone is at least as interesting and instructive as the variability among vertical zones at a given site. However, far less effort has been devoted to documenting and explaining changes in intertidal communities along the shore. Patchiness exists on all scales. Very few species occur everywhere within the SCB, and often a great deal of variation exists in the abundance of a given species among those locations where it is present. Even in local areas where a species is common or abundant, its population density may vary significantly among patches of substrate that are no more than a few meters distant from one another. Finally, most species inhabit particular microhabitats.

Not surprisingly, most is known about small-scale patchiness. Variability within a study site is relatively easy to document and is generally amenable to experimental analysis. Site-to-site differences are much more difficult to study. The best data available for examining large-scale differences in intertidal communities are probably those that resulted from the study sponsored by the Bureau of Land Management in the 1970s (Littler 1979, 1980; Seapy and Littler 1980) (see also chap. 7).

Seapy and Littler (1980) compared communities at Cayucos in central California, as well as at San Diego, not far from the Mexican border, and at nine sites on the offshore islands. They found that the islands most affected by the California Current (Santa Rosa, San Miguel, and San Nicolas) clustered with Cayucos, while those more affected by the California Countercurrent (San Clemente, Santa Catalina, Santa Barbara, and Santa Cruz) clustered with San Diego.

Another well-documented large-scale pattern is a change in the mussel fauna as one goes from exposed to protected sites (Harger 1970a, 1972c). In southern and central California, *Mytilus edulis* is absent from extremely exposed rocks, where *Mytilus californianus* and *Septifer bifurcatus* are common. In contrast, *M. californianus* is not found in quiet bays, where *M. edulis* is abundant. All three species are well represented at sites with intermediate levels of wave energy.

Large differences also exist within single sites. Harger and Landenberger (1971) found that the abundance of mussels varied tremendously from piling to piling among the hundreds of physically similar pilings supporting Ellwood pier near Santa Barbara. Large mussel beds were more likely to be dislodged by storms than were smaller ones, and beds that had suffered seastar predation were at greater risk of storm damage than others. Sousa (1979a,b), working at Ellwood boulder field, found that even within a small area, predictable differences existed in the composition of communities on boulders of different sizes. These differences were a function of the stability of the boulders and the frequency with which they were dislodged and overturned. When small cobbles were glued down, they soon became similar to larger rocks in community composition (Sousa 1979b).

Even on a very small scale, species are generally distributed nonrandomly. At first glance, the several species of limpets inhabiting a rock seem quite mixed. However, a closer look reveals that *Collisella scabra* is most numerous on horizontal surfaces and *Collisella digitalis* tends to congregate on sides of rocks (Haven 1971). *Collisella strigatella* occupies both the tops and the sides (Dixon 1978). Also, *C. scabra* is commonly found among barnacles, whereas *C. digitalis* and *C. strigatella* are more likely to be found on bare rock or in clearings within the barnacle cover (Choat 1977). Stephenson and Stephenson (1972) noted that at La Jolla the hard rocks embedded in soft sandstone tended to have a greater cover of barnacles and a lesser cover of coralline algae than on the surrounding matrix, which seemed to retain water better. Beds of the mussels *Mytilus* spp. are generally on exposed surfaces, whereas *Septifer bifurcatus* is more common in crevices and under rocks (Ricketts et al. 1968). In the laboratory, the *Mytilus* species soon crawl out when buried in gravel, but *Septifer* remains buried and, when placed on the surface, seeks out depressions (Harger 1968). Many species are most common in "cryptic" habitats such as crevices and the undersides of boulders (Ricketts et al. 1968; Swarbrick 1984).

TEMPORAL CHANGES

To judge the significance of natural and man-made perturbations to communities of organisms, it is important to have some understanding of the persistence of the component species and of the normal temporal variability in their population densities. The approximately 3-year duration of most studies is woefully inadequate for these purposes because most of the dominant intertidal organisms are relatively long-lived. Connell (1986) found only a handful of studies in which species had been observed for at least one turnover of the population. Of these, many were qualitative and the temporal variability in population densities could not be estimated.

Ellwood boulder field, west of the University of California at Santa Barbara, has been used repeatedly for ecological studies since 1973 (Sousa 1979a,b; Sousa et al. 1981; Schroeter 1978; Fawcett 1979, 1984; Dean and Connell 1987a,b,c; Swarbrick 1984). During the 15 years for which there are data, a remarkable change in community composition has occurred (S. Swarbrick, S. Schroeter, and J. Connell unpubl. data). From 1973 to 1983, the average abundances of the major species of algae and invertebrates throughout the boulder field appeared to fluctuate seasonally around a relatively constant average level. The fluctuations were smallest for algae, limpets, chitons, and urchins and slightly greater for sea stars, octopuses, and crabs. The greater fluctuations of species in the latter three groups may have been in response to the tidal cycle and storms since most of those species occur in the subtidal as well as the intertidal and may have moved in and out of the boulder field in response to spring and neap tides. Large increases were also noted after severe storms, when individuals were apparently washed into the intertidal and persisted there for several months. Abundances were more variable on the smaller scale of individual boulders.

Much of this variation in community composition was due to disturbances caused by boulders being overturned. When this happened, the algae on the upper surfaces were placed in the dark or in contact with sea urchins, where they either died gradually (Sousa 1979b) or were grazed off (Sousa 1979b; Sousa et al. 1981). The sessile invertebrates on the undersides of such boulders also died because they were exposed to predators or harsh physical conditions (Swarbrick 1984). The newly opened spaces were colonized by invertebrates on the undersides of the boulders and by algae on the tops. After the initial colonization, a regular succession of species of both groups occurred. After about 2 years without further disturbance, the tops of the boulders were dominated by *Gigartina canaliculata* (Sousa 1979b) and the bottoms were covered by the remains of dead individuals and a low cover of living bryozoans and tube-building polychaetes (Swarbrick 1984). Because the attached species colonizing any given boulder kept changing over the 2 years or so before abundances stabilized, the average abundance of any species in the boulder field depended on the frequency of disturbance of the boulders. The important feature of this system was that for both invertebrate and algal assemblages, diversity was highest at intermediate frequencies of disturbance.

This community underwent a profound change in composition and dynamics soon after the series of large storms in early 1983. Wave energy was so high that virtually all boulders were violently tumbled, and all species of algae and sessile and motile invertebrates were driven to low densities. Early recovery was as predicted. The tops of the boulders were heavily colonized by a green alga, *Ulva* sp., and the tube-building polychaete *Phragmatopoma californica*. However, succession did not proceed as expected. Unlike the pattern of the preceding 10 years, *Ulva* and *Phragmatopoma* were not replaced by perennial red algae; they still dominated the boulder tops in 1988, 5 years later. *Phragmatopoma* cemented together boulders and stabilized them. The community on the undersides of boulders was generally as expected in the absence of disturbance. There were few species and few individuals. The cause of this state of change is not yet understood. This change in community structure and dynamics on a scale of tens of years makes apparent the need for long-term studies.

MECHANISMS PRODUCING COMMUNITY PATTERNS ON ROCKY BEACHES

Physical Factors. Species can be constrained in their distributions by inherent physiological or morphological limitations. The assumption of such a mechanism is implicit in discussions of "warm water," "sand-loving," or "desiccation-tolerant" species. However, the basis for such designations is almost always a correlation between the species' abundance and perceived habitat characteristics. Experimental demonstrations of the limiting effects of physical factors are rare (Connell 1972; Underwood 1979).

Much attention has been directed to the effects of high temperatures and desiccation on the high shore (Connell 1972; Wolcott 1973; Underwood 1979). That these factors might be limiting is suggested by the fact that in wet, cool areas many species are found higher on the shore than elsewhere. Vertical ranges are often shifted upward where shade or strong wave action and splash occur (Lewis 1964). Both the littorines and limpets that occupy the highest shore move up and down in the splash zone with changes in wetting conditions (Frank 1965; Bock and Johnson 1967). Following hot, dry periods, as when low tides coincide with Santa Ana winds (hot desert winds characteristic of southern California) or calm seas and hot days, dead and dying animals are often in evidence in the upper intertidal (Frank 1965; Stimson 1968; Sutherland 1970). On Santa Cruz Island, abrupt increases in temperature of as much as 10°C were associated with the onset of Santa Ana winds (Seapy and Littler 1982). During the period of extreme exposure, pronounced declines occurred in the abundance of several middle to upper intertidal macrophytes. *Barnacles* also declined during this period, perhaps as a result of the drying conditions.

It has also been demonstrated for many species that position on the shore correlates with physiological or morphological adaptations to drying conditions (Kensler 1967; Davies 1969; Foster 1971). Such correlations have been reported for species occupying different levels (Hardin 1968; Bock and Johnson 1967; Wolcott 1973; Schroeter 1978) and for high and low individuals of the same species (Segal 1956a,b; Segal and Dehnel 1962; Seapy and Hoppe 1973).

Wolcott (1973) found that low-level species of limpets had lower thermal tolerances than species from the high shore and that they lost water 11 times faster. The species from the high shore, *Collisella scabra* and *Collisella digitalis*, form a mucous sheet between the shell margin and the substrate that prevents water loss. *C. scabra* is less adept at this, but requires less of a seal because its shell fits closely with the home scar that it excavates (Lindberg and Dwyer 1983) and to which it returns between tides (Hewatt 1940). *C. digitalis*, in contrast, can span gaps up to 15% of shell diameter (Wolcott 1973). When this sheet was experimentally removed, *C. digitalis* desiccated at about the same rate as the low-level species. *Collisella strigatella* has also been observed to form mucous sheets under stressful drying conditions in the laboratory and field (Dixon 1978).

Wolcott (1973) also discovered that, although environmental conditions on the high shore often exceeded the tolerances of the low-shore species, their vertical distribution was not constrained by mortality high on the shore. The low-level species simply avoided stressful habitats. Littorines, too, actively select their positions on the shore and, when experimentally displaced, soon return to their original area (Bock and Johnson 1967). In the low intertidal, purple urchins occur in areas of high wave surge, whereas red urchins are rare in such habitats. When urchins were transplanted to habitats with strong surge, the purple urchins remained, but the red urchins quickly moved to sheltered sites (Schroeter 1978).

The fact that distribution at low tide is limited by a behavioral response to drying conditions rather than by increased mortality is, no doubt, important to the individual that is moving rather than dying. However, in the context of community interactions, the significance of behavioral adaptations is less obvious because the distribution of mobile organisms when they are immersed and feeding is poorly documented. Nonetheless, in the case of marine predators, the effect seems clear. As the tide retreats, predaceous snails and starfish seek moist shelter, presumably to avoid lethal drying. Thus, despite their ability to prey on individuals in the low intertidal, they may be prevented from doing so higher on the shore because of lack of time during periods of immersion (Connell 1970; Paine 1974).

Grazers also tend to aggregate in moist areas between tides (Wolcott 1973). Most seek shelter within the zone of their grazing activities. Others move up and down the shore. *Notoacmea fenestrata*, for example, retreats to the bottom of rocks at low tide and nestles into the sand but grazes tens of centimeters higher when immersed (Dixon 1978).

Strong water motion is another stressful physical factor in the intertidal zone that affects community composition (Denny 1985; Denny et al. 1985). Harger (1970a) found that the distribution of three species of mussels in southern California was related to wave force. Both *Septifer bifurcatus* and *Mytilus californianus* inhabit wave-swept shores, but *Mytilus edulis* tends to be common only in more protected areas (see Suchanek 1978). Both *Mytilus* species were common on a horizontal bench near Carpinteria in Santa Barbara County, but after a severe storm most of the *M. edulis* were gone (Harger 1970b).

Harger (1970a) measured the force necessary to pull mussels off the substrate. *Septifer* was most firmly attached and *M. edulis* most easily dislodged. Larger *M. californianus* were more strongly attached than smaller individuals, but the relationship between size and dislodgement force was constant over most of the size range for *M. edulis*. This explains the fact that the maximum size of *M. edulis* was inversely correlated with wave force among study sites.

North (1954) found that *Littorina planaxis* was larger on the extreme high shore and on the protected shoreward side of outcrops. He discovered that large individuals were more easily dislodged when experimentally subjected to strong water motion than were small snails, and he suggested this as the explanation for the distributional patterns.

Besides dislodging organisms, waves cause damage by battering substrates with rocks or debris. In the Pacific Northwest, battering by wave-tossed logs is a significant cause of mortality (Dayton 1971). Within the SCB, damage is more likely to be caused by rocks being slammed together. The probability of impact and its severity depends on the significant wave height, the availability and size of projectiles, and the microhabitat of the organisms (Shanks and Wright 1986). Shanks and Wright found that the number of new dents in aluminum targets was closely correlated with the incidence of new spaces cleared in aggregations of the barnacle *Chthamalus fissus*. Solitary barnacles were at much greater risk than those in dense populations. Shanks and Wright (1986) also discovered that sharp edges of boulders were more likely to be struck than boulder tops and that vertical faces of boulders were impacted least of the three microhabitats. Other evidence suggested that damage was greater in the high intertidal, which was in harmony with the predictions of Denny's (1985) model. Finally, wave damage had a strong seasonal component. Limpets (*Lottia gigantea, Collisella digitalis, Collisella scabra,* and *Collisella limatula*) were chronically damaged by rocks. Damage varied seasonally from near zero to about 0.5% of the populations per day, being highest during the winter stormy period.

Stimson (1968) made similar observations. To quantify shell damage, he glued limpet shells to boards which he then affixed to raised shale slabs at Ellwood Reef. The rate of crushing was highest during winter months when the sand level was low and the underlying cobbles exposed. On the shale reefs, seasonal fluctuations occurred in the number of limpets (mostly *Collisella strigatella*), with an annual decline in the winter. On nearby pier pilings, which probably received less battering, limpet numbers did not show seasonal cycles.

Occasionally, mortality caused by storm-tossed rocks can be catastrophic. During a period of severe storms in February and March 1978, boulders at the mouth of streams near Santa Barbara were overturned and scoured by waterborne cobbles. Barnacle populations were decimated, and the limpets *Collisella digitalis, C. scabra,* and *C. strigatella* became locally extinct on over 50% of the boulders (Dixon 1978). The limpet *Notoacmea fenestrata*, which occupies a more protected microhabitat, also suffered severe mortality but disappeared completely from only 19% of the boulders. Boulders at headlands were much less affected, probably because of a shortage of projectiles.

Fluctuations in sand height can also have important population and community consequences, especially in southern California where many rocky habitats are adjacent to sandy beaches or actually occur in a sand matrix (Littler 1980; Littler et al. 1983; Taylor and Littler 1982). At Ellwood beach, west of Santa Barbara, sand height varies about 60 cm over the year, seasonally covering and exposing large areas of cobbles and raised reefs (Stimson 1968). Fluctuations of a meter or more can occur in a 24-hour period (J. Dixon pers. observ.). A few species, such as the clonal anemone *Anthopleura elegantissima* (Taylor and Littler 1982), can survive extended periods of burial, but most probably cannot. Taylor and Littler (1982) found that the abundance of the tube-building worm *Phragmatopoma californica* was inversely correlated with sand cover and that the worms died after 5 days of burial in the laboratory. At the same site, Littler et al. (1983) found that quadrats that were seasonally inundated by sand tended to be dominated by rapidly colonizing species, whereas pinnacles that were not buried were dominated by *Mytilus californianus* and other long-lived animals.

Biological Interactions. The importance of predators in intertidal systems has long been appreciated (Paine 1966; Kitching and Ebling 1967; Dayton 1971; Connell 1972). In the northeast Pacific, predation by the seastar *Pisaster ochraceus* excludes mussels from the lower intertidal (Paine 1966, 1974). Although extensive mussel beds are less common in the SCB, where they do occur, the lower limit also appears to be determined by predators (Landenberger 1967). Occasionally mussels are found subtidally on pier pilings covered by large aggregations of the clonal anemone *Corynactis californica*. *Corynactis*, which has huge nematocysts, acts as a barrier to *Pisaster*. When the anemones were experimentally removed, starfish soon found and ate the mussels (Landenberger 1967). A similar interaction between *Corynactis* and *Pisaster* has since been observed on the supporting structures of offshore oil platforms (Wolfson et al. 1979), and Vance (1978) found that a diverse group of epibionts protected a sessile clam from detection and predation by seastars.

Barnacles can also be limited in their lower distribution by predation. In Washington, *Balanus glandula* settle over a broader area of shore than they generally occupy as adults. However, all but the individuals highest in the intertidal are eaten by several species of *Nucella* (Connell 1970). *Balanus glandula* never reaches an invulnerable size and so is excluded from the lower shore, where predators are abundant. In contrast, *Semibalanus cariosus* can withstand attacks from predatory snails in most of the intertidal zone if it can survive for the approximately 2 years required to attain large size (Connell 1972; Dayton 1971). However, in the lowest areas, it is never safe (Connell 1972). As a result, size may be a refuge on the high shore but not in the lowest intertidal area (Connell 1972).

Within the SCB, little attention has been directed to barnacle populations. Although it is known that barnacles in this area are eaten by *Pisaster ochraceus* and several species of predaceous snails (Murdoch 1969; Fotheringham 1974), the population consequences have not been investigated. Two species of barnacles, *Megabalanus californicus* and *Tetraclita rubescens*, attain a large size and are found in the lower intertidal. Whether predation controls the density and size distribution of these species is not known. However, work farther north indicates that *Tetraclita* settles mostly in the low intertidal and can attain a size at which it is safe from predators (Hines 1978, 1979).

The distribution of intertidal grazing molluscs is also constrained by predators. Fawcett (1984) found that *Tegula funebralis* occurred lower on the shore where predators (mostly *Octopus* spp., *Pisaster ochraceus*, and crabs of the genus *Cancer*) were less numerous. At sites with many predators, transplanted snails were either killed outright or quickly migrated up the shore. Transplanted individuals that originated at sites where predation was low moved up the shore more slowly and suffered heavier mortality than individuals that came from areas where predation was high. Small individuals moved less and suffered higher mortality.

The distribution of limpets is also affected by predators. Many limpets show strong avoidance and escape responses to *Pisaster ochraceus* (Bullock 1953; Feder 1963; Margolin 1964; Phillips 1975). Phillips (1976) found that *Collisella limatula* and *Notoacmea scutum* moved upward where *Pisaster ochraceus* was caged on the rocks. On Santa Catalina Island, *C. limatula* and *C. scabra* feed only while awash (Wells 1980). Between tides, *C. limatula* moves under rock and *C. scabra* returns to its home scar. Wells (1980) showed that in the laboratory these species suffer higher mortality from *Octopus* spp. if they are forced to remain in the open away from shelter or a home site.

Although biotic interactions may most commonly affect the abundance of populations in the lower parts of the littoral zone, they can also be important on the high shore. Terrestrial predators such as birds and mammals are able to forage longer on the high shore. Frank (1982) found that oystercatchers ate significant numbers of limpets. Individuals in cracks and crevices, under rocks, and on vertical surfaces were less vulnerable than those occupying exposed horizontal surfaces. Oystercatchers are common on the southern California Channel Islands but not on the southern California mainland. Heavy predation on limpets, particularly *Lottia gigantea,* has also been attributed to gulls (Lindberg and Chu 1983).

Competition is another strong interaction that can impart structure to communities. One of the most carefully examined competitive interactions is that between the sea mussels *Mytilus californianus* and *Mytilus edulis* (Harger 1968, 1970a,b,c,d, 1972a,b,c). The case is particularly interesting because of the demonstrated interactions of physical factors and biological attributes. *Mytilus californianus* grows rapidly to a large size, has a thick shell, attaches firmly to the substrate, is not easily dislodged by waves, and thrives on the most exposed coasts. Conversely, *M. edulis* grows more slowly to a smaller size, has a thin shell, has weaker byssal threads, is more easily dislodged, and is rare in very high energy environments. When buried or in mixed clumps, *M. edulis* crawls to the surface, which gives it an advantage in sheltered environments. In such habitats, *M. californianus* is left in the center of clumps, where it is smothered by accumulating silt. Where these mussels occur together in intermediate habitats, juvenile *M. edulis* move quickly to the outside of clumps, where they grow very rapidly. In time, however, *M. californianus* are able to grow through, and if *M. edulis* are incorporated into the matrix of the clump, they are prone to physical damage. Dead *M. edulis,* but not *M. californianus,* are common inside mussel clumps. In these intermediate environments, the two species coexist in proportions that reflect wave exposure and recent storm history.

Competition also occurs between the tube-building worm *Phragmatopoma californica* and the clonal anemone *Anthopleura elegantissima* in the middle intertidal area. When Taylor and Littler (1982) removed *Anthopleura,* an increase first occurred in barnacles, algae, and grazing molluscs. Later, *Phragmatopoma* colonized the clearings, overgrew the barnacles, and became the dominant holder of space. This did not occur in experimental quadrats in the upper intertidal.

Taylor and Littler (1982) also created clearings in the high, middle, and lower intertidal zones. In the high quadrats, *Anthopleura* returned to its original abundance within 2 years and there was no colonization by *Phragmatopoma.* The reverse took place in low clearings. There, the tube worms returned to their original abundance in less than 1 year and no colonization by the anemones occurred. In the middle quadrats, both species recruited successfully. *Anthopleura* returned to its original percentage of cover, and *Phragmatopoma,* which was originally absent, increased to more than 10% cover. Taylor and Littler (1982) suggested that *Phragmatopoma* is the superior competitor but is prevented from dominating the middle region because it suffers severe mortality from periodic sand inundations that *Anthopleura* can tolerate. They speculate that the tube worms are unable to tolerate the harsh drying conditions of the high shore.

Competition has also been well documented among grazing molluscs. Many of the grazers in the middle and upper intertidal areas of eastern Pacific shores rely for food on the microalgal film that grows on rocks (Nicotri 1977) and are potential competitors. Haven (1971) showed that microhabitat differences existed between *Collisella digitalis* and *Collisella scabra*. However, where they occurred together, they reduced one another's rate of growth, presumably as a result of exploitative competition (Haven 1973). Similar competitive relationships have been shown for other pairs of species. Where their distributions overlap, *Notoacmea fenestrata* and *Collisella strigatella* compete for food, as do *C. strigatella* and *C. digitalis* (Dixon 1978). *C. strigatella* are also constrained in their distribution where they occur with *C. digitalis*. When the latter are experimentally removed, *C. strigatella* occur higher on the shore (Choat 1977; Dixon 1978). The mechanism is not clear, but it seems likely that when *C. strigatella* reaches a larger size in the absence of its competitors, it can then tolerate harsher drying conditions and it moves higher to utilize unexploited food resources.

Competition can also take the form of interference. The giant limpet *Lottia gigantea* maintains a territory that it defends against other grazers and recruiting sessile species (Stimson 1970). If *Collisella* spp. are allowed onto the territory by removing the resident *Lottia*, they remove nearly all the algae because smaller individuals graze closer to the rock than do larger ones (Stimson 1973). Therefore, maintenance of a territory allows the development of the thick algal film *Lottia* requires. Under most conditions, this results in increased horizontal patchiness in the abundance of the smaller limpet species. However, in areas where *Lottia* are very abundant, as on the west end of Santa Cruz Island or between Point Conception and Gaviota, the other limpets may be virtually absent from a broad vertical section of the shore except on the shells of *Lottia* (J. Dixon pers. observ.; S. Schroeter pers. observ.). It may be that this is the normal situation in undisturbed habitats.

Interference competition also occurs between urchin species in the low intertidal zone. Red urchins reduce the abundance of purple urchins by prodding nearby individuals with their spines (Schroeter 1978). Red urchins have longer spines and they irritate the epithelial covering of the purple urchins, causing them to give ground. When red urchins were experimentally removed from intertidal habitats, purple urchins rapidly colonized. Both urchin species also occur subtidally, where they appear to have less effect on one another's distribution.

Human Activities. From Gaviota to Point Conception, access to the shore is extremely limited and intertidal communities are probably as pristine as any on the California coast. From Gaviota to Ventura, the human population density is relatively low, many of the rocky shores are not easily accessible, and the intertidal communities appear less perturbed than farther south. From Ventura to San Diego, most rocky shores border densely populated urban areas and are subject to intense collecting and trampling when exposed.

Intertidal invertebrates are collected for food, bait, and specimens. Besides the loss of the individuals actually collected, there are indirect effects of collecting. Mussel clumps are torn apart in the search for bait worms, collectors dislodge and overturn boulders, and enthusiastic clam diggers leave behind amazing excavations in the gravel matrix of boulder fields.

Although the effects of human disturbance are very difficult to measure, they are probably significant. In 1947, when Stephenson and Stephenson (1972) visited La Jolla, both pink and green abalones were common under boulders. Today, the presence of abalones in that area would be a remarkable curiosity. In recent years, the populations of many other species have also changed markedly. In general, the trend seems to be toward fewer and smaller individuals. T. A. Ebert (pers. comm.) has been studying intertidal populations in San Diego County since 1969. Whereas clumps of large mussels, *Mytilus californianus*, and large turban shells, *Astraea undosa*, were once common at his study sites, only small individuals can now be found. In areas near La Jolla where purple urchins were common in 1981 (Russell 1984), they are now rare. In 1986, in 4 hours of searching, only one urchin was found (T. A. Ebert pers. comm.).

Of course, it is not possible, after the fact, to demonstrate that changes in natural populations are due to human depredations. Nonetheless, the changes are suggestive and ominous. Any species that is considered tasty or decorative may be in peril. For these reasons, it is increasingly risky to draw ecological inferences from sampling data obtained from the "urban intertidal." Studies conducted in such areas should explicitly consider the effects of human predation and physical disturbance.

Sandy Beaches

Although sandy beaches are the predominant intertidal habitats in the SCB, we know far less about the invertebrate communities found there than we do about those on rocky substrates. This is probably a reflection of the difficulties involved in conducting ecological studies in sand. Most sand-dwelling organisms are difficult to mark and are very mobile. As a result, individuals or even groups of the same individuals cannot easily be monitored through time. For many species, immigration and emigration rates are high, and this contributes to the large amount of temporal and spatial patchiness in population density that is often observed. In addition, field experiments are very difficult to carry out successfully in an unstable substrate located in a high-energy environment. Despite these daunting problems, some interesting work has been carried out on sandy beaches in the SCB.

VERTICAL ZONATION

Although less obvious to the passerby than it is on rocky shores, vertical zonation of invertebrates also occurs on sandy beaches. However, the animals that live in sand are quite mobile and often change position with changes in water level. For such species, position on the shore is better described relative to the swash zone than to tidal datum.

Despite changes in position in response to tides, it is predictable that some species will be found higher or lower than others. The common invertebrates in the upper intertidal are several species of beach hoppers (amphipods in the genus *Orchestoidea*), the predatory isopod *Excirolana chiltoni*, and three species of polychaetes (fig. 8.4). Of the latter group, the bloodworm *Euzonus mucronata* is most abundant and is a major prey of *Excirolana* (Cubit 1970; Ricketts et al. 1968). Beetles are also common in the high intertidal. On beaches near Santa Barbara, the staphlinid beetle *Thinopinus pictus* is an important predator on beach hoppers (Craig 1968, 1970). Interestingly, in a 10-year study of sandy beaches located throughout the SCB, confirmed records of *Thinopinus* were restricted almost entirely to the area around Santa Barbara (Straughan 1982). However, unidentified staphlinids were recorded at most of the study sites, so the presence of a predaceous beetle may be characteristic of the high intertidal on sandy beaches throughout the SCB.

The middle intertidal is characterized by the sand crabs *Emerita analoga* and *Lepidopa californica*, the polychaete *Nephtys californica*, the snail *Olivella biplicata*, and the clam *Donax gouldi*. *Emerita* is generally the most abundant of the common middle intertidal invertebrates, often comprising over 99% of the individuals found on a given beach (Straughan 1983; Wenner 1988). The two molluscs *Olivella* and *Donax* are sometimes nearly as abundant as *Emerita*, but their importance is probably often underestimated because of their extreme spatial and temporal patchiness.

As is generally true throughout the world, common species in the low intertidal are predominantly polychaetes and nemerteans (Straughan 1982). In addition, the large sand crab *Blepharipoda occidentalis* and the pismo clam *Tivela stultorum* can be found on most southern California beaches throughout the year. However, they are generally not abundant. *Blepharipoda* is several orders of magnitude less dense than the sand crabs *Emerita*

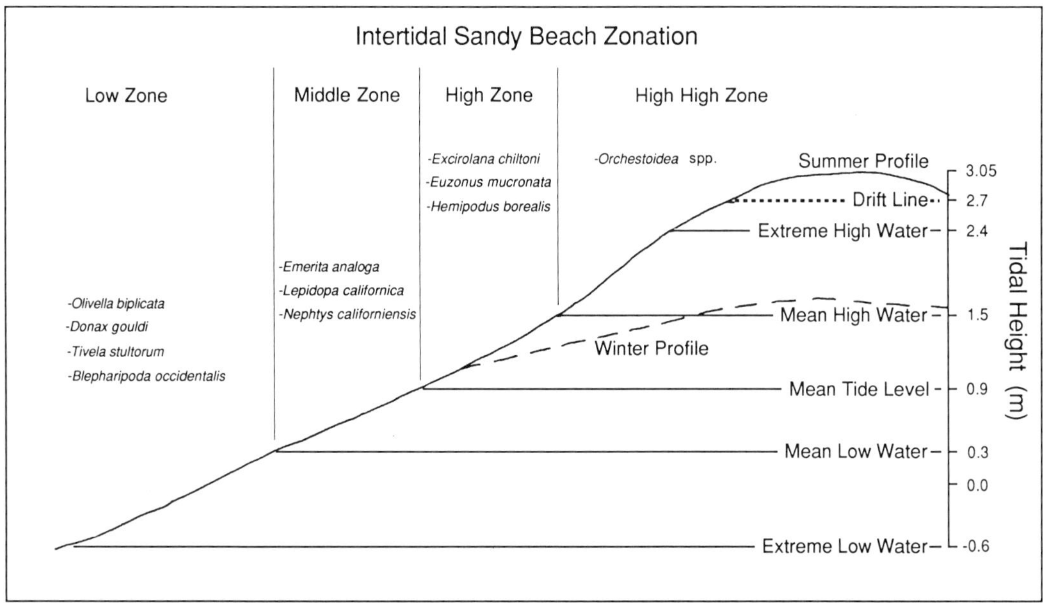

Figure 8.4. Intertidal zonation of a sandy beach in southern California.

and *Lepidopa*, which are found higher on the shore (Straughan 1983). Although *Tivela* tends to be more abundant in the shallow subtidal today (Morin et al. 1985, 1988), this clam was once abundant in the intertidal (Ricketts et al. 1968), and its present rarity is probably the result of overharvesting.

HORIZONTAL PATCHINESS

The mechanisms that produce the observed patchy distribution of species along the shore are best understood for the sand crab *Emerita analoga* and the snail *Olivella biplicata*. Sand crabs show horizontal patchiness on scales from meters to kilometers because of alterations in longshore current patterns caused by natural topographic features and man-made objects such as pier pilings and jetties (Cubit 1968). Both man-made and natural features can concentrate sand crabs by creating convergence areas where the water pools. Another feature that can cause increases in local concentrations is drainage channels of varying permanence.

Onuf (1972) discovered that dense populations of *Olivella* were almost always found within 0.5 km of the lees of headlands and were sparse elsewhere. Where they do occur, they form discontinuous patches in the middle intertidal. A total of 95% of the population was found in 1–2% of the available habitat. Evidence from field experiments indicated that patchiness on both small and large scales resulted from a combination of passive sorting by waves and individual behavioral responses (Onuf 1972).

TEMPORAL CHANGES

The local abundance of invertebrate species on sandy beaches changes significantly over time periods varying from hours to decades. Sand crabs move up and down the beach with the tides and maintain their position in the swash zone. As a result, abundances at any one location in the intertidal roughly follow a curve that is similar to that of the tides (Efford 1965; Cubit 1968; Perry 1980). Beach hoppers (*Orchestoidea* spp.) also migrate up and down the intertidal in rough synchrony with the tides. At low tides, beach hoppers emerge from burrows in the high intertidal, move down the beach, and feed on stranded algae. They then retreat before the incoming tide and return to their burrows at the next high tide (Fawcett 1969). This pattern is more obvious for some species than others, and it appears to be influenced by the lunar phase as well as the tidal cycle (Fawcett 1969).

Seasonal variability in population density is striking in sand crabs. In most years, they tend to be abundant in the summer and fall. Their numbers decline dramatically during most winters, when they are virtually absent from many beaches (Barnes and Wenner 1968; Perry 1980).

The only information regarding long-term variability is for the bean clam *Donax gouldii*. In the first half of the century, this clam was abundant enough to be commercially fished (Coe 1955; MacGinitie and MacGinitie 1949), but its numbers have since declined (Wenner 1988). Studies at La Jolla from 1937 to 1955 documented two spectacular population increases in 1938 and in 1949, when densities increased from less than 20 m^{-2} to more than 10,000 m^{-2} (Coe 1953, 1955). These increases occurred approximately uniformly along a 5-mile stretch of beach, but no data from other beaches in the SCB exist upon which to judge such large-scale spatial patchiness (Coe 1955). Coe's (1955) historical reconstruction indicates aperiodic population explosions at intervals varying from 2 to 15 years from 1894 through 1955. The increases were followed 1 or 2 years later by crashes, which Coe (1955) attributed to disease outbreaks caused by the single-celled organism *Dermocystidium marinum*. No such population outbreaks and crashes have been reported since 1955, but this may be because no one is looking.

MECHANISMS PRODUCING COMMUNITY PATTERNS

Physical Factors. Shifts in the composition of invertebrate assemblages are correlated with changes in beach slope and sand texture. Within a beach, crustaceans and molluscs are most common in the steeper, coarser, and dryer upper intertidal. Polychaetes and nemerteans are the dominant invertebrates in the low intertidal, where beach slope is shallower and the sand is finer and wetter (Wenner 1988; McLachlan and Hesp 1984; Straughan 1982).

Studies of *Olivella biplicata* and *Emerita analoga* illustrate how physical forces can influence the distribution and abundance of sand-dwelling invertebrates. In the middle intertidal, *Emerita* form aggregations (Efford 1965; Cubit 1968; Barnes and Wenner 1968; Perry 1980; Auyong 1981) that move up and down with the tides (MacGinitie 1938; Efford 1965; Cubit 1968). Small male crabs occur highest on the shore, whereas the large females predominate lower down (MacGinitie 1938; Efford 1965; Cubit 1968). Endogenous biological rhythms have been hypothesized to account for both aggregations and their changes in position over the tidal cycle (Efford 1965). However, the weight of observational evidence (Cubit 1968; Barnes and Wenner 1968; Perry 1980), including studies of marked individuals (Dillery and Knapp 1969), suggests that aggregations are formed by the response of crabs to waves and currents interacting with structural features on the beach.

Cubit (1968) showed experimentally that the reactions of sand crabs to physical forces can account for the observed distributional patterns within a beach. Breaking waves suspend sand at the lower end of the swash zone. Sand crabs react by burrowing out of the sand and are then carried up the beach. As the upwash loses momentum, the crabs burrow into the sand again and begin feeding. Uprushing water reaching the relatively well-drained sand at the top of the wash zone moves over and through it, causing it to become thixotropic, or fluid. Crabs respond to this state change by burrowing out, to be carried down the beach until the momentum of the backwash lessens. They then burrow back into the sand (Cubit 1968). The very simple behavior of burrowing out when the beach sand becomes fluid, combined with the physical properties of waves and their effects on beach sands, explains how aggregations of sand crabs are maintained in a swash zone that shifts among tidal levels.

This mechanism may also account for two additional phenomena: (1) the lower density or absence of aggregations on shallow-sloped, fine-grained beaches where waves are small; and (2) the predominance of small sand crabs near the top of the swash zone and large ones near the bottom. The change from thixotropic to nonthixotropic sand is less marked on shallow-sloping beaches because waves are smaller and sand is finer. Small waves suspend less sand at the bottom of the swash zone, and fine sand tends to be saturated with water and does not undergo the shift from a nonfluid to a fluid state. Therefore, the stimuli that cause sand crabs to burrow out of the sand and to be sorted by the wave swash are missing on such beaches. The nonrandom distribution of sizes along the vertical gradient is probably attributable to differences in the behavior of individuals of different sizes. Small sand crabs burrow less deeply in the sand than larger ones and so are more likely to be excavated by waves in the lower wash zone and moved up the shore. In the upper end of the swash zone, they burrow in and are able to maintain their position because wave momentum is less.

Physical mechanisms also account for the horizontal distribution of *E. analoga* on the beach. Several investigators have noted that when cusps form on beaches, aggregations of sand crabs are often found in their bays (Cubit 1968; Dillery and Knapp 1969), although this pattern does not occur invariably (Perry 1980). Cubit (1968) found that sites he termed *convergence areas* were even more likely to have dense aggregations of crabs. A convergence area occurs where water enters at a faster rate than it leaves. Such areas can be created by intersecting wave patterns that are not parallel to the shore or by obstructions such as large rocks or pier pilings. The association of aggregations with cusp bays was strongest when the latter coincided with a convergence area. In addition, sand crabs tend to aggregate along sand channels that are created by seawater draining from temporary or permanent pools (Cubit 1968).

Experimental and observational studies of the snail *Olivella biplicata* provide another example of how an interaction between simple behavior and physical forces accounts for major features of distribution and abundance. *O. biplicata* form aggregations in the middle to lower intertidal on south-facing beaches and in the lees of headlands (Onuf 1972). The large-scale distributional pattern correlates with physical processes. Waves in the northern half of the SCB are predominantly from the west or northwest, creating a gradient from high to low energy as one proceeds northwest to southeast around the headlands and then from low to high energy as one proceeds away from the headlands.

The evidence for passive sorting by size of *O. biplicata* within a beach is twofold: (1) the distribution of snails by size in aggregations is remarkably similar to that of passively sorted pebbles; and (2) *O. biplicata* becomes inactive, and thus pebblelike, when waves are high (Onuf 1972). Once aggregations are formed, they tend to persist because the movements of *Olivella* fall off dramatically as density increases. Experiments have shown that this behavior is simply a response to obstacles and is the same whether the obstacle is another *Olivella* or an inanimate object such as a pebble (Onuf 1972). To our knowledge, no studies to date have directly documented the effects of large disturbances on the biota of sandy beaches in the SCB.

Biological Interactions. Sandy beaches are usually regarded as places where physical factors are more important than biological processes (Straughan 1982; McLachlan and Bate 1983; Wenner 1988). Although there are few pertinent studies, some evidence suggests, however, that biological interactions are also important.

Thinopinus pictus, a staphlinid beetle, preys on beach hoppers in the upper intertidal and is itself limited in its distribution by the presence of prey (Craig 1968). Beach hopper density, however, is apparently not significantly affected by beetle predation (Craig 1968).

Sand crabs are heavily preyed upon by two common shallow subtidal fishes, California corbina (*Menticirrhus undulatus*) and barred surfperch (*Amphistichus argenteus*), comprising over 90% of the diet of the latter (Carlisle et al. 1960; Fitch and Lavenberg 1971). The population consequences of fish predation on sand-dwelling invertebrates in southern California has not yet been investigated.

Straughan (1982) presents distributional data suggesting that interspecific competition may influence the distributions of several species of polychaetes along the SCB or along the intertidal gradient within certain beaches.

Human Activities. Human activities that likely have affected sandy beach invertebrates include the construction of jetties, groins, and sea walls; construction of highways and buildings on sea cliffs; the damming and channelization of streams and rivers; beach maintenance and cleaning; and accidental oil spills. Damming of streams reduces sediment input to downstream beaches and, in some cases, has caused significant erosion that could affect invertebrate communities.

The evidence that these changes affect populations of sandy beach invertebrates is scant and indirect. Straughan (1982) found that a sea wall reduced the slope, average grain size, and intertidal range of a section of sand beach. These changes appeared to select *for* polychaetes and nemerteans and *against* most crustaceans. Sand beaches in the populated central portion of the SCB are regularly cleaned of kelp and other debris (Straughan 1982), thereby reducing the food supply of many of the sand beach invertebrates (e.g., beach hoppers and perhaps sand crabs). In Straughan's bight-wide study, these beaches were consistently depauperate and lacked beach hoppers over a 10-year period. Although confounded by the fact that sand is periodically added to many of the beaches that are cleaned (Straughan 1982), the pattern is suggestive enough to warrant further study.

Rocky Subtidal

PATTERNS OF DISTRIBUTION AND ABUNDANCE

No studies have been done on the depth distribution of sublittoral species comparable to the studies of zonation in the rocky intertidal zone. Such investigations are much more difficult under water because the distance across the depth gradient is usually two or three orders of magnitude greater than the distance across the intertidal gradient, whereas the area one can examine visually from a single site is at least one order of magnitude smaller under water. Exceptions are vertical outcrops and underwater cliffs.

Pequegnat (1961a,b, 1964) studied the communities occupying two submarine hogbacks—folded siltstone outcrops with nearly vertical sides—in water depths of 10 m and 17 m. The pattern of zonation on these reefs was as striking as that seen on intertidal rocks. The tops were dominated by rock oysters, *Chama pellucida,* or mussels, *Mytilus edulis* and *Mytilus californianus.* These species formed thick crusts inhabited by many other species. Lower on the vertical portion of the reefs, the substrate was covered by patches of calcareous bryozoans, gorgonians, stony corals, and rock scallops. Near the bottom of each outcrop were relatively few species and populations were sparse. The most conspicuous organisms were stony corals, gorgonians, sponges, barnacles, and sea urchins. In general, the biomass and numbers of the epifauna decreased from the top to the bottom of an outcrop.

Although the general pattern in the vertical distribution of organisms on the hogback in 10 m of water was similar to that on the 17-m deep hogback, the organisms present at a given depth of water were quite different. The top of the deep reef was located at about the same depth as the bottom of the shallow reef, but the number of species and individuals on the top of the deep reef was most similar to the number on the top of the shallow reef (Pequegnat 1964). Thus, an important gradient apparently occurs from the top to the bottom of submarine outcrops that is independent of absolute depth. Pequegnat argued that the community patterns were reflections of a reduction in both water movement and concentration of plankton with distance down a vertical outcrop. These patterns are reminiscent of those observed on boulders or outcrops located at different heights in the intertidal zone.

A few other examples of depth zonation occur among subtidal species. The various species of abalones occupy different, although broadly overlapping, depth ranges (Cox 1962). Black abalone (*Haliotis cracherodii*) are most abundant in the intertidal zone. As one proceeds into deeper water, the dominant species are, successively, green abalone (*Haliotis fulgens*), pink (*Haliotis corrugata*), red (*Haliotis rufescens*), and finally, white (*Haliotis sorensoni*) and threaded abalones (*Haliotis assimilis*). Different species of sea urchins also tend to occupy different depth ranges. Although both species often occur in dense populations in deeper water (Laur et al. 1986), purple urchins (*Strongylocentrotus purpuratus*) are generally much more abundant in the immediate subtidal than are red urchins (*Strongylocentrotus franciscanus*) (Schroeter 1978). At Santa Catalina Island, red and purple urchins and the diadematid urchin *Centrostephanus coronatus* are all common, but their centers of distribution vary. Purple urchins are most abundant in depths less than about 5 m, red urchins predominate at intermediate depths, and *C. coronatus* is the most numerous species below about 10 m (S. Schroeter and J. Dixon unpubl. data). A similar pattern was observed at Corona del Mar, but at that site, the species replacement occurred over a horizontal distance of about 500 m from very shallow water to a depth of about 18 m (Pequegnat 1964).

MECHANISMS PRODUCING COMMUNITY PATTERNS ON ROCKY BOTTOMS

Physical Factors. Bottom surge produced by passing swells is the physical factor having the most obvious biological effects on the biota in shallow coastal habitats. Large, long-period swells produced by unusually severe storms dislodge both motile and sessile organisms, turn over rocks, and suspend and redistribute sediments, with catastrophic effects on benthic populations (Ebeling et al. 1985; Tegner and Dayton 1987). Even waves associated with more local storms of average intensity suspend sediments and dislodge poorly attached motile invertebrates in depths less than about 15 m. Because wave energy drops off rapidly with increasing depth and varies predictably from place to place along the shore as a result of differences in bottom topography and exposure, it is a factor that has potential for producing many of the observed differences in biological communities.

Relatively persistent differences in community composition from site to site associated with differences in depth or wave exposure have been extensively documented for communities in central California (Foster 1982) and are probably common elsewhere. Unfortunately, to assess the importance of water movement, long-term observations may be necessary because site-to-site differences in the effects of waves during "normal" periods are not necessarily indicative of differences in the effects of unusually severe storms. The effects of storms on giant kelp (*Macrocystis pyrifera*) illustrate this point. In San Diego County, the northern coast is sheltered from north swells by the Channel Islands (Pawka et al. 1984). Nonetheless, during most years, mortality of giant kelp plants (which is largely caused by dislodgement by waves) is similar at Point Loma to the south and at San Onofre to the north (Dayton et al. 1984; Dean et al. 1987). The severe storms of January 1983, however, had devastating effects at Point Loma to depths of 18 m (Dayton and Tegner 1984), but mortality was not unusual at San Onofre (Dean et al. 1987).

The depth zonation of different species of sea urchins reflects, in part, differences in tolerance to wave surge (Lissner 1980, 1983;

Schroeter 1978). Purple urchins can survive greater wave surge than red urchins and are often abundant in very shallow habitats (Schroeter 1978). Red urchins occur shallower than *Centrostephanus*, a circumstance that also reflects differences in ability to survive wave surge (Lissner 1978). A similar mechanism may underlie the distributional pattern of abalone species. Green abalones, which are more likely to inhabit cracks and crevices than are pink and white abalones, survive better than the other species in shallow areas where wave energy is high (Tutschulte 1976).

Water and substrate movements can also influence the behavior and small-scale distribution of motile invertebrates. For example, red and purple urchins will cross patches of sand during calm periods, but not when bottom waters are turbulent (Laur et al. 1986). White urchins, *Lytechinus anamesus* (= *pictus*), respond to water turbulence by covering themselves with shells and small pebbles and by ceasing to move (Lees and Carter 1972). This behavior may result in decreased feeding activity for white urchins (Dean et al. 1984), as it does for *Centrostephanus coronatus*, which does not emerge from shelter holes when bottom surge is strong (Lissner 1980).

Currents may also affect the orientation and distribution of invertebrates, particularly sessile suspension feeders. Grigg (1972) found that colonies of the sea fans *Muricea californica* and *Muricea fruticosa* grow perpendicular to the direction of prevailing currents, which probably increases feeding efficiency.

Biological Interactions. Predation is probably as important in the subtidal as it is in intertidal habitats, but far less information is available on which to assess the population and community consequences. Most studies of predation in shallow waters have been mainly concerned with predatory behavior, dietary preferences, or energetics (Winget 1968; Wobber 1975; Rosenthal and Chess 1972; Rosenthal 1971; Leighton 1971; Gerard 1976; Day and Osman 1981; Tegner and Levin 1983).

The population consequences of predation are difficult to measure and may be apparent only in small areas where feeding is intense. *Octopus* spp. fed on a wide variety of food items, but caused significant reductions only in populations of their preferred prey, small decapod crustaceans, and only very close to their shelter holes (Ambrose 1980). Similarly, the sea urchin *Centrostephanus coronatus* had significant effects on sessile invertebrates only in the immediate vicinity of its shelter holes (Vance 1979; Vance and Schmitt 1979). It is interesting that this urchin is itself limited in its foraging by other predators. *Centrostephanus* that were experimentally prohibited from entering their shelter holes at daylight were eaten by the labrid fish *Semicossyphus pulcher*, the California sheephead (Nelson and Vance 1979).

Larger scale patterns in distribution and abundance may also result from predation. Tegner and Dayton (1977, 1981) and Tegner (1980) suggested that in southern California kelp forests, predators control the habitat use and abundance of red and purple sea urchins and are responsible for the observed size–frequency distributions. Purple urchins appear to be much more susceptible than red urchins to predation by sheephead, lobsters, and starfish. Also, they do not attain a size at which they are safe, and they tend to occupy cryptic habitats (Rosenthal and Chess 1972; Moitoza and Phillips 1979; Tegner and Dayton 1981; Tegner and Levin 1983). The major predators of red urchins also appear to be sheephead and lobster (Tegner 1980). The proportion of urchins in cryptic habitats appears to vary inversely with the density of these predators (Tegner and Dayton 1981). Tegner and Dayton (1981) suggested that the observed bimodal size distribution of urchins at their study sites was also a function of predation. Juvenile red urchins seek shelter beneath the spine canopy of adults, where they are safe from predators (Tegner and Dayton 1977). Urchins larger than about 90 mm in test diameter have long spines and are generally able to ward off predators. Red urchins of intermediate size (50–80 mm) are particularly vulnerable because they are too large to shelter beneath adults but too small to withstand attack by sheephead or lobsters. Thus, a mode of small urchins and a mode of large urchins have survived.

Cowen (1983) tested the importance of sheephead predation by removing sheephead at an experimental site on San Nicolas Island. Compared to an adjacent control site, the density of red urchins increased and their distribution shifted from cryptic to more open microhabitats. However, populations at both experimental sites were unimodal, and no significant changes in size frequency distributions occurred during the 10 months of the study. Thus, although the mechanism proposed by Tegner and Dayton (1981) probably does result in bimodal size distributions, one should bear in mind that the simple observation of such a distribution is neither necessary nor sufficient evidence that predation is important. Predators may have strong effects where the population is unimodal, and bimodal distributions may be caused by other factors such as periodic recruitment and slow growth of large individuals.

The presence of large populations of the small white urchin *Lytechinus anamesus* in areas where sheephead are abundant is a puzzle. White urchins never reach a large size, but they occupy exposed substrates in a great variety of habitats in shallow and deep water. Although these suppositions have never been investigated, it may be that these small urchins are relatively unpalatable, energetically not worth much effort, or both. Alternatively, predation rates may be low in areas with large *L. anamesus* populations.

One of the more intriguing instances of predation on urchins was described by Coyer et al. (1987). At Anacapa Island, *L. anamesus* mobbed red and purple urchins and preyed on them. On some sampling dates, a large fraction of both the red and purple urchin populations showed evidence of attack (primarily chewed spines). The authors speculate that the attacks may have been stimulated by a food shortage caused by overgrazing. We have since observed red and purple urchins with chewed spines where they co-occur with white urchins in northern San Diego County, but the frequency of such spine damage was low (S. Schroeter and J. Dixon pers. observ.).

If a predator elicits a strong escape response, it can affect the distribution of its prey even though it may catch and consume relatively few individuals. For example, the bat star *Asterina* (= *Patiria*) *miniata* is seldom observed to eat white urchins (*Lytechinus anamesus*) in the field. Laboratory experiments indicate that because of a strong escape response, larger urchins can avoid being eaten. Most predation occurs on individuals smaller than about 15 mm, which move slower than the large individuals (Schroeter et al. 1983). In a field experiment where *A. miniata* were transplanted to dense patches of white urchins, they evoked escape responses that cleared 1-m^2 areas of large white urchins within an hour. The authors speculate that the absence of white urchins from large areas of suitable hard substrate was due to a combination of predation on newly settled urchins and avoidance behavior by large urchins. The fact that white urchin abundance increased in many rocky areas after a sea star dieoff associated with the 1982–1984 El Niño (Tegner and Dayton 1987) is in harmony with the predictions of this hypothesis, but does not provide a test.

Similar interactions have been observed with other species. Schmitt (1982, 1985) found that at Santa Catalina Island the sea star *Pisaster giganteus* affected the relative abundance of two turban snails, *Tegula aureotincta* and *Tegula eiseni*, on small patch reefs. In southeast Alaska, the sea star *Pycnopodia helianthoides* apparently influences the small-scale distribution of strongylocentrotid species (Duggins 1983). As in the previous examples, escape responses of the prey, and not direct predation, was the mechanism producing the observed patterns.

Disease has profoundly affected the abundance of echinoderms in southern California.

Populations of virtually all of the shallow-water asteroids and the holothurian *Parastichopus parvimensis* were severely reduced throughout the SCB by a series of disease outbreaks that occurred between 1978 and 1984 (Tegner and Dayton 1987; S. Schroeter and J. Dixon unpubl. data). Sea urchins have also suffered from disease. *Strongylocentrotus franciscanus, Strongylocentrotus purpuratus, Lytechinus anamesus,* and *Centrostephanus coronatus* have all experienced disease outbreaks on Santa Catalina Island and on the mainland coast (S. Schroeter and J. Dixon unpubl. data). On Santa Catalina, an epizootic in late summer and fall of 1984 caused a drastic reduction in the density of shallow-water populations of red and purple urchins.

The causative agent of the disease in the bat star *Asterina miniata* and in probably all the asteroids is a bacterium of the genus *Vibrio* (S. Schroeter, J. Dixon, J. Hose, and R. Smith unpubl. data). A bacterium also appears to be the agent of disease in the case of the urchins (Gilles and Pearse 1986; J. Dixon and S. Schroeter unpubl. data), but the linkage is not as well established as for the sea stars. The possible consequences of these disease outbreaks are great because the affected urchins and sea stars are long-lived and have strong effects on their algal and invertebrate prey.

Although fewer studies have been done on competitive interactions in the subtidal, they are probably important. Schroeter (1978) presented two kinds of evidence that red and purple urchins compete with one another. The first suggested that competition was for food. In cages, red urchins lowered the growth rates of both body and gonadal tissue of purple urchins, while purple urchins had no effect on reds. A second set of experiments showed that fencing with spines, a kind of interference competition, was responsible for differences in depth distributions in the shallow subtidal as well as for microhabitat differences at deeper sites. Purple urchins were restricted to harsher habitats (sites that were shallower or more exposed to wave surge) by interactions with red urchins, which were superior interference competitors. Laboratory experiments suggested that interference with feeding was the mechanism underlying the reduced growth of purple urchins confined with their competitors. However, despite the potential for competitive interactions, purple urchins are often abundant in deeper water (Laur et al. 1986); occasionally they even form mixed grazing aggregations with red urchins. The conditions under which competition constrains the distribution of purple urchins are not yet well understood.

Based on laboratory feeding experiments, Tegner and Levin (1983) suggested that red urchins and red abalone (*Haliotis rufescens*) may compete with one another for drift algae. However, the importance of this competitive interaction in nature and its effects on the distribution and abundance of urchins and abalone has yet to be investigated.

In most shallow seas, sea urchins are the most important herbivores (Lawrence 1975; Foster and Schiel 1985; Harrold and Pearse 1987). In California, there is interest in sea urchins not only because they strongly affect the abundance of macroalgae and sessile invertebrates (Foster and Schiel 1985; Harrold and Pearse 1987) but also because they are commercially fished (Kato 1972; Kato and Schroeter 1985). One of the most striking patterns in the subtidal is the urchin "barrens," an area of hard substrate characterized by very high urchin densities and very low cover of foliose algae and sessile invertebrates. The creation and maintenance of these patterns have received a great deal of attention (North and Pearse 1970; Lawrence 1975; Foster and Schiel 1985; Harrold and Pearse 1987).

Where live kelp is abundant, drift kelp is also abundant (Mattison et al. 1977; Harrold and Reed 1985) and urchins tend to be sedentary, feeding on drift algae they capture with their spines and tube feet. However, when drift is no longer available and urchins starve, they become more mobile, form "fronts," and graze algae and invertebrates off the rocks (Dean et al. 1984; Harrold and Reed 1985). Harrold and Reed (1985) were able to reduce the amount of drift algae in an experimental area that changed urchin behavior from sedentary to moving and grazing.

At Naples Reef in Santa Barbara County, grazing by purple urchins resulted in a barrens after a severe storm destroyed the giant kelp canopy (Ebeling et al. 1985). Kelp did not recruit successfully until after the density of urchins was drastically reduced by mortality during a second storm that churned up the bottom. Although the history of Naples Reef suggests that urchins can prevent kelp recruitment, Harrold and Reed (1985) observed recruitment of kelp in spite of high urchin densities. It is clear that there is still much to learn about the dynamics of the urchin–kelp relationship and its interaction with other environmental factors.

The results of several studies suggest that the settling preferences of larvae may determine the distribution of juveniles and adults. For example, juvenile abalone are particularly abundant in red coralline algal mats. Morse et al. (1980) and Morse and Morse (1984) showed that chemical cues present in the algae trigger settlement of the larvae. A study of bryozoans also demonstrated that settling preferences of larvae affect the distribution of later life stages of subtidal invertebrates. Bernstein and Jung (1979) showed that distributional patterns occurring among the bryozoans found on kelp blades within a kelp bed are largely a product of habitat selection by the larvae.

Cameron and Schroeter (1980), however, showed that red and purple urchin larvae tend to settle on surfaces where bacterial films are abundant, but not near or under adult urchins where juveniles tend to occur (Tegner and Dayton 1977; Schroeter 1978). Therefore, the microhabitat distribution of juveniles must be due to postsettlement events. On a larger scale, adult purple urchins are generally many times more abundant in barren areas than in kelp beds, but newly settled juveniles occur in high numbers in both habitats (Rowley 1989).

In some species, young individuals and adults are found in completely different habitats. For example, young lobsters are found in subtidal beds of surfgrass rather than in the immediately adjacent kelp bed habitat occupied by the adults of the population (Engle 1979). Some sea stars also appear to have nursery grounds that are different from the habitats used by adults. *Pisaster giganteus* recruit on the blades of giant kelp plants (T. Herrlinger, S. Schroeter, J. Dixon, and R. Smith unpubl. data). Although it seems likely that these recruitment patterns reflect the settlement patterns, this has not been demonstrated.

Mutualistic interactions have not received much attention but are probably common and important. Vance (1978) removed the epibionts from shells of the rock jingle *Chama pellucida*. By doing so, he increased predation on the jingle by the sea star *Pisaster giganteus*. Vance (1978) demonstrated that the *Chama* shells were preferred to cleared rock as a settling surface, which suggested that *Chama* and its epibionts have a mutualistic relationship.

Osman and Haugness (1981) discovered a mutualistic interaction between the bryozoan *Celleporaria brunnea* and the hydroid *Zanclea* sp. *Zanclea* polyps on the bryozoan sting small predators and nearby competitors, which increases the survival of *Celleporaria* and helps it to overgrow competing species. Because *Celleporaria* is a substrate on which *Zanclea* can grow, both hydroid and bryozoan occupy more space than they otherwise would.

The importance of physical disturbance and subsequent biological events in structuring subtidal communities is just now becoming appreciated. Studying the encrusting community composed of small sessile invertebrates and algae, Breitburg (1983, 1985) demonstrated that species composition within the community differed greatly, depending on whether the newly exposed rock surfaces were subsequently subject to grazing by urchins. If subject to grazing, the surfaces became dominated by crustose coralline algae which, in turn, limited both the numbers and types of invertebrates that could colonize the space. Mediated by the crustose algae, the effect of grazing was present for significant periods of time, even when the grazer was no longer active. Breitburg (1983, 1985) points out that the eventual species combinations found on rock surfaces are very difficult to predict because of the complex interactions between established populations and recruiting species. The results of these studies of small spatial scale patterns provide some insight into the complex processes that govern the establishment of the larger scale reef-to-reef patterns in community structure.

Sandy Subtidal

VERTICAL ZONATION

In shallow waters, the epifauna on sandy bottoms is well developed and dominated by suspension feeders. In water deeper than about 10 m, the abundance of epifaunal animals declines markedly and most of those present are carnivores and scavengers (Morin et al. 1985, 1988). Apparently, the epifaunal suspension feeders have a partial refuge from predators in the harsh physical environment characterized by strong wave surge, but they suffer intense predation in deeper, lower energy waters (Morin et al. 1985; VanBlaricom 1978, 1982).

Whereas the density of epifaunal species declines with depth, that of the infauna increases (Barnard 1963), probably because of the greater stability of the sediments. There also appears to be a shift from infaunal communities dominated by crustaceans to communities in which polychaetes are predominant (Oliver et al. 1980). Oliver et al. (1980) ascribed this pattern in Monterey Bay to the effects of substrate disturbance. In shallow areas (<14 m), the abundant animals were small deposit-feeding amphipods and ostracods. Few of the species present occupied tubes, although in the deeper polychaete zone (>14 m), most animals established and maintained tubes or burrows that would be destroyed by shifting sediments in shallower water. In San Diego County (Dexter 1978; VanBlaricom 1978), amphipod crustaceans dominated infaunal assemblages at both shallow (<8 m) and moderately deep (17 m) sites, but the densities of common infaunal species were about two orders of magnitude higher at the deep site (table 8.5). Polychaetes dominated the infaunal assemblage in samples from sites deeper than about 20 m.

The zonation along the complex depth gradient that is apparent for higher taxa and trophic groups is also found within genera. There are species replacements with depth among the starfishes *Astropecten* spp. (Davis 1978; Morin et al. 1985, 1988), the amphipods *Paraphoxus* spp. (Barnard 1963; Oliver et al. 1980), and the gastropods *Olivella* spp. (Morin et al. 1988). There are undoubtedly many more examples. The mechanisms underlying these zonation patterns have generally not been investigated. In the case of *Astropecten*, Davis (1978) found that the deeper water species *Astropecten verrilli* was less tolerant of wave surge than was *Astropecten armatus*. However, *A. armatus* was also dominant in agonistic encounters, and one might expect it to occur at both depths. Davis speculated that *A. verrilli* was better at exploiting common food resources, so the distribution of *A. armatus* may ultimately be determined by food supply.

Changes in density with depth are also apparent on a smaller scale. In shallow water at Zuma Beach, sand dollars (*Dendraster excentricus*) formed a dense bed whose boundaries fluctuated seasonally. In the summer, the bed occurred at depths between 6 and 9 m. Its offshore boundary expanded 30–80 m seaward in the winter to depths of 11–12 m (Morin et al. 1985). The most common species inshore of the bed (at depths between 2.6 and 6.4 m) were juveniles of sand dollars and the pennatulacean *Renilla kollikeri*. Pismo clams (*Tivela stultorum*) were also found occasionally. Other common species inshore of the bed were the hermit crab *Holopagurus pilosus*, the snail *Nassarius fossatus*, and the barnacle *Balanus pacificus*. The latter were almost always found fouling the tests of sand dollars and were seen on every survey. The sea stars *Pisaster brevispinus* and *Pisaster gigan-*

Table 8.5. *Common Infaunal Invertebrates in the SCB at a Shallow (<7.5 m) and a Deeper Site (17 m)*[a]

	Imperial Beach (1–7 m)		SIO (17 m)	
Taxonomic Group	Number m^{-2}	% of Total	Number m^{-2}	% of Total
Polychaetes	99	5	2400	16
Crustaceans	219	56	10900	72
Molluscs	42	11	1500	10
Others	34	9	400	3

[a]The shallow site (Imperial Beach) was sampled in 1976 (Dexter 1978), and the deep site (Scripps Institution of Oceanography) from 1974 to 1977 (VanBlaricom 1982).

teus, which feed on juvenile sand dollars, were also found occasionally.

Offshore of the bed, in depths between 9 and 13 m, the most common species was *Stylatula elongata*, a suspension-feeding sea pen. The other common species were the snail *Polinices altus* (a predator and scavenger), the predatory sea stars *Astropecten armatus* and *A. verrilli* (the latter being about five times more abundant than the former), and three brachyuran crabs (*Cancer gracilis, Heterocrypta occidentalis*, and *Loxorhynchus grandis*). The tube-building polychaete *Diopatra ornata* was most abundant seaward of the bed, but was also found inside the outer margin.

The nudibranch *Armina californica*, a major predator on juvenile *Renilla*, had a bimodal distribution (Kastendiek 1975, 1976, 1982). It was rare within the sand dollar bed and abundant inshore and offshore of it. Peak abundances inshore of the bed occurred during the calm summer months (Morin et al. 1985).

HORIZONTAL PATCHINESS

Broad expanses of apparently suitable habitat occur that are devoid of sand dollars (Merrill and Hobson 1970; J. Kastendiek pers. comm.), and data from offshore of Scripps Institution of Oceanography (SIO) provide evidence that a well-developed sand dollar bed can persist for long periods of time then disappear. Analysis of size distributions at Imperial Beach suggests that sand dollar beds may form as a result of recruitment episodes spaced 4 years or longer. Significant asynchrony in these recruitment events occurs on a scale of less than 1 km (Dexter 1978). The patchiness of sand dollar populations is also remarkable because of its effect on a host of other species in the community, as illustrated by the studies at Zuma Beach (Morin et al. 1985).

A striking example of large-scale horizontal patchiness is the dominance of the sea anemones *Harenactis attenuata* and *Zaolutus actius* in shallow water at SIO (table 8.6) (Fager 1968; Davis and VanBlaricom 1978). Although both species were present at Zuma Beach, they were much less abundant and occurred along the seaward edge of the sand dollar bed in depths of 10 m or greater (table 8.6) (Morin et al. 1985).

Fager (1964) described several extremely patchy distributional patterns. At his SIO study site, the polychaete *Owenia fusiformis* occurred mainly in an elliptical patch that measured 150 m by 60 m. Densities were 500–1000 m^{-2} within the patch. The anemone *Zaolutus actius* used the *Owenia* tubes as attachment sites and was present at a density of about one per tube. Together these species stabilized the sandy bottom to such an extent that there were no ripples on the sand surface, little sand was in suspension, and a brown diatom film was often present. Breeding individuals of an isopod, a pycnogonid, and a

Table 8.6. *Abundances of Common Invertebrates in Shallow Sandy Subtidal Habitats*[a]

	Scripps Institution of Oceanography				Zuma Beach 1971–1974[3]		Imperial Beach 1974[4]	
	1957–1963[1]		1974–1975[2]					
Species	Number m^{-2}	(Depth) (m)	Number m^{-2}	(Depth) (m)	Number m^{-2}	(Depth) (m)	Number m^{-2}	(Depth) (m)
Anthozoans								
Harenactis attenuata	6.48	5	3.37	5	P	11	—	—
Zaolutus actius	1.76	5	0.30	5	P	11	—	—
Renilla kollikeri	1.68	5	1.93	5	0.52	4.3	P	—
Stylatula elongata	P	>13	160.0	>13	0.60	11	—	—
Polychaetes								
Owenia fusiformia	750.0	5	0.63	5	—	—	P	—
Diopatra ornata	—	5	—	5	1.12	9.2	—	—
Diopatra splendidissima	P	5	1.5	>13	—	—	—	—
Magelona pitelkai	—	—	—	—	—	—	12.1	4.5
Rhynchospio arenicola	—	—	—	—	—	—	11.5	4.5
Thalanessa spinosa	—	—	—	—	—	—	9.5	4.5
Cirripedia								
Balanus pacificus	P	5	—	—	P	9.2	—	—
Anomurans								
Holopagurus pilosus	0.26	5	04.0	5	0.002	7.5	P	4.5
Blepharipoda occidentalis	P	5	0	5	P	4.5	P	4.5

Brachyurans								
Cancer gracilis	—	5	16.3	5	P	11	P	4.5
Heterocrypta occidentalis	—	5	12.8	5	P	11	—	—
Loxorhynchus grandis	—	5	0	5	P	7.5	—	—
Molluscs								
Bivalves								
Tivela stultorum	—	5	—	5	0.45	4.3	—	—
Gastropods								
Nassarius fossatus	0.07	5	0.003	5	P	7.5	2.6	4.5
Nassarius perpinguis	0.37	5	0.02	5	P	11	P	4.5
Polinices recluzianua	0.10	5	0.003	5	—	—	3.1	4.5
Polinices altus	—	—	—	—	0.03	11	—	—
Armina californica	P	5	0	5	0.05	(a)	—	—
Olivella biplicata	—	—	—	—			1.2	4.5
Echinoderms								
Echinoids								
Dendraster excentricus	(b)	13	0.29	5	930.0	7.5	13.9	4.5
Lovenia cordiformis	—	—	2.38	>13	—	—	—	—
Ophiuroids								
Amphiodia occidentalis	0.65	5	0.37	5	P	11	P	4.5
Amphiodia digitata	—	—	—	—	—	—	1.6	4.5
Asteroids								
Astropecten armatus	0.06	5	0.04	5	0.05	9–13(c)		—
Astropecten verrilli	—	5	—	5	0.25	9–13(c)		—
Pisaster brevispinus	—	5	—	5	P	7.5(c)		—

[a]Densities and depths are for peak abundances for each site by time period combination: (a) = bimodal distribution peaking at 4.3 and 11 m depths; (b) = dense sand dollar bed present, but abundances not estimated; (c) = sampling technique unable to confirm absence; P = present.

Sources: [1] = Fager 1968; [2] = Davis and VanBlaricom 1978; [3] = Morin et al. 1985; [4] = Dexter 1978.

hermit crab species were often seen in the bed but not elsewhere at the site.

TEMPORAL CHANGE

Fager's (1964) observations of *Owenia fusiformis* also provide a striking example of temporal patchiness. When first observed in June 1956, this bed was probably several years old. It disappeared in less than a year, apparently due to predation by the rays *Urolophus halleri* and *Myliobatis californicus*. In October 1956, depressions appeared in the bed and rays were more common in the area. By February 1957, the bed was well broken up, although densities within clumps were still high. However, in a period of about 3 weeks during the following summer, the density plummeted to 2.2 worms per square meter. Twenty years after it was destroyed, it was still absent (Davis and VanBlaricom 1978).

From March 1974 through September 1975, Davis and VanBlaricom (1978) resampled the transects Fager (1964, 1968) had established in 1957 and sampled through 1964. Although most species that Fager observed were still present, their densities had generally declined considerably and the densities on the two transects were no longer similar. In addition, the dense sand dollar bed had disappeared. Populations that appeared temporally stable for 6 years turned out to be temporally patchy over a period of 19 years.

MECHANISMS PRODUCING COMMUNITY PATTERNS

Physical Factors. Many changes in the physical environment are associated with increasing depth. One of the most important, especially in sandy areas, is the decrease in average orbital velocities and, thus, surge with increase in depth. In habitats where the average orbital velocity is low, sand is finer and more stable and is characterized by a reducing layer that is closer to the surface.

The distribution of sand dollars and their orientation to the substrate provide several good examples of the effects of wave and current regimes on distribution. In the protected outer coast of the SCB (including Zuma Beach), the shoreward limit of the sand dollar beds occurs near the breaker line at mean lower low water. Most of the sand dollars at the shoreward edge of the bed are small juveniles that remain buried. Proceeding seaward, sand dollars become progressively larger and more abundant (Merrill and Hobson 1970). The same distributional pattern was observed 10 years later at Zuma Beach (Morin et al. 1985).

Two mechanisms, both related to a response to wave action, seem to set the shoreward distributional limit of sand dollars. Morin et al. (1985) found that during storms or experimental transplants, sand dollars respond to the higher wave action and sand movement inshore by actively migrating to deeper waters. Records of dispersion along the seaward margin of beds after heavy storms (Merrill and Hobson 1970; Morin et al. 1985) also suggest the role of passive sorting.

In addition to setting the shoreward distributional limit, the wave regime also influences the position of individual sand dollars on and in the substrate. In bays and relatively calm beaches along the outer protected coast, sand dollars are either buried or assume a horizontal position on top of the substrate. As waves become bigger and currents stronger, sand dollars assume an upright position with the anterior end in the sand and the oral surface forming an acute angle with the substrate. The mouth is usually oriented toward the prevailing currents or wave surge (Merrill and Hobson 1970; Timko [O'Neill] 1976; O'Neill 1978).

Evidence exists that wave action limits the shoreward distribution of several of the important predators in the sand bottom community. These include the gastropod *Armina californica*, which is the primary predator on juvenile *Renilla* (Kastendiek 1982), and the sea stars *Pisaster brevispinus* and *Pisaster giganteus* (J. Kastendiek pers. comm.) and *Astropecten armatus* and *Astropecten verrilli* (Davis 1978; Kastendiek 1982; Morin et al. 1985). *A. verrilli* is more severely constrained by wave action than *A. armatus* (Davis 1978).

The effects of wave action on the infauna are not as well documented and tend to be based on correlational arguments. Oliver et al. (1980) ascribed the decline in the abundance of tube-building polychaetes in shallow waters in Monterey Bay to increasing substrate disturbance, which is positively correlated with wave action. This hypothesis is supported by observations at Zuma Beach and SIO, where abundances of the tube-building polychaetes *Diopatra ornata* (Morin et al. 1985) and *Diopatra splendidissima* (Davis and VanBlaricom 1978) were greatest in depths greater than 10 m and decreased shoreward. In the case of *D. ornata*, however, the pattern was partly due to preemption of space by sand dollars (Morin et al. 1985), as discussed in the following section.

Biological Interactions. A number of studies performed in shallow subtidal habitats at Zuma Beach clearly demonstrate the important effects of biological interactions on patterns of distribution and abundance. However, in almost every case, these biological interactions are strongly mediated by physical factors, usually those associated with wave action.

Predation clearly affects the distribution and abundance of sand dollars (*Dendraster*). Despite the fact that small sand dollars settle both seaward and shoreward of adult beds, they survive well only in the shoreward areas. Predation is the major factor limiting the seaward distribution of the recruits. *Pisaster brevispinus* and *P. giganteus* prey on all sizes of sand dollars within the bed and along the seaward and shoreward margins. The snail *Balcis rutila* also preys on sand dollars and *Astropecten* (Davis 1978). In turn, *Astropecten* prey heavily on various gastropods such as *Olivella* (Davis 1978) and the sea pansy *Renilla kollikeri* (Kastendiek 1982). They also feed on juvenile sand dollars and thereby limit recruitment to very shallow water inshore of the adult bed. These inshore areas act as refuges for two reasons. First, the bed of adult sand dollars serves as a physical barrier to sea stars (Kastendiek 1982). Second, any stars that manage to cross the barrier find themselves in shallow water, where their foraging activities are constrained by wave action.

Predation mediated by wave action also affects the distribution and abundance of the sea pansy *Renilla kollikeri*. The inshore distributional limits of *R. kollikeri* and its two main predators, *Astropecten armatus* (= *brasiliensis*) and the nudibranch *Armina californica*, are all set by intolerance of the predators to wave action. As was the case for sand dollars and their sea star predators, the sand dollar bed acts as a physical barrier to the sea pansy's predators, and wave action sets a deeper inshore limit for the predators than for the prey, thus providing an inshore refuge (Kastendiek 1982; Morin et al. 1985). Near their inshore limits of distribution, many *Armina* and *Astropecten* bury themselves in the substrate, which prevents them from foraging effectively on *R. kollikeri* (Kastendiek 1982). The proportion of predators burying themselves is positively correlated to wave height (Kastendiek 1982).

The studies at Zuma Beach also provide experimental evidence for the importance of competition. Kastendiek's (1982) experimental removals of sand dollars showed that they physically exclude *R. kollikeri* and *Astropecten*. In addition, observations of shifts in distributions in response to seasonal changes in the sand dollar bed suggest that competition with sand dollars may limit the shoreward distributions of the snails *Polinices altus* and *Megasurcula carpenteriana*, the asteroids *Astropecten verrilli* and *Luidia foliolata*, the crab *Heterocrypta occidentalis*, the tube-building polychaete *Diopatra ornata*, and the sea pen *Stylatula elongata* (Morin et al. 1985). Exclusion does not appear to be as complete for the latter two species, which are small and narrow enough to live among the sand dollars (Morin et al. 1985). This is particularly true for *D. ornata*, which can be found in moderate abundances within the sand dollar bed near its seaward margins (Morin et al. 1985).

Perhaps one of the more intriguing results of the Zuma Beach studies is the evidence for the opposing effects of sand dollars on the distribution and abundance of sea pansies. Although interference competition with sand dollars reduces *R. kollikeri* abundance within the bed, sand dollars also tend to form a protective physical barrier against predators. Kastendiek (1982) speculated that, whereas the immediate effect of removing sand dollars would be to increase the abundance of *R. kollikeri,* populations would eventually decline in response to increased predation following the removal of the barrier to predators.

Compared to the epifaunal sand dollar community, much less is known about biological interactions among infaunal species. Nevertheless, it appears that predation operates here as well and is mediated to some degree by wave action or associated factors. Much of our information on biological interactions comes from studies conducted in 17 m of water offshore of SIO (VanBlaricom 1978, 1982). The abundant predators in this system were the sea star *Astropecten verrilli*, the crab *Portunus xantusii*, the small flatfish *Citharichthys* spp., and the two large rays *Urolophus halleri* and *Myliobatus californicus*. The effect of these predators on the distribution and abundance of the infauna appeared to be indirect. *Citharichthys* spp., the marine analog to cattle egrets or antbirds, prey primarily on crustaceans stirred up by the feeding activities of the larger rays. However, no evidence suggests that their predation affects distribution or abundance. Predation by *A. verrilli* clearly affects the abundance of newly recruited *Cancer gracilis* and *Portunus xantusii* and may ultimately limit populations of these predators. This could result in an indirect effect on the prey of the crabs (VanBlaricom 1982).

The most important indirect effect of predation on infaunal communities resulted from the disturbance caused by the feeding of large rays. A predictable succession of species occurred over a 4–6 week period in the pits created by the feeding rays (VanBlaricom 1978, 1982). Two early waves of recruitment, one peaking 3 days and the other 1–3 weeks after the formation of the pits, were composed of small motile crustaceans, chiefly amphipods but also ostracods, cumaceans, and tanaidids. After 4–6 weeks, the most abundant species were more sedentary crustaceans, a bivalve (*Tellina modesta*), and a polychaete (*Mediomastus acutus*) (VanBlaricom 1982). The rarity or absence of late colonizers in new ray pits seems to be explained by their limited colonizing ability. However, the mechanisms responsible for the temporal replacement of early colonizers have not been clearly demonstrated. VanBlaricom (1982) provides some evidence that food quantity and quality decline as the feeding pits age and fill in with sand, suggesting that when this occurs, resource levels may be too low to support populations of the early colonizers.

Deep Benthic Assemblages

Deep Rocky Substrate Assemblages

Beyond the depths of kelp beds (> 30 m), approximately 3% of the sea floor is rocky outcrops, rubble, and talus inhabited by marine invertebrate assemblages (fig. 8.5a). On the mainland shelf, these rocky areas are generally interspersed with soft substrate such as sand or gravel (fig. 8.6). Offshore, the Channel Island shelves, Santa Rosa–Cortes Ridge, and Tanner and Cortes banks are composed primarily of base rock and rocky outcrops that may be covered with a thin veneer of sediment. Deep, hard substrate assemblages are the least-studied benthic habitats in the region because they exceed SCUBA diving depths and cannot be easily sampled with grab or coring devices or trawls. Hard substrates occur to depths over 500 m in the region and include seamounts and man-made structures. Mainly because of interest in de-

(a)

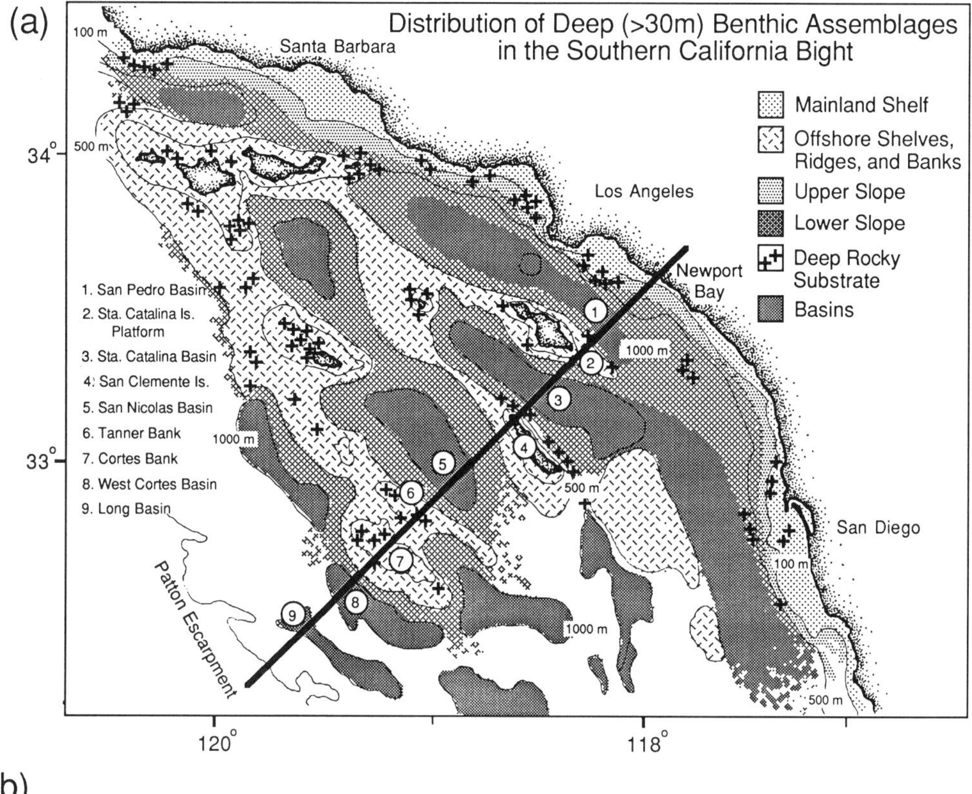

(b)

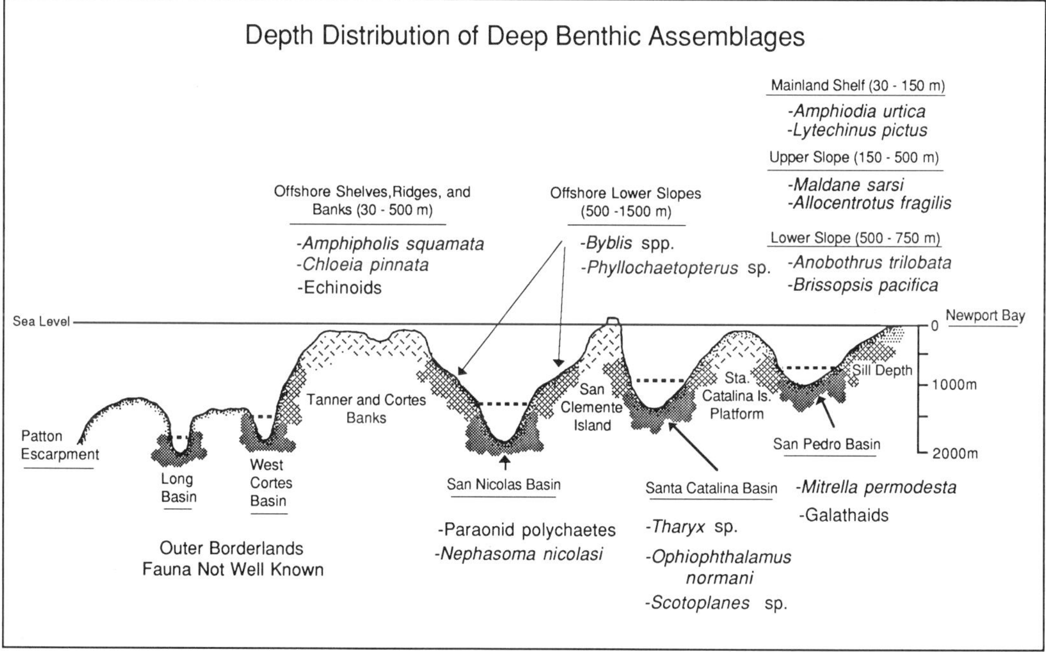

Figure 8.5. (a) Chart of the southern California borderland showing the distribution of the major deep benthic assemblages (deeper than 30 m). Heavy transect line is shown on cross-section in part (b). (b) Cross-section of the southern California borderland showing depth distribution of major deep benthic assemblages and characteristic fauna.

Figure 8.6. Photograph of mixed sandy, rocky benthic assemblage in outer, central Santa Monica Bay (110 m). The regularly spaced ophiuroids are *Ophionereis* sp. (SCCWRP photograph.)

veloping off-shore oil, the first deep rocky samples were collected in the 1970s (S. Smith et al. 1975). This and all subsequent studies were descriptive; they surveyed assemblage composition and structure.

The main sampling device for these habitats is the remote camera. Most early studies were not quantitative. However, quantitative phototransect methodologies are currently being developed (Science Applications International Corp. 1986; Kinnetics Laboratories Inc. 1988). The use of submersibles and remotely operated vehicles (ROVs) has allowed the collection of many invertebrate species, but because these areas have not been well sampled previously, many new taxa are being discovered. Only about 45% of the species collected in the Santa Maria Basin were described taxa (Science Applications International Corp. 1986).

Deep hard substrate assemblages have been studied in two areas of the SCB: (1) the northern Santa Barbara Channel, including the Point Conception–Point Arguello area, and the western end of San Miguel Island; and (2) Tanner and Cortes banks. Other areas are partially known from photographs or incidental samples.

SPECIES COMPOSITION AND ABUNDANCES

In the northern Santa Barbara Channel, deep hard substrate assemblages have different species composition and abundances with increasing water depth and with changes in the relief of the rock substrate (Nekton Inc. and Kinnetics Laboratories Inc. 1983, 1984; Science Applications International Corp. 1986). Three different hard substrate assemblages were described from outer shelf–upper slope depths (105–213 m): (1) a medium low-relief assemblage composed mostly of ophiuroids, brachiopods, and anemones; (2) a high medium-relief assemblage composed mostly of anemones (*Corynactis californica*) and corals (*Lophelia californica*); and (3) a generally distributed assemblage dominated by the crinoid *Florometra serratissima*, the anemone *Metridium senile*, and cup corals. Differences among these assemblages were attributed to tolerances of the taxa to potentially high sediment loads and scour (Science Applications International Corp. 1986).

More recent studies described only two hard bottom assemblages that were depth related (table 8.7). The dominant taxa composed a Komokoiacean (Foraminifera)–hydroid mat that encrusted over 72% of the rock surfaces and was present at all of the 11 sites sampled. These mats nearly disappeared during the 1982 El Niño. Other dominant species included the ophiuroid *Ophiacantha diplasia*, the crinoid *Florometra*, and an undescribed anemone (table 8.7). The most abundant of the taxa listed in table 8.7 occurred at all of the sites, and no one taxon was dominant at any of the study sites. Patches (several individuals) of *Florometra* and the brachiopod *Terebratulina* were observed. Abundances and percentage of cover increased with water depth and vertical relief of the substrate; high relief apparently amplified the effects of depth (Hardin et al. 1988). Depth was the most important factor in determining species composition. In contrast to previous studies (Science Applications International Corp. 1986), however, distinct high- and low-relief assemblages were not apparent.

The northwestern end of the San Miguel Island platform, on the lower slope-sill (290–625 m), was composed of rocky outcrop and small cobbles, boulders, and slabs (<1 m). Some of the outcrops had vertical ledges of >2 m (Parr and Shrake 1989). The bottom was dominated (>80%) by a low-growing turf (possibly the Komokoaicean–hydroid mat), anemones, amphipods, polychaetes, and ectoprocts. Larger species included sponges, gorgonians (*Stenella* sp.), *Florometra*, and the anemone *Stomphia* (fig. 8.7). A high degree of commensalism was evident among the large sponges, which provided structure and habitat for a variety of other animals such as crinoids, shrimp, and ophiuroids. In contrast to hard substrate assemblages in shallower water (105–213 m), cup corals, anemones, brachiopods, and lithotid and galatheid crabs were much less prevalent.

Numerous samples were collected from the Channel Island shelves during the Bureau of Land Management (BLM) baseline and benchmark studies, using a box corer (Fauchald and Jones 1979b, 1983). Grab samples that contained rock or cobble were characterized by the ophiuroids *Ophiopholis bakeri* and *Amphipholis squamata*. One of the most significant discoveries in these samples was of the monoplacophoran *Vema* (*Laevipilina*) *hyaling* from the outer Santa Rosa–Cortes Ridge (McLean 1979).

Deep hard substrate assemblages of Tanner and Cortes banks change with water depth, as do the assemblages in the nearshore areas (S. Smith 1976; S. Smith et al. 1975). At shallow depths (14–40 m), algae such as *Eisenia arborea*, erect corallines, and *Corynactis californica* are dominant. Algae are dominant to about 90 m, below which invertebrates, particularly *Florometra serratissima* are dominant (Lewbel et al. 1981; Lissner and Dorsey 1986).

The hard substrate assemblages of the offshore ridge and banks are similar in species composition to those on the nearshore areas (table 8.7). Depth ranges and abundances of most species increase in the offshore areas, largely due to the lack of sediments from terrestrial sources and increased nutrition from the productivity of the offshore areas. Algae occur deeper where there is increased light penetration and nutrition.

Although few specific feeding studies have been conducted, Hardin et al. (1988) and Minerals Management Service (1988) concluded that most deep hard bottom species are suspension feeders. Sponges, corals, gor-

Table 8.7. *Most Common and Abundant Invertebrates from Deep Hard Substrates*

Location	Species	Taxon	Reference
Point Conception	*Scrupocellaria varians*	Ectoprocta	Science Applications International Corporation 1986
	Crisia maxima	Ectoprocta	
	Micropora coriacea	Ectoprocta	
	Diaperoecia californica	Ectoprocta	
	Anguinella palmata	Ectoprocta	
	Pholoides aspera	Polychaeta	
	Hiatella arctica	Pelecypoda	
	Amphipholis squamata	Echinodermata	
	Perusa papillata	Polychaeta	
	Platidia hornii	Ectoprocta	
	Sponge sp. #75	Porifera	
	Komokoiacea–hydroid mat	—	Mineral Management Service 1988
	Ophiuroidea	—	
	Florometra serratissima	Echinodermata	
	Sabellidae	Polychaeta	
	Anemone sp. #25	Cnidaria	
	Metridium senile	Cnidaria	
	Paracyathus stearnsii	Cnidaria	
Tanner and Cortes banks	*Ophiothrix spiculata*	Echinodermata	Lissner and Dorsey 1986
	Mediaster aequalis	Asteroidea	
	Patiria miniata	Asteroidea	
	Plumarella longispina	Ectoprocta	
	Sponges	Porifera	

gonians, anemones, ophiuroids, and crinoids are all functional suspension feeders. Currents strong enough to erode sediments away and expose base rock probably also carry high suspended food loads and favor suspension feeders. Suspension feeders were more abundant on high-relief features than on low-relief features. This may be the result of adverse effects of sedimentation on low-relief areas or of greater food availability on high-relief substrates.

Few studies of life history and no experimental studies of ecological interactions of deep hard substrate organisms have been done in the region. Most of the information available was summarized by Minerals Management Service (1988). Their summaries for some of the most common and abundant deep hard substrate species are listed in table 8.8.

Several types of disturbances can affect the distribution and abundance of deep hard bottom species. Three sources of disturbance to assemblages of these species include: (1) sedimentation effects; (2) anchoring effects; and (3) hydrocarbon toxicity effects (see chap. 12). Disturbances create space on the substrate that is available for new organisms to inhabit. Which organisms inhabit the space depends on many factors such as type and size of patch, frequency of disturbance, availability of propagules in the plankton, water movement, and larval settlement. Postsettlement processes (succession) within patches

Figure 8.7. Photograph of deep hard substrate benthic assemblage off Point Arguello, 213 m, October 1988. The large, light-colored anemones are *Stomphia didemon* and the smaller ones are *Amphianthus californicus*. Also prominent is the sponge *Leucetta* and the ophiuroid *Ophiacantha*. (Photograph courtesy of Kinnetics Laboratories.)

mitigate the eventual composition of the assemblage (Minerals Management Service 1988).

Recolonization from physical disturbances on high-relief substrates (that is, anchor scars) can occur initially by larvae from nearby species with short-range dispersal (anemones and cup corals) and by vegetative growth of bordering species (sponges and anemones). Motile predator-scavengers (such as echinoderms) as well as larvae of long-range dispersal forms (such as hydroids, ectoprocts, and asteroids) would also colonize. Recolonization of low-relief base rock (after overlying sediments have eroded) can occur initially by larvae from more distant rocky areas, by long-range dispersal forms, and by asexual reproduction of existing species such as anemones. Motile predator-scavengers would then emigrate (Minerals Management Service 1988).

Deep Soft Sediment Assemblages

Extensive areas (over 100,000 km^2) of deep benthic habitats (>30 m) exist in the SCB (fig. 8.5a,b). These areas vary in water depth, sediment type, organic content, influence of terrestrial sedimentation, oceanographic conditions such as upwelling, and contaminant inputs from large metropolitan areas. As a result, the soft substrate benthic assemblages of the region are complex and diverse. However, the degree to which oceanographic factors influence species composition and di-

Table 8.8. *Summary of Life History Information of Marine Benthic Invertebrates Representative of Deep-Water Hard Substrate Communities of the SCB (after Minerals Management Service 1988)*

Taxa	Feeding/ Trophic	Growth/Longevity	Reproduction/Dispersal/ Recruitment	Motility	Competitive Interactions	Potential for Recovery Following Disturbance
Porifera (sponges)	Filter feeder	Variable to spp.	Sexual reproduction includes both internal and external fertilization; some brood; release flagellated larvae which last from hours to a few days, then become crawling larvae and undergo metamorphosis. Asexual reproduction including fragmentation, budding, and formation of gemmules and pseudolarvae.	Sessile	Sponges are preyed on by nudibranchs, chitons, sea stars, and fishes; they contain many chemicals to deter grazing and predation.	
Cnidaria *Allopora* spp.	Passive filter-feeding planktivore	Slow growth, long-lived once established; few colonies survive first year. Colonies 30 cm height at least 20 years old.	Retained tiny medusae produce gametes and brood larvae. Demersal larvae settle within 1 day; limited dispersal.	Once established, permanently attached.	No observed predators; young colonies killed by being overgrown or by sedimentation.	Recovery would probably require many years.
Muricea spp.	Passive filter-feeding planktivore	Upright, branching, grow to large size; slow growth—5 to 10 years to reach sexual maturity in *M. californica*.	Sexes separate; long-lived larvae (30 days) for broad dispersal.	Once established, permanently attached.		
Metridium senile (plumose anemone)	Passive filter-feeding zooplanktivore	Long-lived, medium fast growth days to weeks. Asexual reproduction–pedal laceration.	Sexual reproduction; external fertilization; planula larvae planktonic.	Limited mobility, pedal crawling; adults apparently can detach and drift or roll to new location.	"Catch" tentacles for inter/intraspecific aggression; acontia for interspecific defense; preyed on by nudibranchs and asteroids.	Potential for rapid recovery by adult migration; longer recovery via larval recruitment (years).
Balanophyllia elegans (orange cup coral)	—	Solitary coral; long-lived, slow growth rate.	Short-lived planula larvae, limited dispersal; tolerance for sediments unknown.	Permanently attached; broken individuals not likely to recolonize/ attach.		Dispersal very limited; years to recover.

Echinodermata *Florometra serratissima*	Passive filter/ suspension feeder; ciliary mucus feeder; protozoan and microzoo-planktivore	Long-lived	Sexes separate; external fertilization; larvae short-lived—limited dispersal; regeneration of body parts common.	Attached by cirri—creeping rate 40 m h^{-1}; some individuals may "swim."	Preyed on by some gastropods; adult immigration may be important to colonize new areas.	Limited larval dispersal once established; recovery in a few years.
Mediaster aequalis (red sea star)	In rocky habitats, prefers encrusting sponges and ectoprocts	*Mediaster* 18–20 cm at least 10 years old.		Mobile		
Amphipholis squamata	Suspension/ filter feeder and sediment detritivore	Small species, rapid growth	Protandry—internal fertilization; ovoviviparous, direct development, limited dispersal.	Mobile	Preyed on by fishes.	Rapid recovery once recruited to an area; may raft to new areas on debris, flotsam.
Ophiothrix spiculata (spiny brittle star)	Suspension/ filter feeder and surface deposit feeder		Sexes separate; external fertilization; free-swimming and feeding planktonic larvae.	Mobile	Preyed on by sea stars, fishes, crabs.	
Annelida *Eudistylia polymorpha* (feather-duster worm); *Sabella crassicornis*	Filter feeder	Large, slow-growing, annual to longer-lived species.	Hermaphroditic, short-lived planktonic larvae; limited dispersal; asexual fission.	Sessile; discreetly motile.	Fish predation on anterior portion of worms; regenerate new parts.	
Ectoprocta (bryozoa or moss animals)		Colonial growth forms, either sheet-like encrusting or upright branching forms. Generally short-lived, rapidly overgrown by other species; early colonizers.	Sexual reproduction for new colonies; asexual reproduction to increase colony size. Many different types of larvae all swim or move about, providing for broad dispersal.		Most ectoprocts are overgrown by later succession species.	Rapid recovery for most species, weeks to months for small encrusting forms; 1–2 years for larger growth forms.
Diaperoecia californica (southern staghorn bryozoan)	Active impact suspension feeder	Large upright growth; one of the longer lived species.		Sessile		

versity in deep soft substrate habitats is not understood (Spies 1984).

A considerable amount of information exists about animal–sediment relationships, although most of what is known is correlative and not experimental. Multivariate analyses of patterns in sediment quality indicate that water depth, sediment grain size, and organic content account for most of the variation in the sediment parameters and that these patterns correspond to patterns in benthic species composition and abundances (Smith and Greene 1976; Fauchald and Jones 1979b; Thompson et al. 1987a).

No dynamic studies of the effects of sediment deposition on benthic assemblages have been conducted in the region that comprises the SCB. Studies in other areas have shown that assemblage structure, particularly feeding strategies, may be closely tied to modes and rates of sediment transport (Miller et al. 1984). The presence of benthic animals may also modify sediment and sediment transport. Construction of tubes or mucous-lined burrows, feeding, respiratory, locomotory activities, and fecal pellet production by benthic animals can alter sediment quality, transport, and surface microcirculation (Aller and Yingst 1978; Eckman et al. 1981; Nowell et al. 1981).

Macrobenthic assemblages off southern California can be divided into four major benthic habitat types, each with subassemblages, that generally reflect major physiographic and sedimentologic features in the sea floor of the region: (1) mainland shelves; (2) offshore shelves, ridges, and banks; (3) basin slopes; and (4) basin floors (fig. 8.5a, b) (Fauchald and Jones 1979a, b, 1983).

Assemblages of large, motile megabenthic invertebrates in the region are not as well studied as the macrobenthos, especially in the offshore areas and in most of the basins. As with the macrofauna, nearshore megafauna exist in rather distinct shelf, slope, and basin assemblages (Thompson et al. 1993). Classification into nominal benthic assemblages reflects general patterns of benthic species composition and diversity, but does not necessarily imply that distinct zonation occurs. Rather, species abundances change gradually over space, time, and substrate types.

OUTER MAINLAND SHELF

Species Composition and Abundance. Macrofaunal assemblages on the outer mainland shelf (30–150 m) have been sampled extensively (Allan Hancock Foundation 1965; Jones 1969; Fauchald and Jones 1979a, b, 1983; Word and Mearns 1979; Thompson et al. 1987a). Most muddy areas of the outer shelf are inhabited by the red ophiuroid *Amphiodia urtica*, which is usually numerically dominant. They are most abundant between 40 and 80 m, where they may exist in dense populations (average is about 470 m^{-2}, maximum is 1464 m^{-2}). This large-scale heterogeneous assemblage exists on nearly all of the mainland shelf. Other species in this assemblage are listed in table 8.9.

Several other outer shelf assemblages have been described (Barnard and Hartman 1959; Barnard and Ziesenhenne 1961; Jones 1969), but most of these probably represent temporal or spatial subassemblages of the *A. urtica* assemblage. It is the presence and stability of *A. urtica* that unifies the outer mainland shelf assemblage.

Amphiodia urtica and the polychaete *Paraprionospio pinnata* are about the only species that have been consistently collected in shelf-wide surveys. Composition of the subdominant mainland shelf assemblages appears to change over time periods of 5–10 years. The pelecypod *Cyclocardia* (= *Carditu*) *ventricosa* was considered a codominant of the Santa Barbara shelf benthos based on its contribution to the biomass, but its abundances have decreased from about 72 m^{-2} in the late 1950s (Jones 1969) to less than 35 m^{-2} in the late 1970s (Fauchald and Jones 1979a,b; Word and Mearns 1979). The echiuran *Listriolobus pelodes* was very abundant (92 m^{-2}) on the Santa Barbara shelf in 1959–1960 and was considered to be a separate assemblage (Barnard and Hartman 1959). Its abundances appear to fluctuate considerably (Fauchald 1971; Stull et al. 1986a). A large recruitment of *L. pelodes* was observed in 1975 all along the mainland shelf (Pilger 1980; Fauchald and Jones 1983). By 1976, densities had reached an average of 118 m^{-2}, and by 1977 had decreased to 44 m^{-2} (Fauchald and Jones 1979b). The polychaete *Spiophanes missionensis* has increased in abundance from about 23 m^{-2} (Jones 1969) to 161 m^{-2} (Thompson et al. 1987a), becoming second in abundance to *A. urtica*.

The few studies of annual or seasonal variation in species composition and abundance have shown very little short-term variation (Fauchald and Jones 1979a; Stull et al. 1986b). Only a few species, such as the polychaetes *Pectinaria californiensis* and *Myriochele* sp. M, possibly show significant seasonal fluctuations.

Macrobenthic assemblages in areas of high sand content (e.g., off Imperial Beach and in Santa Monica Bay) are different from those in muddy areas. *Amphiodia urtica* is less abundant or absent, and other species occur instead, such as the pelecypod *Tellina modesta*, the gastropod *Caecum crebricinctum*, and the ophiuroid *Amphipholis hexacanthus* (Barnard and Ziesenhenne 1961; Thompson et al. 1987a). Macrobenthic assemblages near waste discharges also differ dramatically. Again, *A. urtica* does not occur, but the pelecypod *Parvilucina tenuisculpta* and the polychaete *Capitella capitata* are numerically dominant (see chap. 12).

There is little similarity between the megafaunal species collected in trawls and the infaunal species collected in grab or core samples. Megafaunal assemblages on the mainland shelf are much more heterogeneous than macrofaunal assemblages. The mainland shelf megabenthic assemblage ranges between 10 and 137 m. The most common and abundant megafaunal species are the white urchin *Lytechinus pictus*, the ridgeback prawn *Sicyonia ingentis*, and the asteroid *Astropecten verrilli* (table 8.10). *S. ingentis* is much more abundant on the outer shelf (45–315 m) in an apparently transitional assemblage that includes both shelf and slope species.

Megafaunal assemblages on the mainland shelf may be affected by severe pollution, El Niño, or strong storms in shallow water (Thompson et al. 1993). Low-diversity megabenthic assemblages existed off Palos Verdes prior to 1980 while the Los Angeles County sewage outfall was discharging highly contaminated effluent. During the 1982–1983 El Niño, abundances of *S. ingentis* and the galathaid *Pleuroncodes planipes* increased by two orders of magnitude. During that same time, exceptionally strong storm surf disturbed the shallowest shelf sites (18–37 m), producing a low-diversity megabenthic assemblage (table 8.10).

Species Diversity and Biomass. Numbers of macrofaunal species, individuals, and species diversity decrease over shelf depths (Fauchald and Jones 1979a,b; Thompson et al. 1987a). Average diversity on the outer mainland shelf is similar to that on the offshore shelves, ridges, and banks (table 8.11). Biomass is extremely variable (up to 76%) due to chance collections of large motile invertebrates such as holothuroids, echinoids, and echiurans.

Most macrobenthic populations (76–97%) on the mainland shelf are randomly dispersed on the sea floor (scale = 50–1000 m) (table 8.12). They are seldom uniformly dispersed. Species with aggregated dispersions are often the result of recruitment (*Amphiodia urtica* and *Chloeia pinnata*) or other life history phenomena such as adult immigration (*Pectinaria californiensis*) (Fauchald and Jones 1979b). Megafaunal species diversity and biomass increase with shelf depth, opposite to macrofauna (table 8.13) (Thompson et al. 1987a). The moderate to low frequencies of occurrence (table 8.9) indicate the patchy nature of megabenthic populations.

Sediment and Trophic Relationships. About equal proportions of outer shelf macrofaunal species are burrowing (38%) or tube dwelling (35%) (table 8.14). The most common and

Table 8.9. *Most Common and Abundant Benthic Macrofaunal Species in Each Major Deep Benthic Assemblage of the SCB*

		Nearshore				Offshore		
Species	Taxon[a]	Mainland Shelf (30–150 m)	Upper Slope (161–632 m)	Lower Slope (480–851 m)	Basins (627–938 m)	Shelves, Ridges, Banks (17–1160 m)	Lower Slope (541–1768 m)	Basins (1357–2571 m)
Amphiodia urtica	Op	470	+	–	–	51	–	–
Spiophanes missionensis	P	230	+	–	–	+	–	–
Paraprionospio pinnata	P	46	13	–	–	+	–	–
Cyclocardia ventricosa	Pe	26	8	+	–	+	–	–
Euphilomedes spp.	Os	106	14	+	–	74	–	–
Listriolobus pelodes	E	50	+	–	–	+	–	–
Pectinaria californiensis	P	60	23			+		
Maldane sarsi	P	40	113	20	–	+	+	–
Lumbrineris spp.	P	+	16	6	+	27	29	1
Saturnia californica	Pe	–	+	14	–	–	–	–
Anobothrus trilobata	P	–	+	52	–	–	+	–
Aplacophora	M	–	+	24	1	–	+	+
Arhynchite californicus	E	–	5	+	–	+	+	–
Mitrella permodesta	G	–	7	33	9			
Phyllochaetopterus limicolus	P	–	–	+	6	–	7	+
Tharyx spp.	P	39	7	6	4	+	29	25
Listriolobus hexamyotus	E	–	+	4	–	+	+	–

Amphipholis squamata	Op	+	–	–	–	52	23	–
Parvilucina tenuisculpta	Pe	188	–	–	–	157	–	–
Chloeia pinnata	P	91	9	–	–	190	–	–
Photis californica	A	+	+	–	–	58	–	–
Byblis spp.	A	+	+	–	–	+	225	–
Paraonidae	P	+	+	+	2	+	8	22
Ophiura leptoctenia	Op	–	–	+	–	+	8	15

[a] Op = ophiuroid; P = polychaete; Pe = pelecypod; Os = ostracod; E = echiuran; M = molluscan; G = gastropod; A = amphipod; + = present, but not numerically dominant; – = absent.

Data are the average number of organisms per square meter summarized from numerous sources: Fauchald and Jones 1979a,b, 1983; Barnard and Hartman 1959; Barnard and Ziesenhenne 1961; Jones 1969; Thompson et al. 1984a, 1987a; and Hartman and Barnard 1958, 1960.

Table 8.10. *Most Common and Abundant Megafaunal Species from Nearshore Shelf, Slope, and Basins*[a]

Species	Taxon[b]	Contaminated Shelf Palos Verdes pre-1980 (23–137 m)	Storm/ El Niño Shallow Shelf post-1981 (18–37 m)	"Normal" Mainland Shelf (10–137 m)	Outer Shelf– Upper Slope (45–315 m)	Mid-slope (300–490 m)	Lower Slope (478–780 m)	San Pedro and Santa Monica Basins (715–878 m)
Portunus xantusii	C	+ (.03)	2 (.55)	1 (.09)	+ (.03)	0	0	0
Astropecten verrilli	E	+ (.09)	7 (.62)	39 (.79)	33 (.45)	0	0	0
Mursia gaudichaudii	C	5 (.59)	+ (.03)	1 (.33)	11 (.60)	0	0	0
Pleurobranchaea californiensis	M	3 (.47)	+ (.03)	2 (.40)	8 (.68)	1 (.32)	0	0
Lytechinus pictus	E	4 (.05)	85 (.52)	414 (.45)	88 (.23)	0	0	0
Sicyonia ingentis	C	5 (.22)	+ (.01)	74 (.64)	782 (.88)	0	0	0
Pleuroncodes planipes	C	+ (.03)	+ (.07)	16 (.03)	662 (.11)	0	+ (.12)	0
Allocentrotus fragilis	E	0	0	+ (.03)	83 (.36)	391 (.96)	1 (.06)	0
Brissopsis pacifica	E	0	0	+	17 (.06)	3919 (1.0)	1225 (.94)	0
Pannychia sp.	E	0	0	0	+	16 (.60)	49 (.88)	0
Ophiomusium jolliensis	E	0	0	0	0	8 (.60)	82 (.88)	0
Ophioscolex corynetes	E	0	0	0	0	+ (.12)	849 (.24)	0
Porifera	P	+ (.03)	0	+ (.06)	+ (.02)	+ (.08)	+ (.82)	+ (.67)
Munida quadrispina	C	0	0	0	+ (.02)	1 (.16)	11 (.59)	115 (.50)
Munidopsis hystrix	C	0	0	0	<1	0	4 (.47)	177 (.83)

[a] Values given are mean number per trawl followed by frequency of occurrence (in parentheses).

[b] C = crustacean; E = echinoderm; M = molluscan; P = Porifera; + = present <0.5/trawl.

Data after Thompson et al. 1993.

Table 8.11. *Averages for Several Assemblage Structure Parameters of the Major SCB Assemblages*[a]

	Average Number of of Species per Core	Average Number of Individuals per m^2	H′ Species Diversity	C Dominance	Average Biomass (wet g m^{-2})
Mainland shelf	72	3914	3.25	0.10	159
Nearshore upper slope	19	615	2.34	0.15	239
Nearshore lower slope	14	555	1.96	0.21	40
Offshore lower slope	28	1253	2.55	0.13	37
Offshore shelves, ridges	53	3077	3.18	0.10	110
Tanner Bank	41	2038	2.69	0.11	65
San Pedro Basin	4	92	0.92	0.43	6
Santa Cruz Basin	16	454	2.18	0.14	24
San Nicolas Basin	9	214	1.40	0.32	15
Abyssal North Pacific[b]	11	115	—	—	—

[a]Data from Fauchald and Jones (1979b), 0.063-m^2 box corer, 1.0-mm screen. H′ is Shannon-Werner species diversity (log *n*); C is Simpson's index of dominance.

[b]Data after Hessler and Jumars (1974), 1/4-m^2 box corer, 0.3-mm screen.

Table 8.12. *Spatial Dispersion of Macrofaunal Populations*[a]

Macrobenthic Assemblage	Percentage of Species with: Aggregation	Random Dispersion	Uniform
Mainland shelf	7.5	92.1	0.4
Nearshore upper slope	3.7	96.3	0
Nearshore lower slope	5.6	94.4	0
Offshore lower slope	6.3	93.7	0
Island shelves	8.0	91.5	0.5
Tanner Bank	10.0	90.0	0
Santa Cruz Basin	9.0	91.1	0
San Nicolas Basin	0	100	0
San Pedro Basin	0	100	0

[a]Scale = <1000 m; *n* = 4 replicates per site, winter and summer averaged.
Data from Fauchald and Jones (1979b).

abundant benthic species (such as those listed in table 8.9) ingest mostly detrital aggregates at the sediment surface. Only a few macrofaunal species (such as the polychaetes *Glycera capitata* and *Chloeia pinnata*) ingest large quantities of particulate organic material or animal remains. The most abundant megafaunal species also ingest mostly sediment, but *Lytechinus pictus* is omnivorous. It also ingests particulate organic matter, animal remains, and larvae of other benthos (Southern California Coastal Water Research Project unpubl. data). Most asteroids and decapods are predator–scavengers (Thompson 1982; Striplin 1987).

Overall, the mainland shelf assemblage

Table 8.13. *Comparison of Mean Number of Species, Individuals, and Biomass of Megafauna per Trawl among Nearshore Deep Benthic Habitats*[a]

Benthic Habitat	Number of Trawls	Number of Species	Number of Individuals	Biomass (wet kg)
Pre-1980 Palos Verdes shelf	78	5.8 (3.7)	68.3 (200.4)	3.0 (3.9)
Storm–El Niño shelf	29	5.6 (3.7)	101.0 (345.2)	0.9 (1.1)
Normal mainland shelf	658	13.1 (6.5)	577.0 (1539.0)	6.6 (11.3)
Outer shelf–upper slope	343	13.7 (6.0)	1869.5 (5347.4)	30.7 (53.5)
Mid-slope	25	18.0 (5.0)	4944.9 (4626.5)	64.2 (73.8)
Lower slope	17	29.5 (11.5)	3874.2 (6408.0)	48.6 (28.9)
Basins	6	11.0 (2.3)	1027.3 (1624.4)	3.9 (4.7)

[a]25-ft otter trawl; distance ~0.5 km. (Standard deviation given in parentheses.)
Data from Thompson et al. 1993.

Table 8.14. *Life Habits of Deep Benthic Assemblages in the SCB*

Benthic Habitat	Percentage of Assemblage (Abundance)				Percentage of Species Ingesting Foraminifera
	Burrowing	Tube Dwelling	Sessile	Surface Motile	
Mainland shelf	37.5	34.5	12.3	15.7	42.6
Nearshore upper slope	52.6	35.9	5.9	5.4	57.0
Nearshore and offshore lower slopes	40.3	26.1	17.5	14.6	65.0
Basins	12.5	33.3	13.9	40.3	73.6
Island shelf, ridge	33.1	36.3	2.9	27.7	76.4
Tanner Bank	23.4	49.0	1.4	26.2	74.5

After Thompson 1982.

obtains most of its food from detrital aggregates ingested at the sediment surface (table 8.15). Only about 22% of the assemblage feed as suspension feeders. Nearly half of the species ingest Foraminifera (table 8.14), but they do not contribute large volumes to diets except for *Listriolobus pelodes* and *Cyclocardia ventricosa* (Thompson 1982).

Mainland shelf species can ingest mineral particles as large as 350 μm. By size, clay minerals (<4 μm) are always the most abundant particles ingested. By volume, however, the modal particle sizes ingested by each species are usually different. Differences in diets, particle ranges utilized, and feeding stratum (water, sediment surface, or subsurface) may facilitate the coexistence of the numerous deposit-feeding species in these benthic assemblages. In addition, many shelf species can apparently switch their diets to whatever food is most available. Many species (such as *Amphiodia urtica* and *C. ventricosa*) can feed either as surface detrital feeders or as suspension feeders. *A urtica* can be induced to suspension feed if current speed is increased. Because of the large contributions of these species to assemblage abundances and biomass, switches in their diets or feeding stratum may shift the feeding function of their assemblage (Thompson 1982).

Life Histories and Ecological Interactions. Studies of life histories of shelf species indicate no general reproductive patterns or strong sea-

Table 8.15. *Percentages of Each Macrobenthic Assemblage (abundance times biomass) That Use Each Food Stratum Category*[a]

Macrobenthic Assemblage	Food Stratum Category: Detrital Aggregates			Single Mineral Particles			Particulate Organic Matter			Animal Remains			Foraminiferans		
	Sus	Sur	Sub	Sus	Sur	Sub	Sus	Sur	Sub	Sus	Sur	Sub	Sus	Sur	Sub
Mainland shelf	18	45	2	3	27	2	+	1	0	0	1	0	1	1	0
Island shelf	5	28	8	1	24	8	0	2	0	0	11	0	+	8	4
Tanner Bank	2	42	2	+	16	3	0	13	0	0	15	0	0	6	1
Nearshore upper slope	7	47	28	2	13	3	0	1	0	0	1	1	+	1	+
Nearshore lower slope	4	65	18	1	7	4	0	+	+	0	0	0	1	1	0
Offshore lower slope	+	51	14	0	11	7	0	9	+	0	1	0	0	4	3
San Pedro Basin	7	83	0	2	7	0	0	0	0	0	0	0	1	1	0
Santa Cruz Basin	0	21	0	0	8	1	0	63	0	0	8	0	0	0	0
San Nicolas Basin	0	74	0	0	8	0	0	17	0	0	2	0	0	0	0

[a] Sus = suspension feeder; Sur = surface feeder; Sub = subsurface feeder; + = trace, <0.5%. Facultative species were apportioned subjectively among categories.

After Thompson (1982), based on gut contents analyses.

sonality. Most species that have been studied have different reproductive cycles or life history strategies. These differences are no doubt interwoven to produce the complex assemblages observed.

The population size structure and growth of *Amphiodia urtica* were analyzed from Puget Sound by Lie (1968). He showed bimodal size frequency distributions and inferred spring recruitment. He calculated an annual growth rate of less than 2.0 mg dry wt yr^{-1}; individuals can live up to 5 years. In the SCB, juveniles have been collected in both winter and summer samples (Fauchald and Jones 1979b), suggesting nonseasonal or continuous breeding.

The clam *Cyclocardia ventricosa* reproduces year-round and broods its larvae. Juvenile clams can attach to the adult with a byssal thread (Jones 1963; Jones and Thompson 1987a). *Listriolobus pelodes* spawns in late winter through July, but the larvae can settle at any time of the year (Pilger 1980). *Lytechinus pictus* spawns in the spring with an annual late summer larval settlement (Southern California Coastal Water Research Project unpubl. data). *Sicyonia ingentis* is a free-spawning shrimp that undergoes multiple spawns during a prolonged intermolt in the summer. They molt synchronously in the winter and spring (Anderson et al. 1985). The ostracod *Euphilomedes producta* reproduces all year but peaks in September (Baker 1975).

Studies of ecological interactions such as predation and competition among benthic species on the mainland shelf have not been conducted.

OFFSHORE SHELVES, RIDGES, AND BANKS

Species Composition and Abundances. The soft sediments on the shelves and platforms of the Channel Islands, the Santa Rosa–Cortes Ridge, and the Tanner and Cortes banks provide a unique benthic habitat. Unlike those on the mainland shelf, macrobenthic assemblages that inhabit these areas extend to about 500 m on the basin upper slope. These assemblages are much more spatially heterogeneous than on the mainland shelf, and several local subassemblages exist that appear to reflect differences in sediment types (Fauchald and Jones 1979a, b; 1983).

Most areas (63–500 m) are inhabited by the

ophiuroids *Amphipholis squamata* and *Amphiodia urtica,* the pelecypod *Parvilucina tenuisculpta,* the polychaete *Chloeia pinnata,* and the amphipod *Photis* spp. (table 8.9). All of these species also occur on the mainland shelf, but at offshore areas, their abundances are different and they may become locally dominant. At some sites on the southern San Miguel and Santa Rosa Island shelves (mostly 100–300 m), *P. tenuisculpta* (mean abundance 166 m^{-2}, maximum 1872 m^{-2}), and the gastropod *Caecum crebricinctum* (mean abundance 72 m^{-2}) are most abundant. *P. tenuisculpta* is commonly collected throughout the SCB and is most abundant on the mainland shelf near sewage outfalls (up to 3568 m^{-2}) and off Point Conception (up to 2192 m^{-2}) (Jones and Thompson 1984). Shallower areas offshore (30–164 m) were inhabited by the amphipods *Photis californica* and *Photis lacia* (pooled average 511 m^{-2}), and the ophiuroid *A. squamata* occurred in their highest densities (164 m^{-2}). This ophiuroid was the most abundant species on the Santa Rosa Island shelf (67 m^{-2}). Another low-diversity subassemblage associated with coarse, sandy sediments is dominated by the ophiuroid *Ophiopholis bakeri* (110 m^{-2}).

The most abundant species on the southern San Miguel Island shelf is the sipunculan *Nephasoma diaphanes.* It occurs in densities up to 264 m^{-2}, and most of them inhabit the empty distal chambers of the arenaceous foraminiferan *Astrorhyza* sp. The small hydroid (*Perigonimus,* new sp.) often uses the introvert of *N. diaphanes* as a substrate (Ljubenkov and Thompson unpubl. ms). The monoplacophoran *Vema* sp. was collected from the southern Santa Rosa–Cortes Ridge in 1976 (McLean 1979).

Most of the megabenthic invertebrates that inhabit offshore areas also occur on the mainland areas. Otter trawls from the San Miguel Island shelf and Tanner Bank (75–313 m) collected mostly echinoids (*Allocentrotus fragilis* and *Lytechinus pictus*). Other abundant species included the opisthobranch *Philine alba,* the asteroid *Luidia foliolata,* the holothuroid *Parastichopus californicus,* and the siphonophore *Dromalia alexanderi* (Mearns et al. 1978).

As on the mainland shelf, very little information exists about seasonal, annual, or longer term variations in the benthos. Samples collected at various times and areas demonstrate considerable spatial and temporal heterogeneity in the offshore macrobenthic assemblages (Fauchald and Jones 1979a, b, 1983). Significant changes in macrofaunal densities at several offshore sites were caused by recruitment of *Chloeia pinnata.* Average abundances of *C. pinnata* increased from 40 m^{-2} in winter to 606 m^{-2} (maximum 5984 m^{-2}) in summer 1977 because of newly recruited juveniles. This "fireworm" is widely distributed throughout the region (depths 13–819 m) but is most abundant in the offshore areas (Jones and Thompson 1987b).

Assemblage Structure and Diversity. The macrobenthic assemblages of the offshore shelves, ridges, and banks are among the most diverse of the deep benthic habitats in the SCB (table 8.11), particularly, on the southern San Miguel Island shelf ($H' = 3.77$) (Fauchald and Jones 1979b). This high species diversity is probably the result of several factors, including persistent upwelling (which affects the productivity of the area) and the wide range of sediment types. In addition, commensalism, such as that described for *Nephasoma diaphanes,* demonstrates a biological mechanism that contributes to the high diversity of this site.

As on the mainland shelf, offshore macrobenthic assemblages are composed of mostly randomly dispersed populations (scale = 1000 m radius) (table 8.12). However, more aggregated populations occurred in these assemblages than in any other in the region. Where large recruitments of *Chloeia pinnata* occurred, they were highly aggregated (Jones and Thompson 1987b). Other common species such as *Amphipholis squamata, Parvilucina tenuisculpta,* and *Amphiodia urtica* were usually randomly dispersed at the shallower sites, but they were occasionally aggregated at some deeper sites as well (Fauchald and Jones 1979b). Megabenthic assemblages were not markedly more diverse than those on the mainland shelf. Up to 25 species and 18 kg of wet biomass per trawl were collected (Mearns et al. 1978).

Sediment and Trophic Relationships. Possibly because of strong currents and coarser sediments on the offshore shelves, ridges, and banks, more tubicolous species occur than in any other benthic habitat. Fewer burrowing species occur on the offshore areas than on the mainland areas (table 8.14). Using x-radiographs of sediment cores, Edwards (1985) also showed that burrowing and bioturbation in Santa Rosa–Cortes Ridge assemblages were negligible.

Such high-energy, highly productive areas have a high proportion of suspension feeders (36% of species); surface detrital feeding species are equally common (Fauchald and Jones 1983). As on the mainland shelf, the most abundant species on the offshore areas are deposit feeders. *Chloeia pinnata* and *Allocentrotus fragilis* ingest large proportions of particulate organic material and may act as predator–scavengers. The ophiuroids probably function as both deposit and suspension feeders, depending on currents and available food.

Although the offshore assemblages use primarily sediments as food, particulate organic material, animal remains, and Foraminifera (76% of the species) contribute more to their diets than in any other benthic habitat of the region (table 8.15). These differences in diets and life modes from mainland shelf assemblages appear to reflect differences in the sediments and the types of foods that are available in each area (Thompson 1982).

BASIN SLOPES

More of the sea floor in the SCB exists on slopes than on any other habitat. Gradients of decreasing sediment grain size, increasing organic material, and decreasing dissolved oxygen concentrations over increasing slope depth are reflected by gradients in species composition and abundance.

Submarine canyons are features of the slopes that provide a different habitat. Generally, canyons have coarser sediment and more organic material than the adjacent slopes. However, macrofauna in canyons are similar to shelf fauna at upper slope depths, but at deeper slope depths, canyon macrofauna can differ from those on the slope proper by 30–60% (Hartman 1963; Thompson et al. 1984a). Upper and lower slope assemblages are recognizable on all basin slopes studied (Thompson and Jones 1987).

Species Composition and Abundances. Nearshore basin upper slope macrofaunal assemblages occur between 161 and 632 m. On the San Diego Trough, San Pedro Basin, and Santa Monica Basin slopes the most abundant species are the polychaetes *Maldane sarsi, Pectinaria californiensis,* and *Chloeia pinnata* (table 8.9). More abundant on the adjacent mainland shelf, these species decrease in abundance on the slopes but become dominant (Hartman 1966; Fauchald and Jones 1979a, b, 1983). *Chloeia pinnata* is also an important member of the benthos on the offshore shelves, ridges, and banks. The large echiuran *Arhynchite californicus* is most abundant on upper slopes between 83 and 558 m. They exist in densities of about 19 m^{-2} and can contribute up to 88% of the macrofaunal biomass (Thompson 1982).

The assemblage on the Santa Barbara Basin upper slope is different from the other slopes. There, the polychaetes *Heteromastus* sp. (up to 108 m^{-2}) and *Prionospio lobulata* (up to 83 m^{-2}) were most abundant (Thompson and Jones 1987).

Lower slope assemblages of the nearshore basins occur between 480 and 851 m and are composed of species that generally occur only at those depths (table 8.9). On the San Diego Trough and San Pedro Basin lower slopes, the most abundant species is the small pelecypod *Saturnia californica,* which apparently occurs only off southern California at depths of 460–1215 m (Hartman 1966; Fauchald and Jones 1979b, 1983). This species decreases in abundance in the northern nearshore basins, and none were collected from the Santa Barbara Basin lower slope. Instead, the most abundant species on the Santa Barbara Basin lower slope was the snail *Mitrella permodesta*, which is more commonly collected on the nearshore basin floors. Also, several apparently endemic species (such as the amphipod *Byblis barbarensis* and the pelecypod *Cyclocardia barbarensis*) occur on the lower slopes of Santa Barbara Basin (Thompson and Jones 1987).

At least nine species of aplacophoran molluscans also occur on the nearshore basin lower slopes. The echiuran *Listriolobus hexamyotus* is commonly collected at depths of 315–728 m on both nearshore and offshore lower slopes. They often contribute most to macrofaunal sample biomass (up to 66%) (Thompson 1982).

Offshore upper slopes are essentially extensions of the shallower island shelves, ridges, and banks assemblage, to about 500 m (see preceding section).

Offshore lower slope assemblages have been sampled from the Santa Cruz, San Nicolas, and East Cortes basins (Fauchald and Jones 1979b, 1983). These assemblages range from 541 to 1768 m and are inhabited by different macrobenthic species than those found on the nearshore lower slopes, although some species occur in both areas (the polychaete *Prionospio lobulata* and the aplacophoran *Chaetoderma* sp. B). Offshore lower slope assemblages are not as faunally unique as those on the nearshore slopes. Instead, they appear to be transitional assemblages since they include many shallower ridge–bank species such as the ophiuroid *Amphipholis squamata* and the polychaete *Lumbrineris cruzensis*, as well as many offshore basin floor species such as the ophiuroid *Ophiura leptoctenia*.

The most common and abundant megafaunal species on the nearshore mid-slopes are the echinoids *Allocentrotus fragilis* and *Brisaster latifrons*. On the San Diego Trough, San Pedro Basin, and Santa Monica Basin slopes, *A. fragilis* contributes up to 95% of the megafaunal biomass. On the Santa Barbara Basin slope, *B. latifrons* contributes comparable percentages. On the lower slopes of the San Pedro and Santa Monica basins, the dominant megabenthic species (abundance and biomass) is the echinoid *Brissopsis pacifica*. Abundances of these three species of echinoids are different over slope depth and among the basin slopes of the region (fig. 8.8); all of them are often collected in the same trawl catch (Thompson et al. 1987b). About 19 species of asteroids also inhabit the lower slopes (fig. 8.9) (Laughlin and Thompson unpubl. ms.). The benthic siphonophore *Dromalia alexanderi* is an unusual species that is found only in the SCB. It inhabits slopes (130–750 m) in both the nearshore and offshore basins (Pugh 1983).

A glass sponge association consisting of five to seven species of sponges exists near sill depths (459–867 m) (table 8.10) (Hartman 1955; Thompson et al. 1984a). These sponges provide substrate for a variety of ophiuroids, polychaetes, sipunculans, and other benthos (fig. 8.10). The presence of this association near the sills suggests increased suspended particle flux, but this has not been documented.

The megabenthos of the offshore slopes are poorly known. The few recorded samples show *Allocentrotus fragilis* and the shrimp *Pandalus jordani* as the most abundant species (Mearns et al. 1978).

Little seasonal or annual variation has been observed in macrobenthic species composition and abundances at most slope sites (Fauchald and Jones 1979b; Thompson et al. 1984a). Only a few species showed seasonal differences. *Pectinaria californiensis* was significantly more abundant in winter than in summer (1977) samples, probably because of adult immigration. As on the mainland shelf, some long-term temporal changes in abundances have apparently occurred on the nearshore upper slopes. Hartman's (1955, 1966) surveys showed *Chloeia pinnata* as much more abundant and *Maldane sarsi* as much less

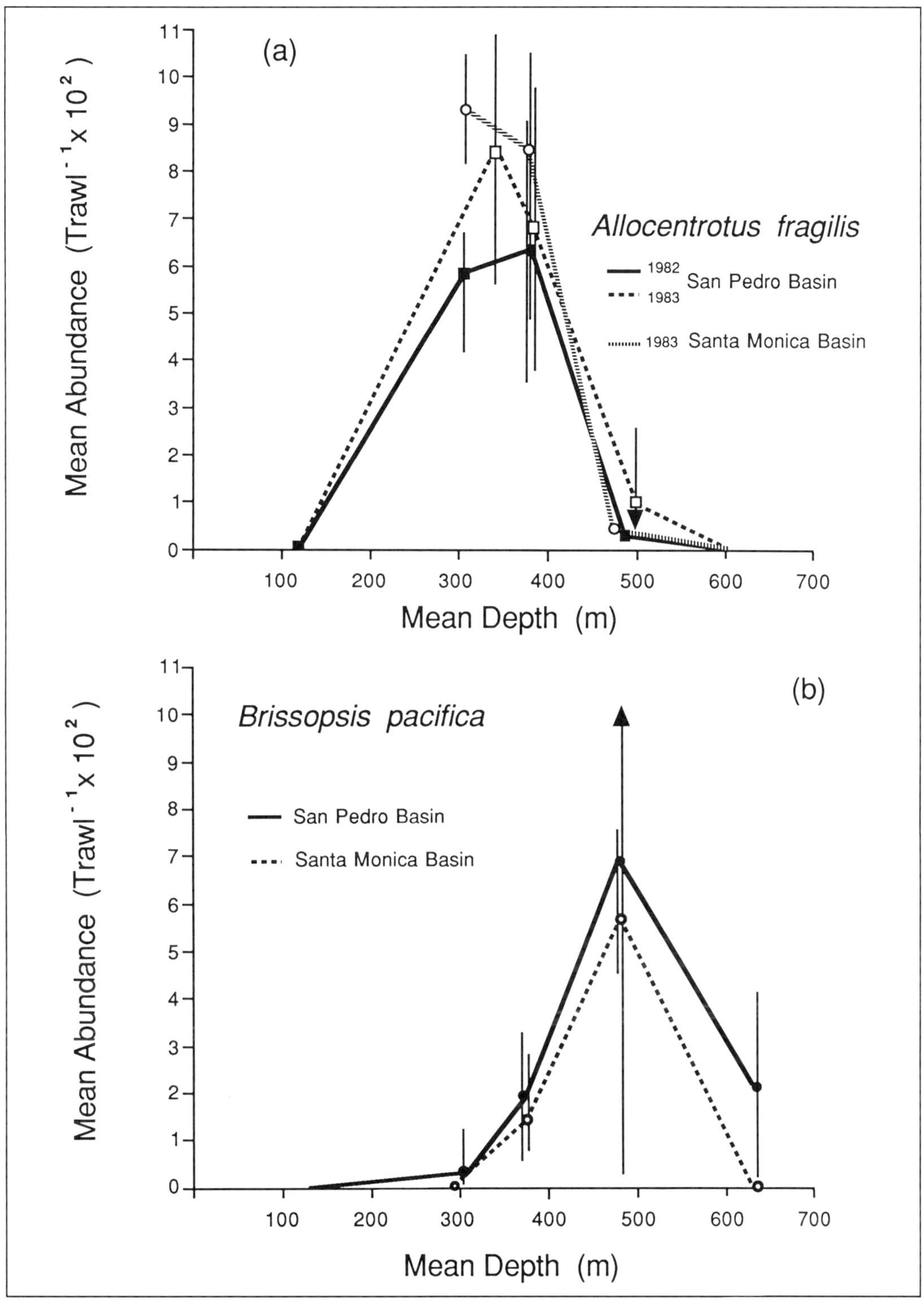

Figure 8.8. Depth distributions of four echinoid species on coastal basin slopes. (a) *Allocentrotus fragilis*. (b) *Brissopsis pacifica*. (c) *Brisaster latifrons*. (d) *Spatangus californicus*. (After Thompson et al. 1987a.)

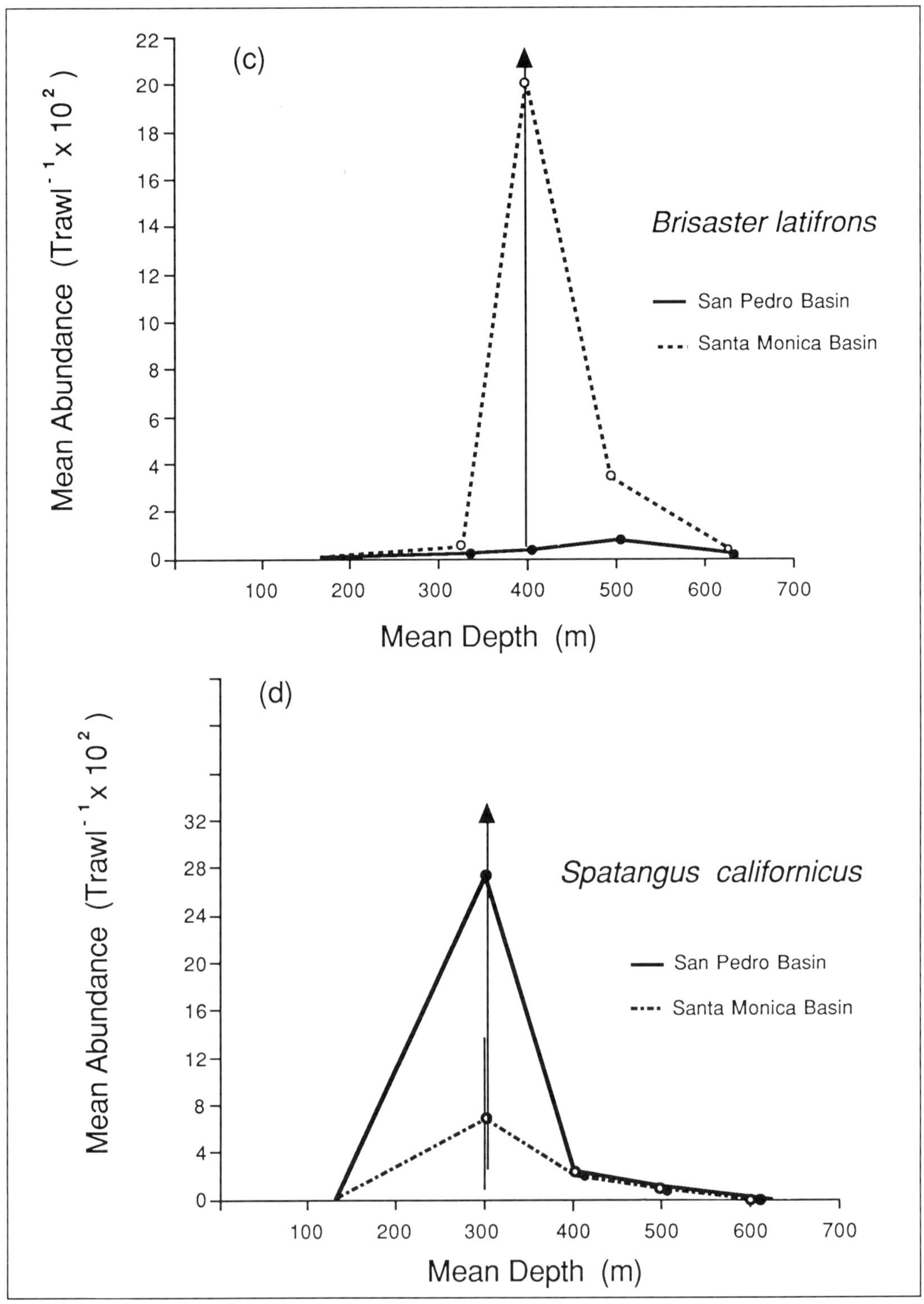

Figure 8.8. (continued)

Figure 8.9. Photograph of the lower slope of San Pedro Basin at 685 m. The ophiuroid *Ophiomusium jolliensis,* the irregular echinoid *Brissopsis pacifica,* and the asteroid *Myxoderma platyacanthum* are common megabenthic inhabitants of this area. (Photograph courtesy of G. F. Jones, University of Southern California.)

abundant than Fauchald and Jones' (1979b, 1983) surveys did.

Diversity and Biomass. Macrobenthic species diversity and biomass decrease over slope depth (table 8.11). These parameters are similar on nearshore and offshore slopes at equivalent depths. On the lower slope, most of these parameters approach their lowest values. Species diversity increases slightly on the lower slope apron in a manner similar to that described off the east coast of the United States and attributed to a variety of causes, including the relative importance of productivity, competition, and predation (Sanders et al. 1965; Rex 1981).

Megafaunal species diversity and biomass increase over slope depth (table 8.13), a trend opposite to that observed for the macrofauna. These measures of megafaunal assemblage structure reach their maximum values on the lower slopes, then decrease in the nearshore basins. The trends in megafaunal diversity and biomass may be due to the elevated organic content (food supply) of lower slope sediments or to the increased number of predators (asteroids) on the lower slope. Below the sills on the basin floors, the organic content of the sediments is even higher; however, low oxygen concentrations and episodes of anoxia may limit the development of the assemblage (see section on Basins).

Slope macrofaunal assemblages are composed mostly of randomly dispersed populations (scale <1000 m) (table 8.12). On the

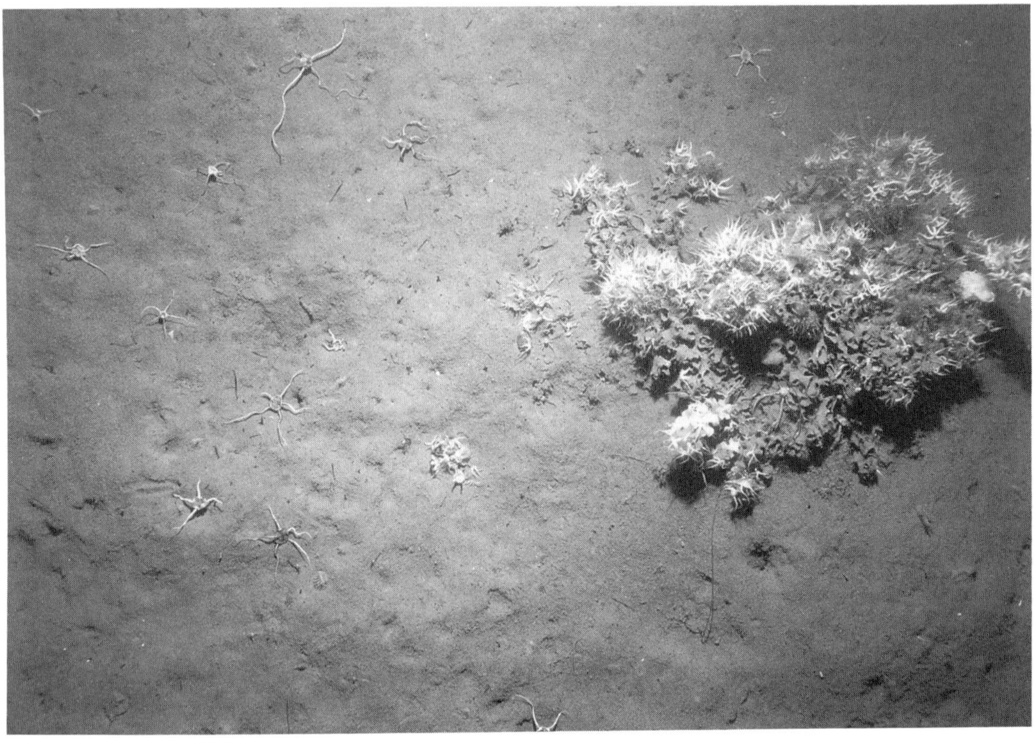

Figure 8.10. Photograph of a glass sponge (on right) taken near the sill of the San Pedro Basin (698 m). These unidentified sponges increase dramatically in frequency at these depths and may provide additional habitats for numerous other organisms. (Photograph courtesy of G. F. Jones, University of Southern California.)

upper slope, *Pectinaria californiensis* and *Chloeia pinnata* were often aggregated due to recruitment (Fauchald and Jones 1979b). Samples collected 2 km apart (49-site grid) on the San Pedro slope showed that the most abundant species were aggregated (Thompson et al. 1984a), demonstrating heterogeneity on a scale of 1–2 km on the slopes. The heart urchins *Brisaster latifrons* and *Brissopsis pacifica* exist in large-scale (kilometer-scale) herds that roam the slopes (Thompson et al. 1987b).

On the lower slopes, slightly more species are aggregated than on the upper slopes or basins. For example, on the East Cortes Basin lower slope, the amphipod *Byblis tannerensis* was collected in densities of 400 m^{-2} at only 1 of 11 sites sampled (Fauchald and Jones 1983).

Sediment and Trophic Relationships. More burrowing species are found among slope macrofauna (up to 53%) than in any other benthic habitat (table 8.14). This may be a reflection of sediment instability on the slopes. Turbidity and mass flows transport sediment downslope (Field and Edwards 1980), creating an unstable habitat to which the benthos are apparently well adapted. More sessile species occur on the nearshore lower slope than on the offshore lower slope, but more surface motile species occur on the offshore lower slopes than on the nearshore lower slope.

The most abundant slope macrofaunal species, *Maldane sarsi* and *Pectinaria californiensis,* are tubicolous burrowing species. Each of the slope echinoids inhabits a slightly different stratum of the sediment (fig. 8.11). The heart

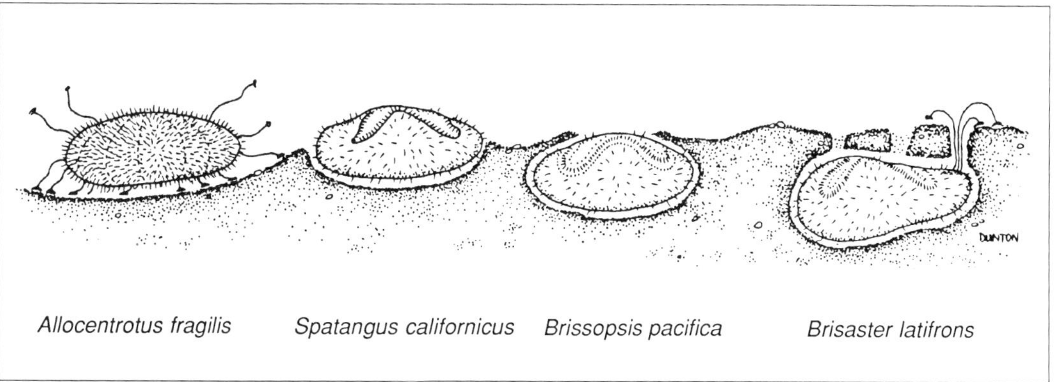

Figure 8.11. Diagram showing burrowing habits of the four slope echinoids. (After Thompson and Laughlin unpubl. ms.)

urchins *Brissopsis pacifica* and *Brisaster latifrons* may have a large effect on slope sediments as they plow through the sediment surface creating disturbed furrows and trails. The latter species can emerge in response to shifts in current direction that resuspend sediments (Nichols et al. 1989).

The importance of bioturbation in slope assemblages is not well understood. Although obvious in photographs, physical evidence of bioturbation in sediment on the Santa Cruz Basin slopes was difficult to obtain because of the noncompacted nature of the sediments (Edwards 1985). Most of the macrobenthic species from slope habitats are subsurface detrital feeders (42–47%). On the lower slope, filter-feeding species were nearly as abundant (Fauchald and Jones 1983).

Slope macrofaunal assemblages obtain most of their food as surface detrital aggregate feeders (table 8.15). Subsurface feeders contribute more to assemblage biomass on the upper slopes than they do in any other benthic habitat in the region. The lower slopes of the Santa Barbara Basin are different from the other lower slopes in that no subsurface detrital aggregate feeders are present. This condition is more characteristic of the basin floor assemblage and may be related to the very low dissolved oxygen values found there (Emery and Hülseman 1962; Cimberg and Smith 1987). The low contribution of single-mineral particles to diets on the slopes is probably due to the reduced percentage of sand in slope sediments, particularly on offshore basin slopes. However, two of the most abundant species on the upper slopes, the polychaetes *Maldane sarsi* and *Pectinaria californiensis*, ingest mostly single-mineral particles. More organic material, animal remains, and Foraminifera were ingested on the offshore lower slopes than on the nearshore slopes. The polychaetes *Glycera capitata* and *Chloeia pinnata* were the most abundant infaunal predator–scavengers collected on the upper slopes, but none of the macrofaunal species examined from the nearshore lower slopes contained animal remains, and less than 1% contained particulate organic material. Instead, the large, motile asteroids that occur on the lower slopes are the primary benthic predator–scavengers. The asteroid *Dipsacaster* sp. ingests *Brissopsis pacifica* (Thompson et al. 1984a).

Between 57 and 65% of slope macrofaunal species consume Foraminifera (table 8.14). However, only a few species, such as the clam *Cyclocardia ventricosa*, the polychaete *Nephtys cornuta*, and the urchin *Brisaster latifrons*, ingest Foraminifera in large enough quantities for them to be considered a major food item. Foraminifera are always found in the guts of *Saturnia californica* (Thompson 1982).

All slope echinoids feed from the sediment

surface and ingest mostly detrital aggregates. However, *Allocentrotus fragilis* also ingests moderate quantities of particulate organic material and animal remains and is considered to be an omnivore. *Brissopsis pacifica* populations on the lower slope ingest about 161 mg dry sediment m^{-2} h^{-1} and consume about 794.7 μg O_2 m^{-2} h^{-1}. *A. fragilis* populations ingest about 2 mg dry sediment m^{-2} h^{-1} and consume about 309 μl O_2 m^{-2} h^{-1}. *A. fragilis,* however, ingests much more nutritious food material (Thompson et al. 1984b). Differences in burrowing strata (fig. 8.11) and sediment components ingested, including mineral particle sizes, probably facilitate the coexistence of the four slope urchin species (Thompson and Laughlin unpubl. ms.).

Slope fish consume a wide range of slope invertebrates. Echinoids are consumed by several fish, such as the short- and longspine combfish and Dover sole. Isopods, amphipods, mysids, the decapod *Pleuronocodes,* and echiuran proboscises were among the most common invertebrates collected from fish stomachs, although some of the prey species were rare in grab samples, which suggests prey selectivity or differential prey availability (Southern California Coastal Water Research Project 1983).

Life Histories and Ecology. Populations of *Allocentrotus fragilis* and *Brissopsis pacifica* are composed of several adult size classes. Although *A. fragilis* probably spawns semiannually (Sumich and McCauley 1973), recruitment was observed only once over 2 years of study (Thompson et al. 1987b). *A. fragilis* is a long-lived species (7–10 years) with a slow growth rate (Sumich and McCauley 1973).

The polychaete *Pectinaria californica* reproduces annually with large recruitments in July (Nichols 1975). Populations usually consist of several adult size classes, which probably can emigrate over kilometer-scale distances, producing population patchiness (Fauchald and Jones 1979a; Thompson et al. 1984a). This species contributes up to 42% of benthic macrofaunal respiration in Puget Sound from the turnover of up to 1 kg dry sediment m^{-2} $month^{-1}$. In contrast, larger but rarer species (*Brisaster latifrons* and the holothuroid *Molpadia intermedia*) contributed very little to benthic metabolism (Nichols 1974, 1975).

BASINS

Species Composition and Abundances. The benthic assemblages from the subsill areas of each basin in the region differ slightly from one another. The greatest differences exist between the nearshore basins (Santa Barbara, Santa Monica, and San Pedro basins and San Diego Trough) and the offshore basins and could be related to differences in proximity to land and sources of sediment, sedimentation rate, sources and frequency of subsill replenishment water, and productivity of overlaying water. However, no specific studies of these relationships have been conducted.

Benthic assemblages in the nearshore basins are not stable or persistent over time periods of several years. In the Santa Barbara Basin, alternating layers of shells, well-mixed sediment, and anoxic sediment collected in cores demonstrated alternating episodes of inhabitation and anoxia (Hülsemann and Emery 1961; Emery and Hülsemann 1962). Samples from the deepest part of the basin were nearly devoid of living organisms and contained elevated sulfide levels, but they contained empty shells of the pelecypod *Macoma* sp. and polychaete fecal pellets. Sites near the base of the slope were inhabited by the gastropod *Mitrella permodesta* and aplacophorans (Hartman and Barnard 1958, 1960).

The ephemeral nature of the benthos in the Santa Barbara Basin is probably due to episodes of flushing of the subsill basin water probably associated with upwelling (Sholkovitz and Gieskes 1971), reoxygenating near bottom water, which allows animals to recolonize the basin floor. The water slowly becomes anoxic within a few months and the animals die. Other factors influencing the benthos in the Santa Barbara Basin include heavy rainfall that causes the adjacent Santa Clara River to flood and discharge large amounts of sediment into the basin (Soutar and Crill 1977), turbidity currents, and mass sediment flows (Sholkovitz and Soutar 1975).

The floor of the Santa Monica Basin is mostly uninhabited by macrofauna. Live macrofaunal organisms were collected from only 26% of the sites (n = 19) sampled by Hartman and Barnard (1958, 1960) and from none of the five sites sampled by Fauchald and Jones (1983). The sediments there are laminated (Savrda et al. 1984) and, possibly because of the absence of macrofauna, are inhabited by unusually high densities of Foraminifera (Abrams 1979).

No molluscs were collected in the San Pedro Basin in 1932 (Natland 1957), and only 43% of the sites (n = 70) sampled in the 1950s contained live macrofauna, thus the basin was termed *impoverished* (Hartman 1955; Hartman and Barnard 1958, 1960). However, all 21 sites sampled by Fauchald and Jones (1979b, 1983) contained live organisms (approximately 15 species) again demonstrating episodes of inhabitation. The most abundant macrofaunal species collected in the San Pedro Basin were the gastropod *Mitrella permodesta* and the polychaetes *Amage* spp. and *Phyllochaetopterus limicolus* (table 8.9). *M. permodesta* lives at nearshore lower slope and basin depths between 421 and 888 m (Fauchald and Jones 1983; Thompson et al. 1984a) and was collected in densities up to 176 m^{-2} in the San Pedro Basin.

Megafaunal animals (approximately eight species) inhabit both the San Pedro and Santa Monica basins. The most abundant species are the galathaid crabs *Munida quadrispina* and *Munidopsis hystrix* (table 8.10). Above the sill on the lower slope, the megafaunal assemblage is dominated by echinoderms, but they do not occur below the sills. The dramatic change in megabenthic assemblages at the sills of the nearshore basins is not related to any apparent discontinuities in sediment characteristics (e.g., Anderhalt and Reed 1978) or oxygen profiles (e.g., Minard 1968). The frequency of flushing in the Santa Barbara Basin can range from seasonal (Sholkovitz and Gieskes 1971) to "several years" (Emery and Hülsemann 1962), but no estimates have been made for the other nearshore basins.

The San Diego Trough is not an enclosed basin, and the benthos there is similar to the benthos in the offshore basins rather than to those in the other nearshore basins. The most common and abundant macrofaunal species from the San Diego Trough is the polychaete *Tharyx* spp. (Jumars 1975a, b). The most abundant megafaunal species are the ophiuroid *Ophiomuseum lymani* and the holothuroid *Scotoplanes* sp. (Barham et al. 1967).

The macrobenthos of the offshore basins have been termed the "*Aricidea uschakovi–Tharyx tessellata*" (polychaetes) assemblage, bur some basins have been considered subassemblages due to the presence of other organisms (Hartman and Barnard 1960). The taxonomy of many basin species has been revised, including the species for whom the assemblage was named (Straughan and Klink 1980). For comparisons, the higher taxon names Paraonidae and *Tharyx* spp. are used in this chapter (table 8.9).

In the Santa Catalina Basin, the polychaete *Tharyx* spp. is the most abundant species, while the pelecypod *Saturnia kennerlyi* is most abundant in the Santa Cruz Basin. This species is related to *Saturnia californica,* which inhabits the nearshore basin lower slopes. The ophiuroids *Ophiura leptoctenia* and *Ophiophthalamus normani* are among the most abundant species in these middle basins, but they decrease in abundance in the outer basins.

The benthos of the Santa Catalina Basin has received more detailed study than that of any other basin in the region. *O. normani* contributes over 95% to megafaunal assemblage abundances and biomass and occurs in average densities of up to 17 m^{-2} (C. Smith and Hamilton 1983; C. Smith 1983).

The San Nicolas Basin assemblage was described as a "sipunculids" subassemblage of the *Aricidea–Tharyx* community (Hartman and Barnard 1960) because the sipunculan *Nephasoma nicolasi* is often among the most abundant species collected in this basin. This species inhabits several of the other outer basins of the region, but it occurs in highest densities (up to 128 m^{-2}) in this basin. The population is spatially and temporally patchy. Members of this species are most abundant at the base of the slope descending into the basin from Tanner Bank, where organic material in the sediments is highest (Thompson 1980).

Other outer basins (San Clemente, Tanner, West Cortes, East Cortes, Long, and Velero) are inhabited by similar macrofaunal assemblages (table 8.9). The existence of the clam *Solemya* sp. in Tanner and San Nicolas basins suggests the presence of anoxic areas, as congeners are commonly collected in sulfide-rich areas near sewage outfalls. The benthic boundary layer of the San Clemente Basin is inhabited by the pelagic holothuroid *Peniagone diaphana* (Barnes et al. 1976).

No Name Basin has never been sampled and only two samples have been collected from the Patton Escarpment (Hartman and Barnard 1960). The macrobenthos from the Patton Escarpment are very similar to those in the outer basins (table 8.9), but how the basin benthos in the region compare to abyssal benthos adjacent to the SCB is not known. Macrobenthos in the central North Pacific include mostly cirratulid polychaetes (e.g., *Tharyx* spp.) (Hessler and Jumars 1974).

Temporal variation in basin benthos has not been studied in detail. The lack of long-term persistence in the nearshore basins was previously discussed, although it is not well understood. Little annual change (winter–summer 1977) occurs in the basin assemblages (Fauchald and Jones 1979b). It is difficult to compare information from various basin surveys conducted at different times due to the many taxonomic changes made in a wide variety of taxa. In addition, Hartman and Barnard's samples were collected over several years (1955–1959), further confounding comparisons.

Diversity and Biomass. The basins generally have lower macrofaunal species diversity and biomass than the other benthic habitats in the region (table 8.11). Some sites in the nearshore basins have no macrofauna; those with macrofauna have very low species diversity (San Pedro Basin, $H' = 0.92$). In the middle basins (Santa Cruz Basin), species diversity and biomass can be much higher than in the nearshore basins.

In contrast to the macrofauna, megafaunal diversity and biomass in the nearshore basins is higher than on the adjacent shelves (table 8.13). Although most of these megafaunal species do not burrow, their presence is not accounted for in currently accepted models of diversity in oxygen-deficient assemblages (Rhoads and Morse 1971; Savrda et al. 1984).

Macrofaunal assemblages in most of the basins studied are composed of mostly randomly dispersed populations (table 8.12). Detailed studies of spatial dispersion patterns of bathyal assemblages in the San Diego Trough and Santa Catalina Basin showed that population aggregation may occur at several spatial scales ranging from kilometers to centimeters (Jumars 1975a, b, 1978; Thistle 1978). In the San Diego Trough, the polychaetes *Polyophthalamus* sp. and *Ceratocephale* sp. have different patterns of dispersion, reflecting their life mode and behavior. Patchiness in their abundances therefore exists at the scale of the individual organism (centimeters) (Jumars 1976).

Experimental nekton falls in the Santa Catalina Basin caused many species to aggregate to feed, both on the carrion and secondarily as predators on the scavengers. These low-intensity disturbances of the sediment caused a reduction in infaunal species diversity, but few species were totally eliminated. The dominant infaunal species *Tharyx* sp. decreased most in abundance (C. Smith 1986).

Sediment and Trophic Relationships. Most basin macrobenthic species are motile and live at the sediment surface (40%), or they are tubicolous (33%) (table 8.14). Anaerobic conditions and high sedimentation rates in the Santa Barbara and Santa Monica basins result in laminated sediments, as there are no mac-

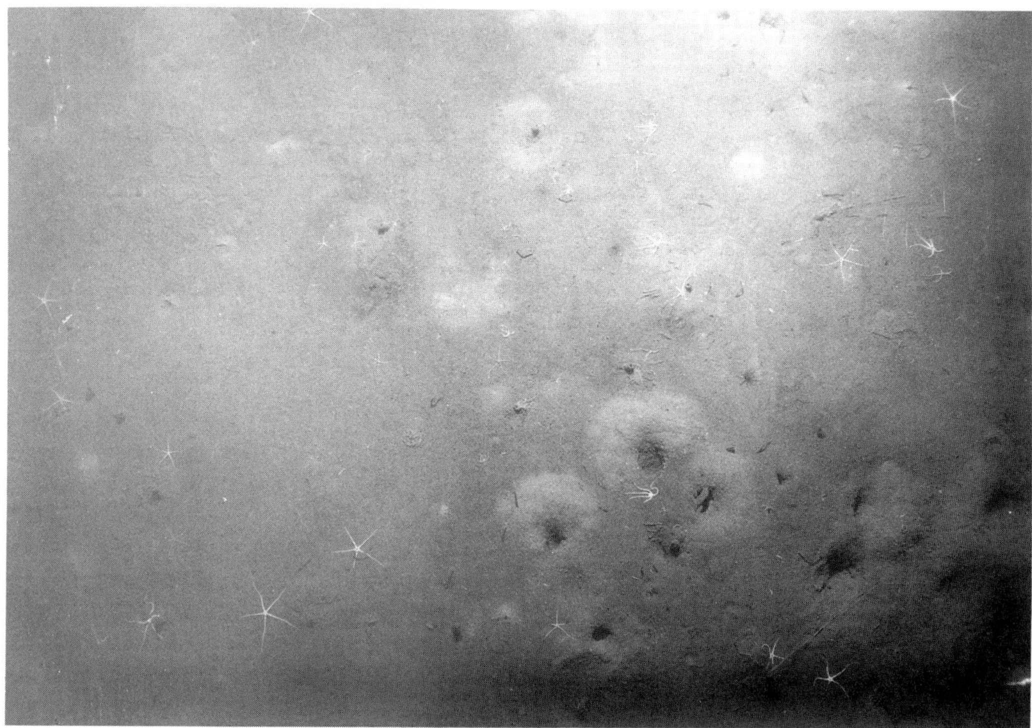

Figure 8.12. Photograph of San Nicolas Basin floor (1754 m). The ophiuroid *Ophiophthalamus normani* is a common inhabitant of the outer basins; the inhabitants of the large burrows are not known. (Photograph courtesy of G. F. Jones, University of Southern California.)

rofaunal animals to burrow and mix the sediments. During dysaerobic flushing periods, colonizing benthic species can mix the surface sediment layers. Savrda et al. (1984) concluded on the basis of x-radiographs and photographs, however, that they do not produce burrows. The offshore basins have higher dissolved oxygen concentrations, and the basin floors may have large burrows and mounds (fig. 8.12); sediments are homogeneously mixed to 40 cm (Edwards 1985). In the Santa Catalina Basin, mounds occur on about 2% of the basin floor and are presumably created by the echiuran *Prometor benthophyla*. Deposition of fecal material and subsurface sediment from burrowing on mounds was measured at 1–2 cm $month^{-1}$, about 1000 times the natural basin sedimentation rate (C. Smith et al. 1986).

The most common and abundant basin species ingest mostly detrital aggregates, but the ophiuroids *Ophiophthalamus normani* and *Ophiura leptoctenia* can ingest large amounts of particulate organic material (POM). These species are attracted to nekton and kelp falls that occur in the middle basins (C. Smith 1986).

Sinking of dead nekton or drift kelp to the basin floors is an important source of food for bathyal megabenthos. The scavenging amphipods *Paralicella caperesca* and *Orchomene* spp. (K. Smith and Baldwin 1982) and the echinoderms *Ophiophthalamus normani* and *Scotoplanes globosa* (C. Smith 1985) can rapidly locate and utilize such food sources. The dominant megabenthic species in the Santa Catalina Basin (*O. normani* and *S. globosa*) account for about 17% of the total benthic boundary layer energy demand (777 cal m^{-2} d^{-1}), which is about three times lower than in the plankton or sediment components (K. Smith 1983). Nekton falls provide about 11% of the benthic assemblage respiratory requirements (C. Smith 1986). Asteroid and lithodid crabs are also attracted to nekton falls to prey on the ophiuroids attracted to nekton falls.

Species that depend on sporadic food sources such as nekton or kelp falls should be expected to withstand long periods of starvation, locate and respond to food rapidly, maximize their feeding rate and quantity ingested, and optimize their assimilation and storage of the food (K. Smith and Baldwin 1982). In contrast, macrobenthic assemblages in the San Pedro and San Nicolas basins ingest mostly surface detrital aggregates (table 8.15). The differences in diets of the benthos in the various basins probably reflect differences in depositional characteristics of these basins or availability of dead nekton that may drop to these basin floors. Benthic suspension feeders and subsurface feeders are rare in most basins, reflecting low current speeds and dysaerobic sediment conditions.

Life Histories and Ecology. Reproduction of the benthos in the San Diego Trough is continuous in most species. Only 2 of 11 species (brachiopod and scaphopod) spawn seasonally (Rokop 1974, 1977). Seasonal settlement of ophiuroid larvae was observed in the San Pedro Basin (Fauchald and Jones 1979b), but adults have not been recorded there.

Colonization of azoic basin sediment (10% of background density in 5 months) (Levin and Smith 1984) and areas disturbed by nekton falls (more than 6 weeks) (C. Smith 1986) was rather slow in the Santa Catalina Basin compared to other areas. However, recolonization of experimental mounds in the same area was much more rapid (up to 85% of background abundances in 50 days) (C. Smith et al. 1986).

Environmentally stable deep-sea areas inhabited by species with large population sizes should favor the acquisition and maintenance of large pools of genetic variation. The enclosed basins of the SCB may function similarly to islands in controlling larval dispersal and gene flow. However, studies of genetic variation in the gastropod *Bathybembix bairdii* in the San Diego Trough and Santa Catalina Basin showed that those populations were genetically homogeneous at levels similar to other bathyal species; this finding suggests unrestricted gene flow among the basins (Siebenaller 1978).

COMPARISONS AMONG DEEP SOFT SUBSTRATE BENTHIC ASSEMBLAGES

The greatest differences in species composition and diversity among deep soft substrate benthic assemblages of the region occur over water depth. This most likely reflects decreasing sediment grain size, increasing organic content, and decreasing dissolved oxygen concentrations that occur over depth.

The largest change in species composition occurs midslope, at about 500 m. This is in contrast to the east coast of the United States where the greatest change in the benthos occurs at the shelf–slope break, about 200 m (Rex 1981; Grassle et al. 1975). Differences in benthic assemblages in the SCB also exist between nearshore and offshore sites of similar depth. These differences appear to be related to increased productivity, water motion (waves and currents), and decreased contribution of terrestrial detrital sediment on the offshore areas. Although many species inhabit both the nearshore and offshore shelf areas, changes in abundance occur between these areas. Species that inhabit both areas can apparently tolerate wide ranges of sediment grain size and organic content and can shift their feeding modes to utilize the different types of foods available in each habitat. Differences between mainland and insular shelf decapod crustacean assemblages are thought to reflect substrate differences. Species adapted to fine sediments occur more frequently on the mainland, and species that prefer coarse sediments and rock occur more frequently on the island shelves (Wicksten 1980).

Nearshore upper slopes are inhabited by many mainland shelf species and represent transitional assemblages between faunally distinct shelf and lower slope assemblages. At similar depths offshore, upper slope assemblages are extensions of the island shelf, ridge, and banks assemblages to about 500 m. Offshore lower slope assemblages are transitional between the shallower ridge bank and basin floor assemblages, as they are inhabited by species from both habitats.

Nearshore and offshore basins are inhabited by quite different assemblages. Nearshore basins have the highest sedimentation rates and can experience episodic anoxia that limits the development of a diverse basin fauna. The benthos of the Santa Barbara basin slopes and floor are unique, and the slope habitat is "compressed" into shallower depths, probably because of the extremely low dissolved oxygen concentrations there. In contrast, offshore basins have higher oxygen concentrations and are inhabited by stable and diverse benthic assemblages.

Each "basin system" (including shelf and slope) is influenced by a different set of oceanographic and sedimentary conditions, so the benthos of each system can therefore be different. Adjacent shelf fauna are most similar, but the fauna of each basin slope differs slightly from that on adjacent basin slopes at the same depth. Lower slope and basin floor assemblages are further divergent and can be quite different.

The nominal macrobenthic assemblages of the SCB have depth ranges similar to those off Oregon (Pereyra and Alton 1972) and other areas of the world (Menzies et al. 1973).

Echinoderms often dominate benthic communities throughout the world (Thorson 1957), and southern California is no exception. All of the deep soft substrate benthic assemblages of the region, except those on the nearshore basin floors, are dominated (abundance and biomass) by echinoderms, usually ophiuroids and echinoids. Their absence from the nearshore basins may be due to low oxygen concentrations found there. Ophiuroids and holothuroids may dominate in the offshore basins.

Most benthic assemblages in the region appear to be temporally stable over annual and biennial time periods. Except for changes in abundances due to recruitment of a few species at shelf and slope depths, large annual fluctuations do not occur in assemblage species composition and structure. However, over longer time periods (5–10 years), evidence shows that abundances of some shelf and slope species change considerably. Fluctuations in nearshore basin assemblages that are attributed to episodes of flushing and anoxia were discussed earlier. Periodic factors such as El Niño and storm surf effects have been observed in mainland shelf and slope megabenthic assemblages (Thompson et al. 1993) and deep rocky assemblages (Minerals Management Service 1988). The general lack of long-term studies and studies of life histories and reproduction of benthic invertebrates precludes a clear understanding of the dynamics of the benthos.

Macrofaunal (grab and core samples) species diversity and biomass decrease over depth from shelf to basins, but megafaunal (trawls) diversity and biomass increase over depth to their maximum values on the slopes. Macrofaunal species diversity is highest on the southern Channel Island shelves and Santa Rosa–Cortes Ridge and is lowest in the nearshore basins. Mechanisms that cause these differences in diversity and biomass are not obvious and have not been studied in the region except in the San Diego Trough and Santa Catalina Basin. Trends in species diversity and biomass on the east coast of the United States are different than in the SCB. Both macro- and megafaunal diversity increase with depth to a maximum on the lower slope of the east coast (2000–3000 m) (Rex 1981). Differences in these trends probably reflect differences in mechanisms that control and maintain diversity in the different regions.

The relationship between species diversity and scales of dispersion has been studied extensively in the basins of the SCB. Species diversity can be influenced considerably by small-scale differences in spatial dispersion of the benthos. Within all deep soft substrate macrobenthic assemblages of the region, most populations are randomly dispersed (all scales) (table 8.12). This suggests that the factors regulating and maintaining assemblage structure are also random in nature (Jumars 1976; Rex 1981). Although in low proportions, more populations are aggregated on the offshore ridges and banks than in any other areas, possibly reflecting the heterogeneity of substrate types. Very few species are uniformly dispersed. Megabenthic species tend to be more patchy in their distributions, particularly the dominant echinoids that occur in kilometer-scale "herds."

Low-intensity disturbances from a variety of biological mechanisms, such as nekton falls, creation of mounds from burrowing, and deposit feeding, occur on the sea floor. These disturbances produce successional patches in the assemblage, each of different structure and diversity, equal to the size of the disturbance ($0.01–1m^2$) (Jumars 1976; Thistle 1978). This pattern of small-scale disturbances creates an areal (within habitat) mosaic of different aged patches that elevates the diversity of the habitat as a whole. Highly motile species can rapidly recolonize after disturbances or are attracted to feed on other colonizers. This further increases the diversity within each patch. The measurement of random dispersion of most populations thus seems in contrast to that in the previous discussion and may reflect the low intensity of many disturbances and the low sampling intensities of most surveys, which may make detection of randomness spurious (C. Smith 1986).

While this scenario has been supported by sampling in the basins, it has not been determined whether it occurs at similar scales of space and time in shallower areas or other basins. Certainly physical disturbances play a larger role in shallower water. On the slopes, the large herds of urchins and unstable sediments may have large-scale effects. In the nearshore basins, flushing events may have much more influence than biological events.

Bioturbation of sediment can occur by burrowing, feeding, or defecation by benthic organisms. The result is generally a thin (1–2 cm), intensely mixed layer of sediment at the surface overlying less mixed sediment as depth increases to about 15 cm (Jumars 1978). This general scheme may be modified in some areas, depending on oceanographic and sedimentary factors. Sediments in the Santa Cruz Basin are homogeneously mixed to about 40 cm (Edwards 1985), and sediments in the nearshore basins, where sedimentation rates are highest, are layered to greater depths (Savrda et al. 1984).

Subsurface feeders and burrowers occur in their highest proportions on the basin slopes, probably reflecting increased sediment transport caused by turbidity and mass sediment flows creating an unstable habitat to which the benthos are apparently well adapted. Burrowing species are least common in the nearshore basins, reflecting the low dissolved oxygen concentrations there. Instead, basins have the highest proportions of surface motile species. Tubicolous species occur in the highest proportions on the offshore shelves, reflecting a high-energy environment that requires an anchoring mechanism. Bioturbation can decrease the abundances of tubicolous species and can increase the abundances of suspension and surface-feeding species (Brenchly 1981). Such interactions demonstrate the dynamic nature of animal–sediment relationships and suggest that feeding and classification schemes alone are not sufficient to understand these relationships.

All benthic assemblages in the region function mainly as deposit feeders. Most species ingest detrital aggregates at the sediment surface (table 8.15). Proportions in this food stratum category are highest in the San Pedro Basin (83% of assemblage), where sedimentation rates are highest, and lowest on island shelves, which are nondepositional (28%).

Unaggregated mineral particles contribute most to diets on the offshore areas where more sand is present and decrease in diets at slope and basin depths where a lack of sand occurs in those sediments. Suspension feeding contributes more to feeding of the mainland shelf assemblage than any other habitat. Particulate organic material, animal remains, and Foraminifera contribute more to diets on the offshore shelves, ridges, and banks than in other habitats. This is different from mainland shelf sites at similar depths and reflects both the lack of detrital sediments on the offshore areas and the high biological productivity of the offshore banks. Animals that inhabit both nearshore and offshore areas can switch their diet to feed on the most available food source. Higher proportions of the benthos (number of species) ingest Foraminifera as water depth increases on the lower slopes, where 74% of the species ingest Foraminifera. However, Foraminifera contribute least to gut content volumes in basin species (Thompson 1982).

Deposit-feeding species can have a large influence on sediment quality. Most deposit feeders utilize particles at the sediment surface. This activity maintains the animal's preferred particle sizes at the sediment surface. Thus, selection and feeding can control the sediment surface texture (Jumars et al. 1981). The presence of animal tracks, fecal pellets, and coils can alter interparticle adhesion, which can change near-bottom current flow and sediment resuspension velocities (Nowell et al. 1981). Tubes constructed by benthic animals can also alter sediment depositional characteristics, causing changes in bacterial colonization, feeding strategies, and larval settlement patterns on the bottom (Eckman et al. 1981; Eckman 1985). Other biologically mediated factors that can influence sediment quality are fecal pellet breakdown rates and the differential accrual, erodability, and burial of fecal pellets versus free sediments (Jumars et al. 1981; Miller and Jumars 1986). Deposit feeding also has a large effect on sediment chemistry by increasing sediment solute transport, reaction rates, and microbial distributions (Aller and Yingst 1985; Hylleberg and Henriksen 1980).

Commercially and Recreationally Important Species

Marine invertebrates have been utilized as food and their shells have served as ornaments, fish hooks, and utensils since before recorded history. The commercial harvesting of molluscs, crustaceans, and sea urchins developed as the population of southern California began to expand in the latter part of the nineteenth century. In recent years, some attempts have been made, with limited success (especially with abalones), to culture some species for use as human food.

The coastal Indians of southern California, such as the Chumash Tribe, used a variety of invertebrates for food and turned their shells to practical as well as ornamental uses (Greenwood and Browne 1969). The knowledge of which species were used by the Indians is obtained by sifting through the mounds of shell fragments called kitchen middens. Pelecypods were primarily used for food, but the more colorful species, especially gastropods, were used as beads or ornaments. The California mussel was shaped into fish hooks. Gastropods and the one species of scaphopod were primarily used as beads or ornaments. Larger shells, such as the rock scallop, were used as utensils. Undoubtedly, other species of marine invertebrates were used for food and other purposes.

The commercially important species of invertebrates in the SCB are listed in table 8.16. The data are presented according to the species, the common name, the amount harvested in kilograms, and the value in 1976 dollars. Most of the species collected are used as food and include species of molluscs, decapod crustaceans, and sea urchins. Decapods and pelecypods are also collected and sold for fish bait. In terms of amount landed, more squid are collected than any other species of invertebrate. Sea urchins are the most valuable species, followed by the California spiny lobster and the red abalone (table 8.16). The California spiny lobster yields the highest dollar return per kilogram.

Presumably, the elevated temperatures caused by the 1982–1983 El Niño affected squid landings in the SCB during the 1982–1985 period. Landings of this pelagic species decreased by as much as two orders of magnitude from 1981. The population apparently recovered in 1986 since the landings that year were the largest since 1982 (table 8.16). Landings of the other commercially important invertebrate species in the SCB apparently were not affected by the El Niño, which may be attributed to their benthic habitat.

The total value in 1976 dollars of the common commercially important invertebrates varied from $3.4 million (1983) to $5.3 million (1986). The dollar value of the landings in recent years would be much higher if inflation were considered. The value of these species will undoubtedly rise in the 1990s because of the interplay of demand, scarcity, and inflation.

The pelecypods and decapods listed in table 8.17 plus many of the species in table 8.16 are collected by people for recreation. They are also used as food or as fish bait. Undoubtedly, population levels and locality affect the collection of many additional species of invertebrates. A valid marine California Department of Fish and Game license is required to collect many of the species listed in tables 8.16 and 8.17; in addition, many of the species have size and quantity limitations. Some of the decapods, especially the bay ghost shrimp, sand crab, and pelecypods, are sold as fish bait. The dollar value of this resource is impossible to determine because many people collect their own bait. No records are kept as to the number of specimens of each species collected annually by individuals. Many of the species listed in table 8.17 are rare (such as *Pecten diegensis*) or not as common as in the past, such as *Ostrea lurida*, because of environmental changes in the estuaries of the SCB.

Summary and Prospectus for Future Research

The SCB is inhabited by many different marine benthic assemblages reflecting the complex physiography and oceanography of the region. The largest differences among these assemblages are related to water depth and substratum type. Two general patterns of distribution and abundance are apparent in all benthic habitats: zonation along gradients of physical stress, and spatial and temporal differences in abundances within those zones.

Zonation occurs in all of the major benthic habitats of the SCB. These zones form recognizable assemblages of invertebrates, each with different species composition and abundances. Zonation is most obvious in habitats with the most extreme physical or chemical gradients, such as in the rocky intertidal. Where these abiotic gradients are more diffuse, zones are broader and less obvious, such as on the basin slopes. Zonation appears to be the result of a complex interplay of gradients of species abundances that reflects chemical, physical, and biological factors that mold assemblages into what we observe.

In the rocky intertidal, the upper distributional limits are set by intolerance to harsh physical conditions, whereas the lower limits are set by predators or competitors. Exceptions to this generalization exist, however. In the shallow sandy subtidal, the physical gradient of wave surge limits shoreward distributions, while competition for space and predation combine to set seaward limits.

Much less is known about the causes of zonation in deep benthic habitats. Zonation in sublittoral and bathyal areas is thought to reflect differences in adaptations to environmental stability, in life histories, and in trophic differences, but the role of each of these factors is unknown.

In addition to species composition, other assemblage characteristics such as species richness and densities are also different among the major benthic assemblages of the SCB. Species richness appears to be lowest in the most unstable habitats such as sandy beaches, subtidal embayments, and basin slopes. Sediment scour, transport, and deposition resulting from wave action and tidal factors in the shallow areas, and mass sediment flows and turbidity currents on the basin slopes may prohibit the development of diverse assemblages. Species richness and densities in nearshore basins can be quite low where anoxia occurs. Species richness and densities appear to be the highest in subtidal and sublittoral rocky assemblages, which probably provide considerable heterogeneity of habitats. Benthic biomass appears to be highest on the basin slopes, possibly because of the presence there of large herds of urchins.

Within benthic assemblages are recognizable patterns in abundances of organisms. These differences are generally ascribed to small-scale (meter-scale) patch-sized dynamics and involve physical and biological components. In all assemblages, scales of patchiness exist that generally reflect the disturbance regime of the habitat and the temporal succession of the assemblage.

In shallow areas most populations exhibit patchiness. In the rocky intertidal, spatial patchiness within a tidal zone appears to be set by the frequency and intensity of physical disturbance. In the sandy intertidal, extensive spatial patchiness within tidal zones also exists. The mechanisms here are not as well understood as in the rocky intertidal, but in several cases, they appear to be the result of responses of individuals to variations in water motion or wave force caused by physical heterogeneity at a variety of spatial scales.

In deep benthic assemblages, most populations are randomly dispersed. Physical conditions are more homogeneous, and disturbances such as nekton falls and predation are either of very low intensity or occur randomly in space and time; thus, the species appear more randomly dispersed. In general, the distributions and abundances of benthic assemblages appear to be persistent and stable in the region. However, phenomena such as disease outbreaks, the effects of unusually severe storms, El Niño, and episodes of flushing and anoxia in the nearshore basins can greatly change species composition and structure.

Anthropogenic effects are another major theme in the ecology of invertebrate assemblages in the SCB. Particularly in embayments and intertidal areas, the effects of humans are making it increasingly difficult to observe the natural patterns. No truly pristine embayments exist in the SCB; they have all been altered directly by dredging and construction or indirectly by the damming of headwaters. Many of the invertebrate species in these embayments are species introduced by international commerce. Intertidal areas, especially south of Santa Barbara County, are subject to chronic perturbation and manipulation. Rocky shores suffer intense trampling and collecting. Sandy beaches are groomed to meet the recreational desires of human populations. Nearshore subtidal areas are subject to heavy pressure from both commercial and sports fishermen who take significant numbers of sea urchins, abalones, lobster, and fin fishes. In addition, kelp harvesters kill large numbers of sea urchins in an effort to protect kelp resources. There is a great need to document and determine natural assemblage patterns and their long-term variability where relatively unaltered assemblages still exist. Furthermore, understanding natural variations will help put man-caused effects into perspective. It is only against these natural patterns that changes can be evaluated.

Although much work has been done in intertidal and shallow rocky habitats, few ecological studies have been conducted in subtidal sandy and muddy habitats. Two-thirds of the SCB embayments have been sampled, but in many instances studies are limited to one benthic survey. Knowledge of the benthos in the outer basins of the SCB is

Table 8.16. *Commercially Important Marine Invertebrates in the SCB*

Species	1976		1980		1981		1982		1983		1984		1985		1986		1987	
	Wt[a]	$[b]	Wt	$	Wt	$	Wt	$	Wt	$	Wt	$	Wt	$	Wt	$	Wt	$
Phylum Mollusca																		
Haliotis corrugata (Pink abalone)	195.6	411.1	55.5	116.7	33.3	69.9	36.5	76.6	26.6	55.9	21.9	46.1	25.3	53.2	18.8	39.5	8.5	17.9
Haliotis cracherodii (Black abalone)	161.8	131.4	234.1	189.6	222.8	180.5	276.5	223.9	208.4	168.8	195.2	158.1	161.7	130.9	117.3	95.0	136.2	110.3
Haliotis fulgens (Green abalone)	54.6	114.6	25.7	54.0	25.0	54.5	31.8	66.8	28.6	60.0	18.1	37.9	—	—	8.8	18.4	8.7	18.3
Haliotis rufescens (Red abalone)	306.3	631.5	197.7	407.2	163.6	336.9	153.6	316.3	83.5	172.0	104.9	216.1	92.8	191.3	88.9	183.1	124.3	252.2
Loligo opalescens (squid)	6932.7	400.9	8218.7	476.7	9903.0	—	5693.6	330.2	861.3	50.0	73.3	4.3	535.6	31.1	1495.7	867.5	1364.4	791.3
Phylum Arthropoda																		
Cancer spp. (Rock crabs)	556.9	342.0	431.0	262.9	510.0	310.8	440.7	268.8	426.5	260.2	569.8	347.6	555.8	339.0	633.4	386.4	591.1	360.6
Pandalus platyceros (Spot prawn)	14.5	0.7	17.2	0.9	113.6	5.7	90.7	4.5	29.2	1.5	8.8	0.4	9.8	0.5	34.6	1.7	30.6	1.5
Panulirus interruptus (California lobster)	132.7	663.2	183.8	919.1	209.3	1046.7	214.0	1070.1	215.7	1078.4	189.8	948.9	185.7	928.4	214.9	1074.6	195.5	977.3
Sicyonia ingentis (Ridgeback prawn)	1.4	2.0	23.0	32.2	86.1	120.5	56.8	79.6	65.0	91.0	265.7	372.0	391.1	547.6	272.7	381.8	105.1	147.2
Phylum Echinodermata																		
Strongylocentrotus spp. (sea urchins)	4994.8	1039.6	9950.0	2069.6	10755.5	2237.3	8390.2	1745.2	7150.2	1487.2	6623.0	1377.6	7719.2	1605.6	1086.4	2259.7	1001.0	2082.0
Totals (kg × 10^3)	13351.3		19336.7		22021.7		15386.4		9095.0		8081.5		9677.0		27210.4		24854.0	
Totals (1976 dollars)		3637.0		4528.9		4910.2		4105.4		3425.0		3509.0		3827.6		5307.7		4758.6

[a] Wt = weight in kg × 10^3.

[b] $ = value in thousands of 1976 dollars.

Summarized from Oliphant (1979) and unpublished records from California Department of Fish and Game.

Table 8.17. *Commonly Occurring Pelecypods and Decapods Collected Recreationally for Food or Fish Bait in the SCB*

Species	Common Name
Argopecten aequisulcatus	Speckled scallop
Callianassa californiensis	Bay ghost shrimp
Chione californiensis	Branded chione
Chione fluctifraga	Smooth chione
Chione undatellum	Wavy chione
Emerita analoga	Sand crab
Hippolysmata californica	Red shrimp
Mytilus spp.	Bay and sea mussels
Nuttallia nuttallii	Purple clam
Ostrea lurida	Native oyster
Pecten diegensis	San Diego scallop
Pododesmus cepio	Abalone jingle
Sanguinolaria nuttalli	Purple clam
Saxidomus nuttallii	Common Washington clam
Tagelus californianus	California jacknife clam
Trachycardium quadragenarium	Spiny cockle
Tresus nuttallii	Gaper

limited to only a few samples and incomplete species lists.

Catalogs of regional biodiversity do not yet exist, since taxonomic knowledge of invertebrates of the SCB is still in development. There are hundreds of known, but undescribed, benthic species in the SCB. Over 50% of the species collected in the most recent (1977–1989) large-scale surveys of the deep benthos were undescribed taxa. Summaries of existing information about each group of invertebrates are needed to support the work of scientists of many disciplines. The problem is further compounded by the virtual absence of training and, therefore, of young scientists entering the field of systematic zoology.

Little is known about the life history and biology of most benthic invertebrates inhabiting the SCB. Episodic, short-, and long-term population fluctuations have been noted in a few instances, but their causes are unknown. One of the most important and difficult areas of research is the factors affecting settlement and early survival of organisms with planktonic larvae. For example, in not one instance has the parental source of a population of such benthic organisms been identified. The length of time in the plankton of SCB benthic species is universally unknown, although estimates have been made based on laboratory observation of a few species. The determinants of mesoscale (kilometer-scale) settlement patterns are not known. For most species, the importance of settlement events to later recruitment is also unknown.

There is virtually no information on interspecific interactions of deep benthic species, their relationships with their substrata, how they modify sediments, benthic–pelagic interactions, or the effects of specific pollutants on the benthos. The basins of the SCB are the ultimate sink for sediments, including contaminants, but the effects of accumulating contamination on basin benthos have not been studied.

The tremendous horizontal and vertical

gradients of physical and chemical conditions, physiography, local upwelling centers, and the presence of 16 different basins in the SCB provide a unique area and exciting possibilities for the study of factors that control the distribution and abundance of benthic organisms and their assemblage structure and function. Advances in deep-ocean technology, such as ROVs and submersibles, provide access to deep-sea areas not accessible until recently.

Finally, benthic invertebrates are important intermediate links in marine food chains, and they are but one part of the marine ecosystem. Little is known about food chain or energy flow relationships in any of the benthic assemblages of the region. The nature and magnitudes of their interactions with other ecosystem components no doubt differ for each benthic assemblage, but these interactions and differences are virtually unknown. The ecology of the benthic invertebrates of the region cannot be understood without consideration of their role in the entire ecosystem.

Literature Cited

Abrams, M. A., 1979. Quantitative analysis of the benthic Foraminifera in Santa Monica Basin, California. Master's Thesis, Univ. of Southern California, Los Angeles. 188pp.

Allan Hancock Foundation, 1965. An oceanographic and biological survey of the southern California mainland shelf. State Water Quality Control Board, Sacramento. Publ. No. 27. Prep. by Allan Hancock Foundation, Univ. of Southern California, Los Angeles. 232 pp.

Allan Hancock Foundation, 1975. Environmental investigations and analysis, Los Angeles Harbor. Allan Hancock Foundation, Univ. of Southern California, Los Angeles. 582pp.

Aller, R. C., and J. Y. Yingst, 1978. Biogeochemistry of tube-dwellings: A study of the sedentary polychaete *Amphitrite ornata*. *J. Mar. Res.* 36:201–254.

Aller, R. C., and J. Y. Yingst, 1985. Effects of the marine deposit-feeders *Heteromastus filiformis* (Polychaeta), *Macoma balthica* (Bivalvia), and *Tellina texana* (Bivalvia) on averaged sedimentary solute transport, reaction rates, and microbial distributions. *J. Mar. Res.* 43:615–645.

Ambrose, R. F., 1980. Importance of octopus predation in a southern California subtidal community. *Am. Zool.* 20(4):884.

Anderhalt, R., and W. E. Reed, 1978. Sedimentary and organic characterization. Southern California Baseline Study, Benthic Year 2. Vol. III, Rep. 15.0. Rep. SAI–77–917-LJ. Prep. by Science Applications, Inc., La Jolla, CA.

Anderson, S. L., L. W. Botsford, and W. H. Clark, Jr., 1985. Size distributions and sex ratios of ridgeback prawns (*Sicyonia ingentis*) in the Santa Barbara Channel (1979–1981). *Calif. Coop. Oceanic Fish. Invest. Rep.* 26:169–174.

Atlantis Scientific, 1972. Environmental impact analysis for the San Elijo Regional Sewerage System. Phase I. Oceanographic and shoreline study. Phase II. Lagoon, creek, and upland study. Final Rep. Prep. for Environmental Development Agency, County of San Diego. Prep. by Atlantis Scientific, Beverly Hills, CA. 221pp.

Auyong, J. S. H., 1981. Comparison of population structure of sand crabs (*Emerita analoga* Stimpson) living at increasing distances from a power plant. Master's Thesis, Univ. of California, Santa Barbara. 79pp.

Baker, J. H., 1975. Distributions, ecology, and life histories of selected Cypridinacea (Myodocopida, Ostracoda) from the southern California mainland shelf. Ph.D. Dissertation, Univ. of Houston, Houston, TX. 185pp.

Bakus, G. J., 1989. The marine biology of southern California. Occas. Pap. (New Ser.) No. 7, Allan Hancock Foundation, Univ. of Southern California, Los Angeles. 61pp.

Barham, E. G., N. J. Ayer, Jr., and R. E. Boyce, 1967. Macrobenthos of the San Diego Trough: Photographic census and observations from bathyscaphe, Trieste. *Deep-Sea Res.* 14:773–784.

Barnard, J. L., 1963. Relationship of benthic Amphipoda to invertebrate communities of inshore sublittoral sands of southern California. *Pac. Nat.* 3:439–467.

Barnard, J. L., 1970. Benthic ecology of Bahia de San Quintin, Baja, California. *Smithson. Contrib. Zool.* No. 44, 60pp.

Barnard, J. L., and O. Hartman, 1959. The sea bottom off Santa Barbara, California: Biomass and community structure. *Pac. Nat.* 1(6):1–16.

Barnard, J. L., and D. J. Reish, 1959. Ecology of Amphidoda and Polychaeta of Newport Bay, California. Occas. Pap. No. 21. Prep. by Allan Hancock Foundation, Univ. of Southern California, Los Angeles. 106pp.

Barnard, J. L., and F. C. Ziesenhenne, 1961. Ophiuroid communities of southern California coastal bottoms.. *Pac. Nat.* 2(2):131–152.

Barnes, A. T., L. B. Quetin, and J. J. Childress, 1976. Deep-sea macroplanktonic sea cucumbers: Suspended sediment feeders captured from deep submergence vehicle. *Science.* 194: 1083–1085.

Barnes, N. B., and A. M. Wenner, 1968. Seasonal variation in the sand crab *Emerita analoga* (Decapoda, Hippidae) in the Santa Barbara area of California. *Limnol. Oceanogr.* 13(3):465–475.

Bernstein, B. B., and N. Jung, 1979. Selective pressures and coevolution in a kelp canopy community in southern California. *Ecol. Monogr.* 49(3):335–355.

Bock, C. E., and R. E. Johnson, 1967. The role of behavior in determining the intertidal zonation of *Littorina planaxis* and *Littorina scutulata* Gould, 1849. *Veliger.* 10:42–54.

Breitburg, D. L., 1983. Residual effects of grazing: Inhibition of competitor recruitment by encrusting coralline algae. *Ecology.* 65:1136–1143.

Breitburg, D. L., 1985. Development of a subtidal epibenthic community: Factors affecting species composition and the mechanisms of succession. *Oecologia (Berl.).* 65:173–184.

Brenchley, G. A., 1981. Disturbance and community structure: An experimental study of bioturbation in marine soft-bottom environments. *J. Mar. Res.* 39:767–790.

Bullock, T. H., 1953. Predator recognition and escape responses of some intertidal gastropods in presence of starfish. *Behaviour.* 5:130–140.

Burnett, B. R., 1981. Quantitative sampling of nanobiota (microbiota) of the deep-sea benthos—III. The bathyal San Diego Trough. *Deep-Sea Res.* 28A(7):649–663.

Butler, A. J., 1979. Relationships between height on the shore and size distributions of *Thais* spp. (Gastropoda:Muricidae). *J. Exp. Mar. Biol. Ecol.* 41:163–194.

Butman, C. A., 1987. Larval settlement of soft-sediment invertebrates: The spatial scales of pattern explained by active habitat selection and the merging role of hydrodynamical processes. *Oceanogr. Mar. Biol. Annu. Rev.* 25:113–165.

California State Coastal Conservancy, 1982. Buena Vista Lagoon watershed enhancement program. Staff Rep. Oakland, CA.

Cameron, R. A., and S. C. Schroeter, 1980. Sea urchin recruitment: Effect of substrate selection on juvenile distribution. *Mar. Ecol. Prog. Ser.* 2:243–247.

Carlisle, J. G., Jr., J. W. Schott, and N. J. Abramson, 1960. The barred surfperch (*Amphisticus argenteus* Agassiz) in southern California. *Calif. Dept. Fish Game Fish Bull.* No. 109. 79pp.

Carlton, J. T., 1985. Transoceanic and interoceanic dispersal of coastal marine organisms: The biology of ballast water. *Oceanogr. Mar. Biol. Annu. Rev.* 23:313–371.

Choat, J. H., 1977. The influence of sessile organisms on the population biology of three species of acmaeid limpets. *J. Exp. Mar. Biol. Ecol.* 26:1–26.

Cimberg, R. L., and R. W. Smith, 1987. Distribution of benthic communities relative to oxygen levels near Pt. Conception, California. Proceedings, Third California Island Symposium. Santa Barbara Museum of Natural History.

Coe, W. R., 1953. Resurgent populations of littoral marine invertebrates and their dependence on ocean currents and tidal currents. *Ecology.* 33:225–229.

Coe, W. R., 1955. Ecology of the bean clan *Donax gouldi* on the coast of southern California. *Ecology.* 36:512–514.

Connell, J. H., 1970. A predator–prey system in the marine intertidal region. I. *Balanus glandula* and several predatory species of *Thais. Ecol. Monogr.* 40:49–78.

Connell, J. H., 1972. Community interactions on marine rocky intertidal shores. *Annu. Rev. Ecol. Syst.* 3:169–192.

Connell, J. H., 1975. Some mechanisms producing structure in natural communities: A model and evidence from field experiments. In: M. L. Cody, and J. M. Diamond, eds. *Ecology and Evolution of Communities.* Belknap Press, Cambridge, MA. pp. 460–490.

Connell, J. H., 1986. Variation and persistence of rocky shore populations. In: P. G. Moore, and R. Seed, eds. *The Ecology of Rocky Coasts.* Columbia Univ. Press, NY. pp. 57–69.

Cowen, R. K., 1983. The effect of sheephead (*Semicossyphus pulcher*) predation on red sea urchin (*Strongylocentrotus franciscanus*) populations: An experimental analysis. *Oecologia.* 58: 249–255.

Cox, K. W., 1962. California abalones, family Haliotidae. *Calif. Dept. Fish Game Fish Bull.* No. 118. 133pp.
Coyer, J. A., J. M. Engle, R. F. Ambrose, and B. V. Nelson, 1987. Utilization of purple and red sea urchins (*Strongylocentrotus purpuratus* Stimpson and *S. franciscanus* Agassiz) as food by the white sea urchin (*Lytechinus anamesus* Clark) in the field and laboratory. *J. Exp. Mar. Biol. Ecol.* 105:21–38.
Craig, P. C., 1968. Natural history of an intertidal sand beetle, *Thinopinus pictus* (Staphylinidae). Master's Thesis, Univ. of California, Santa Barbara. 49pp.
Craig, P. C., 1970. The behavior and distribution of the intertidal sand beetle, *Thinopinus pictus* (Coleoptera:Staphylinidae). *Ecology.* 51:1012–1017.
Cubit, J., 1968. Behavior and physical factors causing migration and aggregation of the sand crab *Emerita analoga* (Stimpson). *Ecology.* 50(1): 118–123.
Cubit, J., 1970. The effects of the 1969 Santa Barbara oil spill on marine intertidal invertebrates. In: R. W. Holmes, and F. A. DeWitt, eds. *Santa Barbara Oil Symposium—Offshore Petroleum Production, an Environmental Inquiry.* Natl. Sci. Found., Div. Grad. Educ. Sci. and Mar. Sci. Inst., Univ. of California, Santa Barbara. pp. 131–136.
Cubit, J. D., 1984. Herbivory and the seasonal abundance of algae on a high intertidal rocky shore. *Ecology.* 65:1904–1917.
Davies, P. S., 1969. Physiological ecology of *Patella.* III. Desiccation effects. *J. Mar. Biol. Assoc. U.K.* 38:291–304.
Davis, N., 1978. Studies of the southern California nearshore sand bottom community. Ph.D. Dissertation, Univ. of California, San Diego.
Davis, N., and G. R. VanBlaricom, 1978. Spatial and temporal heterogeneity in a sand bottom epifaunal community of invertebrates in shallow water. *Limnol. Oceanogr.* 23(3):417–427.
Day, R. W., and R. W. Osman, 1981. Predation by *Patiria miniata* (Asteroidea) on bryozoans: Prey diversity may depend on the mechanisms of succession. *Oecologia (Berl.).* 51(3):300–309.
Dayton, P. K., 1971. Competition, disturbance and community organization: The provision and subsequent utilization of space in a rocky intertidal community. *Ecol. Monogr.* 45:137–159.
Dayton, P. K., and M. Tegner, 1984. Catastrophic storms, El Niño and patch stability in a southern California kelp community. *Science.* 224: 283–285.
Dayton, P. K., V. Currie, T. Gerrodette, B. D. Keller, R. Rosenthal, and D. Ven Tresca, 1984. Patch dynamics and stability of some California kelp communities. *Ecol. Monogr.* 54:253–289.
Dean, R. L., and J. H. Connell, 1987a. Marine invertebrates in an algal succession. 1. Variations in abundance and diversity with succession. *J. Exp. Mar. Biol. Ecol.* 109(3):195–215.
Dean, R. L., and J. H. Connell, 1987b. Marine invertebrates in an algal succession. 2. Tests of hypotheses to explain changes in diversity with succession. *J. Exp. Mar. Biol. Ecol.* 109(3):217–247.
Dean, R. L., and J. H. Connell, 1987c. Marine invertebrates in an algal succession. 3. Mechanisms linking habitat complexity with diversity. *J. Exp. Mar. Biol. Ecol.* 109(3):249–273.
Dean, T. A., S. C. Schroeter, and J. D. Dixon, 1984. Effects of grazing by two species of sea urchins (*Stongylocentrotus franciscanus* and *Lytechinus anamesus*) on recruitment and survival of two species of kelp (*Macrocystis pyrifera* and *Pterygophora californica*). *Mar. Biol.* 78(3):301–313.
Dean, T. A., S. C. Schroeter, J. D. Dixon, L. E. Deysher, and F. R. Jacobsen, 1987. The effects of the operation of the San Onofre Nuclear Generating Station on the giant kelp, *Macrocystis pyrifera*: Background information and the biology of kelp. Final Report to the Marine Review Committee of the California Coastal Commission. December 31, 1987.
Denny, M. W., 1985. Wave forces on intertidal organisms: A case study. *Limnol. Oceanogr.* 30(6):1171–1187.
Denny, M. W., T. L. Daniel, and M. A. R. Koehl, 1985. Mechanical limits to size in wave-swept organisms. *Ecol. Monogr.* 55(1):69–102.
Dexter, D. M., 1978. The infauna of a subtidal, sand-bottom community at Imperial Beach, California. *Calif. Fish Game.* 64(4):268–279.
Dexter, D. M., 1983. Soft bottom infaunal communities in Mission Bay, California. *Calif. Fish Game.* 69(1):5–17.
Dillery, D. G., and L. V. Knapp, 1969. Longshore movements of the sand crab, *Emerita analoga* (Decapoda, Hippidae). *Crustaceana.* 18:233–240.
Dixon, J. D., 1978. Determinants of the local distribution of four closely related species of

herbivorous marine snails. Ph.D. Dissertation, Univ. of California, Santa Barbara. 235pp.
Doering, P. M., and D. W. Phillips, 1983. Maintenance of the shore-level size gradient in the marine snail *Tegula funebralis* (A. Adams): Importance of behavioral responses to light and sea star predators. *J. Exp. Mar. Biol. Ecol.* 67: 159–173.
Duggins, D. O., 1983. Starfish predation and the creation of mosaic patterns in a kelp-dominated community. *Ecology*. 64(6):1610–1619.
Ebeling, A. W., D. R. Laur, and R. J. Rowley, 1985. Severe storm disturbances and reversal of community structure in a southern California kelp forest. *Mar. Biol.* 84:287–294.
Eckman, J. E., 1985. Flow disruption by an animal-tube mimic affects sediment bacterial colonization. *J. Mar. Res.* 43:419–435.
Eckman, J. E., A. Nowell, and P. A. Jumars, 1981. Sediment destabilization by animal tubes. *J. Mar. Res.* 39:361–374.
Edwards, B. D., 1985. Bioturbation in a dysaerobic bathyal basin: California borderland. In: *Soc. Econ. Paleon. Mineral. Special Publ. 35.* pp. 309–331.
Efford, I. E., 1965. Aggregation in the sand crab, *Emerita analoga* (Stimpson). *J. Anim. Ecol.* 34:63–75.
Emery, K. O., and J. Hülsemann, 1962. The relationships of sediments, life and water in a marine basin. *Deep-Sea Res.* 8:165–180.
Engle, J. M., 1979. Ecology and growth of juvenile California spiny lobster, *Panulirus interruptus* (Randall). Ph.D. Dissertation, Univ. of Southern California, Los Angeles.
Fager, E. W., 1964. Marine sediments: Effects of a tube-building polychaete. *Science*. 143:356–359.
Fager, E. W., 1968. A sand-bottom epifaunal community of invertebrates in shallow water. *Limnol. Oceanogr.* 13:448–464.
Fauchald, K., 1971. The benthic fauna in the Santa Barbara Channel following the January 1969 oil spill. In: *Biological and Oceanographical Survey of the Santa Barbara Channel Oil Spill, 1969–1970.* Vol. 1. Biology and bacteriology. Sea Grant Publ. No. 2. Allan Hancock Foundation, Univ. of Southern California, Los Angeles. pp. 61–116.
Fauchald, K., and G. F. Jones, 1979a. A survey of five additional southern California study sites. In: *Southern California Outer Continental Shelf Environmental Baseline Study, 1976/1977 (Second Year) Benthic Program*. Vol. II, Principal Invest. Reps., Ser. 2, Rep. 18. Available from: NTIS, Springfield, VA; PB80–166101. Science Applications, Inc., La Jolla, CA.
Fauchald, K., and G. F. Jones, 1979b. Variation in community structures on shelf, slope and basin macrofaunal communities of the Southern California Bight. In: *Southern California Outer Continental Shelf Environmental Baseline Study, 1976/1977 (Second Year) Benthic Program.* Vol. II, Principal Invest. Reps., Ser. 2, Rep. 19. Available from: NTIS, Springfield, VA; PB80–166101. Science Applications, Inc., La Jolla, CA.
Fauchald, K., and G. F. Jones, 1983. Benthic macrofauna. In: *Southern California Baseline Studies and Analysis, 1975/1976.* Vol. III, Rep. 2.4 (SAI–76–809-LJ). Available from: NTIS, Springfield, VA; PB83–149799. Science Applications, Inc., La Jolla, CA.
Fauchald, K., and P. A. Jumars, 1979. The diet of worms: A study of polychaete feeding guilds. *Oceanogr. Mar. Biol. Annu. Rev.* 17:192–284.
Fawcett, J. J., 1969. Zonation and temporal distribution of three species of beach-dwelling amphipods of the genus *Orchestoidea* (Talitridae). Master's Thesis, Univ. of California, Santa Barbara. 94pp.
Fawcett, M. H., 1979. The consequences of latitudinal variation in predation for some marine intertidal herbivores. Ph.D. Dissertation, Univ. of California, Santa Barbara. 156pp.
Fawcett, M. H., 1984. Local and latitudinal variation in predation on an herbivorous marine snail. *Ecology*. 65(4):1214–1230.
Feder, R. G., 1963. Gastropod defensive responses and their effectiveness in reducing predation by starfish. *Ecology*. 44:505–512.
Field, M. E., and B. D. Edwards, 1980. Slopes of the southern California continental borderland: A regime of mass transport. In: M. E. Field, A. H. Bouma, I. P. Colburn, R. G. Douglas, and J. C. Ingle, eds. *Proceedings of the 4th Pacific Coast Paleogeography Symposium, Quaternary Depositional Environments of the Pacific Coast.* Bakersfield, CA, April 9, 1980. pp. 169–184.
Fitch, J. E., and R. J. Lavenberg, eds., 1971. *Marine Food and Game Fishes of California*. Univ. of California Press, Berkeley, CA. 179pp.
Foster, B. A., 1971. Desiccation as a factor in the intertidal zonation of barnacles. *Mar. Biol.* 8:12–29.
Foster, M. S., 1982. The regulation of macroalgal

associations in kelp forests. In: L. M. Srivastava, ed. *Synthetic and Degradative Processes in Marine Macrophytes*. Walter de Gruyter Co., Berlin. pp. 185–205.

Foster, M. S., and D. R. Schiel, 1985. Ecology of giant kelp forests in California: A community profile. *U.S. Fish Wildl. Serv. Biol. Rep.* 85 (7.2). 152pp.

Fotheringham, N., 1974. Trophic complexity in a littoral boulder field. *Limnol. Oceanogr.* 19(1): 84–91.

Frank, P. W., 1965. The biodemography of an intertidal snail population. *Ecology*. 46:831–844.

Frank, P. W., 1982. Effects of winter feeding on limpets by black oystercatchers, *Haematopus bachmani*. *Ecology*. 63:1352–1362.

Gerard, V. A., 1976. Some aspects of material dynamics and energy flow in a kelp forest in Monterey Bay, California. Ph.D. Dissertation, Univ. of California, Santa Cruz. 173pp.

Gilles, K. W., and J. S. Pearse, 1986. Studies on disease in sea urchins, *Strongylocentrotus purpuratus*: Experimental infection and bacterial virulence. *Dis. Aquat. Organisms*. 1:105–114.

Grassle, J. F., H. L. Sanders, R. R. Hessler, G. T. Rowe, and T. McLellan, 1975. Pattern and zonation: A study of the bathyal megafauna using the research submersible *Alvin*. *Deep-Sea Res.* 22:457–481.

Greenwood, R. S., and R. O. Browne, 1969. A coastal Chumash village: Excavation of Shisholop, Ventura County, California. *Mem. South. Calif. Acad. Sci.* 8:1–79.

Grigg, R. W., 1972. Orientation and growth form of sea fans. *Limnol. Oceanogr.* 17:185–192.

Hardin, D. D., 1968. A comparative study of lethal temperatures in the limpets *Acmaea scabra* and *Acmaea digitalis*. *Veliger*. 11(Suppl.):83–87.

Hardin, D., R. Spies, and T. Parr, 1988. Hard bottom epifaunal assemblages. Chapter 4. In: J. Hyland, and J. Neff, eds. *California OCS Phase II Monitoring Program*. Year One, Annual Rep. Vol. 1. MMS 87–0115. Available from: NTIS, Springfield, VA; PB89–128151/AS. Battelle Ocean Sciences.

Harger, J. R. E., 1968. The role of behavioral traits in influencing the distribution of two species of sea mussel, *Mytilus edulis* and *Mytilus californianus*. *Veliger*. 11(1):45–48.

Harger, J. R. E., 1970a. Comparisons among growth characteristics of two species of sea mussel, *Mytilus edulis* and *Mytilus californianus*. *Veliger*. 13:44–56.

Harger, J. R. E., 1970b. The effect of wave impact on some aspects of the biology of sea mussels. *Veliger*. 12:401–414.

Harger, J. R. E., 1970c. The role of behavioral traits in influencing the distribution of two species of sea mussel *Mytilus edulis* and *Mytilus californianus*. *Veliger*. 11:45–49.

Harger, J. R. E., 1970d. The effect of species composition on the survival of mixed populations of the sea mussels *Mytilus californianus* and *Mytilus edulis*. *Veliger*. 13:147–152.

Harger, J. R. E., 1972a. Competitive coexistence among intertidal invertebrates. *Am. Sci.* 60: 600–607.

Harger, J. R. E., 1972b. Variation and relative 'niche' size in the sea mussel *Mytilus edulis* in association with *Mytilus californianus*. *Veliger*. 14:275–281.

Harger, J. R. E., 1972c. Competitive co-existence: Maintenance of interacting associations of the sea mussels *Mytilus edulis* and *Mytilus californianus*. *Veliger*. 14:387–410.

Harger, J. R. E., and D. E. Landenberger, 1971. The effect of storms as a density dependent mortality factor on populations of sea mussels. *Veliger*. 14:195–201.

Harrold, C., and J. S. Pearse, 1987. The ecological role of echinoderms in kelp forests. *Echinoderm Stud.* 2:137–233.

Harrold, C., and D. C. Reed, 1985. Food availability, sea urchin grazing, and kelp forest community structure. *Ecology*. 66:1160–1169.

Hartman, O., 1955. Quantitative survey of the benthos of San Pedro Basin, southern California, Part I, preliminary results. *Allan Hancock Pacific Expeditions*. 19(1):1–185.

Hartman, O., 1963. Submarine canyons of southern California, Part II, biology. *Allan Hancock Pacific Expeditions*. 27(2):1–424.

Hartman, O., 1966. Quantitative survey of the benthos of San Pedro Basin, southern California, Part II, final results and conclusions. *Allan Hancock Pacific Expeditions*. 19(2):186–456.

Hartman, O., and J. L. Barnard, 1958. The benthic fauna of the deep basins off southern California. *Allan Hancock Pacific Expeditions*. 22(1): 1–67.

Hartman, O., and J. L. Barnard, 1960. The benthic fauna of the deep basins off southern California. *Allan Hancock Pacific Expeditions*. 22(2):69–297.

Haven, S. B., 1971. Niche differences in the intertidal limpets *Acmaea scabra* and *Acmaea digitalis*

(Gastropoda) in central California. *Veliger*. 13: 231–248.

Haven, S. B., 1973. Competition for food between the intertidal gastropods *Acmaea scabra* and *Acmaea digitalis*. *Ecology*. 54:143–151.

Hessler, R. R., and P. A. Jumars, 1974. Abyssal community analysis from replicate box cores in the central North Pacific. *Deep-Sea Res*. 21: 185–209.

Hewatt, W. G., 1940. Observations on the homing limpet, *Acmaea scabra* Gould. *Am. Midl. Nat*. 24:205–208.

Hines, A. H., 1978. Reproduction in three species of intertidal barnacles from central California. *Biol. Bull. (Woods Hole)*. 154:262–281.

Hines, A. H., 1979. The comparative reproductive ecology of three species of intertidal barnacles. In: S. E. Stancyk, ed. *Reproductive Ecology of Marine Invertebrates*. Univ. of South Carolina Press, Columbia, SC. pp. 213–234.

Hülsemann, J., and K.O. Emery, 1961. Stratification in recent sediments of Santa Barbara Basin as controlled by organisms and water character. *J. Geol*. 69:279–290.

Hylleberg, J., and K. Henriksen, 1980. The central role of bioturbation in sediment mineralization and element recycling. *Ophelia*. 1(Suppl.): 1–16.

James, R. C. and D. J. Reish, 1978. The influence of the Los Angeles River on benthic polychaetous annelids. In: D. F. Soule and M. Oguri, eds. *Marine studies of San Pedro Bay, California. Part 14. Biological investigations*. Harbors Environmental Projects, Allan Hancock Foundation, and the Office of Sea Grant Programs Institute of Marine and Coastal Studies, Univ. South. Calif., Los Angeles, CA. pp. 115–134.

Jones, G. F., 1963. Brood protection in three southern California species of the pelecypod *Cardita*. *Wassmann J. Biol*. 21(2):141–148.

Jones, G. F., 1969. The benthic macrofauna of the mainland shelf of southern California. *Allan Hancock Monogr. Mar. Biol*. Vol. 4. 219pp.

Jones, G. F., and B. E. Thompson, 1984. The ecology of *Parvilucina tenuisculpta* (Carpenter, 1864) (Bivalvia: Lucinidae) on the southern California borderland. *Veliger*. 26:188–198.

Jones, G. F., and B. E. Thompson, 1987a. The ecology of *Cyclocardia ventricosa* (Gould 1850) (Bivalvia: Carditidae) on the southern California borderland. *Veliger*. 29(4):374–383.

Jones, G. F., and B.E. Thompson, 1987b. The distribution and abundance of *Chloeia pinnata* Moore, 1911 (Polychaeta: Amphinomidae) on the southern California borderland. *Pac. Sci*. 41:122–131.

Jumars, P. A., 1975a. Methods for measurement of structure in deep-sea macrobenthos. *Mar Biol*. 30(3):245–252.

Jumars, P. A., 1975b. Environmental grain and polychaete species' diversity in a bathyal benthic community. *Mar. Biol*. 30(3):253–266.

Jumars, P. A., 1976. Deep-sea species diversity: Does it have a characteristic scale? *J. Mar. Res*. 34:217–246.

Jumars, P. A., 1978. Spatial autocorrelation with RUM (Remote Underwater Manipulator): Vertical and horizontal structure of a bathyal benthic community. *Deep-sea Res*. 25(7):589–604.

Jumars, P. A., A. R. M. Nowell, and R. F. L. Self, 1981. A simple model of flow-sediment-organisms interactions. *Mar. Geol*. 42:155–172.

Kastendiek, J. E., 1975. The role of behavior and interspecific interactions in determining the distribution and abundance of *Renilla kollikeri* Pfeffer, a member of a subtidal sand bottom community. Ph.D. Dissertation, Univ. of California, Los Angeles.

Kastendiek, J. E., 1976. Behavior of the sea pansy *Renilla kollikeri* Pfeffer (Coelenterata: Pennatulacea) and its influence on the distribution and biological interactions of the species. *Biol. Bull. (Woods Hole)*. 151:518–537.

Kastendiek, J. E., 1982. Factors determining the distribution of the sea pansy, *Renilla kollikeri*, in a subtidal sand–bottom habitat. *Oecologia (Berl.)*. 52:340–347.

Kato, S., 1972. Sea Urchins: A new fishery develops in California. *Mar. Fish. Rev*. 34:23–30.

Kato, S., and S. C. Schroeter, 1985. Biology of the red sea urchin, *Strongylocentrotus franciscanus*, and its fishery in California. *Mar. Fish. Rev*. 47(3):1–20.

Kauwling, T. J., and D. J. Reish, 1975. A quantitative study of the benthic polychaetous annelids of Anaheim Bay and Huntington Harbour, California. In: E. D. Lane, and C. W. Hill, eds. The marine resources of Anaheim Bay. Calif. *Dep. Fish Game Fish Bull*. No. 165. pp. 57–68.

Kensler, C. B., 1967. Desiccation resistance of intertidal crevice species as a factor in their zonation. *J. Anim. Ecol*. 36:391–406.

Kinnetics Laboratories Inc., 1988. Texaco USA site-specific faunal characterization survey for Platform Harvest OCS lease P–0315, Point Conception, California (unpubl. report), Carlsbad, CA.

Kirchman, D., S. Graham, D. Reish, and R. Mitchell, 1982. Lectins may mediate in the settlement and metamorphosis of *Janua (Dexiospira) brasiliensis* Grube (Polychaete: Spirorbidae). *Mar. Biol. Lett.* 3:131–142.

Kitching, J. A., and F. J. Ebling, 1967. Ecological studies at Lough Ine. *Adv. Ecol. Res.* 4:197–291.

Landenberger, D. E., 1967. A study of predation and predatory behavior in the Pacific starfish, *Pisaster*. Ph.D. Dissertation, Univ. of California, Santa Barbara.

Lane, E. D., 1975. Quantitative aspects of the life history of the diamond turbot, *Hypsopsetta guttulata* (Girard), in Anaheim Bay. In: E. D. Lane, and C. W. Hill, eds. *The Marine Resources of Anaheim Bay*. Calif. Dept. Fish Game Fish. Bull. No. 165. pp. 153–173.

Lane, E. D., and C. W. Hill eds., 1975. The marine resources of Anaheim Bay. *Calif. Dep. Fish. Game Fish.* Bull. No. 165. 195pp.

Laughlin, J. D., and B. E. Thompson. Distribution and feeding of asteroids from San Pedro and Santa Monica basin slopes off southern California. Unpubl. manuscript.

Laur, D. R., A. W. Ebeling, and D. C. Reed, 1986. Experimental evaluations of substrate types as barriers to sea urchin (*Strongylocentrotus* spp.) movement. *Mar. Biol.* 93:209–215.

Lawrence, J. M. 1975. On the relationships between marine plants and sea urchins. *Oceanogr. Mar. Biol. Annu. Rev.* 13:213–286.

Lees, D. C., and G. A. Carter, 1972. The covering response to surge, sunlight, and ultraviolet light in *Lytechinus anamesus* (Echinoidea). *Ecology*. 53:1127–1133.

Leighton, D. L., 1971. Grazing activities of benthic invertebrates in southern California kelp beds. *Nova Hedwegia*. 32:421–453.

Levin, L. A., 1981. Dispersion feeding behavior and competition in two spionid polychaetes. *J. Mar. Res.* 39:99–117.

Levin, L. A., 1984. Life history and dispersal patterns in a dense infaunal polychaete assemblage: Community structure and response to disturbance. *Ecology*. 65(4):1185–1200.

Levin, L. A., and C. R. Smith, 1984. Response of background fauna to disturbance and enrichment in the deep sea: A sediment tray experiment. *Deep-Sea Res.* 31:1277–1285.

Lewbel, G. S., A. Wolfson, T. Gerrodette, W. H. Lippincott, J. L. Wilson, and M. M. Littler, 1981. Shallow-water benthic communities on California's outer continental shelf. *Mar. Ecol. Prog. Ser.* 4(2):159–168.

Lewis, J. R., 1964. *The Ecology of Rocky Shores*. The English Universities Press Ltd., London. 323pp.

Lie, U., 1968. A quantitative study of benthic infauna in Puget Sound, Washington, U. S. A., in 1963–1964. *Fiskeridir. Skr. Ser. Havunders.* 14(5):229–556.

Lindberg, D. R., and E. W. Chu, 1983. Western gull predation on owl limpets: Different methods at different localities. *Veliger*. 25:347–348.

Lindberg, D. R., and K. R. Dwyer, 1983. The topography, formation and role of the home depression of *Collisella scabra* (Gould). *Veliger*. 25:229–234.

Lissner, A. L., 1978. Factors affecting the distribution and abundance of the sea urchin *Centrostephanus coronatus* Verrill at Santa Catalina Island. Ph.D. Dissertation, Univ. of Southern California, Los Angeles. 168pp.

Lissner, A. L., 1980. Some effects of turbulence on the activity of the sea urchin *Centrostephanus coronatus* Verrill. *J. Exp. Mar. Biol. Ecol.* 48: 185–193.

Lissner, A. L., 1983. Relationship of water motion to the shallow water distribution and morphology of two species of sea urchins. *J. Mar. Res.* 41:691–709.

Lissner, A. L., and J. H. Dorsey, 1986. Deep-water biological assemblages of a hard-bottom bank-ridge complex of the southern California continental borderland. *Bull. South. Calif. Acad. Sci.* 85:87–101.

Littler, M. M., ed., 1979. The distribution, abundance and community structure of rocky intertidal and tidepool biotas in the Southern California Bight. In: *Southern California Baseline Study, Year Three.* U.S. Dept. of Interior, Bureau of Land Management, Washington, D.C.

Littler, M. M., 1980. Overview of the rocky intertidal systems of southern California. In: D. M. Power, ed. *The California Islands: Proceedings of a Multidisciplinary Symposium*. Haagen Printing, Santa Barbara, CA. pp. 265–306.

Littler, M. M., and D. R. Martz, 1979. Assessments of the distribution, abundance and community structure of rocky intertidal organisms near South Point, Santa Rosa Island. Part 1.14. In : M. M. Littler, ed. *The Distribution, Abundance and Community Structure of Rocky Intertidal and Tidepool Biotas in the Southern California Bight.* U.S. Dept. of Interior, Bureau of Land Management, Washington, D.C.

Littler, M. M., D. R. Martz, and D. S. Littler, 1983. Effects of recurrent sand deposition on rocky intertidal organisms: Importance of substrate heterogeneity in a fluctuating environment. *Mar. Ecol. Prog. Ser.* 11(2):129–140.

Ljubenkov, J., and B. Thompson. A tripartite commensal relationship from southern California with the description of a new hydrozoan. Unpubl. manuscript.

MacDonald, K. B., 1976. Natural resources of Carpinteria Marsh. *Calif. Dep. Fish Game, Coastal Wetlands Ser.* No. 13.

MacGinitie, G. E., 1938. Movements and mating habits of the sand crab, *Emerita analoga. Am. Midl. Nat.* 19:471–481.

MacGinitie, G. E., and N. MacGinitie, 1949. *Natural History of Marine Animals.* MacGraw-Hill, New York. 473pp.

MacKenthum, K. M., and L. P. Parrish, 1968. *San Diego Bay. An evaluation of the benthic environment.* U.S. Dep. Interior, Tidal Water Pollution Control Administration. Tech. Rep.

McLachlan, A., and G. C. Bate, 1983. Sandy beach ecology-workshop report. In: A. McLachlan, and T. Erasmus, eds. *Sandy Beaches as Ecosystems. Developments in Hydrobiology, 19.* Dr. W. Junk Publishers, The Hague, The Netherlands. pp. 569–572.

McLachlan, A., and P. Hesp, 1984. Faunal response to morphology and water circulation of a sandy beach with cusps. *Mar. Ecol. Prog. Ser.* 19:133–144.

McLean, J. H., 1979. A new monoplacophoran limpet from the continental shelf off southern California. *Contrib. Sci. Nat. Hist. Mus. Los Angeles County.* 307:1–19.

Margolin, A. S., 1964. A running response of *Acmaea* to sea stars. *Ecology.* 45:191–193.

Marine Biological Consultants, Inc., 1972. Survey of biological conditions off Aliso Creek and Doheny State Beach, Orange County, California, July 1972. Prep. for Boyle Engineering. Prep. by Marine Biological Consultants, Inc., Costa Mesa, CA.

Marine Biological Consultants, Inc., 1974. Organisms in Goleta and Carpinteria Sloughs. Prep. for the University of California, Santa Barbara. Prep. by Marine Biological Consultants, Inc., Costa Mesa, CA.

Marine Biological Consultants and Southern California Coastal Water Research Project, 1980. Irvine Ranch Water District Upper Newport Bay and Stream Augmentation Program. Final Report. Costa Mesa and Long Beach, CA. 57pp.

Mattison, J. E., J. D. Trent, A. L. Shanks, T. B. Akin, and J. S. Pearse, 1977. Movement and feeding activity of red sea urchins (*Strongylocentrotus franciscanus*) adjacent to a kelp forest. *Mar. Biol.* 39:25–30.

Mearns, A. J., K. V. Pierce, and M. J. Sherwood, 1978. Health, abundance and diversity of bottomfish and shellfish populations at proposed and existing offshore drilling sites in the Southern California Bight. Science Applications Inc., La Jolla, CA. 91pp.

Menge, B. A., and J. P. Sutherland, 1976. Species diversity gradients: Synthesis of the roles of predation, competition and temporal heterogeneity. *Am. Nat.* 110:351–369.

Menge, B. A., and J. P. Sutherland, 1987. Community regulation: Variation in disturbance, competition, and predation in relation to environmental stress and recruitment. *Am. Nat.* 130:730–757.

Menzies, R. J., R. Y. George, and G. T. Rowe, 1973. *Abyssal Environment and Ecology of the World Oceans.* John Wiley & Sons, New York. 488pp.

Merrill, R. J., and E. S. Hobson, 1970. Field observations of *Dendraster excentricus,* a sand dollar of western North America. *Am. Midl. Nat.* 83: 595–624.

Miller, D. C., and P. A. Jumars, 1986. Pellet accumulation, sediment supply, and crowding as determinants of surface deposit-feeding rate in *Pseudopolydora kempi japonica. J. Exp. Mar. Biol. Ecol.* 99:1–17.

Miller, D. C., P. A. Jumars, and A. R. M. Nowell, 1984. Effects of sediment transport on deposit feeding: Scaling arguments. *Limnol. Oceanogr.* 29:1202–1217.

Minard, D., 1968. Water quality in submarine basins off southern California. U.S. Dept. of Interior, Pacific Southwest Region. 31pp.

Minerals Management Service, 1988. Review of recovery and recolonization of hard substrate communities of the outer continental shelf. In: A. Lissner, ed. *MMS Rep. 88–0034.* U.S. Dept. of Interior, Minerals Management Service.

Moitozoa, D. J., and D. W. Phillips, 1979. Prey

defense, predator preference, and on-random diet: The interactions between *Pycnopodia helianthoides* and two species of urchins. *Mar. Biol.* 53:299–304.

Morin, J. G., J. E. Kastendiek, A. Harrington, and N. Davis, 1985. Organization and patterns of interactions in a subtidal sand community on an exposed coast. *Mar. Ecol. Prog. Ser.* 27:163–185.

Morin, J. G., J. E. Kastendiek, A. Harrington, and N. Davis, 1988. Organisms of a subtidal sand community in southern California. *Bull. South. Calif. Acad. Sci.* 87:1–11.

Morse, A. N., and D. E. Morse, 1984. Recruitment and metamorphosis of *Haliotis* larvae induced by molecules uniquely available at the surfaces of crustose red algae. *J. Exp. Mar. Biol. Ecol.* 75:191–215.

Morse, D. E., M. Tegner, H. Duncan, N. Hooker, G. Trevelyan, and A. Cameron, 1980. Induction of settling and metamorphosis of planktonic molluscan (*Haliotis*) larvae 3. Signaling by metabolites of intact algae is dependant on contact. In: D. Mueller-Schwarze, and R. M. Silverstein, eds. *Chemical Signals: Vertebrates and Aquatic Invertebrates.* Plenum Press, New York. pp. 67–86.

Mudie, P. J., B. Browning, and J. Speth, 1976. The natural resources of San Dieguito and Batiquitos Lagoons. *Calif. Dep. Fish Game, Coastal Wetland Ser.* No. 12.

Murdoch, W. W., 1969. Switching in general predators: Experiments on predator specificity and stability of prey populations. *Ecol. Monogr.* 39:335–354.

Natland, M. L. 1957. Paleoecology of West Coast tertiary sediments. *Geol. Soc. Am. Mem.* 67: 543–572.

Nekton Inc. and Kinnetics Laboratories Inc., 1983. Site-specific faunal characterization survey for Platform Harvest, Point Conception, California. Prep. for Texaco, USA. 2–1 to E–18pp.

Nekton Inc. and Kinnetics Laboratories Inc., 1984. An ecological study of discharged drilling fluids on a hard bottom community in the western Santa Barbara Channel, Calif. State Lease PRC–2725.1. Interim Rep. for Texaco USA. II–1 to VII–3pp.

Nelson, B. V., and R. R. Vance, 1979. Diel foraging patterns of the sea urchin *Centrostephanus coronatus* as a predator avoidance strategy. *Mar. Biol.* 51(3):251–258.

Nichols, F. H., 1974. Sediment turnover by a deposit-feeding polychaete. *Limnol. Oceanogr.* 19:945–950.

Nichols, F. H., 1975. Dynamics and energetics of three deposit feeding benthic invertebrate populations in Puget Sound, Washington. *Ecol. Monogr.* 45(1):57–82.

Nichols, F. H., D. A. Cacchione, D. E. Drake, and J. K. Thompson, 1989. Emergence of burrowing urchins from California continental shelf sediments—A response to alongshore current reversals? *Estuarine Coastal Shelf Sci.* 29: 171–182.

Nicotri, M. E., 1977. Grazing effects of four marine intertidal herbivores on the microflora. *Ecology.* 58:1020–1032.

North, W. J., 1954. Size distribution, erosive activities, and gross metabolic efficiency of the marine intertidal snails *Littorina planaxis* and *L. scutulata. Biol. Bull.* 106:185–197.

North, W. J., and J. S. Pearse, 1970. Sea urchin population explosion in southern California coastal waters. *Science.* 167:209.

Nowell, A. R. M., P. J. Jumars, and J. E. Eckman, 1981. Effects of biological activity on the entrainment of marine sediments. *Mar. Geol.* 42:133–153.

Oliphant, M. S., 1979. California marine fish landings for 1976. *Calif. Dept. Fish Game Fish Bull.* No. 170. 56pp.

Oliver, J. S., P. N. Slattery, L. W. Hulberg, and J. W. Nybakken, 1980. Relationships between wave disturbance and zonation of benthic invertebrate communities along a subtidal high-energy beach in Monterey, California. *U.S. Natl. Mar. Fish. Serv. Fish. Bull.* 78:437–454.

O'Neill, P. L., 1978. Hydrodynamic analysis of feeding in sand dollars. *Oecologia (Berl.).* 34: 157–174.

Onuf, C. P., 1972. Aspects of the population biology of the intertidal snail *Olivella biplicata:* Distribution, nutrition and effects of natural enemies. Ph.D. Dissertation, Univ. of California, Santa Barbara. 223pp.

Onuf, C. P., 1987. The ecology of Mugu lagoon, California. An estuarine profile, 85(7.15). Prep. by U.S. Fish and Wildlife Service, National Wetlands Research Center, Washington, D.C. 122pp.

Osman, R. W., and J. A. Haugness, 1981. Mutualism among sessile invertebrates: A mediator of competition and predation. *Science.* 211:846–848.

Paine, R. T., 1966. Food web complexity and species diversity. *Am. Nat.* 100:65–75.

Paine, R. T., 1969. The *Pisaster–Tegula* interaction: Prey patches, predator food preference, and community structure. *Ecology.* 50:950–961.

Paine, R. T., 1974. Intertidal community structure. Experimental studies on the relationship between a dominant competitor and its principal predator. *Oecologia (Berl.).* 15:93–120.

Paine, R. T., 1977. Controlled manipulations in the marine intertidal zone and their contributions to ecological theory. *Spec. Publ. Acad. Nat. Sci. Phila.* 12:245–270.

Parr, T., and J. Shrake, 1989. Benthic habitat characterization survey for POCS lease blocks 0514 (Cicero Northwest), 0515 (Tamora), 0516 (Tamora East) and 0517 (Iago). Kinnetic Laboratories Inc., Carlsbad, CA.

Pawka, S. S., D. L. Inman, and R. T. Guza, 1984. Island sheltering of surface gravity waves: Model and experiment. *Cont. Shelf Res.* 3(1): 35–53.

Pequegnat, W. E., 1961a. New world for marine biologists. *Nat. Hist.* 70:8–17.

Pequegnat, W. E., 1961b. Life in the SCUBA zone. *Nat. Hist.* 70:47–55.

Pequegnat, W. E., 1964. The epifauna of a California siltstone reef. *Ecology.* 45:272–283.

Pereyra, W. T., and M. S. Alton, 1972. Distribution and relative abundance of invertebrates off the northern Oregon coast. In: A. T. Pruter, and D. L. Alverson, eds. *The Columbia River Estuary and Adjacent Ocean Waters, Bioenvironmental Studies.* Univ. of Washington Press, Seattle, WA. pp. 475–537.

Perry, D. M., 1980. Factors influencing aggregation patterns in the sand crab *Emerita analoga* (Crustacea: Hippidae). *Oecologia (Berl.).* 45: 379–384.

Perry, D. M., 1983. Optimal foraging in the predaceous rocky intertidal snail *Acanthina spirata* (Blainville, 1932). Ph.D. Dissertation, Univ. of Southern California, Los Angeles.

Peterson, C. H., 1977. Competitive organization of the soft bottom macrobenthic communities of southern California lagoons. *Mar. Biol.* 43(4):343–360.

Peterson, C. H., and M. L. Quammen, 1982. Siphon nipping: Its importance to small fishes and its impact on growth of the bivalve *Protothaca staminea* (Conrad). *J. Exp. Mar. Biol. Ecol.* 63:249–268.

Phillips, D. W., 1975. Distance chemoreception triggered avoidance behavior of the limpets *Acmaea (Collisella) limatula* and *Acmaea (Notoacmaea) scutum* to the predatory starfish *Pisaster ochraceous. J. Exp. Zool.* 191:199–210.

Phillips, D. W., 1976. The effect of a species-specific avoidance response to predatory starfish on the intertidal distribution of two gastropods. *Oecologia (Berl.).* 23:83–94.

Pilger, J. F., 1980. The annual cycle of oogenesis, spawning, and larval settlement of the echiuran *Listriolobus pelodes* off southern California. *Pac. Sci.* 34(2):129–142.

Pugh, P. R., 1983. Benthic siphonophores: A review of the family Rhodaliidae (Siphonophora, Physonectae). *Philos. Trans. R. Soc. Lond. B.* 301:165–300.

Quammen, M. L., 1982. Influence of subtle substrate differences on feeding by shore birds on intertidal mud flats: *Mar. Biol.* 71(3):339–343.

Quammen, M. L., 1984. Predation by shorebirds, fish, and crabs on invertebrates in intertidal mudflats. An experimental test. *Ecology.* 65: 529–537.

Reish, D. J., 1956. An ecological study of Lower San Gabriel River, California, with special reference to pollution. *Calif. Fish Game.* 42(1): 51–61.

Reish, D. J., 1959. *An Ecological Study of Pollution in Los Angeles–Long Beach Harbors, California.* Allan Hancock Foundation, Los Angeles. 119pp.

Reish, D. J., 1961a. The use of the sediment bottle collector for monitoring polluted marine waters. *Calif. Fish Game.* 47(3):261–272.

Reish, D. J., 1961b. The relationship of temperature and dissolved oxygen to the seasonal settlement of the polychaetous annelid *Hydroides norvegica* (Gunnerus). *Bull. South. Calif. Acad. Sci.* 60(1):1–11.

Reish, D. J., 1963b. Further studies on the benthic fauna in a recently constructed boat harbor in southern California. *Bull. South. Calif. Acad. Sci.* 62(1):23–32.

Reish, D. J., 1964a. Discussion of the *Mytilus californianus* community on newly constructed rock jetties in southern California. *Veliger.* 7(2):95–101.

Reish, D. J., 1964b. A quantitative study of the benthic polychaetous annelids of Catalina Harbor, Santa Catalina Island, California. *Bull. South. Calif. Acad. Sci.* 63(2):86–92.

Reish, D. J., 1971. Seasonal settlement of polychaetous annelids on test panels in Los Angeles–

Long Beach harbors 1950–1951. *J. Fish. Res. Board Can.* 28(10):1459–1467.
Reish, D. J., 1972. *Marine Life of Southern California, Emphasizing Marine Life of Los Angeles–Orange Counties.* Kendall-Hunt Publishing Co., Dubuque, IA. 164pp.
Reish, D. J., 1975. Biological considerations in the development of Shoreline Aquatic Park and Rainbow Lagoon, Long Beach, California. Report to City of Long Beach, 64pp.
Reish, D. J., 1977. *Marine biology of Channel Islands Harbor.* Atlantis Scientific, Beverly Hills, CA. 28pp.
Reish, D. J., 1980. The marine life of Playa Vista, California. Report to the Summa Corporation, Las Vegas, Nevada.
Reish, D. J., 1981. Culture methods for rearing polychaetous annelids through sexual maturity. In: *Laboratory Animal Management, Marine Invertebrates.* National Academy Press, Washington, D.C. pp. 180–198.
Reish, D. J., and H. A. Winter, 1954. The ecology of Alamitos Bay, California, with special reference to pollution. *Calif. Fish Game.* 40(2):105–121.
Rex, M. A., 1981. Community structure in the deep-sea benthos. *Annu. Rev. Ecol. Syst.* 12: 331–353.
Rhoads, D. C., and J. W. Morse, 1971. Evolutionary and ecologic significance of oxygen-deficient marine basins. *Lethaia.* 4:413–428.
Ricketts, E. F., J. Calvin, and J. W. Hedgpeth, 1968. *Between Pacific Tides.* 4th edition. Stanford Univ. Press, Palo Alto, CA. 614pp.
Rokop, F. J., 1974. Reproductive patterns in the deep sea benthos. *Science.* 186(4165):743–744.
Rokop, F. J., 1977. Patterns of reproduction in the deep-sea benthic crustaceans, a reevaluation. *Deep-Sea Res.* 24(7):683–691.
Rosenthal, R. J., 1971. Trophic interaction between the sea star *Pisaster giganteus* and the gastropod *Kelletia kelletii. U. S. Natl. Mar. Fish. Serv. Fish. Bull.* 69:669–679.
Rosenthal, R. J., and J. R. Chess, 1972. A predator–prey relationship between the leather star, *Dermasterias imbricata*, and the purple urchin, *Strongylocentrotus purpuratus. U.S. Natl. Mar. Fish. Serv. Fish. Bull.* 70:205–216.
Rowley, R. J., 1989. Settlement and recruitment of sea urchins (*Strongylocentrotus* spp.) in a sea-urchin barren ground and a kelp bed: Are populations regulated by settlement or post-settlement processes? *Mar. Biol.* 100:485–494.
Russell, M. P., 1984. Life history traits and resource allocation in the purple sea urchin *Strongylocentrotus purpuratus.* Master's Thesis, San Diego State Univ., San Diego.
Sanders, H. L., R. R. Hessler, and G. R. Hampson, 1965. An introduction to the study of deep-sea benthic faunal assemblages along the Gay Head–Bermuda transect. *Deep-Sea Res.* 12:845–867.
Savrda, C. E., D. J. Bottjer, and D. S. Gorsline, 1984. Development of a comprehensive oxygen-deficient marine biofacies model: Evidence from Santa Monica, San Pedro, and Santa Barbara basins, California continental borderland. *Am. Assoc. Pet. Geol. Bull.* 68(9): 1179–1192.
Schmitt, R. J., 1982. Consequences of dissimilar defenses against predation in a subtidal marine community. *Ecology.* 63:1588–1601.
Schmitt, R. J., 1985. Competitive interactions of two mobile prey species in a patchy environment. *Ecology.* 66(3):950–958.
Schreiber, T. C., Jr., 1974. Seasonal distribution of polychaetous annelids in Anaheim Bay, California. Master's Thesis, California State Univ., Long Beach, CA. 73pp.
Schreiber, R. W., ed., 1981. *The biota of Ballona region, Los Angeles County.* Los Angeles County Natural History Museum. 400pp.
Schroeter, S. C., 1978. Experimental studies of competition as a factor affecting the distribution and abundance of purple sea urchins, *Strongylocentrotus purpuratus* Stimpson. Ph.D. Dissertation, Univ. of California, Santa Barbara, CA.
Schroeter, S. C., J. Dixon, and J. Kastendiek, 1983. Effects of the starfish *Patiria miniata* on the distribution of the sea urchin *Lytechinus anamesus* in a southern Californian kelp forest. *Oecologia (Berl.).* 56(2–3):141–147.
Science Applications International Corporation, 1986. Assessment of long-term changes in biological communities in the Santa Maria Basin and Western Santa Barbara Channel. In: Phase I, Vol. II. Synthesis of Findings. U.S. Dept. of Interior, Minerals Management Service, Abstract #14–12–0001–30032. U.S. Dept. of Interior, Bureau of Land Management, Washington, D.C. pp. 116–192.
Seapy, R. R., 1981. Structure, distribution, and seasonal dynamics of the benthic community in Upper Newport Bay, California. *Calif. Dept. Fish Game, Mar. Res. Tech. Rep.* No. 46, 74pp.
Seapy, R. R., and W. J. Hoppe, 1973. Morpho-

logical and behavioral adaptations to desiccation in the intertidal limpet *Acmaea (Collisella) strigatella*. *Veliger*. 16:181–188.

Seapy, R. R., and M. M. Littler, 1979. Assessments of the distribution, abundance, and community structure of rocky intertidal organisms at Cave Canyon, Santa Barbara Island. In: M. M. Littler, ed. *The Distribution, Abundance, and Community Structure of Rocky Intertidal and Tidepool Biotas in the Southern California Bight.* Southern California Baseline Study, Year Three. Vol. II. Chapter 1.15. U.S. Dept. of Interior, Bureau of Land Management, Washington, D.C.

Seapy, R. R., and M. M. Littler, 1980. Biogeography of rocky intertidal macroinvertebrates of the Southern California Islands. In: D. M. Power, ed. *The California Islands: Proceedings of a Multidisciplinary Symposium*. Haagen Printing, Santa Barbara, CA. pp. 307–324.

Seapy, R. R., and M. M. Littler, 1982. Population and species diversity fluctuations in a rocky intertidal community relative to severe aerial exposure and sediment burial. *Mar. Biol.* 71(1): 87–96.

Segal, E., 1956a. Adaptive differences in water-holding capacity in an intertidal gastropod. *Ecology*. 37:174–178.

Segal, E., 1956b. Microgeographic variation as thermal acclimation in an intertidal mollusc. *Biol. Bull.* 111:129–152.

Segal, E., and P. A. Dehnel, 1962. Osmotic behavior in an intertidal limpet, *Acmaea limatula*. *Biol. Bull.* 122:417–430.

Shanks, A. L., and W. G. Wright, 1986. Adding teeth to wave action: The destructive effects of wave-borne rocks on intertidal organisms. *Oecologia (Berl.)*. 69(3):420–428.

Shepard, F. P., 1973. *Submarine Geology*. 3d edition. Harper & Row, New York. 517pp.

Sholkovitz, E. R., and J. M. Gieskes, 1971. A physical-chemical study of the flushing of the Santa Barbara Basin. *Limnol. Oceanogr.* 16(3): 479–489.

Sholkovitz, E. R., and A. Soutar, 1975. Changes in the composition of the bottom water of the Santa Barbara Basin: Effect of turbidity currents. *Deep-Sea Res.* 22(1):13–22.

Shotwell, J. A., 1950. Distribution of volume and relative linear measurement changes in *Acmaea*, the limpet. *Ecology*. 31:51–61.

Siebenaller, J. F., 1978. Genetic variation in deep-sea invertebrate populations of the bathyal gastropod *Bathybembix bairdii*. *Mar. Biol.* 47(3): 265–276.

Smith, C. R., 1983. Enrichment, disturbance and deep-sea community structure: The significance of large organic falls to bathyal benthos in Santa Catalina Basin. Ph.D. Dissertation, Univ. of California, San Diego. 310pp.

Smith, C. R., 1985. Food for the deep sea: Utilization, dispersal, and flux of nekton falls at the Santa Catalina Basin floor. *Deep-Sea Res.* 32:417–442.

Smith, C. R., 1986. Nekton falls, low-intensity disturbance and community structure of infaunal benthos in the deep sea. *J. Mar. Res.* 44(3):567–600.

Smith, C. R., and S. C. Hamilton, 1983. Epibenthic megafauna of a bathyal basin off southern California: Patterns of abundance, biomass, and dispersion. *Deep Sea Res.* 30(9):907–928.

Smith, C. R., P. A. Jumars, and J. DeMaster, 1986. *In situ* studies of megafaunal mounds indicate rapid sediment turnover and community response at the deep-sea floor. *Nature*. 323:251–253.

Smith, K. L., Jr., 1983. Metabolism of two dominant epibenthic echinoderms measured at bathyal depths in the Santa Catalina Basin, Pacific Ocean. *Mar. Biol.* 72(3):249–256.

Smith, K. L., Jr., and R. J. Baldwin, 1982. Scavenging deep-sea amphipods: Effects of food odor on oxygen consumption and a proposed metabolic strategy. *Mar. Biol.* 68(3)287–298.

Smith, R. I., and J. T. Carlton, eds., 1975. *Light's Manual: Intertidal invertebrates of the central California coast.* 3d edition. Univ. of California Press, Berkeley, CA. 716pp.

Smith, R. W., and C. S. Greene, 1976. Biological communities near submarine outfall. *J. Water Pollut. Control Fed.* 48:1894–1912.

Smith, S. H., 1976. Report of the preliminary biological assessment of Tanner Bank and Cortes Bank offshore southern California. U.S. Dept. of Interior, Bureau of Land Management, Los Angeles, CA.

Smith, S. H., H. Davis, and G. Chan, 1975. Preliminary biological survey (subtidal) of Tanner Bank and Cortes Bank, offshore southern California with emphasis on the hydrocoral *Allopora californica*. U.S. Dept. of Interior, Bureau of Land Management, Los Angeles, CA.

Soule, D. F., 1976. Biomass and recolonization studies in the outer Los Angeles Harbor. In: D. F. Soule, and M. Oguri, eds. *Marine Studies of San Pedro Bay, California*. Part 2. Harbors Environmental Projects. Allan Hancock Foundation, Univ. of Southern California, Los Angeles.

Soule, D. F., and M. Oguri, eds., 1975. Marine studies of San Pedro Bay, California. Part 8. Environmental biology. Allan Hancock Foundation, Univ. of Southern California, Los Angeles. 122pp.

Soule, D. F., and M. Oguri, eds., 1980. Marine studies of San Pedro Bay, California. Part 17. The marine environment in Los Angeles and Long Beach harbors during 1978. Univ. of Southern California, Sea Grant Program, Los Angeles. 667pp.

Soule, D. F., and M. Oguri, eds., 1986. Marine studies of San Pedro Bay, California. Part 20B. The marine environment of Marina del Rey, California, in 1985. Allan Hancock Foundation, Univ. of Southern California, Los Angeles. 291pp.

Sousa, W. P., 1979a. Disturbance in marine intertidal boulder fields: The nonequilibrium maintenance of species diversity. *Ecology*. 60(6): 1225–1239.

Sousa, W. P., 1979b. Experimental investigations of disturbance and ecological succession in a rocky intertidal algal community. *Ecol. Monogr.* 49(3):227–254.

Sousa, W. P., S. C. Schroeter, and S. D. Gaines, 1981. Latitudinal variation in intertidal algal community structure: The influence of grazing and vegetative propagation. *Oecologia (Berl.)*. 48:297–307.

Soutar, A., and P. A. Crill, 1977. Sedimentation and climatic patterns in the Santa Barbara Basin during the 19th and 20th centuries. *Geol. Soc. Am. Bull.* 88:1161–1172.

Southern California Coastal Water Research Project, 1973. The ecology of the Southern California Bight: Implications for water quality management. T.R. 104, Chap. 7, coastal fish populations. In: South. Calif. Coastal Water Res. Proj., El Segundo, CA. pp. 193–203.

Southern California Coastal Water Research Project, 1983. A survey of the slope off Orange Co., California: First-year report. South. Calif. Coastal Water Res. Proj., Long Beach, CA. 178pp.

Southern California Ocean Studies Consortium, 1974. A summary of knowledge of the southern California coastal zone and offshore areas. U.S. Dept. of Interior, Bureau of Land Management, Los Angeles, CA.

Spies, R., 1984. Benthic–pelagic coupling in sewage affected marine ecosystems. *Mar. Environ. Res.* 13(3):195–230.

Stein, J. E., and E. G. Kreppert, 1971. *An environmental evaluation of the Bolsa Chica area*. Vol. 3, Dillingham Environmental Co., La Jolla, CA. 117pp.

Stephenson, T. A., and A. Stephenson, 1949. The universal features of zonation between tidemarks on rocky shores. *J. Ecol.* 37:289–305.

Stephenson, T. A., and A. Stephenson, 1972. *Life between tidemarks on rocky shores*. W. H. Freeman and Co., San Francisco. 425pp.

Stimson, J. S., 1968. The population ecology of the limpets *Lottia gigantea* (Gray) and several species of *Acmaea* (Eschsholtz) coexisting on an intertidal shore. Ph.D. Dissertation, Univ. of California, Santa Barbara. 144pp.

Stimson, J. S., 1970. Territorial behavior in the owl limpet, *Lottia gigantea*. *Ecology*. 51:113–118.

Stimson, J. S., 1973. The role of the territory in the ecology of the intertidal limpet *Lottia gigantea* (Gray). *Ecology*. 54:1020–1030.

Straughan, D., 1982. Inventory of the natural resources of sandy beaches in southern California. Allan Hancock Foundation, Los Angeles, CA. 447pp.

Straughan, D. S., 1983. Sandy beach communities exposed to natural oil seepage. In: *Proceedings, 1983 Oil Spill Conference (Prevention Behavior, Control Clean-up)*. Sponsored American Petroleum Institute, Environmental Protection Agency, and U.S.C.G., San Antonio, TX. pp. 485–490.

Straughan, D., and R. W. Klink, 1980. A taxonomic listing of common marine invertebrate species from southern California. Tech. Report No. 3. Prep. by Allan Hancock Foundation, Univ. of Southern California, Los Angeles. 281pp.

Striplin, P. L., 1987. Resource utilization by *Astropecten verrilli* along gradients of organic enrichment. Master's Thesis, California State Univ., Long Beach, CA. 158pp.

Stull, J. K., C. I. Haydock, and D. E. Montagne, 1986a. Effects of *Listriolobus pelodes* (Echiura) on coastal shelf benthic communities and sediments modified by a major California wastewater discharge. *Estuarine Coastal Shelf Sci.* 22(1):1–17.

Stull, J. K., C. I. Haydock, R. W. Smith, and D. E. Montagne, 1986b. Long-term changes in the benthic community on the coastal shelf of Palos Verdes, southern California. *Mar. Biol.* 91(4):539–551.

Suchanek, T. H., 1978. The ecology of *Mytilus edulis* L. in exposed rocky intertidal communities. *J. Exp. Mar. Biol. Ecol.* 31:105–120.

Sumich, J. L., and J. E. McCauley, 1973. Growth of a sea urchin, *Allocentrotus fragilis,* off the Oregon coast. *Pac. Sci.* 27(2):156–167.

Sutherland, J. P., 1970. Dynamics of high and low populations of the limpet *Acmaea scabra* (Gould). *Ecol. Monogr.* 40:169–188.

Swarbrick, S. L., 1984. Disturbance, recruitment, and competition in a marine invertebrate community. Ph.D. Dissertation, Univ. of California, Santa Barbara, CA. 316pp.

Taylor, P. R., and M. M. Littler, 1982. The roles of compensatory mortality, physical disturbance, and substrate retention in the development and organization of a sand-influenced, rocky-intertidal community. *Ecology.* 63:135–146.

Tegner, M. J., 1980. Multispecies considerations of resources management in southern California kelp beds. *Can. Tech. Rep. Fish. Aquat. Sci.* 954:125–143.

Tegner, M. J., and P. K. Dayton, 1977. Sea urchin recruitment patterns and implications of commercial fishing. *Science.* 196:324–326.

Tegner, M. J., and P. K. Dayton, 1981. Population structure, recruitment, and mortality of two sea urchins (*Strongylocentrotus franciscanus* and *S. purpuratus*) in a kelp forest. *Mar. Ecol. Prog. Ser.* 5(3):255–268.

Tegner, M. J., and P. K. Dayton, 1987. El Niño effects on southern California kelp forest communities. *Adv. Ecol. Res.* 17:243–279.

Tegner, M. J., and L. A. Levin, 1983. Spiny lobster and sea urchins: Analysis of a predator–prey interaction. *J. Exp. Mar. Biol. Ecol.* 73:125–150.

Tekmarine, Inc., 1986. Review of concept design material related to coastal features and marsh restoration at the mouth of the Santa Ana River. U.S. Army Corps of Engineers. Prep. by Tekmarine, Inc., Sierra Madre, CA. 63pp.

Thistle, D., 1978. Harpacticoid dispersion patterns: Implications for deep-sea diversity maintenance. *J. Mar. Res.* 36(2):377–397.

Thompson, B. E., 1980. A new bathyl sipunculan from southern California, with ecological notes. *Deep-Sea Res.* 27A:951–957.

Thompson, B. E., 1982. Food resource utilization and partitioning in macrobenthic communities of the southern California borderland. Ph.D. Dissertation, Univ. of Southern California, Los Angeles.

Thompson, B. E., and G. F. Jones, 1987. Benthic macrofaunal assemblages of slope habitats in the southern California borderland. Occ. Pap., New Ser. No. 6, Allan Hancock Foundation, Univ. of Southern California, Los Angeles, 21pp.

Thompson, B. E., and J. D. Laughlin. Feeding and coexistence of four echinoid species from basin slopes off southern California. Unpubl. manuscript.

Thompson, B. E., J. N. Cross, J. D. Laughlin, G. P. Hershelman, R. W. Gossett, and D. T. Tsukada, 1984a. Sediment and biological conditions on coastal slopes. In: W. Bascom, ed. *Biennial Report 1983–1984.* South. Calif. Coastal Water Res. Proj., Long Beach, CA. pp. 37–68.

Thompson, B. E., J. D. Laughlin, and D. T. Tsukada, 1984b. Ingestion and oxygen consumption by slope echinoids. South. Calif. Coastal Water Res. Proj., Long Beach, CA.

Thompson, B. E., J. D. Laughlin, and D. T. Tsukada, 1987a. 1985 reference site survey. Tech. Report C–221, South. Calif. Coastal Water Res. Proj., Long Beach, CA. 50pp.

Thompson, B. E., G. F. Jones, J. D. Laughlin, and D. T. Tsukada, 1987b. Distribution, abundance, and size composition of echinoids from basin slopes off southern California. *Bull. South. Calif. Acad. Sci.* 86(3):113–125.

Thompson, B. E., D. T. Tsukada, and J. D. Laughlin, 1993. Megabenthic assemblages of the coastal shelf, slopes, and basins off southern California. *Bull. South. Calif. Acad. Sci.* 92:25–42.

Thorson, G., 1946. Reproduction and larval development of Danish marine bottom invertebrates, with special reference to the planktonic larvae in the Sound (Oresund). *Medd. Komm. Dan. Fisk. Havunders., Ser. Plankton.* 4:1–523.

Thorson, G., 1957. Bottom communities. *Geol. Soc. Am. Mem.* 67:461–534.

Timko (O'Neill), P. L., 1976. Sand dollars as suspension feeders: A new description of feeding in *Dendraster excentricus*. *Biol. Bull.* 151:247–259.

Turner, C. H., 1970. A survey of the effects of waterway expansion, Channel Islands Harbor. In: Channel Islands Harbor Ecological Study, Appendix A. Moffatt and Nichol Engineers, Long Beach.

Tutschulte, T., 1976. The comparative ecology of three species of abalones. Ph.D. Dissertation, Univ. of California, Santa Barbara, CA.

Underwood, A. J., 1978. A refutation of critical tidal levels as determinants of the structure of intertidal communities on British shores. *J. Exp. Mar. Biol. Ecol.* 33:261–276.

Underwood, A. J., 1979. The ecology of intertidal gastropods. *Adv. Mar. Biol.* 16:111–210.

Underwood, A. J., and E. J. Denley, 1984. Paradigms, explanations, and generalizations in models for the structure of intertidal communities on rocky shores. In: D. R. Strong, D. Simberloff, L. G. Abele, and A. B. Thistle, eds. *Ecological communities: Conceptual Issues and the Evidence.* Princeton Univ. Press, Princeton, NJ. pp. 151–180.

U.S. Fish and Wildlife Service, 1981. Santa Margarita Estuary enhancement plan. Director of Natural Resources, Camp Pendleton, CA.

VanBlaricom, G. R., 1978. Disturbance, predation, and resource allocation in a high-energy sand-bottom ecosystem: Experimental analysis of critical structuring processes for the infaunal community. Ph.D. Dissertation, Univ. of California, San Diego. 328pp.

VanBlaricom, G. R., 1982. Experimental analysis of structural regulation in a marine sand community exposed to oceanic swell. *Ecol. Monogr.* 52(3):283–305.

Vance, R. R., 1978. A mutualistic interaction between a sessile marine clam and its epibionts. *Ecology.* 59(4):679–685.

Vance, R. R., 1979. Effects of grazing by the sea urchin, *Centrostephanus coronatus,* on prey community composition. *Ecology.* 60:537–546.

Vance, R. R., and R. J. Schmitt, 1979. The effect of the predator avoidance behavior of the sea urchin, *Centrostephanus coronatus,* on the breadth of its diet. *Oecologia (Berl.)* 44:21–25.

Vermeij, G. J., 1972. Intraspecific shore-level size gradients in intertidal molluscs. *Ecology.* 53: 693–700.

Ware, R. R., 1979. The food habits of the white croaker *Genyonemus lineatus* and an infaunal analysis near areas of waste discharge in outer Los Angeles Harbor. Master's Thesis, California State Univ., Long Beach, CA. 164pp.

Wells, R. A., 1980. Activity pattern as a mechanism of predator avoidance in two species of acmaeid limpet. *J. Exp. Mar. Biol. Ecol.* 48: 151–168.

Wenner, A. M., 1988. Crustaceans and other invertebrates as indicators of beach pollution, Chap. 9. In: D. F. Soule, and G. S. Kleppel, eds. *Marine Organisms as Indicators.* Springer-Verlag, New York. pp. 200–229.

Wicksten, M. K., 1980. Mainland and insular assemblages of benthic decapod crustaceans of southern California. In: D. M. Power, ed. *The California Islands: Proceedings of a Multidisciplinary Symposium.* Haagen Printing, Santa Barbara, CA. pp. 357–367.

Winget, R. R., 1968. Trophic relationships and metabolic energy budget of the California spiny lobster, *Panulirus interruptus* (Randall). Master's Thesis, San Diego State Univ., San Diego.

Wobber, D. R., 1975. Agonism in asteroids. *Biol. Bull.* 148:483–496.

Wolcott, T. G., 1973. Physiological ecology and intertidal zonation in limpets (*Acmaea*): A critical look at "limiting factors." *Biol. Bull.* 145: 389–422.

Wolfson, A., G. VanBlaricom, N. Davis, and G. S. Lewbel, 1979. The marine life of an offshore oil platform. *Mar. Ecol. Prog. Ser.* 1(1):81–89.

Word, J. Q., and A. J. Mearns, 1979. 60-meter control survey off southern California. Tech. Mem. 229. South. Calif. Coastal Water Res. Proj., El Segundo, CA. 58pp.

Zedler, J. B., and C. S. Nordby, 1986. The ecology of Tijuana Estuary, California: An estuarine profile. U.S. Fish Wildl. Serv. Biol. Rep. 85(7.5). Prep. by San Diego State Univ., Dep. of Biology, San Diego, CA. 104pp.

Chapter 9

Fishes

Jeffrey N. Cross and Larry G. Allen

Introduction

In this chapter, we review information on the ecology of fishes in the major habitats of the Southern California Bight (SCB). The SCB is located in the transition area between Pacific subarctic, Pacific equatorial, and North Pacific central water masses (Sverdrup et al. 1942), and the fish fauna contains representatives from each of these sources. The type and amount of ecological information on fishes are not equable among habitats or among species. What is known reflects various investigator interests, sampling logistics, and economic and political exigencies. For example, research on epipelagic fishes has concentrated on the distribution, abundance, and life-history parameters of commercially exploited species; thus more is known about this small group of fishes than any other group in the SCB. Research on mesopelagic and bathypelagic fishes has emphasized physiological and ecological adaptations to life in the deep sea. Research on shallow-water, soft-bottom fishes has focused on the effects of habitat alterations and pollution.

Because the properties and populations of the ocean are neither uniformly nor randomly distributed, we emphasize spatial and temporal patterns of distribution and abundance and the relationships between physical forcing processes and biological responses. Relevant spatial scales for individuals range from 10^{-2} m for a feeding fish larva to 10^{6} m for the ambit of an adult fish. Spatial scales for populations span a distance of 10^{2} m for some rocky reef fishes to 10^{6} m for epipelagic fishes. Temporal scales extend from one day for larval starvation to several years for population–generation time (Bakun 1984). To the extent that they are known, recruitment and causes of changes in abundance are discussed in this chapter. Recruitment can mean appearance of young fish at the nursery grounds (population recruitment) or appearance of young fish in a particular type of fishing gear (fishery recruitment).

Zoogeography

Of the 554 species and 144 families of California marine fishes (Miller and Lea 1972), 481 species (87%) and 129 families (90%) occur in the SCB (Horn 1974). The greatest number of southern range terminations by temperate and boreal species, and the greatest number of northern range terminations by subtropical and tropical species, occur between 32° and 34°N lat. (Horn and Allen 1978). High species richness is probably due to the complex topography, convergence of several water masses, and changeable environmental conditions. In the northern Channel Islands, the transition from northern species to southern species occurs over a relatively short distance. The ichthyofauna of Santa Cruz Island at the eastern end is typically southern, while the ichthyofauna of San Miguel at the western end is typically northern (Hubbs 1974; Love et al. 1985).

Point Conception (34.5°N), a widely recognized faunal boundary (Briggs 1974) as well as the northern boundary of the SCB, is more important as a barrier to southern species than to northern ones. Northern species cross Point Conception by moving into deeper water off southern California, a phenomenon known as submergence (Hubbs 1948; Briggs 1974) and by occupying upwelling areas on the south side of headlands, especially off Baja California (Hubbs 1948, 1960, 1974). A few southern species find warm-water refugia north of Point Conception in coastal bays and estuaries (Hubbs 1948, 1960, 1974; Horn and Allen 1978) or recruit to southern and central California during years when the southward-flowing California Current weakens (McGleneghan 1973; Cowen 1985).

Seasonal migrations of subtropical and temperate fishes into the SCB (Parrish et al. 1981; Bedford and Hagerman 1983; Cailliet and Bedford 1983) and invasions of tropical fishes during warm-water years associated with El Niño (Radovich 1961; Mearns 1988) illustrate the dynamic nature of the southern California ichthyofauna. Less well known are the invasions of northern fishes during cool years (Hubbs 1948) and even during warm-water events (Phillips 1958; Radovich 1961). The reasons for the extensive migrations of fishes as a result of fluctuating environmental conditions are not well understood, but appear to be related to changes in the total (physical and biological) environment (Lynn 1984; Squire 1987). Warm- and cool-water events affect fish recruitment in the SCB and alter the composition of some fish assemblages for several years (Mearns et al. 1980; Stephens and Zerba 1981; Cowen 1985; Love et al. 1985, 1986).

Life Histories

Life histories are evolutionary responses to the impacts of changing environmental conditions on survival and reproduction of different age classes. Internal constraints (physiology and genetic variation) together with events in the external environment determine the possible combinations of age-specific survival and reproduction. Life histories constitute a balance among the conflicting resource requirements of maintenance, growth, reproduction, storage, and survival (Roff 1984, 1988; Partridge and Harvey 1988; Stearns 1989), and more than one adaptive solution to a particular set of environmental conditions can exist (Southwood 1988). The life history traits of many fishes (age at maturity, maximum age, maximum size, growth rate, and natural mortality) are correlated (Adams 1980; Pauly 1980; Gunderson and Dygert 1988). In general, fishes with low natural adult mortality rates grow slowly, mature late, and have a low annual reproductive effort. Fishes with high natural adult mortality rates grow quickly, mature early, and have a high annual reproductive effort (table 9.1). Correlations between life history traits may change sign over the range of environmental conditions normally encountered (Stearns 1989).

Life history strategies have implications for exploitation and management. Species that grow slowly, mature late, have low natural mortality rates, and long life spans are more sensitive to overfishing than species that grow quickly, mature early, have high rates of natural mortality, and short life spans. Fisheries on short-lived species are more productive because these species can be fished at younger ages and at higher levels of fishing mortality. Given a minimum population size, short-lived species will recover more rapidly from overfishing (Adams 1980). Harvesting strategies that impose high age-specific mortality rates may cause genetic changes in the rates of survival, growth, and reproduction.

Reproduction

The reproductive cycle of species with northern affinities in the SCB generally peaks from winter to spring; for example, Pacific hake, *Merluccius productus* (Bailey et al. 1982), and olive rockfish, *Sebastes serranoides* (Love and Westphal 1981). The reproductive cycle of species with southern affinities generally peaks from spring to summer, for example, kelp bass, *Paralabrax clathratus* (Smith and Young 1966), and queenfish, *Seriphus politus* (DeMartini and Fountain 1981). Some species display gonadal activity throughout the year. These species include splitnose rockfish, *Sebastes diploproa* (Boehlert and Kappenman 1980), and northern anchovy, *Engraulis mordax* (Stauffer and Parker 1980).

Lunar and diel spawning periodicities are superimposed on the annual cycle. Grunion (*Leuresthes tenuis*) spawn in the intertidal zone on sandy beaches on the first few nights following each new and full moon during spring and summer. Spawning occurs 1–3 hours after high tide, and the eggs are deposited in the sand. The next series of high tides about 10 days later exposes them. The eggs hatch 2 or 3 minutes after they are washed out of the sand (Walker 1952; Middaugh et al. 1983). Queenfish spawn from late afternoon to evening, and spawning activity is most intense during the first quarter of the moon. Females may spawn 24 times during the spawning season, expending the equivalent of 114% of their body weight in eggs. Spawning at night may reduce predation on adults and their planktonic eggs, and spawning when tidal exchanges are low may keep the eggs and larvae nearshore (DeMartini and Fountain 1981). Species that attach eggs to a substrate (e.g., garibaldi, *Hypsypops rubicunda*; blacksmith, *Chromis punctipinnis*) (Limbaugh 1964) spawn during the day so they can choose appropriate substrate (Blaxter and Hunter 1982).

Serial, or asynchronous, spawning (iteroparity) is typical of small, short-lived, warm-temperate fishes with planktonic eggs (e.g., queenfish and northern anchovy). Iteroparity may be a reproductive adaptation to variable phytoplankton and zooplankton production cycles, a sort of "bet-hedging" against poor conditions for planktonic larvae (DeMartini and Fountain 1981). Total or synchronous spawning (semelparity) is characteristic of

Table 9.1. *Life-history Parameters for Selected North Pacific Marine Fishes*[a]

	M	GSI	L (cm)	L_∞ (cm)	K	T_m (yr)	$T_{.01}$ (yr)
Northern anchovy	0.92	0.65	11.6	16.6	0.30	1	6
Pacific herring	0.56	0.29	20.9	26.3	0.36	3	10
Pacific cod	0.65	0.16	62.2	86.6	0.33	3	8
English sole	0.26	0.18	38.8	37.2	0.31	4	12
Pacific halibut	0.19	0.11	91.3	109.6	0.13	12	23
Spiny dogfish	0.09	0.04	102.6	112.6	0.06	23	57

[a]M = natural mortality, GSI = ratio of gonad weight to somatic weight, L = average size of mature female, L_∞ = maximum female size from von Bertalanffy growth equation, K = growth parameter (how rapidly fish approach L_∞) from von Bertalanffy growth model, T_m = age when 50% of females are mature, $T_{.01}$ = age at which female abundance is 1% of abundance at T_m.

From Gunderson and Dygert 1988.

species with low adult mortality rates and variable recruitment (e.g., some rockfish, *Sebastes* spp.) and species with high adult mortality rates and stable recruitment (e.g., Pacific salmon, *Oncorhynchus* spp.). Olive rockfish spawn once, and large females produce up to 5 × 10^5 eggs (Love and Westphal 1981). Bocaccio (*Sebastes paucispinis*) spawn twice (Moser 1967a,b) and large females produce up to 2.3 × 10^6 eggs (Phillips 1964). High adult survival and long life enable *Sebastes* spp. to survive periods of poor recruitment. High fecundity permits relatively few individuals to produce a strong year class when environmental conditions become favorable (Warner and Chesson 1985).

Migration

Migrations are common among marine fishes and are usually related to feeding and reproduction. Migratory behavior evolves in concert with other life history traits; the probability of its appearance is increased by factors such as schooling and the trade-off between energy used in migration and energy stored for reproduction (Roff 1988). Dover sole (*Microstomus pacificus*) migrate into deep water during winter for reproduction and into shallow water in the summer for feeding (Hagerman 1952; Cross 1985). Scorpionfish (*Scorpaena guttata*) migrate offshore and aggregate at traditional spawning grounds from May through August. Tagged fish moved up to 42 km to a spawning site in San Pedro Bay (Love et al. 1987). In the fall, Pacific hake (also known as Pacific whiting) migrate from their summer feeding grounds in the California Current off the Pacific Northwest to their winter spawning grounds off southern California and Baja California. Adults return north in the spring and summer, while juveniles remain off California (Bailey 1981; Bailey et al. 1982). Species that migrate long distances are generally larger, grow more rapidly, and mature later in life and at a larger size than nonmigratory species (Roff 1988).

Some species move little during their lives. Kelp bass live up to 30 years and have a home range on the order of 1 km (Young 1963; Quast 1968a). Garibaldi, which live 15–17 years, defend a territory that includes a shelter hole, a grazing area, and—for some males—a nest site. Males have well-defined territories and defend them against most intruders within 1 m of the bottom throughout the year. They maintain thick patches of red algae as nest sites and may use the same site for several years. Females and bachelor males do not defend their territories as vigorously (Clarke 1970).

Splitnose Rockfish and Northern Anchovy

The diversity of life histories in the SCB is illustrated by a comparison of splitnose rockfish and northern anchovy. Splitnose rockfish are loosely aggregating, nektobenthic fishes that live on mud from 60 to 600 m depth (M. J. Allen 1982; Cross 1987). They grow to more than 30 cm in length and may live for 80 years (Bennett et al. 1982). The age of 50% maturity is 7 years for females at 19 cm total length (TL) and 9 years for males at 22 cm TL. Males usually complete spermatogenesis and mate before females complete vitellogenesis. Fertilization and development are internal. Sperm remain viable in the ovaries until ovulation, which often occurs several months after mating (Echeverria 1987). Females develop a vascular network throughout the ovaries for gas exchange and nourishment of the developing embryos during the 1-month gestation (Moser 1967a, b; Boehlert and Yoklavich 1984). Yolk sac larvae are released from February to July (Phillips 1964). Prejuveniles at 10–14 mm standard length (SL) congregate under drifting objects such as kelp patties (Mitchell and Hunter 1970) from August through December. Juveniles at 40–50 mm SL emigrate from the surface waters in April and May (Boehlert 1977), descend to 200–250 m depth, and migrate horizontally until they contact the bottom (Moser and Ahlstrom 1978).

Northern anchovies are epipelagic schooling fishes that live about 5 years and grow rapidly to 18–22 cm. Age at sexual maturity varies between years and may be a density-dependent function of population size. At high population abundances during the mid-1970s, most anchovies were mature at 2 years; at low population abundances in the early 1980s, 90% of age 0–1 fish were mature (Pacific Fishery Management Council 1983). Northern anchovy shed eggs and sperm in the water column and have no parental care. They spawn throughout the year in the SCB, but about 80% of the spawning occurs from winter to spring (Stauffer and Parker 1980).

The stock may consist of several subpopulations adapted to spawning at different times (Hedgecock 1986). During the peak season, adults spawn every 6–8 days between sunset and midnight. Females may spawn 20 times per year. Their ovaries contain the caloric equivalent of two batches of eggs and their fat stores contain the caloric equivalent of 13 batches; the remaining energy must be derived from contemporaneous plankton production (Hunter and Goldberg 1980; Hunter and Leong 1981). Anchovy larvae are the most abundant pelagic fish larvae in the California Current; most (85%) occur in the upper 50 m of the water column (Ahlstrom 1959a, 1965). The larvae begin feeding when they reach 4 mm length and metamorphose at 35 mm after about 2.5 months in the plankton (O'Connell 1981; Hunter 1976). Schooling begins during mid-larval life and develops through the juvenile stage (Blaxter and Hunter 1982).

Commercial and Recreational Fishing

Fishes in the SCB support important commercial and recreational fisheries; more than 100 species appear in the catches. Tunas, mackerel, bonito, and anchovy dominate (>80%) commercial landings, which account for about 4% of the total U.S. catch (approximately 6×10^9 lb, or 2.7×10^9 kg). Los Angeles area ports rank among the top 10 ports in the United States in quantity and value of the commercial catch (U.S. Department of Commerce 1987). Commercial fisheries off California have been studied and monitored since the early part of the twentieth century (MacCall 1986). Recreational fishing began in the late nineteenth century (Young 1969), but monitoring did not begin until the late 1940s (MacCall 1986). Recreational fishermen in the SCB land about 60% of the total recreational catch in California. Fishermen on private and commercial passenger fishing vessels account for more than 80% of the recreational catch (all fish caught).

Recreational landings (those fish kept) in the SCB averaged 14×10^6 fish (20×10^6 lb, or 9×10^6 kg) from 1981 to 1985 and were dominated (>70%) by mackerel, bonito, rockfishes, sea basses, and croakers (Helvey et al. 1987). Recreational landings in the SCB account for about 5% of the total recreational landings in the continental United States (approximately 3×10^8 fish) (Anonymous 1987).

Fishery landing statistics indicate only the availability of fish to the fishery and should be interpreted with caution. Fluctuations in fish landings result from complex interactions between environmental influences on distribution and recruitment, effects of weather on fishing location and success, localized behavior of the fish, and the economics of fishing (Clark et al. 1975).

Subsistence fisheries have existed in the SCB for many centuries, while commercial fisheries date back only to the mid-nineteenth century (Methot 1983a). Mechanization, echo sounding, and the advent of synthetic fibers are responsible for the development of the modern fishing industry (Kristjonsson 1959). Jurisdiction over California's marine resources is complex. Most commercial fish species are managed by the California Legislature; some fish species are managed by the federal Pacific Fishery Management Council; others, including recreational fishes and some commercial species, are managed by the state Fish and Game Commission (FGC). The FGC was established by the state of California in 1870 to preserve and restore the state's fish resources, but efforts to maintain sport fish populations did not really begin until 1945 (Klingbeil 1983; Methot 1983a; Venrick 1983; Huppert 1983; MacCall 1986). Declines in the stocks of some nearshore fishes have resulted in gear restrictions, fish size and bag limits, and fishery closures (MacCall et al. 1976; Fleming 1983; Klingbeil 1983; Methot 1983a; Venrick 1983). Unfortunately, it is often difficult to apportion the reasons for a fishery's demise among overfishing, habitat degradation, pollution, and natural variability of the population (Vojkovich and Reed 1983).

Adaptations of Exploited Species

The California Current is a dynamic system and its inhabitants have evolved life histories to accommodate unpredictability. Exploited species include the following: short-lived, fast-growing, highly fecund, epipelagic schooling fishes, such as northern anchovy, Pacific sardine (*Sardinops sagax*), Pacific mackerel (*Scomber japonicus*), and jack mackerel (*Trachurus symmetricus*); long-lived, slow-growing, moderately fecund, usually demersal fishes, such as rockfishes and flatfishes; and large, fast-growing, migratory fishes, such as tunas, yellowtail (*Seriola lalandei*), and swordfish (*Xiphias gladius*) (Venrick 1983).

The abundance and landings of epipelagic fishes rise and fall rapidly in response to environmental fluctuations. During the 1982–1983 El Niño, commercial landings of anchovy declined 90% and commercial landings of mackerel declined 11% (Smith 1985). The ultimate persistence of epipelagic fishes depends on remnant populations that act as "seed stocks" when conditions become favorable (Venrick 1983; Methot 1983a). Epipelagic fishes also fall prey to other fishes, marine birds, and marine mammals (Huppert 1983). Seabird reproductive success is directly proportional to the abundance of small epipelagic fishes (MacCall 1986).

Long-lived demersal species have forfeited early reproductive maturity for long adult life. Populations of long-lived species tend to remain stable because of the simultaneous occurrence of several reproductive year classes. Their ultimate persistence depends on a long reproductive life (Venrick 1983). Long-lived species are not resilient to heavy fishing pressure; once depressed, populations may take decades to recover (Klingbeil 1983; Venrick 1983).

Migratory species deal with the unpredictability of the ocean by actively seeking favorable areas and avoiding unfavorable conditions (Venrick 1983). Commercial and recreational landings of subtropical species, such as yellowtail, California barracuda (*Sphyraena argentea*), skipjack tuna (*Euthynnus pelamis*), and yellowfin tuna (*Thunnus albacares*), increase dramatically during warm years when large numbers migrate into the SCB from distribution centers off Mexico (Squire 1983a, 1987; Thomson et al. 1985). Commercial landings of skipjack, yellowfin, and albacore (*Thunnus alalunga*) tunas increased 382%, 258%, and 67%, respectively, during the 1982–1983 El Niño (Smith 1985). Landings of albacore declined during the 1958–1959 El Niño when the fish bypassed the SCB for cooler waters farther north (Holts 1985).

Fishes of the Pelagic Zone

The pelagic realm is the largest habitat in the SCB and the home of 40% of the species and 50% of the families of fishes. Nearly 200 species (70 families) have been collected in the California Current (Berry and Perkins 1966), 79 species (30 families) have been collected in the coastal waters (Horn 1974), and 124 species have been collected in the mesopelagic and bathypelagic zones (Lavenberg and Ebeling 1967). Epipelagic fishes are relatively large, active, fast-growing, long-lived fishes that reproduce early and repeatedly. Mesopelagic fishes are relatively small, slow-growing, long-lived fishes that reproduce early and repeatedly. Bathypelagic fishes are relatively large, sluggish, rapid-growing, slightly shorter lived fishes that reproduce late and maybe only once (Childress et al. 1980).

Light penetration, water temperature, and water mass structure define vertical zonation. The epipelagic zone is euphotic, and temperatures fluctuate diurnally and seasonally. It is approximately 50 m deep in turbid nearshore waters and expands offshore in clear oceanic waters. The mesopelagic zone is characterized by steep environmental gradients. It extends from the permanent thermocline, below the compensation depth, to the 6°C isotherm at 500–600 m. The bathypelagic zone is characterized by uniformity and extends nearly to the bottom. It is absent or restricted in the nearshore basins and expands offshore (Lavenberg and Ebeling 1967).

Epipelagic

The epipelagic zone is inhabited by epipelagic fishes, mesopelagic fishes that migrate to the surface waters at night (nyctoepipelagic), bottom-associated species that feed in the water column (nektobenthic) (Berry and Perkins 1966; Horn 1980), and the eggs and larvae of most pelagic and demersal fishes (Loeb et al. 1983a). The fish fauna is dominated by small, planktivorous schooling fishes such as northern anchovy and Pacific mackerel; by predatory schooling fishes such as Pacific bonito (*Sarda chiliensis*) and yellowtail; and by large, solitary predatory fishes such as blue shark (*Prionace glauca*) and swordfish (*Xiphias gladius*) (Mais 1974, 1977; Squire 1983b; Bedford and Hagerman 1983; Cailliet and Bedford 1983). The autecology of the epipelagic planktivores is well known because they support important commercial fisheries; far less is known about the large predatory fishes.

Northern anchovy, Pacific sardine, jack mackerel, Pacific mackerel, and Pacific hake are residents of the California Current system, and their reproductive strategies are adapted to its flow characteristics (Parrish et al. 1981). Northern anchovy is the most abundant epipelagic fish and may be the usually dominant species (MacCall et al. 1976; Squire 1983b). From spring through fall, the SCB is inhabited by Pacific saury (*Cololabis saira*), bluefin tuna (*Thunnus thynnus*), yellowtail, and many large, solitary predators that emigrate from tropical and oceanic areas (Bedford and Hagerman 1983; Cailliet and Bedford 1983).

The California Current is one of four major eastern boundary currents. The others are the Peru Current off the west coast of South America, the Canary Current off the west coast of southern Europe and northern Africa, and the Benguela Current off the west coast of southern Africa. Eastern boundary currents are characterized by equatorward surface flow, coastal upwelling, narrow continental shelf, and temperate climates. The four currents are physically similar and are dominated by a small number of closely related temperate pelagic fishes that can reach large population sizes: anchovy (*Engraulis*), sardine (*Sardinops* or *Sardina*), jack mackerel (*Trachurus*), hake (*Merluccius*), mackerel (*Scomber*), and bonito (*Sarda*) (Parrish et al. 1983).

PRODUCTIVITY AND TROPHIC STRUCTURE

Food chains in the more productive areas of the world's oceans are relatively short (Jones and Henderson 1987) (fig. 9.1). Chemical analyses demonstrate that the epipelagic foodweb in the SCB is more simplified and more structured than nearshore benthic foodwebs (Young et al. 1980; Mearns et al. 1981). Schooling pelagic fishes are important in the recycling of nutrients (McCarthy and Whitledge 1972). Regenerated production accounts for about 65% of total production in the SCB (Eppley and Peterson 1979).

Northern anchovies feed on zooplankton (51% crustaceans and 36% other species) and phytoplankton (11%) (Loukashkin 1970). Depending on the size of food particles available, anchovies feed by filtering or biting (Leong and O'Connell 1969; O'Connell 1972). The average biomass of an anchovy school is about 15 kg m^{-2} (range 0.1–125) (Hewitt et al. 1976). Assuming that the food requirement for a biomass of 15 kg m^{-2} is about 16 g C m^{-2} d^{-1} and that the average primary productivity is about 1 g C m^{-2} d^{-1} in the SCB, a 200-m-diameter school of anchovies with a conversion efficiency of 0.12 would have to swim 21 km per day (about two body lengths per second for a 12-cm fish) to remain in equilibrium with local primary productivity (Smith 1978a).

Anchovy egg production amounted to an estimated 0.9% of primary production in 1957, a year of average larval abundance, and 1.8% in 1966, a year of high larval abundance. If egg production accounts for 10% of food consumption, then anchovies consumed 9% of primary production in 1957 and 18% in 1966 (Smith and Eppley 1982). Changes in primary productivity were correlated with changes in recruitment in some fish stocks (Parrish and MacCall 1978; Fiedler et al. 1986).

SPATIAL VARIABILITY

Schooling is characteristic of fish inhabiting the epipelagic zone (Smith 1981). Within a species, schooling can vary from well-defined, compact aggregations to widespread, scattering layers (Mais 1974). Commercial fishermen recognize more than a dozen different school types among eastern Pacific tunas (Scott 1969; Scott and Flittner 1972). The formation of schools among clupeoids depends on vision, but the maintenance of school structure depends on vision and lateral line stimuli (Blaxter and Hunter 1982). The formation and maintenance of schools is affected by light level. Schooling fish are randomly distributed in darkness; they join groups that form and disperse as light levels rise, then form compact schools as light levels rise still further (Hunter 1968; Hunter and Nicholl 1985). Schooling increases intraspecific competition for food, but the disadvantages must be outweighed by reduction in predation and by facilitation of reproduction (Smith 1978a; Blaxter and Hunter 1982).

Smith (1981 p. 3) observed that "Schooling pelagic fish occupy a hydrographic province whose geographic extent can double or triple in a few years and return to the original as rapidly. . . . [A] pelagic schooling fish occupies a very small portion [0.5%] of its available environment at any instant, but it occupies it very intensely."

Smith (1978a) defined four spatial scales for pelagic schooling fishes: behavioral (scale of aggregation caused by individual behavior, that is, the fish school); hydrographic (scale that attracts and keeps fish in a small geographic area, e.g., upwelling and zooplankton blooms); physiological (distribution of a species determined by its physiological limits); and external (scale at which food or predators enter the environment of a species from outside its area of distribution).

Individual epipelagic fish schools aggre-

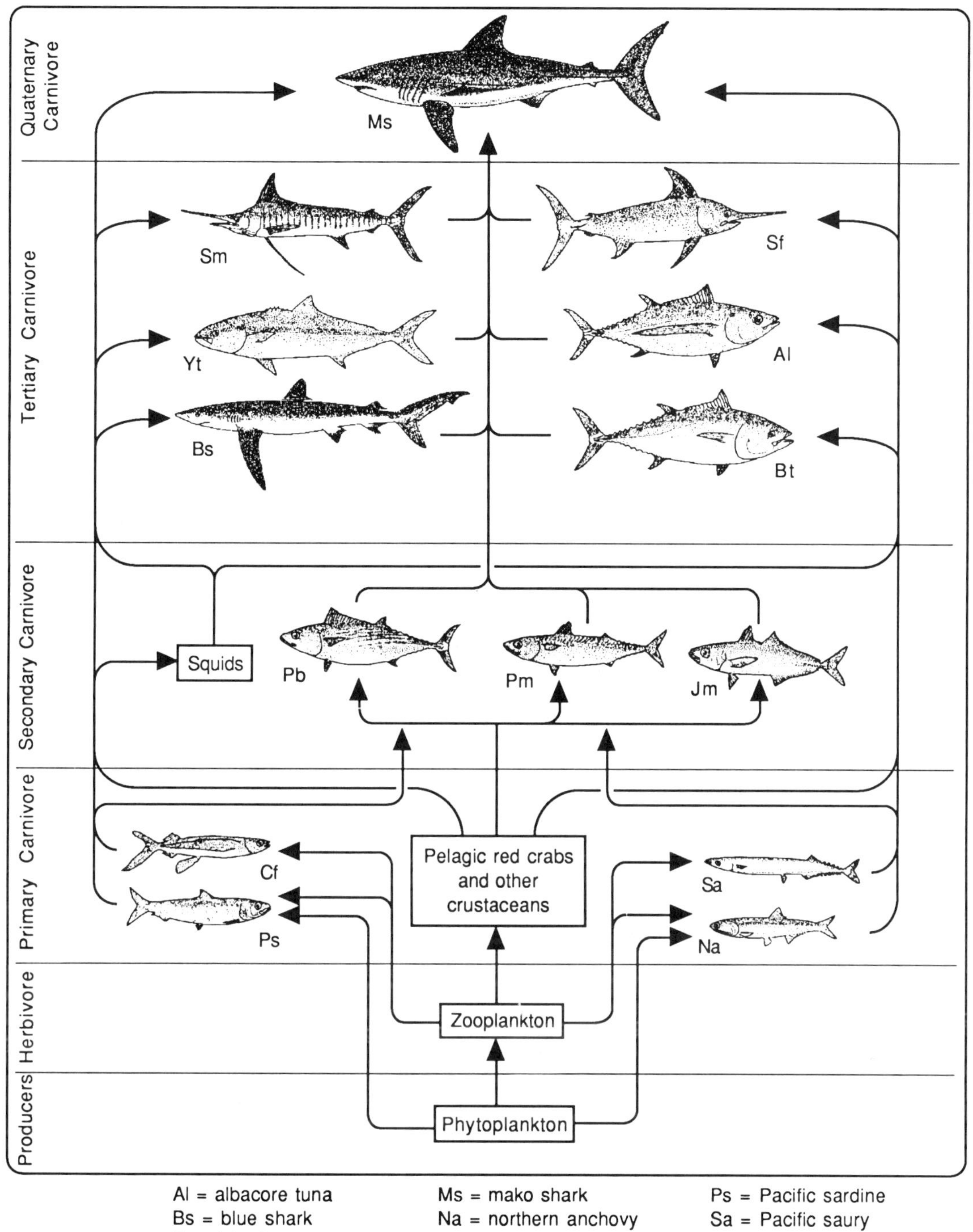

Figure 9.1. Trophic relationships of epipelagic fishes.

gate into school *groups* that occupy areas on the order of 10 km (Smith 1978a,b). The distribution of school groups is often patchy and nonrandom (Mais 1974, 1977). There may be 3500 schools in an area 10 km in diameter; for anchovies, this represents about 1% of the biomass of the total stock. The school group comprises a wider range of anchovy sizes and ages than an individual school. The hydrographic features that control this scale of aggregation have not been identified (Smith 1978a).

The central population of northern anchovy occurs in the SCB. Much of the population occurs inshore in the northern part of the bight during the fall. The fish move offshore and southeast with the onset of spawning in late winter. The largest schools occur within 40 km of the coast over deep water, often over escarpments and submarine canyons (Mais 1974; Pacific Fishery Management Council 1983). Anchovy rarely spawn at temperatures of less than 13.5°C and at phytoplankton pigment levels of less than 0.2 mg m^{-3}. The northern and offshore limits of spawning are determined by cold, upwelled water advected from north of Point Conception into the SCB. The southern limit is determined by low phytoplankton pigment levels (Fiedler 1983). From 1981 to 1982, the cold-water boundary occurred within the SCB (fig. 9.2). During the 1982–1983 El Niño, surface water temperatures of less than 14°C did not occur within the SCB. Anchovy eggs were collected farther north and offshore, an increase in spawning range of 43% (Fiedler et al. 1986). Physical changes such as El Niño are characterized by major displacements of water masses. Many pelagic species respond to geographic changes in the boundaries between water masses by redistributing themselves according to temperature and other attributes of the water (MacCall 1984).

TEMPORAL VARIABILITY

During the day, northern anchovy occur in small, low-density schools near the surface and in large, loosely compacted schools in deep water (110–220 m). Schools rise to the surface at night and disperse into thin scattering layers. Between midnight and dawn, the fish condense into schools and return to deep water (Messersmith et al. 1969; Mais 1974, 1977). Schools of jack mackerel remain near the bottom or under kelp canopies in shallow rocky areas during the day, then venture into deeper surrounding areas at night (Mais 1974). Laboratory experiments suggest that light is sufficient for jack mackerel to maintain schools near the surface on clear moonless nights and to feed effectively near the surface on full moonlight nights (Hunter 1968).

Large epipelagic predatory fishes make diel migrations in response to changes in location and availability of food. When squid (*Loligo opalescens*) are abundant in shallow water in spring, blue sharks migrate from offshore into shallow waters at night and return offshore in the morning. During the summer and fall when squid abundance is low, blue sharks remain offshore throughout the day and night (Sciarrotta and Nelson 1977). Swordfish tracked off Baja California remained during the day near an inshore bank, where they may have preyed on demersal fishes; at night they swam offshore, where they may have preyed on squid or other vertically migrating prey (Carey and Robison 1981).

Epipelagic fish abundances change seasonally. Anchovy schools are more abundant and larger in inshore areas of the northern SCB during summer and fall (table 9.2), probably due to the large number of juveniles produced during the preceding spawning season (Smith 1981). From late winter to spring, the anchovy schools move offshore to the southeast for spawning (Mais 1974, 1977). Yellowtail migrate into the SCB from central and northern Baja California as surface water temperatures warm in the spring. They move offshore to spawn in the summer and return south as surface water temperatures cool in late summer and fall. Yellowtail overwintered in the SCB during the 1957–1959 Cali-

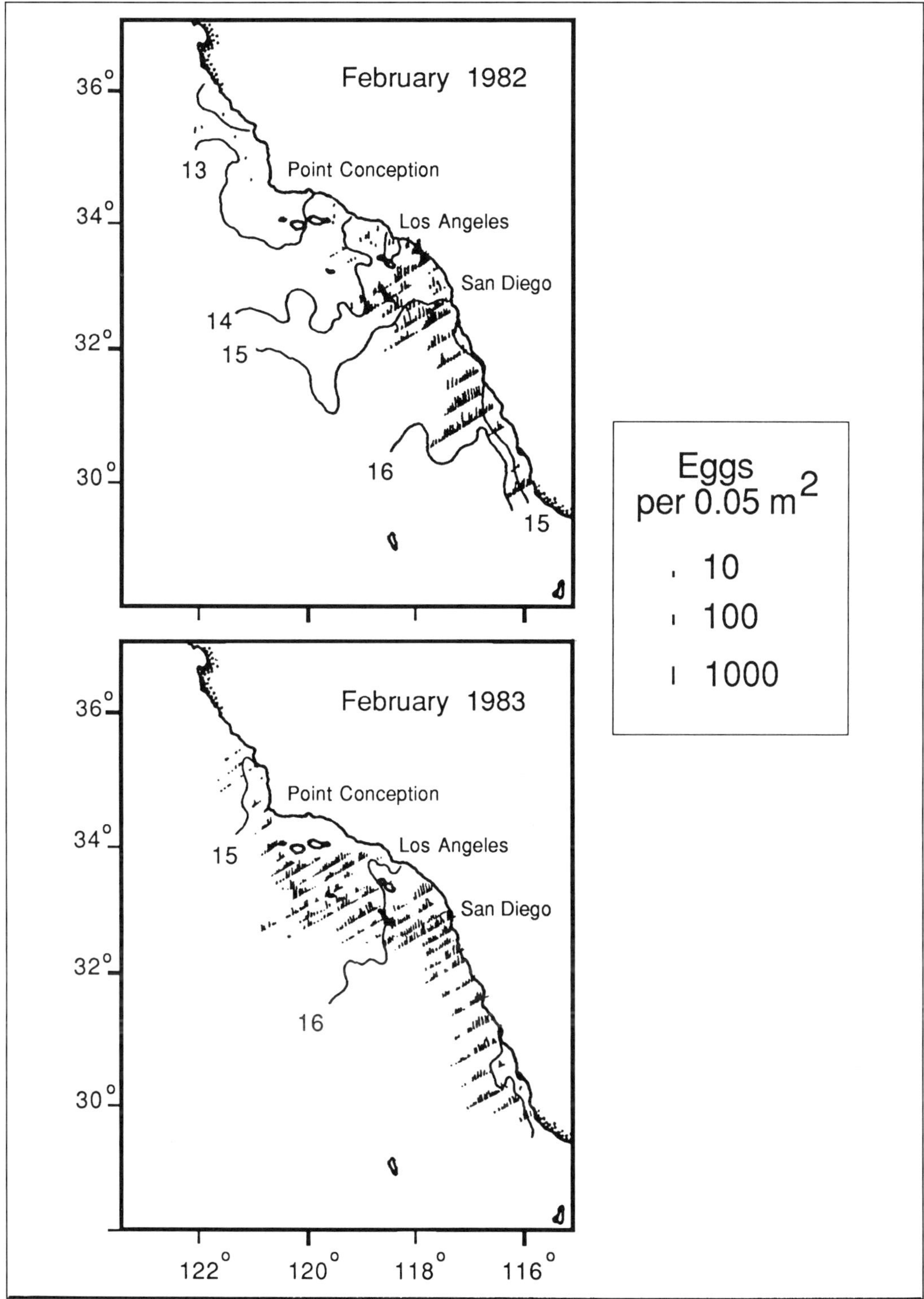

Figure 9.2. Sea surface temperature (°C) isotherms and distribution of northern anchovy eggs. (Redrawn from Fiedler 1983.)

Table 9.2. *Estimated Number and Size of Northern Anchovy Schools by Season*

Season	Schools (number km^{-1})	School Size (m^{-2} school)
Winter	0.79	1151
Spring	0.73	772
Summer	1.22	1327
Fall	1.98	1683

From Smith 1981.

fornia El Niño, when water temperatures were several degrees above average (Baxter 1960; Crooke 1983).

The total biomass of the Pacific mackerel, one of the most thoroughly studied and variable fisheries in the world, exploded twice in the last 60 years (MacCall et al. 1985; Prager and MacCall 1988) and at least once in the 1800s (Soutar and Isaacs 1974). The ecological effects of large biomass are confounded by concomitant environmental and fishery-related changes (MacCall et al. 1985). Large year classes are produced during years of favorable environmental conditions and large spawning biomass. When fishing pressure is heavy, or during a series of environmentally unfavorable years, and especially when both occur together, spawning biomass falls to levels where even optimal environmental conditions do not result in a large year class (Parrish and MacCall 1978).

Soutar and Isaacs (1974) found a generally consistent relationship between scale deposition in the anoxic basins off southern and Baja California and contemporaneous population estimates of Pacific sardine, northern anchovy, and Pacific hake. They assumed that scale deposition was an index of population size and concluded that sardines went through cycles of abundance, whereas anchovy and hake remained relatively stable (fig. 9.3). They estimated the mean biomass of the three species was 2×10^6 t from the 1940s to the 1970s, 8×10^6 t for the entire time series, and 15×10^6 t for the 1980s.

Soutar and Isaacs (1974, p. 270) concluded that ". . . most of man's experience in the waters off the Californias appears to be associated with low pelagic-fish productivity."

THE NEARSHORE ZONE

There are few published studies on pelagic fish assemblages in the neritic zone (within approximately 20 km of the coast) of the SCB. Because of nearshore currents, mesoscale and local upwelling, and tidal and internal wave-induced mixing, it is a different environment from waters farther offshore (Petersen et al. 1986). Northern anchovy is the most abundant species (>80% of all fishes caught) caught within 3 km of the coast; queenfish, white croaker (*Genyonemus lineatus*), Pacific butterfish (*Peprilus simillimus*), and three silversides (Atherinidae) are common. Except for northern anchovy, there is little seasonal variation in total catch. Anchovy abundance increases during the summer due to recruitment of juveniles (L. G. Allen and DeMartini 1983).

Schooling neritic fishes display temporal activity patterns. Queenfish spend the day in inactive schools close to shore (<1.5 km) in shallow water (<10 m) and disperse offshore (>3.5 km) into deeper water (10–20 m) at night to feed on planktonic prey (Hobson and Chess 1976; Hobson et al. 1981; L. G. Allen and DeMartini 1983; DeMartini et al. 1985). Olive rockfish, salema (*Xenistius californiensis*), and walleye surfperch (*Hyperprosopon anale*) also school inactively near shore, reefs, or kelp beds during the day and disperse offshore to feed on planktonic prey in the water column at night (Hobson and Chess 1976; Hobson et al. 1981). Blacksmith are schooling planktivores and one of the most abundant species on nearshore reefs. During the day, they feed in the water column at varying distances from the reef; at dusk, they return to the reef and shelter among the rocks for the night (Ebeling and Bray 1976; Hobson and Chess 1976; Bray 1980). Adult blacksmith import as much as 8 g C m^{-2} yr^{-1} deposited as feces in nocturnal shelters (Bray et al. 1981).

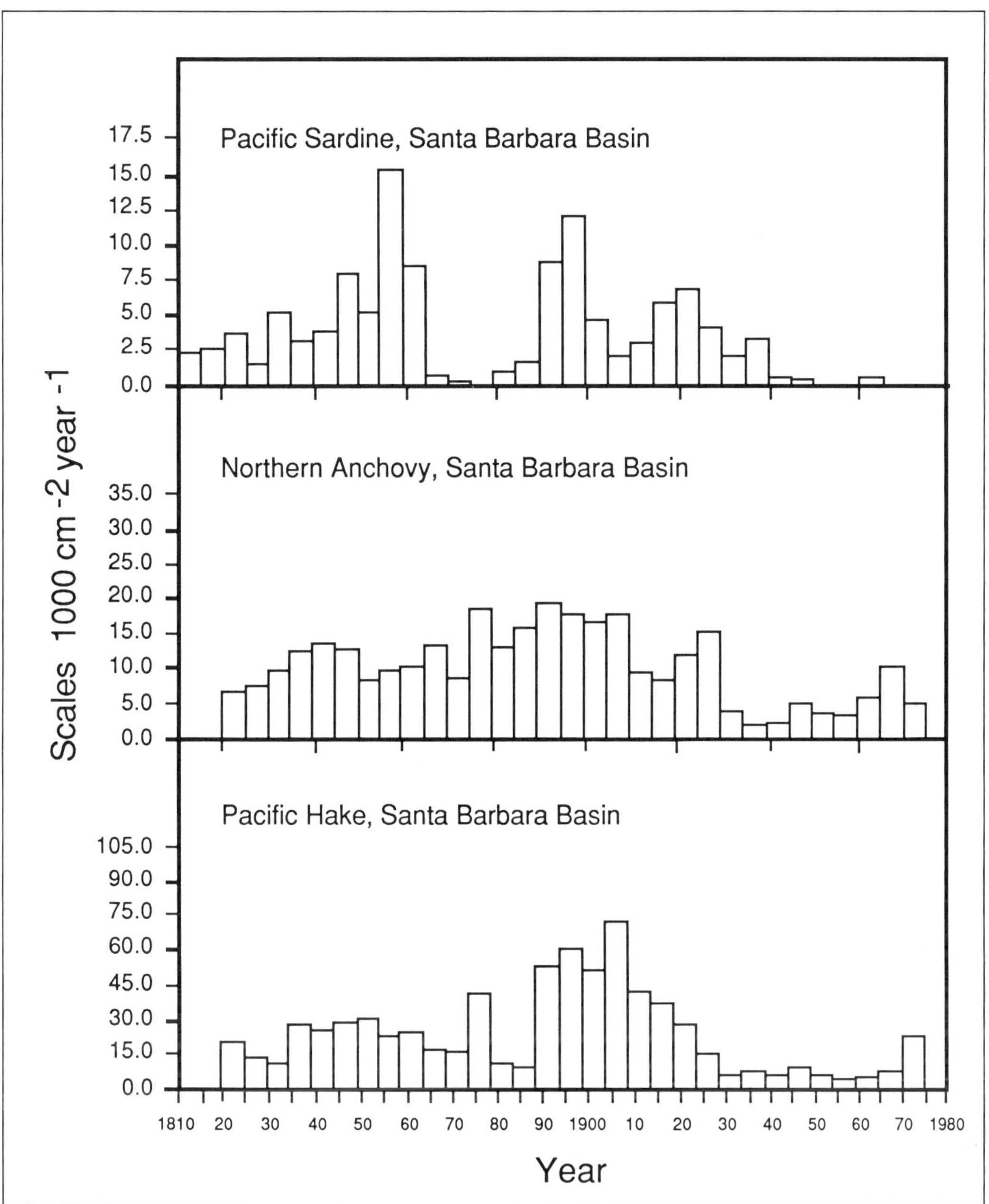

Figure 9.3. Scale deposition rates of Pacific sardine, northern anchovy, and Pacific hake. (Redrawn from Soutar and Isaacs 1974.)

CHANGES IN POPULATION ABUNDANCE

The largest and best known population fluctuations occur among pelagic fishes (Soutar and Isaacs 1969, 1974), especially anchovies, sardines, and herrings, which constitute about 30% of the world fish catch (70 × 10^6 t) (Lasker 1985). Hypotheses advanced to explain changes in the abundance of exploited fish populations can be organized into four categories: intraspecific dynamics, interspecific interactions, environmental fluctuations, and fishing (Shepherd et al. 1984; Skud 1982, 1983; Parrish et al. 1983; MacCall 1984; Lasker 1985; Cury 1988). Most explanations include some combination of these hypotheses because of the complexity of the system (Rothschild 1986).

Fluctuations in Pacific mackerel and northern anchovy biomass are attributable largely to fluctuations in recruitment. Recruitment success in Pacific mackerel has been roughly cyclical since the 1930s (MacCall et al. 1985) (fig. 9.4). Reproductive success in 1976 was about 750 times reproductive success in 1983 (Parrish and MacCall 1978). Anchovy biomass was high in the early 1970s because of favorable environmental conditions, low adult mortality, and above average recruitment (Methot 1982). The subsequent decline in anchovy abundance was attributed to a return to more normal environmental conditions, increased fishing pressure, and predation by an expanding Pacific mackerel population (Mais 1981; Pacific Fishery Management Council 1983; MacCall et al. 1985).

Except for fishing, the causes of fluctuations in abundance can affect eggs, larvae, juveniles, and adults. Eggs are about 1 mm in diameter, and the stage lasts only a few days. Larvae are a few millimeters long, and the stage lasts several weeks. Juveniles a few centimeters long leave the plankton and mature into adults in several years (Rothschild 1986). The number of adults may be four orders of magnitude lower than the number of newly hatched larvae (Bakun 1986). Substantial variation exists in mortality rates among early life history stages (Smith and Moser 1988), and little correlation exists between mortality rates during the planktonic phase and the abundance of recruits (Hunter 1981) or between the abundance of recruits and abundance of their parents. The correlations are better between juvenile and adult abundances and between adult and egg abundances, suggesting that the mortality rate stabilizes during the late larval or early juvenile stage (Rothschild 1986). The number of surviving northern anchovy larvae at day 19 is not correlated to recruitment at age 1; on average, only 0.34% of 19-day-old larvae survive to age 1. The proportion of 9- to 12-month-old fish in the commercial fishery off Mexico is correlated with the abundance of age 1 fish (Peterman et al. 1988).

Intraspecific Dynamics. Northern anchovy, Pacific sardine, and Pacific mackerel undergo great variations in stock size. Density-dependent effects on growth, reproduction, and mortality, if they exist, are weak (Knaggs and Parrish 1973; Parrish and MacCall 1978; MacCall 1979, 1980; Blaxter and Hunter 1982). Some evidence, however, points to density-dependent growth in Pacific mackerel (Prager and MacCall 1988). The degree of density dependence may be a function of fecundity (Cushing 1971; Cushing and Harris 1973) or feeding habits (Ware 1980). Pelagic fishes with low fecundity (such as clupeiforms) lack strong density-dependent population regulatory mechanisms, have highly variable recruitment and population size, and are vulnerable to long-term changes in the environment (Cushing 1975). Low trophic level fishes (clupeiforms) feed on small plankton with short generation times; higher trophic level fishes feed on larger plankton with generation times at least an order of magnitude longer. Predators on longer lived prey can have a more significant and lasting effect on food resources than predators on shorter lived prey (Ware 1980; Blaxter and Hunter 1982).

Interspecific Interactions. Competition, predation, parasitism, and disease can affect pelagic fish population abundance. Little is known about the effects of parasitism and disease, but they are probably not important sources of mortality among clupeiforms (Blaxter and Hunter 1982). Nematodes occur in a high proportion of teleosts off central and southern California. Of the nearly 2300 individuals (68 species, 21 families) examined by Dailey et al. (1981), 42% (42 species) harbored larval *Anisakis* sp. and 6% (17 species) harbored larval *Phocanema* sp. The incidence of infection

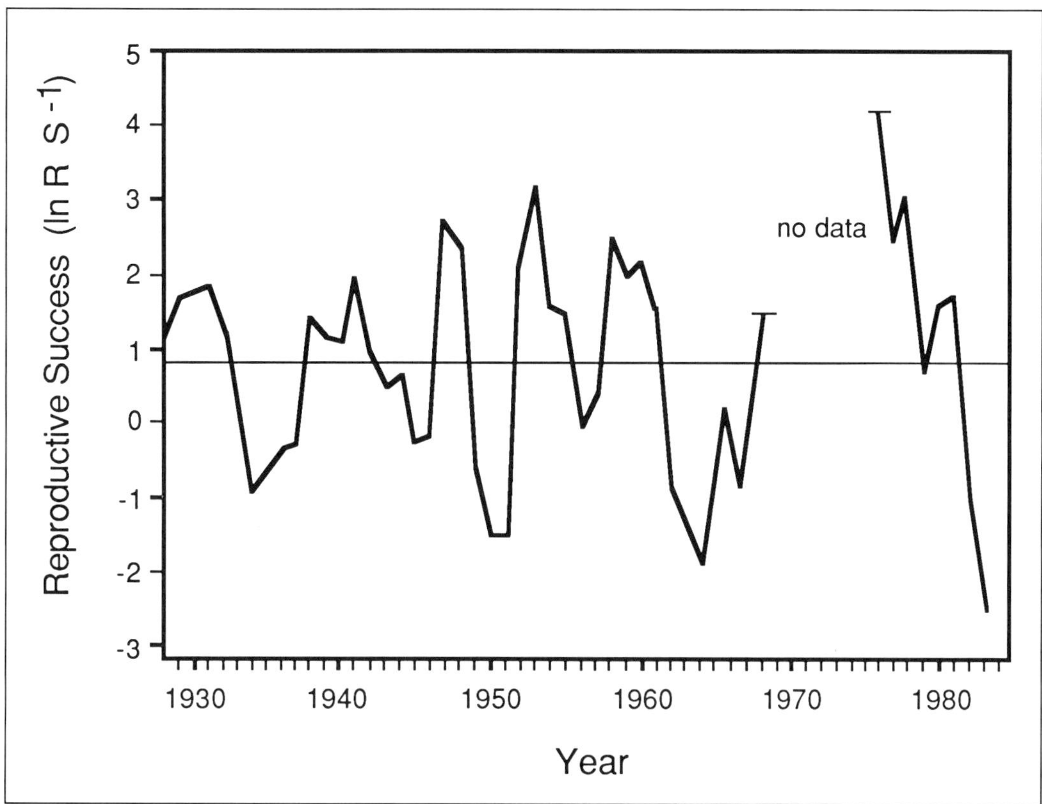

Figure 9.4. Reproductive success of Pacific bonito. R = recruitment in numbers; S = spawning biomass in kilograms. (Redrawn from MacCall et al. 1985.)

of Pacific hake by the myxosporean parasite *Kudoa* spp. can be as high as 90–100%. *Kudoa* invade the white muscle tissue and cause blackening that affects the visual quality of fillets. Following death of the fish, proteolytic enzymes from the parasite liquefy host muscle tissue, further reducing its appeal to consumers (Kabata and Whitaker 1985; Nelson et al. 1985). The effects of the parasite on the fish are unknown. Heavily infected individuals are able to swim because the parasite does not infect the red muscle tissue involved in normal cruising (Kabata and Whitaker 1985). Proximate composition of moderately infected individuals is not altered. Proximate composition of severely affected hake (2–3% of the population) is altered significantly; muscle protein content can decline by 30% (Patashnik et al. 1982; Nelson et al. 1985).

Competition for food between adult Pacific sardine and northern anchovy is often cited as the cause of replacement in this species pair (Murphy 1966; Radovich 1981, 1982). The evidence for competitive replacement, however, is circumstantial (MacCall 1984, 1986). The two species eat similar prey, but their feeding habits are difficult to quantify because their diets are diverse and because they can change feeding modes from filtering to biting with changes in prey size (O'Connell 1972). The increase of available food after the sardine population collapsed may have been responsible for the increased duration of peak spawning and the increased number of spawning batches produced by anchovy (Hunter and Leong 1981). But even if sardine and anchovy diets completely overlapped, competition would not be demonstrated. At high plankton densities, both species could occupy the same feeding niche without competing for food (MacCall 1984). Sardines were abundant from Baja California north to British Columbia while the center of the anchovy population was located south of Point Conception. The fact that anchovies did not occupy habitats north of Point Conception after the sardine declined was taken as evidence against competitive replacement by MacCall (1983). This neglects the importance of genetically distinct populations adapted to slightly different environmental conditions within a species (Hedgecock 1986; Utter et al. 1989). The southern population of sardines may have competed with the anchovy for limited food resources during difficult times.

Little is known about the nature and magnitude of predation on schooling fishes in the California Current (Smith 1978a). The eggs and larvae of northern anchovy are consumed by invertebrate and vertebrate planktivores, including adult anchovies. These life stages are short (2–4 months), but mortality is high. Juvenile and adult anchovies are eaten by many species of fish, nearly all seabirds and many shorebirds, marine mammals, and humans as well (Pacific Fishery Management Council 1983). Piscivorous fishes may affect anchovy natural mortality and spawning biomass (Smith 1978a; Pacific Fishery Management Council 1983). Anchovy natural mortality was low during the early to mid-1970s, a period when Pacific mackerel biomass was also low (7000 t). Mortality increased in the late 1970s and early 1980s when the mackerel population expanded to 166,000 t (Parrish and MacCall 1978; Pacific Fishery Management Council 1983). Brown pelicans (*Pelecanus occidentalis californicus*) survive almost completely on anchovies, and fledgling success is correlated with anchovy biomass (Anderson et al. 1980). On average, predators consume 46% of the initial anchovy spawning biomass per year (Pacific Fishery Management Council 1983).

Albacore tuna migrate eastward from the central North Pacific to the warming coastal waters in spring and early summer. By midsummer, large numbers occur at oceanic fronts along the seaward edge of the coastal water mass, especially in upwelling areas (Laurs and Lynn 1977; Laurs et al. 1984; Fiedler and Bernard 1987). Here they feed on anchovies, sauries, and to a lesser extent, euphausiids, decapods, squids, and other fishes (Pinkas et al. 1971; Bernard et al. 1985; Fiedler and Bernard 1987). Albacore tuna can consume 15% of the anchovy population production; they also may affect year-to-year changes in reproductive success (Smith 1978a). An influx of scombrid predators into the SCB during the 1982–1983 California El Niño may have increased predation on juvenile anchovies and contributed to the low recruitment in 1983 (Fiedler et al. 1986).

Although marine mammals consume 150–300 million Pacific hake annually, they have little impact on the hake population. Half of the fish consumed by marine mammals may belong to one year class (3–5% of the recruits). Northern fur seals (*Callorhinus ursinus*) consume 9,000–12,000 t annually which is less than 10% of the commercial harvest. When hake are abundant, they constitute a large portion of the diets of their marine mammal predators. When hake are not abundant, their predators switch to rockfish and squid (Bailey and Ainley 1982; Antonelis and Perez 1984).

Environmental Processes. The epipelagic ecosystem of the California Current may function permanently under nonequilibrium conditions in which biological processes (competition and predation) are less important than environmental fluctuations in regulating species abundances (McGowan 1974, 1977; Bernal 1981; Parrish et al. 1981). The environment can mediate changes in population abundance through (1) direct physiological effects, (2) feeding effects caused by temporal and spatial variations in food abundance and quality, (3) localized effects of upwelling and storms, and (4) widespread oceanographic events such as El Niño. Direct physiological effects related to temperature, oxygen, and salinity are not important sources of mortality among clupeiforms (Blaxter and Hunter 1982).

Murphy (1966) attributed the 3–5 year cycles of spawning success and failure in Pacific sardine to changes in the environment. The spawning success (reproductive biomass greater than spawning biomass) of a population at carrying capacity should fluctuate around some equilibrium value. Spawning success should be good below carrying capacity and poor above it. The coherence of spawning success and failure in sardine suggests that they were tracking the carrying capacity of the environment rather than fluctuating about some equilibrium (Smith and Moser 1988).

Year-class strengths of widow rockfish (*Sebastes entomelas*) and chilipepper rockfish (*Sebastes goodei*) are correlated with environmental conditions in the California Current. Widow rockfish recruitment is highest in warm years, when transport by the California Current and upwelling are weakest. Chilipepper recruitment is highest in cool years, when transport by the California Current and upwelling are strongest (Norton 1987).

Major periods on the order of 30–60 years exist in the scale deposition rates of pelagic fishes (Soutar and Isaacs 1974). Since generation times are less than 10 years, Smith and Moser (1988) concluded that fluctuations in the environment probably caused the periodicity.

Fishing. The Pacific sardine once supported the largest fishery by weight in the United States. Commercial exploitation began in the nineteenth century and expanded following World War I. Annual landings averaged over 500,000 short tons from 1935 to 1945 (Talbot 1973). The population and the fishery collapsed in the 1950s for reasons that are not clear (Murphy 1966; Talbot 1973; MacCall 1979; Radovich 1981, 1982). However, MacCall (1983) theorizes that overexploitation of a depressed population, coupled perhaps with reduced recruitment from excessive losses of reproductive products caused by increased offshore transport (Cury and Roy 1989), may have led to the demise of the sardine.

Pelagic fishes with low fecundity (such as sardines and herrings) lack strong density-dependent population regulatory mechanisms, have highly variable recruitment and population size, are vulnerable to long-term changes in the environment, and are susceptible to overfishing (Cushing 1975; Blaxter and Hunter 1982).

Lasker (1985, p. 32) states that "A clupeoid collapse can be due to heavy fishing mortality, which reduces the mean age of the population and forces the very young fish to sustain the reproductive load of the population. If the environment intervenes in some way to cause excessive mortality of the eggs, larvae, or even small juveniles, then—at the most—only two successively poor year classes . . . will reduce the population below the critical minimum stock size [when there is also] continued fishing on the adult stock."

The largest scientific problem impeding effective management of living marine resources is the lack of understanding of the interactions among environmental variability, fishery exploitation, and recruitment fluctuations (Bakun 1986).

Population Regulation. Following collapse, populations of epipelagic fishes may remain depressed much longer than their generation times, but then rebound to high levels relatively quickly (MacCall 1983; Tanaka 1983). There are times during the long "extinction" period when climate is probably favorable yet the stocks do not respond (Cury 1988). The causes of collapse have received more attention than the mechanisms that maintain biomass and produce good recruitment at low stock levels. Skud (1982, 1983) and Cury (1988) suggest that populations are regulated by the interplay among stock abundance, interspecific competition, and climate variability. When populations are abundant, most contacts are intraspecific; when populations are reduced, most contacts are interspecific. The decline of one species is often supplanted by the increase of another that then confronts a novel situation: change in climate and reduced dominance by a competitor.

Pelagic fishes have high biotic potential (produce thousands or millions of eggs per year per female) and demographic plasticity (time of reproduction, fecundity, condition, and growth), all of which are attributes that are adaptive in physically and biologically variable environments. Certain genotypes will be optimal in novel circumstances, and "evolutionary novelties" are more likely to be expressed when populations are small (Cury 1988). According to Hedgecock (1986), different spawning times in the anchovy population off central California may be due to different genotypes. The reaction of fish to climate varies with population size (dominance). When sardines were abundant and anchovies were rare, sardine recruitment was positively correlated with salinity and negatively correlated with temperature (factors associated with upwelling). When sardines were rare and anchovies were abundant, the signs of the correlations were reversed (Skud 1982, 1983).

Stocks collapse when abundance declines and climatic conditions are unfavorable. Strong selective pressures (interspecific competition) and the reduced impact of climate lead to a long period of extinction. Stocks rebound when the climate turns favorable and novel evolutionary modifications appear in the population (Cury 1988).

Ichthyoplankton

Most fishes release eggs and sperm in the water column. Fertilization is external, and both eggs and larvae are subject to oceanic diffusion and advection. Even among species that bear live young (e.g., Scorpaenidae and Zoarcidae) or attach their eggs to a substrate (e.g., Cottidae and Pomacentridae), the newly hatched larvae are usually pelagic.

Moser (1981, p. 90) characterized the larval stage as ". . . that period in ontogeny when major portions of a fish fauna come together and share prey, predators, and abiotic variables."

Ichthyoplankton of the California Current are generally well known; more than 200 taxa have been identified (table 9.3). Faunal affinities of the top 10 species in the California Current reflect its transitional nature. Northern anchovy, Pacific hake, Pacific sardine, and jack mackerel are eastern boundary current species. Rockfish (*Sebastes* spp.) and northern lampfish (*Stenobrachius leucopsarus*) are subarctic–transitional species. California smoothtongue (*Leuroglossus stilbius*) is a transitional species that occurs across the Pacific Ocean between the subarctic and central water masses. Sanddabs (*Citharichthys* spp.) occur along the coast from the subarctic to the tropics. Mexican lampfish (*Triphoturus mexicanus*) are found in the southern part of the California Current, and the lightfish (*Vinciguerria lucetia*) occurs in the eastern tropical Pacific (Smith and Moser 1988). Larval fish abundance is greatest in the SCB and off northern Baja California; one-half of the dominant taxa representing more than 90% of the larvae are shared with the California Current (Loeb et al. 1983a; Smith and Moser 1988). Lavenberg et al. (1986) and Walker et al. (1987) identified more than 150 taxa within a few kilometers of the coast in the SCB, and more than 20 have been identified in bays and estuaries (Leithiser 1981; Horn and Allen 1981a, 1985).

Sampling methods and taxonomic problems may distort the results of research on the spatial and temporal patterns of fish larvae and trends in their distribution and abundance. Many kinds of larval fishes can detect and avoid nets during the day (Lenarz 1973; Brewer and Kleppel 1986). Older larval and juvenile anchovies avoid plankton nets (Murphy and Clutter 1972) and require new sampling devices (Methot 1986) for more accurate studies of these fishes, their patterns, and trends. Species that occur primarily in the

Table 9.3. *Relative Abundance of Top 10 Ranking Larval Taxa*[a]

Taxa	Percent	Taxa	Percent
California Current		Southern California Bight	
Northern anchovy	49	Northern anchovy	80
Pacific hake	9	*Sebastes* spp.	6
Vinciguerria lucetia	8	California smoothtongue	4
Sebastes spp.	6	Pacific hake	3
California smoothtongue	4	Mexican lampfish	2
Mexican lampfish	3	Sciaenidae	2
Northern lampfish	3	Popeye blacksmelt	<1
Citharichthys spp.	2	Shortbelly rockfish	<1
Pacific sardine	2	Bocaccio	<1
Jack mackerel	2	Speckled sanddab	<1
SCB <100 km of coast		Upper Newport Bay	
Northern anchovy	83	Arrow goby	27
Sebastes spp.	4	*Anchoa* spp.	19
California smoothtongue	4	Gobiidae	12
Northern lampfish	2	Northern anchovy	10
White croaker	2	Cheekspot goby	9
Pacific hake	2	Shadow goby	8
Citharichthys spp.	<1	Sciaenidae	4
Popeye blacksmelt	<1	Longjaw mudsucker	4
California halibut	<1	White croaker	3
Hypsoblennius spp.	<1	Barred pipefish	2

[a] California Current from 1951 to 1981 (n = 30,296; Smith and Moser 1988); Southern California Bight (CalCOFI Region 7) in 1975 (n = 479; Loeb et al. 1983a, b); Southern California Bight within 100 km of the coast from 1974 to 1976 (n = 112; Gruber et al. 1982); and in Upper Newport Bay from 1978 to 1979 (n = 56; Horn and Allen 1985).

neuston (Gruber et al. 1982) or in the epibenthic layer (Watson 1982; Schlotterbeck and Connally 1982) require special sampling methods (Barnett et al. 1984). Taxonomic problems are greatest among the rockfish (*Sebastes*, >70 species), croakers (Sciaendae, >10 species), sanddabs (*Citharichthys*, 6 species), sea basses (*Paralabrax*, 3 species), blennies (*Hypsoblennius*, 3 species), and some subtropical forms. Taxonomic problems are greatest near the coast, where these groups make up a significant proportion of the ichthyoplankton (Gruber et al. 1982; Loeb et al. 1983a).

Research on ichthyoplankton dynamics in the SCB has focused primarily on Pacific sardine, northern anchovy, and Pacific mackerel (Hunter 1981; Sherman et al. 1983). The collapse of the Pacific sardine fishery in 1947 prompted the formation of the Marine Research Committee (MRC) (Croker 1982; Radovich 1982), which then established the California Cooperative Sardine Research Program to research physical and chemical oceanography, productivity, spawning and recruitment of sardines, and dynamics of the sardine population and fishery. In 1953 the sardine program was expanded to other species and renamed the California Cooperative Oceanic Fisheries Investigations (CalCOFI). CalCOFI's reformulated objective was as follows: to understand the factors that govern the abundance, distribution, and variations of pelagic marine fishes, with emphasis on the oceanic and biological factors affecting sardines and their associates in the California Current system (Talbot 1973; Baxter 1982).

GEOGRAPHICAL DISTRIBUTION

Geographical differences in larval fish distributions in the California Current are related to water mass distributions and to the warm- or cold-water affinities of the fishes (Loeb et al. 1983a; Moser et al. 1987; Smith and Moser 1988). About 80% of fish larvae collected north of Point Conception are mesopelagic fishes and *Sebastes* spp. South of Point Conception, more than 80% of the fish larvae are northern anchovy and Pacific hake. With increasing distance from the coast, anchovy, rockfish, and flatfish larval abundances decrease, while the larval abundances of hake and jack mackerel increase (Loeb et al. 1983a, b).

Within the SCB, jack mackerel, Pacific hake, and mesopelagic fishes are most abundant 10–100 km from the coast. California halibut, turbots (*Pleuronichthys* spp.), sea basses (*Paralabrax* spp.) (Gruber et al. 1982), and blennies (*Hypsoblennius* spp.) (Stevens and Moser 1982) are most abundant within 10 km of the coast. White croaker are most abundant within 4 km of the coast. Clinids (*Gibbonsia* spp.), queenfish, California clingfish (*Gobiesox rhessodon*), gobies, silversides, and diamond turbot (*Hypsopsetta guttulata*) are most abundant within 2 km of the coast (Barnett et al. 1984). Northern anchovy, rockfish (*Sebastes* spp.), and sanddabs (*Citharichthys* spp.) have no apparent horizontal distribution patterns (Gruber et al. 1982).

VERTICAL DISTRIBUTION

Ahlstrom (1959a) identified two groups of fish larvae based on their vertical distributions: those that are most abundant in the upper mixed layer and the upper part of the thermocline (0–125 m depth) (anchovy, sardine, mackerel, rockfish, sanddabs, and lampfish) and those that are most plentiful within or below the thermocline (Pacific hake, California smoothtongue, and snubnose blacksmelt [*Bathylagus wesethi*]). Most fish larvae belong to the first group. Anchovies, sardines, and mackerel usually occur within 50 m of the surface (Ahlstrom 1959a). Saury, grunion, and flying fish larvae (Exocoetidae) are virtually restricted to the neuston by day (Gruber et al. 1982). Near the coast, clinids, white croaker, queenfish, California clingfish, and gobies are most abundant near the bottom; silversides are found in greatest numbers near the surface (Barnett et al. 1984).

Vertical distributions may change during development. Hake larvae less than 8 mm in length are most abundant between 50 and 100 m depth; 8–12 mm larvae are most numerous between 100 and 300 m; larvae greater than 12 mm are most abundant between 200 and 300 m (Bailey 1982). White croaker larvae less than 7 mm are most abundant in the neuston and at mid-depth, while larger larvae occur in greatest numbers in the epibenthic layer (Schlotterbeck and Connally 1982; Brewer and Kleppel 1986).

TEMPORAL VARIABILITY

The larger larvae of most fishes probably avoid plankton nets and are undersampled by daytime tows compared to nighttime tows (Ahlstrom 1959a; Lenarz 1973; Brewer and Kleppel 1986). The night–day catch ratio for northern anchovy increases 0.64 for every millimeter increase in size above 3.5 mm (Ahlstrom 1959a). Net avoidance can be complicated by diel changes in vertical distribution. Anchovy larvae of greater than 12 mm migrate to the surface waters at night to fill their swim bladders (Hunter and Sanchez 1976; Brewer and Kleppel 1986). Blennies and California halibut (*Paralichthys californicus*) occur deeper in the water column during the day than they do at night. California clingfish and reef finspot (*Paraclinus integripinnis*) settle from the water column into the epibenthic layer at night (Barnett et al. 1984).

In the SCB, two peaks in ichthyoplankton abundance occur that have different taxonomic characteristics (Gruber et al. 1982; Loeb et al. 1983a; Walker et al. 1987). In the nearshore zone, the winter–spring assemblage is dominated (69%) by the larvae of fishes with northern range limits from Oregon to Canada, while the summer–fall assemblage is dominated (91%) by the larvae of fishes with northern range limits from Point Conception to Monterey. Cold-water fishes near the southern limits of their range generally spawn during the cold months of the year, while warm-water species near the northern limit of their range generally spawn during the warm months (Walker et al. 1987).

Egg size, and hence larval size, of many winter–spring and spring–summer spawning fishes generally decline during the spawning season (Ware 1975). Eggs of northern anchovy spawned in February and March are larger than eggs spawned during the summer (Hunter 1976). Eggs of Pacific mackerel spawned in April and May are larger than eggs spawned in June and July (Fry 1936). Queenfish eggs declined from 685 μm in March to 580 μm in August (DeMartini and Fountain 1981). Larger larval size at first feeding may be an advantage when the abundance of zooplankton prey is low and when lower water temperatures cause slower growth rates (Ware 1975; Hunter 1976). A strong seasonal signal is not evident for phytoplankton standing stock and productivity in the SCB, but it is evident for zooplankton. Zooplankton abundance peaks in June and July (Mullin 1986).

Interannual environmental changes in the California Current affect the magnitude and seasonal extent of spawning of many species as well as the geographical distribution of their larvae. During the 1958–1959 El Niño, larvae of species with cold-water affinities (California smoothtongue, northern lampfish, *Sebastes* spp., and Pacific hake) declined in abundance in the SCB and off northern Baja California, and their distributional range contracted northward. Larvae of species with warm-water affinities (snubnose blacksmelt, Mexican lampfish, broadfin lampfish [*Lampanyctus ritteri*] and *Vinciguerria lucetia*) increased in abundance in the SCB, and their distributional range expanded northward (Moser et al. 1987). The impact of El Niño on egg and larval physiological tolerance, development time, and metabolism are not known (Bailey and Incze 1985). The larval abundance of many fishes with northern affinities recovered to previous levels soon after the return of normal oceanic conditions in the early 1960s. However, the larval abundance of *Sebastes* spp. did not recover for nearly a decade (MacCall 1986).

THE NEARSHORE ZONE

Concern about nearshore fish populations has increased in the last decade because of the growing importance of recreational fisheries and the increased human modification of the habitat (Gruber et al. 1982; Barnett et al. 1984). The abundance of larvae of coastal fishes increases nearshore (Gruber et al. 1982; Loeb et al. 1983a; Barnett et al. 1984). Older larvae of white croaker, queenfish, and California halibut are concentrated in the epibenthic layer close to shore (Watson 1982; Barnett et al. 1984), and recruitment is greatest in the shallowest water (Love et al. 1986). Young-of-the-year white seabass (*Atractoscion nobilis*) recruit to drift algae and debris outside the surf zone (L. G. Allen and Franklin 1988). Because bays and estuaries in the SCB are small and few in number, the nearshore zone is a nursery area comparable in importance to estuaries along the Atlantic coast (Sherwood 1980; Barnett et al. 1984).

Spawning of nearshore species is limited to the coastline and to offshore islands and banks. Eggs and larvae of nearshore fishes are dispersed by currents, and must either remain nearshore or return to the coast in sufficient numbers to maintain adult populations. Larvae with long development times would be carried offshore, especially during upwelling, so they require a shoreward transport mechanism. The larvae of nearshore fishes may develop rapidly and therefore remain close to shore (Gruber et al. 1982). California halibut larvae settle out of the plankton at 8–12 mm SL after 30 days (L. G. Allen 1988), while northern anchovy larvae recruit at 35 mm SL after 74 days (Hunter 1976). Larvae that end up offshore may be transported onshore by surface slicks. Internal waves cause convergences that compress scattered material and organisms into surface slicks that generally move toward shore (Shanks 1983; Kingsford and Choat 1986).

The nearshore zone is not a preferred spawning habitat for northern anchovy (Brewer and Smith 1982). The seasonal cycle and standing stock of eggs and larvae in the nearshore zone (<44 m water depth) are comparable to those of the SCB as a whole. The nearshore zone constitutes 3.8% of the areal extent of the SCB and contains 3% of the anchovy larvae. Densities of phytoplankton and microzooplankton are often higher in the nearshore zone than farther offshore, but production of young anchovy is not enhanced. Anchovy eggs and young larvae suffer higher mortalities in the nearshore zone than in the SCB as a whole. Cannibalism and predation by planktivores may be higher nearshore. Survival of feeding larvae nearshore improves with time and, by day 30, is comparable to that of the SCB. Anchovy production is enhanced nearshore if improved survival rates occur during the juvenile stage (Hewitt and Brewer 1983).

The importance of the nearshore zone as a nursery area to pelagic fishes may change with population size. When adult Pacific sardine biomass is large (100–200 $\times$ 10^3 t) (MacCall 1979), sardine eggs occur in inshore and offshore areas (Ahlstrom 1959b). When adult biomass is low (5–20 $\times$ 10^3 t), egg and larval sardines are collected only near the coast (Gruber et al. 1982; Wolf and Smith 1986). The resurgence of the sardine was first noted in the inshore waters of the central SCB in 1982, but as the intensity of spawning increased, sardine larvae were collected in inshore waters in other parts of the SCB (Lavenberg et al. 1986). The nearshore zone, with its higher productivity, may be important for the recovery of depleted fish stocks (Gruber et al. 1982).

LARVAL PATCHINESS

Sardine, anchovy, and jack mackerel eggs are aggregated at spawning (Taft 1960; Hewitt 1981; Smith and Hewitt 1985) as a result of parental spawning behavior (Smith 1973). School-sized patches of sardine eggs tens to hundreds of meters in diameter persist for several days (Smith 1973; Smith and Hewitt 1985). Anchovy and jack mackerel larvae are patchy for 4–5 days after spawning. The larvae disperse (5–10 days after spawning for jack mackerel, 10–20 days for anchovy), although never to a completely random pattern, and they eventually become patchy again. The increase in patchiness of older larvae may be a response to food gradients, uneven predator pressure, hydrographic features, or development of social behavior that facilitates schooling (Hewitt 1981). The increase in patchiness in anchovy occurs at about the time schooling begins (Hunter 1981). Hewitt (1981 p. 230) argued that patchiness is ". . . one of a suite of co-evolved characters which together describe the life history of the larval phase." One consequence of larval patchiness is that a large number of samples are required to estimate abundance with a high level of precision (Smith 1981).

LARVAL FEEDING

The larvae of many marine fishes are epipelagic, diurnal particulate planktivores, while the adults occupy different habitats and feed in various specialized ways (Hunter 1981). Northern anchovy eggs and larvae occur in the upper mixed layer of the ocean (Ahlstrom 1959a). The larvae are eel like, with thin oblique muscles, long digestive tracts, and short tails. At hatching, each has a large yolk mass, a beating heart, functional trunk muscles, lateral line neuromasts that are probably functional, and possibly, functional olfactory organs.

Anchovy larvae begin feeding at about 4 mm in length (O'Connell 1981). Like all clupeiforms, they strike at food items once from an S-shaped position. After sighting prey, the anchovy larva sculls toward it, opens its mouth, straightens out, then drives forward, consuming the prey. Larvae that are 6–10 mm in length search 0.1–1.0 l h^{-1}. Once they encounter a dense patch of food, they decrease swimming speed and increase their turns. Prey capture time decreases from 1.5–2.0 seconds at first feeding to 0.6 seconds by the time the larva is 17 mm; success rate increases from 10% at first feeding to 90% in about 3 weeks (Hunter 1972, 1981). Anchovy larvae are diurnal particulate feeders that choose increasingly larger prey as they grow. On laboratory diets, feeding and growth rates increase with increasing concentrations of prey (Theilacker 1987). Significant changes that occur in jaw-branchial apparatus, digestive tract, and locomotor apparatus during metamorphosis (which is complete at 35 mm) are probably due to the change from particulate to filter feeding (Hunter 1976). The wide gape necessary for filtering would be awkward for particulate feeding by the larvae (O'Connell 1981).

Pacific mackerel eggs and larvae also occur in the upper mixed layer (Ahlstrom 1959a). The larvae are stouter than anchovy, have relatively thick muscles, short digestive tracts, and long tails (O'Connell 1981). They represent the more rigid, more agile scombroid type, and strike at prey more than once. The mackerel larvae feeding mode is similar to that of adults: the larvae drive straight forward and bite their prey, and they are piscivorous by 10 mm in length (Hunter 1981). They prey on copepods, cladocerans, oikopleurans, gastropods, invertebrate eggs, diatoms, fecal pellets, and fish larvae. Mackerel larvae grow rapidly, have a high metabolic rate, are fast swimmers, and consume increasingly larger prey, including conspecifics, as they grow. They complete metamorphosis at 15 mm length within 2–3 weeks (Hunter and Kimbrell 1980a). Body depth, head size, mouth size, and prey size increase more rapidly in mackerel than in anchovy (Hunter 1981). Because feeding mode does not change from larvae to adults, development in Pacific mackerel proceeds toward the adult form from early in the larval stage (O'Connell 1981).

Pacific hake larvae occur in colder water within or below the thermocline (Ahlstrom 1959a). Growth and development are slower and metabolic rates and energy requirements are lower for hake larvae than for anchovy or mackerel larvae. First-feeding hake larvae must ingest 0.13 cal d^{-1} at 12°C to satisfy their metabolic and growth costs. Pacific mackerel 3–5 days old must ingest 0.20 cal d^{-1} at 19°C to satisfy their costs (Bailey 1982). Hake larvae have larger mouths than anchovy and mackerel, and they feed on a wide size range of food particles (copepod eggs, copepodites, nauplii, and adults) that changes little with increasing larval size (3–11 mm). Hake larvae probably hunt passively, while anchovy and mackerel larvae hunt actively. Large mouth, slow gut evacuation time, and wide size range of prey is a successful adaptive strategy in the relatively deep, cold strata inhabited by hake larvae (Sumida and Moser 1980; Bailey 1982).

CAUSES OF LARVAL MORTALITY

Larval mortality is determined by internal factors, such as the quality and quantity of yolk, and by external factors, such as transport, turbulence, predator abundance, and food supply. The mechanisms controlling survival change during the larval period, act at different temporal and spatial scales, and may be different in different species (Lasker 1975, 1978; Smith 1981; Hewitt et al. 1985; Fiedler 1986; Peterman and Bradford 1987). The critical time scale for turbulence to induce mortality by dissipating dense patches of food may be several days at most. The critical time scale for offshore transport to remove larvae from a suitable habitat may be a month or more. Differences in timing of oceanographic processes have disparate effects on survival. Seasonal and interannual variations in upwelling affect the average background concentration of food organisms. Upwelling is a favorable factor if it occurs prior to the spawning season and upstream of the spawning grounds; it is unfavorable if it occurs at the time and location of spawning (Bakun and Parrish 1982). If larval survival is controlled by small-scale patchiness, then recruitment is the sum of events occurring in numerous egg and larval patches. Variations in recruitment are caused by changes in the sum of events in the patches and changes in the extent of the area rich enough to feed the larvae (Hunter 1981; Lasker 1981a, b).

The distribution of early Pacific mackerel larvae is dependent on prevailing surface currents; their survival is dependent on the amount and type of plankton and predators in the vicinity of the spawning spot. Older larvae and postlarvae swim sufficiently to mix with survivors from widely separated spawning spots. Their distribution is still dependent on prevailing surface currents, but their survival is dependent on competition for food, cannibalism by survivors of earlier spawnings and older cohorts, and interspecific predation. Environmental components of mortality are largely determined by primary production and surface transport of larvae (Parrish and MacCall 1978).

Larval Transport. Upwelling areas off central and northern California are characterized by strong offshore transport most of the year. Exploitable biomass is dominated by migratory pelagic fishes (hake, sardine, mackerel, and bonito) that spawn in the SCB and off Baja California, where offshore advection of eggs and larvae is minimal. Large recruitment variations among these fishes may be caused by deviations from usual transport conditions. Resident fishes constitute a small portion of the exploitable biomass in the upwelling region. They spawn from winter through early spring, prior to the onset of upwelling, and possess features that reduce the planktonic phase of their early life history: anadromy (salmon), demersal eggs (cottids and greenlings), and viviparity (rockfishes and surfperches). Northern anchovy, because of its small size, does not migrate to the upwelling areas off central and northern California to feed and is resident largely within the SCB. Peak spawning occurs from January through March when winds are weak (Parrish et al. 1981) and may be timed so that the larvae avoid low food availability in winter and high offshore transport in late spring (Methot 1983b).

Stable Ocean. Larval habitat is critical to survival. Successful first feeding by young northern anchovy larvae depends on dense blooms of edible and nutritious plankton that exist only in the absence of wind-generated turbulence. Edibility is determined by size and nutrition is determined by species; some species, such as armored dinoflagellates, are the proper size, but cannot be digested by larval anchovy. Upwelling prior to spawning increases primary production and results in successful copepod reproduction. Storms and upwelling during spawning affect the stability of these patches, disperse the plankton, and reduce concentrations below the level necessary for survival. Fish larvae may find sufficient densities of food organisms only during part of the season. The geographic extent of the area rich enough to feed anchovy larvae can change during a spawning season as well as from year to year (Lasker 1975, 1978, 1981a, b).

The southern (Baja), central (SCB), and northern (Columbia River) subpopulations of anchovy spawn at different times of the year, but the spawning season in each area is characterized by weak coastal upwelling and offshore transport (Husby and Nelson 1982).

The winds that drive upwelling transfer energy to the water column, creating turbulence that may affect the stability of plankton patches. Mortality among 5- to 19-day-old anchovy larvae is high (95% annual average) and is correlated with turbulent mixing. The greater the frequency of high speed wind events during the spawning season, the higher the daily larval mortality rate (Peterman and Bradford 1987). Some degree of turbulent mixing is necessary to get nutrients into the mixed layer from below. If no turbulent mixing occurred, nutrient fluxes would be reduced and so would primary production. If too much mixing occurred (wind speeds $>$5–6 m s^{-1}), fine-scale food strata and anchovy larvae would be dispersed. The most favorable conditions for larval anchovy survival are probably extended periods of low-turbulence "windows" during the critical developmental period. Anchovy have adapted to an optimal level of turbulent mixing in the stable upper layers of the ocean by spawning in different areas at different times of the year (Cury and Roy 1989; Husby and Nelson 1982).

Predation. Predation may be the largest source of larval mortality among clupeiforms (Blaxter and Hunter 1982) and jack mackerel. From 99.5 to 99.95% of jack mackerel larvae are lost to predation between fertilization and yolk sac absorption (Hewitt et al. 1985). Copepods, euphausiids, hyperiid and gammarid amphipods, and chaetognaths can capture or fatally injure larval northern anchovy (Lillelund and Lasker 1971; Theilacker and Lasker 1974; Brewer et al. 1984), although they may be important predators only during the yolk sac stage (Hunter 1981). Cannibalism of eggs and larvae by juvenile and adult clupeiforms and predation by other fishes are probably more important sources of mortality (Hunter 1981; Blaxter and Hunter 1982). Northern anchovy in the SCB contain, on average, 5.1 eggs per stomach and consume an estimated 86 eggs per fish per day. This represents 17% of daily egg production and 32% of the natural egg mortality (Hunter and Kimbrell 1980b).

Starvation. Prior to first feeding, predation is the most important source of larval mortality among jack mackerel (50–80% per day). Starvation is a significant source of mortality when the larvae begin to feed at 6 days of age (45% per day). As the larvae grow, feeding ability improves and starvation rates approach zero at 14 days of age; predation is again the predominant source of mortality (20% per day) (Hewitt et al. 1985; Theilacker 1986). Without feeding, anchovy starve about 4 days after their yolk reserves are gone (Lasker et al. 1970). Hake larvae starve in 6–12 days because they occur deeper in the water column at colder temperatures and have slower growth and developmental rates (Bailey 1982). About 70% of the first-feeding jack mackerel larvae collected in the California Current were starving, while about 12% of the larvae collected near islands and banks in the SCB were starving (Theilacker 1986). About 8% of the 8-mm anchovy larvae in the SCB were starving (O'Connell 1981) for a mortality rate of 35–46% per day (Theilacker 1986).

Parental Deficiencies at Time of Spawning. Both larval size at first feeding and the time a larva can exist before the onset of irreversible starvation are determined largely by egg size and temperature. Larvae from small pelagic eggs can exist about 1–2 days after yolk absorption, while larvae from large demersal eggs can exist for 6 days or more before the onset of irreversible starvation. Survival ability is increased in larvae from larger eggs, but fecundity is lower and incubation times for the stage most vulnerable to predation are longer. Optimal egg size is a trade-off between fecundity and the risks of starvation and predation (Hunter 1981).

Mesopelagic and Bathypelagic

The midwater fish fauna of the SCB contains more than 120 species from the Pacific subarctic, Pacific equatorial, and eastern North Pacific central water masses. The mesopelagic zone is composed of waters transitional between Pacific subarctic and Pacific central water masses and is characterized by steep environmental gradients. Light reaches the top of the mesopelagic zone, where a seasonal thermocline exists, but neither light nor seasonal temperature fluctuations occur at the bottom of the zone. The proportion of Pacific equatorial water increases with depth. The boundary between the mesopelagic and bathypelagic is the depth where equatorial water makes up greater than 50% of the water mass (Lavenberg and Ebeling 1967).

Dominating the SCB fauna are species with centers of distribution in the Pacific equatorial water mass, such as lightfish (*Vinciguerria lucetia*) and bigscale (*Melamphaes acanthomus*), and species with circumequatorial distributions, such as hatchetfish (*Sternoptyx diaphana*) and bristlemouth (*Cyclothone pseudopallida*) (76 species). Most southern species are bathypelagic, few are strong migrators, and they are most common during intrusions of southern water (summer and fall). Northern species (32) have centers of distribution in the Pacific subarctic water mass (robust blacksmelt [*Bathylagus milleri*] and bigscale [*Melamphaes lugubris*]) and in the transitional waters off the west coast (California smoothtongue and northern lampfish). Most northern species inhabit the lower mesopelagic, although a few inhabit the upper mesopelagic and bathypelagic zones. Northern species are most common during intrusions of northern water (late winter to spring). Central species (16) have centers of distribution in the relatively infertile central Pacific water masses (bigeye lightfish [*Danaphos oculatus*] and bristlemouth [*Cyclothone pallida*]). Most central species are mesopelagic and are more common offshore (Lavenberg and Ebeling 1967; Ebeling et al. 1970).

Sampling problems complicate the interpretation of midwater net collections. Midwater fishes have been sampled by horizontal and oblique tows of closing and nonclosing trawls. Nonclosing nets fished in deep water can be contaminated with large catches during descent and ascent (Clarke 1973; Rainwater 1975; Pearcy et al. 1977). Catches from oblique tows are often comparable in composition and abundance to catches from horizontal tows (Robison 1972; Atsatt and Seapy 1974). Towing speed, net size, and mesh size also affect catches (Pearcy and Laurs 1966; Aron and Collard 1969; Clarke 1973; Gartner et al. 1989). Net avoidance is greater for smaller nets (Paxton 1967a). Avoidance may be greater during the day than at night (Pearcy and Laurs 1966) and may be greater on nights with a full moon than on nights with a new moon (Clarke 1973). Escapement of small fishes through the mesh of nets typically used in midwater surveys can be high (Gartner et al. 1989). Food habits studies can be biased by feeding in the net after capture, which varies intra- and interspecifically (Lancraft and Robison 1980).

ECOLOGICAL AND PHYSIOLOGICAL ADAPTATIONS

Mesopelagic fishes generally are countershaded, pinkish or translucent, and have firm flesh and pneumatic or fat-filled swimbladders. Bathypelagic fishes generally are black or red and have soft flesh and fat-filled swimbladders. Young bathypelagic fishes that inhabit the mesopelagic zone are more similar in appearance to mesopelagic fishes than they are to bathypelagic fishes (Lavenberg and Ebeling 1967; DeWitt 1972). Neutral buoyancy may be attained through gas-filled swimbladder, high water content, or high lipid content. Fishes with low lipid and water content that lack a gas-filled swimbladder may have to swim continuously to maintain their position (Neighbors and Nafpaktitis 1982; Neighbors 1988). Northern lampfish that are less than 40 mm SL have gas-filled swimbladders; larger individuals have swimbladders invested with fat (Pearcy et al. 1979). Species living in poorly oxygenated waters have larger gills (Lavenberg and Ebeling 1967; DeWitt 1972).

Shallow-living species are small, grow

slowly and asymptotically, mature relatively early, and reproduce throughout their lives. Deep-living species are large, grow rapidly and nonasymptotically, mature late in life, and may reproduce only once (Childress et al. 1980). The water content of deep midwater fishes increases, and the lipid and skeletal ash content decreases, with increasing depth. Deep species have a lower caloric content per unit weight and therefore require less energy per unit of growth than shallow species (Childress and Nygaard 1973; Childress et al. 1980; Bailey and Robison 1986). Metabolic rate also decreases with increasing depth. The metabolic rates of deep, nonmigratory fishes may be only 5–10% of the rates of vertically migrating species (Childress and Nygaard 1973; Smith and Hessler 1974). Deep fishes have sacrificed muscular strength and mobility for lower metabolic rates (Torres et al. 1979; Childress and Somero 1979; Childress et al. 1980).

The energy conservation characteristics of deep-sea animals are usually attributed to low food availability (Marshall 1971; Childress and Somero 1979; Bailey and Robison 1986). However, bathypelagic fishes, because of their larger size, have a higher lifetime total energy use than vertically migrating mesopelagic fishes. Childress et al. (1980) hypothesized that selective forces requiring greater energy expenditure in surface waters (e.g., predation, water turbulence, and horizontal and vertical migrations) are relaxed in the deep sea, permitting the evolution of energy conservation strategies. However, mesopelagic fishes from the California Current, an environment with relatively high food availability, have higher caloric density than ecologically comparable fishes from the eastern North Pacific Gyre, an environment with low food availability. Mesopelagic fishes from the less productive environment chemically resemble deeper living fishes in the more productive environment, suggesting that midwater communities are food limited and that food availability influences energy conservation strategies (Bailey and Robison 1986).

Many mesopelagic fishes possess photophores (organs that contain luminous bacteria) or produce light through chemical reactions in glandular cells. Light organ patterns are species specific and occasionally sexually dimorphic. Light organs may be involved in species or sexual recognition, prey attraction, or camouflage (Marshall 1972). Photophores in the lightfish (*Vinciguerria lucetia*) appear during metamorphosis; formation and pigmentation are complete by the end of metamorphosis (Ahlstrom and Counts 1958).

VERTICAL MIGRATION

Biomass is low in the deep sea and food gradients are vertical. Horizontal movements to locate food in the deep midwaters are less productive than horizontal movements made by epipelagic fishes (Torres et al. 1979; Childress et al. 1980). Consequently, many deep midwater fishes make diel vertical migrations. Daily round trip vertical migrations of 1000 m require energy expenditure. Vertical migrations take 1–2 hours, and migrators encounter rapid environmental changes. Dissolved oxygen concentrations can change by one or two orders of magnitude, temperature can change by more than 20°C, and hydrostatic pressure can change by 30 to 40 atm. During the day, migrating fishes are inactive and hypoxic or anaerobic at depth (Douglas et al. 1976). The advantage of migrating must outweigh the disadvantages (energy expended); the advantage most commonly cited is abundant food in the epipelagic zone (Marshall 1971; Torres et al. 1979). Not all individuals in a population of a vertically migrating species migrate every night. It is not known whether individual fish are migrators or nonmigrators, or whether individual fish migrate during some nights but not during others (Paxton 1967a; Pearcy et al. 1977).

In general, fishes with swimbladders (myctophids and gonostomatids) make considerable vertical migrations, and fishes without swimbladders (melamphaids and sternoptychids) make limited vertical migrations. Vertical migrators have closed swimbladders (physoclistous): no connection (pneumatic duct) exists between the swimbladder and the gut as in the northern anchovy (physostomous). Gas-filled swimbladders backscatter sound energy and produce traces on shipboard echo sounders (Farquhar 1977). Fishes within scattering layers may be clumped into loose aggregations on the order of 15 m vertically and 1 km horizontally (Greenlaw and Pearcy 1985). Large catches of northern lampfish in single hauls suggest nonrandom distribution that may approach schooling (Paxton 1967a). The extent of migration varies ontogenetically among bristlemouths (*Cyclothone* spp.). Postlarvae have gas-filled swimbladders and live in the upper mixed layer. The swimbladder regresses during metamorphosis and becomes invested with fat. Adults have no swimbladder and live in deeper water (Farquhar 1977).

TROPHIC STRUCTURE

Productivity may determine the standing crop of mesopelagic fishes among the various basins in the SCB (Brown 1974). Midwater fishes prey opportunistically on planktonic organisms, especially crustaceans (fig. 9.5) and, in turn, are preyed on by squids, sharks, salmon, tunas, other midwater fishes, seabirds, and marine mammals (Paxton 1967b). The diets of midwater fishes vary with prey availability and ontogeny. California smoothtongue prey on larvaceans and salps during the productive part of the year (January–June) and on copepods when productivity declines and larvaceans and salps are uncommon (Cailliet 1972). Silvery hatchetfish (*Argyopelecus sladeni*) and slender hatchetfish (*Argyopelecus affinis*) switch from copepods and ostracods to chaetognaths, euphausiids, and salps as they grow (Hopkins and Baird 1977). Robison and Bailey (1982) recognized five feeding guilds among midwater fishes: (1) vertically migrating zooplanktivores (myctophids), (2) nonmigrating zooplanktivores (sternoptychids), (3) stalking predators (stomiatids), (4) ambush predators (ceratioids), and (5) pursuit predators (evermannellids).

Vertically migrating fishes feed on copepods, euphausiids, and sergestid shrimps that also make vertical migrations (Farquhar 1977), and they generally exhibit distinct diel feeding cycles. Nonmigrating mesopelagic species and bathypelagic species that ingest large prey have periodic or acyclic feeding patterns (Holton 1969; Hopkins and Baird 1977). Nonmigratory mesopelagic and bathypelagic lanternfishes eat a wider size range of prey than their shallow-living relatives. Deeper living species have well-developed pharyngeal baskets that concentrate prey as small as 1 mm, and they have larger bodies and mouths that can handle larger prey (Ebeling and Cailliet 1974). Nonmigratory species evolved large body size and mouth structure to take advantage of the small, abundant, high-calorie migratory prey that are lethargic in deep water during the day. Because of their higher caloric content, migratory species may be energetically more important than nonmigratory species in the midwater community than their biomass would suggest (Childress and Nygaard 1973; Torres et al. 1979).

Assimilation efficiency is lower among shallow species than among deep species and lower among shallow species from productive environments than among shallow species from less productive environments. The feces of midwater fishes sink at an average rate of 1 km d^{-1} and contribute about 10% of the organic nutrient transport to sediments in the Santa Barbara Basin (Robison and Bailey 1982).

SPATIAL VARIABILITY

The horizontal and vertical distributions of midwater fishes generally coincide with water mass distributions (Ebeling 1967; Lavenberg and Ebeling 1967; Ahlstrom 1969; Ebeling et al. 1970). The diversity of midwater fishes decreases as one approaches the coast, partly because the various water masses are progressively altered nearshore and partly because of the restricted vertical extent of the habitat in the shallow inshore basins (Laven-

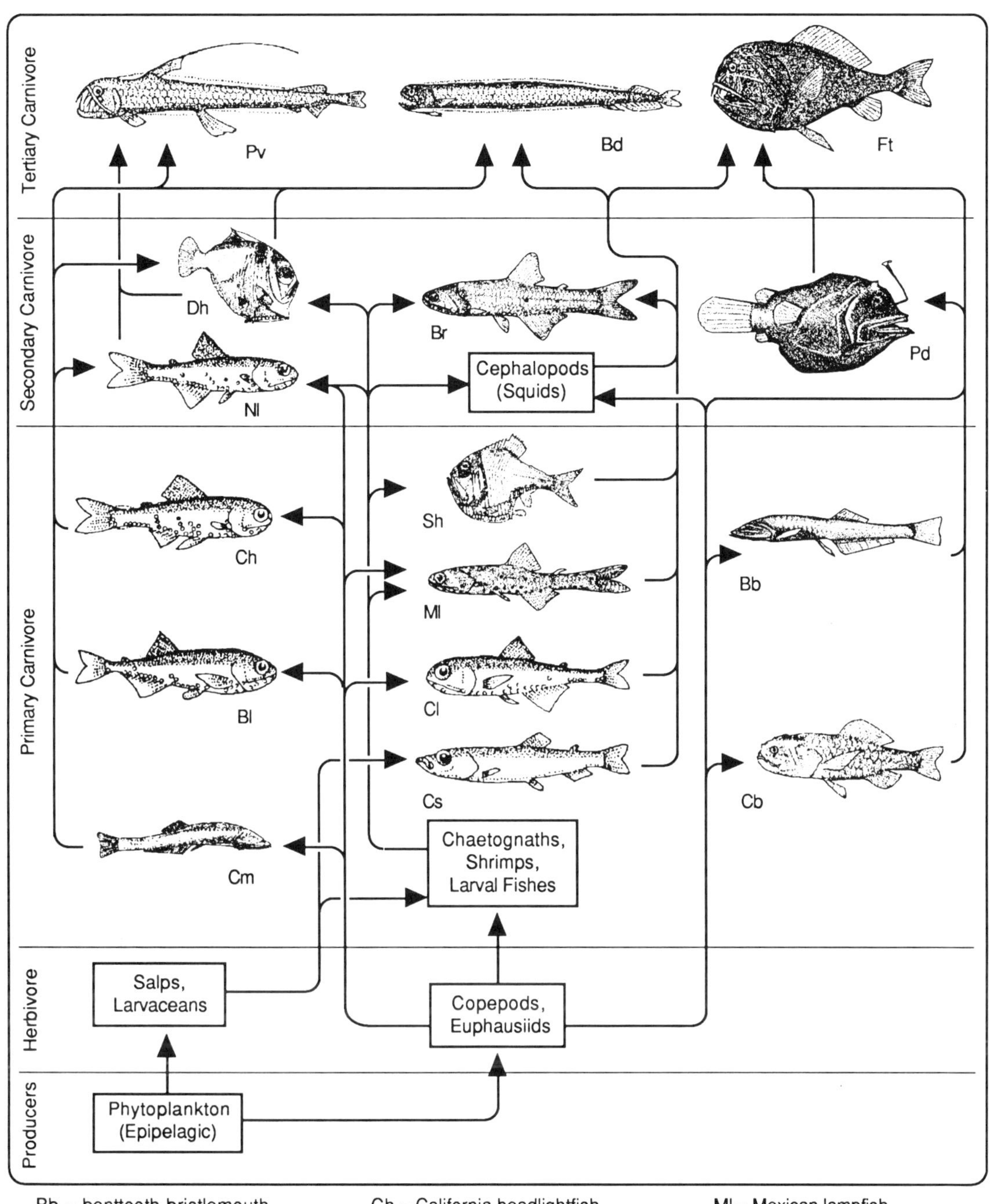

Bb = benttooth bristlemouth
Bd = blackbelly dragonfish
Bl = blue lanternfish
Br = broadfin lampfish
Cb = crested bigscale
Ch = California headlightfish
Cl = California lanternfish
Cm = common bristlemouth
Cs = California smoothtongue
Dh = dollar hatchetfish
Ft = fangtooth
Ml = Mexican lampfish
Nl = northern lampfish
Pd = Pacific dreamer
Pv = Pacific viperfish
Sh = silvery hatchetfish

Figure 9.5. Trophic relationships of mesopelagic and upper bathypelagic fishes.

berg and Ebeling 1967; Ebeling et al. 1970; Brown 1974). Midwater fish abundance is greatest in the more productive nearshore basins (fig. 9.6) (Brown 1974). Abundance and diversity decrease from the mesopelagic to the bathypelagic. The mesopelagic constitutes about one-third of the deep midwaters off southern California, but it contains about 60% of the species (Ebeling et al. 1970; Brown 1974).

Productivity and standing crop may determine the distribution of fishes among the water masses (Ebeling 1967; Barnett 1975), while temperature, light, and dissolved oxygen may determine their vertical distribution. Data are lacking, however, on the effects of predators, prey, and competitors (Paxton 1967a; Robison 1972).

Rainwater (1975) recognized three zones in the Santa Catalina Basin: upper mesopelagic (200–350 m), lower mesopelagic (350–550 m), and bathypelagic (550–800 m). The waters at 200 m in the Santa Catalina Basin are depauperate during the day, but are inhabited by California flashlightfish (*Protomyctophum crockeri*), *Lampanyctus regalis*, California lanternfish (*Symbolophorus californiensis*), and silvery hatchetfish at night. *Cyclothone signata*, northern lampfish, blue lanternfish (*Tarletonbeania crenularis*), California headlightfish (*Diaphus theta*), and slender hatchetfish are the most abundant fishes at 350 m. Mexican lampfish, California smoothtongue, broadfin lampfish, and *Stomias atriventer* migrate from the lower mesopelagic to the surface waters at night. Benttooth bristlemouth (*Cyclothone acclinidens*), *Sternoptyx obscura*, and snubnose blacksmelt remain in the lower mesopelagic day and night. Benttooth bristlemouth, snubnose blacksmelt, *Melanostigma pammelas*, *Poromitra crassiceps*, and *Cyclothone pallida* are the most abundant fishes in the bathypelagic zone and do not migrate (Rainwater 1975).

The bathymetric zone is more distinct in the deep waters offshore, but it merges with the mesopelagic zone in the shallow waters nearshore (Ebeling et al. 1970). The bathymetric zone does not exist in the Santa Barbara Basin, and it is restricted in the San Pedro and Santa Monica basins (600–900 m), where the bathypelagic fish fauna is poorly developed. Sill depths in the San Pedro and Santa Monica basins (approximately 400–700 m) occur in the oxygen minimum layer; the dissolved oxygen concentration in subsill waters is less than 0.2 ml l^{-1}. The bathypelagic fish fauna is more diverse in the Santa Cruz and Santa Catalina basins and in the San Diego Trough. Sill depths occur at approximately 1000 m, and the dissolved oxygen concentration is about 0.5 ml l^{-1} in the oxygen minimum layer. The bathypelagic fish fauna is most diverse in the deep offshore basins, where sill depths occur approximately between 1100 and 1900 m and the dissolved oxygen concentration is about 0.7 ml l^{-1} in the oxygen minimum layer (800 m) (Lavenberg and Ebeling 1967; Brown 1974).

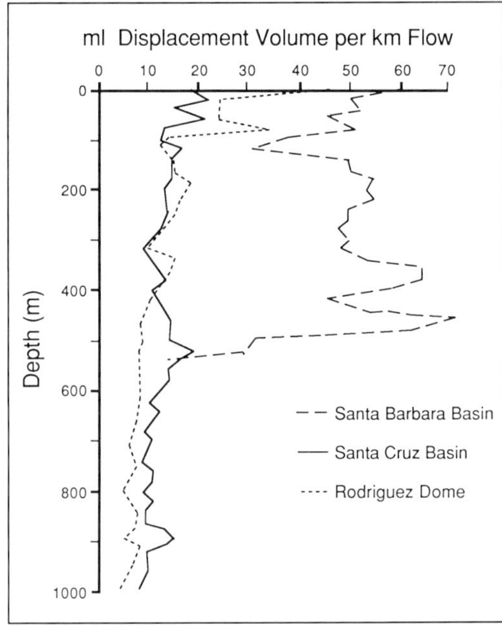

Figure 9.6. Vertical profiles of fish abundance for Santa Barbara Basin, Santa Cruz Basin, and Rodriguez Dome. Displacement volume was standardized by kilometer of linear flow for each collection. (Redrawn from Brown 1974.)

TEMPORAL VARIABILITY

The physical environment of the deep midwaters in the SCB is seasonally variable because of changes in currents and mixing patterns. Northern waters have a greater influence on the upper mesopelagic in the winter and spring, and northern species are usually most abundant at that time. Southern waters have a greater influence in the summer and fall, and southern species are usually most abundant at that time (Paxton 1967a; Ebeling et al. 1970). Lanternfishes are most abundant in the San Pedro Basin in the winter, and Paxton (1967a) theorized that these seasonal fluctuations in catches may be a response to environmental changes (Paxton 1967a). However, Rainwater (1975) concluded that the dominant species in the Santa Catalina Basin were adapted to the changing mixing patterns and that seasonal effects were not significant. The abundance of bathypelagic fishes changes relatively little with season (Ebeling et al. 1970).

RECRUITMENT

Mesopelagic and bathypelagic fishes probably spawn at depth, and the eggs develop as they rise to the surface. Larvae inhabit the productive euphotic zone where they prey on copepods and larval invertebrates. Young fish migrate down to their adult depths during and after metamorphosis (Marshall 1971). From 1955 to 1960, larvae of deep midwater fishes made up 20–40% of the larvae collected in the surface waters of the California Current. Subarctic and transition species were more abundant in years that were colder than average, while equatorial and central Pacific species were more abundant in years that were warmer than average (Ahlstrom 1969). Some of the species that occur in the deep midwaters in the SCB are nonreproductive expatriates from other water masses. All life history stages of the bigscale (*Melamphaes acanthomus*) occur in the equatorial water off Panama; only large, sterile adults occur in the SCB (Ebeling 1962).

DETERMINANTS OF COMMUNITY STRUCTURE

The California Current is a transition zone where different water masses with different flora, fauna, and selection pressures are blended together by large-scale lateral mixing processes. Because the California Current is an open system, biological interactions are obscured by the physical processes associated with large-scale horizontal advection (Barnett 1975; McGowan 1977). It is not surprising that the rank order abundance of mesopelagic fish larvae in collections from the California Current vary from year to year (Ahlstrom 1969).

Oceanic gyres, like the central gyre in the North Pacific Ocean, are physically, chemically, and biologically "monotonous," that is, they have a characteristic climate, seasonality, density, and structure. Lateral transfer rates of allochthonous flora, fauna, and nutrients to gyre centers are small. Because gyres are geologically old and semiclosed, biological interactions (competition and predation) and feedback loops have had a long time to develop, and *in situ* physical and biological processes regulate the state of the systems (Barnett 1975; McGowan 1977). The rank order abundance of mesopelagic fishes and their larvae in collections from the North Pacific Central Gyre are remarkably constant from year to year (Barnett 1975; Loeb 1979).

Fishes of Soft Substrates

Soft substrates are the predominant benthic habitat in the SCB. Mainland and island shelves constitute 11% of the sea surface of the SCB; their sediments are generally sands, silty sands, and sandy silts. Basin and trough slopes constitute 63% and basin and trough floors 17% of the sea surface area; their sediments are generally silty clays and clayey silts (Emery 1960). About 40% of the species and

50% of the families of fishes in the SCB occur on soft substrates along the open coast. At least 71 species (34 families) occur in the surf zone (Carlisle et al. 1960), 126 species (43 families) occur on the mainland shelf (M. J. Allen 1982), and 47 species (22 families) occur on the mainland slope (Cross 1987). The fifteen harbors in the SCB contain more than 25% of the fish species; Chamberlain (1974) collected 132 species and Horn and Allen (1981b) collected 113 species in Los Angeles–Long Beach Harbor. The seven bays and estuaries in the SCB contain about 10% of the fish species; Horn and Allen (1981a, 1985) collected 46 species (28 families) in Upper Newport Bay.

Bays and Estuaries

Compared to the large, river-dominated estuaries common in other parts of the world, bays and estuaries in the SCB are small and few in number. The seven bays and estuaries are physically and chemically variable environments dominated by eurythermal and euryhaline fishes (table 9.4). Bays are nursery areas for several species, including the economically important California halibut (L. G. Allen 1988). Juveniles of noncommercial fishes dominate bay and estuarine fish assemblages (L. G. Allen 1982) and serve as forage fishes for economically important species (Horn 1980). The number of species is variable among bays and is controlled by bay size and mouth width (Horn and Allen 1976). Although 46 species of fish occur in Upper Newport Bay, 7 species account for more than 97% of the individuals (Horn and Allen 1981a, 1985).

About 75% of the bays and estuaries in the SCB have been dredged, filled, and converted to harbors and marinas (Horn and Allen 1985). Fish assemblages of natural bays and man-made harbors are dominated by low trophic level and juvenile fishes with strong seasonal abundance patterns. However, shallow-water *bay* fish assemblages occur nowhere else in the SCB, while *harbor* fish assemblages are indistinguishable from assemblages on the open coast (L. G. Allen 1985). Harbors supplement, but do not substitute for, bays and estuaries as nearshore fish habitats (Horn and Allen 1985).

Bays are the only habitat in the SCB where fishes have been successfully introduced by humans. Striped bass (*Morone saxatilis*), which are native to the Atlantic Coast of the United States, were introduced into Upper Newport Bay during the 1970s, but do not maintain a reproducing population (Horn et al. 1984). Yellowfin gobies (*Acanthogobius flavimanus*), native to Japan, were first collected in Newport Bay in 1978 (Horn and Allen 1981a), but are now one of the most common gobies in the bay. Yellowfin gobies are voracious benthic predators that threaten the native fish populations, especially the longjaw mudsucker (*Gillichthys mirabilis*) and other gobies. The mosquitofish (*Gambusia affinis*), a freshwater fish from the central and eastern United States, was probably introduced in or around Newport Bay for mosquito control.

PRODUCTIVITY AND TROPHIC STRUCTURE

High numerical dominance, high production, and short food chains are characteristic of bays and estuaries (Adams 1976). Total annual production of littoral fishes in Upper Newport Bay was 9.35 g dry wt m^{-2} in 1978, one of the highest recorded values for shallow-water fish assemblages. Most (76%) of the annual production occurred during spring and summer. Topsmelt (*Atherinops affinis*) constituted 77% of individuals, 80% of biomass, and 85% of total production. In 1986–1987, topsmelt constituted 56% of individuals and striped mullet (*Mugil cephalus*) constituted 80% of biomass (L. G. Allen 1982).

Topsmelt and striped mullet are detritivores and herbivores (fig. 9.7), and the proportion of plant material in their diets increases with age (Horn and Allen 1985). Topsmelt less than 50 mm SL eat calanoid, harpacticoid, and cyclopoid copepods while those larger than 50 mm SL eat macroalgae,

Table 9.4. *Relative Abundance of Fishes Collected by Otter Trawl (3.8 m), Gillnet (45.6 m), Bag Seine (15.2 m), and Small Seine (4.6 m) in Upper Newport Bay in 1978–1979*

	Relative Abundance (%)			
Species	Otter Trawl	Gillnet	Bag Seine	Small Seine
Round stingray	<1	<1		
Deepbody anchovy	8	58	1	<1
Northern anchovy	1		<1	
Slough anchovy	7		3	<1
Topsmelt	3	22	91	47
California grunion	<1		<1	
California killifish	<1		2	29
Mosquitofish			<1	15
Barred pipefish	<1		<1	<1
Bay pipefish			<1	<1
Spotted sand bass	1	<1		
Striped bass	1	1		
Barred sand bass	1	<1		
Green sunfish			<1	<1
Yellowfin croaker	2	9		
Queenfish	1	<1		
Shiner surfperch	47	<1	1	
Black surfperch	8	<1		
Striped mullet	<1	6	<1	<1
Arrow goby	<1		<1	7
Cheekspot goby	<1		<1	<1
Shadow goby	<1		<1	<1
Longjaw mudsucker	<1		<1	<1
Diamond turbot	8	<1	<1	<1
California halibut	5	<1	<1	

From Horn and Allen 1981a, 1985.

detritus, and benthic pennate diatoms. Gut length increases during the first year and exceeds standard length at about 50 mm coincident with the shift to a mainly herbivorous diet (L. G. Allen 1980). Primary carnivores are moderately abundant and higher carnivores are uncommon in Upper Newport Bay (L. G. Allen 1982).

SPATIAL VARIABILITY

Upper Newport Bay is the least altered bay–estuary system in the SCB. The principal habitats are channels, littoral, and panne. Channels are deep (>2 m), have the lowest temperatures, and are the least physically variable habitat. Shiner surfperch (*Cymatogaster aggregata*), black surfperch (*Embiotoca jacksoni*), and diamond turbot (*Hypsopsetta guttulata*) are the dominant bottom fishes; deep-body anchovy (*Anchoa compressa*), topsmelt, and striped mullet are the dominant water column fishes. The panne (marsh pools and channels) is shallow (<1 m) and turbid and has the highest temperatures and the most physically variable habitat. California

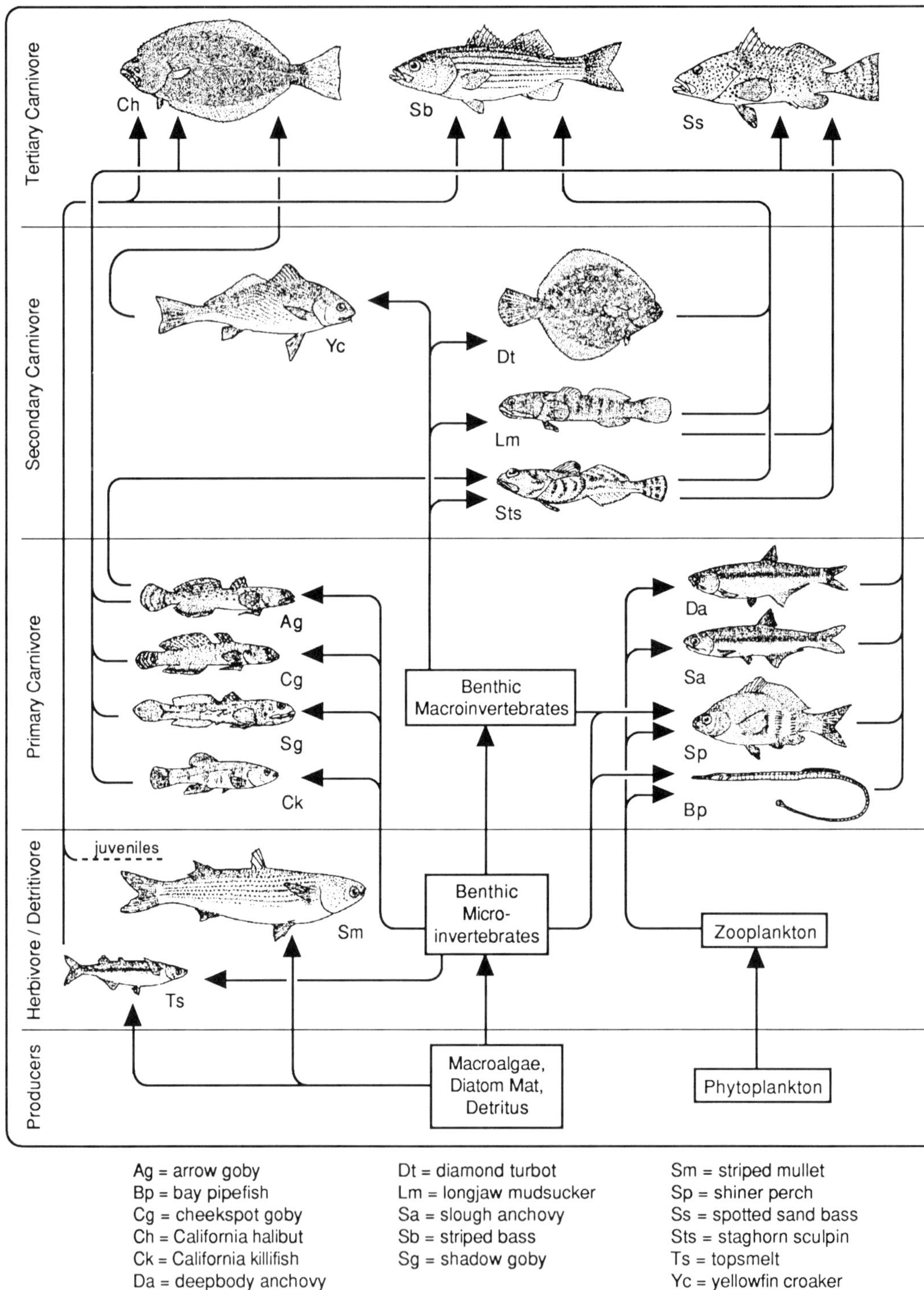

Figure 9.7. Trophic relationships of bay and estuary fishes.

killifish (*Fundulus parvipinnis*) and the mosquitofish are the dominant species. The littoral zone is physically intermediate and produces macroalgal beds seasonally. Topsmelt, mullet, and California killifish are the dominant species (Horn and Allen 1981a).

TEMPORAL VARIABILITY

Temperature and salinity account for 83% of the variation in the abundance of littoral fishes in Upper Newport Bay (L. G. Allen 1982). Temperature and salinity are low during winter storms (January–March), increase after the rainy season, and remain high throughout the summer. During late fall and winter, fish abundance declines and the assemblage is dominated by resident species (fig. 9.8). During the summer, fish abundance (especially resident juveniles) increases and species from the lower bay and outer coast immigrate to the upper bay. Abundance, diversity, and secondary production remain high throughout summer and into fall, when seasonal species and some residents emigrate from the upper bay. Emigration transfers energy from the highly productive littoral zone to the deeper parts of the bay and to nearshore areas outside the bay (Horn and Allen 1981a, 1985; L. G. Allen 1982).

Composition of the fish assemblage in Newport Bay has remained fairly stable since 1974. However, it is difficult to predict which species will be most abundant from one year to the next. The structure of bay and estuarine fish populations is determined largely by variability in reproductive and recruitment success, which are related to oceanographic conditions and to variability in sediment and nutrient loadings, which are in turn related to storm intensity (Horn and Allen 1985). Unseasonable rains in 1978 caused prolonged low salinities in Upper Newport Bay. As a result, resident species declined in abundance and stenohaline periodic species did not enter the upper bay (Horn and Allen 1981a, 1985; L. G. Allen 1982). In Mugu Lagoon, major storms were followed by declines in temperature and salinity and by increases in sediment loads. Topsmelt and shiner surfperch declined in abundance when heavy sedimentation reduced available low-tide habitat (Onuf and Quammen 1983).

Artificial Harbors

The Los Angeles–Long Beach Harbor complex is the largest and most intensively studied harbor in the SCB (Fay and Vallee 1978). Originally, it was a large estuary of the Los Angeles River that contained marshlands, mudflats, and marsh grasses. Since 1872, channels have been dredged, lowlands filled, breakwaters built, and the outlet of the Los Angeles River relocated (Reish et al. 1980). More than 25% of the fish species in the SCB have been collected in the harbor complex. High species richness is credited to the protected nature of the harbor complex, good water circulation, high productivity, abundant food supply, and mixture of substrates (Stephens 1978). Because harbors have increased the available habitats for many nearshore fishes in the last century, they are important nursery areas. Over 80% of the fishes collected near Cabrillo Beach in Los Angeles Harbor were juveniles (L. G. Allen et al. 1983).

Fish populations in Los Angeles Harbor are similar in several ways to populations in Upper Newport Bay, a relatively unaltered estuary. Both habitats are dominated by low trophic level fishes with seasonal peaks in abundance and biomass, and the distribution and abundance of fishes in both are determined largely by physical factors. Temperature and depth account for 76% of the variation of individual species abundances in Los Angeles Harbor. However, species composition differs between the habitats, and the peaks in abundance of ichthyoplankton, juveniles, and adults are not coincident (Horn and Allen 1981b; L. G. Allen et al. 1983). Northern anchovy, white croaker, and queenfish dominate harbor fish assemblages (table 9.5) (Horn and Allen 1981b). The ranking

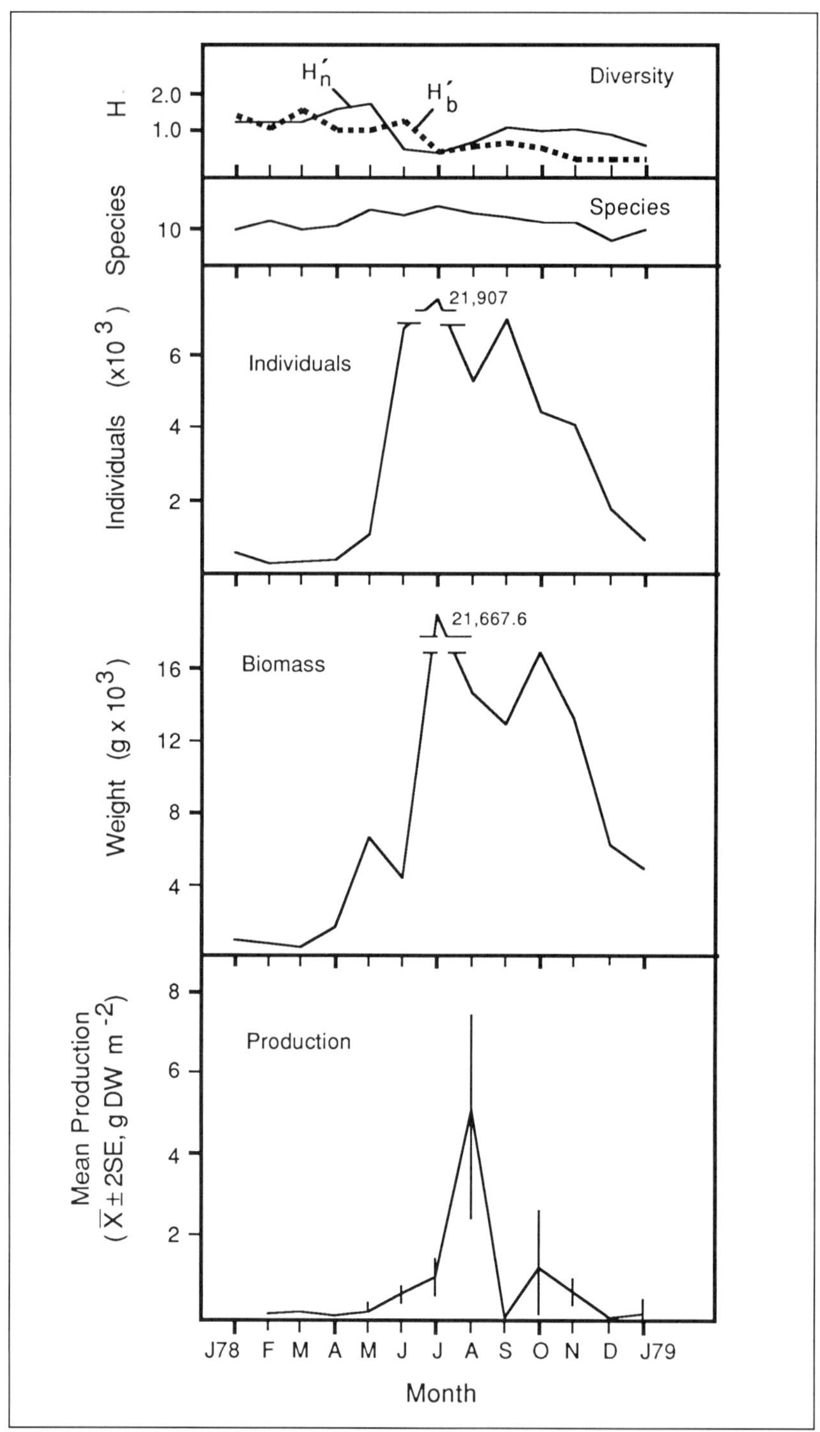

Figure 9.8. Monthly variation in diversity (H'_n for numbers, H'_b for biomass), number of species, number of individuals, biomass, and production ($\bar{X} \pm 2SE$) of fishes caught in the littoral zone of Newport Bay from January 1978 to January 1979. DW = is dry weight. (Redrawn from L. G. Allen 1982.)

Table 9.5. *Relative Abundance of Fishes Collected by Four Sampling Gears in the Los Angeles–Long Beach Harbor Complex*

Species	Relative Abundance (%)			
	Otter Trawl	Gillnet	Purse Seine	Beach Seine
Pacific sardine		<1	12	
Threadfin shad				2
Northern anchovy	15	7	45	24
Jacksmelt		4	<1	<1
Topsmelt		<1	<1	41
California grunion			<1	2
California corbina		1		
Queenfish	9	13	19	4
White croaker	41	25	4	<1
Dwarf surfperch				1
White surfperch	8	16		<1
Shiner surfperch	3	7		<1
Pile surfperch		2		
Walleye surfperch	<1	5		<1
Black surfperch		4		<1
California barracuda		<1	1	
Bay goby	3			<1
Arrow goby				14
Cheekspot goby				7
Jack mackerel		<1	1	
Pacific butterfish	<1	3	2	
Speckled sanddab	5			
California halibut	<1	<1		<1
California tonguefish	7			

Data compiled from Horn and Allen (1981b), L. G. Allen et al. (1983), and Marine Biological Consultants (1984).

and relative abundances of dominant fishes in harbors are nearly identical to their ranking and relative abundances on the open coast (L. G. Allen et al. 1983; L. G. Allen 1985).

PRODUCTION AND TROPHIC STRUCTURE

Total annual production of trawl-caught fishes was 1.7–1.9 g dry wt m^{-2} in Long Beach Harbor (Marine Biological Consultants 1984) and 4.0 g dry wt m^{-2} in Los Angeles Harbor (Stephens et al. 1974), less than half of the estimated fish production in Upper Newport Bay, but within the range of other marine habitats (L. G. Allen 1982). Production estimates for trawl-caught fishes do not include pelagic species, a major component of the harbor ichthyofauna (table 9.5). Planktivores dominate harbor fish assemblages (Horn and Allen 1981b). Food habit data for most harbor species are scarce, however.

SPATIAL VARIABILITY

The predominant habitats in Los Angeles–Long Beach Harbor are soft sediments and open water. Most of the abundant fishes in harbors are also common on the open coast (fig. 9.9). Harbors may also contain hard substrates (riprap, breakwaters, bulkheads, floats, and pilings), subtidal eelgrass and algal beds, and sandy beaches. Each habitat supports a distinctive fish assemblage. Breakwater and riprap fish assemblages are similar to those of natural and artificial rock reefs (L. G. Allen 1985). Algal beds (*Gracillaria* sp.) support a distinct group of fishes, including dwarf surfperch (*Micrometrus minimus*), spotted kelpfish (*Gibbonsia elegans*), giant kelpfish (*Heterostichus rostratus*), and kelp pipefish (*Syngnathus californiensis*) (L. G. Allen et al. 1983). Unfortunately, this unusual habitat was destroyed by dredging operations in Los Angeles Harbor during construction of a new marina in the early 1980s.

TEMPORAL VARIABILITY

Fish abundance and biomass in harbors are highest in summer and early fall (fig. 9.10) due to recruitment of resident fishes (northern anchovy, white croaker, queenfish, and various surfperches) and to the immigration of warm-water and periodic species (Pacific bonito, California barracuda, grey smoothhound [*Mustelus californicus*], and leopard shark [*Triakis semifasciata*]). The fewest number of species and individuals were collected in the winter (Stephens et al. 1974; L. G. Allen et al. 1983).

Open Coast

Fishes of the mainland shelf have been sampled extensively since the 1950s, primarily by monitoring programs associated with municipal wastewater and power plant discharges. However, few studies have collected and analyzed more than 3 years of data. Less is known about fishes of sandy beaches and the mainland slope, and much less is known about fishes of island shelves and slopes, basins, and deep-water habitats farther offshore. Small otter trawls (7–8 m headrope) are the primary sampling gear, and catches are dominated by small sedentary fishes (M. J. Allen 1982). Large mobile fishes that dominate hook-and-line catches (M. J. Allen et al. 1975; Cross 1987), photographic surveys (Southern California Coastal Water Research Project 1973a, 1974; Moore and Mearns 1980a), and observations from submersibles (Clarke et al. 1967) are rare or absent in trawl catches (table 9.6). For example, 47 species of fish were caught by otter trawl and longline between 300 and 600 m. Otter trawls caught 42 species and longlines caught 19 species, but only 14 species (33%) were caught by both types of gear (Cross 1987).

STANDING CROP AND TROPHIC STRUCTURE

Based on daytime tows with a 7.6-m trawl, estimates of abundance of standing crop are 3.5–3.9 wet g m^{-2} on the outer shelf and upper slope and 0.3 wet g m^{-2} in the nearshore basins. The low value for the basins is probably due to low dissolved oxygen concentrations (0.1–0.3 mg l^{-1}) (Cross 1987). The abundance of macrofauna parallels the abundance of oxygen in the basins (Rittenberg et al. 1955). Pearcy et al. (1982) estimated the standing crop of demersal fishes on the upper slope off Oregon to be 2–3 wet g m^{-2} by 3-m beam trawl and 6–11 wet g m^{-2} by 23-m trawl. Differences in net construction, capture efficiency, and trawling methods among studies make it difficult to compare estimates of abundance.

Soft-bottom fishes along the open coast eat a wide variety of prey, although some emphasize particular taxa (fig. 9.11). The diets of most species change during ontogeny, and the diets of some species change with changing food availability (VanBlaricom 1982; Plummer et al. 1983; Cross et al. 1985). Of the 40 dominant species on the mainland shelf, 80% prey primarily on crustaceans, 10% on polychaetes, 5% on fish, 2.5% on echiuroids, and 2.5% on ophiuroids. Of the 32 species that eat crustaceans, 31% prey primarily on gammarids, 16% on mysids, 16% on euphausiids, 13% on calanoids, 13% on crabs, 9% on shrimp, and 3% on ostra-

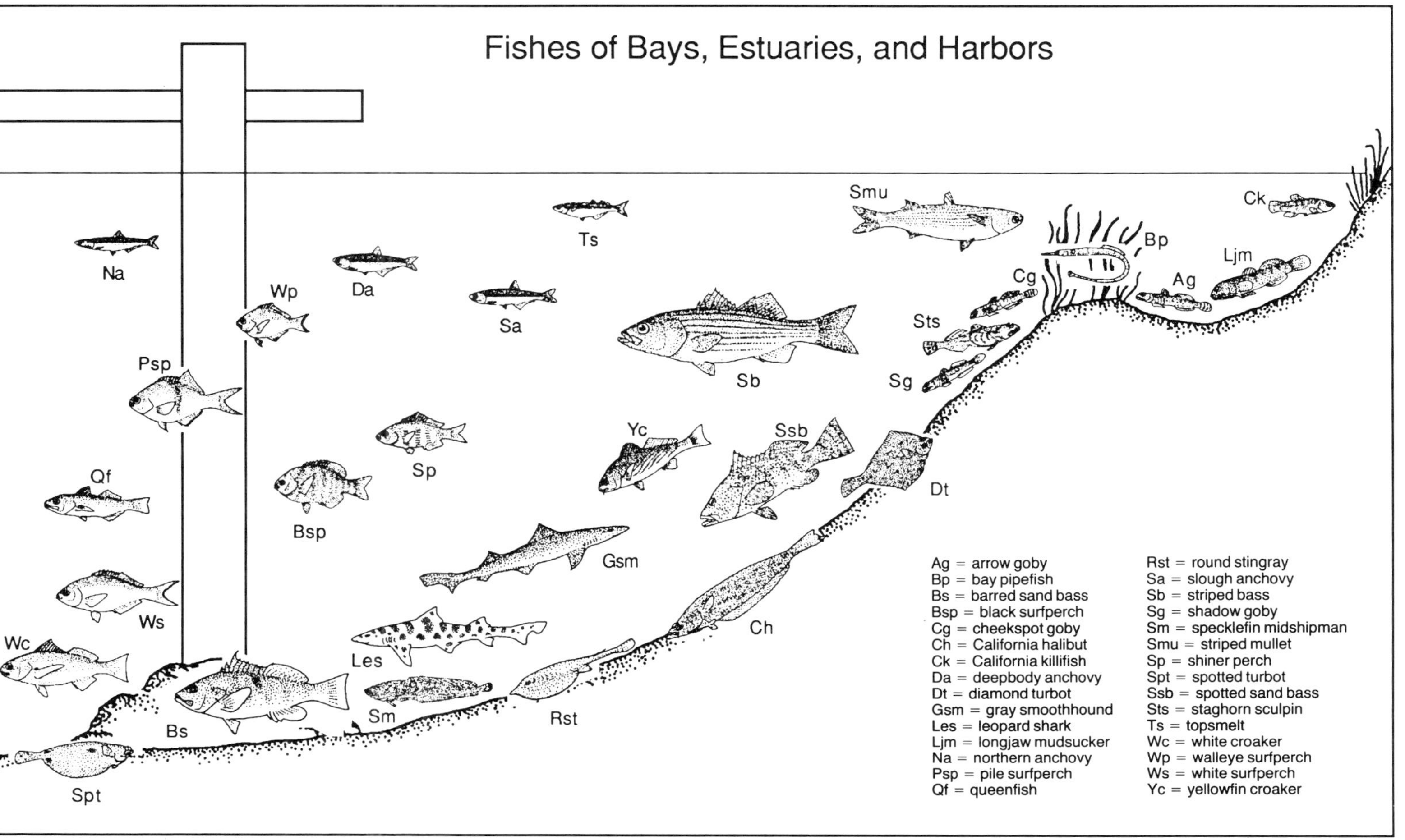

Figure 9.9. Schematic diagram of fishes within bay, estuary, and harbor habitats.

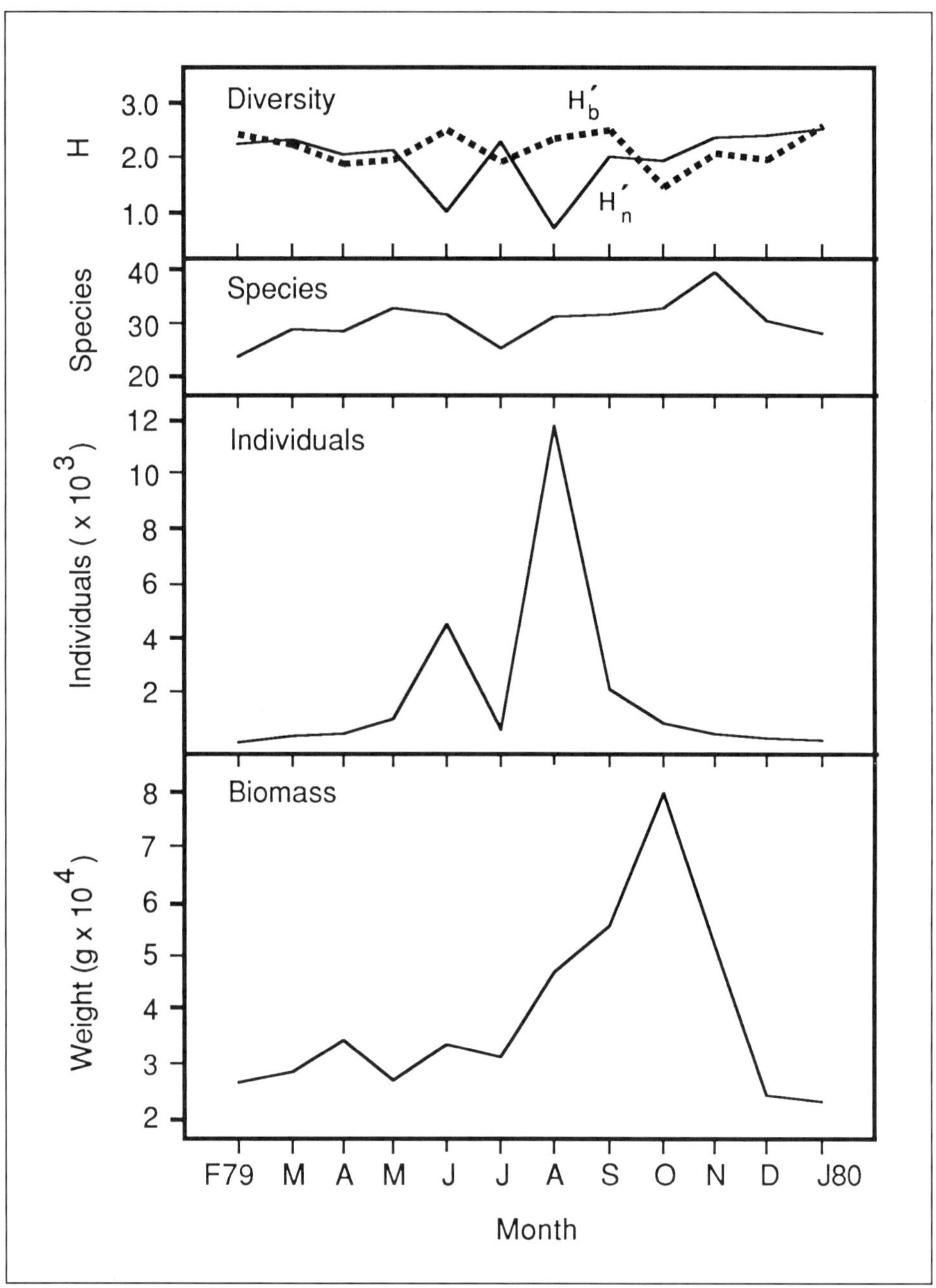

Figure 9.10. Monthly variation in diversity (H_n' for numbers, H_b' for biomass), number of species, number of individuals, and biomass of fishes caught off Cabrillo Beach in Los Angeles Harbor from February 1979 to January 1980. (Redrawn from L. G. Allen et al. 1983.)

Table 9.6. *Rank of Dominant Species Taken on Soft Bottom from 20 to 180 m in Santa Monica Bay by 7.6-m Otter Trawl, 24-m 100-hook Setline, and 6-hook Rockcod Rig Fished by Rod-and-Reel*

	Otter Trawl (n = 346)[a]	Setline (n = 13)	Rod-and-reel (n = 15)
Pacific hagfish		5	
Spotted ratfish		4	
Spiny dogfish		1	
Greenspotted rockfish			5
Vermillion rockfish			3
Bocaccio			1
Stripetail rockfish	1		
Sablefish		2	4
Yellowchin sculpin	5		
White croaker		3	2
Pacific sanddab	4		
Speckled sanddab	3		
Dover sole	2		

[a] n is number of collections.

From M. J. Allen et al. 1975.

cods (M. J. Allen 1982). Allen identified four foraging types: (1) species that eat only pelagic prey (pelagivores, such as queenfish and California halibut); (2) species that eat primarily pelagic prey but also some benthic prey (pelagobenthivores, such as sablefish [*Anoplopoma fimbria*] and slender sole [*Eopsetta exilis*]); (3) species that eat primarily benthic prey but also some pelagic prey (benthopelagivores, such as white croaker and shortspine combfish [*Zaniolepis frenata*]); and (4) species that eat only benthic prey (benthivores, such as spotted cusk-eel [*Chilara taylori*] and Dover sole).

The impact of soft-bottom fishes on their prey rarely has been investigated in the SCB. Round stingrays (*Urolophus halleri*) and bat rays (*Myliobatis californica*) dig pits to expose benthic infauna. Speckled sanddab (*Citharichthys stigmaeus*) are attracted to digging rays and prey on organisms flushed into the water column. A variety of highly mobile invertebrates rapidly colonize the pits and consume the detrital organic matter that accumulates at the sand–water interface. As the pits fill with sand, the quality of the organic material changes as does the pit infauna. On average, about 25% of the bottom in 17 m of water off La Jolla was recovering from ray disturbance. During summer and fall when rays can disturb as much as 5% of the bottom per day, as much as 100% of the bottom was recovering from ray disturbance (VanBlaricom 1982).

SPATIAL VARIABILITY

The composition of fish assemblages on the mainland shelf and slope changes with depth and latitude, although trends with depth are more dramatic than trends with latitude (M. J. Allen 1982). Faunal turnover is most rapid from the surf zone out to about 25 m depth and between 100 and 150 m depth (J. Cross unpubl. data). Schooling water column fishes dominate just outside the surf zone throughout the SCB. Flatfishes constitute an increasing proportion of the catch from north to south and from surf zone to the outer shelf (Love et al. 1986; M. J. Allen 1982). Fish as-

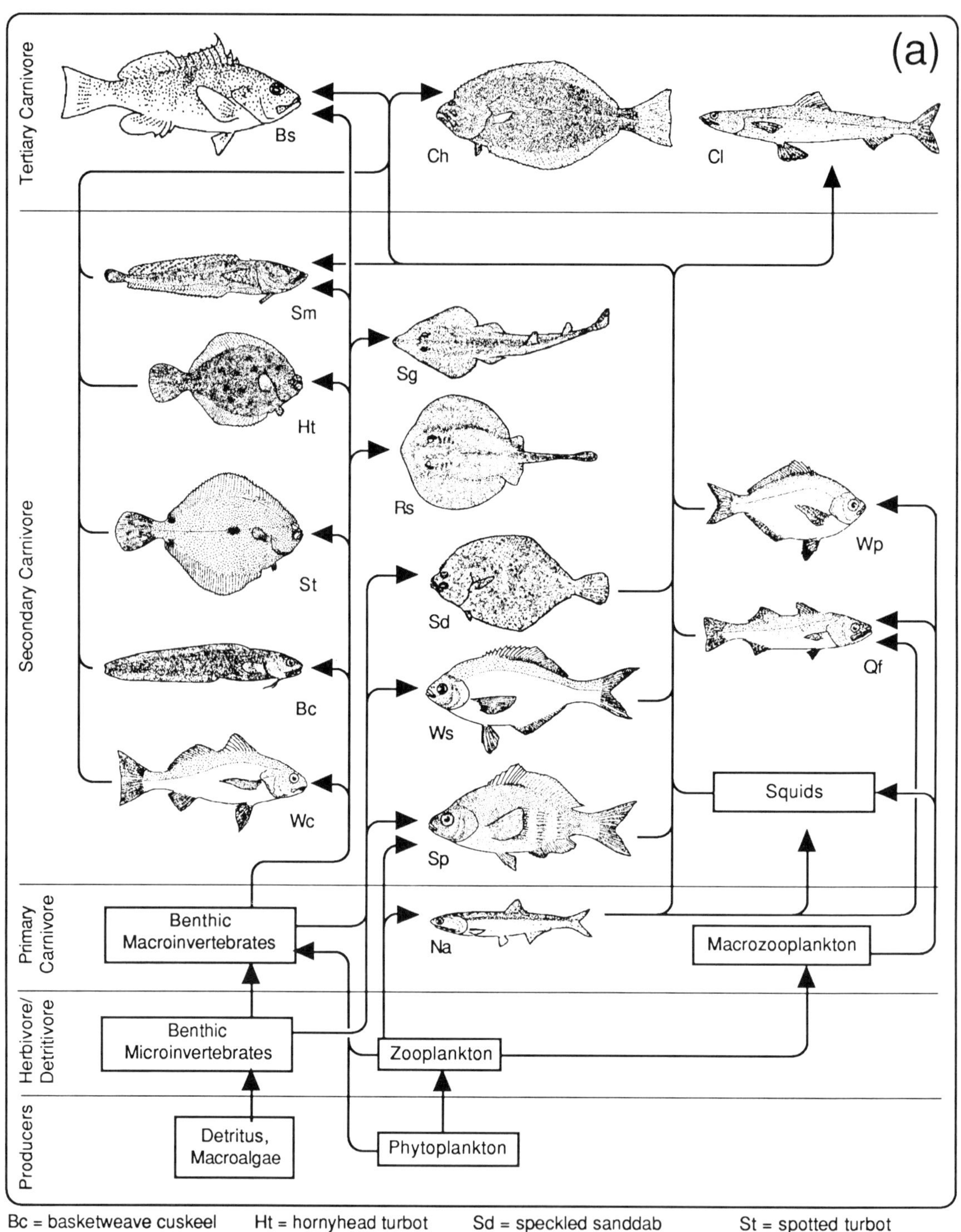

Figure 9.11. (a) Trophic relationships of inner shelf fishes. (b) Trophic relationships of outer shelf fishes.

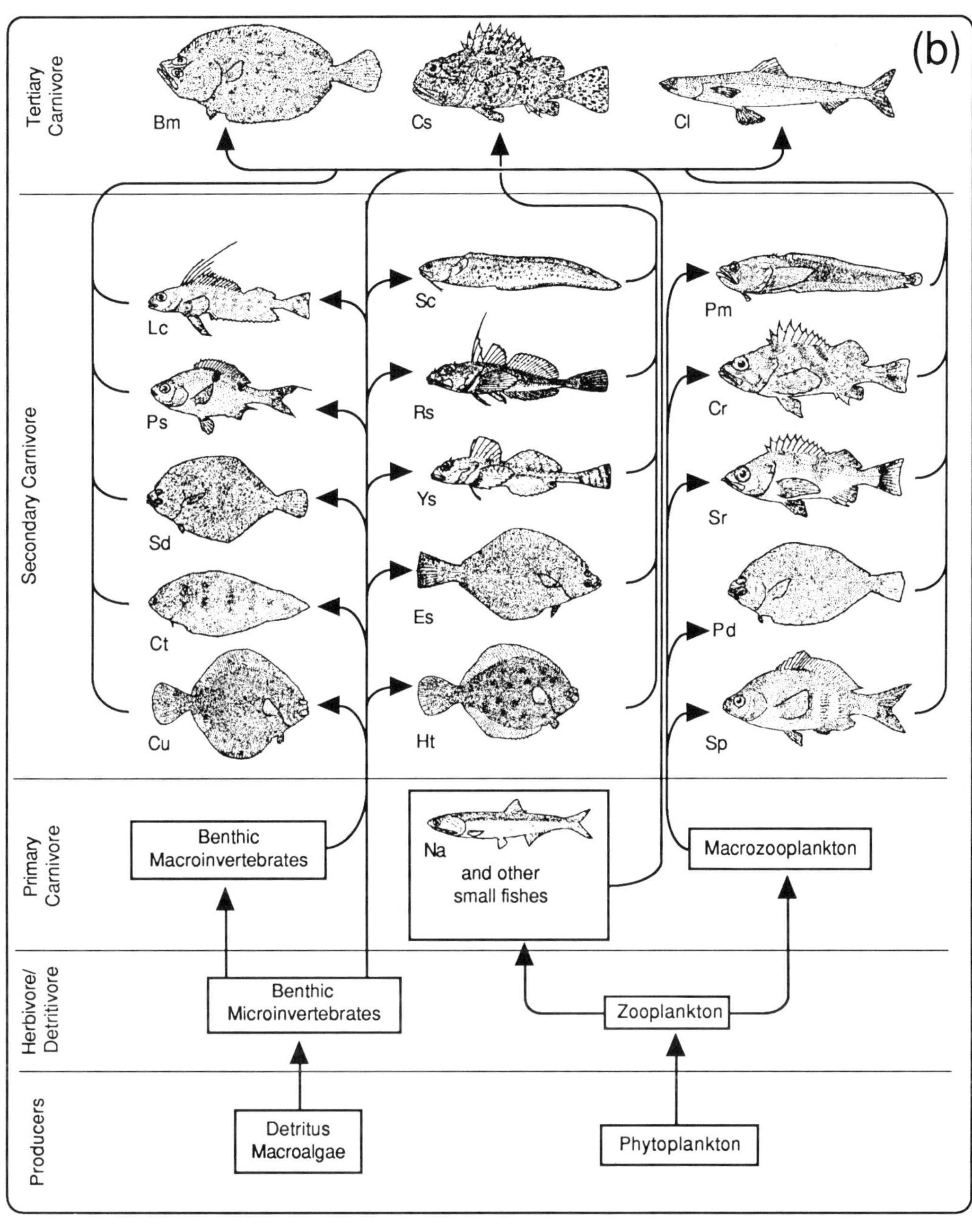

Bm = bigmouth sole
Cl = California lizardfish
Cr = calico rockfish
Cs = California scorpionfish
Ct = California tonguefish
Cu = curlfin turbot
Es = English sole
Ht = hornyhead turbot
Lc = longspine combfish
Na = northern anchovy
Pd = Pacific sanddab
Pm = plainfin midshipman
Ps = pink surfperch
Rs = roughback sculpin
Sc = spotted cuskeel
Sd = speckled sanddab
Sp = shiner perch
Sr = stripetail rockfish
Ys = yellowchin sculpin

Figure 9.11. (continued)

semblages on the mainland slope are similar throughout the SCB (Cross 1987).

Many demersal fishes common on the mainland shelf and slope also occur at the offshore islands and banks, but abundances often differ between nearshore and offshore areas (M. J. Allen 1982; Mearns et al. 1973, 1978). Sediments of the mainland shelf and slope are predominantly silty clays and have a high total organic carbon content; the infauna is dominated by polychaetes and gastropods. Sediments of the offshore banks and slopes are sandier and have a higher biogenic content and a lower total organic carbon content; the infauna is dominated by ostracods, amphipods, and pelecypods (Thompson and Jones 1987). Pacific sanddab (*Citharichthys sordidus*), slender sole, Dover sole, and rex sole (*Glyptocephalus zachirus*) are common along the mainland, but only Pacific sanddab and slender sole are common at the islands. Dover and rex soles prey primarily on sedentary polychaetes, while slender sole and Pacific sanddab prey primarily on mysids, amphipods, and ostracods. Differences in the distribution and abundance of preferred invertebrate prey are probably a determining factor in the distribution and abundance of these fishes (compare Cross et al. 1985).

TEMPORAL VARIABILITY

Nighttime trawl catches on the mainland shelf produce two to four times more individuals, two to three times more biomass, and up to two times more species than daytime catches. Catch estimates from night trawls are less variable than from day trawls, suggesting that nearshore fishes are less contagiously distributed at night. Night-active species (e.g., basketweave cuskeel [*Ophidion scrippsae*], plainfin midshipman [*Porichthys notatus*], and California tonguefish [*Symphurus atricauda*]) are rare or absent in day trawls. Large-bodied elasmobranchs (e.g., shovelnose guitarfish [*Rhinobatus productus*], California skate [*Raja inornata*], and bat ray) are caught more frequently at night. The most abundant and frequently occurring species are caught in similar proportions day and night, however (DeMartini and Allen 1984).

Demersal fishes on the shelf and slope make seasonal bathymetric migrations in response to changes in their physical and biological environment. In Santa Monica Bay, bottom-water temperatures and dissolved oxygen concentrations increase with proximity to shore. Water temperature and dissolved oxygen are lower in summer (8–11°C and 4–6 mg l^{-1}) than in winter (11–14°C and 6–8 mg l^{-1}), as is turbulence. Trawl catches are highest on the shelf and lowest on the slope in summer, but the pattern is reversed in winter (Southern California Coastal Water Research Project 1973b; Love et al. 1986; Cross 1987). California halibut settle out of the plankton in bays and embayments (L. G. Allen 1988). Juveniles migrate to the open coast, then into progressively deeper water as they grow; adults migrate back into shallow water for spawning (Frey 1971; Plummer et al. 1983). In summer, Dover sole migrate from the slope to the shelf to feed, then migrate back into deep water in the winter to reproduce (Hagerman 1952; Alton 1972; Cross 1985). Scorpionfish migrate from shallow water to traditional spawning areas in deep water during early summer; they return to shallow water in early fall (Hartmann 1987; Love et al. 1987).

Significant interannual variability in trawl fish catches is common in the SCB (M. J. Allen and Voglin 1976; Mearns 1978, 1979; Moore and Mearns 1980b) and is correlated with changes in sea surface temperature (fig. 9.12). Otter trawl catches declined in Santa Monica Bay in 1958–1959 (Carlisle 1969) and throughout the SCB in 1983–1984 when surface water temperatures rose as much as 4°C above the long-term mean (Love et al. 1986). Catches off San Diego declined during the summer of 1983 (El Niño), remained low through the summer of 1984, and returned to pre–El Niño levels by winter 1985 (table 9.7). During the El Niño, the mean number of species per trawl declined 16–32%, and the

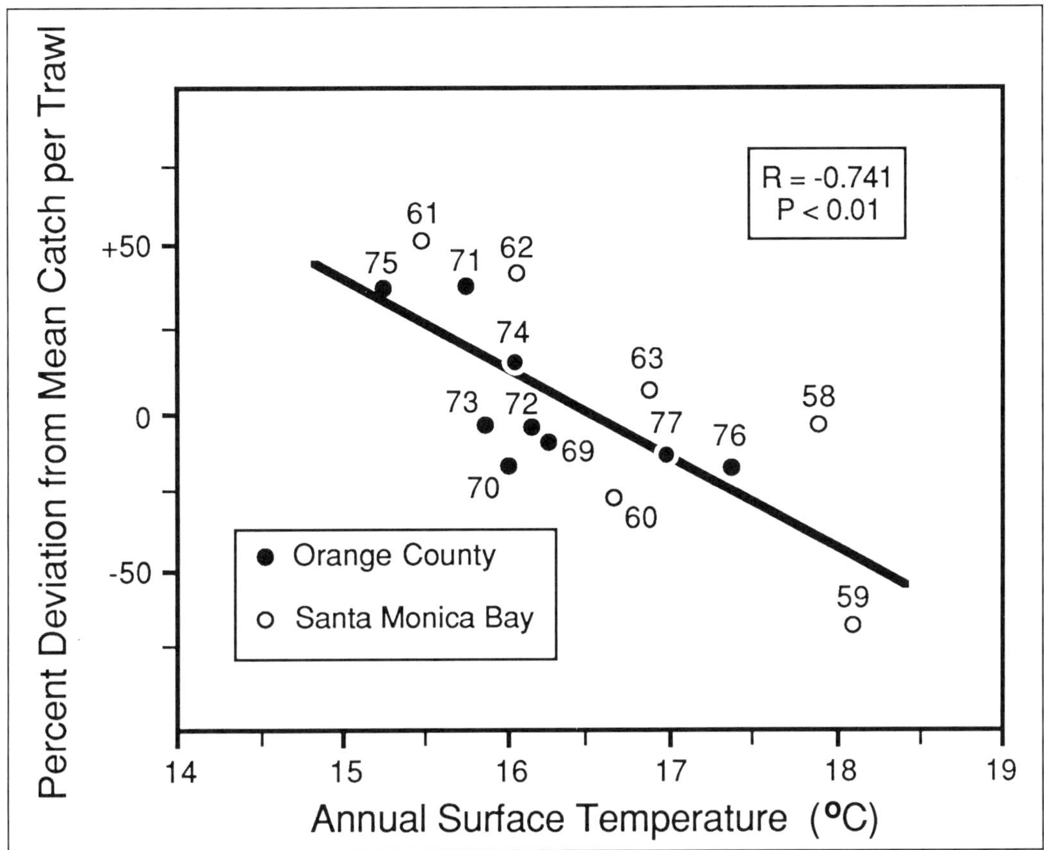

Figure 9.12. Deviation (in percent) of the annual mean catch per trawl from the long-term mean versus annual average sea surface temperature for 1958–1963 and 1969–1978. (Redrawn from Mearns 1978.)

mean number of individuals and mean catch weight declined 57–65%. As water temperatures warmed, many species moved into deeper, cooler water and some species moved northward along the coast.

RECRUITMENT

Recruitment to the mainland shelf occurs throughout the year, but is greatest from winter through spring. The timing and strength of recruitment varies from year to year (fig. 9.13) and may be coherent throughout the SCB (Sherwood 1980; Love et al. 1986). Young fish appeared in abundance on the mainland shelf off Huntington Beach only a few times from 1971 through 1978. For the entire period, young fish averaged 20% of the catch and represented only 13 species. During 1975, however, young fish from more than 20 species constituted 50% of the total catch. Large catches of young fish during that year were correlated with declining water temperature and water clarity; low catches during the other years were correlated with warm, clear water (Mearns 1979).

Changes in recruitment affect the composition of soft-bottom fish assemblages. Stripetail rockfish (*Sebastes saxicola*) was the dominant rockfish from 15 to 150 m throughout the SCB from 1971 to 1975. Significant recruitment occurred in 1971, 1973, and 1975 when minimum winter water temperatures were decreasing and average annual tem-

Table 9.7. *Otter Trawl (7.6 m) Catch Parameters during El Niño (June 1983–July 1984) Compared to Pre- and Post-El Niño (June 1982–January 1983 and January 1985–January 1988) in 60 m near San Diego*

	n[a]	Species		Individuals		Weight (kg)	
		Mean	SD[b]	Mean	SD	Mean	SD
Summer							
Pre- and post-	24	15.2	2.1	201	110	5.5	3.0
El Niño	12	10.3	2.8	73	52	1.9	1.9
Combined	36	13.6	3.3	158	112	4.3	3.2
Winter							
Pre- and post-	30	13.9	3.0	236	133	10.3	7.7
El Niño	6	11.7	3.4	102	94	3.6	3.1
Combined	36	13.5	3.2	213	136	9.2	7.5

[a] n is number of 10-min. trawls.

[b] SD is standard deviation.

From J. Cross, unpubl. data.

peratures were 16°C or less. Significant recruitment of calico rockfish (*Sebastes dallii*) occurred in 1975 and 1977 when minimum winter water temperatures were increasing and average annual temperatures were 17°C or higher. Calico rockfish were the dominant rockfish in trawl catches from 20 to 60 m in the central and southern SCB from 1975 to 1978 (Mearns et al. 1980). The ranges of both species extend south to central Baja California. Calico rockfish occur as far north as San Francisco, but are rare north of Santa Barbara. Stripetail rockfish live in deeper water and occur as far north as Alaska (Eschmeyer et al. 1983). The switch in dominance from stripetail rockfish to calico rockfish spanned 100 km of coast south of Point Dume and was linked to changing oceanographic conditions; fishing was not a factor because both species are small and of no commercial or recreational importance (Mearns et al. 1980).

ANTHROPOGENIC IMPACTS

The coastal waters of the SCB have been used for more than a century as a medium for the dilution, dispersal, and degradation of human and industrial wastes (Schafer 1989) (see chap. 12). Since the 1940s, thousands of metric tons of chlorinated hydrocarbons, tens of thousands of metric tons of trace metals, and millions of metric tons of oil, grease, and nutrients have reached the nearshore waters in municipal wastewater discharge alone. Elevated body burdens of chlorinated hydrocarbons have been detected in fishes hundreds of kilometers from shore (MacGregor 1974; Brown et al. 1986). Power generating stations each year kill billions of fish larvae and hundreds of thousands of juveniles and adults in seawater used for cooling (Herbinson 1981; Connally et al 1982).

More than 80% of the anthropogenic inputs from municipal and industrial discharges, urban runoff, and ocean dumping enters the SCB between Point Dume and Dana Point. Impacts on demersal fishes inhabiting this region include the following: elevated body burdens of chlorinated and petroleum hydrocarbons (Brown et al. 1986; Malins et al. 1987; McCain et al. 1988; Young et al. 1980, 1988); chromosome aberrations (Hose et al. 1987); increased hepatic mixed-function oxidase activities (Spies et al. 1982); enlarged, fatty-vacuolated livers containing degenerating cells and tumors (Pierce et al. 1977; Ma-

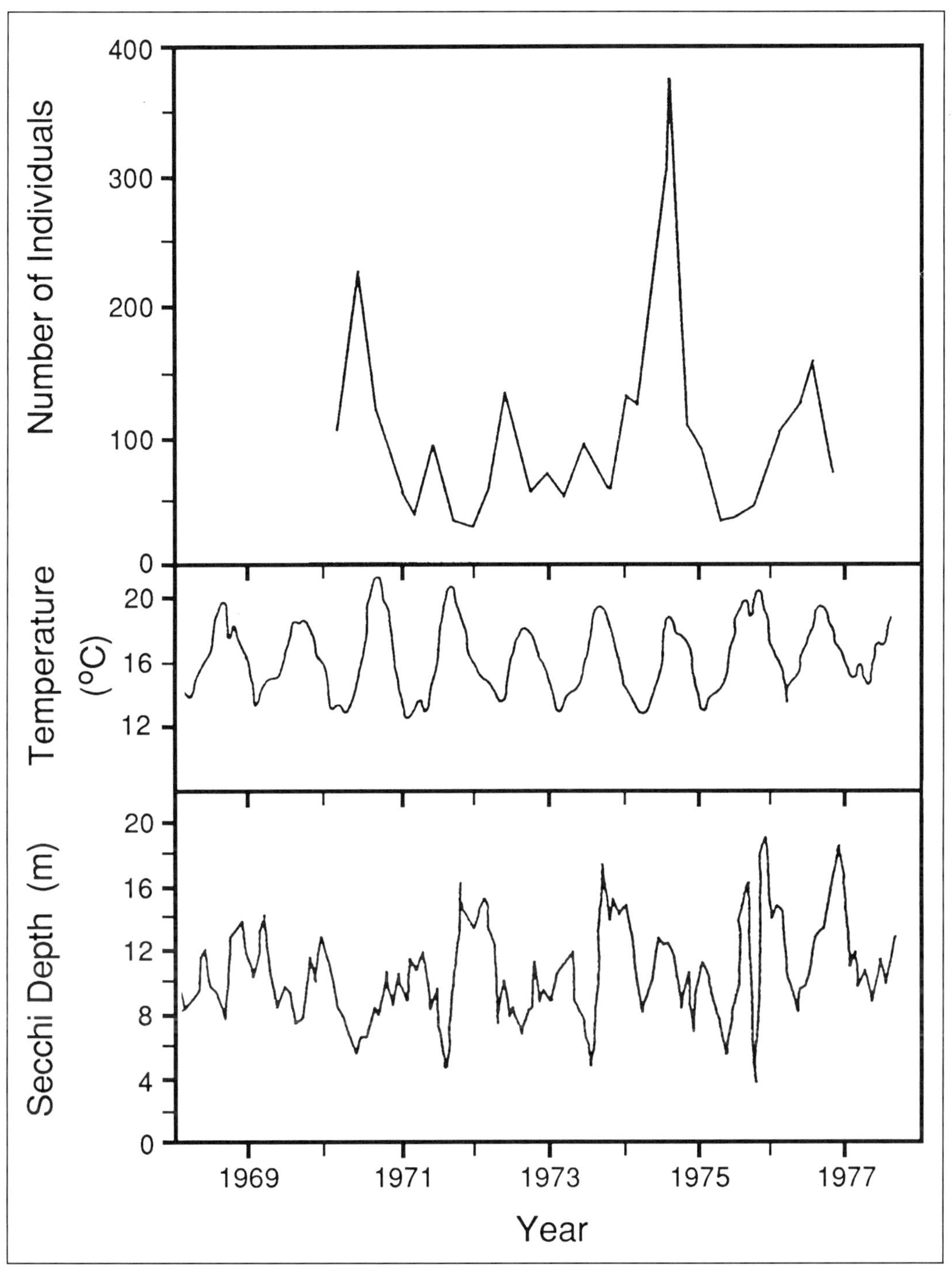

Figure 9.13. Fluctuations in average catch per trawl of fish of less than 56 mm SL in San Pedro Bay, and average monthly sea surface temperature and secchi disk depth in Santa Monica Bay for 1969–1978. (Redrawn from Mearns 1979.)

lins et al. 1987); skeletal abnormalities (Valentine et al. 1973; Valentine 1975); increased incidences of fin erosion, epidermal tumors, and oral papillomas (Mearns and Sherwood 1977; Cross 1988); reduced fecundity and egg viability (Cross and Hose 1988; Hose et al. 1989); altered migratory behavior (Cross 1986); reduced survival rates (Cross 1985); modified food habits and food webs (Cross et al. 1985; Spies et al. 1989); and altered distributions and abundances (Carlisle 1969; Spies 1984; Cross et al. 1985). Some of the most affected species (Hose et al. 1987; Malins et al. 1987; Cross and Hose 1988; Hose et al. 1989) are targets of important commercial and recreational fisheries (Puffer et al. 1982; Love et al. 1984; Stull et al. 1987).

Anthropogenic wastes in the central SCB affect nearshore fishes at all levels of organization. Researchers should take this into account when designing surveys and experiments in the central SCB.

Fishes of Hard Substrates and Kelp Beds

Hard substrates are the least abundant, but among the most important, of fish habitats. About 30% of the species and 40% of the families of fishes in the SCB occur on hard substrates. Shallow (<30 m) reefs may be made of cobble (low relief), bench rock (high relief), or a combination of the two; mixed rock–cobble reefs are probably the most common type in the SCB. Both substrates support kelp beds (Quast 1968b, c). Deep reefs or banks include isolated mounds, the rocky shoulders of submarine canyons, and submerged mountains. Substrates are often mixtures of coarse sand, calcareous organic debris, and rocks (Cross 1987). More than 125 species (>40 families) of fish live in and around shallow rock reefs and kelp beds (Quast 1968b, c; Feder et al. 1974). About 30 species (10 families) occur on deep banks (Cross 1987), and only 6 species are resident in the rocky intertidal (Cross 1982; L. G. Allen 1985). Subtidal reefs are targeted by recreational and commercial fishermen and divers throughout the year.

Rocky Intertidal

The rocky intertidal is a turbulent and dynamic environment where fishes must cope with waves, surge, and physiological stresses imposed by the ebb and flow of the tide (Horn and Gibson 1988). Only six species of fish are residents of the rocky intertidal in the SCB. Wooly sculpin (*Clinocottus analis*), reef finspot, rockpool blenny (*Hypsoblennius gilberti*), spotted kelpfish, and California clingfish spend all but their larval existence in the intertidal and shallow subtidal. Opaleye (*Girella nigricans*) live in the intertidal as juveniles, and dwarf surfperch (*Micrometrus aurora*) often release their young in tidepools. Environmental variability (temperature range of at least 15°C) and unpredictability (low dissolved oxygen concentrations and periodic sedimentation of tidepools) probably account for low fish species richness (Cross 1982; L. G. Allen 1985). Most residents eat amphipods, isopods, copepods, polychaetes, and gastropods (Mitchell 1953; Lacy 1973; Cross 1982). Juvenile opaleye are largely herbivorous; only the smallest fish (<32 mm) eat significant amounts of animal prey (Williams and Williams 1955). Little is known about the spatial and temporal variations in abundance, recruitment, and production of intertidal fish populations in the SCB.

Intertidal fishes tolerate extreme and rapid changes in temperature, salinity, dissolved oxygen, and pH (Graham 1970; Congleton 1980). Juvenile opaleye exhibit strong thermotaxis toward a mean temperature, about 26°C, which may direct their inshore migration to preferred nursery areas in the rocky intertidal and shallow subtidal (Norris 1963). During nighttime low tides, respiration in tidepools can create hypoxic conditions for fish. California clingfish, wooly sculpin, and reef finspot may either breathe air directly or ventilate maximally oxygenated water from directly below the air–water interface during hypoxic conditions (Congleton 1980). Some species can resist desiccation beneath rocks or in algal beds when stranded by the ebbing tide (Horn and Riegle 1981).

Intertidal fishes resist surge and turbulence by inhabiting holes, rock crevices, or algae. California clingfish adhere to rocks by a thoracic suction disc, a modification of the pectoral and pelvic fins. Wooly sculpin, reef finspot, and rockpool blenny grasp the substrate with their paired fins. Wooly sculpin and juvenile opaleye can return to specific tidepools when displaced during high tide (Williams 1957; Richkus 1978; Valle 1989). Homing helps these fishes avoid unfavorable situations at low tide.

Rocky Reefs and Kelp Beds

While more than 125 species of fish live on reefs in the SCB, only 50–60 species are common and only 15–20 are abundant and conspicuous to divers (fig. 9.14) (Quast 1968b, c; Feder et al. 1974). Many kelp bed fishes also occur in subtidal surfgrass (*Phyllospadix* sp.) meadows (DeMartini 1981) and shallow rock reefs devoid of kelp (Stephens et al. 1984; Carr 1989).

Fish abundance on reefs is related to the presence or absence of kelp (*Macrocystis pyrifera*) and substrate relief, although bottom relief greater than 1 m has little effect on fish species diversity and abundance (Patton et al. 1985). Fish are more abundant on cobble reefs with high densities of *M. pyrifera* (23–30 plants per 100 m^2) than on cobble reefs with low densities (8 plants per 100 m^2) (Larson and DeMartini 1984). The abundances of water column fishes (kelp surfperch [*Brachyistius frenatus*], kelp bass, giant kelpfish, and kelp rockfish [*Sebastes atrovirens*]) are directly correlated with kelp density. Kelp surfperch have a threshold kelp density below which they are rare and above which they increase rapidly in abundance (Coyer 1979). Removal of *M. pyrifera* from a reef near San Luis Obispo was followed by a 63% decline in fish biomass and a 16% decline in the number of fish species (Bodkin 1988). Fishes that sheltered in the kelp, foraged on kelp epizooics, or preyed on these species declined in abundance. However, total fish abundance did not change substantially after *M. pyrifera* recruited to a previously kelpless, high-relief reef off Palos Verdes; only kelp bass increased in abundance (Stephens et al. 1984).

Kelp beds are not important spawning areas for fishes (Feder et al. 1974; Stephens et al. 1984); they are, nevertheless, important nursery areas for juvenile fishes. Recruitment and abundance of young-of-the-year and juvenile kelp bass, giant kelpfish, and kelp surfperch are greater in the presence of *M. pyrifera* than in its absence (Carr 1989). Kelp bass approximately 10 mm SL recruit to kelp beds and drift algae after 1 month in the plankton (R. Lavenberg pers. comm.). They are most abundant in the presence of kelp regardless of the type of substrate. Juvenile and adult kelp bass occur in kelp beds and on rocky reefs devoid of kelp. The decreased dependence on kelp as the fish get older is probably due to the decreased vulnerability of larger individuals to predation (Carr 1989). Cryptic fishes, such as *Gibbonsia* spp. and island kelpfish (*Alloclinus holderi*), are more abundant and recruit in greater numbers to areas cleared of *M. pyrifera* because of the increased abundance of understory algae (Carr 1989).

Most reef fish studies are conducted by divers using SCUBA. Sampling is usually nondestructive and restricted to times when water conditions are suitable for counting and filming, usually summer and fall (Ebeling et al. 1980a; Larson and DeMartini 1984). Reef fish abundances have been estimated by several methods: direct visual observations along fixed or temporary belt transects (Quast 1968b), visual observations with cine cameras or videocameras on timed swims along belt transects (Ebeling et al. 1980a), observations at baited (Gotshall 1987) and poison stations (Stephens et al. 1986), and an *in situ* mark-recapture technique (Davis and Anderson 1989). Abundance estimates can be biased,

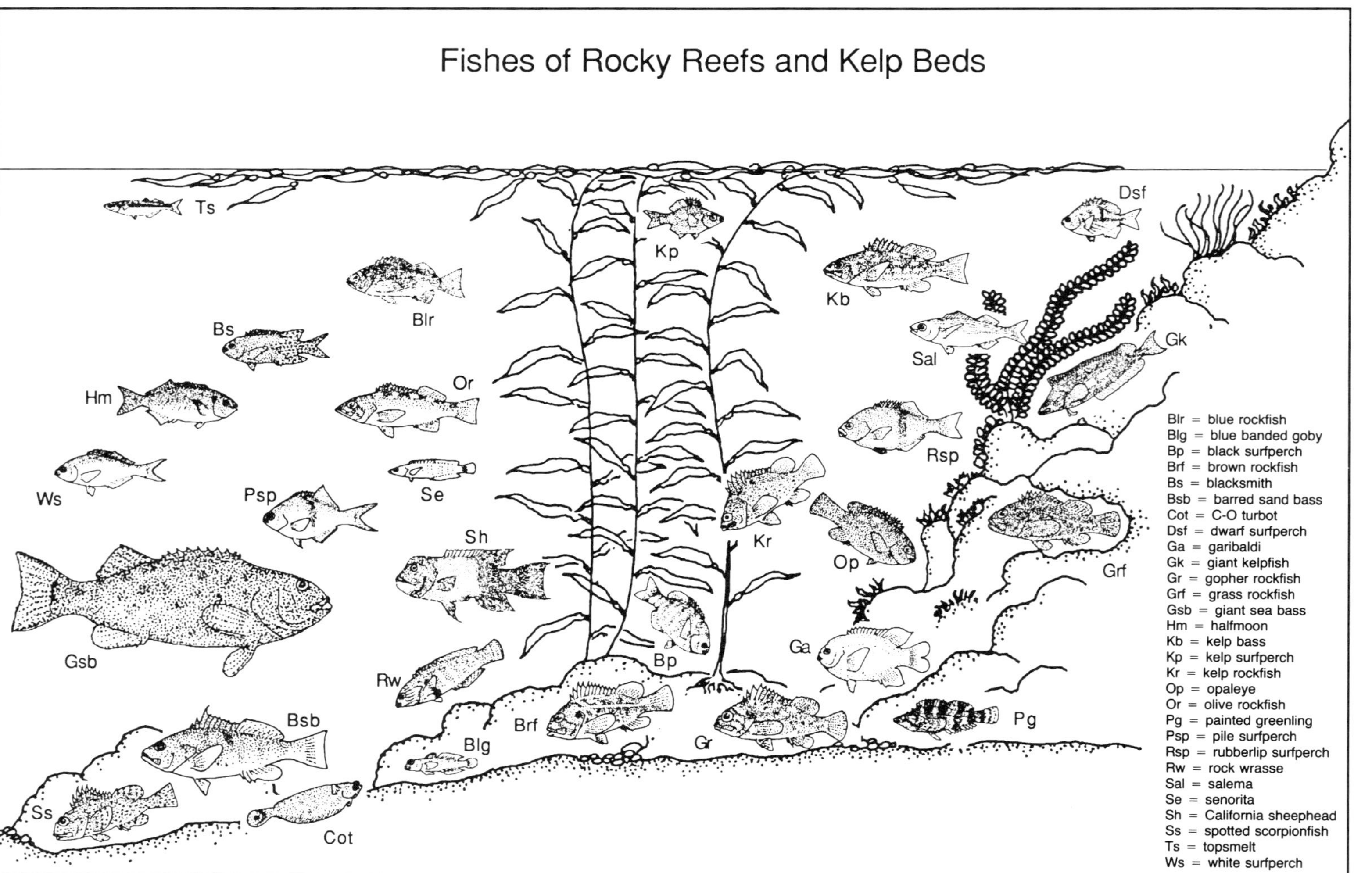

Figure 9.14. Schematic diagram of fishes within rocky reef and kelp bed habitats.

however, by horizontal water visibility, surge, habitat complexity, fish behavior, optical limitations of cameras and films, and inaccurate estimates of area or volume searched (Ebeling et al. 1980a; DeMartini and Roberts 1982; Larson and DeMartini 1984; Gotshall 1987; Davis and Anderson 1989). Visual surveys by divers may also underestimate actual reef fish abundances (Davis and Anderson 1989).

STANDING CROP AND TROPHIC STRUCTURE

Estimates of the standing crop of fishes on high-relief reefs (33–328 g wet m^{-2}) (Quast 1968b; Stephens et al. 1984), low-relief reefs (39–65 wet g m^{-2}) (Larson and DeMartini 1984), and breakwaters (389 g wet m^{-2}) (Stephens et al. 1984) are above the median for coral reefs and freshwater lakes (Horn 1980). The order of magnitude differences among high-relief reefs are at least partially due to sampling techniques that either excluded or included cryptic fishes.

In the northern SCB, the kelp canopy is dominated by planktivores and kelp browsers (e.g., blacksmith, kelp surfperch, blue rockfish [*Sebastes mystinus*], juvenile olive rockfish, and senorita [*Oxyjulis californica*]) (fig. 9.15) (Ebeling et al. 1980a, b). Canopy assemblages are characterized by many individuals distributed unevenly among a few species. The reef itself is dominated by grazers and ambushers (pile surfperch [*Damalichthys vacca*], black surfperch, garibaldi, California sheephead [*Semicossphyus pulcher*], gopher rockfish [*Sebastes carnatus*], and black-and-yellow rockfish [*Sebastes chrysomelas*]). Bottom assemblages are characterized by a more even distribution of individuals among more species. The composition of canopy assemblages is less temporally stable than the composition of bottom assemblages. Planktivores depend on a resource that is strongly affected by local oceanographic conditions, while grazers and ambushers depend on rock and turf infauna that are not affected so immediately by local oceanographic conditions (Ebeling et al. 1980a).

Reef fishes influence the behavior and abundance of their prey and the composition of the reef community as well. For example, California sheephead consume a variety of prey, including sea urchins (*Strongylocentrotus franciscanus*). Although sea urchins rank only seventh in dietary importance, both the presence and absence of sheephead in the reef community have significant impact on the urchin population. Where sheephead are uncommon (0–35 ha^{-1}), 10–30% of the sea urchins occur in exposed positions on the reef; where sheephead are abundant (200–500 ha^{-1}), no urchins are exposed. Moreover, removal of sheephead from a reef resulted in both an increase in sea urchin numbers (26% yr^{-1}) and an increase in the number of exposed individuals (1–2%) (Cowen 1983). Senorita prey on bryozoans, polychaetes, hydroids, barnacles, and bivalves that encrust or settle on blades of *M. pyrifera*. They also prey on gastropods that live on *M. pyrifera* plants and crustaceans that live in the canopy or in midwater. By excluding barnacles and bivalves, senorita affect the diversity of the system, and by grazing on *M. pyrifera* blades, senorita can destroy small kelp beds. Senorita can also stabilize the system by regulating herbivorous isopods that severely damage kelp plants at high densities (Bernstein and Jung 1979).

SPATIAL VARIABILITY

The composition of reef fish assemblages is influenced by the physical characteristics of the reef (Ebeling et al. 1980a, b; Larson and DeMartini 1984) and by water temperatures, particularly as they relate to adult distributions and reproductive success (Quast 1968c; Stephens and Zerba 1981; Stephens et al. 1984) (table 9.8). Shelter-seeking species (e.g., blacksmith, garibaldi, grass rockfish [*Sebastes rastrelliger*], brown rockfish [*Sebastes auriculatus*], and gopher rockfish) are abundant on high-relief reefs, but are rare or absent on low-relief reefs (Larson and DeMartini 1984). Warm-water species dominate reefs in the southern SCB, and cool-water species

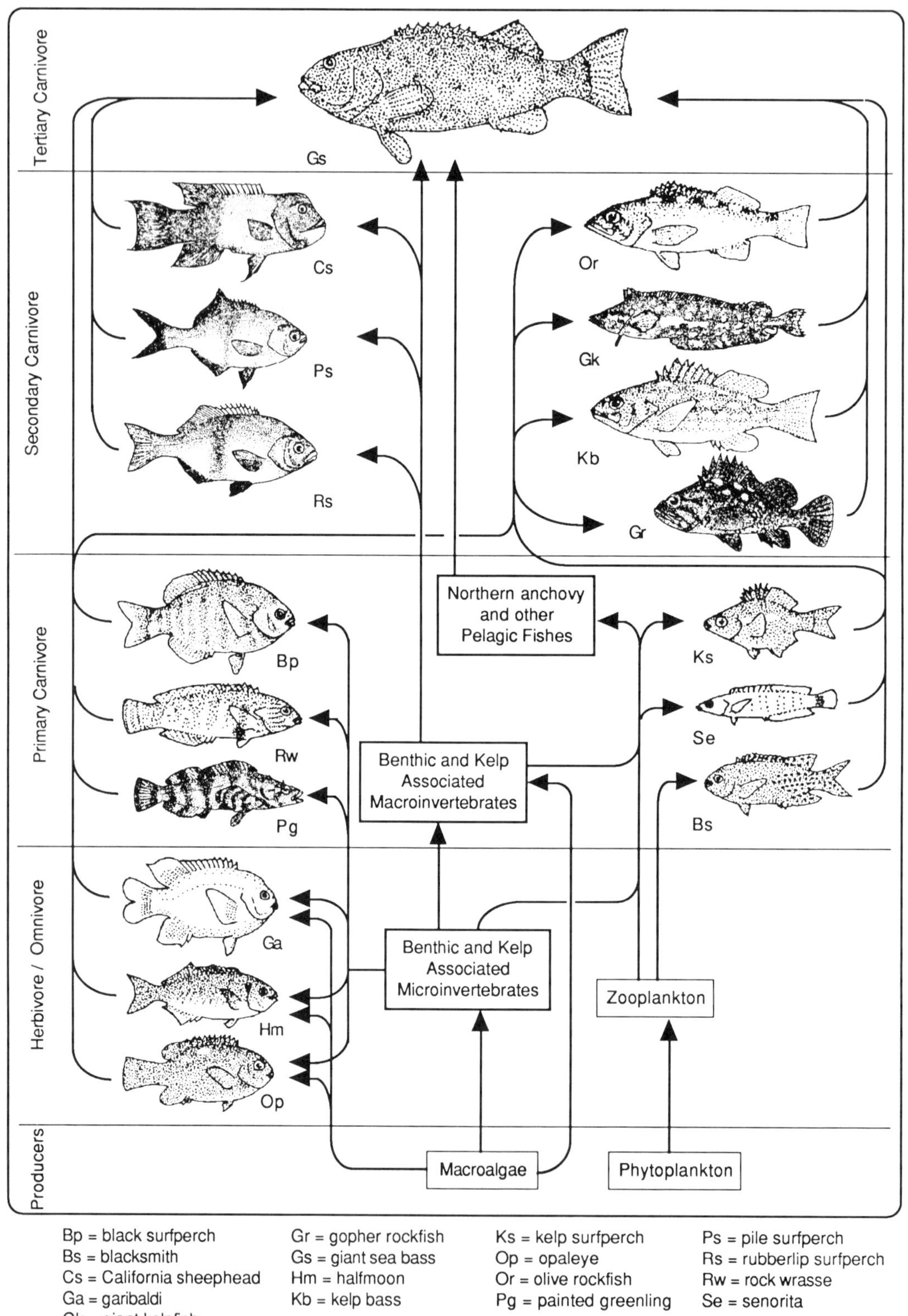

Figure 9.15. Trophic relationships of rocky reef and kelp bed fishes.

Table 9.8. *Relative Abundance (%) of Common, Conspicuous Fishes in the Canopy and near the Bottom of Kelp Beds in the SCB*[a]

Fishes	Santa Barbara	Santa Cruz Island	Palos Verdes	San Onofre
Canopy				
Silverside species	0	0	nd	5
Blue rockfish	41	7	0	0
Kelp rockfish	<1	3	0	0
Olive rockfish	2	9	1	0
Kelp bass	5	3	6	9
Jack mackerel	0	0	7	24
Opaleye	1	3	6	<1
Halfmoon	3	<1	1	2
Kelp surfperch	2	35	1	1
Black surfperch	<1	<1	<1	<1
Pile surfperch	<1	<1	<1	<1
White surfperch	<1	<1	0	9
Rubberlip surfperch	<1	<1	<1	<1
Blacksmith	36	33	27	<1
Senorita	9	5	52	47
California sheephead	<1	<1	0	1
Giant kelpfish	<1	<1	<1	<1
Bottom				
Blue rockfish	7	5	13	0
Rockfish species	3	11	2	<1
Painted greenling	2	1	nd	0
Kelp bass	11	14	5	15
Barred sand bass	<1	0	<1	5
Jack mackerel	0	0	15	0
Salema	0	0	0	8
Black croaker	<1	0	0	10
Opaleye	7	11	3	0
Halfmoon	2	<1	<1	<1
Kelp surfperch	<1	<1	<1	0
Black surfperch	28	8	9	6
Pile surfperch	6	3	1	2
Striped surfperch	4	3	0	0
Rainbow surfperch	8	<1	<1	4
White surfperch	<1	<1	nd	6
Rubberlip surfperch	2	1	<1	1
Blacksmith	7	13	9	0
Garibaldi	1	14	3	<1
Senorita	7	3	33	23
Rock wrasse	0	<1	2	3
California sheephead	3	11	3	16
Giant kelpfish	<1	<1	<1	<1

[a] Reefs at Santa Cruz Island, Santa Barbara (Ebeling et al. 1980a), and off the Palos Verdes Peninsula (Stephens et al. 1984) are high-relief bench rock reefs. The reef at San Onofre (Larson and DeMartini 1984) is a low-relief cobble reef on sand. nd = present but no data.

dominate reefs in the northern SCB. The transition between warm-water and cool-water species occurs in the central SCB, and reefs there contain both groups of fishes. The particular thermal regime in existence determines which group dominates the fauna (Stephens et al. 1984). Some cool-water fishes exist as marginal populations in the SCB and possess localized specializations to their restricted niches (Alevizon 1975a; Haldorson and Moser 1979).

Relatively little is known about the small cryptic fishes that inhabit the understory algae and substrates of shallow reefs (Stephens et al. 1970; Behrents 1987; Hartney 1989). The density of cryptic fishes (283 per 100 m^2) at Santa Catalina Island is four times that of the conspicuous fishes (72 per 100 m^2). Inclusion of cryptic fishes with conspicuous fishes increases diversity by 41%, the number of species by 50%, and biomass by 10%. The small size, rapid growth, and short life span of kelpfishes and gobies may result in relatively high secondary productivity (Bouvier 1987).

TEMPORAL VARIABILITY

The diel behavior of rocky reef fishes is similar to the diel behavior of coral reef fishes (Ebeling and Bray 1976; Hobson and Chess 1976; Hobson et al. 1981). During the day, 65% of diver-observed fishes are in the water column, 25% are within 1 m of the bottom, 7% are resting on the bottom, and 2% are in holes and crevices. Less than half as many fishes are observed at night. Of these, 22% are in the water column, 11% are within 1 m of the bottom, 17% are resting on the bottom, and 50% are in holes and crevices (Ebeling and Bray 1976).

Diurnally active species seek shelter in or near the reef at night. Species with tropical affinities shelter in holes and crevices in the reef (garibaldi, blacksmith, and California sheephead), bury themselves in sand (senorita and rock wrasse, *Halichoeres semicinctus*), or shelter in algal turf (rock wrasse) (Fitch and Lavenberg 1975; Ebeling and Bray 1976; Hobson et al. 1981). Species with temperate affinities (surfperches) descend from the water column at dusk and remain quiescent close to or on the bottom at night (Ebeling and Bray 1976). Pressure from crepuscular and nocturnal predators may be responsible for the twilight transition between diurnally and nocturnally active reef fishes. Spectral sensitivities of their scotopic visual pigments cluster around 500 nm, optimal for vision during twilight when crepuscular predators are active (Hobson and Chess 1976; Hobson et al. 1981).

Diurnal planktivores (blacksmith, kelp surfperch, senorita, and small juvenile olive rockfish) have small, highly modified jaws for feeding on small (<2 mm) zooplankters. They forage in the water column above the reef or at some distance from it (Hobson and Chess 1976). During the day, blacksmith gather at the up current end of the reef where zooplankton densities are greatest; at dusk, they seek shelter in the reef for the night (Bray 1980). Adult blacksmith import as much as 8 g C m^{-2} yr^{-1} deposited as feces in nocturnal shelters (Bray et al. 1981). Nocturnal planktivores (kelp rockfish, large juvenile olive rockfish, walleye surfperch, salema, and queenfish) have large mouths and eyes and feed on large (>2 mm) zooplankters. They spend the day in resting aggregations on or near the reef, but they disperse offshore at night to feed on zooplankters that ascend from the reef and surrounding sand into the water column. At dawn, the fish descend from the water column and form aggregations that migrate back to diurnal resting sites (Hobson and Chess 1976).

Little information exists on seasonal and interannual patterns of distribution and abundance of rocky reef and kelp bed fishes. Some species decline in numbers during winter and spring (Miller and Geibel 1973; Gotshall 1987). Rainbow surfperch (*Hypsurus caryi*) arrive at Naples Reef near Santa Barbara in late spring and leave in the fall (Laur and Ebeling 1983). Giant sea bass (*Stereolepis gigas*) aggregate in shallow water on traditional spawning grounds from June through September (Frey 1971). Between 1974 and 1981, sea surface temperatures increased and kelp became established on a reef off the Palos Verdes Peninsula. Changes in the abundance of one-third of the 24 common species were correlated with water temperature (fig. 9.16). Warm-water species (kelp surfperch, garibaldi, senorita, and California sheephead) increased in abundance. Cold-water species (blackeye goby [*Coryphopterus nicholsii*], painted greenling [*Oxylebius pictus*], pile surfperch, and blue rockfish) decreased in abundance. The remaining species changed little or fluctuated unpredictably (Stephens et al. 1984).

CHANGES IN FISH POPULATION ABUNDANCE

Competition. Interspecific competition influences the bathymetric distribution of adult black-and-yellow rockfish and gopher rockfish. Black-and-yellow rockfish aggressively exclude gopher rockfish from preferred foraging areas in the food-rich shallower parts of the reef. The distributions of the rockfish are determined by preferential settlement of young fish of the two species at different depths and by interference competition through interspecific territoriality among older and larger individuals (Larson 1980a, b).

Studies on five species of surfperches in the northern SCB concluded that the fishes live sympatrically with minimal competition by partitioning available space and food resources. Resource partitioning and character displacement were attributed to competitive interactions in the past (Alevizon 1975a, b; Hixon 1980; Schmitt and Coyer 1982, 1983; Holbrook and Schmitt 1986). Hixon (1980) experimentally demonstrated interspecific competition for foraging space between adult striped surfperch (*Embiotoca lateralis*) and black surfperch. Striped surfperch are more abundant in the food-rich, shallower parts of the reef, while black surfperch are more abundant in food-poor, deeper parts of the reef. Striped surfperch aggressively dominate black surfperch. When striped surfperch were removed, black surfperch became more abundant in shallower parts of the reef. The distributions of juvenile striped and black surfperches overlap, and they often forage together nonaggressively over the shallow parts of the reef (Hixon 1980).

An alternative view is that individualistic responses allow the five surfperches to coexist (Ebeling and Laur 1986). The geographic ranges of the species overlap in the SCB: three species are near the center and two species are near the end of their ranges. Environmental conditions on a particular reef are more favorable for some species than for others, so population densities vary among reefs. Interspecific competition between striped and black surfperches occurs on reefs where both species are abundant, but does not occur on reefs where one species is abundant and the other is rare. Each species uses the reef to its best advantage independently of the other species (Ebeling and Laur 1986).

Hartney (1989) reached a similar conclusion in a study of the sympatric bluebanded goby (*Lythrypnus dalli*) and zebra goby (*Lythrypnus zebra*). The more visible bluebanded goby forages from horizontal surfaces on exposed outcrops and feeds mainly on planktonic crustaceans. The more retiring zebra goby forages vertically on walls of crevices (or upside down beneath rocky overhangs) and feeds mainly on benthic crustaceans. Microspatial preferences do not change when the congener is removed. Microhabitat and foraging differences established over evolutionary time may or may not have been the result of interspecific competition, a hypothesis that is difficult to test experimentally (Connell 1980).

Predation. Little is known about the effects of predation on reef fish populations. Reef fishes recognize their predators, and even young individuals make behavioral adjustments to reduce their risk (Schmitt and Holbrook 1985). The risk of predation influences

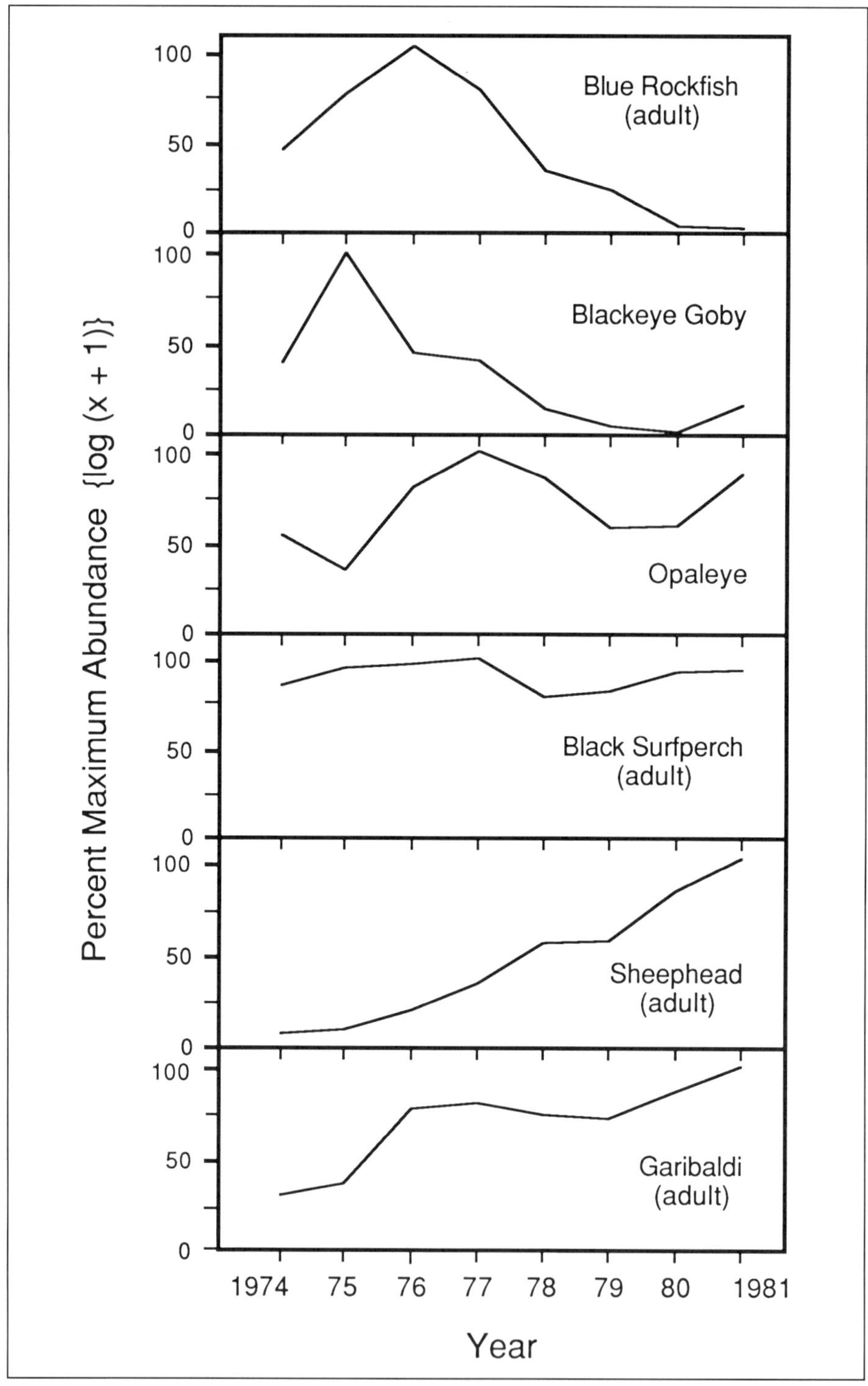

Figure 9.16. Percentage of maximum abundance of some common kelp bed fishes at Palos Verdes Point for 1974–1981. (Redrawn from Stephens et al. 1984.)

microhabitat selection and foraging by juvenile black surfperch (Holbrook and Schmitt 1984, 1988a, b) and may be responsible for the twilight transition between diurnally and nocturnally active reef fishes (Hobson and Chess 1976; Hobson et al. 1981). Fishes that rest on or near the bottom at night are vulnerable to predation by the Pacific electric ray (*Torpedo californica*), a nocturnal predator that immobilizes its prey by electrical discharges (Bray and Hixon 1978). The availability of appropriately sized refugia affects survival of young bluebanded gobies (Behrents 1987). In the last century, populations of some important reef fish predators have decreased (giant sea bass and white seabass), while others have increased (seals and sea lions).

Recruitment. Recruitment of gopher rockfish, black-and-yellow rockfish (Larson 1980b), California sheephead (Cowen 1985), cryptic fishes (Stephens et al. 1986), and kelp bass (Carr 1989) fluctuates from year to year. The presence or absence of *Macrocystis pyrifera* affects the composition of species recruiting from the plankton and may modify the spatial and temporal dynamics of kelp bed fishes (Carr 1989). Some reef fishes, such as the California moray eel (*Gymnothorax mordax*), are common in the SCB, but do not reproduce. The moray population is maintained by larvae that are spawned off Baja California and drift northward into the SCB (McGleneghan 1973).

Interannual variability in California sheephead recruitment at San Nicolas Island is related to changes in the current regime (fig. 9.17). The sheephead, a warm-water fish, is common in the SCB but rare north of Point Conception. Sheephead recruitment is highest during major events, such as El Niño, that last 1–2 years and bring large amounts of warm, larval-rich water to San Nicolas Island. Sheephead recruitment is limited and sporadic during minor events, such as current meanders, that last days or weeks and bring modest amounts of warm water to San Nicolas Island. Because of their length, major events can affect many species. Sheephead, bluebanded and zebra gobies, rock wrasse, and garibaldi—all warm-water species—recruited in high numbers at the northern end of their ranges during the

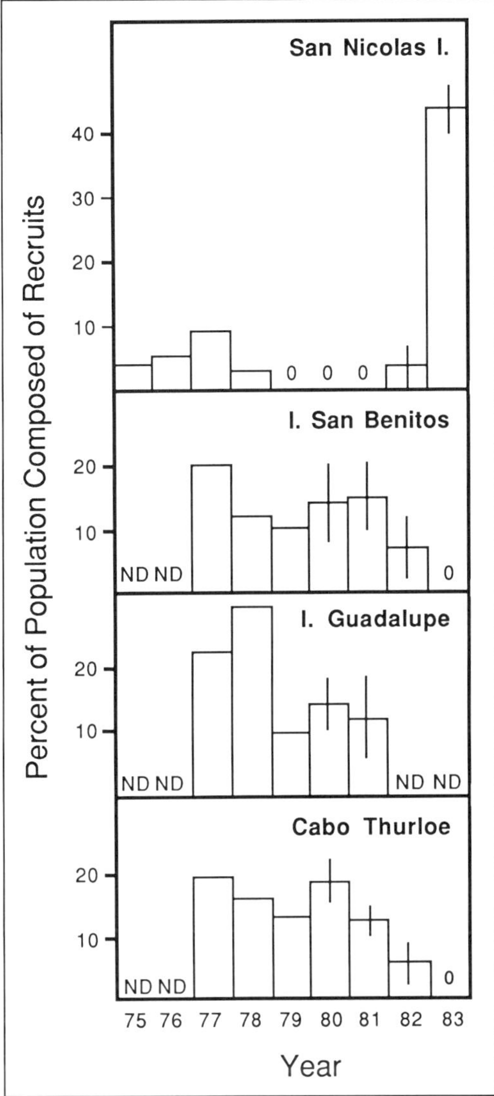

Figure 9.17. Annual recruitment success of California sheephead in southern California and Baja California for 1975–1983. Data for 1975–1979 based on population age structure; data for 1980–1983 based on field counts ($\bar{X} \pm 2SE$). ND = no data; 0 = no recruits. (Redrawn from Cowen 1985.)

1982–1983 El Niño. Minor events benefit species with larvae in the right place at the right time, especially those with long-lived larvae. They probably also account for the occasional recruitment success of one species while other species show no effect (Cowen 1985).

Environmental Processes. Major storms are rare in the SCB, but unusually severe storms occurred in 1980 and 1983 (Dayton and Tegner 1984). Ebeling et al. (1985) described the impact of 6-m-high waves generated by these storms on the kelp bed, sea urchins, and surfperches of Naples Reef off Santa Barbara. The first storm removed the entire kelp canopy and most of the understory kelp. The disappearance of the canopy drastically reduced available drift kelp, the preferred food of sea urchins. Consequently, urchins emerged from shelters in the reef and consumed most of the living plants, including newly recruited kelps and turf algae, creating "barrens." Surfperches declined sharply in abundance because their invertebrate prey lived in turf and juveniles sheltered in the understory kelp. The second storm decimated the urchins, although some survived in deep refugia. *Macrocystis pyrifera* and understory kelps recruited to the reef and persisted where urchins were scarce. Juvenile surfperch reappeared, drift kelp began to accumulate, and urchins resumed living in shelters. Thus, the second storm reversed the effects of the first.

The zoogeographic transition between warm- and cool-water reef fishes occurs in the central SCB. Before 1976, temperate fishes (rockfishes) dominated the breakwater at King Harbor, and tropical fishes (wrasses, damselfishes, and sea basses) dominated nearby reefs off Palos Verdes Peninsula. The gradual loss of cool-water species from King Harbor resulted in increased similarity between the reefs in 1977. After 1979, warm-water species dominated both reefs (Stephens et al. 1984).

Fishing. Reef fishes constitute a major portion of the catches of recreational fishermen and sport divers in the SCB (Frey 1971; Helvey et al. 1987), but little is known about the effects of fishing on reef populations. Olive rockfish is a common reef inhabitant and a dominant species in the sport catch in the northern SCB (Love et al. 1985). Although they are water column fishes, olive rockfish rarely move between reefs. Populations on heavily fished reefs are dominated by juveniles and subadults; large, mature individuals are selectively removed by fishermen. Reproduction on lightly fished reefs may supply recruits to heavily fished areas (Love 1980).

Deep Reefs

Little is known about the fish assemblages of banks and seamounts deeper than the no-decompression SCUBA limits (>50 m) in the SCB. Fishes on deep banks have been photographed by remote cameras (Moore and Mearns 1980a), observed and photographed from submersibles (Lissner and Dorsey 1986), and caught on hook-and-line (Cross 1987; Genin et al. 1988). Deep reefs are targeted by recreational and commercial fishermen, especially during the colder months of the year when warm-water pelagic species are less abundant. Studies reporting fish catches from deep reefs usually lack complete species lists and habitat information (Love 1981; Love et al. 1985; Hartmann 1987).

The fish assemblages of deep reefs are dominated by large, mobile, nektobenthic fishes (rockfish [*Sebastes* spp.], sablefish, Pacific hake, spotted ratfish [*Hydrolagus colliei*], and spiny dogfish [*Squalus acanthias*]). Longline catches on the deepest banks (350–500 m) are smaller and contain fewer species during the summer than the rest of the year. Smaller catches are probably the result of bathymetric movements of some species and latitudinal movements of others (Cross 1987).

Banks and seamounts possess unique physical characteristics that affect local biological processes. They deflect isotherms upward and are the focus of upwelling. This results in increased primary, and perhaps secondary, productivity and attracts pelagic fishes (Genin and Boehlert 1985). Rockfish ascend as much as 50 m into the water column from deep (100–140 m) banks to feed on euphausiids in the morning. Euphausiids spend the day in deep water (>200 m) surrounding the banks and ascend to the surface at night. In the morning, euphausiids descend to their daytime depths, but currents advect some of them over the banks, where they are consumed by rockfish (Genin et al. 1988).

Artificial Reefs

More than 25 artificial reefs have been built in the SCB since 1958 to enhance populations of sport fishes (Carlisle et al. 1964; Turner et al. 1969) and to mitigate the impact of coastal power plants (Grant et al. 1982; Grove 1982). Reefs have been constructed from quarry rock, concrete riprap, automobiles, streetcars, and scuttled ships. Breakwaters, jetties, man-made islands, bridge piers, submarine pipelines, and oil platforms also function as artificial reefs. New structures quickly attract fish and other organisms, and within a few years, they support natural communities similar to other well-established structures (Turner et al. 1969; Davis et al. 1982; Grant et al. 1982). Fish assemblages on artificial reefs in the SCB are indistinguishable from their natural counterparts (Stephens and Zerba 1981; DeMartini and Roberts 1982; L. G. Allen 1985; Ambrose and Swarbrick 1989). More than 100 species of fish have been observed on the breakwater at King Harbor (Stephens and Zerba 1981).

Artificial reefs are colonized by juveniles settling from the plankton or recruiting from other habitats (DeWees and Gotshall 1974), as well as by adults emigrating from nearby natural reefs (Matthews 1985). Despite considerable research worldwide, it is not known whether artificial reefs simply attract and concentrate fish from a large area, permitting more efficient exploitation, or whether they actually increase fish production by increasing successful recruitment (Bohnsack and Sutherland 1985; Bohnsack 1989).

STANDING CROP AND TROPHIC STRUCTURE

The density and biomass of fishes on artificial reefs are often greater than on nearby natural reefs (Stephens et al. 1984: Jessee et al. 1985; Ambrose and Swarbrick 1989), and fishing success can be two or three times greater on artificial reefs (Turner et al. 1969). Estimates of benthic fish biomass were greater on 10 artificial reefs (mean of 45 g m^{-2}, SD of 19) than on 16 natural reefs (mean of 33 g m^{-2}, SD of 24) in the SCB. Artificial reefs have a higher perimeter to area ratio than natural reefs and are usually sited on sand plains isolated from other reefs. Fish densities may be higher on artificial reefs because they attract individuals from a proportionately larger area than natural reefs (Ambrose and Swarbrick 1989) or because artificial reefs are more complex and provide more cover and more diverse habitats than natural reefs (Bohnsack and Sutherland 1985; DeMartini et al. 1989).

Because artificial reefs are smaller than natural reefs, numerical and biomass abundances are far greater on natural reefs. Estimates of fish biomass per unit area were similar on a 1.3-ha artificial reef (40 g m^{-2}) and a 90-ha natural reef (44 g m^{-2}) off San Onofre, but total biomass on the artificial reef (0.4–0.5 t) was much lower than on the natural reef (29–39 t) (DeMartini et al. 1989). Estimates of total biomass on 10 artificial reefs (mean of 0.9 t, SD of 1.0) was much lower than on 16 natural reefs (mean of 45 t, SD of 96) (Ambrose and Swarbrick 1989).

Fishes on an artificial reef off San Diego feed on the reef and in the surrounding sand. On average, 66% of the diet of barred sand bass (*Paralabrax nebulifer*) came from the sand and 34% came from the reef. The diet of white surfperch (*Phanerodon furcatus*) was divided into 43% from the sand, 31% from the reef, and 21% from the plankton. Individual fishes, however, foraged either principally on the reef or principally in the sand. Reef fishes foraging on the sand reduced the density of *Stylatula elongata*, a relatively large, sessile, epifaunal sea pen, but had no discernible effects on small infauna (Davis et al. 1982).

SPATIAL VARIABILITY

The breakwater at King Harbor is the most intensively studied artificial reef in the SCB. The fish fauna is more diverse and more abundant than that of other breakwaters and low-relief, natural reefs (table 9.9). Stephens (1978) attributed the high number of species (>100) at King Harbor to the thermal diversity of the area. In summer, upwelled water from nearby Redondo Canyon enters the harbor and lowers surface temperatures; in winter, power plant thermal effluent discharged into the harbor increases surface water temperatures. A thermocline exists shallower than 9 m during most of the year. Fishes below the thermocline (rockfishes, surfperches, barred sand bass, blackeye goby, and painted greenling) prefer cooler water or the sand–rock interface at the base of the breakwater. Fishes above the thermocline (opaleye and topsmelt) prefer warmer water (Stephens and Zerba 1981).

TEMPORAL VARIABILITY

Adult dwarf surfperch, black surfperch, pile surfperch, rainbow surfperch, and white surfperch immigrate to the breakwater at King Harbor from offshore areas preceding parturition by females. Young-of-the-year surfperches recruit to the breakwater from April through July. Adults of all but the dwarf surfperch emigrate to deeper water in the fall when surface waters warm and the thermocline deepens (Terry and Stephens 1976). The number of species and total abundance of fishes on the breakwater at King Harbor remained relatively stable from 1974 to 1979. The abundance of individual species, however, fluctuated by more than an order of magnitude (Stephens and Zerba 1981). A decline in surfperch abundance was correlated with decreasing recruitment success (Stephens et al. 1986).

RECRUITMENT

Stephens et al. (1986) examined the relationship between the planktonic abundance of larvae and recruitment success by removing all fish from a small reef in King Harbor every month for 16 months. The number of early larvae (post–yolk sac, preflexion) of oviparous species was correlated with the number of adults; the abundance of older larvae was not. Bluebanded goby and roughcheek sculpin (*Artedius creaseri*), the most abundant species, recruited to the reef over many months, reaching a peak from late summer to winter; recruitment was correlated with planktonic abundance. In the absence of predators on the reef, species that were abundant in the plankton had the greatest recruitment success.

Summary and Prospectus for Future Research

In this chapter, we have stressed spatial and temporal patterns in fish distribution and abundance and relationships between environmental processes and biological responses. Here we touch on some of the major points and suggest important research needs.

What we know of marine fishes is based largely on samples collected by remote devices, primarily towed nets. Each sampler has specific characteristics that bias catch parameter estimates because of inter- and intraspecific variability in organism catchability (Wardle 1983, 1986; Glass and Wardle 1989), which affects the accuracy of standing crop and production estimates. Sampling biases are rarely discussed or examined experimentally (for exceptions, see Lenarz 1972; Murphy and Clutter 1972).

Northern anchovy and Pacific mackerel are among the best known fishes in the world. For most fishes in the SCB, however, growth, reproduction, age structure of adults, transport of larvae, recruitment, early life histories, and factors affecting mortality at various stages are unknown.

Life history information is important for management of exploited species. Fishes that grow slowly, mature late, have low natural mortality rates, and have long life spans are

Table 9.9. *Number of Species per Survey and Transect, and Number of Fish per Transect from Diver Surveys of King Harbor and Santa Monica Breakwaters and a Natural Reef off the Palos Verdes Peninsula*

	Species per Survey			Species per Transect			Individuals per Transect		
	Mean	SD[a]	n[b]	Mean	SD	n	Mean	SD	n
King Harbor (leeward)	47.0	6.8	17	13.3	3.0	225	282	120	225
King Harbor (seaward)	No data			11.8	1.3	150	342	150	150
Santa Monica (leeward)	23.0	5.7	2	10.5	1.6	15	162	90	15
Palos Verdes	17.0	5.8	9	8.0	1.5	108	90	75	108

[a] SD is one standard deviation.
[b] n is sample size.
From Stephens 1978.

more sensitive to overfishing than fishes that grow quickly, mature early, have high rates of natural mortality, and have short life spans. Lack of detailed knowledge about rockfish life histories has resulted in overexploitation and population crashes of one species after another, causing boom and bust cycles in their commercial fisheries (Beamish and McFarlane 1983; Gunderson 1984; Leaman 1987).

The number of individuals declines precipitously between the egg and juvenile stages; there is little evidence that variable mortality rates during the planktonic stage determine the strength of recruiting year classes (Hunter 1981). Mortality stabilizes during the late larval or early juvenile stage when there are better correlations between juvenile and adult abundances and between adult and egg abundances (Rothschild 1986). Predation, starvation, and other causes of natural mortality during late larval and juvenile stages, and their correlation with recruitment, require more study.

Few early life history and recruitment studies have been done on nearshore demersal and reef fishes. Nearshore species develop faster and recruit at a smaller size than epipelagic species, so offshore transport may be less of a problem (Gruber et al. 1982). Fluxes in recruitment control the size of some adult reef fish populations and probably account for dominance shifts in soft-bottom fish assemblages. Future studies should focus on interactions between large scale physical and biological factors that affect larval dispersal, small–scale factors that influence larval settlement, and factors that determine juvenile survival after settlement (Cowen 1985).

Population traits, such as diets and growth rates, are usually expressed as means. This can conceal substantial between-individual variability that is biologically and ecologically important. Conspecific surfperches on the same reef have different feeding strategies and diets that result in differential growth and fecundity (Schmitt and Coyer 1982; Holbrook and Schmitt 1986). Individual rockfish of a "territorial" species may hold a territory that includes shelter and feeding sites (territorial), hold one that includes shelter, but forage on a different part of the reef (commuter), or not hold any territory (floater) (Larson 1980a). Long-term studies are needed to examine endogenous variation in individual and population attributes and how these traits respond to environmental fluctuations.

More behavioral studies are needed. Most existing studies deal with feeding, yet we know little about how a predator's feeding strategy and success are affected by its hunger, experience, or adaptation to certain prey types (Bakun et al. 1982). Schooling piscivorous species engage in cooperative hunting, but we do not know if it is commonplace (Schmitt and Strand 1982). Fishes rapidly colonize new artificial reefs, but we do not know if early arrivals differ from conspecifics in their innate tendency for dispersal (Alevizon and Gorham 1989). Observations of reproductive behavior are perhaps the rarest of all (Magnuson and Prescott 1966; Helvey 1982). We do not know how fish identify areas appropriate for reproduction (Bakun et al. 1982).

Much of our knowledge about the link between the physics of the ocean and biological processes was developed in the California Current system, a transition zone where different water masses with different flora, fauna, and selection pressures are blended together by large-scale lateral mixing. Connections between environmental processes and biological responses are difficult to observe and investigate experimentally (Bakun et al. 1982). Relationships may change under different biological conditions and with differences in the timing of environmental processes. When sardines were abundant and anchovies were rare, sardine recruitment was positively correlated with salinity and negatively correlated with temperature. When sardines were rare and anchovies were abundant, the signs of the correlations were reversed (Skud 1982, 1983). Upwelling prior to spawning increases primary production and results in successful copepod reproduction. Storms and upwelling during spawning disperse the plankton and reduce concentrations below the level necessary for fish larval survival (Lasker 1981 a, b).

Large-scale horizontal and vertical gradients in biomass, and probably secondary production, and interannual and interdecadal changes in environmental conditions are well known. Small-scale temporal and spatial patchiness, particularly during the early life histories of epipelagic fishes, are also well documented. However, we know little about the coupling between the large and small scales, such as between dense plankton blooms required by first-feeding anchovy larvae and interannual changes in the extent of the area rich enough to feed them (Hunter 1981; Lasker 1981 a, b).

Large deviations from long-term mean environmental conditions affect fish migrations, feeding, reproductive and recruitment success, and the composition of fish assemblages. Fishes may migrate to, and reproduce in, areas from which they were previously physiologically excluded. Their eggs and larvae may be carried by currents to regions where adults do not occur. New fish assemblages that arise from these invasions may persist through succeeding perturbations because of dynamic processes such as competition and predation. These processes do not confer resistance to every possible perturbation, so the resulting assemblages are only conditionally stable (Sutherland 1974).

The epipelagic ecosystem of the California Current may be permanently under nonequilibrium conditions, where biological processes are less important than environmental fluctuations in regulating species abundances (McGowan 1977). Cycles of spawning success and failure in pelagic fishes suggest that fish are tracking the carrying capacity of the environment rather that fluctuating around some equilibrium (Murphy 1966; Smith and Moser 1988). The ecological effects of large epipelagic fish biomass are poorly known (MacCall et al. 1985).

Predation (including cannibalism) on fish eggs and larvae is a major source of larval mortality for some epipelagic fishes. The effect of vertically migrating mesopelagic predators on the early life stages of fishes in the epipelagic zone is unknown (Smith and Moser 1988). The egg and larval predators are unknown for most species, as is their impact on recruitment. Predation may also be an important source of mortality among juvenile fishes in benthic communities. Predation by established adults can limit the number of recruits or increase postsettlement juvenile mortality. Among reef fishes, the risk of predation affects microhabitat selection and foraging (Holbrook and Schmitt 1988 a, b) and may be the cause of refuge-seeking behavior at dawn and dusk (Hobson et al. 1981).

Fish predation can modify prey assemblages and affect community structure. California sheephead influence the abundance and distribution of sea urchins on shallow reefs (Cowen 1983). Blacksmith reduce the abundance of plankton carried over reefs by currents (Bray 1980). Senorita affect the diversity and stability of kelp beds (Bernstein and Jung 1979). Grazing fishes reduce sea pen densities on soft substrates surrounding artificial reefs (Davis et al. 1982). Digging rays create disturbances in soft sediments that increase the availability of detrital organic material and modify benthic infaunal communities (Van Blaricom 1982).

Primary productivity near the coast in the SCB is determined largely by upwelling (Eppley et al. 1979); farther offshore, it is determined by advection of nutrients. Interannual fluctuations in primary productivity and changes in flow patterns of the California Current affect secondary production (Chelton et al. 1982) and behavior of pelagic and benthic communities. Recruitment in some species is correlated with primary productivity.

Few data exist on fish production in the SCB. Estimates for harbor and bay fishes (1–10 g dry wt m^{-2} yr^{-1}) are within the range of other marine habitats. Fish production is difficult to estimate because it requires knowledge of growth, reproduction, age structure, feeding, metabolism, and behavior of individuals in populations. Standing crop, which has been measured in most SCB habitats, varies over three orders of magnitude (table 9.10). Little information exists on the relationship between production in the water column and production by the benthos in the SCB. The flux of particulate organic matter to the benthos is related to primary production in the water column, lateral advective input, and physical stress on the bottom (Eppley and Peterson 1979; Emerson 1989). The amount and quality of the particulate matter affects benthic community structure, biomass, and metabolism (Smith et al. 1983; Smetacek 1984). Energy is also exchanged between the benthos and the water column by the export of eggs to surface waters by deep-water benthic fishes and by the import of juveniles (Moser 1974).

Little is known about fish assemblages from canyons, deep banks, and seamounts in the SCB, habitats targeted by recreational and commercial fishermen. Nektobenthic rockfishes concentrate on banks and seamounts and along the shelf margin near submarine canyons to intercept nocturnal, vertically migrating euphausiids as they descend to their daytime depths (Genin et al. 1988).

Interannual variations in the abundance of adult mesopelagic and bathypelagic fishes are not well documented. From 1955 to 1960, larvae of deep mid-water fishes made up 20–40% of the larvae collected in the surface waters of the California Current. Subarctic and transition species were more abundant in colder than average years, while equatorial and central Pacific species were more abundant in warmer than average years (Ahlstrom 1969). The horizontal and vertical distributions of mid-water fishes are correlated with productivity, temperature, light, and dissolved oxygen. However, the effects of predators, prey, and competitors have not been examined.

Artificial reefs in the SCB are built to attract juvenile and adult fishes. High colonization rates, fish densities, and catch rates are taken as evidence that artificial reefs increase fish production. However, few data are available to test whether artificial reefs increase fish production or simply attract fish from surrounding habitats. Increased production is most likely to occur on isolated reefs among fishes that are demersal, habitat limited, philopatric, territorial, and obligatory reef dwellers. Attraction is most likely to occur on reefs surrounded by heavily exploited natural reefs and among fishes that are mobile, re-

Table 9.10. *Standing Crop of Fishes in Various Habitats in the SCB*[a]

Habitat	Standing Crop (wet g m^{-2})	Reference
Bay	4.1	L. G. Allen 1982
Harbor	2.8–3.5	L. G. Allen et al. 1983
Outer shelf/upper slope	3.5–3.9	Cross 1987
Basins	0.3	Cross 1987
Kelp bed—low relief	39–65	Larson and DeMartini 1984; DeMartini et al. 1989
Kelp bed—high relief	33–328	Quast 1968b; Stephens et al. 1984
Artificial reef	40–389	Stephens et al. 1984; DeMartini et al. 1989
Epipelagic	31–58	Hewitt et al. 1976
Epipelagic anchovy school	15,000	Hewitt et al. 1976
Mesopelagic (<1000 m)	2.4–3.6[b]	Pearcy and Laurs 1966

[a] Estimates are per m^2 of sea floor except for epipelagic and mesopelagic, where estimates are per m^2 of sea surface.
[b] Off Oregon (no estimates available for the SCB).

cruitment limited, and facultative reef dwellers (Bohnsack 1989). Reefs built to increase larval recruitment, and juvenile survival may be more economical and more effective at increasing net fish production (Bohnsack and Sutherland 1985).

Literature Cited

Adams, P. B., 1980. Life history patterns in marine fishes and their consequences of fisheries management. *U.S. Natl. Mar. Fish. Serv. Fish. Bull.* 78:1–12.

Adams, S. M., 1976. The ecology of eelgrass, *Zostera marina* (L.), fish communities. I. Structural analysis. *J. Exp. Mar. Biol. Ecol.* 22:269–291.

Ahlstrom, E. H., 1959a. Vertical distribution of pelagic fish eggs and larvae off California and Baja California. *U.S. Fish Wildl. Serv. Fish. Bull.* 60(161):107–146.

Ahlstrom, E. H., 1959b. Distribution and abundance of eggs of the Pacific sardine, 1952–1956. *U.S. Natl. Mar. Fish. Serv. Fish. Bull.* 60:185–213.

Ahlstrom, E. H., 1965. Kinds and abundances of fishes in the California Current region based on egg and larval surveys. *Calif. Coop. Oceanic Fish. Invest. Rep.* 10:31–52.

Ahlstrom, E. H., 1969. Mesopelagic and bathypelagic fishes in the California Current region. *Calif. Coop. Oceanic Fish. Invest. Rep.* 13:39–44.

Ahlstrom, E. H., and R. C. Counts, 1958. Development and distribution of *Vinciguerria lucetia* and related species in the eastern Pacific. *U.S. Fish Wildl. Serv. Fish. Bull.* 58(139):363–416.

Alevizon, W. S., 1975a. Comparative feeding ecology of a kelp-bed embiotocid (*Embiotoca lateralis*). *Copeia.* 1975:608–615.

Alevizon, W. S., 1975b. Spatial overlap and competition in congeneric surfperches (Embiotocidae) off Santa Barbara, California. *Copeia* 1975:352–356.

Alevizon, W. S., and J. C. Gorham, 1989. Effects of artificial reef deployment on nearby resident fishes. *Bull. Mar. Sci.* 44:646–661.

Allen, L. G., 1980. Structure and productivity of the littoral fish assemblage of Upper Newport Bay, California. Ph.D. Dissertation, Univ. of Southern California, Los Angeles. 175 pp.

Allen, L. G. 1982. Seasonal abundance, composition, and productivity of the littoral fish assemblage in Upper Newport Bay, California. *U.S. Natl. Mar. Fish. Serv. Fish. Bull.* 80:769–790.

Allen, L. G., 1985. A habitat analysis of the nearshore marine fishes from southern California. *Bull. South. Calif. Acad. Sci.* 84:133–155.

Allen, L. G., 1988. Recruitment, distribution, and feeding habits of young-of-the-year California halibut (*Paralichthys californicus*) in the vicinity of Alamitos Bay–Long Beach Harbor, California, 1983–1985. *Bull. South. Calif. Acad. Sci.* 87:19–30.

Allen, L. G., and E. E. DeMartini, 1983. Temporal and spatial patterns of nearshore distribution and abundance of the pelagic fishes off San Onofre, Oceanside, California. *U.S. Natl. Mar. Fish. Serv. Fish. Bull.* 81(3):569–586.

Allen, L. G., and M. P. Franklin, 1988. Distribution and abundance of young-of-the-year white seabass, *Atractoscion nobilis,* in the vicinity of Long Beach Harbor, California, in 1984–1987. *Calif. Fish Game.* 74:245–248.

Allen, L. G., M. H. Horn, F. A. Edmands II, and C. A. Usui, 1983. Structure and seasonal dynamics of the fish assemblage in the Cabrillo Beach area of Los Angeles Harbor, California. *Bull. South. Calif. Acad. Sci.* 82:47–70.

Allen, M. J., 1982. Functional structure of soft-bottom fish communities of the southern California shelf. Ph.D. Dissertation, Univ. of California, San Diego. 577pp.

Allen, M. J., and R. Voglin, 1976. Regional and local variation of bottom fish and invertebrate populations. In: W. Bascom, ed. *Annual Report.* South. Calif. Coastal Water Res. Proj., El Segundo, CA. pp. 217–221.

Allen, M. J., J. B. Isaacs, and R. M. Voglin, 1975. Hook-and-line survey of demersal fishes in Santa Monica Bay. Tech. Mem. 222. South. Calif. Coastal Water Res. Proj., El Segundo, CA. 23pp.

Alton, M. S., 1972. Characteristics of the demersal fish fauna inhabiting the outer continental shelf and slope off the northern Oregon coast. In: A. T. Pruter, and D. L. Alverson, eds. *The Columbia River Estuary and Adjacent Ocean Waters.* Univ. of Washington Press, Seattle. pp. 583–634.

Ambrose, R. F., and S. L. Swarbrick, 1989. Comparison of fish assemblages on artificial and natural reefs off the coast of southern California. *Bull. Mar. Sci.* 44:718–733.

Anderson, D. W., F. Gress, K. F. Mais, and P. R. Kelly, 1980. Brown pelicans as anchovy stock indicators and their relationships to commercial fishing. *Calif. Coop. Oceanic Fish. Invest. Rep.* 21:54–61.

Anonymous, 1987. United States marine recreational fisheries, 1986. *Mar. Fish. Rev.* 49:166–173.

Antonelis, G. A., Jr., and M. A. Perez, 1984. Estimated annual food consumption by northern fur seals in the California Current. *Calif. Coop. Oceanic Fish. Invest. Rep.* 25:135–145.

Aron, W., and S. Collard, 1969. A study of the influence of net speed on catch. *Limnol. Oceanogr.* 14:242–249.

Atsatt, L. H., and R. R. Seapy, 1974. An analysis of sampling variability in replicated midwater trawls off southern California. *J. Exp. Mar. Biol. Ecol.* 14:261–273.

Bailey, K. M., 1981. Larval transport and recruitment of Pacific hake, *Merluccius productus. Mar. Ecol. Prog. Ser.* 6:1–9.

Bailey, K. M., 1982. The early life history of Pacific hake, *Merluccius productus. U.S. Natl. Mar. Fish. Serv. Fish. Bull.* 80:589–598.

Bailey, K. M., and D. Ainley, 1982. The dynamics of California sea lion predation on Pacific hake. *Fish. Res.* 1:163–176.

Bailey, K. M., and L. S. Incze, 1985. El Niño and the early life history and recruitment of fishes in temperate marine waters. In: W. S. Wooster, and D. L. Fluharty, eds. *El Niño North: Niño Effects in the Eastern Subarctic Pacific Ocean.* Washington Sea Grant Program, Seattle. pp. 143–165.

Bailey, K. M., R. C. Francis, and P. R. Stevens, 1982. The life history and fishery of Pacific whiting, *Merluccius productus. Calif. Coop. Oceanic Fish. Invest. Rep.* 23:81–98.

Bailey, T. G., and B. H. Robison, 1986. Food availability as a selective factor on the chemical compositions of midwater fishes in the eastern North Pacific. *Mar. Biol.* 91:131–141.

Bakun, A., 1984. Report of the working group on environmental studies and monitoring. In: J. Csirke, and G. D. Sharp, eds. *Reports of the Expert Consultation to Examine Changes in Abundance and Species Composition of Neritic Fish Resources.* FAO Fish. Rep./ FAO Inf. Pesca, No. 291, Vol. 1. pp. 41–54.

Bakun, A., 1986. Definition of environmental variability affecting biological processes in large marine ecosystems. In: K. Sherman, and L. M. Alexander, eds. *Variability and Management of Large Marine Ecosystems.* Am. Assoc. Adv. Sci., Selected Symp. 99. Westview Press, Boulder, CO. pp. 80–108.

Bakun, A., and R. H. Parrish, 1982. Turbulence, transport, and pelagic fish in the California and Peru Current systems. *Calif. Coop. Oceanic Fish. Invest. Rep.* 23:99–112.

Bakun, A., J. Beyer, D. Pauly, J. G. Pope, and G. D. Sharp, 1982. Ocean sciences in relation to living resources. *Can. J. Fish. Aquat. Sci.* 39:1059–1070.

Barnett, M. A., 1975. Studies on the patterns of distribution of mesopelagic fish faunal assemblages in the Central Pacific and their temporal persistence in the gyres. Ph.D. Dissertation, Univ. of California, San Diego. 145pp.

Barnett, M. A., A. E. Jahn, P. D. Sertic, and W. Watson, 1984. Distribution of ichthyoplankton off San Onofre, California, and methods for sampling very shallow coastal waters. *U.S. Natl. Mar. Fish. Serv. Fish. Bull.* 82:97–111.

Baxter, J. L., 1960. A study of the yellowtail, *Seriola dorsalis* (Gill). *Calif. Dept. Fish Game Fish Bull.* No. 110. 91pp.

Baxter, J. L., 1982. The role of the marine research committee and CalCOFI. *Calif. Coop. Oceanic Fish. Invest. Rep.* 23:35–42.

Beamish, R. J., and G. A. McFarlane, 1983. The forgotten requirement for age validation in fisheries biology. *Trans. Am. Fish. Soc.* 112: 735–743.

Bedford, D. W., and F. B. Hagerman, 1983. The billfish fishery resource of the California Current. *Calif. Coop. Oceanic Fish. Invest. Rep.* 24: 70–78.

Behrents, K. C., 1987. The influence of shelter availability on recruitment and early juvenile survivorship of *Lythypnus dalli* Gilbert (Pisces: Gobiidae). *J. Exp. Mar. Biol. Ecol.* 107(1):45–59.

Bennett, J. T., G. W. Boehlert, and K. K. Turekian, 1982. Confirmation of longevity in *Sebastes diploproa* (Pisces: Scorpaenidae) from $^{210}Pb/^{226}Ra$ measurements in otoliths. *Mar. Biol.* 71:209–215.

Bernal, P. A., 1981. A review of the low-frequency response of the pelagic ecosystem in the California Current. *Calif. Coop. Oceanic Fish. Invest. Rep.* 22:49–62.

Bernard, H. J., J. B. Hedgepeth, and S. B. Reilly, 1985. Stomach contents of albacore, skipjack, and bonito caught off southern California during summer 1983. *Calif. Coop. Oceanic Fish. Invest. Rep.* 26:175–182.

Bernstein, B. B., and N. Jung, 1979. Selective pressures and coevolution in a kelp canopy community in southern California. *Ecol. Monogr.* 49:335–355.

Berry, F. H., and H. C. Perkins, 1966. Survey of pelagic fishes in the California Current area. *U.S. Natl. Mar. Fish. Serv. Fish. Bull.* 65:625–682.

Blaxter, J. H. S., and J. R. Hunter, 1982. The biology of the clupeoid fishes. In: J. H. S. Blaxter, F. S. Russell, and M. Younge, eds. *Advances in Marine Biology*. Vol. 20. Academic Press, London. pp. 1–223.

Bodkin, J. L., 1988. Effects of kelp forest removal on associated fish assemblages in central California. *J. Exp. Mar. Biol. Ecol.* 117:227–238.

Boehlert, G. W., 1977. Timing of the surface-to-benthic migration in juvenile rockfish, *Sebastes diploproa*, off southern California. *U.S. Natl. Mar. Fish. Serv. Fish. Bull.* 75:887–890.

Boehlert, G. W., and R. F. Kappenman, 1980. Variation of growth with latitude in two species of rockfish (*Sebastes pinniger* and *S. diploproa*) from the northeast Pacific Ocean. *Mar. Ecol. Prog. Ser.* 3:1–10.

Boehlert, G. W., and M. Yoklavich, 1984. Reproduction, embryonic energetics, and the maternal–fetal relationship in the viviparous genus *Sebastes* (Pisces: Scorpaenidae). *Biol. Bull. (Woods Hole)*. 167:354–370.

Bohnsack, J. A., 1989. Are high densities of fishes at artificial reefs the result of habitat limitation or behavioral preference? *Bull. Mar. Sci.* 44: 631–645.

Bohnsack, J. A., and D. L. Sutherland, 1985. Artificial reef research: A review with recommendations for future priorities. *Bull. Mar. Sci.* 37:11–39.

Bouvier, L. S., 1987. The cryptic fish fauna of a shallow rock reef habitat, Santa Catalina Island, California. Master's Thesis, California State Univ., Northridge, CA. 81pp.

Bray, R. N., 1980. Influence of water currents and zooplankton densities on daily foraging movements of blacksmith, *Chromis punctipinnis*, a planktivorous reef fish. *U.S. Natl. Mar. Fish. Serv. Fish. Bull.* 78(4):829–841.

Bray, R. N., and M. A. Hixon, 1978. Night-shocker: Predatory behavior of the Pacific electric ray (*Torpedo californica*). *Science*. 200:333–334.

Bray, R. N., A. C. Miller, and G. G. Geesey, 1981. The fish connection: A trophic link between planktonic and rocky reef communities? *Science*. 214:204–205.

Brewer, G. D., and G. S. Kleppel, 1986. Diel vertical distribution of fish larvae and their prey in nearshore waters of southern California. *Mar. Ecol. Prog. Ser.* 27:217–226.

Brewer, G. D., and P. E. Smith, 1982. Northern anchovy and Pacific sardine spawning off southern California during 1978–1980: Preliminary observations on the importance of the nearshore coastal region. *Calif. Coop. Oceanic Fish. Invest. Rep.* 23:160–171.

Brewer, G. D., G. S. Kleppel, and M. Dempsey, 1984. Apparent predation on ichthyoplankton by zooplankton and fishes in nearshore waters of southern California. *Mar. Biol.* 80(1):17–28.

Briggs, J. C., 1974. *Marine Zoogeography*. McGraw-Hill Book Co., New York. 475pp.

Brown, D. A., R. W. Gossett, G. P. Hershelman, C. F. Ward, A. M. Westcott, and J. N. Cross, 1986. Municipal wastewater contamination in the Southern California Bight: Part 1. Metal and organic contaminants in sediments and organisms. *Mar. Environ. Res.* 18(4):291–310.

Brown, D. W., 1974. Hydrography and midwater fishes of three contiguous oceanic areas off Santa Barbara, California. *Los Angeles Cty. Mus. Contrib. Sci.* 261:1–30.

Cailliet, G. M., 1972. The study of feeding habits of two marine fishes in relation to plankton ecology. *Trans. Am. Microsc. Soc.* 91:88–89.

Cailliet, G. M., and D. W. Bedford, 1983. The biology of three pelagic sharks from California waters, and their emerging fisheries: A review. *Calif. Coop. Oceanic Fish. Invest. Rep.* 24:57–69.

Carey, F. G., and B. H. Robison, 1981. Daily patterns in the activities of swordfish, *Xiphias gladius,* observed by acoustic telemetry. *U.S. Natl. Mar. Fish. Serv. Fish. Bull.* 79(2):277–292.

Carlisle, J. G., Jr., 1969. Results of a six-year trawl study in an area of heavy waste discharge: Santa Monica Bay, California. *Calif. Fish Game.* 55(1):26–46.

Carlisle, J. G., Jr., J. W. Schott, and N. J. Abramson, 1960. The barred surfperch (*Amphisticus argenteus* Agassiz) in southern California. *Calif. Dept. Fish Game Fish Bull.* No. 109. 79pp.

Carlisle, J. G., Jr., C. H. Turner, and E. E. Ebert, 1964. Artificial habitat in the marine environment. *Calif. Dept. Fish Game Fish Bull.* No. 124. 93pp.

Carr, M. H., 1989. Effects of macroalgal assemblages on the recruitment of temperate reef fishes. *J. Exp. Mar. Biol. Ecol.* 126:59–76.

Chamberlain, D. W., 1974. A checklist of fishes from Los Angeles–Long Beach Harbors. In: D. Soule, and M. Oguri, eds. *Marine Studies of San Pedro Bay. Part IV. Environmental Field Investigations.* Allan Hancock Found. Publ. USC-SG–1–74. pp. 43–78.

Chelton, D. B., P. A. Bernal, and J. A. McGowan, 1982. Large-scale interannual physical and biological interactions in the California Current. *J. Mar. Res.* 40(4):1095–1125.

Childress, J. J., and M. H. Nygaard, 1973. The chemical composition of midwater fishes as a function of depth of occurrence off southern California. *Deep-Sea Res.* 20:1093–1109.

Childress, J. J., and G. N. Somero, 1979. Depth-related enzymic activities in muscle, brain and heart of deep-living pelagic marine teleosts. *Mar. Biol.* 52:273–283.

Childress, J. J., S. M. Taylor, G. M. Cailliet, and M. H. Price, 1980. Patterns of growth, energy utilization and reproduction in some meso- and bathypelagic fishes off southern California. *Mar. Biol.* 61:27–40.

Clark, N. E., T. J. Blasing, and H. C. Fritts, 1975. Influence of interannual climatic fluctuations on biological systems. *Nature.* 256:302–305.

Clarke, T. A., 1970. Territorial behavior and population dynamics of a pomacentrid fish, the garibaldi, *Hypsypops rubicunda*. *Ecol. Monogr.* 40:189–212.

Clarke, T. A., 1973. Some aspects of the ecology of lanternfishes (Myctophidae) in the Pacific Ocean near Hawaii. *U.S. Natl. Mar. Fish. Serv. Fish. Bull.* 71:401–434.

Clarke, T. A., A. O. Flechsig, and R. W. Grigg, 1967. Ecological studies during project Sealab II. *Science.* 157:1381–1389.

Congleton, J. L., 1980. Observations on the responses of some southern California tidepool fishes to nocturnal hypoxic stress. *Comp. Biochem. Physiol.* 66A:719–722.

Connally, D. W., R. E. Schlotterbeck, and D. A. Rundstrom, 1982. Marine ichthyoplankton entrainment studies, Vol. II. Analysis and interpretation, August 1979–September 1980. MBC Applied Environmental Sciences, Costa Mesa, CA. 62pp.

Connell, J. H., 1980. Diversity and the coevolution of competitors, or the ghost of competition past. *Oikos.* 35:131–138.

Cowen, R. K., 1983. The effect of sheephead (*Semicossyphus pulcher*) predation on red sea urchin (*Strongylocentrotus franciscanus*) populations: An experimental analysis. *Oecologia (Berl.).* 58:249–255.

Cowen, R. K., 1985. Large-scale pattern of recruitment by the labrid, *Semicossyphus pulcher*: Causes and implications. *J. Mar. Res.* 43:719–742.

Coyer, J. A., 1979. The invertebrate assemblage associated with *Macrocystis pyrifera* and its utilization as a food source by kelp forest fishes. Ph.D. Dissertation, Univ. of Southern California, Los Angeles. 346pp.

Croker, R. S., 1982. An iconoclast's view of California fisheries research, 1929–1962. *Calif. Coop. Oceanic Fish. Invest. Rep.* 23:29–34.

Crooke, S. J., 1983. Yellowtail, *Seriola lalandei* Valenciennes. *Calif. Coop. Oceanic Fish. Invest. Rep.* 24:84–87.

Cross, J. N., 1982. Intertidal fishes of southern California. In: W. Bascom, ed. *Biennial Report 1981–1982.* South. Calif. Coastal Water Res. Proj., Long Beach, CA. pp. 111–117.

Cross, J. N., 1985. Fin erosion among fishes collected near a southern California municipal wastewater outfall (1971–82). *U.S. Natl. Mar. Fish. Serv. Fish. Bull.* 83(2):195–206.

Cross, J. N., 1986. Epidermal tumors in *Microstomus pacificus* (Pleuronectidae) collected near a municipal wastewater outfall in the coastal waters off Los Angeles (1971–1983). *Calif. Fish Game.* 72(2):68–77.

Cross, J. N., 1987. Demersal fishes of the upper continental slope off southern California. *Calif. Coop. Oceanic Fish. Invest. Rep.* 28:155–167.

Cross, J. N., 1988. Fin erosion and epidermal tumors in demersal fish from southern California. In: D. A. Wolfe, and T. P. O'Connor, eds. *Ocean Processes in Marine Pollution. Vol. 5. Urban Wastes in the Coastal Marine Environments.* Krieger Publishing Co., Malabar, FL. pp. 57–64.

Cross, J. N., and J. E. Hose, 1988. Evidence of impaired reproduction in white croaker (*Genyonemus lineatus*) from contaminated areas off southern California. *Mar. Environ. Res.* 24:185–188.

Cross, J. N., J. N. Roney, and G. S. Kleppel, 1985. Fish food habits along a pollution gradient. *Calif. Fish Game.* 71(1):28–39.

Cury, P., 1988. Pressions sélectives et nouveautés évolutives: Une hypothèse pour comprendre certains aspects des fluctuations à long terme des poissons pélagiques côtiers. *Can. J. Fish. Aquat. Sci.* 45:1099–1107.

Cury, P., and C. Roy, 1989. Optimal environmental window and pelagic fish recruitment success in upwelling areas. *Can. J. Fish. Aquat. Sci.* 46:670–680.

Cushing, D. H., 1971. The dependence of recruitment on parent stock in different groups of fishes. *J. Cons. Cons. Int. Explor. Mer.* 33:340–362.

Cushing, D. H., 1975. *Marine Ecology and Fisheries.* Cambridge Univ. Press, Cambridge, MA. 278pp.

Cushing, D. H., and J. G. K. Harris, 1973. Stock and recruitment and the problem of density dependence. *Rapp. P.-V. Réun. Cons. Int. Explor. Mer.* 164:142–155.

Dailey, M. D., L. A. Jensen, and B. W. Hill, 1981. Larval anisakine roundworms of marine fishes from southern and central California, with comments on public health significance. *Calif. Fish Game.* 67:240–245.

Davis, G. E., and T. W. Anderson, 1989. Population estimates of four kelp forest fishes and an evaluation of three *in situ* assessment techniques. *Bull. Mar. Sci.* 44:1138–1151.

Davis, N., G. R. VanBlaricom, and P. K. Dayton, 1982. Man-made structures on marine sediments: Effects on adjacent benthic communities. *Mar. Biol.* 70(3):295–303.

Dayton, P. K., and M. J. Tegner, 1984. Catastrophic storms, El Niño, and patch stability in a southern California kelp community. *Science.* 224:283–285.

DeMartini, E. E., 1981. The spring–summer ichthyofauna of surfgrass (*Phyllospadix*) meadows near San Diego, California. *Bull. South. Calif. Acad. Sci.* 80:81–90.

DeMartini, E. E., and L. G. Allen, 1984. Diel variation in catch parameters for fishes sampled by a 7.6-m otter trawl in southern California coastal waters. *Calif. Coop. Oceanic Fish. Invest. Rep.* 25:119–134.

DeMartini, E. E., and R. K. Fountain, 1981. Ovarian cycling frequency and batch fecundity in the queenfish, *Seriphus politus*: Attributes representative of serial spawning fishes. *U.S. Natl. Mar. Fish. Serv. Fish. Bull.* 79:547–560.

DeMartini, E. E., and D. Roberts, 1982. An empirical test of biases in the rapid visual technique for species–time censuses of reef fish assemblages. *Mar. Biol.* 70:129–134.

DeMartini, E. E., L. G. Allen, R. K. Fountain, and D. Roberts, 1985. Diel and depth variations in the sex-specific abundance, size composition, and food habits of queenfish, *Seriphus politus*

(Sciaenidae). *U.S. Natl. Mar. Fish. Serv. Fish. Bull.* 83:171–185.

DeMartini, E. E., D. A. Roberts, and T. W. Anderson, 1989. Contrasting patterns of fish density and abundance at an artificial rock reef and a cobble-bottom kelp forest. *Bull. Mar. Sci.* 44:881–892.

DeWees, C. M., and D. W. Gotshall, 1974. An experimental artificial reef in Humboldt Bay, California. *Calif. Fish Game.* 60:109–127.

DeWitt, F. A., Jr., 1972. Bathymetric distributions of two common deep-sea fishes, *Cyclothone acclinidens* and *C. signata*, off southern California. *Copeia.* 1972:88–96.

Douglas, E. L., W. A. Friedl, and G. V. Pickwell, 1976. Fishes in oxygen-minimum zones: Blood oxygenation characteristics. *Science.* 191:957–959.

Ebeling, A. W., 1962. Melamphaidae. I. Systematics and zoogeography of the species in the bathypelagic fish genus *Melamphaes* Gunther. *Dana-Rep.* 58:1–164.

Ebeling, A. W., 1967. Zoogeography of tropical deep-sea animals. *Stud. Trop. Oceanogr. (Miami).* 5:593–613.

Ebeling, A. W., and R. N. Bray, 1976. Day versus night activity of reef fishes in a kelp forest off Santa Barbara, California. *U.S. Natl. Mar. Fish. Serv. Fish. Bull.* 74(4):703–717.

Ebeling, A. W., and G. M. Cailliet, 1974. Mouth size and predator strategy of midwater fishes. *Deep-Sea Res.* 21:959–968.

Ebeling, A. W., and D. R. Laur, 1986. Foraging in surfperches: Resource partitioning or individualistic responses? *Environ. Biol. Fishes.* 16: 123–133.

Ebeling, A. W., R. M. Ibara, R. J. Lavenburg, and F. J. Rohlf, 1970. Ecological groups of deep-sea animals off southern California. *Bull. Los Ang. Cty. Mus. Nat. Hist. Sci.* 6:1–43.

Ebeling, A. W., R. J. Larson, W. S. Alevizon, and R. N. Bray, 1980a. Annual variability of reef-fish assemblages in kelp forests off Santa Barbara, California. *U.S. Natl. Mar. Fish. Serv. Fish. Bull.* 78(2):361–377.

Ebeling, A. W., R. J. Larson, and W. S. Alevizon, 1980b. Habitat groups and island–mainland distribution of kelp-bed fishes off Santa Barbara, California. In: D. M. Powers, ed. *The California Islands, Proceedings of a Multidisciplinary Symposium.* Santa Barbara Museum of Natural History, pp. 403–431.

Ebeling, A. W., D. R. Laur, and R. J. Rowley, 1985. Severe storm disturbances and reversal of community structure in a southern California kelp forest. *Mar. Biol.* 84:287–294.

Echeverria, T. W., 1987. Thirty-four species of California rockfishes: Maturity and seasonality of reproduction. *U.S. Natl. Mar. Fish. Serv. Fish. Bull.* 85:229–250.

Emerson, C. W., 1989. Wind stress limitation of benthic secondary production in shallow, soft-sediment communities. *Mar. Ecol. Prog. Ser.* 53:65–77.

Emery, K. O., 1960. *The Sea Off Southern California.* John Wiley & Sons, New York. 366pp.

Eppley, R. W., and B. J. Peterson, 1979. Particulate organic matter flux and planktonic new production in the deep ocean. *Nature.* 282:677–680.

Eppley, R. W., E. H. Renger, and W. G. Harrison, 1979. Nitrate and phytoplankton production in southern California coastal waters. *Limnol. Oceanogr.* 24:483–494.

Eschmeyer, W. N., E. S. Herald, and H. Hammann, eds., 1983. *A Field Guide to Pacific Coast Fishes of North America.* Houghton Mifflin Co., Boston. 336pp.

Farquhar, B. B., 1977. Biological sound scattering in the oceans: A review. In: N. R. Andersen, and B. J. Zahuranec, eds. *Oceanic Sound Scattering Prediction.* Plenum Press, New York. pp. 493–527.

Fay, R. C., and J. A. Vallee, 1978. Some biological components of marine harbors and bays in southern California. In: *The Urban Harbor Environment.* Tech. Rep. No. 1, South. Calif. Ocean Studies Consortium, Long Beach, CA. pp. 61–71.

Feder, H. M., C. H. Turner, and C. Limbaugh, 1974. Observations on fishes associated with kelp beds in southern California. *Calif. Dept. Fish Game Fish Bull.* No. 160. 144pp.

Fiedler, P. C., 1983. Satellite remote sensing of the habitat of spawning anchovy in the Southern California Bight. *Calif. Coop. Oceanic Fish. Invest. Rep.* 24:202–209.

Fiedler, P. C., 1986. Offshore entrainment of anchovy spawning habitat, eggs, and larvae by a displaced eddy in 1985. *Calif. Coop. Oceanic Fish. Invest. Rep.* 27:144–152.

Fiedler, P. C., and H. J. Bernard, 1987. Tuna aggregation and feeding near fronts observed in satellite imagery. *Cont. Shelf Res.* 7:871–881.

Fiedler, P. C., R. D. Methot, and R. P. Hewitt, 1986. Effects of California El Niño 1982–1984 on the northern anchovy. *J. Mar. Res.* 44(2): 317–338.

Fitch, J. E., and R. J. Lavenberg, 1975. *Tidepool and Nearshore Fishes of California*. Univ. of California Press, Berkeley, CA. 156pp.

Fleming, E. R., 1983. Factors influencing management of offshore fisheries resources. In: R. H. Stoud, ed. *Marine Recreational Fisheries*. Vol. 8, Sport Fishing Institute, Washington, D.C. pp. 149–160.

Frey, H. W., ed., 1971. California's living marine resources and their utilization. *Calif. Dept. Fish Game*. 148pp.

Fry, D. H., Jr., 1936. A description of the eggs and larvae of the Pacific mackerel. *Calif. Fish Game*. 22:28–29.

Gartner, J. V., Jr., W. J. Conely, and T. L. Hopkins, 1989. Escapement by fishes from midwater trawls: A case study using lanternfishes (Pisces: Myctophidae). *U.S. Natl. Mar. Fish. Serv. Fish. Bull.* 87:213–222.

Genin, A., and G. W. Boehlert, 1985. Dynamics of temperature and chlorophyll structures above a seamount: An oceanic experiment. *J. Mar. Res.* 43:907–924.

Genin, A., L. Haury, and P. Greenblatt, 1988. Interactions of migrating zooplankton with shallow topography: Predation by rockfishes and intensification of patchiness. *Deep-Sea Res.* 35(2A):151–175.

Glass, W. W., and C. S. Wardle, 1989. Comparison of the reactions of fish to a trawl gear, at high and low light intensities. *Fish. Res.* 7:249–266.

Gotshall, D. W., 1987. The use of baited stations by divers to obtain fish relative abundance data. *Calif. Fish Game*. 73:214–229.

Graham, J. B., 1970. Temperature sensitivity of two species of intertidal fishes. *Copeia*. 1970: 49–56.

Grant, J. J., K. C. Wilson, A. Grover, and H. A. Togstad, 1982. Early development of Pendleton artificial reef. *Mar. Fish. Rev.* 44(6–7):53–60.

Greenlaw, C. F., and W. G. Pearcy, 1985. Acoustical patchiness of mesopelagic organisms. *J. Mar. Res.* 43:163–178.

Grove, R. S., 1982. Artificial reefs as a resource management option for siting coastal power stations in southern California. *Mar. Fish. Rev.* 44(6–7):24–27.

Gruber, D., E. H. Ahlstrom, and M. M. Mullin, 1982. Distribution of ichthyoplankton in the Southern California Bight. *Calif. Coop. Oceanic Fish. Invest. Rep.* 23:172–179.

Gunderson, D., 1984. The great widow rockfish hunt of 1980–82. *N. Am. J. Fish. Manage.* 4:465–468.

Gunderson, D. R., and P. H. Dygert, 1988. Reproductive effort as a predictor of natural mortality rate. *J. Cons. Cons. Int. Explor. Mer.* 44:200–209.

Hagerman, F. B., 1952. The biology of Dover sole, *Microstomus pacificus* (Lockington). *Calif. Dept. Fish Game Fish Bull.* No. 85. 48pp.

Haldorson, L., and M. Moser, 1979. Geographic patterns of prey utilization in two species of surfperch (Embiotocidae). *Copeia*. 1979:567–572.

Hartmann, A. R., 1987. Movement of scorpionfishes (Scorpaenidae: *Sebastes* and *Scorpaena*) in the Southern California Bight. *Calif. Fish Game*. 73:68–79.

Hartney, K. B., 1989. The foraging ecology of two sympatric gobiid fishes: Importance of behavior in prey type selection. *Environ. Biol. Fishes*. 26:105–118.

Hedgecock, D., 1986. Recognizing subpopulations in California's mixed pelagic fish stocks. In: D. Hedgecock, ed. *Identifying Fish Subpopulations*. Calif. Sea Grant Program Publ. Rep. No. T-CSGCP-013, La Jolla. pp. 26–30.

Helvey, M., 1982. First observations of courtship behavior in rockfish, genus *Sebastes*. *Copeia*. 1982:763–770.

Helvey, M., S. J. Crooke, and P. A. Milone, 1987. Marine recreational fishing and associated state-federal research in California, Hawaii, and the Pacific Island Territories. *Mar. Fish. Rev.* 49(2):8–14.

Herbinson, K. T., 1981. 316(b) fish impingement inventory. Res. Develop. Ser. Southern California Edison, Rosemead, CA. 178pp.

Hewitt, R. P., 1981. The value of pattern in the distribution of young fish. *Rapp. P.-V. Réun. Cons. Int. Explor. Mer.* 178:229–236.

Hewitt, R. P., and G. D. Brewer, 1983. Nearshore production of young anchovy. *Calif. Coop. Oceanic Fish. Invest. Rep.* 24:235–244.

Hewitt, R. P., P. E. Smith, and J. C. Brown, 1976. Development and use of sonar mapping for pelagic stock assessment in the California Current area. *U.S. Natl. Mar. Fish. Serv. Fish. Bull.* 74:281–300.

Hewitt, R. P., G. H. Theilacker, and N. C. H. Lo, 1985. Causes of mortality in young jack mackerel. *Mar. Ecol. Prog. Ser.* 26:1–10.

Hixon, M. A., 1980. Competitive interactions between California reef fishes of the genus *Embiotoca. Ecology.* 61:918–931.

Hobson, E. S., and J. R. Chess, 1976. Trophic interactions among fishes and zooplankters nearshore at Santa Catalina Island, California. *U.S. Natl. Mar. Fish. Serv. Fish. Bull.* 74(3):567–598.

Hobson, E. S., W. N. McFarland, and J. R. Chess, 1981. Crepuscular and nocturnal activities of Californian nearshore fishes, with consideration of their scotopic visual pigments and the photic environment. *U.S. Natl. Mar. Fish. Serv. Fish. Bull.* 79(1):1–30.

Holbrook, S. J., and R. J. Schmitt, 1984. Experimental analysis of patch selection by foraging black surfperch (*Embiotica jacksoni* Agazzi). *J. Exp. Mar. Biol. Ecol.* 79:39–64.

Holbrook, S. J., and R. J. Schmitt, 1986. Food acquisition by competing surfperch on a patchy environmental gradient. *Environ. Biol. Fishes.* 16:135–146.

Holbrook, S. J., and R. J. Schmitt, 1988a. The combined effects of predation risk and food reward on patch selection. *Ecology.* 69:125–134.

Holbrook, S. J., and R. J. Schmitt, 1988b. Effects of predation risk on foraging behavior: Mechanisms altering patch choice. *J. Exp. Mar. Biol. Ecol.* 121:151–163.

Holton, A. A., 1969. Feeding behavior of a vertically migrating lanternfish. *Pacific Sci.* 23:325–331.

Holts, D., 1985. Recreational albacore, *Thunnus alalunga,* fishery by U.S. West Coast commercial passenger fishing vessels. *Mar. Fish. Rev.* 47(3):48–53.

Hopkins, T. L., and R. C. Baird, 1977. Aspects of the feeding ecology of oceanic midwater fishes. In: N. R. Andersen, and B. J. Zahuranec, eds. *Oceanic Sound Scattering Prediction.* Plenum Press, New York. pp. 325–360.

Horn, M. H., 1974. Fishes. In: M. D. Dailey, B. Hill, and N. Lansing, eds. *A Summary of Knowledge of the Southern California Coastal Zone and Offshore Areas. Vol. II, Biological Environment.* Prepared for Div. Mar. Minerals, Bur. Land Manage., U.S. Dept. of Interior, Contract No. 08550-CT4–1. South. Calif. Ocean Studies Consortium, Long Beach, CA. pp. 1–11 to 11–124.

Horn, M. H., 1980. Diversity and ecological roles of noncommercial fishes in California marine habitats. *Calif. Coop. Oceanic Fish. Invest. Rep.* 21:37–47.

Horn, M. H., and L. G. Allen, 1976. Number of species and faunal resemblance of marine fishes in California bays and estuaries. *Bull. South. Calif. Acad. Sci.* 75:159–170.

Horn, M. H., and L. G. Allen, 1978. A distributional analysis of California coastal marine fishes. *J. Biogeogr.* 5:23–42.

Horn, M. H., and L. G. Allen, 1981a. Ecology of fishes in Upper Newport Bay, California: Seasonal dynamics and community structure. *Calif. Dept. Fish Game, Mar. Res. Tech. Rep.* No. 45, 102pp.

Horn, M. H., and L. G. Allen, 1981b. A review and synthesis of ichthyofaunal studies in the vicinity of Los Angeles and Long Beach Harbors, Los Angeles County, California. Final Rep. to U.S. Fish Wildl. Serv., Ecolog. Serv., Laguna Niguel, CA, 96pp.

Horn, M. H., and L. G. Allen, 1985. Fish community ecology in southern California bays and estuaries. In: A. Yáñez-Aranciba, ed. *Fish Community Ecology in Estuaries and Coastal Lagoons: Towards an Ecosystem Integration.* DR (R) UNAM, Mexico. pp. 169–190.

Horn, M. H., and R. N. Gibson, 1988. Intertidal fishes. *Sci. Am.* 256:64–70.

Horn, M. H., and K. C. Riegle, 1981. Evaporative water loss and intertidal vertical distribution in relation to body size and morphology of stichaeoid fishes from California. *J. Exp. Mar. Biol. Ecol.* 50:273–288.

Horn, M. H., L. G. Allen, and F. D. Hagner, 1984. Ecological status of striped bass, *Morone saxatilis,* in Upper Newport Bay, California. *Calif. Fish Game.* 70:180–182.

Hose, J. E., J. N. Cross, S. G. Smith, and D. Diehl, 1987. Elevated circulating erythrocyte micronuclei in fishes from contaminated sites off southern California. *Environ. Res.* 22:167–176.

Hose, J. E., J. N. Cross, S. G. Smith, and D. Diehl, 1989. Reproductive impairment in a fish inhabiting a contaminated coastal environment off southern California. *Environ. Pollut.* 57:139–148.

Hubbs, C. L., 1948. Changes in the fish fauna of western North America correlated with changes in ocean temperature. *J. Mar. Res.* 7(3):459–482.

Hubbs, C. L., 1960. The marine vertebrates of the outer coast. *Syst. Zool.* 9:134–147.

Hubbs, C. L., 1974. Review, marine zoogeography by John C. Briggs. *Copeia*. 1974:1002–1005.

Hunter, J. R., 1968. Effects of light on schooling and feeding of jack mackerel, *Trachurus symmetricus*. *J. Fish. Res. Board Can.* 25:393–407.

Hunter, J. R., 1972. Swimming and feeding behavior of larval anchovy *Engraulis mordax*. *U.S. Natl. Mar. Fish. Serv. Fish. Bull.* 70:821–838.

Hunter, J. R., 1976. Culture and growth of northern anchovy. *Engraulis mordax*, larvae. *U.S. Natl. Mar. Fish. Serv. Fish. Bull.* 74:81–88.

Hunter, J. R., 1981. Feeding ecology and predation of marine fish larvae. In: R. Lasker, ed. *Marine Fish Larvae*. Washington Sea Grant Program, Seattle, WA. pp. 33–77.

Hunter, J. R., and S. R. Goldberg, 1980. Spawning incidence and batch fecundity in northern anchovy, *Engraulis mordax*. *U.S. Natl. Mar. Fish. Serv. Fish. Bull.* 77(3):641–652.

Hunter, J. R., and C. A. Kimbrell, 1980a. Early life history of Pacific mackerel, *Scomber japonicus*. *U.S. Natl. Mar. Fish. Serv. Fish. Bull.* 78:89–101.

Hunter, J. R., and C. A. Kimbrell, 1980b. Egg cannibalism in the northern anchovy, *Engraulis mordax*. *U.S. Natl. Mar. Fish. Serv. Fish. Bull.* 78:811–816.

Hunter, J. R., and R. Leong, 1981. The spawning energetics of female northern anchovy, *Engraulis mordax*. *U.S. Natl. Mar. Fish. Serv. Fish. Bull.* 79(2):215–230.

Hunter, J. R., and R. Nicholl, 1985. Visual threshold for schooling in northern anchovy, *Engraulis mordax*. *U.S. Natl. Mar. Fish. Serv. Fish. Bull.* 83:235–242.

Hunter, J. R., and C. Sanchez, 1976. Diel changes in swim bladder inflation of the larvae of the northern anchovy, *Engraulis mordax*. *U.S. Natl. Mar. Fish. Serv. Fish. Bull.* 74:847–855.

Huppert, D. D., 1983. Forage fish management: Experience with California's anchovy fishery. In: R. H. Stoud, ed. *Marine Recreational Fisheries*. Vol. 8, Sport Fishing Institute, Washington, D.C. pp. 173–183.

Husby, D. M., and C. S. Nelson, 1982. Turbulence and vertical stability in the California Current. *Calif. Coop. Oceanic Fish. Invest. Rep.* 23:113–129.

Jessee, W. N., J. W. Carter, and A. L. Carpenter, 1985. Density estimates of five warm-temperate reef fishes associated with an artificial reef, a natural reef and a kelp forest. In: F. M. Ditri, ed. *Artificial Reefs: Marine and Freshwater Applications*. Lewis Publishing, Chelsea, MI. pp. 383–400.

Jones, R., and W. W. Henderson, 1987. The dynamics of energy transfer in marine food chains. *S. Afr. J. Mar. Sci.* 5:447–465.

Kabata, Z., and D. J. Whitaker, 1985. Parasites as a limiting factor in exploitation of Pacific whiting, *Merluccius productus*. *Mar. Fish. Rev.* 47(2):55–59.

Kingsford, M. H., and J. H. Choat, 1986. Influence of surface slicks on the distribution and onshore movement of small fish. *Mar. Biol.* 91: 161–171.

Klingbeil, R. A., 1983. Is marine recreational fisheries management possible? In: R. H. Stoud, ed. *Marine Recreational Fisheries*. Vol. 8, Sport Fishing Institute, Washington, D.C. pp. 205–216.

Knaggs, E. H., and R. H. Parrish, 1973. Maturation and growth of Pacific mackerel, *Scomber japonicus* Houttuyn. *Calif. Fish Game*. 59:114–120.

Kristjonsson, H., 1959. Introduction: Modern trends in fishing. In: H. Kristjonsson, ed. *Modern Fishing Gear of the World*. Fish. News Int., London. pp. xxv–xxxi.

Lacy, S. B., 1973. Food relationships and population ecology of the clingfish *Gobiesox rhessodon* (Smith). Master's Thesis, San Diego State Univ., San Diego. 77pp.

Lancraft, T. M., and B. H. Robison, 1980. Evidence of postcapture ingestion by midwater fishes in trawl nets. *U.S. Natl. Mar. Fish. Serv. Fish. Bull.* 77:713–715.

Larson, R. J., 1980a. Territorial behavior of the black and yellow rockfish and gopher rockfish (Scorpaenidae, *Sebastes*). *Mar Biol.* 58:111–122.

Larson, R. J., 1980b. Competition, habitat selection, and the bathymetric segregation of two rockfish (*Sebastes*) species. *Ecol. Monogr.* 50: 221–239.

Larson, R. J., and E. E. DeMartini, 1984. Abundance and vertical distribution of fishes in a cobble-bottom kelp forest off San Onofre, California. *U.S. Natl. Mar. Fish. Serv. Fish. Bull.* 82(1):37–53.

Lasker, R., 1975. Field criteria for survival of anchovy larvae: The relation between inshore chlorophyll maximum layers and successful first feeding. *U.S. Natl. Mar. Fish. Serv. Fish. Bull.* 73(3):453–462.

Lasker, R., 1978. The relation between oceanographic conditions and larval anchovy food in the California Current: Identification of factors contributing to recruitment failure. *Rapp. P.-V. Réun. Cons. Int. Explor. Mer.* 173:212–230.

Lasker, R., 1981a. Factors contributing to variable recruitment of the northern anchovy (*Engraulis mordax*) in the California Current: Contrasting years, 1975 through 1978. *Rapp. P.-V. Réun. Cons. Int. Explor. Mer.* 178:375–388.

Lasker, R., 1981b. The role of a stable ocean in larval fish survival and subsequent recruitment. In: R. Lasker, ed. *Marine Fish Larvae: Morphology, Ecology, and Relation to Fisheries.* Univ. of Washington Press, Seattle, WA. pp. 80–87.

Lasker, R., 1985. What limits clupeoid population? *Can. J. Fish. Aquat. Sci.* 42(Suppl. 1):31–38.

Lasker, R., H. M. Feder, G. H. Theilacker, and R. C. May, 1970. Feeding, growth, and survival of *Engraulis mordax* larvae reared in the laboratory. *Mar. Biol.* 5:345–353.

Laur, D. R., and A. W. Ebeling, 1983. Predator–prey relationships in surfperches. *Environ. Biol. Fishes.* 8:217–229.

Laurs, R. M., and R. J. Lynn, 1977. Seasonal migration of North Pacific albacore, *Thunnus alalunga,* into North American coastal waters: Distribution, relative abundance, and association with transition zone waters. *U.S. Natl. Mar. Fish. Serv. Fish. Bull.* 75(4):795–822.

Laurs, R. M., P. C. Fiedler, and D. R. Montgomery, 1984. Albacore tuna catch distributions relative to environmental features observed from satellites. *Deep-Sea Res.* 31(9A):1085–1099.

Lavenberg, R. J., and A. W. Ebeling, 1967. Distribution of midwater fishes among deep water basins of the southern California shelf. In: R. N. Philbrick, ed. *Proceedings, Symposium on the Biology of the California Islands.* Santa Barbara Botanic Garden, Santa Barbara. pp. 185–201.

Lavenberg, R. J., G. E. McGowen, A. E. Jahn, J. H. Petersen, and T. C. Sciarrotta, 1986. Abundance of southern California nearshore ichthyoplankton: 1974–1984. *Calif. Coop. Oceanic Fish. Invest. Rep.* 27:53–64.

Leaman, B. M., 1987. Incorporating reproductive value into Pacific Ocean perch management. In: *Proceedings, International Rockfish Symposium.* Alaska Sea Grant Rep. No. 87–2. pp. 355–368.

Leithiser, R. M., 1981. Distribution and seasonal abundance of larval fishes in a pristine southern California salt marsh. *Rapp. P.-V. Réun. Cons. Int. Explor. Mer.* 178:174–175.

Lenarz, W. H., 1972. Mesh retention of larvae of *Sardinops caerulea* and *Engraulis mordax* by plankton nets. *U.S. Natl. Mar. Fish. Serv. Fish. Bull.* 70:839–848.

Lenarz, W. H., 1973. Dependence of catch rates on size of fish larvae. *Rapp. P.-V. Réun. Cons. Int. Explor. Mer.* 164:270–275.

Leong, R. J. H., and C. P. O'Connell, 1969. A laboratory study of particulate and filter feeding of the northern anchovy (*Engraulis mordax*). *J. Fish. Res. Board Can.* 26:557–582.

Lillelund, K., and R. Lasker, 1971. Laboratory studies on predation by marine copepods on fish larvae. *U.S. Natl. Mar. Fish. Serv. Fish. Bull.* 69:655–667.

Limbaugh, C., 1964. Notes on the life history of two Californian pomacentrids: Garibaldis, *Hypsypops rubicunda* (Girard), and blacksmiths, *Chromis punctipinnis* (Cooper). *Pacific Sci.* 18: 41–50.

Lissner, A. L., and J. H. Dorsey, 1986. Deepwater biological assemblages of a hard-bottom bank-ridge complex of the southern California continental borderland. *Bull. South. Calif. Acad. Sci.* 85:87–101.

Loeb, V. J., 1979. Larval fishes in the zooplankton community of the North Pacific Central Gyre. *Mar. Biol.* 53:173–191.

Loeb, V. J., P. E. Smith, and H. G. Moser, 1983a. Geographical and seasonal patterns of larval fish species structure in the California Current area, 1975. *Calif. Coop. Oceanic Fish. Invest. Rep.* 24:132–151.

Loeb, V. J., P. E. Smith, and H. G. Moser, 1983b. Ichthyoplankton and zooplankton abundance patterns in the California Current area, 1975. *Calif. Coop. Oceanic Fish. Invest. Rep.* 24:109–131.

Loukashkin, A. S., 1970. On the diet and feeding behavior of the northern anchovy, *Engraulis mordax* (Girard). *Proc. Calif. Acad. Sci.* 37:419–458.

Love, M. S., 1980. Isolation of olive rockfish, *Sebastes serranoides,* populations off southern California. *U.S. Natl. Mar. Fish. Serv. Fish. Bull.* 77:975–983.

Love, M. S., 1981. Evidence of movements of some deepwater rockfishes (Scorpaenidae: genus *Sebastes*) off southern California. *Calif. Fish Game.* 67:246–249.

Love, M. S., and W. V. Westphal, 1981. Growth, reproduction, and food habits of olive rockfish, *Sebastes serranoides,* off central California. *U.S. Natl. Mar. Fish. Serv. Fish. Bull.* 79:533–545.

Love, M. S., B. Axell, P. A. Morris, R. Collins, and A. Brooks, 1987. Life history and fishery of the California scorpionfish, *Scorpaena guttata,* within the Southern California Bight. *U.S. Natl. Mar. Fish. Serv. Fish. Bull.* 85:99–116.

Love, M. S., G. E. McGowen, W. Westphal, R. J. Lavenberg, and L. Martin, 1984. Aspects of the life history and fishery of the white croaker, *Genyonemus lineatus* (Sciaenidae), off California. *U.S. Natl. Mar. Fish. Serv. Fish. Bull.* 82:179–198.

Love, M. S., J. S. Stephens, Jr., P. A. Morris, M. M. Singer, M. Sandhu, and T. C. Sciarrotta, 1986. Inshore soft substrata fishes in the Southern California Bight: An overview. *Calif. Coop. Oceanic Fish. Invest. Rep.* 27:84–106.

Love, M. S., W. Westphal, and R. A. Collins, 1985. Distributional patterns of fishes captured aboard commercial passenger fishing vessels along the northern Channel Islands, California. *U.S. Natl. Mar. Fish. Serv. Fish. Bull.* 83:243–251.

Lynn, R., 1984. Measuring physical–oceanographic features relevant to the migration of fishes. In: J. D. McCleave, G. P. Arnold, J. J. Dodson, and W. H. Neill, eds. *Mechanisms of Migration in Fishes.* Plenum Press, New York. pp. 471–486.

McCain, B. B., D. W. Brown, M. M. Krahn, M. S. Myers, R. C. Clark, Jr., S.-L. Chan, and D. C. Malins, 1988. Marine pollution problems, North American West Coast. *Aquat. Toxicol.* 11:143–162.

MacCall, A. D., 1979. Population estimates for the waning years of the Pacific sardine fishery. *Calif. Coop. Oceanic Fish. Invest. Rep.* 20:72–82.

MacCall, A. D., 1980. Population models for the northern anchovy (*Engraulis mordax*). *Rapp. P.-V. Réun. Cons. Int. Explor. Mer.* 177:292–306.

MacCall, A. D., 1983. Variability of pelagic fish stocks off California. In: J. Csirke, and G. D. Sharp, eds. *Reports of the Expert Consultation to Examine Changes in Abundance and Species Composition of Neritic Fish Resources.* FAO Fish. Rep./FAO Inf. Pesca, No. 291, Vol. 2. pp. 101–112.

MacCall, A. D., 1984. Report of the working group on resources study and monitoring. In: J. Csirke, and G. D. Sharp, eds. *Reports of the Expert Consultation to Examine Changes in Abundance and Species Composition of Neritic Fish Resources.* FAO Fish. Rep./ FAO Inf. Pesca, No. 291. Vol. 1. pp. 9–39.

MacCall, A. D., 1986. Changes in the biomass of the California Current ecosystem. In: K. Sherman, and L. M. Alexander, eds. *Variability and Management of Large Marine Ecosystems.* Am. Assoc. Adv. Sci., Selected Symp. 99. Westview Press, Boulder, CO. pp. 33–54.

MacCall, A. D., R. A. Klingbeil, and R. D. Methot, 1985. Recent increased abundance and potential productivity of Pacific mackerel (*Scomber japonicus*). *Calif. Coop. Oceanic Fish. Invest. Rep.* 26:119–129.

MacCall, A. D., G. D. Stauffer, and J.-P. Troadec, 1976. Southern California recreational and commercial marine fisheries. *Mar. Fish. Rev.* 38(1):1–32.

McCarthy, J. J., and T. E. Whitledge, 1972. Nitrogen excretion by anchovy (*Engraulis mordax* and *E. ringens*) and jack mackerel (*Trachurus symmetricus*). *U.S. Natl. Mar. Fish. Serv. Fish. Bull.* 70:395–402.

McGleneghan, K., 1973. The ecology and behavior of the California moray eel, *Gymnothorax mordax* (Ayres 1859), with descriptions of its larva and leptocephali of some other east Pacific Muraenidae. Ph.D. Dissertation, Univ. of Southern California, Los Angeles. 228pp.

McGowan, J. A., 1974. The nature of oceanic ecosystems. In: C. B. Miller, ed. *The Biology of the Oceanic Pacific.* Oregon State Univ. Press, Corvallis, OR. pp. 9–28.

McGowan, J. A., 1977. What regulates pelagic community structure in the Pacific? In: N. R. Andersen, and B. J. Zahuranec, eds. *Oceanic Sound Scattering Prediction.* Plenum Press, New York. pp. 423–443.

MacGregor, J. S., 1974. Changes in the amount and proportions of DDT and its metabolites, DDE and DDD, in the marine environment off southern California, 1949–72. *U.S. Natl. Mar. Fish. Serv. Fish. Bull.* 72:275–293.

Magnuson, J. J., and J. H. Prescott, 1966. Courtship, locomotion, feeding, and miscellaneous behavior of Pacific bonito (*Sarda chiliensis*). *Anim. Behav.* 14:54–67.

Mais, K. F., 1974. Pelagic fish surveys in the California Current. *Calif. Dept. Fish Game Fish Bull.* No. 162. 79pp.

Mais, K. F., 1977. Acoustic surveys of northern

anchovies in the California Current system, 1966–1972. *Rapp. P.-V. Réun. Cons. Int. Explor. Mer.* 177:287–295.
Mais, K. F., 1981. Age–composition changes in the anchovy, *Engraulis mordax,* central population. *Calif. Coop. Oceanic Fish. Invest. Rep.* 22:82–87.
Malins, D. C., B. B. McCain, D. W. Brown, M. S. Myers, M. M. Krahn, and S.-L. Chan, 1987. Toxic chemicals, including aromatic and chlorinated hydrocarbons and their derivatives, and liver lesions in white croaker (*Genyonemus lineatus*) from the vicinity of Los Angeles. *Environ. Sci. & Technol.* 21(8):765–770.
Marine Biological Consultants, 1984. Outer Long Beach Harbor–Queensway Bay biological baseline survey. Prepared for the Port of Long Beach, Division of Port Planning.
Marshall, N. B., 1971. *Explorations in the Life of Fishes.* Harvard Univ. Press, Cambridge, MA. 204pp.
Marshall, N. B., 1972. *The Life of Fishes.* Universe Books, New York. 402pp.
Matthews, K. R., 1985. Species similarity and movement of fishes on natural and artificial reefs in Monterey Bay, California. *Bull. Mar. Sci.* 37:252–270.
Mearns, A. J., 1978. Variations in coastal physical and biological conditions, 1969–78. In: W. Bascom, ed. *Annual Report.* South. Calif. Coastal Water Res. Proj., El Segundo, CA. pp. 147–156.
Mearns, A. J., 1979. Abundance, composition, and recruitment of nearshore fish assemblages on the southern California mainland shelf. *Calif. Coop. Oceanic Fish. Invest. Rep.* 20:111–119.
Mearns, A. J., 1988. The "odd fish": Unusual occurrences of marine life as indicators of changing ocean conditions. In: D. F. Soule, and G. S. Kleppel, eds. *Marine Organisms as Indicators.* Springer-Verlag, New York. pp. 137–176.
Mearns, A. J., and M. J. Sherwood, 1977. Distributions of neoplasms and other diseases in marine fishes relative to the discharge of waste water. *Ann. N.Y. Acad. Sci.* 298:210–224.
Mearns, A. J., J. Allen, M. D. Moore, and M. J. Sherwood, 1980. Distribution, abundance, and recruitment of soft-bottom rockfishes (Scorpaenidae: *Sebastes*) on the southern California mainland shelf. *Calif. Coop. Oceanic Fish. Invest. Rep.* 21:180–190.
Mearns, A. J., J. Allen, and M. J. Sherwood, 1973. An otter trawl survey off the Palos Verdes Peninsula and Santa Catalina Island, May–June, 1972. Tech. Mem. 204. South. Calif. Coastal Water Res. Proj., El Segundo, CA. 21pp.
Mearns, A. J., K. V. Pierce, and M. J. Sherwood, 1978. Health, abundance and diversity of bottom-fish and shellfish populations at proposed and existing offshore drilling sites in the Southern California Bight. Final Report on Contract No. 696–20. Science Applications, Inc., La Jolla, CA. 91pp.
Mearns, A. J., D. R. Young, R. J. Olsen, and H. A. Schafer, 1981. Trophic structure and the cesium–potassium ratio in pelagic ecosystems. *Calif. Coop. Oceanic Fish. Invest. Rep.* 22:99–110.
Messersmith, J. D., J. L. Baxter, and P. M. Roedel, 1969. The anchovy resources of the California Current region off California and Baja California. *Calif. Coop. Oceanic Fish. Invest. Rep.* 13:32–38.
Methot, R. D., 1982. Age-specific abundance and mortality of northern anchovy. NOAA-NMFS-SWFC Admin. Rep. LJ–82–31. Southwest Fish. Center, La Jolla, CA.
Methot, R. D., 1983a. Management of California's nearshore fishes. In: R. H. Stoud, ed. *Marine Recreational Fisheries.* Vol. 8, Sport Fishing Institute, Washington, D.C. pp. 161–172.
Methot, R. D., 1983b. Seasonal variation in survival of larval northern anchovy, *Engraulis mordax,* estimated from the age distribution of juveniles. *U.S. Natl. Mar. Fish. Serv. Fish. Bull.* 81(4):741–750.
Methot, R. D., 1986. Frame trawl for sampling pelagic juvenile fish. *Calif Coop. Oceanic Fish. Invest. Rep.* 27:267–278.
Middaugh, D. P., H. W. Kohl, and L. E. Burnett, 1983. Concurrent measurement of intertidal environmental variables and embryo survival for the California grunion, *Leuresthes tenuis,* and the Atlantic silverside, *Menida menida* (Pisces: Atherinidae). *Calif. Fish Game.* 69:89–96.
Miller, D. J., and J. J. Geibel, 1973. Summary of blue rockfish and lingcod life histories; a reef ecology study; and giant kelp, *Macrocystis pyrifera,* experiments in Monterey Bay, California. *Calif. Dept. Fish Game Fish Bull.* No. 158. 137pp.
Miller, D. J., and R. N. Lea, 1972. Guide to the coastal marine fishes of California. *Calif. Dept. Fish Game Fish Bull.* No. 157. 249pp.

Mitchell, C. T., and J. R. Hunter, 1970. Fishes associated with drifting kelp, *Macrocystis pyrifera,* off the coast of southern California and northern Baja California. *Calif. Fish Game.* 56:288–297.

Mitchell, D. F., 1953. An analysis of stomach contents of California tide pool fishes. *Am. Midl. Nat.* 49(3):862–871.

Moore, M. D., and A. J. Mearns, 1980a. Photographic survey of benthic fish and invertebrate communities in Santa Monica Bay. In: W. Bascom, ed. *Biennial Report, 1979–1980.* South. Calif. Coastal Water Res. Proj., Long Beach, CA. pp. 139–147.

Moore, M. D., and A. J. Mearns, 1980b. Changes in bottom fish population off Palos Verdes, 1970–1980. In: W. Bascom, ed. *Biennial Report, 1979–1980.* South. Calif. Coastal Water Res. Proj., Long Beach, CA. pp. 21–33.

Moser, H. G., 1967a. Seasonal histological changes in the gonads of *Sebastodes paucispinis* Ayres, an ovoviviparous teleost (Family Scorpaenidae). *J. Morphol.* 123:329–354.

Moser, H. G., 1967b. Reproduction and development of *Sebastodes paucispinis* and comparison with other rockfishes off southern California. *Copeia.* 1967(4):773–797.

Moser, H. G., 1974. Development and distribution of larvae and juveniles of *Sebastolobus* (Pisces; family Scorpaenidae). *U.S. Natl. Mar. Fish. Serv. Fish. Bull.* 72:865–884.

Moser, H. G., 1981. Morphological and functional aspects of marine fish larvae. In: R. Lasker, ed. *Marine Fish Larvae.* Washington Sea Grant Program, Seattle, WA. pp. 90–131.

Moser, H. G., and E. H. Ahlstrom, 1978. Larvae and pelagic juveniles of blackgill rockfish, *Sebastes melanostomus,* taken in midwater trawls off southern California and Baja California. *J. Fish. Res. Board Can.* 35:981–996.

Moser, H. G., P. E. Smith, and L. E. Eber, 1987. Larval fish assemblages in the California Current region, 1954–1960, a period of dynamic environmental change. *Calif. Coop. Oceanic Fish Invest. Rep.* 28:97–124.

Mullin, M. M., 1986. Spatial and temporal scales and patterns. In: R. W. Eppley, ed. *Lecture Notes on Coastal and Estuarine Studies, Vol. 15. Plankton Dynamics of the Southern California Bight.* Springer-Verlag, Berlin. pp. 216–273.

Murphy, G. I., 1966. Population biology of the Pacific sardine (*Sardinops caerulea*). *Proc. Calif. Acad. Sci.* 24:1–84.

Murphy, G. I., and R. I. Clutter, 1972. Sampling anchovy larvae with a plankton purse seine. *U.S. Natl. Mar. Fish. Serv. Fish. Bull.* 70:789–798.

Neighbors, M. A., 1988. Triacylglycerols and wax esters in the lipids of deep midwater teleost fishes of the Southern California Bight. *Mar. Biol.* 98:15–22.

Neighbors, M. A., and B. G. Nafpaktitis, 1982. Lipid compositions, water contents, swimbladder morphologies and buoyancies of nineteen species of midwater fishes (18 myctophids and 1 neoscopelid). *Mar. Biol.* 66:207–215.

Nelson, R. W., H. J. Barnett, and G. Kudo, 1985. Preservation and processing characteristics of Pacific whiting, *Merluccius productus. Mar. Fish. Rev.* 47(2):60–74.

Norris, K. S., 1963. The function of temperature in the ecology of the percoid fish *Girella nigricans* (Ayres). *Ecol. Monogr.* 33:23–62.

Norton, J., 1987. Ocean climate influences on groundfish recruitment in the California Current. In: *Proceedings, International Rockfish Symposium.* Alaska Sea Grant Rep. No. 87–2. pp. 73–98.

O'Connell, C. P., 1972. The interrelation of biting and filtering in the feeding activity of the northern anchovy (*Engraulis mordax*). *J. Fish. Res. Board Can.* 29:285–293.

O'Connell, C. P., 1981. Development of organ systems in the northern anchovy, *Engraulis mordax,* and other teleosts. *Am. Zool.* 21:429–446.

Onuf, C. P., and M. L. Quammen, 1983. Fishes in a California Coastal lagoon: Effects of major storms on distribution and abundance. *Mar. Ecol. Prog. Ser.* 12:1–14.

Pacific Fishery Management Council, 1983. Northern anchovy fishery management plan. Pacific Fishery Management Council, Portland, OR.

Parrish, R. H., and A. D. MacCall, 1978. Climatic variation and exploitation in the Pacific mackerel fishery. *Calif. Dept. Fish Game Fish Bull.* No. 167. 110pp.

Parrish, R. H., A. Bakun, D. M. Husby, and C. S. Nelson, 1983. Comparative climatology of selected environmental processes in relation to eastern boundary current pelagic fish reproduction. In: J. Csirke, and G. D. Sharp, eds. *Proceedings of the Expert Consultation to Examine Changes in Abundance and Species Composition of Neritic Fish Resources.* FAO Fish. Rep./FAO Inf. Pesca, No. 291, Vol. 3. pp. 731–777.

Parrish, R. H., C. S. Nelson, and A. Bakun, 1981.

Transport mechanisms and reproductive success of fishes in the California Current. *Biol. Oceanogr.* 1:175–203.

Partridge, L., and P. H. Harvey, 1988. The ecological context of life history evolution. *Science.* 241:1449–1455.

Patashnik, M., H. S. Groninger, Jr., H. Barnett, G. Kudo, and B. Koury, 1982. Pacific whiting, *Merluccius productus:* I. Abnormal muscle texture caused by myxosporidian-induced proteolysis. *Mar. Fish. Rev.* 44(5):1–12.

Patton, M. L., R. S. Grove, and R. F. Harman, 1985. What do natural reefs tell us about designing artificial reefs in southern California? *Bull. Mar. Sci.* 37(1):279–298.

Pauly, D., 1980. On the relationships between natural mortality, growth parameters, and mean environmental temperature in 175 fish stocks. *J. Cons. Cons. Int. Explor. Mer.* 39:175–192.

Paxton, J. R., 1967a. A distributional analysis for the lanternfishes (Family Myctophidae) of the San Pedro Basin, California. *Copeia.* 1967:422–440.

Paxton, J. R., 1967b. Biological notes on southern California lanternfishes (Family Myctophidae). *Calif. Fish Game.* 53:214–217.

Pearcy, W. G., and R. M. Laurs, 1966. Vertical migration and distribution of mesopelagic fishes off Oregon. *Deep-Sea Res.* 13:153–165.

Pearcy, W. G., E. E. Krygier, R. Mesecar, and F. Ramsey, 1977. Vertical distribution and migration of oceanic micronekton off Oregon. *Deep-Sea Res.* 24:223–245.

Pearcy, W. G., H. V. Lorz, and W. Peterson, 1979. Comparison of the feeding habits of migratory and non-migratory *Stenobrachius leucopsarus* (Myctophidae). *Mar. Biol.* 51:1–8.

Pearcy, W. G., D. L. Stein, and R. S. Carey, 1982. The deep-sea benthic fish fauna of the northeastern Pacific Ocean on Cascadia and Tufts abyssal plains and adjoining continental slopes. *Biol. Oceanogr.* 1:375–428.

Peterman, R. M., and M. J. Bradford, 1987. Wind speed and mortality rate of a marine fish, the northern anchovy (*Engraulis mordax*). *Science.* 235:354–356.

Peterman, R. M., M. J. Bradford, N. C. H. Lo, and R. D. Methot, 1988. Contribution of early life stages to interannual variability in recruitment of northern anchovy (*Engraulis mordax*). *Can. J. Fish. Aquat. Sci.* 45(1):8–16.

Petersen, J. H., A. E. Jahn, R. J. Lavenberg, G. E. McGowen, and R. S. Grove, 1986. Physical–chemical characteristics and zooplankton biomass on the continental shelf off southern California. *Calif. Coop. Oceanic Fish. Invest. Rep.* 27:36–52.

Phillips, J. B., 1958. Southerly occurrences of three northern species of fish during 1957, a warm water year on the California coast. *Calif. Fish Game.* 44:349–350.

Phillips, J. B., 1964. Life history studies on ten species of rockfish (Genus *Sebastodes*). *Calif. Dept. Fish Game Fish Bull.* No. 126. 70pp.

Pierce, K., B. McCain, and M. J. Sherwood, 1977. Histology of liver tissue from Dover sole. In: W. Bascom, ed. *Annual Report, 1977.* South. Calif. Coastal Water Res. Proj., El Segundo, CA. pp. 207–212.

Pinkas, L., M. S. Oliphant, and I. L. K. Iverson, eds. 1971. Food habits of albacore, bluefin tuna, and bonito in California waters. *Calif. Dept. Fish Game Fish Bull.* No. 152. 105pp.

Plummer, K. M., E. E. DeMartini, and D. A. Roberts, 1983. The feeding habits and distribution of juvenile-small adult California halibut (*Paralichthys californicus*) in coastal waters off northern San Diego County. *Calif. Coop. Oceanic Fish. Invest. Rep.* 24:194–210.

Prager, M. H., and A. D. MacCall, 1988. Revised estimates of historical spawning biomass of the Pacific mackerel, *Scomber japonicus. Calif. Coop. Oceanic Fish. Invest. Rep.* 29:81–90.

Puffer, H. W., S. P. Azen, and M. J. Duda, 1982. Sportfishing activity and catches in polluted coastal regions of metropolitan Los Angeles. *N. Am. J. Fish. Manage.* 2:75–79.

Quast, J. C., 1968a. Observations on the food and biology of the kelp bass, *Paralabrax clathratus,* with notes on its sport fishery at San Diego, California. In: W. J. North, and C. L. Hubbs, eds. *Utilization of Kelp-Bed Resources in Southern California.* Calif. Dept. Fish Game Fish Bull. No. 139. pp. 35–55.

Quast, J. C., 1968b. Estimates of the populations and the standing crop of fishes. In: W. J. North, and C. L. Hubbs, eds. *Utilization of Kelp-Bed Resources in Southern California.* Calif. Dept. Fish Game Fish Bull. No. 139. pp. 57–79.

Quast, J. C., 1968c. Fish fauna of the rocky inshore zone. In: W. J. North, and C. L. Hubbs, eds. *Utilization of Kelp-Bed Resources in Southern California.* Calif. Dept. Fish Game Fish Bull. No. 139. pp. 109–142.

Radovich, J., 1961. Relationships of some marine

organisms of the northeast Pacific to water temperature, particularly during 1957 through 1959. *Calif. Dept. Fish Game Fish Bull.* No. 112. 62pp.

Radovich, J., 1981. The collapse of the California sardine fishery: What have we learned? In: M. H. Glantz, and J. D. Thompson, eds. *Resource Management and Environmental Uncertainty. Advances in Environmental Science and Technology*. Vol. 11. John Wiley & Sons, New York. pp. 107–136.

Radovich, J., 1982. The collapse of the California sardine fishery: What have we learned? *Calif. Coop. Oceanic Fish. Invest. Rep.* 23:57–78.

Rainwater, C. L., 1975. An ecological study of midwater fishes in Santa Catalina Basin, off southern California, using cluster analysis. Ph.D. Dissertation, Univ. of Southern California, Los Angeles. 159pp.

Reish, D. J., D. F. Soule, and J. D. Soule, 1980. The benthic biological conditions of Los Angeles–Long Beach harbors: Results of 28 years of investigations and monitoring. *Helgol. Meeresunters.* 34:193–205.

Richkus, W. A., 1978. A quantitative study of intertidepool movement of the wooly sculpin *Clinocottus analis*. *Mar. Biol.* 49:271–284.

Rittenberg, S. C., K. O. Emery, and W. L. Orr, 1955. Regeneration of nutrients in sediments of marine basins. *Deep-Sea Res.* 3:23–45.

Robison, B. H., 1972. Distribution of the midwater fishes of the Gulf of California. *Copeia.* 1972:448–461.

Robison, B. H., and T. G. Bailey, 1982. Nutrient energy flux in midwater fishes. In: G. M. Cailliet, and C. A. Simenstad, eds. *Gutshop '81, Fish Food Habits Studies.* Washington Sea Grant Program, Seattle, WA. pp. 80–87.

Roff, D. A., 1984. The evolution of life history parameters in teleosts. *Can. J. Fish. Aquat. Sci.* 41:989–1000.

Roff, D. A., 1988. The evolution of migration and some life history parameters in marine fishes. *Environ. Biol. Fishes.* 22:133–146.

Rothschild, B. J., 1986. *Dynamics of Fish Populations.* Harvard Univ. Press, Cambridge, MA. 277pp.

Schafer, H. A., 1989. Improving southern California's coastal waters. *J. Water Pollut. Control Fed.* 61:1395–1401.

Schlotterbeck, R. E., and D. W. Connally, 1982. Vertical stratification of three nearshore southern California larval fishes (*Engraulis mordax, Genyonemus lineatus,* and *Seriphus politus*). *U.S. Natl. Mar. Fish. Serv. Fish. Bull.* 80:895–902.

Schmitt, R. J., and J. A. Coyer, 1982. The foraging ecology of sympatric marine fish in the genus *Embiotoca* (Embiotocidae): Importance of foraging behavior in prey size selection. *Oecologia (Berl.).* 55:369–378.

Schmitt, R. J., and J. A. Coyer, 1983. Variation in surfperch diets between allopatry and sympatry: Circumstantial evidence for competition. *Oecologia (Berl.).* 58:402–410.

Schmitt, R. J., and S. J. Holbrook, 1985. Patch selection by juvenile black surfperch (Embiotocidae) under variable risk: Interactive influence of food quality and structural complexity. *J. Exp. Mar. Biol. Ecol.* 79:39–64.

Schmitt, R. J., and S. W. Strand, 1982. Cooperative foraging by yellowtail, *Seriola lalandei* (Carangidae) on two species of fish. *Copeia.* 1982:714–717.

Sciarrotta, T. C., and D. R. Nelson, 1977. Diel behavior of the blue shark, *Prionace glauca,* near Santa Catalina Island, California. *U.S. Natl. Mar. Fish. Serv. Fish. Bull.* 75:519–528.

Scott, J. M., 1969. Tuna schooling terminology. *Calif. Fish Game.* 55:136–140.

Scott, J. M., and G. A. Flittner, 1972. Behavior of bluefin tuna schools in the eastern North Pacific Ocean as inferred from fishermen's logbooks, 1960–67. *U.S. Natl. Mar. Fish. Serv. Fish. Bull.* 70:915–927.

Shanks, A. L., 1983. Surface slicks associated with tidally forced internal waves may transport pelagic larvae of benthic invertebrates and fishes shoreward. *Mar. Ecol. Prog. Ser.* 13:311–315.

Shepherd, J. G., J. G. Pope, and R. D. Cousens, 1984. Variations in fish stocks and hypotheses concerning their links with climate. *Rapp. P.-V. Réun. Cons. Int. Explor. Mer.* 185:255–267.

Sherman, K., R. Lasker, W. Richards, and A. W. Kendall, Jr., 1983. Ichthyoplankton and fish recruitment in large marine ecosystems. *Mar. Fish. Rev.* 45:1–25.

Sherwood, M. J., 1980. Recruitment of nearshore demersal fishes. In: W. Bascom, ed. *Biennial Report, 1979–1980.* South. Calif. Coastal Water Res. Proj., Long Beach, CA. pp. 319–333.

Skud, B. E., 1982. Dominance in fishes: The relationship between environment and abundance. *Science.* 216:144–149.

Skud, B. E., 1983. Interactions of pelagic fishes and the relation between environmental factors and abundance. In: J. Csirke, and G. D. Sharp, eds. *Reports of the Expert Consultation to Examine Changes in Abundance and Species Composition of Neritic Fish Resources.* FAO Fish. Rep./ FAO Inf. Pesca, No. 291, Vol. 2. pp. 1133–1140.

Smetacek, V., 1984. The supply of food to the benthos. In: M. J. Fasham, ed. *Flows of Energy and Materials in Marine Ecosystems: Theory and Practice.* Plenum Press, New York. pp. 517–547.

Smith, C. L., and P. H. Young, 1966. Gonad structure and the reproductive cycle of the kelp bass, *Paralabrax clathratus* (Girard), with comments on the relationships of the serranid genus *Paralabrax. Calif. Fish Game.* 52:283–292.

Smith, K. L., Jr., and R. R. Hessler, 1974. Respiration of benthopelagic fishes: *In situ* measurements at 1230 meters. *Science.* 184:72–73.

Smith, K. L., Jr., M. B. Laver, and N. O. Brown, 1983. Sediment community oxygen consumption and nutrient exchange in the central and eastern North Pacific. *Limnol. Oceanogr.* 28: 882–898.

Smith, P. E., 1973. The mortality and dispersal of sardine eggs and larvae. *Rapp. P.-V. Réun. Cons. Int. Explor. Mer.* 164:282–292.

Smith, P. E., 1978a. Biological effects of ocean variability: Time and space scales of biological response. *Rapp. P.-V. Réun. Cons. Int. Explor. Mer.* 173:117–127.

Smith, P. E., 1978b. Precision sonar mapping for pelagic fish assessment in the California Current. *J. Cons. Int. Explor. Mer.* 38:33–40.

Smith, P. E., 1981. Fisheries of coastal pelagic schooling fish. In: R. Lasker, ed. *Marine Fish Larvae.* Washington Sea Grant Program, Seattle, WA. pp. 1–31.

Smith, P. E., 1985. A case history of an anti El Niño to El Niño transition on plankton and nekton distribution and abundances. In: W. S. Wooster, and D. L. Fluharty, eds. *El Niño North: Niño Effects in the Eastern Subarctic Pacific Ocean.* Washington Sea Grant Program, Seattle. pp. 121–142.

Smith, P. E., and R. W. Eppley, 1982. Primary production and the anchovy population in the Southern California Bight: Comparison of time-series. *Limnol. Oceanogr.* 27:1–17.

Smith, P. E., and R. P. Hewitt, 1985. Anchovy egg dispersal and mortality as inferred from close-interval observations. *Calif. Coop. Oceanic Fish. Invest. Rep.* 26:97–110.

Smith, P. E., and H. G. Moser, 1988. CalCOFI time series: An overview of fishes. *Calif. Coop. Oceanic Fish. Invest. Rep.* 29:66–77.

Soutar, A., and J. D. Isaacs, 1969. History of fish populations inferred from fish scales in anaerobic sediments off California. *Calif. Coop. Oceanic Fish. Invest. Rep.* 13:63–70.

Soutar, A., and J. D. Isaacs, 1974. Abundance of pelagic fish during the 19th and 20th centuries as recorded in anaerobic sediment off the Californias. *U.S. Natl. Mar. Fish. Serv. Fish. Bull.* 72(2):257–273.

Southern California Coastal Water Research Project, 1973a. Baited camera observations of demersal fish. Tech. Mem. 207. Prep. by South. Calif. Coastal Water Res. Proj., El Segundo, CA. 10pp.

Southern California Coastal Water Research Project, 1973b. The ecology of the Southern California Bight: Implications for water quality management. TR-104. South. Calif. Coastal Water Res. Proj., El Segundo, Ca. 531pp.

Southern California Coastal Water Research Project, 1974. Observations with a baited movie camera. In: W. Bascom, ed. *Annual Report.* South. Calif. Coastal Water Res. Proj., El Segundo, CA. pp. 45–52.

Southwood, T. R. E., 1988. Tactics, strategies and templets. *Oikos.* 52:3–18.

Spies, R. B., 1984. Benthic–pelagic coupling in sewage affected marine ecosystems. *Mar. Environ. Res.* 13(3):195–230.

Spies, R. B., J. S. Felton, and L. Dillard, 1982. Hepatic mixed function oxidases in California flatfishes are increased in contaminated environments and by oil and PCB ingestion. *Mar. Biol.* 70(2):117–127.

Spies, R. B., H. Kruger, R. Ireland, and D. W. Rice, Jr., 1989. Stable isotope ratios and contaminant concentrations in a sewage-distorted food web. *Mar. Ecol. Prog. Ser.* 54:157–170.

Squire, J. L., Jr., 1983a. Warm water and southern California recreational fishing: A brief review and prospects for 1983. *Mar. Fish Rev.* 45(4–6): 27–34.

Squire, J. L., Jr., 1983b. Abundance of pelagic resources off California, 1963–78, as measured by an airborne fish monitoring program. NOAA Tech. Rep. NMFS SSRF–762, U.S. Dept. of

Commerce, Natl. Mar. Fish. Serv., Scientific Publications Office, Seattle, WA, 75pp.

Squire, J. L., Jr., 1987. Relation of sea surface temperature changes during the 1983 El Niño to the geographical distribution of some important recreational pelagic species and their catch temperature parameters. *Mar. Fish. Rev.* 49(2):44–57.

Stauffer, G., and K. Parker, 1980. Estimate of the spawning biomass of the northern anchovy central subpopulation for the 1978–79 fishing season. *Calif. Coop. Oceanic Fish. Invest. Rep.* 21:12–16.

Stearns, S. C., 1989. Trade-offs in life-history evolution. *Funct. Ecol.* 3:259–268.

Stephens, J. S., Jr., 1978. Breakwaters and harbors as productive habitats for fish populations: Why are fishes attracted to urban complexes? In: M. D. Dailey, S. N Murray, and E. Segal, eds. *Proceedings, First Southern California Ocean Studies Consortium Symposium. The Urban Harbor Environment.* Tech. Pap. No. 1. South. Calif. Ocean Studies Consortium, California State Univ., Long Beach, CA. pp. 49–60.

Stephens, J. S., Jr., and K. E. Zerba, 1981. Factors affecting fish diversity on a temperate reef. *Environ. Biol. Fish.* 6:111–121.

Stephens, J. S., Jr., R. K. Johnson, G. S. Key, and J. E. McCosker, 1970. The comparative ecology of three sympatric species of California blennies of the genus *Hypsoblennius* Gill (Teleostomi: Blenniidae). *Ecol. Monogr.* 40:213–233.

Stephens, J. S., Jr., G. A. Jordan, P. A. Morris, M. M. Singer, and G. E. McGowen, 1986. Can we relate larval fish abundance to recruitment or population stability? A preliminary analysis of recruitment to a temperate rocky reef. *Calif Coop. Oceanic Fish. Invest. Rep.* 27:65–83.

Stephens, J. S., Jr., P. A. Morris, K. Zerba, and M. Love, 1984. Factors affecting fish diversity on a temperate reef: The fish assemblage of Palos Verdes Point, 1974–1981. *Environ. Biol. Fishes.* 11:259–275.

Stephens, J. S., Jr., C. Terry, S. Subber, and M. J. Allen, 1974. Abundance, distribution, seasonality and productivity of the fish populations in Los Angeles Harbor, 1972–73. In: D. Soule, and M. Oguri, eds. *Marine Studies of San Pedro Bay. Part IV. Environmental Field Investigations.* No. USC-SG—6–72. Allan Hancock Foundation, Univ. of Southern California, Los Angeles, CA. pp. 1–42.

Stevens, E. G., and H. G. Moser, 1982. Observations on the early life history of the mussel blenny, *Hypsoblennius jenkinsi,* and the bay blenny, *Hypsoblennius gentilis,* from specimens reared in the laboratory. *Calif. Coop. Oceanic Fish. Invest. Rep.* 23:269–275.

Stull, J. K., K. A. Dryden, and P. A. Gregory, 1987. A historical review of fisheries statistics and environmental and societal influences off the Palos Verdes Peninsula, California. *Calif. Coop. Oceanic Fish. Invest. Rep.* 28:135–154.

Sumida, B. Y., and H. G. Moser, 1980. Food and feeding of Pacific hake larvae, *Merluccius productus,* off southern California and northern Baja California. *Calif. Coop. Oceanic Fish. Invest. Rep.* 21:161–166.

Sutherland, J. P., 1974. Multiple stable points in natural communities. *Am. Nat.* 108:859–873.

Sverdrup, H. U., M. W. Johnson, and R. H. Fleming, 1942. *The Oceans, Their Physics, Chemistry, and General Biology.* Prentice-Hall, Inc., New York. 1087pp.

Taft, B. A., 1960. A statistical study of the estimation of abundance of sardine (*Sardinops caerulea*) eggs. *Limnol. Oceanogr.* 5:254–264.

Talbot, G. B., 1973. The California sardine-anchovy fisheries. *Trans. Am. Fish. Soc.* 102(1): 178–187.

Tanaka, S., 1983. Variation of pelagic fish stocks in waters around Japan. In: J. Csirke, and G. D. Sharp, eds. *Reports of the Expert Consultation to Examine Changes in Abundance and Species Composition of Neritic Fish Resources.* FAO Fish Rep./FAO Inf. Pesca, No. 291, Vol. 2. pp. 17–36.

Terry, C., and J. S. Stephens, Jr., 1976. A study of the orientation of selected embiotocid fishes to depth and shifting seasonal vertical temperature gradients. *Bull. South. Calif. Acad. Sci.* 75:170–183.

Theilacker, G. H., 1986. Starvation induced mortality of young sea-caught jack mackerel, *Trachurus symmetricus,* determined with histological and morphological methods. *U.S. Natl. Mar. Fish. Serv. Fish. Bull.* 84(1):1–17.

Theilacker, G. H., 1987. Feeding ecology and growth energetics of larval northern anchovy, *Engraulis mordax. U.S. Natl. Mar. Fish. Serv. Fish.* Bull. 85:213–228.

Theilacker, G., and R. Lasker, 1974. Laboratory studies of predation by euphausiid shrimps on fish larvae. In: J. H. S. Blaxter, ed. *The Early*

Life History of Fish. Springer-Verlag, New York. pp. 287–299.

Thompson, B. E., and G. F. Jones, 1987. Benthic macrofaunal assemblages of slope habitats in the southern California borderland. *Allan Hancock Found. Occas. Pap. (New Ser.)* No. 6. 21pp.

Thomson, C., A. Grover, and W. L. Craig, 1985. Status of the California coastal pelagic fisheries. *U.S. Natl. Mar. Fish Serv., Southwest Fish. Center, Admin. Rep.* SWR 85–1. 27pp.

Torres, J. J., B. W. Belman, and J. J. Childress, 1979. Oxygen consumption rates of midwater fishes as a function of depth of occurrence. *Deep-Sea Res.* 26A:185–197.

Turner, C. H., E. E. Ebert, and R. R. Given, 1969. Man-made reef ecology. *Calif. Dept. Fish Game Fish Bull.* No. 146. 221pp.

U.S. Department of Commerce, 1987. Fisheries of the United States, 1986. U.S. Dept. of Commerce, Natl. Oceanic Atmos. Adm., Washington, D.C. 119pp.

Utter, F., G. Milner, G. Stahl, and D. Teel, 1989. Genetic population structure of chinook salmon, *Onchorhynchus tshawytscha,* in the Pacific Northwest. *U.S. Natl. Mar. Fish. Serv. Fish. Bull.* 87:239–264.

Valentine, D. W., 1975. Skeletal anomalies in marine teleosts. In: W. E. Ribelin, and G. Migaki, eds. *The Pathology of Fishes.* Univ. of Wisconsin Press, Madison, WI. pp. 695–718.

Valentine, D. W., M. E. Soule, and P. Samollow, 1973. Asymmetry analysis in fishes: A possible statistical indicator of environmental stress. *U.S. Natl. Mar. Fish. Serv. Fish. Bull.* 71:357–370.

Valle, C. F., 1989. The homing behavior and intertidal movements of the opaleye, *Girella nigricans* (Pisces: Kyphosidae). Master's Thesis, California State Univ., Long Beach, CA. 80pp.

VanBlaricom, G. R., 1982. Experimental analyses of structural regulation in a marine sand community exposed to oceanic swell. *Ecol. Monogr.* 53:283–305.

Venrick, E. L., 1983. The marine recreational fisheries of the California Current. In: R. H. Stoud, ed. *Marine Recreational Fisheries.* Vol. 8, Sport Fishing Institute, Washington, D.C. pp. 13–24.

Vojkovich, M., and R. J. Reed, 1983. White seabass, *Atractoscion nobilis,* in California–Mexican waters: Status of the fishery. *Calif. Coop. Oceanic Fish. Invest. Rep.* 24:79–83.

Walker, B. W., 1952. A guide to the grunion. *Calif. Fish Game.* 38:409–420.

Walker, H. J., Jr., W. Watson, and A. M. Barnett, 1987. Seasonal occurrence of larval fishes in the nearshore Southern California Bight off San Onofre, California. *Estuarine Coastal Shelf Sci.* 25:91–109.

Wardle, C. S., 1983. Fish reactions to towed gears. In: A. MacDonald, and I. G. Priede, eds. *Experimental Biology at Sea.* Academic Press, New York. pp. 167–195.

Wardle, C. S., 1986. Investigating the visual reaction of fish in towed otter trawls. *Prog. Underwater Sci.* 11:95–99.

Ware, D. M., 1975. Relation between egg size, growth, and natural mortality of larval fish. *J. Fish Res. Board Can.* 32:2503–2512.

Ware, D. M., 1980. Bioenergetics of stock and recruitment. *Can. J. Fish. Aquat. Sci.* 37:1012–1024.

Warner, R. R., and P. L. Chesson, 1985. Coexistence mediated by recruitment fluctuations: A field guide to the storage effect. *Am. Nat.* 125:769–787.

Watson, W., 1982. Development of eggs and larvae of the white croaker, *Genyonemus lineatus* Ayres (Pisces: Sciaenidae), off the southern California coast. *U.S. Natl. Mar. Fish. Serv. Fish. Bull.* 80:403–417.

Williams, G. C., 1957. Homing behavior of the California rocky shore fishes. *Univ. Calif. Publ. Zool.* 59:249–284.

Williams, G. C., and D. C. Williams, 1955. Observations on the feeding habits of the opaleye, *Girella nigricans. Calif. Fish Game.* 41:203–208.

Wolf, P., and P. E. Smith, 1986. The relative magnitude of the 1985 Pacific sardine spawning biomass off southern California. *Calif. Coop. Oceanic Fish. Invest. Rep.* 27:25–31.

Young, D. R., R. W. Gossett, and T. C. Heesen, 1988. Persistence of chlorinated hydrocarbon contamination in California marine ecosystem. In: D. A. Wolfe, and T. P. O'Connor, eds. *Ocean Processes in Marine Pollution. Vol. 5. Urban Wastes in Coastal Marine Environments.* Krieger Publishing Co., Malabar, FL. pp. 33–41.

Young, D. R., A. J. Mearns, T.-K. Jan, T. C. Heesen, M. D. Moore, R. P. Eganhouse, G. P. Hershelman, and R. W. Gossett, 1980. Trophic structure and pollutant concentrations in marine ecosystems of southern California. *Calif. Coop. Oceanic Fish. Invest. Rep.* 21:197–206.

Young, P. H., 1963. The kelp bass (*Paralabrax clathratus*) and its fishery, 1947–1958. *Calif. Dept. Fish Game Fish Bull.* No. 122. 67pp.
Young, P. H., 1969. The California partyboat fishery 1947–1967. *Calif. Dept. Fish Game Fish Bull.* No. 145. 91pp.

Chapter 10

Birds

Patricia Herron Baird

Introduction

More than 195 species of birds use coastal or offshore aquatic habitats in the Southern California Bight (SCB). Population numbers are not accurately documented, but breeding birds number in the thousands and migratory populations number in the millions. The SCB is the only California breeding location for three seabird species and is also the northern or southern limit of breeding ranges for many other species.

Although the SCB covers a definite geographic area, birds are mobile, of course, and do not restrict their ranges to its boundaries. Most of the avifauna discussed in this chapter regularly travel in and out of the SCB in their feeding forays, migrate through the area, or use only a part of it. Some use only the ocean-influenced marshes and estuaries.

This chapter summarizes much of the published and unpublished literature on distribution, abundance, migration, breeding biology, and feeding ecology of marshbirds (herons, rails, cranes, and ibises), waterbirds (ducks, geese, coots, and grebes), shorebirds, and seabirds (birds that spend a portion of their lives—or reside—on or near the coastal or offshore habitats). Each of these species assemblages is discussed separately under each of the following categories: distribution and abundance, nesting and breeding ecology, food habits and foraging ecology, and, where known, energy and nutrient cycling. The discussion of species includes only those with large population numbers, those of "special concern" (Remsen 1978) or endangered, those at the northern or southern limit of their ranges or whose California populations breed only in the SCB, and those whose habitats are threatened. Passerines, for example, often remain in the marshes to nest and feed, but are not addressed here because of their minimal contact with the marine environment.

Loss of coastal habitat (and with it, loss of species diversity) is one of the most important environmental issues in the SCB. The cumulative effects of incremental habitat degradation have become especially noticeable in the latter part of this century. Originally, natural diversity in the SCB was high due to climate, geographical location, and ocean currents, although the number of wetlands or ocean-influenced areas was limited. Before the twentieth century, less than 10 wetlands had extensive areas of lagoons, river mouth estuaries, and salt and freshwater marshes. These wetlands were extremely important, however, as overwintering or staging areas for the many birds that came to southern California from their northern or southern breeding grounds. One of the largest wetlands lay between what is now the Redondo–Hermosa Beach area and Long Beach, part of the changing delta of the Los Angeles River. However, the same factors that gave rise to diverse habitats along the coast also attracted a large human population, which in turn led to the destruction of natural habitats, especially coastal wetlands (California Nature Conservancy 1987).

Marshes in the SCB are few and all are threatened, but the bays and river mouths, that is, the coastal wetlands, are particularly threatened. Most of these have been channeled, filled, dredged, diked, or converted to farmland, pastures, harbors, marinas, cities, or garbage dumps (California Nature Conservancy 1987). Today the only remaining salt marshes of substantial size are the Tijuana River mouth, Anaheim Bay, Upper Newport Bay, and Bolsa Chica Lagoon (fig. 10.1).

The extent of the discussion of each SCB avifauna group reflects the amount of time and in what number each group is present in the SCB. Marshbirds are relatively scarce in the SCB, compared to their abundance in other parts of the world, so their discussion is brief. Since the majority of species of waterfowl and shorebirds found in the SCB breed elsewhere, their discussion is limited and addresses abundance, distribution, and feeding ecology during winter residence and migratory periods. Seabirds occupy the SCB year-round, with various groups of species breeding, overwintering, and migrating through the SCB. Because they have the greatest numbers of species and biomass of all SCB avifauna, discussion of seabirds is most extensive.

To understand the interaction of marine birds with other biota in the SCB and their placement in the food web, the following aspects of their ecology are addressed: species composition and seasonal abundance in the SCB, locations of species concentrations, annual reproductive success of breeding populations, varieties and amounts of prey species taken, and energy or nutrient cycling. Finally, human effects on marine birds are discussed, with suggestions on how best to mitigate these effects.

Methods Used in Data Collection

Accurate estimates of numbers of birds occupying wetlands can be obtained by regularly scheduled complete counts of all birds seen at each wetland throughout the year. A knowledge of the daily and annual patterns of the species that inhabit wetlands is required to obtain the most accurate counts. Birds that

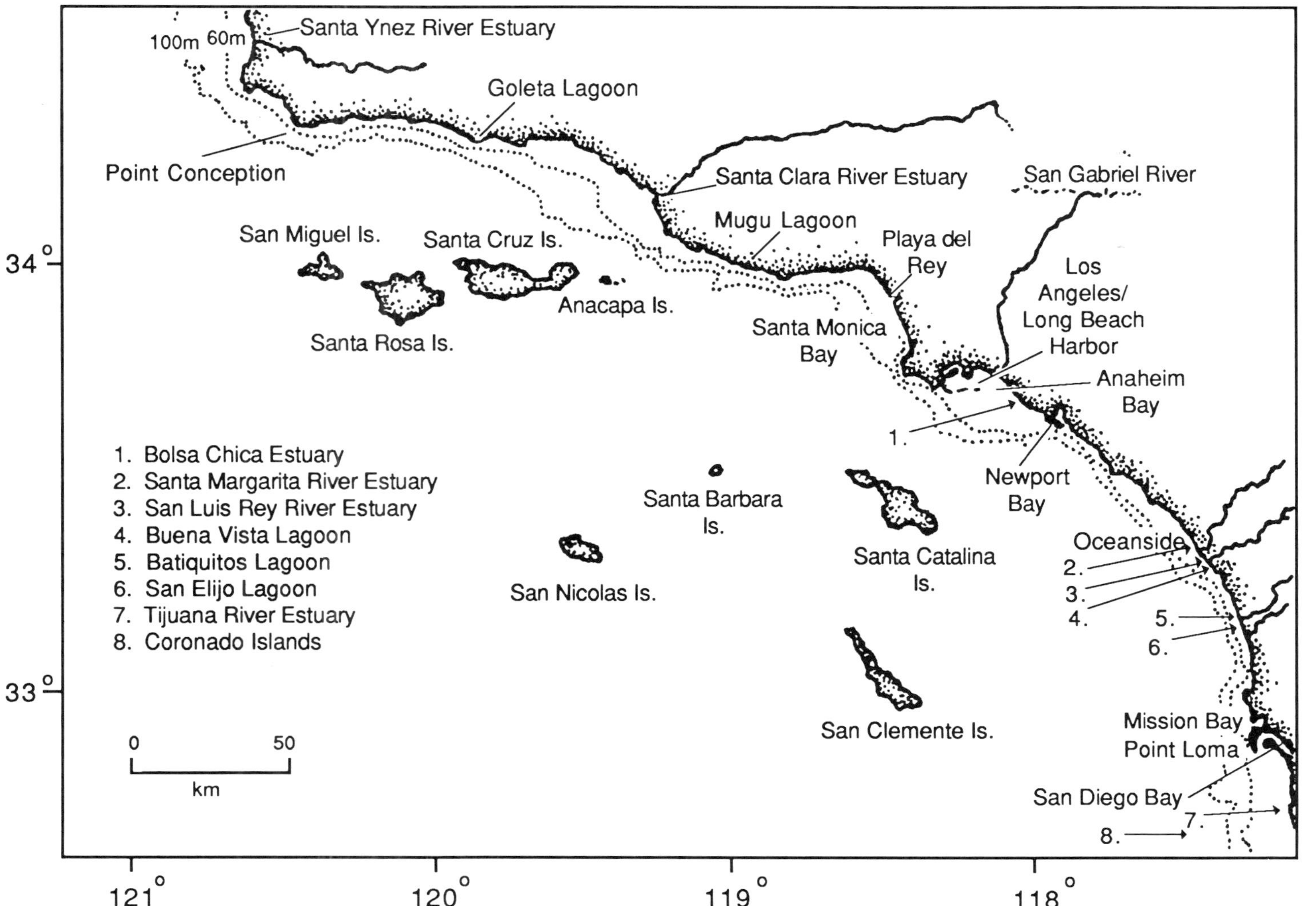

Figure 10.1. Location of areas in the SCB important for marshbirds, waterbirds, shorebirds, and seabirds.

feed at sea are more difficult to count because the area over which they forage is vastly larger than wetlands and because seabirds are almost always moving so that any count is a rough estimate of true population numbers. Since cyclic events in the ocean affect the abundance and distribution of fauna over time, at-sea surveys need to be conducted over a period of several years to account for variations in populations that accompany these events.

Determination of habitat requirements is important for management programs because seabirds, waterfowl, herons, cranes, and rails usually feed and nest in different habitats. Also, knowledge of the chronology of the feeding cycle can help in species management. Reproductive success is determined by recording the number of eggs laid per breeding pair (laying success), the number of eggs hatched per egg laid (hatching success), or the number of chicks fledged per mated pair.

The most accurate way to determine types of prey taken is to collect the stomach contents of the adult or to collect food destined for the chick by regurgitation or by bill-taping (Baird 1986). Prey requirements can also be estimated by observing foraging birds or adults feeding chicks. Number and frequency of occurrence of prey per food sample or stomach have been the two standard measurements in food habit comparisons in addition to weight and length of whole prey (Ashmole and Ashmole 1968; Diamond 1983).

Estimates of intercolonial movement, species longevity, and population growth can be obtained by marking a sample in a population and recording their movements or presence over a number of years.

Distribution and Abundance

Habitats

Natural habitats are important not only for breeding birds but also as temporary resting or staging areas for migrants and visitors. Although some habitats do not seem to be used often by birds, short periods of use may be very important or even critical for maintaining a species population in the area. Habitats that have been the most severely reduced in the SCB are coastal salt marshes and kelp beds. Other habitats that have been destroyed, either directly by conversion to other land uses or indirectly by erosion or pollution, include island beaches, island cliffs, and island soil habitats (burrowing areas for murrelets, petrels, and auklets). To maintain a species in an area, it is necessary to preserve foraging areas as well as nesting sites, for without a habitat in which to forage, a species will not remain in the area. Wetlands in some bays or estuaries of the SCB have recently been enhanced by increasing tidal influences and by construction of islands. These enhancements have attracted greater numbers and more varieties of marine birds (especially marshbirds, shorebirds, and terns) than were previously present (Burkett 1988).

The Channel Islands are especially important as habitats because they offer nesting sites to some seabird species for which suitable habitats are very scarce elsewhere in the SCB (Bender et al. 1974).

Marshbirds

Because of the natural scarcity of wetlands in southern California, historical populations of marshbirds were never large, and because of land conversion, their numbers are even smaller today. Except in the case of the endangered light-footed clapper rail (*Rallus longirostris levipes*), precise seasonal counts of marshbirds have never been conducted, either earlier in the century or in more recent years. However, we can note general population trends from anecdotal counts. For example, we know that the only remaining important areas used by herons today are Goleta Marsh and vicinity, Santa Clara Estuary, Mugu Lagoon, Ballona Wetlands, Bolsa Chica Lagoon, Anaheim Bay, Upper Newport Bay, Santa Margarita River Estuary, several lagoons in coastal San Diego County, Mission Bay, San Diego Bay, and the Tijuana River Estuary. These habitats are used year-round, but they are used most heavily in winter.

Wetland use varies among species. Marshbirds often fly a short distance inland to roost or nest, but they return each day to the wetlands to feed. For this reason, protection of nesting habitat alone is not sufficient to maintain present population levels.

The greatest numbers of marshbirds in the SCB (16 heron species and 8 rail species) occur during the nonbreeding seasons (fig. 10.2), and most of these species can be found at all of the marshes shown in figure 10.1 and listed in table 10.1. Only 9 heron and 6 rail species breed in the SCB, including the light-footed clapper rail, which is of particular concern because of its endangered species status. Marshbirds do not breed every year. The most common nesters are the great blue heron (*Ardea herodias*), snowy egret (*Egretta thula*), green-backed heron (*Butorides striatus*), and black-crowned night heron (*Nycticorax nycticorax*).

Species whose populations have decreased the most over this century are the least bittern (*Ixobrychus exilis*), great blue heron, white-faced ibis (*Plegadis chihi*), wood stork (*Mycteria americana*), and light-footed clapper rail (Willet 1933; Garrett and Dunn 1981). The American coot (*Fulica americana*) and cattle egret (*Bubulcus ibis*) are the only marshbirds whose populations have remained stable or have increased (Garrett and Dunn 1981). Estimates of relative abundance and SCB distribution of the most common species are shown in fig. 10.2. Accounts of other heron, egret, rail, and ibis species can be found in Garrett and Dunn (1981).

Great blue herons, among the most common and visible herons, are found in a variety of aquatic habitats, especially in estuaries. Snowy egrets are found in marsh and estuarine habitats and along river courses and irrigation channels in winter. During migration, small numbers can be found on rocky shorelines.

Cattle egrets occur in coastal estuaries mainly in fall and winter (September–early November) and are most common in the San Diego area. They also forage in agricultural fields. Their numbers have been steadily increasing since 1964 when they first arrived in the SCB, and the increase reflects part of a global pattern of expansion of this highly adaptable egret.

The black-crowned night heron is the more abundant of the two night herons found in the SCB (the yellow-crowned night heron, *Nyctanassa violacea,* is rare). It is most numerous in San Diego County, occupying a variety of water-edge habitats where trees or dense marshes are located nearby for roosting or nesting. It forages in coastal estuaries, including lake shores and freshwater marshes and channels, but is often found on pilings and piers or on the shores of bays and harbors.

The white-faced ibis is listed in the highest priority category of "Bird Species of Special Concern" by the California Department of Fish and Game (Remsen 1978), and it is on the "Blue List," the list of species showing signs of noncyclical population decline or range contraction (table 10.3). This ibis used to breed all along the coastal area (Grinnell and Miller 1944) and could be found in the winter throughout all coastal marshes (Remsen 1978). Now, however, only three wintering areas remain in California that have more than 25 birds per site (McCaskie 1976, 1977). Wintering areas with largest population numbers in the SCB are Point Mugu (approximately 25 birds) and Oceanside (approximately 35 birds) (Remsen 1978). Because extensive, shallow, and grassy marshes are required for breeding, the destruction of southern California marshes (many of which were allowed to dry up for mosquito and cattail control) was responsible for the population decrease of the white-faced ibis. DDE, a breakdown product of DDT, also contributed to the demise of these ibises by thinning the shells of their eggs (Remsen 1978). Currently, foraging birds can be found in flooded fields, marshes, ditches, and—less commonly—in

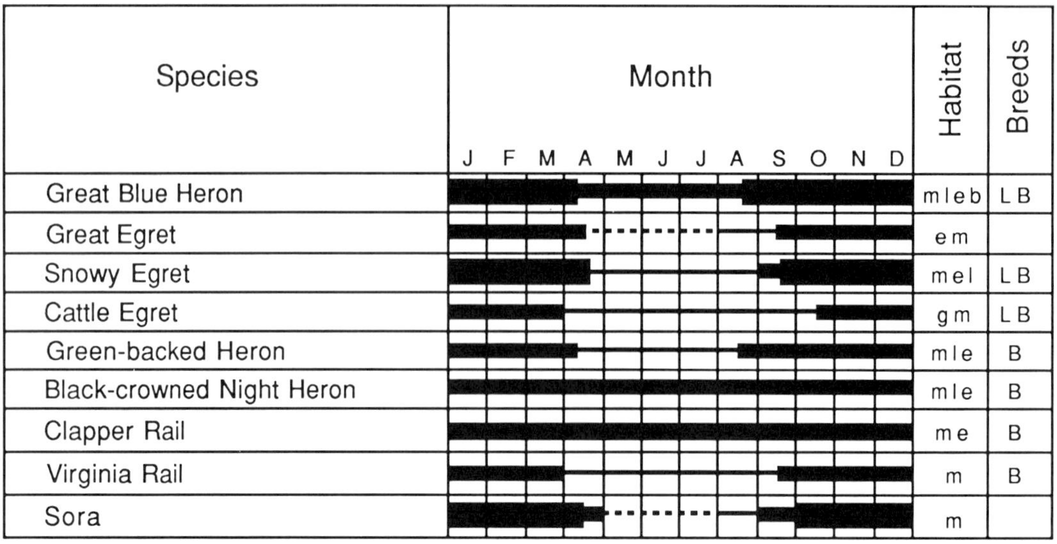

Species	Month (J F M A M J J A S O N D)	Habitat	Breeds
Great Blue Heron		mleb	LB
Great Egret		em	
Snowy Egret		mel	LB
Cattle Egret		gm	LB
Green-backed Heron		mle	B
Black-crowned Night Heron		mle	B
Clapper Rail		me	B
Virginia Rail		m	B
Sora		m	

Key

Common to Abundant: Species always present, in fairly large numbers. Can refer either to a generalist occupying most areas in moderate numbers or a specialist occupying specific habitats in large numbers.

Fairly Common: Species usually found in its proper habitat throughout, but in low numbers.

Uncommon: Species found only in a few of its proper habitats, always in low numbers.

Rare: One or two sightings of the species found almost every year in its habitat in low numbers. (A regular migrant in low numbers.)

Casual: Species not regularly found in its habitat, but when found, always in low numbers.

A **Accidental:** Individual record. Not enough numbers found to detect a pattern. Species sighted ten or fewer times in the region.

Irregularly Common: An invasionary species, occurring regularly and unpredictably, but in moderate to high numbers.

Abundance Tables

Common unless stated

B = breeds

I = intermittent

L = local

r = rare

re = record

U = uncommon

Habitat

o = open ocean; > 1 km offshore

i = inshore ocean waters, bays, and harbors ; < 1 km offshore

b = beaches; sandy and rocky shores, jetties

e = estuaries; mudflats, tidal lagoons and brackish water

g = grassland; field and agriculture

m = marshes; fresh and/or salt water

l = lakes, reservoirs, large streams, and rivers

Figure 10.2. Seasonal distribution and relative abundance of marshbirds (herons, rails, and egrets) in coastal regions of the SCB. (Adapted from Garrett and Dunn 1981, with additions.)

Table 10.1. *Important Marshes of the Light-Footed Clapper Rail*

County	Marsh	Numbers (1984)
Santa Barbara	Carpinteria Marsh	52
Orange	Anaheim Bay–Seal Beach	48
Orange	Upper Newport Bay	224
San Diego	San Elijo Lagoon	20
San Diego	Kendell–Frost Reserve	48
San Diego	Sweetwater Marsh	28
San Diego	Tijuana Marsh (Oneonta Lagoon)	76

From Zembal and Massey 1985.

Table 10.2. *Population Numbers and Densities of the Principal Species of Seabirds in the SCB*

Species	Annual or Highest Population Size	Highest Density (birds km^{-2})
Pacific loons	40,000–46,000	0.4–1.8
Western and Clark's grebes	27,000	no value[a]
Scoters	125,000	1–5
Pink-footed shearwaters	40,000–400,000	5–8
Sooty shearwaters	2.7–4.7 million	no value
Black-vented shearwaters	20,000–30,000	80
Northern fulmars	120,000–300,000	2–3
Leach's storm-petrels	150,000	2.6
Black storm-petrels	100,000	2.0
Least storm-petrels	200,000	no value
Brown pelicans	6,000–90,000	no value
Red and red-necked phalaropes	925,000	7
Bonaparte's gulls	300,000	10
Heermann's gulls	45,000	no value
California gulls	5,000	5–8
Herring gulls	32,500	no value
Western gulls	25,000–50,000	no value
Black-legged kittiwakes	50,000–300,000	2–3
Common and arctic terns	30,000–50,000	no value
Common murres	20,000–30,000	no value
Cassin's auklets	50,000–100,000	100
Rhinoceros auklets	100,000–300,000	2–5

[a]No value = density not determined.
From Briggs et al. 1987.

Table 10.3. *Marine Bird Species on the Federal Endangered or Rare List, the Blue List, or the California List of Species of Special Concern*[a]

Species	Limited Range SCB	Endangered List	Rare List	Blue List	Special Concern	Populations in Decline
Common loon					1	
Black-footed albatross	LR					X
Short-tailed shearwater	LR					X
Black storm-petrel	LR				3	
Ashy storm-petrel					3	
Fork-tailed storm-petrel	LR					X
Brown pelican	LR	E			3	
Double-crested cormorant					2	X
Pelagic cormorant	LR					
Whistling swan						X
Canada goose						X
Greater white-fronted goose						X
Snow goose						X
Ross's goose						X
Ruddy duck						X
Great blue heron						X
Least bittern						X
Wood stork						X
White-faced ibis				B	1	X
Sandhill crane						X
Virginia rail						X
Sora						X
Black rail			R[b]			X
Light-footed clapper rail		E				X
Common moorhen						X
American avocet				B	2	X
Snowy plover				B	2	X
Common snipe						X
California least tern		E				
Caspian tern						X
Forster's tern						X
Elegant tern					3	X
Royal tern						X

Table 10.3. *Marine Bird Species on the Federal Endangered or Rare List, the Blue List, or the California List of Species of Special Concern*[a] (continued)

Species	Limited Range SCB	Endangered List	Rare List	Blue List	Special Concern	Populations in Decline
Skimmer					3	
Pigeon guillemot	LR					
Xantus' murrelet	LR					
Common murre						X
Tufted puffin						X
Rhinoceros auklet					3	

[a]Codes for Species of Special Concern: 1 = highest priority; 2 = second priority; 3 = third priority. Codes are defined in text.

[b]Listed as rare in California.

estuaries. To increase numbers of this species, it is proposed that we allow wet areas to remain wet and become overgrown by marsh vegetation and purposely flood existing grassy areas (Remsen 1978).

Least bitterns, once numerous, are now uncommon but are found occasionally in most of the wetlands in the SCB.

The most dramatic population reductions of rails have been those of the light-footed clapper rail and the black rail (*Laterallus jamaicensis*), although populations of Virginia rails (*Rallus limicola*), soras (*Porzana carolina*), and common moorhens (*Gallinula chloropus*) have also suffered declines (Willet 1933; Garrett and Dunn 1981). The light-footed clapper rail is the only federally listed endangered marsh bird in the SCB, and its listing is the result of severe population decline in recent decades. The decline has been most obvious in large estuaries dominated by the salt marsh vegetation *Salicornia* and *Spartina*.

Numbers of breeding light-footed clapper rails in the SCB have decreased from 500–750 breeding birds in 1974 to 346 in 1981 and to 254 in 1984 (Zembal and Massey 1985). Approximately 88% of California's clapper rail population is found in the SCB, and Upper Newport Bay alone contains about 48% of the total statewide population (Zembal and Massey 1985).

Regular annual decreases in clapper rail population numbers coincide with heavy winter and early spring storms (Zembal and Massey 1985). These storms drive tides high into rail habitat, and subsequent runoff destroys some nesting areas. These storms are not new to the SCB, of course. But their destructive effects in tandem with other habitat losses, along with predation by the introduced red fox, have caused a steep decline in the clapper rail population. Predation on both chicks and eggs by red foxes seems to have had a major impact on the population at Anaheim Bay (G. Hunt pers. comm.), and it appears to be difficult for this population to recover.

A recovery plan designed to protect, restore, create, and manage marsh habitats for the light-footed clapper rail is being implemented. The plan proposes to improve existing habitats as well as create and protect a minimum of 20 marshes measuring approximately 4087 ha, an area capable of supporting 800 pairs of rails.

Virginia rails are common winter visitors; in summer, their breeding populations are small but ubiquitous (<20 birds per site) (Garrett and Dunn 1981). During winter, Virginia rails can be found both inland and on salt marshes (Garrett and Dunn 1981; Unitt 1984). The sora rail is a common winter visitor in freshwater marshes, along the marshy borders of small ponds, and in lakes and salt marshes of the SCB region; small nonbreeding populations occur in the summer. Because the sora used to breed here, it is possible that the species may reestablish a breeding population in the SCB. However, no current reported breeding records exist for the sora in the area.

Black rails are listed as "rare" by the state of California because of the recent decline in their numbers. Historically, they occupied the coastal salt marshes in large numbers (California Department of Fish and Game 1980); the remaining marshes now occupied are shown in figure 10.3. The most abundant rail is the American coot. Since it acts ecologically like a duck, and is usually found with them, it is discussed with waterfowl.

The sandhill crane (*Grus canadensis*), which breeds in the subarctic, is a winter visitor to the SCB and is rare along the coast (Garrett and Dunn 1981). It was formerly more widespread during the winter. Today, only a few records document the occurrence of small flocks along the coast in Santa Barbara and Los Angeles counties (Garrett and Dunn 1981).

Waterbirds

Few suitable habitats for waterbird feeding and nesting occur in the SCB. Ducks, geese, coots, and grebes inhabit the limited number of brackish lagoons, estuaries, and bays along the coast, but larger numbers of these waterbirds occupy freshwater ponds and lakes in the area. The greatest numbers are present from October to March (fig. 10.4), and the majority migrate northward to breed during the summer. Numerous wintering species include the American wigeon (*Anas americana*), green-winged teal (*Anas crecca*), northern pintail (*Anas acuta*), Canada goose (*Branta canadensis*), brant (*Branta bernicla*), and American coot (a rail). A few waterfowl species and the American coot breed yearly in the region in small numbers, perhaps 20–50 birds for each species. Because most of these birds do not occupy the marine environment, their treatment here is minimal.

Historically, populations of the tundra swan (*Cygnus columbianus*), greater white-fronted goose (*Anser albifrons*), Canada goose, snow goose (*Chen caerulescens*), and Ross's goose (*Chen rossii*) were more common along the coast; only the snow and Ross's goose populations were also more abundant on the Channel Islands than they are today. At present, populations of these species are low, and some have moved inland to the Salton Sea. In contrast, the ring-necked duck (*Aythya collaris*), rare before 1930, is one of the few species besides the mallard (*Anas platyrhynchos*) that is increasing in numbers. It is believed that the ring-necked duck's increase is due to the increase in artificial lakes.

Twenty-five species of ducks, five species of geese, two species of swans, four species of grebes, and the American coot have been recorded in the SCB (fig. 10.5). Areas supporting the largest numbers include Upper Newport Bay, San Elijo Lagoon, Bolsa Chica Lagoon, and the Santa Clara River mouth. Enhancement of wetlands in some bays or estuaries along the coast of the SCB by increasing tidal influence and by construction of islands attracts a greater number and variety of waterbirds to these wetlands (Burkett 1988). Accounts of the most abundant waterbirds are given in the paragraphs that follow; the months of greatest abundance and concentration of the more common waterbirds are shown in figure 10.4. Detailed descriptions and distributions of the less common species can be found in Unitt (1984) and Garrett and Dunn (1981).

The northern pintail, with populations in individual lagoons numbering in the thousands, is the most abundant dabbling duck in fall and winter (Unitt 1984). In the middle to late August peak, large flocks are seen over the open ocean.

The American wigeon is an abundant winter visitor, especially on fresh water lakes, lagoons, river mouths, and bays (Unitt 1984). They are rarely found on the Channel Islands, except on Santa Rosa Island, where they are a common winter visitor (Garrett and Dunn 1981).

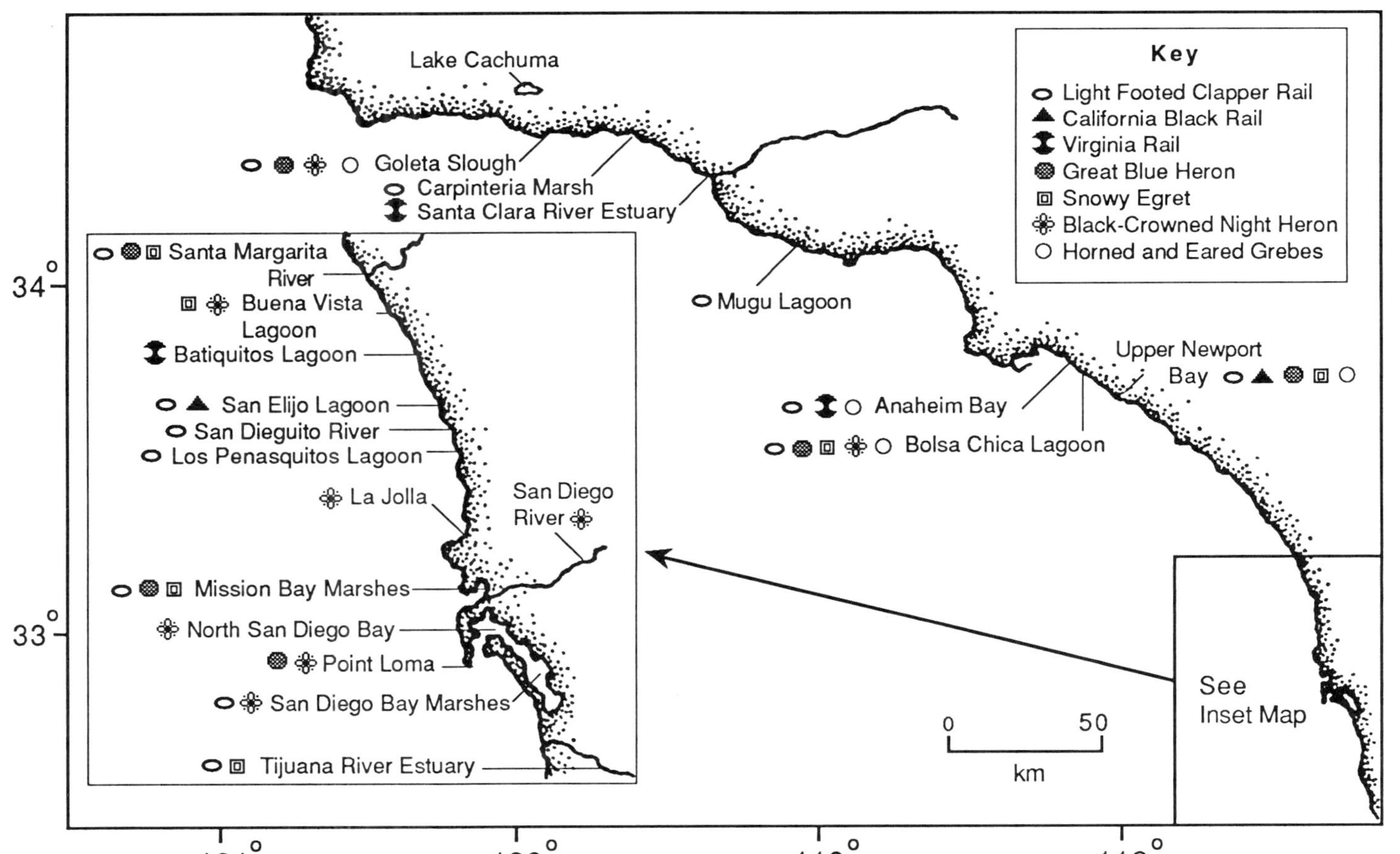

Figure 10.3. Location of marshes most heavily used by the principal species of marshbirds in the SCB. (From Garrett and Dunn 1981; Unitt 1984.)

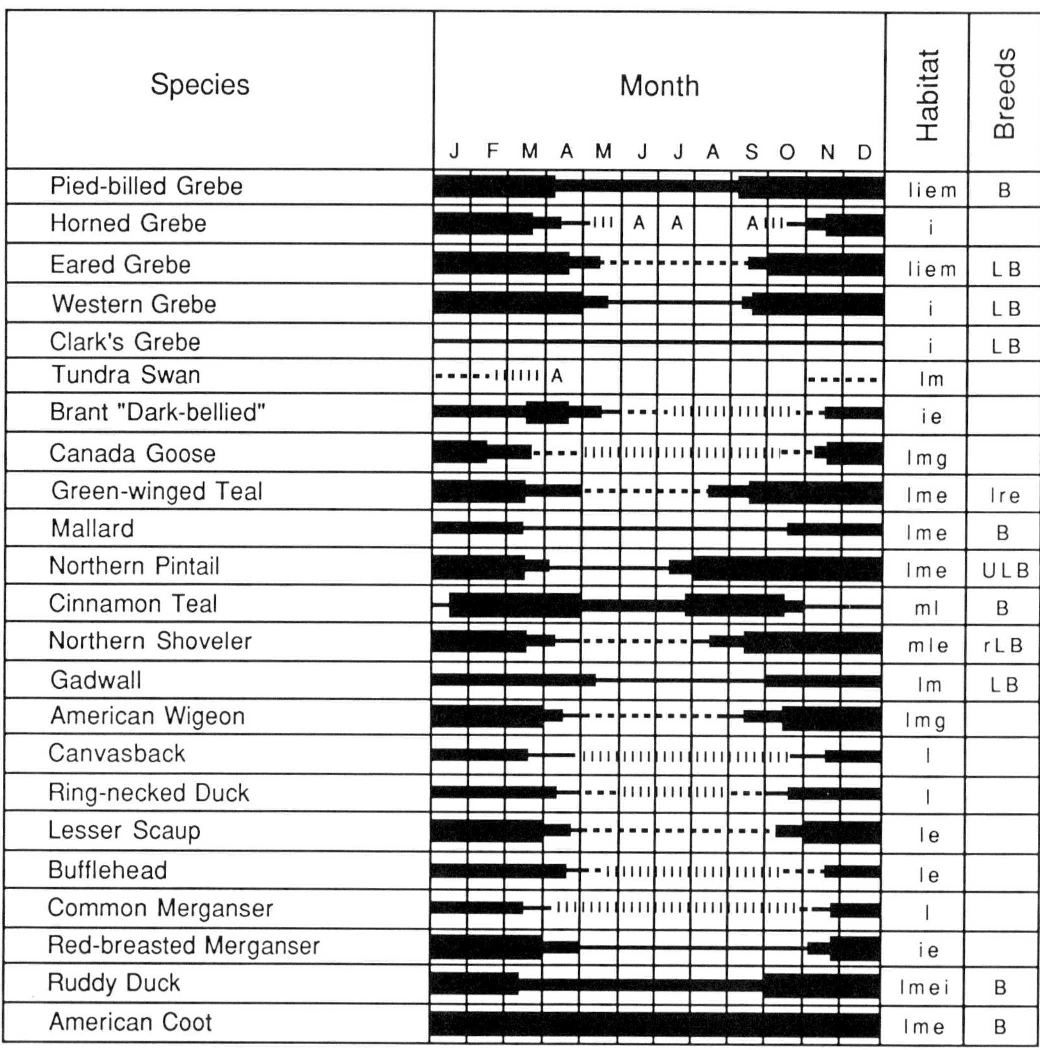

Figure 10.4. Seasonal distribution and relative abundance of waterbirds (swans, geese, ducks, grebes, and coots) in coastal regions of the SCB. (Adapted from Garrett and Dunn 1981, with additions.)

Green-winged teals, whose populations number in the thousands, are one of the most numerous winter dabblers; they occur in a variety of shallow freshwater habitats and in brackish coastal estuaries throughout the SCB (Garrett and Dunn 1981; McCaskie 1980). During migration, small numbers (such as 10) are found in saltwater bays (Unitt 1984).

Northern shovelers (*Anas clypeata*) are abundant migrants and winter visitors (numbering in the thousands), but they are rare in the summer (Garrett and Dunn 1981). In winter, they are found widespread along the mainland coast, occurring in greatest numbers in lagoons, estuaries, and lakes and where extensive shallows are available for foraging (Garrett and Dunn 1981; Unitt 1984).

Ruddy ducks (*Oxyura jamaicensis*) are also abundant as migrants and winter visitors (Unitt 1984) and are common throughout the summer, when they breed at most of the salt water and freshwater marshes in the SCB. They frequent any body of freshwater or salt water deep enough to accommodate their

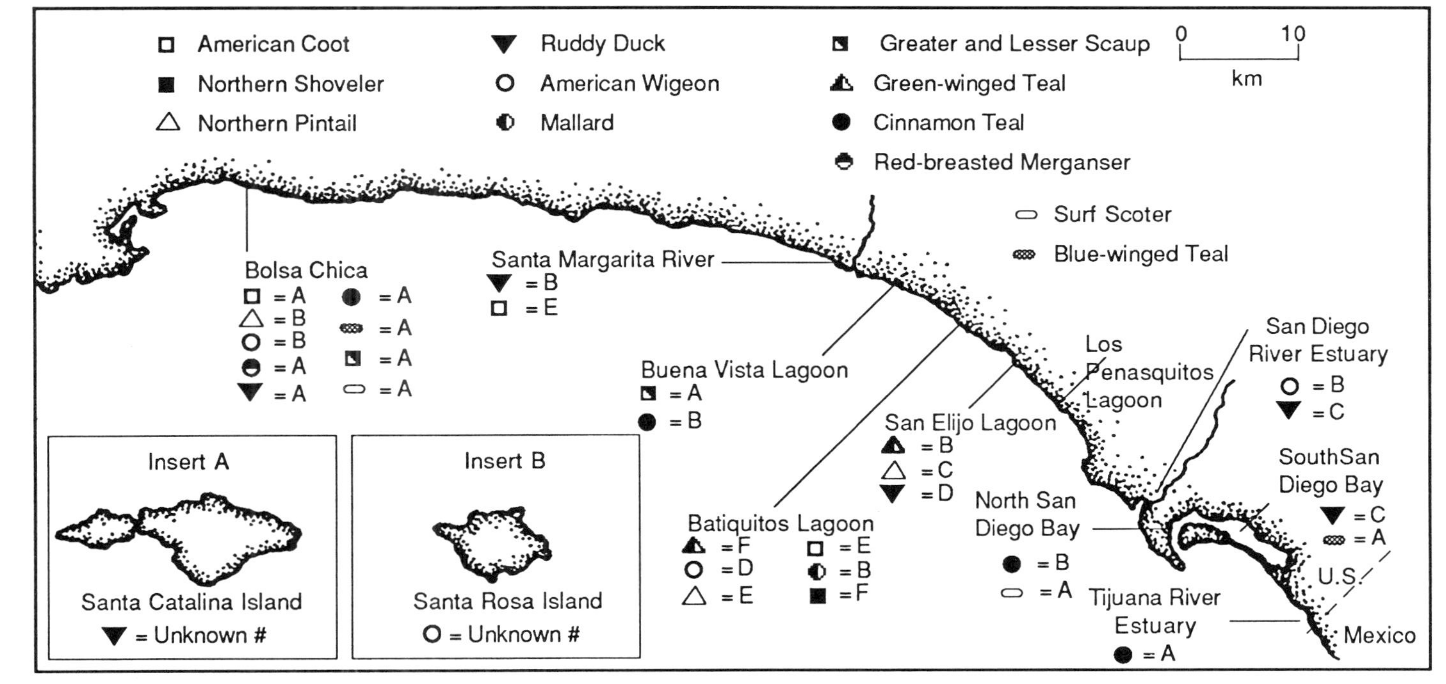

Figure 10.5. Areas of greatest concentrations of waterbirds in the SCB. Numbers given are maximum sightings in 1 day. (From Unitt 1984; American Birds 1986.)

diving habits. However, Unitt (1984) reports that their populations have decreased in this century.

The wild mallard population's abundance and distribution is difficult to measure because of the widespread establishment of feral birds, some of which are difficult to distinguish from the wild population (Garrett and Dunn 1981). Mallards are found in lagoons, estuaries, ponds, and lakes, but they nest in marshes.

Cinnamon teals (*Anas cyanoptera*) are common during all seasons, but are particularly abundant during spring migration in San Diego Bay and offshore. Summer individuals occur throughout the SCB and are occasionally found on the Channel Islands. In winter, these waterbirds frequent lagoons, ponds, and lakes (Unitt 1984).

Lesser scaups (*Aythya affinis*) are numerous from October to April, when their populations number in the thousands. They are rare in the SCB in summer, however. During the winter, most of these birds occupy estuaries, lagoons, bays, lakes, and ponds along the coast, but they are also found occasionally in low numbers in the Channel Islands (Garrett and Dunn 1981; Unitt 1984).

The American coot is particularly adapted to a human-modified environment, and it is the rail least affected by marshland destruction. It reaches peak abundance in winter when breeders arrive from the north. Coots inhabit virtually any nonmarine aquatic habitat, and they have been seen by the thousands in coastal estuaries, harbors, and shallow bays during the winter season (e.g., San Elijo and Batiquitos lagoons) (Garrett and Dunn 1981; Unitt 1984). Breeders are common but occur in low numbers (20–35) in most of the wetlands in the SCB.

Eared (*Podiceps nigricollis*) and horned (*Podiceps auritus*) grebes, unlike the western (*Aechmophorus occidentalis*) and Clark's (*Aechmophorus clarkii*) grebes, are usually found close to shore. Horned grebes make up less than 5% of all grebes seen on open coastal waters. They are present in and around the Channel Islands as well as in mainland waters. High counts near the islands in the Santa Barbara Channel occur in January and February, while lowest counts are recorded in late winter. Populations begin increasing in mid-spring as birds migrate in from the Gulf of California and other southern areas (Briggs et al. 1987).

Shorebirds

Shorebirds are most abundant in the SCB in the winter (fig. 10.6). Twenty-one species inhabit many of the SCB bays and estuaries. The majority of these birds feed in shallow water, but some also frequent irrigated fields and lawns.

The black-necked stilt (*Himantopus mexicanus*) winters along coastal Orange and San Diego counties and at Point Mugu. It also breeds at Point Mugu as well as in Goleta Slough, Santa Clara River Estuary, Bolsa Chica Lagoon, San Joaquin marsh, and all the estuaries in San Diego County. The American avocet (*Recurvirostra americana*), which is closely related to the stilt, winters in coastal areas of Orange and San Diego counties and at Playa del Rey. Once more widespread in the SCB, this species now breeds in estuaries of the Santa Clara River, in Bolsa Chica Lagoon, and in all estuaries in San Diego County (Garrett and Dunn 1981).

The black-bellied plover (*Pluvialis squatarola*) is one of the most numerous shorebirds overwintering in the SCB. It prefers the estuarine mudflats but is also found on sandy beaches and rocky shores. Although it once was widespread, the snowy plover (*Charadrius alexandrinus*) is now less common in the SCB than the black-bellied plover (Garrett and Dunn 1981). The snowy plover's preferred habitats are wide sandy beaches, but development and occupation of these beaches by humans in the twentieth century have destroyed the habitats and severely reduced the numbers of this species. It is now found mainly in San Diego County, on Vandenburg Air Force Base in Santa Barbara County, and on San Miguel, Santa Rosa, and San Nicolas islands, although small numbers do still breed at other SCB estuaries (Garrett and Dunn

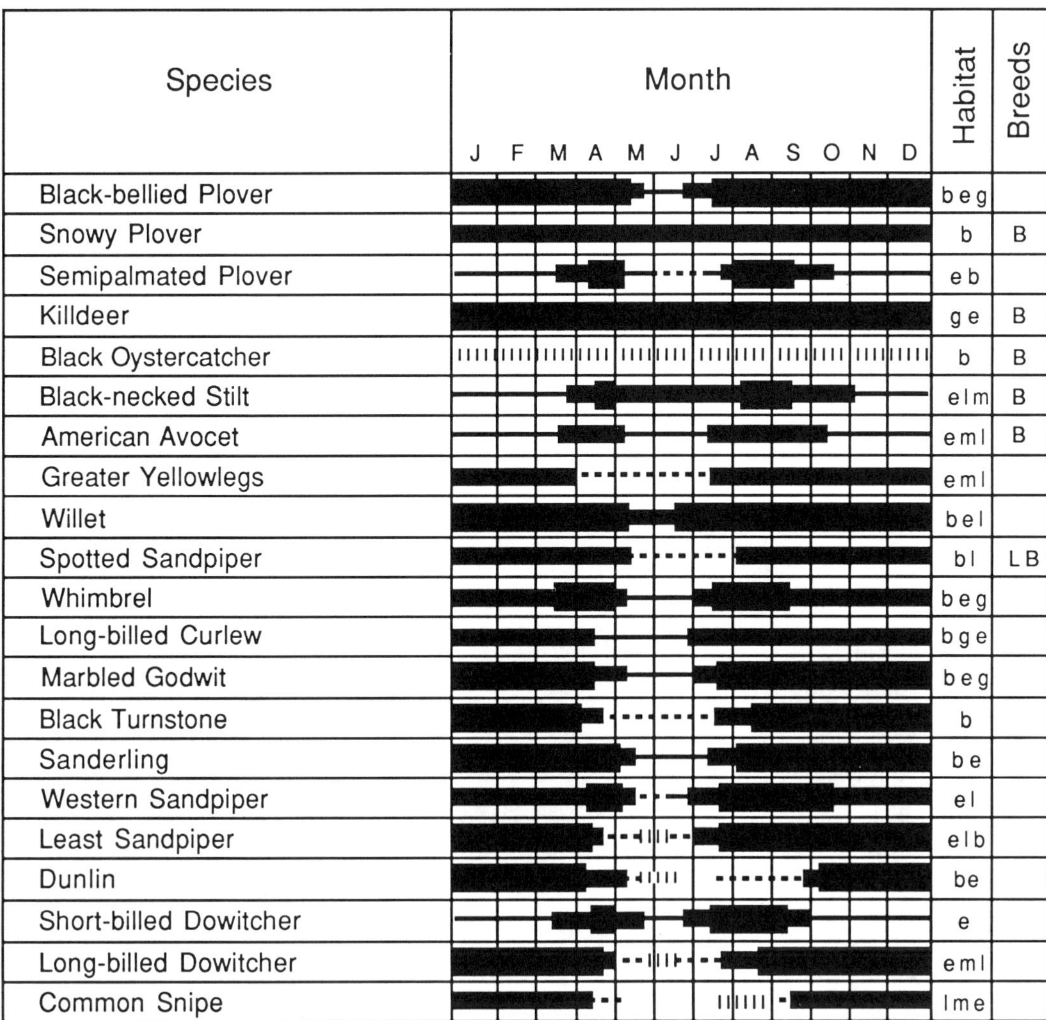

Figure 10.6. Seasonal distribution and relative abundance of shorebirds in coastal regions of the SCB. (Adapted from Garrett and Dunn 1981, with additions.)

1981). The snowy plover is now included in the second priority list of Bird Species of Special Concern in California (Remsen 1978) and has been added to the Blue List, which includes species showing signs of noncyclical population declines or range contractions (Arbib 1977). The semipalmated plover (*Charadrius semipalmatus*) prefers mudflats of coastal estuaries in which to overwinter, and it also is not found in great abundance in the SCB.

Killdeer (*Charadrius vociferus*) inhabit wetlands and fields throughout the SCB and are common breeders near water, irrigated fields, and lawns. Long-billed dowitchers (*Limnodromus scolopaceus*) are another species commonly found in flooded fields, although they also frequent borders of river channels during their spring and fall migrations.

Greater and lesser yellowlegs (*Tringa melanoleuca* and *T. flavipes*) are sometimes found together, with the greater yellowlegs being more widespread inland. Lesser yellowlegs are found mainly in coastal estuaries.

Spotted sandpipers (*Actitis macularia*) are present mainly in the winter and breed primarily along the Santa Ynez, Santa Clara, and

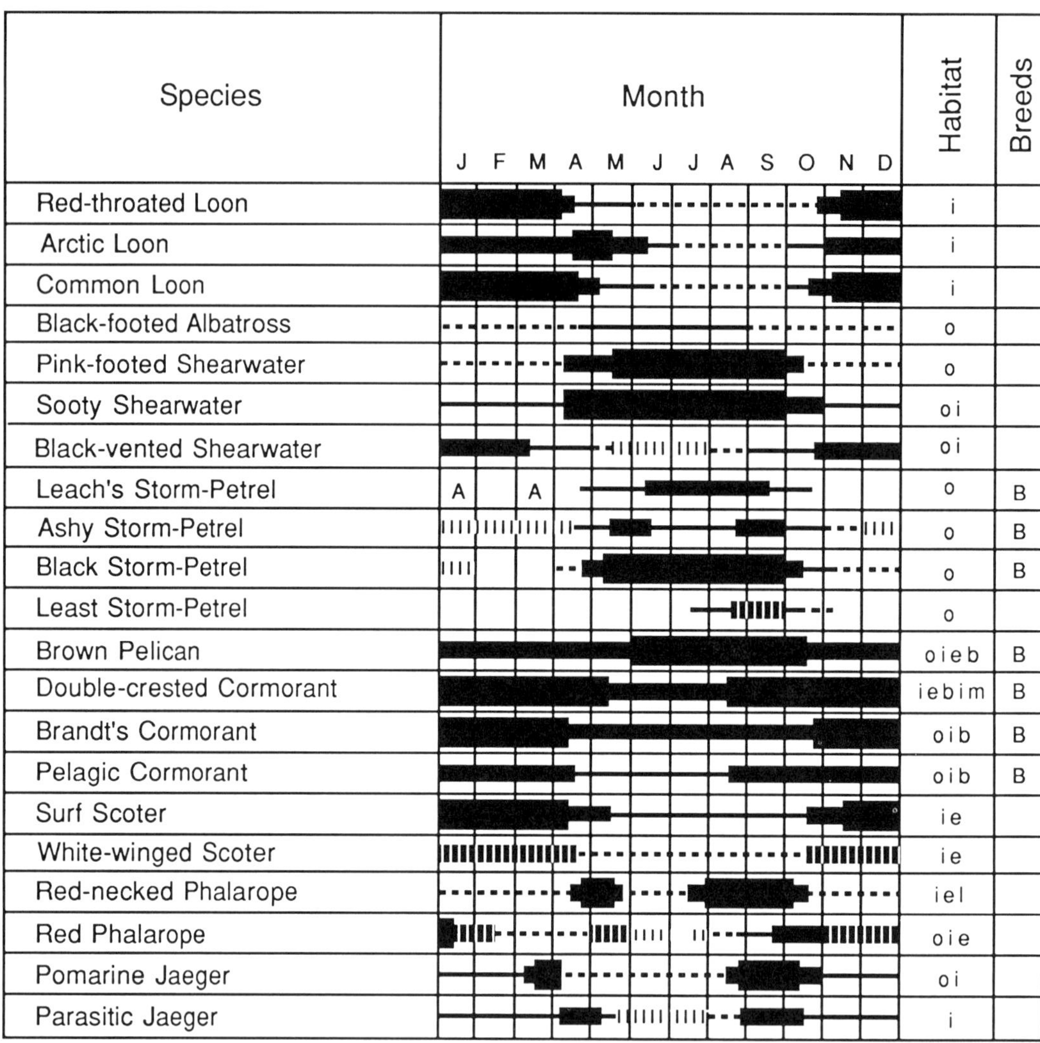

Figure 10.7. Seasonal distribution and relative abundance of seabirds, scoters, and phalaropes in coastal regions of the SCB. (Adapted from Garrett and Dunn 1981, with additions.)

Ventura rivers. Willets (*Catoptrophorus semipalmatus*) are common along beaches and on the Channel Islands in the winter, but they do not breed in the SCB (Bent 1927). They feed on sandy or rocky coasts, estuaries, and flooded fields.

Long-billed curlews (*Numenius americanus*) are winter season birds. They prefer coastal estuaries, especially with *Salicornia* vegetation, and occasionally are found in irrigated fields. Areas where they are most likely to be found are Point Mugu, Anaheim Bay, Newport Bay, San Diego Bay, and Tijuana River Estuary.

The marbled godwit (*Limosa fedoa*) is a wintering bird of open beaches, estuaries, and mudflats. They especially prefer *Salicornia* marshes, but occasionally occupy irrigated fields. Black turnstones (*Arenaria melanocephala*) are found only on rocky coasts during the winter and are thus absent from most of the SCB because of a lack of this habitat. However, they are found on the Channel Islands and in San Diego Bay (Garrett and Dunn 1981).

Sanderlings (*Calidris alba*) overwinter along sandy beaches, feeding at the water's edge. This feeding behavior makes them

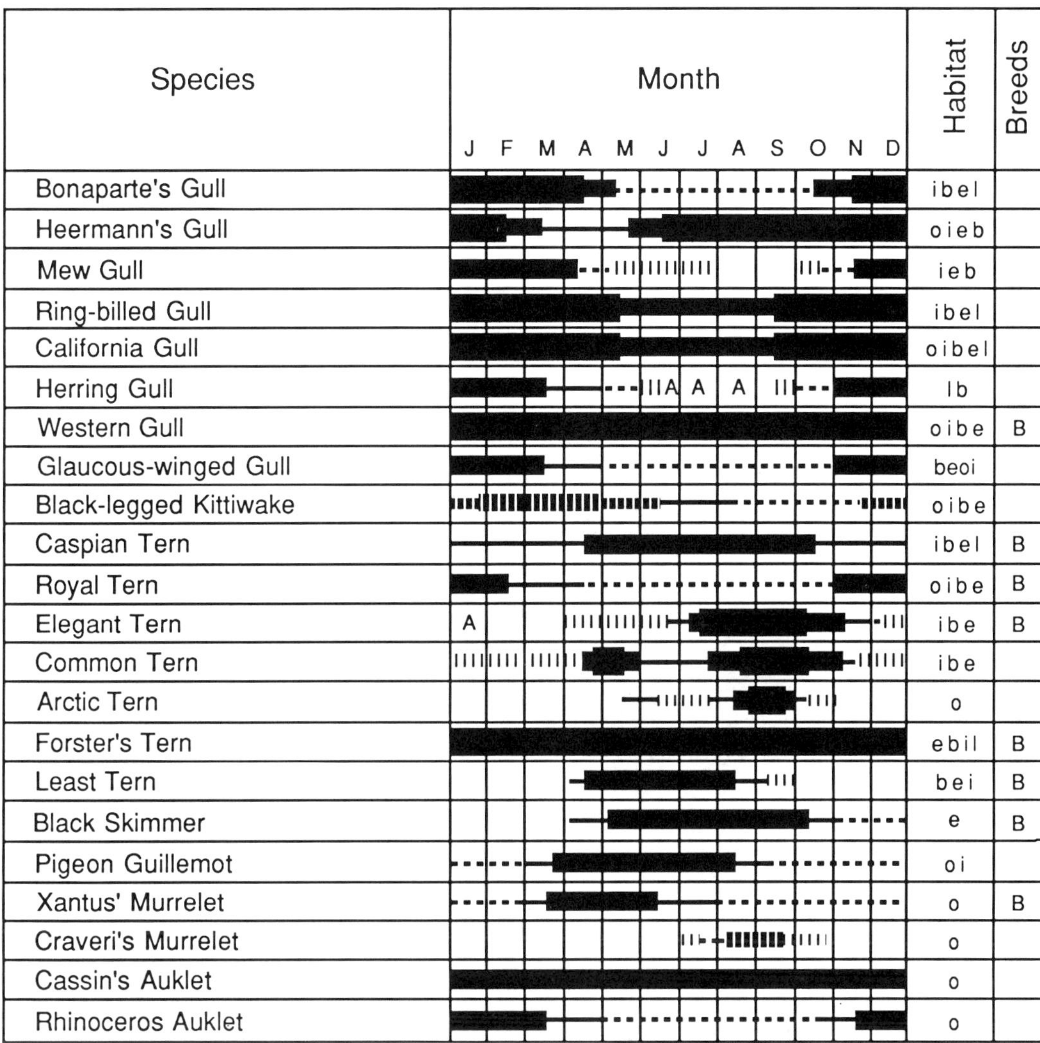

Figure 10.7. (continued)

especially vulnerable to oil spills. They are found throughout the SCB and on the Channel Islands. Western sandpipers (*Calidris mauri*) overwinter in large estuaries, especially San Diego Bay, Newport Bay, and Point Mugu, yet occur sparingly throughout other wetlands of the SCB. Least sandpipers (*Calidris minutilla*) are found both in winter and summer in habitats similar to those of western sandpipers. Summer birds are southbound migrants. They do not nest in the SCB. Dunlin (*Calidris alpina*) are winter birds, usually inhabiting tidal mudflats and the Channel Islands. Common snipe (*Gallinago gallinago*) winter along the coast and on Santa Catalina Island. They formerly bred inland in the SCB, but no longer do so (Garrett and Dunn 1981).

Seabirds

Seabirds, together with sea ducks (scoters), loons, and western grebes, constitute the greatest biomass of avifauna that use the SCB. Of the seabirds, the shearwaters, storm-petrels, phalaropes, gulls, terns, and auklets are most numerous (table 10.2). A total of 43 species of seabirds use the SCB, and of these, 20 species predominate (table 10.2). Their use is year-round, with 17 species breeding, 10 additional species overwintering, and 20 others migrating through (fig. 10.7). These visitors or migrants are numerous, their populations commonly numbering in the thousands. In spring the visitors are mainly austral breeders, in fall they are subtropical breeders, and in winter they are mainly Alaskan breeders (Briggs and Chu 1987). All seabirds that breed in the SCB, except the terns and skimmers, nest on the Channel Islands (12 species and almost 49,000 birds, 23,400 of which nest on Prince Island alone) (Hunt et al. 1981).

Some seabird species that use the SCB are considered particularly important because of their large population numbers, their limited ranges, the rapid decrease of their populations, or their use of critical or unique habitats.

POPULATIONS WITH RESTRICTED RANGES

Because Point Conception and the northern Channel Islands mark the boundary zone between warm and cold waters in California (Hunt et al. 1980), these areas are also the northern or southern breeding limits of a few species whose major concentrations are either in warmer or colder waters. For example, the black storm-petrel (*Oceanodroma melania*), brown pelican (*Pelecanus occidentalis*), and Xantus' murrelet (*Synthliboramphus hypoleucus*) breed only as far north as the Channel Islands; these species are more abundant in Mexico. In contrast, pelagic cormorants (*Phalacrocorax pelagicus*) and pigeon guillemots (*Cepphus columba*) are most numerous north of Point Conception, and the northern Channel Islands are the southern limit of their breeding range. Other seabirds with a limited foraging range in the SCB include the black-footed albatross (*Diomedea nigripes*), short-tailed shearwater (*Puffinus tenuirostris*), and fork-tailed storm-petrel (*Oceanodroma furcata*).

POPULATIONS WHOSE SIZES HAVE DECREASED IN THIS CENTURY

The black-footed albatross was found in large numbers in the SCB before this century, but because of the destruction of its colonies in the mid-Pacific, its numbers have decreased dramatically worldwide as well as in the SCB. However, this species is beginning to recover in numbers. The short-tailed shearwater and the fork-tailed storm-petrel were also more numerous throughout California earlier in this century (Garrett and Dunn 1981). In the early 1900s, common murres (*Uria aalge*) and tufted puffins (*Fratercula cirrhata*) bred as far south as the Channel Islands, but their ranges have been severely restricted and they no longer breed in the SCB.

SPECIAL HABITATS

Habitats of special concern include all wetlands (and land bordering wetlands) and the Channel Islands. The Channel Islands are especially important for seabirds because their position offshore makes them more influenced by oceanic waters than the mainland. However, the shallow waters of the continental shelf surrounding the islands also give them nearshore properties. These islands are also influenced less by human development than the mainland. The entire California breeding populations of black storm-petrels (150 birds), Xantus' murrelets (5000 birds), and brown pelicans (2690 birds) nest on the Channel Islands, as do the entire southern California populations of ashy and Leach's storm-petrels (*Oceanodroma homochroa* and *Oceanodroma leucorhoa*) (Hunt et al. 1981). These species are thus the focus of particular concern because their entire breeding populations could be adversely affected, even destroyed, by environmental perturbations such as oil spills.

Seabirds that forage in wetlands in the SCB (gulls, terns, and skimmers) are also important because all wetlands in southern California are rare or threatened by development (California Nature Conservancy 1987). Birds such as terns and skimmers that use these threatened areas exclusively are considered to be especially important species because of their dependency on these habitats.

ABUNDANCE AND DISTRIBUTION: AT-SEA AND ON COLONIES

The SCB, in comparison to areas such as the Gulf of Alaska, has very few colonies of seabirds and these colonies are smaller, their population numbers being one or two orders of magnitude less (Sowls et al. 1980; Baird and Gould 1983). Reasons for lower numbers include less abundant food during the breeding season and fewer preferred nesting sites than in more northern areas. Prince, Middle Anacapa, Santa Barbara, and San Nicolas islands have the largest seabird colonies, ranging from about 1000 to 7000 birds, except for Prince, which supports more than 23,000 birds (Hunt et al. 1980). Most colonies, however, have fewer than 100 birds (fig. 10.8). Offshore rocks, inaccessible cliff faces, or mainland slopes are preferred nest sites, most likely because they are inaccessible to predators. About 51,350 seabirds, constituting 13.6% of California's breeding population, breed in the SCB (Sowls et al. 1980; Massey 1988; C. Collins pers. comm.). However, seabird populations could be much larger than they are now were it not for human activity, the greatest single force causing changes in seabird populations in the SCB (Hunt et al. 1980). Introduction of exotic animals such as predators (cats and rats) and destroyers of the habitat (sheep, goats, mules, and rabbits) has contributed significantly to population declines. Likewise, specimen egg collectors of the early 1900s helped to decimate annual reproductive output. Also, from 1942 to 1965, San Miguel Island and the northwest end of San Nicolas Island were used by the U.S. Navy as an aerial bombing range and missile target. Finally, short-sighted commercial fishing policies allowed heavy decreases in prey fish, negatively affecting seabird population numbers (Hunt et al. 1981).

This summary of at-sea distributions of birds in the SCB is from a series of surveys conducted over a number of years and covering all four seasons (Briggs et al. 1987). The contrasts between species composition of avifauna in southern California and those in central and northern California are more dramatic than contrasts between those of central and northern California because oceanographic conditions change the most radically within the SCB, exchanging warm San Diego waters for cooler waters north of Point Conception.

Predominant in the avifauna of California, in terms of numbers and biomass, are those high-latitude nesters (both north and south) that attain greatest abundance in cool waters of the upwelling zone (Briggs et al. 1987). Likewise, in warmer waters seaward of the upwelling zone in the eastern half of the SCB (south of the cool California Current), bird numbers are greatest in winter when these high-latitude birds visit the SCB. Birds associated with cooler (northern or upwelling-influenced) water or warmer (southern) water are listed in table 10.4.

Diversity of seabirds is lowest from May to August, with highest values in fall to early spring, reflecting the arrival of species that nest elsewhere (Briggs et al. 1987). The greatest numbers of species appear during fall and spring migration; lowest numbers are observed in June and July. The proportion of avifauna in the SCB that are nesting residents is quite small compared to populations in central and northern California (Briggs et al. 1987). Annual declines in species diversity in late spring and early summer reflect dominance by shearwaters and phalaropes. These two species account for half or more of the individuals present not only in the SCB but in the entire state of California as well (Briggs et al. 1987).

Only the more important seabird species are discussed here with respect to nesting and foraging distributions. Typical seabird habitats and seasonal numbers or densities are given when known (table 10.5), and distribution of colonies with numbers of breeding pairs are displayed in figure 10.8. Figure 10.9 shows areas of greatest foraging concentration for the most important seabird species throughout all seasons. Accounts of other seabird species that inhabit the SCB can be found in Garrett and Dunn (1981).

LOONS

Loons are discussed with seabirds because they occupy the same habitat. Three species

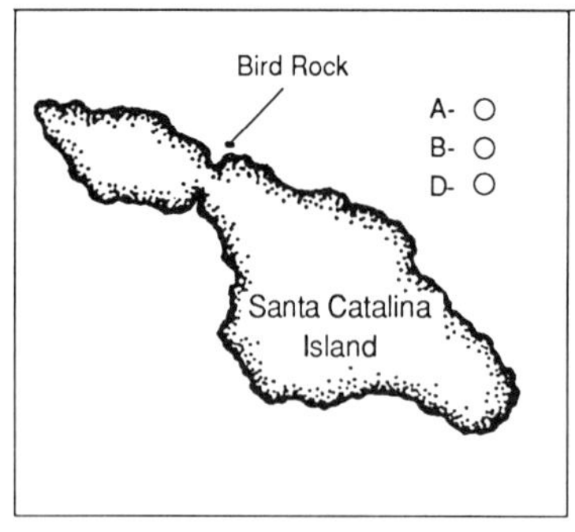

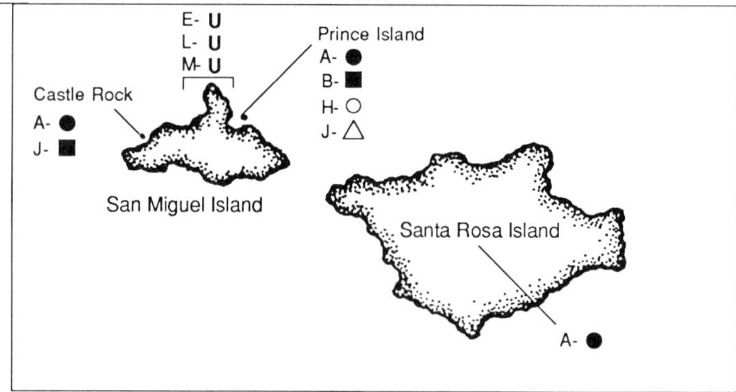

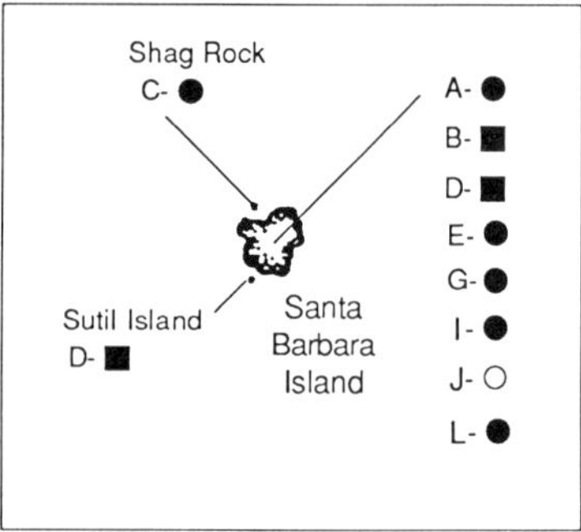

Breeding Colonies of Sea Birds

A- Brandt's Cormorant
B- Western Gull
C- Pigeon Guillemot
D- Xantus' Murrelet
E- Double-crested/Pelagic Cormorant
F- California Least Tern
G- Black Storm-Petrel
H- Least Storm-Petrel
I- Brown Pelican
J- Cassin's Auklet
K- Rhinoceros Auklet
L- Ashy Storm-Petrel
M- Leach's Storm-Petrel

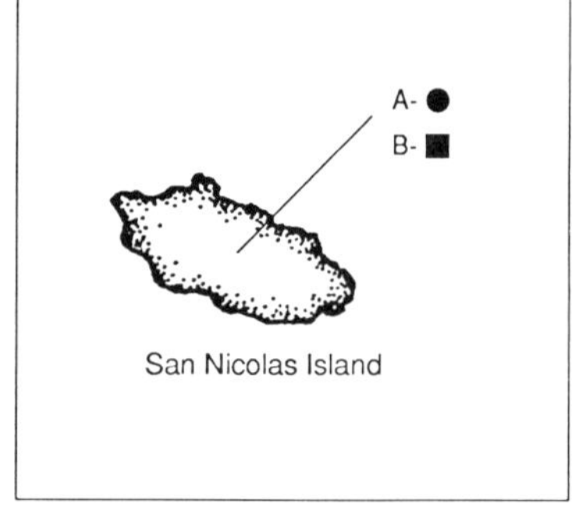

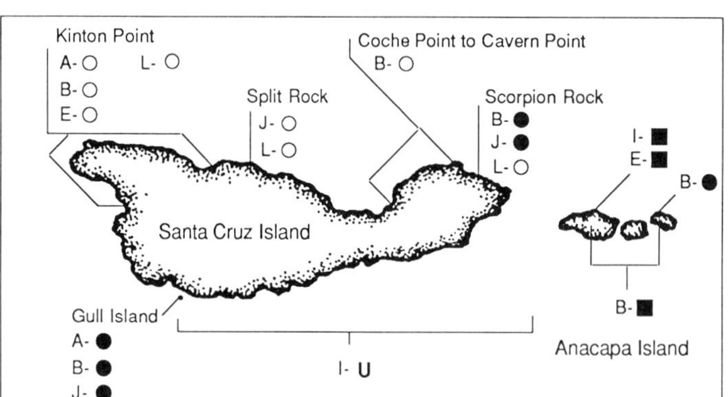

Colony Number

△ = 10,000 - 1,000,000
■ = 1,000 - 10,000
● = 100 - 1,000
○ = <100
U = Unknown

Figure 10.8. Abundance and distribution of seabirds at breeding colonies in the SCB. (From Briggs et al. 1987; K. Garrett pers. comm.)

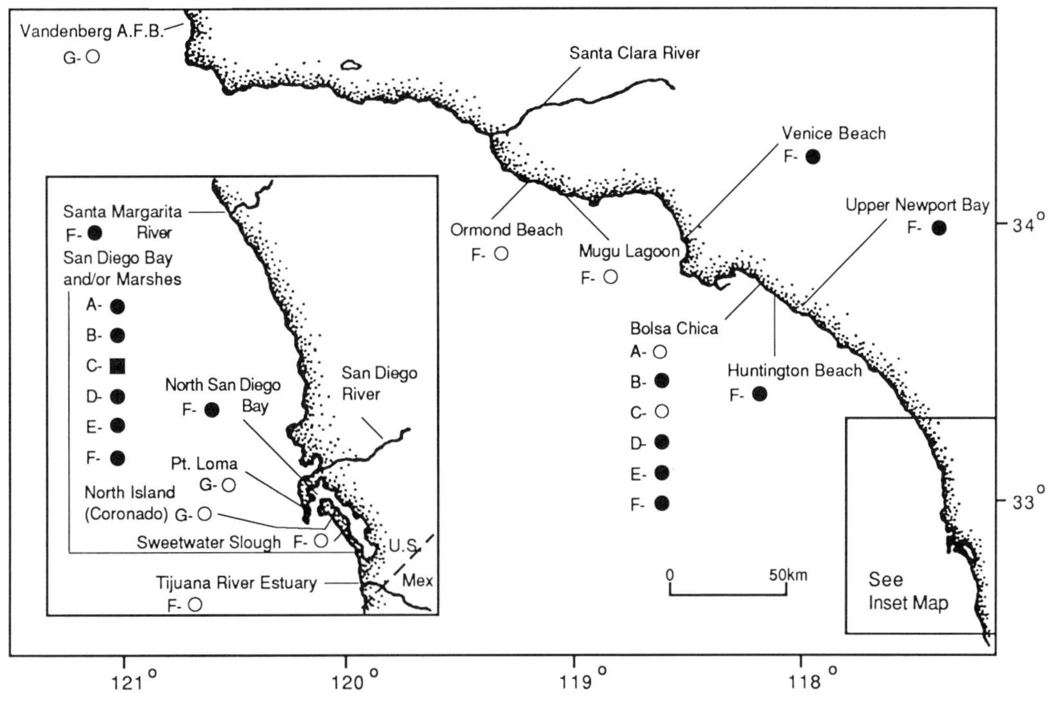

Figure 10.8. (continued)

Table 10.4. *Birds Indicative of Water Temperatures in the SCB*

Species Indicating Warmer Water	Species Indicating Cooler or Upwelling Water
Least storm-petrel	Black-footed albatross
Ashy storm-petrel	Sooty shearwater
Black storm-petrel	Cassin's auklet
Black-vented shearwater	Western and Clark's grebes
Pink-footed shearwater	Western gull
Heermann's gull	California gull
Brown pelican	
Elegant tern	
Royal tern	
Xantus' murrelet	

From Briggs et al. 1987.

Table 10.5. *Numbers, Densities, and Special Habitats Used by Seabirds in the SCB*[a]

Species	Numbers	Densities (birds km^{-2})	Habitats
Pacific loon[b]	M: 40,000–60,000 W: 5000 (Mar–Dec)	0.4–1.8 Max = 80 in Santa Barbara Channel	
Surf scoter[b,c]	T: 125,000 M: late Mar–early May and late Oct; SU: hundreds W: 5000; peak in Dec	1–5 (Dec–Mar)	On shelf and slope. Northern Channel islands important for M. Water temperatures: 10–13°C Unacceptable: >14.5°C Sandy substrate, lee of promontories
Black-footed albatross[d,e,f]	F and W: lowest numbers	S: 0.05–0.1 (May–Jun)	Coolest and most productive waters
Pink-footed shearwater[b]	Max: 60,000–400,000 (May–Sep) peaks: May–Jun; Aug–Sep	5–8	Continental shelf and slope, cool shallow regions; water 15–20°C
Sooty shearwater[b]	2.7–4.7 million statewide		Continental shelf, surface water 10–15°C; prefer water 11–14°C
Black-vented shearwater[b,g]	F–W: 20,000–30,000 statewide (Sep–Dec)	Max = 80 near Oceanside	Present all months except April Nearshore 75 km north from San Diego
Leach's storm-petrel[b]	SU: 150,000 (late Aug)	SU–F: 0.5–2.6 (Jun–Oct); W–S: 0.05–0.5 (Dec–May)	Seaward of continental slope, California Current in summer
Black storm-petrel[b]	SU–F: 100,000 (Aug–Sep)	SU–F: 2.0 W–S: <0.1	In highest surface temperatures
Ashy storm-petrel[b]	1400 Point Buchon south		
Least storm-petrel[b]	F: 200,000		Warmest waters
Brown pelican[b]	56,000–80,000 (May–Dec) 10,000 on land peak: Sep–Oct 5,000–6,000 (Dec–Mar)		At-sea greatest abundance just after period of maximum surface temperatures. Prefer fronts with sharp thermal gradients but not the warmest water.
Brandt's cormorant[b]	20,000 SU to S (80,000 all Calif.) 5500 breeding	Highest densities late SU to late S	Forage in shallow water

Phalaropes[b]	925,000 (3.7 million all Calif.)	S: 70 F: 4.5	W: seaward of shelf; S: near mainland coast near islands, especially Santa Barbara Island kelp beds; M: outermost shelf and upper continental slope
Pomarine jaeger[b]	All seasons except S Peak in F 60,000 statewide W: 1000		Aug: Santa Barbara Channel; W: rest of SCB; F migration: cool water, especially 40 km seaward of Santa Rosa–Cortes Ridge
Bonaparte's gull[b]	S M: 300,000 at one sighting W: 15,000	Nov: 10 Apr–Jun: 0.1	Within 40 km of mainland or on Channel Islands
Heermann's gull[b]	S: 8,000–10,000 (Jun–Jul) F: 15,000 At-sea: several 1000's		Nest in Mexico; most numerous gull on beaches, Santa Barbara Channel to San Diego County
Ring-billed gull[b]	W: 10,000		Protected bays and estuaries; seldom >1 km from shore
California gull[b]	F–W: 5000 beaches Peak M: 150,000 (Oct–Mar)	5–8 Highest within 50 km of mainland	Neritic water
Herring gull[b]	F–W: 2500 beaches 30,000 at-sea	Peak Jan–Mar	San Diego dump; Santa Rosa–Cortes Ridge; Santa Barbara Channel
Western gull[b]	S–SU: 5,000–10,000 (breeding) W: 25,000–50,000 (Jan–Feb peak)		Generally not >25 km seaward of shelf break. Numbers decrease with oceanic warming
Black-legged kittiwake[b]	W: invasionary, irregular Up to 50,000–300,000	Up to 2–3	Offshore
Royal tern	435 around Channel Islands especially San Miguel, Santa Rosa, and San Clemente		Rare >1 km at sea
Elegant tern	S: 450 Bolsa Chica Several thousand along mainland beaches		

Table 10.5. *Numbers, Densities, and Special Habitats Used by Seabirds in the SCB*[a] (continued)

Species	Numbers	Densities (birds km^{-2})	Habitats
Common and arctic terns[b]	F M: 30,000–50,000 S M: peak Apr–May S densities lower than F		Many migrate seaward of SCB. Common terns more numerous within 25 km of mainland on coastal beaches, estuaries. Arctic terns more numerous 25 km offshore. Both occur throughout SCB; concentrated over the slope primarily in clear waters, outside upwelling zone. Migration west of Santa Barbara Channel to Cortes Bank.
Forster's tern	Late S: 500 on beaches + 80 Bolsa Chica W: none		Mainland beaches, coastal bays and estuaries, up to 15 km from shore.
Caspian tern	900 breeders San Diego 200 at Bolsa Chica		Nests at San Diego Bay and Bolsa Chica Lagoon
Black skimmer	150 Bolsa Chica		
Cassin's auklet[b]	Breeding: 22,000 20,000 at San Miguel W: 500,000–1,000,000 statewide F: 250,000–500,000 50–100,000 SCB.	>100 (near colonies at San Miguel S and SU)	Mid-shelf seaward to 150 km offshore. Concentrated over continental shelf and slope San Miguel to Point Buchon in central and western portions of SCB. SU–F: outer shelf, coastal upwelling (May–Oct); W–S: deeper water, >2000 m (Nov–Apr).
Rhinoceros auklet[b]	W: peak 100,000–300,000 (Feb–Mar) After Apr: few birds		W: abundant offshore. Rest of year, western SCB shelf break, Santa Barbara Channel islands.

[a]Because of different foraging behavior (solitary or group), some species with small population numbers may have higher localized densities than do species with larger populations. M = migration; S = spring; SU = summer; F = fall; W = winter; T = total; I = instantaneous counts.
[b]Briggs et al. 1987. [c]Unitt 1984. [d]Sanger 1974. [e]Ainley 1976. [f]Briggs and Chu 1987. [g]K. Garrett, pers. comm.

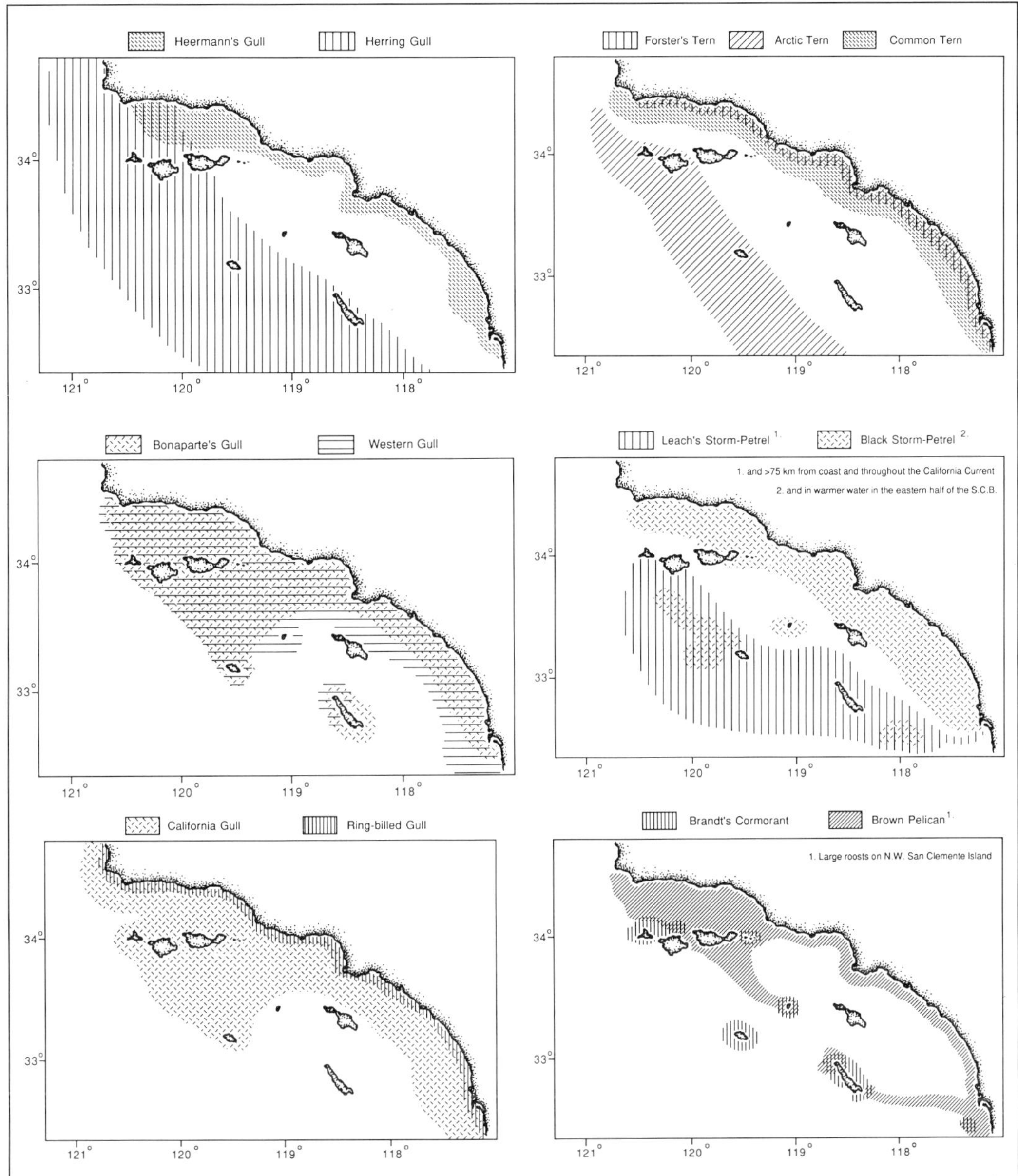

Figure 10.9. Most heavily used foraging areas for selected seabirds in the SCB.

of loons (red-throated [*Gavia stellata*], Pacific [*Gavia pacifica*], and common [*Gavia immer*]) inhabit the SCB; the most abundant and widely distributed of the three is the Pacific loon (Small 1974; Ainley 1976). It is the most common loon found within 40 km of the mainland and is also found offshore in high density (at 75 km, 1.0 bird km^{-2}) (Briggs et al. 1987). This species prefers winter habitats in sheltered areas along the mainland and island coasts.

Common loons are on the highest priority list of Bird Species of Special Concern (Remsen 1978), meaning that they face immediate extirpation of their entire California breeding population if current trends continue. They

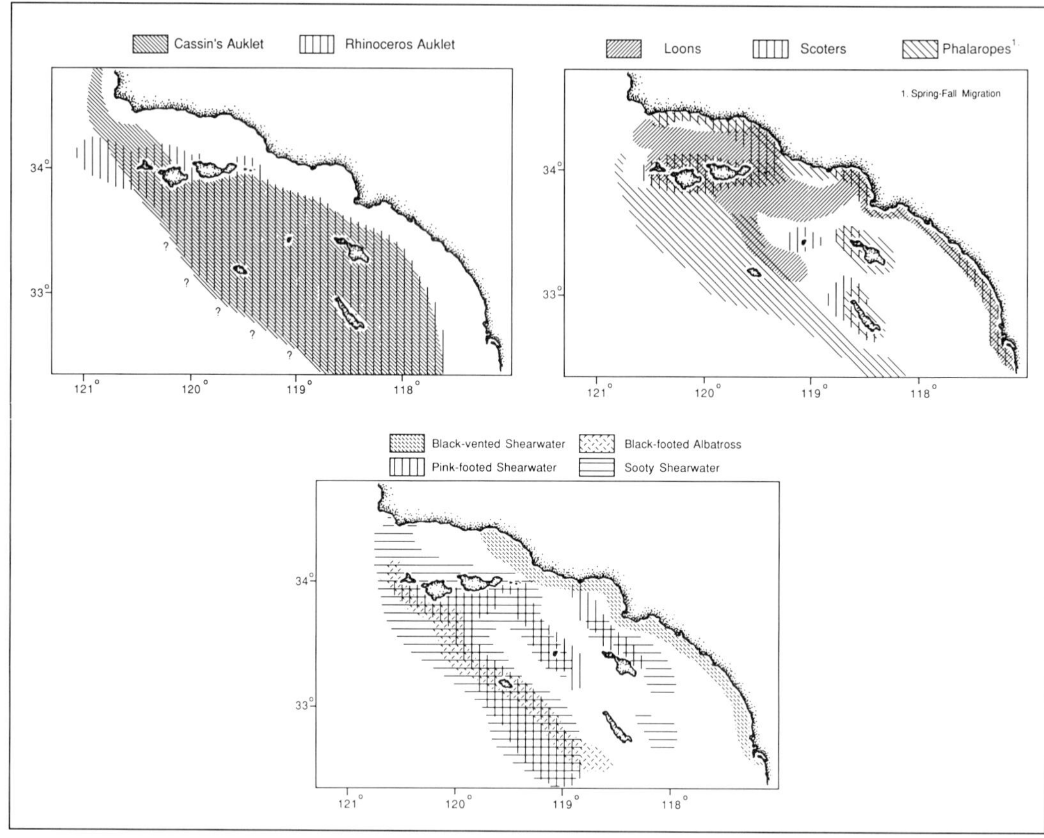

Figure 10.9. (continued)

are more numerous in northern California, but in the SCB they number fewer than 1000 birds from December through March, with spring migration numbers at 5000–10,000 (Briggs et al. 1987).

WESTERN AND CLARK'S GREBES

The western grebe (including the similar appearing Clark's grebe) is a predominant species in waters within 0.5 km of the mainland coast from October through May and is included with seabirds because it occupies the same habitat (Briggs et al. 1987). Its preferred foraging habitat is over sandy bottoms less than 10 m deep, especially downwind from major headlands. The shallow waters at the eastern end of the Santa Barbara Channel are known to support 2000–27,000 (average of 10,000 over three surveys) of these grebes. In summer, however, the counts along the coast are low (500–800 birds), although this may be an underestimation (Briggs et al. 1987).

ALBATROSSES

The black-footed albatross is the most numerous albatross in the SCB (and in all U.S. coastal waters as well), but it is two to ten times more common north of Point Conception (Sanger 1974). In the SCB, this species is more numerous far offshore in the California Current proper than within 100 km of the coast (Briggs et al. 1987). Peak sightings have been documented in the coolest, most productive southern California waters within 25 km of the Santa Rosa–Cortes Ridge, especially near San Miguel Island and the Tanner–Cortes Bank area (Briggs et al. 1987).

FULMARS

The northern fulmar (*Fulmarus glacialis*) is one of the more numerous species in the SCB, numbering from 120,000 to 300,000, with highest densities of 2–3 birds per square kilometer occurring from October through March or April (K. Briggs pers. comm.). These birds are most numerous seaward of the middle of the continental shelf (Briggs et al. 1987) and are reported to be associated with regions of cool surface temperatures and high surface salinity (Ainley 1976). This may explain their abundance near upwellings at Point Conception and San Miguel Island. Their population numbers decline during environmental warming (Ainley 1976), and their numbers can vary by an order of magnitude (Vermeer and Rankin 1984).

SHEARWATERS

The pink-footed shearwater (*Puffinus creatopus*) is more abundant off southern California than it is north of Point Conception, but it is associated with cool (15–20°C), shallow regions of the SCB (Ainley 1976; Briggs and Chu 1987).

Sooty shearwaters (*Puffinus griseus*) are one of the most numerous and common species found in the SCB. The greatest numbers occur over the continental shelf in May when large concentrations appear near or downstream from stable waters of upwelling, especially when a thermohaline front forms at the seaward edge of the upwelling (Briggs et al. 1987). These shearwaters can be as dense as 1 per square kilometer over the shelf, and slope abundances of greater than 10 birds per square kilometer are fairly common in June. Maximum densities have been 18 birds per square kilometer (Briggs and Chu 1987).

Buller's and black-vented shearwaters (*Puffinus bulleri* and *Puffinus opisthomelas*) are not as numerous as sooty shearwaters; however, their populations number in the thousands. Buller's shearwaters occur mainly on the seaward side of the shelf break on the warmer sides of temperature fronts and are least common in California coastal waters during years of warm temperatures (Briggs et al. 1987). Black-vented shearwaters occur mainly nearshore, from the San Diego area northward about 75 km (Briggs et al. 1987). The shearwater extends its range northward during years of high fall temperatures at San Diego (which usually occur when the Davidson current develops strongly over the shelf) (Ainley 1976; Briggs et al. 1987).

STORM-PETRELS

Leach's storm-petrel is a highly pelagic species found year-round off southern California. This petrel nests in small numbers on Prince Island, but the majority of California nesters occur in the region just south of the SCB in the Coronado Islands. This species is always more abundant seaward of the central continental slope than over the shelf and is always the most numerous petrel in the California Current (Briggs et al. 1987). In midsummer, many nonbreeding Leach's storm-petrels migrate into an area more than 75 km from the mainland. Later, they move to the northwest, where they are concentrated in California Current waters seaward of the outermost islands (Briggs et al. 1987). The few petrels that remain in the winter are found more than 100 km seaward of Point Conception (Crossin 1974).

Black storm-petrels are listed in the third priority category of Bird Species of Special Concern (Remsen 1978) because of their small numbers in California. Although they are not in danger of extinction, they are vulnerable should a threat arise. They are found during all months in the SCB, with a peak in late summer to fall. They are known to nest only at Santa Barbara and Sutil islands in small numbers (Pitman and Speich 1976).

Ashy storm-petrels are not abundant in the SCB, but concentrations of this species occur around San Miguel Island and along the shelf break to 25 km seaward along the Santa Rosa Ridge during the May–September nesting season (Briggs et al. 1987). Their population numbers are small, and for this reason, they are placed on the third priority list of Bird Species of Special Concern (Remsen 1978). Threats to ashy storm-petrels are human disturbance, introduction of small mammals on breeding sites, and oil spills in their foraging areas. Except for the Farallon Islands, Prince Island and Castle Rock (off San Miguel Island) have the largest colonies of these storm-petrels (Sowls et al. 1980).

Least storm-petrels are fairly abundant (200,000) fall visitors to the SCB. They have been sighted primarily in the warmest water and are more abundant south of the SCB, especially in the Gulf of California (Briggs et al. 1987).

PELICANS

The brown pelican is on both the California and the Federal Endangered Species lists, although the species has begun to recover from its lowest points reached in the 1960s and early 1970s. At the time of maximum numbers (September–October), about 70–80% of the statewide population occurs south of Point Conception, including 56,000–80,000 birds at-sea and another 10,000 on land (Briggs et al. 1987). Anacapa Island was the only breeding site used by large numbers from 1968 through 1979 (Remsen 1978; Briggs et al. 1981), and today it is the northernmost nesting colony. Scorpion Rock at Santa Cruz Island and Santa Barbara Island were recolonized recently, but breeding there has been sporadic (Remsen 1978; Briggs et al. 1981). Prince Island, a historical colony, has not been recolonized. Annual nesting populations vary from a few hundred to more than 6000 pairs (Briggs et al. 1987).

Brown pelicans roost on land at night at a small number of sites. The greatest numbers of birds roosting in the SCB are found on Santa Barbara Island, where over 6000 pelicans have been counted. The Channel Islands are the most important diurnal and nocturnal roosting areas, and the use of sites is variable, changing seasonally and yearly. The largest numbers of pelicans forage in shallow, warm water within 20 km of the coast; densities at sea have been found to be highest just after the period of maximum surface temperature (August–October). However, birds are not necessarily found in the warmest waters (Briggs et al. 1987). Peak populations have been recorded at 65,000–94,000 from September to October. During migration in August and December, pelicans have no affinity for any particular environmental variable (Briggs et al. 1981).

Low reproductive output linked with pollution by chlorohydrocarbons (especially DDT) is mainly responsible for the population decline of brown pelicans (Risebrough et al. 1971; Jehl 1973; Anderson and Anderson 1976; Anderson and Keith 1980; Anderson et al. 1982). Anchovies are the preferred food of brown pelicans, and they have been found to contain high levels of DDE, the breakdown product of DDT (Anderson et al. 1975). The banning of DDT and protection of brown pelican nesting areas from human disturbance during the breeding season seem to be encouraging population increases again (Anderson et al. 1975; Anderson and Keith 1980). However, some members of the population spend part of the year outside the United States, so the species is still subject to harm from DDT, for this chemical has not been banned in some other countries. Population numbers are also sensitive to fluctuations in anchovy abundance. Therefore, management of these fish stocks at relatively high levels, combined with worldwide banning of DDT, will help to ensure adult pelican survival.

CORMORANTS

Cormorants are not as abundant in the SCB as they are in central and northern California. Brandt's cormorant (*Phalacrocorax penicillatus*), the most abundant species in the SCB, occurs in highest densities between late summer and late spring, a pattern that reflects an influx of northern birds after their breeding season is over (Osborne 1973; DeSante and Ainley 1980).

Brandt's cormorant is usually found within 25 km of mainland or island roosts and colonies, generally not more than 10 km from shore (fig. 10.8) (Briggs et al. 1987). This cormorant nests on islands or on inaccessible mainland bluffs. About 5500 nesting birds are present in the SCB, but the largest colonies of 1400–1800 birds each are at the Santa Barbara Channel Islands (Hunt et al. 1981). Colonies of all species of cormorants often shift location from year to year.

Double-crested cormorants (*Phalacrocorax auritus*) used to breed on the majority of the coastal cliffs and offshore islands in the SCB, but they are rare at present (Remsen 1978). They have declined in numbers throughout their North American range and are on the second highest priority list of Bird Species of Special Concern (Remsen 1978). This listing means that they are "definitely on the decline in a large portion of their range in California, but populations are still sufficiently substantial that danger is not immediate." Habitat destruction, human disturbance, fluctuating water levels in freshwater nesting areas, and DDE thinning of eggshells have contributed to nesting failures and subsequent decline of both the double-crested and the Brandt's cormorant (Gress et al. 1973; Hunt et al. 1980).

SCOTERS

Scoters feed alongside seabirds. They spend the majority of their lives on or near the ocean, staying in shallow nearshore waters and in bays and estuaries. They are among the most numerous waterbirds of inshore waters and harbors during winter and early spring, and are more abundant near the Channel Islands than near the mainland (Garrett and Dunn 1981). Because of their choice of nearshore habitats, they are vulnerable to inshore oil spills that often end up at the beach. The surf scoter (*Melanitta perspicillata*) is the most abundant scoter off southern California, composing 90–95% of all scoters south of the Santa Barbara Channel Islands (white-winged scoters [*Melanitta fusca*] compose the other 5–10%). The black scoter (*Melanitta nigra*) is unimportant in the SCB.

PHALAROPES

Red and red-necked phalaropes (*Phalaropus fulicaria* and *Phalaropus lobatus*) occur together in large numbers as they migrate through the SCB (fig. 10.9), although red phalaropes predominate. The red phalarope is more often found greater than 50 km from the mainland; the red-necked phalarope usually appears close to shore. Migratory populations of red phalaropes peak approximately one month later than those of red-necked phalaropes. However, because these two species often occur together, they are displayed together in the tables and figures of this chapter. During spring migration, phalarope numbers are highest in neritic waters near coasts and islands, for example, more than 35,000 in kelp beds in the Santa Barbara Channel Islands (Briggs et al. 1987). Spring and fall migrants concentrate over the outermost shelf and upper continental slope, and phalarope numbers are higher seaward of the shelf than close to shore. Densities are highest in winter.

JAEGERS

Pomarine jaegers (*Stercorarius pomarinus*) are present in the SCB in small flocks or singly, except during summer when only a few individuals are present; they are the most numerous jaeger found offshore. On the southward migration from their arctic nesting grounds, they reach the SCB in August, concentrating near the Santa Barbara Channel before dispersing into the rest of the SCB in mid-September (Briggs et al. 1987). Highest numbers are found in cool waters within 40 km of the Santa Rosa–Cortes Ridge (Briggs et al. 1987).

Parasitic jaegers (*Stercorarius parasiticus*) dominate inshore waters within 15 km of the mainland and inland shore. They number about one order of magnitude less than the pomarine jaeger.

GULLS

Most gulls in the SCB are present in greatest numbers only during the winter, and only one species, the western gull (*Larus occidentalis*), breeds in the SCB. Bonaparte's gulls (*Larus philadelphia*) are more abundant at sea during their migration period than any other gull. Their overwintering population is centered in southern California, where they remain along the coast from December through March. Up to 300,000 birds were counted at one sighting during migration in March through May (Briggs et al. 1987).

Heermann's gull (*Larus heermanni*) is the most numerous gull on the beaches along the Santa Barbara Channel and in San Diego County (Briggs et al. 1987). It does not, however, appear in large numbers at sea near the Channel Islands or over open water between the islands and the coast, and it usually forages no more than a few kilometers offshore.

Ring-billed gulls (*Larus delawarensis*) are found in protected bays and estuaries of mainland beaches and are seldom more than 1 km from shore during the winter. In late spring, they migrate inland to their nesting areas in the Rocky Mountain states (Baird 1976). California gulls (*Larus californicus*) are among the most abundant gulls in neritic waters in the fall and winter, although they also forage at sea (Briggs et al. 1987). Numbers peak from January through March; in spring, most California gulls leave for their inland nesting grounds.

Herring gulls (*Larus argentatus*) peak in January through March and can number up to 2500 birds on the mainland and islands (Briggs et al. 1987). They are the most numerous on the beaches of eastern Santa Barbara Channel and the San Diego area. Total populations can reach 30,000 in winter, and they are most numerous west of the Santa Rosa–Cortes Ridge (Briggs et al. 1987).

Western gulls occur in neritic waters throughout the year and stay close to their colonies during April through August (Briggs et al. 1987). They are rarely found greater than 25 km seaward of the shelf break. Declines in their population numbers over a period of years have been associated with periods of ocean warming and storminess (Briggs et al. 1987).

The black-legged kittiwake (*Rissa tridactyla*) is found mainly in northern California, although they occasionally migrate into the SCB in large numbers during the winter (fig. 10.7) (Briggs et al. 1987).

TERNS

All species of terns are sensitive to human disturbance and habitat destruction. Their colonies are found only on the mainland, either on sandy beaches or on islands in estuaries or lagoons. Because they nest in close proximity to populated areas or areas targeted for development, their nesting habitat is slowly diminishing; their populations have been decreasing since the beginning of this century. For these reasons, all species of terns are discussed here, even though their populations may be small.

In the nonbreeding season, royal terns (*Sterna maxima*) are found in low numbers (less than 450 birds) around the shores of San Miguel, Santa Rosa, and San Clemente islands, rarely more than 1 km offshore. They nest at San Diego Bay and at Bolsa Chica Lagoon. Peak numbers occur in September, and the majority of their California population occurs in the SCB (B. Massey pers. comm.).

Elegant terns (*Sterna elegans*) number several thousand and are found along the mainland beaches of southern California, rarely more than 4 km from the coast. They are in the third priority category of Bird Species of Special Concern (Remsen 1978) because of their low numbers and their vulnerability to extirpation. Their breeding population numbers have been slowly increasing, but in some areas, more rapidly. For example, in Bolsa Chica Lagoon in 1988, there were 900 birds nesting, up from 30 in 1987 (Burkett 1988). They also nest at the south end of San Diego Bay (900 birds), and they court at Camp Pendleton at the Santa Margarita River estuary (McCaskie 1974). In late summer and fall, they migrate northward to northern California.

Most common and arctic terns (*Sterna hirundo* and *S. paradisaea*) migrate seaward of the SCB from their northern breeding grounds on their way to overwintering in South America. Of those found in the SCB, common terns are most numerous within 25 km of the mainland, while arctic terns predominate seaward of 25 km from shore (Briggs et al. 1987). Numbers peak in April and May and are 10 times more numerous than in the fall. Southern California numbers account for only 30–50% of those present in northern California (30,000–50,000).

Forster's terns (*Sterna forsteri*) are common along mainland beaches, coastal bays, and estuaries along the entire coast, but they are also found up to 15 km from shore. About 500 birds occur scattered along the beaches in the summer.

The Caspian tern (*Sterna caspia*) nests in Bolsa Chica Lagoon and in south San Diego Bay, where breeders number about 900 birds (California Department of Fish and Game 1980). The black skimmer (*Rynchops niger*), a recent breeder in the SCB since 1984, nests in Bolsa Chica Lagoon (>300 birds), Anaheim Bay, and south San Diego Bay (70–80 birds; W. Schew pers. comm.). This species is on the third priority list of Bird Species of Special Concern (Remsen 1978).

The California least tern (*Sterna antillarum browni*) is listed on both the California and Federal Endangered Species lists, and its nesting sites are protected. Most colonies are small (<50 birds), ranging from the Santa Ynez estuary in the north to the Tijuana River estuary in the south (Massey 1988). Largest colonies (as of 1988) of 150 to more than 300 birds were at Point Mugu, Venice Beach, Anaheim Bay, Bolsa Chica Lagoon, Huntington Beach State Park, Upper Newport Bay, Santa Margarita River estuary, and San Diego's Lindbergh Field (Massey 1988). Many colonies of this species are now protected behind chain link or electric fences. Predator control around one of the most successful least tern colonies, Santa Margarita River at Camp Pendleton, has helped to lower egg and chick mortality. Populations of this tern have been steadily increasing since the mid-1970s when mitigation measures were first initiated. The breeding season of 1988 found approximately 2670 adults, an increase of approximately 300 from 1987 (Baird 1989) and one of the highest breeding populations since the El Niño in 1982–1983 (Massey 1988). In 1988 the population returned to its 1983 level, indicating recovery from El Niño effects.

ALCIDS

Common murres usually inhabit the Santa Barbara Channel in the winter (20,000–30,000 birds) and are most common in water less than 150 m deep. They often dive to 100 m (T. DeGange pers. comm.), and because of this foraging characteristic, one of their greatest threats comes from the gill net industry (see section on effects of human activities). Thousands of murres drown yearly in gill nets (H. Carter pers. comm., in Sowls et al. 1980). The El Niño–Southern Oscillation (ENSO) of 1982–1983 also caused failure of northern nesters and an overall decrease by two-thirds in densities at sea in midwinter in northern and central California (Briggs et al. 1987). Murres are a species vulnerable to oil spills because they sit on the water for long periods of time after feeding.

Xantus' murrelets nest both in the SCB and in Mexico. Their main nesting area in the SCB is Santa Barbara Island (5000 birds) (Hunt et al. 1981), and they remain close to the colony from March through May (Briggs et al. 1987). Adults escort their dependent young to sea in May, and at this time, the SCB population disperses from San Diego to Rodriguez Dome (Briggs et al. 1987).

The breeding population of Cassin's auklet (*Ptychoramphus aleuticus*) is concentrated in two places in California: in the Farallon Islands (approximately 100,000), and on Prince Island, off San Miguel Island (approximately 20,000) (Briggs et al. 1987). Total southern California breeding populations are now about 22,000, but they were formerly much more common at Santa Barbara Island (Hunt et al., in Power 1980). Cassin's auklets are one of the most numerous species found from midshelf seaward to 150 km offshore. From August through October, however, in the postnesting period, they disperse throughout the SCB (Briggs et al. 1987). They are nocturnal on their colonies, feed diurnally at sea, and nest in burrows on offshore islands or in shallows north of Prince Island (Hunt et al. 1980).

Numbers of both Xantus' murrelets and Cassin's auklets at Santa Barbara Island first declined then later increased because of changes in predation pressure (introduction and then removal of terrestrial predators as well as disappearance of peregrine falcons as a breeding species there). Causes of population fluctuations of auklets on Prince Island are not known (Hunt et al. 1980).

Rhinoceros auklets (*Cerorhinca monocerata*) constitute up to 30% of all seabirds off southern California in the winter (100,000–300,000 birds), composing one of the most important elements of the wintering fauna south of Monterey (Briggs et al. 1987). They are included on the third priority list of Bird Species of Special Concern (Remsen 1978). Their SCB populations are particularly at risk for two reasons: (1) like murres, they remain on the water after feeding and are therefore vulnerable to oil spills; and (2) the majority of their eastern Pacific nesting population is located off California in February and March and could be decimated by environmental perturbations at this time (Briggs et al. 1987). These auklets are abundant on offshore waters the length of the state during winter and are most often found seaward of the shelf break.

Breeding and Reproductive Ecology

Reproductive success of marine birds depends on various biotic and abiotic factors. There is growing evidence that ocean temperatures, rainfall, and wind can influence abundance, distribution, and catchability of prey in some oceanic areas (Solomonsen 1955; Pugh 1962; Dunn 1973; Birkhead 1976; Vermeer et al. 1979; Vermeer 1980; Ingraham 1982; Baird 1990). Prey abundance and availability, in turn, influence reproductive success (Vermeer et al. 1979; Vermeer 1980; Anderson et al. 1982; Baird 1991). Except for pelicans (Anderson et al. 1980, 1982) and western gulls (Hunt and Butler 1980), few studies to date have described this association for seabirds of the SCB. Presumably, however, the same factors that influence marine bird populations elsewhere influence those in the SCB. High levels of variation in annual food supply can influence high survival variation in reproductive output of bird populations.

Correct timing of egg laying is important for reproductive success because if chicks hatch during periods of abundant prey, the probability of their survival is greatest. However, too early or too late egg laying produces chicks that are faced with a decreased food supply. Greater chick mortality, lower fledging weights, and reduced fledging success have all been noted for seabird chicks that hatched either early or late (Coulson and White 1961; Vermeer 1979; Baird and Gould 1983).

Availability of preferred foods in adequate amounts is known to influence fledging surges in some species (Boersma 1978; Harris and Hislop 1978; Baird 1979, 1990; Hunt and Butler 1980; Anderson et al. 1982). Different species react to food shortages in different ways. Some abandon the breeding grounds, others lay reduced clutches, and others allow the chicks to starve.

Temporary food abundances can also influence fledging success. Populations of scavengers, such as gulls, often remain stable or even increase, partially because of their scavenging food habits, because they can usually find enough for the young to eat (Hunt and Hunt 1975; LFPPO 1989). Whereas fish prey populations for many other species of birds have decreased, garbage dumps and dumping of offal at sea have increased in this century, enabling gulls and other scavenging populations to thrive (Hunt and Hunt 1975).

Because of their highly visible and stable colonies, seabirds can be used as indicators of the health of the ocean. Observations of annual population numbers, and of annual reproductive success, can reveal patterns of highs and lows that are often correlated with food-rich or food-poor years, respectively (Boersma 1978; Baird 1979, 1990; Anderson et al. 1982). Prey abundance, in turn, can be correlated with abiotic changes in the oceanic environment (Dunn 1973; Birkhead 1976; Vermeer et al. 1979; Vermeer 1980; Ingraham 1982; Baird 1990). Detection of changes in abiotic factors or in populations of fish or invertebrates is more difficult than detection of changes in numbers or success of seabirds, and thus seabirds can be used as biological indicators of these changes.

Marshbirds

Known nesting locations of marshbirds are shown in figure 10.3, although some small active rookeries may exist elsewhere (Garrett and Dunn 1981). The breeding ranges of great blue herons and Virginia rails have been substantially reduced in this century because of development along the coast and destruction of coastal wetlands (Garrett and Dunn 1981). In contrast, cattle and snowy egrets, which nest colonially with other herons and ibises in both salt water and freshwater habitats, are expanding their breeding ranges.

Virtually nothing is known about green-backed heron nesting biology and distribution in the SCB. Although black-crowned night herons often nest with other species of herons or ibises, their breeding distribution in the SCB is also poorly known (Unitt 1984). Small colonies of this species, other than those displayed in figure 10.3, probably occur in all wetlands of the SCB. At least three former nesting sites of black-crowned night herons are known (Imperial Beach and Balboa Park in San Diego County, and Inglewood in Los Angeles County) (Garrett and Dunn 1981; Unitt 1984).

Waterbirds

The SCB is relatively unimportant as a breeding area for most waterbirds. Breeding ranges of most duck and goose species lie northward of the SCB, and of the 33 species of waterfowl present in winter, only 9 have been recorded breeding in the coastal region. Of these 9, only 4 species commonly breed in the SCB. The remaining 5 breed in only a few areas, nest in extremely low numbers in many wetlands, or have been found breeding only once or twice in the SCB. These 5 uncommon or rare nesters include green- and blue-winged (*Anas discors*) teals, northern pintails, northern shovelers, and redheads (*Anas americana*).

The four species of ducks that are widespread breeders in the SCB (i.e., they breed in larger numbers in most wetlands) include mallards, cinnamon teals, gadwalls (*Anas strepera*), and ruddy ducks.

Shorebirds

Few shorebirds breed in the SCB. American avocets, black-necked stilts, snowy plovers, spotted and least sandpipers, willets, black oystercatchers (*Haematopus bachmani*), and killdeer are the only species that breed in any great number (Palmer 1976a; Garrett and Dunn 1981). Information on their breeding biology from the SCB is poorly known, and the breeding parameters listed in table 10.6 are from studies made elsewhere in the United States. These species can be found breeding at all the marshes shown in figure 10.1.

Table 10.6. *Reproductive Output of Selected Seabird Species in the SCB*[a]

Species	Mean Clutch Size	Mean Hatch Rate (Chicks per Egg)	Chicks per Nest	References
Leach's storm-petrel	1.0	0.8	0.7	Boersma and Wheelwright 1979; Boersma et al. 1980; Baird and Gould 1983; Ainsley and Atkinson 1937
Fork-tailed storm-petrel	1.0	0.8	0.7	Hunt et al. 1981; Baird and Gould 1983; Boersma et al. 1980
Ashy storm-petrel	1.0	0.8	0.7	Hunt et al. 1981; Bent 1922; Grau et al. 1977
Black storm-petrel	1.0	0.8	0.7	Bent 1922; Hunt et al. 1981
Brown pelican	2.0–3.0	?	0.6	Ainley et al. 1974; Anderson et al. 1975, 1980
Double-crested cormorant	2.3 (2–7)	?	0.8	Ainley and Whitt 1974; Ainley et al. 1974; Frame 1972; Ayers 1975
Brandt's cormorant	2.4 (2–6)	?	0.8	Ainley and Whitt 1974; Sowls et al. 1980
Pelagic cormorant	2.2–3.7 (3–7)	(0.3–0.7)	(0.5–0.9)	Ainley and Whitt 1974; Baird and Gould 1983
Western gull	3.0 (1–3)	0.8	1.8	Hunt 1972; Hunt et al. 1981
Forster's tern	3.0 (1–3)	0.8	1.0	Bent 1921; Hunt et al. 1981
Caspian tern	3.0 (1–3)	0.8	1.2	Bent 1921; Hunt et al. 1981
California least tern	2.0 (1–4)	0.9	0.7–1.1	Massey 1989; D. Statlander, J. Tutton pers. comm.
Pigeon guillemot	1.7 (1–2)	0.9	1.3	Hunt et al. 1981; Drent 1965; Kuletz 1983
Xantus' murrelet	1.7 (1–2)	0.4	?	Sowls et al. 1980
Cassin's auklet	1.0	0.5	0.4	Manuwal 1974

[a]Some data are taken from reproductive studies on these same species outside of the SCB. Parentheses indicate range.

Seabirds

The majority of seabirds that inhabit the SCB migrate north or south for their breeding season; 17 species, however, do breed there. The entire California populations of 3 of these breeders (black storm-petrels, Xantus' murrelets, and brown pelicans) are found in the SCB. Table 10.7 shows breeding chronology, mean incubation, and nestling periods of some of the more important seabirds.

Seabirds are long lived but they mature slowly; many do not reproduce until they are 4 years of age or older, and their reproductive rate is low. The breeding cycle incorporates a series of easily identifiable stages leading to production of young, and loss at any of these stages results in lowered productivity. Some species lay only one egg per season; most others lay two to three (table 10.6) (Sowls et al. 1980; Baird and Gould 1983).

Hatching rates seem to be fairly high

Table 10.7. *Breeding Chronology, Incubation, and Nestling Periods of Selected Seabirds in the SCB*

Species	Laying Dates	Hatching Dates	Fledging Dates	Mean Incubation Duration (days)	Nestling Duration (days)
Ashy storm-petrel	1/01–>6/06	<2/15–>7/20	<5/05–10/10	42	76
Leach's storm-petrel	3/25–7/05	5/05–8/25	7/15–11/02	42	66 (63–70)
Brown pelican	2/05–7/05	3/05–8/05	6/05–11/02	30	52
Brandt's cormorant	2/20–6/20	>3/25–>7/15	3/12–>8/20	28–32[b]	40–42
Pelagic cormorant	4/15–>6/20	5/18–>7/20	7/10–>9/03	31	49 (42–58)
Western gull	4/08–6/08	5/08–7/08	6/18–8/25	29	42–45
California least tern	4/25–7/25	5/16–8/15	6/10–9/01	21	20
Caspian tern	4/10–6/16	6/02–7/12	6/15–8/15	25	36–38
Elegant tern	4/20–6/02	6/02–7/15	6/15–8/15	25	32–35
Forster's tern	5/26–6/15	6/02–7/15	—[a]	25	—[a]
Black skimmer	6/02–9/01	6/29–9/28	—[a]	23	34–36
Xantus' murrelet	3/08–6/20	4/15–8/03	4/15–8/05	41	2
Cassin's auklet	>3/05–>6/10	>4/15–>7/15	>5/20–>8/25	38	41–45

[a] Information not available.

[b] Range.

Sources: Sowls et al. 1980; Baird and Gould 1983; Kirven 1969; Massey and Atwood 1981; B. Massey, W. Schew, E. Burkett, K. Keane, and C. Collins pers. comm.

(>80%) in all but the alcids, although fledging rates (chicks fledged per breeding pair) of all species are usually less than one (table 10.6). Because of this low reproductive output, adult populations of seabirds are very important. The success of these species depends on the longevity of the adult, and any decreases in adult numbers due to displacement or mortality, especially reduction in the breeding populations, can severely affect the success and abundance of the entire population.

A few species have been studied in depth in the SCB. Brown pelicans are reported to be making a recovery since DDT was banned (Anderson et al. 1975). Abundance and surface availability of anchovies, together with absence of human disturbances, correspond with pelican breeding success (Anderson and Keith 1980; Anderson et al. 1980, 1982).

Breeding populations of California least terns have been studied intensively at Venice Beach since the mid-1970s and at Santa Margarita Estuary since 1987, and brief surveys of adults have been conducted statewide since the early 1970s (Massey 1988; Baird 1989). These terns lay reduced clutches in years of poor food availability (Baird 1989), and in particularly poor food years, three adults have been found attending one nest (Baird 1989). Adults are also known to move to other colonies or fail to breed at all if food is poor (Baird 1989).

Hunt and Butler (1980) studied western gulls at Santa Barbara Island for a number of years. They have reported an unusual attribute of these gulls: females sometimes pair with other females, laying "supernormal" clutches in one nest. Western gulls also often respond to decreases in availability of schooling fish by failing to breed (Hunt and Butler 1980).

Foraging Ecology and Food Habits

Where a species forages gives an indication of how vulnerable it is to natural or artificial perturbations of the environment. How a bird feeds (e.g., from the surface or by diving) also gives some indication of how long it might be exposed to pollution or any kind of environmental change at the feeding site. The habitat in which a species forages also may be rare or threatened, and knowledge of this foraging habitat may help in management of the species' population in the SCB.

To forage, marshbirds, waterfowl, and shorebirds concentrate in the few remaining estuaries, but these estuaries are often severely impacted by winter storms that destroy prime feeding areas. Because so few wetland habitats remain in the SCB, once a particular foraging area is temporarily destroyed, it is difficult for these birds to relocate to new and uncrowded habitats.

Knowledge of the types, quantities, proportions, and weights of prey a bird species eats helps to place avifauna within the food web, aids in studies of energy and nutrient cycling through the ecosystem, and can be used to help manage avifauna populations. For example, if prey species are known, then any change in prey populations can be predicted to influence the known predator bird populations (Ashmole and Ashmole 1968; Baird 1979, 1990; Vermeer and Westerheim 1982). Although the predominant prey of a species is important, prey found consistently, yet in low numbers, may also be extremely valuable in sustaining a population when the more abundant prey decreases (Ashmole 1963).

Marshbirds

Food habits of marshbirds in the SCB are not well known. Prey items described here for the various species are from more general accounts in California. Areas where marshbirds forage are known from National Audubon Society's annual counts, from intermittent surveys by the California Department of Fish and Game, and from anecdotal accounts. Adult food and typical foraging areas are listed in table 10.8.

The great blue heron feeds in both salt and

Table 10.8. *Food Habits and Foraging Areas of Marshbirds*[a]

Species	Food Eaten (with percentages)	Foraging Areas	References
Great blue heron	68% fish (43% "nongame"), 8% crayfish, 4% amphibians, 5% mice and shrews, 8% insects, rabbits, birds, small lizards	Salt and fresh shallow water, shorelines, wetlands, tidal flats, sandbars, surf, wet meadows, dry pastures, ocean kelp beds	Palmer 1976a,b; K. Briggs pers. comm.; Packard 1943
Great egret	Fish, frogs, salamanders, snakes, snails, crustaceans, insects, small mammals	Freshwater marshes and ponds; feed occasionally with double-crested cormorants	Palmer 1976b
Snowy egret	Small fishes, frogs, lizards, snakes, crustaceans, worms, snails, insects, small rodents, especially *Microtus*	Fresh, brackish, or salt water, lentic tule marshes	Palmer 1976b
Cattle egret	Insects, especially grasshoppers and fly larvae	Less aquatic than other herons. Willows, fields, shorelines of lakes and ponds	Palmer 1976b
Black-crowned night heron	>51% fish (shad, herring, suckers), 6% frogs, 22% crustaceans, 16% aquatic insects, shrimp, tadpoles, snakes, salamanders, molluscs, marine annelids, small birds, and mammals (3% rats and mice)	Diverse habitats: tule marshes, boat marinas	Palmer 1976b
Clapper rail	Decapods (shrimp, crayfish, crabs), small molluscs, aquatic insects, beetles, snails, small fish, clam worms, dipteran larvae		Martin et al. 1951
Virginia rail	Beetles, snails, spiders, dipteran larvae, bugs, seeds of marsh plants (bulrush)		Martin et al. 1951
Sora rail	Beetles and other insects, snails, spiders, crustaceans		Martin et al. 1951
White-faced ibis	Crayfish, insects (beetles, Orthoptera), earthworms	Freshwater ponds with rushes	Palmer 1976b
American bittern	20.3% fish (eels, catfish, bullheads, pickerels, sunfish, suckers, killifishes, sticklebacks, yellow perch), 19.0% crayfish, 23.1% insects, 20.6% frogs and salamanders, 9.6% mice, shrews, voles, 5.2% garter snakes, 2.2% crabs, spiders		Howell 1932; Palmer 1976a,b; Ingraham 1941
Least bittern	40% fish (topminnow, mud minnows, sunfish, yellow perch), 10% crayfish, 21% dragonflies, 12% aquatic bugs, frogs, tadpoles, salamanders, leeches, slugs, crustaceans, occasionally small mammals	Freshwater, especially cattail marshes or aquatic or semiaquatic vegetation. Marshes with bushes or woody growth, tussocks, tall grass, sedgy bogs	Palmer 1976a,b; Howell 1941

[a]Information is from western sites, not necessarily the SCB.

fresh shallow water, on shorelines of wetlands, and on tidal flats or sandbars. Occasionally it feeds in the surf or in meadows, in pastures, on shoulders of roads, or while standing on kelp a few hundred meters offshore (Palmer 1976b; K. Briggs pers. comm.). It has been found in fields on Año Nuevo Island in central California and feeding on baby brush rabbits in grassy fields (K. Briggs pers. comm.).

Black-crowned night herons are found in diverse habitats from tule marshes to boat marinas, where they often perch on masts as hunting or roosting platforms. Adults often hunt at night; young are fed shrimp and fish. Adult food consists of several species of crustaceans, squids, and fishes (Briggs and Chu 1987; Chu 1984). Great egrets (*Casmerodius albus*) fly away from their inland roosts to feed, primarily in freshwater marshes and ponds. They occasionally feed with double-crested cormorants (Christman 1957) and may also alight on the water to feed (Palmer 1976b).

Cattle egrets are less aquatic than are other herons; they eat insects in fields and along bodies of water. Snowy egrets hunt in fresh, brackish, or salt water in lentic areas, often preferring tule marshes (Palmer 1976b). In contrast, white-faced ibises forage in freshwater ponds with rush vegetation. Clapper rails eat mainly aquatic food, especially decapods, small molluscs, aquatic insects, small fish, and clam worms, while Virginia rails eat terrestrial species, mainly beetles, snails, spiders, hemipterans, and dipteran larvae (Martin et al. 1951). Sora rails differ from other rails in that they consume a large quantity of plant food.

Waterbirds

Most waterfowl and grebes forage in wetlands, preferring freshwater marshes and ponds, yet also frequenting brackish salt water estuaries. Exceptions are western grebes, which spend most of their time foraging along the ocean coast, and Canada geese, which are usually found foraging in fields. The areas of greatest concentration of waterbirds are shown in figure 10.5.

Studies on food habits of waterfowl in the SCB, either in winter or summer, are few. Most ducks eat freshwater plants and forage mainly in freshwater, and information on their food habits in these freshwater environments can be found in Martin et al. (1951) and Palmer (1976a). The information presented in table 10.9 is from summaries of food habits of the most common waterbirds in marine-influenced areas. Although some of these studies are from other west coast sites, it is assumed that food eaten there is similar to food eaten in the SCB.

Shorebirds

Shorebirds in the SCB, present mainly in the winter, obtain a large part of their daily energy requirements from intertidal habitats, especially estuaries and beaches. If there is a loss of, or a decrease in, their feeding grounds in one estuary, many birds will move to another estuary. In this way, the displaced population may become distributed over several wetlands and the total population is not reduced. However, when wetlands are as scarce and as isolated as they are in southern California, this normal redistribution may be impossible. Because only so much food is available in each marsh or intertidal area, the removal of just part of a feeding area may mean that the affected population will not be able to move to an already occupied habitat and, therefore, may move away from the area entirely.

Shorebirds also display behavioral responses to feeding densities of their own species. The number of birds feeding in preferred habitats reaches an upper limit and does not increase (Goss-Custard 1979). "Overflow" birds will shift to less preferred areas to feed, but will not crowd together. If less preferred areas are food depauperate, birds pushed into these areas will often leave. The amount and local distribution, and sometimes the absolute density, of prey biomass in estuarine or tidal-influenced habitats are major determining factors of shorebird abundance and distribution (Baird et al. 1985).

The period of greatest competition among shorebirds for prey is in midwinter (Quammen 1981, 1982). There are more shorebirds present at this time crowding onto the few estuaries in the SCB, with a concomitant decline in the abundance of their invertebrate prey (Goss-Custard 1979). Various factors are responsible for this decline in prey abundance. First, the actual prey biomass decreases from summer to fall because of several factors: (1) little replacement growth from the smaller size classes (Goss-Custard 1979; Quammen 1981), (2) mortality of the preferred size class through shorebird consumption or by emigration, and (3) a decrease in biomass of the individual prey species (some bivalves lose up to 40% of their weight during the winter) (Baird et al. 1985). Second, prey is less accessible to shorebirds and harder to capture during the winter because the prey may burrow more deeply or become less active; up to 80% of the prey in a given area can be too deep (Reading and MacGrorty 1978). However, the one attractive factor for shorebirds that can offset lower prey abundance is that in the SCB, the greatest daytime minus tides occur in winter.

Short-term changes also occur in numbers of accessible prey whose concentrations vary with mud temperature (Goss-Custard 1979). Additionally, preferred habitat may be lost in the winter because of storms, tides (the greatest plus and minus tides occur in winter), or overgrowth of vegetation. In southern California, the main growing season for much ground cover is in the winter, and grasses, *Spartina*, or other vegetation may grow over and cover much of the supralittoral zone. When this happens, prey availability sharply decreases and shorebirds may no longer feed in the area.

The mechanical interference of substrates is another factor that influences the shorebird's choice of feeding location. Often, soil resistance to bill movement, rather than prey density, will determine the spatial distribution of these birds in the environment (Quammen 1982). Also, sand grains may interfere with the detection and capture of prey that are similar in size to sand grains. Preferred substrates for shallow-feeding shorebirds (dowitchers, western sandpipers, dunlins, and American avocets) are those with less mud and a moderate amount of sand (42–80%). The preferred prey are several species of small polychaetes and oligochaetes (Quammen 1982). Prey densities in these sandier areas are not reduced. Although prey numbers are reduced in mud substrates (8% sand), shallow-feeding birds often spend more time on them.

Table 10.10 lists preferred prey of some of the more common shorebirds just south of the SCB (Reish and Barnard 1990), and Table 10.11 lists prey for birds found in the SCB (Bent 1927; Martin et al. 1951; Palmer 1976a).

The black oystercatcher forages intertidally on rocks. Its principal prey is the California mussel (*Mytilus californianus*) (30–40%); the remainder of the oystercatcher diet consists of limpets, chitons, barnacles, marine worms, and assorted crustaceans (Martin et al. 1951). Young oystercatchers are primarily fed crabs (Hartwick 1976; Helbing 1977). Very young chicks (<1 week old) are fed mussels, nemerteans, polychaete worms, and tenebrionid beetles, while older chicks are fed mostly limpets (Goss-Custard 1979). In winter, black oystercatchers are known to forage on mudflats at low tides and in grassy fields at high tide where they consume earthworms (Hartwick and Blaylock 1979). The typical feeding habitat of the oystercatcher is an area where many surface pollutants, especially oil, accumulate due to onshore drift. Oil can contaminate not only their feathers but also their prey.

Curlews forage on shores of lakes, rivers, salt marshes, and sandy beaches; they are considered to be more upland foragers than are other shorebirds (Palmer 1976a). Their terrestrial diet includes toads, grasshoppers, and

Table 10.9. *Food of Waterfowl, Grebes, and Coots*

Species	Predominant Food (with percentages)	Comments	References
Canada goose	25–50% pondweed and sago; 10–25% barley and hardstem bulrush; 5–10% wheat, wild barley, bromegrass	Forage in fields.	Martin et al. 1951
Green-winged teal	10-25% pondweed and bulrush; 5–10% windmillet, smartweed, and spikens	Surface feeders; pirate from other ducks. Seeds important.	Martin et al. 1951
Cinnamon teal	10–25% pondweed and bulrush; 5–10% saltgrass, sedges, and horned pondweed	Pirate from other ducks. Food similar to that of green-winged teal.	Martin et al. 1951
Northern shoveler	75% plants; 10–25% pondweed; 5–10% bulrush, spikebush, and saltgrass; animal prey (freshwater molluscs, aquatic insects, crustaceans)	Surface feeders. Do not dive, yet reach bottom in shallow areas. Sift organic material from muddy bottoms with bill.	Martin et al. 1951
Ring-necked duck	75% plants; 10–25% pondweed, muskgrass	Feeds on seeds of watershield and water lily. Divers. Found in rafts of scaup ducks.	Martin et al. 1951
Lesser scaup	Animal food primarily; 67% molluscs (snails) and dragonfly and damselfly nymphs, larval caddisflies, midges, and other dipterans, crustaceans, water boatmen; 5–10% wigeongrass and pondweed	Forage on open water in coastal bays or tide-influenced mouths of large rivers. Divers.	Martin et al. 1951
Pied-billed grebe	46.3% insects; 27% crayfish; 24.2% fish (those found in shallow areas, such as carp, sculpin, catfish, sticklebacks)	Associated with coots at edges of duck flocks. Winter: coastal salt water, lakes, ponds with abundance of emergent vegetation, marshes with open water, and marshy inlets and bays.	Palmer 1976a; Wetmore 1925
Western grebe	81% fish (cyprinids, chubs, catfish, smelt, herring, mullet, sacramento perch); crustaceans; 17% insects (grasshoppers, mayflies, aquatic beetles)	The most marine grebe species. Highly susceptible to oiling.	Palmer 1976a; Lawrence 1950
Eared grebe	Damselflies, dragonflies, grasshoppers, mayflies, caddisflies, bugs, beetles, moths, butterflies, midges		Palmer 1976a
Horned grebe	43% fish (cyprinids, darters, anchovies, sculpin, silversides, yellow perch, cabezon, sea perch, squaw fish); 46% insects (water beetles, caddisfly larvae, mayflies, crustaceans, small frogs, and salamanders)		Martin et al. 1951; Wetmore 1925
American coot	Submerged aquatic vegetation especially duckweed (25–50%), wigeongrass and muskgrass (10–25%); alga; insects (aquatic beetles and bugs, dragonflies)	Forage along coasts and in fresh water or brackish ponds. Surface feeders and divers.	Martin et al. 1951

Black brant	25–50% eelgrass	Forage in bays.	Martin et al. 1951; Cottam et al. 1944
Red-breasted merganser	Fish (86% "nongame")	Forage in marine waters.	Martin et al. 1951
Common merganser	Fish (40% "nongame")	Forage in inland waters.	Martin et al. 1951
Bufflehead	Mainly insects, especially caddisfly larvae, dragonfly nymphs, larval and adult beetles, water boatman, crustaceans, molluscs (univalves)	Divers. Inhabits lakes, bays, large streams, ponds.	Martin et al. 1951
Great, or tule, white-fronted goose	10–25% barley, wheat; 5–10% cattail root and bulrush rootstock	Grain fields, tule marshes.	Martin et al. 1951
Ross's goose	83% oats (seeds and vegetation); 2-5% barley seeds and vegetation		Martin et al. 1951
Barrow's goldeneye	Animal food dominates (molluscs especially, insects, crustaceans)		Martin et al. 1951
Mallard	10–25% pondweed, bulrush; 5–10% barley	Inland ponds, shallow lakes, rivers, upland areas, grain fields.	Martin et al. 1951
Ruddy duck	Animal food: 33% insects (midges, horsefly and caddisfly larvae, water boatmen, molluscs, crustaceans); 10–25% pondweed, bulrush; 5–10% wigeongrass	Bays, estuaries.	Martin et al. 1951
Northern pintail	10–25% pondweed, bulrush; 5–10% windmillet		Martin et al. 1951
Gadwall	10–25% pondweed, alga (muskgrass); 5–10% wigeongrass, horned pondweed	Bays, open ocean, coasts.	Martin et al. 1951
American wigeon	10–25% pondweed, wigeongrass; 5–10% spikebush, alga, alfalfa	Resemble coots in feeding ecology—mainly eat aquatic plants.	Martin et al. 1951; Hochbaum 1944
Canvasback	53% pondweed vegetation and seeds	Large rafts found in bays and other open water.	Martin et al. 1951
Tundra swan	25–50% grasses; 10–25% pondweed, sago (vegetation and seeds), horsetail, burreed (vegetation and seeds)		Martin et al. 1951
Trumpeter swan	60% pondweed, sago (vegetation and seeds); 5–10% waterbutter-cup, arrowhead		Martin et al. 1951

Table 10.10. *Percentage of Prey Species from Shorebirds at Bahia de San Quintin*[a]

Prey Species	Shorebird Species						
	Willet (9/10)	Snowy Plover (3/3)	Western Sandpiper (7/9)	Dunlin (14/17)	Least Sandpiper (17/30)	Short-billed Dowitcher (10/11)	Marbled Godwit (13/15)
Polychaetes							
Boccardia uncata						<1	
Glycera americana					2.0		
Marphysa sanguinea							51.8
Neanthes arenaceodentata				60.2	2.0	14.3	24.1
Onuphis microcephala							18.0
O. pigmentata	1.3						
Prionospia malmgreni						<1	
Scoloplos spp.						3.6	
Spionid						<1	
Eunicid, unidentified	1.3						
Sipunculids, unidentified				<1	2.0		
Ostracods, unidentified		60.7					
Myodocope, unidentified					2.0	8.0	
Isopods							
Anthurid, unidentified		3.6					
Cirolanid, unidentified				11.7			
Ligia spp.	9.2				4.1		
Paracerceis spp.	22.4					<1	
Rocinela cf. *Tuberculosa*						<1	
Isopods, unidentified		7.1					
Amphipods							
Allorchestes spp.					10.2		
Amphithoe plumulosa				<1			

Amphithoe spp.					4.1		
Amphithoid, unidentified						1.8	
Corophium baconi		3.6					
Corophium spp.		14.3				33.0	<1
Hyale frequens		7.1					
Hyale spp.	32.9						
Lembos spp.		3.6					
Orchestia traskiana	19.7						
Paraphoxus spinosus							<1
Phoxocephalid, unidentified						1.8	
Pontogenia spp.					2.0		
Decapods, unidentified	9.2						2.6
Chironomid larvae	4.0		92.6	8.6	71.4	16.1	1.3
Gastropods							
Acteocina inculta				12.5		<1	
Cystiscus politulus			1.9	3.1		16.1	
Odostomia spp.			5.6	1.6			
Turbonella spp.				<1			
Fish, unidentified							1.3
Total number of prey items	76	28	54	128	49	112	228

[a] Numbers in parentheses indicate number of birds with food in stomach per number of birds examined. Crustaceans identified by J. L. Barnard, other species by D. J. Reish. Percentages based on total number of prey items present in each species of bird.

Data from Reish and Barnard 1990.

Table 10.11. *Food Items of the Most Common Shorebirds in the SCB*

Species	Most Common Food (with percentages)	Comments (habitat, spp. associations)
Black-bellied plover	Insects, marine worms, small crustaceans, molluscs	Small flocks along beaches
Snowy plover	Insects, marine worms, small crustaceans, molluscs	Small flocks along beaches
Semipalmated plover	Insects, marine worms, small crustaceans, molluscs	Small flocks along beaches
Killdeer	Insects (beetles, caterpillars, ants, grasshoppers, bugs)	Grassy areas, fields
Black oystercatcher	Marine worms, barnacles, mussels, other crustaceans, limpets	Rocky marine coast
Black-necked stilt	Insects, especially beetles, nymphs of dragon- and caddisflies, water bugs, mayflies, molluscs, small fish, crustaceans	Marshes, tidal-influenced wetlands
American avocet	Insects, crustaceans, molluscs (snails), clam worms, and 10–25% pondweed	Marshes, tidal-influenced wetlands
Greater yellowlegs	Fish, insects, snails, worms, crustaceans	Shallow estuaries
Willet	Aquatic insects, marine worms, small crabs, molluscs, fish	Shallow ponds, estuaries
Long-billed curlew	Snails, other small molluscs, insects, crayfish, crabs, other crustaceans, worms, spiders	Shorelines
Marbled godwit	50% plant food (winter), 24% (spring), 45% (summer), 65% (fall). Also insects, especially fly larvae, beetles, 25–50% pondweed, sago (seeds and vegetation)	
Spotted sandpiper	Nymphs of caddisflies, dragonflies, beetles, ants, small crustaceans, molluscs, marine worms	Lakes, streams
Ruddy turnstone	Small crustaceans, molluscs, insects	Rocky shorelines
Black turnstone	Small crustaceans, molluscs, insects	Rocky shorelines

From Martin et al. 1951; Bent 1927; Palmer 1976a.

beetles, and their aquatic food consists of crayfish, small crabs, and snails.

Killdeers are also shorebirds of grassy areas, and 98% of their food consists of insects. Killdeers are most often seen foraging in fields, on irrigated land, along river banks, and on manicured lawns in commercial and industrial complexes. Their most common insect prey are beetles, weevils, grasshoppers, caterpillars, ants, true bugs, and caddis flies (Palmer 1976a).

Avocets and black-necked stilts are among the more common large shorebirds in southern California marshes. Avocets forage in sun-baked flats and marshes bordering shallow lakes. Stilts also forage in the shallow muddy borders of alkaline lakes and in brackish ponds of salt marshes. Both species eat aquatic food, especially insect nymphs and larvae.

Willets forage in salt and brackish marshes, in saline or broad mudflats, or on sand flats in bays and estuaries. They also forage along muddy banks of creeks, in pond holes in salt marshes, and along rocky shores (Palmer 1976a). Their preferred food consists mainly of aquatic insects, marine worms, small crabs, fishes, and molluscs.

Snowy plovers feed in compact groups in the wet sandy areas of beaches at the surf line. Their food consists mainly of small crustaceans and marine worms. They occasionally feed at inland sites, especially along muddy shores of lakes, where their primary prey are beetles and flies (Palmer 1976a).

Seabirds

Seabirds most commonly eat fishes, squid, and crustaceans. Oceanic productivity determines the abundance and distribution of seabird prey and, thus, the abundance and distribution of seabirds. Highest ocean productivity is in waters overlying the continental shelves; in areas of upwelling, where cold, rich, deep water replaces surface water; and in areas of convergence, where water of different properties comes together and mixing and turbulence occur (Evans 1980). As can be seen from figure 10.9, these are the areas where the majority of seabirds forage.

Upwelling, a localized phenomenon along the SCB coast (see chap.2), is extremely important for seabird foraging. During late winter, large numbers of dinoflagellates concentrate in the upwelling areas, furnishing food for larvae of northern anchovies, which spawn in the zone of upwelling at this time (Briggs et al. 1987) (see chap. 9). The northern anchovy is one of the most common prey items of seabirds in the SCB. In spring and early summer, rockfishes and flatfishes spawn in the upwelling zones and are found in the majority of seabird diets at this time (Briggs et al. 1987). Months of greatest extent and persistence of upwelling off the SCB are April through early June. Point Arguello-Point Conception is one of the major centers of upwelling, where waters are coolest and winds are strongest and persist in directions favorable for upwelling. Upwelling fronts and high temperature gradients seem to be the most important factors isolating different seabird species groups (Briggs et al. 1987). Storms can influence the survival of larval prey fish such as anchovies and thus can negatively affect seabird foraging success, productivity, and chick growth (Lasker 1981; Baird 1990).

Another important hydrographic feature of the SCB is the Southern California Eddy (a recurvature of the California Current), which forms south and east of Point Conception. This eddy affects hydrographic patterns over a large part of the SCB (Briggs et al. 1987). The west part of this eddy is a cool advected mass that is contiguous with the major upwelling at Point Conception, and contrasts sharply with the warmer subtropical waters east of the Santa Rosa–Cortes Ridge as well as the cool upwelling (Briggs et al. 1987). The boundary between these water types is just east of San Nicolas Island. These varied water types and the transport of nutrients to the surface at these zones attract a variety of birds typical of both cool, northern and warm, subtropical waters, making the SCB an area for diverse avifauna (Hunt et al. 1980; Owen 1980; Parrish et al. 1981).

A branch of cool, subarctic water descends as far south as San Diego, 200–300 km from shore, where warmer, saltier water of the Southern California Countercurrent is near the coast (Chelton 1980). When this countercurrent is less developed, or when there is mixing by storms, birds normally associated with cool western SCB waters (kittiwakes and fulmars) move eastward toward the coast (Briggs et al. 1987). Large variations in avian population numbers can occur from year to year because of such changing ocean conditions.

Seabirds are among the top consumers in the SCB, along with marine mammals. They are generally planktivores or piscivores, and they exploit the majority of the oceanic habitats by using diverse feeding methods. Some seabirds are pelagic feeders, foraging far offshore; some remain within sheltered bays to feed, while others feed along the shore. Some feed on the surface of the ocean, barely touching it; others dive to great depths.

Table 10.12 summarizes the foraging areas and feeding methods used by the major seabird groups in the SCB. Figure 10.9 shows areas of greatest foraging concentration of selected seabird species. Foraging methods are important to mention, for they indicate the amount of time a bird is in contact with the ocean and, thus, its vulnerability to contamination of the water. In addition, knowledge of the general area where a species group usually feeds will help managers make decisions regarding action after oceanic contamination of any kind.

Diving species predominate in regions where upwelling and other processes maintain high standing stocks of phytoplankton, making the water turbid (Ainley 1977; Briggs et al. 1987). Cassin's auklets, sooty shearwaters, and phalaropes (nondivers) abound where water clarity is relatively low (Briggs et al. 1987). Diving species that also surface seize (e.g., sooty shearwaters) and gulls (that surface seize) also have high abundance in the turbid upwelling zones. Other diving species that are abundant in these areas are Pacific loons, western and Clark's grebes, Brandt's cormorants, scoters, and rhinoceros auklets (*Cerorhinca monocerata*). Leach's storm-petrel is the only surface-feeding species found in high abundance in clear, blue offshore waters (Briggs et al. 1987).

Fish (e.g., sauries [*Cololabis* spp.], mackerel, flatfish [*Citharichthys* spp.], and tuna) and planktonic marine invertebrates (euphausiids [*Thysanoessa* spp.] and copepods [*Calanus pacificus*]) or squid (*Gonatus* spp., *Onychotenthis boreal-japonicus*) are common prey species of seabirds. Squid (*Loligo opalescens*), euphausiids (*Euphausia pacifica*), anchovies, and rockfish make up 60–70% of seabirds' total annual consumption. Consumers of fish and squid spend most of their time feeding in neritic areas away from upwelling, while plankton eaters tend to forage seaward of the shelf (Briggs and Chu 1987). Mobile prey such as fish and squid may be easier to capture in neritic areas because wind-induced turbulence is less (Ainley 1977; Briggs et al. 1983a). Zooplankton, especially euphausiids, are also a major prey for Cassin's auklets, scoters, and shearwaters, whose populations number in the thousands. Squids are also a preferred prey of shearwaters.

Most albatrosses, shearwaters, storm-petrels, gulls, and terns feed in the upper few meters of the ocean. Most of these, with the exception of terns and storm-petrels, also spend much time sitting on the water after feeding and, therefore, are quite vulnerable to surface contaminants.

Cormorants, loons, grebes, scoters, and alcids pursue their prey underwater, often to great depths, feeding throughout the water column. Grebes, scoters, and alcids often remain on the water after diving bouts. It has been demonstrated that in years when fish or invertebrate stocks may be stratified at greater depths, the pursuit divers tend to catch more of the preferred prey because they can sample the entire water column. It is also known that their reproductive output during these years remains at levels similar to those in good food years (Vermeer 1980; Baird 1991).

The ashy storm-petrel forages in the California Current just off the continental shelf, but Leach's storm-petrel forages over a much broader pelagic area (Ainley et al. 1974). Food of storm-petrels consists of small invertebrates and fish captured at the surface (Baird and Gould 1983). The surface-feeding behavior of petrels makes them vulnerable not only to oil slicks but to any kind of surface pollution. They have been known to concentrate crude oil in their digestive tracts (D. Boersma pers. comm.). However, they do not normally feed in flocks and are usually widely dispersed throughout the SCB, so the chances of large numbers of this group becoming contaminated by surface pollutants is small.

Brown pelicans eat fish, mainly the north-

Table 10.12. *Foraging Methods and Areas of Most Important Seabirds of the SCB*[a]

Species	Foraging Methods	Foraging Areas	Season	Geographic and Hydrographic Preferences	Maximum Depth (m)
Black-footed albatross	Surface seize, surface dip	Surface pelagic	Spring, summer	Seaward of coastal upwelling zone (CUZ)	1
Northern fulmar	Surface seize minor: pursuit plunge, surface dip	Surface pelagic	Winter	Seaward of CUZ	1
Pink-footed shearwater	Pursuit plunge surface seize	Surface midwater	Summer	11–14°C, neritic water	
Buller's shearwater	Surface seize minor: surface filch, surface-dive moderate: dip (flight feed)	Surface, water column			
Sooty shearwater	Surface dive pursuit plunge moderate: shallow plunge minor: surface seize hydroplane	Water column, surface and upper layers, subsurface, pelagic	Spring and early summer	11–14°C, neritic water	5
Leach's storm-petrel	Surface seize, flight dip, patter	Surface, pelagic	All	California Current (CC); seaward of CUZ	Few cm
Ashy storm-petrel	Unknown	Surface	All	Outer CUZ	
Black storm-petrel	Surface seize	Surface			
Brown pelican	Surface plunge aerial plunge	Surface, settle on water, groups or singly	Late summer and fall	Nearshore fresh-water, 13–17°C translucent neritic water	
Brandt's cormorant	Underwater pursuit	Bottom midwater subsurface singly	All	Nearshore, fresh-water, neritic above rocky and sandy sub-strate, on sur-facing fish schools in open water	70

Table 10.12. *Foraging Methods and Areas of Most Important Seabirds of the SCB*[a] (continued)

Species	Foraging Methods	Foraging Areas	Season	Geographic and Hydrographic Preferences	Maximum Depth (m)
Double-crested and pelagic cormorants	Underwater pursuit	Bottom midwater flocks, singly	All	Neritic, nearshore, freshwater, littoral-benthic zone, pelagic feeding over rocky bottoms	20 (DCC) 36 (PC)
Cassin's auklet	Pursuit dive	Entire water column	All	Feed over shelf break 10–14°C in CUZ	40
Rhinoceros auklet	Pursuit dive	Entire water column	Winter	CC and Davidson Current (DC)	60
Scoter spp.	Pursuit dive	Benthos	All, especially fall, winter, spring	CUZ and coastal shallow zones	
Phalarope spp.	Spin dip	Surface	Spring, fall	Convergences bordering upwelling	Few cm
Pomarine jaeger	Piracy	Surface	Fall, spring	All	
Arctic loon	Pursuit dive	Subsurface	Fall, spring	Coastal shallow	
Black oystercatcher	Probing	Rocky coasts	All	Intertidal	
Western grebe	Pursuit dive	Subsurface	Winter	Coastal shallow	
Western gull	Subsurface seize, dip	Surface	All	Neritic water	1
Herring gull	Subsurface seize, dip	Surface	Winter	Neritic water	1
California gull	Subsurface seize, dip	Surface	Fall, spring	Neritic water	1
Heermann's gull	Surface seize, dip	Surface	Spring, fall	Coastal	1
Black-legged kittiwake	Surface dip, surface plunge	Surface	Winter	CC, DC	1–2
Arctic and common terns	Aerial plunge	Surface	Fall	CC seaward of CUZ	0.5
Coastal tern spp.	Aerial plunge	Surface	Summer	Coastal, shallow lagoons	

[a] See text for references.

ern anchovy (Palmer 1976a; Anderson et al. 1975, 1980), and pelican population numbers fluctuate directly with anchovy abundance (Anderson et al. 1980). In fact, their abundance correlates so well with anchovy abundance that brown pelicans can be used as indicators of anchovy stock (Anderson et al. 1980).

It appears that pelicans are more susceptible than other birds to contamination by

DDE because they, more than other species, produce thin-shelled eggs as a direct result of this DDT metabolite (Anderson et al. 1975). The weight of the parent often crushes these eggs and the unhatched chicks die. Although DDT dumping into the SCB was halted in 1970, residues of DDE still remain. Because pelicans sit on the water for a long time after feeding, they are also vulnerable to contamination by surface pollution.

Cormorants, unlike other seabirds, must dry their feathers after wetting them during their dives. Thus, they remain in the water less time than do other seabirds. But since they do not float very well, oiling of their feathers would most likely make them sink and drown (K. Briggs pers. comm.). Likewise, since cormorants cannot store energy as body fat, any disruption in daily feeding activities due to oiling of the ocean or of themselves would be detrimental to their well-being (K. Briggs pers. comm.). Double-crested cormorants forage nearshore, often singly, and those nesting on the Channel Islands may even fly to the mainland coast to feed (Ainley pers. comm., in Sowls et al. 1980). Brandt's and pelagic cormorants feed in flocks, often among feeding assemblages of other seabirds.

The food of western gulls is mainly anchovies, rockfish, and Pacific sauries, although they, like all gulls, are omnivores. They spend much time on the water after feeding in flocks, thus are also vulnerable to sea surface pollutants.

Xantus' murrelets dive for food, sitting on the surface after they eat. Their chicks accompany their parents to the ocean when they are 1–2 days old and are fed by their parents until they are old enough to fly (Murray et al. 1983). This species is most vulnerable to oil spills or surface pollutants because both age classes—chicks and adults—remain on the water during the entire chick stage; the chicks cannot fly away from a contaminated area, and the parents will not leave the chicks.

Cassin's auklets, one of the more numerous seabirds in the SCB (with up to 23,000 breeding in the Channel Islands), forage pelagically in large flocks and, like other alcids, often remain on the ocean surface after they feed. Because of their large numbers, dense feeding assemblages, and postfeeding behavior, they are very vulnerable to oil spills or surface pollutants.

The avifauna of the SCB over the shelf and shelf break are similar to those of the offshore regions in central and northern California because of the upwellings and cooler waters in these areas (Briggs et al. 1987). The topography of the ocean floor is complex in the SCB. Shallow banks and deep basins occur both offshore and near the mainland. Upwelling occurs both at the western edge of the SCB and along the mainland, a phenomenon that does not exist in central or northern California where depth, temperature, and distance from land are cross-correlated (Briggs et al. 1987).

Anchovies, a preferred seabird prey in the SCB, occur in spawning schools in late spring. In summer, postspawning anchovies occasionally form large near-surface schools in the daytime near the coast, peaking in abundance in October (Briggs et al. 1987). This coastal concentration attracts birds that would normally feed offshore (fig. 10.9). Rockfish and squid, preferred prey for many species, are also found nearshore in summer; juvenile rockfish and euphausiids are abundant in the northwest SCB in the spring (Brinton 1976; Parrish et al. 1981). Distribution and abundance of seabirds reflect these concentrations of prey (fig. 10.9).

Concentrations of avifauna over the shelf and upper continental slope rather than in offshore waters may also result from the more stable, vertically stratified California Current waters, where concentrations of phyto- and zooplankton tend to be found near or below the thermocline by day (30–100 m), making them more difficult to locate (Briggs and Chu 1986). Also, surface convergences are usually more common near the shelf break than offshore, and these convergent fronts often concentrate macroplankton and their fish predators, thus attracting seabirds. Briggs and Chu (1986) found that upwelling fronts mark the offshore limits of distribution for some coastal species, such as western gulls, and also mark the near-coast limits for typically offshore species, such as Leach's storm-petrel and Buller's shearwater (*Puffinus bulleri*). Briggs et al. (1983a,b 1984) found that phalaropes, pelicans, Cassin's auklets, and sooty shearwaters were associated with the frontal zone.

Waters off California, including those in the SCB, have been sampled for various physical parameters that, in turn, have been associated with seabird abundance. Briggs et al. (1987) performed a principal component analysis of these parameters and seabird associations and found that the most important components to explain variation in density of seabirds were (1) distance from shore paired with depth, (2) variation in temperature latitudinally, and (3) thermal gradients.

Densities of common murres, Cassin's auklets, western gulls, and sooty shearwaters were inversely related to distance from shore and depth and positively related to temperature gradient. Densities of rhinoceros auklets, black-legged kittiwakes, and sometimes northern fulmars and black-footed albatrosses were positively associated with the latitude–temperature gradient (high latitude, cooler temperatures) and usually with higher temperature gradients also. Densities of Leach's storm-petrels and phalaropes (and sometimes common and Arctic terns and Buller's shearwaters) varied in density as distance from shore and depth increased.

Estimated Food Consumption

The knowledge of energy flow through the ecosystem can be used as a diagnostic tool in research and management. Measurements of prey consumption by marine birds and shorebirds can be used as tools to monitor the health of the oceanic ecosystem (Boersma 1978; Baird 1979, 1990, 1992). Information on energy or nutrient flow in avifauna in the SCB is scanty to nonexistent. However, it is important to understand the amount of biomass removed from the ecosystem by seabirds. Thus, the following values have been taken from other studies and can be applied to birds in the SCB. Because of their relative unimportance in the SCB and because of lack of studies, marshbirds and waterbirds are not discussed in this section.

Large-bodied birds require a smaller amount of food per gram than do small-bodied birds, but the total biomass of smaller birds may be greater and thus they are as important in energy and nutrient cycling. For example, a 3000-g gannet requires 0.21 kcal g^{-1} of body weight per day, while in contrast, a 35-g Wilson's storm petrel requires 0.72 kcal g^{-1} per day (Powers and Brown 1987). Thus, numbers of birds (abundance) must be combined with their allometric scaling when one is calculating energy transfer rates (Lasiewski and Dawson 1967; Kendeigh 1970). This information must then be combined with the number of days various species remain in the SCB. Transfer rates can then be used to show the relative importance of various prey groups (Powers and Brown 1987).

Shorebirds

Shorebirds are one of the major paths of energy flow from intertidal benthic invertebrates (Goss-Custard et al. 1979; Baird et al. 1985). They reportedly have removed up to 90% of the standing crop of prey, such as large *Hydrobia* or *Nereis*, during a single winter (Evans et al. 1979). A more conservative estimate is probably 35–60% (Goss-Custard 1977a,b; Baird et al. 1985). The distribution of some species of shorebirds may be closely related to variations in the absolute density of their preferred prey (Baird et al. 1985), although the simple annual productivity of an area will not predict the number of birds found in any one location. The energy transfer from invertebrate prey to bird predator varies widely from place to place, and no absolutely clear relationship seems to exist between productivity of prey and prey consumption by birds.

Numerous studies have been done of energy cycling by shorebirds from Europe, but none exist for the SCB. Literature from European shorebird communities suggests that shorebirds (primarily zoobenthos feeders) take 8–44% of the total annual productivity from estuaries, with an average range of about 17–20%. Studies have shown, for example, that areas of 1300 km^2 (i.e., Wadden Sea in Holland) can support 1.5–2.5 million shorebirds, which are estimated to cycle 77.4–90.0 kJ m^{-2} through the ecosystem, consuming about 17% of the total annual production (Baird et al. 1985). Even in smaller areas, shorebirds can take a large percentage of the invertebrate population from a mudflat. For example, in the Tees Estuary in northeastern England, an intertidal mudflat of 1.4 km^2, shorebirds consume about 44% of the annual production (total prey production of 1192×10^6 kJ yr^{-1}). In the winter, about 79,000 shorebirds consume approximately 20% of the invertebrate population productivity (Baird et al. 1985).

A perusal of the data reveals that energy cycling through the marine ecosystem is more efficient than that found in terrestrial ecosystems. It is safe to say that shorebirds consume from 17 to 40% of all invertebrate annual production on their wintering grounds (Baird et al. 1985). Thus, birds must be included in any summary of energy and nutrient transfer in estuarine ecosystems.

Seabirds

Seabirds are important members of the upper trophic levels and are responsible for removing anywhere from 14–29% of various fish stocks (table 10.13) (Schaefer 1970; Robertson 1972; Furness 1978; Furness and Cooper 1982; Furness and Ainley 1984). Recent estimates of food requirements of pelagic birds have shown that bird populations are capable of removing prey at rates sufficient to have an impact on local prey stocks (Wiens and Scott 1975; Furness 1978) and at rates comparable to those of marine mammals (Briggs and Chu 1987).

Table 10.13. *Three-Year Average Seabird Biomass*

Biomass (kg km^{-2})	Location
55.2 ± 8.7	Northern California
68.3 ± 5.9	Central California
8.38 ± 0.9	SCB (includes continental slope)
10.87 ± 1.8	Northern California slope
13.34 ± 1.9	Central California slope

From Briggs et al. 1987.

Average density and biomass of seabirds are low in the SCB compared to other areas of California (Briggs et al. 1987). Although seabird distribution is very patchy, average biomass is a loose estimate of kilograms per area and can be calculated for any given area. Average biomass in the SCB is estimated to be 8.38 kg km^{-2}. This is comparable to that estimated for cold core eddies of the Gulf Stream, 0.1–7.9 kg km^{-2} (Haney 1986), and for the oceanic South Pacific, 0.9–10.2 kg km^{-2} (Briggs et al. 1987). For comparison, the Ross Sea in late summer has a biomass of 44 kg km^{-2}, and a variety of terrestrial ecosystems have biomasses ranging from 2 to 78 kg km^{-2} (Szaro and Balda 1979; Briggs et al. 1987). Biomass of seabirds in the neritic zone is much greater than offshore (Briggs and Chu 1987).

Studies of ENSO episodes (in southern California June 1975–January 1977) reveal that biomass, but not density, is significantly higher on the central California slope than in southern California. No difference exists between northern and southern slopes in either biomass or density (Briggs et al. 1987).

Density peaks of seabirds in neritic waters can occur at various times over a period of years. In the SCB these have occurred, for example, in June and December 1975, May 1976, and June and October 1977 (Briggs et al. 1987). Some areas of highest biomass have been found in the Santa Barbara Channel and near the Santa Rosa–Cortes Ridge, together with aggregate bird density exceeding 50 birds per square kilometer, equivalent to biomass values of 34–46 kg km^{-2} (Briggs et al. 1987).

The year-round center of highest carbon flux in the SCB begins just north of Point Conception at Point Burgess and extends southeast for 300 km to the Cortes Bank. Carbon flux in this area is about 3 mg C m^{-2} d^{-1} (Briggs and Chu 1987). Especially in the Santa Barbara Channel, the maximum monthly fluxes are about 100 times higher than those of coastwide averages. Waters around the Channel Islands also have periodic monthly maximum fluxes (Briggs and Chu 1987).

SCB waters support far fewer numbers of birds than those north of Point Conception. A small number of species also dominate the avian biomass in the SCB. Food consumption by birds is lowest in January and February and in June and July. Two species that probably cycle the most carbon through the ecosystem are the sooty shearwaters and the brown pelicans. Shearwater numbers are highest from late spring to midsummer, and at this time, they are estimated to consume about 0.13 mg C m^{-2} d^{-1}. Their annual consumption is about 18,000 t wet weight (Briggs and Chu 1987). Brown pelicans take about 0.02 mg C m^{-2} d^{-1} in the early fall. Their annual consumption is about 16,500–18,000 t (Briggs and Chu 1987).

Total daily consumption of fish and squid in the SCB has been estimated as 200–300% more than that of plankton. In all California, the estimate has been 75,000–300,000 t yr^{-1} for all seabirds (extrapolated from Briggs and Chu 1987). For comparison, total commercial and sportfishing catches (for all of California) have been estimated at 300,000–600,000 t yr^{-1} (Green 1978; Briggs and Chu 1987).

Effects of Human Activities on Birds

Seabirds in California waters are seriously threatened by at least two human activities: (1) petroleum leaks or discharges and (2) gill netting for fish. Stenzel et al. (1988) have shown that oiling is the most common cause of death in bird carcasses found washed up on California beaches. After large oil spills, the numerous dead species found on the beaches are grebes and loons and, secondarily, cormorants. Recent evidence shows that the grebes and loons probably died prior to being oiled (Varoujean 1982). Cormorants that died in these spills, however, were determined to have died from oiling.

Oiling is most damaging to diving birds whose oil-escape pattern is to dive, therefore becoming exposed to more oil. These species do not usually rest on land, and resting on the water after oiling occurs subjects them to greater thermal stress than that suffered by birds that remain on land.

Sublethal oil toxicity may affect the behavior of birds, having a narcotic effect that either hampers their ability to further avoid oil or produces an escape response (Varoujean 1982). The hypothesis that previous exposure to oil may produce future avoidance patterns is corroborated by finding juvenile pelicans and gulls making contact with oil more than do adults. Likewise, residents familiar with permanent oil seeps tend to avoid these more than migrants do (Varoujean 1982). The overriding factors determining whether a diving seabird will avoid oil are related to the bird's ability to detect oil presence before contacting it and to avoid oil successfully once it is encountered (Varoujean 1982).

Under conditions of low physiological stress, birds may survive being lightly oiled or ingesting a small quantity of oil (Holmes and Cronshaw 1977; Holmes et al. 1978). However, when birds are breeding, light oiling may lead to a decrease in reproductive success, such as reduced clutch size or hatching failure (Ainley et al. 1979, 1981; King and Lefever 1979; Fry et al. 1985, 1986), or it might cause the birds to change breeding sites or abandon nests (Fry et al. 1985).

Research suggests that toxic components of oil ingested by seabirds are absorbed by the intestinal tract and exported to the liver and ovary, where they may be deposited in the yolk of forming eggs (Grau et al. 1977). These petroleum components have been found to inhibit sodium and water absorption by intestinal mucosa, thus disturbing sodium and potassium metabolism for both the adult bird and the forming embryo. Less than normal amounts of yolk are deposited, cracks occur in the yolk ring, and eggs have thinner shells (Grau et al. 1977). Oil-contaminated Cassin's auklets have exhibited dissociation of liver cells, and oiled auklets and common murres show necrosis of kidney tubules (Fry and Lowenstine 1982). Hematocrits of oiled birds are depressed, with packed cell volumes as low as 15–20%, and many of these birds have hemolytic anemia (Fry and Lowenstine 1982, 1985).

Despite intensified efforts by the petroleum industry to reduce the frequency of oil spills, a single major spill might have disastrous consequences for some species. For example, flocks of tens of thousands of sooty shearwaters often feed close to the shore of Santa Barbara and San Luis Obispo counties (Briggs and Chu 1987; California Cooperative Oceanic Fisheries Investigations 1988), and a major oil spill near one of these preferred feeding areas would clearly affect large numbers of birds. Other species of birds that occur in large numbers in the SCB and spend considerable time on the water's surface, and thus would be most affected by oil spills, include Cassin's auklet and red and red-necked phalaropes.

The commercial fishing industry is equally as dangerous a threat to seabirds in a variety of ways. The use of gill nets for commercial fishing has increased dramatically in recent years, and they catch and kill large numbers of diving birds when they are set for fishing periods (DeGange and Newby 1980; King 1984). Drifting gill nets that are lost or discarded may also entangle diving birds as they forage for prey (DeGange and Newby 1980). King (1984) estimated that the annual kill of seabirds by the Japanese fisheries in the North Pacific and Bering Sea is 250,000–750,000 birds per year, with sooty shearwaters comprising 60% and alcids 33% of birds killed. The nearshore halibut fishery in Monterey Bay has killed 10,000 birds per month since 1979 (King 1984), about 80% of these birds being sooty shearwaters and common murres. The birds most susceptible to mortality from gill nets in the SCB are those that dive to depths to which the gill nets are set; these species are mainly large alcids (tufted puffins and common murres) and sooty shearwaters. Takekawa et al. (1987) stated that recent dramatic decreases in the southern California populations of common murres are probably due to mortality in gill nets. In northern California, where there is no gill netting fishery, common murre populations are stable (Takekawa et al. 1987).

In addition to killing birds outright, commercial fishing operations often crop 50–70% of fish production so that little is left for natural predators (Furness and Ainley 1984). Fishing operations not only place their nets in preferred seabird foraging areas but also take certain fish species similar in age (size) to prey consumed by seabirds (Schaffner 1986), thus directly competing with the birds. Numerous studies have shown a correspondence between high stocks of certain fish species and overall breeding success of seabirds (Baird 1991; Vermeer 1980; Anderson et al. 1982; Schaffner 1986).

Overfishing of seabird prey stock is known to decimate seabird populations, some permanently (Ainley and Lewis 1974; Idyll 1973; Lid 1981; Furness 1982; MacCall 1984; Barrett and Vader 1984). Commercial fishing has not been well regulated, therefore overfishing in some areas, especially of anchovies, has occurred (Hunt et al. 1980). Anchovies are particularly important to brown pelican and elegant tern reproduction success (Schaffner 1986). Over 86% of chick regurgitations of elegant terns at some sites contained anchovies. On the Pacific coast, the northern anchovy spawning biomass and recruitment continue to decline (Schaffner 1986).

Another human-caused problem affecting seabirds is the dumping of plastics either at sea or into flood channels. Plastics are found throughout the ocean and are known to be consumed by seabirds (Day 1979). At least 60 species of Pacific Ocean seabirds ingest plastics while foraging (Fry 1987), and several species regurgitate plastics to chicks. The possible deleterious effects of plastics on chick growth and survival are unknown, although Fry (1987) reports that ingested plastics appear to reduce survival of Laysan albatross chicks. However, Sievert et al. (1988) report that the effect of plastic ingestion on growth and survival of Laysan (*Diomedea immutabilis*) and black-footed albatross chicks appears to be insignificant. Similarly, the deleterious effects of plastic ingestion on adult birds are unknown. Fry (1987) suggests smaller alcids may be at a higher risk because of an inability to regurgitate plastics.

Human disturbance of colonies is another effect that had become worse in recent years but seems to be fairly well-regulated today. San Nicolas and San Miguel islands are no longer used for target practice by the U.S. Navy. Other bird nesting areas have been improved, created, and protected by additional regulations and tighter controls over human use of the Channel Islands. Many California least tern colonies on busy public beaches have been fenced and posted, and new marshlands have been created at the mouth of the Santa Ana River. Acreage at Bolsa Chica and Ballona wetlands has been ceded for wildlife refuges instead of development. These are just a few examples of the most recent efforts to mitigate previous wildlife and habitat loss.

Summary and Prospectus for Future Research

This chapter covers the most important birds using the SCB, with importance being based on high numbers, use of threatened habitats, a species' threatened status, or some other characteristic that makes the species unique. At least 195 species of marshbirds, waterfowl, shorebirds, and seabirds use coastal or offshore aquatic habitat in the SCB. The majority of these migrate through the area, and population numbers of each transiting species can range as high as a million individuals.

Of all marine species, marshbird numbers are lowest throughout the SCB, with most species numbering only in the hundreds. Waterfowl populations are more abundant, with the most common species numbering in the thousands. Of the 39 species of waterfowl found in the SCB, only 4 of the 9 breeding species are common. Shorebirds are common only in the winter in the SCB where they are concentrated in bays and estuaries. Populations can number in the thousands.

Seabirds, together with scoters, loons, and western grebes, contribute the greatest avifaunal biomass in the SCB. Twenty-three species predominate. Seabirds use the SCB year-round, and some of the migrants can constitute the largest biomass of seabirds at any one instant in the SCB. Seabird densities can be as great as 70 birds km^{-2} for migrants (e.g., phalaropes), and up to 1000 birds km^{-2} for breeders near their colonies (Cassin's auklets). Spring visitors are mainly austral breeders, fall migrants are subtropical breeders, and winter migrants are mainly Alaskan breeders. Seabird populations number in the thousands to tens of thousands, with some reaching hundreds of thousands of individuals. Shearwaters and phalaropes account for half or more of the seabirds found in the SCB; rhinoceros auklets account for a third of the seabirds in the winter.

Seventeen species of seabirds totaling more than 51,000 individuals breed in the SCB. Breeding habitat for seabirds in the SCB, except for terns and skimmers, is located entirely in the Channel Islands. Prince Island, off San Miguel Island, ranks highest with 23,400 seabirds breeding there.

Some of the most important species in the SCB are the California least tern and the light-footed clapper rail (both endangered due to habitat destruction), the brown pelican (endangered due to DDT use and overfishing of their prey), and Cassin's auklets (nesting in only two known sites in California;

one colony off San Miguel Island has over 20,000 birds). Low reproductive output, limited breeding sites, and DDT have been responsible for population decline of the brown pelican. Rhinoceros auklets are also an important species because most of their eastern Pacific nesting population is located off the coast of California in February and March. An environmental disaster in the SCB at this time could severely decrease their population numbers. Also, because of their low numbers and destruction of their habitat, all duck, rail, heron, egret, and shorebird species, except the coot and the mallard, are considered important. Because the entire or nearly entire California breeding populations of black storm-petrels, Xantus' murrelets, and brown pelicans nest in the Channel Islands, their populations are of particular concern because each species' California breeding population could be exterminated or severely affected by environmental perturbations such as oil spills.

The few wetland areas of the SCB have gained importance for ocean-influenced birds because 90% of coastal marshes have been eliminated in the last century. All are wholly or partially protected, and new marshes are slowly being developed. Coastal wetlands are most important for the 25 species of herons and rails (including the light-footed clapper rail) and the 36 species of waterfowl occurring in the SCB. Greatest use of the wetlands occurs from October to March. These wetlands, which include bays, estuaries, marshes, and lagoons, are critically important overwintering habitat for waterfowl and marshbirds as well as shorebirds. The shorebird species that frequent the beaches of the SCB during the winter are most vulnerable to oil spills. Some seabirds are dependent on the threatened coastal wetlands for foraging and nesting and thus are also considered vulnerable because of their dependency on these rare habitats.

The introduction of exotic animals (predators such as cats and rats and habitat destroyers such as goats and rabbits) and aerial bombing of important seabird habitat at San Miguel and San Nicolas islands have seriously decreased many breeding seabird populations. Overfishing by commercial fleets continues to decimate seabird prey populations, thus lowering breeding success and population numbers. Also, in all seasons, foraging seabirds are drowned when caught in fishing nets. The widespread use of DDT led to eggshell thinning in many species of birds, especially in larger birds that ate larger fish in which biomagnification of this pesticide had occurred.

Recent control or elimination of predators and habitat-destroying species, creation of artificial lakes and nesting islands, enhancement of wetlands, and elimination of some pesticides have aided the increase in population numbers for some species, especially ducks, pelicans, light-footed clapper rails, and California least terns. However, pesticide poisoning and habitat destruction still remain a major threat for species that overwinter south of the U.S. border where there are fewer environmental controls.

Effluent from storm drains has made many wetlands and shorelines unusable or dangerous for shorebirds, marshbirds, and waterfowl. Also, the explosion of the human population in southern California with concomitant demands for space has resulted in avian habitat destruction or fragmentation, further decreasing population numbers of all birds that use the wetland or shoreline habitats. The increase of shipping with the expansion of the ports of Los Angeles and Long Beach has increased the probability of a major oil spill in the SCB. In short, numbers of nearly all ocean-influenced birds have declined drastically, some critically, and not enough is being done to rectify this situation. The major culprit is habitat fragmentation and loss.

Seabirds mature slowly and are long-lived; many do not reproduce until they are 4 years of age or older. Reproductive rates are low; many species lay only one egg per season, and the rest only two or three. Fledging rates are also low, usually less than one chick per breeding pair. Thus, any calamity that affects the adult birds, such as oil spills or habitat loss, can have extremely deleterious effects.

Breeding success, as well as distribution and abundance of foraging marine birds, is directly dependent on food availability. For example, anchovy abundance and availability have been directly related to brown pelican and California least tern breeding success, and the commercial overfishing of the anchovy could result in successive years of low reproductive success for these two endangered species. In other ecosystems, such as colonies off the coast of Norway, entire seabird breeding populations have disappeared because of overfishing of the major prey item (in this case, herring).

Because ocean productivity determines the abundance and distribution of seabirds, many seabirds can be found over water that overlies continental shelves where cold, rich, deep water upwells, as well as in areas of convergence and mixing, which occurs in localized areas lasting for several weeks. The greatest extent and persistence of upwelling occurs in April through early June. The Point Arguello–Point Conception area is one of the major upwelling centers in the SCB. Because of the mixing of different types of waters, the SCB harbors a variety of prey and thus a variety of marine birds. A great diversity of birds typical of both cool northern and warm subtropical waters can be found at boundaries between the cool Southern California Eddy (contiguous with the major upwelling at Point Conception) and the warmer subtropical waters east of the Santa Rosa–Cortes Ridge, just east of San Nicolas Island.

Knowledge of birds' prey as well as foraging methods and energy and nutrient cycling helps in the management of important commercial species, such as anchovy, smelt, squid, and euphausiids, by using the birds as biological indicators of the ocean's health.

Although studies on food habits of all species of SCB avifauna are not available, some of the most important species' diets are summarized in this chapter (e.g., sooty shearwaters, brown pelicans, western gulls, Cassin's auklets, Xantus' murrelets, and various shorebirds). Anchovies, sauries, mackerel, rockfish, tuna, squid, and euphausiids are the main seabird prey. Additionally, foraging methods and habitats are listed for most of the avifauna because this knowledge (1) indicates how vulnerable certain species may be to ocean-carried pollution, and (2) reveals those habitats in the SCB that are most important for birds. For example, fish and squid consumers spend most of their time in neritic areas away from upwelling, while plankton eaters (shearwaters and Cassin's auklets) tend to forage seaward of the shelf. Rockfish and squid are found nearshore in summer, juvenile rockfish and euphausiids are abundant in the northwest SCB in spring, and anchovy abundance peaks in October near the coast.

Energy and nutrient cycling through the avifauna is presented in this chapter because they can be used as diagnostic tools in research and management. For example, shorebirds constitute one of the major paths of energy flow from intertidal benthic invertebrates and have been reported to remove 35–60% of the standing crop of prey during a single winter. Thus, they can have a major impact on invertebrate populations as well as wetland ecosystems.

Seabirds are known to take 14–29% of fish stocks, rates comparable to those for marine mammals, and sooty shearwaters and brown pelicans are thought to cycle the most carbon through the ecosystem annually (up to 18,000 t each). They often concentrate near year-round centers of highest carbon flux from just north of Point Conception or Point Buchon to 300 km southeast to the Cortes Bank and the Santa Barbara Channel. Total daily consumption of fish and squid in the SCB has been estimated to be 200–300% more than that of plankton. Yet marine birds take much less fish (75,000–300,000 t yr^{-1}) than do commercial and sport fishermen (300,000–600,000 t yr^{-1}). Needed research includes long-term studies of breeding biology on banded populations to help identify mortality rates; population growth, decline, or stability; and population age structure.

After abundance and distribution of SCB

avifauna have been reasonably determined, the next most important data to be collected are the types of prey consumed, the proportion of the meals these prey comprise, and the numbers in which different prey species are taken, as well as energy and nutrient content. Answers to these questions, coupled with abundance data for birds, will help to indicate the amount of biomass consumed and, ultimately, the amount of energy and nutrients that avifauna are cycling through the system. Only after these basic questions are answered can other important ecosystem questions be addressed and further measures taken to regulate land development and commercial fishing that negatively affect all species.

Acknowledgments

I would like to thank Ken Briggs not only for his prolific publications on seabirds in California, without which this chapter would have been impossible to write, but also for his very helpful and detailed review of the chapter and suggestions for improvement. I thank George Hunt, Jr., and Kimball Garrett for their many useful comments on the chapter.

I thank Don Reish and Dee Dee Rypka for hunting down obscure manuscripts that were germane to the chapter synthesis, and I thank Charlie Collins, Barbara Massey, and Don Reish for helpful suggestions for improving the chapter. Finally, I thank J. D. Herron, Murray Dailey, and the many members of the MMS staff and Quality Review Board for their encouragement and support throughout this project.

Literature Cited

Ainley, D. G., 1976. The occurrence of seabirds in the coastal region of California. *West. Birds*. 7:33–68.

Ainley, D. G., 1977. Feeding methods of seabirds: A comparison of polar and tropical nesting communities in the eastern Pacific Ocean. In: G. A. Llano, ed. *Adaptations Within Antarctic Ecosystems*. Gulf Publishing Co., Houston, TX. pp. 669–686.

Ainley, D. G., and T. J. Lewis, 1974. The history of the Farallon Islands: marine birds populations, 1854–1972. *Condor*. 76:432–446.

Ainley, D. G., and M.C. Whitt, 1974. Numbers of marine birds breeding in Northern California. *West. Birds*. 4:65–70.

Ainley, D. G., S. Morrell, and T. J. Lewis, 1974. Patterns in the life histories of storm petrels on the Farallon Islands. *Living Bird*. 13:295–312.

Ainley, D. G., R. J. Boekelheide, T. J. Lewis, H. R. Huber, C. S. Strong, and T. A. Wooton, 1979. Influence of petroleum on egg formation and embryonic development in seabirds. Final Report. Natl. Oceanic Atmos. Adm., Juneau, AK.

Ainley, D. G., C. R. Grau, T. E. Roudybush, S. H. Morrell, and J. M. Utts, 1981. Petroleum ingestion reduces reproduction in Cassin's auklets. *Mar. Pollut. Bull.* 12(9):314–317.

Ainsley, J. A., and R. Atkinson, 1937. On the breeding habits of Leach's and fork-tailed petrels. *Br. Birds*. 30:234–238.

American Birds, 1986. Eighty-sixth Christmas bird count. Count numbers 1407, 1410, 1424, 1425, 1429, 1443, 1449, 1459. *Am. Birds* 40(4).

Anderson, D. W., and I. T. Anderson, 1976. Distribution and status of brown pelicans in the California Current. *Am. Birds*. 30:3–12.

Anderson, D. W., and J. O. Keith, 1980. The human influence on seabird nesting success: Conservation implications. *Biol. Conserv*. 18(1):65–80

Anderson, D. W., J. R. Jehl, R. W. Risebrough, L. A. Woods, Jr., L. R. De Weese, and W. G. Edgecomb, 1975. Brown pelicans: Improved reproduction off the southern California coast. *Science*. 190:806–808.

Anderson, D. W., F. Gress, K. F. Mais, and P. R. Kelly, 1980. Brown pelicans as anchovy stock indicators and their relationships to commercial fishing. *Calif. Coop. Oceanic Fish. Invest. Rep.* 21:54–61.

Anderson, D. W., F. Gress, and K. F. Mais, 1982. Brown pelicans *Pelecanus occidentalis californicus*: Influence of food supply on reproduction. *Oikos*. 39(1):23–31.

Arbib, R., 1977. The Blue List for 1978. *Am. Birds*. 31:1087–1096.

Ashmole, M. J., and N. P. Ashmole, 1968. The use of food samples from seabirds in the study of seasonal variation in the surface fauna of tropical oceanic areas. *Pac. Sci.* 22:1–10.

Ashmole, N. P., 1963. The regulation of numbers of tropical ocean birds. *Ibis*. 103:458–473.

Ayers, D., 1975. Reproductive performance of the

double-crested cormorant in Humboldt Bay, California. Master's Thesis, California State Univ., Humboldt, Arcata, CA.

Baird, O., P. R. Evans, H. Milne, and M. W. Pienkowski, 1985. Utilization by shorebirds of benthic invertebrate production in intertidal areas. *Oceanogr. Mar. Biol. Annu. Rev.* 23:573–597.

Baird, P., 1976. Comparative ecology of California and ring-billed gulls. Ph.D. Dissertation, Univ. of Montana, Missoula, MT.

Baird, P., 1979. Reproductive success and breeding biology of seabirds at Sitkalidak Strait, Kodiak Island, Alaska. In: *Annu. Reps. of Principal Investigators, Vol. 2.* NOAA Environ. Res. Lab., Boulder, CO. pp. 107–186.

Baird, P., 1986. A method for collecting food from seabird chicks. *Wilson Bull.* 98:169–170.

Baird, P., 1989. Population biology of the California least tern at the Santa Margarita River estuary. Unpublished report. Dept. of Biology, California State Univ., Long Beach.

Baird, P., 1990. Influence of abiotic factors and prey distribution on diet and reproductive success of three seabird species in Alaska. *Ornis Scand.* 21:224–235.

Baird, P., 1991. Optimal foraging and intraspecific competition: A diet comparison among breeders, nonbreeders and chicks in the tufted puffin. *Condor* 93:503–515.

Baird P., 1992. Seabirds as indicators of the oceanic environment. In: P. M. Grifman and S. E. Yoder, eds. *Perspectives on the Marine Environment.* Sea Grant Program Special Publ. USCSG-TR-01-92, Univ. of Southern California, Los Angeles.

Baird, P., and P. Gould, 1983. The breeding biology and feeding ecology of marine birds in the Gulf of Alaska. In: P. Baird, and P. Gould, eds. Natl. Oceanic Atmos. Adm., OCSEAP Final Reps. 45(1986):121–505.

Barrett, R. T., and W. Vader, 1984. Seabird conservation in Norway. In: J. P. Croxall, P. G. H. Evans, and R. W. Schreiber, eds. *Status and Conservation of the World's Seabirds.* Int. Comm. Bird Preserv. Tech. Publ. 2. pp. 323–333.

Bender, K. E., C. T. Collins, and S. L. Warter, 1974. Marine and shore birds. In: M. D. Dailey, B. Hill, and N. Lansing, eds. *A Summary of Knowledge of the Southern California Coastal Zone and Offshore Areas. Vol. 2, Biological environment.* Southern California Ocean Studies Consortium, Long Beach, CA. pp. 13-1–13-103.

Bent, A. C., 1921. Life histories of North American gulls and terns. Vol. 113, Smithson. Inst., *U.S. Natl. Mus. Bull.*, Washington, D.C. 345pp.

Bent, A. C., 1922. Life histories of North American petrels, pelicans, and their allies. Vol. 121, Smithson. Inst., *U.S. Natl. Mus. Bull.*, Washington, D.C. 343pp.

Bent, A. C., 1927. Life histories of North American shore birds. Parts I and II, Vols. 142 and 146, Smithson. Inst., *U.S. Natl. Mus. Bull.*, Washington, D.C. 420pp. and 412pp.

Birkhead, T. R., 1976. Effects of sea conditions on rates at which guillemots feed chicks. *Br. Birds.* 69:490–492.

Boersma, P. D., 1978. Breeding patterns of Galapagos penguins as an indicator of oceanographic conditions. *Science.* 200:1481–1483.

Boersma, P. D., and N. T. Wheelwright, 1979. The costs of egg neglect in the Procellariformes: Reproductive adaptations in the fork-tailed storm petrel. *Condor.* 81:157–165.

Boersma, P. D., N. T. Wheelwright, M. K. Nerini, and E. S. Wheelwright, 1980. The breeding biology of the fork-tailed storm petrel (*Oceanodroma furcata*). *Auk.* 97:268–282.

Briggs, K. T., and E. W. Chu, 1986. Sooty shearwaters off California: Distribution, abundance, and habitat use. *Condor.* 88:355–364

Briggs, K. T., and E. W. Chu, 1987. Trophic relationships and food requirements of California seabirds: Updating models of trophic impacts. In: J. P. Croxall, ed. *Seabirds: Feeding Ecology and Role in Marine Ecosystems.* Cambridge Univ. Press, London. pp. 297–304

Briggs, K. T., E. W. Chu, D. B. Lewis, W. B. Tyler, R. L. Pitman, and G. L. Hunt, Jr., 1981. Distribution, numbers, and seasonal status of seabirds of the Southern California Bight. In: *Summary of Marine Mammal and Seabird Surveys of the Southern California Bight Area, 1975–1978.* Investigators' reports, book I, part III. Rep. to U.S. Dept. of Interior, Bureau of Land Management. Available from: NTIS, Springfield, VA; PB81–248205. pp. 1–399.

Briggs, K. T., W. B. Tyler, D. B. Lewis, and K. F. Dettman, 1983a. Seabirds of central and northern California 1980–1983. Status, abundance and distribution. Available from: Natl. Tech. Info. Serv. Rep.; PB85–183846. Springfield, VA. 246pp.

Briggs, K. T., W. B. Tyler, D. B. Lewis, P. R. Kelly, and D. A. Croll, 1983b. Brown pelicans

in central and northern California. *J. Field Ornithol.* 54(4):353–373.
Briggs, K. T., K. F. Dettman, D. B. Lewis, and W. B. Tyler, 1984. Phalarope feeding in relation to autumn upwelling off California. In: D. N. Nettleship, G. A. Sanger, and P. F. Springer, eds. *Marine Birds, Their Feeding Ecology and Commercial Fisheries Relationships.* Canadian Wildlife Service Special Publ., Ottawa. pp. 51–62.
Briggs, K. T., W. B. Tyler, D. B. Lewis, and D. R. Carlson, 1987. Bird communities at sea off California: 1975–1983. *Studies in Avian Biol. No. 11.* Cooper Ornithological Society. 74pp.
Brinton, E., 1976. Population biology of *Euphausia pacifica* off southern California. *U.S. Natl. Mar. Fish. Serv. Fish. Bull.* 74:733–762.
Burkett, E., 1988. Abundance, distribution, and status of birds at Bolsa Chica. Unpublished Tech. Rep. Calif. Dept. of Fish and Game, Sacramento, CA.
California Cooperative Oceanic Fisheries Investigations, 1988. Physical, chemical and biological data. Cruises 8709 and 8711. In: *Data Report SIO Ref. 88–8.* Scripps Institution of Oceanography, La Jolla, CA. pp. 91–94.
California Department of Fish and Game, 1980. At the crossroads: A report on the status of California's endangered and rare fish and wildlife. *Calif. Fish Game.* 147pp.
California Nature Conservancy, 1987. Sliding towards extinction: The state of California's natural heritage. San Francisco, CA. 123pp.
Chelton, D. B., Jr., 1980. Low-frequency sea level variability along the west coast of north America. Ph.D. Dissertation, Scripps Institution of Oceanography, La Jolla, CA. 212pp.
Christman, G., 1957. Some interspecific relations in the feeding of estuarine birds. *Condor.* 59: 343–344.
Chu, E. W., 1984. Sooty shearwaters off California: Diet and energy gain. In: D. N. Nettleship, G. A. Sanger, and P. F. Springer, eds. *Marine Birds, Their Feeding Ecology and Commercial Fisheries Relationships.* Can. Wildl. Serv., Ottawa. pp. 64–71.
Cottam, C., J. J. Lynch, and A. L. Nelson, 1944. Food habits and management of American sea brant. *J. Wildl. Manage.* 8:36–56.
Coulson, J. C., and E. White, 1961. Analysis of the factors influencing the clutch size of kittiwakes. *Proc. Zool. Soc. Lond.* 136:2017–2217.
Crossin, R. S., 1974. The storm-petrels (Hydrobatidae). In: W. B. King, ed. *Pelagic Studies of Seabirds in the Central and Eastern Pacific Ocean.* Smithson. Contrib. Zool. Washington, D.C. pp. 154–205.
Day, R., 1979. The accumulation of plastic particles in seabirds in Alaska. Master's Thesis, Univ. of Alaska, Fairbanks, AK.
DeGange, A. R., and T. C. Newby, 1980. Mortality of seabirds and fish in a lost salmon driftnet. *Mar. Pollut. Bull.* 11:322–323.
DeSante, D. F., and D. G. Ainley, 1980. The avifauna of the South Farallon Islands, California. *Stud. Avian Biol. No. 4.* Cooper Ornithological Society. 104pp.
Diamond, A. W., 1983. Feeding overlap in some tropical and temperate seabird communities. *Stud. Avian Biol.* 8:24–46.
Drent, R. H., 1965. Breeding biology of the pigeon guillemot, *Cephhus columba. Ardea.* 53: 99–160.
Dunn, E. K., 1973. Changes in fishing ability of terns associated with windspeed and sea surface conditions. *Nature (Lond.).* 224:520–521.
Evans, P. R., D. M. Herdson, P. J. Knights, and M. W. Pienkowski, 1979. Short-term effects of reclamation of part of Seal Sands, Teesmouth, on wintering waders and shelduck. I. Shorebird diets, invertebrate densities and the impact of predation on the invertebrates. *Oecologia* 41: 183–206.
Evans, R. M., 1980. Development of behavior in seabirds: An ecological perspective. In: J. Burger, B. L. Olla, and H. E. Winn, eds. *Behavior of Marine Animals, Vol. 4: Marine Birds.* Plenum Press, New York.
Frame, M. A., 1972. Cormorant nesting, San Luis Obispo County, California, 1972. Calif. Fish Game Spec. Wildl. Invest. Prog. Rep. WS4R4.
Fry, D. M., 1987. Plastics ingestion by seabirds: Estimating species at risk. *Pac. Seabird Group Bull.* 14(1):27.
Fry, D. M., and L. J. Lowenstine, 1982. Insults to alcids: Injuries caused by food, by burrowing, and by oil contamination. *Pac. Seabird Group Bull.* 9(2):78–79.
Fry, D. M., and L. J. Lowenstine, 1985. Pathology of common murres and Cassin's auklets exposed to oil. *Arch. Environ. Contam. Toxicol.* 14:725–737.
Fry, D. M., R. Boekelheide, J. Swenson, A. Kang, J. Young, and C. R. Grau, 1985. Long-term responses of breeding seabirds to oil exposure. *Pac. Seabird Group Bull.* 12(1):22.

Fry, D. M., J. Swenson, L. A. Addiego, C. R. Grau, and A. Kang, 1986. Reduced reproduction of wedge-tailed shearwaters exposed to weathered Santa Barbara crude oil. *Arch. Environ. Contam. Toxicol.* 15:453–463.

Furness, R. W., 1978. Energy requirements of seabird communities: A bioenergetics model. *J. Anim. Ecol.* 47(1):39–53.

Furness, R. W., 1982. Competition between fish and seabird communities. *Adv. Mar. Biol.* 20: 225–307.

Furness, R. W., and D. G. Ainley, 1984. Threats to seabird populations presented by commercial fisheries. In: J. P. Croxall, P. G. H. Evans, and R. W. Schreiber, eds. *Status and Conservation of the World's Seabirds.* Int. Comm. Bird Preserv. Tech. Publ. 2. pp. 701–708.

Furness, R. W., and J. Cooper, 1982. Interaction between breeding seabird populations and fish populations of the southern Benguela region, South Africa. *Mar. Ecol. Prog. Ser.* 8:243–250.

Garrett, K., and J. Dunn, 1981. *Birds of Southern California: Status and Distribution.* Los Angeles Audubon Society, Los Angeles. 408pp.

Goss-Custard, J. D., 1977a. The energetics of prey selection by redshank *Tinga totanus* (L.) in relation to prey density. *J. Anim. Ecol.* 46:1–19.

Goss-Custard, J. D., 1977b. Responses of redshank *Tringa totanus* to the absolute and relative densities of two prey species. *J. Anim. Ecol.* 46:867–874.

Goss-Custard, J. D., 1979. Effect of habitat loss on the numbers of overwintering shorebirds. *Stud. Avian Biol.* 2:167–177.

Goss-Custard, J. D., P. R. Evans, H. Milne, and M. W. Pienkowski, 1979. Utilization by shorebirds of benthic invertebrate production in intertidal areas. *Oceanogr. Mar. Biol. Ann. Rev.* 23:573–597.

Grau, C. R., T. Roudybush, J. Dobbs, and J. Wathen, 1977. Altered yolk structure and reduced hatchability of eggs from birds fed single doses of petroleum oils. *Science.* 195:779–781.

Green, K. A., 1978. Ecosystem description of the California Current. Rep. to U.S. Marine Mammals Commission, MMC–77/11. Contract MM7 AC026. 73pp.

Gress, F., R. W. Risebrough, D. W. Anderson, L. F. Kiff, and J. R. Jehl, 1973. Reproductive failures of double-crested cormorants in southern California and Baja California. *Wilson Bull.* 85:197–208.

Grinnell, J., and A. H. Miller, 1944. Distribution of birds of California. *Pac. Coast Avifauna.* 27:1–608.

Haney, J. C., 1986. Seabird segregation at Gulf Stream frontal eddies. *Mar. Ecol. Prog. Ser.* 28:279–285.

Harris, M. P., and J. R. G. Hislop, 1978. The food of young puffins, *Fratercula arctica. J. Zool. (Lond.).* 185:213–236.

Hartwick, E. B., 1976. Foraging strategy of the black oystercatcher (*Haematopus bachmani* Audubon). *Can. J. Zool.* 54:142–155.

Hartwick, E. B., and W. Blaylock, 1979. Winter ecology of a black oystercatcher population. *Stud. Avian Biol.* 2:207–215.

Helbing, G. C., 1977. Maintenance activities of the black oystercatcher, *Haematopus bachmani* Audubon, in northwestern California. Master's Thesis, California State Univ., Humboldt, Arcata, CA.

Hochbaum, H. A., 1944. *The Canvasback on a Prairie Marsh.* The American Wildlife Institute, Washington, D.C. 201pp.

Holmes, W. N., and J. Cronshaw, 1977. Biological effects of petroleum on marine birds. In: D. C. Malins, ed. *Effects of Petroleum on Arctic and Subarctic Marine Environments and Organisms.* Vol. 2. Academic Press, New York. pp. 359–398.

Holmes, W. N., J. Cronshaw, and J. Gorsline, 1978. Some effects of ingested petroleum on seawater-adapted ducks (*Anas platyrhynchos*). *Environ. Res.* 17(2):177–190.

Howell, A. H., 1932. Florida bird life. Tech. Rep. Florida Dep. Game Fresh Water Fish, Tallahassee, FL.

Howell, J. C., 1941. Early nesting at Cape Sable, Florida. *Auk.* 58:105–106.

Hunt, G. L., Jr., 1972. Influence of food distribution and human disturbance on the reproductive success of herring gulls. *Ecology.* 53: 1051–1061.

Hunt, G. L., Jr., and J. L. Butler, 1980. Reproductive ecology of western gulls and Xantus' murrelets with respect to food resources in the Southern California Bight. *Calif. Coop. Oceanic Fish. Invest. Rep.* 21:62–67.

Hunt, G. L., Jr., and M. W. Hunt, 1975. Reproductive ecology of the western gull: The importance of nest spacing. *Auk.* 92:270–279.

Hunt, G. L., Jr., R. L. Pitman, and H. L. Jones, 1980. Distribution and abundance of seabirds breeding on the California Channel Islands. In D. M. Power, ed. *The California Islands: Proceedings of a Multidisciplinary Symposium*. Santa Barbara Museum of Natural History, Santa Barbara, CA. pp. 443–460.

Hunt, G. L., Jr., R. L. Pitman, M. Naughton, K. Winnet, A. Newman, P. R. Kelly, and K. T. Briggs, 1981. Distribution, status, reproductive ecology, and foraging habits of breeding seabirds. In: *Summary of Marine Mammals and Seabird Surveys of the Southern California Bight Area 1975–1978*. Vol. III, investigators' reports, Part III, Book II, seabirds. Rep. to U.S. Dept. of Interior, Bureau of Land Management. Available from: NTIS, Springfield, VA; PB81–248205.

Idyll, C. P., 1973. The anchovy crisis. *Sci Am*. 228:22–29.

Ingraham, W. J., 1982. The anomalous surface salinity minima area across the northern Gulf of Alaska and its relation to fisheries. *Mar. Fish. Rev*. 41:8–19.

Ingraham, W. M., 1941. American bittern eats garter snake. *Auk*. 58:253.

Jehl, J. R., Jr., 1973. Studies of a declining population of brown pelicans in northwest Baja California. *Condor*. 75:69–79.

Kendeigh, S. C., 1970. Energy requirements for existence in relation to size of bird. *Condor*. 72:60–65.

King, J. G., and C. A. Lefever, 1979. Effects of oil transferred from incubating gulls to their eggs. *Mar. Pollut. Bull*. 10:319–321.

King, W. B., 1984. Incidental mortality of seabirds in gill nets in the northern Pacific. In: J. P. Croxall, P. G. H. Evans, and R. W. Schreiber, eds. *Status and Conservation of the World's Seabirds*. Int. Comm. Bird Preserv. Tech. Publ. 2. pp. 709–731.

Kirven, M., 1969. The breeding biology of Caspian terns (*Hydroprogne caspia*) and elegant terns (*Thalasseus elegans*) at San Diego Bay. Master's Thesis, San Diego State Univ., San Diego, CA. 110pp.

Kuletz, K., 1983. Mechanisms and consequences of foraging behavior in a population of breeding pigeon guillemots (*Cepphus columba*). Master's Thesis, Univ. of California, Irvine, CA.

Lasiewski, R. C., and W. R. Dawson, 1967. A reexamination of the relation between standard metabolic rate and body weight of birds. *Condor*. 69:13–23.

Lasker, R., 1981. Factors contributing to variable recruitment of the northern anchovy (*Engraulis mordax*) in the California Current: Contrasting years, 1975 through 1978. *Rapp. P.-V. Réun. Cons. Int. Explor. Mer*. 178:375–388.

Lawrence, G. E. 1950. The diving and feeding activity of the western grebe on the breeding grounds. *Condor*. 52:3–16.

LFPPO, 1989. *Ligue Française pour la Protection des Oiseaux*. Summary Research Notes, G. Bentz, compiler. Station Ornithologique de l'Ile Grande, Pleumeur-Bodou, France. 5pp.

Lid, G., 1981. Reproduction of the puffins on ROST in the Lofoten Islands in 1964–1980. *Fauna Norv. Ser. C. Cinclus*. 4:30–39.

MacCall, A., 1984. Seabird fisheries trophic interactions: Interaction in the eastern Pacific corridor currents, California and Peru. In: D. N. Nettleship, G. A. Sanger, and P. F. Springer, eds. *Marine Birds, Their Feeding Ecology and Commercial Fisheries Relationships*. Canadian Wildlife Service Special Publ., Ottawa. pp. 136–148.

McCaskie, G., 1974. The nesting season: Southern Pacific coast region. *Am. Birds*. 28:948–951.

McCaskie, G., 1976. The fall migration: Southern Pacific coast region. *Am. Birds*. 30:125–130.

McCaskie, G., 1977. The nesting season: Southern Pacific coast region. *Am. Birds*. 31:372–376.

McCaskie, G., 1980. Southern Pacific coast regional report. *Am. Birds*. 34(3):305–309.

Manuwal, D. A., 1974. The natural history of Cassin's auklet (*Ptychoramphus aleuticus*). *Condor*. 76:421–431.

Martin, A. C., H. S. Zim, and A. L. Nelson, 1951. *American Wildlife and Plants, a Guide to Wildlife Food Habits*. McGraw Hill Book Co., New York. 500pp.

Massey, B. W., 1988. California least tern study: 1988 breeding season. Annu. Rep. to Calif. Dept. of Fish and Game, Long Beach, CA.

Massey, B. W., 1989. California least tern study: 1989 breeding season. Annu. Rep. to Calif. Dept. of Fish and Game, Long Beach, CA.

Massey, B. W., and J. L. Atwood, 1981. Second-wave nesting of the California least tern: Age composition and reproductive success. *Auk*. 98(3):596–605.

Murray, K. G., K. Winnett-Murray, Z. Eppley, G. L. Hunt, Jr., and D. B. Schwartz, 1983. Breeding biology of the Xantus' murrelet. *Condor*. 85:12–21.

Osborne, T., 1973. Recent nesting of the rhinoceros auklet in California. *Condor.* 75:463–464.

Owen, R. W., Jr., 1980. Eddies of the California Current system: Physical and ecological characteristics. In: D. M. Power, ed. *The California Islands: Proceedings of a Multidisciplinary Symposium.* Santa Barbara Museum of Natural History, Santa Barbara, CA. pp. 237–263.

Packard, F. M., 1943. Predation upon a Wilson's phalarope by Treganza's heron. *Auk.* 60:97.

Palmer, R. S., ed., 1976a. *Handbook of North American Birds. Vols. 2 and 3: Waterfowl.* Yale Univ. Press, New Haven, CT. 521pp. and 560pp.

Palmer, R. S., ed., 1976b. *Handbook of North American Birds. Vol. 1: Loons through Flamingos.* Yale Univ. Press, New Haven, CT. 567pp.

Parrish, R. H., C. S. Nelson, and A. Bakun, 1981. Transport mechanisms and reproductive success of fishes in the California Current. *Biol. Oceanogr.* 1(2):175–203.

Pitman, R. L., and S. M. Speich, 1976. Black storm petrel breeds in the United States. *West. Birds.* 7:71.

Power, D. M., ed., 1980. *The California Islands: Proceedings of a Multidisciplinary Symposium.* Santa Barbara Museum of Natural History, Santa Barbara, CA. 787pp.

Powers, K. D., and R. G. B. Brown, 1987. Seabirds. In: R. H. Backus, and D. W. Bourne, eds. *Georges Bank.* The MIT Press, Cambridge, MA. pp. 359–371.

Pugh, J. R., 1962. Effect of certain electrical parameters and water resistivities in mortality of fingerling silver salmon. *U.S. Fish Wildl. Serv. Fish. Bull.* 62(208):223–234.

Quammen, M. L., 1981. Use of enclosures in studies of predation by shorebirds on intertidal mudflats. *Auk.* 98:812–817.

Quammen, M. L., 1982. Influence of subtle substrate differences on feeding by shore birds on intertidal mud flats. *Mar. Biol.* 71(3):339–343.

Reading, C. J., and S. MacGrorty, 1978. Seasonal variations in the burying depth of *Macoma balthica* (L.) and its accessibility to wading birds. *Estuarine Coastal Mar. Sci.* 6:135–144.

Reish, D. J., and J. L. Barnard, 1990. Marine invertebrates as food for the shorebirds of Bahia de San Quintin, Baja California. In: M. Dailey and H. Bertsch, eds. *Proceedings, VIII International Marine Biology Symposium,* Ensenada, Mexico. June 4–8, 1990. pp. 1–6.

Remsen, J. V., Jr., 1978. Bird species of special concern in California: An annotated list of declining or vulnerable bird species. Calif. Dept. Fish Game Rep. 78–1. 54pp.

Risebrough, R. W., F. C. Sibley, and M. N. Kirven, 1971. Reproductive failure of the brown pelican on Anacapa Island in 1969. *Am. Birds.* 25:8–9.

Robertson, I., 1972. Studies of fish eating birds and their influence on stocks of the Pacific herring in the Gull Islands of British Columbia. Herring Invest., Pac. Biol. Station, Nanaimo, BC.

Sanger, G. A., 1974. Black-footed albatross (*Diomeda nigripes*). In: W. B. King, ed. *Pelagic Studies of Seabirds in the Central and Eastern Pacific Ocean.* Smithson. Contrib. Zool. 158:96–128.

Schaefer, M. B., 1970. Men, birds, and anchovies in the Peru Current: Dynamic interaction. *Trans. Am. Fish. Soc.* 99(3):461–467.

Schaffner, F. C., 1986. Trends in elegant tern and northern anchovy populations in California. *Condor.* 88(3):347–354.

Sievert, P. O., L. Sileo, and S. I. Fefer, 1988. Prevalence and effect of plastic ingestion in Hawaiian seabirds. *Pac. Seabird Group Bull.* 15(1):38.

Small, A., 1974. *The Birds of California.* Winchester Press, New York. 310pp.

Solomonsen, F., 1955. The food production in the sea and the annual cycle of Faeroese marine birds. *Oikos.* 56:92–100.

Sowls, A. S., A. R. DeGange, S. W. Nelson, and G. S. Lester, 1980. Catalog of California seabird colonies. U.S. Fish Wildl. Serv., Biol. Serv. Program, Coastal Ecosys. Proj. FWS/OBS–80/37. Washington, D.C.

Stenzel, L. E., G. W. Page, M. R. Carter, and D. G. Ainley, 1988. Chronic oiling of marine birds found on beached bird surveys in California. *Pac. Seabird Group Bull.* 15(1):39.

Szaro, R. C., and R. P. Balda, 1979. Bird community dynamics in a ponderosa pine forest. *Stud. Avian Biol. No. 3,* Cooper Ornithol. Soc. 66pp.

Takekawa, J. E., T. E. Harvey, and H. R. Carter, 1987. Population decline of the common murre in central California. *Pac. Seabird Group Bull.* 15:40.

Unitt, P., 1984. *The Birds of San Diego County.* San Diego Society of Natural History, San Diego, CA. 276pp.

Varoujean, D. H., 1982. Seabird–oil spill behavior study. *Pac. Seabird Group Bull.* 9(2):78.

Vermeer, K., 1979. Nesting requirements, food,

and breeding distribution of rhinoceros auklet *Cerorhinca monocerata* and tufted puffin *Lundha cirrhata*. *Ardea*. 67:101–110.

Vermeer, K., 1980. The importance of timing and type of prey to reproductive success of rhinoceros auklets, *Cerorhinca monocerata*. *Ibis*. 122:343–350.

Vermeer, K., and L. Rankin, 1984. Pelagic seabird populations in Hecate Strait and Queen Charlotte Sound: Comparisons with the west coast of the Queen Charlotte Islands. *Can. Tech. Rep. Hydrogr. Ocean Sci*. 52:1–40.

Vermeer, K., and S. J. Westerheim, 1982. Fish changes in diets of nestling rhinoceros auklets and their implications. In: D. N. Nettleship, G. A. Sanger, and P. F. Springer, eds. *Marine Birds, Their Feeding Ecology and Commercial Fisheries Relationships*. Canadian Wildlife Service Special Publ., Ottawa. pp. 96–105.

Vermeer, K., L. Cullen, and M. Porter, 1979. A provisional explanation of the reproductive failure of tufted puffins, *Lunda cirrhata*, on Triangle Island, British Columbia. *Ibis*. 121:348–354.

Wetmore, A., 1925. Food of American phalaropes, avocets, and stilts. *U.S. Dept. of Agriculture Bull*. 1359.

Wiens, J. A., and J. M. Scott, 1975. Model estimation of energy flow in Oregon coastal populations. *Condor*. 77:439–452.

Willet, G., 1933. A revised list of the birds of southwestern California. *Pac. Coast Avifauna*. 21:1–204.

Zembal, R., and B. W. Massey, 1985. Distribution of the light-footed clapper rail in California, 1980–1984. *Am. Birds*. 39:135–137.

Chapter 11

Marine Mammals

Michael L. Bonnell and Murray D. Dailey

Introduction

The Southern California Bight (SCB) has one of the largest and most diverse marine mammal populations in the world. It includes all but one of the baleen whale species found in the eastern North Pacific; more than a dozen species of porpoises, dolphins, and other toothed whales; six species of pinnipeds; and the sea otter (table 11.1). Some species are migrants that pass through the SCB on their way to calving or feeding grounds located elsewhere. Some are seasonal visitors that remain for only a few weeks to exploit a particular food resource. Other species have resident populations in the area for many months or year-round.

In some seasons, the combined abundance of marine mammals in the SCB may reach 150,000 animals, representing as many as 30 different species. In part, this extraordinarily great use of the SCB is due to the large ranges occupied by many marine mammal species. Within its range, a species may undergo seasonal migrations to reach breeding and feeding grounds and to accommodate changing environmental conditions (Gaskin 1982). These movements bring many species into or through the SCB. Also, the SCB is a region of overlap of faunas characteristic of both warm temperate (San Diegan) and

Table 11.1. *Marine Mammals of the Eastern North Pacific and Status of Species in the SCB*

Species	Historical Population Size[a] (North Pacific)	Present Population Size[b] (North Pacific)	Status and Size of Population in SCB
Order Cetacea			
Baleen Whales (Suborder Mysticeti)			
Blue whale[E] (*Balaenoptera musculus*)	5000	1400–1900[c]	Migratory population. Reaches peak numbers in summer as population moves northward from subtropical wintering grounds.
Fin whale[E] (*Balaenoptera physalus*)	42,000–45,000	14,620–18,630[d]	Migratory population. A few are present in the SCB year-round; population reaches peak numbers in summer.
Sei whale[E] (*Balaenoptera borealis*)	45,000–50,000	22,000–37,000[e]	Migratory population. Seen only in summer months, primarily in offshore waters; uncommon in SCB.
Bryde's whale (*Balaenoptera edeni*)	20,000–30,000	14,600–39,900	Rare; represented in SCB by single sighting near San Diego.
Minke whale (*Balaenoptera acutorostrata*)	—	13,500	Migratory population. Common in SCB throughout the year, but reaches peak numbers in spring and summer.
Humpback whale[E] (*Megaptera novaeangliae*)	15,000	1200	Migratory population. Uncommon in SCB; peak abundance in summer and autumn.
Gray whale[E] (*Eschrichtius robustus*)	15,000	21,113[f,g]	Most of world population passes through SCB in winter and spring.
Northern right-whale[E] (*Balaena glacialis*) (Also referred to as *Eubalaena glacialis*)	15,000	200[h]	Occasional visitor. Represented in SCB by two sightings; rare.
Toothed Whales (Suborder Odontoceti)			
Sperm whale[E] (*Physeter macrocephalus*)	>307,300[i]	472,000	Occasional visitor; typically inhabits offshore waters. Uncommon in SCB.
Common dolphin (*Delphinus delphis*)	—	>900,000[j]	Year-round resident; mean population of 57,000 in summer and autumn. Common.
Northern right-whale dolphin (*Lissodelphis borealis*)	—	—[f]	Seasonal resident population in winter and spring (<16,000); common.

Table 11.1. *Marine Mammals of the Eastern North Pacific and Status of Species in the SCB* (continued)

Species	Historical Population Size[a] (North Pacific)	Present Population Size[b] (North Pacific)	Status and Size of Population in SCB
Pacific white-sided dolphin (*Lagenorhynchus obliquidens*)	—	30,000–50,000[f,j]	Year-round resident; population in SCB may reach 12,000 animals. Common.
Risso's dolphin (*Grampus griseus*)	—	—	Year-round resident; peak population of about 4000 in summer and autumn. Common.
Dall's porpoise (*Phocoenoides dalli*)	—	1.4–2.8 million[f,k]	Year-round resident; present throughout the year, but is at peak abundance in SCB in autumn and winter (<1000). Common.
Bottlenose dolphin (*Tursiops truncatus*; also referred to as *T. gilli*	—	—	Year-round resident; two populations may be present in the SCB (<1000). Common.
Harbor porpoise (*Phocoena phocoena*)	—	—	Occasional visitor. Stranding records from Santa Barbara to Los Angeles; rare in SCB.
Short-finned pilot whale (*Globicephala macrorhynchus*; also referred to as *G. scammonii*)	—	—	Year-round resident population of about 400 animals; increases to about 600 in winter. Common prior to 1982.
Killer whale (*Orcinus orca*)	—	—	Occasional visitor; most often present in summer and winter. Uncommon.
False killer whale (*Pseudorca crassidens*)	—	—[f]	Occasional visitor; possible mass stranding represented by skeletal remains; several sightings in SCB and offshore waters; widely distributed in eastern North Pacific. Rare.
Cuvier's beaked whale (*Ziphius cavirostris*)	—	—[f]	Occasional visitor; known in the SCB from sightings and strandings. Uncommon.
Baird's beaked whale (*Berardius bairdii*)	—	—[f]	Occasional visitor; several sightings in SCB or pelagic waters. Rare.
Hubb's beaked whale (*Mesoplodon carlhubbsi*)	—	—[f]	Occasional visitor; represented in SCB by strandings (several sightings at sea of *Mesoplodon* sp. may be this species). Uncommon.
Ginkgo-toothed beaked whale (*Mesoplodon ginkgodens*)	—	—	Possible visitor; represented in SCB by single stranding record.

Table 11.1. *Marine Mammals of the Eastern North Pacific and Status of Species in the SCB* (continued)

Species	Historical Population Size[a] (North Pacific)	Present Population Size[b] (North Pacific)	Status and Size of Population in SCB
Hector's beaked whale (*Mesoplodon hectori*)	—	—	Occasional visitor; represented in SCB by strandings and probable sightings. Rare.
Blainville's beaked whale (*Mesoplodon densirostris*)	—	—[f]	Possible visitor; typically found in central North Pacific.
Bering Sea beaked whale (*Mesoplodon stejnegeri*)	—	—[f]	Possible visitor; typically found in Bering Sea and Gulf of Alaska.
Dwarf sperm whale (*Kogia simus*)	—	—[f]	Possible visitor; known from strandings in central California and Mexico.
Pygmy sperm whale (*Kogia breviceps*)	—	—[f]	Occasional visitor; known from strandings only. Rare.
Striped dolphin (*Stenella coeruleoalba*)	—	>2,300,000[j]	Occasional visitor; known from strandings and sightings. Rare.
Spinner dolphin (*Stenella longirostris*)	—	>900,000[j]	Possible visitor; eastern tropical Pacific species.
Spotted dolphin (*Stenella attenuata*)	—	>2,200,000[j]	Possible visitor; eastern tropical Pacific species.
Rough-toothed dolphin (*Steno bredanensis*)	—	—	Possible visitor; eastern tropical Pacific species.
Order Carnivora Pinnipeds (Suborder Pinnipedia)			
California sea lion (*Zalophus californianus*)	—	157,000	Year-round resident; peak summer population of about 87,000. Abundant.
Northern (Steller's) sea lion (*Eumetopias jubatus*)	—	95,000–122,000[f]	Occasional visitor; no longer breeds in SCB, but a few males usually present in summer. Uncommon.
Northern fur seal (*Callorhinus ursinus*)	—	1,151,000[f]	Year-round resident. San Miguel colony reaches peak of 4000 in summer; pelagic population of about 5000 off SCB in winter and spring. Common.
Guadalupe fur seal[T] (*Arctocephalus townsendi*)	—	1600[f]	Occasional visitor. Presently breeds only on Isla de Guadalupe, Mexico; winter pelagic range includes SCB. Uncommon.

Table 11.1. *Marine Mammals of the Eastern North Pacific and Status of Species in the SCB* (continued)

Species	Historical Population Size[a] (North Pacific)	Present Population Size[b] (North Pacific)	Status and Size of Population in SCB
Northern elephant seal (*Mirounga angustirostris*)	—	>100,000[f]	Year-round resident; winter breeding population of about 27,000 on land. Common on land; uncommon at sea.
Pacific harbor seal (*Phoca vitulina*)	—	312,000	Year-round resident; population of up to 5000 in SCB during early summer (1987 est.). Common.
Fissipeds (Suborder Fissipedia)			
Sea otter[T] (*Enhydra lutris*)	300,000[l]	100,000–168,000[f,m]	Year-round resident. Locally abundant in nearshore waters around San Nicolas Is.; a few wanderers near mainland and other islands. Total of 40–60 in SCB (1987 est.).

Note: Dash indicates that population estimate cannot be made from existing data.

[E]Designated as endangered under the Endangered Species Act of 1973 (ESA). The National Marine Fisheries Service has proposed delisting of the eastern North Pacific (California) stock of gray whales based on the determination that it has recovered to its preexploitation size and is no longer in danger of extinction (Nov. 22, 1991; Federal Register 56:58869).

[T]Designated as threatened under the Endangered Species Act of 1973. Sea otters in the SCB are afforded the protections of a listed species only within a "translocation zone" of waters extending 10–19 nautical miles from the 15-m isobath around San Nicolas Island. Sea otters in the SCB found outside the San Nicolas Island translocation zone are treated, for purposes of the ESA, as if belonging to a species proposed for listing (Public Law 99–625 of 1986). Therefore, outside the translocation zone, incidental take of sea otters in the course of otherwise lawful activities is not considered a violation of either the Endangered Species Act or the Marine Mammal Protection Act.

[a]Food and Agriculture Organization of the United Nations (FAO) 1978; other sources as cited. Historical size estimated for population prior to modern mechanized whaling of this century; may include more than one management stock.

[b]U.S. Department of Commerce 1987; other sources as cited.

[c]Braham 1984a.

[d]Ohsumi and Wada 1974.

[e]Ohsumi and Fukuda 1975.

[f]World population limited to North Pacific area.

[g]Breiwick et al. 1988.

[h]Braham and Rice 1984.

[i]Estimate includes only exploitable adults.

[j]Estimate includes only eastern North Pacific stock.

[k]Jones et al. 1987.

[l]Johnson 1982.

[m]Estes 1980; Calkins and Schneider 1985; Rotterdam and Simon-Jackson 1988.

cool temperate (Montereyan) divisions of the California zoogeographic province, and perhaps other provinces as well (Hubbs 1960). Several marine mammal species reach the southern limit of their ranges in the SCB, while other species are at their northern range limits.

In examining the status of marine mammals in the SCB, it is important to remember that most species have been heavily exploited in the last two centuries for oil, pelts, and other products. Populations may still be recovering. As a measure of recovery, present abundance of a few species can be compared with their historical population. However, the size of marine mammal populations prior to exploitation is often poorly known. Much of the historical data is based on accounts of whaling, sealing, or sea otter harvests. Thus, only rough estimates are available for a few commercially valuable species (table 11.1). Some species that were nearly brought to extinction appear to have recovered, but populations of other species remain at precariously low levels. While numbers are uncertain, there is little doubt that populations of most marine mammals were much larger in the past than they are today.

Recognition that stocks of many marine mammal species remain at much reduced levels has led to passage of laws regulating human activities where marine mammals might be adversely affected. The most specific of these laws is the Marine Mammal Protection Act of 1972 (MMPA), which prohibits, in part, the intentional taking, import, or export of any marine mammal without a permit. "Taking" includes any attempt to harass, hunt, capture, or kill a marine mammal. A second law protecting certain species is the Endangered Species Act of 1973 (ESA), as amended by Public Law 97–304 (the 1982 amendments to the ESA). Marine mammal species listed by the ESA as endangered or threatened are identified in table 11.1. The prohibitions of the ESA are similar to those of the MMPA in that no listed species may be subject to intentional or attempted take, importation, exportation, or other commerce. Some unintentional take or take resulting from research or recovery actions may be authorized under the ESA. However, the MMPA is more restrictive. Any species listed under the ESA is considered depleted under the MMPA, that is, to be beneath its optimum sustainable population. Unintentional take of a depleted species can be allowed by permit only when a particular human activity is determined to have a negligible impact. Intentional take of a depleted species is allowed only under the terms of a scientific research permit.

Through migrations and seasonal concentrations in parts of their range, marine mammals can exert great influence on the ecosystem. Most marine mammal species are very mobile and can go where food is plentiful. They are efficient feeders and utilize a wide variety of food resources, including zooplankton, large invertebrates, fish, and even other marine mammals. As a group, marine mammals affect energy flow among several trophic levels. However, the ecological role of marine mammals in the SCB is still not well understood. Free-ranging marine mammals are difficult to study. Not only do they remain underwater and out of sight a significant portion of the time but many species are uncommon in the SCB and rarely observed. Even the most abundant species are wary and often restrict their activities to offshore waters and remote islands.

To understand how marine mammals interact with their environment, several kinds of information are needed. First, we must know what species use the SCB, their annual cycles of abundance, and patterns of distribution. Further, we need to know what they eat, how their diets vary seasonally, and the energetic costs of life in the SCB. Much of this information is available. Systematic aerial and vessel surveys have revealed the seasonal patterns of abundance and distribution of marine mammals at sea and on land. Substantial progress has also been made in characterizing the diet and estimating the energetic needs of at least a few marine mammal species. From this information, it is possible to estimate some of the impacts of marine mammals on their food resources and to anticipate the ecological consequences of changes in marine mammal populations and their environment.

Description of the Marine Mammal Fauna

The marine mammal fauna of the SCB includes at least 34 species that have been identified from sightings or strandings. Some species are known to be common in the SCB, while others are extremely rare. Because marine mammals are difficult to observe, the status of some rare species is unclear. Such species may occur in very small numbers on a regular basis or may enter the SCB infrequently during periods of unusual oceanographic conditions. Generally, use of the SCB by marine mammals is fairly well described compared to other parts of the eastern North Pacific. A great deal of new information has been gathered in the last decade through the sponsorship of the Minerals Management Service Environmental Studies Program (U.S. Department of the Interior), as well as under the auspices of the National Marine Fisheries Service (U.S. Department of Commerce).

Generalizing from even the most intensive survey data has risks. Marine mammals are very mobile. Population centers may shift on large spatial scales (>100 km) over small time scales (days or weeks) as marine mammals exploit short-lived resources of food (e.g., patches of small planktonic crustaceans or spawning concentrations of fish and squid). The availability of food varies greatly with oceanographic conditions, and corresponding changes in the distribution of marine mammals may occur as well.

Baleen Whales

Three taxonomic families of baleen whales inhabit the SCB: the rorquals (Balaenopteridae), the gray whale (Eschrichtiidae), and the right whale (Balaenidae). The rorquals, a group that includes the blue, fin, sei, Bryde's, minke, and humpback whales, have characteristic pleated throats that expand to take in water, which is then strained outward through the baleen. All of the baleen whales have extensive ranges in the North Pacific, extending from high-latitude feeding grounds occupied in the summer to subtropical calving grounds occupied in the winter (fig. 11.1). The mating season typically begins during the southbound migration and extends from mid-autumn through winter. Gestation lasts 11–12 months for blue, fin, sei, Bryde's, and humpback whales (Laws 1961; Johnson and Wolman 1984; Mizrock et al. 1984a). The gestation period of gray whales lasts approximately 13.5 months (Rice 1983). For all of these species, females become sexually mature at about 7–8 years of age, and thereafter, they calve every 2–3 years. Calves are usually weaned on the feeding grounds at 6–9 months of age. The smaller minke whales have a gestation period of about 10 months, and females may calve each year (Mitchell 1978).

The blue whale (*Balaenoptera musculus*), which is the largest of the rorquals, has a worldwide distribution in circumpolar and temperate waters. Blue whales in the North Pacific are presently considered to be a single stock, although little is known about their range and movements. The summer feeding grounds are generally believed to be in the Gulf of Alaska and off the eastern Aleutian Islands (Berzin and Rovnin 1966). Catch records show that the peak abundance of blue whales in Alaskan waters occurs in June and July (Rice 1974). The whales migrate southward in the autumn, reaching waters off Baja California in October; the actual location of their wintering grounds is unknown, but calving may occur in subtropical waters farther to the south or offshore (Rice 1974). Following the mating and calving season, blue whales again migrate northward. The general migration pathway through the SCB follows the continental slope (fig. 11.2). Blue whales

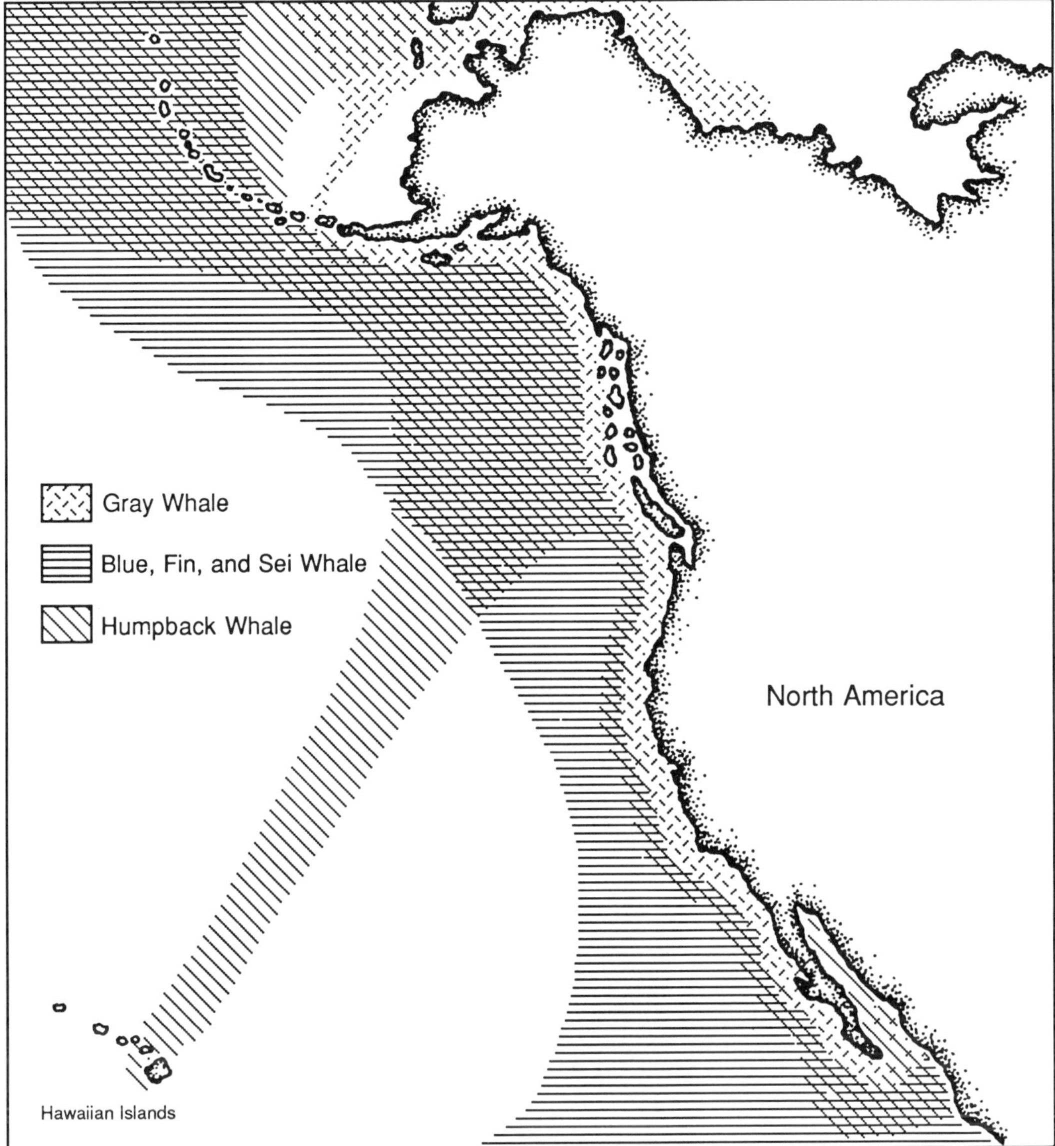

Figure 11.1. Approximate distribution and pathways of seasonal migration for blue, fin, sei, humpback, and gray whales in the eastern North Pacific. (Data sources described in text.)

are gone from Baja California waters by early summer and reach peak abundances in the SCB in June. On ship and aerial surveys in the SCB, 97% of all sightings were recorded between June and October (Dohl et al. 1981). Blue whales are rarely present in the SCB after October. Line transect analysis of aerial survey data indicates that blue whales in the SCB number about 50 animals in June and half that in October. These whales appear to have a summer population in the SCB of at least 20 animals that are concentrated between July and September in highly productive waters of the western Santa Barbara Channel offshore to about Rodriguez Seamount.

Fin whales (*Balaenoptera physalus*) have a worldwide distribution, with at least two dis-

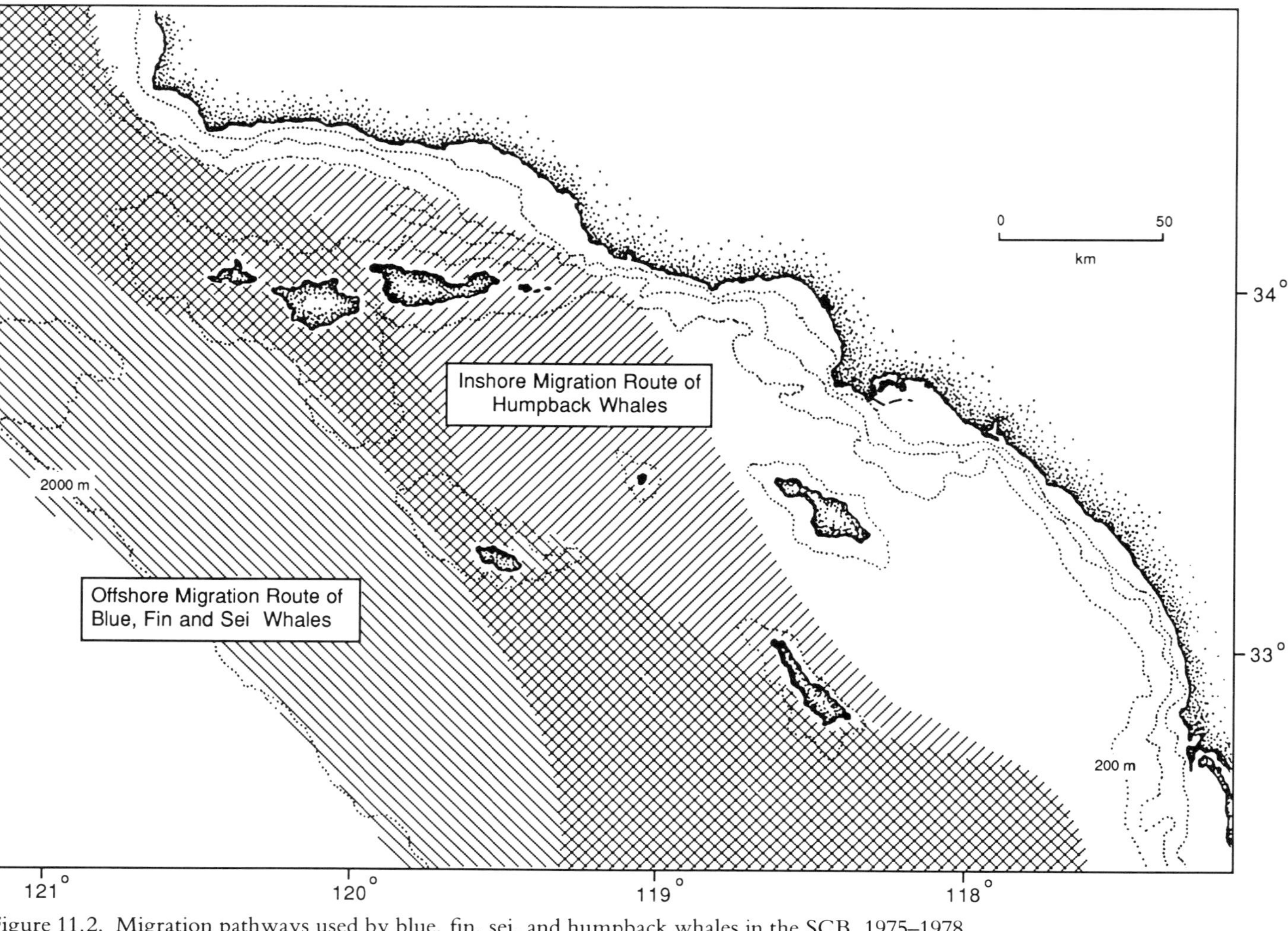

Figure 11.2. Migration pathways used by blue, fin, sei, and humpback whales in the SCB, 1975–1978.

tinct populations in the North Pacific. Fin whales in the Asian stock are found in subtropical and temperate waters of the western North Pacific in winter (Nemoto 1959); in spring and summer, they migrate northward to the Chukchi Sea (Berzin and Rovnin 1966). Animals of the eastern stock migrate northward from subtropical wintering grounds (presumably located off Mexico) to the Gulf of Alaska and the Bering and Chukchi seas (Brueggeman et al. 1987). On surveys in the SCB, fin whales have been seen in all seasons; however, 87% were observed from March through October (Dohl et al. 1981). In June, about 70 animals may be present as these whales migrate northward through the area. Their summer population is estimated at 28 animals, increasing to about 50 in October when southbound migrants reach peak numbers. Their summer distribution includes the Santa Rosa-San Nicolas Ridge and inshore waters toward Anacapa and Santa Catalina islands. Like the blue whale, the migration pathway of fin whales through the SCB follows the continental slope (fig. 11.2).

Sei whales (*Balaenoptera borealis*) are found in temperate waters worldwide, but are considered more boreal in distribution than the other balaenopterids. Three stocks probably exist in the North Pacific (Mizrock et al. 1984b). Perhaps because of small population size or their offshore distribution, sei whales are uncommon in the eastern North Pacific. Only three sightings of sei whales in California waters have been recorded, including one over Cortes Bank (Dohl et al. 1981). The closely related Bryde's whale (*Balaenoptera edeni*) has only recently been distinguished from the sei whale. Though more tropical than other whales of this genus, it is occasionally found in the SCB. A stranding of a Bryde's whale occurred near La Jolla in 1954 (Leatherwood et al. 1982).

Minke whales (*Balaenoptera acutorostrata*), which are abundant worldwide, are still hunted in the Southern Hemisphere, where the population is estimated to be 258,300 (U.S. Department of Commerce 1987). In the eastern North Pacific, minke whales appear to be broadly distributed between Baja California, Mexico, and the Chukchi Sea in the summer (Rice 1974). The winter range is more compressed, extending from Islas Revillagigedo, Mexico, to central California (Rice 1974). Only a few minke whales have been noted in winter months in central and northern California (Dohl et al. 1983); however, in the SCB, newborn or very young minke whales are sometimes present in the winter (Norris and Prescott 1961; Dohl et al. 1981). From these sightings, it is likely that the SCB is within the calving range of this species.

The cycle of abundance in the SCB shows small but distinct peaks in June and November, suggesting that a few whales migrate into and—probably through—the SCB in the spring, with return movement southward in the autumn. The summer population of minke whales has a relatively compact distribution that includes the western Santa Barbara Channel; the Santa Rosa-San Nicolas Ridge; the island shelves south of San Miguel, Santa Rosa, and Santa Cruz islands; and the east side of San Nicolas Island (fig. 11.3). Sightings are also clustered near the Anacapas and southward over the eastern rim of the Santa Cruz Basin. The size of the summer population is difficult to estimate, but surveys have consistently recorded 8–16 animals (Dohl et al. 1981). Line transect analysis of data collected on aerial surveys during the spring and autumn peaks shows the migrant population to range from about 31 whales (spring) to 67 whales (autumn). The distribution of minke whales in autumn broadens to include most inshore waters. However, minke whales are most abundant along the Santa Rosa–Cortes Ridge near San Miguel and Santa Rosa islands and in waters between Santa Catalina Island and Forty-Mile Bank southeast of San Clemente Island. Very few minkes are present in the SCB in winter months, but the population rapidly increases in late spring. From May through July, minke whales are seen over an area stretching from the San Diego Basin all the way into central

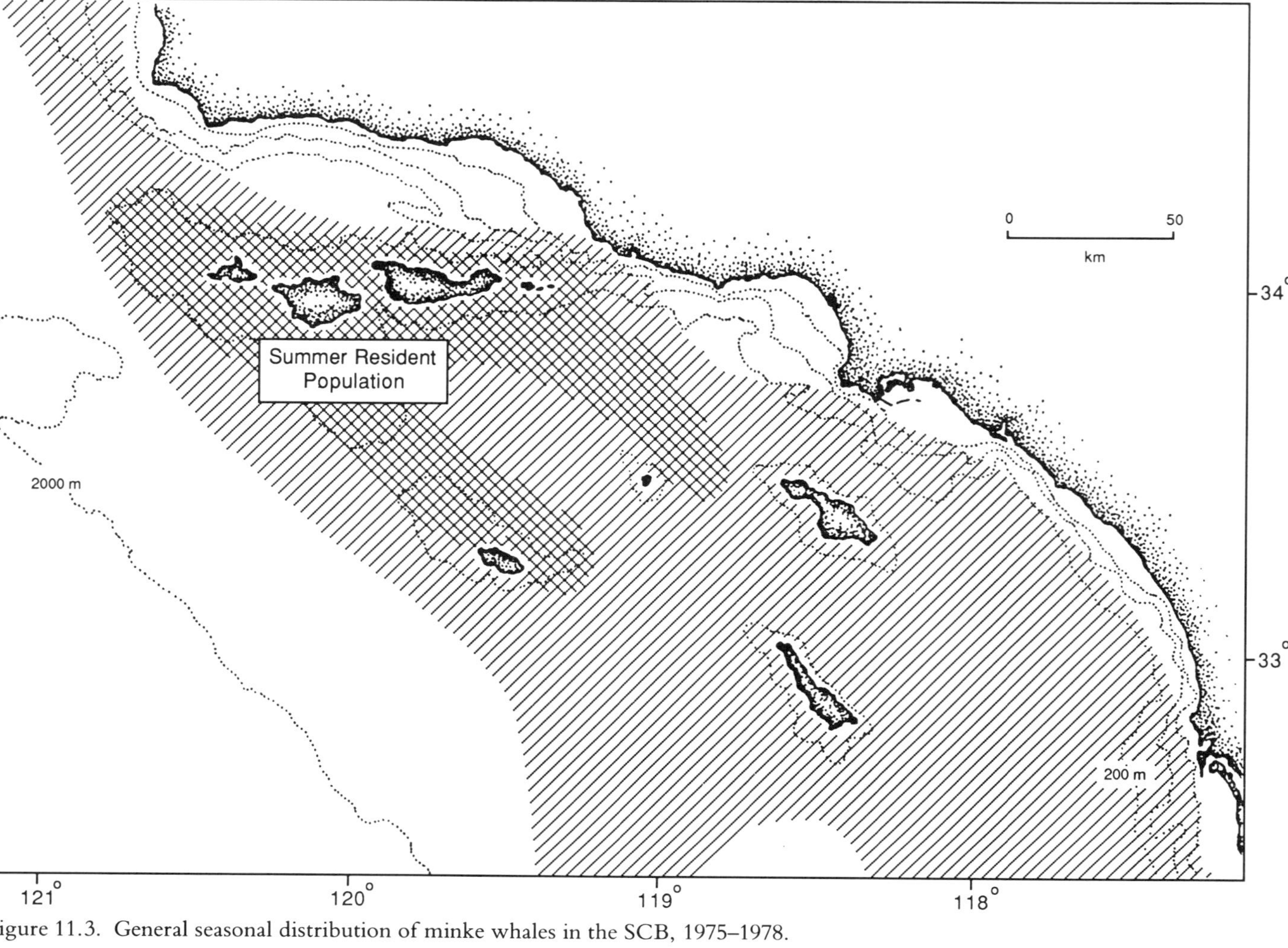

Figure 11.3. General seasonal distribution of minke whales in the SCB, 1975–1978.

waters of the SCB, especially in the region of Lasuen Knoll, east of Santa Catalina Island in the San Pedro Channel. In contrast with the summer–autumn distribution, minke whales are virtually absent during spring over the Santa Cruz Basin and most of the Santa Rosa–Cortes Ridge.

Humpback whales (*Megaptera novaeangliae*) also occur worldwide. In the North Pacific, they range in the summer months from Arctic waters south to Japan and central California. Three major wintering areas exist, including waters off Mexico (Rice 1974); Hawaii (Baker et al. 1986); the Marianas, Bonin, and Ryukyu islands; and Taiwan (Nishiwaki 1959). On the basis of discovery markers and photographs of fluke patterns, it appears that whales from wintering grounds south of Japan and off Hawaii and Mexico intermingle during summer months in the waters of southeast Alaska, Prince William Sound, and western Gulf of Alaska (Baker et al. 1986). Combining estimates from Rice and Wolman (1982), Baker et al. (1985), and their own surveys, Brueggeman et al. (1987) estimated the total population in the summer feeding areas of Alaska to be 1007 whales. Photoidentification studies indicate that some interchange exists between the Hawaiian and Mexican wintering grounds (Darling and McSweeney 1985). The eastern North Pacific stock apparently has a summer feeding range extending from central California to the waters off southeastern Alaska and the eastern Aleutian Islands. Peak numbers off central and northern California occur from July through November; the population during these months has been estimated at about 338 animals (Dohl et al. 1983).

Wintering grounds for humpbacks off Mexico are known to include the outer coast of southern Baja California, the mainland coast of Mexico, and the Islas Revillagigedo (Rice 1974; Urbán-R. and Aguayo-L. 1987). A few humpbacks are also present throughout the year in the Gulf of California (Urbán-R. and Aguayo-L. 1987). The northward migration begins in March or April and is led by newly pregnant females and immatures (Wolman 1986). Humpback whales are present in the SCB from March through June and again from September through December. In these months, sightings are sparse and widespread within the central waters of the SCB. No summer resident population of humpback whales occurs in the area; neither do these whales appear to linger into the winter breeding season. Line transect analysis of data collected on aerial surveys shows the migrant population in the SCB to range from about 25 to 35 whales (present in the SCB over two 4-day survey periods in June and October). Migrants passing through the SCB appear to follow a more inshore path than blue, fin, or sei whales (fig. 11.2).

The migration of gray whales (*Eschrichtius robustus*) between calving lagoons in Baja California, Mexico, and feeding grounds in the Bering Sea is one of the most significant events in coastal waters of the eastern North Pacific. Gray whales are found only in the North Pacific (an Atlantic form is extinct). Two populations exist, although the western population is still severely depleted and may be unable to recover (Brownell 1977). The eastern stock, known as the California gray whale, has a growing population that has now reached preexploitation levels. At present, the population is estimated at more than 21,000 animals (Breiwick et al. 1988; Reilly 1991).

California gray whales are typically found in summer months in feeding grounds in the Bering Sea and southern Chukchi Sea (Rice and Wolman 1971; Braham 1984b), although a few gray whales pass the summer at locations along the coast between Baja California and southeastern Alaska (Patten and Samaras 1977; Sprague et al. 1978; Darling 1979; Dohl et al. 1983; Gill and Hall 1983; Sullivan et al. 1983; Sumich 1984). Departure from feeding grounds in Alaska begins in October and November led by pregnant females and followed by females that have recently ovulated, adult males, then immature whales of both sexes (Rice and Wolman 1971). Southbound gray whales pass the Oregon coast beginning in December and continuing through February, with a peak in late December and early January (Herzing and Mate 1984). The southbound migration through the SCB also begins in December and extends through February. However, the northbound migration is more protracted, lasting from February though at least May. In the SCB, gray whales are typically absent from August through November.

Along most of the coast, gray whales are known to migrate very close to land (Herzing and Mate 1984; Reilly 1984; Rice et al. 1984; Rugh 1984; Sund and O'Connor 1974). The distance offshore of migrating gray whales appears to vary according to the nature of coastal morphology. In central California, for example, many gray whales take a direct path farther offshore across the Farallon Basin, Monterey Bay, and Estero Bay (Dohl et al. 1983; Poole 1984). Data collected at several locations along the migratory route indicate that southbound migrants are generally found at greater distances from shore than those returning north (Dohl et al. 1983; Braham 1984b; Herzing and Mate 1984; Poole 1984).

Migration through the SCB by way of an offshore pathway has been noted by Rice (1965) and Leatherwood (1974). More recently, Dohl et al. (1981) described the major features of gray whale migration through the SCB; they based their descriptions on 392 sightings of more than 900 gray whales made on aerial and vessel surveys. The migration pathway through the SCB is broad and rather diffuse (fig. 11.4). Gray whales are seen up to 200 km offshore following three general routes. However, over 50% of all sightings occur within 15 km from the mainland shore. For convenience of description, we refer to these routes as nearshore, inshore, and offshore. The nearshore pathway closely follows the mainland shore most of the way between Point Conception and Point Vicente, although it takes a more direct line off the coast from Santa Barbara to Ventura and across Santa Monica Bay. Near Point Vicente or Point Fermin, part of the migrant population appears to veer southward, creating a notable clustering of sightings around Santa Catalina Island on the southbound migration.

Some of these whales may return to the nearshore route in the vicinity of Newport Beach; others join migrants using an inshore route, streaming southeastward toward Point Loma and Punta Descanso. The inshore route includes the northern chain of Channel Islands, especially Santa Cruz Island and the Anacapas, and then follows the eastern rim of the Santa Cruz Basin to Santa Barbara Island and Osborn Bank. From Santa Barbara Island, gray whales travel either east to Santa Catalina Island or southeast to San Clemente Island. A direct open water route is then taken from Santa Catalina Island to Point Loma or Punta Descanso, or from San Clemente Island to the vicinity of Punta Banda. The offshore route follows the undersea ridge from Santa Rosa Island, past San Nicolas Island, and over Tanner and Cortes banks into Mexican waters. This route would reach the mainland shore of Baja California much farther to the south.

Line transect analysis of aerial survey data indicates that the January population in the SCB may reach about 820 animals present during a 4-day sampling period. Radio-tagging studies show that gray whales on southbound migration remain in the vicinity of the northern island chain for a period from 1 to 4 days (minimum estimate based on loss of signal) (Jones and Swartz 1986). A few calves have been observed among southbound gray whales in central and northern California and in the SCB (Dohl et al. 1981, 1983; Jones and Swartz 1986).

Wintering areas for the California gray whale include the west coast of Baja California and parts of the east coast of the Gulf of California. Gray whales with calves also have been seen near oceanic Isla de Guadalupe (Gilmore 1955). Although some descriptions of the species' range indicate that San Diego Bay was historically used as a calving ground,

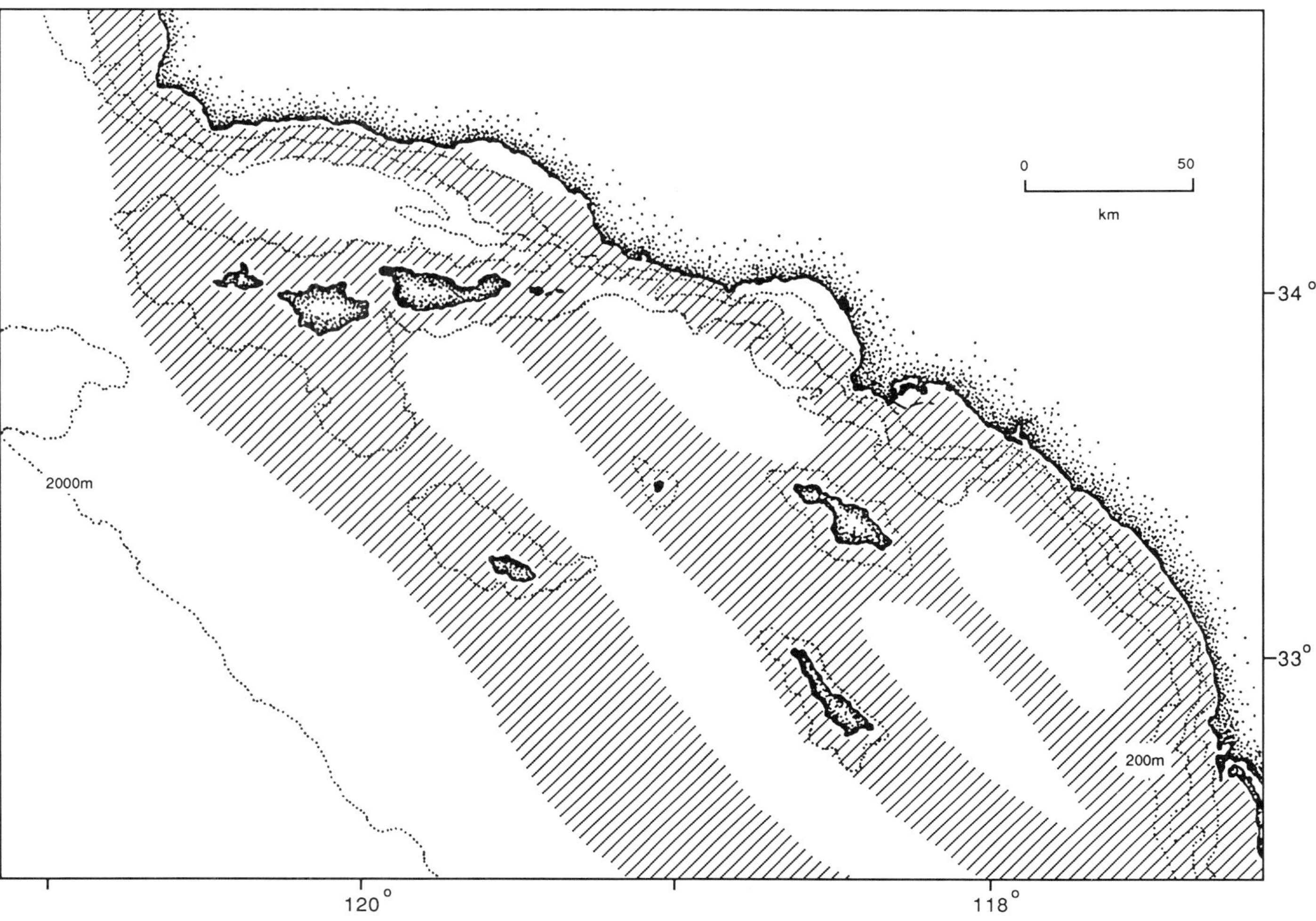

Figure 11.4. Migration pathways used by gray whales passing through the SCB, 1975–1978.

supporting evidence is lacking (Henderson 1972, cited in Rice et al. 1984). At present, approximately 85% of observed numbers of gray whale calves in Baja California are born each winter in lagoons along the outer coast of Baja California, especially Laguna Guerrero Negro, Laguna Ojo de Liebre (Scammon's Lagoon), Laguna San Ignacio, and Estero Soledad (Rice et al. 1981). A few calves are also born in open waters of Bahia Magdalena, Bahia Almejas, and Bahia Santa Maria. Departure from the lagoons begins in mid-February and occurs in two phases. The first gray whales to leave include pregnant females, anestrous females, adult males, and immatures; the second, later phase of migration includes cows with calves (Rice et al. 1984).

Northward-migrating gray whales appear in the SCB from February through May, and the peak of the northward migration occurs in March (Leatherwood 1974). The total number of gray whales present in the SCB during a 4-day survey period in March was estimated at 570 animals. Their northbound migration path does not differ notably from the southbound path. About one-half of the northbound migrants remain close to the mainland shore, while others, including some mother–calf pairs, follow the inshore or offshore routes. While most mother–calf pairs appear to remain fairly close to land, a few have been observed far offshore over the Cortes Banks and the Santa Rosa Ridge (Dohl et al. 1981). In central California, Poole (1984) found independent gray whales moving through the area in the months of February and March. Mothers with dependent calves were observed from mid-April through mid-May, predominantly in nearshore waters. Dohl et al. (1983) found mother–calf pairs in central and northern California remaining within 9.25 km from shore except when crossing the Gulf of the Farallones, offshore of San Francisco Bay.

Historically, right whales (*Eubalaena glacialis*) inhabited temperate to subarctic waters in both hemispheres. In the eastern North Pacific, right whales occurred from the Bering Sea to central Baja California, Mexico, but these whales were extensively hunted and stocks are now depleted. The present population in the North Pacific is thought to number no more than about 200 animals (Braham and Rice 1984). Like other baleen whales, right whales appear to migrate from high-latitude feeding grounds toward more temperate waters in the autumn and winter. The location of their calving grounds is unknown, but sightings have been noted off Baja California (Rice and Fiscus 1968). Summer feeding grounds may generally stretch across the North Pacific from about 50° to 63°N latitude (Omura 1958). A recent sighting in the Bering Sea was reported by Brueggeman et al. (1984). The paucity of records along the western seaboard of North America suggests that right whales migrate to summer grounds from the western North Pacific or pelagic waters of the eastern or central North Pacific (Braham and Rice 1984). In this century, there have been only two sightings and one stranding of right whales in the SCB (Gilmore 1956; Woodhouse and Strickley 1982; Scarff 1986).

Odontocete (Toothed) Whales

The sperm whale (*Physeter macrocephalus*) is rare over the continental shelf, but it is considered abundant in that part of the eastern North Pacific directly offshore of the SCB, which is referred to as the "Eastern Grid" on surveys of the Smithsonian Institution's Pacific Ocean Biological Survey Program (Brownell and DeLong 1968). The sperm whale is the largest of the toothed whales and is found predominantly in temperate to tropical pelagic waters of both hemispheres. Schools of females and immature whales remain year-round in waters south of about 45°N latitude (Gosho et al. 1984). Adult males utilize northern feeding grounds in areas where extreme bottom topography promotes concentration of food resources (Berzin and Rovnin 1966). Sperm whales are common year-round in deep waters off central and northern California (Dohl et al. 1983). A few sightings over San Juan Seamount, west of the SCB, have been reported by Dohl et al. (1981).

Common dolphins (*Delphinus delphis*) appear in temperate to tropical waters worldwide. In the eastern North Pacific, they range from the equator to at least central California, although their presence north of the SCB is usually associated with warm water years. Sightings off central California have been reported along the Big Sur coast, Monterey Bay, and the Farallon Basin (Brownell 1964; Fiscus and Niggol 1965; Christmann 1983; Dohl et al. 1983). The northernmost record is a stranding in British Columbia (Guiguet 1954). According to Evans (1982), the general range of the species in the eastern Pacific extends from 40°S to 36°N latitude and offshore to about 132°W longitude. Within this area, four stocks have been described. In the SCB, both a neritic (inshore) form and a pelagic (offshore) form occur. Common dolphins breed seasonally, with calving peaks in spring and fall. Females reach sexual maturity at 3–5 years of age and bear a single calf every 2–3 years (Leatherwood and Reeves 1978). In the SCB, schools with young animals have been seen most often in offshore waters (Dohl et al. 1981).

The common dolphin is by far the most abundant cetacean species in the SCB and accounts for 57–84% of the total seasonal cetacean population in the area (Dohl et al. 1981). The pattern of occupation in the SCB by common dolphins is strongly seasonal. These dolphins are most abundant in the summer and autumn; they withdraw southward in the winter. A vector analysis shows that movements within the area from December through June were predominantly toward the north, northwest, and west, while movements from July through November were in the opposite direction (Evans 1982). Dohl et al. (1986) carried out an analysis of common dolphin sightings recorded on aerial surveys conducted in the SCB from 1975 to 1978. In winter and spring, common dolphins were found almost exclusively in the eastern parts of the SCB. Total abundance in the SCB from January through June averaged 15,400 individuals (coefficient of variation, C.V., 36%). Through the summer, the number of common dolphins east of Santa Catalina and San Clemente islands remained high, but the distribution expanded westward to include the central part of the SCB. From July through December, total area abundance averaged 57,300 animals (C.V. 17%). The summer and autumn distribution included most of the SCB inshore of the Patton Escarpment (fig. 11.5).

A portion of the population in the SCB consists of a pelagic form typically found farther offshore (Brownell and DeLong 1968; Dohl et al. 1981, 1986). In summer and autumn, mixed schools of the paler, neritic form and the larger, more brilliantly marked pelagic form occur in central waters of the SCB over the Santa Rosa–Cortes Ridge and along the south side of the northern island chain. Schools farther offshore over the Patton Escarpment and San Juan Seamount are comprised exclusively of the pelagic form (Dohl et al. 1986).

Northern right-whale dolphins (*Lissodelphis borealis*) are found only in the North Pacific, where they occur off the west coast of the United States and off Japan. Leatherwood and Walker (1979), reviewing records of sightings, concluded that the northern right-whale dolphin is distributed in the temperate eastern North Pacific from British Columbia to about San Diego. The species occurs in the SCB primarily in the winter and spring when it is distributed over and inshore of the Santa Rosa–Cortes Ridge (Dohl et al. 1981). They are also observed around Santa Catalina and San Clemente islands, where their presence may be associated with the spawning of squid (Dohl et al. 1981; Oliver and Jackson 1987). The species is rarely seen in pelagic waters west of the SCB; however, Brownell and DeLong (1968) considered it to be abundant in northern parts of the eastern

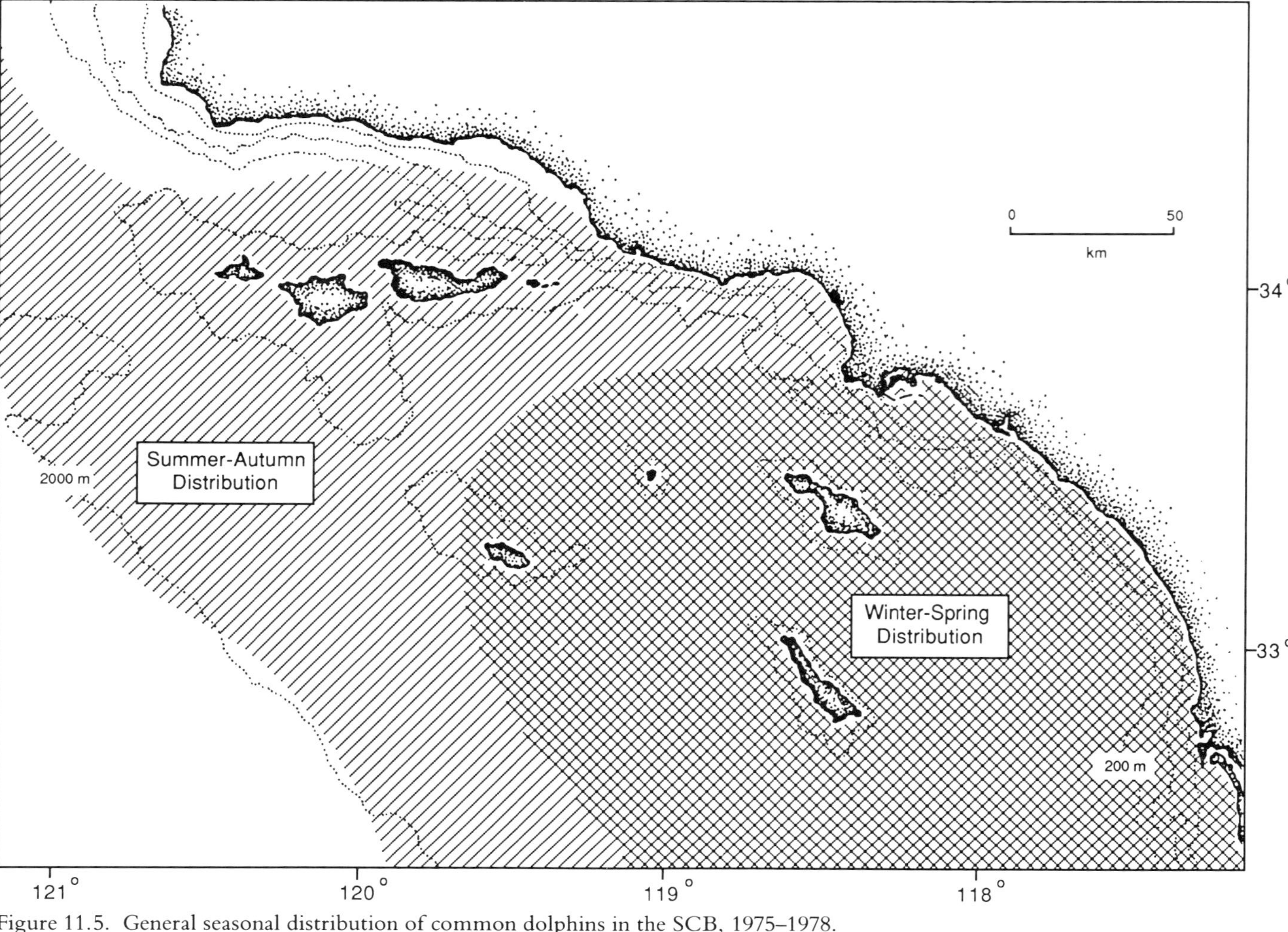

Figure 11.5. General seasonal distribution of common dolphins in the SCB, 1975–1978.

grid. The general pattern of movement of the population appears to be southward and inshore with cooling of water temperature in the late autumn and winter, and northward and offshore as temperatures increase in the late spring and summer (Leatherwood and Walker 1979).

In the SCB, northern right-whale dolphins are common in the winter and spring, but become rare after June. The mean population from January through May is about 6500 ± 2400 (standard error of estimate, SE), generally increasing through these months as northern right-whale dolphins enter the area from the north. As numbers increase, the area of occupation expands southward in inshore waters which include the island shelves (fig. 11.6). By May, northern right-whale dolphins are observed around the northern chain of Channel Islands, over the Santa Rosa–Cortes Ridge to the south, and over the Catalina and San Clemente escarpments. During the warmer water months from July through November, most of the population is found farther north. Peak numbers off central and northern California in these months are estimated at more than 61,000 animals (Dohl et al. 1983). Newborns and immatures have been observed in the SCB and throughout central and northern California waters (Dohl et al. 1981, 1983). Although sightings of immatures have been recorded in all seasons in central and northern California, more than 70% of the sightings occurred in winter. In California waters, most calving occurs in the SCB and within about 50 km of the central California coast (Dohl et al. 1981, 1983).

Pacific white-sided dolphins (*Lagenorhynchus obliquidens*) are found only in temperate waters of the eastern and western North Pacific. While seasonally present in southern parts of the Gulf of Alaska, they rarely occur very far north in these waters (Hall 1979; Consiglieri and Braham 1982). Occurrence of the species in subarctic waters is probably associated with periods of mild oceanographic conditions (Leatherwood et al. 1982). The southern range limit appears to be off the tip of the Baja California peninsula and southern waters of the Gulf of California. Pacific white-sided dolphins are present year-round in California waters, but appear to reach peak numbers in the autumn when most schools are seen south of Monterey Bay. The population in central and northern California is estimated to range from 26,000 to 33,500 animals through most of the year, but reaches 86,000 animals in the autumn (Dohl et al. 1983).

In the SCB, Pacific white-sided dolphins are present year-round, with a mean population of about 6000 animals. The pattern of use of the area changes seasonally as animals move southward in the autumn and northward in the spring (fig. 11.7). Peak numbers, exceeding 10,000 animals, are reached in the area from about September through November. During these months, abundance increases first in northern waters of the SCB, especially in the western Santa Barbara Channel, over the San Miguel Island shelf, and along the Santa Rosa–San Nicolas Ridge (Dohl et al. 1981). By October and November, Pacific white-sided dolphins are fairly widespread in the area at slightly lower densities. By November, sightings in central and eastern waters of the SCB begin to increase, indicating a general dispersal southward. Sightings reported by observers with the National Marine Fisheries Services (NMFS) Dolphin–Tuna Program indicate that the occurrence of Pacific white-sided dolpins in Mexican waters from about 25° to 30°N latitude also increase in the autumn months (Leatherwood et al. 1984). Numbers of immatures in the SCB are greatest in late summer (Norris and Prescott 1961; Dohl et al. 1981).

By winter, numbers of Pacific white-sided dolphins sharply decline in both California and Mexican waters. During January in the SCB, Dohl et al. (1981) recorded sightings only in offshore waters over the Santa Rosa–Cortes Ridge, suggesting that the reduction in numbers in the area results from an offshore shift in the species distribution. In the spring, distribution in the SCB shifts to inshore waters, where Pacific white-sided dol-

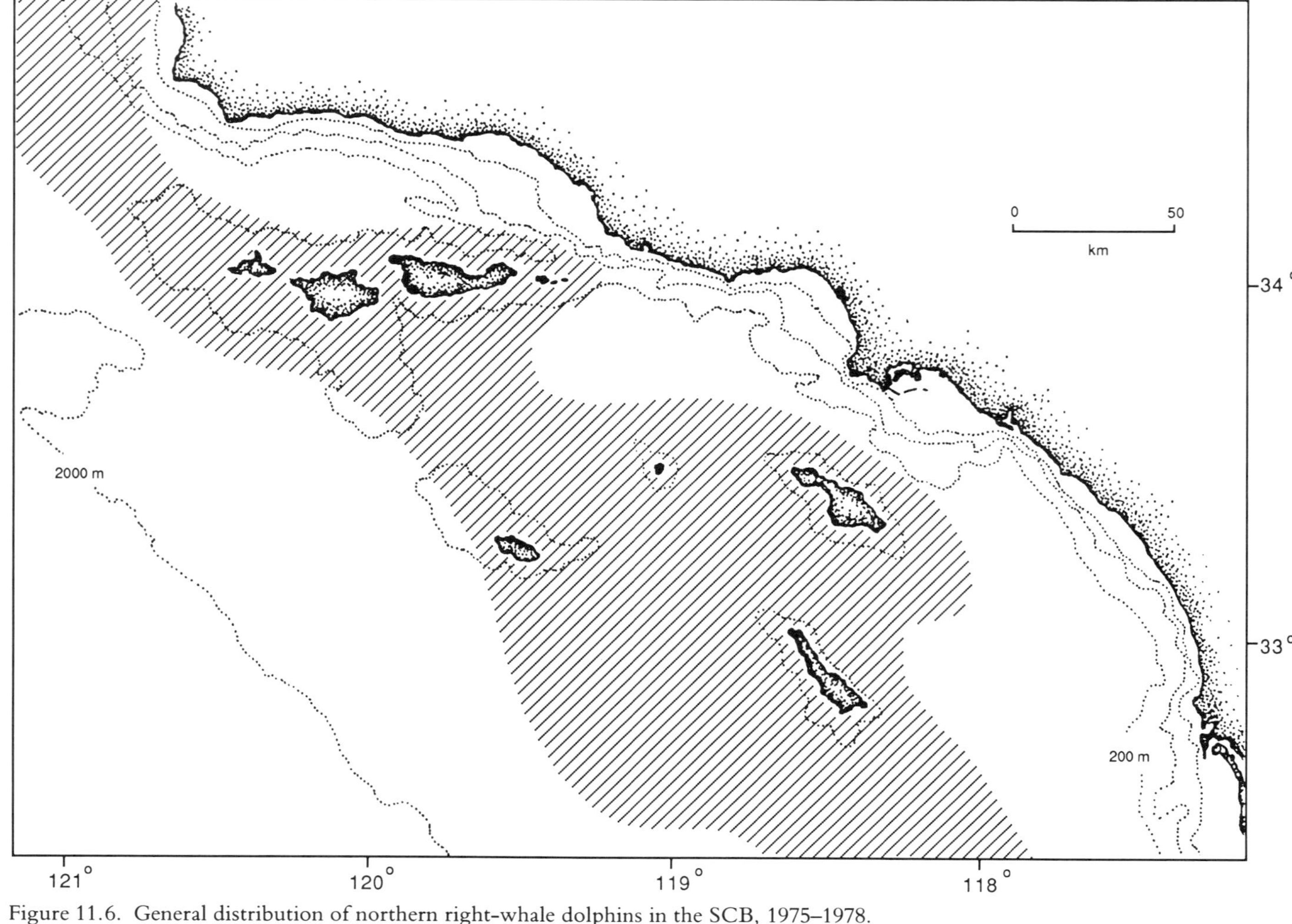

Figure 11.6. General distribution of northern right-whale dolphins in the SCB, 1975–1978.

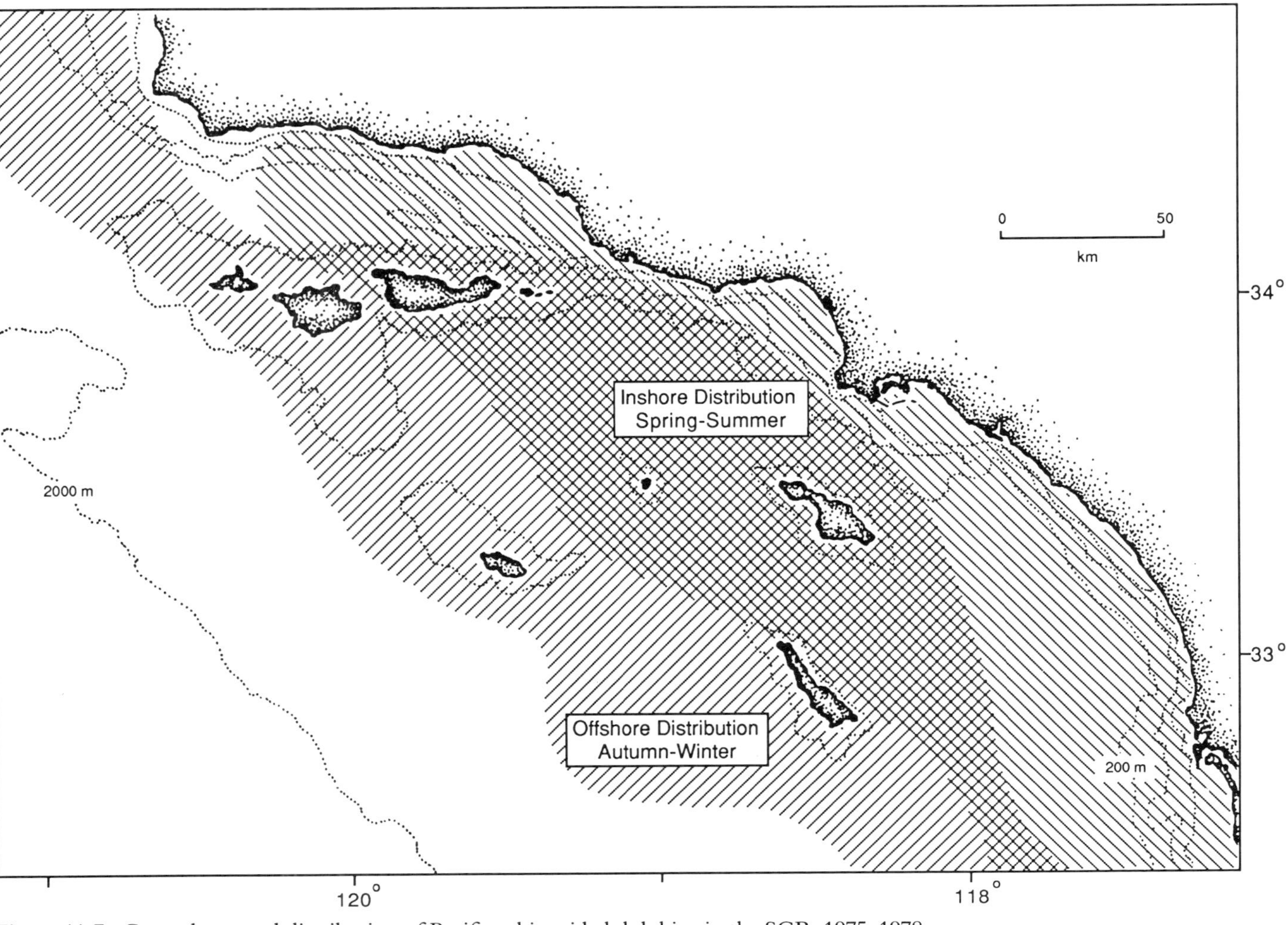

Figure 11.7. General seasonal distribution of Pacific white-sided dolphins in the SCB, 1975–1978.

phin sightings are typically clustered within 30 km of shore. Dohl et al. (1981) continued recording a few sightings close inshore through June and July. The significance of the winter–spring shift in the population from offshore to inshore waters is unclear, but may simply be the expression of an age or sex segregation of the population that persists into late summer.

Risso's dolphins (*Grampus griseus*) are widespread in tropical and warm temperate waters of both the Pacific and Atlantic oceans. In the eastern North Pacific, they range from the equator as far north as British Columbia (Guiguet and Pike 1965). The pattern of sightings and strandings suggests that Risso's dolphins are currently extending their range northward as water temperatures increase (Leatherwood et al. 1980, 1982). In the SCB, Risso's dolphins are present year-round, with a mean annual abundance of about 1400 animals. Peak numbers in the area occur from September through November (Dohl et al. 1981). Through summer and autumn months, Risso's dolphins in the SCB are distributed inshore of the Santa Rosa–Cortes Ridge. Through winter and spring, the population shifts offshore except in the vicinity of the northern chain of Channel Islands. Risso's dolphins are found predominantly over and seaward of the Santa Rosa–Cortes Ridge and over the San Clemente Escarpment and associated offshore banks (fig. 11.8). The population in the SCB reaches about 2770 ± 620 (SE) during September through November. For the remainder of the year, the mean population averages about 850 animals. In contrast, Dohl et al. (1983) estimated population size in central and northern California to be about 13,000 animals in the summer, increasing to 30,000 in the winter. Immature Risso's dolphins in California have been seen in all seasons, although 60–66% have been observed from December through February (Dohl et al. 1983).

The Dall's porpoise (*Phocoenoides dalli*) is found only in the North Pacific where it is considered to be a cool water species. Its patterns of movement in California are probably temperature related, but are not well understood. Dall's porpoises are rarely found in waters warmer than about 17°C (Morejohn 1979; Leatherwood et al. 1982). In the eastern North Pacific, Dall's porpoises range from subpolar waters of the Bering Sea and the Gulf of Alaska southward to about 30°N latitude in cool temperate waters off Baja California (Leatherwood and Fielding 1974; Leatherwood et al. 1982). They can be found in California year-round, but are most common in the winter months (Morejohn 1979). In the SCB, Dall's porpoises are most numerous in cold water years and are rarely found in inshore waters except between late autumn and late spring (Leatherwood et al. 1982). The population in the SCB numbers about 50–150 animals from March through October, but increases to over 1000 in January and February. In adjacent waters of central and northern California, the population ranges from 3500 to about 8750 (Dohl et al. 1983). The summer range of the Dall's porpoise in the SCB includes the western Santa Barbara Channel, the waters around the northern chain of Channel Islands, and the Santa Rosa–San Nicolas Ridge (fig. 11.9). In the autumn, the population expands southward and, by winter, occupies most of the inshore waters of the SCB, including the shelves of the Santa Catalina, San Clemente, and Santa Barbara islands. Newborn animals have not been observed in the SCB. Off central and northern California, Dohl et al. (1983) recorded 95% of all immature animals during the months of June through September.

Bottlenose dolphins (*Tursiops* sp.) in the eastern North Pacific range from the equator northward into central California. The northern range limit probably depends on sea surface temperature and other oceanographic conditions that influence the abundance of food (Leatherwood and Reeves 1978). Norris and Prescott (1961) proposed that distinct coastal and offshore populations exist in the SCB and that coastal and offshore forms differ slightly in size, coloration, and behavior

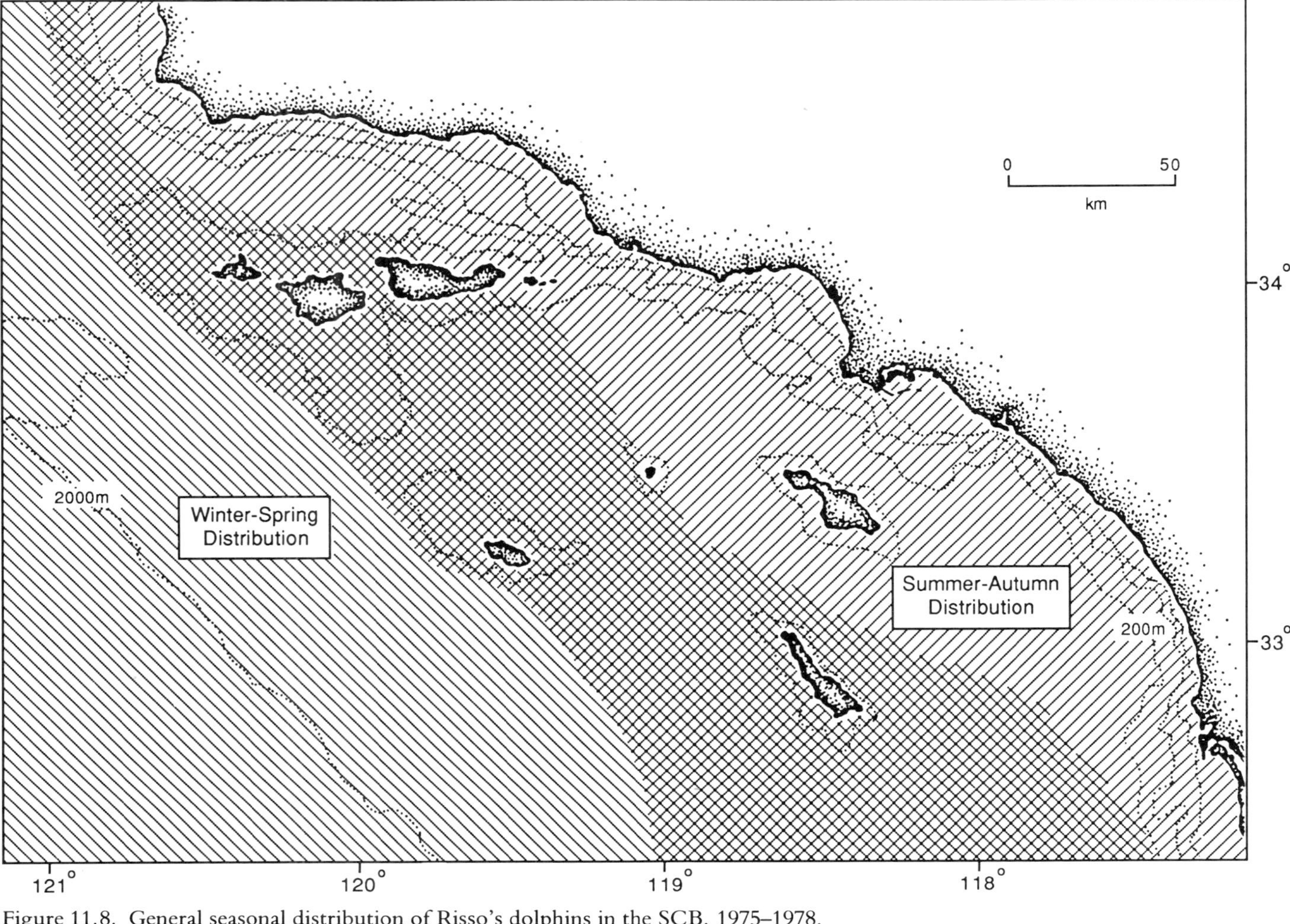

Figure 11.8. General seasonal distribution of Risso's dolphins in the SCB, 1975–1978.

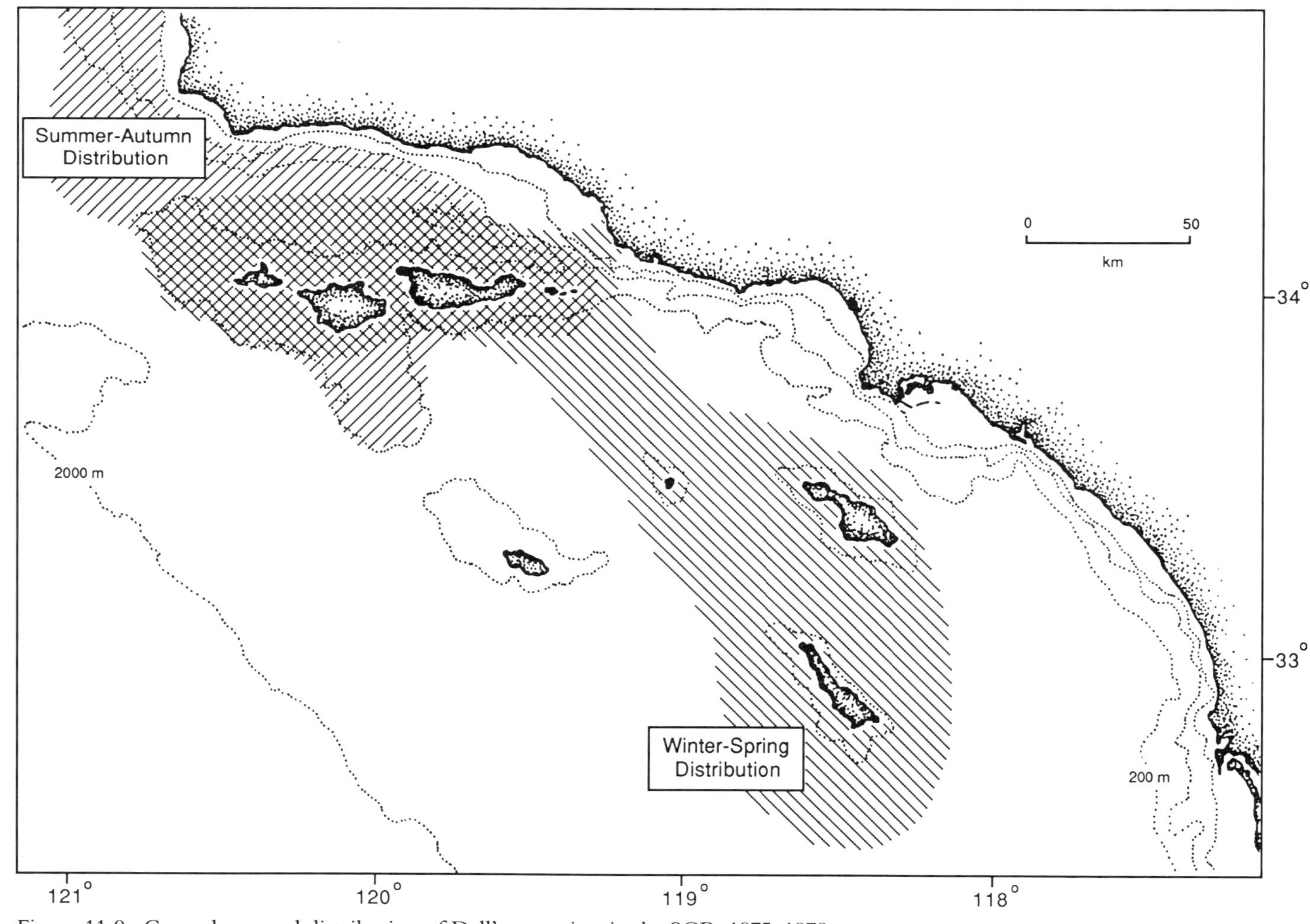

Figure 11.9. General seasonal distribution of Dall's porpoises in the SCB, 1975–1978.

as well as in distribution. No information is available on the degree of intermixing, but it has been noted that specimens from the two populations have different parasites, which suggests that some degree of isolation exists (Walker 1975).

The coastal population of bottlenose dolphins typically inhabits waters within 1 km of the mainland shore, while the offshore population appears to move between nearshore waters around several southern California islands (Dohl et al. 1981). Studies of coastal populations show that these dolphins appear to form small resident groups that occupy relatively stable home ranges (see Shane et al. 1986 for review). Generally, these home ranges are discrete and little overlap of populations occurs. However, extensive movements of small groups outside the population's home range have also been noted (Wells et al. 1983; Dohl et al. 1983).

Possible expansion of home range, resulting in relocation of a population center, is reported by Dohl (1987) and Wolf et al. (1987). Bottlenose dolphins remain in the SCB year-round (fig. 11.10), although the population size and distribution do change seasonally. Local coastal populations exist along the mainland shore, especially in the eastern part of the SCB. Population centers may shift from year to year as dictated by changes in abundance of prey species.

A coastal population of about 240 bottlenose dolphins near San Diego has been described by Hansen (1983). Movements of many individuals in this population between Orange County and Ensenada have been reported by Defran et al. (1987), indicating that the population's home range extends at least that distance. Recently, bottlenose dolphins have been seen in increasing numbers in the northern SCB along the mainland shore of Ventura and Santa Barbara counties (Wolf et al. 1987) and along the central California coast to Morro Bay (Dohl 1987). Photographs document the interchange of bottlenose dolphins between the Santa Barbara Channel and the San Diego–Ensenada parts of the range. These observations suggest that, at present, the coastal population inhabits this entire area. During the El Niño episode of 1982–1983, a group of at least 25 bottlenose dolphins ventured north past Monterey Bay (Wells et al. 1983).

A relatively large population of bottlenose dolphins also exists in offshore waters of the SCB, centering around Santa Catalina Island most of the year (Dohl et al. 1981; Oliver and Jackson 1987). Most sightings of bottlenose dolphins near this island were recorded in winter months, when coastal and offshore populations may combine. From data collected by Dohl et al. (1981) in 1975–1978, the population in the SCB was estimated to be about 186 ± 20 (SE) in January, at which time bottlenose dolphins were distributed only in nearshore waters of Santa Catalina and San Clemente islands. The summer population (June through September) in the SCB numbered about 550 ± 70 (SE) and was clearly greater than that of the winter. However, the population in summer was also much more widely distributed, occupying waters near Santa Catalina, San Clemente, Santa Barbara, Anacapa, and San Nicolas islands as well as the mainland shore.

Harbor porpoises (*Phocoena phocoena*) inhabit cool temperate to subarctic waters in both the North Pacific and North Atlantic. In the eastern North Pacific, they are found from about 34°N latitude northward into Alaska waters. Harbor porpoises typically occur in small groups in coastal waters within a few km of shore (Dohl et al. 1983). The California, Oregon, and Washington population is estimated at 49,800 animals, of which fewer than 1% are found south of Monterey Bay (Barlow 1987). A few harbor porpoises commonly occur off Point Arguello just north of the SCB (Dohl et al. 1983; Oliver and Jackson 1987), and strandings have been recorded along the mainland coast between Santa Barbara and Los Angeles (Leatherwood et al. 1982).

Pilot whales (*Globicephala* sp.) range widely from the Gulf of Alaska to the eastern tropical

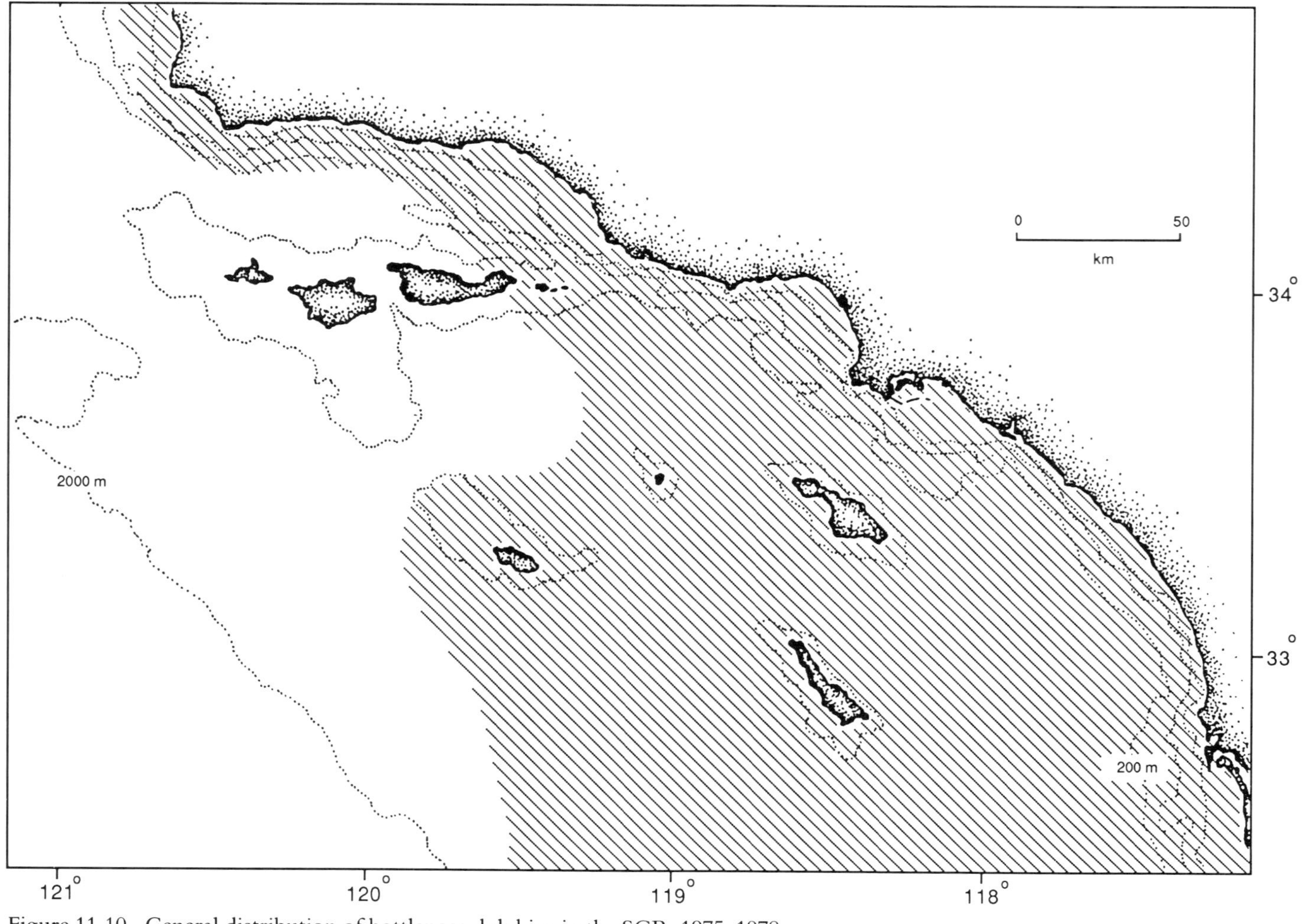

Figure 11.10. General distribution of bottlenose dolphins in the SCB, 1975–1978.

Pacific. More than one species may be present in the eastern North Pacific, but taxonomic questions have yet to be resolved. The life history of pilot whales in the Pacific is presumed to be similar to that of pilot whales in the North Atlantic. Female pilot whales occupying waters off Newfoundland require 15–16 months for gestation, and most of their calves are born over a 6-month period; young are weaned after about 22 months (Sergeant 1962). The species is thought to be polygynous, with competition for females occurring among males. Reproductive herds in Newfoundland contain twice as many females as males in the summer, but during the remainder of the year, males are geographically segregated from herds of females and immatures.

In the SCB, pilot whales have a resident population of about 370 animals (Dohl et al. 1981). From March through about mid-November, pilot whales are found predominantly in groups of less than 20 animals in the vicinity of the northern chain of Channel Islands and the Palos Verdes Peninsula (fig. 11.11). In the winter, the resident population is joined by seasonal visitors entering the SCB, and group size increases to about 35 animals. The population reaches a peak in January, when nearly all sightings typically occur within about 20 km of Santa Catalina Island and the adjacent Palos Verdes Peninsula. The concentration of pilot whales within this single small area leads us to believe that winter is a season in which mixing of segregated groups occurs (probably males joining groups of females and immatures) and breeding takes place. Because of the concentration in this area, population status can be assessed with relative ease by direct counts rather than line transect sampling of the population. Counts in December and January can total as high as 430 to 540 animals. However, high counts such as these were obtained in a year of relatively typical oceanographic conditions (1975 through early 1976); subsequent counts have been less.

The status of the pilot whale population in the SCB may be affected by changes in water temperature which, in turn, may affect the abundance of squid. January counts on aerial surveys declined sharply from 1976 to 1978 as sea surface temperature increased significantly above the 20-year mean. By 1981, with the return of more typical oceanographic conditions, the population of pilot whales around Santa Catalina Island reached a size comparable to that of 1976 (Oliver and Jackson 1987). However, numbers of pilot whales counted in January of subsequent years have been very low, which suggests that the more severe El Niño–Southern Oscillation (ENSO) temperature anomaly of 1982–1983 may have stalled the recovery (Shane 1984; Oliver and Jackson 1987).

Killer whales (*Orcinus orca*) are found worldwide, but appear to be most common in colder waters within about 800 km of land (Heyning and Dahlheim 1988). In the eastern North Pacific, killer whales are found from the Beaufort, Chukchi, and Bering seas (Braham and Dahlheim 1982; Brueggeman et al. 1984) to the tropical eastern Pacific (Dahlheim et al. 1982). In California waters, strandings of killer whales occur infrequently (Dahlheim et al. 1982). Sightings at sea are relatively common, with 93% recorded from Monterey Bay northward (Dohl et al. 1983). A few sightings have been recorded in summer and winter in the SCB (Dohl et al. 1981).

The false killer whale (*Pseudorca crassidens*) occurs in tropical to warm temperate waters worldwide. In the eastern North Pacific, sightings north of about 30°N latitude are rare (Fiscus and Niggol 1965; Sullivan and Houck 1979; Leatherwood et al. 1982). These whales are considered relatively common in pelagic waters of the eastern tropical Pacific, but only a few sightings in the SCB have been reported (Norris and Prescott 1961; Mitchell 1965; Leatherwood et al. 1982).

Several species of beaked whales occur in the SCB, but all are uncommon. The Cuvier's beaked whale (*Ziphius cavirostris*) may be the most abundant. This species is widely distributed worldwide in tropical and warm tem-

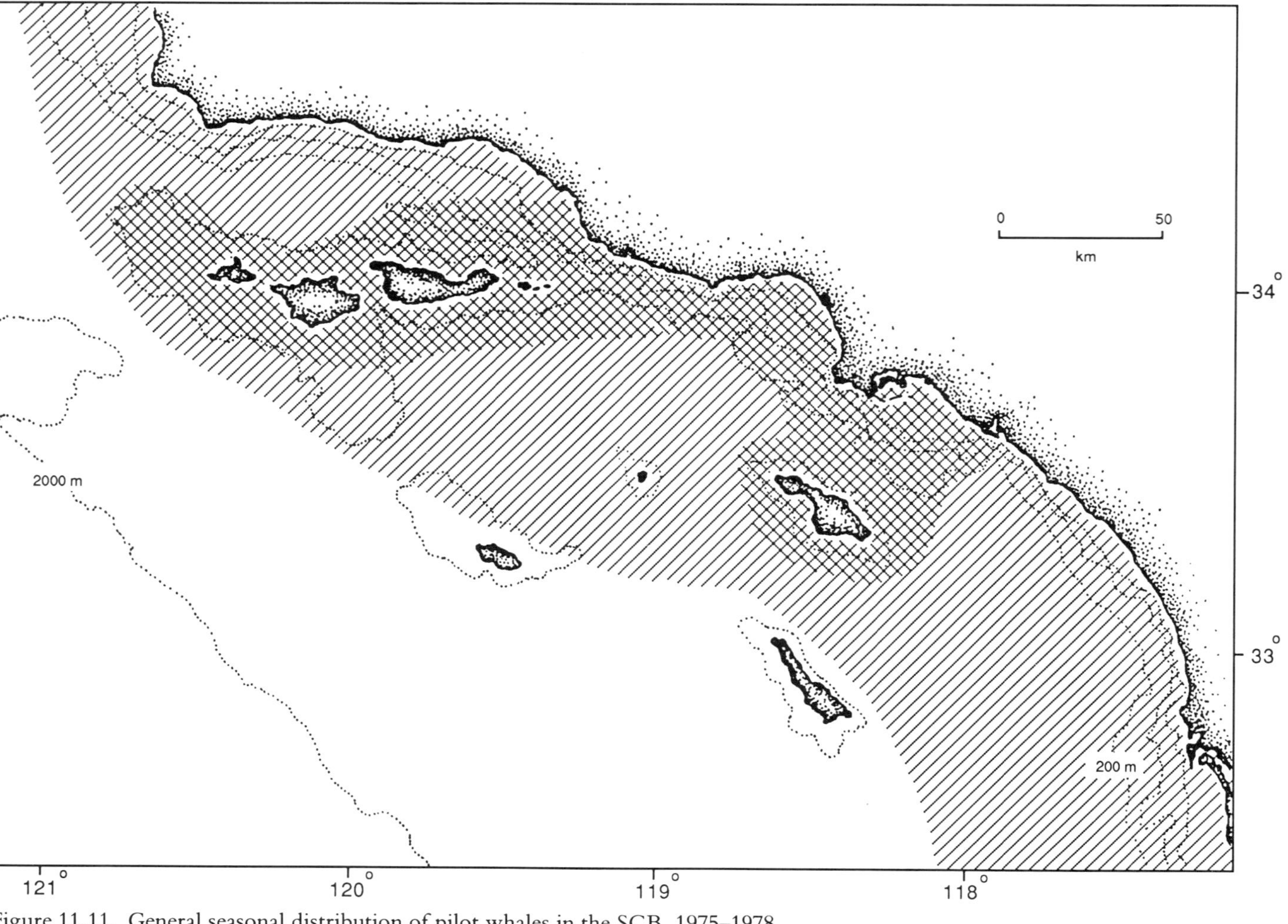

Figure 11.11. General seasonal distribution of pilot whales in the SCB, 1975–1978.

perate waters. From the stranding record, it appears that Cuvier's beaked whales are distributed in the eastern North Pacific nearly continuously from Alaska to Baja California. Of the 41 known strandings, 12 have occurred in the SCB (Mitchell 1968). Three strandings have also been recorded near Morro Bay in central California (Roest 1970). Though not common, sightings of several Cuvier's beaked whales were recorded on surveys of California waters (Dohl et al. 1981, 1983).

Five species of beaked whales of the genus *Mesoplodon* have also been described in the eastern North Pacific, and other undescribed species may exist as well (Leatherwood et al. 1982; Mead et al. 1982; Pitman et al. 1987). All are uncommon and, even when seen, can be identified as to species only by close examination. Strandings of *Mesoplodon carlhubbsi, M. ginkgodens,* and *M. hectori* have occurred in the SCB. Dohl et al. (1981) report sightings of 16 *Mesoplodon* beaked whales in the SCB between 1975 and 1978. In addition, a pair of beaked whales seen near Santa Catalina Island were tentatively identified as *M. hectori* (Mead 1981). Baird's beaked whales (*Berardius bairdii*) may occasionally enter the SCB, though no sightings have been reported (Dohl et al. 1981). This species is endemic to the North Pacific and is found predominantly in cool temperate to Arctic waters (Dohl et al. 1983).

The pygmy sperm whale (*Kogia breviceps*) is distributed worldwide in tropical and warm temperate waters (Leatherwood et al. 1982). Sightings and strandings have occurred in California, including several in the eastern part of the SCB (Hubbs 1951; Roest 1970; Sullivan and Houck 1979). A sighting near Santa Catalina Island was reported by Oliver and Jackson (1987). The dwarf sperm whale (*Kogia simus*) is known to occur in the eastern tropical Pacific (Scott and Cordaro 1987) and may also occasionally be present in the SCB.

Striped, spinner, and spotted dolphins of the genus *Stenella* are found mainly in the eastern tropical Pacific, where they are abundant and often form large mixed schools (Polacheck 1987). A few strandings and sightings of the striped dolphin, *Stenella coeruleoalba,* have been reported in the SCB (Norris and Prescott 1961; Hubbs et. al. 1973). The rough-toothed dolphin, *Steno bredanensis,* which is typically found in tropical and warm temperate waters (Perrin and Walker 1975), may also occur occasionally in the SCB.

Pinnipeds

The California sea lion, *Zalophus californianus,* is the most abundant pinniped in the SCB, representing 50–93% of all pinnipeds on land and about 95% of all sightings at sea (Bonnell et al. 1981; Bonnell and Ford 1987). Prior to the 1930s, California sea lions were rare in the area and the distribution of the eastern Pacific population was largely restricted to Mexican waters surrounding the Baja California peninsula. Small populations, which have now disappeared, also existed on the Galapagos Islands and the islands of northern Japan. The historical record is not well documented for this species, but Scammon (1874) noted that sea lions were very abundant in the mid–1800s along the coast of California and Baja California. Extensive hunting of sea lions so reduced the populations that by 1927, fewer than 800 could be counted in the entire state (Bonnot 1928). Population growth and range expansion led to recolonization of islands in the SCB in the 1930s (Bartholomew and Boolootian 1960). Counts of pups on rookeries show the population in the SCB to be increasing at a rate of about 7% per year (fig. 11.12). The world population of the species is estimated to number over 157,000 animals (U.S. Department of Commerce 1987). California and Mexico have approximately equal sea lion populations (Mate 1977; Le Boeuf et al. 1983; Bonnell et al. 1983).

California sea lions presently range from Acapulco, Mexico (Gallo-R. and Ortega-O. 1986), to British Columbia, Canada (Bigg 1973). The northernmost rookery is on San

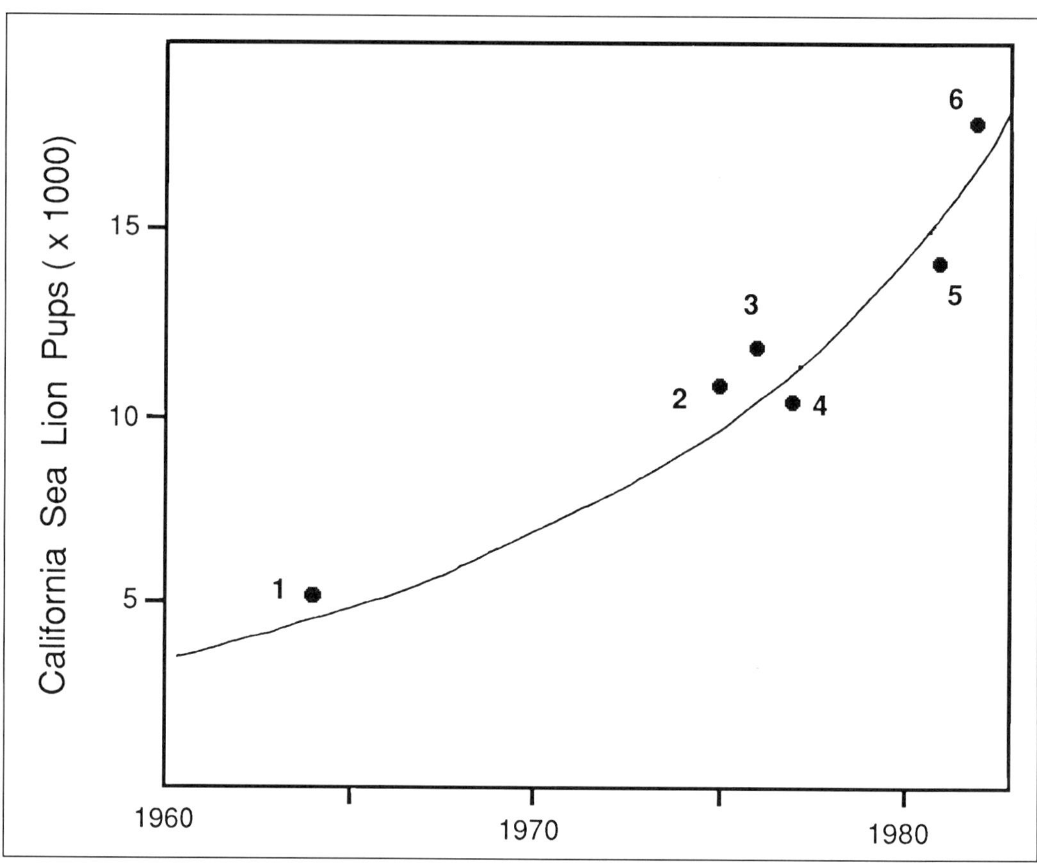

Figure 11.12. Growth of the California sea lion population in the SCB as represented by total number of pups counted on rookery islands. Curve is plotted from the expression: $N = 52.15e^{0.07t}$; $R^2 = 0.95$. Data sources: (1) Odell 1971; (2–4) Bonnell et al. 1981; (5–6) Bonnell 1982a. (Redrawn from Bonnell et al. 1983.)

Miguel Island in the SCB, although a few pups are occasionally born on islands off central California (Pierotti et al. 1977; Keith et al. 1984). In the SCB, large breeding colonies exist on San Miguel and San Nicolas islands, and smaller breeding colonies are found on Santa Barbara and San Clemente islands. A few locations on other islands are used for resting by foraging animals (Bonnell et al. 1981). Peak numbers on land occur in late June during the summer breeding season. During the breeding season of 1982, more than 51,000 California sea lions were counted on land in the SCB. Approximately 52% inhabited San Miguel Island (Bonnell 1982a) and 40% occupied San Nicolas Island (Stewart and Yochem 1984).

Including animals foraging at sea, the total SCB population during the summer breeding season of 1982 was estimated at about 63,000 (Bonnell and Ford 1987). Since 1983, the California population has been growing at almost 12% per year and was estimated to contain about 87,000 animals in 1986 (Boveng 1988). In both summer and autumn seasons, the distribution is widespread, but is generally centered in the SCB around the rookery islands (fig. 11.13). Following the breeding season, most adult males and many juveniles depart and spend the autumn and winter in waters farther north. During the winter, the distribution in the SCB shifts eastward to the waters around Santa Catalina and San Clemente islands and southward to Tanner and

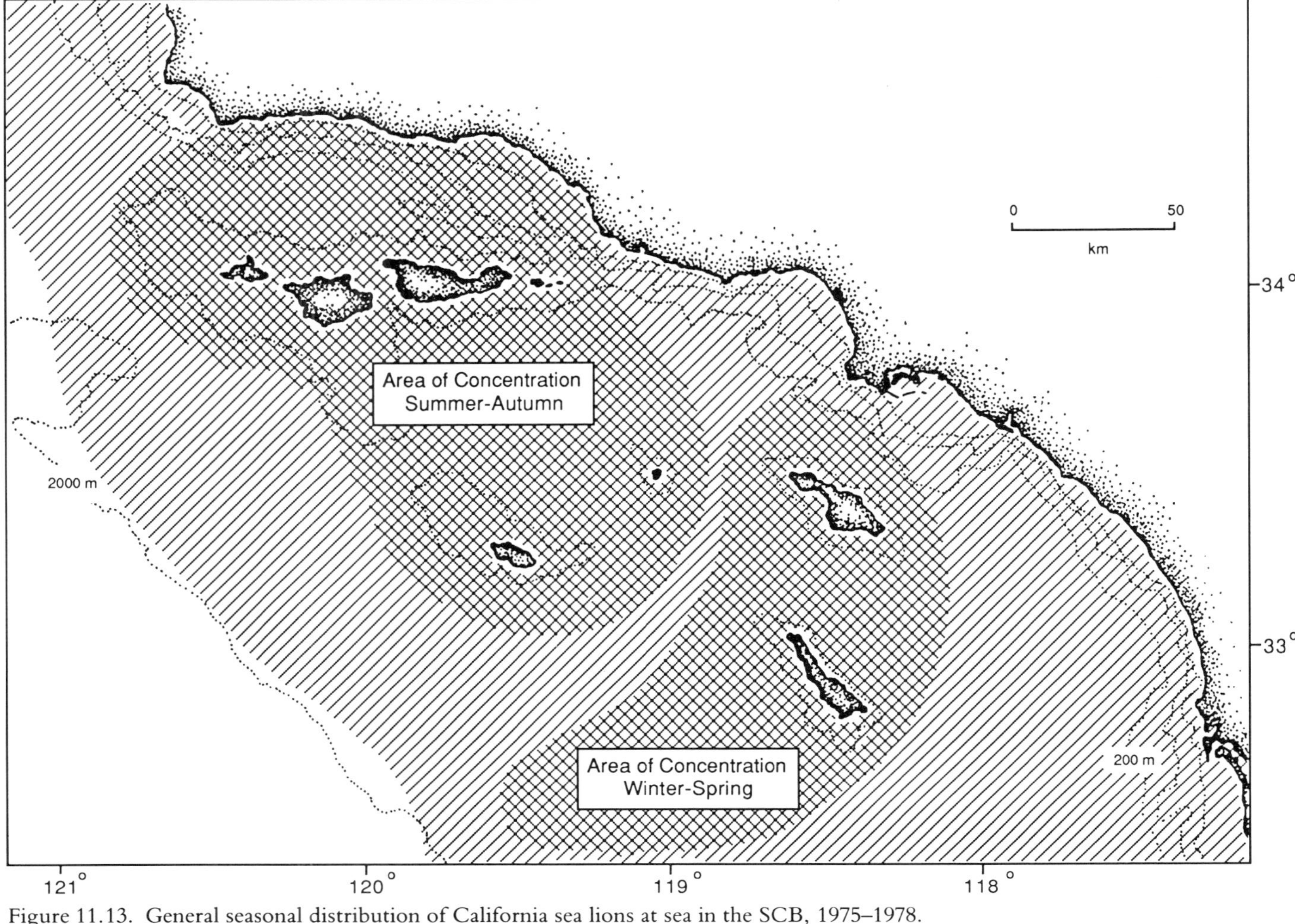

Figure 11.13. General seasonal distribution of California sea lions at sea in the SCB, 1975–1978.

Cortes banks. The mean population in the SCB from January through March is estimated at 43,000 (Bonnell and Ford 1987).

Steller's sea lions (*Eumetopias jubatus*), once the most abundant pinniped in the SCB, are now a straggler in the area. Numbers have declined precipitously in the last several decades, but the causes of the decline are not well understood (Bartholomew 1967; Le Boeuf and Bonnell 1980). A few adult or subadult males typically occupy territories on relict rookeries at the west end of San Miguel Island and adjacent rocks in the summer months. Steller's sea lions are also occasionally seen at sea in southern California waters on ship or aerial surveys (Bonnell et al. 1981, 1983). The last reported Steller's sea lion pups on San Miguel Island were observed in the summer of 1980 (Stewart 1980). The SCB is the southern extreme of the historical breeding range of the species, and at present, 96% of the world population is found in Alaska or Siberian waters (Loughlin et al. 1984). Decline in numbers of Steller's sea lions in the SCB was documented by area-wide censuses in the 1960s and 1970s (Bartholomew and Boolootian 1960; Odell 1971; Bonnell et al. 1981).

The near disappearance of the species in the SCB over a 30-year period was assumed to represent a probable northward shift in the range, perhaps caused by competition with the burgeoning population of California sea lions in the southern California area. However, recent censuses of Steller's sea lions in Alaska show a significant decline in the size of that population as well (Braham et al. 1980; Merrick et al. 1987). In fact, the present population of Steller's sea lions in Alaska may be less than half that of 30 years ago (Merrick et al. 1987). The cause of the decline is not yet known, but several possibilities have been suggested. These include disease, entanglement mortality, and limitations in availability of food (Merrick et al. 1987).

The northern fur seal, *Callorhinus ursinus*, is the most abundant otarlid in the Northern Hemisphere, with a world population estimated at over 1.1 million (U.S. Department of Commerce 1987). Most of the population is associated with rookery islands in the Bering Sea and the Sea of Okhotsk. Following the summer breeding season, females and juveniles migrate southward, while adult males remain in the northern parts of the range (Lander and Kajimura 1976). During most of the year, few animals haul out on land. The pelagic range of northern fur seals extends southward to about 30°N latitude and includes offshore waters of the SCB (Kenyon and Wilke 1953; Lander and Kajimura 1982), but only a small percentage of the total population in pelagic waters of the eastern North Pacific occurs over the continental shelf (< 200 m depth).

Antonelis and Perez (1984) estimated that over 300,000 northern fur seals are present in waters off California, Oregon, and Washington in the winter and spring. As might be expected, 80% of sightings in or near the SCB have been recorded offshore of the continental slope, especially in waters west of San Miguel Island (Bonnell et al. 1981). Most northern fur seals have been observed over the Santa Rosa–Cortes Ridge, the San Nicolas Basin, and the Tanner and Cortes banks (fig. 11.14). Almost 97% of all fur seals seen on surveys of the SCB were recorded in the winter and spring. Bonnell (1985) estimated that the mean abundance of fur seals in the SCB and offshore waters to about 122° 30′W longitude during the winter and spring was 5000 ± 465 (SE).

In the mid–1960s, northern fur seals colonized San Miguel Island in the SCB (Peterson et al. 1968). The San Miguel population numbered less than 100 animals when it was discovered; since then it has increased rapidly. Through 1981, pup production at San Miguel Island grew at a rate of 10.25% or more per year, apparently because of the immigration of parturient females from the Bering Sea population (Bonnell et al. 1983). The breeding population peaked in 1982 at more than 3600 animals on land, but then plummeted to less than 1500 in 1983 during the El Niño temperature anomaly (U.S. Depart-

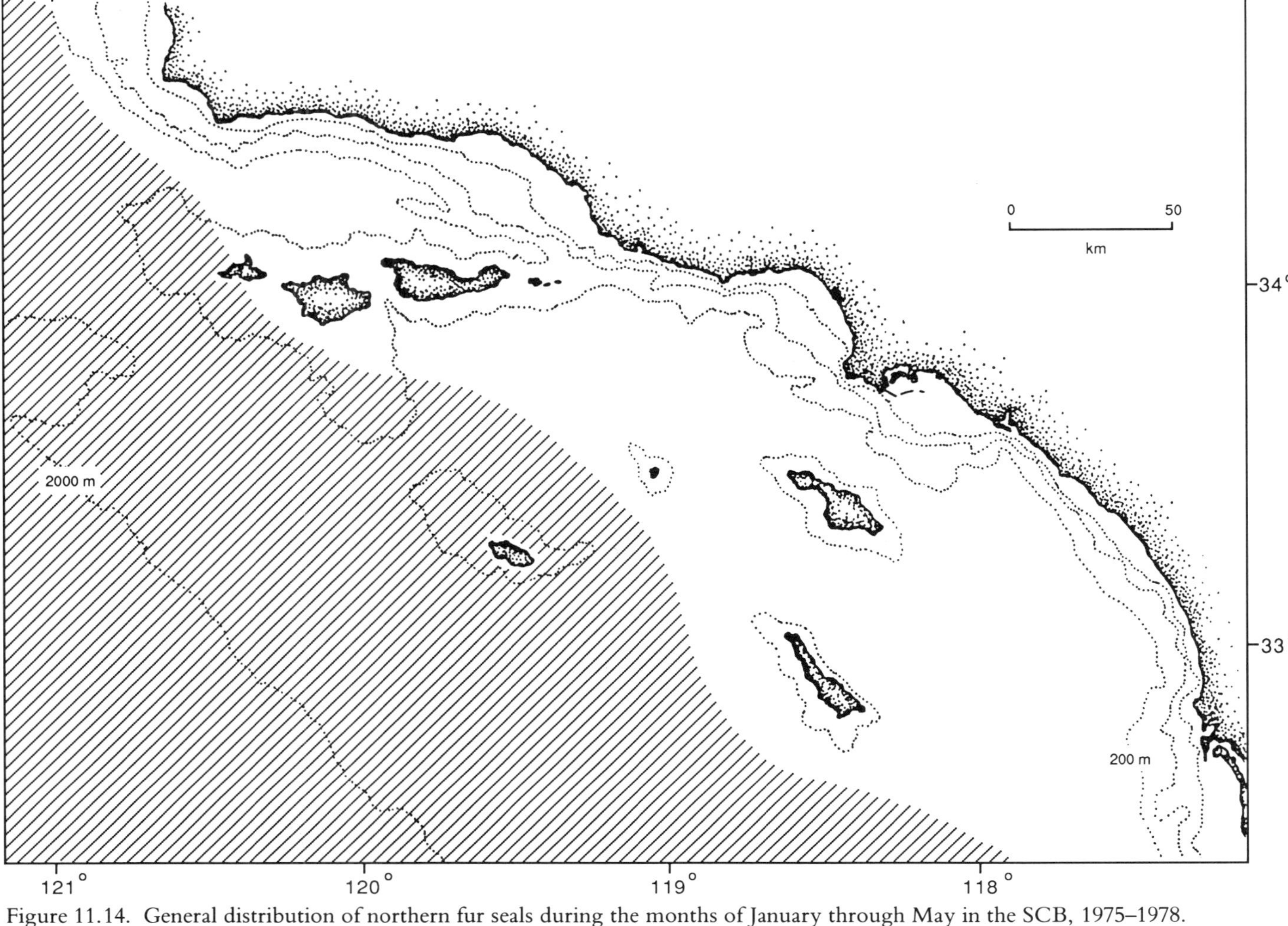

Figure 11.14. General distribution of northern fur seals during the months of January through May in the SCB, 1975–1978.

ment of Commerce 1986). Although some recovery was noted in 1984 (an increase to about 2000 animals), it was concluded by NMFS biologists that the El Niño episode might have had a long-term effect on pup production because of relocation of females or increased female mortality (Antonelis and DeLong 1986).

Northern fur seals on San Miguel Island have an annual cycle of numbers on land similar to the large Pribilof Island population. The breeding season begins in late May and continues through September or October; the population on land declines rapidly and few fur seals are found on land past November (Bonnell et al. 1981). The foraging range of female northern fur seals from the San Miguel colony in the summer and autumn is not clearly understood. However, most sightings from June through November have been recorded in an area to the west of San Miguel Island, over Rodriguez Seamount and northward at a distance of about 24–67 km offshore to an area near Point Sal (Bonnell et al. 1981, 1983).

Northern fur seals have been harvested commercially for two centuries (Lander 1980). Under the terms of the North Pacific Fur Seal Convention, the participating countries (the United States, Canada, Japan, and Russia) conduct the harvest, and the population is managed to achieve maximum sustained yield (York 1987). In the late 1970s, the population on the Pribilof Islands began to decline at about 5% per year, apparently due to entanglement in lost or discarded fishing nets and net fragments (Fowler 1982, 1985). More recently, entanglement mortality has decreased concurrent with a shift of commercial fishing effort to deeper waters, and the number of pups born each year has stabilized (York 1990).

The Guadalupe fur seal, *Arctocephalus townsendi*, is an occasional visitor to the SCB, but presently breeds only on Isla de Guadalupe, a volcanic island 260 km west of Baja California (Pierson 1987). Historically, Guadalupe fur seals probably ranged from Islas Revillagigedo, Mexico (18°N) to the Farallon Islands off San Francisco (Repenning et al. 1971). Archaeological evidence from Indian middens reveals that Guadalupe fur seals may once have been very abundant in southern California (Lyon 1937; Walker and Craig 1979), but the species was hunted intensively throughout the 1800s and was considered extinct by the turn of the century. Sightings of a few animals in California waters during the last 40 years or so rekindled hope for the recovery of the species (Bonnot et al. 1938; Bartholomew 1950). A careful search of remote Isla de Guadalupe in 1954 led to the discovery of a small remnant breeding population (Hubbs 1956). Since then, the population appears to have grown at a rate of 10% per year, and presently numbers at least 2000 animals (Fleischer 1987; Pierson 1987).

In the last 20 years, there have been 50 sightings of Guadalupe fur seals in the SCB (Stewart et al. 1987). However, there appears to be little trend in the frequency of sightings. Most occurred after researchers initiated long-term observational studies on San Miguel Island and San Nicolas Island in 1969 and 1980, respectively; therefore, increased sightings may represent increased observation effort. Single sightings have been recorded on Santa Barbara Island and San Clemente Island as well. There have also been three sightings of Guadalupe fur seals at sea (Brownell and DeLong 1968; Bonnell et al. 1981). Because recognition of individuals depends on observation of characteristic scars or marks (Stewart 1981b), the total number recorded in the SCB probably includes repeat sightings of the same individual on different islands or in different seasons or years. The actual number of individual Guadalupe fur seals hauling out on islands of the SCB each year is not known. Territoriality and sexual behavior have been observed in male Guadalupe fur seals on San Nicolas and San Miguel islands in the summer months (Stewart et al. 1987). It is thus likely that breeding will eventually occur on these islands if the population continues to grow and expand its range into the SCB.

To provide protection for this recovering species, the Guadalupe fur seal was listed as a Threatened species in December 1985 (50 FR 51251–51258).

Historically, northern elephant seals (*Mirounga angustirostris*) were found by the thousands along much of the coast of California and Baja California and were known to breed from Cabo San Lazaro, Mexico, to Point Reyes (Scammon 1874). Years of commercial harvest so reduced their numbers that, by about 1890, the species is believed to have consisted of a single herd of fewer than 100 animals on Isla de Guadalupe (Townsend 1899; Bartholomew and Hubbs 1960). When hunting ceased, elephant seals began to increase in numbers, and by the 1930s, they recolonized other islands of Baja California and the SCB (Huey 1930a; Bartholomew and Hubbs 1960). At present, northern elephant seals have a breeding range from Isla Natividad in Baja California to Point Reyes near San Francisco (Le Boeuf and Mate 1978; Allen 1987) and the general direction of range expansion is northward. A single pup has been observed on Castle Rock off Point Saint George in northern California (Bonnell et al. 1983).

Most northern elephant seals are presently associated with rookery islands in the SCB. The largest population in this part of the range is on San Miguel Island, which accounts for more than 75% of the total numbers on land in the SCB. About 23% of the breeding population is found on San Nicolas Island, and small colonies exist on Santa Barbara and San Clemente islands. Maximum numbers on land occur during the winter when northern elephant seals congregate on rookery islands to breed, and in spring when most females and juveniles haul out to molt. Counts obtained at the peak of the breeding season in late January 1982 totaled nearly 20,000 animals on land in the SCB (Bonnell 1982b; Stewart and Yochem 1984). However, counts underestimate breeding population size because the breeding season extends over a 3-month period and each female is on the rookery for an average of only 34 days (Le Boeuf et. al. 1972). Furthermore, a few pups leave the rookery before others are born. Bonnell et al. (1981), using the Le Boeuf–Sylvan statistical model, estimated that the breeding population of northern elephant seals in the SCB totaled about 27,500 in 1982.

Overall, the world population of northern elephant seals has increased rather steadily, at a rate of about 7% per year from at least the mid–1970s through the present. However, growth of the Mexican portion of the population may have leveled off between 1977 and 1978 (years of the last full-range census); counts on Mexican rookeries show almost no increase in pup production between these years. While approximately 44% of the total pup production of this species occurred on islands of the SCB in 1978 (Bonnell et al. 1981), this figure increased to nearly 55% by 1982 because of continued immigration from the Mexican colonies. Growth estimates of the northern elephant seal population in the SCB, based on counts of pups, are graphed in figure 11.15. Total population size, which is a combination of the breeding population plus estimated numbers of juveniles not present on rookeries, can be calculated by using information on life history and age-specific survival (Le Boeuf et al. 1974; Reiter et al. 1978). The world population of northern elephant seals was estimated at 77,000 in 1982 (Bonnell et al. 1983) and currently numbers about 114,000 animals.

Northern elephant seals fast for the duration of their stay on land. When they go to sea, some of the animals spend 2 weeks or more feeding almost continuously at great depths (Le Boeuf et al. 1988). During this period, northern elephant seals may spend as much as 85% of their time underwater and, consequently, are infrequently seen on aerial or vessel surveys. Northern elephant seals disperse widely in waters from Baja California to the Gulf of Alaska following the breeding and molting seasons (Condit and Le Boeuf 1984). Although observed at sea throughout the year, northern elephant seals are never

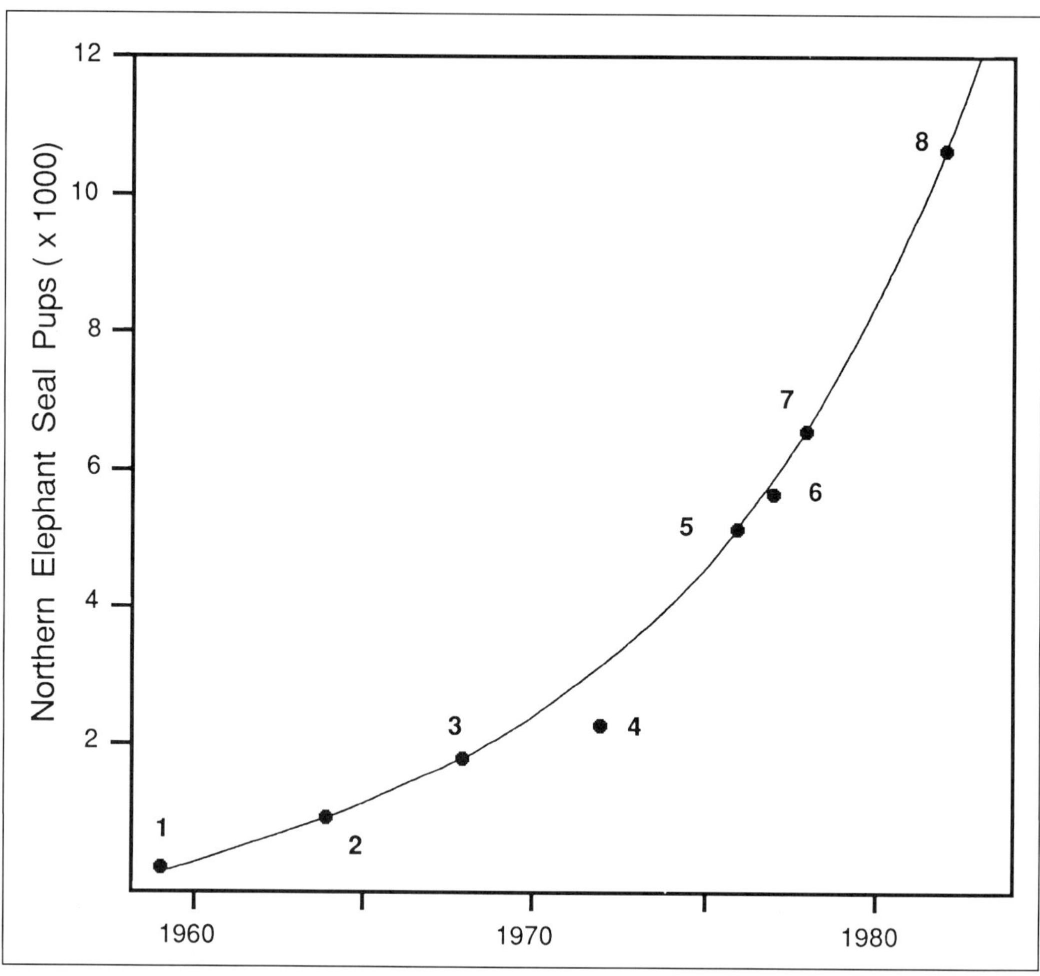

Figure 11.15. Growth of northern elephant seal population in the SCB expressed as total number of pups counted on rookery islands; curve plotted through points. Data sources: (1) Bartholomew and Boolootian 1960; (2) Odell 1971; (3) Peterson et al. 1968; (4) Antonelis et al. 1981; (5–7) Bonnell et al. 1981; (8) Bonnell 1982b.

very abundant in the SCB (Bonnell et al. 1981). On surveys conducted between 1975 and 1978, 83 sightings of northern elephant seals were recorded for a mean annual abundance of about 490 ±18 (SE) (Bonnell et al. 1981).

Harbor seals (*Phoca vitulina*), unlike other pinnipeds of the SCB that are endemic to the North Pacific, are found in the North Atlantic as well. Pacific harbor seals (*P. v. richardsi*) range from Herschel Island in Alaska to Baja California, but only 14% are found south of Alaska (Shaughnessy and Fay 1977; U.S. Department of Commerce 1987). The population in the SCB is near the southern limit of the subspecies' range. Scheffer (1958) suggests that harbor seals probably do not range south of central Baja California because of a rise in sea surface temperature. Historical information on population size in the SCB is lacking. Bartholomew (1967) estimated the southern California population during the early 1960s to be only 500 animals. Odell (1971) counted 645 on the island shores in

June 1964, and counts made between 1975 and 1977 increased from 1200 to 1800 (Bonnell et al. 1981).

More recently, counts of harbor seals on the islands and mainland shore of the SCB have been conducted under the auspices of the National Marine Fisheries Service (NMFS) (Stewart 1982; Miller et al. 1983a; Hanan 1986). The largest count of 4106 animals was recorded in late May and early June of 1982. Of the total, 75% were counted on the northern island chain (San Miguel, Santa Rosa, Santa Cruz, and Anacapa). Basing their work on counts of harbor seals on land, NMFS researchers calculated that the population in the SCB has been increasing at a rate of 10% annually for the last two decades. However, uncertainty exists about what portion of the population may be on land at any given time of day. While observational studies have demonstrated a strong diurnal hauling pattern (Stewart 1984), telemetry studies show that individual harbor seals may not haul out each day, and all harbor seals hauling out on a particular day may not be on land during the same period of time (Yochem et al. 1987). For these reasons, counts of animals on land probably underestimate true population size.

Harbor seals on the islands of the SCB exhibit a distinctive annual cycle of abundance. Numbers increase sharply through the spring as adults haul out to breed and pups are born. Peak counts of harbor seals on land are obtained in late May or early June (Stewart 1981a, 1982). Numbers on land remain high through early summer, when adults undergo their annual molt. In late July, numbers plummet to less than one-third of the summer population and remain at this low level through the autumn and winter (Stewart and Yochem 1984). It is unlikely, however, that most harbor seals actually leave the SCB during seasons of minimum abundance on land. Populations in central and northern California, as well as elsewhere in the species' range, exhibit a similar pattern of abundance on land (Sullivan 1980; Bonnell et al. 1983; Johnson and Jeffries 1977).

Harbor seals are typically seen in coastal waters and apparently forage relatively close to shore. In the SCB, 71% of all harbor seals seen at sea have been within 10 km of land; greatest numbers were seen during autumn months, following the breeding and molting seasons (Bonnell et al. 1981). While 80% of the sightings occur in relatively shallow water (<200 m), it is known that harbor seals can dive quite deeply. Off the Big Sur coast, a harbor seal was recovered from a sablefish trap set at a depth of more than 500 m (Kolb and Norris 1981).

Sea Otters

Historically, the range of the sea otter, *Enhydra lutris,* extended from the northern islands of the Japanese Archipelago northeast to include the Commander, Pribilof, and Aleutian islands, and southward along the Alaska Peninsula and the west coast of North America to about Morro Hermoso, Baja California, Mexico (Kenyon 1969). There is little doubt that sea otters inhabited the nearshore waters of the SCB, and some studies suggest they were quite abundant in the area (Lyon 1937; Ogden 1941; Rozaire 1959; Walker and Craig 1979). Commercial exploitation of sea otters began about 1741 with the discovery of the Commander Islands by the Bering expedition. Hunters harvested sea otters relentlessly throughout their range for 170 years, causing the world population to shrink to an estimated 1000–2000 animals (Kenyon 1969). Protection was extended to sea otters in 1911 under the International Fur Seal Treaty. By then, remnant populations of sea otters persisted only in the Kurile Islands and Kamchatka Peninsula, in the Commander and Aleutian islands, in southeastern Alaska, in the Queen Charlotte Islands in British Columbia, near Point Sur in California, and at Islas San Benito, Mexico (Kenyon 1969). Although no longer hunted, several of these remaining populations, including those in southeastern Alaska, British Columbia, and

Mexico, continued to decline and became extinct.

The sea otter population from the Kurile Islands to Prince William Sound has apparently recovered and now occupies most of the historical habitat in this part of the range. The northern population may include 150,000–300,000 animals (Kenyon 1969; Johnson 1982). In California, the remnant population near Point Sur was estimated in 1914 to include about 50 animals and was known to only a few local residents and interested naturalists (Bryant 1915). The population was brought to public attention with the opening of the Pacific Coast Highway in 1938 (Bolin 1938) and was then estimated to include 150–300 animals (Fisher 1939; Boolootian 1961). By 1960, the population occupied nearshore waters between Monterey Peninsula and Point Piedras Blancas and was believed to include 1000 animals (California Department of Fish and Game 1976). Growth continued at a rate of about 5% per year through the mid–1970s (Ralls et al. 1983). In 1976, the population reached a peak of 1500–1800 independent animals (Estes and Jameson 1983; Geibel and Miller 1984); it subsequently underwent a slow decline that lasted through 1982, at which time it stabilized at about 1200 independent animals.

It is now generally agreed that the reduction in numbers of sea otters from 1976 through 1985 resulted from entanglement mortality in set-nets for halibut and rockfish. The California Department of Fish and Game estimates that an average of 80 sea otters drowned each year in set-nets (Wendell et al. 1985). Legislation prohibiting the use of such nets in the sea otter range was enacted in 1985 and 1986, and population growth appears to have resumed. In 1987, the U.S. Fish and Wildlife Service (USFWS) estimated the population of sea otters to be about 1650 animals.

In 1977, the California population of sea otters was listed as threatened under the Endangered Species Act. A plan for the recovery of the population was developed by the U.S. Fish and Wildlife Service (1982). The recovery plan called for an effort to establish a second population of California sea otters at another location within historical habitat as a hedge against catastrophic impact of an oil spill in the central California range. The slow growth of the central California population was viewed as an impetus for action to implement the recovery plan. Under the Endangered Species Act, the proposed translocation of sea otters could be undertaken solely as a recovery action, but existing authorities of the Marine Mammal Protection Act did not allow for establishment and long-term management of a new colony of a depleted species. For the recovery action to proceed, a new law specifically authorizing the translocation of California sea otters was enacted by Congress in November 1986 (Public Law 99–625). After a lengthy review process, 69 sea otters were translocated to San Nicolas Island in the autumn of 1987. Of the total number initially translocated, about 20 were still present at the end of the first year. A few died in gill nets or were found dead from other causes. Several returned to the mainland and were taken back to their central California range, and others were unaccounted for following a winter storm. A few sea otters are occasionally present elsewhere in the SCB. These otters may be wanderers from the central California population or from the new San Nicolas Island colony.

Diet and Feeding of Marine Mammals

Study of diet and food consumption of marine mammals presents obvious difficulties resulting from their life in a marine environment, yet a great deal is known from a variety of sources. During years of pelagic whaling, the diets of most large whales were described from animals brought aboard ships or taken to shore stations for processing. The diet of the abundant northern fur seal was also investigated directly by collecting a sample of the population on its pelagic foraging grounds (Kajimura 1985; Perez and Bigg 1986). But for most species, knowledge of feeding habits relies on the infrequent discovery of stranded specimens or direct observations of foraging behavior at sea.

Identification of stomach contents from stranded or collected animals depends to some degree on the state of digestion of food. However, the presence of hard parts (such as squid beaks and fish otoliths) has permitted the identification of some prey species even in partially or fully digested remains (Fitch and Brownell 1968; Murie 1987). The diets of some pinniped species have been described from scats or vomitus collected on haul-out sites (Antonelis et al. 1984). Lavage of stomach contents from chemically immobilized animals has been used to describe the diet of northern elephant seals (Antonelis et al. 1987).

Knowledge of diet can sometimes provide important clues to where and how marine mammals feed (Gaskin 1982). The general habitats of many fishes and squid can be differentiated into neritic and pelagic realms and further identified by the depth at which the prey species is found—that is, near the bottom or in midwater. Thus, it may be possible to determine the general foraging habitat of marine mammals based on the distribution of their prey species.

The quantity of food consumed can be estimated in a variety of ways. Often, the amount sufficient to maintain the weight of captive animals in aquaria is used to approximate the food consumption of free-ranging animals (Keyes 1968; Sergeant 1969; Ridgway 1972). However, the energy that a marine mammal can obtain from food differs with prey species, and total consumption in the natural habitat may be less or more than that in captivity (Brodie 1975). Another approach to estimation of food consumption utilizes measurement of preformed water ingested in the prey of marine mammals (Nagy and Costa 1980). This technique, which can be used only for animals that do not ingest seawater, relies on the measured dilution of a dose of isotopically labeled water over the period of time during which a marine mammal forages (Costa 1987, 1988). From water flux, food consumption can be estimated if the following parameters are known: metabolic efficiency of the diet, rate of metabolic water produced per gram of food consumed, and the energy and water content of probable prey species (Shoemaker et al. 1976; Costa 1988). Estimates of the quantity of food required to meet the energy expenditure of subject animals can also be made using the doubly labeled water method (Costa 1987). Because use of these techniques requires the recapture of subject animals, it has been applied among marine mammals only for pinnipeds that return to land (Costa and Gentry 1986; Costa et al. 1986).

Baleen Whales

Since no studies of baleen whale feeding and food habits have ever been conducted in the SCB, some assumptions must be made based on the findings of studies in other parts of their range (Nemoto 1959; Nemoto and Kawamura 1974; Kawamura 1982). All species exhibit a northward movement in the eastern North Pacific in the spring or summer to exploit more abundant food resources in highly productive subarctic waters. However, feeding in the SCB and other locations off California probably occurs on both northward and southward migrations, and a few individuals of some species, especially fin and minke whales, are present in the SCB throughout the summer months when most feeding typically occurs.

Blue whales are considered virtually monophagous, consuming only a variety of euphausiid crustaceans (table 11.2). Off California, blue whales are known to consume mostly *Euphausia pacifica* and the slightly larger *Thysanoessa spinifera* (Rice 1986; Johnson 1987; Schoenherr 1987). Rice (1986) calculates that, to sustain growth and reproduction, a blue whale weighing about 68,000 kg needs to consume more than 3400 kg of

Table 11.2. *Potential Food Items of Blue, Fin, Sei, Minke, and Humpback Whales*[a]

Food Item	Blue Whale	Fin Whale	Sei Whale	Minke Whale	Humpback Whale
Crustaceans					
Euphausiids	X	X	X	X	X
Pelagic red crabs	X				
Copepods		X	X		X
Fish					
Anchovy		X	X	X	X
Saury		X	X		X

[a]Not all food items may typically be consumed in the SCB.

Sources: Nemoto 1959; Mitchell 1978; Kawamura 1982; Mizrock et al. 1984a, b; Rice 1986; Wolman 1986; Johnson 1987; Schoenherr 1987.

euphausiids per day (about 5% of its body mass) during the 6-month period when it feeds. Blue whales also have been observed to feed on dense shoals of pelagic red crabs (*Pleuroncodes planipes*) off Baja California (Rice 1986).

Fin whales have a broader diet that includes euphausiids and copepods, as well as small schooling fish when they are present in sufficient concentrations. Sei whales are known to prefer copepods, but take a wide variety of prey, including euphausiids and fish. The diet of minke whales, which are present year-round in the SCB, is not well known, but may be very diverse. From several studies, minke whales are known to take both euphausiids and fish (Mitchell 1978). Their distribution in the SCB suggests that minke whales may exploit some specific but not yet identified inshore resources. Humpback whales feed on euphausiids, copepods, and schooling fish (Wolman 1986). Sergeant (1969) estimated the feeding rate of adult rorquals at about 4% of their body mass per day during that period of the year when feeding occurs.

Unlike other baleen whales, gray whales feed predominantly on benthic organisms (Rice and Wolman 1971; Oliver et al. 1983). The mechanism of feeding is described by Nerini (1984) from several reported observations. As the gray whale swims slowly forward, it rolls to one side and passes within a few centimeters of a benthic food source while creating suction in its open mouth to draw in the food and sediment. The whale retains food while pushing the slurry of water, silt, and sand outward through the baleen. Observation by divers and side-scan sonar shows that the activity leaves shallow depressions on the sea floor which can be used to identify feeding grounds (Nerini et al. 1980; Nelson et al. 1983; Oliver et al. 1983). Gray whales are also capable of taking pelagic prey such as schools of bait fish by a skimming or engulfing behavior (Gilmore 1961; Sund 1975). Wellington and Anderson (1978) and Swartz and Jones (1982) reported interesting observations of juvenile gray whales and mothers with calves skimming in dense beds of kelp or eel grass in one of the breeding lagoons in Mexico and along parts of the island and mainland shores of the SCB.

On feeding grounds in shallow waters of the Bering and Chukchi seas, gray whales primarily consume several species of amphipods; polychaetes are also a principal prey species in some areas (Bogoslovskaya et al. 1981). A variety of other invertebrates, and occasionally fish, are ingested incidentally to benthic feeding on primary food types. Nerini (1984) reviewed the prey species known to be consumed by gray whales and pointed out that the list is extensive because of the gray whale's opportunistic approach to feeding and the nonselective nature of the feeding mechanism. Summer populations of gray whales also exist from northern California to southeastern Alaska. Observations of apparent feeding have been described for four locations off northern California by Dohl et al. (1983) and for Vancouver Island, British Columbia, by Murison et al. (1984). Other unpublished observations are reviewed by Nerini (1984).

Generally, gray whales rarely feed during the migrations between summer feeding areas and winter calving areas (Rice and Wolman 1971). Nonetheless, a few observations of feeding have been reported. On the southbound migration, feeding on bait fish by gray whales passing Monterey Bay was observed by Sund (1975). Observations of feeding in the SCB have been reported for nearshore waters around the northern island chain, in the Santa Barbara Channel, near Point Mugu, and near San Diego (Pike 1962; Wellington and Anderson 1978; Jones and Swartz 1986). Benthic feeding was noted by M. Poole for waters in the vicinity of Piedras Blancas, by A. Roest for waters near San Luis Obispo (unpubl. observ. cited in Nerini 1984), and by Dohl et al. (1983) for mother-calf pairs in Carmel Bay.

From the foregoing evidence, there is little doubt that feeding along the migration path occasionally occurs as gray whales encounter exploitable concentrations of food resources. However, Rice et al. (1984) stated that gray whales without calves lose between 0.21 and 0.37% of their body weight per day on migration to and from breeding grounds and that this weight loss is sufficient to account for energy needs in the winter. A comparable figure of 0.2% loss of body mass per day over the migration period has been calculated by Sumich (1983). Thus, feeding during migration for these animals is probably not obligatory. Lactating females, however, are subject to greater energetic demands and may make greater use of available food encountered during their slower passage northward. Data are unavailable to estimate the food consumption of gray whales passing through the SCB, but it is likely that their impact on stocks of prey species is quite small.

Odontocete (Toothed) Whales

Common dolphins are believed to feed at night on deep-scattering layer organisms. Using radiotelemetry, Evans (1971) showed that common dolphins in the SCB make deeper dives at night as many species of fish and squid rise into epipelagic waters (<200 m). Large resting schools, or possibly food-searching schools (Norris and Dohl 1980a), form in daylight hours and disperse into smaller feeding schools after dark. Common dolphins feed on a variety of schooling fish, including Pacific whiting, northern anchovy, myctophids (lanternfish), and squid. Two common dolphins from the SCB that were examined by Fitch and Brownell (1968) contained mostly anchovy and Pacific whiting. A few myctophids and other deep-water species were also present, and one animal had consumed Pacific saury (an epipelagic species). The diet suggests that common dolphins may feed from the surface down to about 200–250 m; however, Evans's (1971) study indicates that night dives to about 100 m are more typical. The autumn and winter diet probably consists mostly of squid and anchovy, while in spring and summer it includes deep-sea smelt, lanternfish, and Pacific whiting (Leatherwood et al. 1982). Information on the diet of common dolphins is summarized in table 11.3.

Northern right-whale dolphins apparently feed on both mesopelagic fish and squid and forage at depths of 200–250 m (Norris and Prescott 1961). Few strandings have occurred and little is known about their feeding habits. Sullivan and Houck (1979) found squid beaks of the family Gonatidae in the stomach of a northern right-whale dolphin stranded in northern California. The stomach of a northern right-whale dolphin stranded in the SCB contained squid beaks and otoliths of lan-

Table 11.3. *Reported Prey Species of Common Pinnipeds and Toothed Whales That Inhabit the SCB*[a]

Prey Species	Pinnipeds				Toothed Whales					
	Z.c.[b]	*M.a.*	*C.u.*	*P.v.*	*D.d.*	*L.b.*	*L.o.*	*G.g.*	*P.d.*	*G.m.*
Teleost Fish										
Pacific whiting (*Merluccius productus*)	X	X	X	X	X	X	X		X	
Rockfish (*Sebastes* sp.)	X	X	X	X			X		X	
Blacksmith (*Chromis punctipinnis*)	X			X						
Surfperches (Embiotocidae)	X	X		X						
Greenlings and lingcod (Hexagrammidae)	X			X						
Northern anchovy (*Engraulis mordax*)	X	X	X	X	X		X		X	
Lanternfish (Myctophidae)	X		X		X	X	X	X	X	
Pinpoint lampfish (*Lampanyctus regalis*)									X	
North. lampfish (*Stenobrachius leucopsarus*)	X									
Medusafish (*Icichthys lockingtoni*)	X				X		X			
Cusk eels (Ophidiidae)	X	X		X	X				X	
Midshipman (*Porichthys notatus*)	X	X		X	X		X			
Flounders and sole (Pleuronectidae)	X	X		X						
White croaker (*Genyonemus lineatus*)				X						
Pacific sandab (*Citharichthys sordidus*)		X		X			X		X	
Salmon (*Oncorhynchus* sp.)	X		X							
Pacific saury (*Cololabis saira*)			X		X		X		X	
Jack mackerel (*Trachurus symmetricus*)	X		X				X		X	
Pacific (chub) mackerel (*Scomber japonicus*)	X		X							
Pacific tomcod (*Microgadus proximus*)	X			X					X	
Smelts (Osmeridae)	X			X	X	X	X	X	X	
Eulachon (*Thaleichthys pacificus*)			X				X			
Pacific herring (*Clupea harengus*)	X		X						X	
Calif. smoothtongue (*Bathylagus stilbius*)	X				X				X	
Pacific pompano (*Peprilus simillimus*)	X								X	

Queenfish (*Seriphus politus*)	X						X			
Sablefish (*Anoplopoma fimbria*)	X	X	X	X					X	
Snailfish (*Liparis* sp.)									X	
Eelpouts (Zoarcidae)		X		X					X	
Grenadier (Macrouridae)									X	
Eels (Anguilliformes)									X	
Elasmobranchs										
Brown cat shark (*Apristurus brunneus*)		X								
Ratfish (*Hydrolagus colliei*)		X								
Stingray (*Urolophus halleri*)		X								
Blue shark (*Prionace glauca*)		X								
Angel shark (*Squatina californica*)		X								
Skate (*Cephaloscyllium ventriosum*)		X								
Dogfish (*Squalus acanthias*)		X								
Swell shark (*Raja* sp.)		X								
Horn shark (*Heterodontus francisci*)		X								
Thornback (*Platyrhinoidis triseriata*)		X								
Egg cases (various species)		X								
Cyclostomes										
Lamprey (*Lampetra tridentata*)	X	X		X						
Hagfish (*Eptatretus* sp.)		X		X						
Egg cases		X								
Cephalopods										
Unid. squid					X	X	X		X	
Market squid (*Loligo opalescens*)	X	X	X		X		X		X	X
Octopoteuthidae							X	X	X	
Octopoteuthis deletron		X								
Onychoteuthis borealijaponicus	X	X	X		X		X	X	X	
Moroteuthis robusta		X								X

Table 11.3. *Reported Prey Species of Common Pinnipeds and Toothed Whales That Inhabit the SCB*[a] (continued)

	Pinnipeds				Toothed Whales					
Prey Species	*Z.c.*[b]	*M.a.*	*C.u.*	*P.v.*	*D.d.*	*L.b.*	*L.o.*	*G.g.*	*P.d.*	*G.m.*
Histioteuthis sp.		X								X
Dosidicus gigas								X		
Gonatidae	X	X	X			X	X	X	X	
Berryteuthis magister		X								
Gonatopsis sp.		X					X		X	
Galiteuthis sp.		X								
Gonatus sp.	X	X			X			X	X	
Taningia danae		X								
Taonius pavo		X								
Ocythoe tuberculata		X								
Abraliopsis felis		X					X		X	
Chiroteuthis calyx		X					X	X		
Vampyroteuthis internalis		X								
Chanchidae		X					X		X	
Cranchia scabra		X								
Octopus sp.	X	X		X			X		X	
Cuttlefish (*Rossia pacifica*)		X								

Crustaceans		
Pelagic red crabs (*Pleuroncodes planipes*)	X	X
Pasiphaea pacifica		X
Euphausia sp.		X
Notostomus japonicus		
Hemisquilla californiensis		X

[a] List of food types includes only species known to occur off California; not all prey species listed occur in southern California waters. Some rarely consumed food items have been omitted.

[b] Species abbreviations and sources as follows:

Z.c. = California sea lion (*Zalophus californianus*). Sources: Lowry et al. 1987; Antonelis et al. 1984; Roffe and Mate 1984 (lampreys); Heath and Francis 1983; Jones 1981; Briggs and Davis 1972 (salmonids).

M.a. = Northern elephant seal (*Mirounga angustirostris*). Sources: Antonelis et al. 1987; Hacker 1986; Condit and Le Boeuf 1984; Nagorsen and Stewart 1983; Jones 1981; Antonelis and Fiscus 1980; Morejohn and Baltz 1970; Pike and MacAskie 1969; Freiberg and Dumas 1954; Huey 1930b.

C.u. = Northern fur seal (*Callorhinus ursinus*). Sources: Perez and Bigg 1986; Antonelis and Perez 1984; Stroud et al. 1981; Spalding 1964; Scheffer 1958.

P.v. = Harbor seal (*Phoca vitulina*). Sources: Haaker et al. 1984; Jones 1981; Morejohn et al. 1977.

D.d. = Common dolphin (*Delphinus delphis*). Sources: Jones 1981; Fitch and Brownell 1968; Fiscus and Niggol 1965.

L.b. = Northern right-whale dolphin (*Lissodelphis borealis*). Sources: Fitch and Brownell 1968; Sullivan and Houck 1979; Dohl et al. 1983.

L.o. = Pacific white-sided dolphin (*Lagenorhynchus obliquidens*). Sources: Stroud et al. 1981; Jones 1981; Fitch and Brownell 1968; Ridgway 1966; Fiscus and Niggol 1965; Best 1963; Houck 1961; Norris and Prescott 1961; Brown and Norris 1956.

G.g. = Risso's dolphin (*Grampus griseus*). Source: Jones 1981.

P.d. = Dall's porpoise (*Phocoenoides dalli*). Sources: Jones 1981; Stroud et al. 1981; Loeb 1972; Fiscus and Niggol 1965; Norris and Prescott 1961; Brown and Norris 1956; Scheffer 1953.

G.m. = Short-finned pilot whale (*Globicephala macrorhynchus*). Sources: Seagars and Henderson 1985; Hacker 1986.

ternfish (Fitch and Brownell 1968). Lanternfish are deep-scattering layer organisms found during daytime at depths of 250 m or more, suggesting that northern right-whale dolphins may feed mostly at night as lanternfish rise closer to the surface.

The diet of the Pacific white-sided dolphin also includes both fish and squid. Fitch and Brownell (1968) found mostly Pacific whiting (*Merluccius productus*) and anchovies (*Engraulis mordax*), along with squid, in stomachs of two stranded Pacific white-sided dolphins from the SCB. Brown and Norris (1956) and Ridgway (1966) indicated the presence of market squid (*Loligo opalescens*) in stomach contents, and Fiscus and Niggol (1965) found *Gonatus* sp. in the stomachs of Pacific white-sided dolphins collected off California. Based on size of prey, Pacific white-sided dolphins probably feed in both epipelagic (0–200 m) and upper mesopelagic (>200 m) zones. Demersal fish have been found among stomach contents, suggesting that these dolphins may occasionally feed near the bottom in continental shelf waters. On the basis of data obtained from captive animals fed a diet of herring and mackerel, Sergeant (1969) and Ridgway (1972) calculate food consumption to be about 7–8% of body mass per day.

The diet of Dall's porpoises is among the most diverse for any cetacean (table 11.3). Most of the known food items of the species were identified in a study of Dall's porpoise in Monterey Bay (Loeb 1972). A great many prey species were noted, but the most abundant were Pacific whiting, juvenile rockfish, and market squid *Loligo opalescens*. Other cephalopods found included *Abraliopsis, Gonatus* sp., *Onychoteuthis* sp., and *Octopus* sp. Dall's porpoises probably feed primarily in epipelagic and mesopelagic zones and occasionally take bottom-dwelling prey. In southern California waters, Brown and Norris (1956) noted that Dall's porpoises consume anchovies. Norris and Prescott (1961) reported that one Dall's porpoise from the SCB had consumed whiting, jack mackerel, and squid.

The diet of bottlenose dolphins in the SCB has not been completely described. In the Gulf of Mexico, bottlenose dolphins are known to consume a wide variety of fishes plus squid, shrimp, and crabs (Gunter 1942; Kemp 1949, cited in Leatherwood 1975). Association of bottlenose dolphins with shrimp boats in the Gulf of California has been observed by Norris and Prescott (1961) and Leatherwood (1975). In coastal waters of the SCB, bottlenose dolphins are believed to feed opportunistically on a wide variety of fish, cephalopods, and crustaceans (Norris and Prescott 1961; Leatherwood et al. 1982).

Pilot whales in the SCB are presumed to feed almost entirely on squid, but data on stomach contents are scant. One clue to the squid-eating habit of pilot whales is the anterior emphasis in the distribution of teeth in both jaws (Gaskin 1982). Stomachs of the few pilot whales found stranded in the SCB have contained only squid, including *Loligo opalescens, Moroteuthis robusta,* and *Histioteuthis dofleini* (Hall et al. 1971; Seagars and Henderson 1985; Hacker 1986). Pilot whales in the North Atlantic, which also feed primarily on squid, occasionally supplement their diet with fish (Sergeant 1962). The rate of consumption of food is estimated to be 4–5% of body weight per day, which amounts to about 13.6 kg for an average-sized adult (Kritzler 1952; Sergeant 1962).

Risso's dolphins also probably eat only squid, but little is known of their diet from strandings. Sergeant and Fisher (1957) pointed out that, like the pilot whale, Risso's dolphins lack teeth in the upper jaw and probably consume only soft invertebrates. The stomach of a single Risso's dolphin found stranded off the coast of central California was found to contain only remains of the pelagic squid *Dosidicus gigas* (Orr 1966). Farther north, Stroud (1968) reported that analyses of the stomach contents of a Risso's dolphin stranded in Washington revealed six species of squid (*Onychoteuthis, Octopodoteuthis, Chiroteuthis, Gonatus,* and two unidentified gonatid squid).

Killer whales have a varied diet that includes fish, birds, seals, and even other whales (Rice 1968). Stomach contents vary with the season or food resource being exploited. Primarily fishes and squids, but also porpoises and seals, were found in stomachs of killer whales in one feeding study. In another, mostly marine mammals were found (reviewed by Scheffer 1978). Observations of killer whales attacking gray whales and other baleen whales have been reported along the U.S. west coast (Hancock 1965; Morejohn 1968; Baldridge 1972).

Pinnipeds

The diet of California sea lions in the SCB has been described from analyses of scats collected on rookery islands (Heath and Francis 1983; Antonelis et al. 1984; Lowry et al. 1987). In terms of frequency of occurrence, Pacific whiting, market squid, rockfish, anchovy, mackerel, octopus, and several pelagic squid appear to predominate in the diet (table 11.3). Pacific whiting, market squid, juvenile rockfish, and northern anchovy represented about 80% of all prey species identified from scats collected on San Miguel Island in the spring and summer of 1978 and 1979 (Antonelis et al. 1984). Composition of the diet varies from place to place and from season to season. The relative contribution of different food types in the diet was also found to change during the 1983 El Niño temperature anomaly that apparently affected the availability of prey (DeLong and Antonelis 1991). Costa (1986) used the water flux method to determine the total food consumption of several lactating female sea lions in 1983 and 1984. Food consumption during foraging bouts ranged from 11.6% of the body mass per day in 1983 to about 10.4% in 1984. While food consumption remained about the same, both the energy expenditure required to obtain food and the length of the foraging bout were significantly greater during the El Niño temperature anomaly of 1983 (Costa 1986; Costa et al. 1991).

Northern elephant seals appear to have one of the most varied diets of any marine mammal (table 11.3). While they feed mostly on cephalopods and Pacific whiting, elephant seals are also known to consume a variety of small sharks and rays (Hacker 1986; Antonelis et al. 1987). On the basis of stomach contents of a specimen stranded near San Diego, Huey (1930b) first suggested that elephant seals probably forage near the bottom in deep water. Morejohn and Baltz (1970) also found benthic or epibenthic species in a dead elephant seal found in Monterey Bay. Hacker (1986) examined the stomach contents of 21 elephant seals from the SCB and found 20 species of fish and 11 species of cephalopods, crustaceans, and tunicates. In that study, cusk eel, whiting, ratfish, and midshipman were the most frequently occurring fish, while market squid and octopus accounted for most of the cephalopod remains. Antonelis et al. (1987) identified no less than 20 different species of cephalopods by lavage of stomach contents from healthy elephant seals on San Miguel Island. Pacific whiting and pelagic red crabs (*Pleuroncodes planipes*) also appeared frequently in samples. Elephant seals fast throughout their breeding or molting seasons and may disperse northward over a 1000 km^2 area between periods of time on land (Condit and Le Boeuf 1984). Thus, the stomach contents of northern elephant seals on land in the SCB probably reflect feeding success in other parts of their range.

Costa et al. (1986) investigated the energetic costs of lactation by nursing female northern elephant seals on their rookery islands. They found that females lose an average of 42.2% of their initial body mass over the breeding season while nursing a single pup. Costa et al. (1986) calculate that energy obtained by a female elephant seal over the year averages about 143 MJ d^{-1}, 10% of which exceeds maintenance requirements and is stored as fat to support fasting on land and lactation. On a diet of squid, this food consumption would total only about 5.7% of the body mass per day (mean mass of 425 kg from Costa et al. 1986). Part of this remarkable efficiency is due to the temporal separation of nursing on land from feeding at sea. Costa et al. (1986) pointed out that a female, unencumbered by her pup (weaned after 28 days), can forage most efficiently. Furthermore, the cost of lactation is amortized over the 300 days per year spent at sea.

The diet of northern fur seals during winter and spring in pelagic waters of the North Pacific has been described by Kajimura (1985) and Perez and Bigg (1986) and includes a variety of fish and squid (table 11.3). Off southern California, these seals consume Pacific whiting and anchovy in nearly equal quantities; combined, these two foods represent about two-thirds of prey taken (Perez and Bigg 1986). In oceanic waters, the pelagic squid *Onychoteuthis* sp. makes up most of the remainder of the diet. Northern fur seals over the continental shelf also consume jack mackerel, market squid, and rockfish.

Food consumption in captive fur seals subsisting on a diet of herring and mackerel averaged 7.25% ± 2.3% (SE) of body mass per day (Spotte and Adams 1981). Food consumption was significantly greater during the winter months of lowest water and air temperature. Bigg et al. (1978) found similar food consumption in captive fur seals in an earlier study. Through analysis of stomach contents, Perez and Mooney (1986) calculate that lactating females have a daily feeding rate about 1.6 times greater than nonlactating animals of similar body mass. Using the doubly labeled water method, Costa and Gentry (1986) developed an index of foraging efficiency for northern fur seals based on the ratio of the net gain in energy obtained through foraging and the metabolic energy expended. Lactating northern fur seals, to compensate for decreased availability of food from 1981 to 1982, increased their foraging intensity but not the length of feeding trips. Net energy gain for these years remained about the same, but the foraging efficiency declined by half (Costa and Gentry 1986).

Little is known about the diet and prey consumption of harbor seals resident in the SCB, although the diet of this species has been well studied in northern waters (Spalding 1964; Pitcher 1980; Brown and Mate 1983). A wide variety of prey has been identified in central California (Jones 1981; Morejohn et al. 1977). Most are fish found in neritic waters; the only cephalopod reported is *Octopus* sp. In the SCB, chance observations of foraging behavior indicate that harbor seals consume blacksmith and some surfperches (Haaker et al. 1984). The habitats of prominant prey species suggest that harbor seals forage mostly on benthic or epibenthic fish in relatively shallow water. Using the weight of the seals' stomach contents as a basis, Spalding (1964) estimated the seals' food consumption to be about 7% of body mass per day (observed maximum of 11%).

Sea Otters

Sea otters are not abundant in the SCB, and the translocated population at San Nicolas Island (U.S. Fish and Wildlife Service 1987) is never expected to exceed about 300 animals at carrying capacity. However, the small colony may have a major local impact on the nearshore community at this island. Sea otters eat a diet of nearshore macroinvertebrates almost exclusively and, depending on water temperature and other factors, may consume more than 25% of their body weight per day in food (Costa 1978, 1982). This high rate of food consumption is necessary to support the elevated metabolic rate that serves, along with the insulation provided by the pelage, to maintain body temperature in cold waters (Costa and Kooyman 1984). The diet includes most species found in waters to about 30 m depth (the approximate diving limit of sea otters), including sea urchins, abalone, crabs, snails, limpets, mussels, clams, chitons, squid, and octopus (U.S. Fish and Wildlife Service 1982). VanBlaricom and Estes (1988) described the community ecology of sea otters.

Trophic Impacts of Marine Mammals

A general diagram of the food web of marine mammals in the SCB is shown in figure 11.16. Intermediate trophic level fish and

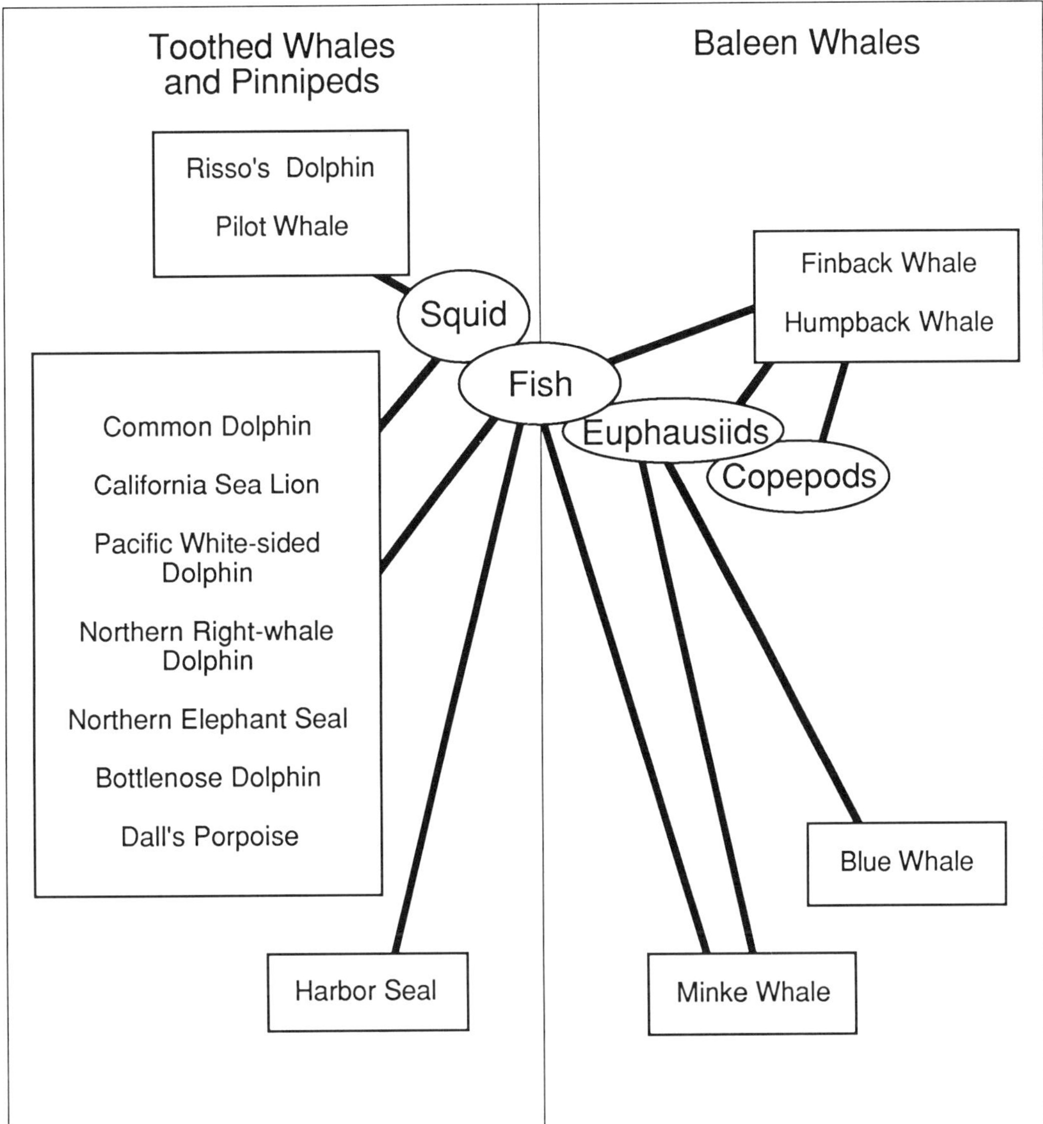

Figure 11.16. Food resources of common marine mammal species that feed in the SCB.

squid support large populations of pinnipeds, porpoises, and dolphins. Dense schools of small fish are also consumed by fin, minke, and humpback whales, but these species, along with blue whales, are predominantly sustained by patches of euphausiid crustaceans. The resulting impact of marine mammals on stocks of fish, squid, and crustaceans in the SCB is principally a function of how many marine mammals are present in the area. The abundance of each marine mammal species varies seasonally as animals enter or leave the SCB. Annual cycles, described from intensive survey studies, show some interannual variation in timing or duration of peak numbers. However, the general features of annual cycles appear to be quite consistent from year to year.

Species of baleen whales have similar annual cycles of abundance in the SCB, with large numbers of northbound migrants entering the area in late spring. A few blue, fin, and minke whales linger through the summer, while humpback whales appear to simply pass through the SCB. In the autumn, another peak in abundance occurs as southbound migrants enter the SCB. Typically, only gray whales and a few minke whales are present in the winter months. Annual cycles of the common toothed whales are shown in figure 11.17. Populations of most porpoises and dolphins occupy the SCB year-round and exploit the greater abundance of fish and squid found in these productive waters. However, numbers vary from season to season, which suggests that their pattern of use of the area is tied to predictable changes in abundance or availability of prey species.

Cool-water species, such as the northern right-whale dolphin and Dall's porpoise, have their greatest impact on local food resources in the winter and spring. Pilot whales also reach peak numbers in the SCB in the winter, though perhaps for different reasons. Common dolphins, Pacific white-sided dolphins, and Risso's dolphins all reach peak abundances in the SCB in the summer and autumn months.

Pinnipeds have relatively complex patterns of annual cycles of abundance (fig. 11.18). California sea lions, which have the largest foraging population, are clearly the most significant predator on stocks of fish and squid in the SCB. The greatest numbers of foraging California sea lions are present during and immediately after the breeding season in middle to late summer. The population thins by late autumn, but a second smaller peak may occur in the winter when some California sea lions return to the SCB to exploit spawning concentrations of anchovy, whiting, and squid. On average, California sea lions account for more than one-half of the total number of foraging marine mammals in the SCB.

Northern elephant seals are also very abundant in the area, especially during their winter and spring breeding and molting seasons. Unlike sea lions, northern elephant seals fast throughout their stay on land. Following departure from the islands, they disperse widely throughout their range and have relatively little impact on stocks of fish and squid in the SCB. Northern fur seals have a small breeding population in the area during summer and autumn, but most foraging occurs in waters over or beyond the shelf break. From December through May, many fur seals from the Bering Sea overwinter off the SCB. However, in these months as well, impacts of fur seals on stocks of fish and squid largely occur in pelagic waters rather than in the SCB. Harbor seals are present year-round in the SCB despite counts showing few animals on land in autumn and winter. While numbers are relatively small, harbor seals may have a significant impact on fish populations in benthic and epibenthic communities of the shallow island shelves.

Consumption of food resources by marine mammals can be calculated from data on the size of the foraging population of each species and from the estimated daily food consumption of individual animals (table 11.4). Food consumption varies through the year as the size and composition of the fauna change. Overall, the average food consumption of marine mammals in the SCB is slightly more than 1000 t d^{-1}. From a winter minimum of about 650 t d^{-1}, food consumption increases through the spring to a maximum of more than 1400 t d^{-1} in June (fig. 11.19).

Of all marine mammal species using the SCB, California sea lions have by far the greatest impact on local food resources, accounting for an average of about 38% of all food consumption. Pinnipeds as a group account for nearly one-half of the annual food consumption by marine mammals in the SCB. Common dolphins are the second most important predator, accounting for more than 15% of the annual food consumption. Toothed whales as a group account for one-third of all food consumption by marine mammals in the SCB. Baleen whales have their greatest impact on food resources in late

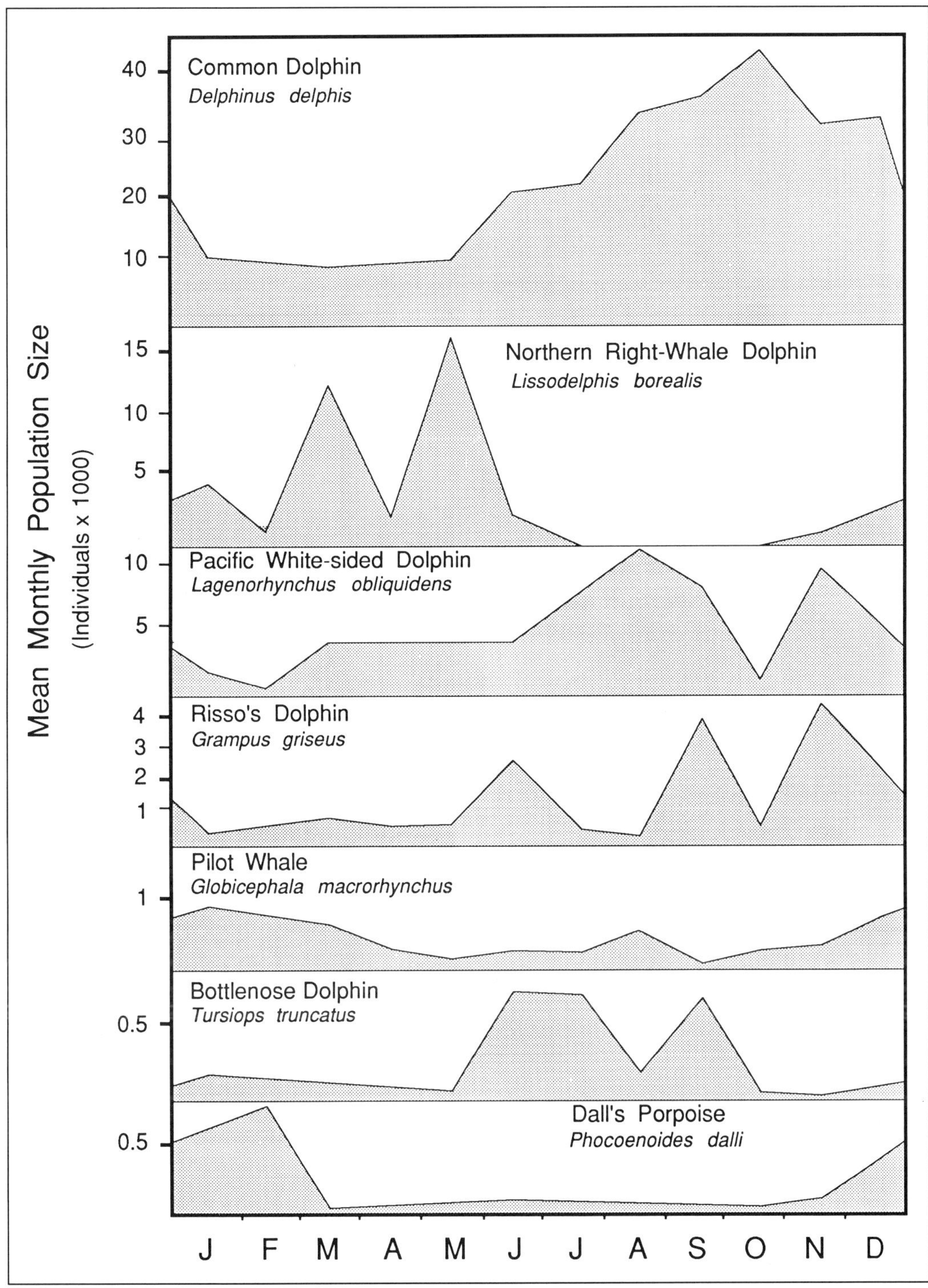

Figure 11.17. Relative abundance of common odontocete cetaceans in the SCB. Annual cycles are plotted from mean monthly population estimates obtained from line transect analysis of MMS-OCS data collected on aerial surveys, 1975–1978; extrapolation area = 6×10^4 km^2.

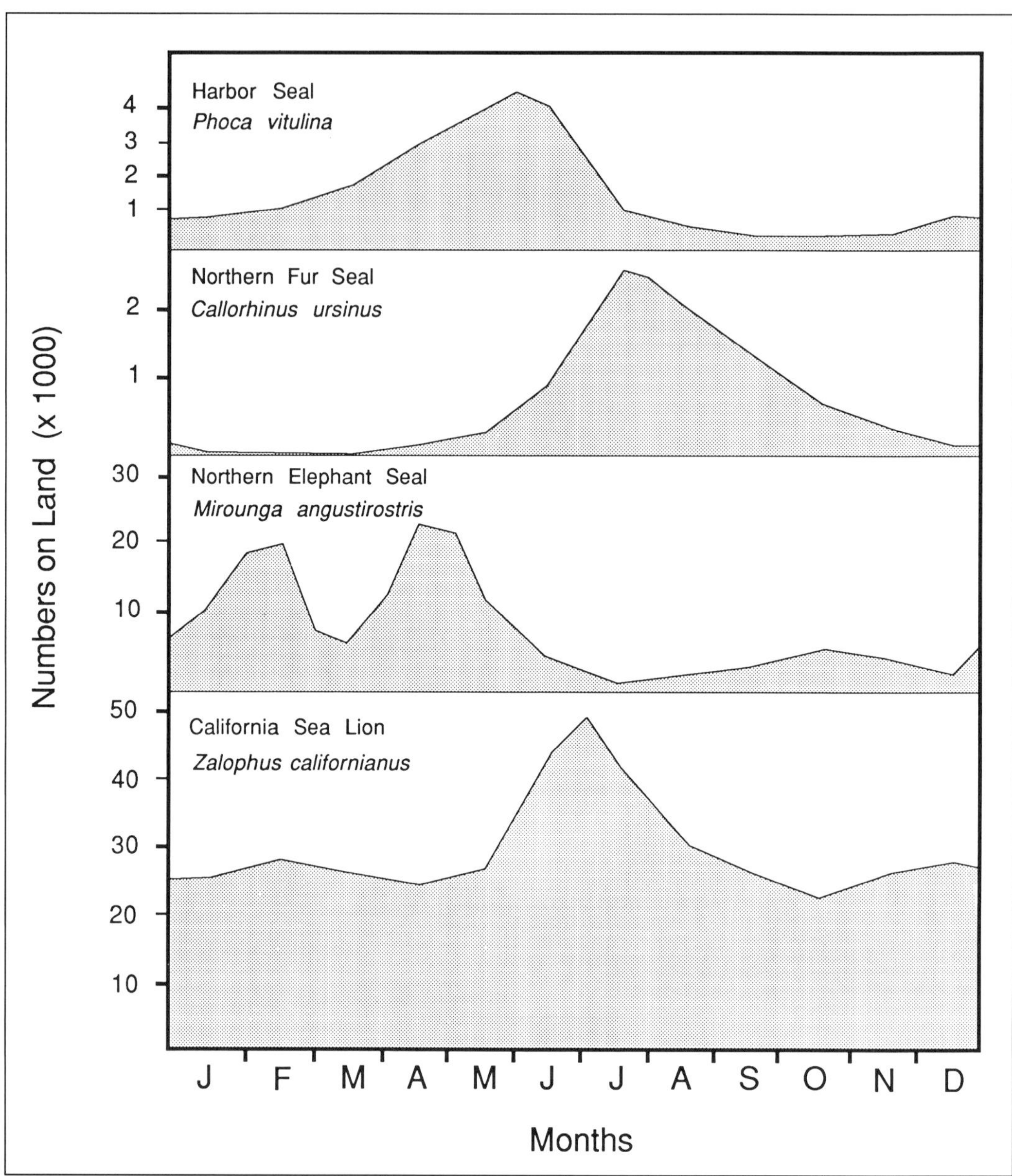

Figure 11.18. Annual cycle of abundance of pinnipeds on land in the SCB. (Redrawn from Bonnell et al. [1981] and adjusted to 1982 breeding population.)

spring and autumn as migrants occupy SCB waters. Overall, feeding by baleen whales accounts for 18.5% of the total food consumption by marine mammals in the SCB.

The proportion of various food items in the diet of marine mammals in the SCB is still poorly known, especially for cetacean species. However, we can estimate the approximate impacts on general groups of prey species. For our calculations, we assume that blue whales consume only euphausiids, while small schooling fish constitute 33% of the diet of fin, minke, and humpback whales. The diet of common toothed whales, except

Table 11.4. *Monthly Food Consumption of Common Marine Mammals in the SCB*[a]

Species	Amount Consumed (metric tons per day)											
	Jan	Feb	Mar	Apr	May	Jun	Jul	Aug	Sep	Oct	Nov	Dec
Blue whale	0	0	0	52.8	110.0	220.0	88.0	88.0	88.0	110.0	57.2	0
Fin whale	0	0	0	73.6	147.2	224.0	89.6	89.6	89.6	160.0	80.0	0
Minke whale	24.0	24.0	24.0	30.0	36.0	42.0	36.0	36.0	36.0	80.4	39.6	24.0
Humpback whale	0	0	28.0	56.0	56.0	78.4	0	0	56.0	78.4	56.0	28.0
Common dolphin	83.2	80.2	77.2	84.0	90.7	152.2	153.7	252.7	278.2	302.2	228.7	236.3
Northern right-whale dolphin	27.6	7.8	69.6	14.4	91.8	15.0	0	0	0	0	3.0	15.0
Pacific white-sided dolphin	20.6	0.5	38.5	37.6	37.6	36.7	72.4	108.1	86.5	10.3	85.5	52.6
Risso's dolphin	26.2	1.6	47.2	31.5	36.7	126.0	15.7	5.3	346.5	42.0	341.2	210.0
Dall's porpoise	7.9	10.5	0.5	0.9	0.9	1.3	1.3	1.3	1.3	1.3	2.6	5.2
Pilot whale	21.6	14.4	14.4	14.4	7.2	14.4	10.8	14.4	14.4	10.8	14.4	21.6
Bottlenose dolphin	5.8	4.9	3.7	3.1	3.1	21.5	20.9	6.1	20.3	0.9	1.8	3.7
California sea lion	390.4	404.8	388.0	371.2	388.0	322.4	463.2	344.0	250.4	358.4	358.4	388.0
Northern elephant seal	24.0	82.2	161.4	82.2	24.1	161.4	125.3	89.6	52.0	13.6	13.6	13.6
Northern fur seal	2.1	2.1	2.1	2.1	2.1	3.5	3.5	3.5	3.5	2.1	2.1	2.1
Harbor seal	24.4	24.4	24.4	24.4	24.4	24.4	24.4	24.4	24.4	24.4	24.4	24.4
Totals	657.8	657.4	879.0	878.2	1055.8	1443.2	1104.8	1063.0	1347.1	1194.8	1308.5	1024.5

[a] Estimates calculated as product of mean monthly population within 60,000 km^2 area inshore of the 2000-m isobath, average individual body mass, and food consumption rate expressed as percentage of body mass per day; for pinnipeds, separate estimates have been combined for lactating and nonlactating components in foraging population. Values of body mass and food consumption rate taken from published estimates (see text).

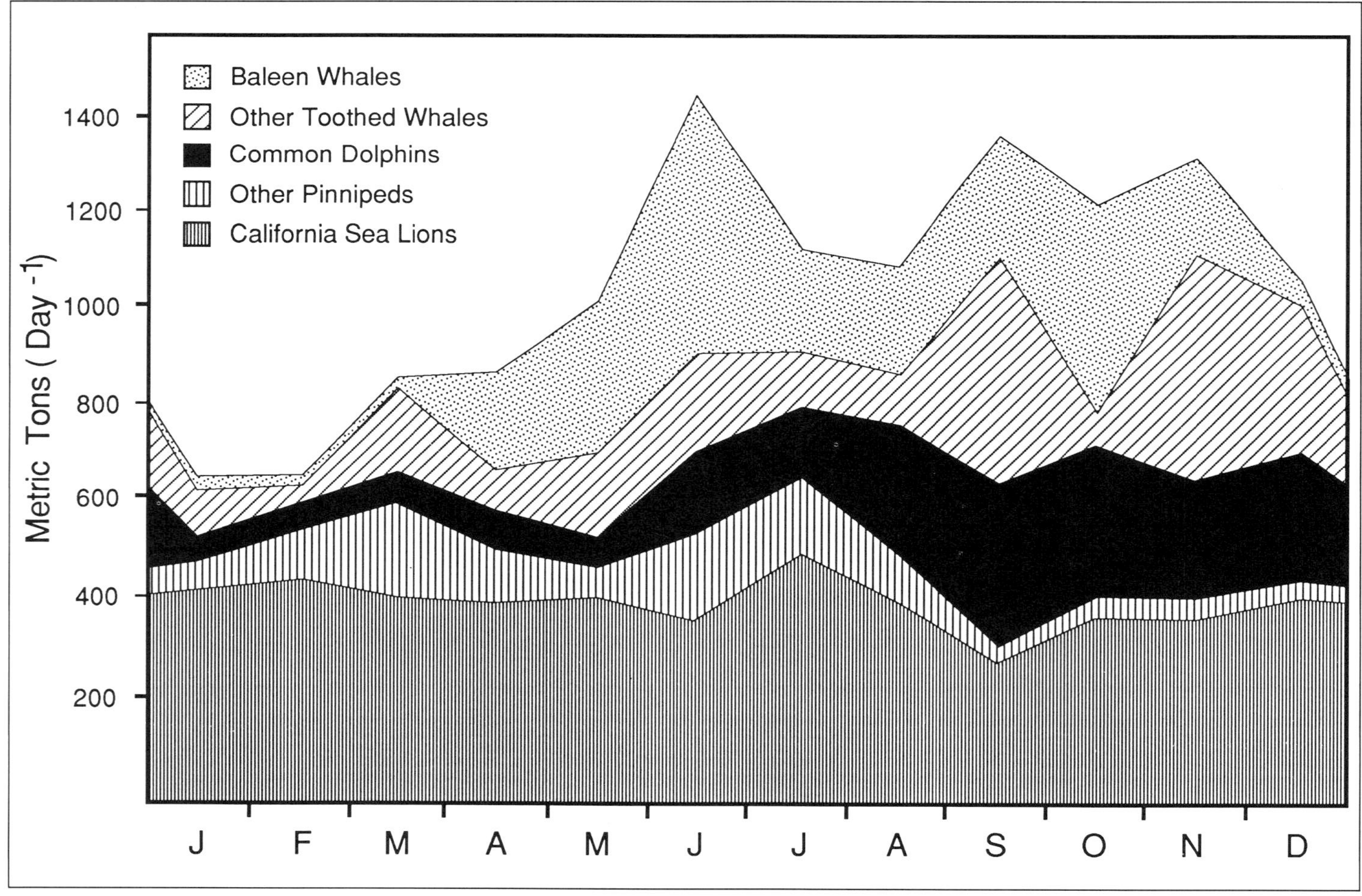

Figure 11.19. Food consumption of marine mammals in the SCB. Upper line represents total food consumption by all groups combined (metric tons per day).

for pilot whales and Risso's dolphins which consume only squid, are assumed to consist of 50% fish and 50% squid. The diet of California sea lions includes all three groups, with approximately equal proportions of fish and squid and 5% pelagic red crabs. Northern elephant seals are assumed to consume equal proportions of fish and squid, while harbor seals consume 90% fish and only 10% cephalopods (octopus). Northern fur seals consume 65% fish and 35% pelagic squid. These estimates are only reasonable approximations based on data presented in the previous section; it is likely that proportions of food types in the diet vary from place to place and from season to season. The impact of foraging marine mammals on crustaceans, cephalopods, and fish in the SCB is shown in figure 11.20.

While some overlap exists, fish and cephalopods are assumed to represent a higher trophic level than crustaceans. Consumption of fish, which include anchovies, Pacific whiting, rockfish, mackerel, lanternfish, and others, remains quite constant through the year. Cephalopods, which include mostly squid (both neritic and pelagic species), are consumed throughout the year, with only a slight increase during the winter. Two exceptional peaks in consumption of squid occur in the autumn as Risso's dolphins reach their maximum abundance in inshore waters of the SCB. Consumption of crustaceans (mostly euphausiids, copepods, and pelagic red crabs) shows peaks in the spring and autumn during migration of whales through the SCB.

Natural Mortality and Environmental Concerns

Much of the information gathered about natural mortality in marine mammals comes from stranded or beach cast animals. Six species of pinnipeds strand along the California coast, and of these species, California sea lions, northern elephant seals, and harbor seals comprise over 90% of the total. In 1983 and 1984, a total of 1775 stranded pinnipeds were reported in the SCB (Seagars et al. 1986). The largest number of these were from Los Angeles (44%), Orange (29%) and San Diego (16%) counties. The California Marine Mammal Stranding Network (CMMSN) lists the various causes of pinniped strandings as disease, human related (clubbed, shot, etc.), incidental fishing mortality, and miscellaneous.

Explanations of cetacean strandings have been primarily a combination of fact and speculation. It is commonly assumed that both single and multiple strandings are caused by disease, impairment, social or environmental disturbance, or accident (Cowan et al. 1986). Cetacean strandings along the California coast are nearly all single in nature, whereas multiple strandings are more common on the east coast of the United States. From 1983 to 1987 the CMMSN reports 429 cetacean strandings in California. The network lists 23 reported species, with harbor porpoises, common dolphins, gray whales, bottlenose dolphins, and Pacific white-sided dolphins making up 83% of the total (D. Seagars pers. comm.).

For 1983 and 1984, Seagars et al. (1986) listed 80 stranded animals in the SCB out of a total 225 for the entire state. These authors list the type of stranding occurrences as unknown, human related, and gill net, but no necropsies were performed and histopathology was not examined. However, Cowan et al. (1986) carried out extensive necropsy work on 51 cetaceans collected from Point Mugu to Huntington Beach. They conclude that many of the diseases found in these animals are compatible with life and that only those lesions in certain critical areas would cause problems and result in stranding. Obviously, stranded animal samples would preclude certain catastrophic diseases as the cause of immediate death, for example, massive myocardial infarct and massive brain hemorrhage in critical areas. These animals would not survive long enough to reach the beach.

Given the large number of factors working against the survival of marine mammals

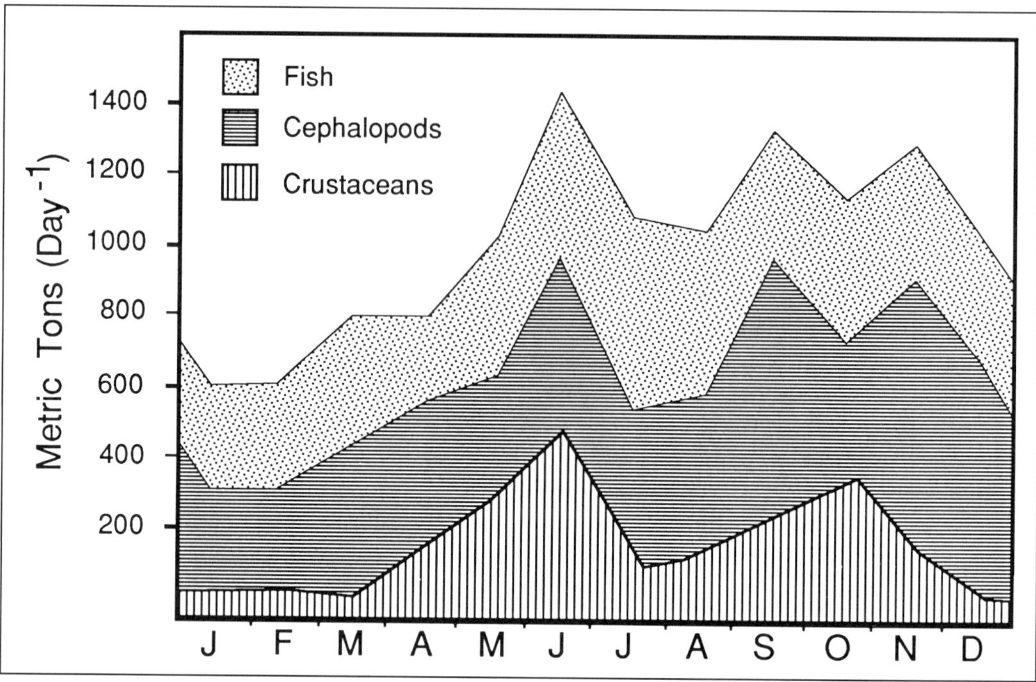

Figure 11.20. Food resources needed to sustain the marine mammal population of the SCB. Upper line represents total food consumption per day (metric tons).

throughout their lives, (predators, accidents, humans, contamination, and disease), it is logical to assume that high natural mortality plays an important part in the control of populations.

Parasitism and Disease

CETACEANS

The causes of disease in cetaceans can be attributed to several major groups of organisms: viruses, fungi, bacteria, and parasites (protozoan and metazoan). Little is known of the role viruses play in the mortality of free-swimming cetaceans. However, fungi, bacteria, and parasites have all been reported from both free-ranging and stranded animals (Dailey 1985; Cowan et al. 1986).

According to Sweeney et al. (1976), systemic fungal disease has been reported as the cause of death from a variety of cetacean species. The fungi appear to be introduced into the nearshore water through soil runoff, and they are presumed to be a normal part of the fauna of ocean waters throughout the world. Fungal organisms are opportunists whose presence can lead to a symptomatic condition when the host is compromised by other conditions. Given these facts, it is unlikely that systemic mycotic diseases play a primary role in mortality of cetacean populations.

Bacterial problems, however, do appear to be a major cause of serious diseases found in cetaceans. Howard et al. (1983) listed bacterial infections of the skin, respiratory system, and urogenital system; bacteria was also implicated as the cause of septicemia and abscess. In their study, the largest number of isolates of bacteria from cetaceans was 27 from pneumonic lungs, followed by 13 from septicemias, and 3 from abscesses. *Pseudomonas* sp. was the most common bacterial type isolated from cetaceans with bacterial pneumonia. A list of bacterial isolates from cetacea is given by Dailey (1985).

Parasitism by helminths as a factor in the natural mortality of cetaceans seems to be limited to the nematodes and trematodes (Walker et al. 1984). Perrin and Powers (1980) reported quantification of mortality related to the nematode *Crassicauda* sp. in spotted dolphins (*Stenella attenuata*) in the eastern tropical Pacific. The authors stated that estimates of natural mortality of the spotted dolphin in the eastern tropical Pacific range from 7 to 9% per year. By using the dentinal layer deposition as an estimate of age (one layer per year), the rounded-off estimates of nematode-induced annual mortality are 1% per year for animals with eight or more dentinal layers and 3% per year for those with more than five layers. They concluded that these estimates represent 11–14% of the natural mortality of the population, and they supported the concept that parasitism constitutes a major factor in natural mortality of small cetaceans. Geraci et al. (1978) also concluded that worms of the genus *Crassicauda* may cause reproductive failure in Atlantic white-sided dolphins (*Lagenorhynchus acutus*) through invasion of mammary glands (fig. 11.21).

The occurrence of trematode parasites in the brain of cetaceans as a probable cause of mortality in southern California was first reported by Ridgway and Dailey in 1972. Subsequent studies in southern California (Schroeder et al. 1973; Dailey and Walker 1978; Cowan et al. 1986) concluded that the genus *Nasitrema* is implicated in the pathology found in the inner ear and brain. The damage to the cetacean nervous system caused by the migrating parasite is found in both the cerebrum and cerebellum. Fresh lesions were found to be convoluted burrows, in which eggs had been deposited (Cowan et al. 1986). The destructive lesions in and around the acoustic nerve, both inside the dura in the regions of the auditory ossicles and immediately inside the nerve, are considered an indication of where the parasite enters the brain (Cowan et al. 1986). Morimitsu et al. (1987) report *Nasitrema* sp. and *Nasitrema gondo* as the cause of mass strandings in the pilot whale and false killer whale, respectively. They state that parasitogenic eighth cranial neuropathy was directly responsible for two live mass strandings at Miyazaki and Iki, Japan.

PINNIPEDS

Several viruses have been found in southern California pinnipeds. Serotypes of calciviruses (San Miguel Sea Lion Virus, or SMSV) have been isolated from vesicular lesions and aborted fetuses of California sea lions, northern fur seals, and elephant seal pups (Smith et al. 1973). In addition to causing lesions in pinnipeds, the SMSV has been implicated as a probable cause of premature parturition in California sea lions. Twenty percent of the California sea lion pups born on San Miguel Island die due to premature parturition. Specimens collected (Gilmartin et al. 1976) from premature partus animals resulted in the recovery of SMSV and contaminants. The DDE levels of the prematurely parturient cows' blubber and livers were 7.6 and 4.8 times greater, respectively, than corresponding tissue concentrations in full-term animals. Polychlorinated biphenyl residues were 4.4 and 3.8 times greater in the blubber and livers of aborting animals, respectively, than they were in the same tissues of full-term sea lions. Premature partus females had tissue imbalances of mercury, selenium, cadmium, and bromine. Gilmartin et al. (1976) suggested an interrelationship of disease agents and environmental contaminants as the cause of premature parturition.

Britt and Howard (1983) reported on viral hepatitis from five California sea lions stranded along the coast of Los Angeles County. They indicated that this viral disease is lethal to marine mammals, but just what role it is playing in the population is still unknown.

Bacterial infections of the respiratory tract and lungs are a major health problem in pinnipeds found in southern California and constitute a source of significant mortality (Sweeney 1973). In diseased California sea lions stranded on southern California beaches, bacterial infections appear to affect mainly

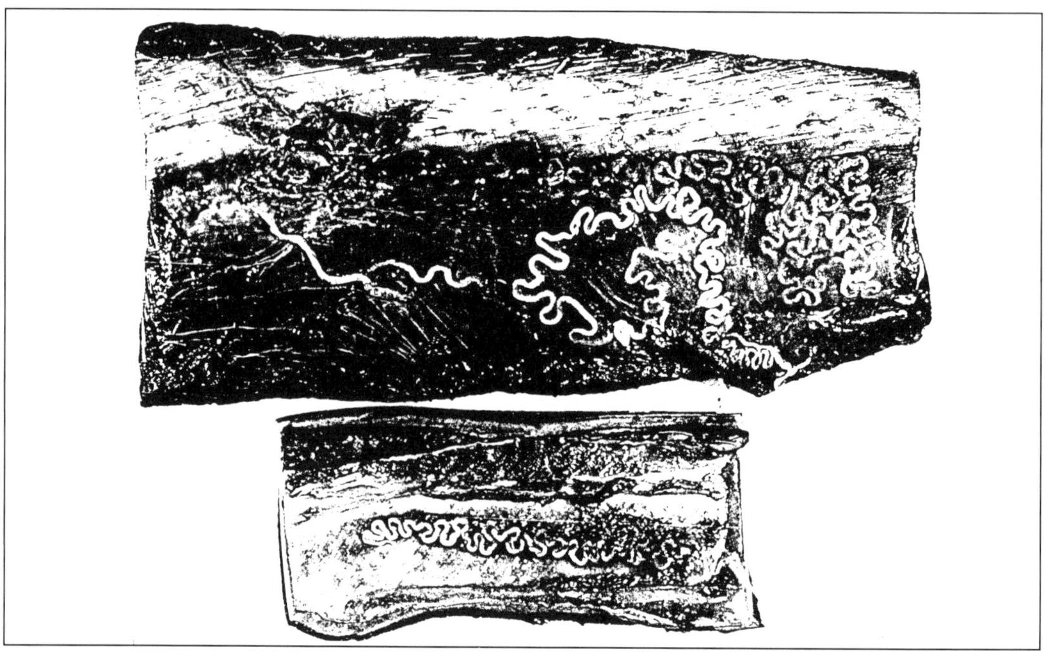

Figure 11.21. *Crassicauda* sp. in tissue of *Lagenorhynchus acutus*. (Photograph: courtesy of New England Aquarium and J. R. Geraci.)

young animals. During an examination of 51 stranded sea lions, Sweeney and Gilmartin (1974) found pneumonia–pneumonitis in 39 cases. In sea lions under 2 years of age, bacterial pneumonia is frequently associated with lungworm infection (Lauckner 1985). At least 11 different strains of bacteria have been isolated from lungs of California sea lions suffering from a lungworm infection of the nematode *Parafilaroides decorus*. The most common of these were *Escherichia coli* and *Klebsiella pneumoniae* (Lauckner 1985).

Girella nigricans (opaleye), a fish found in tidepools at the Channel Islands rookeries, facilitates the transmission of *P. decorus*. These fish feed upon sea lion excrement that contains the larval parasite, then are themselves fed upon by young sea lions learning to catch fish (Dailey 1970) (fig. 11.22). The larvae of these nematodes induce the cells that line the trachea and bronchi to secrete such large amounts of mucus that the young sea lions have difficulty breathing. In the lungs, these worms inhabit the alveolar sacs, where they may cause the formation of verminous granulomas. The combination of these respiratory problems plus the stress of rookery life appears to be one reason why large numbers of emaciated, beach cast California sea lions are found along the beaches of the SCB each year.

Stranded (beached) phocids, such as the northern elephant seals, appear to suffer primarily from stomach worms. These nematodes normally attach to the stomach mucosa, often in clusters that cause large ulcers. The pathology produced is considerable, often resulting in gastritis, enteritis, diarrhea, dehydration, and anemia (Blessing and Peitz 1970; Ridgway et al. 1975). Hookworm (*Uncinaria* sp.), heartworm (*Depetalonema spirocauda*), and another respiratory worm (*Otastrongylus circumlitus*) also have been implicated as possible problems for the pinniped population of the western United States. Although all these worms are found in pinnipeds in the SCB, the effect they may have on their populations is not known at this time.

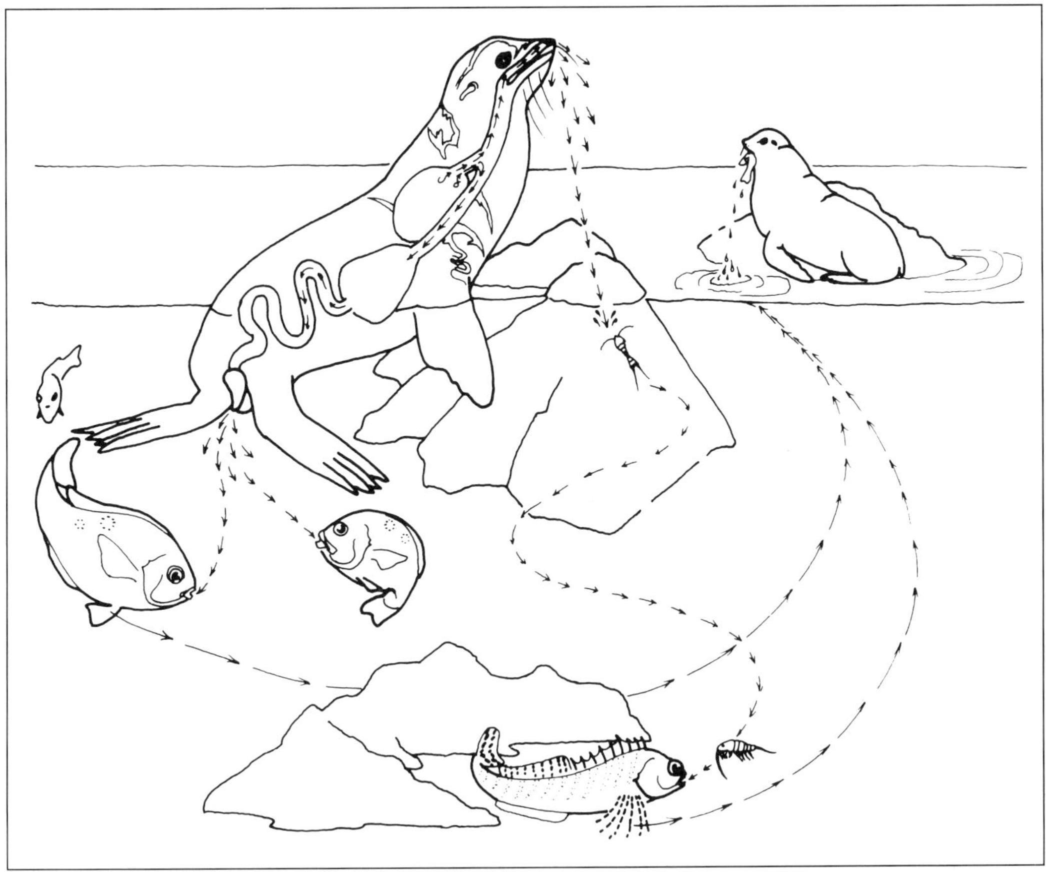

Figure 11.22. Life cycle of the sea lion lungworm *Parafilaroides decorus*. (Drawing by M. D. Dailey.)

SEA OTTERS

Pathological changes found in sea otters dying with symptoms of hemorrhagic enteritis show severe hyperaemia of the villi and desquamation of the mucosal epithelium. The otters become weak and cannot stand or walk and they are semicomatose or very lethargic. This is primarily caused by *Clostridium* sp. (thought to be present in all otters), which becomes virulent when the animal is subjected to stress such as starvation or nervous tension (Kenyon 1969). Rausch (1953) termed the disease (gastroenteritis) "idiopathic" (arising spontaneously). Otter populations in California also have been found affected by the fungus *Coccidioides immitis* (Sweeney 1978; Cornell et al. 1979).

A number of parasites infect sea otters, but only a few have been rated as having a pathogenic effect. A trematode (*Micropallus pirum*) and larval nematodes (*Pseudoterranova decipiens*) have been identified as the cause of death in these animals by Rausch (1953) and Fay et al. (1978). Hennessy and Morejohn (1977) blame another parasite in the order Acanthocephala (*Polymorphus kenti*) for causing necrotic ulcerations that perforate the intestinal wall and contribute to the death of the otter. Since the acanthocephalans are the most common parasite in the otters translocated to San Nicolas Island, we believe that these parasites undoubtedly present a possible problem for the newly introduced population.

Predation

Among cetaceans, predation appears to play a relatively minor role in regulating population size. While cetaceans have few predators, the most notable are sharks, killer whales, and false killer whales (Evans 1987). Studies indicate that shark attacks have been evident on 3–18% of cetaceans examined (Norris and Dohl 1980a, b). Perryman and Foster (1980) report that the false killer whale was the predator most often seen chasing or attacking dolphins during fishing operations in the eastern tropical Pacific. Dwarf sperm whales may also prey on dolphins in nets (Scott and Cordaro 1987). Since attacks on dolphins by small whales are not commonly reported, it may be that the unique situations created by the purse-seine fishery cause this form of predation. Killer whales have been observed attacking gray, minke, and humpback whales, but instances of predation appear to be relatively uncommon (Hancock 1965; Baldridge 1972; Whitehead and Glass 1985). Sharks and killer whales prey upon pinnipeds (Rice 1968; Le Boeuf et al. 1982). In central California, white sharks (*Carcharodon carcharias*) are clearly the most significant predator of northern elephant seals and probably harbor seals as well (Ainley et al. 1981; Le Boeuf et al. 1982). In the SCB, white sharks are known to prey on California sea lions (Scholl 1983). Attacks by white sharks may account for 9–15% of all mortality to the sea otter population in central California (Ames and Morejohn 1980).

Fisheries Conflicts

Marine mammals share the marine environment with humans, and in the SCB, perhaps more than anywhere else in the eastern North Pacific, both compete for the same resources. Thus far, marine mammal populations continue to grow and do not appear to be food limited in years of normal ocean temperature conditions. However, some evidence suggests that little surplus of food exists, and prey switching by marine mammals cannot fully compensate for reductions in abundance of their principal prey species. During El Niño, reduced availability of food may have caused some cetacean species to undergo shifts in distribution. It also seriously affected foraging success and pup survival for breeding populations of pinnipeds that could not relocate. At a scale similar to El Niño, human fishing activities may result in reductions in abundance of food available to marine mammals, with corresponding impacts. Conversely, release of some fish stocks from predation pressure by fisheries may have a positive effect on population growth of some marine mammal species (Ainley et al. 1982).

The interrelationship of fisheries and marine mammal populations is not well understood, and any investigation of the relationship may require construction of a dynamic systems model. What we do know is that the balance is rather delicate. The food needs of present marine mammal populations are great, and the possibility exists that unregulated fisheries can place sufficient strain on the system to cause some populations in the SCB to become food limited. Viewed from another perspective, populations of some marine mammal species may be greater than desirable, given other uses of the environment by humans.

Direct impacts of fisheries on marine mammals appear to result from accidental entanglement in nets and shooting or harassment of marine mammals entering areas where fishing is underway. DeMaster et al. (1985) reviewed conflicts between marine mammals and fisheries off California (see also DeMaster et al. 1982). Most of the take of marine mammals each year results from the shooting of California sea lions in the commercial salmon trolling fishery (about 300 sea lions in 1981) and the drowning of probably 600-1200 California sea lions per year in the shark-drift gill-net fishery (Miller et al. 1982). Many harbor seals, harbor porpoises, and seabirds are taken each year in the halibut gill-net fishery off central California (Miller et al. 1982). In the SCB, substantial take of sea lions and pilot whales occurs in the squid round-haul fishery near Santa Catalina Island. Miller et al. (1983b) estimated the take of pilot whales to be 4–12% of the population per year. Gill-net mortality of sea otters in their central California range is estimated to have exceeded annual recruitment over most of the last decade (Wendell et al. 1985). The result was a net decline in numbers of sea otters until gill-net fishing in sea otter habitat was prohibited by law in 1985.

Another fisheries conflict clearly exists when marine mammals remove fish from nets or long lines or take anadromous fish entering rivers or locks. Taking of fish caught in nets or on hook and line is almost exclusively perpetrated by California sea lions. California sea lions and harbor seals are the predominant predators of anadromous fish entering rivers and locks. While many methods have been used to discourage this foraging behavior (e.g., "seal bombs" to startle pinnipeds, and agents placed in dead fish to create an aversion response in animals that eat them), none have been very effective. Ultimately, the few animals that are engaging in this activity may either have to be captured for aquaria or otherwise removed from the population.

Environmental Contaminants

Much attention has been focused on the presence of environmental contaminants in the marine environment and the potential harm they might cause to marine mammals (Holden and Marsden 1967; Risebrough et al. 1968). While some tenuous conclusions have been drawn on the basis of effects on other animals, we have not yet clearly demonstrated the impacts of most environmental contaminants on marine mammals. This does not mean, of course, that effects postulated by researchers are not likely to occur. What it does mean is that the effects of environmental contaminants are difficult to distinguish from natural variation in numbers. Death and stranding of adults, juveniles, and pups, together with abortion of fetuses, are the most visible signs of serious problems for marine mammals. Decline in population size (e.g., declines in northern fur seal, Steller's sea lion, and California sea otter populations) may slowly continue over a decade or more before we become convinced that the problem is real and begin to search for a solution.

Marine mammals incorporate environmental contaminants through their food. Generally, the higher up the food chain a marine mammal feeds, the greater the concentrations of contaminants to which it will be exposed through its food and the greater the concentrations that are likely to be found in the animal's tissues. To reduce the concentrations of contaminants and forestall problems that may occur later, efforts must be made to minimize the input of contaminants into the SCB ecosystem. Unlike the open coast, where currents can dilute and diffuse pollutants, the SCB has a residence time for water that is on the order of 2–3 months, which allows contaminants to become increasingly concentrated (Jones 1971). Gaskin (1982) noted that if one protects the habitat, animal populations can show remarkable resilience in the face of other external pressures.

Three major groups of environmental contaminants exist that may presently affect marine mammals in the SCB: chlorinated hydrocarbons such as DDT and PCB, heavy metals such as mercury, and oil pollution. A few heavy metals occur naturally in sediments, and marine mammals may have evolved mechanisms to partially detoxify some of them (Piotrowski et al. 1977). Natural seeps are known to occur in the Santa Barbara Channel of the SCB and in the Beaufort Sea. Natural seeps have not been connected with mortality of marine mammals, but it must be assumed that this is because most resident marine mammals have learned to avoid them.

Most input of pollutants into the marine environment comes from agricultural runoff, industrial waste, sewage outfalls, pumping the bilges of ships, and other anthropogenic sources. A wide variety of chlorinated hydrocarbons have been identified in marine mammal tissues, including DDT, DDE, DDD, dieldrin, endrin, heptachlor, isomers of chlordane, and toxaphene (Holden 1975; O'Shea et al. 1980). Polychlorinated biphenyls (PCB) are also known to be widespread in the marine environment (Risebrough 1971). Researchers have found high concentrations of chlorinated hydrocarbons in tissues of marine mammals of the SCB, including common dolphins, Pacific white-sided dolphins, Dall's porpoises, pilot whales, bottlenose dolphins, harbor seals, sea otters, and California sea lions (Le Boeuf and Bonnell 1971; Shaw 1971; DeLong et al. 1973; MacGregor 1974; O'Shea et al. 1980). The highest reported concentrations of DDT in any marine mammal occurred in tissues of bottlenose dolphins inhabiting coastal waters of the SCB (O'Shea et al. 1980).

The principal source of DDT contamination in the SCB was traced to local release into the Los Angeles sewer system from a major pesticide manufacturing plant (see chap. 3). Discharge from this plant was halted in 1970, although residues continued to wash out of the system in decreasing amounts for years. MacGregor (1974) described the time course of contamination of waters of the SCB. During this same period, levels of DDT and metabolites in gray whales were quite low compared to those found in resident marine mammals; however, the disparity in levels may be attributable to the fact that little feeding by gray whales occurs during migration through California and Mexican waters (Wolman and Wilson 1970).

All chlorinated hydrocarbons are lipid soluble and become stored in fats of marine mammals. Compared to other marine organisms, marine mammals are long-lived; therefore, accumulation of chlorinated hydrocarbons in the fat and other tissues may be a function of age (Gaskin 1982). Chlorinated hydrocarbons have been linked to reproductive failure in California sea lions in the SCB (DeLong et al. 1973; Gilmartin et al. 1976), ringed seals in Sweden (Helle et al. 1976), and several other species of mammals and birds.

Heavy metals such as mercury, cadmium, and lead vary in toxicity. Some are bound in the form of metalloproteins or present as less toxic salts (e.g., mercurous chloride); others are very toxic (e.g., mercuric chloride and methyl-mercuric chloride). Unlike the chlorinated hydrocarbons, heavy metals are not deposited in fat and tend to find their way into the liver and kidneys. The concentrations of heavy metals among marine mammals are highly variable according to the characteristics of the waters the animals inhabit. Most studies have been conducted elsewhere, and little is known concerning the levels of heavy metals in marine mammals of the SCB. Levels of heavy metals in California sea lions have been measured, however, and a possible association with premature births reported (Braham 1973; Buhler et al. 1975; Martin et al. 1976). Sea otters in central California have been found to have relatively high levels of cadmium, but no deleterious effects have been observed (Martin 1979; Risebrough 1984).

The presence of oil and oil products in the marine environment presents different problems. Oil can directly affect marine mammals by contact with the skin, fur, or baleen. Furthermore, the potential exists for ingestion of oil as a fur seal or sea otter grooms its fur—or as a pinniped nurses its pup. Englehardt (1983) and Geraci and St. Aubin (1988) provided reviews of potential effects of oil on marine mammals. The most notable effects on cetacean species would probably be the fouling of baleen plates or the inhalation of oil or volatile components (Geraci and St. Aubin 1988). Some information on potential dangers can be obtained from laboratory studies. Such work indicates that fouling of baleen can occur, but that clearance of oil on baleen plates takes place quite rapidly (Geraci and St. Aubin 1985a, b). A greater threat to the health of a whale or dolphin may result from irritation of the lungs or incorporation of hydrocarbons into tissues following inhalation of vapors above an oil slick (Geraci and St. Aubin 1988).

Adverse effects of oil on northern fur seals and sea otters have been demonstrated under laboratory conditions, where oiling over 25–30% of the body surface resulted in severe hypothermia and an elevated metabolic rate in typical water temperatures. These fur-bearing species are the only marine mammals that depend on the integrity of an air layer trapped in the fur to provide insulation and buoyancy. Researchers concluded that the oil contamination would cause death of the animal unless the fur was cleaned and restored to its original insulative capability (Kooyman et al. 1976, 1977; Costa and Kooyman 1980, 1982). The Exxon *Valdez* oil spill, in which over 1000 sea otters died in Prince William Sound, leaves little doubt about the sensitivity of these animals to oil contamination.

Acute toxic effects of oil ingestion during grooming have been observed in polar bears (Oritsland et al. 1981) and river otters, *Lutris lutris* (Baker et al. 1981). Presumably, other marine mammals that groom their fur, such as fur seals and sea otters in the SCB, could experience similar effects. Variable reactions to ingestion of oil, ranging from relatively rapid recovery to sickness and death, have been reported in common (harbor) seals, gray seals, and ringed seals (Duguy and Babin 1976; Davis and Anderson 1976; Geraci and Smith 1976; Englehardt 1982). Geraci and St. Aubin (1988) conclude that marine mammals subject to other kinds of internal or environmental stress might be most susceptible to effects of oil exposure.

Summary and Prospectus for Future Research

At least 20 species of marine mammals regularly use the SCB, although several are migratory and simply pass through the area. An additional 14 species have been seen infrequently in the SCB or are known from stranding records. The only true migrants that pass through the area are baleen whales, the most common of which are the blue, fin, humpback, minke, and gray whales. While a few of each species may be present at any time of the year, most occur in the SCB during migrations between high-latitude feeding grounds and subtropical calving grounds. The balaenopterid whales (blue, fin, sei, and minke) typically migrate along the shelf break over the 2000 m isobath, while humpback whales move through the SCB closer inshore over the continental shelf. The gray whale migration pathway extends offshore to the continental slope, but about one-half of the total migrant population typically remains within 15 km of shore.

Peak numbers of blue, fin, sei, minke, and humpback whales occur in the area during migration in June and October. However, small numbers of blue, fin, and minke whales spend the summer in the SCB, feeding on euphausiid crustaceans and small schooling fish over the shelf break. Minke whales occur in the winter as well, and the SCB may be part of their calving range. Gray whales are present continuously from about December through May, appearing about the same time as southbound migrants passing through the SCB en route to the calving lagoons of Baja California. Later, in two phases, gray whales migrate northward to feeding grounds in the Bering Sea. Most northbound gray whales pass through the SCB in February and March; mothers and calves come through the area later in the spring. Gray whales feed sparingly in the SCB.

Seven species of odontocetes or toothed whales regularly inhabit the SCB on a year-round or seasonal basis. Because of their numbers, the most ecologically significant species in the SCB are the common dolphin, the Pacific white-sided dolphin, the northern right-whale dolphin, Risso's dolphin, and Dall's porpoise. None is endemic to the SCB; they use the area as part of a larger North Pacific range. Most reach peak abundance in the SCB in the autumn months of highest water temperature. Common dolphins, which are found from equatorial waters to southern California, are by far the most abundant in the area. With a population of more than 60,000 animals present in summer and autumn, common dolphins account for nearly three-fourths of all cetaceans in the SCB.

Six species of pinnipeds occur in the SCB, including the Guadalupe fur seal (still rare) and the Steller's sea lion (straggler of a declining population). Two others, the northern fur seal and northern elephant seal, breed on the outermost islands of the SCB, but have vast foraging ranges in waters of the California Current. Their impact on the more coastal waters of the SCB is probably quite small. California sea lions have a population of more than 63,000 in the SCB during the summer and autumn. Pacific harbor seals have a small but growing population of about 5000 animals. California sea lions are probably the single most important predator on fishes and squid in the SCB, accounting for an estimated 38% of all food consumption in the area by marine mammals.

Over the year, food consumption by marine mammals totals about 1000 t d^{-1}. The baleen whales represent about 40% of all marine mammal biomass in the SCB, but account for only about 18% of the total impact of marine mammals on local food resources. Toothed whales as a group, which represent 18% of all marine mammal biomass in the SCB, account for 33% of all food consumption. Their diet is composed of a wide variety of fish and squid. Pinnipeds represent about 42% of the total marine mammal biomass of the SCB. However, they account for nearly one-half of the total food consumption by marine mammals in the area.

Through the use of satellite imagery, the seaward boundary of upwelled inshore waters can often be recognized as a color-temperature front (Bernstein et al. 1977; Smith and Baker 1982). Smith et al. (1986) used satellite maps of sea surface chlorophyll to describe some habitat characteristics of cetaceans observed on NMFS surveys of the California Current area. They found that cetaceans occurred to a significantly greater extent in upwelled coastal waters along the central California coast than in the less productive oceanic waters. Though still exploratory in nature, the work of Smith et al. (1986) points the way toward a characterization of marine mammal habitat using remote sensing of sea surface features.

Most species of dolphins and porpoises exhibit offshore–inshore shifts in different seasons and cannot be easily classified as belonging to either domain. The common dolphin, for example, is typically considered a neritic form, but during autumn the inshore population is joined by pelagic dolphins that otherwise do not occur in the SCB. Many odontocetes may have population segments using both inshore and offshore waters to optimize exploitation of food resources. Among pinnipeds, California sea lions and harbor seals appear to be predominantly neritic forms, although California sea lions are occasionally seen as far west as Rodriguez Seamount.

Among the inshore or coastal fauna, there are a few species that are found almost exclusively in nearshore waters. The most notable are the sea otters and harbor porpoises. Bottlenose dolphins and pilot whales also have populations or at least population components in the SCB that appear to be nearshore in distribution.

While we know a great deal about the habits of marine mammals using the SCB, there is much yet to be learned. Surveys conducted in the 1970s provide an excellent picture of how marine mammals were distributed in a time of apparently normal oceanic conditions. But scant data suggest that some populations may have changed their patterns of use of the SCB since that time. Given events on the scale of El Niño, we really have little understanding of the range of natural variability that can exist in the abundance and distribution of marine mammals populations. Probably the single most important information need that future research must address is this: how do marine mammal populations respond to changes in oceanographic conditions? To develop any useful description of marine mammal use of the SCB, it will be necessary to monitor marine mammal populations in the area over a period of time sufficiently long to encompass the cyclical changes in environmental conditions that can be reasonably expected to occur (on the order of 10 years).

The development of an understanding of the impacts of marine mammals on their food resources in the SCB may be more remote. While considerable progress has been made studying the food habits and energetics of pinnipeds, very little is known about the diet and energetic requirements of dolphins and porpoises. These data can probably be obtained only by capture of small cetaceans and lavage of stomach contents. And last, we still have an incomplete understanding of the causes of mortality in marine mammal populations that live in an environment increasingly modified by human activities. At a minimum, continued monitoring of levels of environmental contaminants in stranded specimens is essential to alert us to impacts that may be difficult to halt or reverse.

Literature Cited

Ainley, D. G., C. S. Strong, H. R. Huber, T. J. Lewis, and S. H. Morrell, 1981. Predation by sharks on pinnipeds at the Farallon Islands. *U.S. Natl. Mar. Fish. Serv. Fish. Bull.* 78: 941–945.

Ainley, D. G., H. R. Huber, and K. M. Bailey, 1982. Population fluctuations of California sea lions and the Pacific whiting fishery off central California. *U.S. Natl. Mar. Fish. Serv. Fish. Bull.* 80:253–258.

Allen, S. G., 1987. Harbor seals and northern elephant seals in Point Reyes. In: *Proceedings of the Symposium on Current Research Topics in the Marine Environment.* Presented by the Gulf of the Farallones National Marine Sanctuary and Point Reyes National Seashore, and the Environmental Action Committee of West Marin. pp. 16–19.

Ames, J. A., and G. V. Morejohn, 1980. Evidence of white shark (*Carcharodon carcharias*) attacks on sea otters (*Enhydra lutris*). *Calif. Fish Game.* 66:196–209.

Antonelis, G. A., Jr., and R. L. DeLong, 1986. Population and behavioral studies, San Miguel Island, California (Adams Cove and Castle Rock). In: P. Kozloff, ed. *Fur Seal Investigations.* NOAA Tech. Memo. NMFS F/NWC-97. pp. 49–53.

Antonelis, G. A., Jr., and C. H. Fiscus, 1980. The pinnipeds of the California current. *Calif. Coop. Oceanic Fish. Invest. Rep.* 21:68–78.

Antonelis, G. A., Jr., and M. A. Perez, 1984. Estimated annual food consumption by northern fur seals in the California Current. *Calif. Coop. Oceanic Fish. Invest. Rep.* 25:135–145.

Antonelis, G. A., Jr., S. Leatherwood, and D. K. Odell, 1981. Population growth and censuses of the northern elephant seal, *Mirounga angustirostris,* on the California Channel Islands, 1958–78. *U.S. Natl. Mar. Fish. Serv. Fish. Bull.* 79:562–567.

Antonelis, G. A., Jr., C. H. Fiscus, and R. L. DeLong, 1984. Spring and summer prey of California sea lions, *Zalophus californianus,* at San Miguel Island, California, 1978–79. *U.S. Natl. Mar. Fish. Serv. Fish. Bull.* 82(1):67–76.

Antonelis, G. A., Jr., M. S. Lowry, D. P. DeMaster, and C. H. Fiscus, 1987. Assessing northern elephant seal feeding habits by stomach lavage. *Mar. Mamm. Sci.* 3:308–322.

Baker, C. S., L. M. Herman, A. Perry, W. S. Lawton, J. M. Straley, and J. H. Straley, 1985. Population characteristics and migration of summer and late-season humpback whales, *Megaptera novaeangliae,* in southeastern Alaska. *Mar. Mamm. Sci.* 1(4):304–323.

Baker, C. S., L. M. Herman, A. Perry, W. S. Lawton, J. M. Straley, A. A. Wolman, G. D. Kaufman, H. E. Winn, J. D. Hall, J. M. Reinke, and J. Ostman, 1986. Migratory movement and population structure of humpback whales (*Megaptera novaeangliae*) in the central and eastern North Pacific. *Mar. Ecol. Prog. Ser.* 31:105–119.

Baker, J. R., A. M. Jones, T. P. Jones, and H. C. Watson, 1981. Otter (*Lutris lutris*) mortality and marine oil pollution. *Biol. Conserv.* 20:311–321.

Baldridge, A. J., 1972. Killer whales attack and eat a gray whale. *J. Mammal.* 53(4):898–900.

Baldridge, A. J., 1974. Migrant gray whales with calves and sexual behavior of gray whales in the Monterey area of central California, 1967–73. *U.S. Natl. Mar. Fish. Serv. Fish. Bull.* 72(2): 615–618.

Barlow, J., 1987. Abundance estimation for harbor porpoise (*Phocoena phocoena*) based on ship surveys along the coasts of California, Oregon and Washington. NOAA-NMFS-SWFC Admin. Rep. No. LJ–87–05. 36pp.

Bartholomew, G. A., 1950. A male Guadalupe fur seal on San Nicolas Island, California. *J. Mammal*. 31:175–180.

Bartholomew, G. A., 1967. Seal and sea lion populations of the California Islands. In: R. N. Philbrick, ed. *Proceedings, Symposium on the Biology of the California Islands*. Santa Barbara Botanic Garden, Santa Barbara, CA. pp. 229–244.

Bartholomew, G. A., and R. A. Boolootian, 1960. Population structure of the pinnipeds of the Channel Islands, California. *J. Mammal*. 41: 366–375.

Bartholomew, G. A., and C. L. Hubbs, 1960. Population growth and seasonal movements of the northern elephant seal, *Mirounga angustirostris*. *Mammalia*. 24:313–323.

Bernstein, R. L., L. L. Breaker, and R. Whritner, 1977. California Current eddy formation: Ship, air, and satellite results. *Science*. 195(4276):353–359.

Berzin, A. A., and A. A. Rovnin, 1966. The distribution and migrations of whales in the northeastern part of the Pacific, Chukchi, and Bering Seas. *Izvestia TINRO*. 58:179–207.

Best, E. A., 1963. Contributions to the biology of Pacific hake, *Merluccius productus* (Ayres). *Calif. Coop. Oceanic Fish. Invest. Rep*. 9:51–56.

Bigg, M. A., 1973. Census of California sea lions on southern Vancouver Island, British Columbia. *J. Mammal*. 54:285–287.

Bigg, M. A., I. B. MacAskie, and G. Ellis, 1978. Studies on captive fur seals. Progress Report No. 2. Can. Fish. Mar. Serv. Manuscr. Rep. No. 1471, 21pp.

Blessing, M. H., and G. Peitz, 1970. Anisakisbefall in Gefangenschaft gehaltener mariner Sauger. In: Verk. II. Int. Symp. uber die Enkrankungen der Zootiere (Budapest, 1970). pp. 283–286.

Bogoslovskaya, L. S., L. M. Votrogov, and T. N. Semenova, 1981. Feeding habits of the gray whale off Chukotka. *Rep. Int. Whaling Comm*. 31:507–510.

Bolin, R. L., 1938. Reappearance of the southern sea otter along the California coast. *J. Mammal*. 19:301–303.

Bonnell, M. L., 1982a. Flight report: Coastal survey 3–11, Supplemental, 5 July 1982. Institute of Marine Sciences, Univ. of California, Santa Cruz. Unpubl., 5pp.

Bonnell, M. L., 1982b. Flight report: Coastal survey 3–09, Supplemental, 25 January 1982. Institute of Marine Sciences, Univ. of California, Santa Cruz. Unpubl., 5pp.

Bonnell, M. L., 1985. Abundance and distribution of northern fur seals in the pelagic waters of California, 1980–1983. In: *Sixth Biennial Conference on the Biology of Marine Mammals*, Vancouver, B.C.

Bonnell, M. L., and R. G. Ford, 1987. California sea lion distribution: A statistical analysis of aerial transect data. *J. Wildl. Manage*. 51(1):13–20.

Bonnell, M. L., B. J. Le Boeuf, M. O. Pierson, D. H. Dettman, G. D. Farrens, and C. B. Heath, 1981. Pinnipeds of the Southern California Bight. Summary Report, 1975–1978, Part I. Vol. III, Principal Investigator's Reports, Marine Mammal and Seabird Survey of the Southern California Bight Area. Available from: NTIS, Springfield, VA; PB 81–248–71, 535pp.

Bonnell, M. L., M. O. Pierson, and G. D. Farrens, 1983. Pinnipeds and sea otters of central and northern California, 1980–1983: Status, abundance, and distribution. Final Report to U.S. Dept. of Interior, Minerals Management Service, Contract 14–12–0001–29090. OCS Study MMS 84–0044, 220pp.

Bonnot, P., 1928, The sea lions of California. *Calif. Fish Game*. 14:1–16.

Bonnot, P., G. H. Clark, and S. R. Hatton, 1938. California sea lion census for 1938. *Calif. Fish Game*. 24:415–419.

Boolootian, R. A., 1961. The distribution of the California sea otter. *Calif. Fish Game*. 47:287–292.

Boveng, P., 1988. Status of the California sea lion population on the U.S. west coast. NOAA-NMFS-SWFC Admin. Rep. No. LJ–88–07. NOAA, Natl. Marine Fish. Serv., Southwest Fish. Center, La Jolla, CA.

Braham, H. W., 1973. Lead in the California sea lion (*Zalophus californianus*). *Environ. Pollut*. 5(4):253–258.

Braham, H. W., 1984a. The status of endangered whales: An overview. In: J. M. Breiwick, and H. W. Braham, eds. *The Status of Endangered Whales. Mar. Fish. Rev*. 46:2–6.

Braham, H. W., 1984b. Distribution and migration of gray whales in Alaska. In: M. L. Jones, S. L. Swartz, and S. Leatherwood, eds. *The Gray Whale, Eschrichtius robustus*. Academic Press, Orlando, FL. pp. 241–266.

Braham, H. W., and M. E. Dahlheim, 1982. Killer whales in Alaska documented in the Platforms of Opportunity Program. *Rep. Int. Whaling Comm.* 32:643–646.

Braham, H. W. and D. W. Rice, 1984. The right whale, *Balaena glacialis*. In: J. M. Breiwick, and H. W. Braham, eds. *The Status of Endangered Whales. Mar. Fish. Rev.* 46:38–44.

Braham, H. W., R. D. Everitt, and D. J. Rugh, 1980. Northern sea lion population decline in the eastern Aleutian Islands. *J. Wildl. Manage.* 44(1):25–33.

Breiwick, J. M., D. J. Rugh, D. E. Withrow, M. E. Dahlheim, and S. T. Buckland, 1988. Preliminary population estimate of gray whales during the 1987/88 southward migration. *Rep. Int. Whaling Comm.* Paper IWC SC/40/PS12, San Diego, CA.

Briggs, K. T., and C. W. Davis, 1972. A study of predation by sea lions on salmon in Monterey Bay. *Calif. Fish Game.* 58(1):37–43.

Britt, J. O., and E. B. Howard, 1983. Virus diseases. In: E. B. Howard, ed. *Pathobiology of Marine Mammals.* Vol. 1. CRC Press, Boca Raton, FL. pp. 47–67.

Brodie, C. F., 1975. Cetacean energetics, an overview of intraspecific size variation. *Ecology.* 56:152–161.

Brown, D. H., and K. S. Norris, 1956. Observations of captive and wild cetaceans. *J. Mammal.* 37:311–326.

Brown, R. F., and B. R. Mate, 1983. Abundance, movements, and feeding habits of harbor seals, *Phoca vitulina,* at Netarts and Tillamook bays, Oregon. *U.S. Natl. Mar. Fish. Serv. Fish. Bull.* 81(2):291–301.

Brownell, R. L., Jr., 1964. Observations of odontocetes in central California waters. *Nor. Hvalfangst-Tid.* 3:60–66.

Brownell, R. L., Jr., 1977. Current status of the gray whale. *Rep. Int. Whaling Comm.* 27:209–211.

Brownell, R. L., Jr., and R. L. DeLong, 1968. Marine mammals of the eastern grid area. Report, Pacific Ocean Biological Survey Program. Smithsonian Institution, Washington, D.C. 10pp.

Brueggeman, J. J., R. A. Grotefendt, and A. W. Erickson, 1984. Endangered whale abundance and distribution in the Navarin Basin of the Bering Sea during the ice-free period. *Alaska Sea Grant Rep.* 84:201–236.

Brueggeman, J. J., G. A. Green, R. A. Grotefendt, and D. G. Chapman, 1987. Aerial surveys of endangered cetaceans and other marine mammals in the northwestern Gulf of Alaska and southeastern Bering Sea. Report of OCSEAP Research Unit 673 to the Minerals Management Service, Alaska OCS Region, and NOAA Office of Oceanography and Marine Assessment Office, 156pp.

Bryant, H. C., 1915. Sea otters near Point Sur. *Calif. Fish Game.* 1:134–135.

Buhler, D. R., R. R. Claeys, and B. R. Mate, 1975. Heavy metals and chlorinated hydrocarbon residues in California sea lions, *Zalophus californianus. J. Fish. Res. Board Can.* 32(12): 2391–2397.

California Department of Fish and Game, 1976. A proposal for sea otter protection and research, and request for return of management to the State of California. Unpubl. Report, January 1976.

Calkins, D. G., and K. B. Schneider, 1985. The sea otter (*Enhydra lutris*). In: J. J. Burns, K. J. Frost, and L. F. Lowry, eds. *Marine Mammal Species Accounts. Wildl. Tech. Bull. No.* 7. Alaska Dept. Fish Game. pp. 37–45.

Christmann, J., 1983. Common dolphins, *Delphinus delphis* in Monterey Bay. *Calif. Fish Game.* 69(1):56–58.

Condit, R., and B. J. Le Boeuf, 1984. Feeding habits and feeding grounds of the northern elephant seal. *J. Mammal.* 65(2):281–290.

Consiglieri, L. D., and H. W. Braham, 1982. Seasonal distribution and relative abundance of marine mammals in the Gulf of Alaska. Final Report of Research Unit 68, OCS Environmental Assessment Program, NOAA, 204pp.

Cornell, L. H., K. G. Osborn, J. E. Antrim, and J. G. Simpson, 1979. Cacciordomycosis in a California sea otter (*Enhydra lutris*). *J. Wildl. Dis.* 15:373–378.

Costa, D. P. 1978. The ecological energetics, water, and electrolyte balance of the California sea otter, *Enhydra lutris.* Ph.D. Dissertation, Univ. of California, Santa Cruz.

Costa, D. P., 1982. Energy, nitrogen, and electrolyte flux and sea-water drinking in the sea otter, *Enhydra lutris. Physiol. Zool.* 55:35–44.

Costa, D. P., 1986. Assessment of the impact of the California sea lion and the northern elephant seal on commercial fisheries. In: Final Report for California Sea Grant. Biennial Report of Completed Projects 1984–1986. Calif. Sea Grant College Prog., Univ. of California, La Jolla, CA. Publ. No. R-CSGCP–024. pp. 36–43.

Costa, D. P., 1987. Isotopic methods for quantifying material and energy intake of free-ranging marine mammals. In: A. C. Huntly, D. P. Costa, G. A. J. Worthy, and M. A. Castellini, eds. *Approaches to Marine Mammal Energetics.* Soc. Mar. Mammal., Spec. Publ. No. 1.

Costa, D. P. 1988. Methods for studying the energetics of freely diving animals. *Can. J. Zool.* 66:45–52.

Costa, D. P., and R. L. Gentry, 1986. Free-ranging energetics of northern fur seals. In: R. L. Gentry, and G. L. Kooyman, eds. *Fur Seals: Maternal Strategies on Land and at Sea.* Princeton Univ. Press, N.J. pp. 79–109.

Costa, D. P., and G. L. Kooyman, 1980. Effects of oil contamination on the sea otter *Enhydra lutris.* Final Report of Research Unit No. 71, OCS Environmental Assessment Program, NOAA, 18pp.

Costa, D. P., and G. L. Kooyman, 1982. Oxygen consumption, thermoregulation, and the effect of fur oiling and washing on the sea otter, *Enhydra lutris. Can. J. Zool.* 60:2761–2767.

Costa, D. P., and G. L. Kooyman, 1984. Contributions of specific dynamic action to heat balance and thermoregulation in the sea otter *Enhydra lutris. Physiol. Zool.* 57:199–203.

Costa, D. P., B. J. Le Beouf, A. C. Huntley, and C. L. Ortiz, 1986. The energetics of lactation in the northern elephant seal, *Mirounga angustirostris. J. Zool.* (*Lond.*). 209:21–33.

Costa, D. P., G. A. Antonelis, Jr., and R. L. DeLong, 1991. Effects of El Niño on the foraging energetics of the California sea lion. In: F. Trillmich, and K. Ono, eds. *Pinnipeds and El Niño: Responses to Environmental Stress.* Springer-Verlag, Heidelberg. pp. 156–165.

Cowan, D. R., W. A. Walker, and R. L. Brownell, Jr., 1986. Pathology of small cetaceans stranded along southern California beaches. In: M. M. Bryden, and R. Harrison, eds. *Research on Dolphins.* Oxford Univ. Press, Oxford. pp. 323–367.

Dahlheim, M. E., S. Leatherwood, and W. F. Perrin, 1982. Distribution of killer whales in the warm temperate and tropical eastern Pacific. *Rep. Int. Whaling Comm.* 32:647–653.

Dailey, M. D., 1970. The transmission of *Parafilaroides decorus* (Nematoda: Melastrongyloidea) in the California sea lion (*Zalophus californianus*). *Proc. Helminthol. Soc. Wash.* 37:215–222.

Dailey, M. D.,1985. Diseases of Mammalia: Cetacea. In: O. Kinne, ed. *Diseases of Marine Animals.* Vol. IV, part 2. *Reptilia, Aves, Mammalia.* Biologische Anstalt Helgoland, Hamburg, FRG. pp. 805–847.

Dailey, M. D., and W. A. Walker, 1978. Parasitism as a factor (?) in single strandings of southern California cetaceans. *J. Parasitol.* 64: 593–596.

Darling, J. D., 1979. Summer abundance and distribution of gray whales along the Washington and British Columbia coast. Unpubl. report prepared for Northwest and Alaska Fisheries Center, Natl. Mar. Fish. Serv.

Darling, J. D., and D. J. McSweeney, 1985. Observations on the migrations of North Pacific humpback whales (*Megaptera novaeangliae*). *Can. J. Zool.* 63:308–314.

Davis, J. E., and S. S. Anderson, 1976. Effects of oil pollution on breeding gray seals. *Mar. Pollut. Bull.* 7:115–118.

Defran, R. H., D. L. Kelly, G. M. Shultz, A. C. Weaver, and M. A. Espinoza, 1987. Occurrence and movements of the bottlenose dolphin in the Southern California Bight (abstract). In: Seventh Biennial Conference on the Biology of Marine Mammals, December 1987, Miami, FL.

DeLong, R. L., and G. A. Antonelis, Jr., 1991. Effects of the 1982–1983 El Niño on the northern fur seal population at San Miguel Island, California. In: F. Trillmich, and K. Ono, eds. *Pinnipeds and El Niño: Responses to Environmental Stress.* Springer-Verlag, Heidelberg. pp. 75–83.

DeLong, R. L., W. G. Gilmartin, and J. G. Simpson, 1973. Premature births in California sea lions: Associated with high organochlorine pollutant residue levels. *Science.* 181:1168–1169.

DeMaster, D. P., D. J. Miller, D. Goodman, R. L. DeLong, and B. S. Stewart, 1982. Assessment of California sea lion fishery interactions. In: K. Sabol, ed. *Transactions of the 47th North American Wildlife and Natural Resources Confer-*

ence, March 26–31, 1982, Portland OR. Wildlife Manage. Inst., Washington, D.C. pp. 253–264.

DeMaster, D. P., D. J. Miller, J. R. Henderson, and J. M. Coe, 1985. Conflicts between marine mammals and fisheries off the coast of California. In: J. R. Beddington, R. J. H. Beverton, and D. M. Lavigne, eds. *Marine Mammals and Fisheries.* George Allen & Unwin, London, pp. 111–118.

Dohl, T. P., 1987. Atypical distribution, movement and seasonality of five cetacean species in California waters (abstract). In: Seventh Biennial Conference on the Biology of Marine Mammals, December 1987, Miami, FL.

Dohl, T. P., K. S. Norris, R. C. Guess, J. D. Bryant, and M. W. Honig, 1981. Cetacea of the Southern California Bight. Part II of Investigator's Reports: Summary of Marine Mammal and Seabird Surveys of the Southern California Bight Area, 1975–1978. Available from: NTIS, Springfield, VA; PB 81–248–189. 414pp.

Dohl, T. P., R. C. Guess, M. L. Duman, and R. C. Helm, 1983. Cetaceans of central and northern California, 1980–1983: Status, abundance, and distribution. U.S. Dept. of Interior, Minerals Management Service, Pacific OCS Region. OCS Study MMS 84–005. 284pp.

Dohl, T. P., M. L. Bonnell, and R. G. Ford, 1986. Distribution and abundance of common dolphin, *Delphinus delphis,* in the Southern California Bight: A quantitative assessment based upon aerial transect data. *U.S. Natl. Mar. Fish. Serv. Fish. Bull.* 84:333–343.

Duguy, R., and P. L. Babin, 1976. Intoxication aigine par les hydrocarbures observee chex un phoque veaumarin (*Phoca vitulina*). *Ann. Soc. Sci. Nat. Charente-Marit.* 6:194–196.

Englehardt, F. R., 1982. Hydrocarbon metabolism and cortisol balance in oil exposed ringed seals, *Phoca hispida. Comp. Biochem. Physiol.* 72C:133–136.

Englehardt, F. R., 1983. Petroleum effects on marine mammals. *Aquat. Toxicol.* 4:199–217.

Estes, J. A., 1980. *Enhydra lutris. Mamm. Species.* 133:1–8.

Estes, J. A., and R. J. Jameson, 1983. Summary of available population information on California sea otters. POCS Tech. Pap. No. 83–11. Prepared by the U.S. Fish and Wildlife Service for the Minerals Management Service, 29pp.

Evans, P. G. H., 1987. *The Natural History of Whales and Dolphins.* Facts on File Publications, New York. 342pp.

Evans, W. E., 1971. Orientation behavior of delphinids: Radiotelemetric studies. *Ann. N.Y. Acad. Sci.* 188:142–160.

Evans, W. E., 1982. Distribution and differentiation of stock of *Delphinus delphis* Linnaeus in the northeastern Pacific. In: *Mammals of the Seas: Small Cetaceans, Seals, Sirenians, and Otters.* FAO Fisheries Series No. 5, Vol. IV. Food and Agriculture Organization of the United Nations, Rome. pp. 45–66.

Fay, F. H., R. A. Dieterich, L. M. Shults, and B. P. Kelly, 1978. Morbidity and mortality of marine mammals. In: *Environmental Assessment of the Alaskan Continental Shelf.* NOAA/ERL Annual Reports of Principal Investigators for the year ending March 1978. Vol. 1. Mammals–Birds. pp. 39–79.

Fiscus, C. H., and K. Niggol, 1965. Observations of cetaceans off California, Oregon, and Washington. *U.S. Fish Wildl. Serv. Spec. Sci. Rep. Fish.* No. 498, 27pp.

Fisher, E. M., 1939. Habits of the southern sea otter. *J. Mammal.* 20:21–36.

Fitch, J. E., and R. L. Brownell, Jr., 1968. Fish otoliths in cetacean stomachs and their importance in interpreting feeding habits. *J. Fish. Res. Board Can.* 25:2561–2578.

Fleischer, L. A., 1987. Guadalupe fur seal, *Arctocephalus townsendi.* In: J. P. Croxall, and R. L. Gentry, eds. *Status, Biology, and Ecology of Fur Seals.* Proceedings of International Symposium and Workshop, Cambridge, U.K., 23–27 April 1984. NOAA Tech. Rep. NMFS 51. pp. 43–51.

Food and Agriculture Organization of the United Nations, 1978. Mammals in the seas. FAO Fisheries Series No. 5. Report of the FAO Advisory Committee on Marine Resources Research, Working Party on Marine Mammals.

Fowler, C. W., 1982. Interactions of northern fur seals and commercial fisheries. *Trans. N. Am. Wildl. Nat. Resour. Conf.* 47:278–292.

Fowler, C. W., 1985. An evaluation of the role of entanglement in the population dynamics of northern fur seals on the Pribilof Islands. In: R. S. Shomura, and H. O. Yoshida, eds. *Proceedings of a Workshop on the Fate and Impact of Marine Debris,* 16–29 Nov. 1984, Honolulu, HI. U.S. Dept. of Commerce, NOAA Tech. Memo. NOAA-TM-NMFS-SWFC–54. pp. 291–307.

Freiberg, R. E., and P. C. Dumas, 1954. The elephant seal (*Mirounga angustirostris*) in Oregon. *J. Mammal.* 35:129.

Gallo-R., J. P., and A. Ortega-O., 1986. The first report of *Zalophus californianus* in Acapulco, Mexico. *Mar. Mamm. Sci.* 2(2):158.

Gaskin, D. E., 1982. *The Ecology of Whales and Dolphins.* Heinemann Educational Books Ltd., London. 459 pp.

Geibel, J. J., and D. J. Miller, 1984. Estimation of sea otter, *Enhydra lutris,* population with confidence bounds, from air and ground counters. *Calif. Fish and Game.* 70:225–233.

Geraci, J. R., and T. G. Smith, 1976. Direct and indirect effects of oil on ringed seals (*Phoca hispida*) of the Beaufort Sea. *J. Fish. Res. Board Can.* 33:84.

Geraci, J. R., and D. J. St. Aubin, 1985a. Effects of offshore oil and gas development on marine mammals and turtles. In: D. F. Boesch, and N. N. Rabalais, eds. *The Long-term Effects of Offshore Oil and Gas Development: An Assessment and a Research Strategy.* Final Report to NOAA, Natl. Mar. Pollut. Prog. Off., for Interagency Committee on Ocean Pollution, Research, Development and Monitoring. Louisiana Univ. Mar. Consortium, Chauvin, LA. pp. 12.1–12.34.

Geraci, J. R., and D. J. St. Aubin, 1985b. Study of the effects of oil on cetaceans. Final Report. U.S. Dept. of the Interior, Bureau of Land Management, Washington, D.C. Contract No. AA551-CT9–29, 274pp.

Geraci, J. R., and D. J. St. Aubin, 1988. Synthesis of effects of oil on marine mammals. In: J. R. Geraci, and D. J. St. Aubin, eds. Final Report prepared for the Minerals Management Service, U.S. Dept. of the Interior. Contract No. 14–12–0001–30293. Battelle Memorial Inst., Ventura, CA.

Geraci, J. R., M. D. Dailey, and D. J. St. Aubin, 1978. Parasitic mastitis in the Atlantic white-sided dolphin *Lagenorynchus acutus* as a probable factor in herd productivity. *J. Fish. Res. Board Can.* 35:1350–1355.

Gill, R. E., Jr., and J. D. Hall, 1983. Use of nearshore and estuarine areas of the southeastern Bering Sea by gray whales (*Eschrichtius robustus*). *Arctic.* 36:275–281.

Gilmartin, W. G., R. L. DeLong, A. W. Smith, J. C. Sweeney, B. W. DeLappe, R. W. Risebrough, L. A. Griner, M. D. Dailey, and D. B. Peakall, 1976. Premature parturition in the California sea lion. *J. Wild. Dis.* 12:104–115.

Gilmore, R. M., 1955. The return of the gray whale. *Sci. Am.* 192:62–74.

Gilmore, R. M., 1956. Rare right whale visits California. *Pac. Discovery.* 9:20–25.

Gilmore, R. M., 1961. *The Story of the Gray Whale,* 2nd ed. Pioneer Printers, San Diego, CA.

Gosho, M. E., D. W. Rice, and J. M. Breiwick, 1984. The sperm whale, *Physeter macrocephalus.* In: J. M. Breiwick, and H. W. Braham, eds. *The Status of Endangered Whales. Mar. Fish. Rev.* 46:54–64.

Guiguet, C. J., 1954. A record of Baird's dolphin (*Delphinus bairdi* Dall) in British Columbia. *Can. Field Nat.* 68:136.

Guiguet, C. J., and G. C. Pike, 1965. First specimen record of the gray grampus or Risso dolphin, *Grampus griseus* (Cuvier) from British Columbia. *Murrelet.* 46:16.

Gunter, G., 1942. Contributions to the natural history of the bottle-nose dolphin, *Tursiops truncatus* (Montague), on the Texas coast, with particular reference to food habits. *J. Mammal.* 23:267–276.

Haaker, P. L., D. O. Parker, and K. C. Henderson, 1984. Observations of harbor seal, *Phoca vitulina richardsi,* feeding in southern California waters. *Bull. South. Calif. Acad. Sci.* 83:152–153.

Hacker, E. S., 1986. Stomach content analysis of short-finned pilot whales (*Globicephala macrorhynchus*) and northern elephant seals (*Mirounga angustirostris* from the Southern California Bight. NOAA-NMFS-SWFC Admin. Rep. No. LJ–86–08C, 34pp.

Hall, J. D., 1979. A survey of cetaceans of Prince William Sound and adjacent vicinity: Their numbers and seasonal movements. Final Report, OCS Environmental Assessment Program, NOAA, 83pp.

Hall, J. D., W. G. Gilmartin, and J. L. Mattson, 1971. Investigation of a Pacific pilot whale stranding on San Clemente Island. *J. Wildl. Dis.* 7(4):324–327.

Hanan, D. A., 1986. California Department of Fish and Game, Coastal Marine Mammal Study, Annual Report for the Period July 1, 1983–June 30, 1984. NOAA-NMFS-SWFC Admin. Rep. No. LJ–86–16. Prep. by NOAA, Natl. Mar. Fish. Serv., Southwest Fish. Center, La Jolla, CA.

Hancock, D., 1965. Killer whales kill and eat a minke whale. *J. Mammal.* 46:341–342.

Hansen, L. J., 1983. Population biology of the coastal bottlenose dolphin (*Tursiops truncatus*) of southern California. Master's Thesis, California State Univ., Sacramento. 104pp.

Heath, C. B., and J. M. Francis, 1983. California sea lion population dynamics and feeding ecology, with applications for management. NOAA-NMFS-SWFC Admin. Rep. LJ–83–04C. Prep. by NOAA, Natl. Mar. Fish. Serv., Southwest Fish. Center, La Jolla, CA. 47pp.

Helle, E., M. Olsson, and S. Jensen, 1976. DDT and PCB levels and reproduction in ringed seal from the Bothnian Bay. *Ambio.* 5:188–189.

Hennessy, S. C., and G. V. Morejohn, 1977. Acanthocephalan parasites of the sea otter, *Enhydra lutris,* off coastal California. *Calif. Fish Game.* 63:268–272.

Herzing, D. L., and B. R. Mate, 1984. Gray whale migration along the Oregon coast, 1978–1981. In: M. L. Jones, S. L. Swartz, and S. Leatherwood, eds. *The Gray Whale, Eschrichtius robustus.* Academic Press, Orlando, FL. pp. 289–307.

Heyning, J. E., and M. E. Dahlheim, 1988. *Orcinus orca.* Mammalian Species No. 304, the American Society of Mammalogists, 9pp.

Holden, A. V., 1975. The accumulation of oceanic contaminants in marine mammals. *Rapp. P.-V. Réun. Cons. Int. Explor. Mer.* 169:353–361.

Holden, A. V., and K. Marsden, 1967. Organochlorine pesticides in seals and porpoises. *Nature.* 216:1274–1276.

Houck, W. J., 1961. Notes on the Pacific striped porpoise. *J. Mammal.* 42:107.

Howard, E. B., J. O. Britt, G. K. Matsumoto, R. Itahara, and C. N. Nagano, 1983. Bacterial diseases. In: E. B. Howard, ed. *Pathobiology of Marine Mammal Diseases.* Vol. 1. CRC Press, Boca Raton, FL. pp. 70–117.

Hubbs, C. L., 1951. Eastern Pacific records and general distribution of the pygmy sperm whale. *J. Mammal.* 32:403–410.

Hubbs, C. L., 1956. Back from oblivion. Guadalupe fur seal: Still a living species. *Pac. Discovery.* 9:14–21.

Hubbs, C. L., 1960. The marine vertebrates of the outer coast. *Syst. Zool.* 9:134–147.

Hubbs, C. L., W. F. Perrin, and K. C. Balcomb, 1973. *Stenella coeruleoalba* in the eastern and central North Pacific. *J. Mammal.* 54:549–552.

Huey, L. M., 1930a. Past and present status of the northern elephant seal with a note on the Guadalupe fur seal. *J. Mammal.* 11:188–194.

Huey, L. M., 1930b. Capture of an elephant seal off San Diego California, with notes on stomach contents. *J. Mammal.* 11:229–231.

Johnson, A. M., 1982. The sea otter, *Enhydra lutris.* In: *Mammals in the Seas.* FAO Fisheries Series No. 5. Vol. IV. Food and Agriculture Organization of the United Nations, Rome. pp. 521–525.

Johnson, J. H., and A. A. Wolman, 1984. The humpback whale, *Megaptera novaeangliae.* In: J. M. Breiwick, and H. W. Braham, eds. *The Status of Endangered Whales. Mar. Fish. Rev.* 46:30–37.

Johnson, M. L., and S. J. Jeffries, 1977. Population evaluation of the harbor seal (*Phoca vitulina richardsi*) in the waters of the State of Washington. Final Report to the U.S. Marine Mammal Commission. Contract No. MM5AC019, 33pp.

Johnson, T. M., 1987. Blue whale (*Balaenoptera musculus*) feeding on euphausiids (*Thysanoessa spinifera*) in the Gulf of the Farallones, California, 1986 (abstract). In: Seventh Biennial Conference on the Biology of Marine Mammals, December 1987, Miami, FL.

Jones, J. H., 1971. General circulation and water characteristics in the Southern California Bight. Rep. TR 101. South. Calif. Coastal Water Res. Proj., El Segundo, CA. 37pp.

Jones, L. L., G. C. Bouchet, and B. J. Turnock, 1987. Comprehensive report on the incidental take, biology, and status of Dall's porpoise. Submitted to the Ad Hoc Committee on Marine Mammals, Int. North Pacific Fish. Comm. 79pp.

Jones, M. L., and S. L. Swartz, 1986. Radiotelemetric study and aerial census of gray whales in the Channel Islands National Marine Sanctuary during the southward migration, January 1986. Draft Final Report prepared for the Northwest and Alaska Fisheries Center, Natl. Mar. Fish. Serv., and the Sanctuary Programs Division, NOAA. Contract No. 50-ABNF–6–00067.

Jones, R. E., 1981. Food habits of smaller marine mammals from northern California. *Proc. Calif. Acad. Sci.* 42:409–433.

Kajimura, H., 1985. Opportunistic feeding by the northern fur seal (*Callorhinus ursinus*). In: J. R.

Beddington, R. J. H. Beverton, and D. M. Lavigne, eds. *Marine Mammals and Fisheries*. George Allen & Unwin, London. pp. 300–318.
Kawamura, A., 1982. Food habits and prey distributions of three rorqual species in the north Pacific Ocean. *Sci. Rep. Whales Res. Inst.*, Tokyo. 34:59–91.
Keith, E. O., R. S. Condit, and B. J. Le Boeuf, 1984. California sea lions breeding at Año Nuevo Island, California. *J. Mammal.* 65:695.
Kenyon, K. W., 1969. The sea otter in the eastern Pacific Ocean. North Am. Fauna No. 68. Bur. Sport Fish. Wildl. Prep. by U.S. Gov. Printing Office, Washington, D.C. 352pp.
Kenyon, K. W., and F. Wilke, 1953. Migration of the northern fur seal, *Callorhinus ursinus*. *J. Mammal.* 34(1):86–98.
Keyes, M. C., 1968. The nutrition of pinnipeds. In: R. J. Harrison, ed. *The Behavior and Physiology of Pinnipeds*. Appleton-Century-Crofts, New York. pp. 359–395.
Kolb, P. M., and K. S. Norris, 1981. A harbor seal, *Phoca vitulina richardsi,* taken from a sablefish trap. *Calif. Fish Game*. 67:123–124.
Kooyman, G. L., R. L. Gentry, and W. B. McAllister, 1976. Physiological impact of oil on pinnipeds. NWAFC Proc. Rep., Natl. Mar. Fish. Serv., Seattle, Washington, 23pp.
Kooyman, G. L., R. W. Davis, and M. A. Castellini, 1977. Thermal conductance of immersed pinniped and sea otter pelts before and after oiling with Prudhoe Bay crude. In: D. A. Wolfe, ed. *Fate and Effects of Petroleum Hydrocarbons in Marine Ecosystems and Organisms*. Pergamon Press, Oxford, U.K. pp. 151–157.
Kritzler, H., 1952. Observations on the pilot whale in captivity. *J. Mammal.* 33:321–334.
Lander, R. H., ed., 1980. Summary of northern fur seal data and collection procedures: Land data of the United States and Soviet Union (excluding tag and recovery records). NOAA Tech. Memo. NMFS F/NWC–3. U.S. Dept. of Commerce, NOAA, Natl. Mar. Fish. Serv. Vol. 1, 315pp.
Lander, R. H., and H. Kajimura, 1976. Status of northern fur seals. Report to the Advisory Committee on Marine Resources Research. Food and Agriculture Organization of the United Nations, Rome. 50 pp.
Lander, R. H., and H. Kajimura, 1982. Status of northern fur seals. In: *Mammals in the Seas*. FAO Fisheries Series No. 5. Vol. IV. Food and Agriculture Organization of the United Nations. pp. 319–345.
Lauckner, G., 1985. Diseases of Mammalia: Pinnipedia. In: O. Kinne, ed. *Diseases of Marine Animals*. Vol. IV, Part 2. *Reptilia, Aves, Mammalia*. Biologische Anstalt Helgoland, Hamburg, FRG. pp. 683–793.
Laws, R. M., 1961. Reproduction, growth, and age of southern fin whales. *Discovery Rep.* 31: 327–486.
Leatherwood, J. S., 1974. Aerial observations of migrating gray whales, *Eschrichtius robustus,* off southern California, 1969–72. *Mar. Fish. Rev.* 36(4):45–49.
Leatherwood, S., 1975. Some observations of feeding behavior of bottle-nosed dolphins *Tursiops truncatus* in the northern Gulf of Mexico and *Tursiops* cf. *T. gilli* off southern California, Baja California, and Nayarit, Mexico. *Mar. Fish. Rev.* 37(9):10–16.
Leatherwood, S., and M. R. Fielding, 1974. A summary of distribution and movements of Dall porpoises *Phocoenoides dalli* off southern California and Baja California. Working Paper No. 42. FAO Advisory Committee on Marine Resources Research. Food and Agriculture Organization of the United Nations, Rome.
Leatherwood, S., and R. R. Reeves, 1978. Porpoises and dolphins. In: D. Haley, ed. *Marine Mammals of Eastern North Pacific and Arctic Waters*. Pacific Search Press, Seattle, WA. pp. 96–111.
Leatherwood, S., and W. A. Walker, 1979. The northern right-whale dolphin, *Lissodelphis borealis* Peale, in the eastern North Pacific. In: H. W. Winn, and B. L. Olla, eds. *Behavior of Marine Animals*. Vol. 3. Plenum Press, New York. pp. 85–141.
Leatherwood, S., W. F. Perrin, V. L. Kirby, C. L. Hubbs, and M. E. Dahlheim, 1980. Distribution and movements of Risso's dolphin, *Grampus griseus,* in the eastern North Pacific. *U.S. Natl. Mar. Fish. Serv. Fish. Bull.* 77(4): 951–963.
Leatherwood, S., R. R. Reeves, W. F. Perrin, and W. E. Evans, 1982. Whales, dolphins, and porpoises of the eastern North Pacific and adjacent Arctic waters: A guide to their identification. U.S. Dept. of Commerce, NOAA Tech. Rep., Natl. Mar. Fish. Serv. Circ. 444, 245pp.
Leatherwood, S., R. R. Reeves, A. E. Bowles,

B. S. Stewart, and K. R. Goodrich, 1984. Distribution, seasonal movements, and abundance of Pacific white–sided dolphins *Lagenorhynchus obliquidens* in the eastern North Pacific. *Sci. Rep. Whales Res. Inst.,* Tokyo. 35:129–158.

Le Boeuf, B. J., and M. L. Bonnell, 1971. DDT in California sea lions. *Nature*. 234:108–110.

Le Boeuf, B. J., and M. L. Bonnell, 1980. Pinnipeds of the California islands: Abundance and distribution. In: D. M. Power, ed. *The California Islands: Proceedings of a Multidisciplinary Symposium*. Haagen Printing, Santa Barbara, CA. pp. 475–493.

Le Boeuf, B. J., and B. R. Mate, 1978. Elephant seals colonize additional Mexican and California islands. *J. Mammal*. 59:621–622.

Le Boeuf, B. J., J. Whiting, and R. F. Gantt, 1972. Perinatal behavior of northern elephant seal females and their young. *Behaviour*. 43:121–156.

Le Boeuf, B. J., D. G. Ainley, and T. J. Lewis, 1974. Elephant seals on the Farallones: Population structure of an incipient breeding colony. *J. Mammal*. 55(2):370–385.

Le Boeuf, B. J., M. Riedman, and R. S. Keyes, 1982. White shark predation on pinnipeds in California coastal waters. *U.S. Natl. Mar. Fish. Serv. Fish. Bull*. 80(4):891–895.

Le Boeuf, B. J., D. Aurioles, R. Condit, C. Fox, R. Gisiner, R. Romero, and F. Sinsel, 1983. Size and distribution of the California sea lion population in Mexico. *Proc. Calif. Acad. of Sci*. 43(7):77–85.

Le Boeuf, B. J., D. P. Costa, A. C. Huntley, and S. D. Feldkamp, 1988. Continuous deep diving in female northern elephant seals *Mirounga angustirostris*. *Can. J. Zool*. 66(2):446–458.

Loeb, V. J., 1972. A study of the distribution and feeding habits of the Dall porpoise in Monterey Bay, California. Master's Thesis, San Jose State College, San Jose, CA. 62pp.

Loughlin, T. R., D. J. Rugh, and C. H. Fiscus, 1984. Northern sea lion distribution and abundance: 1956–80. *J. Wildl. Manage*. 48(3):729–740.

Lowry, M. S., C. W. Oliver, and J. B. Wexler, 1987. The feeding habits of the California sea lion (*Zalophus californianus*) at San Clemente Island, California, from September 1981 through September 1985 (abstract). In: Seventh Biennial Conference on the Biology of Marine Mammals, December 1987, Miami, FL.

Lyon, G. M., 1937. Pinnipeds and a sea otter from the Point Mugu shell mound of California. *Univ. Calif. Publ. Biol. Sci*. 1:133–168.

MacGregor, J. S., 1974. Changes in the amount and proportions of DDT and its metabolites, DDE and DDD, in the marine environment off southern California, 1949–72. *U.S. Natl. Mar. Fish. Serv. Fish. Bull*. 72(2):275–293.

Martin, J. H., 1979. Bioaccumulation of heavy metals by littoral and pelagic marine organisms. U.S. Environ. Protec. Agency Rep. No. 600/3–79–039, 56pp.

Martin, J. H., P. D. Elliot, V. C. Anderlini, D. Girvin, S. A. Jacobs, R. W. Risebrough, R. L. DeLong, and W. G. Gilmartin, 1976. Mercury–selenium–bromide imbalance in premature parturient California sea lions. *Mar. Biol*. 35:91–104.

Mate, B. R., 1977. Aerial censusing of pinnipeds in the eastern Pacific for assessment of population numbers, migratory distributions, rookery stability, breeding effort, and recruitment. Final Report to U.S. Marine Mammal Commission. Contract No. MM5AC001. Available from: NTIS, Springfield, VA; PB–265–859-9 VA; PB–265–859–9, 67pp.

Mead, J. G., 1981. First records of *Mesoplodon hectori* (Ziphidae) from the Northern Hemisphere and a description of the adult male. *J. Mammal*. 62:430–432.

Mead, J. G., W. A. Walker, and W. J. Houck, 1982. Biological observations on *Mesoplodon carlhubbsi* (Cetacea: Ziphiidae). *Smithson. Contrib. Zool*. 344:1–25.

Merrick, R. L., T. R. Loughlin, and D. G. Calkins, 1987. Decline in abundance of the northern sea lion, *Eumetopias jubatus,* in Alaska, 1956–86. *U.S. Natl. Mar. Fish. Serv. Fish. Bull*. 85(2):351–365.

Miller, D. J., M. J. Herder, and J. P. Scholl, 1982. California marine mammal–fishery interaction study, 1979–1981. Final Report prep. by Calif. Dept. of Fish and Game for NOAA-NMFS-SWFC No. 79-ABC–00149.

Miller, D. J., M. Herder, J. Scholl, and P. Law, 1983a. Harbor seal, *Phoca vitulina,* censuses in California. In: D. J. Miller, ed. *Coastal Marine Mammal Study,* Annual Report for the Period of July 1, 1981–June 30, 1982. NOAA-NMFS-SWFC Admin. Rep. LJ–83–21C. Prep. by NOAA, Natl. Mar. Fish. Serv., Southwest Fish. Center, La Jolla, CA.

Miller, D. J., M. Herder, and J. Scholl, 1983b.

California marine mammal-fishery interaction study, 1979–1981. NOAA-NMFS-SWFC, Admin. Rep. LJ–83–13C. Prep. by NOAA, Natl. Mar. Fish. Serv., Southwest Fish. Center, La Jolla, CA.

Mitchell, E. D., 1965. Evidence for mass strandings of the false killer whale, *Pseudorca crassidens,* in the eastern North Pacific Ocean. *Norwegian Whaling Gazette.* 54(8):172–177.

Mitchell, E. D., 1968. Northeast Pacific stranding distribution and seasonality of Cuvier's beaked whale, *Ziphius cavirostris. Can. J. Zool.* 46:265–279.

Mitchell, E. D., 1978. Finner whales. In: D. Haley, ed. *Marine Mammals of Eastern North Pacific and Arctic Waters.* Pacific Search Press, Seattle, WA. pp. 36–45.

Mizrock, S. A., D. W. Rice, and J. M. Breiwick, 1984a. The blue whale, *Balaenoptera musculus.* In: J. M. Breiwick, and H. W. Braham, eds. *The Status of Endangered Whales. Mar. Fish. Rev.* 46:15–19.

Mizrock, S. A., D. W. Rice, and J. M. Breiwick, 1984b. The fin whale, *Balaenoptera physalus.* In: J. M. Breiwick, and H. W. Braham, eds. *The Status of Endangered Whales. Mar. Fish. Rev.* 46:20–24.

Morejohn, G. V., 1968. A killer whale–gray whale encounter. *J. Mammal.* 49:327–328.

Morejohn, G. V., 1979. The natural history of Dall's porpoise in the north Pacific Ocean. In: H. E. Wynn, and B. L. Olla, eds. *Behavior of Marine Animals.* Vol. 3: *Cetaceans.* Plenum Press, New York. pp. 45–83.

Morejohn, G. V., and D. M. Baltz, 1970. Contents of the stomach of an elephant seal. *J. Mammal.* 51:173–174.

Morejohn, G. V., J. T. Harvey, R. C. Helm, and R. L. Cross, 1977. Feeding habits of harbor seals, *Phoca vitulina,* in Elkhorn Slough, Monterey Bay, California (unpubl. ms.). 29pp.

Morimitsu, T., T. Nagai, M. Ide, H. Kawano, A. Narchuu, M. Koono, and A. Ishii, 1987. Mass stranding of Odontoceti caused by parasitogenic eighth cranial neuropathy. *J. Wildl. Dis.* 23:586–591.

Murie, D. J., 1987. Experimental approaches to stomach content analyses of piscivorous marine mammals. In: A. C. Huntley, D. P. Costa, G. A. J. Worthy, and M. A. Castellini, eds. *Approaches to Marine Mammal Energetics.* Soc. Mar. Mammal., Spec. Publ. No. 1. pp. 147–163.

Murison, L. D., D. J. Murie, K. R. Morin, and J. de Silva Curiel, 1984. Foraging of the gray whale along the west coast of Vancouver Island, British Columbia. In: M. L. Jones, S. L. Swartz, and S. Leatherwood, eds. *The Gray Whale, Eschrichtius robustus.* Academic Press, Orlando, FL. pp. 451–463.

Nagorsen, D. W., and G. E. Stewart, 1983. A dwarf sperm whale (*Kogia simus*) from the Pacific coast of Canada. *J. Mammal.* 64:505–506.

Nagy, K., and D. P. Costa, 1980. Water flux in animals: Analysis of potential errors in the tritiated water technique. *Am. J. Physiol.* 238: 454–465.

Nelson, C. H., K. R. Johnson, and H.-L. Mitchell, 1983. Assessment of gray whale feeding grounds and sea floor interaction in the northeastern Bering Sea. NOAA, OCS Environmental Assessment Program. Draft Final Report.

Nemoto, T., 1959. Food of baleen whales with reference to whale movements. *Sci. Rep. Whales Res. Inst.,* Tokyo. 14:149–290.

Nemoto, T., and A. Kawamura, 1974. Characteristics of food habits and distribution of baleen whales with special reference to the abundance of north Pacific sei and Bryde's whale. In: *Rep. Int. Whaling Comm.* Spec. Issue No. 1. pp. 80–87.

Nerini, M. K., 1984. A review of gray whale feeding ecology. In: M. L. Jones, S. L. Swartz, and S. Leatherwood, eds. *The Gray Whale, Eschrichtius robustus.* Academic Press, Orlando, FL. pp. 423–450.

Nerini, M. K., L. Jones, and H. W. Braham, 1980. Gray whale feeding ecology. Final Report of Research Unit 593 to NOAA, OCS Environmental Assessment Program. Prep. by Natl. Mar. Fish. Serv., Contract No. R7120828.

Nishiwaki, M., 1959. Humpback whales in Ryukyuan waters. *Sci. Rep. Whales Res. Inst.,* Tokyo. 14:49–87.

Norris, K. S., and T. P. Dohl, 1980a. The structure and functions of cetacean schools. In: L. M. Herman, ed. *Cetacean Behavior: Mechanisms and Processes.* John Wiley & Sons, New York. pp. 211–261.

Norris, K. S., and T. P. Dohl, 1980b. The behavior of the Hawaiian spinner porpoise, *Stenella longirostris. U.S. Natl. Mar. Fish. Serv. Fish. Bull.* 77:821–849.

Norris, K. S., and J. H. Prescott, 1961. Observations of Pacific cetaceans in California and Mexican waters. *Univ. Calif. Publ. in Zool.* 63:291–402.

Odell, D. K., 1971. Censuses of pinnipeds breeding on the California Channel Islands. *J. Mammal.* 152(1):187–190.

Ogden, A., 1941. *The California Sea Otter Trade, 1784–1848.* Univ. of California Press, Berkeley, CA. 251pp.

Ohsumi, S., and Y. Fukuda, 1975. On the estimates of exploitable population size and replacement yield for the Antarctic sei whale by use of catch and effort data. *Rep. Int. Whaling Comm.* 25:102–105.

Ohsumi, S., and S. Wada, 1974. Status of whale stocks in the North Pacific, 1972. *Rep. Int. Whaling Comm.* 24:114–126.

Oliver, C. W., and T. D. Jackson, 1987. Occurrence and distribution of marine mammals at sea from aerial surveys conducted along the U.S. west coast between December 15, 1980, and December 17, 1985. NOAA-NMFS-SWFC Admin. Rep. LJ–87–19. Prep. by NOAA, Natl. Mar. Fish. Serv., Southwest Fish. Center, La Jolla, CA.

Oliver, J. S., P. N. Slattery, M. A. Silberstein, and E. F. O'Connor, 1983. A comparison of gray whale *Eschrichtius robustus* feeding in the Bering Sea and Baja California Mexico. *U.S. Natl. Mar. Fish. Serv. Fish. Bull.* 81(3):513–522.

Omura, H., 1958. North Pacific right whale. *Sci. Rep. Whales Res. Inst.,* Tokyo. 13:1–52.

Oritsland, N. A., F. R. Englehardt, F. A. Juck, R. J. Hurst, and P. D. Watts, 1981. Effects of crude oil on polar bears. Dept. of Indian Affairs and Northern Development, Canada. Publ. No. QS–8283–020-EE-A1. Environmental Studies No. 24, 268pp.

Orr, R. T., 1966. Risso's dolphin on the Pacific coast of North America. *J. Mammal.* 47(2):341–343.

O'Shea, T. J., R. L. Brownell, Jr., D. R. Clark, Jr., W. A. Walker, M. L. Gay, and T. G. Lamont, 1980. Fish, wildlife, and estuaries: Organochlorine pollutants in small cetaceans from the Pacific and south Atlantic oceans Nov. 1968–June 1976. *Pestic. Monit. J.* 14(2):35–46.

Patten, D. R., and W. F. Samaras, 1977. Unseasonable occurrences of gray whales. *Bull. South. Calif. Acad. Sci.* 76:205–208.

Perez, M. A., and M. A. Bigg, 1986. Diet of northern fur seals, *Callorhinus ursinus,* off western North America. *U.S. Natl. Mar. Fish. Serv. Fish. Bull.* 84:957–971.

Perez, M. A., and E. E. Mooney, 1986. Increased food and energy consumption of lactating northern fur seals, *Callorhinus ursinus. U.S. Natl. Mar. Fish. Serv. Fish. Bull.* 84:371–381.

Perrin, W. F., and J. E. Powers, 1980. Role of a nematode in natural mortality of spotted dolphins. *J. Wildl. Manage.* 44:960–963.

Perrin, W. F., and W. A. Walker, 1975. The rough-toothed porpoise, *Steno bredanensis,* in the eastern tropical Pacific. *J. Mammal.* 56:905–907.

Perryman, W. L., and T. C. Foster, 1980. Preliminary report on predation by small whales, mainly the false killer whale, *Pseudorca crassidens,* on dolphins (*Stenella* spp. and *Delphinus delphis*) in the eastern tropical Pacific. Natl. Mar. Fish. Serv., Southwest Fish. Center Admin. Rep. LJ–80–05. 9pp.

Peterson, R. S., B. J. Le Boeuf, and R. L. DeLong, 1968. Fur seals from the Bering Sea breeding in California. *Nature (Lond.).* 219:899–901.

Pierotti, R. J., D. G. Ainley, T. J. Lewis, and M. C. Coulter, 1977. Birth of a California sea lion on southeast Farallon Island. *Calif. Fish Game.* 63(1):64–66.

Pierson, M. O., 1987. Breeding behavior of the Guadalupe fur seal, *Arctocephalus townsendi.* In: J. P. Croxall, and R. L. Gentry, eds. *Status, Biology, and Ecology of Fur Seals.* Proceedings of International Symposium and Workshop, Cambridge, U.K., 23–27 April 1984. NOAA Tech. Rep. NMFS 51. pp. 83–94.

Pike, G. C., 1962. Migration and feeding of the gray whales (*Eschrichtius robustus*). *J. Fish. Res. Board. Can.* 19:815–838.

Pike, G. C., and I. B. MacAskie, 1969. Marine mammals of British Columbia. *Bull. Fish. Res. Board Can.* No. 171.

Piotrowski, J. K., E. M. Berm, and A. Werner, 1977. Cadmium and mercury binding to metallothionein as influenced by selenium. *Biochem. Pharmacol.* 26:2191–2192.

Pitcher, K. W., 1980. Food of the harbor seal, *Phoca vitulina richardsi,* in the Gulf of Alaska. *U.S. Natl. Mar. Fish. Serv. Fish. Bull.* 78(2): 544–549.

Pitman, R. L., A. Aguayo-L., and J. Urbán-R., 1987. Observations of an unidentified beaked whale (*Mesoplodon* sp.) in the eastern tropical Pacific. *Mar. Mamm. Sci.* 3:345–352.

Polacheck, T., 1987. Relative abundance, distribution and interspecific relationship of cetacean schools in the eastern tropical Pacific. *Mar. Mamm. Sci.* 3:54–77.

Poole, M. M., 1984. Migration corridors of gray

whales along the central California coast, 1980–1982. In: M. L. Jones, S. L. Swartz, and S. Leatherwood, eds. *The Gray Whale, Eschrichtius robustus*. Academic Press, Orlando, FL. pp. 389–407.

Ralls, K., J. Ballou, and R. L. Brownell, 1983. Genetic diversity in California sea otters: Theoretical considerations and management implications. *Biol. Conserv*. 25:209–232.

Rausch, R. C., 1953. Studies on the helminth fauna of Alaska. XIII. Disease in the sea otter, with special reference to helminth parasites. *Ecology*. 34:584–604.

Reilly, S. B., 1984. Assessing gray whale abundance: A review. In: M. L. Jones, S. L. Swartz, and S. Leatherwood, eds. *The Gray Whale, Eschrichtius robustus*. Academic Press, Orlando, FL. pp. 203–223.

Reilly, S. B., 1991. Population biology and status of eastern Pacific gray whales: recent developments. In: D. McCullough and R. Barrett, eds. *Wildlife 2001: Populations*. Elsevier, New York.

Reiter, J., N. L. Stinson, and B. J. Le Boeuf, 1978. Northern elephant seal development: The transition from weaning to nutritional independence. *Behav. Ecol. Sociobiol*. 3:337–367.

Repenning, C. A., R. S. Peterson, and C. L. Hubbs, 1971. Contributions to the systematics of the southern fur seals, with particular reference to the Juan Fernandez and Guadalupe species. In: W. H. Burt, ed. *Antarctic Pinnipedia. Antarct. Res. Ser. 18*. American Geophysical Union, Washington, D.C. pp. 1–34.

Rice, D. W., 1965. Offshore southward migration of gray whales off southern California. *J. Mammal*. 46:504–505.

Rice, D. W., 1968. Stomach contents and feeding behavior of killer whales in the eastern North Pacific. *Nor. Hvalfangst-Tid*. 57:35–38.

Rice, D. W., 1974. Whales and whale research in the eastern North Pacific. In: W. E. Schevill, ed. *The Whale Problem: A Status Report*. Harvard Univ. Press, Cambridge, MA. pp. 170–195.

Rice, D. W., 1983. Gestation period and fetal growth of the gray whale. *Rep. Int. Whaling Comm*. 33:539–544.

Rice, D. W., 1986. Blue whale. In: D. Haley, ed. *Marine Mammals of the Eastern North Pacific and Arctic Waters*. 2nd ed. Pacific Search Press, Seattle, WA. pp. 40–45.

Rice, D. W., and C. W. Fiscus, 1968. Right whales in the southeastern North Pacific. *Nor. Hvalfangst-Tid*. 57:105–107.

Rice, D. W., and A. A. Wolman, 1971. The life history and ecology of the gray whale (*Eschrichtius robustus*). *Am. Soc. Mammal. Spec. Publ*. No. 3, 142pp.

Rice, D. W., and A. A. Wolman, 1982. Whale census in the Gulf of Alaska, June to August, 1980. *Rep. Int. Whaling Comm*. 32:491–498.

Rice, D. W., A. A. Wolman, D. E. Withrow, and L. A. Fleischer, 1981. Gray whales on the winter grounds in Baja California. *Rep. Int. Whaling Comm*. 33:539–544.

Rice, D. W., A. A. Wolman, and H. W. Braham, 1984. The gray whale, *Eschrichtius robustus*. In: J. M. Briewick, and H. W. Braham, eds. *The Status of Endangered Whales. Mar. Fish. Rev*. 46(4):7–14.

Ridgway, S. H., 1966. Dall porpoise, *Phocoenoides dalli* (True): Observations in captivity and at sea. *Nor. Hvalfangst-Tid*. 55:97–110.

Ridgway, S. H., 1972. Homeostasis in the aquatic environment. In: S. H. Ridgway, ed. *Mammals of the Sea: Biology and Medicine*. C. C. Thomas Publ., Springfield, IL. pp. 590–747.

Ridgway, S. H., and M. D. Dailey, 1972. Cerebral and cerebellar involvement of trematode parasites in dolphins and their possible role in stranding. *J. Wildl. Dis*. 8:33–43.

Ridgway, S. H., J. R. Geraci, and W. Medway, 1975. Diseases of pinnipeds. *Rapp. P.-V. Réun. Cons. Int. Explor. Mer*. 169:327–337.

Risebrough, R. W., 1971. Chlorinated hydrocarbons. In: D. W. Hood, ed. *Impingement of Man on the Oceans*. Wiley Interscience, New York. pp. 259–286.

Risebrough, R. W., 1984. Accumulation patterns of heavy metals and chlorinated hydrocarbons by sea otters, *Enhydra lutris*, in California. Draft report submitted to U.S. Marine Mammal Commission, Washington, D.C.

Risebrough, R. W., P. Reiche, S. G. Herman, D. B. Peakall, and M. N. Kirven, 1968. Polychlorinated biphenyls in the global ecosystem. *Nature*. 220:1098–1102.

Roest, A. I., 1970. *Kogia sinus* and other cetaceans from San Luis Obispo County, California. *J. Mammal*. 51(2):410–417.

Roffe, T. J., and B. R. Mate, 1984. Abundances and feeding habits of pinnipeds in the Rogue River, Oregon. *J. Wildl. Manage*. 48(4):1262–1274.

Rotterdam, L. J., and T. Simon-Jackson, 1988. Sea

otter, *Enhydra lutris*. In: J. W. Lentfer, ed. *Selected Marine Mammals of Alaska: Species Account with Research and Management Recommendations*. Marine Mammal Commission, Washington, D.C. pp. 237–275.

Rozaire, C. E., 1959. Archeological investigations at two sites on San Nicolas Island, California. *Master Key*. 33:129–152.

Rugh, D. J., 1984. Census of gray whales at Unimak Pass, Alaska: November–December 1977–1979. In: M. L. Jones, S. L. Swartz, and S. Leatherwood, eds. *The Gray Whale, Eschrichtius robustus*. Academic Press, Orlando, FL. pp. 225–248.

Scammon, C. M., 1874. *The Marine Mammals of the Northwest Coast of North America*. J. H. Carmony & Company, San Francisco, CA. 319pp.

Scarff, J. E., 1986. Historic and present distribution of the right whale (*Eubalaena glacialis*) in the eastern North Pacific south of 50 degrees N and east of 180 degrees W. In: *Rep. Int. Whaling Comm. Spec. Issue 10*. pp. 43–63.

Scheffer, V. B., 1953. Measurements and stomach contents of eleven delphinids from the northeast Pacific. *Murrelet*. 34:27–30.

Scheffer, V. B., 1958. *Seals, Sea Lions, and Walruses*. Stanford Univ. Press, Palo Alto, CA. 179pp.

Scheffer, V. B., 1978. Killer whale. In: D. Haley, ed. *Marine Mammals of Eastern North Pacific and Arctic Waters*. Pacific Search Press, Seattle, WA. pp. 120–127.

Schoenherr, J. R., 1987. The feeding ecology of blue whales in Monterey Bay, California, during fall 1986 (abstract). In: Seventh Biennial Conference on the Biology of Marine Mammals, December 1987, Miami, FL.

Scholl, J. P., 1983. Skull fragments of the California sea lion (*Zalophus californianus*) in stomach of a white shark (*Carcharodon carcharias*). *J. Mammal*. 64:332.

Schroeder, C. R., C. A. Delli Quadri, R. W. McIntyre, and W. A. Walker, 1973. Marine mammal disease surveillance program in Los Angeles County. *J. Am. Vet. Med. Assoc*. 163: 580–581.

Scott, M. D., and J. G. Cordaro, 1987. Behavioral observations of the dwarf sperm whale, *Kogia simus*. *Mar. Mamm. Sci*. 3:353–354.

Seagars, D. J., and J. R. Henderson, 1985. Cephalopod remains from the stomach of a short-finned pilot whale collected near Santa Catalina Island, California. *J. Mammal*. 66:777–779.

Seagars, D. J., J. H. Lecky, J. J. Slawson, and H. S. Stone, 1986. Evaluation of the California marine mammal stranding network as a management tool based on records for 1983–1984. Natl. Mar. Fish. Serv., Southwest Fish. Center Admin. Rep. SWR–86–5. 34pp.

Sergeant, D. E., 1962. The biology of the pilot or pot-head whale *Globicephala melaena* (Traill) in Newfoundland waters. *Bull. Fish. Res. Board Can*. 132:1–84.

Sergeant, D. E., 1969. Feeding rates of cetacea. *Fiskeridir. Skr. Ser. Havunders*. 15:246–258.

Sergeant, D. E., and H. D. Fisher, 1957. The smaller cetacea of eastern Canadian waters. *J. Fish. Res. Board Can*. 14:83–115.

Shane, S. H., 1984. Pilot whales and other marine mammals at Santa Catalina Island, California, in 1983–84. NOAA-NMFS-SWFC Admin. Rep. LJ–84–28C.

Shane, S. H., R. S. Wells, and B. Wursig, 1986. Ecology, behavior and social organization of the bottlenose dolphin: A review. *Mar. Mamm. Sci*. 2:34–63.

Shaughnessy, P. D., and F. H. Fay, 1977. A review of the taxonomy and nomenclature of North Pacific harbor seals. *J. Zool. (Lond)*. 182(3):385–420.

Shaw, S. B., 1971. Chlorinated hydrocarbon pesticides in California sea otters and harbor seals. *Calif. Fish Game*. 57:290–294.

Shoemaker, V. H., K. A. Nagy, and W. R. Costa, 1976. Energy utilization and temperature regulation of jack rabbits in the Mojave Desert. *Physiol. Zool*. 49:364–375.

Smith, A. W., T. G. Akers, S. H. Madin, and N. A. Vedros, 1973. San Miguel sea lion virus isolation, preliminary characterization, and relationship to vesicular exanthema of swine virus. *Nature (Lond.)*. 244:108–110.

Smith, R. C., and K. S. Baker, 1982. Oceanic chlorophyll concentrations as determined by satellite (*Nimbus*–7 coastal zone color scanner). *Mar. Biol*. 66(3):269–280.

Smith, R. C., P. Dustan, D. Au, K. S. Baker, and E. A. Dunlap, 1986. Distribution of cetaceans and sea surface chlorophyll concentrations in the California Current. *Mar. Biol*. 91(3):385–402.

Spalding, D. J., 1964. Comparative feeding habits of the fur seal, sea lion, and harbour seal on the British Columbia coast. *Bull. Fish. Res. Board Can.*, No. 146, 52pp.

Spotte, S., and G. Adams, 1981. Feeding rate of

captive adult female northern fur seals, *Callorhinus ursinus*. *U.S. Natl. Mar. Fish. Serv. Fish Bull.* 79:182–184.

Sprague, J. G., N. B. Miller, and J. L. Sumich, 1978. Observations of gray whales in Laguna de San Quintin, northwest Baja California, Mexico. *J. Mammal.* 59(2):425–427.

Stewart, B. S., 1980. Historical and present populations of pinnipeds in the Channel Islands. In: J. R. Jehl, Jr., and C. F. Cooper, eds. Potential Effects of Space Shuttle Sonic Booms on the Biota and Geology of the California Channel Islands: Research Reports. Technical Report 80–1, prepared by the Center for Marine Studies, San Diego State Univ., San Diego, CA, for the U.S. Air Force, Space Division (Contracts F–04701–78-C–0060 and F–04701–78-C–0178). pp. 45–98.

Stewart, B. S., 1981a. Aerial censuses of harbor seals (*Phoca vitulina richardsi*) on the southern California Channel Islands, 27–29 June 1981. Hubbs/Sea World Res. Inst. Tech. Rep. 81–129. 19pp.

Stewart, B. S., 1981b. The Guadalupe fur seal (*Arctocephalus townsendi*) on San Nicolas Island, California. *Bull. South. Calif. Acad. Sci.* 80:134–136.

Stewart, B. S., 1982. Peak 1982 aerial census of harbor seal populations on the southern California Channel Islands: Final report. Sea World Research Institute–Hubbs Marine Research Center Technical Report #82–143. San Diego, CA. 8pp.

Stewart, B. S., 1984. Diurnal hauling patterns of harbor seals at San Miguel Island, California. *J. Wildl. Manage.* 48(4):1459–1461.

Stewart, B. S., and P. K. Yochem, 1984. Seasonal abundance of pinnipeds at San Nicolas Island, California, 1980–1982. *Bull. South. Calif. Acad. Sci.* 83(3):121–132.

Stewart, B. S., P. K. Yochem, R. L. DeLong, and G. A. Antonelis, Jr., 1987. Interactions between Guadalupe fur seals and California sea lions at San Nicolas and San Miguel Islands, California. In: J. P. Croxall, and R. L. Gentry, eds. *Status, Biology, and Ecology of Fur Seals.* Proceedings of International Symposium and Workshop, Cambridge, U.K., 23–27 April 1984. NOAA Tech. Rep. NMFS 51. pp. 103–106.

Stroud, R. K., 1968. Risso's dolphin in Washington State. *J. Mammal.* 49:347–348.

Stroud, R. K., C. H. Fiscus, and H. Kajimura, 1981. Food of the Pacific white-sided dolphin, *Lagenorhynchus obliquidens*, Dall's porpoise, *Phocoenoides dalli,* and northern fur seal, *Callorhinus ursinus,* off California and Washington. *U.S. Natl. Mar. Fish. Serv. Fish. Bull.* 78:951–959.

Sullivan, R. M., 1980. Seasonal occurrence and haul-out use in pinnipeds along Humboldt County, California. *J. Mammal.* 61:754–760.

Sullivan, R. M., and W. J. Houck, 1979. Sightings and strandings of cetaceans from northern California. *J. Mammal.* 60(4):828–833.

Sullivan, R. M., J. D. Stack, and W. J. Houck, 1983. Observations of gray whales (*Eschrichtius robustus*) along north California. *J. Mammal.* 64:689–692.

Sumich, J. L., 1983. Swimming velocities, breathing patterns, and estimated costs of locomotion in migrating gray whales, *Eschrichtius robustus. Can. J. Zool.* 61(3):647–652.

Sumich, J. L., 1984. Gray whales along the Oregon coast in summer, 1977–1980. *Murrelet.* 65: 33–40.

Sund, P. N., 1975. Evidence of feeding during migration and of an early birth of the California gray whale *Eschrichtius robustus. J. Mammal.* 56(1):265–266.

Sund, P. N., and J. L. O'Connor, 1974. Aerial observations of gray whales during 1973. *Mar. Fish. Rev.* 36(4):51–55.

Swartz, S. L., and M. L. Jones, 1982. Demographic studies and habitat assessment of gray whales, *Eschrichtius robustus,* in Laguna San Ignacio, Baja California Mexico. Final Report prep. by Cetacean Research Associates, San Diego, CA, for the Marine Mammal Commission, Washington, D.C. Available from: NTIS, Springfield, VA; PB82–123373, 56pp.

Sweeney, J. C., 1973. Management of pinniped diseases. *Am. Assoc. Zoo Vet. Annu. Proc.* pp. 141–171.

Sweeney, J. C., 1978. Marine mammals (Cetacea, Pinnipedia, and Sirenia): Non-infectious diseases. In: M. E. Fowler, ed. *Zoo and Wild Animal Medicine.* Saunders, Philadelphia. pp. 596–599.

Sweeney, J. C., and W. G. Gilmartin, 1974. Survey of diseases in free-living California sea lions. *J. Wildl. Dis.* 10:370–376.

Sweeney, J. C., G. Migaki, P. M. Vainik, and R. H. Conklin, 1976. Systemic mycosis in marine mammals. *J. Am. Vet. Med. Assoc.* 169: 946–948.

Townsend, C. H., 1899. Notes on the fur seals of

Guadalupe, the Galapagos and Lobos Islands. In: *The Fur Seals and Fur Seal Islands of the North Pacific Ocean.* Rep. of fur seal invest., 1896–1897. Part 3. U.S. Gov. Printing Office, Washington, D.C. pp. 265–274.

U.S. Department of Commerce, 1986. Fur Seal Investigations, 1984. NOAA Tech. Mem. NMFS F/NWC–97.

U.S. Department of Commerce, 1987. Marine Mammal Protection Act of 1972, Annual Report 1986/87. NOAA, Natl. Mar. Fish. Serv.

U.S. Fish and Wildlife Service, 1982. Southern sea otter recovery plan. USFWS Office of Endangered Species, Ventura, CA. 66pp.

U.S. Fish and Wildlife Service, 1987. Translocation plan for southern sea otters. Appendix B. Final Environmental Impact Statement: Translocation of Southern Sea Otters. Prep. by U.S. Fish Wildl. Serv., Sacramento, CA, and the Institute of Marine Studies, Univ. of California, Santa Cruz, CA.

Urbán-R., J., and A. Aguayo-L., 1987. Spatial and seasonal distribution of the humpback whale, *Megaptera novaeangliae,* in the Mexican Pacific. *Mar. Mamm. Sci.* 3:333–344.

VanBlaricom, G. R., and J. A. Estes, eds., 1988. *The Community Ecology of Sea Otters. Ecological Studies: Analysis and Synthesis.* Vol. 65. Springer-Verlag, New York. 237pp.

Walker, P. L., and S. Craig, 1979. Archeological evidence concerning the prehistoric occurrence of sea mammals at Point Bennett, San Miguel Island. *Calif. Fish Game.* 65:50–54.

Walker, W., 1975. Review of the live capture fishery for smaller cetaceans taken in southern California waters for public display. 1966–73. *J. Fish. Res. Board Can.* 32:1197–1211.

Walker, W. A., F. G. Hochbert, and E. S. Hacker, 1984. The potential use of the parasites *Crassicauda* (Nematoda) and *Nasitrema* (Platyhelminths) as biological tags and their role in the natural mortality of common dolphins, *Delphinus delphis,* in the eastern North Pacific. Natl. Mar. Fish. Serv., Southwest Fish. Center Admin. Rep. LJ–84–08C. 31pp.

Wellington, G. M., and S. Anderson, 1978. Surface feeding by a juvenile gray whale *Eschrichtius robustus. U. S. Natl. Mar. Fish. Serv. Fish. Bull.* 76(1):290–293.

Wells, R. S., T. P. Dohl, L. J. Hansen, A. B. Baldridge, and D. L. Kelly, 1983. Extraordinary movements of bottlenose dolphins (*Tursiops* sp.) along the coast of California (abstract). Am. Soc. Mammal. Ann. Meet., Humboldt State Univ., Arcata, CA.

Wendell, F. E., R. A. Hardy, and J. A. Ames, 1985. Assessment of the accidental take of sea otters, *Enhydra lutris,* in gill and trammel nets. Unpubl. Rep., Calif. Dept. Fish Game, Mar. Resour. Branch. 30pp.

Whitehead, H., and C. Glass, 1985. Orcas (killer whales) attack humpback whales. *J. Mammal.* 66:183–185.

Wolf, J. M., R. H. Defran, and G. M. Shultz, 1987. Photographic surveys of the bottlenose dolphin (*Tursiops truncatus*) along the Ventura–Santa Barbara coastline of southern California (abstract). In: Seventh Biennial Conference on the Biology of Marine Mammals, December 1987, Miami, FL.

Wolman, A. A., 1986. Humpback whale. In: D. Haley, ed. *Marine Mammals of Eastern North Pacific and Arctic Waters.* 2nd ed. Pacific Search Press, Seattle, WA. pp. 57–63.

Wolman, A. A., and A. J. Wilson, 1970. Occurrence of pesticides in whales. *Pestic. Monit. J.* 4:8–10.

Woodhouse, C. D., Jr., and J. Strickley, 1982. Sighting of northern right whale in the Santa Barbara Channel. *J. Mammal.* 63:701–702.

Yochem, P. K., B. S. Stewart, R. L. DeLong, and D. P. DeMaster, 1987. Diel haul-out patterns and site fidelity of harbor seals (*Phoca vitulina richardsi*) on San Miguel Island, California, in autumn. *Mar. Mamm. Sci.* 3(4):323–332.

York, A. E., 1987. Northern fur seal, *Callorhinus ursinus,* eastern Pacific population (Pribilof Islands, Alaska, and San Miguel Island, California). In: J. P. Croxall, and R. L. Gentry, eds. *Status, Biology, and Ecology of Fur Seals.* Proceedings of International Symposium and Workshop, Cambridge, U.K., 23–27 April 1984. NOAA Tech. Rep. NMFS 51. pp. 9–21.

York, A. E., 1990. Trends in numbers of pups born on St. George and St. Paul Islands, 1973–88. In: H. Kajimura, ed. Fur Seal Investigations, 1987–1988. NOAA Tech. Mem. NMFS-F/NWC–180. Natl. Marine Fish. Serv. pp. 31–37.

Chapter 12

Human Impacts

Jack W. Anderson, Donald J. Reish, Robert B. Spies, Michael E. Brady, and Elbert W. Segelhorst

Introduction

History of Contamination

The coastal region of southern California contains one of the most rapidly expanding urban complexes in the United States. Approximately two-thirds of the population of this area reside in Los Angeles and Orange counties (fig. 12.1). Intense development within the region has directly influenced the coastal waters, traditionally a repository for human wastes. The first inventory of contaminant inputs to the Southern California Bight (SCB) was compiled by the Southern

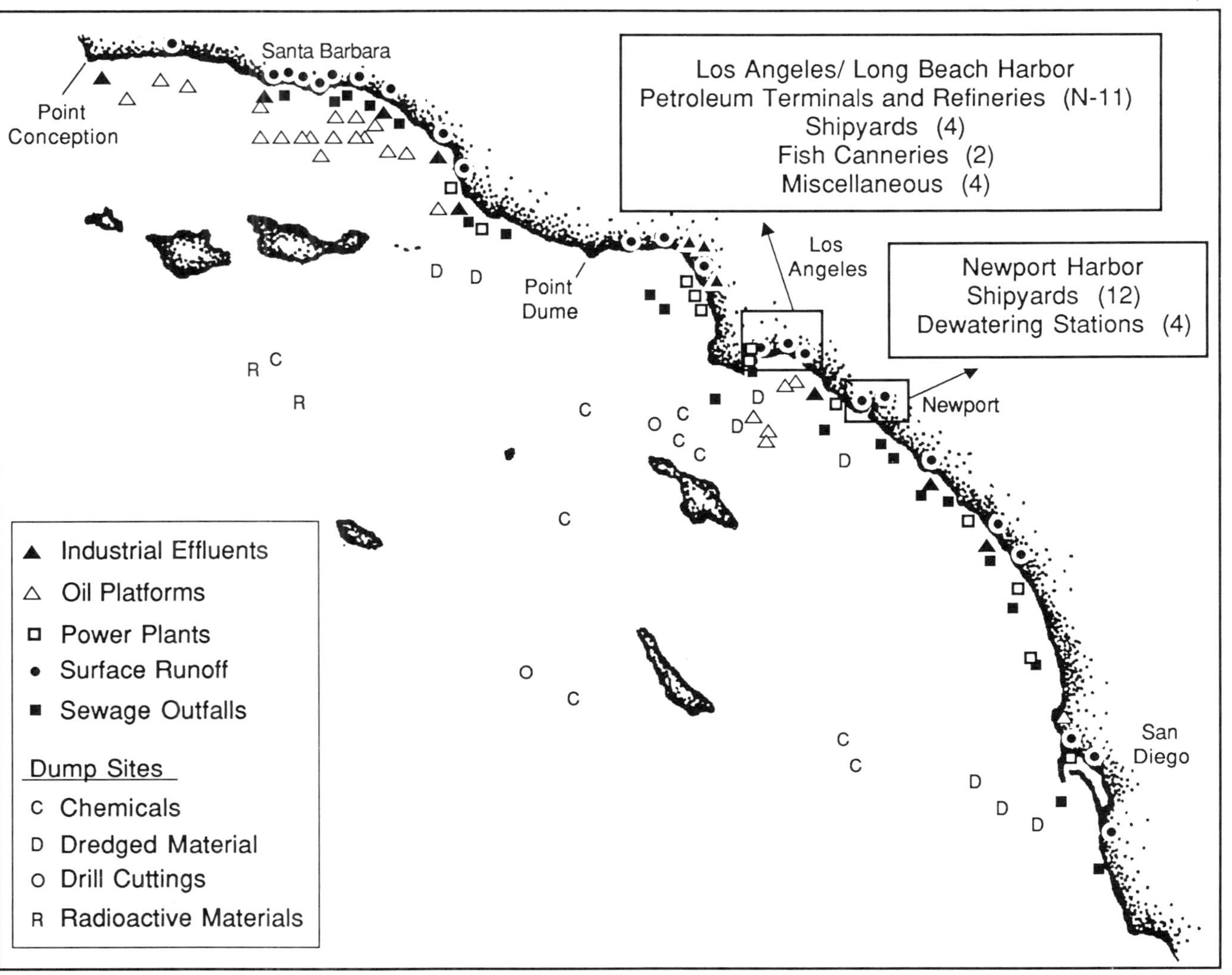

Figure 12.1. Location of major contaminant inputs to the SCB.

California Coastal Water Research Project (SCCWRP) in 1971–1972 (Southern California Coastal Water Research Project 1973). To the extent possible, SCCWRP examined and quantified "constituents . . . introduced through man's physical contact with the waters and [through] exploitation of marine resources" (p. 55). No new comprehensive inventory has been accomplished since that time.

Over the years, local, state, and federal agencies and universities have published many reports on contamination in the SCB. In addition to published reports, a large body of valuable unpublished information exists on contaminant inputs collected by both public agencies and private interests. But these data are largely unknown beyond the agencies that either collect them or monitor their collection. So that these published reports and the more obscure data collections might contribute to this synthesis, the authors initiated a thorough and systematic search, compilation, and summary of the best available data.

Evidence of high levels of DDT compounds (including DDE and other degradation products) in sediments on the Santa Monica and San Pedro coastal shelves and basins in 1973 came from MacGregor (1976), and both trace metals and organochlorines in sediments were described by Chen and Lu (1974). Risebrough (1969) and Klein and Goldberg (1970) first reported the occurrence of organochlorine compounds and mercury, respectively, in marine fish or macroinvertebrates from sites within the SCB. MacGregor (1974) analyzed DDT and polychlorinated biphenyls (PCBs) in museum specimens of midwater fish collected from the SCB as early as 1940. In a study of contaminant levels in commercial fishes collected off the west coast, Stout and Beezhold (1981) found the highest concentrations of DDT in fish from the waters off Los Angeles. Chlorinated hydrocarbon body burdens declined steadily from southern California to Alaska, except for isolated "hot spots" in San Francisco Bay and Puget Sound. Burnett (1971) found a similar trend for DDT in mole crabs collected from San Francisco to the Mexican border. Matta et al. (1985) found that fish and invertebrates from the coastal waters off Los Angeles had the highest levels of DDT and the third highest levels of PCBs (after Puget Sound and Vancouver, B.C.) along the west coast. Nearly two decades after the source of DDT was removed, fish livers from the Los Angeles region (Point Dume to Los Angeles Harbor) still contained the highest levels of DDT in the United States (Mearns et al. 1988). Bertine and Goldberg (1977) have observed that trace metal concentrations in coastal sediments have decreased from the high levels found up to about 1970.

A draft report from NOAA (Mearns et al. 1991), entitled *Contaminant Trends in the SCB: Inventory and Assessment,* provides a comprehensive review of the past and present levels of metals, chlorinated organics, and fossil fuel hydrocarbons in the sediments and organisms of the SCB. This chapter cannot begin to cover the depth and range of findings contained within the report, but it will present a few of the conclusions. For example, there appears to be a pattern of distribution of DDT, PCBs, chlordane, and tin in fish livers that approximates the levels found in nearby sediment sources and mussels. Unlike tin, most trace metals are elevated in livers of fish collected in sites remote from sources of input. Low or possibly depressed levels of these metals are found in fish livers from sites known by other measures of contamination to be polluted. Data sets are available for about 10 years on mussels at Royal Palms, near the Los Angeles County Sewer Discharge (LACSD) outfall, and at Oceanside. At Royal Palms from about 1977 to 1985, concentrations of DDT, PCBs, chromium, and lead in mussels declined by factors of two to ten as sewage inputs of these and other trace metals decreased by 50–70%. In contrast, levels of silver, cadmium, copper, mercury, and zinc varied at Royal Palms, with no clear trends. Mussels at Oceanside exhibited variable concentrations of the contaminants that were always lower than those at Royal Palms. Since sediment concentrations followed the decreases in input at Royal Palms, it appears that mussel levels at this site are independent of sewage inputs of cadmium, copper, mercury, and zinc.

The report concludes that all pollutants measured have accumulated, over reference site values, in sediments at one site (Palos Verdes) and most of these have also accumulated in mussels. Far fewer contaminants have shown accumulation in other macroinvertebrates and fish. The only strong evidence of increasing pollutant concentrations with increasing trophic level (biomagnification) is for mercury and chlorinated organics (DDT, PCBS, and chlordane).

Sources of Contamination

The SCB receives nutrients and contaminants from a wide variety of sources that differ in their composition and modes of timing and delivery to the area (table 12.1). While there are nearly 200 discrete sources, these sources generally fall into five major categories: waste outfalls, river and storm runoff, atmospheric fallout, ocean dumping, and current advection. An additional mechanism for classifying inputs is to consider these three characteristics: (1) space—where and how contaminants are introduced, (2) time—the timing and duration of the input event and input route, and (3) contaminant gradient—the concentration difference between the input and the receiving waters. Evaluation of each class of input in this scheme (disregarding data availability for the moment) provides a time series for contaminant inputs that is likely to provide the best information.

The inputs most readily evaluated would have four characteristics in common: (1) spatially localized with few discrete sources, (2) temporally continuous with long durations and low variability, (3) concentrated compared to the receiving environment, and (4) regularly monitored for compliance. Municipal wastewater discharges and discrete industrial discharges possess most of these characteristics.

Some inputs that are very difficult to estimate characteristically are (1) spatially extensive with many diffuse sources, (2) temporally intermittent (especially aperiodic) with short durations and high variability, (3) dilute compared to the receiving environment, and (4) infrequently or never sampled or analyzed. Current advection and atmospheric fallout exhibit most of these characteristics. Most classes of contaminant inputs in the SCB fall somewhere between these extremes (table 12.1).

The concern of this chapter is with constituents having fluxes from anthropogenic inputs that are significant in comparison to the natural mobilization rates in the SCB. Current advection is not considered because data on contaminant levels in the California Current and the volume of flow into various regions of the SCB are sparse. We recognize from a recent report (Mearns et al. 1991) that the distribution of several trace metals, particularly cadmium, in biota collected throughout the SCB over several years may not be from land sources but from ocean currents. However, industrial inputs to the California Current upstream from the SCB are small compared to the major municipal and industrial inputs within the SCB.

The main transport paths of contaminants from the land to the SCB are waste outfalls, river and storm runoff, atmospheric fallout, and ocean dumping. Inputs of metals and oil and grease from waste outfalls exceed inputs from river and storm runoff by a factor of three (table 12.2). Most of the water used for domestic and industrial purposes (together with some agricultural waters) enters municipal wastewater treatment plants and eventually empties into the ocean (Southern California Coastal Water Research Project 1973). In 1980, 17 wastewater treatment plants (fig. 12.1) discharged approximately 1.6×10^9 m^3 yr^{-1} into the SCB (table 12.2).

Other sources of contaminants include direct industrial waste discharge to the SCB

Table 12.1. *Categorization of Contaminant Inputs to the SCB*[a]

	Spatial			Temporal			
Input	Location	Loci	Dispersion	Flow	Duration (Event/System)	Variability	Mass Intensity (Concentration)
Terrestrial							
Waste discharge							
Municipal outfalls	Bottom nearshore	Localized few	Discrete	Continuous	Long/long	Low	Concentrated
Power plants	Bottom nearshore	Localized few	Discrete	Continuous	Long/long	Low	Dilute
Ocean dumping	Surface near-/offshore	Localized few	Discrete	Intermittent aperiodic	Long/short	High	Concentrated
Industrial outfalls	Surface/bottom nearshore	Localized many	Discrete	Continuous	Long/long	Unknown	Concentrated
Surface runoff							
Storm	Surface near-shore	Localized many	Discrete	Intermittent periodic/aperiodic	Long/short	High	Concentrated
Dry	Surface near-shore	Localized many	Discrete	Continuous	Long/long	Unknown	Concentrated
Atmospheric fallout							
Rain	Surface near-/offshore	Extensive	Diffuse	Intermittent periodic/aperiodic	Long/short	Unknown	Dilute
Dry	Surface	Extensive	Diffuse	Continuous	Long/long	Unknown	Dilute
Marine							
Oil seeps	Bottom near-/offshore	Localized few	Discrete	Continuous aperiodic	Long/long	Low	Concentrated
Oil platforms	Bottom near-/offshore	Localized few	Discrete	Continuous	Long/long	Unknown	Concentrated
Shipping and boating	Surface near-/offshore	Extensive many	Discrete	Intermittent aperiodic	Short/short	Low	Concentrated
Current advection	Surface to bottom near-/offshore	Extensive	Diffuse	Continuous/intermittent	Long/short	Unknown	Dilute

[a] After Southern California Coastal Water Research Project (1973) and Eganhouse (1982).

Table 12.2. *Volume of Contaminant Inputs in the SCB in 1970–1971 and 1980*

	Volume (m^3 yr^{-1})	
	1971–1972	1980
Municipal wastewaters[a]	1.5×10^9	1.6×10^9
Industrial outfalls[a]	0.25×10^9	0.46×10^9
Returned cooling waters[a]	7.7×10^9	14.9×10^9
Surface runoff[a]	0.57×10^9	0.56×10^9
Chemical ocean dumping[b]	3.79×10^2	—
Oil seeps[c]	$5.8–58.0 \times 10^3$	$2.3–35.0 \times 10^3$

Sources:

[a] Southern California Coastal Water Research Project (1973).

[b] Off Los Angeles: Chartrand et al. (1985).

[c] From Point Conception to Long Beach: Wilson et al. (1973), Fischer (1978).

and power plant thermal effluents (table 12.2). In 1980, there were more than 64 permitted industrial discharges to the SCB, 17 electrical generating plants, and 26 offshore oil platforms. Discrete industrial outfalls, although numerous, individually discharge low volumes of wastes; in total, they discharge about one-fourth the volume of municipal outfalls. Most industrial discharges are associated with the petroleum industry (oil production, transport, and processing). The discharges from oil platforms vary over the lives of the platforms. Discharges are greatest during the construction and drilling phases or the first 3 to 4 years of the 30-year average life span. Produced water generated from platforms in the SCB is either reinjected to the reservoir from which it came or transported to the coast by pipeline for treatment and subsequent discharge nearshore.

Power plants discharge 10 times the volume of municipal wastewater treatment plants (table 12.2). Thermal effluents are returned ocean waters that carry waste heat and a small volume of contaminants compared to municipal wastewaters. SCCWRP (1973) estimated that the waste heat loss for 1971–1972 from the 14 plants operating at that time was 1.96×10^7 kW or 4.1×10^{14} g-cal d^{-1}. With a net solar and atmospheric radiation of 0.1 g-cal cm^{-2} d^{-1}, the amount of waste heat from the power plants was equivalent to the absorbed radiation of 20 km^2 of coastal water. Waste heat is not likely to affect coastal fish populations significantly. However, the impingement of juvenile and adult fish and invertebrates and the entrainment of fish eggs and larvae in cooling water intakes during normal operations and heat treatments do have impact. The use of biocides (chlorine) may also have an impact in the receiving waters, but estimates of the effects of chlorine alone on biota of the SCB are not well understood.

Surface runoff is composed of storm and dry weather flows that differ in contaminant concentrations, timing, and duration. Surface runoff is approximately one-third the volume of municipal wastewater discharge (table 12.2). Eighteen major watersheds draining 3.3×10^4 km^2 through more than 200 rivers, streams, and storm drains discharge into the SCB. Estimates made in 1973 suggest that storms account for approximately 70% of the surface runoff (Southern California Coastal Water Research Project 1973) and probably carry more contaminants than dry weather runoff to the SCB. More than 60% of the input of the extractable organic compounds from the Los Angeles River to marine waters occurs during storm runoff (Eganhouse et al. 1981). The Los Angeles River accounts for approximately 28% of the volume of surface runoff to the SCB. Urban runoff of petroleum compounds from this river represents about 1% of the petroleum compounds discharged by runoff to marine waters throughout the world (Eganhouse and Kaplan 1981).

A variety of wastes either have been or are currently being dumped from vessels into the SCB, including industrial (bulk and containerized) and radioactive wastes, cuttings and mud from oil drilling, dredged materials, refuse and garbage, and military explosives. SCCWRP (1973) found very little in the public record about specific contaminants in these wastes. They concluded that only oil refinery and chemical wastes contributed significant amounts of contaminants. Industrial wastes were dumped in the San Pedro Channel off Los Angeles from the 1930s to 1975 (table 12.2). Between 1961 and 1972, more than 11.3×10^3 m^3 of such wastes were dumped in the channel by two salvage companies (Chartrand et al. 1985). Most ocean dumping of dredged material was discontinued during the mid-1970s, but in recent years, dumping has again become a major input of particulates to the coastal environment. Sediments dredged from bays and harbors that are not used for landfill or are unsuitable for beach replenishment are dumped at one of three sites off Los Angeles (LA 2), Newport Beach (LA 3), and San Diego (LA 5) (designated D on fig. 12.1). Between 1977 and 1986, 3.4×10^6 m^3 of material was discharged at the Los Angeles and San Diego sites. Historical records for the Newport Beach site could not be found, but in 1987, 0.84×10^6 m^3 of material dredged from Upper Newport Bay was dumped at this site. Although some limited sediment chemistry and bioassays are required before a discharge permit is issued, the data are generally not adequate to estimate contaminant emissions.

The magnitude of contaminant input into the SCB from the atmosphere is not well understood. Atmospheric contributions are difficult to quantify because of the behavior of particles in air, the diffuse sources of particles, and the high concentration gradients near densely populated areas (Southern California Coastal Water Research Project 1973; Huntzicker et al. 1975). Dry aerial fallout is probably a minor source of most metals in the coastal waters near densely populated areas in southern California. With the exception of lead, dry aerial inputs of toxic metals (excluding manganese and iron, which are relatively nontoxic) and chlorinated hydrocarbons were found to be one to two orders of magnitude lower than those of municipal wastewaters (Southern California Coastal Water Research Project 1973). The importance of aerial fallout of contaminants would not change much even if major fires burned year-round (Young and Jan 1976).

Mass Emissions

SCCWRP (1973) identified the major inputs as municipal wastewater discharge, surface runoff, vessel coating, ocean dumping, and rainfall, and they estimated the annual mass emissions of eleven metals and two chlorinated hydrocarbons in 1971–1972 (table 12.3). Mass emissions of silver (Ag), cadmium (Cd), chromium (Cr), copper (Cu), nickel (Ni), and DDT were highest in municipal wastewater discharges; mass emission of iron (Fe) was highest in surface runoff; mass emissions of cobalt (Co) and PCBs were highest in ocean dumping; and mass emissions of mercury (Hg), lead (Pb), zinc (Zn), and manganese (Mn) were highest in storm runoff.

Discrete sources (178) of contaminant and nutrient input to the SCB from Point Conception to the Mexican border have been identified (fig. 12.1). These sources include municipal wastewater outfalls, industrial outfalls, power plant cooling flows, surface runoff, oil platforms, and dump sites. Twenty-six discrete input sources are located in the area

Table 12.3. *Percentage Contribution and Total Mass of Trace Metal and Chlorinated Hydrocarbon Inputs by Five Types of Ocean Discharges to the SCB in 1970–1971*

	Municipal Wastewater (%)	Surface Runoff (%)	Vessel Coating (%)	Ocean Dumping (%)	Rainfall (%)	Annual Total (t)
Ag	84	7	—	9	—	18
Cd	78	1	<1	20	—	69
Co	14	23	—	64	—	22
Cr	92	4	<1	4	—	703
Cu	41	1	28	2	29	1,399
Hg	18	<1	24	9	47	17
Ni	79	4	—	7	10	398
Pb	18	8	<1	3	71	1,399
Zn	40	2	4	1	52	4,201
Fe	19	80	—	<1	<1	32,320
Mn	13	23	—	4	61	793
DDT	54	<1	—	40	6	35
PCB	23	<1	9	64	5	44

From Southern California Coastal Water Research Project (1973).

from Point Conception to Point Dume (table 12.4). In this area, the largest freshwater inputs, and probably the largest sources of nutrients and contaminants, are the Santa Clara and Ventura rivers and the municipal wastewater treatment plant at Oxnard. Seventy distinct input sources are located in the area between Point Dume to San Mateo Point (southern Orange County). The largest sources of freshwater inputs are three municipal wastewater treatment plants (Los Angeles County, the City of Los Angeles, and Orange County) and the Los Angeles and San Gabriel rivers. This region contains the largest industrial discharges (from Champlin, Chevron, and Union Oil refineries) and the greatest number of power-generating stations in the SCB. Twenty-four discrete input sources are located between San Mateo Point to the Mexican border (table 12.4). The largest source of freshwater is the municipal wastewater treatment plant at Point Loma, which discharges more than twice the volume of the three largest rivers combined. This area contains three electrical power-generating stations. Fifty-seven unrelated input sources were identified in the offshore waters, which are dominated by dump sites and oil platforms (table 12.4).

By far, the majority of discrete inputs to the SCB are located between Point Dume and San Mateo Point. Approximately 82% of municipal wastewater effluents, 95% of the discrete industrial wastes, 70% of power plant–returned cooling waters, and 71% of surface runoff enter the coastal waters in this area (table 12.4). The majority of contaminants from aerial fallout also probably originate in this region since it contains about 70% of the total southern California population. The highest levels of contamination in sediments and animals in the SCB are also found in this region (Katz and Kaplan 1981; Brown et al. 1984). For these reasons, we will emphasize the findings regarding the distribution of anthropogenic contaminants and the impacts on the biota reported for sites within Los Angeles and Orange counties.

Table 12.4. *Summary of Inputs from Discrete Sources to the SCB by Area*

Area	Class	Flow (million m^3 yr^{-1})	Number
Santa Barbara and Ventura counties	Sewage effluents	52.9	8
	Power plants	2,543.9	2
	Industrial wastes	1.7	6
	Surface runoff	59.4	10
Los Angeles and Orange counties	Sewage effluents	1,348.7	8
	Power plants	11,125.5	8
	Industrial wastes	450.5	44
	Surface runoff	398.9	10
San Diego County	Sewage effluents	231.5	4
	Power plants	2,161.8	5
	Industrial wastes	19.7	12
	Surface runoff	80.6	3
Offshore	Sewage effluents	0.4	1
	Power plants	4.5	1
	Dump sites	899.6	31
	Oil platforms	79.5	24

Municipal Wastes

Reish (1984) reviewed available information on the treatment of domestic wastes, methods of disposal, quantitation of contaminants, and the fate and effects of these inputs. Some of the information contained in that review, which primarily drew upon data from the SCB, is discussed and updated here.

Inputs

The concentrations of contaminants in municipal wastewater entering the SCB have been efficiently measured by the dischargers for many years, and SCCWRP has been compiling them since 1971. In a recent review, Schafer (1989) described the trends in these effluents between 1971 and 1988. Seven of the 19 municipalities bordering the SCB, including the four largest that discharge to these marine waters, have compiled a history of discharge data. Table 12.5 lists the 1986 and 1987 concentrations of numerous nutrients and contaminants discharged from the four largest of these municipal facilities. Volumes discharged to the SCB range from about 57×10^6 l d^{-1} for the small plants to about 1470×10^6 l d^{-1} for the city and county of Los Angeles. The Hyperion 7-mile outfall in Santa Monica Bay discharged diluted, digested sludge at the head of Santa Monica Canyon from September 1957 until November 1987. While this discharge represented only a small volume (15×10^6 l d^{-1}), the mass emissions of suspended solids were about equal to that of the 5-mile outfall, which exhibits 100 times the fluid volume.

A comparison of mass emissions from the combination of the seven outfalls (two in Santa Monica Bay) during the period of 1971–1987 is shown in table 12.6. Total flows have increased from 3.5×10^9 l d^{-1} to 4.5×10^9 l d^{-1} over this period, a time of rapid population expansion in southern California. During this time, both DDT and PCBs were banned in the United States, and water quality regulations became more stringent as a result of the 1972 enforcement of the Clean Water Act by the California State Water Resources Control Board and the Environmental Protection Agency. Table 12.6 dem-

Table 12.5. *Calculated Annual Mass Emissions (in t yr^{-1}, except as noted) for 1986 and 1987*

	L.A. County		L.A. City (5-mile)		L.A. City (7-mile)[a]		Orange County	
Constituent	1986	1987	1986	1987	1986	1987	1986	1987
Flow								
l (10^9 yr^{-1})	503	506	542	518	5.4	4.7	329	348
MGD[b]	364	366	392	375	3.9	3.4	238	252
Susp. solids	41,200	36,900	41,700	30,000	70,200[c]	59,200[c]	16,100	16,400
BOD	50,300	54,600	80,200	60,000	—	—	25,000	26,500
Oil and grease	5,130	5,610	11,500	7,900	3,710	3,260	3,320	4,500
NO_3-N	261	253	149	7	—	—	—	—
NO_2-N	10	15	60	135	—	—	—	—
NH_3-N	20,100	19,200	8,670	8,400	2,810	2,280	8,220	8,700
Organic-N	4,020	3,740	4,080	3,330	2,760	2,600	—	—
Total P	4,070	3,790	2,930	2,280	1,300	1,290	—	—
MBAS[d]	1,410	1,670	2,390	2,200	—	—	—	—
Cyanide	10	10	9	14	0.4	0.4	0.2	<7
Phenol	805	1,010	21	21	0.8	0.5	—	—
Ag	4	4	9	5	2	1	4	3
As	4	4	5	4	2	2	1	1
Cd	2	1	5	3	4	3	1	1
Cr	29	26	19	7	25	17	9	7
Cu	26	21	65	30	71	44	27	26
Hg	0.2	0.1	0.2	0.1	0.2	0.1	0.1	0.1
Ni	30	26	61	29	14	7	11	10
Pb	28	23	28	22	18	8	7	3
Se	7	7	0.5	<3	0.3	0.2	—	—
Zn	80	61	118	109	85	63	30	24
Total DDT (kg yr^{-1})	35	30	3	<10	0.8	<1	12	14
Total PCB (kg yr^{-1})	ND[e]	ND	<54	4	26	2	454	250

[a]Terminated November 1987.
[b]Million gallons per day.
[c]Total solids.
[d]MBAS = methylene blue activated substances.
[e]ND = not detected.
From Southern California Coastal Water Research Project (1988).

onstrates the significant reduction in the mass emissions of these two chlorinated organics to present (1992) relatively low levels. During the period 1971–1987, all trace metals except silver decreased significantly. Although silver is one of the better indicators of municipal waste discharge, its various sources are not well known. These sources are likely numerous and difficult to regulate, however.

The pattern of change in population size,

Table 12.6. *Combined Mass Emissions (in t yr^{-1}, except as noted) for 1971–1987 from Seven Municipal Outfalls*

Constituent	1971	1972	1973	1974	1975	1976	1977	1978
Flow								
1 (10^9 yr^{-1})	1,286	1,274	1,319	1,336	1,361	1,419	1,335	1,402
MGD	931	922	955	967	985	1,027	966	1,015
Total Susp. Solids	288,000	279,000	270,000	264,000	287,000	288,000	244,000	256,000
BOD[d]	283,000	250,000	217,000	222,000	237,000	259,000	244,000	237,000
Oil and Grease	63,500	60,600	57,400	54,700	57,400	59,100	49,000	49,000
NH_3-N	56,600	39,900	45,900	37,000	36,600	37,400	41,200	39,500
Ag	18	21	29	22	26	20	34	32
As	—	—	—	21	12	11	14	15
Cd	57	34	49	55	50	45	42	45
Cr	676	673	695	690	580	593	366	280
Cu	559	485	509	575	511	507	412	417
Hg	—	—	—	3	2	3	3	2
Ni	339	273	318	314	124	307	264	320
Pb	243	226	180	199	196	191	152	219
Se[e]	—	—	—	18	17	22	23	23
Zn	1,880	1,210	1,360	1,320	1,142	1,064	837	905
Total DDT[f] (kg yr^{-1})	21,700	6,600	4,120	2,120	1,990	1,670	920	1,110
Total PCB[f] (kg yr^{-1})	8,730	9,830	4,620	9,390	6,010	4,310	2,180	2,510

[a] SERRA and Encina data first included.

[b] Discharge from Hyperion 7-mile outfall was terminated in November 1987.

[c] Four largest treatment plants (JWPCP, Hyperion, Orange Co., and Point Loma). From SCCWRP (1990).

[d] Hyperion 7-mile outfall data excluded.

[e] Data include only JWPCP, Hyperion 5- and 7-mile outfalls and Point Loma.

[f] Values for 1971–1975 are from SCCWRP's final report to the U.S. Environmental Protection Agency for Grant Nos. 801153 and R803707.

From Southern California Coastal Water Research Project (1988).

flow rates, and mass emission of total solids is shown in figure 12.2. Over the period when population and flows were constantly increasing (fig. 12.2a), mass emissions of suspended particulates decreased from 288,000 to 205,000 t yr^{-1} (fig. 12.2b). This reduction was the result of enhanced source control and improved particulate removal by advanced primary treatment and increases in secondary treatment, which together resulted in a one-third decrease in the mass of contaminants entering the ocean. Figure 12.3 provides a view of the rate of decrease in mass emissions of several contaminants (Ag, Cd, Zn, Cr, Cu, oil and grease, DDTs, and PCBs). Dramatic decreases, except for Ag, were exhibited throughout the period between 1971 and 1988. Elimination of a waste stream from the largest manufacturer of DDT in the country produced the rapid reduction in DDT after 1972.

Effects

An extensive field survey of surface sediment contamination in Santa Monica Bay and

1979	1980	1981	1982[a]	1983	1984	1985	1986	1987[b]	1988[c]	1989[c]
1,456	1,516	1,516	1,567	1,611	1,622	1,644	1,691	1,702	1,632	1,658
1,054	1,097	1,097	1,134	1,166	1,174	1,190	1,224	1,231	1,178	1,200
3,000	233,000	226,000	227,000	247,000	198,000	205,000	187,000	162,000	97,000	83,400
6,000	260,000	264,000	269,000	256,000	230,000	255,000	184,000	169,000	168,800	161,100
5,000	39,000	37,000	31,900	36,300	30,200	34,300	29,300	26,600	25,300	22,600
1,200	42,000	41,000	44,000	40,600	40,800	44,200	43,900	45,600	44,600	45,500
42	31	28	26	26	25	27	22	15	11	11
15	11	12	9	10	18	16	12	12	8.9	7.4
42	40	33	21	24	16	17	15	10	3.4	1.9
237	275	187	203	164	140	110	88	60	29	22
359	336	339	286	247	252	240	205	135	76	68
3	2	2	1	1	1	1	1	<1	0.4	0.4
256	224	167	169	165	134	120	129	78	63	54
223	175	130	123	99	94	120	106	64	50	27
8	11	15	9	10	9	13	8	7	6.7	7.6
724	730	540	549	505	374	377	345	276	151	146
760	640	470	290	220	310	58	50	53	26	20
1,190	1,130	1,250	860	1,440	1,340	820	480	250	ND	ND

offshore of the Palos Verdes Peninsula was conducted by SCCWRP between 1975 and 1978 (Southern California Coastal Water Research Project 1978). The sampling grid (fig. 12.4a) included numerous stations along isobaths from 15 to 300 m. Concentration contours have been prepared for several contaminants, including DDT, PCBs, and several trace metals. Figure 12.4b shows the distribution of silver in the region sampled, a pattern that is very similar to that of the other contaminants measured. Some small areas near the outfalls from Los Angeles County and the city of Los Angeles (sludge, 7-mile outfall) exhibit concentrations of greater than 20 mg kg^{-1}, or nearly three orders of magnitude greater than levels found at similar depths (60 m) in reference sites sampled in 1985 (Thompson et al. 1987). The distribution of cadmium (fig. 12.4c) shows that in 1978, most of Santa Monica Bay and the inshore portion of the shelf off Palos Verdes had less than 2 mg kg^{-1} in the surface sediments. Contours indicating concentrations of contaminants around the outfalls in Santa Monica Bay (5-mile and 7-mile) and off Palos Verdes decreased in surface area as the ranges increased from about 5 mg kg^{-1} to greater than 50 mg kg^{-1}.

To get a historical perspective, and to see the success of regulatory actions taken in 1972, one needs only to follow the changes that occurred in the influent and effluent quality of a major wastewater treatment plant. The Joint Water Pollution Control Plant (JWPCP) of the County Sanitation Districts of Los Angeles recorded changes in the concentrations of several contaminants in both incoming and outgoing waste streams over the period 1975–1985 (fig. 12.5). The influent

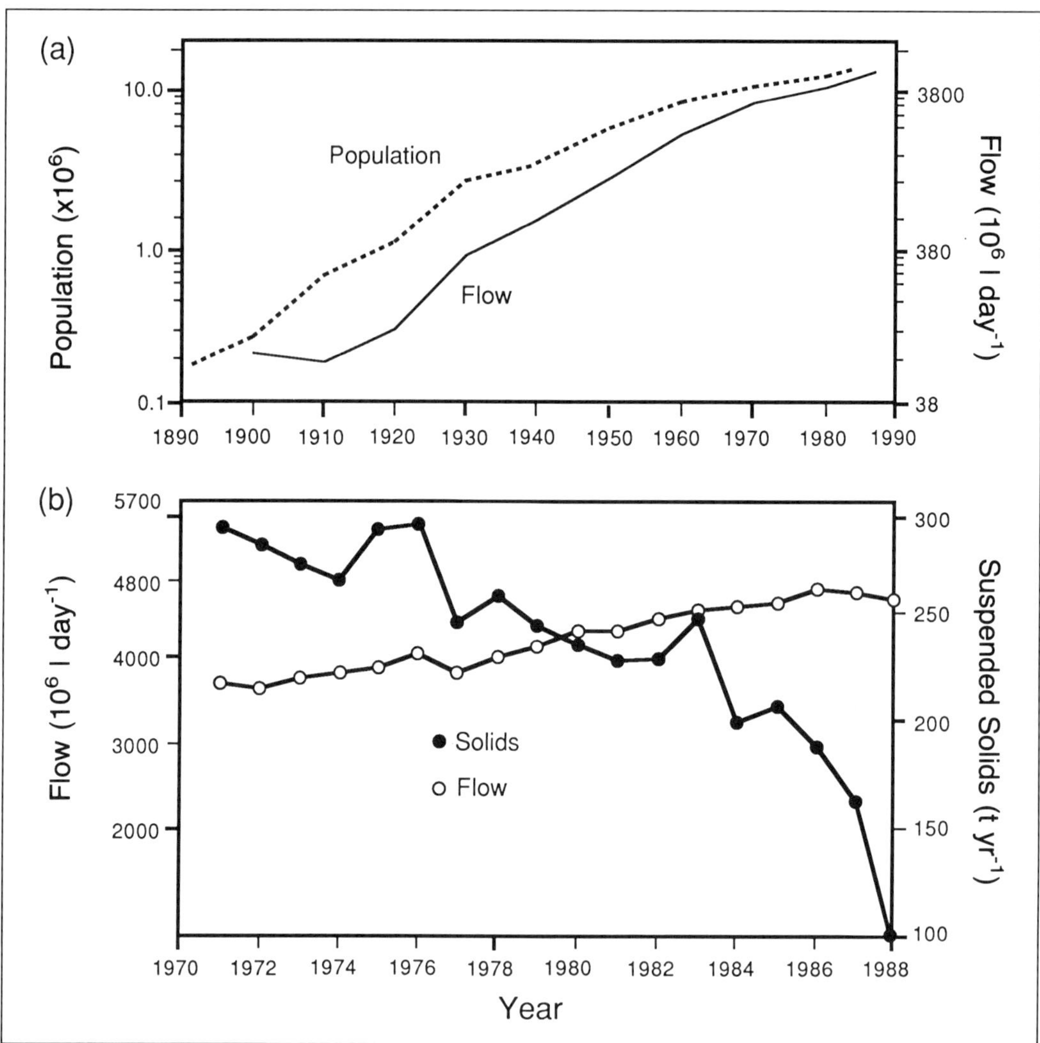

Figure 12.2. Population growth, municipal wastewater flows, and solids emissions for the SCB.

mass of oil and grease (fig. 12.5a) that entered the treatment plant from numerous sources decreased little over the 10-year period, but improved treatment technology (particularly suspended solids removal) produced a significant decrease in the amounts entering the ocean. The record for chromium (fig. 12.5b) provides evidence of the success of source regulations because the influent reductions are substantial and the final effluent reductions have followed closely. Zinc data (fig. 12.5c) give an indication that both source control and improved treatment procedures contributed to the reduction in effluent mass emissions.

Measurements in the environment near the effluent discharge provide evidence that the ecology of the area has improved as the concentrations of contaminants in the surface sediments have decreased. Zinc, which exhibited a distribution pattern similar to that of silver in 1978 (fig. 12.4b), was found in 1985 to be present at high concentrations in only small "pockets" very near the discharge (fig. 12.6a). The depth profile for zinc at three stations at varying distances (2, 8, and 21 km

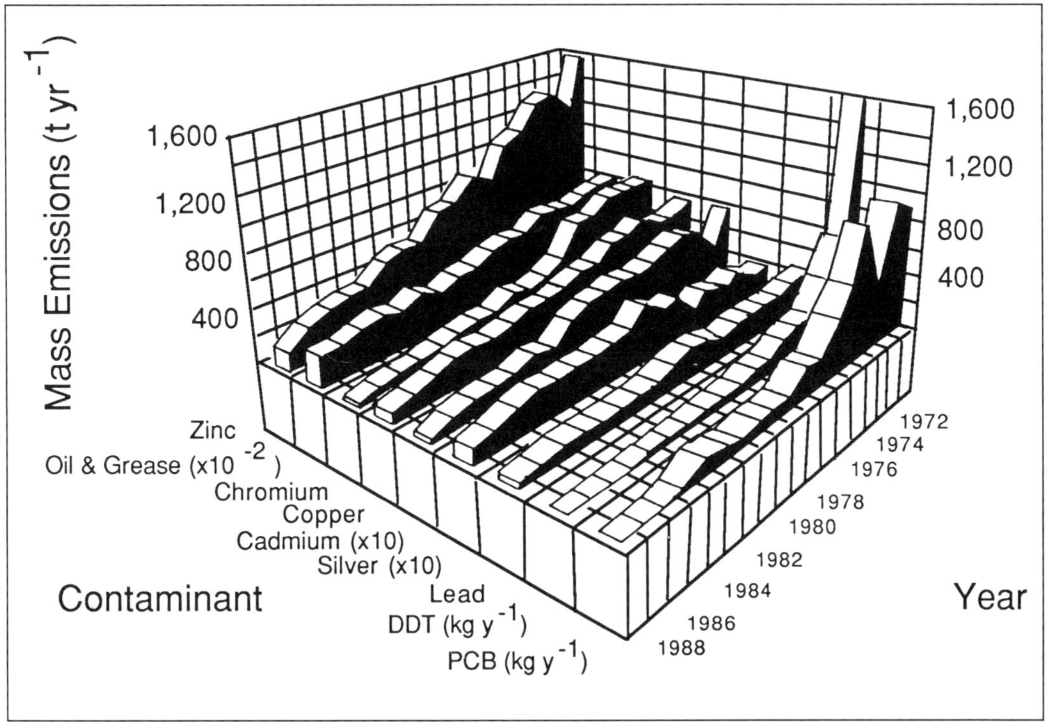

Figure 12.3. Trends in annual mass emissions of contaminants from municipal wastewater to California coastal water from 1971 to 1988. Quantities are in metric tons per year unless shown otherwise.

northwest) from the outfall (station 6C the nearest) shows a cover of about 25 cm of cleaner sediment in the region of greatest deposition (fig. 12.6b). This same pattern is shown in figure 12.6c for DDT; however, the great similarity between the profiles for 1985 and those for 1983 seems to indicate that the rate of deposition is slowing (fig. 12.6d).

Some of the changes observed in the biota of the region are illustrated in figures 12.6e and f. Beginning in about 1978, the kelp canopy of the Palos Verdes shelf showed dramatic improvement, likely resulting from decreased mass emissions of suspended particulates (fig. 12.6e). Incidence of fin erosion in Dover sole collected near the outfall had decreased significantly since 1972 to a very low percentage in 1986 (fig. 12.6f). Recent analyses of muscle tissue from white croaker (J. Stull unpubl. data) showed that concentrations of DDT have decreased to about 5 mg kg^{-1} from about 20 mg kg^{-1} in the 1970s at stations off Palos Verdes. Fish collected in Santa Monica Bay at the Venice and Hermosa beach piers are considerably lower in DDT (83 and 287 $\mu g\ kg^{-1}$, respectively).

Swartz et al. (1985, 1986) provided other evidence of improved ecological conditions near Palos Verdes. From collections made in 1980 and 1983 along a transect to the northwest of the JWPCP outfall, these authors found decreases in the levels of contaminants in the surface sediments and a reduction in toxicity for amphipods. Sediments obtained from locations near the outfall in 1980 produced significant reductions in amphipod survival, but tests with samples from the same stations in 1983 showed no significant reduction in survival. Chemical analyses of contaminants conducted for both collections showed that the highest correlations with biological effects were for total oil and grease, hydrocarbon oil and grease, and lead. At the most contaminated stations, a 71% reduc-

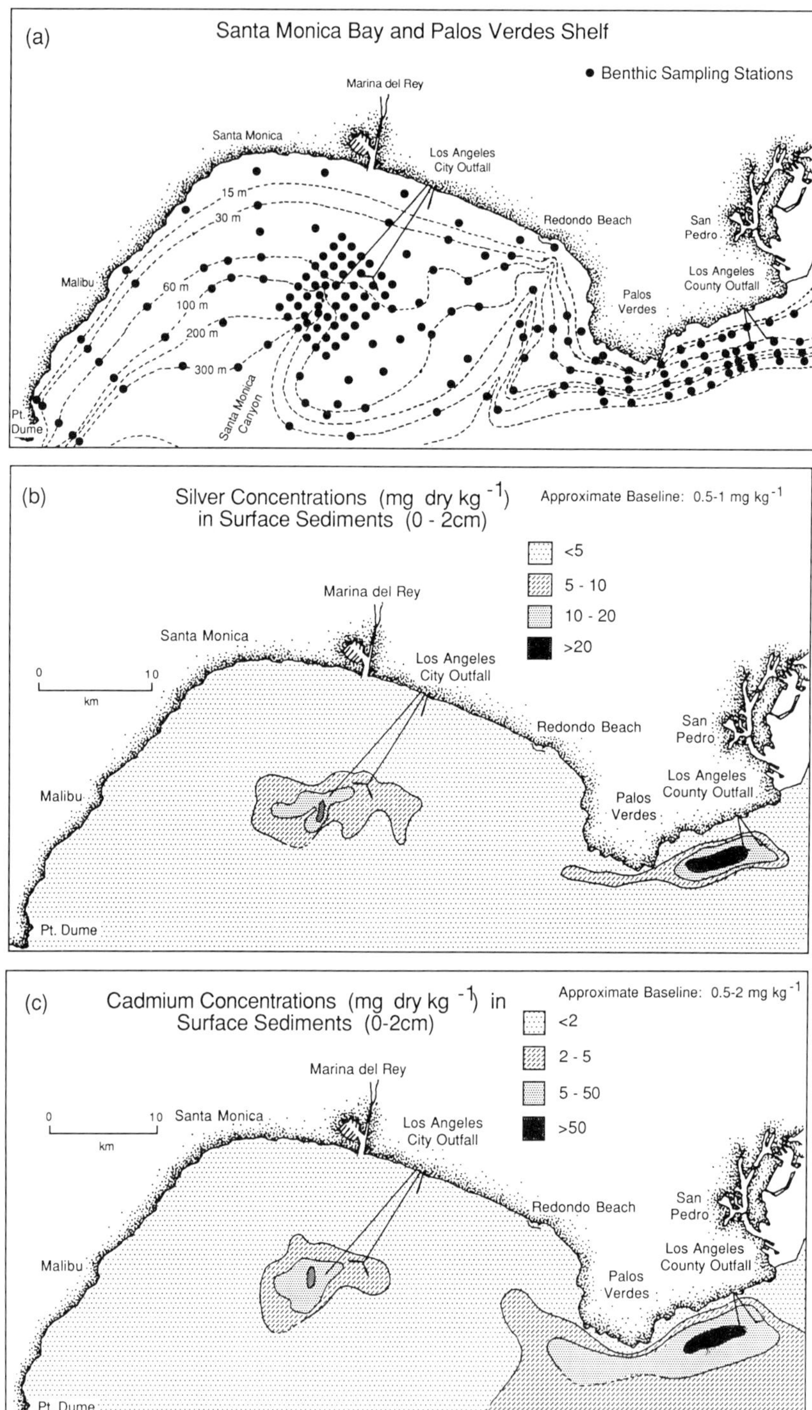

Figure 12.4. (a) Sampling grid for a 1978 benthic survey in Santa Monica Bay and off Palos Verdes. Depth contours and municipal wastewater ocean outfalls are also shown. (b) Concentration contours for silver on surface sediments (top 2 cm) sampled in 1978. (c) Concentration contours for cadmium on surface sediments sampled in 1978.

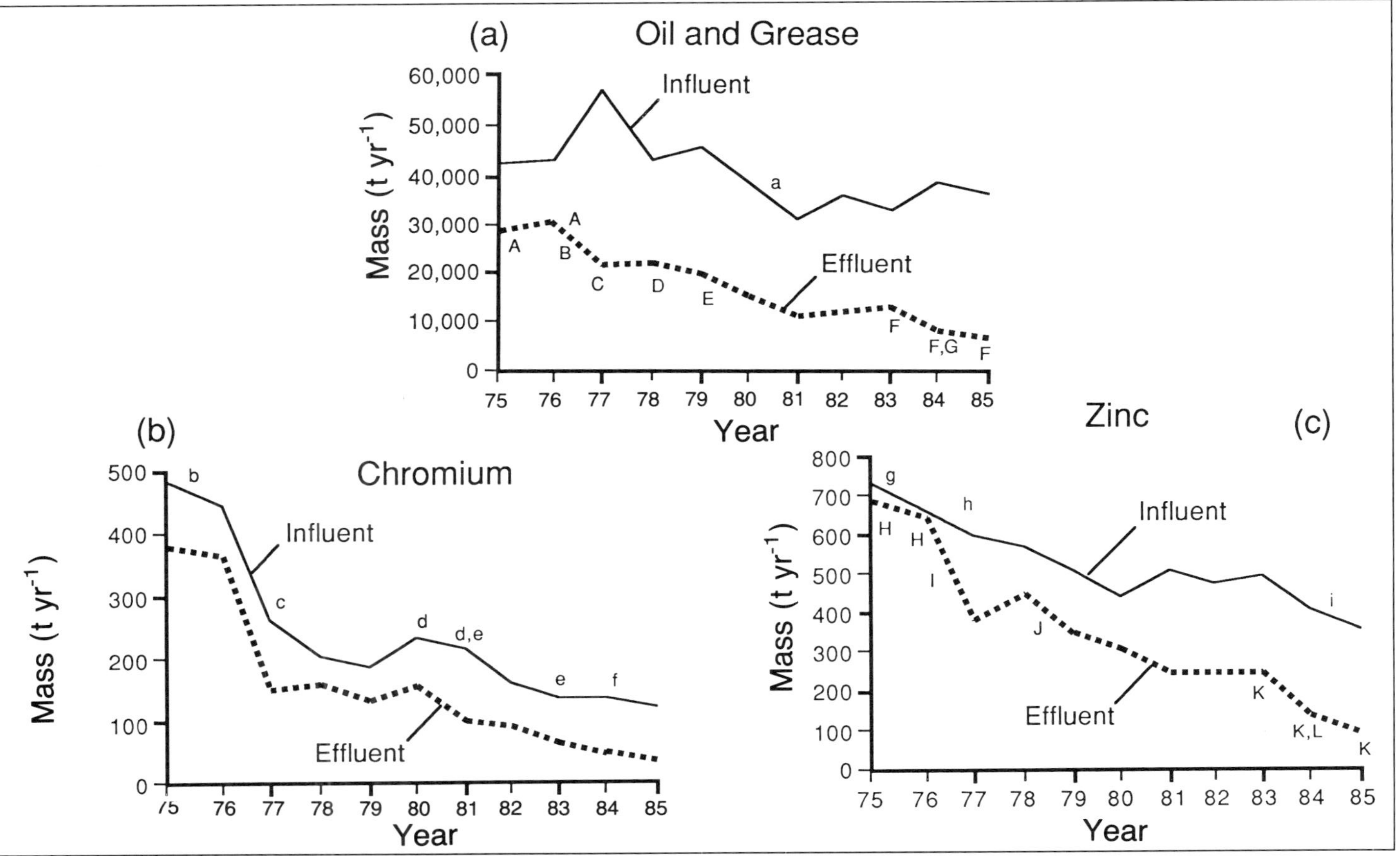

Figure 12.5. Impacts of source control and improved wastewater treatment at the Joint Water Pollution Control Plant (JWPCP) of Los Angeles County Sanitation Districts on annual mass emissions between 1975 and 1985. (a) Changes in oil and grease emissions to the ocean as a result of these factors: a = greater enforcement and service charges; A = construction; B = digester cleaning facility; C = basket centrifuges; D = polymer, more digesters; E = effluent screens; F = secondary treatment; G = scroll centrifuges. (b) Changes in chromium emissions to the ocean as a result of these factors: b = phase I limits announced; c = limits enforced; d = violations; e = enforcement; f = EPA categorical pretreatment standards. (c) Changes in zinc emissions to the ocean as a result of same factors. H = construction; I = basket centrifuges; J = polymer; K = secondary treatment; L = scroll centrifuges; g = phase I limits announced; h = limits enforced; i = EPA categorical pretreatment. (Modified from Stull and Haydock 1988.)

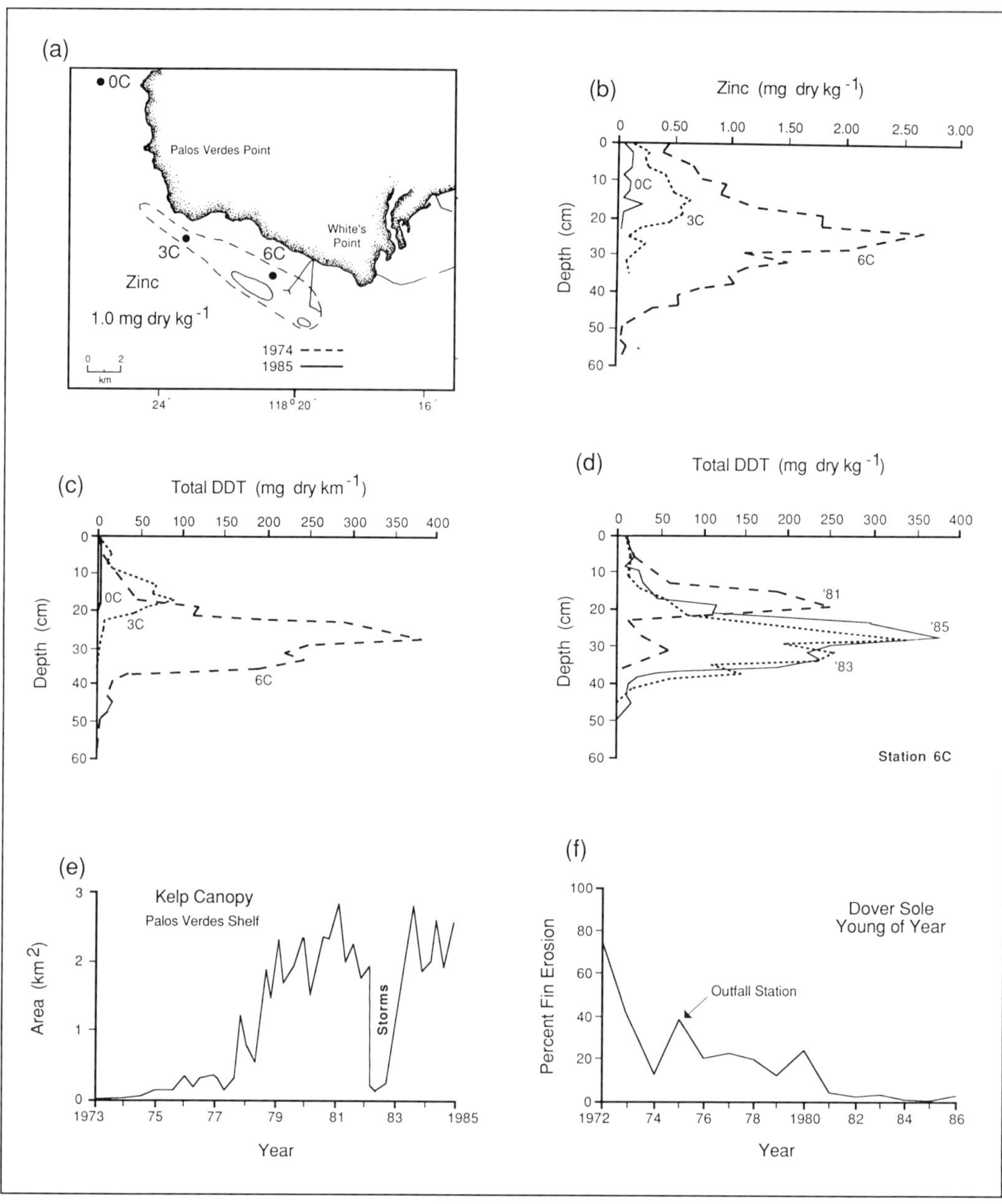

Figure 12.6. Horizontal and vertical distribution of contaminants in sediments off Palos Verdes and associated biological responses. (a) Surface distribution of a 1 mg kg^{-1} concentration of zinc in 1974 and 1985. Station locations shown at 60 m (0C, 3C, and 6C) are used in figures 12.6b–d. (b) Vertical distribution of zinc at the three 60-m station locations in 1985. (c) Vertical distribution of DDT at the same time and locations as in part b. (d) Vertical distribution of DDT at station 6C (nearest outfall) from cores taken in 1981, 1983, and 1985. (e) Area of kelp canopy on the Palos Verdes shelf from 1973 to 1985. (Data from California Department of Fish and Game.) (f) Fin erosion of Dover sole collected near the JWPCP outfall from 1972 to 1986. (From Stull and Haydock 1988.)

tion occurred in biological oxygen demand (BOD) and a 67% decrease occurred in oil and grease between 1980 and 1983. The decrease in effluent oil and grease during the period before and after 1980 corresponds to the changes observed in the sediments near the outfall (fig. 12.5a).

Other studies conducted in the SCB related to the impacts of municipal waste outfalls have examined sediments near the Orange County and San Diego wastewater discharges. Mearns and Greene (1975) described differences in sediment type at depths from about 15–300 m off the Orange County outfalls located 1 mile (old) and 5 miles (new in 1971) offshore between Huntington Beach and Newport Beach. In the spring of 1975, these authors found sewage sludge type particles around both outfalls. Later, a zone very near the 5-mile outfall was found to exhibit copper concentrations 1.5–2.0 times background values (fig. 12.7a) (Greene 1976). That same year (1975), only 4.5 years after initiation of the 5-mile outfall, influence of the discharged waste was observed in the distribution of benthic species (fig. 12.7b). West of the outfall, a reduction in the number of species occurred, but both biomass and the number of individuals showed an enhancement near the diffuser. Extensive monitoring conducted in recent years by the Orange County Sanitation Districts shows that the area of altered benthic community has contracted to a small zone (<1 m^2) near the outfall as inputs of solids and contaminants have decreased.

Extensive and intensive benthic sampling covering the 60-m contour of the SCB was conducted by the SCCWRP staff in 1977. These data were described in several papers included in the 1978 SCCWRP Annual Report. Word (1978) first demonstrated the usefulness of what he called the "infaunal trophic index" (ITI) to define the environmental conditions of sectors of the SCB. The application of this approach was described by Reish (1980) in his discussion of the effects of domestic wastes on benthic marine communities of southern California. The index is based on the selection of four groups of infaunal species separated by types of feeding habits (fig. 12.8). Group I is dominated by suspension feeders, Group II is a combination of suspension and surface detritus feeders, Group III is composed of surface deposit feeders, and Group IV is dominated by subsurface detritus feeders (Word 1978). Detritus, as used here, is separated from deposited substrate by recent settling, at the same time recognizing that "fresh" detritus may eventually become part of the permanent sediment. It is important to note that these group designations are somewhat arbitrary, and they primarily represent assemblages found along a pollution gradient at the 60-m contour in the SCB. The ITI is then calculated by a formula using the numbers of individuals within each group. Problems with this approach include variation of infaunal communities with depth, the formula used for the ITI, and assessment of feeding type and taxonomy. Figure 12.8 shows the separation of groups produced by the treatment of the data from 300 samples in the SCB. Using the ITI, Word (1978) showed the reductions in normal infaunal indices produced by major wastewater discharges in Santa Monica Bay, at Palos Verdes, at Orange County, and at Point Loma in 1977 (fig. 12.9). Taking a closer look at these latter two sites (fig. 12.10), it is possible to observe the gradual changes in the ITI at distances away from the diffusers. Also included in the annual report are maps of most of the SCB showing the distribution of stations, ITI, infaunal biomass, and data on the physical and chemical characteristics of the sediments.

For evaluating the degree of apparent impact from wastewater discharges, Word (1978) defined normal infauna as having an ITI of 60–100, while "changed" was 30–60 and "degraded" was 0–30. From the data collected in 1977 and 1978, Word (1978) defined the areas surrounding five of the major ocean outfalls in the SCB (out to 200 m) as normal, changed, or degraded. Figure 12.11 illustrates

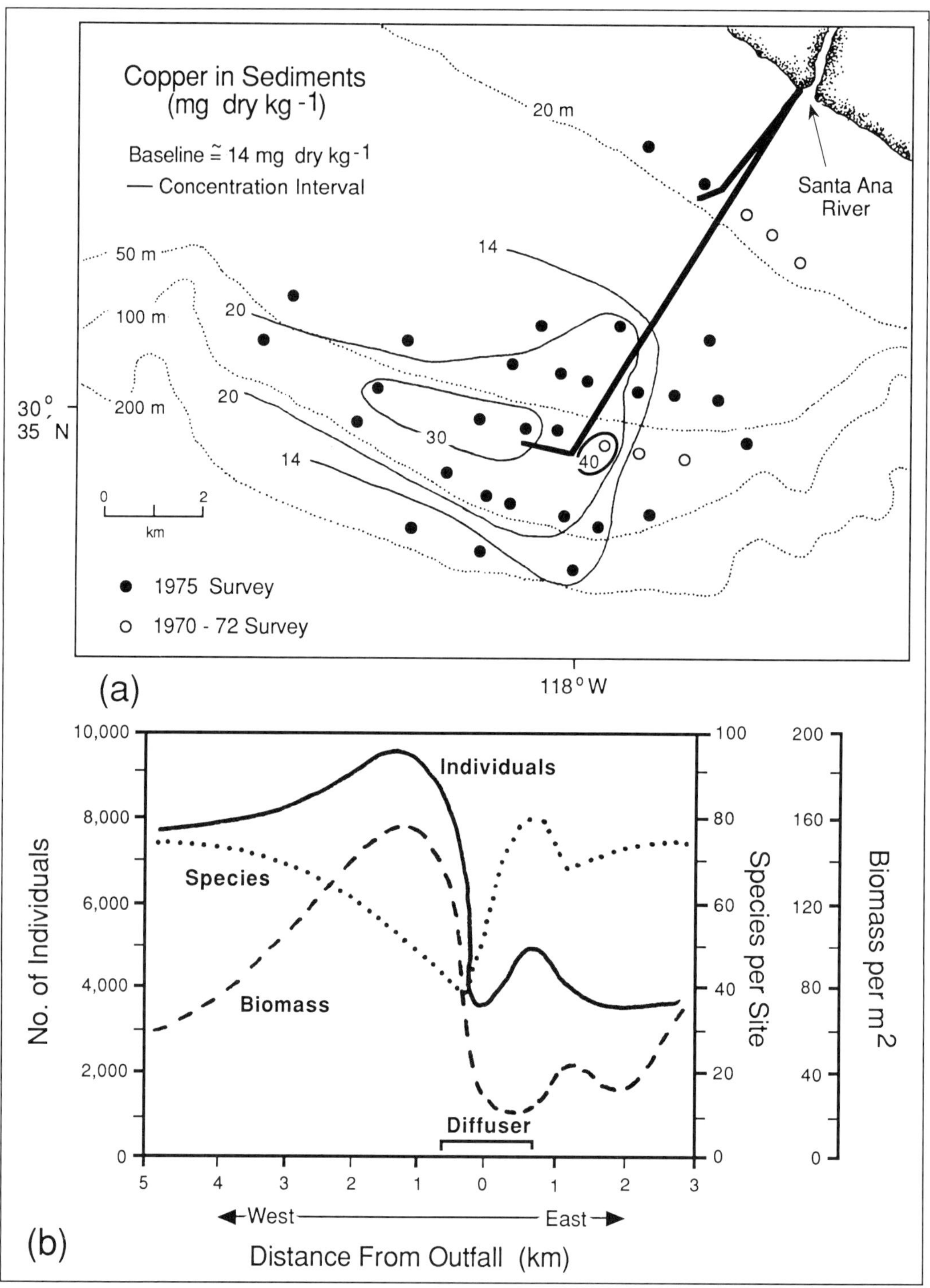

Figure 12.7. Distribution of infauna and contaminants near the Orange County outfall in 1975 (Greene 1976). (a) Sediment copper concentrations. (b) The distributions of individual organisms and biomass per square meter and species per site along the 55-m (diffuser) depth contour. (From Southern California Coastal Water Research Project 1978.)

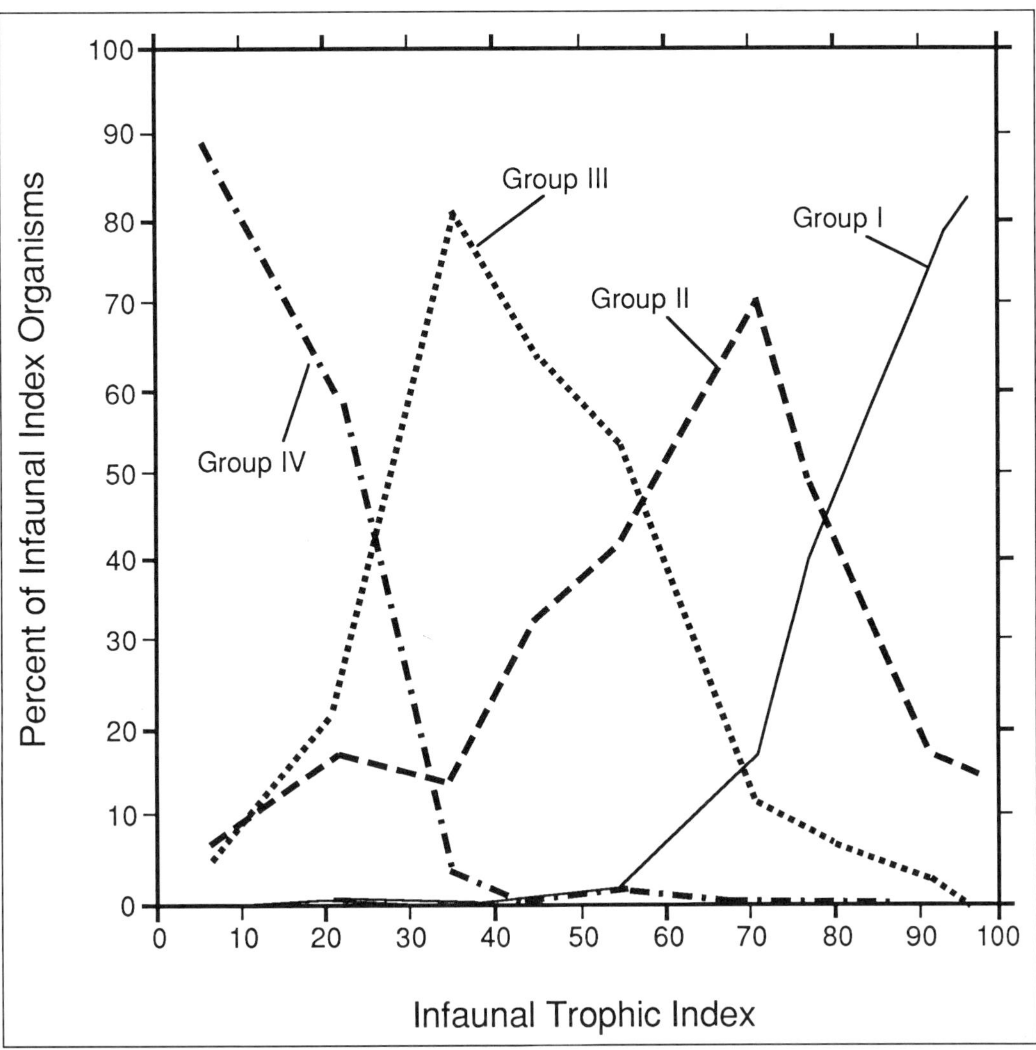

Figure 12.8. Dominance of infaunal trophic index (ITI) groups in 300 samples from southern California stations 20–200 m in depth. Although individuals from all groups are present over the range of ITI values, each group dominates a specific interval of values. (From Southern California Coastal Water Research Project 1978.)

the areas of the zones (in square kilometers) associated with these five municipal outfalls. San Diego and Orange county outfalls had small zones of changed infauna but no degraded zones. These findings were the same in 1988 as in 1978. As would be expected, the two largest volume discharges in Santa Monica Bay and at Palos Verdes were associated with the largest zones of changed infauna, and zones of degraded infauna were observed near the discharges. Overall, Word (1978) estimated that about 4.6% or 168 km^2 of the 3640 km^2 of mainland continental shelf of southern California was changed as a result of deep-water sewage discharges. Only about 12 km^2 (0.3%) of this region was considered "degraded" on a basis of the criteria used in the ITI approach. Data from 1988 were obtained from the dischargers' technical staff (J. Dorsey, J. Stull, T. Gerlinger, and P. Vai-

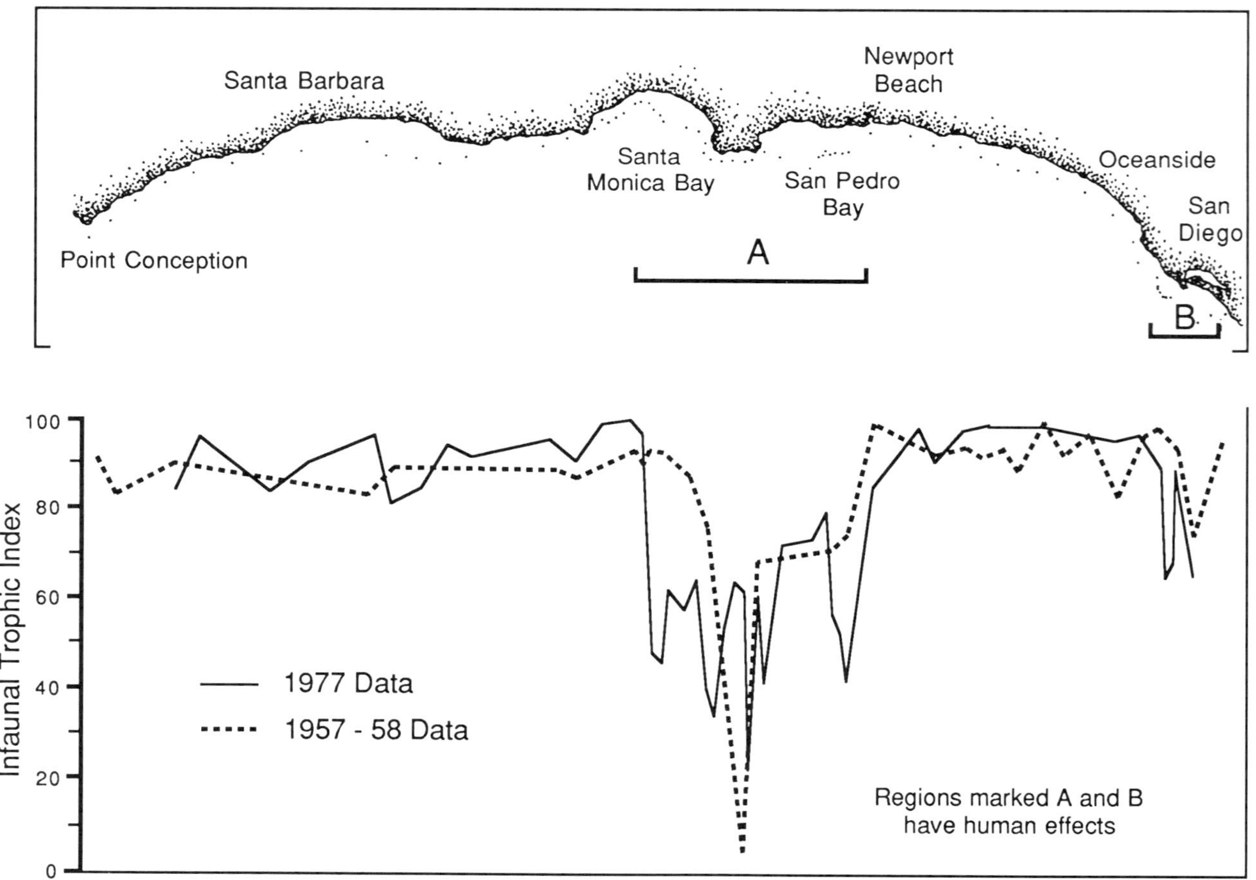

Figure 12.9. Infaunal trophic index (ITI) values calculated from data collected by the Southern California Coastal Water Research Project in 1977 and from data collected by the Allan Hancock Foundation in 1957–1958. (From Southern California Coastal Water Research Project 1978.)

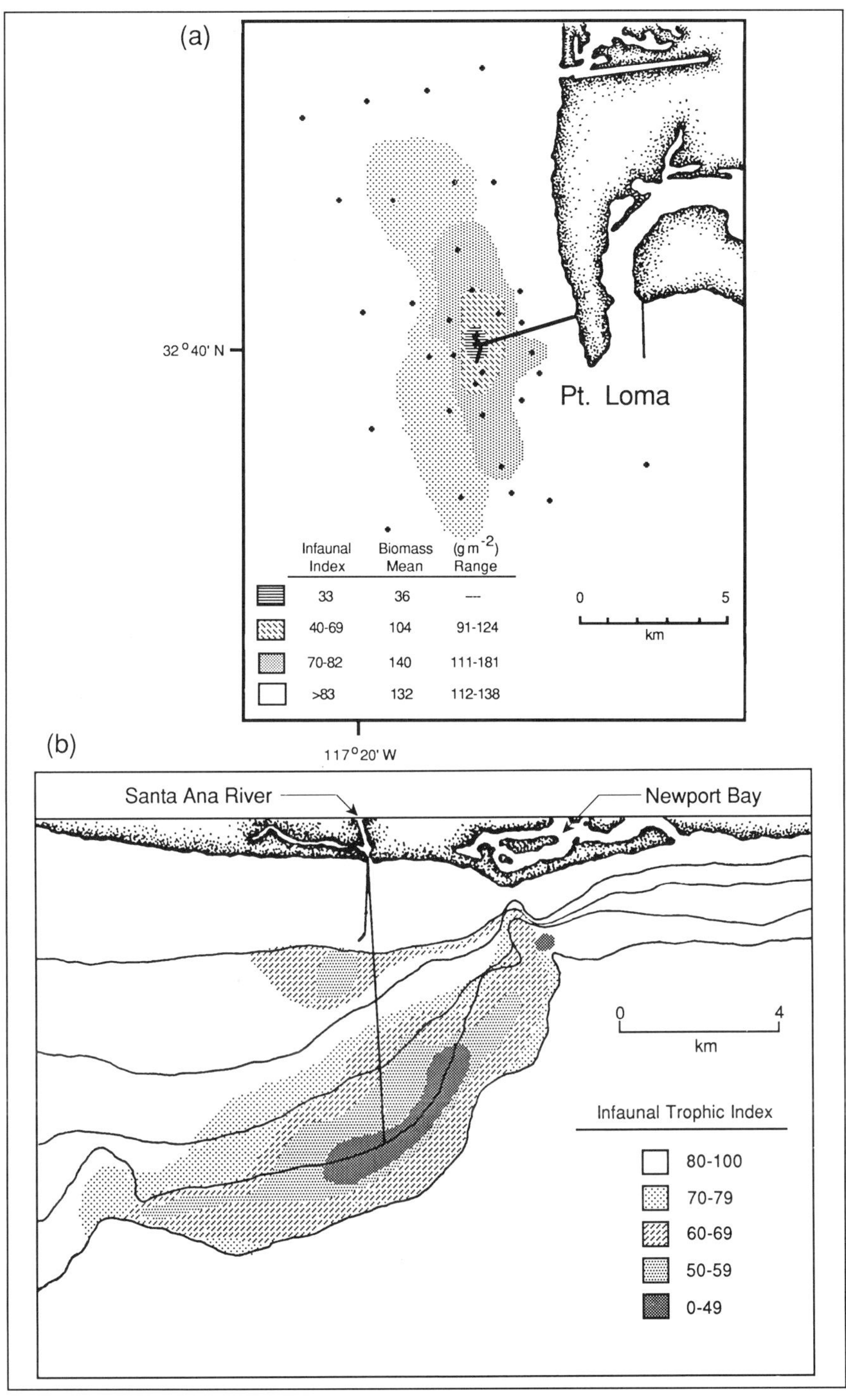

Figure 12.10. (a) Patterns of infauna response around Point Loma outfall, summer 1978. Infaunal trophic index (ITI) values equal to or greater than 83 indicate control conditions; biomass values above the maximum found in control areas (112 g m^{-2}) indicate enhancement of natural benthic infaunal populations. (b) ITI values for Orange County outfall. These values provide a geographically continuous record of sediment conditions in the area sampled. (From Southern California Coastal Water Research Project 1978.)

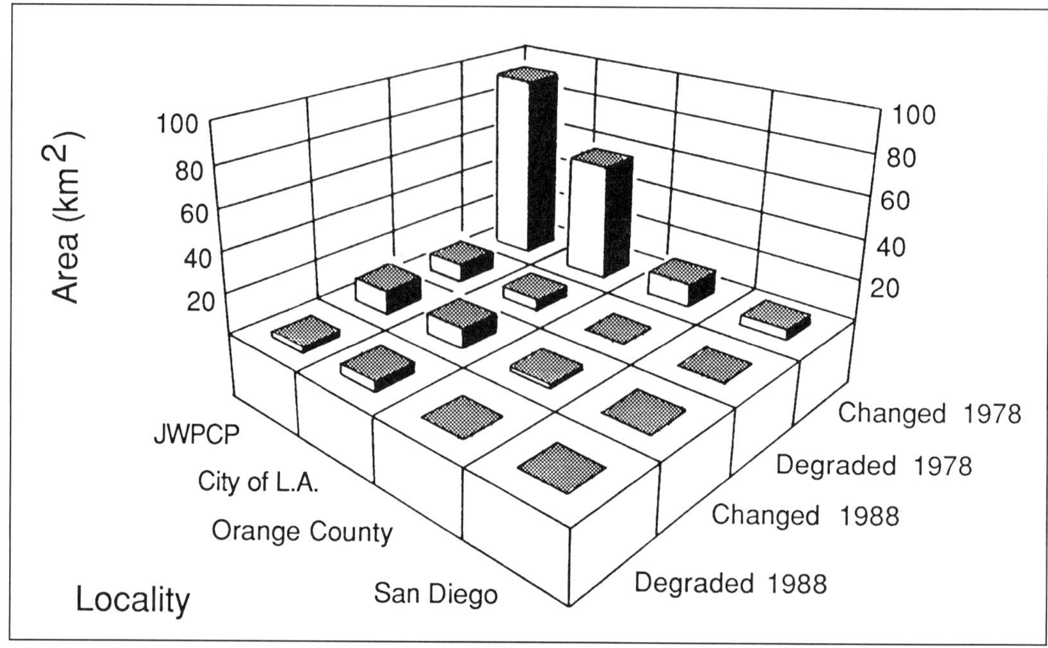

Figure 12.11. Altered coastal areas of southern California. Major municipal wastewater outfall areas are shown that were "changed" and "degraded" in 1978 and 1988. (From Southern California Coastal Water Research Project 1978 and unpubl. data.)

nik). The area changed in 1988 is 28% of that in 1978, and nearly a 50% decrease has occurred in the area degraded. The largest portion of the degraded area is around the sludge outfall in Santa Monica Bay (5 km^2), and some recovery of this area has already been noted since termination of the outfall in November 1987.

Surveys of the benthic infauna in the SCB at locations distant from outfalls had not been conducted since the control stations were analyzed in the 1977–1978 project (Word and Mearns 1979). Thompson et al. (1987), however, collected and analyzed data from 13 sites between Point Conception and the Mexican border that were designated as controls in the earlier survey. Samples from three depths (30, 60, and 150 m) were taken at each location to provide both a comparison to the earlier 60-m survey and to gain more information regarding changes with depth. Included in this "reference" study were analyses of trawled fish and invertebrates, infauna species, sediment characteristics (TVS, TOC, and so on), seven trace metals, chlorinated organics, and polynuclear aromatic hydrocarbons (PAHs). Comparisons of the data with the 1977 60-m survey showed in general that few changes had occurred in either organisms or sediment chemistry. Differences, both up and down, were quite small, with the exception of an order of magnitude decrease in silver concentrations in sediments (means of 0.2 to 0.03 mg kg^{-1}). The authors pointed out the usefulness of the data in relationship to determining differences between stations near outfalls and other sets of stations within the SCB.

Johnson and Roney (1988) described the assemblages of macroinvertebrate and demersal fish caught by trawls in Santa Monica Bay near the two outfalls. *Astropecten verrilli* (sand star) and *Genyonemus lineatus* (white croaker) were the dominant species at each of the outfalls but not in the remainder of the bay. *Lytechinus pictus* (white sea urchin) was dominant at 100 m depths in the northern part of the bay but was absent from the outfall areas. Fish livers often contained levels of

DDT and PCBs near or above FDA standards (DDT, 5 mg kg^{-1}; PCB, 2 mg kg^{-1}), but muscle tissue was generally one to two orders of magnitude less concentrated. Only 16 (1.5%) of the 1066 Dover sole (*Microstomus pacificus*) collected from six upper slope sites showed evidence of epidermal tumors.

Regarding the prevalence of diseased fish, Cross (1986) found a significant geographical difference in the incidence of tumors in Dover sole measuring less than 120 mm in length collected along transects of the shelf off Palos Verdes near the JWPCP outfall. In addition to these epidermal tumors, the incidence of fin erosion in Dover sole was enhanced near outfalls (Cross 1985). Significantly more Dover sole individuals were afflicted with both diseases than predicted by chance ($X^2 = 45.4$, $p < .001$). Mearns and Sherwood (1976), however, concluded that epidermal tumors in Dover sole populations in southern California were not enhanced by municipal waste water discharge. While it is recognized that the occurrence of tumors in pleuronectid fish may be caused by a protozoan parasite, the possibility remains that contaminant-induced stress may make newly recruited fish more susceptible to epidermal tumors.

Brown et al. (1986) collected scorpionfish (*Scorpaena guttata*) from stations as far north as Anacapa Island and south to Ensenada to determine the levels of DDT and PCB in the liver tissue (fig. 12.12). While there was only a slight rise in the amounts of PCBs in fish from the Palos Verdes–Santa Monica Bay area, a major increase in DDT levels was evident at these latter locations, particularly at Palos Verdes. More recently, the staff of Los Angeles County Sanitation Districts (J. Stull unpubl. data) analyzed young (1–3 cm) Dover sole from trawl stations off Palos Verdes and found concentrations of DDE (the primary contaminant from historical DDT input) as high as 8 mg kg^{-1}. Highest concentrations were found northwest of the outfall (T1–T4) at both 60 and 135 m depths (fig. 12.13). The reference station (T0), which has low values, is located around the Palos Verdes Peninsula toward the northeast and extending into Santa Monica Bay. In two recent papers, Cross and Hose (1989) and Hose et al. (1989) described reproductive impairment in white croaker and kelp bass, using a kelp bed near the JWPCP outfall and an area of Los Angeles Harbor near the Palos Verdes Peninsula as the contaminated sites. As found in studies with salmonids (Burdick et al. 1972), it appears that ovarian DDT levels of about 4 mg kg^{-1} are correlated with inhibition of spawning. Since PAHs and trace metals are also high in the contaminated sites, DDT cannot be identified as the only toxicant contributing to the effects on fish reproduction.

One means of defining changes in benthic assemblages around municipal outfalls was presented by Thompson (1982) and by SCCWRP's 1986 annual report concerning the area around the Santa Monica Bay sludge outfall. Data recorded in the 60-m survey (Word and Mearns 1979) were used to map the distribution of three key species that characterized the extent and degree of contamination from discharged sludge. As shown in figure 12.14, the normal areas of the bay contained brittle stars (*Amphiodia urtica*) in densities of greater than 100 m^{-2}. A broad transition zone containing the clam *Parvilucina tenuisculpta* at densities greater than 1000 m^{-2} was found to surround the 5-mile and 7-mile outfalls, except at stations very near the pipes. This species feeds at and below the sediment surface on detritus and deposits and, through endosymbiosis, utilizes sulfur for energy (Jones and Thompson 1984). The stations near the outfalls were defined as the contaminated zone and were characterized by the polychaete *Capitella capitata* at densities greater than 1000 m^{-2}. *Capitella capitata* feeds on deposits and is an opportunistic species, expanding rapidly in numbers where predation and competition are reduced by organic enrichment of the sediments. These three species represent groups I, III, and IV, respectively, in the trophic index classification of Word (1978).

Using the data just discussed from collections made in 1983–1985, Dorsey (1988) concluded that the distribution patterns of these

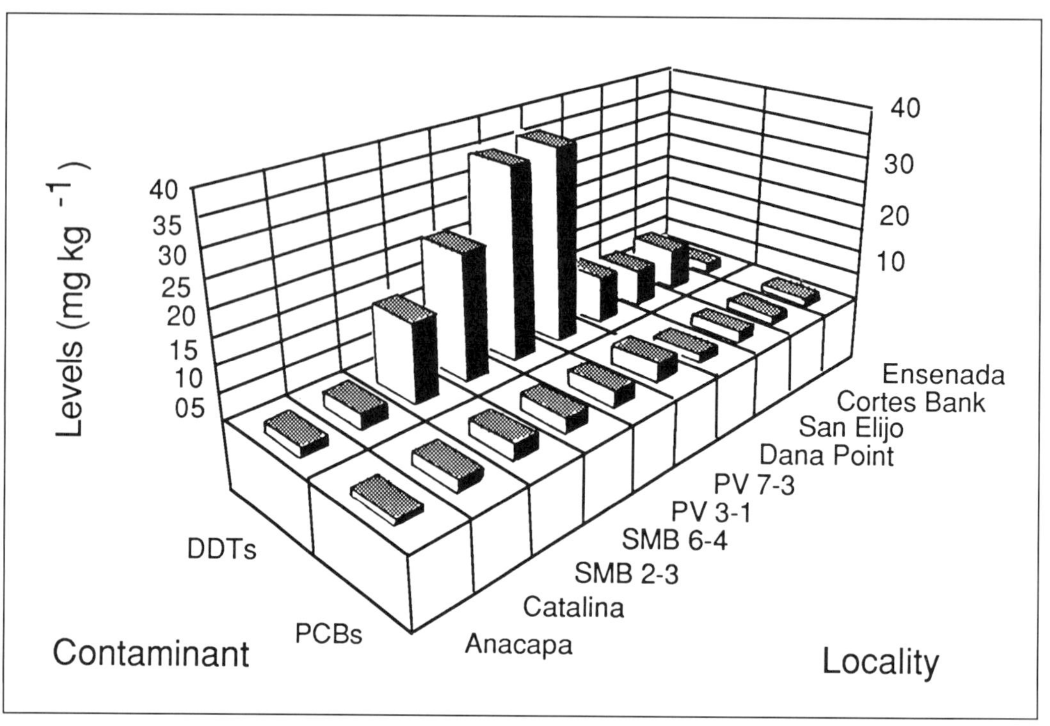

Figure 12.12. Levels of DDT and PCBs (mg kg^{-1}) in scorpionfish livers in 1985. (From Brown et al. 1986.)

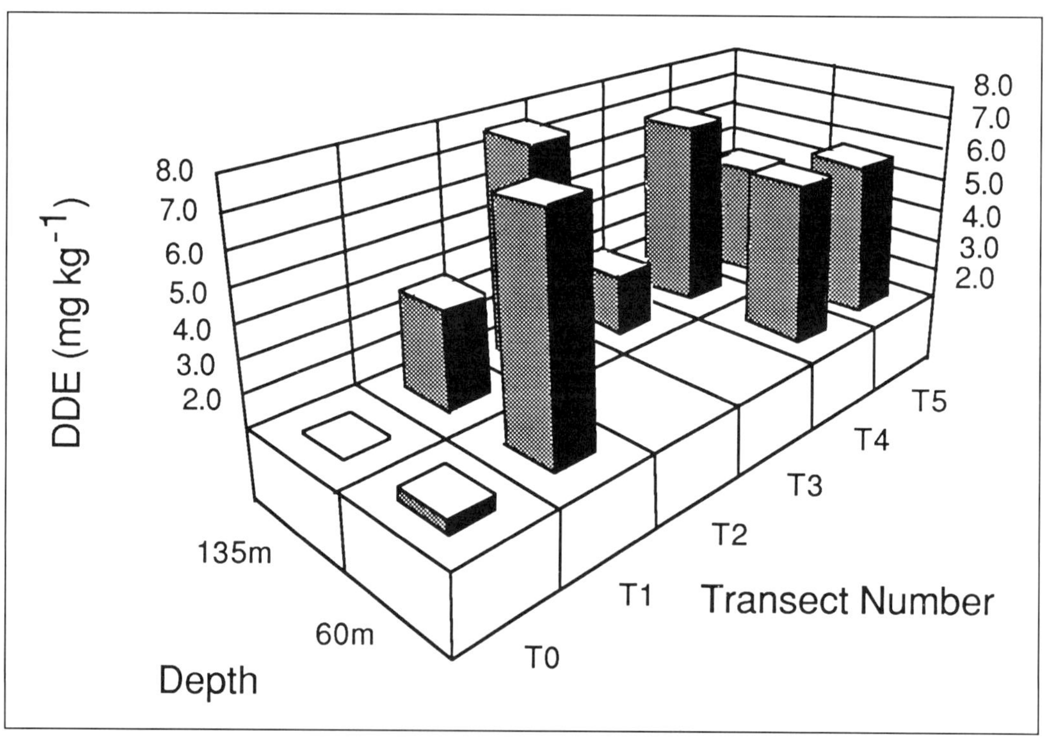

Figure 12.13. DDE in Dover sole from the Palos Verdes shelf (J. Stull unpubl. data). Transect number decreases with distance from outfall.

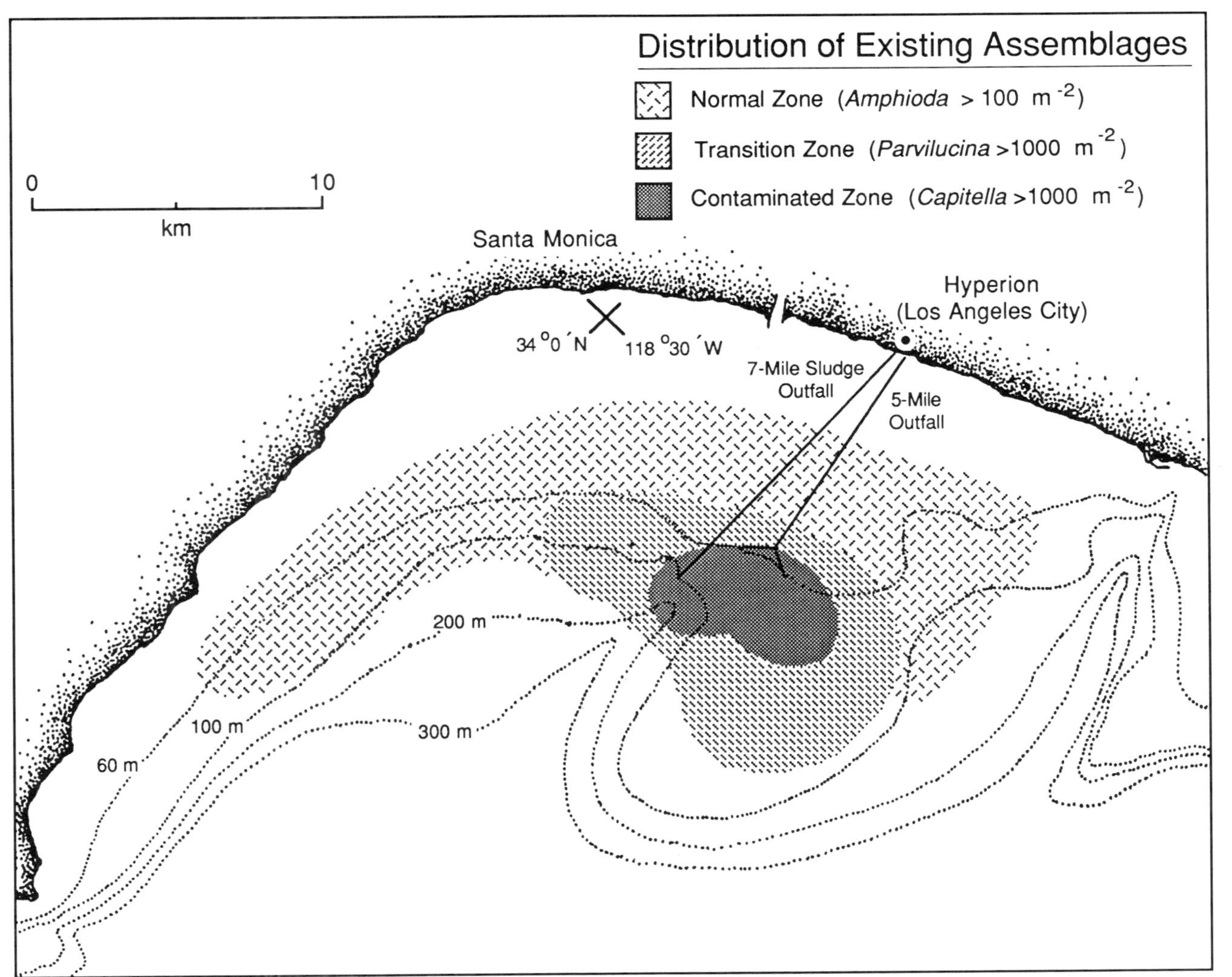

Figure 12.14. Distribution of key indicator species in Santa Monica Bay.

three species appear to have changed very little in 7 years. The only difference seems to be a reduction and condensation of the dense populations of *C. capitata* in a region between the two outfalls. Abundances and distributions of these key species were very similar in 1985 to those in 1977 and 1978.

While the outfalls from the city and county of Los Angeles represent the largest volumes of discharge and have produced impacts to the infauna, other SCB municipal waste discharges are significant. In their 1987 Annual Report, the County Sanitation Districts of Orange County presented a massive amount of data compiled in the ocean monitoring program under their 301h National Pollution Discharge Elimination System (NPDES) permit. Figure 12.15 shows the depth contours around the outfall. Data from this report and the criteria described by Thompson (1982) were used to prepare a distribution map for *Amphiodia urtica, Parvilucina tenuisculpta,* and *Capitella capitata* (fig. 12.15). As discussed earlier, the normal area is defined by high (>100 m^{-2}) concentrations of *A. urtica* at appropriate depths. It is interesting that the high densities of *P. tenuisculpta* and *C. capitata* overlap and are not separated as they were in Santa Monica Bay. One likely explanation for this overlap is that the levels of contamination surrounding this outfall are not high enough to force populations of *P. tenuisculpta* to the outer edge of the more resistant *C. capitata* populations.

Power Plants

Cooling water discharged from coastal power plants represents the largest volume of inflow to the SCB. In 1970–1971, the average annual flow from the 15 major electrical generating stations in southern California was 7.7×10^{12} l yr^{-1}, which was about seven times the corresponding rate for municipal wastewater discharges (Young et al. 1977). In 1977 the influents and effluents of eight of Southern California Edison (SCE) Company's coastal generating plants were sampled for the soluble and particulate (>0.4 μm) concentrations of six trace metals (Cd, Cr, Cu, Ni, Pb, and Zn). Table 12.7 from Young et al. (1977) shows the median influent concentrations of these metals and the increase (decrease for soluble Cr) found in the effluents of the eight generating plants (dissolved and particulate). The authors also placed this contribution to the coastal waters in perspective by preparing table 12.8, which includes inputs of metals from three other sources. Cooling waters in 1977 contributed only about 0.1–0.7% of the total metals known to enter the ocean from anthropogenic inputs. Calculations with more recent data would probably show an increase in the percentage contribution since municipal waste mass emissions have significantly decreased.

One possible impact of coastal generating plants on marine biota is the introduction of chlorine to seawater because it is used in the form of sodium hypochlorite twice a day for about 20 minutes in the generating plants to prevent microbial fouling of condenser tubes. Periodic heat treatments, not chlorine, are used to prevent the fouling of cooling system conduits by marine organisms. The chlorination process was studied by Grove (1983) at seven SCE generating stations. He analyzed total residual oxidants (TRO) before and after the injection of chlorine passed the condenser and when it reached the receiving water. Of the 36 readings at outfall stations (within the "boil"), 21 were found to be below the limit of detection (0.08 mg l^{-1}) during the chlorination process. During 19 successful surveys over a 3-year period, the San Onofre study showed that the TRO concentrations at the condenser inlet were below 1.0 mg l^{-1}, and at the outfall boil, concentrations were below 0.1 mg l^{-1} in all but four surveys. Dilution factors obtained as part of these studies ranged between 5.2 and 19. An example of the TRO values near the outfall during the peak in a normal chlorination cycle was at Huntington Beach, where the maximum TRO at the outfall was 0.11 mg l^{-1} and

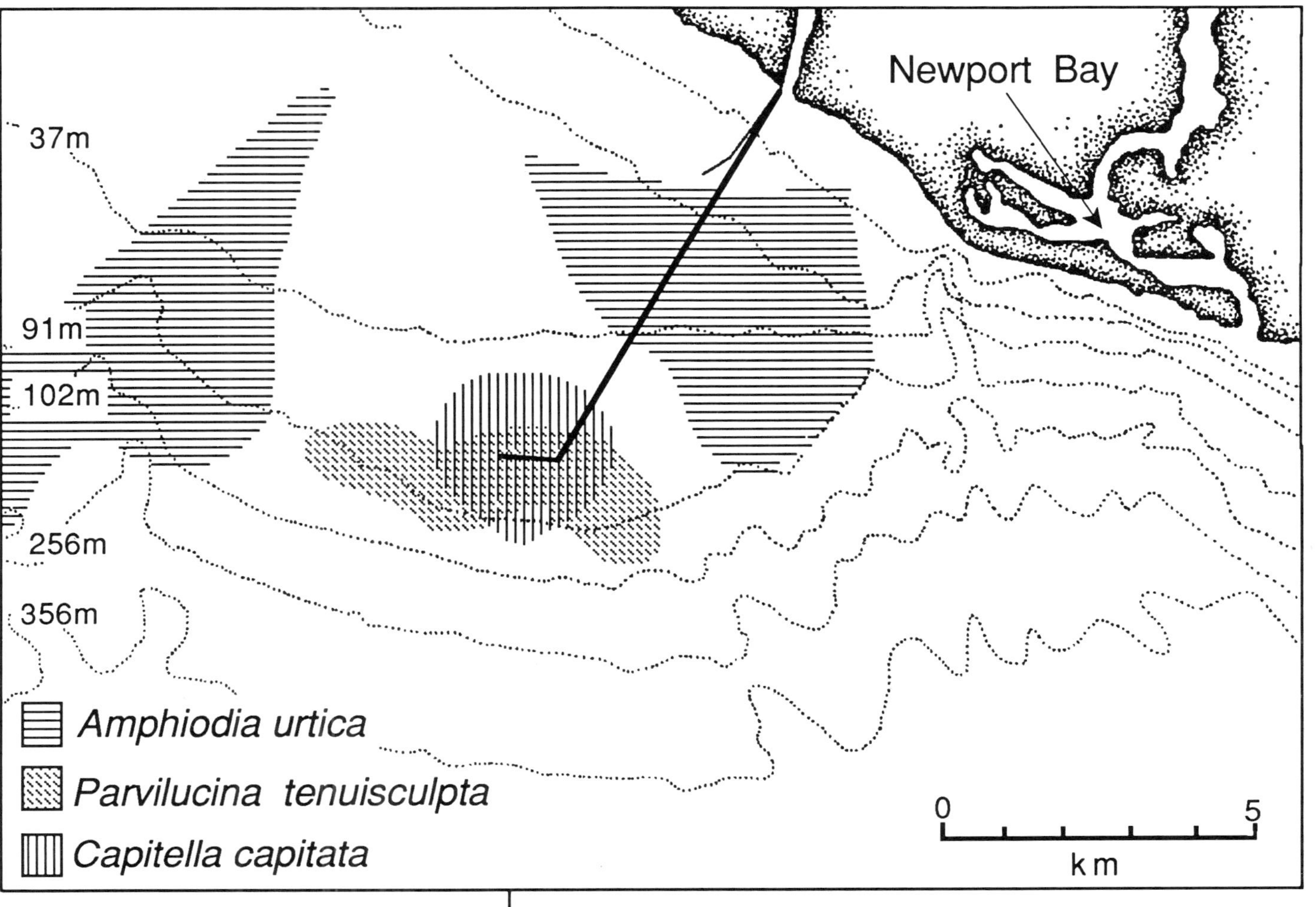

Figure 12.15. Distribution of key indicator species off Orange County, 1986. (From County Sanitation Districts of Orange County data.)

Table 12.7. *Median Concentrations of Six Metals in the Influents of Eight SCE Generating Plants and Median Difference Between Effluent and Influent Concentrations*

Element	Median Influent Concentration ($mg\ kg^{-1}$)		Median Difference Between Effluent and Influent Concentrations ($mg\ kg^{-1}$)	
	Dissolved	Particulate	Dissolved	Particulate
Cadmium	0.06	0.006	0.034	0.005
Chromium	0.16	0.20	(0.01)	0.097
Copper	0.80	0.32	0.21	0.10
Nickel	0.44	0.16	0.10	0.004
Lead	0.14	0.24	0.04	0.07
Zinc	<0.2	0.48	0.09	0.17

From Young et al. (1977).

Table 12.8. *Estimated Annual Inputs ($t\ yr^{-1}$) of Six Metals to the SCB from Eight SCE Generating Stations and Three Other Sources*

Metal	Municipal Wastewater[a]	Dry Aerial Fallout[b]	Storm Runoff[c]	SCE Cooling Waters[d]	Sum	SCE Cooling Waters (% of Sum)
Cadmium	41	0.8	1	0.3	43	0.7
Chromium	576	6.6	25	0.6	608	0.1
Copper	472	31	18	2.1	523	0.4
Nickel	307	12	17	0.7	337	0.2
Lead	161	240	90	0.8	492	0.2
Zinc	1010	150	100	1.8	1260	0.1

[a] Based on 1976 data on total metals in effluents of the Oxnard, Hyperion (Los Angeles City), JWPCP (Los Angeles County), and Orange County treatment plants.

[b] The estimated input in a 100- by 100-km zone off Los Angeles and Orange counties in 1975 (from Young and Jan 1976).

[c] The estimated input of dissolved and particulate metals to the study area during 1971, an abnormally dry year (from Young et al. 1977).

[d] Based on sum of dissolved and particulate metals concentrations.

From Young et al. (1977).

a flux weighted average dilution of 7.5 resulted in a calculated average receiving water dilution to a TRO of 0.015 mg l^{-1}.

Potential effects of chlorine in the marine environment were discussed by Goldman (1979) in a review article describing some of the chlorinated organic compounds created during drinking water chlorination. He summarized some possible effects of uncombined (TRO) chlorine on marine zooplankton and larvae. Data on effects were derived from Capuzzo et al. (1977), whose work included tests with phytoplankton, zooplankton, larval oysters and lobsters, and juvenile fish. Effects were observed when these species were exposed for short periods to about 1.0 mg l^{-1} TRO and survival was measured 48 hours later.

Thatcher (1978) conducted 96-hour bioassays with chlorine on eight species of northwest marine fish and reported LC_{50} values ranging from 0.032 to 0.167 mg l^{-1}. Dinnel et al. (1981) found sea urchin and sand dollar sperm bioassays to be very sensitive to chlorinated sewage and seawater. Values between 2 and 20 ng ml^{-1} TRO were found to be the 50% effective concentrations (EC_{50}). Eppley et al. (1976) measured phytoplankton productivity in the vicinity of the cooling system outfall from the San Onofre Nuclear Generating Station (SONGS). Peak TRO concentrations of 20 to 40 mg ml^{-1} were correlated with a 70–80% reduction in the rate of photosynthesis in the discharge waters. They estimated that this amount of chlorination at SONGS may have resulted in a loss in primary production on the order of 15–30 kg C d^{-1}. After conducting behavioral studies on 13 species of marine fish, Hose et al. (1983) reported avoidance thresholds that averaged 0.20 mg l^{-1} TRO. This value was remarkably similar to other field and laboratory data on both marine and freshwater fish that found the avoidance threshold to be approximately 50% of the median lethal concentration.

Extensive oceanographic and marine biological investigations were conducted in 1979 in the vicinity of SONGS. This site, located between the cities of San Clemente and Oceanside, was studied to assess the effects of the operation of SONGS unit 1 and the construction of units 2 and 3 and to provide a baseline for future assessment of effects of units 2 and 3 when they became operational. The projects included studies of water temperature, nutrients, turbidity, local sedimentology, intertidal and subtidal benthos, kelp, and larval and adult fish (Southern California Edison Company 1979). While thermal changes were judged insignificant compared to natural, temporal, and spatial variations, unit 1 had a perceptible influence on local turbidity. Construction and dredging activities for units 2 and 3 also affected water clarity, but this was short term and spatially limited. Bottom sediment heavy metal concentrations did not exhibit accumulation over the 5-year study period (1975–1979). Other than temporary impacts of the construction pads for units 2 and 3, effects on sedimentation, rocky intertidal, and benthic infaunal communities were not observed.

Queenfish, anchovy, and white croaker were the most abundant larval fishes both offshore of San Onofre and entrained annually at unit 1. Estimates indicated that about 900 million fish larvae are entrained annually at unit 1. The water entrained by unit 1 is biologically representative of midwater, with occasional strong influences from the epibenthos. The discharge of cooling water from unit 1 had no apparent effect on the distribution or abundance of the fish community. Queenfish numerically dominated impingement at unit 1, but walleye surfperch were the dominant species caught during heat treatment of the pipes. The fish community was apparently not adversely affected by the discharge of unit 1, although the reproductive condition of queenfish and white croaker near the discharge was lower than other areas, presumably because of earlier spawning.

Studies of the San Onofre kelp beds indicated that construction activities caused turbidity and sedimentation rates to increase in the upcoast section of the beds. The kelp canopy did not change as a result of dredging activity, but there was deterioration associated with midwinter storms. Recruitment of kelp (*Macrocystis*) was observed in the upcoast section of the study from September to December in 1978 and 1979.

A more recent report (Southern California Edison Company 1987) on SONGS contains information collected in 1986. Trace metal concentrations in the water at stations near SONGS and at a downcoast reference station were all less than the limits set in the NPDES operating permit. Sediment concentrations of chromium, copper, and iron were lower at the downcoast station but often very similar to those at SONGS. Possible impact of SONGS on the San Onofre kelp bed appears to be restricted to a region just downcoast of the unit 3 diffuser. Kelp canopy was reduced in all southern California beds from severe storms in the winter of 1985–1986. Urchin grazing (white, red, and purple urchins) was affecting many of the rocky areas of the San Onofre kelp beds, but the plant and animal assemblages appeared normal and typical of other beds in the SCB.

Body and organ weights of kelp bass and weight of sand bass collected from the area of the cooling water diffusers were heavier than those of control bass, indicating that those near diffusers were in better condition. As noted by SCE in 1979, queenfish, white croaker, and northern anchovy were the most abundant species near SONGS. Queenfish and anchovy represented 91.4% of the fish impinged and 96% of fish returned through the units 2 and 3 fish return system. The estimated annual impingement loss of unit 1 was 56,720 fish (1799 kg). At units 2 and 3, about 70% of the fish entrapped (59,969 kg) were returned to the ocean, leaving 23,092 kg impinged at the plant. Total commercial landings of northern anchovy in the SCB for 1986 were 174,453 kg; the weight of fish impinged represented about 2% of this value. Queenfish loss represented 11,288 kg, but comparison of catch statistics is not feasible because there is little, if any, commercial fishery for this species.

The Marine Review Committee (MRC) was established by the California Coastal Commission in 1974 under Proposition 20 as a condition of the permit for units 2 and 3 of SONGS. The MRC was charged with designing and implementing a broad study of the effects on the marine environment, including pre- and postconstruction and postoperational monitoring and impact analysis. In the 15 years between 1974 and 1989, the MRC staff, through the MRC studies program, has spent $46.5 million to investigate potential effects of the discharges. In an interim report, dated April 18, 1988, and a synopsis presented to the Coastal Commission, some results have been made available. Adverse effects observed include

1. A large decline in midwater fish (particularly small white croaker) out to at least 3 km from SONGS. In normal years of fish abundance, SONGS would be expected to take in and kill about 150,000 white croaker (1.5 t). The average annual catch of this species in southern California has been 150 t by commercial fisheries and 350 t by sport fisheries.
2. The kelp bed and associated community have sustained damage from increased turbidity and a 29.1-hectare cohesive sediment layer.
3. Anchovy larvae within 1–3 km of SONGS have declined about 30%.
4. Large numbers of fish larvae are lost in the intake, and some species of benthic fish have decreased.

SONGS has also affected the nearshore environment by apparently increasing the abundance of many benthic fish species and the meroplankton (temporary plankton) as a whole. Details of these studies and subsequent publications by the scientists involved will be valuable contributions to our understanding of the impacts of coastal nuclear power plants.

River and Storm Runoff

For a number of years, the staff of SCCWRP has been sampling the Los Angeles River at one or more sites to determine the concentrations of contaminants in the runoff water. Examples of the data at three sites along this river are shown in figure 12.16. High levels of oil and grease, zinc, lead, and copper are present in this storm water by the time it reaches Willow Street in Long Beach. This station, which is just north of the point where the river empties into Long Beach Harbor near the Queen Mary, has been sampled since 1971. Major variations in the flow rate and mass emissions of solids and contaminants have been observed at this site, as shown in table 12.9. High-flow volumes are certainly related to increased mass emissions, but the

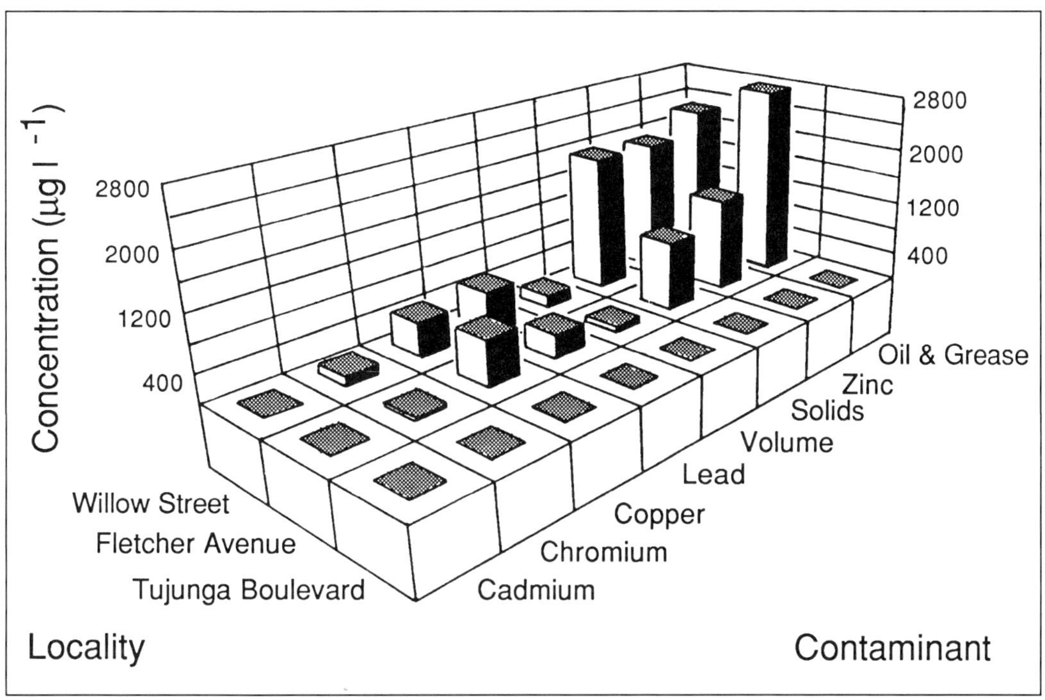

Figure 12.16. Maximum concentrations of contaminants in Los Angeles River runoff. Volume of flow is in $m^3 s^{-1}$, solids are in $\mu g\ l^{-1}$ ($\times\ 2\times10^{-4}$), oil and grease are in $\mu g\ l^{-1}$ ($\times\ 10^{-1}$), and other contaminants are in $\mu g\ l^{-1}$.

general trend in decreasing emissions of DDT and PCBs indicates that most of the sources have been depleted.

As of July 1990, data from the collection and analysis of stormwater by SCCWRP had been entered in computer files for subsequent evaluations and comparisons. The 1989–1990 SCCWRP annual report (SCCWRP 1990) provided estimated contaminant input from the Los Angeles River to the SCB. Except for cadmium, the concentrations of all constituents measured in the Los Angeles River were positively correlated with river flow and suspended solids. Other than cadmium, 70–95% of the estimated annual constituent loads were discharged during high flow days. During high flow days, it was estimated that 33–50% of the cadmium load was discharged. About 40% of the total low flow discharge of the Los Angeles River was composed of sand-filtered secondary effluents from water reclamation plants operated by the city of Los Angeles. Except for cadmium and nickel, the combined mass emissions from these plants amounted to less than 30% of the estimated loads delivered to the ocean by the Los Angeles River. SCCWRP (1990) concluded that annual mass loading estimates of cadmium, copper, lead, zinc, and total DDT were comparable between the Los Angeles River and the Los Angeles County Joint Water Pollution Control Plant (JWPCP). Annual mass emission estimates of chromium and nickel were greater for JWPCP, but those of suspended solids and PCBs were greater for the Los Angeles River.

When considering the significance of runoff, a comparison to the daily flow from a major outfall is useful. In terms of metric tons discharged, a 3-day storm (September 23–25, 1986) introduced to the coastal waters about the same amount of suspended solids and zinc as one month of discharge from the city of Los Angeles outfall (5-mile pipe) in Santa

Table 12.9. *Listing of the Mass Emission Estimates for the Los Angeles River at Willow Street on a Per Storm Basis*

Date (yr:mo:day)	Volume (10^9 l)	Suspended Solids (10^3 kg)	Oil and Grease (10^3 kg)	Cd (kg)	Cr (kg)	Cu (kg)	Fe (kg)	Ni (kg)	Pb (kg)	Zn (kg)	Total DDT (g)	Total PCB (g)
71:12:22	6.7	3,667	108	69	593	1,156	159,129	590	8,621	11,604	8,436	12,411
72:12:04	10.5	10,959	NS[a]	NS	NS	NS	NS	NS	NS	NS	6,458	21,488
73:02:27	8.2	11,081	NS	NS	NS	NS	NS	NS	NS	NS	12,514	46,643
73:03:06	8.0	23,626	NS	NS	NS	NS	NS	NS	NS	NS	6,340	15,317
73:03:11	4.3	6,286	NS	NS	NS	NS	NS	NS	NS	NS	2,536	14,782
79:11:07	2.9	10,334	NS	6	550	433	304,255	293	253	3,170	641	532
80:01:08	22.6	47,820	NS	13	3,279	2,425	1,456,410	1,794	5,576	15,651	9,181	12,288
80:01:10	15.7	23,633	NS	24	752	638	394,657	483	2,624	3,366	1,498	1,758
85:02:09	14.5	9,004	1,466	16	538	743	195,591	268	1,670	4,170	1,020	4,660
86:09:23	11.4	5,805	380	63	507	2,046	NS	523	5,111	7,926	933	3,162
87:01:04	11.5	4,871	69	36	468	1,218	NS	323	1,923	4,906	603	1,365
88:01:17	21.3	19,765	96	27	913	1,367	NS	535	2,082	7,355	1,508	1,510

[a]NS = not sampled.

Monica Bay. The 1973 report by SCCWRP noted that except for lead, zinc, and oil and grease, storm runoff contributed about one order of magnitude less to coastal waters than did municipal wastewater. However, major reductions have occurred in the mass emissions of contaminants from waste treatment plants in the late 1970s and 1980s, suggesting that the relative contribution of storm runoff has increased. A major problem in assessing mass emissions from runoff is the lack of consistency in sampling low-flow (summer) and high-flow periods in the major storm channels. To produce the type of total assessment that would be of most value, it is necessary to understand the following variations in input: (1) with time during a storm, (2) between storms, (3) between low-flow and high-flow, (4) between storm channels, and (5) between years. Using the mean values and the variances, one could provide a reasonably good estimate of annual inputs from storm runoff that could be compared to other sources of pollution to the SCB. Since all of this information is not available from the present database, the estimates are not as accurate as one would like.

Recent developments in Los Angeles, Orange, Riverside, and San Diego counties may provide the level of effort required to obtain the necessary data over the next few years. A process called the "Early Permit" is moving forward in these regions, where there is a commitment on the part of the agencies responsible for the quality of stormwater runoff to begin assessing the factors contributing to contamination and possible solutions. There is, in fact, no clear-cut evidence of which components in stormwater represent a hazard to humans or marine species. While evidence from Santa Monica Bay (J. Dorsey unpubl. data) and other locations that high levels of fecal coliform bacteria on beaches are a result of stormwater inputs, the health impacts on swimmers have not been documented (see chap. 4). There are substantial amounts of chemical contaminants entering coastal waters from storm runoff, but the effects on biota in the sediments or water near these discharges have not been well defined. Similar to the problem of municipal waste inputs, scientists have not yet been able to determine the exact levels of single contaminants on sediments or mixtures that represent a clear hazard to marine species. It is even less clear whether or not the introduction of these chemicals on particulates to coastal waters represents a direct threat to humans. Of primary concern is the introduction of chlorinated organics, such as DDT or PCBs, which accumulate in fish and provide a health risk to individuals who ingest a significant amount of the contaminated species.

Petroleum

Natural Seeps

Natural petroleum seepage has been reported from many areas of the SCB. Most known seeps occur on the mainland shelf, but others have been reported around many of the northern and southern Channel Islands and offshore banks and ridges. On the mainland shelf there are four main seepage zones: Point Conception, Coal Oil Point, Santa Barbara–Rincon, and Santa Monica Bay. Within most of these areas are several recognizable seep trends, often associated with structural features such as anticlines or faults. Detailed maps and descriptions of seeps are available for the Santa Barbara Channel (Fischer 1978) and for Santa Monica Bay (Heyney et al. 1975) (fig. 12.17).

GEOLOGIC SETTING

Marine petroleum seepage in southern California, as elsewhere, results from a combination of geologic circumstances that bring hydrocarbon-bearing source rocks to—or close to—the sea floor. Typically, seepage originates from relatively young (Miocene) sedimentary deposits and is most common along basin margins. In the case of Coal Oil Point,

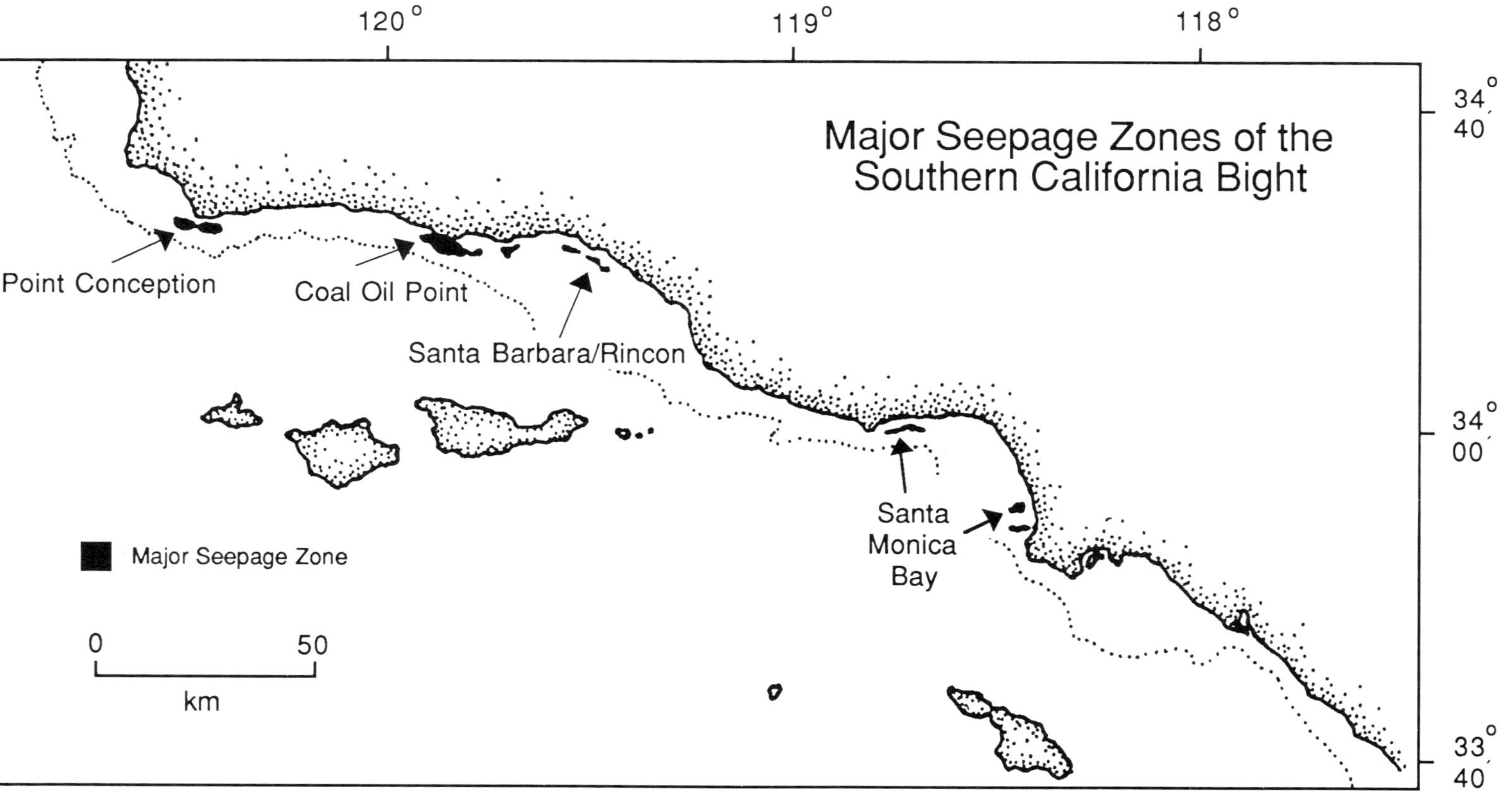

Figure 12.17. Natural oil seep zones along the SCB.

whose bedrock formations are a seaward extension of oil-rich formations of the Ventura–Santa Barbara Basin, later Quaternary or early Holocene uplifting led to the eventual thinning of unconsolidated Quaternary sediments, then to exposure of bedrock on the sea floor (fig. 12.18). Erosion and faulting of these formations subsequently led to petroleum seepage. Specifically, seepage is associated with eroded anticlines and the interactions of faults with cross or tear faults, particularly where sediment overburden is thin or absent, because sediment presents an obstacle to the vertical migration of petroleum fractions that are otherwise less dense than seawater.

Fischer (1978) has proposed an evolutionary sequence of seep trends that is based on the geologic features and types of hydrocarbons emitted by present-day seepage in the Santa Barbara Channel. This scheme serves to illustrate the variety of seepage phenomena in the SCB. Early, or stage I, seeps are characterized by outer fold trends and deep Pliocene–Quaternary overburden; gas seepage may range from none to prolific, with little associated oil. Mature, or stage II, seeps are distinguished by outer fold trends that enclose Miocene Sisquoc Formation in their centers and have little or no Quaternary overburden. Approximately equal amounts of oil and gas emissions, with minor amounts of tar, characterize these seeps. Advanced, or stage III, seeps are characterized by inner fold trends that bring Miocene shale into proximity with the sea floor. These seeps have approximately equal proportions of gas, liquid oil, and tar emissions. Finally, deep erosion of the oil-bearing formations (Monterey shale) and oozing tar mounds mark inactive (stage IV) seeps. Because evidence exists for the deposition of thermally mature petroleum hydrocarbons (triterpenoids) in the unconsolidated Pleistocene sediments of the Santa Barbara basin (Simoneit and Kaplan 1980), there appears to have been sufficient time for such an evolution. Since natural seepage is associated with faults, it is not surprising that seismic activity can affect emissions. For example, following the large earthquake in the Los Angeles area in 1971, active gas boils were seen for about 5 days afterward just offshore of Malibu (Wilkinson 1972).

The Coal Oil Point seep area is the focus of most of the detailed geologic, chemical, and biological studies of seeps in southern California. Both the Coal Oil Point seep and the Isla Vista seep have been described in some detail (Allen et al. 1970). The Coal Oil Point seep is a thin, narrow seep that tends south–southeast from Coal Oil Point for approximately 1 km, extending from a depth of 13 m to a depth of 30 m. The Isla Vista seep lies approximately 1 km due east of the shoreward extent of the Coal Oil Point seep in about 10–16 m of water. Both of these areas are estimated to be about 1000 m^2 in area. The shoreward portions of these areas consist of broken shale infilled to a varying extent with fine sand, but the overburden is generally very thin. The sediment overburden thickens in the offshore portions of these seeps. In some cases, only gas bubbles can be seen to escape from the thicker sediments, often through biogenic structures, such as polychaete tubes (of *Diopatra*). In the Isla Vista seep, much of the seepage is mixed oil and gas, particularly in areas where it is intense. The gas escapes as discrete bubbles that range from only several millimeters to more than 1 cm in diameter. Oil-coated gas bubbles leave a small sheen of oil as they pass through the water surface. Oil may also emerge from separate vents in small droplets several millimeters in diameter, but it is estimated that the largest volumes come from areas of intense seepage where up to 100 point sources for droplets can be seen per square meter, mixed usually with large amounts of gas.

In these intense seepage areas, the oil and gas often appear to be coming from petroleum-soaked sands. It is not known if these sands represent the original geologic sources or are secondarily deposited over the fractured Miocene shale bedrock. The liquid oil has a specific gravity slightly less than that of

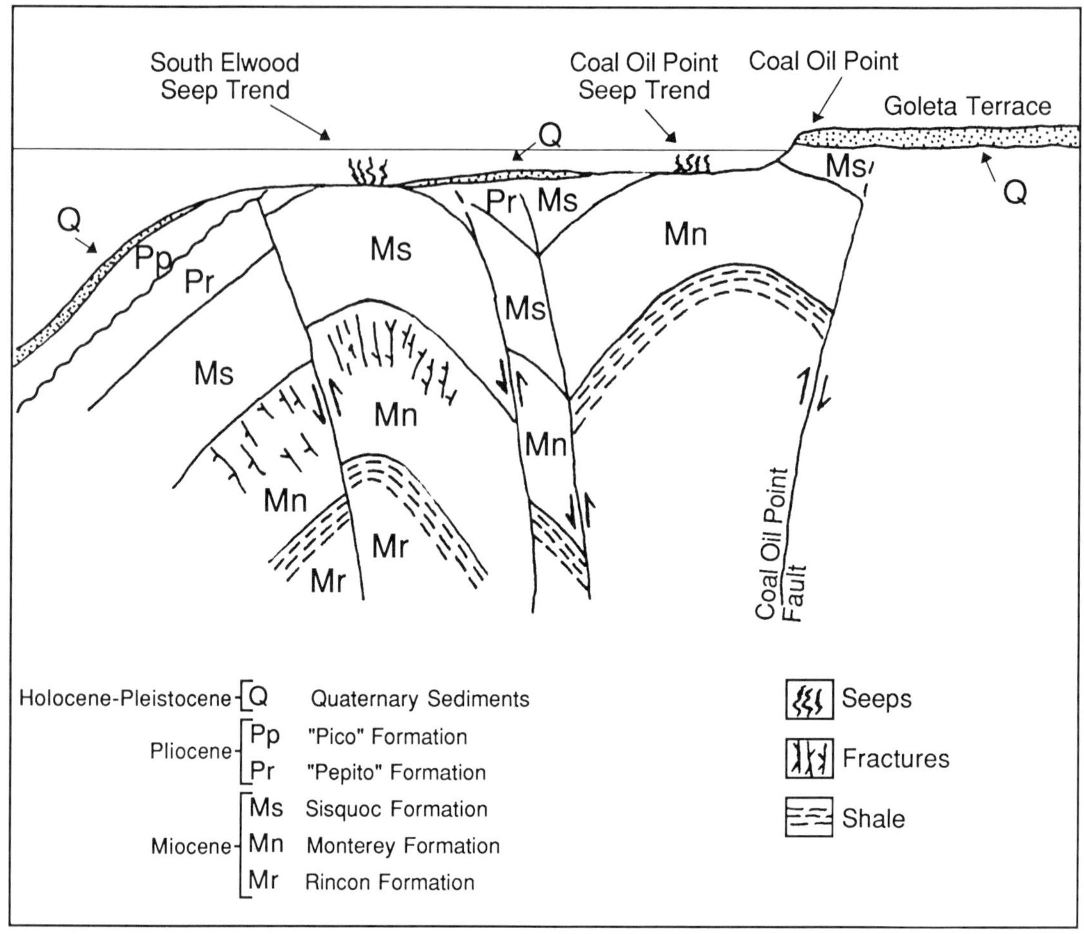

Figure 12.18. A geologic section from Coal Oil Point off the mainland shelf showing major faults, anticlines, and geologic ages of rocks and sediments. (After Fischer 1978.)

water and forms into small, buoyant droplets several millimeters in diameter. The average volume of such droplets from the Coal Oil Point seep is 2.5 ml (Allen et al. 1970). The droplets may migrate slowly into the water column with a long tail attached to the sediment that eventually breaks, allowing the droplets to reach the sea surface.

On sunny and windy days, the oil droplets quickly spread out on the surface of the water and much of their volatile content evaporates. The remaining oil either contributes to the thin slicks that can be seen moving out of the area, usually with the prevailing westerly currents, or sinks to the bottom. Oil tends to accumulate during calm or overcast conditions, especially if surface currents are weak. Under these conditions, oil does not evaporate quickly, making it more likely that sufficient oil will accumulate to make a more persistent, thick slick. Within the Isla Vista seep are numerous areas of intense seepage that appear to account for most of the seepage that emerges from the overall seep. Observations from a submersible in a particularly active seep located at a depth of about 65 m at the eastern edge of the South Elwood trend (about 1600 m east-southeast of Platform Holly) documented more extensive alterations of the sea floor associated with seepage. Prior to 1982, this seep could easily be located from the air by a large boil of natural gas on the sea surface. In September 1982, however, ARCO placed a pyramid baffle system over it to trap natural gas and oil. Large amounts of gas and oil are coming from several discrete vents on the sea floor. Some of the vents were observed to be as large as 5 m in diameter and appeared to be at least 2 m deep.

VOLUME AND COMPOSITION

As yet, no accurate estimate exists of the volume of seepage in the SCB, although perhaps the best estimates for any area in the world are those made for the Santa Barbara Channel (Allen et al. 1970; Fischer 1978). SCCWRP has estimated a rate of oil seepage of approximately 3 ml cm^{-2} h^{-1} from an active area on the outer edge of the Isla Vista seep (unpubl. data). In an industry-sponsored study of gas seepage rates at Coal Oil Plant, at More Mesa, and off Gaviota in the Santa Barbara Channel, measurements of gas seepage rates from individual events ranged over four orders of magnitude, from 0.0004 to 1.3 ml s^{-1} (Vetter and Johnson 1982). In early 1988, ARCO was capturing 42,500 m^3 d^{-1} of gas and approximately 20–35 l d^{-1} of oil in their seep containment system south-southeast of Platform Holly, off Coal Oil Point (D. Keen, ARCO, pers. comm.).

Using a combination of underwater observations, aerial surveillance, surface sampling, and estimates of wind and current speed, Allen et al. (1970) estimated a range of seepage from the Coal Oil Point seep trend of 8000–11,000 l d^{-1} of oil. This was apparently independent of the gas seepage in the area, which is also considerable. Basing his conclusion on seismic profiling data, Fischer (1978) estimated that seepage from the Coal Oil Point seepage trend ranges from 4000 to 64,000 l d^{-1} and that for the entire Santa Barbara Channel ranges from 6400 to 107,200 l d^{-1}. Estimates made from the seismic profiling records may be inaccurate because of disproportionately strong resonance signals from certain sizes of gas bubbles.

The composition of gases from the few southern California seeps that have been analyzed suggests that most of these have a petrogenic origin, that is, they contain <95% methane, with the remaining portion being C_2–C_6 hydrocarbons. By contrast, biogenic hydrocarbon gas is >99% methane. Table 12.10 summarizes the composition of gases from two seepage areas in the Santa Barbara Channel. Like other liquid crude oils, petroleum seep oil is composed of thousands of different compounds. Three classes of compounds usually can be operationally defined by chemical separation, using silica gel chromatography. These result from successive elutriations by hexane, benzene (or toluene), and methanol or from a similar series of solvents of increasing polarity. Gravimetric determination of these fractions in samples from the Carpinteria seep and from three seepage areas in the vicinity of Coal Oil Point yielded similar results for all four oils: 20–28% in the hexane fraction, 43–63% in the benzene fraction, and 8–13% in the methanol fraction (Reed and Kaplan 1977). Further separation of the hexane fraction by molecular sieves to obtain the normal alkanes revealed that, in contrast to many other crude oils, the normal alkanes make up approximately 2–3% of the oil. This is characteristic of other seep oils and is thought to result from microbial degradation (Bailey et al. 1973; Sassen 1980).

The Isla Vista seep oil is, in fact, dominated by other alkylated cycloalkanes and branched alkanes (Spies et al. 1980), which is typical of other seep oils that are strongly naphthenic (Philippi 1977). The remaining fractions of seep oils have not been extensively analyzed, but it is known that a high proportion of aromatic hydrocarbons elute in the benzene fraction. Toluene, alkylated benzenes, indan, naphthalenes, and phenanthrenes have been identified in the oil, water-soluble fractions or sediment, and in water samples from the Isla Vista seep (Spies et al. 1980; Stuermer et al. 1982). Seep oils also have a high sulfur content (3–5%) compared to many other crude oils. A whole series of sulphur-containing compounds have been identified in samples from the Isla Vista petroleum seep (Stuermer et al. 1982). The low molecular weight compounds consist of thiophenes and 5- and 6-membered saturated heterocyclic ring structures with various alkyl substitutions. The higher molecular weight sulfur compounds include many benzothiophenes and dibenzothiophenes as well as their alkylated derivatives.

ECOLOGY

Oil and gas seepage have a profound effect on microbial populations, productivity, and metabolic activities of sediments. Studies in the Isla Vista petroleum seep show that bacterial numbers and productivity in sediments are several times greater than in nearby areas without seepage (Montagna et al. 1986). Benthic oxygen flux, whose biological component is usually dominated (>80%) by microbial contributions, is also several-fold greater in the seep (Bauer et al. 1988). Hydrocarbon degradation and sulfate reduction rates are also much higher in sediments of the Isla Vista petroleum seep (Montagna et al. 1986). These activities appear to be directly related to elevated concentrations of sulfide and eventual depletion of sulfate as well as high alkalinity in pore water.

Basing their measurements on bacterial densities, cell biovolumes, and frequency of dividing cells, Montagna et al. (1986) estimated annual average bacterial productivity in sediments (0–2.7 cm) to be 1010 mg cm^{-2} d^{-1} in an area of intense seepage (Station A) within the Isla Vista petroleum seep and 330 mg cm^{-2} d^{-1} in an area of less intense seepage (Station B), compared to a rate of 230 mg cm^{-2} d^{-1} in a nearby area without seepage (Station C).

Hydrocarbon-degrading microbes often constitute a large fraction of the microflora in sediments with high concentrations of petroleum. They also are frequently found in association with sulfate reducers (Ross 1983). It is not surprising, therefore, that increased capacity for hydrocarbon utilization was found, as assayed by anthracene degradation, at these same three stations within the Isla Vista petroleum seep (Bauer et al. 1988). Similar patterns were observed for naphthalene and hexadecane degradation (Bauer et al. 1988). Recent experiments also indicate that the seep sediments contain microbes capable of rapid oxidation of methane (R. Spies and J. Bauer unpubl. data).

The distribution and abundance of macrofauna and meiofauna in and around the Isla Vista petroleum seep have been thoroughly described. Stations A, B, and C were samples at 2-week intervals for a year (1984–1985) for abundances of meiofauna (Montagna et al. 1987). Sampling revealed that nematodes had significantly greater mean annual abundances at Station A (2.4×10^6 m^2) than at the two stations having lesser amounts of oil (1.3×10^6 m^2 at B and 1.4×10^6 m^2 at C). Juvenile and adult harpacticoid copepods had the highest densities at Station B (1.17×10^6 m^2), but abundances at Station A (0.996×10^6 m^2) and Station C (1.08×10^6 m^2) were not significantly different. These two groups formed 96% of the meiofauna, while 19 other groups comprised the remaining 4%. The 19 groups included ciliates, gastrotrichs, turbellarians, polychaetes, Foraminifera, bivalve juveniles, ostracods, amphipod juveniles, kinorynchs, cumaceans, gastropods, ophiuroid juveniles, halicarids, leptostracans, and tanaidaceans. More of these fauna occurred at Stations B (0.181×10^6 m^2) and C (0.196×10^6 m^2)—which were not significantly different statistically from one another—than at Station A (0.134×10^6 m^2).

The macrofauna community is part of the gray-sand fauna predominant at depths of 12–21 m from Point Conception to Ventura (Barnard and Hartman 1959). It has been classified as the *Nothria–Tellina* assemblage (Jones 1969). The macrofauna at Stations A and B were compared over a 26-month period (1975–1978) (Spies and Davis 1979;

Davis and Spies 1980), during which significantly higher overall densities of macrofauna were found on all dates at Station B. Macrofauna classified as deposit feeders, mainly polychaetes, accounted for much of this difference. Therefore, a somewhat similar conclusion to that reached for the meiofauna also applies to the macrofauna. Diversity and dominance were found to be very similar, however. A later study of the macrofauna at Station A revealed a dramatically reduced macrofauna in and around the active oil intrusions at this station (Spies et al. 1980). Occasional blooms of the polychaete *Capitella* spp. also have been observed at this station. The patterns observed in the macrofauna therefore generally agree with those predicted from the theory of organic enrichment: reduced numbers of individuals and numbers of species near the source of organic matter input, increases in numbers of some species above background at some intermediate level of influence, followed by a return to a more normal community structure at some distance from the source of organic matter (Pearson and Rosenberg 1978).

Spies et al. (1988) performed a set of field experiments to compare directly the effects of oil in sediment to those of a contemporary source of organic matter—kelp (fig. 12.19). The purpose of these experiments was to distinguish the effects of oil toxicity from those that were merely the expected results of a surplus organic matter supply. The addition of 0, 0.1, 1, and 5% of each constituent was made to the sediment, and the treatments were placed in the environment (at Station C) for 35 days for colonization and subsequent development of the fauna. The results showed that the total meiofauna, dominated by nematodes, responded more vigorously to kelp, but the general shapes of the abundance versus treatment curves were similar for each treatment, each producing a maximum abundance response at 0.1% organic addition. The macrofauna responded strongly to both sources of organic matter; still, peak responses were at the lower concentrations of each constituent. Analysis of the carbon stable isotope ratios of the *Capitella* spp. that settled in 9 of the 16 treatments revealed a correlation of $\delta^{13}C$ with the amounts of oil (–24.1 ‰) and kelp (–14.8 ‰) in these treatments (Spies et al. 1988). It was apparent, then, that liquid petroleum was being utilized by the deposit-feeding organisms colonizing these experimental treatments. Recent experiments with multicellular seep organisms indicate rapid utilization of ^{14}C-label from $^{14}CH_4$ (R. Spies and J. Bauer unpubl. data).

RESPONSES OF ORGANISMS

Because harpacticoid copepods are reported to be sensitive to petroleum exposure, the 10 most common species collected during 1984–1985 at Stations A, B, and C in and around the Isla Vista seep were analyzed for a number of measures of reproductive success. These data imply that reproductive development of harpacticoid copepods is enhanced at the seep, but fewer juveniles were produced per gravid female than at the reference station C (Kinnetic Laboratories, Inc. 1989). This suggests a toxic effect on reproductive success. However, another possible interpretation is offered by evidence from the previously discussed field experiment, during which mature females, nauplii, and juvenile harpacticoids were analyzed in each of the treatments. The percent of gravid females and the nauplii to female ratio was found to increase continuously as a function of the percentage of kelp in the treatment sediments, whereas oil had no effect. However, the juvenile to nauplii ratio decreased as kelp concentrations increased. It is thus possible that, in some of the cases observed, decreased juvenile to gravid female ratios could be a manifestation of the processes that tend to result in fewer numbers of organisms and species with increasing food supply (organic loading).

It is clear, however, that petroleum is eliciting a biochemical response in fishes, as the hepatic mixed function oxygenases (MFO) are induced to higher levels of activity in

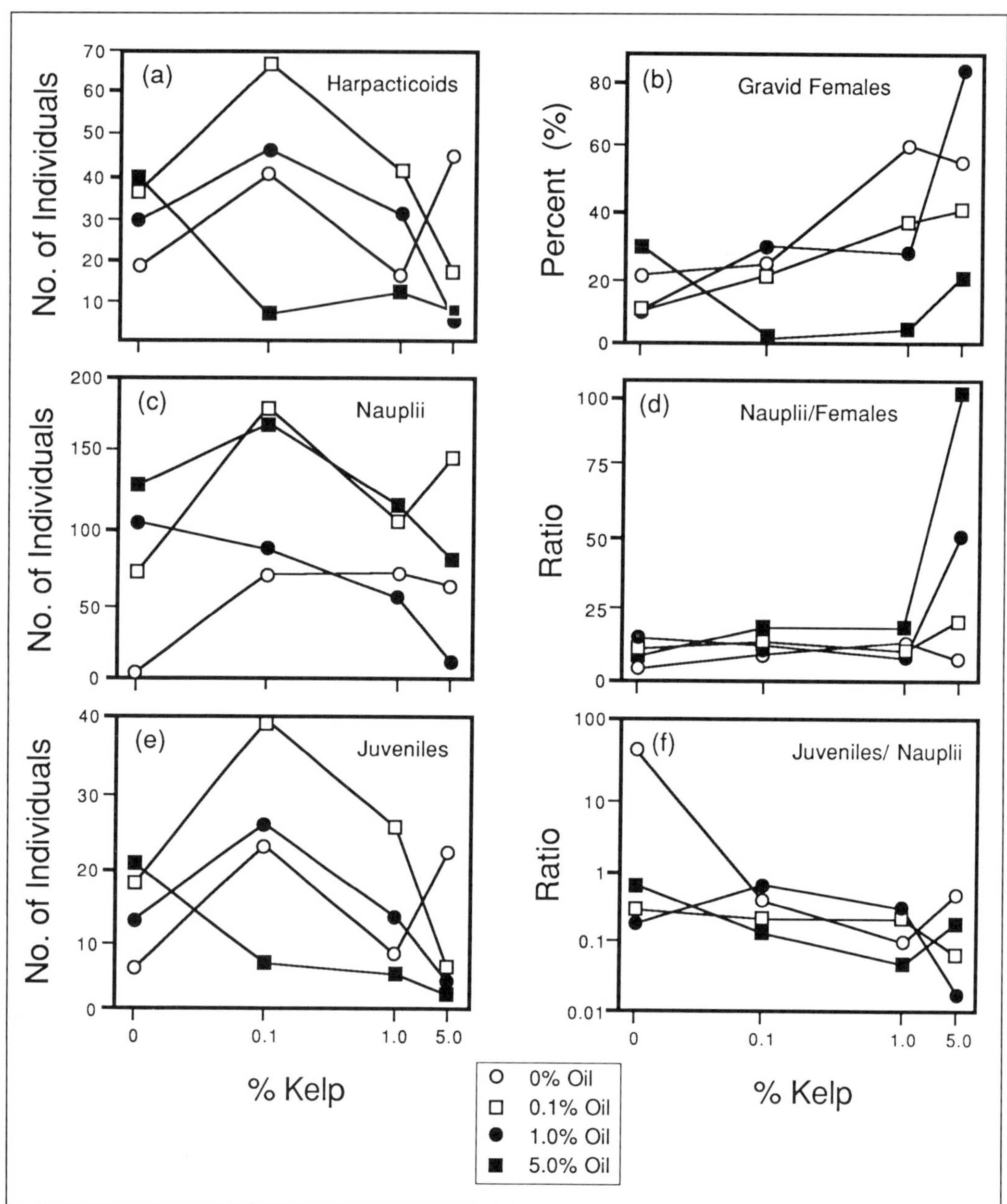

Figure 12.19. Harpacticoid copepods in each of the 16 treatments after 35 days. (a) Total number of harpacticoid copepods (adults and copepodites). (b) Percentage of the females that were gravid. (c) Total number of harpacticoid nauplii. (d) Ratio of nauplii to females. (e) Abundance of copepodites. (f) Ratio of copepodites to nauplii. (After Spies et al. 1988.)

Citharichthys stigmaeus, Citharichthys sordidus, and *Phanerodon furcatus* relative to several comparison areas (Spies et al. 1980, 1982). These enzymes facilitate the elimination of hydrocarbons from tissues by oxidative conversion to more water-soluble compounds. Some products of polynuclear aromatic hydrocarbons, however, are converted to potentially cytogenetic and teratogenic compounds. These appear to exert their initial action by binding covalently to DNA and possibly other large molecules. The elevation of MFO activity appears to be a short-term adaptation to living in these seep environments.

Water-soluble fractions (WSF) of seep oil have been shown to be acutely toxic and to have effects on the growth and proportion of abnormalities in starfish embryos (Spies and Davis 1982). Significant effects were generally seen at 50% WSF and greater, which corresponds approximately to 4.8 mg l^{-1} of low molecular weight aromatic hydrocarbons (Stuermer et al. 1982). This concentration is approximately three orders of magnitude higher than those measured in the water from the seep environment, although pore water samples taken in an active oil intrusion had approximately 1 mg l^{-1} of these compounds (Stuermer et al. 1982).

Oil Spills

To date, two spills of significant magnitude have occurred in the SCB that have been studied to determine biological impacts. On February 7, 1990, a spill of Alaskan crude oil from the tanker *American Trader* occurred off Huntington Beach, California. The impacts of that spill are still being investigated and are not available for inclusion in this review.

Most accidental oil spills occur when tankers run aground and subsequently lose their cargoes through ruptured hulls. In the case of the *American Trader,* it was the anchor that punctured the hull. Prior to February 7, 1990, only two oil spills in the SCB had been of sufficient magnitude or significance to warrant extensive studies of their biological impacts, and both were uncharacteristic spills in that they did not involve grounded vessels. The first of these two SCB spills resulted from a well blowout in the Santa Barbara Channel in 1969. (Since then, three other blowouts of offshore wells have occurred in other parts of the world: Gulf of Mexico, North Sea, and Bay of Campeche.) The second incident, in 1976, was most unusual because it involved the explosion of a tanker (the *Sansinena*) in Los Angeles Harbor and the sinking of heavy Bunker C fuel oil.

Two reports (Straughan 1970, 1971), as well as individual publications, describe the Santa Barbara oil spill and the findings of subsequent studies. The blowout on Platform A in the Santa Barbara Channel began on January 28, 1969, and by February 8, 1969, the flow of oil was substantially reduced. The estimate of oil released during the first 100 days was more than 1.2×10^7 l at a rate of about 1×10^6 l d^{-1} (Mead and Sorensen 1970). Several subsequent oil discharges of smaller rates and volumes occurred up to about December 1969. The total volume of the spill was about one order of magnitude less than the 1.14×10^8 l spill of the *Torrey Canyon* incident in the English Channel in 1967. The spill from the blowout hit about 48.3 km of the California coast; the *Torrey Canyon* mishap affected about 225 km of English and French beaches.

Effects on birds from the spilled oil offer the best documented impacts of the Santa Barbara Channel blowout (Straughan 1971). The California Department of Fish and Game (CDFG) estimated that more than 3600 birds were killed, but the species composition was observed for only 432 dead birds. Western grebes suffered the highest mortality and 253 out of the 432 were loons or grebes. CDFG surveys similar to those conducted since 1965 showed that all fish appeared healthy and there were no indications of impairment in the food chains or damage to fish populations. The total commercial fish catch from the Santa Barbara Channel Islands did not change from 1968, but a decrease occurred in the months of February and March. This reduction was likely the result of the reluctance of fishermen to use their boats or gear where they might be subject to oiling. Ebeling et al. (1972) confirmed this conclusion and also discussed the difficulties of attempting to determine the impacts of the oil spill on fish communities.

A report from the Allan Hancock Foundation stated that no proof of damage to plankton, benthos, or marine mammals could be detected (Straughan 1971). There was mortality among the upper intertidal barnacle *Chthamalus fissus,* probably caused by smothering from oil. *Balanus glandula,* which was larger and generally protruded above the oil layer, survived well. The alga *Hesperophycus harveyanus* and the surfgrass *Phyllospadix* sp. were killed in some localities, but by August 1969, recovery had begun.

Comparisons between the results of this oil spill to spills in England (*Torrey Canyon*) and Massachusetts (*Florida*) led to some hypotheses (Straughan 1970). For example, the much greater impacts from crude oil on the English coast were correctly judged to be the result of very heavy application of a toxic chemical dispersant. The significant impacts on West Falmouth Bay in Massachusetts were considered to be the result of a major difference in the type of oil (high aromatic No. 2 fuel oil) and the physical nature of the coastal environment.

In the second SCB spill, the *Sansinena,* a 70,000-ton tanker, was moored at a dock adjacent to Cabrillo Beach in outer Los Angeles Harbor. On December 17, 1976, the tanker exploded and burned, spilling approximately 4.8×10^6 l (32,000 barrels) of Bunker C fuel oil. The majority of this heavy oil, from which lighter fractions were likely consumed in the fire, sank to the bottom of the harbor. Benthic stations as close as possible to the sunken wreckage were sampled a few days after the explosion, and SCUBA dive transects from the ship were made immediately after the accident and at monthly intervals thereafter for 1 year. A thick oil residue (about 2.5 m) pooled beneath the ship and covered the nearby benthic habitat. One year later, after efforts by salvage crews, the layer of residue was reduced to 2 cm. Severe storms following this observation apparently removed the remaining oil layer in a period of 2 months (Soule and Oguri 1978a). Benthic samples taken initially at 23 stations were reduced to 12 stations on a quarterly basis through November 1977 (Reish et al. 1980).

Apparently the spill had not yet affected the benthos when the first samples were taken a few days after the explosion, since a mean of 31.7 species was identified. One month later, this mean had decreased to 12.4, but stimulation of the surviving species apparently occurred, producing an increase in the numbers of individuals. By the next collection in April 1977, the mean number of species increased to 27.3 and the number of individuals decreased by an order of magnitude. Partial recovery was observed in July, and by November of that same year, the populations appeared normal. Long-term examination of changes was not possible because the area was dredged following salvage and cleanup of oil residue.

In September 1987, an unusual spill that did not involve an oil tanker occurred 20 km southwest of Point Conception when the *Pac Baroness,* carrying finely powdered copper concentrate (71% chalcopyrite), collided with a larger vessel in the fog and sank in 430 m of water. Hyland et al. (1988) reported elevated levels of hydrocarbons on bottom sediments in the vicinity of the sunken vessel, and the hydrocarbon pattern matched that of the fuel oil from the vessel. Analyses of macroinfaunal samples from the same vicinity showed significant reductions in mean number of species, mean number of individuals (combining all species), and combined abundance of amphipod species. Total polyaromatic hydrocarbon (PAH) levels at the sediment stations most impacted by the wreck were from 10 to 3500 times higher than the average background concentration (0.04 mg kg^{-1}).

FATE OF SPILLED OIL

Although there have been few actual case studies concerning oil spills in California waters, many of the studies conducted in subarctic and temperate marine environments provide information that is directly applicable to addressing the fate of oil spilled in California waters, as well as the fate of PAHs in petroleum. The fate of oil spilled onto the surface of the water is depicted in figure 12.20. This generalized schematic addresses oil spilled on the water's surface prior to landfall. A variety of processes can act on the oil and partition it into various compartments in the marine environment, that is, water column, sediments, biota.

Each of these processes may act differentially on the various components of oil. As a result, the chemical composition of hydrocarbons in the seawater, exclusive of the droplets themselves, is distinctly different from that of the whole oil droplets. The water-soluble and more toxic aromatics dominate the dissolved fraction. Animals exposed to the water mass will initially acquire these compounds. However, the total concentration of these dissolved aromatics seldom exceeds 1 mg l^{-1} and usually is in the 0.01–0.1 mg l^{-1} range. The instantaneous concentration of aromatics and of the whole oil droplets themselves is directly dependent on the physical energy of dispersion, that is, the result of wind- and wave-induced turbulence.

The worst case of water column exposure is actually represented by either (1) the subsurface discharge of oil from an oil well blowout (Brooks et al. 1981; Boehm and Fiest 1982) or (2) a chemically dispersed oil plume. In these cases, maximum petroleum hydrocarbon levels seldom exceed 50 mg l^{-1}, and these concentrations are rapidly diluted by a factor of 10^3–10^5 as they are mixed with seawater. A comprehensive review of chemical dispersant use and effects was recently completed by the National Research Council (NRC 1989). The report recommends that dispersants be considered, along with other response options, as a potential first response to oil spills. Major benefits are reduction in effects on birds and fur-bearing mammals and the prevention of oil stranding on sensitive shorelines. However, contingency planning and prior approval of application are necessary for effective use of dispersants.

The potentially more serious environmental effects associated with petroleum hydrocarbon transport mechanisms (specifically, aromatic hydrocarbons) relate to: (1) the sorption of oil droplets on suspended particulate material and transport to the bottom; and (2) the landfall, or beaching, of an oil slick, followed by sorption onto beach sediments and subsequent erosion of these sediments and deposition in the nearshore zone. Studies of oil spills have shown that for significant quantities of oil to reach the bottom, oil must sorb to suspended sediment and sink. The Santa Barbara oil spill exemplifies this process (Kolpack et al. 1971). During the spill, transport and deposition of large quantities of flood runoff material occurred in channel waters near the spill.

Although the sorption-sinking sequence appears to be the major transport route of oil to the benthos, other mechanisms can become significant in certain cases. These are, most notably, fecal pellet transport (Johanssen et al. 1980; Boehm et al. 1982), sinking due to Langmiur circulation, or direct sinking of dense (cold, weathered) oil in areas of low-density water (freshwater input at river mouths or near ice melting). Studies by Juszko et al. (1983) have suggested that the weathering of oil, typically of an initial density of 0.9 g ml^{-1}, can create a residue having a density of 1.02–1.03 g ml^{-1}. Once oil achieves this density, it can sink. The behavior of oil, once it sinks, is determined by the vertical density structure of the water column.

BIOLOGICAL EFFECTS OF OIL SPILLS

Numerous published scientific papers and technical reports deal with the biological im-

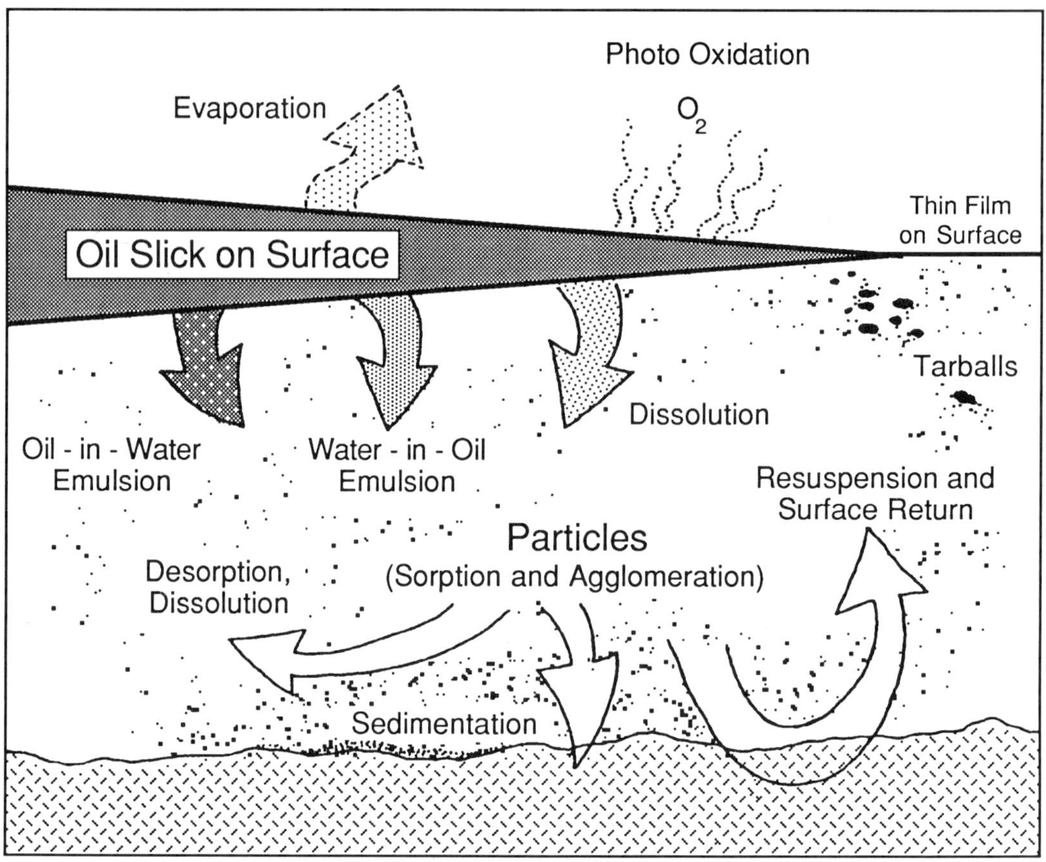

Figure 12.20. Transport and interaction path of oil and suspended particulate matter.

pacts of oil spills. Results of many of these investigations have been summarized in reviews by Clark and Finley (1977), Teal and Howarth (1984), the National Research Council (1985), and Boesch and Rabalais (1987). The evidence is that massive kills of shallow and intertidal benthic organisms occurred whenever and wherever large amounts of fresh crude or refined oil came ashore. Massive immediate kills either did not occur or were impossible to document in spills and blowouts that occurred well offshore and resulted in little or no stranding of fresh oil (*Argo Merchant, Chevron, Ekofish,* and *Ixtox I*) (National Research Council 1985). In the *Florida* diesel fuel oil spill, massive kills occurred in heavily oiled sites where sediment hydrocarbon concentrations exceeded about 130 mg kg^{-1}. Following the *Amoco Cadiz* spill, there were massive reductions of some species, such as ampeliscid amphipods, where sediment hydrocarbon concentrations exceeded 50–100 mg kg^{-1}. Ampeliscid amphipod populations also were severely damaged in the *Florida* and *Tsesis* spills. These amphipods apparently are among the most sensitive dominant benthic species in many temperate and subarctic marine environments.

In most cases, littoral macroalgae appear to be fairly tolerant of oiling, although, as in the *Torrey Canyon* spill, they may be smothered by massive deposits of oil. In some cases, a sensitive species (e.g., *Ascophyllum*) was eliminated and was replaced rapidly by a more tolerant species (*Fucus*). In most cases, the macrofauna associated with the macroalgae was decimated wherever significant amounts of oil came ashore. In the case of the

Torrey Canyon spill, much of the damage to the rocky intertidal fauna was attributed to the use of large amounts of chemical dispersants containing aromatic solvents to clean the shore (Southward and Southward 1978).

Intertidal salt marches, particularly those containing peaty, high-organic soils, tend to be quite sensitive to oil spills, particularly if the oil is fresh when it reaches the marsh and the marsh grasses are in the active growth phase. Coastal marshes of Buzzards Bay, Massachusetts, which were impacted by diesel fuel from the *Florida* spill, were severely damaged, and recovery was incomplete even 8–12 years after the spill (Sanders et al. 1980). Several marshes, especially that at Ile Grande, were severely impacted by the *Amoco Cadiz* spill, and were subsequently destroyed during the cleanup operation. Following the *Arrow* spill, Nova Scotia marshes damaged by the Bunker C residual oil experienced a die-off of *Spartina alterniflora* 1 year after the spill, but substantial recovery was observed 2 years later (Thomas 1973).

Damage to commercially important shellfish has been variable. There were large losses of intertidal clams (*Mya arenaria*) following the *Florida* and *Arrow* spills of refined oil products. Following the *Amoco Cadiz* spill, oysters, shrimp, and lobsters became heavily contaminated with oil, but did not experience high mortalities. In fact, numbers of commercial shrimp actually increased along the north coast of Brittany in the 2 years following the *Amoco Cadiz* spill, possibly because of decreased predation or increased microbial and algal production. Abundance of some microfaunal and meiofaunal species may increase in moderately oiled sediments, providing increased food for benthic deposit feeders such as the clam *Macoma* and the polychaete *Arenicola*.

Very few impacts, other than tainting, have been observed in phytoplankton and zooplankton communities following oil spills. Following the *Tsesis* spill in the Baltic Sea, phytoplankton and planktonic bacterial biomass increased, possibly due to a decrease in the abundance or feeding activity of the larger zooplankton. Some evidence of contamination and sublethal stress in zooplankton populations was reported following the *Arrow, Argo Merchant,* and *Amoco Cadiz* spills. These impacts apparently were of short duration and areal extent.

Adverse impacts on commercially important pelagic and demersal fish were reported following the *Argo Merchant, Tsesis,* and *Amoco Cadiz* spills. A year after the *Tsesis* spill, spawning frequency and hatching success of herring *Clupea harengus* were reduced. For up to 3 years after the *Amoco Cadiz* spill, plaice *Pleuronectes platessa* collected from oil-contaminated estuaries exhibited a variety of biochemical changes and histopathological lesions. In female plaice from the contaminated estuaries, there was evidence of delayed or suppressed ovarian development. Growth rate and fecundity of the fish were reduced.

Effects of Chronic Discharges

The long-term environmental impacts of chronic low-level point-source effluents containing PAHs are quite different from those of massive oil spills. There is usually no massive dieoff of marine fauna and flora. Instead, there may be a gradient of impact, characterized by altered community structure, abundance, and diversity extending around the pollutant source. Several investigations of impacts of refinery effluents and oil tanker terminals have been done on the coastal marine environment, as well as a few studies of impacts of produced-water discharges (reviewed by Dicks and Hartley 1982; Neff 1985).

DRILLING FLUIDS AND PRODUCED WATERS

The production of offshore oil requires the drilling of deep wells, and this process produces a variety of liquid, solid, and gaseous wastes, some of which are discharged to the marine environment. Drilling an offshore well can generate approximately 200–1000 t of drilling fluids and an equal amount of cuttings. When the production of oil and gas begins, fossil ("produced") water from the reservoir must be separated from the oil, treated, and discharged or reinjected. Within the SCB, discharge of drilling fluids is restricted. Produced waters are transported by pipeline to shore for treatment and subsequent nearshore discharge at two sites near Santa Barbara and one in Huntington Beach.

Studies of the effects of drilling fluids strongly implicate the major contribution of petroleum components, particularly diesel fuel, to the toxicity observed in laboratory tests on marine species (Neff and Anderson 1981; Conklin et al. 1983; U.S. Environmental Protection Agency unpubl. data). Conklin et al. (1983) and the ongoing EPA study reported statistically significant inverse relationships between the 96 LC_{50} values for shrimp and mysids, and the concentrations of aromatic hydrocarbons in the fluids similar to those in diesel fuel found in the drilling fluid. The results of acute and sublethal toxicity testing and field studies on both drilling fluids and produced waters were extensively reviewed by the National Academy of Sciences (1983) and Neff (1987). Most 96-hour bioassays with a variety (62 species) of marine species have resulted in LC_{50} values of about 1×10^4 mg of drilling fluid per liter of water. Water column organisms are never exposed to whole unfractionated drilling fluids or even to the lighter liquid or suspended particulate fractions of drilling fluids for long enough to elicit lethal or sublethal responses. Benthic organisms or communities can suffer localized effects where the fluids and cuttings settle out and accumulate on the bottom. The areal extent is likely to be only a few hundred meters, and the rate of recovery depends on the rate of resuspension and sea bed transport. The types of benthic communities and the energy of the environment also govern the degree and duration of impact.

Produced water discharged to the ocean from petroleum operations offshore or on the coast can contain elevated salinity, altered ion ratios, low dissolved oxygen, petroleum hydrocarbons and other organics, and heavy metals. Salinities can be as high as 170 ‰, or about five times oceanic seawater. Unfortunately, little is known about the chronic or sublethal effects of produced water, and most evaluations are inferred from the literature on the chronic and acute effects of petroleum hydrocarbons and heavy metals. Only one published study on the uptake of petroleum hydrocarbons from produced water shows that clams (*Rangia cuneata*) accumulate aromatic compounds when suspended from a platform discharging produced water in shallow (10 m) water (Fucik et al. 1977).

The most comprehensive investigations of the impacts of produced water and related production discharges from oil platforms were those conducted in Trinity Bay in Texas and the Forties Field in the North Sea. Trinity Bay is a shallow, low-salinity estuary. During the 20-month time course of the Trinity Bay study, produced water with a mean total hydrocarbon concentration of 15 mg l^{-1} was discharged from the separator platform through an outfall 1 m above the bottom at a rate of 0.6×10^6–1.6×10^6 l d^{-1} (Armstrong et al. 1979). Hydrocarbons were diluted nearly 2500-fold in the water column within 15 m of the outfall. Bottom sediments were heavily contaminated with medium molecular weight alkanes (C_{10}–C_{28} *n*-paraffins) and aromatics (C_3 benzenes–trimethylphenanthrenes).

The Trinity Bay investigators found a gradient of decreasing naphthalene concentrations from a mean of about 21 mg kg^{-1} at a distance of 15 m from the outfall to the background 500–4800 m from the outfall, depending on direction. There was an inverse gradient of numbers of organisms and numbers of species of benthic infauna with distance from the outfall. Within 15 m of the outfall, the bottom was almost devoid of organisms. Outward from the outfall, benthic faunal abundance was significantly reduced for a distance of approximately 150 m in all directions. Armstrong et al. (1979) estimated that a nominal concentration greater than about 2 mg kg^{-1} total naphthalenes on sediment was necessary to significantly reduce benthic infaunal populations in Trinity Bay.

Results of these investigations should be extrapolated to offshore situations such as in the SCB with extreme caution, however. The shallow, turbid nature of the receiving water at Trinity Bay is unlike most situations encountered offshore. Where water depth is greater and suspended sediment concentrations are lower than those encountered in Trinity Bay, a much smaller fraction of the hydrocarbons in the discharged produced water will be deposited in bottom sediments near the outfall, and adverse effects on the benthos will be much less severe.

Hartley and Ferbrache (1983) reported the results of a benthic monitoring study in the Forties Field, which is located in the British sector of the North Sea in 100–125 m of water, approximately 177 km northeast of Aberdeen, Scotland. They found the benthic fauna to be rich and diverse throughout the study area. However, directly beneath Platform C and in an area extending about 100 m to the west, the benthic macrofauna were severely depressed. There was also evidence of hydrocarbon contamination (305–470 mg kg^{-1} aliphatics), possibly from diesel fuel. Overall, it is apparent that impacts of oil production activities from the Forties Field have been localized, primarily within 450 m of the platforms, and are of low magnitude.

Toxicity of Petroleum Hydrocarbons

The fact that oils, and especially their water-soluble fraction (WSF), vary in the proportion of monoaromatics and higher molecular weight diaromatics and triaromatics led researchers to investigate the toxicity of specific petroleum compounds to a variety of marine organisms. The general trend is an increase in acute toxicity (lower LC_{50} value) as molecular weight increases to the highest molecular weight compound (fluoranthrene) found to produce acute toxicity. Unfortunately, no complete data set exists for any one species, but data from a number of studies verify this general trend (Caldwell et al. 1977; Benville and Korn 1977; Neff et al. 1976; Thomas and Rice 1979; Young 1977; Rossi and Neff 1978; Ott et al. 1978; Lee and Nicol 1978).

The findings presented in many studies over the past 15 years lead to the conclusion that naphthalenes are perhaps the most toxic components in the various oils, but that a portion of the toxicity is contributed by some of the monoaromatic compounds present in initially high concentrations in extracts of crude oil. Other important contributors to the toxicity of some crude and refined petroleums are the phenanthrenes and dibenzothiophenes. A review of the literature can be found in the National Academy of Sciences volume, *Oil in the Sea: Inputs, Fates, and Effects* (National Research Council 1985) and also in Boesch and Rabalais (1987).

To determine the differences in the tolerance to petroleum hydrocarbons at selected early life stages, four species of invertebrates were tested with precisely the same extract of No. 2 fuel oil under the same exposure conditions (Neff et al. 1976; Rossi and Anderson 1976). For *Palaemonetes pugio,* there was an increase in sensitivity with increase in size and age. However, both *Penaeus aztectus* and *Neanthes arenaceodentata* adults were considerably more sensitive to the petroleum extracts than younger stages, and sensitivity decreased with increasing size.

Karinen and Rice (1974) reported impaired molting by crustaceans exposed to crude oil. Mecklenburg et al. (1977) compared animals in various stages of molt. This work and various other observations during toxicity testing have shown that crustaceans are more sensitive at the molt stage than at intermolt periods. Mecklenburg et al. (1977) found that coonstripe shrimp larvae (*Pandalus hypsinotus*) are approximately eight times more sensitive during molting than during intermolt. Obviously, mechanical and physiological stresses must be overcome during molting; also, an uptake of the surrounding water would bring high doses of available contaminants.

Rice et al. (1984) summarized the numerous studies conducted on eggs and larvae of salmon. They noted that eggs and alevins are more tolerant to short-term exposures (96 hours) of petroleum hydrocarbons than are fry or juvenile salmon. Studies on embryos, larvae, and adults of kelp shrimp (*Eualus suckleyi*) and coonstripe shrimp (*Pandalus hypsinotus*) showed that eggs were more tolerant to petroleum hydrocarbons than the adult females. Even during long-term exposures (28 days) in a flowing system, the eggs of shrimp survived if the female carrying them survived. Toxicity values for eggs, larvae, and adults (96-h and 28-d LC_{50}) were between 0.5 and 1.4 mg l^{-1} (or greater) total aromatics. These concentrations are very near the 96-h LC_{50} shown for white shrimp (*Penaeus setiferus*). In other studies, stage I larvae of king crab (*Paralithodes camtschatica*) exhibited 96-h LC_{50} values that were about one-half of the values for juveniles (Brodersen et al. 1977). They also reported an increase of about one order of magnitude (0.2–2.0 mg l^{-1}) in the tolerance of coonstripe shrimp between stage I and stage VI larvae.

Flow-through systems provide consistency of exposure so that either sublethal or lethal responses of organisms can be closely related to specific components and concentrations. Vanderhorst et al. (1976) indicated that flowing exposures can reduce 96-h LC_{50} values to about one-half of static LC_{50} values. Standard static acute tests are also difficult to interpret because time is rather short (4 days) and time of exposure is not allowed to vary. At least in mortality tests of relatively short duration (9 days or less), Anderson et al. (1980) found that the time of exposure is as important as concentration. In this study, a 50% mortality of three crustacean species, *Pandalus danae, Hippolyte clarki*, and *Neomysis awatschensis*, was predicted by extrapolation of a log-log plot of exposure time (up to 9 days) and concentration of exposure. There was a consistent total hydrocarbon exposure (concentration $\times$ time) in mg l^{-1} d^{-1} that produced 50% mortality for these crustacean species. This relationship has been confirmed using *Pandalus danae* and chemically dispersed oil in constant and diluting exposures (Anderson et al. 1981). Anderson (1984), utilizing *Pandalus danae,* also demonstrated a predictable total exposure producing 50% mortality, regardless of the rate of exposure. Obvious limitations to this approach are very short exposures, which do not provide sufficient time for death (a few hours), or very long exposures to low concentrations (over 9 days), which may allow some animals to accommodate to the petroleum components. Flow-through studies ensure that concentration and composition of the exposures are consistent so that the tolerance of a species can be accurately defined. The toxicity for some species under realistic spill conditions can be predicted from a knowledge of dilution rates, movement of the organisms, and alterations of the hydrocarbons under specific field conditions.

Wolfe (1977) summarized much of the available information in symposium proceedings on the fate and effects of petroleum hydrocarbons in marine ecosystems, with emphasis on arctic and subarctic environments. In the same volume, Anderson (1977) reviewed studies on the physiological, histological, behavioral, and reproductive effects on marine species of different oil components of several crude and refined oils. While total hydrocarbon values found to produce abnormal responses varied from 0.1 to 16 mg l^{-1}, there was closer agreement for both total naphthalenes and total aromatics. A relatively small range of total naphthalenes, 0.1–0.6 mg l^{-1}, adversely affected the growth and reproduction of polychaetes, crustaceans, and fish over exposure periods of 1–60 days. Results of sublethal effect studies, compiled in a review by the National Research Council (1985), are in general agreement with these concentration ranges.

The effects of the water-soluble fraction (WSF) of Santa Barbara crude oil on embryos, larvae, and adults of California halibut, northern anchovy, and California mussels were tested by Kanter et al. (1983). This project used flowing concentrations between about 10 and 400 mg l^{-1} WSF in tests to examine bioaccumulation, effects on survival and growth of larvae, and histopathology of adults. As demonstrated in other studies, the organisms accumulated petroleum hydrocarbons, primarily in gill, liver, and digestive glands (mussels). The mussels were least sensitive, and the only effects were on larval growth rate (at 14–559 $\mu g\ l^{-1}$) and the rate of byssal thread production (at 48–413 mg l^{-1}). Survival of larval and adult halibut was inversely proportional to both exposure concentration (between 10 and 600 mg l^{-1}) and duration. Growth rates of anchovy larvae were reduced at 30 mg l^{-1}, and survival of adults was reduced at 177 mg l^{-1} WSF. All exposures were relatively long—from about 2 weeks to about 3 months—so that researchers could attempt to simulate possible chronic exposures in the field.

Attempts to determine the threshold levels of petroleum exposure for marine species may be most productive when expressed in terms of the total exposure (hours or days) to specific hydrocarbons (milligrams per liter or micrograms per liter total aromatics). It appears that exposures to constant concentrations of oil extracts producing a total exposure of about 10 mg l^{-1} h of soluble aromatic hydrocarbons will result in adverse sublethal responses in marine species. Ten hours of exposure to 1 mg l^{-1} should produce the same effect on the organisms as 100 hours of exposure to 0.100 mg l^{-1}. Depending on the metabolic capability of the species, accommodation to very low levels of exposure may take place after about 10 days. McAuliffe (1987) reviewed many biological studies and field measurements of hydrocarbons in seawater, then converted all the information to exposures in milligrams per liter-hours. He found that likely maximum exposures to soluble aromatic hydrocarbons under an oil slick would be 0.002 mg l^{-1} h. If oil were dispersed chemically in the field, the maximum exposure would be about 0.060 mg l^{-1} h. From his evaluation of field measurements and laboratory toxicity tests, McAuliffe (1987) concluded that significant effects on even sensitive life stages of fish and invertebrates in the water column are very unlikely.

Dredging, Filling, and Offshore Disposal of Sediments

In 1972, the Marine Protection, Research, and Sanctuaries Act (Public Law PL 92–532) was established to prevent the disposal of sediments that would have an unacceptable impact on the marine environment at both dredging and disposal sites in ocean waters off the United States. As mandated by the act, the Environmental Protection Agency (EPA) published regulations and criteria governing the dredging and disposal of marine sediments (U.S. Environmental Protection Agency 1977). Guidelines for the implementation of the act were jointly published by the EPA and the U.S. Army Corps of Engineers (1977).

Dredging activities in the SCB have been more extensive in Los Angeles–Long Beach Harbor and San Diego Bay than elsewhere. Maintenance dredging is done less frequently in such localities as Marina del Rey, the mouth of the Los Angeles River, Anaheim Bay, and Mission Bay. Offshore dredging of sand for beach nourishment is done as needed at Surfside, as well as other sandy beach localities in the SCB.

Dredged material is now disposed by barging to approved or interim marine disposal sites, used for landfill, or transported to an approved land disposal site. Dredged material is disposed in water 100–182 m deep off Los Angeles–Long Beach Harbor (LA 2), Newport Bay (LA 3), and San Diego (LA 5).

Public Law 92–532 stimulated considerable research and development, which was largely sponsored by the Waterways Experiment Station in Vicksburg, Mississippi. In turn, research has resulted in many publications dealing with all aspects of the problem of contaminated sediments. The topic has been the subject of symposia (Kester et al.

1983; Landin 1988; Soule and Walsh 1983) and a multidisciplinary field verification program (Johns and Gutjahr-Gobell 1988), as well as bibliographies (Reish 1981; Dillon 1984; Olsen and Adams 1984).

Dredging and disposal of sediments can affect the marine environment and biota through (1) the physical removal of the material at the site; (2) increased water turbidity at the site; (3) toxicant release from the sediments to the water column; (4) toxicant removal from the water column, benthos, and possible uptake by marine organisms; and (5) burial of nearby benthic populations. At the marine disposal site, the existing benthic population can be buried as well as affected by the introduction of differently sized sediment particles and contaminated sediments. Construction of artificial islands and landfill with dredged material removes the area from an existing marine ecosystem.

The potential effects of dredging on the marine environment of outer Los Angeles Harbor was the subject of a review by Soule and Oguri (1976). Other studies have been concerned with the effect of contaminated sediments at the disposal site (Tetra Tech, Inc., and MBC Environmental Services, Inc. 1985a, b; U.S. Environmental Protection Agency 1988). As a requirement under PL 92–532, bioassays must be performed on contaminated sediments prior to dredging and disposal. Bioassay procedures have been outlined by the EPA and the U.S. Army Corps of Engineers (1977), and a manual for the Los Angeles Districts U.S. Army Corps of Engineers has been written (Reish and LeMay 1988). Many such bioassays have been conducted with contaminated sediments from Los Angeles–Long Beach Harbor and San Diego Bay by consulting firms, such as Marine Biological Consultants (1980).

Effects of Dredging

Field experiments were devised to determine if differences would occur in the recolonization of infaunal species between existing sediment and sediment recently exposed by dredging in outer Los Angeles Harbor (Soule and Oguri 1976). Sediments from dredged and nondredged sites, which had been frozen to render them azoic, were placed in containers, transported to the bottom by divers, and left there for varying periods up to 24 weeks. The recolonizing populations were then compared. Over 140 taxa (100 of which were polychaetes) were identified from the containers. There was little difference between the newly exposed sediment and the older sediment. Most of the species that dominated marine communities in outer Los Angeles Harbor were present as early as 6 weeks, but they were not necessarily in dominating proportions. Colonizing populations were less diverse than established populations, and it was believed that 2–3 years would be required for the community to stabilize. This time requirement was similar to that Reish (1961c) found for the initial colonization of the benthos in newly constructed marinas. These experiments by Soule and Oguri indicated that while a period of time is required for the benthos to recover to previous conditions, it will recover from dredging, and assuming that no additional environmental change occurs, the area will gradually return to previous population levels.

Populations in subtidal offshore sandy sediments, in contrast, recovered rapidly from dredging because they exist in a high-energy environment where currents move sediments into the dredged area. Benthic populations from dredged and undredged areas were indistinguishable from one another with regard to species composition, but there was a reduced number of specimens in the dredged area off Seal Beach 3 years after dredging. It was concluded that offshore dredging activities for beach nourishment do not have any long-term effect on the benthic population (Reish 1982).

Effects of Ocean Disposal

Two primary sites for ocean disposal of dredged material are located in the SCB. LA 2 is located off Point Fermin in 100 m depth, and LA 5 is located off San Diego in 182 m depth (U.S. Environmental Protection Agency 1987, 1988). A secondary site, LA 3, is located off Newport Beach near the discharge pipeline of the Sanitation Districts of Orange County. This site received 746,000 m^3 of sediment from Upper Newport Bay in 1987. None of these dump sites were characterized physically or biologically prior to the first dumping of dredged material. A quarterly study was initiated in 1983 at LA 2 and LA 5 for sediment grain size analysis, chemical analyses of both sediments and animal tissue, and benthic fauna (Tetra Tech, Inc. 1985a, b; Tetra Tech, Inc., and MBC Environmental Services, Inc. 1985a, b; U.S. Environmental Protection Agency 1988). The results were compared to nearby reference sites at similar depths and similar benthic conditions. The results of these studies were similar for both sites. However, since the differences between the disposal site and the reference site were greater for LA 2, only the results from this site are discussed here.

Sediments had a greater percentage of sand at the disposal site compared to the reference site. The concentration of metals at the disposal and reference sites is summarized as means and standard deviations in table 12.11. The means were higher at the disposal site for all seven metals and were significantly different for arsenic and zinc. In the absence of historical data, it is impossible to state whether these different sediment characteristics and elevated metal concentrations are the result of the disposal of contaminated dredged materials or the result of natural variability. The tissue of the ridgeback prawn *Sicyonia ingentis* was analyzed for metals from both the disposal and reference sites of LA 2. The results for this epibenthic species are summarized as means in table 12.11; there was no difference in the concentrations of the seven metals in either population.

The benthic infaunal population was sampled quarterly over a period of 1 year at the disposal site LA 2 and a nearby reference site. The numerical results of this quantitative study are presented in table 12.12 as the number of species, number of specimens, and species diversity (H′). All three measures indicate a difference in the benthic population between the two sites. According to the offshore environmental measure of Thompson (1982), which separates populations into three categories (central, transitional, and contaminated based on species comparison and number of species), the species composition at LA 2 is indicative of a transitional zone or moderate zone. The species composition at disposal site LA 2 is dominated by the transitional zone indicators *Mediomastus* sp. (polychaete), *Axinopsida serricata* (pelecypod), *Parvilucina tenuisculpta* (pelecypod), and *Euphilomedes producta* (ostracod). However, species characteristic of the clean environment and contaminated zone were also present at the LA 2 disposal site, but not in dominant numbers.

Comparison of the sediment particle size, sediment chemistry, and benthic infaunal analyses at the LA 2 disposal site with the reference area indicate trends in differences and, in some cases, significant differences. The differences between the two areas suggest that disposal of dredged material was the cause; however, because of the lack of data before dumping commenced, it is impossible to be certain. With the 1983–1984 database as a reference (Tetra Tech, Inc., and MBC Environmental Services, Inc. 1985a), future sampling may determine whether disposal of dredged material or normal variability is the cause of these changes.

Biological Characteristics of Marine Structures

Many of the protected bodies of water within the SCB have been altered to accommodate shipping (Los Angeles–Long Beach harbors), recreational boating and housing (Marina del Rey and Newport Bay), U.S. Navy bases (Long Beach Harbor and San Diego Bay), and commercial fisheries (Los Angeles Harbor and San Diego Bay). All these uses required the construction of jetties, pilings,

Table 12.10. *Chemical Composition of the Gas Phase of Hydrocarbon Seeps in the Santa Barbara Channel*

Component	Percentage Composition	
	Coal Oil Point	Carpinteria
Methane	94.8	87.7
Ethane	0.4	3.4
Propane	2.27	5.0
i-Butane	1.03	1.2
n-Butane	0.92	1.7
2,2-Dimethyl-propane	0.01	0.00
2-Methylbutane	0.54	0.70
n-Pentane	0.01	0.30
Hexanes	0.07	0.12
$\delta^{13}C$ (CH_4)	–38.7 ‰	–40.3‰

From Reed and Kaplan (1977).

Table 12.11. *Concentration of Metals in Sediment and in the Ridgeback Prawn* Sicyonia ingentis *from the Proposed Dredged Material Disposal Site (LA 2) (mean and SD in mg kg^{-1}) and a Reference Site, August 1983*

Metal	Disposal Site		Reference Site	
	Sediment	Tissue	Sediment	Tissue
Arsenic	2.3 ± 1.46	0.38 ± 0.2	0.4 ± 0.16	0.32 ± 0.2
Cadmium	1.8 ± 0.8	0.3 ± 0.1	1.3 ± 0.2	0.37 ± 0.05
Chromium	39.8 ± 20.8	0.39 ± 0.2	32.7 ± 4.0	0.38 ± 0.2
Copper	54.0 ± 132.9	23.5 ± 7.9	16.5 ± 3.5	22.7 ± 7.1
Lead	43.0 ± 35.5	2.36 ± 0.7	13.8 ± 5.9	2.5 ± 0.9
Mercury	0.4 ± 0.4	0.1 ± 0.07	0.04 ± 0.007	0.12 ± 0.08
Zinc	207.6 ± 185.9	51.5 ± 10.5	63.8 ± 16.6	56.2 ± 10.4

Compiled from data in Tetra Tech, Inc., and MBC Environmental Services, Inc. (1985a).

docks, and buildings to carry out a particular activity. Offshore construction includes fishing piers, oil islands, oil platforms, and subtidal artificial reefs. Regardless of the type of construction involved, the environment is altered from its original state. Previous biological communities may be altered, changed, or eliminated; however, new communities do develop as a result of these changes in the habitat. Construction materials usually consist of rock, concrete, and wood, and construction often involves dredging of channels to accommodate ships and boats.

The entrances of all harbors and marinas of the SCB are protected by rock jetties or breakwaters to prevent siltation and damage to structures and boats from heavy seas. The largest of these structures is the breakwater that protects the entrance to the Los Angeles–Long Beach Harbor. The breakwater is built of rock in three sections, the first of which was completed in 1912 and the remaining two in 1928; it is 13.81 km long (McQuat 1951). Other bodies of water, such as Alamitos Bay and Newport Bay, have upcoast and downcoast jetties that extend for about 1–2 km.

Table 12.12. *The Number of Species, Specimens, and Species Diversity (H′) as Means for Benthic Animals from the Proposed Disposal Site (LA 2) and Reference Area, 1983–1984*

Stations	Total Number of Species	Total Number of Specimens	H′
Dredged Disposal Site			
1	90.5[a]	1038	4.52
2	91	1048	4.26
3	75.8	575	4.68
4	55.8	382	4.34
5	48	433	3.55
Mean and SD	72.2 ± 19.7	488 ± 378	4.27 ± 0.4
Reference Site			
1	101.3	830	5.40
2	94	745	5.11
3	69	400	4.70
Mean and SD	88 ± 16.9	658 ± 227	5.07 ± 0.4

[a] Each station mean value based on four replicates.

Compiled from data in Tetra Tech, Inc., and MBC Environmental Services, Inc. (1985a).

Comparison with the biota of rocky shores is made in Chapter 8.

Earlier construction of docks and piers utilized Douglas fir for pilings. As a protection against penetration by marine wood borers, the outer 2–4 cm of the pilings were impregnated with creosote under pressure. More recently, concrete pilings have replaced wooden ones as the primary material for pilings.

Subtidal artificial reefs have been constructed on sandy beaches at many localities in the SCB. Earlier reefs in the SCB made use of abandoned streetcar and automobile bodies (Carlisle et al. 1964); more recent ones use quarry rock (Davis 1985). The primary purpose of these reefs is to provide a hard substrate for the attachment of algal and animal communities that will enhance recreational fishing in the area (see Chapter 9).

Oil islands are limited in number because (1) they can be constructed only in shallow water and (2) they are subject to the laws and restrictions for coastal zone construction. All 6 islands and 24 platforms in marine waters of California are located in the SCB (Boesch and Robillard 1987). Construction of these islands enables the oil industry to drill many oil wells from one locality. The oldest of these is Rincon Island located in Ventura County. The city of Long Beach, together with a consortium of five oil companies, constructed four islands in the shallow water off Long Beach in 1965–1966. The islands are constructed of rock perimeters and a sand-filled interior core. Rock for these islands was barged from Santa Catalina Island, and sand was obtained from the nearby ocean floor. The working area of each island is 1–2 m lower than the perimeter as a precautionary measure in the event of an oil spill. Each island is constructed in a depth of 12 m, extends 5 m above mean low water, and covers a surface area of 25 ha.

The use of oil platforms for offshore drilling in the SCB began in the late 1950s. Oil platforms number 24 in the SCB and are concentrated off Santa Barbara and Orange counties (fig. 12.1). They are located from 2.2 to 17.8 km offshore (Marine Biological

Consultants 1987). The legs of the platforms provide a substrate for the attachment of marine organisms which attract fish (Allen et al. 1976).

Concrete outfall pipes are used for the discharge of domestic wastes and for the intake and discharge of seawater as cooling waters for electrical generating stations. These pipes are generally laid on soft bottom sediments, and they represent a different type of substrate in the area.

The study of biological communities on the various man-made structures is limited. The community structure attached to the legs of oil platforms has been detailed by Mearns and Moore (1976), Wolfson et al. (1979), and Reish (1980); submarine pipelines by Allen et al. (1976); and submarine artificial habitats by Carlisle et al. (1964). The intertidal invertebrates present on the Long Beach oil islands have been described (Southern California Ocean Studies Consortium 1977). Successional studies on the development of intertidal communities on rock jetties and boat floats were described by Reish (1964a, b, c). Artificial reefs have been studied by Carlisle et al. (1964), Carter et al. (1985), and Davis (1985) and reviewed by Bohnsack and Sutherland (1985).

Growth on the oil platform legs is characterized by a heavy mass up to 1.2 m in thickness and is composed of a diverse fauna that includes the mussel *Mytilus californianus,* sea anemones, coral, starfish, bryozoans, and other invertebrate species (Mearns and Moore 1976). A qualitative distribution of the flora and fauna was determined at 1.5-m intervals from the surface to a depth of 12.2 m by Reish (1980) (fig. 12.21). A total of 89 species was collected: 27 crustaceans, 22 polychaetes, 17 molluscs, and 23 species divided among 10 other phyla. Zonation was evident on the platform leg. The mussels *Mytilus californianus* and *M. edulis* and the barnacle *Pollicipes polymerus* were the dominant species from the surface to a depth of 1.5 m. The mussels continued to be dominant to at least a depth of 12.2 m. The barnacles *Balanus tintinnabulum* and *B. trigons* replaced *P. polymerus* and were important elements of the community from 1.5 to 4.5 m depth. The white plume anemone *Metridium senile* was present in large numbers attached to the mussels from 4.5 to at least 12.2 m. The majority of the organisms were either attached to the surface of the mussels or lived in the interstitial spaces. Wolfson et al. (1979) reported the thick growth of *M. californianus* and *M. edulis* extending the entire length of the platform legs, except for the last few meters above the sea floor, where they were replaced by the sea anemone *Corynactis californica.*

Allen et al. (1976) conducted a photographic survey of the organisms associated with the two discharge pipelines in Santa Monica Bay originating from the Hyperion Sewage Treatment Plant. They recorded at least 100 different species that were either attached to the pipes or were present in the immediate vicinity. The anemone *Corynactis* sp. was the dominant animal on the pipes from 10 to 50 m depth. The white plume anemone *Metridium senile* was the most commonly photographed species from 60 to 100 m depth. The majority of the species attached to the pipe fed primarily upon the fine particulate matter suspended in the water column.

Marinas

The settlement of fouling organisms on floating boat docks in Alamitos Bay was studied over a 3-year period to determine whether or not the establishment of the community dominated by the bay mussel *Mytilus edulis* is a seasonal progression or true succession (Reish 1964a). The biological process was similar to that observed on recently constructed rock jetties. The initial inhabitants were determined not by a successional pattern, but by the time of year the float was originally submerged in seawater (Reish 1964c). For example, if the float was initially exposed to seawater during the winter or spring months, then *Mytilus edulis* would set-

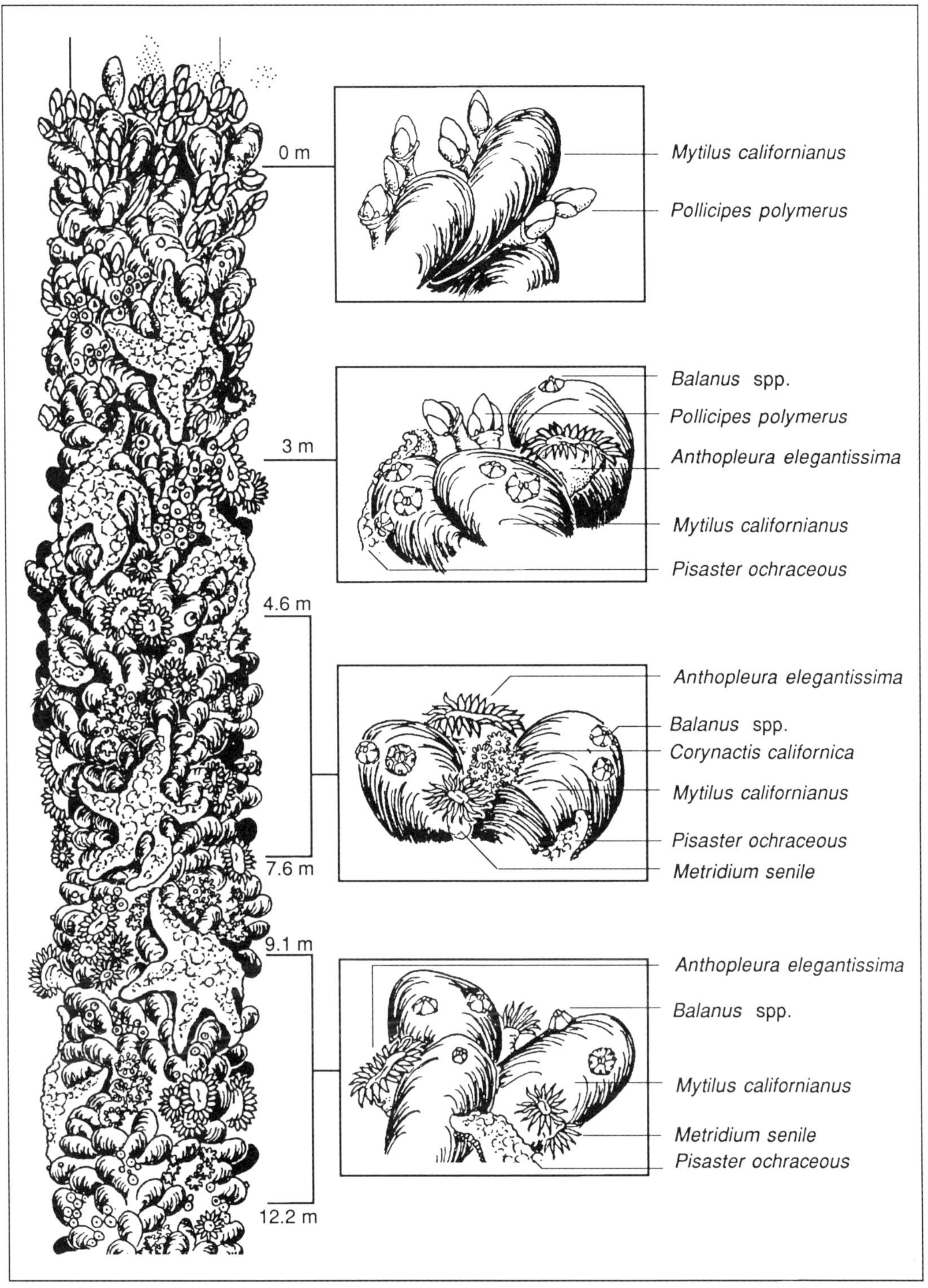

Figure 12.21. Distribution of marine invertebrates on an offshore platform piling. (After Reish 1980.)

tle shortly thereafter and the mussel community was established rapidly. If, however, the initial exposure of the floats was in the second half of the year, other organisms such as bryozoans and tunicates would be the early macroscopic colonizers; *M. edulis* would not settle until the following March. The maximum number of *M. edulis* per surface area was reached in 3 months, at which time each mussel occupied 1.0 cm^2 of surface area. As the individual mussel grew the ensuing 2 years, the number of specimens per unit area decreased and at 2 years each mussel occupied 3.5 cm^2 of surface area.

The *M. edulis* community on the floating boat docks was rich, varied, and dependent upon the time of year (Reish 1964b). The species composition consisted primarily of algae, polychaetes, crustaceans, molluscs, ectoprocts, and tunicates. The number of species and specimens, especially amphipods and polychaetes, was related to the water temperature (fig. 12.22). Population peaks were reached during the summer months when the water temperature was 20°C or greater.

The invertebrates within the *M. edulis* community were either herbivores feeding on algae (amphipods, polychaetes, and gastropods) or suspension feeders (*M. edulis*, tunicates, and ectoprocts). Some polychaetes (i.e., *Halosydna johnsoni* and *Nereis grubei*) were omnivores and may have played a minor role in feeding upon amphipods. While the population of associated mobile invertebrates declined with the decrease in water temperature in the fall months, it is not known whether they were eaten or died.

As previously described, two types of materials are used in piling construction: concrete and creosote impregnated timbers. The population is generally more diverse on concrete pilings and less diverse on timber pilings because of the toxic nature of creosote. Intertidal zonation is evident on a piling present in Alamitos Bay Marina (fig. 12.23). The splash and high-tide zone are dominated by different species of acorn barnacles. The bay mussel dominates the piling community from the mid-tide zone to the sediment line. Many animals are associated with the mussels; the tunicates *Styela plicata* and *S. montereyensis* are the most conspicuous of these. Small crustaceans and polychaetes are present within the interstitial spaces of the mussels. Red algae, such as *Antithamnion* sp., is often present in clearer waters of marinas.

Comparisons between the pilings present in the open coast with those in the protected waters of bays or harbors show both similarities and differences (fig. 12.21 and 12.23). Acorn barnacles, which belong to the genera *Balanus* and *Chthalmus,* dominate the high intertidal zone in both environments. Mussels become the dominant organism throughout the remaining length of the piling, except for the last few meters above the sea floor. *Mytilus californianus* dominates the open coast oil platform legs, where its growth reaches 1.0 m in thickness (Mearns and Moore 1976). In contrast, *M. edulis* is the dominant mussel on bay pilings, where its growth is generally 10 cm or less in thickness. Sea stars and sea urchins are prevalent on open coast pilings but absent on bay pilings. Both associations provide protective spaces for smaller animals such as flatworms, polychaetes, gastropods, pelecypods, crustaceans, bryozoans, brittle stars, and tunicates; however, species diversity is greater within the *M. californianus* community. A band of sea anemones (*Corynactis californica*) is present at the base of the open coast pilings (Wolfson et al. 1979), but the base of bay pilings is devoid of marine life (D. J. Reish pers. observ.). The biological difference between the two piling communities is attributed to the high-energy condition of the open coast in contrast to the low-energy bay environment.

Wooden pilings are subjected to the burrowing habits of five different species of wood borers in the SCB; these are the isopods *Limnoria tripunctata* and *Limnoria quadripunctata,* the amphipod *Chelura terebrans,* and the pelecypods *Lyrodus diegensis* and *Bankia setacea* (Menzies et al. 1963). Creosote is ef-

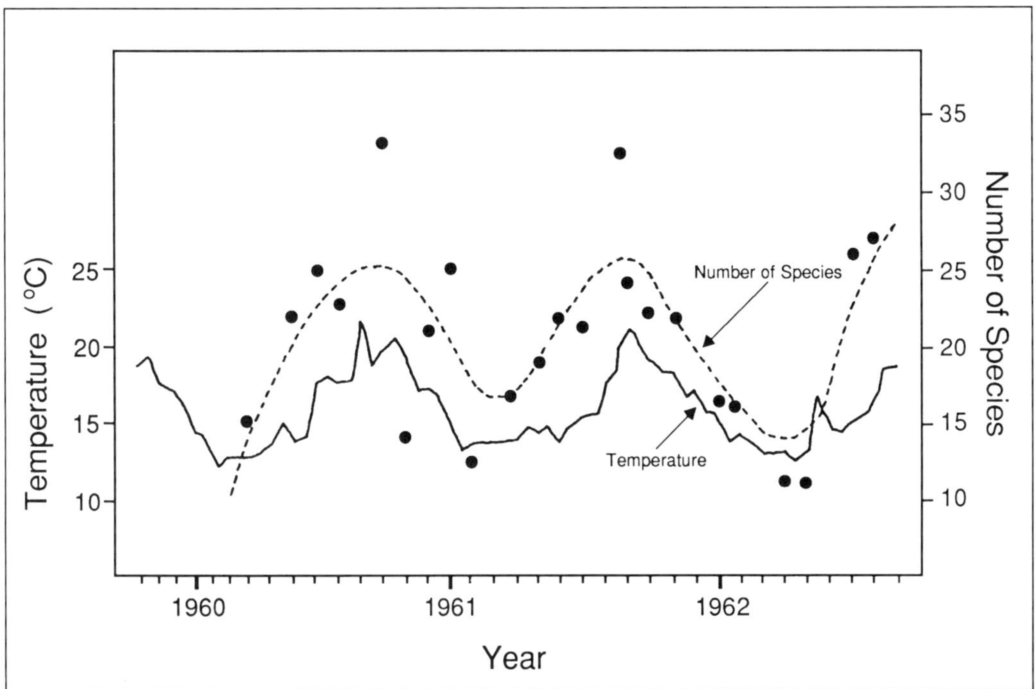

Figure 12.22. The relationship between the number of associated species in the *Mytilus edulis* community and water temperature in Alamitos Bay Marina. (From Reish 1964b.)

fective in preventing the settlement of all species except *L. tripunctata* (Menzies 1951). This isopod has been observed in creosote-impregnated pilings in bays, harbors, and offshore structures. With the initiation of the pollution abatement program in Los Angeles Harbor in 1968, which improved the dissolved oxygen concentration (Reish 1971a), these pilings were subsequently subjected to attacks by *L. tripunctata*. As a preventive measure, each piling was individually wrapped with plastic sheeting to exclude the isopod and to maintain an environment free of dissolved oxygen between the liner and the piling. This technique has proven to be an effective measure for protecting existing wooden pilings (C. M. Wakeman pers. comm.).

Man-made structures in the marine environment present a variety of environments for marine organisms. Because these structures increase the different kinds of habitats, the flora and fauna are frequently more extensive in a marina or harbor than in the original estuarine environment, largely as a result of the invasion of widely distributed or cosmopolitan species (see Chapter 8).

Radioactive Waste Disposal

Low-level radioactive wastes were dumped at six sites off the coast of California between 1946 and 1970. Contaminated material consisted primarily of trace levels of radioactivity on paper towels, rags, clothing, glassware, and laboratory equipment. Most of the contaminated material was embedded in concrete and sealed in 55-gallon steel drums. A total of 55,029 drums were discarded off California, with the majority of these dumped at 900 and 1700 m depths near the Farallon Islands off San Francisco (table 12.13) (U.S. Environmental Protection Agency 1983). Southern California sites were located offshore of Point Hueneme, Los Angeles, and San Diego (table 12.13 and figure 12.1). The number of

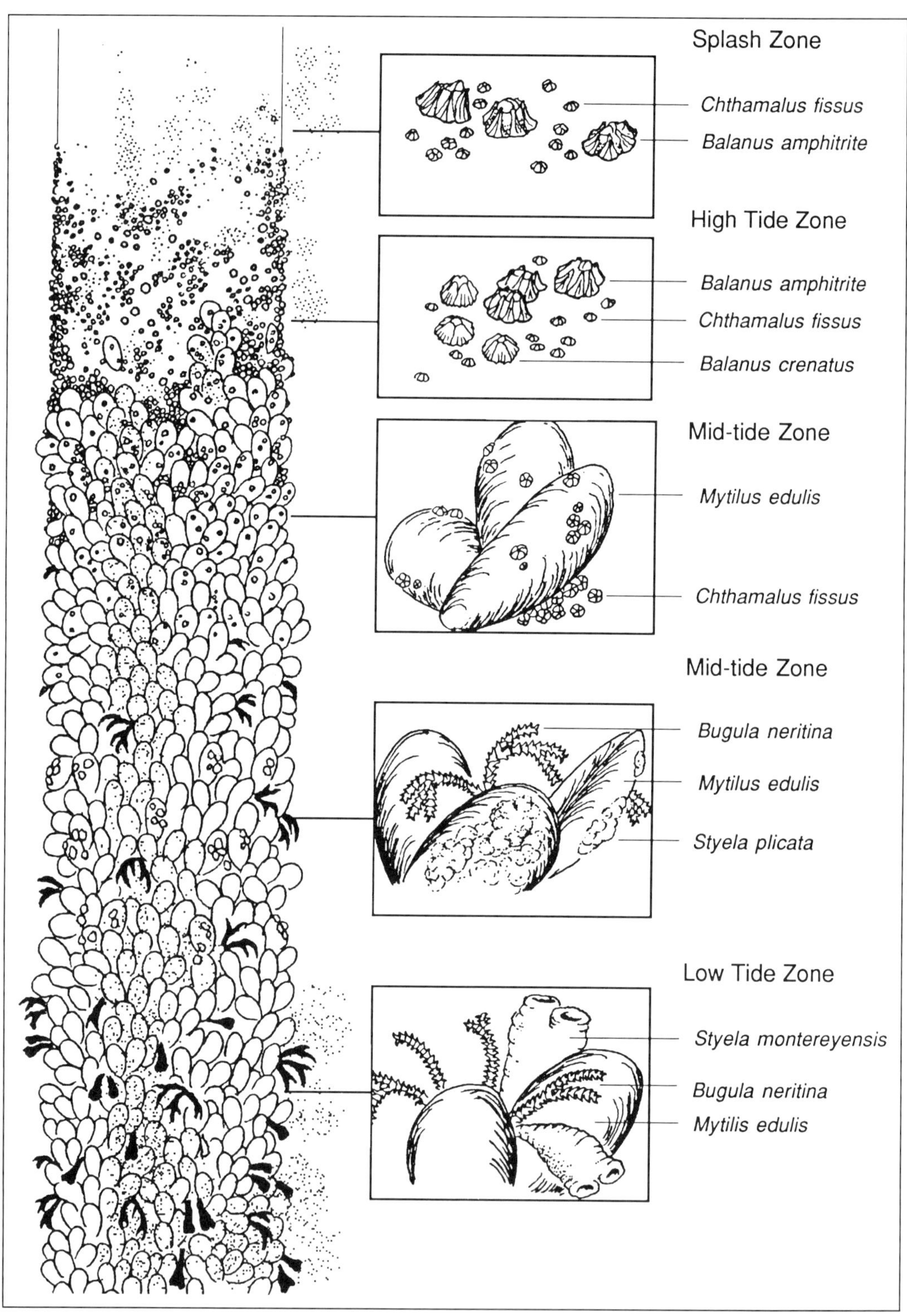

Figure 12.23. Distribution of marine invertebrates on a piling in Alamitos Bay Marina. (After Reish 1982.)

Table 12.13. *Location of the Radioactive Waste Disposal Sites in California*

Site	Location	Depth (m)	Number of Containers	Curies
Farallon Islands	37°38′ N 128°08′ W	900	3,500	1,100
Farallon Islands	37°37′ N 123°17′ W	1,700	44,000	13,400
Cape Mendocino	40°07′ N 139°06′ W	1,990	29	No data
Port Hueneme	33°39′ N 119°28′ W	1,940	3,100	108
Los Angeles	30°43′ N 139°06′ W	4,570	No data	No data
San Diego	32°00′ N 121°30′ W	2,210	4,400	34

From U.S. Environmental Protection Agency (1983).

curies at the time of disposal is indicated in table 12.13. Radioactive decay has diminished the total count in the years following disposal, but the percentage of decay is unknown (U.S. Environmental Protection Agency 1983).

All ocean dumping of radioactive wastes by the United States was terminated in 1970. The EPA was given the responsibility for developing regulations and issuing permits for future disposal of all wastes under PL 92–532. This act also prohibited ocean disposal of high-level radioactive waste and radiological warfare agents. The National Oceanic and Atmospheric Administration (NOAA) and the EPA were directed under the 1983 act to coordinate their oceanographic activities for evaluating ocean dumping, with an option for future disposal of low-level radioactive wastes.

A review by Woodhead (1984) and symposium proceedings edited by Park et al. (1983) summarized much of the knowledge regarding the effects of radioactivity on the ocean environment and its inhabitants. The most extensive studies of disposal sites have been sponsored by the EPA; the focus of much of their research has been on the Pacific Coast off the Farallon Islands because of the greater number of drums dumped at those sites. The EPA sponsored studies at the two Farallon Islands sites in 1974, 1975, 1977, and 1985, as well as a preliminary study at the San Diego site in 1974. Concentrations of ^{239}Pu and ^{240}Pu in the sediment in the vicinity of the drums at the Farallon Islands sites were 2 to 20 times above the expected fallout level (Schell and Sugai 1980). Populations of benthic animals were similar at the dump site and reference stations, and they apparently have not been affected by the elevated levels of these radionuclides (Reish 1983). Since radionuclides have been leaking from the drums, presumably most of them are buried in the sediments either by bioturbation or sedimentation (Schell and Sugai 1980). However, as suggested by Robison and Lancraft (1984), the possibility exists for hazardous material to be transported upward into the pelagic environment by negatively buoyant fish eggs developing on the bottom and moving upward during the larval stages. Whether radionuclides move upward into the pelagic environment in this way has not been demonstrated.

Pollution of Protected Waters

It was not until the 1960s that both the private and public sector of the population realized that the oceanic waters of the world cannot assimilate an infinite amount of waste. Initial studies focused on harbor waters (Reish 1955); as marine pollution expanded into offshore waters, studies were conducted there as well (Duedall et al. 1983). This concern for the marine environment has resulted in the establishment of specialized journals such as *Marine Pollution Bulletin,* in-depth reviews

(Kinne 1984a, b), symposia proceedings (Kinne and Bulnheim 1980), and annual reviews (Reish et al. 1989).

Many different measures are used to determine the severity of marine pollution. Each scientific discipline has its own preferred method of determining whether or not degradation of the environment has occurred and, if so, what is the degree of degradation. However, in spite of such differences of opinion, a consensus exists that the benthic environment yields the greatest amount of data on the effects of pollution. A large percentage of the pollutants entering the system either settle directly to the bottom or become associated with particulates that ultimately settle to the bottom. Benthic studies include biological analyses for micro- and macrobiota, chemical analyses of sediment and organisms for contaminant concentrations, and grain size analyses of sediments.

Los Angeles and Long Beach Harbors

This two-harbor complex is actually one harbor oceanographically (fig. 12.24), and it has been studied more thoroughly and for a longer period of time than any other similar body of water in the world (Anon. 1952; Reish 1959, 1986; Allan Hancock Foundation 1975). The first comprehensive investigation of the environmental conditions in this harbor was made in 1951 (Anon. 1952) and included analyses of the water, sediment, and benthic populations. Approximately 235 outfalls were located throughout the harbor at that time; they carried wastes from different industries, domestic wastes, and stormwaters (Reish 1959). A more detailed seasonal study was conducted in 1954 that emphasized benthic conditions. It was possible to divide the harbor into five different zones on the basis of bottom conditions, species composition, and water characteristics. The results of one of these surveys are presented in figure 12.25a. In general, the polluted zone (characterized by the presence of the polychaete *Capitella capitata*) and the very polluted zone (which lacked macroscopic animal life) were present within the majority of the blind-ending slips or basins located in the inner harbor area, in portions of Long Beach Naval Base, in Fish Harbor, and at the discharge point of the Terminal Island Sewage Treatment Plant. The two intermediate or semihealthy zones, one characterized by the polychaete *Dorvillea articulata* (= *Stauronereis rudolphi* or *Schistomeringos longicornis*) and by *Polydora paucibranchiata* (= *Pseudopolydora paucibranchiata*) and the other by *Cirriformia luxuriosa*, were located in the inner harbor area between the polluted or very polluted zones and the healthy zone of the outer harbor. Healthy zones, characterized by the polychaetes *Tharyx parvus*, *Cossura candida*, and *Nereis procera*, were present in the outer harbor and outer portions of the main channels.

The dissolved oxygen concentrations in the water column were higher (3.5 mg l^{-1}) in the polluted zone, where *Capitella capitata* occurred, than in the two semihealthy zones (3.2 mg l^{-1}). Furthermore, laboratory experiments conducted to determine the tolerance of *C. capitata* and *Dorvillea articulata* to reduced dissolved oxygen concentrations showed that *D. articulata* was more tolerant than *C. capitata* (28-d LC_{50} of 1.5 mg l^{-1} compared to 0.65 mg l^{-1}) (Reish 1970a). Since it is apparent that *C. capitata* required a higher dissolved oxygen content under laboratory experimental conditions, its presence in more polluted areas is presumably related to some other environmental condition.

Ecological conditions remained similar for the ensuing 15 years. In 1968, the state of California issued an order prohibiting the discharge of oil refinery wastes into Dominguez Channel, which had reached 7.5×10^7 l d^{-1}. These wastes entered Los Angeles Harbor through Consolidated Slip. Dissolved oxygen was usually absent, and the bottom lacked macroscopic life (Reish 1959). Compliance with the order was completed in September 1970. A month later, the dissolved oxygen content ranged from 3.8 to 5.2 mg l^{-1}. The improvement in water quality was reflected

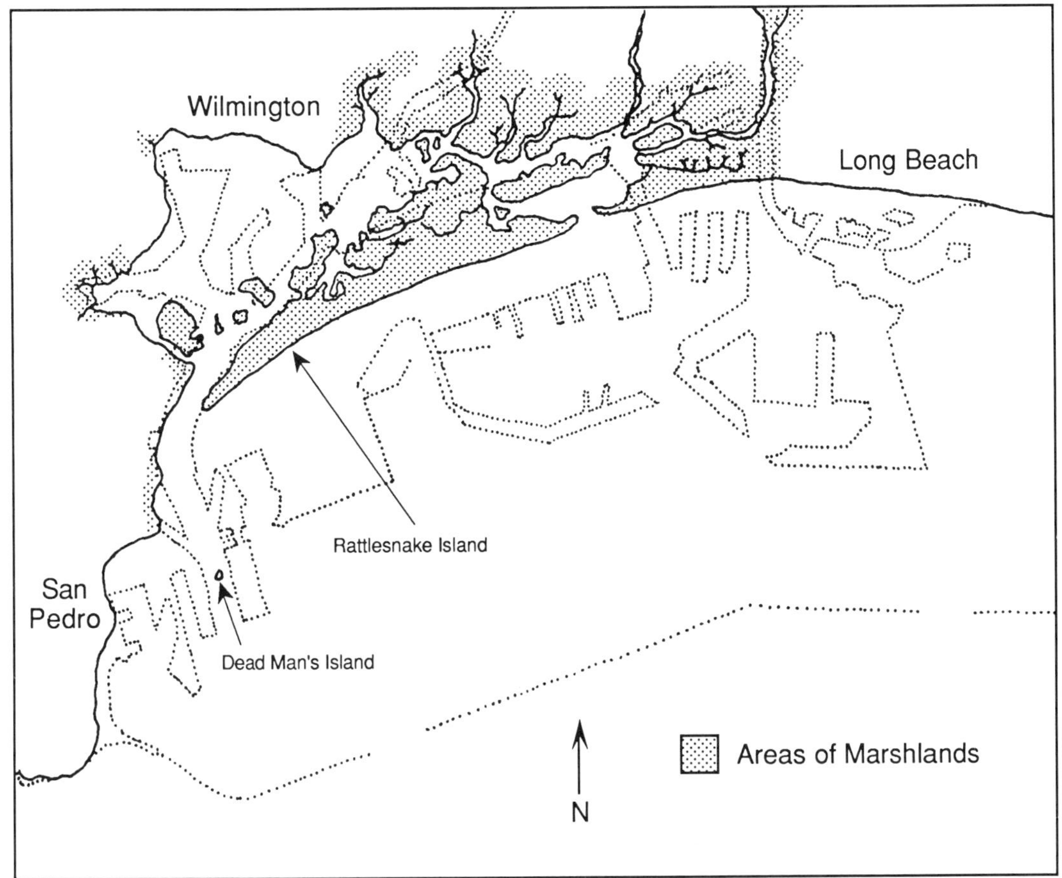

Figure 12.24. Map of San Pedro Bay in 1872 with current map (dotted outline) of Los Angeles–Long Beach Harbor superimposed upon it. (After Reish et al. 1980.)

in the biological conditions. Eighteen species were collected from the benthos where none had been found previously. Prior to the initiation of pollution abatement, the organisms present on boat floats in the Consolidated Slip–East Basin area were composed of a blue-green alga, the green alga *Enteromorpha* sp., and an oligochaete. By October 1970, 26 different species were present on the boat floats, including red and green algae, cnidarians, polychaetes, gastropods, pelecypods including *Mytilus edulis,* barnacles and other crustaceans, and ectoprocts. Of particular interest and importance was the invasion of the wood-boring isopod *Limnoria tripunctata,* which had previously been excluded from the area.

Since this was the first documented case of the effects of a pollution abatement program, it is significant to note the rapidity of recovery once the source of pollutants was removed. One factor contributing to the rapid recovery, which may not be applicable at all localities, is that many of the local species have extended reproductive periods or reproduce throughout the year (Reish 1961a). This reproductive capability of the local population undoubtedly played an important role in the rapid recovery of this polluted body of water.

The pollution abatement program that began in 1968 in Dominguez Channel was extended throughout the harbor area in the 1970s. Scientific evidence for the success of

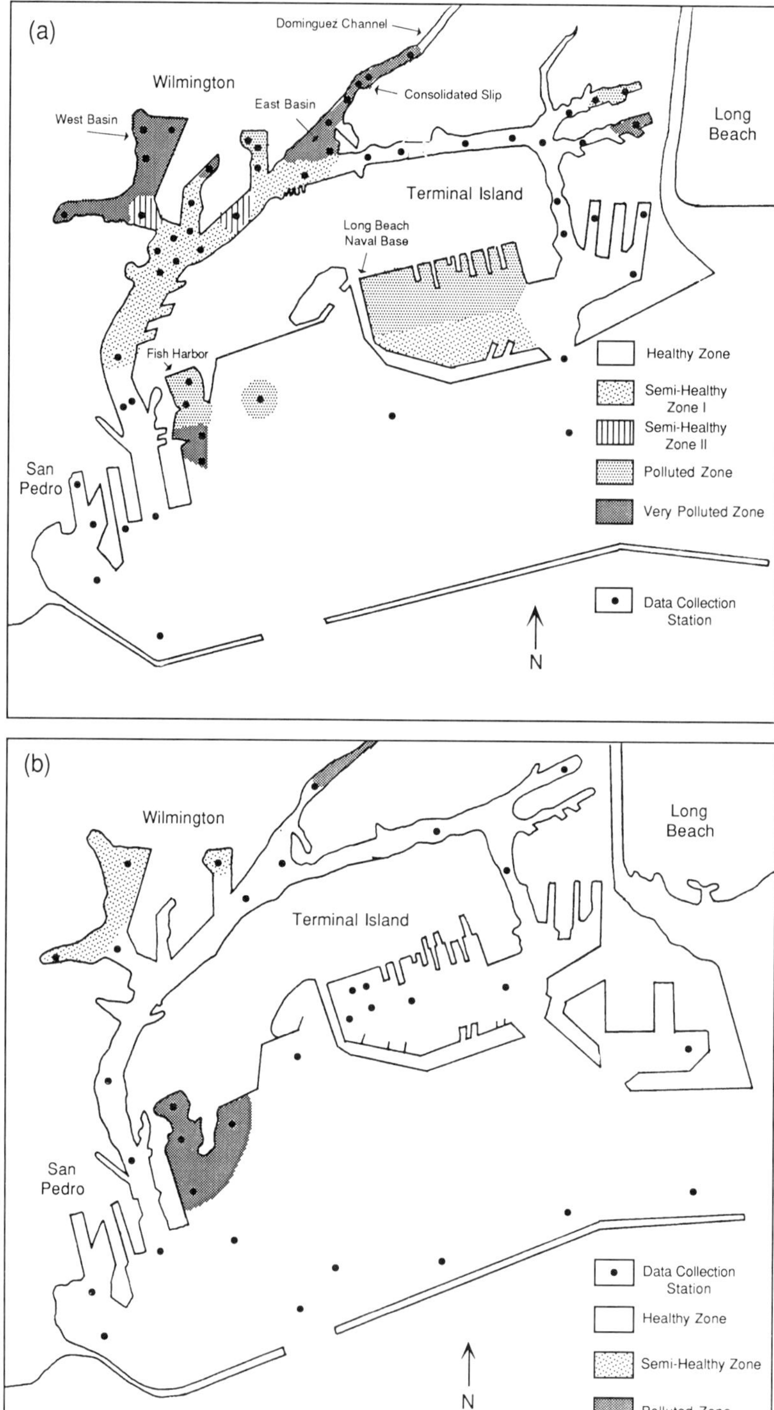

Figure 12.25. Benthic biological conditions in Los Angeles–Long Beach Harbor. (a) Data for Los Angeles–Long Beach Harbor collected in 1954; data for the Long Beach Naval Base collected in 1971. (b) Data for Los Angeles–Long Beach Harbor collected in 1973; data for the Long Beach Naval Base collected in 1978.

this improvement appeared in a series of publications edited by Soule and Oguri (Allan Hancock Foundation 1975; Soule and Oguri 1974, 1975, 1976, 1978b, 1979, 1980). Improvement of benthic biological conditions can be noted by comparing figure 12.25a with figure 12.25b. The very polluted zone characteristic of the inner harbor slips in 1954 no longer occurred in the early 1970s. Some macroscopic life was present at all benthic localities. By 1978, the polluted zone was limited to the blind-ending slips. The main channels and the outer harbor were inhabited by the animals of the healthy zone. The diversity of the benthic fauna increased, with the number of species more than doubling from 1954 to 1973, largely as a result of invasion of offshore species (table 12.14). The species dominance also changed: the polychaetes *Cossura candida* and *Mediomastus californiensis* comprised over 50% of the number of specimens in 1978 compared to 7.6% in 1954 and 28% in 1973–1974. The greater diversity also is indicated by the number of species necessary to comprise over 80% of the population: 8 in 1954, 10 in 1973–1974, and 13 in 1978.

The dissolved oxygen conditions improved throughout the harbor area following the initiation of the pollution abatement program in 1968. Where conditions were at or near zero in the inner harbor in 1954, readings in 1978 were never below 3.5 mg l^{-1} and were generally above 5.0 mg l^{-1} in the inner harbor (Soule and Oguri 1980).

The total organic carbon content of the sediments dropped dramatically with the initiation of the pollution abatement program. In 1954, the amount ranged from 0.2 to 10.7%, with the high values located in Consolidated Slip, which received oil refinery wastes. After pollution abatement began, the amount of organic carbon dropped from a range of 0.86–1.83% in 1973 to 0.26–1.67% in 1978 (Reish 1959; Allan Hancock Foundation 1975; Soule and Oguri 1980).

Chemical analyses were made of the sediments throughout the harbor area in 1973–1974 and 1978 but, unfortunately, not in 1954. A representative sample of these analyses is summarized as means and standard deviations in table 12.15. In general, some pollutant levels decreased in 1978 over the conditions prevailing in 1973–1974. The levels of cadmium, chromium, nickel, lead, and PCBs were less. However, arsenic, zinc, and DDT levels were higher in 1978. The concentrations of mercury and copper were unchanged. Los Angeles Harbor, especially in the Dominguez Channel–East Basin, had the highest levels of most of the pollutants. Arsenic was highest in the vicinity of Pier J in Long Beach Harbor. DDT was highest at that station located outside the Los Angeles breakwater and nearby stations in the harbor. The higher concentrations in the sediments were undoubtedly the result of the high levels of DDT off the Palos Verdes Peninsula (Swartz et al. 1986).

As part of the California State Mussel Watch program, resident specimens of *Mytilus edulis* were collected from different bays and harbors from the SCB and analyzed for contaminants. The data for heavy metals and synthetic organic compounds are summarized in table 12.16 (California State Water Resources Control Board 1982). No one bay or harbor in the SCB could be characterized as having the highest or lowest concentration of metals or synthetic organic compounds. Overall, Anaheim Bay, Newport Bay, and Los Angeles–Long Beach Harbor characteristically had higher values than Alamitos Bay San Diego Bay, and Mission Bay. However, the highest concentration of copper was measured in San Diego Bay. DDT and PCBs were highest in Newport Bay. The concentration of the different contaminants in the mussels is believed to be related to the amount of surface runoff, the water circulation at the site of the mussel collection, and the relationship of the site to commercial activity or recreational boating (California State Water Resources Control Board 1982).

In 1972, 1981, and 1986, benthic sediment samples were collected from areas immediately adjacent to boatyards in Rhine Channel

Table 12.14. *Summary of the Biological, Chemical, and Physical Characteristics of the Three Ecological Areas of Los Angeles–Long Beach Harbors*

Parameter	Outer Harbor	Channel Stations	Inner Harbors
Indicator species	*Cossura candida*	*Euchone limnicola*	*Armandia bioculata*
	Haploscoloplos elongatus	*Nephtys franciscana*	*Capitella capitata*
	Notomastus tenuis	*Cryptomya californica*	*Ophiodromos pugettensis*
	Prionospio pinnata	*Callianassa californiensis*	*Polydora ligni*
	Prionospio pygmaeus		*Pseudopolydora paucibranchiata*
	Tharyx ? parvus		*Schistomeringos longicornis*
	Macoma acolasta		*Theora lubrica*
	Tellina modesta		
Number of species	72	70	45
Dissolved oxygen (mg l^{-1})[a]	5.9	5.2	4.5
Salinity (‰)[a]	33.45	33.57	33.52
Cadmium (mg kg^{-1})[a]	3.0	3.5	4.6
Chromium (mg kg^{-1})[a]	60.7	86.1	157.7
Copper (mg kg^{-1})[a]	116.0	81.0	206.0
Lead (mg kg^{-1})[a]	87.0	102.0	173.0
Mercury (mg kg^{-1})[a]	0.45	0.86	1.7
Zinc (mg kg^{-1})[a]	133.0	160.0	362.0

[a] Data as means.

Modified from Allan Hancock Foundation (1975).

Table 12.15. *Chemical Characteristics of Sediments in Los Angeles–Long Beach Harbors 1973–1974 and 1978*

Chemical	Concentration (mean ± SD) 1973–1974	1978
Arsenic	4.7 ± 3.7	56.5 ± 31.8
Cadmium	2.8 ± 1.3	0.72 ± 0.4
Chromium	80.4 ± 42.5	73.7 ± 46.4
Copper	91.5 ± 74	91.8 ± 63.7
Mercury	0.72 ± 0.82	0.71 ± 0.76
Nickel	56.4 ± 25.5	36.8 ± 22.7
Lead	107.7 ± 81.2	31.7 ± 62.4
Zinc	192.2 ± 114.9	223 ± 210
Total DDT	0.088 ± 0.3	0.1 ± 0.2
Total PCB	2.13 ± 1.4	0.14 ± 0.2

Calculations based on the data from Allan Hancock Foundation (1975) and Soule and Oguri (1980).

in Newport Bay and analyzed for chromium, copper, lead, mercury, and zinc (Liu and Schneider 1988). By 1975, waste discharges from all boatyards were regulated by the Regional Water Quality Control Board. During the 14-year period when the samples were collected, the concentration of copper in the sediment was reduced by more than an order of magnitude from 6220–11,000 $\mu g\ g^{-1}$ to 530–610 $\mu g\ g^{-1}$ dry wt. By 1986, the concentrations for the remaining four metals were only one-half of the concentrations measured in 1972.

San Gabriel River

The mouth of the San Gabriel River forms the boundary at the coast between Los Angeles and Orange counties. Originally this river constituted one of the major rivers of the Los Angeles Basin. However, construction of flood control dams in the San Gabriel Mountains reduced the freshwater flow in the lower reaches of the river so that any significant amount of freshwater in the river occurs only during periods of rainfall. During much of the year, the salinity in the lower San Gabriel River is 33–35 ‰.

The San Gabriel River in 1952 was a shallow stream that received waste discharges from several sewage plants and a chemical plant, brines from oil well production, and cooling waters from an electrical generating plant (Reish 1956). At this time, no macroinvertebrate life was present at any of the 14 stations sampled between the river mouth and the Seventh Street bridge in Long Beach. A few intertidal species were present along the sides of the channel (Reish 1956). In late 1952, the lower San Gabriel River was dredged as a flood control measure. By mid–1954, 12 different species of invertebrates were collected from the mid-channel region; *Capitella capitata* comprised 46% of the population. While benthic conditions had improved as a result of the deepening of the channels by dredging, the same wastes were being discharged into the river. However, all of the agencies that were emptying wastes into the river ceased to do so between 1954 and 1966, when Turner and Strachan (1969) conducted the next benthic study. Turner and Strachan reported collecting 19 species from the benthos of the river in 1966. Although the original power plant no longer functioned, two new electrical generating stations were built, and by 1988, these two plants discharged 8.4×10^9 l d^{-1} of heated cooling waters.

Surveys of the benthic fauna were conducted during the summers of 1970 and 1971 (Reish 1970b, 1971b). The benthic conditions continued to improve; 25 species were present in 1970 and 34 in 1971. *Capitella capitata* comprised only 13.3% of the benthic fauna for 1970 and 3.5% for 1971.

Four sewage treatment plants were constructed along the freshwater region of the river in the 1970s. By 1988, these plants were discharging 3.8×10^8 l d^{-1} of tertiary treated wastes. Benthic diversity of marine invertebrates has declined since 1971, but the decline is attributed to the accidental introduction of the fish *Tilapia mossambica* into the river. This fish is omnivorous, feeding on the benthic fauna and the algae present in the area (J. Stull pers. comm.).

Table 12.16. *Concentration of Metals (in mg kg^{-1}), DDT, and PCB (in μg kg^{-1}) in* Mytilus edulis *Collected from Bays and Harbors in the SCB*

	LA-LB Harbors	Alamitos Bay	Anaheim Bay	Newport Bay	Mission Bay	San Diego Bay
Cadmium	2.8 ± 2.2	2.0 ± 0.3	4.9 ± 1.4	8.4 ± 3.7	1.6 ± 0.4	4.9 ± 2
Chromium	3.5 ± 7.8	0.97 ± 0.15	1.8 ± 0.6	1.8 ± 0.5	1.1 ± 0.1	1.6 ± 0.6
Copper	11.7 ± 4.1	9.0 ± 0.9	9.3 ± 3.3	20 ± 15	7.4 ± 1.2	20.2 ± 17.5
Mercury	0.16 ± 0.1	0.2 ± 0.01	0.18 ± 0.08	0.27 ± 0.1	0.08 ± 0.01	0.15 ± 0.05
Lead	14.3 ± 11.4	1.7 ± 0.5	14.7 ± 12.3	9.5 ± 8.3	3.0 ± 1.6	6.1 ± 2.7
Zinc	291 ± 116	170 ± 38	328 ± 166	240 ± 146	167.5 ± 9.6	237 ± 100
DDT	1373.4 ± 454.7	1146	1859 ± 354	2558 ± 865	—	—
PCB	771 ± 530	760	730 ± 128	1898 ± 2700	210	1733 ± 2347

Data from California State Water Resources Control Board (1982).

Deterioration of a bay or harbor is generally first noted in a blind-ending slip such as those that occur in the inner Los Angeles–Long Beach Harbor. Limited water circulation coupled with an industrial or domestic waste discharge leads to the degradation of the environment. The decline of environmental conditions in these slips occurs in the benthos since many of the wastes eventually sink to the bottom. By the time the benthic invertebrate population is characterized by a polluted assemblage of animals, other parameters of the environment have changed. The dissolved oxygen content of the water is low, especially near the bottom. The sediment appears dark, possesses a sulfide odor, and has elevated levels of contaminants.

Pollution abatement leads to an improvement of environmental conditions, as indicated by any one of many measurable parameters. Repopulation of the benthos is rapid since many of the invertebrate species in the SCB have extended reproductive seasons. The pollution abatement program in Los Angeles–Long Beach Harbor led to a shift in the benthic population, with the polluted and semihealthy assemblage of species moving into formerly very polluted areas and offshore species extending their range into the outer harbor. The improved biological and chemical characteristics of the harbors provide evidence that the initiation of a pollution abatement program can lead to better environmental conditions and can do so in a relatively short period of time.

Potential Impacts of World Shipping

History

The early history of shipping in the SCB has been summarized by Emery (1960). The first ship to visit the SCB was the *Leila Byrd,* which sailed from Boston Harbor in 1808. Early ships anchored in the San Pedro area in what is now known as the Los Angeles–Long Beach Harbor. By the latter part of the nineteenth century, the number of ocean-going vessels making port calls in the SCB reached 500 per year. The development of the Los Angeles–Long Beach Harbor into a modern-day facility began in 1850 with construction of docks and piers in the estuary formed at the mouth of the Los Angeles River (figs. 12.24, 12.25). The number of ships entering the harbor increased sharply after the first section of the present-day breakwater was constructed in 1912 and the final two sections completed in 1928 (McQuat 1951). The completion of the breakwater resulted in many changes in the configuration of the harbor area (U.S. Army Corps of Engineers 1984). The number of ships entering the harbor increased to 8000 per year. And while the number of ships making port calls in Los Angeles–Long Beach Harbor has remained unchanged in recent years (table 12.17), the size of the ships, their draft weights, and the depth of the water required have increased greatly in the past 40 years (table 12.18). Larger ships have increased the productivity, which results in a decrease in the cost of shipping per ton.

The demand for recreational boats in the SCB has continued to expand, especially in the past three decades. This expansion is the result of rapid increases in population, leisure time, and level of affluence in southern California. Locations of the marinas of southern California are listed in table 8.1. The number of boat slips available in the SCB in 1979 is listed by county in table 12.19. These data indicate a nearly 100% occupancy of the boat slips in 1979; the demand is expected to increase in the foreseeable future. An additional 3482 slips have since been added. However, both the construction of additional marinas and the expansion of existing ones will become increasingly more difficult because of fewer suitable locations, use of proposed sites for other purposes, and environmental considerations.

Major Factors

The effects of shipping and boating on the marine environment are many and varied.

Table 12.17. *Summary of Vessel Arrivals at Los Angeles–Long Beach Harbors, 1978–1982*

Vessel Type	Number of Vessels by Year				
	1978	1979	1980	1981	1982
General cargo	3148	3420	3652	3499	4114
Liquid bulk	2715	2013	1686	1269	1559
Cargo passenger	303[a]	151[a]	136[a]	201[a]	140[a]
Bunkers only	1152[a]	2155	2533	3315	2495
Totals	7318	7739	8007	8284	8308

[a] Data for Los Angeles Harbor only; data for these categories included in other vessel types.
Data from U.S. Army Corps of Engineers (1984).

Table 12.18. *Tanker Size Trend, 1945–1979*

Year	Ship Class	Draft Weight (t)	Draft Depth (m)
1945	T-2	16,250	9.14
1955	Veedol	50,800	11.6
1960	Jacques Carrier	70,100	13.7
1965	San Juan Explorer	107,700	15.5
1968	Universal Ireland	331,200	25.0
1972	Globtik Tokyo	484,600	28.0
1976*	Seawise Giant	571,000	30.5
1979*	Prairial	563,800	30.5

From Segelhorst (1974).
*Data projection made in 1974.

Table 12.19. *Recreational Boat Slips in the SCB in 1979*

County	Number of Slips	Percentage Occupancy
Santa Barbara	1,008	98.6
Ventura	3,259	95.9
Los Angeles	14,265	99.6
Orange	10,282	99.9
San Diego	7,788	96.1
Total	36,602	—
Average percentage occupancy	—	98.6

Data from Williams-Kuebelbeck and Associates, Inc. (1979).

They include (1) construction, dredging, and filling operations; (2) oil spills; (3) collisions; (4) contamination by marine paint and other preservatives; and (5) the introduction of marine organisms from other geographical regions.

The increase in the size of ships and the draft depth required has necessitated dredging harbors deeper and disposing of the sediments elsewhere. The effects of dredging on the marine environment have been previously addressed in this chapter.

Oil spills related to shipping activity occur as a result of collisions, overboard washing of tanker cleanings and oily ballast water, and careless transfer of oil from ship to shore or vice versa (table 12.20). As more and larger ships congregate in coastal waters for longer periods of time waiting for a docking berth, the chance of collision increases. While the number of accidents worldwide had increased slightly in 1971–1972 over the previous 2 years, the number of accidents involving larger oil spills doubled during this time. The effects of oil spills as well as other effects caused by petroleum hydrocarbons have already been described in this chapter.

Projections on future vessel activities have been completed based on numbers of commercial vessels using the northern area of the SCB (Point Conception to Long Beach). These studies (John J. Mullen Associates, Inc. 1977; National Maritime Research Center 1981; California Maritime Academy 1984; County of Santa Barbara Energy Division 1989) were able to project activities to the year 2000 and beyond. The most recent County of Santa Barbara study (1989) reported 8458 vessels, or 23.3 trips per day, for the year 1987 (table 12.21). Cargo vessels and oil tankers accounted for 80% and 16% of this total, respectively. In the same study, earlier projections were analyzed and it was predicted that the maximum number of vessel trips per day would increase from 23.3 in 1987 to 43.2 (15,864 per year) in the year 2000 (table 12.22). Projected vessel size has also been examined. In 1987, nearly 50% of the vessels using the SCB were in the 1000–1999 TEU (twenty-ft equivalent units) size class. This size is projected to diminish to only 5% of the total fleet by the year 2020, with the larger 4000–4999 TEU size class being the dominant replacement (fig. 12.26). The expected tonnage is expected to increase from 75 million t in 1980 to 202 million t by the year 2020 (U.S. Army Corps of Engineers 1984). From these analyses, it appears to be a

Table 12.20. *Location of Oil Spills in Los Angeles–Long Beach Harbor 1969–1972*

Location	1969–1970	1971–1972	Totals
Sea	52	68	120
Coastal	60	109	169
Entrance	59	44	103
Harbor	45	95	140
Pier	43	57	100
Unknown	7	3	10
Totals	266	376	642

Data from Keith and Porricelli (1973) and J. J. Henry Co. (1973).

Table 12.21. *Commercial Vessel Traffic for 1987 for the Northern Area of the SCB*

Port	Number of Vessels
Los Angeles–Long Beach	
Cargo	6,582
Tanker	933
Passenger–Cargo	60
Barge	183
Lumber	108
Port Hueneme	
Cargo	212
Santa Barbara	
Tanker	380
Total	8,458

Data from County of Santa Barbara Energy Division (1989).

Table 12.22. *Commercial Vessel Traffic and Projections for the Northern SCB, 1980–2000*

Year	Number of Projected Vessels	
	Per Day	Per Year
1980	24.8–31.4	8,972–11,368
1990	27.8–38.6	10,130–14,110
2000	28.8–43.5	10,496–15,864

Data from National Maritime Research Center (1981).

foregone conclusion that the probability of vessel accidents will increase as will the chances of significant environmental damage (County of Santa Barbara Energy Division 1989).

Boats and ships exposed to seawater present an additional habitat for the attachment of larvae of marine organisms. These organisms, referred to as "fouling organisms," consist primarily of algae, tube-dwelling polychaetes, barnacles, mussels, bryozoans, and tunicates. Settlement of larvae and larval growth rate is rapid, especially during the warmer months of the year (Reish 1961b). The growth of fouling organisms presents a drag to a ship's movement, causing the ship to require a longer time and additional fuel to reach its destination. Marine paint is used to prevent or slow the attachment of fouling organisms to the sides of ships. Antifouling paint contains a biocide, such as cupric oxide or organotin (tributyl tin, TBT), to prevent the attachment of larvae. Three types of antifouling paints have been used: (1) one in which the biocide is not attached to any molecule and is released into the environment by leaching from the paint; (2) a paint that sloughs off, exposing the fresh surface layer (usually containing a copper compound) of biocide; and (3) a self-polishing copolymer paint in which the biocide (in this case TBT) is released by chemical hydrolysis. The use of TBT has the advantage of a 5–7 year life compared to a 1–2 year life for the former two types (Champ 1986).

Organotins began to be widely used in marine paints in the early 1970s because of the known toxicity of the copper-based antifouling paints and the long life of TBT-containing paints. By 1980, it was suspected that TBT was adversely affecting shell growth in oysters. Because this was later shown to be true, France became the first country to ban the use of TBT in marine paints in 1982 (Alzieu 1986). The use of TBT in marine paints has since been banned in the United States on nonaluminum vessels less than 25 m in length. Research on the effects of TBT on the marine environment in the SCB is limited. Stephenson et al. (1986) demonstrated abnormal shell deposition in oysters and mussels that were suspended in bags near known areas of high TBT contamination in San Diego Bay.

Carlton (1985) summarized the various ways in which marine organisms, especially invertebrates, algae, fish, and marine angiosperms, have been introduced into new geographical areas by way of human active or passive ways. It is impossible to ascertain how many species have been introduced into the SCB through human activities. It is reasonable, however, to assume that the majority of these species were transported to the new geographical area by ships since most of the introduced species have been reported from harbors or shallow water (Carlton 1987). Shipping was minimal in the Pacific Ocean until the discovery of gold in California in 1849, but it increased greatly following this discovery and has continued to increase to the present time (table 12.22). Thus, the introduction of exotic species into the SCB is a recent occurrence.

Carlton (1985) stated that marine species are transported by ships either as fouling organisms attached to the sides or in ballast seawater. To a lesser extent, marine organisms have been transported by relocation of oil platforms from one geographical area to another and through mariculture, especially with oyster spat. The principal routes for the introduction of exotic species to the west coast of North America have been by way of Australasia, the southern Asia–southwestern Pacific, and the northwestern Pacific. Most

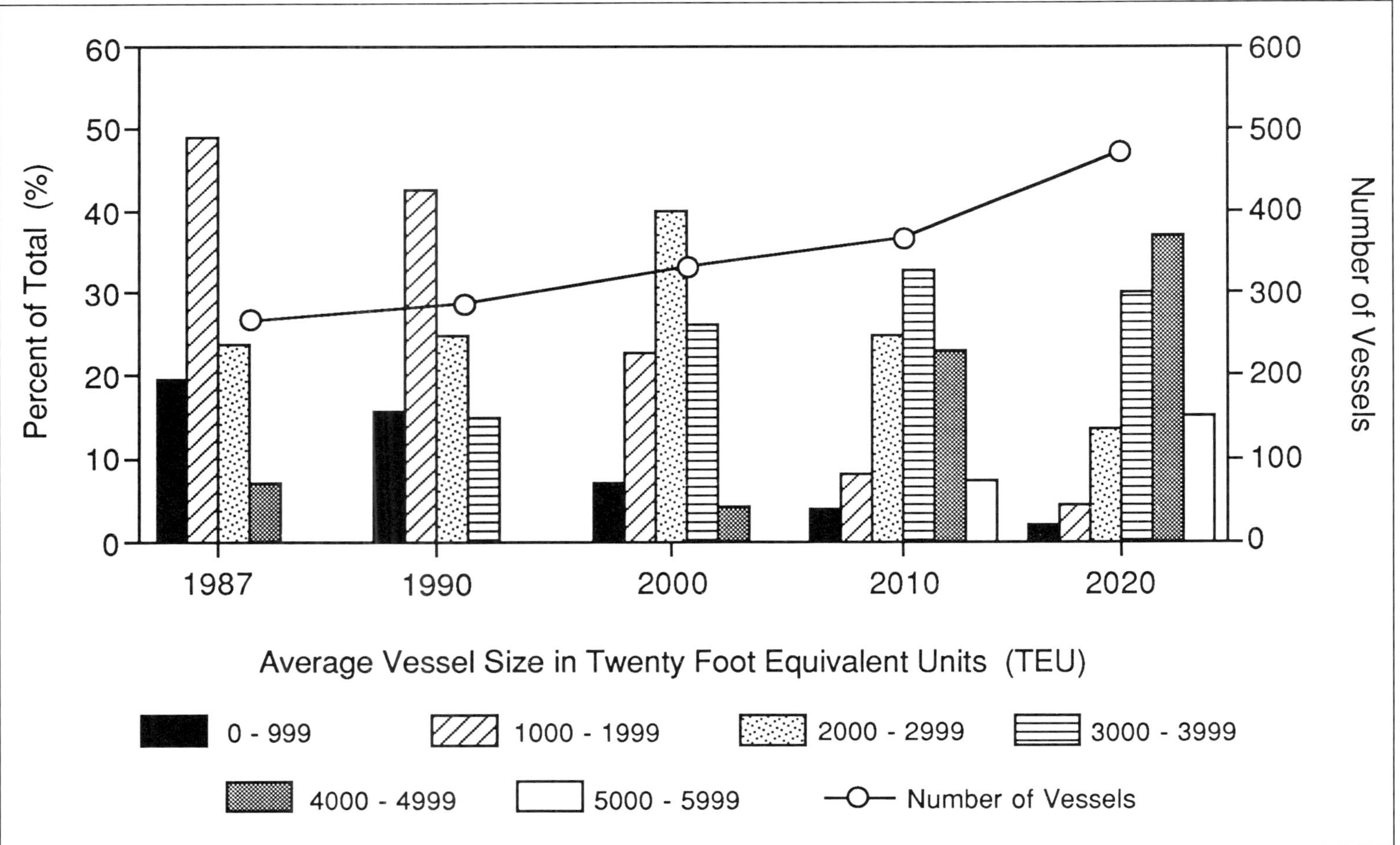

Figure 12.26. A comparison and projected size of commercial vessels entering Los Angeles–Long Beach Harbor, 1987–2020. (After Temple, Barker and Sloane 1988.)

of the reports of exotic species from the west coast have been from harbor areas. Since it is reasonable to assume that the majority of these introduced species were transported by ships, species from harbor areas of other geographical regions would be more tolerant than offshore species and could conceivably survive the voyage and inhabit the new harbor. San Francisco Bay and Los Angeles–Long Beach Harbor have recorded the greatest number of introduced species (Carlton 1985); these two ports are also the busiest on the Pacific Coast.

Shipping and boating will continue to play an important role in the SCB. The emergence of Pacific Rim countries into a dynamic economic force will increase shipping in the Pacific Ocean. Ships will become larger, and harbors, especially Los Angeles–Long Beach Harbor and San Diego Bay, will continue to be modified by construction and dredging to accommodate this growth. Boating as a leisure activity will continue to rise, creating an ever-increasing pressure for additional marine facilities. Not only will the marine life and marine environment be affected by this growth but also the offshore waters as well.

Summary and Prospectus for Future Research

Research studies on the impacts of municipal wastes attest to the fact that the magnitude of effects observed in the late 1980s has been less than that found in the 1970s, and in some cases there have been major improvements. This type of dramatic recovery was also observed in the Los Angeles–Long Beach Harbor area between 1968 and 1970. Our understanding of the impacts of oil discharges, including accidental spills, is that few chronically impacted offshore areas exist and that none have been identified in the SCB. Effects from petroleum in the region have stemmed from a few major spills, and the area impacted has been quite localized. Apparent recovery from two of these has been rapid, but the more recent spills of the *Pac Baronnes* and the *American Trader* have not been followed for sufficient time to evaluate recovery.

A review by Mearns et al. (1991) is extremely valuable in determining the present status of sediments and organisms in the SCB. Table 12.23 summarizes the significance of the massive data base examined by these authors. They conclude that no evidence exists to indicate that levels of any chemical pollutant are increasing, but recent data are weak for silver, lead, tin, selenium, and dieldrin. The only contaminants shown to be biomagnified (in the foodweb) are mercury, PCBs, DDT, and chlordane. While fish tissues have been useful in demonstrating the presence of high levels of organics in the environment, trace metals enrichment in the environment is more likely to be observed in tissues of mussels or other invertebrates.

There is still a great deal to learn regarding the effects of contaminants bound to sediment from various sources. At the federal and state level, research programs investigate the threshold concentrations of specific chemical pollutants alone and in combination with others that produce acute or chronic effects on sensitive marine organisms. Long and Morgan (1990) have reviewed the data available from many different types of sediment effects studies and have provided their estimates of the low and median effects range for many sediment-bound chemicals. This is a critical area of research, deserving one of the highest priorities, since regardless of the source of the contaminated particle (waste, runoff, dredged material, etc.), benthic organisms in the receiving water will be impacted. This is an extremely complex field of research because sediments seldom, if ever, contain only one pollutant, and studies should not only involve toxicologists but geochemists and microbiologists as well. Neither laboratory nor field studies alone will provide adequate answers to numerous questions regarding the factors controlling the partitioning of chemical mixtures between sediments, water, and benthic organisms. Predictions

Table 12.23. *Conceptual Summary of Patterns of Contamination in the SCB*

Contaminant	Accumulation in: Sediment?	Mussels?	Macro-invertebrates?	Fish?	Biomagnification?	Increasing?	Number Yes
Silver	Yes	Yes	Yes	No	No	[?]	3+
Arsenic	Yes	No	No	No	No	No	1
Cadmium	Yes	[?]	Yes	No	No	[?]	2+
Chromium	Yes	Yes	Yes	No	No	No	3
Copper	Yes	[?]	Yes	No	No	No	2+
Mercury	Yes	Yes	No	No	Yes	No	3
Lead	Yes	Yes	[?]	[?]	No	[?]	2+
Selenium	Yes	[?]	[?]	[?]	[?]	[?]	1+
Tin	Yes	Yes	[?]	Yes	[?]	[?]	3+
Zinc	Yes	No	No	No	No	No	1
PAH	Yes	Yes	[?]	Yes	No	No	3+
PCB	Yes	Yes	Yes	Yes	Yes	No	5
DDT	Yes	Yes	Yes	Yes	Yes	No	5
Chlordane	Yes	Yes	[?]	Yes	[?]	No	3+
Dieldrin	Yes	Yes	[?]	Yes	[?]	[?]	3+
Yes	15	10	6	6	4	0	
Uncertain	0	3	6	2	3	6	
No	0	2	3	7	8	9	

From Mearns et al. 1991.

based on the results of short- and long-term exposures in the laboratory will need to be verified in the field. This research requires a higher level of funding to produce the information needed to guide appropriate regulations on the discharge of safe concentrations of sediment-bound chemicals to coastal waters.

Related to contamination on particles that have settled to the bottom and become part of the permanent substrate is the flux of near-bottom fine particulates. Hendricks (in SCCWRP 1988) described the relationship between benthic community structure (using the infaunal trophic index, ITI) at three municipal waste outfalls and the magnitude of effluent particle resuspension and redeposition measured by sediment traps. At all three locations (Palos Verdes, Newport Beach, and Point Loma), the ITI was inversely related to the flux of effluent material in near-bottom waters. At Newport Beach and Point Loma, the ITI decreased by about 12–25 units for each 100 mg cm^{-2} per year increase in near-bottom effluent particle transport flux. It may be that the flux of fine organic particles from effluents has a major influence on benthic communities by both altering the composition of assemblages and contributing to the contamination of organisms. This fine material is not captured in either sediment grabs or water samples, and few chemical analyses of sediment trap material have been conducted. Some evidence shows that areas of the coast not exhibiting high concentrations of chlorinated organics in the sediments do contain fish with higher than expected tissue concentrations. While migration may be the explanation, the undersampled fine near-bottom particulates may represent the source of contamination. Further investigation of the potential effects of this material on benthic communities and foodweb contamination is certainly warranted.

A major recommendation heard at the local, state, and national levels is that a regional monitoring program be developed for the SCB. Many millions of dollars are presently being spent on the compliance monitoring of numerous municipal and industrial effluents. There is no consistency in the types of data collected, and it is not possible to interpret the significance of the massive database. Much emphasis has been placed on collecting data but not nearly enough on developing a better understanding of what to measure and how to identify a problem. It is necessary for public interest groups, dischargers, regulators, and scientists to get together to agree on (1) the usefulness of regional monitoring, (2) funding and administration, and (3) the most meaningful measurements and the criteria for each. In what is one of the wealthiest and most highly populated coastal areas of the world, not one facility is properly equipped or funded to conduct marine environmental research and compile and coordinate the massive database that will be produced in a regional monitoring program. The monitoring plan, the facility, and the staff are all necessary if we are to succeed in keeping in touch with the condition of the marine environment in the SCB.

Literature Cited

Allan Hancock Foundation, 1975. Environmental investigations and analysis, Los Angeles Harbor. Report to the U.S. Army Corps of Engineers, Harbor Environment Project. Prep. by Allan Hancock Foundation, Univ. of Southern California, Los Angeles. 582pp.

Allen, A. A., R. S. Schlueter, and P. J. Mikolaj, 1970. Natural oil seepage at Coal Oil Point, Santa Barbara, California. *Science*. 170:974–977.

Allen, M. J., H. Pecorelli, and J. Word, 1976. Marine organisms around outfall pipes in Santa Monica Bay. *J. Water Pollut. Control Fed.* 48(8): 1881–1893.

Alzieu, C., 1986. TBT detrimental effects on oyster culture in France: Evaluation since antifouling paint regulation. *Proceedings, Oceans '86.* 4:1130–1134.

Anderson, J. W., 1977. Responses to sublethal levels of petroleum hydrocarbons: Are they sensitive indicators and do they correlate with tissue contamination? In: D. A. Wolfe, ed. *Fate and Effects of Petroleum Hydrocarbons in Marine Ecosystems and Organisms*. Pergamon Press, New York. pp. 95–114.

Anderson, J. W., 1984. Development of a diluting concentration method for testing the acute toxicity of chemically dispersed oil. Final Report to the American Petroleum Institute, Washington, D.C. 58pp.

Anderson, J. W., S. L. Kiesser, and J. W. Blaylock, 1980. The cumulative effect of petroleum hydrocarbons on marine crustaceans during constant exposure. *Rapp. P.-V. Réun. Cons. Int. Explor. Mer.* 179:62–70.

Anderson, J. W., S. L. Kiesser, R. M. Bean, R. G. Riley, and B. L. Thomas, 1981. Toxicity of chemically dispersed oil to shrimp exposed to constant and decreasing concentrations in a flowing system. In: *Proceedings, 1981 Oil Spill Conference*. American Petroleum Institute, Washington, D.C. pp. 69–75.

Anonymous, 1952. Los Angeles–Long Beach Harbor pollution survey. Los Angeles Regional Water Pollution Control Board, Los Angeles. 43pp.

Armstrong, H. W., K. Fucik, J. W. Anderson, and J. M. Neff, 1979. Effects of oilfield brine effluent on sediments and benthic organisms in Trinity Bay, Texas. *Mar. Environ. Res.* 2:55–69.

Bailey, N. J. L., A. M. Jobson, and M. A. Rogers, 1973. Bacterial degradation of crude oil: Comparison of field and experimental data. *Chem. Geol.* 11:203–221.

Barnard, J. L., and O. Hartman, 1959. The sea bottom off Santa Barbara, California: Biomass and community structure. *Pac. Nat.* 1(6):1–16.

Bauer, J. E., P. A. Montagna, R. B. Spies, M. C. Prieto, and D. Hardin, 1988. Microbial biogeochemistry and heterotrophy in sediments of a marine hydrocarbon seep. *Limnol. Oceanogr.* 33:1493–1513.

Benville, P. E., Jr., and S. Korn, 1977. The acute

toxicity of six monocyclic aromatic crude oil components to sriped bass (*Morone saxatilis*) and bay shrimp (*Crago franciscorum*). *Calif. Fish Game*. 63:204–209.

Bertine, K. K., and E. D. Goldberg, 1977. History of heavy metal pollution in southern California coastal zone-reprise. *Environ. Sci. & Technol.* 11:297–299.

Boehm, P. D., and D. L. Fiest, 1982. Subsurface distributions of petroleum from an offshore well blowout. The *Ixtoc 1* blowout, Bay of Campeche. *Environ. Sci. & Technol.* 16:67–74.

Boehm, P. D., J. E. Barak, D. L. Fiest, and A. A. Elskus, 1982. A chemical investigation of the transport and fate of petroleum and hydrocarbons in littoral and benthic environments: The *Tsesis* oil spill. *Mar. Environ. Res.* 6:157–188.

Boesch, D. F., and N. N. Rabalais, eds., 1987. *Long-Term Environmental Effects of Offshore Oil and Gas Development*. Elsevier Applied Science, London. 708pp.

Boesch, D. F., and G. A. Robillard, 1987. Physical alteration of marine and coastal habitats resulting from offshore oil and gas development activities. In: D. F. Boesch, and N. N. Rabalais, eds. *Long-Term Environmental Effects of Offshore Oil and Gas Development*. Elsevier Applied Science, London. pp. 619–650.

Bohnsack, J. A., and D. L. Sutherland, 1985. Artificial reef research: A review with recommendations for future priorities. *Bull. Mar. Sci.* 37(1):11–39.

Brodersen, C. C., S. D. Rice, J. W. Short, T. Z. Mecklenburg, and J. F. Karinen, 1977. Sensitivity of larval and adult Alaskan shrimp and crabs to acute exposures of the water-soluble fraction of Cook Inlet crude oil. In: *Proceedings, 1977 Oil Spill Conference (Prevention, Behavior, Control, Cleanup)*. American Petroleum Institute, Washington, D.C. pp. 575–578.

Brooks, J. M., D. A. Wiesenburg, R. A. Burke, Jr., and M. G. Kennicutt, 1981. Gaseous and volatile hydrocarbon inputs from a subsurface oil spill in the Gulf of Mexico. *Environ. Sci. & Technol.* 15:951–959.

Brown, D. A., R. W. Gossett, G. P. Hershelman, C. F. Ward, and J. N. Cross, 1984. Metal and organic contaminants in sediments and animals. In: W. Bascom, ed. *Biennial Report 1983–1984*. South. Calif. Coastal Water Res. Proj., Long Beach, CA. pp. 179–193.

Brown, D. A., R. W. Gossett, G. P. Hershelman, C. F. Ward, A. M. Westcott, and J. N. Cross, 1986. Municipal wastewater contamination in the Southern California Bight: Part 1. Metal and organic contaminants in sediments and organisms. *Mar. Environ. Res.* 18(4):291–310.

Burdick, G. E., H. J. Dean, E. J. Harris, J. Skea, R. Karchers, and C. Frisa, 1972. Effect of rate and duration of feeding DDT on the reproduction of salmonid fishes reared and held under controlled conditions. *N.Y. Fish Game J.* 19: 97–115.

Burnett, R., 1971. DDT residues: Distribution of concentrations in *Emerita analoga* (Stimpson) along coastal California. *Science*. 174:606–608.

Caldwell, R. S., E. M. Caldarone, and M. H. Mallon, 1977. Effects of a seawater-soluble fraction of Cook Inlet crude oil and its major aromatic components on larval stages of the Dungeness crab, *Cancer magister*. In: D. P. Wolfe, ed. *Fate and Effects of Petroleum Hydrocarbons in Marine Ecosystems and Organisms*. Pergamon Press, New York. pp. 210–220.

California Maritime Academy, 1984. Santa Barbara vessel traffic study. Prep. for California Coastal Commission, San Francisco, CA.

California State Water Resources Control Board, 1982. California state mussel watch 1980–81. Trace metals and synthetic organic compounds in mussels from California's coast, bays and estuaries. Water Quality Monitoring Report 81–11 TS. Parts I, II, and III. State Water Resources Control Board, Division of Technical Services, Surveillance and Monitoring, Sacramento, CA (not sequentially numbered).

Capuzzo, J. M., J. C. Goldman, J. A. Davidson, and S. A. Lawrence, 1977. Chlorinated cooling waters in the marine environment: Development of effluent guidelines. *Mar. Pollut.* 8:161–163.

Carlisle, J. G., Jr., C. H. Turner, and E. E. Ebert, 1964. Artificial habitat in the marine environment. *Calif. Dep. Fish Game Fish Bull.* 124:1–93.

Carlton, J. T., 1985. Transoceanic and interoceanic dispersal of coastal marine organisms: The biology of ballast water. *Oceanogr. Mar. Biol. Annu. Rev.* 23:313–371.

Carlton, J. T., 1987. Patterns of transoceanic marine biological invasions in the Pacific Ocean. *Bull. Mar. Sci.* 41:452–465.

Carter, J. W., A. L. Carpenter, M. S. Foster, and W. N. Jessee, 1985. Benthic succession on an artificial reef designed to support a kelp reef community. *Bull. Mar. Sci.* 37(1):86–113.

Champ, M. A., 1986. Organotin symposium: Introduction and overview. *Proceedings, Oceans '86.* 4:1093–1100.

Chartrand, A. B., S. Moy, A. N. Safford, T. Yoshimura, and L. A. Schinazi, 1985. Ocean dumping under Los Angeles Regional Water Quality Control Board Permit: A review of past practices, potential adverse impacts, and recommendations for future action. Unpubl. Rep., California Regional Water Quality Control Board.

Chen, K. Y., and J. C. S. Lu, 1974. Sediment compositions in Los Angeles–Long Beach harbors and San Pedro Basin. In: D. F. Soule, and M. Oguri, eds. *Marine Studies of San Pedro Bay, California, Part 7. Sediment Investigations.* Harbor Environmental Projects, Allan Hancock Foundation, Univ. of Southern California, Los Angeles.

Clark, R. C., Jr., and J. S. Finley, 1977. Effects of oil spills in arctic and subarctic environments. In: D. C. Malins, ed. *Effects of Petroleum on Arctic and Subarctic Marine Environments and Organisms. Vol. II. Biological Effects.* Academic Press, New York. pp. 411–456.

Clean Water Act, 1972. Public Law PL 92–532, The Marine Protection, Research and Sanctuaries Act.

Conklin, P. J., D. Drysdale, D. G. Doughtie, K. R. Rao, J. P. Kabareba, T. R. Gilbert, and R. F. Shokes, 1983. Comparative toxicity of drilling fluids: Role of chromium and petroleum hydrocarbons. *Mar. Environ. Res.* 10:105–125.

County of Santa Barbara, Energy Division, 1989. Marine emergency management study. Technical assistance provided by Science Applications International Corp., La Jolla, CA.

County Sanitation Districts of Orange County, 1987. 1987 Annual Report, County Sanitation Districts of Orange County, Fountain Valley, CA.

Cross, J. N., 1985. Fin erosion among fishes collected near a southern California municipal wastewater outfall (1971–1982). *U.S. Natl. Mar. Fish. Serv. Fish. Bull.* 83(2):195–206.

Cross, J. N., 1986. Epidermal tumors in *Microstomus pacificus* (Pleuronectidae) collected near a municipal wastewater outfall in the coastal waters off Los Angeles (1971–1983). *Calif. Fish Game.* 72(2)68–77.

Cross, J. N., and J. E. Hose, 1989. Reproductive impairment in two species of fish from contaminated areas off southern California. *Proceedings, Oceans '89.* Seattle, WA.

Davis, G. E., 1985. Artificial structures to mitigate marina construction impacts on spiny lobster, *Panulirus argus. Bull. Mar. Sci.* 37(1):151–156.

Davis, P. H., and R. B. Spies, 1980. Infaunal benthos of a natural petroleum seep: Study of community structure. *Mar. Biol.* 59(1):31–41.

Dicks, B., and J. P. Hartley, 1982. The effects of repeated small oil spillages and chronic discharges. *Phil. Trans. Roy. Soc. London.* B297: 285–307.

Dillon, T. M., 1984. Biological consequences of bioaccumulation in aquatic animals: An assessment of the current literature. U.S. Army Engineers Waterways Experiment Station, Vicksburg, MS. 69pp.

Dinnel, P. A., Q. J. Stober, and D. H. DiJulio, 1981. Sea urchin sperm bioassay for sewage and chlorinated seawater and its relation to fish bioassays. *Mar. Environ. Res.* 5:29–39.

Dorsey, J., 1988. Wastewater discharge in Santa Monica Bay. In: *Proceedings, Symposium Managing Inflows to California's Bays and Estuaries.* November 13–15, 1986, Monterey, CA. The Bay Institute, Sausalito, CA. pp. 27–31.

Duedall, I. W., B. H. Ketchum, P. K. Park, and D. R. Kester, eds., 1983. *Wastes in the Ocean. Vol. 1: Industrial and Sewage Wastes in the Ocean.* John Wiley & Sons, New York, 431pp.

Ebeling, A. W., F. A. DeWitt, Jr., W. Werner, and G. M. Cailliet, 1972. Santa Barbara oil spill; fishes. In: *Proceedings, Santa Barbara Oil Symposium.* Sponsored by the National Science Foundation, Division Graduate Education Science and Marine Science Institute, Univ. of California, Santa Barbara, CA. pp. 137–164.

Eganhouse, R. P., 1982. Organic matter in municipal wastes and storm runoff: Characterization and budget to the coastal waters of southern California. Ph.D. Dissertation. Univ. of California, Los Angeles. 248pp.

Eganhouse, R. P., and I. R. Kaplan, 1981. Extractable organic matter in urban stormwater runoff. 1. Transport dynamics and mass emission rates. *Environ. Sci. & Technol.* 15:310–315.

Eganhouse, R. P., B. R. T. Simoneit, and I. R. Kaplan, 1981. Extractable organic matter in urban stormwater runoff. 2. Molecular characterization. *Environ, Sci. & Technol.* 15:315–326.

Emery, K. O., 1960. *The Sea off Southern Cali-*

fornia. John Wiley & Sons, Inc., New York. 366pp.

Eppley, R. W., E. H. Renger, and P. M. Williams, 1976. Chlorine reactions with seawater constituents and the inhibition of photosynthesis of natural marine phytoplankton. *Estuarine Coastal Mar. Sci.* 4(2):147–161.

Fischer, P. J., 1978. Natural gas and oil seeps, Santa Barbara Basin. In: *California Gas, Oil, and Tar Seeps*. The State Lands Commission, Long Beach, CA. pp. 1–62.

Fucik, K. W., H. W. Armstrong, and J. M. Neff, 1977. Uptake of naphthalenes by the clam *Rangia cuneata* in the vicinity of an oil separator platform in Trinity Bay, Texas. In *Proceedings, 1977 Oil Spill Conference (Prevention, Behavior, Control, Cleanup)*. American Petroleum Institute, Washington, D.C.

Goldman, J. C., 1979. Chlorine in the marine environment. *Oceanus*. 22(2):36–43.

Greene, C. S., 1976. Response and recovery of the benthos at Orange County. In: *Annual Report, 1976*. South. Calif. Coastal Water Res. Proj., El Segundo, CA. pp. 197–203.

Grove, R. S., 1983. Dispersion of chlorine at seven southern California coastal generating stations. Chapter 23. In: R. L. Jolley, W. A. Brungs, J. A. Cotruvo, R. B. Cumming, J. S. Mattice, and V. A. Jacobs, eds. *Water Chlorination: Environmental Impact and Health Effects. Vol. 4, Book 1, Chemistry and Water Treatment*. Ann Arbor Science Publishers, Ann Arbor, MI.

Hartley, J. P., and J. Ferbrache, 1983. Biological monitoring of the Forties Oilfield (North Sea). In: *Proceedings, 1983 Oil Spill Conference*. American Petroleum Institute, Washington, D.C. pp. 407–414.

Heyney, T. L., T. R. Nardin, and B. A. Nardin, 1975. Oil and tar seep studies on the shelves off southern California. II. Santa Monica Bay. Univ. of Southern California Geophysics Lab. Tech. Rep. No. 57–3. 46pp.

Hose, J. E., J. N. Cross, S. G. Smith, and D. Diel, 1989. Reproductive impairment in a fish inhabiting a contaminated coastal environment off southern California. *Environ. Pollut.* 57: 139–148.

Hose, J. E., T. D. King, K. E. Zerba, R. J. Stoffel, J. S. Stephens, Jr., and J. A. Dickinson, 1983. Does avoidance of chlorinated seawater protect fish against toxicity? Laboratory and field observations. In: R. L. Jolley, W. A. Brungs, J. A. Cotruvo, R. B. Cumming, J. S. Mattice, and V. A. Jacobs, eds. *Water chlorination: Environmental Impact and Health Effects. Vol. 4, Book 2, Environment, Health, and Risk*. Ann Arbor Science Publishers, Ann Arbor, MI. pp. 967–982.

Huntzicker, J. J., S. K. Friedlander, and C. I. Davidson, 1975. Material balance for automobile-emitted lead in Los Angeles Basin. *Environ. Sci. & Technol.* 9:448–457.

Hyland, J., J. Kennedy, J. Campbell, S. Williams, P. Boehm, A. Uhler, and W. Steinhauer, 1988. Initial effects of the *Pac Baroness* oil and copper spill: Results of hydrocarbon and macrofaunal analyses. Rep. to U.S. Dept. of Commerce, Minerals Management Service, Pacific OCS Region, and EPA, Region 9. 25pp.

J. J. Henry Co., 1973. An analysis of oil outflows due to tanker accidents. U.S. Coast Guard, Washington, D.C.

Johanssen, S., V. Larsson, and P. D. Boehm, 1980. The *Tsesis* oil spill impact on the pelagic ecosystem. *Mar. Pollut. Bull.* 11:284–293.

John J. Mullen Associates, Inc., 1977. Environmental impact report for Point Conception LNG import terminal-draft vessel traffic analysis. Prep. for California Public Utilities Commission, San Francisco, CA.

Johns, D. M., and R. Gutjahr-Gobell, 1988. Bioenergetic effects of black rock harbor dredged material on the polychaete *Nephtys incisa*, a field verification. Tech. Rep. D-88-3, U.S. EPA for U.S. Army Engineers Waterways Experiment Station, Vicksburg, MS. 119pp.

Johnson, S., and J. Roney, 1988. Wastewater discharge in Santa Monica Bay: II. In: *Proceedings, Symposium Managing Inflows to California's Bays and Estuaries*. November 13–15, 1986, Monterey, CA. The Bay Institute, Sausalito, CA. pp. 32–36.

Jones, G. F., 1969. The benthic macrofauna of the mainland shelf of southern California. *Allan Hancock Monogr. Mar. Biol.* Vol. 4. 219pp.

Jones, G. F., and B. E. Thompson, 1984. The ecology of *Parvilucina tenuisculpta* (Carpenter, 1864) (Bivalvia: Lucinidae) on the southern California borderland. *Veliger*. 26:188–198.

Juszko, B. A., D. R. Green, and M. F. Fingas, 1983. Sinking of oil: Water density considerations. In: *Proceedings, Sixth Arctic Marine Oil Spill Program Technical Seminar*. Environmental Protection Service, Ottawa, Canada. pp. 9–13.

Kanter, R. G., R. C. Wingert, W. H. Vick, M. L. Sowby, and C. J. Foley, 1983. California commercial/sport fish and shellfish oil toxicity study. Vol. 2. Synthesis of findings. Final report to Minerals Management Service. Available from: NTIS, Springfield, VA. No. PB84–167220. Prep. by MBC Applied Environmental Sciences, Inc., Costa Mesa, CA. 278pp.

Karinen, J. F., and S. D. Rice, 1974. Effects of Prudhoe Bay crude oil on molting tanner crabs, *Chionocetes bairdi*. *Mar. Fish Rev.* 36(7):31–37.

Katz, A., and I. R. Kaplan, 1981. Heavy metals behavior in coastal sediments of southern California: A critical review and synthesis. *Mar. Chem.* 10(4):261–299.

Keith, V. F., and J. D. Porricelli, 1973. Prevention and control of oil spills. In: *Proceedings, An Analysis of Oil Outflows due to Tanker Accidents.* American Petroleum Institute, U.S. EPA, and U.S. Coast Guard, Washington, D.C.

Kester, D. R., B. H. Ketchum, I. W. Duedall, and P. K. Park, 1983. *Wastes in the Ocean. Vol. 2: Dredged Material Disposal in the Ocean.* John Wiley & Sons, New York. 299pp.

Kinne, O., ed., 1984a. *Marine Ecology*. Vol. 5, part 3. John Wiley & Sons, New York. 527pp.

Kinne, O., ed., 1984b. *Marine Ecology*. Vol. 5, part 4. John Wiley & Sons, New York. 379pp.

Kinne, O., and H.-P. Bulnheim, eds., 1980. Protection of life in the sea. 14th Eur. Mar. Biol. Symp. on Protection of Life in the Sea, September 23–29, 1979, Helgoland (FRG). *Helgol. Meeresunters.* 33(1–4):772.

Kinnetic Laboratories, Inc., 1989. Adaptations of marine organisms to chronic hydrocarbon exposures. MMS 87–0089. Two volumes. U.S. Dept. of Interior, Minerals Management Service, Los Angeles, CA. Prep. by Kinnetic Laboratories, Inc., Carlsbad, CA. 353pp. and 413pp.

Klein, D. H., and E. D. Goldberg, 1970. Mercury in the marine environment. *Environ. Sci. & Technol.* 4(9):765–768.

Kolpack, R. L., J. S. Mattson, J. B. Mark, Jr., and T. C. Tu, 1971. Hydrocarbon content of Santa Barbara Channel sediments. In: R. L. Kolpack, ed. *Biological and Oceanographic Survey of the Santa Barbara Channel Oil Spill 1969–1970. Vol. II. Physical, Chemical and Geological Studies.* Allan Hancock Foundation, Univ. of Southern California, Los Angeles.

Landin, M. C., ed., 1988. Beneficial uses of dredged material. U.S. Army Corps of Engineers, Baltimore District, 231pp.

Lee, W. Y., and J. A. C. Nicol, 1978. Individual and combined toxicity of some petroleum hydrocarbons to the marine amphipod, *Elasmopus pectenicrus*. *Mar. Biol.* 48:215–222.

Liu, E. H., and J. E. Schneider, 1988. Managing inputs to Newport Bay; boatyard discharge. In: *Proceedings, Symposium Managing Inflows to California's Bays and Estuaries.* November 13–15, 1986, Monterey, CA. The Bay Institute, Sausalito, CA. pp. 20–26.

Long, E. R., and L. G. Morgan, 1990. The potential for biological effects of sediment-sorbed contaminants tested in the National Status and Trends Program. NOAA Tech. Mem. NOS OMA 52. Natl. Oceanic Atmos. Admin., Seattle, WA. 175pp.

MacGregor, J. S., 1974. Changes in the amount and proportions of DDT and its metabolites, DDE and DDD, in the marine environment off southern California, 1949–72. *U.S. Natl. Mar. Fish. Serv. Fish. Bull.* 72(2):275–293.

MacGregor, J. S., 1976. DDT and its metabolites in the sediments off southern California. *U.S. Natl. Mar. Fish. Serv. Fish. Bull.* 74(1):27–35.

Marine Biological Consultants, 1980. Biological investigations relating to the proposed ocean disposal of dredged sediments from Berths 80–88 and Berths 62–67. Report to the Port of Long Beach, CA. 35pp.

Marine Biological Consultants, 1987. Ecology of oil/gas platforms offshore California. OCS Study MMS 86–0094. U.S. Dept. of Interior, Minerals Management Service, Pacific OCS Region, Los Angeles, CA. Prep. by Marine Biological Consultants, Inc., Costa Mesa, CA. 92pp.

Matta, M. B., A. J. Mearns, and M. F. Buchman, 1985. Trends in DDT and PCBs in U.S. West Coast fish and invertebrates: A National Status and Trends Program report. Pacific Office, Coastal and Estuarine Assessments Branch, Ocean Assessments Division, Natl. Oceanic Atmos. Admin., Seattle, WA.

McAuliffe, C. D., 1987. Organism exposure to volatile/soluble hydrocarbons from crude oil spills: A field and laboratory comparison. In: *Proceedings, 1987 Oil Spill Conference.* American Petroleum Institute, Washington, D.C. pp. 275–288.

McQuat, H. W., 1951. History of Los Angeles Harbor. In: *First Conference on Coastal Engineering.* Long Beach, CA., 1950. pp. 259–270.

Mead, W. J., and P. E. Sorensen, 1970. The

economic cost of the Santa Barbara oil spill. In: *Proceedings, Santa Barbara Oil Symposium*. December 16–18, 1970. Univ. of California, Santa Barbara, CA. pp. 183–226.

Mearns, A. J., and C. S. Greene, 1975. Benthic response at Orange County outfalls. In: *Annual Report, 1975*. South. Calif. Coastal Water Res. Proj., El Segundo, CA. pp. 85–88.

Mearns, A. J., and M. D. Moore, 1976. Biological study of oil platforms Hilda and Hazel, Santa Barbara Channel, California. Final Rep., Univ. of California, San Diego Research Contract 5L–21174. Institute of Marine Resources, Scripps Institution of Oceanography, La Jolla, CA.

Mearns, A. J., and M. J. Sherwood, 1976. Fin erosion prevalence and environmental changes. In: *Annual Report, 1976*. South. Calif. Coastal Water Res. Proj., El Segundo, CA. pp. 139–141.

Mearns, A. J., M. B. Matta, D. Simecek-Beatty, M. F. Buchman, G. Shigenaka, and W. A. Wert, 1988. PCB and chlorinated pesticide contamination in U.S. fish and shellfish: A historical assessment report. NOAA Tech. Mem. NOS OMA 39. Natl. Oceanic Atmos. Admin., Seattle. 140pp.

Mearns, A. J., M. Matta, G. Shigenaka, D. MacDonald, M. Buchman, H. Harris, J. Golas, and G. Lauenstein, 1991. Contaminant trends in the Southern California Bight: Inventory and Assessment. NOAA Tech. Memo. NOS ORCA 62. Natl. Oceanic Atmos. Admin., Seattle, WA

Mecklenburg, T. A., S. D. Rice, and J. F. Karinen, 1977. Molting and survival of king crab (*Paralithodes camtschatica*) and coonstripe shrimp (*Pandalus hypsinotus*) larvae exposed to Cook Inlet crude oil water-soluble fraction. In: D. A. Wolfe, ed. *Fate and Effects of Petroleum Hydrocarbons in Marine Organisms and Ecosystems*. Pergamon Press, New York. pp. 221–228.

Menzies, R. J., 1951. *Limnoria* and the premature failure of creosoted marine structures in North America. In: *Proceedings, Marine Borer Conference*. U.S. Naval Civil Engineering, Research and Evaluation Laboratory, Port Hueneme, CA. pp. M-1–M-7.

Menzies, R. J., J. Mohr, and C. M. Wakeman, 1963. The seasonal settlement of wood-borers in Los Angeles–Long Beach harbors. *Wasmann J. Biol.* 21:97–120.

Montagna, P. A., J. E. Bauer, M. C. Prieto, D. Hardin, and R. B. Spies, 1986. Benthic metabolism in a natural coastal petroleum seep. *Mar. Ecol. Prog. Ser.* 34:31–40.

Montagna P. A., J. E. Bauer, J. Toal, D. Hardin, and R. B. Spies, 1987. Temporal variability and the relationship between benthic meiofaunal and microbial populations of a natural coastal petroleum seep. *J. Mar. Res.* 45(3):761–789.

National Academy of Sciences, 1983. *Drilling Discharges in the Marine Environment*. National Academy Press, Washington, D.C. 180pp.

National Maritime Research Center, 1981. Santa Barbara Risk Management Program. Prep. for California Coastal Commission, San Francisco, CA.

National Research Council, 1985. *Oil in the Sea: Inputs, Fates, and Effects*. National Academy Press, Washington, D.C. 601pp.

National Research Council, 1989. *Using Oil Spill Dispersants on the Sea*. National Academy Press, Washington, D.C. 335pp.

Neff, J. M., 1985. Biological effects of drilling fluids, drill cuttings and produced waters. In: *An Assessment of the Long-Term Environmental Effects of U.S. Offshore Oil and Gas Development Activities*. Federal Interagency Committee on Pollution Research, Development and Monitoring (COPRDM), U.S. Dept. of Commerce, Washington, D.C.

Neff, J. M., 1987. Biological effects of drilling fluids, drill cuttings, and produced waters. In: D. F. Boesch, and N. N. Rabalais, eds. *Long-Term Environmental Effects of Offshore Oil and Gas Development*. Elsevier Applied Science, London. pp. 469–523.

Neff, J. M., and J. W. Anderson, 1981. *Response of Marine Animals to Petroleum and Specific Petroleum Hydrocarbons*. Halstead Press, New York. 177pp.

Neff, J. M., J. W. Anderson, B. A. Cox, R. B. Laughlin, Jr., S. S. Rossi, and H. E. Tatem, 1976. Effects of petroleum on survival, respiration, and growth of marine animals. In: *Sources, Effects and Sinks of Hydrocarbons in the Aquatic Environment*. American Institute of Biological Sciences, Arlington, VA. pp. 515–539.

Olsen, L. A., and K. Adams, 1984. Effects of contaminated sediment on fish and wildlife: Review and annotated bibliography. U.S. Fish Wildl. Serv. FWS/OBS–82/66. 103pp.

Ott, F. S., R. P. Harris, and S. C. M. O'Hara, 1978. Acute and sublethal toxicity of naphthalene and three methylated derivatives to the estuarine copepod, *Eurytemora affinis*. *Mar. Environ. Res.* 1:49–58.

Park, P. K., D. R. Kester, I. W. Duedall, and B. W. Ketchum, eds., 1983. *Wastes in the Ocean. Vol. 3: Radioactive Wastes in the Ocean*. John Wiley & Sons, New York. 522pp.

Pearson, T. H., and R. Rosenberg, 1978. Macrobenthic succession in relation to organic enrichment and pollution of the marine environment. *Mar. Biol. Annu. Rev.* 16:229–311.

Philippi, G. T., 1977. On the depth, time and mechanism of the origin of the heavy to medium-gravity naphthenic crude oils. *Geochem. Cosmochim. Acta*. 41:33–52.

Reed, W. E., and I. R. Kaplan, 1977. The chemistry of marine petroleum seeps. *J. Geochem. Explor.* 7(2):255–293.

Reish, D. J., 1955. The relation of polychaetous annelids to harbor pollution. *Allan Hancock Foundation Contrib. 164. Public Health Reps.* 70(12):1168–1174.

Reish, D. J., 1956. An ecological study of Lower San Gabriel River, California, with special reference to pollution. *Calif. Fish Game*. 42(1):51–61.

Reish, D. J., 1959. An ecological study of pollution in Los Angeles-Long Beach harbors, California. *Allan Hancock Found. Occas. Pap.* No. 22. 119pp.

Reish, D. J., 1961a. The use of the sediment bottle collector for monitoring polluted marine waters. *Calif. Fish Game*. 47(3):261–272.

Reish, D. J., 1961b. The relationship of temperature and dissolved oxygen to the seasonal settlement of the polychaetous annelid *Hydroides norvegica* (Gunnerus). *Bull. South. Calif. Acad. Sci.* 60(1):1–11.

Reish, D. J., 1961c. A study of benthic fauna in a recently constructed boat harbor in southern California. *Ecology*. 42(1):84–91.

Reish, D. J. 1964a. Studies on the *Mytilus edulis* community in Alamitos Bay, California: I. Development and destruction of the community. *Veliger*. 6(3):124–131.

Reish, D. J., 1964b. Studies on the *Mytilus edulis* community in Alamitos Bay, California: II. Population variations and discussion of the associated organisms. *Veliger*. 6(4):202–207.

Reish, D. J., 1964c. Discussion of the *Mytilus californianus* community on newly constructed rock jetties in southern California. *Veliger*. 7: 95–101.

Reish, D. J., 1970a. The effects of varying concentrations of nutrients, chlorinity, and dissolved oxygen on polychaetous annelids. *Water Res.* 4:721–735.

Reish, D. J., 1970b. Biological conditions in the Lower San Gabriel River. Report to Los Angeles County Sanitation District No. 2. 21pp.

Reish, D. J., 1971a. Effect of pollution abatement in Los Angeles Harbor. *Mar. Pollut. Bull.* 2(5):71–74.

Reish, D. J., 1971b. Biological survey of Lower San Gabriel River. Report to Los Angeles County Sanitation District No. 2. 18pp.

Reish, D. J., 1980. Marine biological characteristics of Platform Holly, Santa Barbara County, California. Report to ARCO. Prep. by Atlantis Scientific, Beverly Hills, CA. 100pp.

Reish, D. J., 1981. Biological investigations on bioassay procedures and test oganisms. U.S. Army Corps of Engineers, Los Angeles District. 237pp.

Reish, D. J., 1982. Benthic survey of the marine invertebrates off Sunset Beach, California. Report to U.S. Army Corps of Engineers, Los Angeles District. 36pp.

Reish, D. J., 1983. Survey of the marine benthic infauna collected from the United States radioactive waste disposal sites off the Farallon Islands, California. U.S. EPA, Office of Radiation Programs. 54pp.

Reish, D. J., 1984. Domestic wastes. In: O. Kinne, ed. *Marine Ecology. Vol. V. Ocean Management*. John Wiley & Sons, New York. pp. 1711–1767.

Reish, D. J., 1986. Benthic invertebrates as indicators of marine pollution: 35 years of study. Vol. 3. Monitoring Strategies Symposium. In: *Oceans '86 Conference Record*. Sponsored by IEEE Ocean Engineering Society and Marine Technology Society. Marine Technology Society, Washington, D.C. pp. 885–888.

Reish, D. J., and J. A. LeMay, 1988. Bioassay manual for dredged sediments. U.S. Army Corps of Engineers, Los Angeles District. 162pp.

Reish, D. J., P. S. Oshida, A. J. Mearns, and P. Ginn, 1989. Effects on salt water organisms. *J. Water Pollut. Control Fed.* 61:1042–1054.

Reish, D. J., D. F. Soule, and J. D. Soule, 1980. Benthic biological conditions of Los Angeles–Long Beach harbors: Results of 28 years of investigations and monitoring. *Helgol. Meeresunters.* 34(2):193–205.

Rice, S. D., D. A. Moles, J. F. Karinen, S. Korn, M. G. Carls, C. C. Brodersen, J. A. Gharrett, and M. M. Babcock, 1984. Effects of petroleum hydrocarbons on Alaskan aquatic organisms: A comprehensive review of all oil-effects research on Alaskan fish and invertebrates conducted by the Auke Bay Laboratory, 1970–81. NOAA Tech. Mem.NMFS F/NWC–67. Natl. Oceanic Atmos. Admin., Natl. Mar. Fish. Serv., Auke Bay, Alaska. 128pp.

Risebrough, R. W., 1969. Chlorinated hydrocarbons in marine ecosystems. In: M. W. Miller, and G. G. Berg, eds. *Chemical Fallout*. C. C. Thomas, Springfield, IL. pp. 5–9.

Robison, B. H., and T. M. Lancraft, 1984. An upward transport mechanism from benthos. *Naturwissenschaften*. 71(6):322–324.

Ross, D., 1983. Ecological studies on sulphate-reducing bacteria in offshore oil storage systems. Ph.D. Dissertation, Heriot-Watt Univ., Edinburgh.

Rossi, S. S., and J. W. Anderson, 1976. Toxicity of water-soluble fractions of No. 2 fuel oil and south Louisiana crude oil to selected stages in the life history of the polychaete, *Neanthes arenaceodentata*. *Bull. Environ. Contam. Toxicol.* 16:18–24.

Rossi, S. S., and J. N. Neff, 1978. Toxicity of polynuclear aromatic hydrocarbons to the marine polychaete, *Neanthes arenaceodentata*. *Mar. Pollut. Bull.* 9:220–223.

Sanders, H. L., J. F. Grassle, G. R. Hampson, L. S. Morse, S. Garner-Price, and C. C. Jones, 1980. Anatomy of an oil spill: Long-term effects of the grounding of the barge *Florida* off West Falmouth, Massachusetts. *J. Mar. Res.* 38:265–380.

Sassen, R., 1980. Biodegradation of crude oil and mineral deposition in a shallow Gulf Coast salt dome. *Org. Geochem.* 2:153–156.

Schafer, H., 1989. Historical trends in municipal wastewater emissions to southern California coastal waters. *J. Water Pollut. Control Fed.* 61:1395–1401.

Schell, W. R., and S. Sugai, 1980. Radionuclides at the U.S. radioactive waste disposal site near the Farallon Islands. *Health Phys.* 39:475–496.

Segelhorst, E. W., 1974. Transportation. In: M. D. Dailey, B. Hill, and N. Lansing, eds. *A Summary of Knowledge of the Southern California Coastal Zone and Offshore Areas. Vol. III, Social and Economic Elements.* Southern California Ocean Studies Consortium, Long Beach, CA. pp.17-1–17-110.

Simoneit, B. R. T., and I. R. Kaplan, 1980. Triterpenoids as molecular indicators of paleoseepage in recent sediments of the Southern California Bight. *Mar. Environ. Res.* 3:113–128.

Soule, D. F., and M. Oguri, eds., 1974. Marine studies of San Pedro Bay, California. Part 4. Environmental field investigations. Harbor Environmental Projects, Allan Hancock Foundation, Univ. of Southern California, Los Angeles. 101pp.

Soule, D. F., and M. Oguri, eds., 1975. Marine studies of San Pedro Bay, California. Part 8. Environmental biology. Harbor Environmental Projects, Allan Hancock Foundation, Univ. of Southern California, Los Angeles. 122pp.

Soule, D. F., and M. Oguri, ed., 1976. Marine studies of San Pedro Bay, California. Part 11. Potential effects of dredging on the biota of outer Los Angeles Harbor. Toxicity, bioassay, and recolonization studies. Harbor Environmental Projects, Allan Hancock Foundation, Univ. of Southern California, Los Angeles. 227pp.

Soule, D. F., and M. Oguri, eds., 1978a. Marine studies of San Pedro Bay, California. Part 15. The impact of the *Sansinena* explosion and Bunker C spill on the marine environment of outer Los Angeles Harbor. Harbor Environmental Projects, Allan Hancock Foundation, Univ. of Southern California, Los Angeles. 273pp.

Soule, D. F., and M. Oguri, eds., 1978b. Marine studies of San Pedro Bay, California. Part 14. Biological investigations. Harbor Environmental Projects, Allan Hancock Foundation, Univ. of Southern California, Los Angeles. 164pp.

Soule, D. F., and M. Oguri, eds., 1979. Marine studies of San Pedro Bay, California. Part 16. Ecological changes in outer Los Angeles–Long Beach harbors following initiation of secondary waste treatment and cessation of fish cannery waste effluent. Harbor Environmental Projects, Allan Hancock Foundation, Univ. of Southern California, Los Angeles. 611pp.

Soule, D. F., and M. Oguri, eds., 1980. Marine studies of San Pedro Bay, California. Part 17. The marine environment in Los Angeles and Long Beach Harbors during 1978. Harbor Environmental Projects, Allan Hancock Foundation, Univ. of Southern California, Los Angeles. 667pp.

Soule, D. F., and D. Walsh, eds., 1983. *Water Dis-*

posal in the Ocean: Minimizing Impact, Maximizing Benefits. Westview Press, Boulder, CO. 296pp.
Southern California Coastal Water Research Project, 1973. The ecology of the Southern California Bight: Implications for water quality management. Rep. TR 104. South. Calif. Coastal Water Res. Proj., El Segundo, CA. 531pp.
Southern California Coastal Water Research Project, 1978. *Annual Report, 1978.* Prep. by South. Calif. Coastal Water Res. Proj., El Segundo, CA.
Southern California Coastal Water Research Project, 1986. *Annual Report, 1986.* Prep. by South. Calif. Coastal Water Res. Proj., Long Beach, CA.
Southern California Coastal Water Research Project, 1988. *Annual Report, 1987.* South. Calif. Coastal Water Res. Proj., Long Beach, CA.
Southern California Coastal Water Research Project, 1990. *Annual Report, 1989–1990.* South. Calif. Coastal Water Res. Proj., Long Beach, CA.
Southern California Edison Company, 1979. Annual report San Onofre Nuclear Generating Station Unit 1. No. 80-RD–100, vols. 4 and 5. South. Calif. Edison Co., Rosemead, CA.
Southern California Edison Company, 1987. Marine environmental analysis and interpretation. San Onofre Nuclear Generating Station. Report of 1986 data, 787-RD–31. South. Calif. Edison, Rosemead, CA.
Southern California Ocean Studies Consortium, 1977. Long Beach downtown marina, environmental impact report. South. Calif. Ocean Studies Consortium, Long Beach, CA. 369pp.
Southward A. J., and E. C. Southward, 1978. Recolonization of rocky shores in Cornwall after use of toxic dispersants to clean up the *Torrey Canyon* oil spill. *J. Fish. Res. Board Can.* 35(5): 682–706.
Spies, R. B., and P. H. Davis, 1979. The infaunal benthos of a natural oil seep in the Santa Barbara Channel. *Mar. Biol.* 50(3):227–237.
Spies, R. B., and P. H. Davis, 1982. Toxicity of Santa Barbara seep oil to starfish embryos: Part 3. Influence of parental exposure and the effects of other crude oils. *Mar. Environ. Res.* 6:3–11.
Spies, R. B., P. H. Davis, and D. H. Stuermer, 1980. Ecology of a submarine petroleum seep off the California coast. In: R. A. Geyer, ed. *Marine Environmental Pollution: 1. Hydrocarbons.* Elsevier Oceanogr. Ser., Elsevier Sci. Publ. Co., Amsterdam (Netherlands). pp. 229–263.
Spies, R. B., J. S. Felton, and L. Dillard, 1982. Hepatic mixed function oxidases in California flatfishes are increased in contaminated environments and by oil and PCB ingestion. *Mar. Biol.* 70(2):117–127.
Spies, R. B., D. Hardin, and J. Toal, 1988. Organic enrichment or toxicity? A comparison of the effects of kelp and crude oil in sediments on the colonization and growth of fauna. *J. Exp. Mar. Biol. Ecol.* 124:261–282.
Stephenson, M. D., D. R. Smith, J. Goetzl, G. Ichikawa, and M. Martin, 1986. Growth abnormalities in mussels and oysters from areas with high levels of tributyltin in San Diego Bay. *Proceedings, Oceans '86.* 4:1245–1251.
Stout, V. F., and F. L. Beezhold, 1981. Chlorinated hydrocarbon levels in fishes and shellfishes of the northeastern Pacific Ocean, including the Hawaiian Islands. *Mar. Fish. Rev.* 43(1): 1–12.
Straughan, D., 1970. Ecological effects of the Santa Barbara oil spill. In: *Santa Barbara Oil Symposium.* December 16–18, 1970. Univ. of California, Santa Barbara, CA. pp. 173–182.
Straughan, D., ed., 1971. Oil pollution and seabirds. In: *Biological and Oceanographic Survey of the Santa Barbara Channel Oil Spill 1969–1970, Vol. I: Biology and Bacteriology.* Allan Hancock Foundation, Univ. of Southern California, Los Angeles. pp. 307–312.
Stuermer, D. H., R. B. Spies, P. H. Davis, D. J. Ng., C. J. Morris, and S. Neal, 1982. The hydrocarbons in the Isla Vista marine seep environment. *Mar. Chem.* 11(5):413–426.
Stull, J. K., and C. I. Haydock, 1988. Wastewater discharges and environmental responses: The Palos Verdes case. In: *Proceedings, Symposium Managing Inflows to California Bays and Estuaries.* November 13–15, 1986, Monterey, CA. The Bay Institute, Sausalito, CA. pp. 44–49.
Swartz, R. C., F. A. Cole, D. W. Schults, and W. A. Deben, 1986. Ecological changes in the Southern California Bight near a large sewage outfall: Benthic conditions in 1980 and 1983. *Mar. Ecol. Prog. Ser.* 31:1–13.
Swartz, R. C., D. W. Schults, G. R. Ditsworth, W. A. Deben, and F. A. Cole, 1985. Sediment toxicity contamination and macrobenthic communities near a large sewage outfall. In: T. P. Boyle, ed. *American Society for Testing and Mate-*

rials Special Tech. Publ. 865. Symposium on Validation and Predictability of Laboratory Methods for Assessing the Fate and Effects of Contaminants in Aquatic Ecosystems; Grand Forks, ND, August 8, 1983. American Society for Testing and Materials, Philadelphia, PA. pp. 152–175.

Teal, J. M., and R. W. Howarth, 1984. Oil spill studies: A review of ecological effects. *Environ. Manager.* 8:27–44.

Temple, Barker and Sloane, 1988. Forecast of the dry-bulk tanker and container fleets serving Los Angeles–Long Beach, 1987–2020. Prep. for U.S. Army Corps of Engineers, Exhibit 411. Prep. by Temple, Barker and Sloane, Lexington, MA.

Tetra Tech, Inc., 1985a. Environmental assessment for final designation for LA 2 ocean dredged material disposal site. Pasadena. 190pp.

Tetra Tech, Inc., 1985b. Environmental assessment for final designation of LA 5 ocean dredged material disposal site. Pasadena. 179pp.

Tetra Tech, Inc., and MBC Environmental Services, Inc. 1985a. Environmental assessment for final designation of LA 2 ocean dredged material disposal site. Pasadena. 190pp.

Tetra Tech, Inc., and MBC Environmental Services, Inc., 1985b. Environmental assessment for final designation of LA 5 ocean dredged material disposal site. Pasadena. 179pp.

Thatcher, T. O., 1978. The relative sensitivity of Pacific Northwest fishes and invertebrates to chlorinated sea water. In: R.Jolley, H. Gorchev, and D. Hamilton, Jr., eds. *Water Chlorination: Environmental Impact and Health Effects, Vol. 2.* Ann Arbor Science, Ann Arbor, MI. pp. 341–350.

Thomas, M. L. H., 1973. Effect of Bunker C oil on intertidal and lagoonal biota in Chedabucto Bay, Nova Scotia. *J. Fish Res. Board Can.* 30(1):83–90.

Thomas, R. E., and S. D. Rice, 1979. The effect of exposure temperatures on oxygen consumption and opercular breathing rates of pink salmon fry exposed to toluene, naphthalene, and water-soluble fractions of Cook Inlet crude oil and No. 2 fuel oil. In: F. J. Vernberg, W. B. Vernberg, and A. Calabrese, eds. *Marine Pollution: Functional Processes.* Academic Press, New York.

Thompson, B. E., 1982. Variations in benthic assemblages, 1981–1982. In: *Annual Report, 1982.* South. Calif. Coastal Water Res. Proj., Long Beach, CA. pp. 45–58.

Thompson, B. E., J. D. Laughlin, and D. T. Tsukada, 1987. 1985 reference site survey. In: *Tech. Rep. 221.* South. Calif. Coastal Water Res. Proj., Long Beach, CA.

Turner, C. H., and A. R. Strachan, 1969. The marine environment in the vicinity of the San Gabriel River mouth. *Calif. Fish Game.* 55:53–68.

U.S. Army Corps of Engineers, 1984. Environmental impact report/environmental impact statement for landfill development. Draft report. U.S. Army Corps of Engineers, Los Angeles District. 272pp.

U.S. Environmental Protection Agency, 1977. Ocean dumping: Final revisions of regulations and criteria. *Federal Register.* 42(7).

U.S. Environmental Protection Agency, 1983. Survey of the marine benthic infauna collected from the United States radioactive waste disposal sites off the Farallon Islands, California. EPA 520/1-83-006. Revised edition. U.S. EPA, Office of Radiation Programs, Washington, D.C. 54pp.

U.S. Environmental Protection Agency, 1987. Environmental impact statement for San Diego (LA 5) ocean dredged material disposal site. Site designation. U.S. EPA, Region 9, San Francisco, CA.

U.S. Environmental Protection Agency, 1988. Final Environmental Impact Statement (EIS) for the Los Angeles/Long Beach (LA 2) ocean dredged material disposal site designation. U.S. EPA, Region 9, San Francisco, CA. 339pp.

U.S. Environmental Protection Agency and U.S. Army Corps of Engineers, 1977. Ecological evaluation of proposed discharge of dredge material into ocean waters. Environmental Effects Laboratory, U.S. Army Engineers Waterways Experiment Station, Vicksburg, MS. 122pp.

Vanderhorst, J. R., C. I. Gibson, and L. J. Moore, 1976. Toxicity of No. 2 fuel oil to coonstripe shrimp. *Mar. Pollut. Bull.* 7(6):106–108.

Vetter, T. G., and J. A. Johnson, 1982. A scientific investigation of natural marine seeps by seismic systems and manned submersible (unpubl. rep.). Prep. by Nekton, Inc., San Diego, CA.

Wilkinson, E. D., 1972. California offshore oil and gas seeps. California Division of Oil and Gas, Publ. No. TR 08. 11pp.

Williams-Kuebelbeck and Associates, Inc., 1979. Survey of southern California marinas. In: *The Official Statement, City of Long Beach $34,500,000 1980 Marine Revenue Bonds*. Williams-Kuebelbeck and Associates, Inc., Redwood City, CA.

Wilson, R. D., P. H. Monaghan, A. Osanik, L. C. Price, and M. A. Rogers, 1973. Estimate of annual input of petroleum to the marine environment from natural marine seepage. *Gulf Coast Assoc. Geol. Soc.* 23:182–193.

Wolfe, D. A., ed., 1977. *Fate and Effects of Petroleum Hydrocarbons in Marine Organisms and Ecosystems*. Pergamon Press, New York. 478pp.

Wolfson, A., G. VanBlaricom, N. Davis, and G. S. Lewbel, 1979. The marine life of an offshore oil platform. *Mar. Ecol. Prog. Ser.* 1(1): 81–89.

Woodhead, D. S., 1984. Contamination due to radioactive materials. In: O. Kinne, ed. *Marine Ecology*. Vol. 5, part 3. John Wiley & Sons, New York. pp. 1111–1287.

Word, J. Q., 1978. The infaunal trophic index. In: *Annual Report, 1978*. South. Calif. Coastal Water Res. Proj., El Segundo, CA. pp. 19–39.

Word, J. Q., and A. J. Mearns, 1979. 60-meter control survey off southern California. In: *Tech. Mem. 229-TR*. South. Calif. Coastal Water Res. Proj., El Segundo, CA. 58pp.

Young, D. R., and T.-K. Jan, 1976. Aerial fallout of metals during a brushfire. In: W. Bascom, ed. *Annual Report, 1976*. South Calif. Coastal Water Res. Proj., El Segundo, CA. pp. 43–47.

Young, D. R., T.-K. Jan, and M. D. Moore, 1977. Metals in power plant cooling water discharges. In: *Annual Report, 1977*. South. Calif. Coastal Water Res. Proj., El Segundo, CA. pp. 25–31.

Young, G. P., 1977. Effects of naphthalene and phenanthrene on the grass shrimp *Palaemonetes pugio* (Holthius). Master's thesis, Graduate College, Texas A&M Univ. College Station, TX. 67pp.

Chapter 13

Governance

David W. Fischer

Governance: An Introduction

Governance refers both to the *exercise* of authority and to the *system* through which power is displayed. Thus, as Kelman (1987) writes, governance answers such questions as who is granted formal authority to make political choices, what procedures will be required in making such choices, and in what ways others attempt to influence those choices. While governance may anticipate experience, policy, the result of governance, does not. Indeed, in the Southern California Bight (SCB), policy is often shaped by the adverse experiences of some users who make a political issue of their experiences in an attempt to influence the political choices within the system of governance available. Scientists are but one group among many interested in influencing policy concerning the SCB.

However, as Caldwell (1984, p. 15) notes, "Critical issues as categorized by science do not always correspond to the perceptions and priorities prevailing in governments. . . . An issue becomes prospectively critical to a government if it becomes sufficiently critical to its political constituency." Of course, what is critical to one group is not necessarily critical to other groups, regardless of the factual information base available.

The large and heterogeneous set of uses found within the SCB cannot be rationed and conflicts cannot be resolved simply by providing and distributing more scientific information on the impacts of such uses on the ecology of the SCB. Moreover, human behavior does not always change in response to increased information alone. Rather, the large number of users must be induced through governance to forego, reduce, or otherwise coordinate their level of use. Even the passage of a new law cannot guarantee that conflicts will be reduced to some level acceptable to all interests involved.

Instead, users respond to the combined hierarchy of incentives generated by the myriad laws, rules, organizations, and known information in relationship to their own preferences, incomes, and perceptions of the resource base. Each user group is a stakeholder in the political arena represented by the SCB, and these user groups rarely agree on the goals to be pursued for the region. As yet, no one organization is responsible for the entire SCB, and few would even agree on which one organization should govern—that is, one group through which goals could be formed and implemented.

In the absence of shared goals and a governance structure that is accountable for such goals, decision making in the SCB takes place through organizations and governments completely disassociated from one another. For example, the many cities adjacent to the SCB each pursue their individual goals separately without an ability to account for the impacts of their decisions beyond their separate jurisdictional boundaries.

Thus, the lack of a single governing organization, as well as the lack of a constituency for the SCB as a whole, means that no one group can represent the SCB in political choices concerning the area (Hershman and Evans 1979). As well, without one governmental authority to speak for the SCB, there is no way to elevate a scientific problem to the status of a political choice in an agreed upon political forum. Instead, a complex web of competing choices and influences permeates a variety of organizations and levels of government.

Management Versus Governance

It is important to distinguish *management* from *governance*. Governance provides the framework and sets the rules for management actions. Governance operates through markets, votes, executive decisions, and judicial findings to affect policy. Management, in contrast, refers to the work of an agency established to implement policy set forth through the governance process. Responsibility for management begins with an already agreed upon goal and generates the means for accomplishing it. Sound management cannot exist without a consensus on goals, nor can it exist when conflicts in values remain unresolved.

It is important to recognize that governance builds the framework for resource management, and for governance to work, it is necessary to achieve consensus among all levels of government and the private sector on the social, economic, and political processes involved. Scientists concerned with the ecology of the SCB too often assume that the consensus they share is also shared by others. However, no consensus exists on which social and economic values should be diminished, enhanced, or otherwise rationed in the SCB. (An early work by Warren et al. [1972] on the concept of the coastal zone called this scientific concern "the management syndrome.") Thus, management typically involves a hierarchical agency that allocates resources within a single use and otherwise implements a policy based on shared values predetermined for that use through the governance process. In contrast, governance provides the basis for making the political choices that created the agency and its specialized management mandate. Overall management of the SCB's resources and uses cannot commence until the values surrounding user conflicts are reconciled through governance.

The Need for Governance

Including the water column and irregular seabed, the SCB encompasses an immense three-dimensional area (see chap. 1). This area is home to the nation's largest urban concentration: nearly 14 million people live, work, and play in five counties and more than 120 cities adjacent to the SCB. In addition, numerous special purpose districts exist throughout the shoreland area (U.S. Department of Commerce 1987; International City Management Association 1987).

The SCB contains a variety of natural and man-made features, including a diverse shoreline with urban and rural land uses, natural and man-made harbors and islands, and a vast water domain. Each feature is identified with a set of associated uses. This broad range of uses subjects the SCB to varying patterns of impacts as the area serves southern California's demand for food, transportation, communication, waste disposal, energy, recreation, and education, among others. An increasing population generates even more urbanization and its consequent demands on the SCB. As the costs of land, construction, and urbanization rise, new and intensified uses of the SCB emerge. With a greater variety of ocean uses and a greater number of users, the SCB is subjected to a growing scarcity of unused spaces and to increasingly incompatible uses that create conflicts. Thus, a vast area once freely used by all falls victim to escalating use and its corollary: mounting costs for all.

As Eckert (1979, p. 8) points out,

> Up to now, most ocean space has been "owned" either communally or not at all. Individual persons or nations have usually had no responsibility to make sure that the value of ocean resources is maintained rather than depleted, that the seas are kept clean rather than polluted, or that areas of particular scarcity are allocated to society's highest-valued purpose. As a result, the oceans . . . tend to be inefficiently dirty and overcongested.

This problem is seen as the "common property" issue in which no one user has the incentive to use ocean spaces efficiently, meaning that values of the water are altered through overuse in such a way that other users' interests are affected negatively.

Because the "costs" thus imposed on *other* users are not borne by offending users, *all* suffer. For example, proponents of marine mammal protection do not take into account the effect on fishermen who, in turn, suffer reduced catches. If the fishermen were compensated by the protectionists, they would presumably be neutral to protection. However, protectionists have no incentive to compensate fishermen's losses, so society's cost of enforcing protection is higher (Hardin and Baden 1977). Even so, in some areas of the SCB, certain uses have been accommodated with seemingly low levels of conflict. For example, the Long Beach area readily mixes oil and gas production and the traffic of a busy port with one of the SCB's largest marinas and other growing marine recreation uses.

As demands on the SCB increase, more costs and conflicts can be expected. Unlike natural systems, which tend toward self-regulating and self-maintaining mechanisms that promote a stable long-term balance, human use tends to involve single purpose additions to and deletions from a natural system. The SCB, at some point, may not absorb all uses and subsequent impacts placed upon it. Its limits could be exceeded, and then the costs of over use would rise dramatically. Such pending scarcity makes the issue of governance inescapable: some system of authority or control is necessary to allocate the SCB's resources to those uses judged to have higher social value (Cicin-Sain and Knecht 1985).

All uses involve an array of private and public governance. Goods shipped through the ports of Los Angeles and Long Beach arrive at a public entity on private ships. Within the ports, public cranes transfer private goods to private land transport on public roads. The U.S. Army Corps of Engineers and private contractors maintain the dredging of the harbor, and the U.S. Coast Guard maintains navigational aids. Operations of the ports are governed in part by the need to conform to the myriad rules laid down by the cities of Los Angeles and Long Beach, the state of California, and the U.S. government.

Oil and gas recovered from city, state, and federal lands or seabeds in the SCB are pumped ashore from private wells on artificial islands and platforms. Private pipelines traverse federal, state, city, and private lands to private refineries, all of which are subject to private, city, state, and federal rules covering environmental impact, air quality, endangered species, production rate, safety, navigation, water quality, and impact on other activities.

Recreation in the SCB consists of private boating, jet skiing, water skiing, sail boarding, surfing, swimming, SCUBA diving, and fishing—plus the interactions among all of these activities. Private boating is supported by public marinas, launch ramps, and parking areas. Private, local, state, and federal organizations govern and facilitate boating. Each of these uses of the SCB can be identified with one or more conflicts. For example, recreational boating includes conflicts with sailboating and motorboating as well as with commercial shipping, the presence of oil platforms, and military use. Oil production conflicts with environmental quality because it causes air and water pollution and requires industrial siting onshore. Ports may conflict with marina and residential use as well as with certain recreational uses.

Just as in plant and animal systems, where each species occupies an ecological niche, each user group also occupies a niche. This niche consists of the sanctioning agency as well as the using group; they work together to expand their niche, as opposed to that of other groups, in a competitive struggle with such groups for increased resources and access to the SCB. A typical niche would consist of the congressional subcommittee, the federal agency, the corresponding state entities, and the constituency—each mutually supporting the other in maintaining and even enlarging its political niche.

Thus, as noted earlier, the SCB has a problem in governance. No one organization or niche can internalize the collective impacts from the SCB's diverse uses. To implement attempts to enhance governance requires the agreement of *all* organizations and user groups (niches) involved. One organization, however, can thwart proposed change through litigation or by simply refusing to participate. Furthermore, organizations designed to manage single resources of the SCB, such as the California Department of Fish and Game, are not capable of resolving conflicting user demands of nonfishing groups because they lack the means to identify and reconcile or accommodate other user groups.

The Complexity of Governance

Each use of the SCB can be identified with some organization that sanctions it, and each organization, in turn, can be identified with some level of government or the private sector. Table 13.1 lists major organizations that sanction or influence use in the SCB.

Federal Level

The federal level contains a wide array of organizations that are directly controlled by decision making in the executive branch, congressional committees, the courts, and independent commissions. No one group in this multiorganizational system has final authority. Rather, the U.S. Constitution provides for split authority in which each federal branch is constrained by the others. Even within the executive branch, where the federal organizations in table 13.1 are located, the statutory laws authorizing these units impose constraints as well as set policy characteristics among them. Each federal organization listed is successful in its sphere only to the extent that it satisfies the multiplicity of interests surrounding it. An organization cannot provoke its constituency or violate rules of other executive organizations operating within its policy field (Bish 1982). Thus, the checks and balances operate among federal organizations to approach some level of consensus on federal use and regulation of the SCB. The problem is in the short-lived nature of the consensus, which changes as federal political priorities, budgets, rules, and personnel change and when changes occur in groups outside of the federal sphere.

Walsh (1981, pp. 73–74) has described this problem as each policy field having

Table 13.1. *Organizations Directly Influencing Uses in the SCB*

International Level

International Maritime Cooperative Organization

Federal Level

Geological Survey
Corps of Engineers
Minerals Management Service
National Science Foundation
Navy
National Marine Fisheries Service
Nuclear Regulatory Commission
National Oceanic and Atmospheric Administration
Coast Guard
Environmental Protection Agency
National Park Service
Fish and Wildlife Service
State Department
Food and Drug Administration
Department of Energy

Federal–State Level

Pacific Fishery Management Council

State Level

Office of Planning and Research
Parks and Recreation
Fish and Game
Air Resources Board
Business, transportation, energy, and housing agencies
Boating and Waterways
State Lands Commission
Coastal Commission
Coastal Conservancy
State universities

Local Level

Ports of Los Angeles and Long Beach
Five counties (Santa Barbara, Ventura, Los Angeles, Orange, San Diego)
120 cities and their departments of planning, public works, and recreation

Private

Oil companies
Fishing companies
Trucking companies
Residents and businesses
Cousteau Society
American Fisheries Society
South Central Coastwatch
Seashore Environmental Alliance
Save the Whales
San Pedro Planning Alliance
Oceanic Society
Natural Resources Defense Council
The Laguna Greenbelt, Inc.
Friends of the Earth
Get Oil Out, Inc./Get Oil Out-Two
Santa Catalina Island Conservancy
National Coalition for Marine Conservation
Friends of Santa Monica Mountains and Seashore
Sailing clubs
Shipping companies
Southern California Edison
Sierra Club
Amigos de Bolsa Chica
American Cetacean Society
Stop Pollution
Sea and Sage Audubon Society
Save Our Coastline
Orange County Marine Institute
Santa Monica Bay Audubon Society
Marine Technology Society
Friends of the Sea Otter
Friends of Anaheim Bay
Friends of Coastal Area Preservation
Ecology Center of Southern California
Scenic Shoreline Preservation, Inc.
Malibu Citizens for Good Community Planning

its own non-governmental constituencies, a complement of involved federal agencies, and congressional infrastructure. But in and among the [fields] there is really very little crossover, coordination, and cooperation. No one is in charge. Thus, while each ocean activity is promoted by well-meaning, hard-working, and often zealous professionals, the net result, in the sense of real national interest, often can be zero.

Precisely because of this situation at the federal level, there have been calls for a "Department of Oceans" or similar reorganizations of federal marine functions. Beginning with the Stratton Commission's finding in 1969 that a cabinet-level oceans agency be created, several proposals for federal reorganization have been made, all of which are described by Bowen (1981).

Nevertheless, no federal proposal to integrate all ocean uses into one organization has been possible. Indeed, as Finn (1980, p. 390) notes, "It is unlikely that governmental reorganization, or adoption of a new institutional theory and structure for marine resource management, such as management by a single entity for multiple-use objectives, could achieve similar advantages" found in regular, formal interagency coordinated decision making. Rather than follow this approach, the Reagan administration consistently demonstrated hostility to the existing federal role in the oceans by attempting to sell ocean information services to the private sector and eliminate coordinating mechanisms such as the National Advisory Committee on Oceans and Atmosphere and the Office of Coastal Resource Management.

However, political change in the oceans arena is difficult. Even a popular president failed to carry out his desire to reduce significantly the federal role in oceans management. An enlargement of the federal role is equally difficult. The sources of these difficulties include the following (Walsh 1981):

1. The oceans do not provide a suitable political focus because of their diffuseness, size, and remoteness.
2. Ocean issues are complex and connect the fields of science, technology, economics, and politics.
3. Ocean issues are insidious and develop over time without easy political recognition.
4. Federal ocean personnel include executive and congressional officials who have short terms of office, which voids the continuity needed for long-term effort.

Another source of political difficulty in ocean policy making is the role of the Congress in spearheading ocean policy. As Cicin-Sain and Knecht (1985) note, Congress spawned a host of subcommittees covering a variety of ocean uses, each of which allowed easy access for a variety of interest groups. They point out (p. 296) that

single-issue groups representing different value positions (some favoring development, others conservation) utilized these Congressional multiple points of access and the growing number of Congressional subcommittees to prevail in separate battles. Nearly every group got some legislative authority to either protect or develop its own preferred aspect of the marine environment.

Thus, the serial passage of single-purpose acts legitimized new interest groups within single ocean sectors, and no governance mechanism was created to develop the basis for political choices among these sectors.. "Use-by-use" decision making thwarted a cohesive and holistic approach because each user group became more adept at political manipulation.

Indeed, Cicin-Sain and Knecht (1985, p. 300) go on to note that

Congress piled upon NOAA [National Oceanic and Atmospheric Administration], like a "stack of bologna," management responsibilities for a large number of management and regulatory—not research—programs . . . the major agency charged with implementing a significant portion of the new management-oriented legislation could by no means be considered as a coherent oceans agency, but rather as an amalgam of separate "fiefdoms," populated with veterans from other agencies. . . .

Experience in creating new organizations demonstrates that these entities are often amalgams of other organizations with little loyalty to the goals of the new organization. Time is required to forge a new nongovernmental constituency and a new organizational identity, both of which are required to achieve political clout.

Attempts to change the system must be routed through Congress, where legislative initiatives are sent to a myriad of congressional subcommittees (each with its separate fiefdom) overseeing a particular executive agency and policy arena. This arrangement diverts change and promotes the status quo. Attempting change through the courts cannot achieve a positive direction but can only restrain an agency from acting. Also, the courts lack substantive knowledge of marine issues and tend to favor decisions based on due process or other points of law. Cicin-Sain and Knecht (1985) summarized the difficulties of the courts as follows: (1) the inability of a single judge to grasp the complexities involved; (2) the exclusion of relevant interests from the limited lawsuit; (3) the inability to address potential conflicts; and (4) the resultant delay as the courts are used to address issues for which they were not designed.

Other Levels

Thus, if one looks solely to the federal government to resolve the governance issue for the oceans, disappointment is bound to follow. Other levels of government also exist that have important roles in oceans governance. When one combines the complex and divided federal decision-making system with the similar complexities and divisions at the state level, an even more diverse picture emerges. Beyond the state level are the scores of separate local government entities that have no authority to coordinate beyond their jurisdictions. Finally, the myriad private profit and nonprofit entities pursue their objectives constrained only by their ability to achieve consent from the complex governmental array woven about them.

Table 13.1 lists only representative selections from a great variety of private nonprofit organizations. If the table listed all private organizations using the SCB, it would encompass many more pages. When private agencies are combined with governmental organizations, their number, variety, degrees of authority, and the extent of their overlapping jurisdictions are staggering. Even if one organization could be created and dedicated to govern the SCB, it would find itself competing with all of the organizations listed in table 13.1 and more. Thus, market transactions, administrative decisions, voting, legal decisions, and bargaining agreements among these organizations all combine to allocate uses throughout the SCB, with or without one organization having jurisdiction over the SCB.

Having noted this built-in limitation on governance, we must add that government actually does exercise total (or nearly total) control over uses in the SCB. Figure 13.1 shows the range of governmental authority from least in the onshore to total in the offshore. The geographic scope of this high degree of governmental control includes that part of the SCB which encompasses both the Territorial Sea and the Exclusive Economic Zone. Generally, the federal government exercises dominion over the latter, while the state has exclusive jurisdiction over the former. The intertidal zone includes both state and local government control, while the shoreland (except for that which is federally or state owned) is left under local government jurisdiction. In the upland areas, private ownership becomes dominant, but is constrained by rules emanating from all levels of government.

However, in that part of the SCB over which the federal government exercises total or nearly total control, it does so through a variety of laws, proclamations, and other provisions. As partially shown in figure 13.2, this lineup of rules generates overlapping ju-

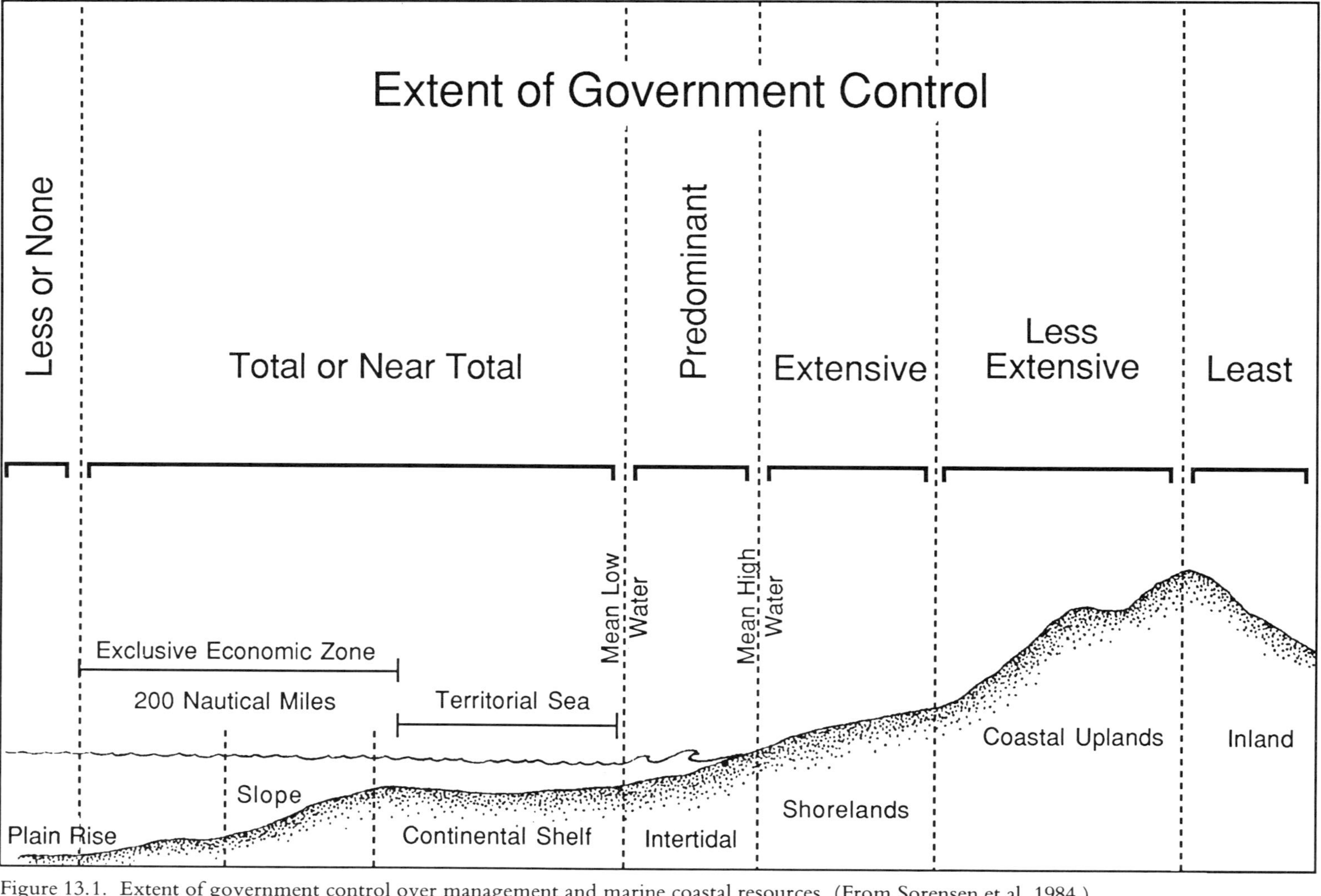

Figure 13.1. Extent of government control over management and marine coastal resources. (From Sorensen et al. 1984.)

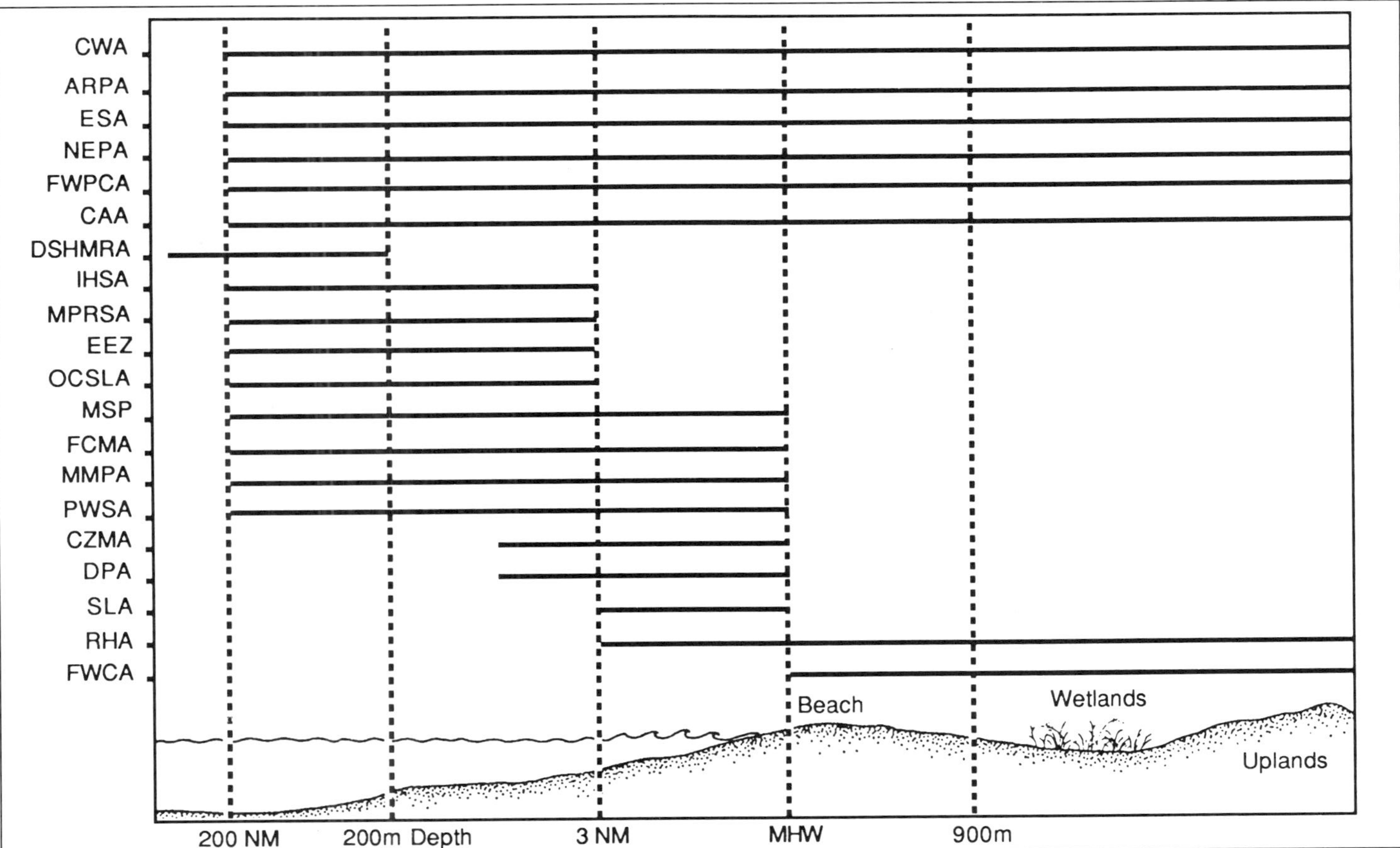

Figure 13.2. Areal coverage of selected federal laws and proclamations in the SCB. Key: ARPA = Archaeological Resources Protection Act; ESA = Endangered Species Act; NEPA = National Environmental Policy Act; FWPCA = Federal Water Pollution Control Act; CAA = Clean Air Act; DSHMRA = Deep Seabed Hard Minerals Resource Act; IHSA = Intervention on the High Seas Act; MPRSA = Marine Protection, Research and Sanctuaries Act; EEZ = Exclusive Economic Zone; OCSLA = Outer Continental Shelf Lands Act; MSP = Marine Sanctuaries Provision; FCMA = Fishery Conservation and Management Act; MMPA = Marine Mammals Protection Act; PWSA = Ports and Waterways Safety Act; CZMA = Coastal Zone Management Act; DPA = Deepwater Ports Act; SLA = Submerged Lands Act; RHA = Rivers and Harbors Act; FWCA = Fish and Wildlife Coordination Act; CWA = Clean Water Act.

risdictions. In addition, a different federal agency is associated with each rule under which it exercises its authority. Thus, the degree of federal authority or jurisdiction in the SCB equals the sum of the number of rules covering its areal extent.

The recent desire to clean up Santa Monica Bay is an example of attempting a more holistic management base by designating the bay as part of the National Estuary Program. This program would affirm a national interest in more effective protection and use of the bay; the hope is that the new rules would be so well coordinated that the bay will be protected. However, the urban watershed surrounding the bay has no direct connection to this federal initiative; therefore the bay will continue to be influenced more by the millions of urbanites than by this one federal program.

Approaches to Governance

Table 13.2 lists selected federal laws and the chief mechanisms used by each to enhance governance in the SCB. The roles of states, other federal agencies, and the public are crucial to federal governance. Indeed, the state of California is to be consulted, to review for sufficiency, to be certified to administer, to plan for federal spending, and to otherwise participate in federal rule making. These formal methods of incorporating state governance in the SCB supplement such procedures as informal coordination and bargaining among federal and state officials.

Each organization holding a stake in the SCB expects some role in the decision process. Thus, each federal agency governing a specific use in this area expects the major role in any conflict surrounding that use. The state of California also has strong economic and environmental interests in the SCB and seeks to determine the outcome of decisions that affect the region. Beyond these central jurisdictions are the governing bodies of coastal cities and counties and numerous regional and special use authorities, each having a perceived share in any conflicts over a proposed decision.

Governance can be achieved in many ways. For example, land ownership allows the Santa Catalina Island Conservancy to set whatever rules it chooses, while others have little recourse to the decision-making process that governs the area. Leasing is another approach; the city of Long Beach, for example, leases shorelands and a water area for occupancy by the *Queen Mary,* and it leases marina slips to private boat users. Direct use or management of recreation occurs when sailing clubs set and enforce sailing rules for their members. Regulation is used when a city zones beach and water areas for certain uses. Planning and coordination integrate recreation and other uses to ensure cohesive use of an area. Citizen participation in decision making is prominent in land use decisions affecting wetlands and beaches. Disgruntled groups can use litigation to block a planned recreation use. Spending is used by government to fund recreation projects and generate information for recreation planning.

Both the National Environmental Policy Act and the Coastal Zone Management Act have been used to affect federal and state governance of the SCB. Each relies on coordination, public participation, spending, and litigation approaches to bring about the desired levels of governance. For example, petroleum use requires the governing federal agency to consult and coordinate this use with the state of California, to share the results of its environmental impact studies with the public, to spend funds to study petroleum impacts, and to defend this use in the federal courts.

Coastal Zone Concept

In an attempt to integrate these various uses and their authorizing laws in the SCB, as elsewhere, the coastal zone concept was introduced and promoted by the federal Coastal Zone Management Act (CZMA) as well as

Table 13.2. *Selected Federal Acts and Their Governance Mechanisms*

Act	Mechanism
National Environmental Policy Act	Interagency, intergovernmental, and public review
Coastal Zone Management Act	State review
Outer Continental Shelf Lands Act	State consultation
Administrative Procedure Act	Public rule-making process
Archaeological Resources Protection Act	Permits
Fish and Wildlife Coordination Act	Interagency preemption
Marine Mammal Protection Act	Prohibition and consultation
Marine Sanctuaries Provision	Public review, interagency state approval
Endangered Species Act	Public listing process and state agreements
Fishery Conservation and Management Act	Foreign agreements and federal–state management plans
Rivers and Harbors Act	Spending and consultation
Federal Water Pollution Control Act	Spending and state certification
Deepwater Port Act	State approval
Clean Air Act	State planning and certification
Ports and Waterways Safety Act	State consultation
Hazardous Liquid Pipeline Safety Act	State certification
Ocean Dumping Act	Preemption

by California's own coastal act. This twin thrust included common acceptance of a prescribed land–water interface as the most appropriate vehicle for coastal governance.

The concept of a coastal zone is a construct often used to summarize the complex ecological relationships found within the land–water interface. As such, it can be described in objective terms that suitably define the ecological systems in a scientific manner on the basis of certain subsystems, including ocean basins, circulatory patterns, longshore circulation cells, estuarine areas, and geomorphic units as well as the human use systems found in the area. Conversely, the coastal zone can be pictured in terms of human uses such as recreation, fishing, housing, and shipping as well as the resources themselves, including fish, wildlife, beaches, surf, and scenery. Planning officials use both "images." More often, however, the coastal zone is viewed as the interface between terrestrial and marine domains where explicit integration of the ecosystems, resources, and uses provides a basis for more informed decision making.

The irony of the Coastal Zone Management Act is found in its set of purposes, which includes the mutually conflicting goals of preservation and development. This conflict precludes use of the act as a basis for choosing between preservation and development goals (Coastal Zone Management Act 1972).

The national Coastal Zone Management Act is based on an attempt to define the coastal zone by its inherent physiographic factors, such as lands, waters, proximity, islands, tidal areas, marshes, wetlands, and beaches. Inland, the definition emphasizes control over uses that physically impact the coastal waters. Surprisingly, federal or nationally owned lands are excluded from the national coastal zone. The state of California defines the coastal zone as ranging from 3 miles seaward to about 1000 yards inland from high tide, landward of any drainage basin flowing into the coastal zone. California includes in its definition all significant coastal resources (natural, man-made, or recreational) and areas where development could directly or cumulatively affect public access to the coast.

The setting of coastal zone boundaries was a highly political process, and the definitions just described reflect this process. However the decisions were made, depending on state and local politics, the national coastal zone became (and still is) a compilation of state-defined zones.

Conceptual Basis

An integrating policy mechanism is one that unifies disparate policies and generates a single, harmonious, mutually supporting outcome. To accomplish this aim, Underdal (1980) has suggested three basic requirements: comprehensiveness, aggregation, and consistency. He defines *comprehensiveness* across several dimensions, but the one that is of particular interest here is the spatial dimension, in which the coastal zone covers all of the consequences recognized as relevant to the decisions at hand. *Aggregation* is defined as the broad coastal zone perspective by which individual use alternatives are assessed, and *consistency* is seen as a coastal area where various uses complement one another. He goes on to note (p. 163) that the purpose of policy integration is to "improve outcomes" through "internalization of externalities."

While Underdal makes clear that, *prima facie,* a set of policies based on narrow interests does not always lead to poor results, he believes that ocean space requires an integrated approach for sound decision making. However, he cautions that marine policy (or an SCB policy) is not by definition more useful than any other construct such as energy, transportation, or minerals policies and that "a common physical characteristic, like 'salt water,' is in itself a poor basis for distinguishing policy areas."

Limitations of the Concept

A coastal zone by itself has significance only as the physical margin between the land and the ocean as influenced by the forces of nature. So long as development is limited strictly to activities that do not adversely affect this marine regimen (such as fishing or beach recreation), there may be no serious conflicts of boundaries used to demarcate a multiple use area. However, when a national perspective drives development (such as bolstering the U.S. economy), the coastal zone soon begins to lose its unique relevance as the basis for governance because the desired impacts and activities have less connection to the ocean. For example, while minerals impacts may affect the marine environment, the issues surrounding minerals development go far beyond coastal considerations to include national security, government revenues, trade balance, and employment.

Another limitation of the coastal zone concept as a basis for planning is the absence within it of effective devices and institutions for raising and settling value, goal, and policy issues. Because coastal zones overlap existing political boundaries, public agency jurisdictions, and private ownership patterns, just as they do economic sectors, they are affected by various separate political bodies that may or may not desire effective coordination. The absence of political and administrative institutions with areal jurisdiction corresponding to a coastal zone spatial arrangement hampers the effectiveness of decision making there. In the case of offshore petroleum development, a virtual ongoing stalemate has occurred, with congressmen, state and local coastal officials, and environmental organizations opting for moratoria on such development rather than participating with the U.S. Department of the Interior in an integrated approach leading to sound multiple use practices (Ciotti 1987).

It has been argued that the ability to focus on a single physical entity and to gather around it a wide group of disciplines will aid materially in the planning process. While theoretically valid, this arrangement has not been workable because of the intransigence of certain groups central to the decision-making process, the cross-cutting economic sectors, and the presence of political jurisdictions that complicate orderly planning and budgeting around one spatial configuration with no political basis. The absence of a constituted government for the SCB makes it difficult to fix responsiveness, control, and accountability throughout the area and to decide among proposed uses when questions of permitting, engineering, maintenance, and financial responsibilities arise.

Summary and Prospectus for Future Research

Governing the SCB can be summarized as an attempt to resolve conflicts. The economic and environmental stakes of the area's many interests propel them to interact with one another, and the process of trying to resolve the myriad conflicts that arise is the essence of governance. Although effective governance is thwarted when conflicts occur, there is no appropriate and agreed upon forum for hearing and resolving these differences. This problem increases the likelihood that any decisions made will be challenged. The area management concept has not served to lessen conflict in the SCB. For example, consultation and coordination have led to litigation and moratoria lasting years rather than to the development of effective decision-making mechanisms. The concern of the governed tends to be related to access to the governors, but even with guaranteed access, accommodation is not always reached.

The prescribed communication process between interests in a conflict situation is at the heart of governance. Federal and state agency representatives are usually technical staff charged with solving a technical problem. Technical questions rest on a set of ethical value judgments about who should govern how. Such basic ethical values are not addressed by the agency's technical experts. Instead, they focus on the logic of the imperative thrust upon them by the source of authority under which they operate. This approach only heightens disagreements between interests operating under separate spheres of authority, especially since laws embody the very language of the conflicting values that are embedded in compromises necessary to winning approval.

Nevertheless, explicitly addressing governance issues is a first step toward resolving them. It is to be expected that newer uses of the SCB must go through several phases on their way to broader acceptance. For example, ocean waste disposal generates public concern, controversy, and opposition. However, as alternatives are studied and safeguards proposed, opposition can be expected to give way to compromise and eventual acceptance. Explicitly dealing with the values embedded in this use, as well as with the technical parameters, can go far in meshing ocean waste disposal with other desired uses of the SCB.

What the SCB lacks is an explicit political arena charged with weighing the conflicting values associated with collective use. Instead, each level of government and each of its agencies acts separately in limited technical spheres with selected aspects of uses that, in concert, impact the SCB. While a regulatory SCB agency may seem an ideal way of resolving conflicts on an area-wide basis, such a quasi-judicial body would compete with the great array of other agencies that have jurisdictional claims over the SCB. In this situation, the opponent to any proposed use merely selects the most appropriate arena conducive to achieving the goal of blocking the use. The more narrow the point of law, the easier to halt a proposed use for which the stakes are high. Regulatory hearings emphasize fact finding and procedural cautions designed to promote acceptance of the decision-making process more than the outcome itself. Thus, fundamental conflicts over ethical values are not weighed in a forum appropriate for exploration of alternatives and trade-offs.

Scientists wishing to improve decision outcomes in the SCB should consider working with their social science counterparts in establishing and promoting parameters of stewardship in the complex arena of shared decision making. As Kelman (1987, p. 16) notes, "Choices about institutional design have an impact on what substantive policies are selected, first of all, because different rules give relative advantages and disadvantages to different participants in the process." For scientists not to be disadvantaged requires a well-coordinated multidisciplinary approach to the issue of institutional complexity because no one group can speak for the SCB.

The lack of published research on the governance of any marine water area is reflected in the paucity of references supporting this chapter. The most frequent research done is on individual resources (e.g., fisheries or oil) rather than on seas or bays, which require an integrated focus from several scientific disciplines. However, if there is one overarching research need it is for holistic, integrated analysis of ocean areas.

It may seem curious that, while all could agree on the usefulness of governance studies, so few are actually done. This phenomenon occurs through lack of an institutional base for such research. University departments are organized by discipline rather than subject, and government agencies are organized by sector or resource rather than area. Research arms of these two types of organization are funded by those who have more incentive to continue extending past effort than to shift to research which, by definition, has strong economic, organizational, and political overtones. Moreover, holistic, integrated research is costly and may be controversial in its conclusions. For example, my book (Fischer 1981) describing and comparing the British and Norwegian approaches to governing the North Sea was opposed by the British.

Nevertheless, a research paradigm can be suggested for the study of governance of any marine area. It would include successive rounds of interviews with all relevant users, regulators, experts, and representatives from all levels of government. The questions needing responses from each group interacting in the area include the following:

1. What is the source of authority for each activity?
2. Is this authority recognized by others?
3. Who is (and who should be) involved in decision making for each activity?
4. What dollar and other values are involved in each activity?
5. What are the goals and objectives of each activity?
6. How is the area-wide problem or conflict defined by each?
7. What information is used by each for decision making?
8. What information flows exist between decision makers for each activity?
9. What alternatives are considered by each?
10. How are decisions made by each?
11. What controls or policy instruments are used to implement and coordinate decisions?
12. What overlaps and gaps exist with other activities?
13. What monitoring is done to determine the "fit" of decisions?
14. What resources (e.g., budget and personnel) and research are available to support each activity?
15. When conflicts between activities arise, how are trade-offs made?
16. What is the role of science in trade-offs?
17. What would be an acceptable forum for area-wide decision making?

These types of questions approach the governance issue for *any* marine area. If all major marine research groups included social scientists for the purpose of linking scientific analysis to governance analysis, one result would be the emergence of a stronger analytical base suitable for the integration of all findings in a truly ecological perspective. In this future, we are all stakeholders.

Literature Cited

Bish, R. L., 1982. *Governing Puget Sound.* Puget Sound Books, Seattle, WA. 137pp.

Bowen, R. E., 1981. The major United States federal government marine organization proposals. In: F. W. Hoole, R. L. Friedheim, and T. M. Hennessy, eds. *Making Ocean Policy: The Politics of Government Organization and Management.* Westview Press, Boulder, CO. pp. 51–70.

Caldwell, L. K., 1984. *International Environmental Policy: Emergence and Dimensions.* Duke Univ. Press, Durham, NC. 367pp.

Cicin-Sain, B., and R. W. Knecht, 1985. The problem of governance of U.S. ocean resources and the new Exclusive Economic Zone. *Ocean Devel. Intl. Law.* 15(3/4):289–320.

Ciotti, P., 1987. Oil war II. *Los Angeles Times Magazine.* 3(42):8–15; 26–29.

Coastal Zone Management Act, 1972. 15 CFR 923–930.

Eckert, R. D., 1979. *The Enclosure of Ocean Resources: Economics and the Law of the Sea.* Hoover Institution Press, Stanford, CA. 390pp.

Finn, D. P., 1980. Interagency relationships in marine resource conflicts: Shore lessons from OCS oil and gas leasing. *Harv. Environ. Law Rev.* 4:359–390.

Fischer, D. W., 1981. *North Sea Oil: The Environmental Interface.* The Univ. of Norway Press, Bergen, Norway. 356pp.

Hardin, G., and J. Baden, 1977. *Managing the Commons.* W. H. Freeman, San Francisco. 294pp.

Hershman, M. J., and N. Evans, 1979. Building support for coastal zone management without an interest group constituency. *Coast. Zone Mgt. J.* 5(3):155–160.

International City Management Association, 1987. *The Municipal Year Book, 1987.* Interagency City Management Association, Washington, D.C.

Kelman, S., 1987. *Making Public Policy: A Hopeful View of American Government.* Basic Books, New York. 332pp.

Sorensen, J. C., S. T. McCreary, M. J. Hershman, 1984. *Institutional Arrangements for Management of Coastal Resources.* Research Planning Institute, Columbia, S.C. 165pp.

Underdal, A., 1980. Integrated marine policy: What? Why? How? *Mar. Policy.* July:159–169.

U.S. Department of Commerce, 1987. Census of governments: Governmental units in 1987. Bureau of Census, U.S. Govt. Printing Office, Washington, D.C.

Walsh, D., 1981. National organization for ocean management: Centralization vs. functionalization. In: F. W. Hoole, R. L. Friedheim, and T. M. Hennessy, eds. *Making Ocean Policy: The Politics of Government Organization and Management.* Westview Press, Boulder, CO. pp. 71–87.

Warren, R., R. L. Bish, L. E. Craine, and M. L. Moss, 1972. Allocating coastal resources: Tradeoff and rationing processes. In: B. H. Ketchum, ed. *The Water's Edge: Critical Problems of the Coastal Zone.* MIT Press, Cambridge, MA. pp. 212–245.

Chapter 14

Ecosystem Interrelationships

Donald W. Hood

Introduction

The Southern California Bight (SCB) is one of the most heavily utilized marine resource areas on earth. This use evolves largely from dependence of some 15 million southern Californians on it for marine recreation, transportation, waste discharge, building materials, and fishery products. It is also used to support broader U.S. national interests in the form of naval bases, international and domestic shipping, and commercial enterprises such as offshore oil and gas development. Oceanographically, the SCB is an exciting place to examine because of its complexity, heavy dependence on broad oceanographic phenomena, and the diverse uses of its resources by humans.

Few places have been studied as intensely as the SCB in attempts to understand how it functions naturally and to determine the impact of human uses on the system. Previous chapters of this book summarized our present knowledge and understanding of the natural science of the SCB, considered the many human inputs (200 identified sources of contaminant inputs), and addressed the status of specific ecosystems that function together to make the SCB a very varied and productive ecosystem. In this chapter, it is the author's purpose to attempt to interrelate the various ecosystems discussed in earlier chapters. This effort involves constructing an energy flow budget for the various components of the system and comparing these findings to those of other well-studied marine ecosystems. In this way, similarities between ecosystems in various parts of the ocean can be shown and those differences that might be unique to the SCB can be identified. This treatment also allows us to apply the generic ecosystem relationships obtained in other regions to the specific ecosystems of the southern California coastal regime.

Previously, few attempts have been made to develop a food web budget for any part of the SCB. One such budget developed for the California Current (Green 1978) emphasized marine mammals; one for Santa Monica Basin (Small et al. 1989) emphasized plankton. Jackson et al. (1989) developed a carbon budget for the upper layer of Santa Monica Bay and San Pedro Basin. Efforts to describe energy transfer budgets for the SCB have been met with difficulty owing to the complexity of the system and the lack of data on the biomass and productivity of biological components. In this book, the authors have synthesized the available information into process-oriented quantitative relationships to the extent possible. The author uses these local data in this chapter, along with relevant generic information from elsewhere, in an attempt to produce a preliminary energy flow budget for the SCB.

Approach

Satellite imagery (Eppley 1986, fig. 1.3), the only truly synoptic method available for estimating phytoplankton biomass, shows highest chlorophyll concentrations in the northern bight and along the coast (<100-m depth). South of Santa Catalina and San Clemente islands and extending north over San Pedro Basin, the concentration of chlorophyll is lower, indicating a different productivity regime for this region. In a more extensive treatment of satellite imagery, Peláez and McGowan (1986) examined 34 months of Coastal Zone Color Scanner (CZCS) data for the California Current that show widely different seasonal patterns in the SCB, but a fairly consistent interannual recurrence of the seasonal patterns. The existence of recurring patterns implies the existence of recurring physical driving forces responsible for their generation.

To recognize that phytoplankton pigments are not a conservative property and to understand the reasons these patterns exist, as well as the factors that change them, requires that considerable detailed interdisciplinary information be obtained. Such information is not yet available. Irregularity in chlorophyll concentration does, however, imply variability in the water properties and flow patterns. The swirls, plumes, and eddylike features observed in the CZCS photographs are a visual manifestation of the complexity of the surface circulation and biological variability found in the SCB. Heterogeneity in the SCB is clearly indicated, and wide variability in the system at all trophic levels is to be expected.

Heterogeneity leads to patchiness and nonequilibrium of the chemical and biological regimes in the SCB. These features, as in other areas, are caused by disturbances in the natural environment and are now regarded as the rule in the natural world rather than the exception, although they vary in degrees of intensity (Lewin 1986; Wolfe and Kjerfve 1986).

If they are significant, disturbances change the stage of succession (species composition) of an ecosystem regardless of whether they are natural or human induced. Disturbance affects both community structure as well as ecosystem processes. The successional path of an ecosystem is altered, hence infrequent or rare events such as El Niño, long-term changes in temperature or water properties, and storms can be large factors in determining the subsequent nature of the system (Odum 1981). In the SCB, which is heavily impacted by human activity, we have an outstanding opportunity to examine what influence we, as humans, have on the natural system and thereby improve our ability to use ocean resources wisely.

Lord Byron (1788–1824) wrote: "Roll on, thou deep and dark blue ocean—roll! Ten thousand fleets sweep over thee in vain; Man marks the earth with ruin—his control stops with the shore." In Byron's time, it was not perceived that humans could disturb the ocean enough to significantly impact its processes. Today, in the SCB and in many other areas of the world, there is concern by many scientists (and near panic by some) that human manipulation and contamination of the environment have exceeded the ocean's capacity to assimilate these impacts, and that such activity may lead to long-term or permanent change in ocean ecosystems. This fear has led to an inherent difficulty in managing ocean resources.

As humans, we tend to seek to preserve natural systems that are destined to change. In fact, we structure systems for monitoring the impact of human activities on the environment to see if change has occurred, then when change does occur, as in Port Valdez, Alaska (Shaw and Hameedi 1988), we are often at a loss to determine cause and effect. It is not a realistic management goal to preserve an ecosystem as it was, but it *is* realistic to attempt to maintain the aggregate anthropogenic disturbance at a level low enough to ensure that natural changes are dominant.

In this chapter, the approach is to examine the flow of energy through the main ecosystems of the SCB as accurately as possible with available data and to evaluate how natural changes would effect this system. Examination of the disturbances caused by humans can then be related to natural disturbances to determine if the SCB is being adversely influenced by human activities and if the natural system prevails in spite of these activities. Comparisons with other well-studied ecosystems of continental shelves are made frequently to give perspective to considerations dealing specifically with the SCB.

Oceanographic Characteristics

The physical features of the SCB were described in chapters 1 and 2. The California Current breaks away from the coast at Point Conception, the northern extremity of the SCB, and its eastern edge flows over Santa Rosa–Cortes Ridge, which lies about 200 km west of San Diego (chap. 2). This movement of water offshore is not unlike that occurring in the Gulf Stream, which originates in the Gulf of Mexico, flows through the Florida Straits, turns east and north, and then swings offshore at Cape Hatteras, North Carolina, as it follows the east coast of the United States. It then passes along the outer edge of the Mid-Atlantic Bight, which encompasses the New York Bight and Georges Bank, two well-studied ecosystems. Off Newfoundland's Grand Banks, the Gulf Stream bends sharply east toward Europe.

As the California Current approaches the Mexican border (32°N latitude), it bifurcates. The eastern branch turns north to form the California Countercurrent, which is the major source water for the SCB. In some ways, this is similar to the flow of the English Channel waters into the southern North Sea, providing the source water for this region and serving to greatly decrease the influence of freshwater input from the highly urbanized (16 cities with populations in excess of 500,000 [Lancelot et al. 1987]) and industrialized countries of western Europe.

The SCB is characterized by many oceanographic features that clearly differentiate it from other continental shelf regions, and yet it has similarities to many well-studied ecosystems with which it can be compared to provide a comprehensive overview of processes important to its function. While the systems chosen for comparison are distributed widely, have different climates, different dominant organisms, and great differences in oceanographic features, they all have been extensively studied in a quantitative way during the past two decades, and they all have features in common that make for interesting and valuable comparisons. A dominant feature, common to all systems discussed, is dependence on the supply of nitrate-N (fixed nitrogen is usually the limiting nutrient to growth) to the euphotic zone, which supports primary production. Consideration of how this nutrient is supplied and the dynamics of its replenishment are fundamental to any ecosystem study. The details of nutrient supply to a given ecosystem are specific, but the processes providing this supply are more generic and, therefore, can be usefully viewed from a wider perspective. Other aspects, including how other systems handle human impacts and ecosystem dynamics, are studied here to provide a broad view that gives support and balance to this analysis. A list of the ecosystems chosen for comparison (or contrast) and an explanation of the rationale used in their choice follow.

Southeastern Bering Sea Shelf. This pristine ecosystem off the west coast of Alaska may be as well understood oceanographically as any in the world. Processes that occur on this broad continental shelf (600 km average width) are probably not unlike those occurring on other shelves (Coachman 1986), but because they occur over longer distances, recognition of the features is easier. This system contains well-established oceanographic fronts that permit the existence of both pelagic and benthic ecosystems on the same shelf. The processes transferring nutrients to the euphotic zone in the summer season are similar to those of the SCB.

Southern North Sea. This region has been an extreme example of anthropogenic influence throughout history. It has been studied intensively by the many nationalities on its borders and has been carefully analyzed to determine the impact of human activity on the natural system. Historically, this region and the connecting English Channel have been areas that exemplify many concepts of how physics, chemistry, and biological processes function and interact in an ecosystem (Cooper 1933). The input of English Channel water from the south, which has different characteristics than that derived from the North Atlantic, somewhat parallels the input from the California Countercurrent to the SCB and the input of waters having other characteristics from the west and north. Many other studies, in particular those concerning nutrient dynamics and particulate transport, are relevant to the SCB.

Southern Benguela Pelagic Ecosystem. The Benguela is one of the four major eastern boundary current regions of the World Ocean. The oceanography of the western coast of South Africa south of about 15°S latitude, like that off California, Peru, and northwest Africa, is dominated by a coastal upwelling system. The SCB does not experience upwelling as intense as in the region north of Point Conception on the California coast, but intermittent upwelling plays a major role in supplying nutrients to this ecosystem. Small shoal fish, mainly anchovies, dominate the pelagic ecosystem in both the Benguela and the SCB.

New York Bight. This region on the eastern border of the United States receives large quantities of freshwater (mainly from the Hudson River) carrying effluents from the principal urban and industrial regions of North America. The major transport of water through the area is from the north to the open ocean (Gulf Stream), where it exits near Cape Hatteras. This system, operating in the western Atlantic, is similar to that in the SCB, except that the flow is opposite and the scale slightly larger.

Georges Bank. Located on the continental shelf east of the New York Bight, this feature appears, superficially at least, to function differently than the SCB. It is not within a major current system, but it is surrounded by an eddy. On the Bank, the water tends to be nonstratified, and a large portion of its nutrient nitrogen is derived from regeneration to ammonium in the sediments. Its primary production is at least twice that of the SCB. It is included here for contrast and also because it has been favored with intense studies that have led to advanced understanding of many ecosystem processes.

Texas–Louisiana Shelf. Broad and shallow, this shelf is heavily influenced by freshwater input from the Mississippi River, which contains a complex mixture of effluents from the heartland of North America. Offshore oil and gas production was first begun in this region by American developers, and the area has sustained its worldwide leadership. While its physical regime is clearly different from that of the SCB, it is chosen here for comparison of the common uses and many similar human impacts it shares with the SCB.

Many features greatly influence the dynamics of the various ecosystems within the SCB and, therefore, energy transfer within and between them. To establish a background for comparison of the SCB ecosystem with others in the world, one needs a comparison of some of the important oceanographic features. A summary of some of the physical characteristics of the previously discussed ecosystems is given in table 14.1.

Physiographic Setting

Off southern California and northern Baja California, continued overriding of the North American plate has produced a much more complex continental slope than is typical of other parts of the world. The margin—designated by Shepard and Emery (1941) as the California Continental Borderland—is wide (up to 300 km) and is composed of ridges and depressions aligned in a northwest–southeast direction. Figure 14.1a compares the bottom topography of a typical cross-section off the coast of southern California with that of a longitudinal section off the Texas Gulf coast. The SCB contains both narrow shelf regions of less than 5 km and broader shelves of 20–40 km, such as at Santa Monica and San Pedro. Most of the world's continental shelves are broad shallow ledges (e.g., Bering Sea, approximately 600 km; off New York Bight, 200 km; Texas Gulf coast, 250 km; North Sea, 600 km) and are typified by the northern Gulf of Mexico (fig. 14.1b).

The California Borderland is a complex of basins (reaching depths of 600–3000 m), rises (at depths of 200–500 m), and islands having a major northwest–southeast trend and a secondary east–west trend (Emery 1960). These features greatly alter the dynamics of the SCB (Peláez and McGowan 1986).

Circulation

The Pacific Ocean bordering the most southwestern coast of the United States (SCB) represents an environment significantly different from that usually associated with an eastern boundary current system. A sudden easterly break in the direction of the coastline at Point Conception causes a decrease in coastal wind stress of more than an order of magnitude in the SCB; thus the effects of wind-forcing mechanisms that dominate north of Point Conception differ significantly from those in the SCB (see chap. 2). The California Current, with a flow of approximately 15 cm s^{-1} and a volume transport of 12×10^6 m^3 s^{-1}, or 12 Sv (Sverdrups) (Pavlova 1966), hugs

Table 14.1. *Some Physical Characteristics of Continental Shelf Ecosystems*

Ecosystem	Area (km^2)	Major Current	Volume (Sv)	Speed ($cm\ s^{-1}$)	Flushing (months)	Salinity Mixed Layer (‰)	Temperature Range (°C)	References
Bering Sea Outer Domain	6×10^4	Bering Slope		1–10	Approx. 3	31.7–32.7	0–10	Coachman (1986)
North Sea Southern	1×10^5	English Channel	0.14	4–5	6–24	33–34 (29–32[a])	6–25	Eisma (1987); Rutgers van der Loeff (1980); Lancelot et al. (1987)
Benguela	4×10^4	Benguela		—	—	34.5–34.9	12–22	Bailey and Chapman (1985)
Mid-Atlantic Bight	8.5×10^4	New England Countercurrent		5–15	3–4	31–32.5	4–17	Walsh (1981b)
Georges Bank	3.3×10^4	Circular Drift		2–8	2–3	32.5–33.3	5–14	Butman and Beardsley (1987); Csanady and Magnell (1987)
Texas–Louisiana	1.5×10^5	Central and West Gulf		11–22	—	32–36.5	15–30	Gallaway (1981)
SCB	3×10^4	California Countercurrent	0.8–1.8	12–18	2–3	33.4–33.6	14–19	See chap. 2

[a] Near German shore, summer and winter.

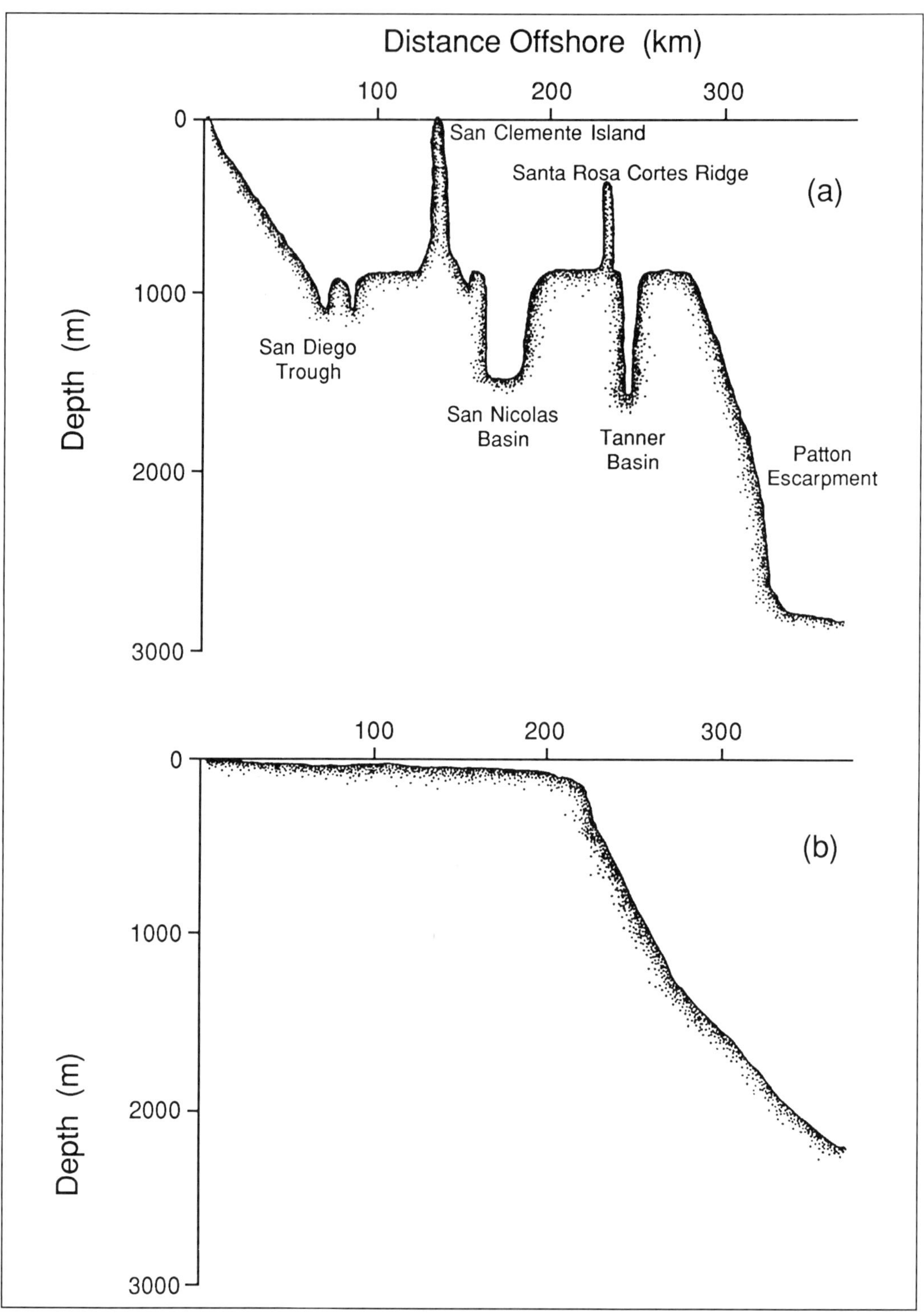

Figure 14.1. (a) Schematic diagram of bottom topography at 33°N off the coast of California (vertical exaggeration, 100×). (b) The same at 94°30′W longitude off the Texas–Gulf of Mexico coast.

the coast north of Point Conception during most of the year. However, from November through February, the coastal winds shift to the southeast, forcing the southern-flowing California Current to move offshore. This produces an inshore surface countercurrent (Davidson Current), which remains until the northwesterly winds resume in the spring (Jones 1971).

Flow in the SCB between and inshore of the Channel Islands, with a speed of 12–18 cm s^{-1} and a volume of 0.8–1.8 Sv, is to the northwest throughout the year, establishing a nearly permanent Southern California Countercurrent or Eddy. Along the coast, over both narrow and broad shelves, there is a persistent equatorward flow, strongest in the winter (approximately 5–10 cm s^{-1}) at a depth of about 30 m. Within the SCB, the speed of surface currents is usually less than 25 cm s^{-1}, although speeds as high as 100 cm s^{-1} have been observed in the shoal waters of the interisland passages. The circulation within the SCB (described in detail in chap. 2) is complex, variable, and undergoes significant year-to-year fluctuations (speeds of 2–4 cm s^{-1}); the greatest differences occur nearshore. The estimated mean residence time of water in the SCB is 2–3 months (Jones 1971) (chap. 2), depending on the seasonal variability of local winds.

Comparison of the flow characteristics of the SCB with the other ecosystems we are examining can be made from data presented in table 14.1. The southeastern Bering Sea has a mean subtidal flow on the order of 5 cm s^{-1} along the shelf toward the northwest. The residence time on the outer shelf for a property introduced near Unimak Island is approximately 3 months (Coachman 1986). The southern North Sea, with an input volume from the English Channel of only 0.14 Sv, flushes every 6–24 months. The current velocities and flushing times of the other systems are about the same as the SCB; that is, a property added near New York Harbor would travel off the shelf at Cape Hatteras in 80–140 days, depending mainly on wind conditions. Similarly, a property placed off San Diego should pass by Point Conception in 60–90 days.

Temperature

The average maximum surface temperature at California Cooperative Oceanic Fisheries Investigations (CalCOFI) stations in the SCB is about 19°C, which occurs in August and September when the surface mixed layer depth is about 10 m. The average lowest surface temperature is about 14°C and occurs from December through March when the mixed layer depth is about 30 m. The 10°C isotherm depth ranges between 80 and 130 m, with the shallowest being in May and June (Eber 1977). Elevation of surface temperatures by 7°–10°C may occur during an El Niño event as a result of the changing strengths of the California Current and the Equatorial Countercurrent (chap. 1).

Surface temperatures of the other ecosystems show greater variability than those in the SCB (table 14.1). The southeastern Bering Sea shelf is much colder (0°–10°C), and the Texas–Louisiana coast is slightly warmer (15°–30°C). High values for the other systems generally fall within the range that is characteristic of the SCB. In the southeastern Bering Sea, mixed layer depths vary widely depending on the frequency and direction of storm tracks (Sambrotto et al. 1986). Annual variation between May 1976 (a cold year) and May 1981 (a warm year) was about 3.5°C. In the North Sea, the average seasonal variation in sea surface temperature for the period 1948–1983 was lowest (6°C) in February and March and highest in August at 15°C. Year-to-year variations are less than 2°C (Colebrook 1985). Georges Bank waters reach a minimum of about 5°C in winter when the waters are isothermal from top to bottom (40–100 m depth). In summer, temperatures may reach 20°C on the eastern edge, but are isothermal at about 15°C over the main part of the Bank. Long-term fluctuations of temperature for shelf waters near Georges Bank have a range of about 2°C and occur on a 10- to 15-year cycle (Flagg 1987). On the northwest Gulf of Mexico shelf, surface temperatures vary from 8–16°C in the winter to 30°C in summer. Winter temperatures show a high correlation with coastal air temperature (Temple et al. 1977).

Salinity

The surface waters of the SCB (40 m inshore; 100 m offshore) are relatively isohaline; salinity peaks in summer (July) at 33.6‰, decreasing to 33.4–33.5‰ in late winter. Gradients are small between the upper layer and a depth of 200 m, where values between 34.0 and 34.2‰ in the nearshore area are found. Most of the stratification of the SCB is derived from temperature differences in the water column.

Salinity of the mixed layer in the SCB is much more stable than in other systems (table 14.1). In the southeastern Bering Sea, from the 50 m (inner front) to 150 m contour (shelf break front), the salinity increases from 31.8 to 33.2‰. The system has three well-defined permanent fronts (three easily distinguishable ecosystems) over a distance of 500 km (Kinder and Coachman 1978).

In winter, the Bering Sea shelf is well mixed to the bottom contour of about 150 m. The coastal domain (<40 m water depth) is usually mixed top to bottom through the year and has a salinity of about 31.8‰. The middle domain (>40 to <100 m) is well mixed to the bottom in the winter, but stratifies with increased insolation in the spring. It is bordered at an inner front with a salinity gradient of 9.3×10^{-3} g kg^{-1} km^{-1} (31.7–32.4‰) and at the outer front by the same gradient (32.4–32.7‰) (Coachman and Charnell 1979). This is a clear example of a system with fronts maintained by different salinity regimes.

The surface salinity in the central part of the North Sea ranges from 34.8 to 35.1‰, and the bottom salinity is 35.0–35.1‰. Salinity along the German shore shows a large range from 29‰ in the summer to 32‰ in the winter. In the southern part of the North Sea, as on Georges Bank, little stratification occurs at any time of the year. The Benguela, Mid-Atlantic Bight, and Texas–Louisiana coastal waters are highly stratified, as is the SCB. The Mid-Atlantic Bight and Texas–Louisiana coast are heavily influenced by freshwater runoff from large rivers. Along the Gulf of Mexico coast (Gallaway 1981), the influence of the Atchafalaya and Mississippi rivers is sometimes seen (salinities of 30–32‰) beyond the 100-m contour as far west as Galveston. Mixing of freshwater has been found as deep as 50 m. In summer, high-salinity water (36‰) flows from the southwest Gulf, forcing the freshwater lense to the Louisiana–Gulf Coast area. It is apparent that freshwater runoff is a dominant factor controlling Gulf Coast ecosystems, which contain the highly prized and commercially valuable penaeid shrimp.

Dissolved Oxygen

Dissolved oxygen in the SCB varies greatly with depth, reflecting the properties of source water coming from the south. In the mixed zone—above 30 m—dissolved oxygen is saturated (5.5–6 ml l^{-1}) throughout the year. At about 100 m (σ_t of 25.8), the values drop to 3.4–3.8 ml l^{-1}; at 200–300 m (σ_t of 26.6), only 1.2–1.4 ml l^{-1} is found. Anaerobic conditions are found at the water–sediment interface in many of the deep basins (Jackson 1986). In open coastal regions, anoxia usually occurs in upwelling situations, such as on the Peruvian shelf, where intense carbon production, sinking of organic matter, and subsequent oxygen demands may overwhelm the oxygen replacement processes (Walsh 1981a). Biological events that cause anoxic conditions are usually associated with dinoflagellate blooms, often referred to as "red tides." Red tides caused extensive fish kills in Santa Monica Bay in 1945 (Somner and Clark 1946), in the North Sea in 1975 (a *Ceratium tripos* bloom) (Reid 1975), and in the New York Bight in 1976 (a bloom of *C. tripos*) (Falkowski et al.

1980). These outbreaks have been attributed to a variety of causes, including domestic pollution (Pra Kash 1975) and natural causes (Stoddard and Walsh 1986).

In the outer domain of the southeast Bering Sea shelf, oxygen concentrations may become as much as 130% saturated (9.15 ml l^{-1}) during the spring bloom in June. Values during the rest of the year to depths of 200 m, deeper than any part of the Bering Sea shelf, rarely fall below 90% saturation (6.8 ml l^{-1}). In depths greater than this (deeper than the shelf break front), low-oxygen water from the oxygen minimum layer of the Bering Sea moves over the shelf break (Codispoti et al. 1986; Niebauer et al. 1981).

The southern North Sea is well mixed to the bottom (50–80 m) most of the year because of the strong tidal action (3-knot surface velocities). Despite the heavy loads of organic matter discharged by the many rivers (Eisma et al. 1982), the open waters remain above 90% saturation (5.1–6.3 ml l^{-1}) throughout the year (Postma 1973). At times, under natural physical forcing and biological responses at climatic time scales, anoxic conditions may occur.

Like other upwelling regions (e.g., off the coasts of Peru, Northwest Africa, and Oregon–California), the South Benguela system in the summer has a strong oxycline associated with the thermocline in the inshore region. Oxygen-deficient water of less than 2 ml l^{-1} is found in water upwelled from >150 m depth, with a temperature of <9.8°C and a σ_t of >26.8.

In the northwestern Gulf of Mexico, the oxygen concentration in the mixed zone is usually near saturation (5.2 ml l^{-1} in winter to 4.5 ml l^{-1} in summer). In 1979, Harper and McKinney (1980) observed hypoxia offshore of the Texas coast in bottom waters 10–30 m deep and greater. Similar hypoxic water (<2 ml l^{-1}) exists along the Louisiana coast between the Mississippi River and Grande Isle, Louisiana. Both conditions are attributed to freshwater runoff with high organic loads and nutrient supplies, which causes stratification of the water column, followed by the input of high particulate loads to the isolated bottom waters.

Unlike the other shelf ecosystems previously discussed, the SCB receives a constant flow of hypoxic water from the south that originates in the oxygen minimum layer of the eastern Pacific Ocean. All waters below 100 m are less than 50% saturated (3 ml l^{-1}) throughout the year.

Nutrients

Physical processes supply nutrients to the euphotic zone and thereby establish the level of primary productivity. The motions of the earth that control incident energy, sea surface temperature, and stability of the water column, coupled with winds, tides, and currents, control the rate of nutrient supply to the euphotic zone, thereby determining the level of productivity a system can support. These processes manifest themselves in different ways in different ecosystems of the world.

The vertical distribution of the nutrients NO_3, SiO_3, PO_4, and O_2 in the SCB is shown in chapter 3. These properties vary widely in concentration from lower values in the southern part of the SCB to higher values in the northern part (fig. 14.2 for NO_3) (see also chap. 3). Ammonium and NO_2 occur almost exclusively in the upper 100 m of the water column. There is almost always an NO_2 maximum, and usually an NH_4 maximum at the nitracline (bottom of the euphotic zone) where the concentration of nitrate rises rapidly—usually starting at the depth of the 1% light level. The O_2 depletion at depth, accompanied by a decline in the NO_3 concentration below the sill depth in deep basins, is indicative of heterotrophic utilization of O_2 followed by anaerobic denitrification. The horizontal distribution of the nutrient isopleths shows increases in concentrations shoreward from 25 km (see chap. 3), probably resulting from shoaling of deeper nearshore nutrient-rich water (Mullin 1986).

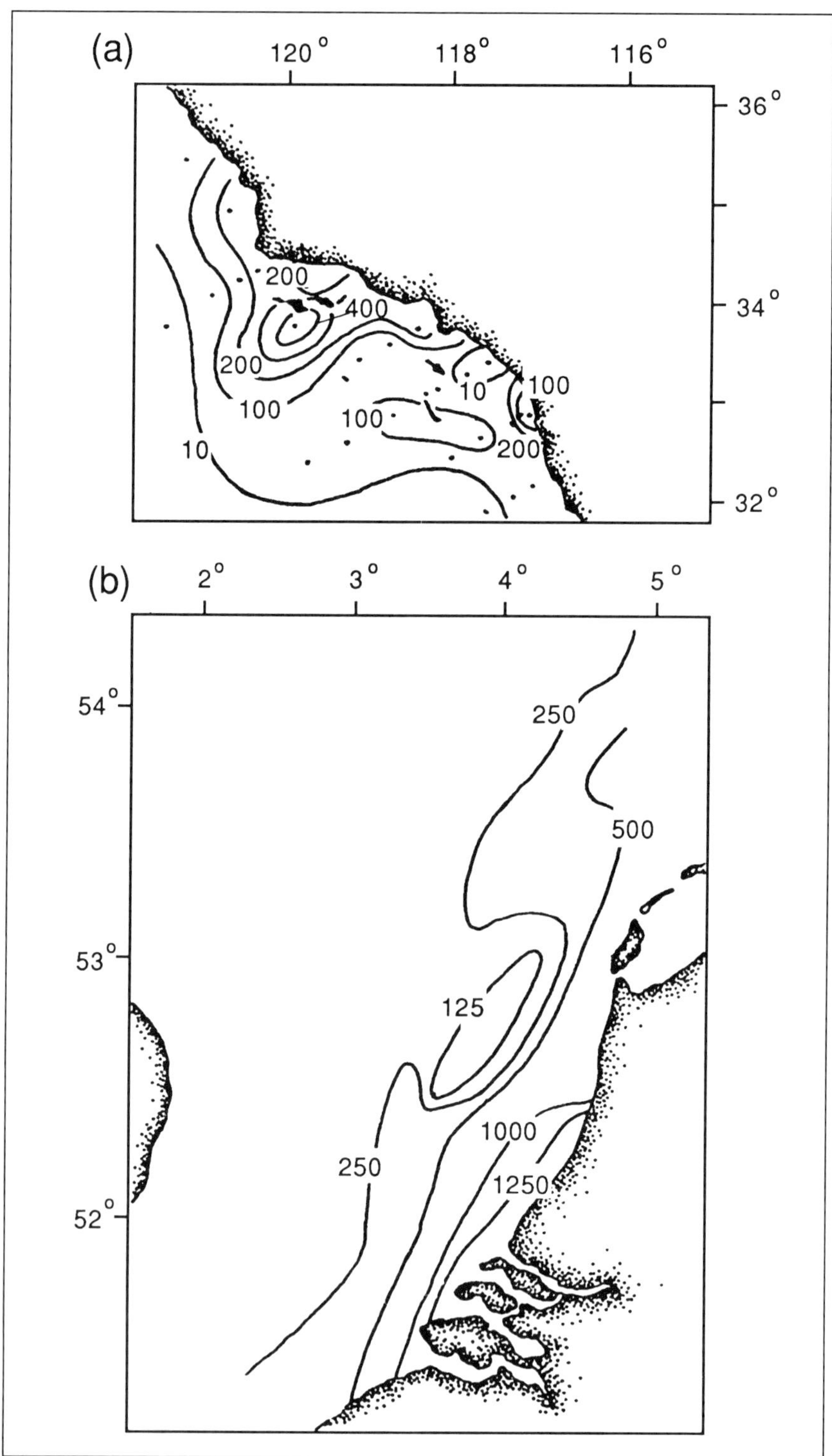

Figure 14.2. Nitrate-N concentrations. (a) SCB, integrated from 0–50 m (CalCOFI Cruise 6902, February–March 1968). (b) Southern North Sea, integrated from 0–25 m. (From Postma 1973.)

The concentration of NO_3 in the mixed zone (20–25 m depth of homogeneous surface properties) is near zero, indicating that this nutrient is utilized by phytoplankton in photosynthesis to the extent it is available (see chap. 3). Any additional productivity would have to rely on the availability of other forms of nitrogen (NH_4, NO_2, and remineralized organic N). In this region, the euphotic zone (that depth region where sunlight is 1% of the surface value) is a two-layered system with a nutrient-poor, light-rich area in the mixed layer and a light-limited, nutrient-rich area at the bottom of the euphotic zone. The top of the nitracline, the depths of rapidly changing NO_3 concentrations, constitutes the boundary between these layers.

The physical processes that bring NO_3 to the euphotic zone are the driving forces for primary production, along with seasonal changes in insolation and stability of the water column. The regenerative processes that lead to primary production from NH_4 are in proportion to the production resulting from the utilization of nitrate (Eppley et al. 1979). Dugdale and Goering (1967) coined the term *new production* to represent that production resulting from nitrate utilization (see chap. 5).

The euphotic zone is the region for production of particulate matter through the growth of phytoplankton and zooplankton and for production of fecal pellets that occurs in this zone. Losses arise from heterotrophic consumption and decomposition and particle sinking. In theory, for a state of equilibrium to exist, the nitrogen lost through particle sinking must balance "new production" from nitrate transferred from deep water. Data from the Santa Monica and San Pedro basins (Jackson et al. 1989) indicate a particle loss from both the upper and lower euphotic zone of 216 mg C m^{-2} d^{-1}, which contained 28 mg N m^{-2} d^{-1}. This compares with measured new production for phytoplankton of 33 mg N m^{-2} d^{-1}. This production rate requires a vertical transport of 10–12 t NO_3-N km^{-2} yr^{-1} to the euphotic zone in the Santa Monica–San Pedro basin area. Such transfer would require, on the average, a vertical transport rate of about 1×10^{-4} cm s^{-1}, assuming that the water is transported from a 200-m depth containing a concentration of 25 μM NO_3. If we assume that this transport rate is representative of the SCB (7.8×10^4 km^2), then 7.8–9.4×10^5 t NO_3-N would be transferred from deeper waters to the euphotic zone each year.

Sholkovitz and Gieskes (1971) computed an eddy diffusion coefficient of 3.9–5.8 cm^2 s^{-1}, which agrees well with the measured value of Berelson et al. (1982) of 3.4 cm^2 s^{-1}, based on ^{222}Rn experiments. With a gradient in NO_3 of 6 μM between the depths of 32 and 42 m (see chap. 3), the vertical transport to the euphotic zone would be about 2.4×10^{-5} μM l^{-1} s^{-1} (using 4 cm^2 s^{-1} diffusivity coefficient) or 8.7×10^2 t NO_3-N km^{-2} d^{-1}. This amounts to a transfer of 9.4×10^5 t NO_3-N to the euphotic zone in the SCB (7.8×10^4 km^2) each year. The remarkable agreement between these two estimates of vertical transport may be fortuitous, but it is satisfying since so many rough estimates were used in the calculations.

Ammonium and urea (0–0.4 μM each) are two other forms of nitrogen used extensively by phytoplankton in the SCB (McCarthy 1972). These compounds are not commonly found in deep ocean water, but are the metabolic products of organisms. They are transient, thermodynamically unstable forms that recycle from heterotrophs to phytoplankton in time scales of minutes to a few days (Azam 1986) (see chap. 4). Ammonium and NO_2 occur almost exclusively in the upper 100 m of the water column (see chap. 3) at concentrations of 0–1 μM. Urea shows no correlation with NH_4 except when blue shark infestations occur. While other marine animals excrete urea (anchovies and zooplankton) (McCarthy and Whitledge 1972), the blue shark appears to be dominant in the SCB (Williams 1986). A NO_2 maximum coinciding with the nitracline is nearly always present in the SCB. This maximum is thought to result from autotrophic oxidation of NH_4 by NH_4-oxidizing bacteria (Olson 1981).

The rates of utilization and regeneration of

nitrogen by the biological communities in the SCB have been examined extensively (Eppley and Holm-Hansen 1986). An estimated two-thirds of the nitrogen demand is supplied by recycled nitrogen, which is derived from ammonium and urea produced by animal and bacteria metabolism in the euphotic zone, and about one-third from NO_3.

There is an additional significant input of NH_4 to the deeper waters of the SCB from the mass emissions of seven municipal outfalls amounting to an aggregated total of about 4.5×10^4 t yr^{-1}. On a bight-wide basis, this represents an equivalent of about 5% of the NO_3 and 2.5% of the NH_4 utilized in photosynthesis. It is, however, improbable that much of this allochthonous NH_4 reaches the euphotic zone. High near surface concentrations have occurred near outfalls (Thomas and Carsola 1980) (e.g., 155μM at a depth of 27 m, 2 km from the White Point outfall), but these invasions of the surface are localized and rarely observed.

The complexity of the physical processes of the SCB (see chap. 2), caused by intermittent upwelling, variable wind patterns, tidal mixing, and reversing main currents, makes predicting the concentration of chemical parameters in the mixing zone very difficult. However, as indicated in figure 14.3 (Reid 1965), the concentration of PO_4-P in the mixed zone, on the average, has a consistent annual cycle. The wide variation in values for phosphate, and probably other nutrients as well, may be attributed to the presence of the coastal countercurrent (low nutrients) along with periods of upwelling (rich in nutrients), year-to-year variations, and consumption by plants.

By comparison, the southeastern Bering Sea shelf becomes mixed top to bottom during the winter and begins the spring bloom period with NO_3 concentrations between 15 and 20 μM. After the establishment of a mixed layer and increased insolation in April, the surface concentrations are reduced to near zero by the beginning of June (Codispoti et al. 1986). Because of seasonal NO_3 depletion on the shelf, the eddy diffusion flux increases the NO_3 content of the impoverished euphotic zone during the spring bloom (Walsh and McRoy 1986). This nitrate is supplied from offshore flux in the 80–100 m depth across the shelf break front and enters the mixed zone by vertical transport. The observed biological uptake rate in the mixed layer of the outer domain was approximately 0.7 μM NO_3 d^{-1} during April and May of 1979–1981. At the end of May, transition to postbloom water conditions occurred, and productivity depended upon resupply of nutrients from deep water and regeneration in the mixed layer, a process essentially the same as that for the SCB throughout the year.

Coachman and Walsh (1981) used microlayer fine structure and salinity gradients of the southeast Bering Sea shelf to estimate a vertical diffusivity for the outer domain of 10^{-1} cm^2 s^{-1}. Using this eddy diffusivity rate and a gradient of 16 μg-at l^{-1} over a 10-m layer beneath the mixed zone or 16×10^{-7} μg-at l^{-1} s^{-1}, one obtains an accumulation rate of 0.14×10^{-2} μM l^{-1} d^{-1}, or about 35.0 g m^{-2} yr^{-1} of NO_3-N to the mixed layer over the 180-day growing period in the Bering Sea. This is about the same order of supply (33.12 g m^{-2} yr^{-1}) required to support the biological uptake rate of 0.7 μM NO_3 l^{-1} d^{-1} obtained by direct $^{15}NO_3$ uptake experiments (Sambrotto and Goering 1979). In the outer shelf domain of the southeast Bering Sea, high concentrations of NH_4 occur at approximately 30 m, or where the density gradient is greatest and light intensity is 0.5% of that at the surface. Nutrient data collected by the PROBES (Processes and Resources of the Bering Sea Shelf) program from 1978 to 1982 (Whitledge et al. 1986) indicated that NH_4 concentrations are inversely correlated with NO_3 concentrations. Thus, when NO_3 concentrations are high in the surface layer, NH_4 concentrations in the subsurface layers are low. Nitrate is the basic source of nitrogen for phytoplankton growth and is also the source of NH_4 through regeneration. The loss of NO_3 from the euphotic zone due to phyto-

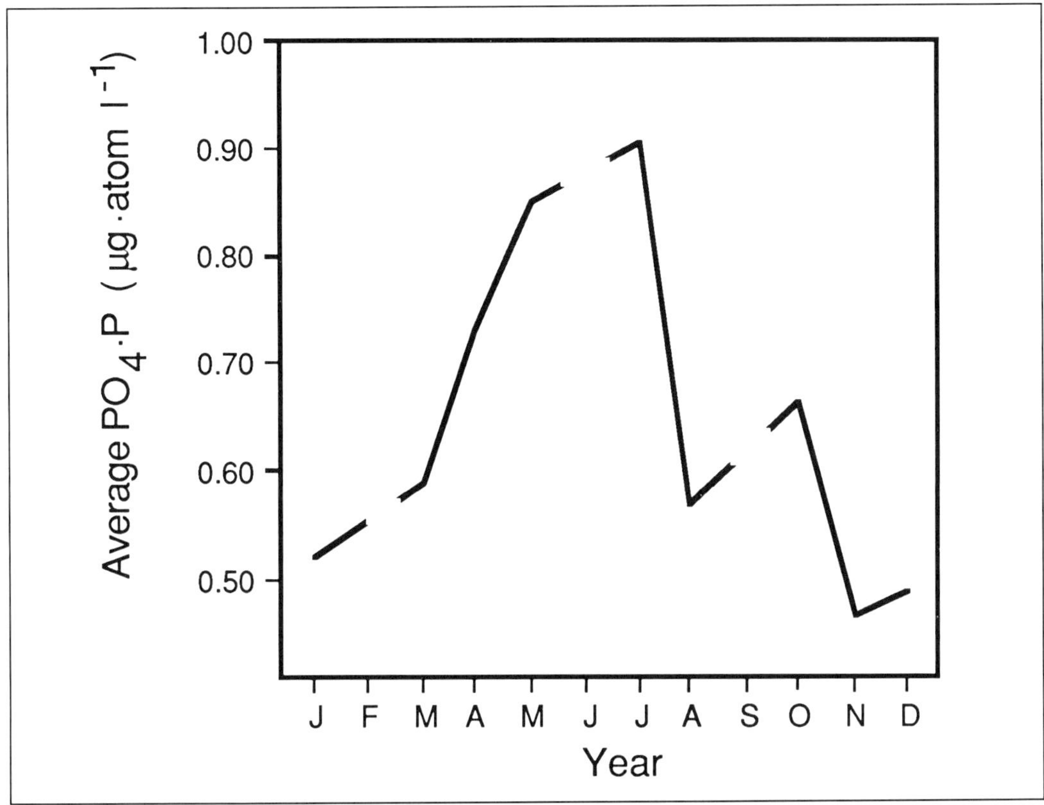

Figure 14.3. Average phosphate-P (μg·atoms l^{-1}) at sea surface 29 km southwest of Point Arguello. (Data collected in various years during the period from 1949 to 1964; Reid 1965.)

plankton growth (largely diatoms) leads to severe changes in the ambient concentrations of the forms of nitrogen-containing nutrients. For example, NH_4 and urea increase, resulting in changes in phytoplankton composition and growth rates (Goering and Iverson 1979). The ratio of NO_3 to NH_4 use varies from very high (>10) during the bloom to approximately 1 after the bloom during the months of June to August (Sambrotto and Goering 1979).

The pattern of nutrient dynamics in the southern part of the North Sea is similar to that in the outer domain of the southeastern shelf of the Bering Sea, except for the high nutrient enrichment from the influx of river water near the coasts of Belgium, the Netherlands, and Great Britain (fig. 14.2). Since 1950, the NO_3 content of the Rhine River has increased from 50 $\mu M\ l^{-1}$ to >300 $\mu M\ l^{-1}$ in 1980 (Van Bennekom and Salamons 1981), and other rivers emptying into the lower North Sea Bight have increased similarly. Outside the coastal areas, phytoplankton growth is limited by depletion of nutrients from the euphotic zone, a result of the slowing of resupply by the seasonal thermocline. The shortage is not totally replaced in the winter because most of the incoming water is from shallow sources and the residence time is too long for complete renewal from nutrient-rich North Atlantic water during the winter season. In the North Sea ecosystem, a very important part of nutrient cycling occurs in the benthic phase rather than in the water column (Billen and Lancelot 1988). It has been estimated that 30–50% of the required nitrogen needed to support phytoplankton production in the offshore southern North Sea Bight is supplied from the sediments. This process appears to be more important in low-energy, shallow coastal waters than it is in the deeper waters of the continental shelves.

Eutrophication caused by anthropogenic inputs (exemplified by the southern North Sea Bight) is also readily observed on the coast of the United States at the mouth of major rivers such as the Hudson and Mississippi. Pristine river content is estimated to be 5–10 μM N, but the Hudson River content is now >50 μM N and the Mississippi is >100 μM N. These high concentrations are not seen north of the Hudson Canyon or on Georges Bank (Walsh et al. 1987a), or east of the Mississippi River, which is a clear indication of the influence of circulation on the source of nutrient supply. (Major flow direction is south along the Mid-Atlantic Bight and to the west along the Louisiana coast.) The mixing regimes on Georges Bank and the Mid-Atlantic Bight are quite different: Georges Bank is well mixed and the Mid-Atlantic Bight is stratified. Between April and November, the entire water column of Georges Bank has an NO_3 concentration of 1μM or less, while the New York Bight is <1 μM in the euphotic zone but 3–5 μM at a depth of 40 m.

Influx of nutrients to Georges Bank depends upon exchange through the fronts surrounding this feature and the regeneration of ammonium in sediments and the deeper water column. In winter, when the water column is isothermal, concentrations of 3.5–5.0 μM NO_3 (present both on the Mid-Atlantic Bight and on Georges Bank) are depleted to a depth of 30 m by April on Georges Bank. Assuming a C:N atomic ratio of 5.3 in plankton tissue (based on plant tissue composition with C:N:P ratio of 80:15:1), a stock of 5 μM NO_3 in the water column would provide for approximately 22 g C of the annual primary production. The remaining requirements (approximately 65 g N m^{-2} yr^{-1}) must be derived from vertical and horizontal transport of nitrogen to the banks.

Because of the vigorous tidal mixing on Georges Bank (tidal velocities of 50–110 cm s^{-1}), Falkowski (1983) estimated the effective rate of vertical mixing by tides to be 1 cm s^{-1}, or 10^4 times the vertical mixing rate in the SCB (1×10^{-4} s^{-1}). Unlike the SCB, where an abundant supply of nutrients is available below the euphotic zone, the shallow Bank (60–80 m depth) depends on regeneration in the sediments and the water column and cross-shelf transport. Between July and September, 90% of the nitrogen-containing nutrients on the Bank are in the form of NH_4. Regeneration accounts for approximately 48.4 g N m^{-2} yr^{-1} on Georges Bank (as compared to 32.9 g N m^{-2} yr^{-1} for the area off New York) (Walsh et al. 1987a), which supports a production rate of approximately 293 g C m^{-2} yr^{-1}.

The dynamics of some environmental parameters that support the energetics of several continental shelf ecosystems is presented in table 14.2. The relationship between the NO_3-N plus NH_4-N supply to the euphotic zone and primary production (to be discussed later) is of particular interest in summarizing this discussion on nitrogen supply. In the other systems examined, the ratio of primary production to nitrogen supply to the euphotic zone through transport of NO_3 and NH_4 ranges between 5.3–8.7. The reasons for the high ratio of 11.8 found in the SCB are not clear, and the difference, while interesting, may not indicate radical differences in the system dynamics because of the difficulty of making these measurements. Ammonium supply to the euphotic zone from below the pycnocline, however, is a significant source of nitrogen, in nonstratified systems (e.g., the southern North Sea, Georges Bank, and many shallow coastal systems). Mineralization in the sediments and deeper water column can become an important source of nitrogen. In the SCB, vertical transport of NH_4 to the euphotic zone does not appear to be an important factor in primary production. Much of the data presented in table 14.2 are rough approximations because rates of vertical transfer are estimates of the average rates of widely differing oceanographic conditions. Also, concentration gradients are averages of a changing hydrographic structure, and primary production data are averages of rates obtained from patchy phytoplankton populations.

Particulates

Particles have long been recognized as very important in the exchange processes that move materials through the water column and eventually into the sediments. Fine-grained organic particles are the primary carriers of heavy metals and other xenobiotic compounds in the sea. It is in this group of materials that many of the by-products of human activity occur. As recently as 1974, Menzel (1974) wrote "particularly lacking is information on the rates of input, utilization, and decomposition of organic matter to and in the deep sea [as well as] to [what] extent . . . these processes influence the chemistry of sea water." The steady increase of knowledge of mid-water communities has increased the awareness of the importance of the rain of detrital material in providing energy to the pelagic, benthypelagic, and benthic communities (Angel and Baker 1982). This led to an extensive expansion in analysis of particulate matter, made possible by expanding collection methods to include direct pumping (Bishop and Edmond 1976) and sediment traps (Knauer and Martin 1981; Fellows et al. 1981) as well as the classic bottle technique. Much information has now emerged on the fluxes of detrital organic matter through water column ecosystems (Angel 1984; Williams 1986) (see also chap. 3), and it now appears that rapidly sinking particles, forms not sampled by techniques used in earlier analyses, account for most of the flow of energy from the euphotic zone to deeper layers of the water column and benthos.

The distribution of the various organic and inorganic materials in the water column is usually described in terms of three categories: dissolved particulate, suspended particulate, and sinking particulate. In practice, these three categories of materials are separated by size schemes. *Dissolved* is that portion which passes through a porous filter, with a pore size of 10–100 nm. It includes colloidal material but attempts to exclude living forms (picoplankton and bacteria). The particulates separated by this technique include both organic and inorganic constituents present in surface and deep waters. Because we deal here with relatively small parcels of water, the residue collected is composed mostly of suspended particulates.

SUSPENDED PARTICULATES

The materials composing this fraction, which consists of clay minerals, organic detritus, and most of the phytoplankton, bacteria, and other biota small enough to be buoyant, sink at a rate of <1 m d^{-1} (McCave 1975). The organic constituents, which include particulate organic carbon (POC), particulate organic nitrogen (PON), and particulate organic phosphorus (POP), have been studied extensively in the SCB. A typical profile for POC and PON for Santa Monica Basin (and probably elsewhere in the SCB) is shown in figure 14.4. The ratio (by atoms) of C:N:P is 100:5:1.1 near the surface and falls to 100:6.8:0.75 at depth, indicating the preferential utilization and recycling of the less refractory organic matter in the surface waters. The residence time of POC in the surface waters of the SCB varies from 3 to over 100 days between seasons and years (Eppley et al. 1983) and is dependent on the amount of new production, the recycling rate within the euphotic zone, and the sinking flux of the POC. The recycling time of POC in the SCB is about 15 days under average conditions. The short residence times are associated with upwelling and the long residence times with oligotrophic surface waters derived from the central Pacific.

The inorganic composition of the suspended particulate fraction collected from the

Table 14.2. *The Rates of Some Processes in the Euphotic Zones of Several Continental Shelf Ecosystems*

	Bering Sea (Outer Domain)	North Sea	Benguela (Southern)	Mid-Atlantic Bight	Georges Bank	Peru	Oregon	SCB
Range in temperature (°C) (mixed layer)	10/yr	19/yr	14/yr	20/yr	10/yr	5/yr	5/yr	—
Nitrate supply to euphotic zone (NO_3–N m^{-2} yr^{-1})	38.5 g[a]	? g[b]	120 g[c]	23 g[d]	35 g[e]	98 g[f]	32.7 g[g]	12 g[h]
Primary production (g C m^{-2} yr^{-1})	200	200	1000	200	293	750	192	142
Particulate matter removed (g m^{-2} yr^{-1})	—	—	—	150	200	—	—	90
Ammonium addition by vertical transport (NH_4–N m^{-2} yr^{-1})	0	15.8 g[i]	0	0	17.4 g[j]	0	0	0
Ratio of primary production to N supply to euphotic zone	5.2	—	8.3	8.7	5.6	7.6	5.9	11.8
Fisheries yield t km^{-2} yr^{-1}	15–20	4.7[k]	35[l]	10	30	100	10	4.5

[a] March concentration (20 μM) plus that accumulated by vertical diffusivity (K_z of 10^{-1} cm^2 s^{-1} for 180-day growing season (Coachman and Walsh 1981).

[b] Not definable because of uncertain lateral inputs.

[c] Upwelling of 3 m d^{-1} of subthermal water containing 25 μM NO_3 l^{-1} for 120 days per year (Bailey and Chapman 1985).

[d] Vertical transport of 0.3 m d^{-1}; source water of 15 μM NO_3 at 100-m depth (Csanady and Hamilton 1987).

[e] K_z of 5 cm^2 s^{-1}; a horizontal coefficient (K_x) of 3×10^6 cm^2 s^{-1}; gradient of 0.2 μM NO_3 over 10-m depth for 150 days and gradient of 0.04 μM NO_3 for 180 days (Walsh et al. 1987a).

[f] Walsh 1981b.

[g] Vertical transport of 3×10^{-2} cm s^{-1}. Source water of 25 μM NO_3 at 200-m depth.

[h] K_z of 3.4 cm^2 s^{-1} and a 7-μM gradient of NO_3 from 30 to 40 m depth. Water column concentration above 30 m was estimated at zero (chap. 3).

[i] A flux of 9.9 nM m^{-2} s^{-1} of NH_4 from the sediment into unstratified water (Rutgers van der Loeff 1980).

[j] Contribution by regeneration in nonstratified water column: 4.8 g N m^{-2} for benthos; 4.3 g N m^{-2} for zooplankton; 8.3 g N m^{-2} for microplankton (Walsh et al. 1987a).

[k] Cohen and Grosslein (1987).

[l] Shannon and Field (1985).

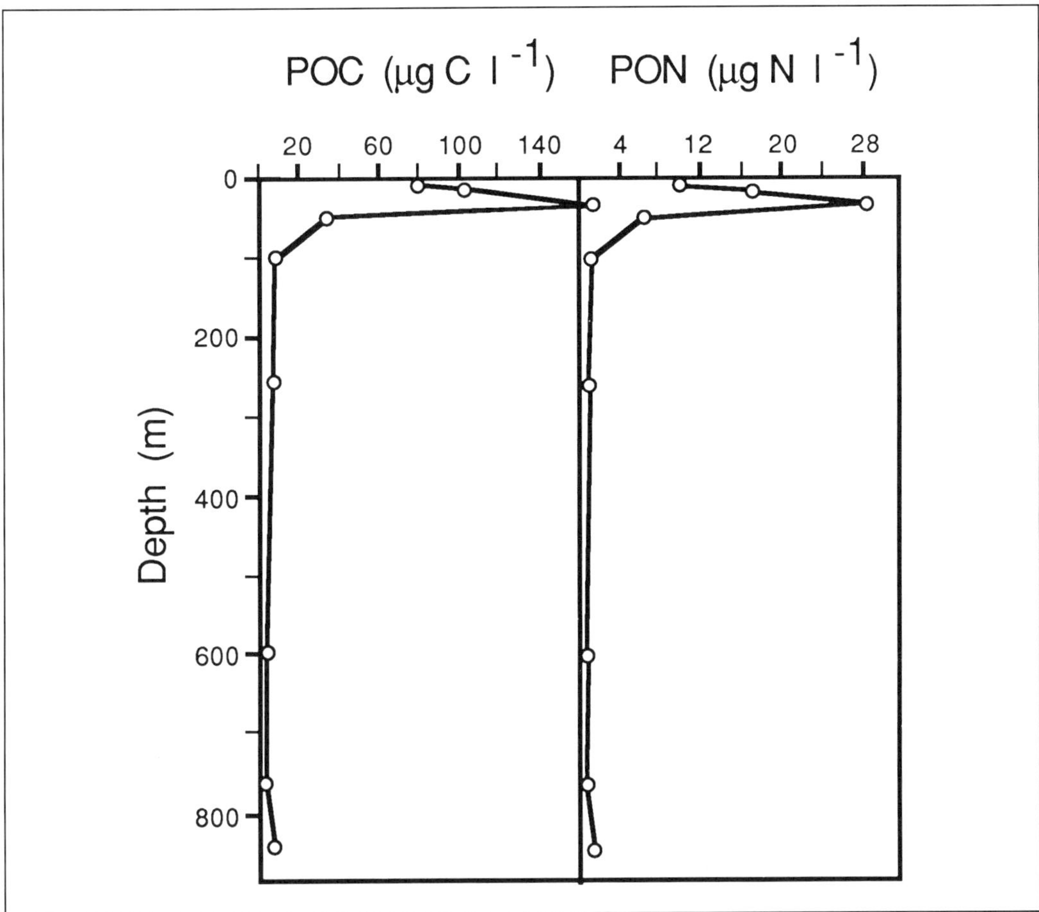

Figure 14.4. Profiles of particulate organic carbon (POC) and nitrogen (PON) in Santa Monica Basin. (From Williams 1986.)

SCB has been examined in detail by Bruland et al. (1974) for the heavy metals Mn, Fe, Ni, Cr, Zn, Co, V, Cu, and Pb and by Shiller (1982) for Mg, Al, Si, P, Cl, K, Cu, Ti, Mn, and Fe to determine the relative importance of terrigenous and biogenic components in sediment sources. These authors found that the concentration levels of most of these metals were controlled by the natural variation in material sources and transport processes and that only occasionally (near sewage outfalls) could high values be attributed to anthropogenic contributions.

The Bering Sea is apparently one of the most particle-rich areas of the world oceans. Loder (1971) found POC values as high as 700 μg l^{-1} in surface waters (<20 m), which decreased to 200 μg l^{-1} at the depth of the 1% light level and further decreased to the relatively high value of 20 μg l^{-1} at the bottom of the water column (500 m) in the Unimak Pass area.

Since many studies do not clearly distinguish between suspended and sinking particulate matter and since recent data emphasize the sinking fraction, further discussion of the particulate fraction is included under sinking particulates.

SINKING PARTICULATES

The "sinking" particulates are those materials collected by sediment traps deployed from the surface to points near the bottom for various time periods. It is the materials included in this fraction that are largely responsible for mass flux from the surface to deeper waters and the sediments. Included in the material of the SCB are cellular and amorphous organic matter, fecal pellets, clay, quartz, opaline shells, and calcareous and siliceous tests (Williams 1986). The sinking rate of material caught in sediment traps averages 100–350 m d^{-1} (Lorenzen et al. 1983) in the subarctic Pacific. In the Santa Monica Basin it is estimated to be 100–200 m d^{-1} (Williams 1986). The total mass flux in San Pedro Basin, measured at two different times in May 1983 with sediment traps deployed at 100, 300, and 500 m, gave values between 100 and 400 mg m^{-2} d^{-1}, depending on time and depth. Between the two deployments, there was evidence for lateral movement of sinking particulates within the water column that supplemented the flux from the surface waters. The C:N ratio varied from 7.3 to 9.5 in the trapped material, and the organic flux represented about 5–6% of the euphotic zone productivity (Williams 1986). This rapid loss of daily fixed carbon is further evidence for the importance of organic carbon recycling and utilization in the euphotic zone. The relationship between the production of organic matter in the euphotic zone and its decomposition and fractionation during sinking is discussed in detail in chapter 3.

The California Basin Study (CaBS), designed to evaluate the processes that control cycling and vertical transport of energy-related materials off the southern California coast (Jackson et al. 1989), has emphasized particle transport and transformation, particularly in the San Pedro and Santa Monica basins. The three purposes of the CaBS program are the determination of (1) regional currents, (2) carbon and nitrogen fluxes through phytoplankton and zooplankton in the euphotic zone and their downward export, and (3) inorganic and organic fluxes into and out of the sediments, as well as characterization of compounds and elements that serve as tracers for terrestrial and anthropogenic inputs. Using diverse and comprehensive techniques, including sediment traps, the CaBS group has been able to provide reasonable mass balances for the movement of carbon through the study area. Results for the euphotic zone—that surface layer illuminated by more than 1% of the incident radiant energy—are given in figure 14.5. These data clearly show that the two-layered system (discussed in the section on nutrients in this chapter) exists in the SCB. Of the particle biomass found in the mixed layer, 53% is bacterial whereas only 31% is composed of phytoplankton. The bacteria mainly make use of dissolved organic carbon (DOC), which appears to be derived from the particulate biomass, at a rate of nearly one-half of the primary productivity. In the deeper water column, the total flux of sinking particulate matter of about 500 mg m^{-2} d^{-1} is composed of about 50 mg C m^{-2} d^{-1} and between 5 and 10 mg N m^{-2} d^{-1}. The rest is inorganic material, which includes siliciclastics from land areas, mineral matter from biogenic processes, and anthropogenic materials from the highly urbanized megalopolis of Los Angeles and San Diego.

Studies of suspended particulate matter in the Bering Sea are more limited than those in the SCB, and few sediment trap data are available. Handa and Tanoue (1981) carefully examined the total particulate load in the water column on a section between Nome, Alaska, and the western Aleutian Islands and related it to the percentage of carbon present (fig. 14.6). In the outer domain, near the Pribilof Islands, the carbon content was 20–25% of the total (36–45% organic matter; factor of 1.8 used to convert carbon to organic matter) (Gordon 1970). Feely et al. (1981) found a particulate concentration of <1 mg l^{-1} for the outer domain containing 35 ± 19% C and a C:N ration of 7.4. The flux rate of particulate matter through the water column has not been evaluated for the eastern Bering Sea shelf. However, on the basis of measurements of the distribution of suspended matter, Baker (1981) estimated a vertical velocity of 10^{-2}–10^{-4} cm s^{-1} (8.64 m d^{-1}–8.6 cm d^{-1}).

The North Sea, specifically its southern

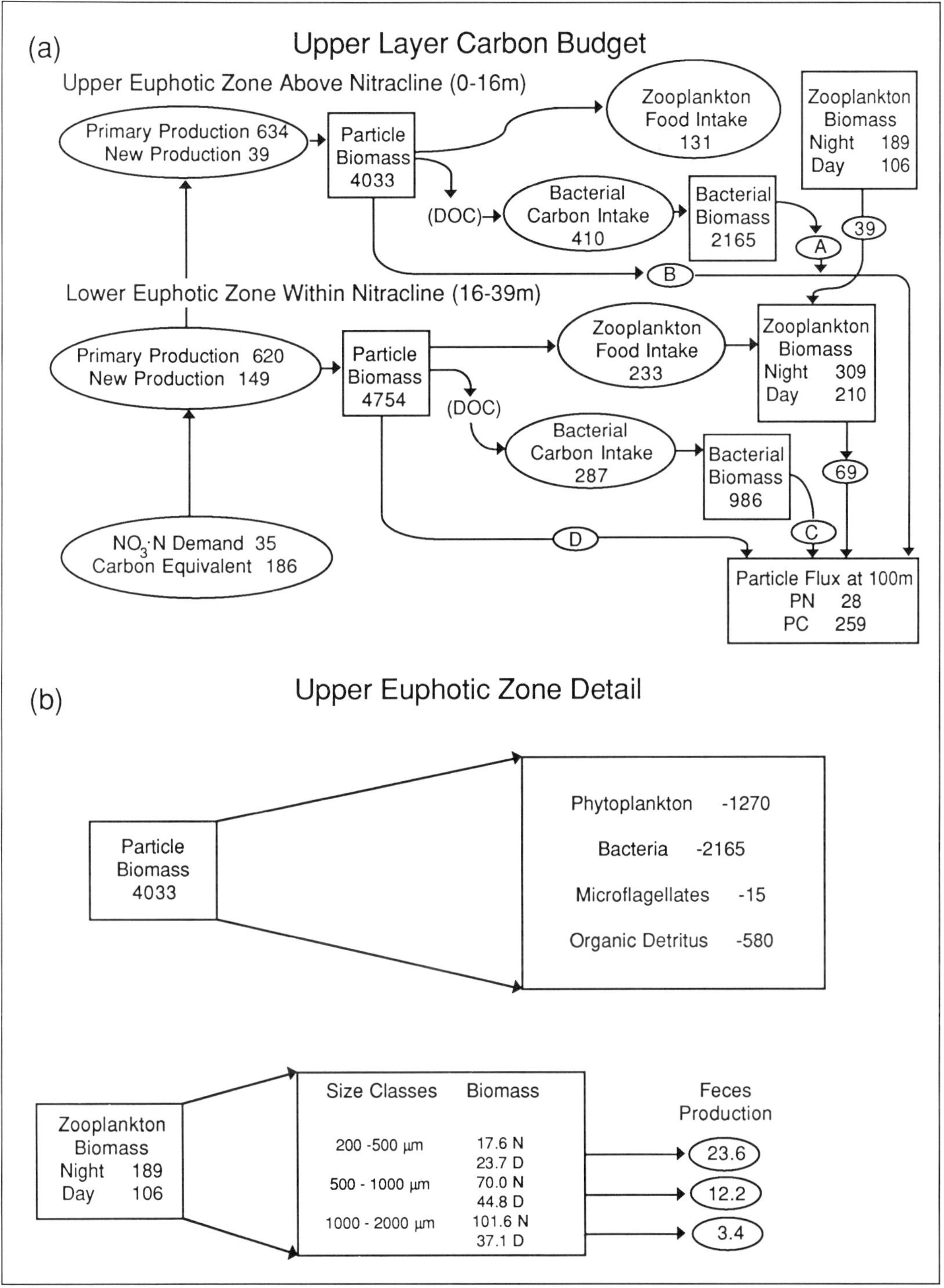

Figure 14.5. Upper layer carbon budget, May 1986. The photic zone is divided into a layer above the bottom nitracline (0–16 m) and a layer within the nitracline (16–39 m) to illustrate relationships among processes within these two layers. Numbers within oval symbols represent rates (mg C m^{-2} d^{-1}), except for NO_3-N demand and particulate nitrogen (PN) particle flux, which are in mg N m^{-2} d^{-1}. Numbers within rectangles are biomasses (mg C m^{-2}). (From Jackson et al. 1989.)

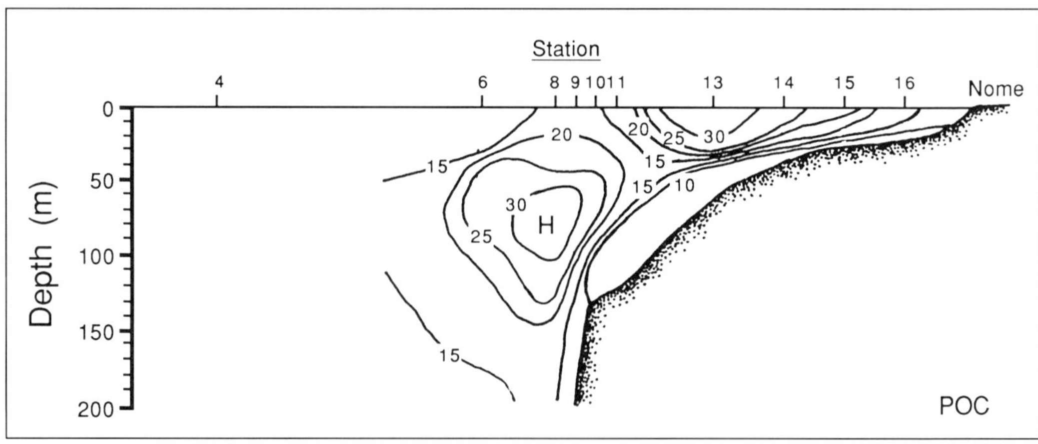

Figure 14.6. Percentage particulate organic carbon (POC) of total particulate matter in the eastern Bering Sea. (From Handa and Tanoue 1981.)

part (south of 54°N), is surrounded by one of the most densely populated and most heavily exploited parts of the world's oceans. The contribution of debris, especially organic material, and nutrients from the land has become a factor of steadily increasing importance to the ecology of the southern North Sea. Extensive studies of the area have been made to relate the importance of contributions of organic matter from land to that formed by *in situ* primary production (Postma 1973; Eisma and Kalf 1987a, b; Eisma 1981; Eisma et al. 1982; Weichart 1985; Joiris et al. 1982). On the basis of suspended matter concentrations in the winter, three regions of the North Sea can be distinguished: (1) the English–Scottish and continental coastal waters, where suspended matter concentrations exceed 10 mg l^{-1} (20% organic matter); (2) the remaining part of the southern North Sea, which has concentrations of 1–5 mg l^{-1} (20% organic matter); and (3) the northern North Sea, where concentrations of 0.1–1.0 mg l^{-1} (40–60% organic matter) are found.

Supply of organic matter to the southern North Sea, which has a surface area of 1×10^5 km^2 and a volume of 2500 km^3, is estimated to be 0.6×10^6 t yr^{-1} from the English Channel, 0.4×10^6 t yr^{-1} from the northern North Sea, and 2.5×10^6 t yr^{-1} from rivers, for a total of 3.5×10^6t yr^{-1}. This is equivalent to an input of 35 g m^{-2} organic matter or about 14 g C m^{-2} yr^{-1}. The average primary production for the southern North Sea is estimated to be nearly 200 g C m^{-2} yr^{-1}. It is evident from these data that production of organic matter *in situ* is nearly 15 times higher than the supply from outside.

In the New York Bight and the greater Mid-Atlantic Bight, large quantities of anthropogenic effluents containing high loads of particulate matter reach the sea. In this case, the sources are river runoff, as in the North Sea, and offshore dumping on the outer continental shelf, which (like the situation in the SCB) places the material below the pycnocline. In the Mid-Atlantic Bight, along-shelf transport is predominantly southwestward, with potential for resuspension of near-bottom particles by a number of wind events during a year-long 1000-km trajectory of individual water parcels (Chapman et al. 1986). This longshore advective process terminates with the entrainment of shelf waters at 35–36°N by the Gulf Stream. Meanwhile, the cross-shelf processes of random dispersion move particles only a few tens of kilometers seaward to waters overlying the continental slope, where they may settle out in a quiescent bottom-current environment. Estimates suggest that total particle exchange across the shelf edge by dispersive processes might be as important in the particle mass balance of this continental shelf as the southward advection (Walsh et al. 1987b).

The shelf waters of the northwestern Gulf of Mexico are characterized, at least seasonally, by a stratified water column with surface (0–5 m) salinities of 30‰ mid-column (5–12 m) salinities of 30.5–32.5‰ and bottom column 12–20 m) salinities of 32.5–34‰, with concentration bands (800–1500 $\mu g\ l^{-1}$) of suspended materials collected at the interfaces. Of these, the bottom layer is most persistent, being detected in all seasons from Louisiana to South Texas (Griffin 1979; Causey 1969). The source of these suspended materials in the turbid layers is freshwater runoff from the land, primarily from the Mississippi River drainage system. The river discharge emanates from the passes as a distinct surface plume containing high concentrations of suspended matter. Because of the distinct differences between the plume and the underlying clear mid-depth water, mixing is limited and the plume may persist well into the Texas coastal waters. The plume is slowly eroded as it moves west, and as small quantities of suspended materials interact with higher salinity water, they settle into the mid-column turbid layer. Eventually, particles reach the bottom where, through transport processes, they accumulate as a nepheloid cloud of fine particles. Of the systems reviewed here, the Texas–Louisiana coast ecosystem appears to be most influenced by river runoff and associated processes.

Flux of particulate matter, both vertically and horizontally, and the processes that control it are critical considerations in determining the flow of energy through the ecosystem. While organic matter, largely fixed in the euphotic zone by phytoplankton, carries energy through the trophic levels to the many biological consumers, transport and transformation of the organic matter to the place—and in the form—for utilization are generally governed by chemical and physical processes. As seen in table 14.3, the flux of particulates sinking out of the euphotic zone has been estimated for only a few ecosystems. This important aspect of energy transfer dynamics, however, has been clearly demonstrated (Jackson et al. 1989) in the SCB and will doubtless receive greater attention in other systems in future research.

Biological Productivity

The marine life of the SCB is abundant and diversified because of the multiplicity of habitats, coupled with unusually stable environmental conditions. The interplay of the physiography, current systems, surface winds, and anthropogenic inputs contribute to the richness of this area. In this discussion, we attempt to answer three questions. First, how high is production within trophic levels in the SCB, and how does it compare to the other continental shelf ecosystems described? Second, can the level of production in the SCB be explained on the basis of ecological conditions found there? Third, can the flow of energy through the ecosystem be reasonably balanced, using available data?

Productivity within the various trophic levels, as treated here, is an estimate of the conversion of food into growth and energy. It is based on energy relationships described by Sorokin (1983) and Sissenwine (1987). Historically, comparisons of productivity among ecosystems have been limited because of the lack of adequate data. However, Cohen and Grosslein (1987) have made a comparison of Georges Bank with other ecosystems in which they reveal the utility of this approach in obtaining a broader understanding of continental shelf ecosystems. A similar comparison is made here for the SCB.

Data used for the SCB is largely obtained from papers included in this work; data for other systems is obtained from the literature. Given the variety of sources of information, many differences exist in data collection and in methodology. The values are, at best, estimates; many uncertainties remain. Confidence level estimations are not possible.

Table 14.3. *Inputs of Organic Carbon and Energy to the SCB*

Source	Area (km^2)	Carbon Input ($g\ C\ m^{-2}\ yr^{-1}$)	Carbon Input ($t\ C\text{-}SCB\ yr^{-1} \times 10^3$)	Coefficient	Energy Input ($kcal\ SCB\ yr^{-1} \times 10^{11}$)	Percentage of Total kcal	References
Autochthonous							
Phytoplankton	7.8×10^4	142	11,100	11.2	1,240	92.5	Chapter 5; Eppley and Holm-Hansen (1986)
Phytoplankton exudates	7.8×10^4	26	2,000	4.0	80	5.9	Iturriaga (1981)
Salt marshes	50	260	13	11.2	1.5	0.08	Chapter 7; Zedler (1982)
Rocky intertidal	60	900	20	11.2	22	1.2	Chapter 7; Littler et al. (1979)
Kelp forests	88	1,000	88	11.2	99	5.7	Chapter 7; Mann (1982)
Allochthonous							
Municipal outfalls	—	—	140	4.0	5.6	0.32	Chapter 12
Natural seeps	—	—	33[a]	11.3	37	0.22	Chapter 12
River runoff	—	—	32	4.0	13	0.07	Chapter 3
Shale erosion	—	—	6	11.3	0.68	0.045	Chapter 3
Dry fallout	7.8×10^4	1.7	13	4.0	0.17	0.030	Chapter 3
Total inputs			13,445		1,498		

[a] Maximum value estimate.

Despite these uncertainties, it is felt that considerable insight into how energy is produced, utilized, and distributed in the SCB can be gained from this type of treatment. These results should be considered preliminary, to be built onto as better and more sophisticated data become available.

Autochthonous Energy Inputs or Primary Production

Phytoplankton in the SCB, as in most ocean environments, are responsible for most of the primary production. Macrophytes, although abundant, account for approximately 1% of the total energy contributed by *in situ* photosynthesis.

PHYTOPLANKTON PARTICULATE PRODUCTION

Phytoplankton show a high variability in growth in SCB, which is related to the flow of the California Current and the intensity of upwelling. A time series for the years 1920–1979 for the primary production in the SCB (Eppley and Holm-Hansen 1986, fig. 5.14) shows a high correlation between southward transport of the California Current and primary production. These data show an annual average production of nearly 150 g C m^{-2}, with a fluctuation around the mean of ± 75 g C yr^{-1}.

Using data from figure 14.5, we estimate that approximately 33% of the particulate biomass in the euphotic zone of the San Pedro and Santa Monica basins area is made up of living phytoplankton. This would give a biomass of 2.9 g C m^{-2} and a production to biomass ratio (P:B) for the area of about 50, using 142 g C m^{-2} yr^{-1} as the primary production (table 14.2) (see also chap. 5). Beers (1986, fig. 4.6) shows the high variability of phytoplankton carbon at a series of stations taken in the vicinity of La Jolla. If 100 μg C l^{-1} is selected as the average value for the top 30 m of the water column, this would give a biomass of 3 g C m^{-2}, which agrees well with the previous computation.

If 142 g C m^{-2} yr^{-1} is used for the rate of primary production for all of the SCB (7.8×10^4 km^2), this would convert to 1.1×10^{13} g of particulate carbon synthesized in the SCB annually. Phytoplankton carbon, converted to an equivalent amount of energy, can be calculated using a factor of 1 g C = 11.2 kcal (Platt and Irwin 1973), thus yielding 1.24×10^{14} kcal of energy, in particulate form, produced by phytoplankton in the SCB each year.

PHYTOPLANKTON EXUDATE PRODUCTION

In the processes of photosynthesis and respiration, phytoplankton excrete to the environment a variable portion of the fixed energy as soluble organic matter. This material, together with that formed during metabolic activities of all of the biomass, constitutes the important fraction known as dissolved organic carbon (DOC), which is particularly important in the microbial food web (chap. 4; Azam 1986). Iturriaga (1981) estimated that 18–19% of the carbon assimilated by phytoplankton was excreted to the water as compounds with a molecular weight of <1000. This would indicate that about 2×10^{12} g C yr^{-1} is fixed in the form of low molecular weight compounds that are not normally included in primary production computations, assuming that the methodology used looks only at the particulate fraction. If we use a C:kcal factor of 10 (kcal in 1 g of metabolites) to convert this to an equivalent amount of energy, then 20×10^{12} kcal are contributed to the system from this fraction for biological consumption (mostly bacteria) each year.

BENTHIC MACROPHYTES

Benthic macrophyte communities are most important contributors to the productivity budgets of the shallow coastal seas. Mann (1982) estimates that production by coastal phytoplankton in areas not influenced by upwelling is between 50 and 250 g C m^{-2} yr^{-1}, whereas communities of marine flowering plants may produce 300–1000 g C m^{-2} yr^{-1}

and seaweed communities may produce 400–1900 g C m^{-2} yr^{-1}. Consequently, knowledge concerning the productivity of the macrophytes is important to energy generation and distribution for the SCB. A total of 492 species of red (347), brown (86), and green (59) macrophytes occur in the SCB. The seaweed flora south of Point Conception are greater in diversity but less in biomass than that found north of Point Conception. Fortunately, our knowledge of macrophyte biology and productivity for the SCB is well developed (chap. 7) and contributes significantly to our understanding of the role of these communities in the overall problem of marine bioenergetics.

SALT MARSH HABITATS

There are 30 salt marshes in the SCB that occupy approximately 50 km^2 (5000 ha). Approximately 27 species of macroalgae characteristic of quiet water (dominated by green algae) inhabit the salt marsh communities. According to Zedler (1982) (chap. 7), these marshes produce between 100 and 1100 g dry wt m^{-2} yr^{-1}. If we assume that 40% of this material is carbon, then the total primary production would be 2–22 $\times$ 10^9 g C yr^{-1} or 2.5–25 $\times$ 10^{10} kcal of energy contributed to the SCB system.

ROCKY INTERTIDAL

Of the approximately 1000 km of coastline in the SCB, about 350 km adjoin the mainland, which is about 15% rocky intertidal. The remainder adjoins the southern California Channel Islands, which are about 72% rocky intertidal. In total, this represents approximately 600 km of coastline that is inhabited by about 100 seaweed species with a wide range of abilities to occupy habitats with varying stability in the substratum and stresses produced by different oceanographic conditions (chap. 7).

The annual productivity of the rocky intertidal community in the SCB is estimated by Littler et al. (1979) to be 900 mg C m^{-2} d^{-1}. To convert this productivity value to an areal-wide contribution to the energy budget is difficult because the area covered by these communities has not been measured. To make a very rough estimate, it is assumed that, on the average, the rocky intertidal community extends 100 m from the shore. With a total linear distance of 600 km and an estimated width of 100 m, the areal coverage would be 60 km^2, or a total productivity of 2 $\times$ 10^{10} g C yr^{-1}, representing 2.2 $\times$ 10^{11} kcal of energy contributed to the SCB.

SUBTIDAL KELP FORESTS

Extensive stands of giant kelp (*Macrocystis*) are conspicuous and well-known features of the SCB. The total canopy coverage is approximately 88 km^2, with about 50% occurring along the mainland and the other 50% associated with the island groups. The kelp forests in the SCB are dynamic over both the short and long term. They are especially vulnerable to heavy storms and changes in nutrient supply (such as those brought on by El Niño events). Productivity within these unusual ecosystems also varies widely. Jackson (1987), in considering the many factors that influence growth in a kelp forest, estimated a net production of about 3 g C m^{-2} d^{-1} for the "typical" forest, or 1095 g C m^{-2} yr^{-1}. Mann (1982) roughly estimated an average productivity of 900–1100 g C m^{-2} yr^{-1}. If a mean value of 1000 g C m^{-2} yr^{-1} is used, a yield of 8.8 $\times$ 10^{10} g C or 9.9 $\times$ 10^{11} kcal energy is calculated to be fixed by the kelp forests each year in the SCB.

Allochthonous Inputs

Many sources of energy in the form of fixed carbon are transported from outside the immediate area (chap. 12) and contributed to the SCB each year. These represent a source of energy that is either utilized, deposited, or transported out of the system. In 1986, municipal outfalls contributed 1.87×10^5 t of solid wastes, which—on the basis of biological oxygen demand (BOD) values—appears to be about 75% carbon. With this assumption, 1.4×10^{11} g C is derived from this source, which is assumed to be usable by indigenous organisms. If we use a conversion factor of 4.0 (kcal g^{-1} of glucose is 3.98), this would yield 5.6×10^{11} kcal of energy to the SCB budget to be used by the biota.

NATURAL PETROLEUM SEEPS

Major oil seep areas occur at Point Conception, Coal Oil Point, Santa Barbara–Rincon (offshore in the vicinity of the city of Santa Barbara and Rincon Beach), and Santa Monica Bay in the SCB. While these hydrocarbons influence the ecology (mostly by increasing hydrocarbon-utilizing bacteria populations) (chap. 12) in the vicinity of the seeps, this carbon eventually finds its way into the food chain as a source of energy. Fischer (1978) estimates that the total seepage for the Santa Barbara Channel is between 6.4×10^3 and 1.1×10^5 l d^{-1}. This is equivalent to a maximum yield of 9×10^7 g C d^{-1} or 3.3×10^{10} g C yr^{-1}. With a conversion factor of 11.3 (heat of combustion of 1 g decane), this yields a maximum energy of 3.7×10^{11} kcal energy to the SCB ecosystem.

MISCELLANEOUS INPUTS

Additional organic energy inputs are known to occur in the SCB. Runoff from rivers, a major source of organic material in many continental shelf ecosystems, is of less importance in the SCB, where it is low. The flow from rivers in the SCB is estimated at 2.42×10^8 m^3 yr^{-1}, which is about 25% of that discharged through municipal waste outfalls (1.6×10^9 m^3 yr^{-1}). However, this runoff contains about 3.2×10^{10} g C, or an energy input of 1.3×10^{11} kcal, 23% of that discharged through municipal waste outfalls (see chap. 3).

Shale erosion adds another 6×10^9 g C, and dry fallout from the atmosphere adds about the same amount of carbon to the SCB annually. These two sources contribute about 4.8×10^{10} kcal of additional usable energy. Other sources of energy are even less significant to the total budget and are neglected in this consideration.

Table 14.3 summarizes the information on fixed carbon inputs to the SCB. The total carbon contributed to the system is estimated to be 1.4×10^7 t, which contains 1.5×10^{14} kcal of energy. Of this amount, over 92% is derived from phytoplankton productivity.

Zooplankton Production

Zooplankton vary in size from those that can pass through a 250-μm mesh net to those that can pass through a 2000-μm net (fig. 14.5). The metabolic activities of the size classes vary greatly, as evidenced by the rates of feces production. The small organisms (microzooplankton) are composed of micrometazoans and protozoans, the latter being able to reproduce by simple asexual binary fission and thus respond quickly to environmental change. The biomass of the microzooplankton is estimated to be only 20–25% of the total zooplankton both inshore and offshore in the SCB (Beers and Stewart 1970) (fig. 14.5). However, this fraction produced 60% of the fecal pellets found in sediment traps placed at the bottom of the mixed layer. These data indicate a threefold greater activity (feeding rate) for the microzooplankton than for a composite of the larger forms.

MICROZOOPLANKTON

On the average, Beers and Stewart (1970) found that 2.4–5.9 mg C m^{-3} of microzooplankton biomass occurred in the upper 100 m of the water column off La Jolla during the summer of 1967. Banse and Mosher (1980) have suggested a P:B ratio for the microzooplankton of 25. If we assume that the average biomass is 400 mg C m^{-2}, then the productivity of this group of organisms would be 10 g C m^{-2} yr^{-1}.

Another estimate of microzooplankton production can be made by partitioning POC into nanoplankton and netplankton and making the following assumptions: (1) microzooplankton graze only on nanoplankton and macrozooplankton only on netplankton (Parsons and Le Brasseur 1970), (2) about 50% of available POC is grazed in shelf ecosystems (Dagg and Turner 1982), and (3) the transfer efficiency for herbivorous zooplankton is 32% (Steele 1974). Primary production for the SCB is estimated to be 70% nanoplankton and 30% netplankton (chap. 5), as are the estimates for the Gulf of Maine and the Mid-Atlantic Bight. Using these data and the primary production value of 142 g C m^{-2} yr^{-1} for the SCB, a production value of 11 g C m^{-2} yr^{-1} (very close to the previous value of 10 g m^{-2} yr^{-1}) for the microzooplankton is obtained. This yields an annual production of 8.6×10^{11} g C, or 9.6×10^{12} kcal of energy from this source to the SCB.

MACROZOOPLANKTON

Data obtained on the CalCOFI line 90, which transected the SCB during the four seasons of 1965–1969, gave an average value of 27 ml m^{-2} wet vol, or 1.22 g C m^{-2} (5% wet vol = carbon for phytoplankton). Jackson et al. (1989) indicated a zooplankton biomass of 500 mg C m^{-2} in the euphotic zone, which is held by a 500-μm or greater size net, in Santa Monica and San Pedro basins. Steele (1974) estimated a P:B ratio for the macrozooplankton of 7. If we use the CalCOFI data and 7 for the P:B ratio, this yields an annual production rate of 8.50 g C m^{-2}, or 6.6×10^{11} g C, equivalent to 6.6×10^{12} kcal energy to the SCB annually. If the alternate method of calculation used for microzooplankton is used here, then the production of macrozooplankton would be 7.81 g C m^{-2} yr^{-1}. This productivity value would yield 6.1×10^{11} g C, or 6.1×10^{12} kcal energy annually to the Bight.

Bacterial Production

The microbiology of the SCB has been carefully discussed in chapter 4. Much of the early work on bacteriology of the sea was done at the Scripps Institution of Oceanography (ZoBell and Feltham 1934; ZoBell 1941). Personnel at this same institution, working in the SCB, have contributed heavily to the establishment of the "microbial loop" as an important component of the marine food web (Azam et al. 1983).

An average bacterioplankton biomass of 9 μg C l^{-1} was observed during all seasons, down to a depth of 40 m, 15 km offshore in San Pedro Channel (Krempin and Sullivan 1981). Bacterial biomass tends to decrease with distance from shore (17 μg C l^{-1} near Los Angeles Harbor to 3 μg C l^{-1} near Santa Catalina Island). Bacterioplankton constitute between 3.3 and 8.8% of the total plankton biomass in the SCB. Bacterioplankton production has been estimated to be 1 μg C l^{-1} d^{-1} in the offshore waters of the SCB and 2–14 μg l^{-1} d^{-1} in the inshore waters (Fuhrman and Azam 1980, 1982), as determined by the thymidine incorporation method. Using a biomass of 3 μg C l^{-1} and a productivity of 1 μg C l^{-1} d^{-1} for the offshore region of the SCB, we obtain a P:B ratio of 91. For the inshore region, an average value of 8 μg C l^{-1} d^{-1} productivity and 9μg C l^{-1} for biomass appears reasonable. This yields a P:B of 325.

Estimating the energy produced by the bacterioplankton for all of the SCB is tenuous because of the difficulty of converting the available data to acceptable values adequately representing all portions of the bight. A conservative estimate of bacterial production can be made by assuming an average production rate of 1 μg l^{-1} extending from the surface to an average functional depth of 100 m. If 1 g of carbon is equivalent to 10 kcal of energy in these organisms, then the estimated production for bacteria would be 38 g C m^{-2} yr^{-1}, or 380 kcal m^{-2} annually. For all of the SCB, 3.0×10^{12} g C and 5.7×10^{13} kcal would be converted from dissolved organic carbon

(DOC) to particulate carbon in the upper water column by the bacterioplankton.

The Benthos

MACROBENTHOS

An average biomass for the macrobenthos was computed for the SCB using data from chapter 8 and estimating the contribution of each bottom feature to the whole: 20% to mainshelf and upper slope, 20% to basins, and the rest to other features. Averaging the biomass data within the features gives a value of 83.2 g wet wt m^{-2}. This is equivalent to 4.16 g C m^{-2} (using the factor 5% wet wt to convert to carbon).

Benthic productivity data for the SCB are not available except for a few isolated species (chap. 8). Estimates of the P:B ratio, taken from other locations, is about 1.5 (Stoker 1981; Steele 1974; Crisp 1975), which would yield a productivity of 6.25 g C m^{-2} yr^{-1}. Using a conversion factor of 1 g wet wt = 0.5 kcal of energy (Cohen and Grosslein 1987) gives a value of 62.5 kcal of energy m^{-2} yr^{-1}. For the SCB, this amounts to 4.88 × 10^{12} kcal of energy annually.

BENTHIC MICROBIOLOGICAL

Bacterial biomass of the sediments on the slope off Orange County in waters 300–600 m deep averaged 58 μg C cm^{-3} (Thompson et al. 1984). Off Newport Beach, 46 μg cm^{-3} were found. If we assume an average value of 52 μg C cm^{-3} at the surface and 1% of that at 20 cm (Rittenberg 1940), this would give an integrated biomass value of 2.7 μg C cm^{-3} for the top 20 cm of sediment, or 540 mg C m^{-2}. For the Orange County slope, this represented 10–36% of total microbial biomass, the rest being protozoan and meiofaunal metazoans (chap. 6). If we assume that 26% is an average value for bacterial carbon, the total biomass represented by this group of organisms would be 2.1 g C m^{-2}. It is probable that only about one-half of these are actively producing (Fuhrman and Azam 1982).

Productivity data for the benthic microbial organisms do not exist; however, Fuhrman and Azam (1982) estimated a 1–4 day doubling time for waterborne bacteria. If we assume that the benthic organisms have similar activity and that about three-fourths of the biomass is composed of the slower growing protozoans and metazoans with a P:B ratio of about 30 (Banse and Mosher 1980), then the ratio of P:B for this component of the benthos would be about 31. Assuming that one-half the benthic microbial biomass is active, then these estimates of P:B would yield a productivity of 8.4 g C m^{-2} yr^{-1}, or 6.5 × 10^{12} kcal yr^{-1} for the SCB.

Metabolic rates for the remaining components have not been determined for the SCB.

Fish

Fish production estimates for the various habitats of the SCB are not available in the literature, and the type and amount of fisheries ecological information for the habitats are not easily comparable (chap. 9). To estimate productivity of fish and thereby try to establish their contribution to energy transfer dynamics, it is necessary to use many sources of information. Most biomass, major prey components, and consumption data for six major groups of fishes were obtained from data contained in chapter 9. (Variances to this are indicated in table 14.4, which also summarizes other data relevant to energy flow.) Consumption rates for demersal soft-bottom fish and mackerel in percentage of body weight per day were gleaned by combining information from chapter 9 with the extensive database assembled for similar groups of fish that inhabit Georges Bank (Sissenwine 1987). For the hard-bottom and mesopelagic fish, data offered by Daan (1973) were used; for large carnivores, our source was Mallicoate and Parrish (1981). The consumption rate for anchovies is a compromise between that suggested by Smith (1978) (1.7% wet wt d^{-1}) and

Table 14.4. *Estimate of Biomass, Consumption, and Productivity of Fishes in the SCB*

	Biomass			Consumption			Productivity	
Classification	($g\ m^{-2}$ wet wt)	(t C-SCB[b] × 10^3)	Prey[a]	(rate % body wt per day)	($g\ m^2\ d^{-1}$ wet wt)	Percentage Growth Efficiency	(t C-SCB × 10^3)	(kcal-SCB[c] × 10^9)
Demersal—soft-bottom[d]	3.8	37.0	c,f,po	1.3	0.046	12	19.5	195
Hard-bottom—reef-kelp[e]	100[f]	1.75	mi,ma,c,ce,f	2.0	2.0	12	1.5	15
Pelagic								
Anchovies[g]	35	340	p,z,de	3.0	1.1	12	470	4,700
Mackerel[h,i]	2.1	2.0	a	1.2	0.025	12	11	106
Large carnivores[j]	8.0	7.8	a,d,eu,ce	1.5	0.096	10	34	340
Mesopelagic[k]	3.0	2.9	co,eu,ch,sa	2.0	0.06	10	21	210
Totals		410			1.35		557	5,566

[a] a = anchovies; c = crustaceans; ce = cephalopods; ch = chaetognaths; co = copepods; d = decapods; de = detritus; eu = euphausiids; f = fish; ma = macrobenthos; mi = microbenthos; p = phytoplankton; sa = salps; po = polychaetes; z = zooplankton.

[b] Wet wt to carbon is obtained by dividing by 8 (Shannon and Field 1985).

[c] For all classifications, fish production was calculated as 1 gram wet wt = 1.25 kcal of energy, or 1 g C = 10 kcal.

[d] Chapter 9; Sissenwine (1987).

[e] Chapter 9; Daan (1973).

[f] Area covered in SCB is estimated to be 140 km^2 (chap. 7).

[g] Chapter 9; Shannon and Field (1985).

[h] Chapter 9; Sissenwine (1987).

[i] Based on 1.6 × 10^5 t mackerel in the SCB (Parrish and MacCall 1978).

[j] Brownell et al. (1976); Mallicoate and Parrish (1981).

[k] Chapter 9; Pearcy and Laurs (1966).

the findings reported by Shannon and Field (1985) (10% wet wt d^{-1}). The compromised value of 3% wet wt d^{-1} was chosen to reflect the lower carbon content of the food prey of anchovy (about one-third phytoplankton, which have a carbon content of 6% versus 12.5% for other biota).

For the demersal soft-bottom fish and mackerel group, growth efficiency estimates were derived from the consumption to biomass and productivity to biomass data of Sissenwine (1987) for Georges Bank. By dividing the C:B ratio by that for P:B, a growth efficiency percentage is obtained. Hard-bottom fishes and anchovies were given the same growth efficiency values as the soft-bottom fishes. Values of 10% were assigned to those groups for which data were not available.

Multiplying the food consumed in wet weight per day by growth efficiency yields productivity values. Conversion to metric tons of carbon for the SCB was accomplished by multiplying the product of consumption in wet weight per day and growth efficiency by 3.56×10^6, a factor derived from the percentage carbon in wet weight, area of the SCB, and the days of production in a year. The estimated total fish production for the SCB is about 5.7×10^5 t yr^{-1}, containing 5.7×10^{12} kcal of energy for distribution in the ecosystem.

Marine Mammals

The rates of consumption and productivity for the most abundant species of marine mammals in the SCB are summarized in table 14.5. A total of 5.3×10^4 cetaceans and 6.2×10^4 pinnipeds are estimated to inhabit the SCB on an average year-round basis. The most abundant cetacean is the common dolphin, which also represents the largest biomass (8000 t) actively feeding in the region. The blue whale represents about the same in biomass, and the gray whale probably exceeds it, but the latter apparently does not feed or feeds very little while migrating through these waters. The California sea lion is the most abundant pinniped (5×10^4 individuals) and represents by far the largest biomass of this group at the present time (1×10^4 t). These organisms consume a wide variety of prey, ranging from zooplankton by the filter-feeding whales to fish and pinnipeds by the large carnivores. Rates of consumption vary from about 2% of body weight per day for the large whales to as much as 40% of body weight per day for the only mustelid of the group, the sea otter. On an annual basis, the marine mammals consume about 675×10^3 t of prey per year that contains about 880×10^9 kcal of energy. On the basis of a 7.8×10^4 km^2 area for the SCB and an energy equivalence where 1 g wet wt contains 1.25 kcal, estimated annual consumption is 4.5 kcal m^{-2} for the cetaceans and 6.75 kcal m^{-2} for the pinnipeds and sea otters.

Growth efficiency of marine mammals at a rate of 10% of consumption (wet wt) yields a productivity of 3.5×10^{11} kcal yr^{-1} for cetaceans and 6.4×10^{11} kcal yr^{-1} for the pinnipeds and sea otters. Scott et al. (1983) estimated from energy requirement considerations that cetaceans of the northeastern shelf of the United States had a transfer efficiency of 23%. Applied to the cetaceans of SCB, this efficiency would yield a productivity of 8.4×10^{11} kcal yr^{-1}.

Birds

The SCB is in the migratory pathway of millions of marine birds. More than 195 species use the area to breed in summer or winter or to pass through each year. Precise numbers of birds using the area at a given time are not accurately documented; however, the foraging habits of the most important species are described in detail in chapter 10. Two species that probably cycle the most energy through the ecosystem, the sooty shearwater and the brown pelican, have been studied extensively. Shearwaters, when present in highest numbers during spring and midsummer, consume about 0.13 mg C m^{-2} d^{-1}. If a 240-day feeding period at this rate is assumed for each

Table 14.5. *Estimate of Biomass, Composition, and Productivity of Marine Mammals in the SCB*

Species	Biomass			Prey[a]	Consumption			Percentage Growth Efficiency	Productivity		References[b]
	Number (average)	kg each	Total (t × 10^3)		(kg d^{-1} wet wt)	(t yr^{-1} × 10^3 wet wt)	(kcal yr^{-1} × 10^{10})		(t C × 10^3)	(kcal × 10^{10})	
Cetaceans											
Blue whale	20	4 × 10^5	8	eu	8,000	58.5	7.3	10–23	0.75–1.7	0.75–1.7	Rice 1986
Fin whale	50			eu, co, ce, f	1,250	23[c]	2.9	10–23	0.29–0.66	0.29–0.66	Mitchell 1986
Minke whale	50			a, s, eu	700	12.8	1.6	10–23	0.16–0.36	0.16–0.36	Calkins 1987
Humpback whale	15			eu, f	1,100	6.0	0.75	10–23	0.075–0.36	0.075–0.36	Calkins 1987
Gray whale[d]	600			am							
Pilot whale	500	900	0.45	sq	35	6.4	0.8	10–23	0.08–0.19	0.08–0.19	Reilly and Shane 1986; Winn et al. 1987
Common dolphin	4 × 10^4	200	8	a, h, my, sq	8	110	13.8	10–23	1.4–3.1	1.4–3.1	Leatherwood and Reeves 1986
Bottlenose dolphin	375	275	0.1	sq, mo, cr	10	0.14	0.017	10–23	0.02–0.04	0.02–0.04	Dohl et al. 1986
White-sided dolphin	6,000	200	1.2	f, a, ce	15[e]	33	4.2	10–23	0.4–0.9	0.4–0.9	Leatherwood and Reeves 1986; Calkins 1987
Right-whale dolphin	3,500	225	0.79	sq, my	15[e]	19.2	2.4	10–23	0.2–0.55	0.2–0.55	Leatherwood and Reeves 1986

Dall's porpoise	250	175	0.044	sq,my,h	12[e]	1.1	0.14	10–23	0.01–0.03	0.01–0.03	Leatherwood and Reeves 1986
Risso's dolphin	1,400	300	4	sq	22[e]	11.2	1.4	10–23	0.14–0.3	0.14–0.3	Leatherwood and Reeves 1986
Cetacean totals	5.3×10^4					252	35.3		3.5–8.4	3.5–8.4	
Pinnipeds											
California sea lion	5×10^4	200	10	f,sq	20	370	46.2	12	5.6	5.6	Mate and DeMaster 1986
Northern fur seal	3,000	115	0.35	f,sq	11	12	1.5	12	0.18	0.18	Fiscus 1986
Elephant seal	500	425	0.35	ce,cr,f,tu	40	9.1	1.14	12	0.13	0.13	Costa et al. 1986
Harbor seal	7,000	115	0.80	f,ce,cr	11	28	3.5	12	0.4	0.4	Calkins 1987
Sea otters	1,200	35	0.04	mo,f,se	7	3.1	0.4	10	0.04	0.04	Calkins 1987

[a] a = anchovies; am = amphipods; ce = cephalopods; co = copepods; c = crustaceans; eu = euphausiids; f = fish; h = hake; mo = molluscs; m = myctophids; se = sea urchins; sq = squid; tu = tunicates.

[b] Chapter 11.

[c] Consumption of 1000–1500 kg d^{-1} (Lockyer 1976).

[d] Gray whales have no record of consumption in the SCB.

[e] Actual values are not available; inferred from data for comparable species.

year, then consumption by this species alone would be about 2.4×10^3 t yr^{-1} in all of the SCB (7.8×10^4 km^2). Brown pelicans consume about 1.7×10^3 t yr^{-1} (Briggs and Chu 1987). Together these species represent about one-half the total number of birds in the SCB (see chap. 10). If it is assumed that these two species also represent one-half the biomass of 8.38 kg km^{-2} (see chap. 10) and the prey consumption rate for the SCB, the total biomass would be 1.31×10^9 g wet wt, or 1.6×10^8 g C. The consumption rate would then be 7.4×10^9 g wet wt yr^{-1}, or 9.25×10^8 g C yr^{-1}. This represents a consumption to biomass ratio (C:B) of 5.8, or a consumption rate of about 18% of body weight per day.

Productivity of marine birds in the SCB has not been measured, nor has it been examined in other shelf areas. Estimates of production for the SCB are made by assuming a 10% growth efficiency on food consumed (wet wt). A consumption rate of 9.25×10^8 g C yr^{-1} would then yield a productivity of 9.25×10^7 g C, or 9.5×10^8 kcal energy annually (0.012 kcal m^{-2} yr^{-1}).

Comparison of Trophic Level Productivity among Areas

Table 14.6 lists estimates of production at various trophic levels for the SCB, the outer domain of the Bering Sea, the North Sea, Georges Bank, the Mid-Atlantic Bight, and the southern Benguela off the southeastern coast of Africa. The SCB is compared with each of the other shelf ecosystems in all categories except macrophytes, cetaceans, and pinnipeds. Satisfactory values for the meiobenthos of the SCB could not be estimated.

The SCB has the lowest primary production of particulate energy (1774 kcal m^{-2}) of any of the shelf ecosystems examined, even though macrophyte production (174 kcal m^{-2}) was included for this system only. The outer domain of the southeastern Bering Sea shelf is slightly higher (1824 kcal m^{-2}), followed by the North Sea (2280 kcal m^{-2}), Georges Bank, and the Mid-Atlantic Bight, the latter two of which are both nearly twice as productive as the SCB (3342 and 3103 kcal m^{-2}, respectively). The southern Benguela, which is one of the world's most active upwelling areas, is over three times more productive (11×10^3 kcal m^{-2}) than the next highest area compared (Georges Bank).

Bacterial production in the SCB is about 80% that of Georges Bank, the only other comparative system for which data exist. However, both locales have production values comparable to that for other open sea situations (Williams 1984). Bacterial productivity is much higher in coastal and estuarine areas.

Microzooplankton productivity for the SCB is less than 60% of that in the outer domain of the Bering Sea, the North Sea, and the Mid-Atlantic Bight, which are about equivalent at nearly 215 kcal m^{-2} each. Georges Bank is about 25% higher (285 kcal m^{-2}). Data for the southern Benguela do not separate the micro- from the macrozooplankton; therefore, the combined value is more than twice that of any other system in this comparison.

Macrozooplankton varies significantly between the six ecosystems compared. The SCB production is by far the lowest (85 kcal m^{-2}), only 60% that of the next lowest, the Mid-Atlantic Bight (137 kcal m^{-2}). The outer domain of the eastern Bering Sea, known to be a pelagic food web (McRoy et al. 1986), has about 90% of the productivity (170 kcal m^{-2}) of the North Sea (186 kcal m^{-2}) at this trophic level, which in turn has about 90% that of Georges Bank (202 kcal m^{-2}).

Productivity data for the meiobenthic community are not available for the SCB, disallowing comparison of this component. Macrobenthos data are available, however. The lowest productivity in the macrobenthos category is in the outer domain of the southeastern Bering Sea, which has a low biomass invertebrate community (5 g C m^{-2} vs. 25 g C m^{-2}) (Stoker 1981) as compared to other

Table 14.6. *Annual Production Estimates (in kcal m^{-2}) at Different Trophic Levels of Some Shelf Ecosystems*

Trophic Level	SCB	Bering Sea Outer Domain[a]	North Sea[a]	Georges Bank[a]	Mid-Atlantic Bight[a]	Benguela
Primary production	1,774	1,824	2,280	3,342	3,103	11,400
Phytoplankton	1,600					
Macrophytes	174					
Bacteria	730	—	—	900	—	—
Microzooplankton	120	204	214	285	220	900[b]
Macrozooplankton	85	170	186	202	137	—
Meiobenthos	—	—	25	13	30	—
Macrobenthos	63	41	100	98	180	—
Fish	72	56	24	52	25	90[c]
Birds	0.012	0.1	—	0.20	—	—
Mammals	1.3–1.9[d]	1.3	0.4	1.0	—	—
Cetaceans	0.5–1.1					
Pinnipeds	0.81					
Average annual landings[e]	2.8	2.5	5.9	8.4	3.7 (7.3)[f]	12.5

[a] Data from Cohen and Grosslein (1987).

[b] Includes both microzooplankton and macrozooplankton (Shannon and Field 1985).

[c] Shannon and Field (1985).

[d] 0.55–1.1 g C m^{-2} yr^{-1} from cetaceans; 0.81 g C m^{-2} yr^{-1} from pinnipeds.

[e] SCB landings data obtained from the National Marine Fisheries Service records, Terminal Island Facility, California, and provided by J. Cross; Bering Sea landings data from Bakkala and Low (1982) for the period 1965–1981; North Sea landings data from ICES, Bullitins Statistiques, for the period 1961–1980; Georges Bank landings data from Cohen and Grosslein (1987); Mid-Atlantic Bight landings data from ICNAF for the period 1968–1982.

[f] Value in parentheses includes menhaden landings for period 1963–1967 (Schaff 1980).

similar environments. Next lowest is the SCB (63 kcal m^{-2}) followed by the North Sea and Georges Bank, which have nearly equal values (100 and 98 kcal m^{-2}, respectively), and the Mid-Atlantic Bight, with the highest value (180 kcal m^{-2}).

Fish production in the ecosystems compared varies widely with time (86 kcal m^{-2} in the 1960s vs. 52 kcal m^{-2} in the 1970s for Georges Bank), reflecting the influence of fishing activities in these areas. Data presented here (as indicated in the footnotes to table 14.6) appear in information published in the 1970s and 1980s. Somewhat surprising is the conclusion that the North Sea, long known as a major fishing area, has the lowest production of the group (24 kcal m^{-2}), followed closely by the Mid-Atlantic Bight (25 kcal m^{-2}). Georges Bank and the outer domain of the eastern Bering Sea, with production rates of 52 and 56 kcal m^{-2}, respectively, yield values about twice the level per unit area of the North Sea and Mid-Atlantic Bight. The SCB, even with its relatively low primary productivity, is an efficient producer of fish. Fish production in the SCB is nearly 80% of fish production in the strong upwelling area of the Benguela, which produces at the rate of 90 kcal m^{-2} yr^{-1}. Both the SCB and the Benguela probably owe their high production rates to the predominance of the pelagic school fish, anchovy, and sardines in the fish biomass. These fish are partially herbivores, feeding on phytoplankton, and thus have a short food chain and fairly high P:B ratio.

Bird production in the SCB (0.012 kcal m^{-2}) is only about 12% of that in the eastern Bering Sea (0.1 kcal m^{-2}) and only 6% of that in Georges Bank (0.2% kcal m^{-2}). The unusually low biomass indicated for birds in the SCB (chap. 10), compared to other areas, makes this truly a distinct feature of the SCB.

The marine mammal productivity estimates for the SCB (1.6 kcal m^{-2}, on average) are higher than for any other area under comparison. The eastern Bering Sea values are lower (1.3 kcal m^{-2}), even though this area supports a huge walrus herd not present in other areas. Georges Bank production is two-thirds as much (1.0 kcal m^{-2}), and the North Sea is only 25% that of the SCB.

Fish landings can be used as an indicator of fish productivity if it is assumed that the fisheries catch is within the bounds of the designated area and that fishing activity does not fluctuate widely from year to year. Landing data for the groups of organisms taken in southern California are given in table 14.7. Because the location of a catch is not indicated in the landing data, it is difficult to know what portion might have been taken on the high seas outside the SCB. Of the areas examined (table 14.6), the highest yield, 12.5 kcal m^{-2}, is from the Benguela (a fishery based largely on schooling herbivores), followed by Georges Bank, with a yield of 8.4 kcal m^{-2}. The Mid-Atlantic Bight is next (7.5 kcal m^{-2}), followed by the North Sea (5.9 kcal m^{-2}), the SCB (4.5 kcal m^{-2}), and finally the eastern Bering Sea, with the lowest value of 2.5 kcal m^{-2}.

Ratios of Production at Different Trophic Levels

Ratios of annual productivity estimates for different trophic levels are shown in table 14.8. These ratios, converted to percentages, are taken from data in table 14.6.

The ratio of macrozooplankton to phytoplankton in the SCB is intermediate to that of the other areas. Its value of 5.3% is greater than that for the Mid-Atlantic Bight (4.4%) and about the same for Georges Bank (6%), but it is much lower than either the North Sea (8.2%) or the outer domain of the eastern Bering Sea Shelf (9.3%). The outer shelves of the Bering Sea and (to a lesser extent) the North Sea are dominated by a pelagic food web that appears to account for these high ratios of large zooplankton to phytoplankton. The microzooplankton to phytoplankton ratios are higher than the macrozooplankton to phytoplankton ratios by 62–86% in the different ecosystems, their order being the same as for the macrozooplankton. If all the zooplankton are considered, the SCB is still intermediate, but the Benguela (data only available for the total) is the lowest, at 7.8%, followed by the Mid-Atlantic Bight (11.5%), the SCB (12%), Georges Bank (14.6%), the North Sea (17.5%), and the outer domain of the Bering Sea (20.5%). Overall, among the systems examined here, these ratios indicate an intermediate level of secondary production for the SCB. The relatively low value for the Benguela probably results from a high fallout of ungrazed phytoplankton in this upwelled area, and the high values on the outer Bering Sea shelf result from a physical regime that permits large, overwintering zooplankton to be present in abundance at the time of the spring bloom (Cooney and Coyle 1982).

The macrobenthos versus phytoplankton production estimates indicate that the SCB is intermediate to the other systems. The highest values for the Mid-Atlantic Bight (5.8%) and North Sea (4.4%) are indicative of systems with a high organic particle flux to the sediments, while Georges Bank (2.9%) and the outer domain of the eastern Bering Sea shelf (2.2%) reflect systems with a high consumption of particulate organic matter in the water column, particularly the euphotic zone.

The production rates of marine mammals versus phytoplankton are 30% higher in the SCB than in other systems examined, reflect-

Table 14.7. *Selected Commercial Fishery Landings (in t × 10^2) in Southern California Averaged over the Years 1980–1987*[a]

Category	Santa Barbara	Los Angeles	San Diego	Totals	Percentage of Total Landing
Abalone	3.8	1	0.64	39.6	2.2
Squid	2.8	45.4	0.03	73.4	4.0
Higher crustaceans	3.6	3.4	2.0	9.0	0.5
Sea urchins	63.6	17.9	6.7	88.2	4.8
Anchovies and sardines	36	127.7	0.01	163.7	9.0
Mackerel	45.4	431.3	0.18	476.9	26.2
Large carnivores	2.4	720	207	929.4	51.0
Soft-bottom fish	3.4	13.1	1.9	18.4	1.0
Hard-bottom fish	5.9	6.7	7.5	20.1	1.1
Totals	164.1	1,325.5	226	1,819	99.8

[a]Data provided by J. Cross and obtained from the National Marine Fisheries Service, Terminal Island Facility, California.

ing the richness of this system in marine mammal production. Conversely, marine birds, not listed in table 14.7, are very unproductive in the SCB (<12% that of the eastern Bering Sea and 6% that of Georges Bank). The reasons for low productivity are not clear, but there appears to be a low abundance of birds that occupy the SCB on a year-round basis, as compared to other areas, even when including the area immediately north of Point Conception (Briggs et al. 1981).

The macrobenthos versus macrozooplankton productivity is a comparison between two groups of organisms that represent a large portion of the secondary production of the ecosystem. The macrozooplankton feed mostly in the water column, while macrobenthos feed near the sediments, but both feed closely coupled to primary production. A high ratio (>50%) implies that a high proportion of the products of primary production are not grazed in the water column, but sink to the bottom where they support a benthic food web. Conversely, a low ratio (e.g., 22% for the outer domain of the Bering Sea) implies a pelagic food web where most of the primary production is grazed before sinking to the bottom. According to this analysis, the Mid-Atlantic Bight (131%) and the SCB (74%) have a highly productive benthic biota, the North Sea (53%) and Georges Bank (48.5%) are intermediate, and the outer domain of the eastern Bering Sea shelf has a low-producing benthic system.

More dramatic evidence of the differences in ecosystem dynamics, with respect to pelagic versus benthic food webs, can be seen on the Bering Sea shelf. The outer domain (as previously indicated) has a ratio of macrobenthos to macrozooplankton of 22, strongly indicating a pelagic food web. The middle domain, separated from the outer domain by a permanent oceanographic front (Coachman 1986), has a ratio of 294; not shown here (see Cohen and Grosslein 1987) is a clear example of a benthic food web. The ecological reasons for the great differences in these two systems geographically adjacent to each other lie in how the spring bloom is coupled to the zooplankton community (Hood 1986). The outer domain is coupled, but the middle is decoupled in the presence of zooplankton, which leads

Table 14.8. *Ratios (in %) of Annual Production at Different Trophic Levels for Several Ecosystem Components*

Ratio of Estimated Production	SCB	Bering Sea Outer Domain	North Sea	Georges Bank	Mid-Atlantic Bight	Benguela
Macrozooplankton/ phytoplankton	5.3	9.3	8.2	6.0	4.4	—
Microzooplankton/ phytoplankton	7.5	11.2	9.3	8.5	7.1	—
All Zooplankton/ phytoplankton	12.0	20.5	17.5	14.6	11.5	7.8
Macrobenthos/ phytoplankton	3.9	2.2	4.4	2.9	5.8	—
Mammals/ phytoplankton	0.1	0.07	0.02	0.03	—	—
Macrobenthos/ macrozooplankton	74	22	53	48.5	131	—
Total zooplankton plus macrobenthos/ phytoplankton	20.6	22.8	21.9	17.5	17.2	
Fish/phytoplankton	4.0	3.1	1.1	1.5	0.8	0.8
Fish/ macrozooplankton	75	33	13	26	18.2	—
Fish/all zooplankton	31	15	6	10.6	7.0	10
Fish landing/ phytoplankton	0.17	0.14	0.26	0.25	0.23	0.11
Fish landing/ fish production	3.8	4.5	24	16	29	13.8

to vastly different ecosystems and processes of energy transfer.

If we assume that phytoplankton productivity is largely consumed by the secondary producers (represented by all the zooplankton and the macrobenthos) and not transported from the system unused, then the ratio of all zooplankton plus macrobenthos to phytoplankton should be fairly constant in similar ecosystems. As seen in table 14.8, this ratio is 17.2–22.8% for the systems compared. These ratios are remarkably close considering the many measurements required and assumptions made and that no allowance is made for oceanographic differences in the systems.

Fish production in the SCB is higher than in any other area compared except the Benguela (table 14.6). Likewise, the fish versus phytoplankton ratio is higher (4.0%) than in any other system—as much as fivefold that of the Mid-Atlantic Bight (0.8%) and Benguela (0.8%). The outer domain of the eastern Bering Sea shelf is about 80% (3.1%) that of the SCB (4.0%). The ratio of fish production to other trophic levels (i.e., fish to macrozooplankton or fish to all zooplankton) is also much higher than for other systems. This discrepancy leads one to suggest that fish production for the SCB is overestimated. The most abundant fish species of the SCB is the anchovy, which according to Hewitt et al.

(1976) has an average biomass of 31–55 g m^{-2} in the SCB. Based on a biomass of 35 g m^{-2}, generic data regarding consumption rates (3%), and growth efficiency (12%), estimates of the productivity of this species are 57 kcal m^{-2} yr^{-1}, or more than 90% of the total fish production. The high ratios observed clearly result from the high production rate assigned to the anchovy. Catch statistics since 1976 indicate that the anchovy populations have apparently decreased and sardines have increased. If the combined biomass has not changed, these calculations would not be greatly affected.

The ratio of fish landings to phytoplankton production divides into two groups within the six systems compared. The Benguela, Bering Sea, and the SCB have only 0.11–0.17% of phytoplankton production landed as fish, while the North Sea, Georges Bank, and the Mid-Atlantic Bight have 0.25–0.28% of their primary production landed. The interesting ratio of fish landings to fish production places the SCB on the low side (3.8%) just below the Bering Sea (4.5%), while the Mid-Atlantic Bight is high (29%), followed by the North Sea (24%), Georges Bank (16%), and the Benguela (13.8%). If fish production in the SCB and the Bering Sea were estimated high, part of the differences in catch efficiency might be explained. However, many other factors such as fishing effort, higher marine mammal production, and foraging habits of large carnivorous fish are involved in creating these differences.

Ecosystem Food Chains and Energy Transfer

The structure and functioning of the many ecosystems of the SCB are the central theme of many presentations in this book. In this section, we attempt to summarize the information presented in this work and elsewhere on the predator–prey relationships of the major species of organisms that inhabit the SCB and to indicate, to the extent possible, the flow of energy through the many ecosystems. This effort was initiated by preparing table 14.9, which indicates the food habits of the major groups of predators according to the relative amounts of prey consumed. There are notable deficiencies in this table because of the lack of quantitative data for several groups listed (higher crustaceans, hard-bottom fish, and birds) and because almost no data are available for others (notably, pelagic molluscs [other than cephalopods], pelagic coelenterates, tunicates, and the meio- and microfauna). Because of these deficiencies, the path for energy flow cannot be complete and an exercise that involves balancing the energy flow for the total SCB lacks some detail. Despite these difficulties, many of the paths of energy flow are reasonably well established and can be used as a valuable tool in analyzing how the SCB functions as a system and in helping to identify areas of research needing greater emphasis.

One purpose of examining predator–prey relationships—that is, what and how much each predator consumes—is to gain insight into how energy flows through the system. By combining the information in table 14.9 with that on biomass and consumption rates discussed in previous sections, it is possible to estimate the energy transfer from prey to predator groups. Accordingly, table 14.10 reveals a number of relationships important to understanding how the SCB functions as an energy distributing system.

Microorganisms are the dominant group in consumption and transport of energy and, closely coupled to phytoplankton, occupy a position at least twice the magnitude (955×10^{11} kcal yr^{-1}) of the microzooplankton (450×10^{11} kcal yr^{-1}), the next most dominant group. Composed mostly of bacteria, freely suspended organisms are the exclusive users of the abundant supply of dissolved organic matter (DOM) excreted by phytoplankton and all other organisms, while other microorganisms adhere to particulate material as some of it is converted to microbial biomass. This group of organisms, either in

Table 14.9. *Food Habits of Major Consumers Found in the SCB*

Consumer	Food Items (% of total)	References
Microorganisms	100% DOC plus detritus	Chapter 4
Microzooplankton	90% phytoplankton; 10% microorganisms	Beers and Stewart 1970; Beers 1986
Macrozooplankton	40% phytoplankton; 50% microzooplankton; 10% ichthyoplankton	Kleppel et al. 1988; Brewer et al. 1984
Higher crustaceans	Detritus; macrozooplankton; ichthyoplankton; meiobenthos	
Cephalopods	25% microorganisms and detritus; 50% crustaceans; 15% pelagic fish; 10% macrobenthos	Bakus 1989; Paxton 1967; Nixon 1987
Fish		
Soft-bottom	80% crustaceans; 5% soft-bottom fish; 5% macrobenthos; 10% meiobenthos	Chapter 9; Cross et al. 1985; Allen 1982
Hard-bottom	Macrophytes; crustaceans; cephalopods; pelagic fish; macrobenthos	Chapter 9
Pelagic		
Anchovy	50% phytoplankton; 30% microorganisms; 15% macrozooplankton; 5% pelagic fish	Chapter 9; Mearns 1982; Bakus 1989; Hewitt et al. 1976
Mackerel	90% anchovy; 10% other	Chapter 9
Large carnivores	35% crustaceans; 25% cephalopods; 40% anchovies and other fish	Brownell et al. 1976; Mallicoate and Parrish 1981
Mesopelagic	50% macrozooplankton; 40% euphausiids; 10% chaetognaths and salps	Chapter 9; Pearcy and Laurs 1966
Meio- and microbenthos	90% microorganisms and detritus; 10% other	Chapter 4; Feder and Jewett 1987
Macrobenthos	50% microorganisms and detritus; macrophytes; crustaceans; meio- and microbenthos; macrobenthos	Chapter 8; Laughlin 1982; Feder and Jewett 1987
Sea birds	Macrozooplankton; crustaceans; cephalopods; anchovies	Chapter 10; Schneider et al. 1986
Cetaceans	10% macrozooplankton; 40% crustaceans; 25% cephalopods; 10% large carnivores; 10% pinnipeds; 5% other	Chapter 11; Calkins 1987
Pinnipeds	10% crustaceans; 50% cephalopods; 20% anchovies; 10% mackerel; 10% large carnivores	Chapter 11; Lavigne et al. 1982
Landings	See table 14.7	

Table 14.10. *Estimated Energy Transferred (kcal × 10^{11}) from Prey Groups to Predators in the SCB*[a]

Predators	Prey[b]																	
	Ph + Ex	Mph	Mio + De	Miz	Maz	Hcr	Cep	Fsb	Fhb	Anc	Mac	Lca	Mep	Mib	Mab	Mbi	Cet	Pin
Microorganisms	860		95															
Microzooplankton	405		45															
Macrozooplankton	88		46	85														
Higher crustaceans		X	X		X									X				
Cephalopods			X		X	X							X		X			
Fishes, soft-bottom						10.5		0.6						1.3	0.6			
Fishes, hard-bottom		X			X	X	X				X			X	X			
Fishes, pelagic																		
Anchovy	196		118			50	11.8							19.6				
Mackerel										8.9								
Large carnivores						12	8.5			14								
Mesopelagic					13	8.6												
Meio- and micro-benthos			X															
Macrobenthos			245			X		X						X	X			
Sea birds					X	X	X			X								
Cetaceans					0.30	1.2	0.75			0.3		0.30						0.3
Pinnipeds						0.50	2.30			0.9	0.46	0.46						
Landings						0.01	0.1	0.02	0.03	0.2	0.6	1.2			0.04			

[a] X indicates that only qualitative data are available.

[b] Ph = phytoplankton; Ex = exudates; Mph = macrophytes; Mio = microorganisms; De = detritus; Miz = microzooplankton; Maz = macrozooplankton; Hcr = higher crustaceans; Cep = cephalopods; Fsb = fishes, soft-bottom; Fhb = fishes, hard-bottom; Anc = anchovies; Mac = mackerel; Lca = large carnivores; Mep = mesopelagic; Mib = meio- and microbenthos; Mab = macrobenthos; Mbi = marine birds; Cet = cetaceans; Pin = pinnipeds.

the water column or in the intestines of other living forms, is largely responsible for the mineralization of nutrients. Thus, along with physical processes, this group supplies recycled nutrients to the euphotic zone to support primary production. The biomass produced by microorganisms is used heavily by microzooplankton and the meiofauna groups, but these alone do not account for all the predation observed (Azam 1986).

The benthos also contains a large biomass of microorganisms (540 mg C m^{-2}) that has a high rate of productivity (8.4 g C m^{-2} yr^{-1}). Probably only 25% of these are bacteria, the rest being protozoan and metazoan grazers (see section on the benthos). Because the function of this group in energy flow is not clear, it is left as "unknown," along with the other members of the meiofauna, in the energy flow schemes to follow. Because the major source of food for this group is thought to be detritus, furnished by all the other groups, its omission from the scheme has minimal impact on the other components.

Microzooplankton, consuming mostly phytoplankton and detritus, are next to microorganisms in importance to the flow of energy through the system. This secondary producer—coupled with that portion of macrozooplankton (e.g., anchovy) and macrobenthos that feed directly on phytoplankton, macrophytes, or detritus—constitutes the main body of herbivores in the community. Energy accumulated at this level in the form of biomass constitutes the foundation of the food web for all carnivores. The next trophic level, or those organisms that feed mainly on zooplankton, herbivorous fish, and macroinvertebrates, are represented, in part, by the higher crustaceans and the cephalopods, for which little quantitative data are available. These deficiencies, and a complete lack of data on other large groups occupying the same trophic level, represent the greatest obstacle in calculating an energy balance for the SCB. While an energy-balanced flow scheme would be highly desirable, our level of understanding of any ecosystem in the world is still insufficient to accomplish that end. The aim, then, is to use the best information available to approximate an energy flow scheme that might eventually be developed into a balanced energy budget.

The data for prey–predator relationships for the higher trophic levels of the SCB is reasonably complete, as indicated in table 14.10. The flow of energy is highest through the pelagic fish, represented here by anchovy, which is primarily a herbivore. Because of their large biomass, anchovy consume large quantities of phytoplankton and detritus. Mackerel feed largely on anchovy. Large carnivores are heavy consumers of large crustaceans and cephalopods, and the mesopelagic organisms feed on large crustaceans and macrozooplankton. The cetaceans and pinnipeds, most directly in competition with humans, extract more energy from the higher trophic levels of prey than do humans, with the exception of the large carnivores (represented mostly by tuna) and mackerel, of which commercial fisheries may extract slightly more.

Scheme for Annual Energy Flow

Information provided in tables 14.8, 14.9, and 14.10, accompanied by biomass and productivity data presented in earlier sections, provides the elements necessary for estimating the flow of energy through the SCB ecosystem. Table 14.11 is the product of such an exercise in which those groups of organisms for which quantitative information is available (either from this work or from the literature) were used to calculate the energy flow involved in production, respiration, assimilation, consumption, and excretion through the food web. The scheme of energy flow based on the data given in table 14.11, supplemented with data from tables 14.8, 14.9, and 14.10, is presented in figure 14.7. The detritus and dissolved organic matter component produced and utilized appears to be in balance, except for the unknown amount of energy represented by "x."

Table 14.11. *Average Biomass (B) and Annual Production (P) of Ecosystem Components in the SCB and Elements of Annual Energy Balance*[a]

Components of Ecosystem	Data for Calculating Energy Balance						Elements of Energy Balance × 10^{12} kcal yr^{-1}				
	B (10^5 t C)	P (10^5 t C)	P/B	Coeff. C to kcal	K_2 (P/A)	I (A/Co)	P	R $P(1-K_2)/K_2$	A (P + R)	Co	F A/I
Phytoplankton	2.3	110	48	11.3	—	—	124	—	—	—	—
Exudates	—	20	—	10	—	—	20	—	—	—	—
Macrophytes	—	12	—	10	—	—	12	—	—	—	—
Bacteria	0.94	28.5	30	10	0.30	1	28.5	66.5	95	95[b]	—
Microzooplankton	0.31	8.6	28	10	0.24	0.8	8.6	27	36[c]	45[d]	9
Macrozooplankton	0.40	6.1	6.5	10	0.38	0.73	6.1	10	16[e]	22[f]	6
Macrobenthos	3.24	4.9	1.5	10	0.14	0.70	4.9	30	34	49[g]	15
Fish	4.1	5.0	1.2	10	0.16	0.65	5.0	26	31	48	17[h]
Birds	0.008	0.009	1.1	10	0.15	0.67	0.009	0.05	0.06	0.09	0.03[i]
Cetaceans	—	0.06	—	10	0.27	0.73	0.06	0.16	0.22	0.3	0.05[j]
Pinnipeds	—	0.065	—	10	0.17	0.82	0.065	0.32	0.38	0.46	0.08[j]
Landings	0.23	—	—	—	—	—	0.23	—	—	—	—

[a] R = metabolic loss; A = assimilated food; Co = consumption; F = nonassimilated part of consumption. All data are for the entire SCB (7.8 × 10^4 km^2).

[b] Based on 30% growth efficiency (Sorokin 1983).

[c] Respiration taken as 15% of body weight per day (McAllister 1972).

[d] Consumption estimated at 40% body weight per day.

[e] Respiration taken as 7% of body weight per day (McAllister 1972).

[f] Consumption estimated at 15% body weight per day.

[g] Macrobenthos is assumed to have 10% growth efficiency (Feder and Jewett 1987).

[h] Value for excretion in fish varies widely with age, species, and food availability. An average value of 25% was chosen.

[i] Based on a consumption:assimilation ratio of 1.33 (Schneider et al. 1986).

[j] Excretion rate of 17% for pinnipeds (Lavigne et al. 1982) is also used for cetaceans for which no data were found.

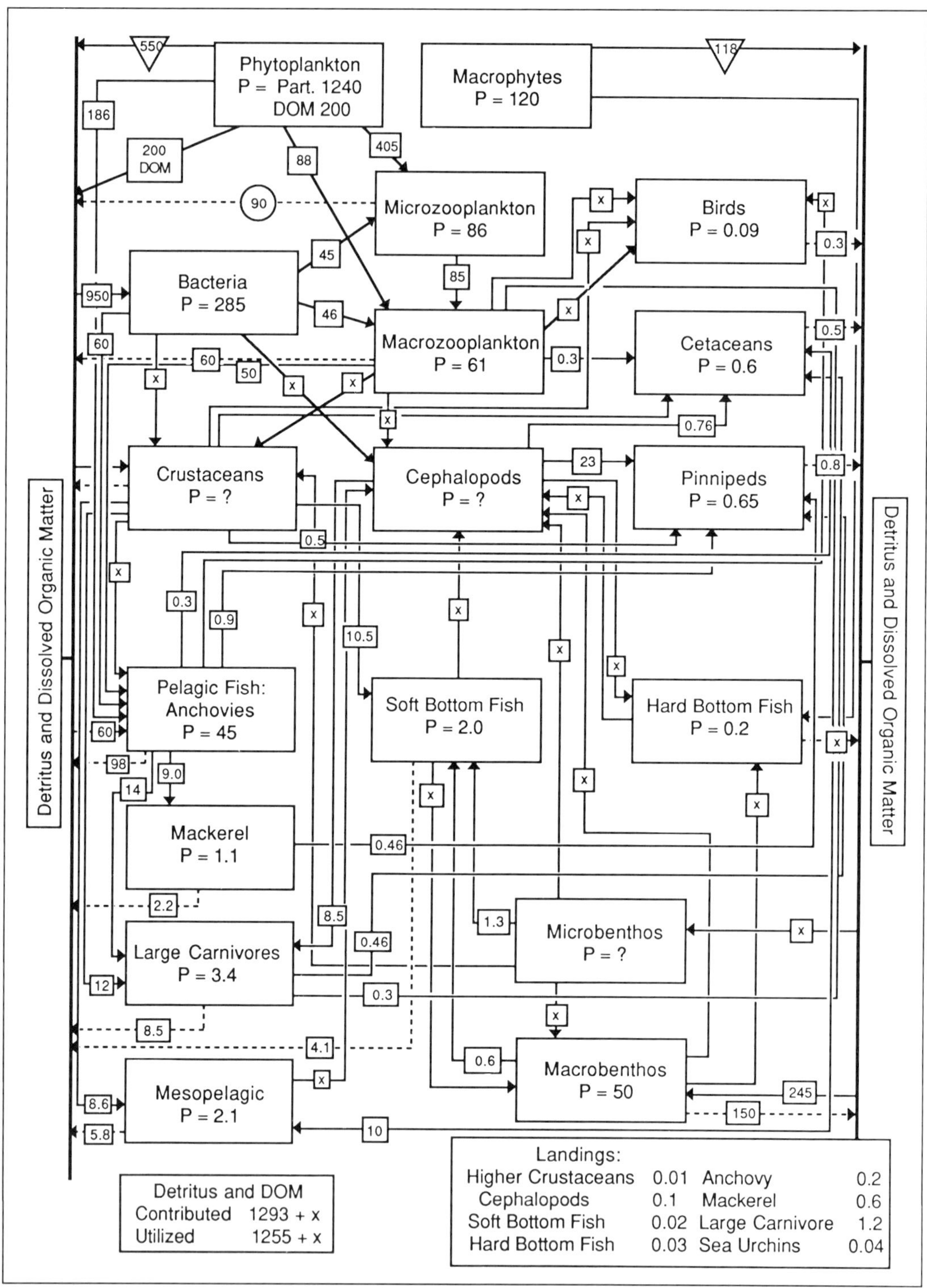

Figure 14.7. Scheme for energy flow in the SCB, in accordance with data given in tables 14.8 and 14.9. Units are in kcal $\times$ 10^{11} yr^{-1}. P = productivity; numbers in triangles = nongrazed production; numbers in squares = consumption by subsequent member of the trophic chain; numbers in circles = nonassimilated food.

The hard-bottom fish, higher crustaceans, and cephalopods contribute an unknown amount of detritus to the pool, and meio- and microfauna and higher crustaceans make use of an unknown amount of the detritus pool. If these rates of consumption and utilization are about the same, then this "catch-all" detritus pool would be balanced. If balance occurs, then no detritus remains to provide for the input losses to the sediment that are known to occur. Also, there is no allowance for allochthonous inputs of energy (see table 14.3), which amount to 11.3×10^{11} kcal yr^{-1} or about 0.7% of the total organic energy input. The allochthonous organic matter enters the system either as dissolved or detrital material and, as such, would enter the pool containing the very large component of photosynthetic products. The relatively small amount derived from this source would be undetectable in the scheme used here to follow energy flow through the entire SCB. Some of these inputs, which are discharged from point sources, could be important in local areas.

Summary and Prospectus for Future Research

It has been the intention here to develop a scheme for the flow of energy through the main pathways in the ecosystem under natural conditions, to look for changes that might occur as a result of natural disturbances, and to compare these to disturbances that might be imposed by human usage of the resource. The purpose of this exercise is to see if natural processes are adversely influenced by human activities or if natural processes prevail—the assumption being that if natural processes dominate, any perturbation to the system induced by humans would be reversible and transitory.

Energy is expressed here as calories fixed by biological processes into an organic form that is used by organisms for metabolism and growth. Once initiated, these processes are largely biological, but the physical and chemical environment necessary for the growth and development of biological systems is controlled by physical and chemical processes (nutrient supply, water column stability, and light) that usually operate on the microscale (centimeters to meters) and mesoscale (meters to 100 kilometers), yet are often affected by large-scale phenomena (>100 kilometers), such as El Niño events or weather systems. It follows that ecosystems thrive because of contributions made by nonbiological processes working in consort to establish a hospitable environment for growth.

The SCB ecosystem benefits from physical and chemical features that provide an environment for growth comparable to that of other major coastal ecosystems. Annual variations of salinity (33.4–33.6%) and temperature (14–19°C) are the smallest of any of the systems examined, although the volume and speed of the dominating current (California Countercurrent) are relatively high and the flushing time rapid (2–3 months).

The continental shelf of southern California is vastly different from other shelves examined. The nearshore waters are deeper and the bottom topography is much rougher, resulting in habitats of rapidly changing depths, an abundance of hard- and soft-bottom regimes, and many deep basins, some interconnected at variable depths, but all with sills. It is these features plus the many island outcrops that clearly distinguish the SCB from the other systems. From the perspective of energy flow, these distinguishing features would show up mostly in the benthic portion of the ecosystem. Because of the difficulty of diving to the depths involved as well as sampling problems, quantitative data for the deep water habitats is lacking, preventing detailed analysis of the influence of these distinguishing features on the ecosystem.

The dissolved oxygen regime of the SCB is similar to those of other continental shelf ecosystems to about 100 m depth (nearly saturated). However, below 100 m, the oxygen concentrations fall to less than 50% saturated

($3 \text{ ml } l^{-1}$) due to the input of hypoxic intermediate Pacific water transported by the California Countercurrent. This feature is typically uncommon to continental shelves except under upwelling conditions, (e.g., Peru and Benguela). Its importance to energy flow, however, is probably minimal since indigenous biota appear to pass reasonably freely through these waters in their diel migrations to find food in surface layers.

As in the other ecosystems examined, the amount of energy generated by the photosynthetic process is limited by the supply of nutrients (usually nitrogen) to the euphotic zone. Much recycling (which supports growth of the plant community) occurs in the euphotic zone, but because of the loss of energy to deeper waters through the sinking of ungrazed (unused) particulate organic matter, a portion of the nutrients must be replaced from below (the most important is usually nitrate). This *rate* of transfer appears to be the most important feature that determines the *amount* of energy that is fixed by coastal ecosystems. The SCB, because of its physics, has a nitracline lying just below the surface mixed zone. This results in a sharp increase of nitrate at the bottom of the euphotic zone but below the mixed zone. Transfer of nitrate from deeper, nutrient-rich layers to the bottom of the euphotic zone is controlled by the rate of vertical eddy diffusivity. On the average, the coefficient of eddy diffusivity for the SCB is about $4 \text{ cm}^2 \text{ s}^{-1}$, which amounts to the transfer of 9.4×10^{11} g NO_3-N yr^{-1} to the euphotic zone of the SCB. If we assume a C:N ratio of 6.5 and g C to kcal of 11.3 for these waters, this would yield 69×10^{12} kcal of energy, or about 48% of the total primary production (144×10^{12} kcal). Conversely, this would indicate that 48% of the primary production would sink ungrazed below the euphotic zone, a value close to that observed (40%) in the California Basin Study, especially if the occurrence of preferential recycling of N in the euphotic zone is considered. The SCB appears to have the lowest rate of vertical transport of nutrients and also the lowest rate of primary production of any of the continental shelf ecosystems examined here.

Microorganisms, especially the bacteria, play a major role in the flow of energy in the SCB. Consuming dissolved organic matter and detritus, this group of organisms, in the water column alone, produces about 40% as much energy to be passed up the food web as do the primary producers. This level of activity appears to be about the same as at Georges Bank, the only other system examined.

The calculated productivity of the micro- and macrozooplankton in the SCB is very low compared to other systems. The estimates are based on limited data and some general assumptions that may have led to these low results. However, predation on these groups of organisms is great, which may limit their biomass and thus their productivity. It is not clear why differences exist between the SCB and other systems.

Quantitative data for the intermediary trophic levels in the system are insufficient to allow estimates of levels of biomass, consumption, or productivity of this important group on which many apex predators feed. The situation for the fishes and marine mammals is better understood and is reflected in the energy flow scheme of figure 14.7.

Clearly, changes have occurred in the SCB over the last decade. These changes probably are best illustrated in the changing composition of commercial landings between 1980 and 1987. As shown in table 14.12, the landings of many of the fisheries species changed radically. Abalone declined at all ports and squid increased nearly eightfold, but the most dramatic changes took place in the decline of the northern anchovy and the increase in sardines. Overall yields have declined in both poundage and in value. What may have caused these changes? The analyses presented here are not sufficiently rigorous to answer this question fully, although some observations are relevant. Since 1920, primary production has varied by as much as twofold between years, but on the average has remained

Table 14.12. *Changes in Landings (landings × 10^{-3} lbs) in the SCB Between 1980 and 1987*

Species Landed	1980	1987	Percentage of 1980
Abalone			
Santa Barbara	796	498	62
Los Angeles	207	98	47
San Diego	127	16	13
Squid			
Santa Barbara	2,413	18,653	773
Los Angeles	15,695	11,418	63
San Diego	11	6	56
Higher crustaceans			
Santa Barbara	734	12,992	187
Los Angeles	376	955	253
San Diego	344	512	149
Northern anchovy			
Santa Barbara	22,817	336	1.4
Los Angeles	64,175	253	0.4
San Diego	—	—	—
Sardine			
Santa Barbara	0	336	—
Los Angeles	115 (1981)	699	610
San Diego	—	—	—
Mackerel			
Santa Barbara	9,858	12,992	131
Los Angeles	99,481	104,819	105
San Diego	17	24	143
Total landings			
Santa Barbara	55,892	52,211	93
Los Angeles	383,336	203,162	52
San Diego	202,308	7,710	3.8

nearly constant. Zooplankton standing stocks between the years 1951 and 1966 fluctuated by about sixfold, but showed no long-term trends. Between 1951 and 1979, the anchovy biomass fluctuated about 30-fold but also showed no long-term trends (Smith and Eppley 1982). Unfortunately, time series data are not available for the decade of the 1980s to compare with the landing data. Nevertheless, even though these natural fluctuations within the lower trophic populations are large, the absence of long-term trends does not explain the rather consistent decrease in landings since 1980.

Other factors may cause declines in landings, but alone they seem inadequate to pinpoint cause and effect. Among these are the increased loading of the system by anthropogenic inputs that may inhibit the growth of specific species. Such inputs are local, or regional at most, except for those that are airborne. Local inputs would be expected to show significant regional differences, but this does not appear to be the case for the landings; also these effects should be seen in the long-term trends in populations of the lower trophic levels, which is not the case.

Harvesting at rates exceeding the sustainable yield is often the cause of decreased landings in many fisheries, but it is usually identifiable with a single species. Decrease in the catch of anchovy, with the increase in the sardine catch, might be indicative of such a problem, which is similar to that experienced in the upwelling region off Peru. However, adding the 87 million pounds of anchovy caught in 1980 to the total catch of 1987 would not alter the 1987 catch appreciably. Besides, the greatest decrease in total landings was in San Diego, an area in which anchovy were not harvested in quantity.

The increased take of squid and higher crustaceans, which are important prey for many of the landed species, appears to be a factor in the decreased catch. However, the energy removed by the fishery is small compared to that consumed by the marine mammals. The increase in population of the marine mammals during recent years (largely the result of efforts to protect them) is implicated as one of the causes of the decline in landings. Determination of the extent of this influence must await the availability of an extended time series of quantitative data.

The ecology of the SCB, despite heavy usage by humans, clearly appears to be controlled by natural processes. Oceanic inputs, dominated by the flow of the California Countercurrent, are apparently great enough to overpower effects caused by human activity.

Opportunities to contribute to understanding of how marine ecosystems function exist in every aspect of marine science, but none has more promise than the quantification of ecological rates and processes (Longhurst 1984). In the analysis undertaken here, the greatest difficulty encountered was in efforts to obtain values for production, respiration, consumption, excretion, and assimilation—all related to the physiology of organisms. Without reliable values for these processes, little can be done to develop a complete energy flow scheme for even a small segment of the system, much less the whole. Other physiologically related measurements, such as growth efficiency and caloric content of prey, are important. Published results for these values differ widely.

Another area of needed research is the acquisition of time series data designed to give information on the variability in physical and chemical conditions as well as on the numbers and distribution of animals. Before any predictions are attempted about the potential effects of human-caused changes, we need a reasonably long time series (10 years or more) of data to determine the nature and effects of natural fluctuations in the SCB ecosystem.

Great difficulty was encountered in estimating the kinds of prey and the proportions of each prey consumed by predators. This probably imposed more uncertainty in calculated energy flow in the SCB than any other consideration. Reliable quantitative estimates of rates of consumption of each prey organism by specific predators is required to evaluate the effects of changes in predator populations on the path of energy flow.

As difficult as it is to define a quantitative energy flow for a ecosystem, the exercise has rewards. At least it more clearly defines the problem and allows one to focus efforts on filling deficiencies in our information and understanding. We hope this chapter has made some progress in this regard.

Acknowledgments

The author wishes to acknowledge the help the contributors to this book gave so freely in providing information needed for preparation of this chapter. Special thanks goes to Dee Dee Rypka for helping in so many ways in the physical preparation of the manuscript; to Dr. David Menzel, Director of Skidaway Institute of Oceanography, for reviewing the manuscript; to the editors for selecting me to write the analysis chapter; to the advisory review board for their encouragement; and to the Marine Minerals Management Service for their partial financial support.

Literature Cited

Allen, L. G., 1982. Seasonal abundance, composition, and productivity of the littoral fish assemblage in Upper Newport Bay, California. *U.S. Natl. Mar. Fish. Serv. Fish. Bull.* 80(4): 769–790.

Angel, M. V., 1984. Detrital organic fluxes through pelagic ecosystems. In: M. J. R. Fasham, ed. *Flows of Energy and Materials in Marine Ecosystems.* Plenum Press, New York. pp. 475–516.

Angel, M. V., and A. C. Baker, 1982. Vertical distribution of the standing crop of plankton and micronekton at three stations in the Northeast Atlantic. *Biol. Oceanogr.* 2(1):1–30.

Azam, F., 1986. Nutrient cycling and food web dynamics in the Southern California Bight: The microbial food web. In: R. W. Eppley, ed. *Lecture Notes on Coastal and Estuarine Studies. Vol. 15. Plankton Dynamics of the Southern California Bight.* Springer-Verlag, Berlin. pp. 274–288.

Azam, F., T., Fenchel, J. G. Field, J. S. Gray, L. A. Meyer-Reil, and F. Thinkstad, 1983. The ecological role of water column microbes in the sea. *Mar. Ecol. Prog. Ser.* 10:257–263.

Bailey, G. W., and P. Chapman, 1985. The nutrient status of the St. Helen Bay Region in February 1979. In: L. V. Shannon, ed. *South African Ocean Colour and Upwelling Experiment.* Sea Fisheries Research Institute, Cape Town. pp. 125–145.

Baker, E. T., 1981. North Aleutian shelf transport experiment. U.S. Dept. of Commerce, NOAA (Natl. Oceanic Atmos. Admin.), OCSEAP. Annual Report 6:329–330.

Bakkala, R., and L. L. Low, eds., 1982. Condition of the groundfish resources of the eastern Bering Sea and Aleutian Islands region in 1982. Natl. Mar. Fish. Serv., Northwest and Alaska Fish. Center, Seattle, WA. Unpubl. Report. 187pp.

Bakus, G. J., 1989. The marine biology of southern California. Allan Hancock Found. Occas. Pap. (New Ser.) No. 7, Univ. of Southern California, Los Angeles. 61pp.

Banse, K., and S. Mosher, 1980. Adult body mass and annual production/biomass relationships of field populations. *Ecol. Monogr.* 50:355–379.

Beers, J. R., 1986. Organisms and the food web. In: R. W. Eppley, ed. *Lecture Notes on Coastal and Estuarine Studies. Vol. 15. Plankton Dynamics of the Southern California Bight.* Springer-Verlag, Berlin. pp. 84–175.

Beers, J. R., and G. L. Stewart, 1970. Numerical abundance and estimated biomass of microzooplankton. In: J. D. H. Strickland, ed. The ecology of plankton off La Jolla, California, in the period April through September 1967. *Bull. Scripps Inst. Oceanogr. Univ. Calif.* 17:67–87.

Berelson, W. M., D. E. Hammond, and C. Fuller, 1982. Radon–222 as a tracer for mixing in the water column and benthic exchange in the southern California borderland. *Earth Planet. Sci. Lett.* 61:41–54.

Billen, G., and C. Lancelot, 1988. Modeling benthic nitrogen cycling in temperate coastal ecosystems. In: T. H. Blackburn, and J. Sorensen, eds. *Nitrogen Cycling in Coastal Marine Environments.* John Wiley & Sons, Ltd., Chicester. pp. 341–378.

Bishop, J. K. B., and J. M. Edmond, 1976. A new large volume filtration system for the sampling of ocean particulate matter. *J. Mar. Res.* 34(2): 181–198.

Brewer, G. D., G. S. Kleppel, and M. Dempsey, 1984. Apparent predation on ichthyoplankton by zooplankton and fishes in nearshore waters of southern California. *Mar. Biol.* 80:(1):17–28.

Briggs, K. T., and E. W. Chu, 1987. Trophic relationships and food requirements of California seabirds: Updating models of trophic impacts. In: J. P. Croxall, ed. *Seabirds: Feeding Ecology and Role in Marine Ecosystems.* Cambridge Univ. Press, Cambridge. pp. 297–304.

Briggs, K. T., E. W. Chu, D. B. Lewis, W. B. Tyler, R. L. Pitman, and G. L. Hunt, Jr., 1981. Distribution, numbers, and seasonal status of seabirds of the Southern California Bight. In: Marine mammal and seabird survey of the Southern California Bight Area, Summary Report 1975–1978, Book 3, Vol. III. Available from: NTIS, Springfield, VA; PB–81–248–197. pp. 1–212.

Brownell, R. L., Jr., L. Eber, A. R. Longhurst, W. F. Perrin, and P. E. Smith, 1976. A preliminary description of the California Current ecosystem including marine mammals. NOAA, Natl. Mar. Fish. Serv., Southwest Fish. Center, La Jolla, CA.

Bruland, K. W., K. Bertine, M. Koide, and E. D. Goldberg, 1974. History of metal pollution in southern California coastal zone. *Environ. Sci. & Technol.* 8(5):425–432.

Butman, B., and R. C. Beardsley, 1987. Physical oceanography. In: R. H. Backus, and D. W. Bourne, eds. *Georges Bank*. MIT Press, Cambridge, MA. pp. 88–98.

Calkins, D. G., 1987. Marine mammals. In: D. W. Hood, and S. T. Zimmerman, eds. *The Gulf of Alaska: Physical Environment and Biological Resources*. U.S. Goverment Printing Office, Washington, D.C. pp. 527–558.

Causey, B. D., 1969. The fish of Seven and One-half Fathom Reef. Master's Thesis, Texas A & I Univ., Kingsville, TX. 110pp.

Chapman, D. C., J. A. Barth, R. C. Beardsley, and R. G. Fairbanks, 1986. On the continuity of mean flow between the Scotian Shelf and the Middle Atlantic Bight. *J. Phys. Oceanogr.* 16: 758–772.

Coachman, L. K., 1986. Circulation, water masses, and fluxes on the southeastern Bering Sea shelf. *Cont. Shelf Res.* 5:23–108.

Coachman, L. K., and R. L. Charnell, 1979. On lateral water mass integration: A case study of Bristol Bay, Alaska. *J. Phys. Oceanogr.* 9:279–297.

Coachman, L. K., and J. J. Walsh, 1981. A diffusion model of cross-shelf exchange of nutrients in the southeastern Bering Sea. *Deep-Sea Res.* 28(8A):819–846.

Codispoti, L. A., G. E. Frederich, and D. W. Hood, 1986. Variability in the inorganic carbon system over the southeastern Bering Sea shelf during spring 1980 and spring–summer 1981. *Cont. Shelf Res.* 5:133–160.

Cohen, E. B., and M. B. Grosslein, 1987. Production on Georges Bank compared with other ecosystems. In: R. H. Backus, and D. W. Bourne, eds. *Georges Bank*. MIT Press, Cambridge, MA. pp. 383–391.

Colebrook, J. M., 1985. Sea surface temperature and zooplankton, North Sea, 1948 to 1983. *J. Cons. Cons. Int. Explor. Mer.* 42(2):179–185.

Cooney, R. T., and K. O. Coyle, 1982. Trophic implications of cross-shelf copepod distributions in the southeastern Bering Sea. *Mar. Biol.* 70:187–196.

Cooper, L. H. N., 1933. Chemical constituents of biological importance in the English Channel, November 1930 to January 1932. *Mar. Biol Assoc. (UK).* 18:677–753.

Costa, D. P., B. J. Le Beouf, A. C. Huntley, and C. L. Ortiz, 1986. The energetics of lactation in the northern elephant seal, *Mirounga angustirostris*. *J. Zool. (Lond.).* 209:21–33.

Crisp, D. J., 1975. Secondary production in the sea. In: D. E. Reichle, J. F. Franklin, and D. W. Goodal, eds. *Productivity of the World Ecosystem*. National Academy of Sciences, Washington, D.C. pp. 71–90.

Cross, J. N., J. N. Roney, and G. S. Kleppel, 1985. Fish food habits along a pollution gradient. *Calif. Fish Game.* 71(1):28–39.

Csanady, G. T., and P. Hamilton, 1987. Circulation of slope water. *Cont. Shelf. Res.* 8(5–7):565–622.

Csanady, G. T., and B. A. Magnell, 1987. Mixing processes. In: R. H. Backus, and D. W. Bourne, eds. *Georges Bank*. MIT Press, Cambridge, MA. pp. 163–169.

Daan, N., 1973. A quantitative analysis of the food intake of North Sea cod, *Gadus morhua*. *Neth. J. Sea Res.* 6(4):479–517.

Dagg, M. J., and J. T. Turner, 1982. The impact of copepod grazing on the phytoplankton of Georges Bank and the New York Bight. *Can. J. Fish. Aquat. Sci.* 39(7):979–990.

Dohl, T. P., M. L. Bonnell, and R. G. Ford, 1986. Distribution and abundance of common dolphin, *Delphinus delphis*, in the Southern California Bight: A quantitative assessment based upon aerial transect data. *U.S. Natl. Mar. Fish. Serv. Fish. Bull.* 84:333–343.

Dugdale, J. L., and J. J. Goering, 1967. Uptake of new and regenerated forms of nitrogen in primary production. *Limnol. Oceanogr.* 12(2):196–206.

Eber, L. E., 1977. Contoured depth–time charts (0 to 250 m, 1950–1966) of temperature, salinity, oxygen and sigma-t at 23 CalCOFI stations in the California Current. *Calif. Coop. Oceanic Fish. Invest. Atlas.* No. 25, 231 charts.

Eisma, D., 1981. The mass-balance of suspended matter and associated pollutants in the North Sea. *Rapp. P.-V. Réun. Cons. Int. Explor. Mer.* 181:7–14.

Eisma, D., 1987. The North Sea: An overview. *Trans. R. Soc. Lond. B.* 316:461–485.

Eisma, D., and J. Kalf, 1987a. Distribution, organic content and particle size of suspended matter in the North Sea. *Neth. J. Sea Res.* 21(4): 265–285.

Eisma, D., and J. Kalf, 1987b. Dispersal, concentration and deposition of suspended matter in the North Sea. *J. Geol. Soc. (Lond.).* 144(1): 161–178.

Eisma, D., G. C. Cadee, and R. Laane, 1982. Supply of suspended matter and particulate and dissolved organic carbon from the Rhine to the coastal North Sea. *Mitt. Geol.-Palaeont. Inst. Univ. Hamburg SCOPE/UNEP Sonderb.* 52: 483–505.

Emery, K. O., 1960. *The Sea off Southern California.* John Wiley & Sons, New York. 366pp.

Eppley, R. W., ed. 1986. *Lecture Notes on Coastal and Estuarine Studies. Vol. 15. Plankton Dynamics of the Southern California Bight.* Springer-Verlag, Berlin. 373pp.

Eppley, R. W., and O. Holm-Hansen, 1986. Primary production in the Southern California Bight. In: R. W. Eppley, ed. *Lecture Notes on Coastal and Estuarine Studies. Vol. 15. Plankton Dynamics of the Southern California Bight.* Springer-Verlag, Berlin. pp. 176–215.

Eppley, R. W., E. H. Renger, and W. G. Harrison, 1979. Nitrate and phytoplankton production in southern California coastal waters. *Limnol. Oceanogr.* 24(3):483–494.

Eppley, R. W., E. H. Renger, and P. R. Betzer, 1983. The residence time of particulate organic carbon in the surface layer of the ocean. *Deep-Sea Res.* 30(3A):311–323.

Falkowski, P. G., 1983. Light-shade adaptation and vertical mixing of marine phytoplankton: A cooperative field study. *J. Mar. Res.* 41(2): 215–237.

Falkowski, P. G., T. S. Hopkins, and J. J. Walsh, 1980. An analysis of factors affecting oxygen depletion in the New York Bight. *J. Mar. Res.* 38(3):479–506.

Feder, H. M., and S. C. Jewett, 1987. The subtidal benthos. In: D. W. Hood, and S. T. Zimmerman, eds. *The Gulf of Alaska: Physical Environment and Biological Resources.* U.S. Government Printing Office, Washington, D.C. pp. 347–396.

Feely, R. A., G. J. Massoth, A. J. Paulson, M. F. Lamb, and E. A. Martin, 1981. Distribution and elemental composition of suspended matter in Alaska coastal waters. *NOAA Tech. Mem. ERL PMEL–27.* 129pp.

Fellows, D. A., D. M. Karl, and G. A. Knauer, 1981. Large particle fluxes and the vertical transport of living carbon in the upper 1500 m of the northeast Pacific Ocean. *Deep-Sea Res.* 28A(9):921–936.

Fischer, P. J., 1978. Natural gas and oil seeps, Santa Barbara Basin. In: *The State Lands Commission 1977, California Gas, Oil, and Tar Seeps.* Sacramento, pp. 1–62.

Fiscus, W. H., 1986. Northern fur seal. In: D. Haley, ed. *Marine Mammals.* 2d ed. Pacific Search Press, Seattle, WA. pp. 175–181.

Flagg, C. N., 1987. Hydrographic structure and variability. In: R. H. Backus, and D. W. Bourne, eds. *Georges Bank.* MIT Press, Cambridge, MA. pp. 108–124.

Fuhrman, J. A., and F. Azam, 1980. Bacterioplankton secondary production estimates for coastal waters of British Columbia, Antarctica, and California. *Appl. Environ. Microbiol.* 39: 1085–1095.

Fuhrman, J. A., and F. Azam, 1982. Thymidine incorporation as a measure of heterotrophic bacterioplankton production in marine surface waters: Evaluation and field results. *Mar. Biol.* 66(2):109–120.

Gallaway, B. J., 1981. An ecosystem analysis of oil and gas development on the Texas–Louisiana continental shelf. FWS/OBS–81/27. U.S. Fish Wildl. Serv., Office of Biological Serv., Washington, D.C. 89pp.

Goering, J. J., and R. L. Iverson, 1979. Plankton distribution of the southeast Bering Sea shelf. In: PROBES Prog. Rep. Vol. II. Inst. Mar. Sci., Univ. of Alaska, Fairbanks, AK. pp. 225–259.

Gordon, D. C., 1970. Some studies of the distribution and composition of particulate organic carbon in the North Atlantic Ocean. *Deep-Sea Res.* 17(2):233–243.

Green, K. A., 1978. Ecosystem description of the California Current. Rep. to U.S. Marine Mammals Commission, MMC–77/11. Contract MM7 AC026. 73pp.

Griffin, G. M., 1979. Evaluation of the effects of oil production platforms on the turbidity of Louisiana shelf waters. In: C. H. Ward, M. E. Bender, and D. J. Reish, eds. Effects of oil drilling and production in the coastal environment. Rice Univ. Stud.. 65. 589pp.

Handa, N., and E. Tanoue, 1981. Organic matter in the Bering Sea and adjacent areas. In: D. W. Hood, and J. A. Calder, eds. *The Eastern Bering Sea Shelf: Oceanography and Biological Resources,* vol. 1. Univ. of Washington Press, Seattle, WA. pp. 359–381.

Harper, D. E., Jr., and L. D. McKinney, 1980. Evaluation of brine disposal from the Bryan Mound site of the Strategic Petroleum Reserve Program. In: R. W. Hann, Jr., and R. E. Randall, eds. *Final Report,* vol. 1, chap. 5. Texas A&M Research Foundation, College Station, TX. p. 79.

Hewitt, R. P., P. E. Smith, and J. C. Brown, 1976. Development and use of sonar mapping for pelagic stock assessment in the California Current area. *U.S. Natl. Mar. Fish. Serv. Fish. Bull.* 74:281–300.

Hood, D. W., 1986. Processes and resources of the Bering Sea shelf (PROBES). *Cont. Shelf Res.* 5(1–2):291.

Iturriaga, R., 1981. Phytoplankton photo-assimilated extracellular products; heterotrophic utilization in marine environments. *Kiel. Meeresforsch. Sonderh.* 5:318–412.

Jackson, G. A., 1986. Physical oceanography of the Southern California Bight. In: R. W. Eppley, ed. *Lecture Notes on Coastal and Estuarine Studies. Vol. 15. Plankton Dynamics of the Southern California Bight.* Springer-Verlag, Berlin. pp. 13–52.

Jackson, G. A., 1987. Modeling the growth and yield of giant kelp, *Macrocystis pyrifera,* off southern California. *Limnol. Oceanogr.* 22:979:995.

Jackson, G. A., F. Azam, A. F. Carlucci, R. W. Eppley, B. Finney, D. S. Gorsline, B. Hickey, C.-A. Huh, R. A. Jahnke, I. R. Kaplan, M. R. Landry, L. F. Small, M. I. Venkatesan, P. M. Williams, and K. M. Wong, 1989. Elemental cycling and fluxes off the coast of southern California. *EOS, Trans. Am. Geophys. Union.* 70:146–155.

Joiris, C., G. Billen, C. Lancelot, M. H. Daro, J. P. Mommaerts, A. Bertels, M. Bossicart, J. Nijs, and J. H. Hecq, 1982. A budget of carbon cycling in the Belgian coastal zone: Relative roles of zooplankton, bacterioplankton and benthos in the utilization of primary production. *Neth. J. Sea Res.* 16:260–275.

Jones, J. H., 1971. General circulation and water characteristics in the Southern California Bight. Report TR 101. South. Calif. Coastal Water Res. Proj., El Segundo, CA. 37pp.

Kinder, T. H., and L. K. Coachman, 1978. The front overlying the continental shelf in the eastern Bering Sea. *J. Geophys. Res.* 83:4551–4559.

Kleppel, G. S., D. Frazel, R. E. Pieper, and D. V. Holliday, 1988. Natural diets of zooplankton off southern California. *Mar. Ecol. Prog. Ser.* 49:231–241.

Knauer, G. A., and J. H. Martin, 1981. Primary production and carbon–nitrogen fluxes in the upper 1500 m of the northeast Pacific. *Limnol. Oceanogr.* 26(1):181–186.

Krempin, D. W., and C. W. Sullivan, 1981. The seasonal abundance, vertical distribution, and relative microbial biomass of chroococcoid cyanobacteria at a station in southern California coastal waters. *Can. J. Microbiol.* 27:1341–1344.

Lancelot, C., G. Billen, A. Sournia, T. Weisse, F. Colijn, M. J. W. Veldhuis, A. Davies, and P. Wassman, 1987. *Phaeocystis* blooms and nutrient enrichment in the continental coastal zones of the North Sea. *Ambio.* 16(1):38–46.

Laughlin, J. D., 1982. Gut analysis of polychaetes and amphipods. In: W. Bascom, ed. *Biennial Report 1981–1982.* South. Calif. Coastal Water Res. Proj., Long Beach, CA. pp. 59–66.

Lavigne, D. M., W. Barchard, and N. A. Oritsland, 1982. Pinniped bioenergetics. In: *Mammals in the Sea,* Vol. IV. Food and Agriculture Organization of the United Nations, Rome. pp. 191–265.

Leatherwood, S., and R. R. Reeves, 1986. Porpoises and dolphins. In: D. Haley, ed. *Marine Mammals of the Eastern North Pacific and Arctic Waters,* 2d ed. Pacific Search Press, Seattle, WA. pp. 110–131.

Lewin, R., 1986. In ecology, change brings stability. *Science.* 234:1071–1073.

Littler, M. M., S. N. Murray, and K. E. Arnold, 1979. Seasonal variations in net photosynthetic performance and cover of intertidal macrophytes. *Aquat. Bot.* 7:35–46.

Lockyer, C., 1976. Growth and energy budget of large baleen whales from the southern hemisphere. Advisory Committee on Marine Resource Research, Food and Agriculture Organization of the United Nations, Rome. ACMRR/MM/SC/38. 33pp.

Loder, T. C., 1971. Distribution of dissolved and particulate organic carbon in Alaska polar, subpolar, and estuarine waters. Ph.D. Dissertation, Univ. of Alaska, Fairbanks, AK. 236pp.

Longhurst, A., 1984. Importance of measuring rates and fluxes in marine ecosystems. In: M. J. R. Fasham, ed. *Flow of Energy and Materials in Marine Ecosystems.* Plenum Press, New York. pp. 3–22.

Lorenzen, C. J., N. A. Welschmeyer, and A. E. Coping, 1983. Particulate organic carbon flux in the sub-arctic Pacific. *Deep-Sea Res.* 30(6A): 639–643.

Mallicoate, D. L., and R. H. Parrish, 1981. Seasonal growth patterns of California stocks of northern anchovy, *Engraulis mordax,* Pacific mackerel, *Scomber japonicus,* and jack mackerel, *Trachurus symmetricus. Calif. Coop. Oceanic Fish. Invest. Rep.* 22:69–81.

Mann, R. G., 1982. *Ecology of Coastal Waters: A Systems Approach*. Univ. of California Press, Berkeley. 322pp.

Mate, B., and D. P. DeMaster, 1986. California sea lion. In: D. Haley, ed. *Marine Mammals of the Eastern North Pacific and Arctic Waters,* 2d ed. Pacific Search Press, Seattle, WA. pp. 196–201.

McAllister, C. D., 1972. Estimates of the transfer of primary production to secondary production at Ocean Station P. In: A. Y. Takenouti, ed. *Biological Oceanography of the North Pacific Ocean*. Idemitsu Shoten, Tokyo. pp. 573–579.

McCarthy, J. J., 1972. The uptake of urea by natural populations of marine phytoplankton. *Limnol. Oceanogr.* 17(5):738–748.

McCarthy, J. J., and T. C. Whitledge, 1972. Nitrogen excretion by anchovy (*Engraulis mordax* and *E. ringens*) and mackerel (*Trachurus symmetricus*). *U.S. Fish Wildl. Serv. Fish. Bull.* 70: 395–401.

McCave, I. N., 1975. Vertical flux of particles in the ocean. *Deep-Sea Res.* 22(7):491–502.

McRoy, C. P., D. W. Hood, L. K. Coachman, J. J. Walsh, and J. J. Goering, 1986. Processes and resources of the Bering Sea shelf (PROBES): The development and accomplishments of the project. *Cont. Shelf Res.* 5:5–22.

Mearns, A. J., 1982. Assigning trophic levels to marine animals. In: W. Bascom, ed. *Biennial Report 1981–1982*. South. Calif. Coastal Water Res. Proj. Long Beach, CA. pp. 125–141.

Menzel, D. W., 1974. Primary productivity, dissolved and particulate organic matter and the sites of oxidation of organic matter. In: E. D. Goldberg, ed. *The Sea,* vol. 5. John Wiley & Sons, New York. pp. 659–678.

Mitchell, C., 1986. California outer continental shelf fisheries database. Annotated bibliography: Fisheries species and oil and gas platforms offshore California. POCS Study No. 86–0092. U.S. Dept. of Interior, Minerals Management Service, Pacific OCS Region, Los Angeles.

Mullin, M. M., 1986. Spatial and temporal scales and patterns. In: R. W. Eppley, ed. *Lecture Notes on Coastal and Estuarine Studies. Vol. 15. Plankton Dynamics of the Southern California Bight*. Springer-Verlag, Berlin. pp. 216–273.

Niebauer, H. J., C. P. McRoy, and J. J. Goering, 1981. Hydrographic data: May 1976; May, June and September–October 1977. *RV Acona* cruises 227, 242, 243, 250, and 251. PROBES data report PDR 81–003. Inst. Mar. Sci., Univ. of Alaska, Fairbanks, AK. 205pp.

Nixon, M., 1987. Cephalopod diets. In: P. R. Boyle, ed. *Cephalopod Life Cycles*. Academic Press, London. pp. 201–219.

Odum, E. P., 1981. The effects of stress on the trajectory of ecological succession. In: G. W. Barrett, and R. Rosenberg, eds. *Stress Effects on Natural Ecosystems*. John Wiley & Sons, Ltd., Sussex, U.K. pp. 43–47.

Olson, R. J., 1981. Nitrogen–15 tracer studies of the primary nitrite maximum. *J. Mar. Res.* 39(2):203–226.

Parrish, R. H., and A. D. MacCall, 1978. Climatic variation and exploitation in the Pacific mackerel fishery. *Calif. Dept. Fish Game Fish Bull.,* no. 167. 110pp.

Parsons, T. R., and R. J. Le Brasseur, 1970. The availability of food to different trophic levels in the marine food chain. In: J. H. Steele, ed. *Marine Food Chains*. Oliver and Boyd, Edinburgh, U.K. pp. 325–343.

Pavlova, Y. V., 1966. Seasonal variations of the California Current (English translation). *Oceanology*. 6:806–814.

Paxton, J. R., 1967. Biological notes on southern California lanternfishes (Family Myctophidae). *Calif. Fish Game*. 53(3):214–217.

Pearcy, W. G., and R. M. Laurs, 1966. Vertical migration and distribution of mesopelagic fishes off Oregon. *Deep-Sea Res.* 13(2):153–165.

Peláez, J., and J. A. McGowan, 1986. Phytoplankton pigment patterns in the California Current as determined by satellite. *Limnol. Oceanogr.* 31(5):927–950.

Platt, T., and B. Irwin, 1973. Caloric content of phytoplankton. *Limnol. Oceanogr.* 18(2):306–310.

Postma, H., 1973. Transport and budget of organic matter in the North Sea. In: E. D. Goldberg, ed. *North Sea Science*. MIT Press, Cambridge, MA. pp. 326–334.

Pra Kash, A., 1975. Dinoflagellate blooms: An overview. In: V. R. Lo Cicero, ed. *Proceedings of the First International Conference on Toxic Dinoflagellate Blooms*. Mass. Sci. Tech. Foundation, Wakefield, MA. pp. 1–6.

Reid, J. L., 1965. Physical oceanography of the region near Point Arguello. IMR 65–19, Inst. Mar. Resour., Univ. of California, La Jolla, CA.

Reid, P. C., 1975. Large-scale changes in North Sea phytoplankton. *Nature (Lond.)*. 257:217–219.

Reilly, S. B., and S. H. Shane, 1986. Pilot whale.

In: D. Haley, ed. *Marine Mammals of the Eastern North Pacific and Arctic Waters,* 2d ed. Pacific Search Press, Seattle, WA. pp. 132–139.

Rice, D. W., 1986. Blue whale. In: D. Haley, ed. *Marine Mammals of the Eastern North Pacific and Arctic Waters,* 2d ed. Pacific Search Press, Seattle, WA. pp. 40–45.

Rittenberg, S. C., 1940. Bacteriological analysis of some long cores of marine sediments. *J. Mar. Res.* 3:191–201.

Rutgers van der Loeff, M. M., 1980. Nutrients in the interstitial waters of the Southern Bight of the North Sea. *Neth. J. Sea Res.* 14(2):144–171.

Sambrotto, R. N., and J. J. Goering, 1979. Changes in nutrient chemistry, cell size, and nitrogen cycling in a diatom population grazed by macrozooplankton. In: PROBES Prog. Rep. Vol. II. Inst. Mar. Sci., Univ. of Alaska, Fairbanks, AK. pp. 198–223.

Sambrotto, R. N., H. J. Niebauer, J. J. Goering, and R. L. Iverson, 1986. Relationships among vertical mixing, nitrate uptake, and phytoplankton growth during the spring bloom in the southeast Bering Sea middle shelf. *Cont. Shelf Res.* 5:161–198.

Schaff, W. E., 1980. An analysis of the dynamic population response of Atlantic menhaden, *Brevoortia tyrannus,* to an increase of fishery. In: A. Saville, ed. *The Assessment and Management of Pelagic Fish Stocks.* Rapp. P.-V. Réun. Cons. Int. Explor. Mer. 177:243–251.

Schneider, D. C., G. L. Hunt, Jr., and N. M. Harrison, 1986. Mass and energy transfer to sea birds in the southeastern Bering Sea. *Cont. Shelf Res.* 5:241–257.

Scott, G. P., R. D. Kenney, T. J. Thompson, and H. E. Winn, 1983. Functional roles and ecological impacts of the cetacean community in waters of the northeastern U.S. continental shelf. *Cons. Int. Explor. Mer.* 12:1–32.

Shannon, L. V., and J. G. Field, 1985. Are fish stocks food-limited in the southern Benguela pelagic ecosystem? *Mar. Ecol. Prog. Ser.* 22:7–19.

Shaw, D. G., and M. J. Hameedi, eds., 1988. *Environmental Studies of Port Valdez, Alaska.* Springer-Verlag, New York. 423pp.

Shepard, F. P., and K. O. Emery, 1941. Submarine topography off the California coast: Canyons and tectonic interpretations. *Geol. Soc. Am. Spec. Pap.,* no. 31. 171pp.

Shiller, A. M., 1982. The geochemistry of particulate major elements in the Santa Barbara Basin and observations on the calcium carbonate–carbon dioxide system in the ocean. Ph.D. Dissertation, Univ. of California, San Diego, CA. 197pp.

Sholkovitz, E. R., and J. M. Gieskes, 1971. A physical–chemical study of the flushing of the Santa Barbara Basin. *Limnol. Oceanogr.* 16(3): 479–489.

Sissenwine, M. P., 1987. Fish and squid production. In: R. H. Backus and D. W. Bourne, eds. *Georges Bank.* MIT Press, Cambridge, MA. pp. 347–355.

Small, L. F., M. R. Landry, R. W. Eppley, F. Azam, and A. F. Carlucci, 1989. Role of plankton in the carbon and nitrogen budgets of Santa Monica Basin, California. *Mar. Ecol. Prog. Ser.* 56:57–74.

Smith, P. E., 1978. Biological effects of ocean variability: Time and space scales of biological response. *Rapp. P.-V. Réun. Cons. Int. Explor. Mer.* 173:117–127.

Smith, P. E., and R. W. Eppley, 1982. Primary production and the anchovy population in the Southern California Bight: Comparison of time-series. *Limnol. Oceanogr.* 27(1):1–17.

Somner, H., and F. N. Clark, 1946. Effect of red water on marine life in Santa Monica Bay, California. *Calif. Fish Game.* 32(2):100–101.

Sorokin, Y. I., 1983. The Black Sea. In: B. H. Ketchum, ed. *Ecosystems of the World. Vol. 26. Estuaries and Enclosed Seas.* Elsevier Scientific Publishing Company, New York. pp. 253–289.

Steele, J. H., 1974. *The Structure of Marine Ecosystems.* Harvard Univ. Press, Cambridge, MA. 128pp.

Stoddard, A., and J. J. Walsh, 1986. Modeling oxygen in the New York Bight: The water quality impact of a potential increase of waste inputs. In: D. A. Wolfe, ed. *Urban Wastes in Coastal Marine Environments.* NOAA Prof. Pap.

Stoker, S., 1981. Benthic invertebrate macrofauna of the Eastern Bering/Chukchi continental shelf. In: D. W. Hood, and J. A. Calder, eds. *The Eastern Bering Sea Shelf: Oceanography and Biological Resources.* Univ. of Washington Press, Seattle, WA. pp. 1069–1090.

Temple, R. F., D. L. Harrington, and J. A. Martin, 1977. Monthly temperature and salinity measurements of waters of the northwestern Gulf of Mexico, 1963–1965. NOAA Tech. Rep. NMFS 55RF–707.

Thomas, W. H., and A. J. Carsola, 1980. Ammonium input to the sea via large sewage outfalls 1. Tracing sewage in southern California waters. *Mar. Environ. Res.* 3(4):277–289.

Thompson, B. E., J. D. Laughlin, and D. T. Tsukada, 1984. Ingestion and oxygen consumption by slope echinoids. In: W. Bascom, ed. *Biennial Report 1983–1984.* South. Calif. Coastal Water Res. Proj., Long Beach, CA. pp. 93–107.

Van Bennekom, A. J., and W. Salamons, 1981. Pathways of nutrient and organic matter from land to oceans through rivers. In: *River Inputs to Ocean Systems.* UNESCO, Geneva, Switz. pp. 35–51.

Walsh, J. J., 1981a. A carbon budget for overfishing off Peru. *Nature (Lond.).* 290:300–304.

Walsh, J. J., 1981b. Shelf-sea ecosystems. In: A. R. Longhurst, ed. *Analysis of Marine Ecosystems.* Academic Press, New York. pp. 159–196.

Walsh, J. J., and C. P. McRoy, 1986. Ecosystem analysis in the southeastern Bering Sea. *Cont. Shelf Res.* 5:259–288.

Walsh, J. J., T. E. Whitledge, J. E. O'Reilly, W. C. Phoel, and A. F. Draxler, 1987a. Nitrogen cycling on Georges Bank and the New York Shelf: A comparison between well-mixed and seasonally stratified waters. In: R. H. Backus, and D. W. Bourne, eds. *Georges Bank.* MIT Press, Cambridge, MA. pp. 234–246.

Walsh, J. J., D. A. Dieterle, and M. B. Meyers, 1987b. A simulation analysis of the fate of phytoplankton within the Mid-Atlantic Bight. *Cont. Shelf Res.* 8(5–7):757–787.

Weichart, G., 1985. High pH values in the German Bight as an indication of intensive primary production. *Dtsch. Hydrogr. Z.* 38(3):93–117.

Whitledge, T. E., W. S. Reeburgh, and J. J. Walsh, 1986. Seasonal inorganic nitrogen distributions and dynamics in the southeastern Bering Sea. *Cont. Shelf Res.* 5:109–132.

Williams, P. J. L., 1984. The emperor's new suit of clothes? In: M. J. R. Fasham, ed. *Flow of Energy and Materials in Marine Ecosystems.* Plenum Press, New York. pp. 271–299.

Williams, P. M., 1986. Chemistry of the dissolved and particulate phases in the water column. In: R. W. Eppley, ed. *Lecture Notes on Coastal and Estuarine Studies. Vol. 15. Plankton Dynamics of the Southern California Bight.* Springer-Verlag, Berlin. pp. 53–83.

Winn, H. E., J. H. W. Hain, M. A. M. Hyman, and G. P. Scott, 1987. Whales, dolphins and porpoises. In: R. H. Backus, and D. W. Bourne, eds. *Georges Bank.* MIT Press, Cambridge, MA. pp. 375–382.

Wolfe, D. A., and B. Kjerfve, 1986. Estuarine variability: An overview. In: D. A. Wolfe, ed. *Estuarine Variability.* Academic Press, New York. pp. 3–17.

Zedler, J. B., 1982. The ecology of southern California coastal salt marshes: A community profile. Report No. FWS/OBS–81/54. Prep. by U.S. Fish and Wildlife Service, Biological Services Program, Washington, D.C. 110pp.

ZoBell, C. E., 1941. Studies on marine bacteria. 1. The cultural requirements of heterotrophic aerobes. *J. Mar. Res.* 4:42–75.

ZoBell, C. E., and C. B. Feltham, 1934. Preliminary studies on the distribution and characteristics of marine bacteria. *Bull. Scripps Inst. Oceanogr.* 3:279–296.

Index

NOTES: Localities, genera and species, and subjects from the main text are indexed in detail. Material in summary and future-research prospectus subsections is indexed in less detail. Acknowledgment and bibliography subsections are not indexed. Information about genera and species should be sought under *both* scientific and common names for completeness, although some overlap does exist for the two kinds of entries. Material in maps, charts, graphs, illustrations, line drawings, and photographs is denoted by **boldface** page numbers; material in tables is denoted by *italic* page numbers.

Designer: UC Press Staff
Compositor: Prestige Typography
Text: 10/12 Bembo
Display: Bembo
Printer: Edwards Brothers, Inc.
Binder: Edwards Brothers, Inc.